LECTURE NOTES ON
FIELD THEORY IN
**CONDENSED
MATTER PHYSICS**

LECTURE NOTES ON FIELD THEORY IN CONDENSED MATTER PHYSICS

Christopher Mudry
Paul Scherrer Institut, Switzerland

World Scientific

NEW JERSEY • LONDON • SINGAPORE • BEIJING • SHANGHAI • HONG KONG • TAIPEI • CHENNAI

Published by

World Scientific Publishing Co. Pte. Ltd.
5 Toh Tuck Link, Singapore 596224
USA office: 27 Warren Street, Suite 401-402, Hackensack, NJ 07601
UK office: 57 Shelton Street, Covent Garden, London WC2H 9HE

Library of Congress Cataloging-in-Publication Data
Mudry, Christopher, 1962– author.
Lecture notes on field theory in condensed matter physics / Christopher Mudry, Paul Scherrer Institut, Switzerland.
pages cm
Includes bibliographical references and index.
ISBN 978-9814449090 (hardcover : alk. paper) -- ISBN 978-9814449106 (softcover : alk. paper)
1. Quantum field theory. 2. Condensed matter. I. Title.
QC174.45.M83 2014
530.4'1--dc23
2013047940

British Library Cataloguing-in-Publication Data
A catalogue record for this book is available from the British Library.

In-house Editor: Song Yu

Printed in Singapore

To my mother Sally, my father Louis, and brother Yves who nurtured my love for physics and to my wife Cristy and son André who sometimes had to compete with it.

Preface

Reading the books from Baym [1], Messiah [2], and Dirac [3], while an undergraduate student at ETHZ, I learnt about how enjoyable and useful it is to learn quantum mechanics from different perspectives. This impression was reinforced as I got exposed to statistical physics and to the diversity of approaches to it found in the books from Becker [4], Callen [5], Huang [6], and Feynman [7].

My initiation to quantum field theory was different. In those days, there seemed to be two separate communities doing many-body physics. I took a proseminar from Prof. Klaus Hepp on the theory of renormalization, who told me that the only book on quantum field theory relevant to his class was that of Itzykson and Zuber [8]. Although I already had taken a proseminar on Fermi liquid theory and had started reading Kittel's *Quantum theory of solid* [9], I had not realized the close connection among the applications of many-body physics to high energy physics, statistical physics, and condensed matter physics. At the time, the few books on quantum field theory for high-energy physics were obsessed with propagators, Feynman diagrams, causality and positivity, and how to make sense of ultraviolet divergences. The venerable books on many-body physics in condensed matter physics were deceptive [10]-[13]. Although they claimed to do many-body physics, single-particle physics was soon enough resuscitating as a mean-field approximation or with particles with diminutive names (quasiparticles). Moreover, these books were full of approximations with mysterious acronyms such as the random phase approximation, involving some magical circular logic.

This cultural divide, manifest in the books published prior to the 80's, was however in the process of disappearing. As I was moving to UIUC to start my PhD, the standard model had established itself as the fundamental theory for particle physics and a steady supply of books devoted to it were being published by the late 80's. The renormalization group had been applied to explain asymptotic freedom of quantum chromodynamics (QCD), to solve the Kondo problem, and had a profound impact on statistical physics. Lattice gauge theory, an approach to solve QCD in the strong coupling problem, was turning into a discipline of its own, bridging relativistic quantum field theory to statistical physics. Algebraic topology, an arcane discipline of mathematics to most physicists, had shown its value to classify defects in the vacua of quantum field theories and the order parameters of symmetry broken phases in condensed matter physics or statistical physics. Algebraic topology could explain the quantization of the quantized Hall effect and the existence of collective excitations obeying exchange statistics evading the "spin-statistics theorem". Integrable models from statistical physics were used to explain the low-energy properties of spin chains, impurity models, and quasi-one-dimensional metals. The second remarkable discovery of the 80's in condensed matter physics was of course that of high T_c superconductivity, a class of materials that defy a solution using perturbation theory to this date.

The first books that I am aware of aiming at overcoming the cultural differences between high-energy physics, statistical physics, and condensed matter physics of the 60's are those of Polyakov [14], Parisi [15], and Negele and Orland [16], whose authors had made seminal contributions to the revolution brought upon by the application of the renormalization group to theoretical physics during the late 60's and 70's. Another generation of authors came along in the 90's with the intent to explain how the machinery of quantum field theory should be applied to condensed matter physics [17]-[22].

Since the turn of the 21st century, concepts and techniques have been shared from condensed matter theory to string theory. The breadth of topics in condensed matter physics makes it impossible to cover all applications of quantum field theory to condensed matter physics in a single book. Correspondingly, the number of books applying quantum field theory to condensed matter physics is steadily increasing and getting more specialized. A student has now the luxury of picking his favorite book and taking advantage of a variety of viewpoints.

This book is the result of teaching the class "*Quantum field theory in condensed matter physics*" at ETHZ. My aim was to demystify some of the condensed matter jargon used in seminars in condensed matter physics for a student at the level of a master degree in physics from ETHZ. I also wanted a student attending my class to obtain a hands-on experience of concepts such as spontaneous symmetry breaking, mean-field theory, random phase approximation, screening, quantum fluctuations, renormalization group flows, critical points, phase transitions driven by topological defects, bosonization, etc. Many books on quantum field theory devote space to the machinery of quantum field theory before solving problems with it. I wanted my teaching to do the reverse, i.e., to develop the needed methodology one problem at a time. I also did not want quantum field theory to become the primary interest. It had to remain a tool to explain as economically as possible fundamental principles of condensed matter physics. I am of the opinion that the most efficient technique for this purpose is to systematically use the path integral representation of quantum mechanics. Path integrals are thus pervasive in this book. However, I assume no more prior knowledge than familiarity with quantum mechanics, at the level of Baym's book say.

The book is organized in two parts. The first part deals with bosons, the second with fermions. In condensed matter physics, this organization principle is not as obvious as would be implied by the standard model of high-energy physics. The fundamental boson of condensed matter physics is the photon. The fundamental fermions of condensed matter physics are the electron and the proton, the charged constituents of the atoms from the periodic table. On the relevant energy and length scales of condensed matter physics, these elementary constituents interact through the rules of quantum electrodynamics at a non-vanishing density of fermionic matter in the ground state. This is the main difference with quantum field theory aiming at explaining high-energy scattering experiments, for which the ground state (the vacuum before and after scattering) has a vanishing density of (fermionic) matter. This difference is of a fundamental nature. The atomic nucleus has a much larger mass than the electrons orbiting around it. In a material, positive charge is localized in position space on the sites of a crystal at low temperatures. As a result, the fermionic nature of the ionic constituents becomes irrelevant. What matters greatly however is that the normal modes of this crystal are phonons, collective excitations obeying Bose-Einstein statistics. The same can happen with electrons. They can localize in position space, in which case the material is called an insulator. Some localized electrons can still interact through their internal spin-1/2 degree of freedom. It is often the case that the collective excitations resulting from the interactions between the spins of localized electrons are collective excitations obeying Bose-Einstein statistics. They are called magnons. Electrons need not be localized in a material, which is then called a metal. In a metal, the mobile electrons exchange photons with each other; they interact through the Coulomb interaction; they interact with the localized positive charge of the crystal, a one-body potential for the electrons; they interact with the phonons, and they might interact with some of the electrons that are localized around the crystalline sites. Solving this many-body problem from the Schrödinger equation is and will be impossible. The Hilbert space is simply too large. Instead, effective theories motivated by phenomenology and simplicity have been the bread and butter of theoretical condensed matter physics. In these models, the elementary local constituents might be bosons, fermions, or more complicated objects of which the simplest examples are quantum spin degrees of freedom. The partition of this book into a part devoted to bosons and a part devoted to fermions refers to the situations when some of the low-energy collective excitations can be shown to obey the Bose-Einstein or Fermi-Dirac statistics, respectively. Even then, we shall show that bosons and fermions emerging from some interacting models on a one-dimensional lattice are interchangeable under the rule of bosonization. The four chapters on bosons cover phonons (as a way to introduce a quantum field theory), superfluidity, restoration of a continuous symmetry by fluctuations at the lower critical dimension, and the Kosterlitz-Thouless phase transition, respectively. The five chapters on fermions cover non-interacting fermions, the random phase approximation in the jellium model, superconductivity, dissipative Josephson junction (an example of dissipative quantum mechanics), and bosonization, respectively. Each chapter ends with a section in which material is presented as a sequence of exercises. Each chapter also comes with an appendix. Some appendices provide distracting intermediary steps. Most appendices contain learning material.

The books of Naoto Nagaosa in Ref. [20] and Mike Stone in Ref. [21] have been very influential while preparing my lectures. I am indebted to their authors for their inspiring books.

Christopher Mudry
Gipf-Oberfrick, July 2013

Acknowledgments

I must start thanking Donald E. Knuth for developing TeX. I typed my master thesis and had I needed to do the same for a book, I would have never written one.

I am grateful to my home institution, the Paul Scherrer Institut (PSI), in the persons of Kurt Clausen who has been supportive of this endeavor and Joël Mesot who has been a steady and reliable advocate of the condensed matter theory group. Since 1999 to this date, I benefited at PSI from a great colleague, Rudolf Morf.

I am indebted to my mentors Eduardo Fradkin (my PhD adviser), Xiao-Gang Wen (my host at MIT), and Bertrand Halperin (my host at Harvard) for shaping my taste in physics. I am also indebted to my friends and long term collaborators Claudio Chamon, Akira Furusaki, Piet Brouwer, and Shinsei Ryu who have been so influential on my understanding of physics.

I have had the good fortune of directing the thesis of three talented students. Andreas Schnyder, Sebastian Guerrero, and Titus Neupert. They were all teaching assistants of my class and made important contributions to the exercises. Titus had also the kindness and patience for converting my figures into artworks.

In the last six months Maurizio Storni has helped me polish my lecture notes into this book. He even shares my compulsive obsession with using TeX for baroque notation! It has been my privilege to benefit from his dedicated and critical reading. Maurizio has been the fairy-godmother of Cinderella for my lecture notes. I only hope there is no midnight deadline. Of course, as convention dictates, all remaining embarrassing mistakes are my responsibility.

Contents

PART 1
Bosons

Chapter 1

The harmonic crystal

Outline

The classical equations of motion for a finite chain of atoms are solved within the harmonic approximation. In the thermodynamic limit, an approximate hydrodynamical description, i.e., a one-dimensional classical field theory, is obtained. Quantization of the finite harmonic chain is undertaken. In the thermodynamic limit, phonons in a one-dimensional lattice are approximated by a quantum hydrodynamical theory, i.e., a one-dimensional quantum field theory.

1.1 Introduction

To illustrate the transition from the one-body to the many-body physics, the harmonic excitations of a crystal are derived classically and quantum mechanically. The thematic of crystallization, i.e., of spontaneous-symmetry breaking of translation symmetry in position space, is addressed in section 1.5 from the point of view of an application of the Mermin-Wagner theorem.

1.2 Classical one-dimensional crystal

1.2.1 *Discrete limit*

For simplicity, we shall consider a one-dimensional world made of N point-like objects (atoms) of mass m and interacting through a potential V. We assume first that the potential V depends only on the coordinates $\eta_n \in \mathbb{R}$, $n = 1, \cdots, N$, of the N atoms,

$$V = V(\eta_1, \cdots, \eta_N). \tag{1.1}$$

Furthermore, we assume that V has a non-degenerate minimum at

$$\bar{\eta}_n = n\,\mathfrak{a}, \qquad n = 1, \cdots, N, \tag{1.2}$$

where $\mathfrak{a}$ is the lattice spacing. For example,

$$\begin{aligned} V(\eta_1,\cdots,\eta_N) = & \left(\frac{\mathfrak{a}}{2\pi}\right)^2 \kappa \sum_{n=1}^{N-1}\left[1-\cos\left(\frac{2\pi}{\mathfrak{a}}\left(\eta_{n+1}-\eta_n\right)\right)\right] \\ & + \left(\frac{\mathfrak{a}}{2\pi}\right)^2 m\,\Omega^2 \sum_{n=1}^{N}\left[1-\cos\left(\frac{2\pi}{\mathfrak{a}}\eta_n\right)\right] \\ & + \text{boundary terms.} \end{aligned} \tag{1.3}$$

The physical interpretation of the real-valued parameters κ and Ω is obtained as follows. For small deviations $\delta\eta_n$ about minimum (1.2), it is natural to expand the potential energy according to

$$\begin{aligned} V(\bar\eta_1+\delta\eta_1,\cdots,\bar\eta_N+\delta\eta_N) = & V(\bar\eta_1,\cdots,\bar\eta_N) + \sum_{n=1}^{N-1}\frac{\kappa}{2}\left(\delta\eta_{n+1}-\delta\eta_n\right)^2 \\ & + \frac{1}{2}m\,\Omega^2\sum_{n=1}^{N}\left(\delta\eta_n\right)^2 + \cdots + \text{boundary terms.} \end{aligned} \tag{1.4}$$

The dimensionful constant κ is the elastic or spring constant. It measures the strength of the linear restoring force between nearest-neighbor atoms. The characteristic frequency Ω measures the strength of an external force that pins atoms to their equilibrium positions (1.2). To put it differently, $m\,\Omega^2$ is the curvature of the potential well that pins an atom to its equilibrium position. Terms that have been neglected in $\cdots$ are of several kinds. Only terms of quadratic order in the nearest-neighbor relative displacement $\delta\eta_{n+1}-\delta\eta_n$ have been accounted for, and all interactions beyond the nearest-neighbor range have been dropped. We have also omitted to spell out what the boundary terms are. They are specified once boundary conditions have been imposed. In the limit $N\to\infty$, the choice of boundary conditions should be immaterial since the bulk potential energy should be of order $L\equiv N\,\mathfrak{a}$, whereas the energy contribution arising from boundary terms should be of order $L^0=1$.

To minimize boundary effects in a finite system, one often imposes periodic boundary conditions

$$\eta_{n+N}=\eta_n, \qquad n=1,\cdots,N. \tag{1.5}$$

An open chain of atoms turns into a ring after imposing periodic boundary conditions. Furthermore, imposing periodic boundary conditions endows the potential with new symmetries within the harmonic approximation defined by[1]

$$V_{\text{har}}(\bar\eta_1+\delta\eta_1,\cdots,\bar\eta_N+\delta\eta_N) := \sum_{n=1}^{N}\left[\frac{\kappa}{2}\left(\delta\eta_{n+1}-\delta\eta_n\right)^2+\frac{1}{2}m\,\Omega^2\left(\delta\eta_n\right)^2\right]. \tag{1.7}$$

[1] Without loss of generality, we have set the classical minimum of the potential energy to zero,

$$V(\bar\eta_1,\cdots,\bar\eta_N)=0. \tag{1.6}$$

First, changing labels according to

$$n \to n + m, \qquad n = 1, \cdots, N, \qquad \forall m \in \mathbb{Z}, \tag{1.8}$$

leaves Eq. (1.7) invariant. Second, translation invariance is recovered in the absence of the pinning potential,

$$\begin{aligned} &\text{Eq. (1.7) with } \Omega = 0 \Longrightarrow \\ &V_{\text{har}}(\eta_1, \cdots, \eta_N) = V_{\text{har}}(\eta_1 + x\,\mathfrak{a}, \cdots, \eta_N + x\,\mathfrak{a}), \end{aligned} \tag{1.9}$$

for any real-valued x.

The kinetic energy of an open chain of atoms is simply given by

$$T(\bar{\eta}_1 + \delta\eta_1, \cdots, \bar{\eta}_N + \delta\eta_N) = \frac{1}{2} m \sum_{n=1}^{N} \left(\frac{\mathrm{d}\delta\eta_n}{\mathrm{d}t} \right)^2 \equiv \frac{1}{2} m \sum_{n=1}^{N} \left(\delta\dot{\eta}_n \right)^2 . \tag{1.10}$$

As was the case for the potential energy, the choice of boundary conditions only affects the kinetic energy by terms of order L^0. It is thus natural to choose periodic boundary conditions if one is interested in extensive properties of the system.

The classical Lagrangian $\mathfrak{L}$ in the harmonic approximation and with periodic boundary conditions is defined by subtracting from the kinetic energy (1.10) the potential energy (1.7),

$$\mathfrak{L} := \sum_{n=1}^{N} \frac{1}{2} \left[m \left(\delta\dot{\eta}_n \right)^2 - \kappa \left(\delta\eta_{n+1} - \delta\eta_n \right)^2 - m\,\Omega^2 \left(\delta\eta_n \right)^2 \right] . \tag{1.11}$$

The classical equations of motion follow from Euler Lagrange equations of motion

$$\frac{\mathrm{d}}{\mathrm{d}t} \left(\frac{\partial \mathfrak{L}}{\partial\, \delta\dot{\eta}_n} \right) = \left(\frac{\partial \mathfrak{L}}{\partial\, \delta\eta_n} \right), \qquad n = 1, \cdots, N. \tag{1.12}$$

They are

$$m\delta\ddot{\eta}_n = \kappa \left(\delta\eta_{n+1} + \delta\eta_{n-1} - 2\delta\eta_n \right) - m\,\Omega^2 \delta\eta_n, \qquad n = 1, \cdots, N, \tag{1.13}$$

with the complex-valued and traveling-wave solutions

$$\delta\eta_n(t) \propto e^{\mathrm{i}(kn - \omega t)}, \qquad \omega^2 = 2\frac{\kappa}{m}\left(1 - \cos k\right) + \Omega^2. \tag{1.14}$$

Imposing periodic boundary conditions allows to identify the normal modes. These are countably-many traveling waves with the frequency to wave-number relation

$$\omega_l = \sqrt{2\frac{\kappa}{m}\left(1 - \cos k_l\right) + \Omega^2}, \qquad k_l = \frac{2\pi}{N} l, \qquad l = 1, \cdots, N. \tag{1.15}$$

The most general real-valued solution of Euler-Lagrange equations (1.13) obeying periodic boundary conditions is

$$\delta\eta_n(t) = \sum_{l=1}^{N} \left[A_l\, e^{+\mathrm{i}(k_l n - \omega_l t)} + A_l^{*} e^{-\mathrm{i}(k_l n - \omega_l t)} \right], \qquad n = 1, \cdots, N. \tag{1.16}$$

Here, the complex-valued expansion coefficient A_l remains arbitrary as long as initial conditions on $\delta\eta_n$ and $\delta\dot{\eta}_n$ have not been specified.

To revert to the Hamilton-Jacobi formalism of classical mechanics, one introduces the canonical momentum $\delta\pi_n$ conjugate to $\delta\eta_n$ through

$$\begin{aligned}\delta\pi_n(t) &:= \frac{\partial \mathfrak{L}}{\partial \delta\dot{\eta}_n} \\ &= -\,\mathrm{i}m\sum_{l=1}^{N}\omega_l\left[A_l\, e^{+\mathrm{i}(k_l n-\omega_l t)} - A_l^*\, e^{-\mathrm{i}(k_l n-\omega_l t)}\right], \quad n=1,\cdots,N,\end{aligned} \tag{1.17}$$

and construct the Hamiltonian

$$\mathfrak{H} = \sum_{n=1}^{N}\frac{1}{2}\left[\frac{(\delta\pi_n)^2}{m} + \kappa\left(\delta\eta_{n+1}-\delta\eta_n\right)^2 + m\,\Omega^2(\delta\eta_n)^2\right] \tag{1.18}$$

from the Lagrangian (1.11) through a Legendre transformation. Hamilton-Jacobi equations of motion are then

$$\delta\dot{\eta}_n = +\frac{\partial\mathfrak{H}}{\partial\delta\pi_n} = \{\delta\eta_n,\mathfrak{H}\}, \qquad \delta\dot{\pi}_n = -\frac{\partial\mathfrak{H}}{\partial\delta\eta_n} = \{\delta\pi_n,\mathfrak{H}\}, \qquad n=1,\cdots,N, \tag{1.19}$$

where $\{\cdot,\cdot\}$ stands for the Poisson brackets.[2]

In the long wave-number limit $k_l \ll 1$, the dispersion relation reduces to

$$\omega_l^2 = \frac{\kappa}{m}k_l^2 + \Omega^2 + \mathcal{O}(k_l^4). \tag{1.21}$$

The pinning potential characterized by the curvature Ω of the potential well has opened a *gap* in the spectrum of normal modes. No solutions to Euler-Lagrange equations (1.13) can be found below the characteristic frequency Ω. By switching off the pinning potential, $\Omega = 0$, the dispersion relation simplifies to

$$\omega_l^2 = \frac{\kappa}{m}k_l^2 + \mathcal{O}(k_l^4). \tag{1.22}$$

The proportionality constant $\sqrt{\kappa/m}$ between frequency and wave number is interpreted as the velocity of propagation of a sound wave in the one-dimensional harmonic chain in units for which the lattice spacing $\mathfrak{a}$ has been set to unity.

1.2.2 *Thermodynamic limit*

The thermodynamic limit $N \to \infty$ emerges naturally if one is interested in the response of a one-dimensional solid to external perturbations as can be induced, say, by compressions. If the characteristic wavelength of a perturbation applied to a solid is much larger than the atomic separation, then the (elastic) response from this solid to this perturbation is dominated by normal modes with arbitrarily small

[2] The Poisson bracket $\{f,g\}$ of two functions f and g of the canonical variables $\delta\eta_n$ and $\delta\pi_n$ is defined by

$$\{f,g\} := \sum_{n=1}^{N}\left(\frac{\partial f}{\partial\delta\eta_n}\frac{\partial g}{\partial\delta\pi_n} - \frac{\partial f}{\partial\delta\pi_n}\frac{\partial g}{\partial\delta\eta_n}\right). \tag{1.20}$$

wave numbers $k \to 0$. If so, it is then much more economical not to account for the discrete nature of this solid as is done in the Lagrangian (1.11) when computing the (elastic) response. To this end, Eq. (1.11) is first rewritten as

$$\begin{aligned}\mathfrak{L} &= \sum_{n=1}^{N} \mathfrak{a}\frac{1}{2}\left[\frac{m}{\mathfrak{a}}\left(\delta\dot{\eta}_n\right)^2 - \kappa\mathfrak{a}\left(\frac{\delta\eta_{n+1}-\delta\eta_n}{\mathfrak{a}}\right)^2 - \frac{m}{\mathfrak{a}}\Omega^2\left(\delta\eta_n\right)^2\right] \\ &=: \sum_{n=1}^{N} \mathfrak{a}\,L_n.\end{aligned} \tag{1.23}$$

We interpret

$$\mu := \frac{m}{\mathfrak{a}}, \qquad Y := \kappa\,\mathfrak{a}, \qquad \xi := \frac{\delta\eta_{n+1}-\delta\eta_n}{\mathfrak{a}}, \qquad \text{and} \qquad L_n, \tag{1.24}$$

as the mass per unit length, the Young's modulus,[3] the elongation per unit length, and the *local* Lagrangian per unit length (the Lagrangian density), respectively. Then, we write

$$\begin{aligned}\mathfrak{L} &= \int_0^L \mathrm{d}x\,\frac{1}{2}\left[\mu\left(\frac{\partial\varphi}{\partial t}\right)^2 - Y\left(\frac{\partial\varphi}{\partial x}\right)^2 - \mu\Omega^2\varphi^2\right] \\ &=: \int_0^L \mathrm{d}x\,\mathcal{L},\end{aligned} \tag{1.26}$$

whereby the following substitutions have been performed.

[1] The discrete sum $\sum_n$ has been replaced by the integral $\int \mathrm{d}x/\mathfrak{a}$ over the semi-open interval $]0, L]$.

[2] The relative displacement $\delta\eta_n$ at time t has been replaced by the value of the real-valued function φ at the space-time coordinate (x,t) obeying periodic boundary conditions in position space,

$$\varphi(x+L,t) = \varphi(x,t), \qquad x \in]0,L], \qquad \forall t \in \mathbb{R}. \tag{1.27}$$

[3] The time derivative of the relative displacement $\delta\eta_n$ at time t has been replaced by the value of the time derivative $(\partial_t\varphi)$ at the space-time coordinate (x,t).

[4] The discrete difference $\delta\eta_{n+1} - \delta\eta_n$ at time t has been replaced by the lattice constant times the value of the derivative $(\partial_x\varphi)$ at the space-time coordinate (x,t).

[5] The integrand $\mathcal{L}$ in Eq. (1.26) is called the Lagrangian density. It is a real-valued function of space and time. From it, one obtains the continuum limit of Euler-Lagrange equations (1.12) according to

$$\partial_t \frac{\delta\mathcal{L}(x,t)}{\delta(\partial_t\varphi)(y,t)} + \partial_x \frac{\delta\mathcal{L}(x,t)}{\delta(\partial_x\varphi)(y,t)} = \frac{\delta\mathcal{L}(x,t)}{\delta\varphi(y,t)}. \tag{1.28}$$

[3] For an elastic rode obeying Hooke's law, the extension ξ of the rode per unit length is proportional to the exerted force F with the Young's modulus Y as the proportionality constant,

$$F = Y\xi. \tag{1.25}$$

Here, the symbol $\delta\mathcal{L}(x,t)$ is to be interpreted as the infinitesimal functional change of $\mathcal{L}$ at the given space-time coordinates (x,t) induced by the Taylor expansion

$$\begin{aligned}\delta\mathcal{L} =& \mathcal{L}[\varphi+\delta\varphi,(\partial_x\varphi)+\delta(\partial_x\varphi),(\partial_t\varphi)+\delta(\partial_t\varphi)] - \mathcal{L}[\varphi,(\partial_x\varphi),(\partial_t\varphi)] \\ =& \frac{\partial\mathcal{L}}{\partial\varphi}\delta\varphi + \frac{\partial\mathcal{L}}{\partial(\partial_x\varphi)}\delta(\partial_x\varphi) + \frac{\partial\mathcal{L}}{\partial(\partial_t\varphi)}\delta(\partial_t\varphi) + \cdots .\end{aligned} \tag{1.29}$$

One must keep in mind that φ, $(\partial_x\varphi)$, and $(\partial_t\varphi)$ are independent "variables". Moreover, one must use the rule

$$\frac{\delta\varphi(x,t)}{\delta\varphi(y,t)} = \delta(x-y) \Longrightarrow \int\limits_0^L \mathrm{d}x\,\frac{\delta\varphi(x,t)}{\delta\varphi(y,t)} = 1, \qquad y\in]0,L], \tag{1.30}$$

that extends the rule

$$\frac{\partial\eta_m}{\partial\eta_n} = \delta_{m,n} \Longrightarrow \sum_{m=1}^{N}\frac{\partial\eta_m}{\partial\eta_n} = 1, \qquad n=1,\cdots,N, \tag{1.31}$$

to the continuum. Otherwise, all the usual rules of differentiation apply to $\delta\cdot/\delta\varphi$.

[6] Equations of motion (1.13) become the one-dimensional sound wave equation

$$\left(\partial_t^2 - v^2\partial_x^2 + \Omega^2\right)\varphi = 0, \qquad v := \sqrt{\frac{Y}{\mu}}, \tag{1.32}$$

after replacing the finite difference

$$\delta\eta_{n+1} + \delta\eta_{n-1} - 2\delta\eta_n = +\left(\delta\eta_{n+1} - \delta\eta_n\right) - \left(\delta\eta_n - \delta\eta_{n-1}\right) \tag{1.33}$$

by $\mathfrak{a}^2$ times the value of the second-order space derivative $(\partial_x^2\varphi)$ at the space-time coordinate (x,t).

The Hamiltonian $\mathfrak{H}$ in the continuum limit follows from Eq. (1.26) with the help of a (functional) Legendre transform or directly from the continuum limit of Eq. (1.18),

$$\begin{aligned}\mathfrak{H} =& \int\limits_0^L \mathrm{d}x\,\frac{1}{2}\left[\frac{\pi^2}{\mu} + Y\left(\frac{\partial\varphi}{\partial x}\right)^2 + \mu\Omega^2\varphi^2\right] \\ =:& \int\limits_0^L \mathrm{d}x\,\mathcal{H},\end{aligned} \tag{1.34a}$$

where the field π is the canonically conjugate to φ,

$$\pi(x,t) := \int\limits_0^L \mathrm{d}y\,\frac{\delta\mathcal{L}(y,t)}{\delta(\partial_t\varphi)(x,t)} = \mu(\partial_t\varphi)(x,t). \tag{1.34b}$$

Probing the one-dimensional harmonic crystal on length scales much larger than the lattice spacing $\mathfrak{a}$ blurs our vision to the point where the crystal appears as an elastic continuum. Viewed without an atomic microscope, the relative displacements

$\delta\eta_n$, $n = 1, \cdots, N$, become a field $\varphi(x,t)$ where x can be any real-valued number provided N is sufficiently large.[4]

The mathematics that justifies this blurring or coarse graining is that, for functions f that vary *slowly on the lattice scale*,

$$\sum_n f(n\,\mathfrak{a}) \longrightarrow \int \frac{\mathrm{d}x}{\mathfrak{a}} f(x). \tag{1.35}$$

In particular,

$$f(m\,\mathfrak{a}) = \sum_n \delta_{m,n} f(n\,\mathfrak{a}) = \sum_n \mathfrak{a}\,\frac{\delta_{m,n}}{\mathfrak{a}} f(n\,\mathfrak{a}) \longrightarrow f(x) = \int \mathrm{d}y\, \delta(x-y) f(y), \tag{1.36}$$

justifies the identification

$$\frac{\delta_{m,n}}{\mathfrak{a}} \longrightarrow \delta(x-y). \tag{1.37}$$

Equation (1.37) tells us that the divergent quantity $\delta(x=0)$ in position space should be thought of as the inverse, $1/\mathfrak{a}$, of the lattice spacing $\mathfrak{a}$. In turn, the number $1/\mathfrak{a}$ can be interpreted as the spacing of normal modes in reciprocal space per unit volume $2\pi/N$ in *wave-number space* by the following argument,

$$\frac{1}{\mathfrak{a}} = \frac{k_{l+1}-k_l}{\mathfrak{a}} \times \frac{1}{2\pi/N}, \qquad k_l := \frac{2\pi}{N} l. \tag{1.38}$$

How does one go from a discrete Fourier sum to a Fourier integral? Start from an even number N of sites for which

$$\sum_{l=1}^{N} e^{\mathrm{i}k_l(m-n)} = N\delta_{m,n}, \qquad k_l := \frac{2\pi}{N} l. \tag{1.39}$$

Multiply both sides of this equation by the inverse of the system size $L = N\,\mathfrak{a}$,

$$\frac{1}{L}\sum_{l=1}^{N} e^{\mathrm{i}k_l(m-n)} = \frac{\delta_{m,n}}{\mathfrak{a}}. \tag{1.40}$$

Since the right-hand side should be identified with $\delta(x-y)$ in the thermodynamic limit $N \to \infty$, the left-hand side should be identified with

$$\frac{1}{L}\sum_{l=1}^{N} e^{\mathrm{i}\frac{k_l}{\mathfrak{a}}(m-n)\,\mathfrak{a}} \longrightarrow \int_0^{2\pi/\mathfrak{a}} \frac{\mathrm{d}k}{2\pi} e^{\mathrm{i}k(x-y)} \approx \int_{-\infty}^{+\infty} \frac{\mathrm{d}k}{2\pi} e^{\mathrm{i}k(x-y)}, \tag{1.41}$$

whereby

$$\frac{k_l}{\mathfrak{a}} \longrightarrow k, \qquad (m-n)\,\mathfrak{a} \longrightarrow x-y. \tag{1.42}$$

---

[4] In mathematics, a (real-valued scalar) field φ is a mapping $\varphi : \mathbb{R}^{d+1} \to \mathbb{R}, (\boldsymbol{r},t) \mapsto \varphi(\boldsymbol{r},t)$. In physics, a field is often abbreviated by the value $\varphi(\boldsymbol{r},t)$ it takes at the point $(\boldsymbol{r},t)$ in $(d+1)$-dimensional (position) space and time.

To see this, recall first that the periodic boundary conditions tell us that $l = 1, \cdots, N$ could have equally well be chosen to run between $-N/2+1$ and $+N/2$. Hence, it is permissible to adopt the more symmetrical rule

$$\frac{1}{L}\sum_{l=1}^{N} \tilde{f}(k_l) \longrightarrow \int\limits_{-\pi/\mathfrak{a}}^{+\pi/\mathfrak{a}} \frac{\mathrm{d}k}{2\pi}\,\tilde{f}(k), \tag{1.43}$$

to convert a finite summation over wave numbers into an integral over the *first Brillouin zone* (reciprocal space) $]-\pi/\mathfrak{a}, +\pi/\mathfrak{a}]$ as the thermodynamic limit $N = L/\mathfrak{a} \to \infty$ is taken. Now, if $f(x)$ is a slowly varying function on the lattice scale $\mathfrak{a}$, its Fourier transform $\tilde{f}(k)$ will be essentially vanishing for $|k| \gg 1/\mathfrak{a}$. In this case, the limits $\pm\pi/\mathfrak{a}$ can safely be replaced by the limits $\pm\infty$ on the right-hand side of Eq. (1.43). We then arrive at the desired integral representation of the delta function in position space,

$$\delta(x-y) = \int\limits_{-\infty}^{+\infty} \frac{\mathrm{d}k}{2\pi}\, e^{\mathrm{i}k(x-y)}. \tag{1.44}$$

Observe that factors of 2π appear in an asymmetrical way in integrals over position (x) and reciprocal (k) spaces. Although this is purely a matter of convention when defining the Fourier transform, there is a physical reasoning behind this choice. Indeed, Eq. (1.43) implies that $\mathrm{d}k/(2\pi)$ has the physical meaning of the number of normal modes in reciprocal space with wave number between k and $k+\mathrm{d}k$ per unit volume L in *position space.* Correspondingly, the divergent quantity $2\pi\,\delta(k=0)$ in reciprocal space has the physical meaning of being the divergent volume $L \to \infty$ of the system as is inferred from

$$\delta(x) = \int\limits_{-\infty}^{+\infty} \frac{\mathrm{d}k}{2\pi}\, e^{\mathrm{i}kx} \Longleftrightarrow 2\pi\delta(k) = \int\limits_{-\infty}^{+\infty} \mathrm{d}x\, e^{-\mathrm{i}kx}. \tag{1.45}$$

1.3 Quantum one-dimensional crystal

1.3.1 *Reminiscences about the harmonic oscillator*

We now turn to the task of giving a quantum-mechanical description for a *non-dissipative* one-dimensional harmonic crystal. One possible route consists in the construction of a Hilbert space with the operators acting on it and whose expectation values can be related to measurable properties of the crystal.[5] In this setting, the time evolution of physical quantities can be calculated either in the Schrödinger or in the Heisenberg picture. We will begin by reviewing these two approaches in

[5] Another route to quantization is by means of the path-integral representation of quantum mechanics as is shown in appendix A.

the context of a single harmonic oscillator. The extension to the harmonic crystal will then follow in a very natural way.

The classical Hamiltonian that describes a single particle of unit mass $m = 1$ confined to a quadratic well with curvature ω^2 is

$$\mathfrak{H} := \frac{1}{2}\left(p^2 + \omega^2 x^2\right). \tag{1.46}$$

Hamilton-Jacobi equations of motion are

$$\begin{aligned}\frac{\mathrm{d}x(t)}{\mathrm{d}t} &= \{x, \mathfrak{H}\} = +\frac{\partial \mathfrak{H}}{\partial p} = p(t),\\ \frac{\mathrm{d}p(t)}{\mathrm{d}t} &= \{p, \mathfrak{H}\} = -\frac{\partial \mathfrak{H}}{\partial x} = -\omega^2 x(t).\end{aligned} \tag{1.47}$$

Solutions to these classical equations of motion are

$$\begin{aligned}x(t) &= A\cos(\omega t) + B\sin(\omega t),\\ p(t) &= \omega\left[-A\sin(\omega t) + B\cos(\omega t)\right].\end{aligned} \tag{1.48}$$

The energy E of the particle is a constant of the motion that depends on the choice of initial conditions through the two real-valued constants A and B,

$$E = \frac{1}{2}\left(A^2 + B^2\right)\omega^2. \tag{1.49}$$

In the Schrödinger picture of quantum mechanics, the position x of the particle and its canonical conjugate p become operators $\hat{x}$ and $\hat{p}$ that (i) act on the Hilbert space of twice-differentiable and square-integrable functions $\Psi : \mathbb{R} \to \mathbb{C}$ and (ii) obey the *canonical commutation relation*

$$[\hat{x}, \hat{p}] := \hat{x}\,\hat{p} - \hat{p}\,\hat{x} = \mathrm{i}\hbar. \tag{1.50}$$

The time evolution (or dynamics in short) of the system is encoded by Schrödinger equation

$$\mathrm{i}\hbar\,\partial_t \Psi(x,t) = \hat{H}\,\Psi(x,t), \tag{1.51a}$$

where the quantum Hamiltonian $\hat{H}$ is given by

$$\hat{H} = \frac{1}{2}\left(\hat{p}^2 + \omega^2 \hat{x}^2\right). \tag{1.51b}$$

The time evolution of the wave function $\Psi(x,t)$ is unique once initial conditions $\Psi(x, t = 0)$ are given.

Solving the time-independent eigenvalue problem

$$\hat{H}\,\psi_n(x) = \varepsilon_n \psi_n(x) \tag{1.52a}$$

is tantamount to solving the time-dependent Schrödinger equation through the Ansatz

$$\Psi(x,t) = \sum_n c_n\,\psi_n(x)\,e^{-\mathrm{i}\varepsilon_n t/\hbar}. \tag{1.52b}$$

The expansion coefficients $c_n \in \mathbb{C}$ are time independent and uniquely determined by the initial condition, say $\Psi(x, t=0)$.

As is well known, the energy eigenvalues ε_n are given by

$$\varepsilon_n = \left(n + \frac{1}{2}\right) \hbar\omega, \qquad n = 0, 1, 2, \cdots . \tag{1.53}$$

The energy eigenfunctions $\psi_n(x)$ are Hermite polynomials multiplying a Gaussian,

$$\begin{aligned}
\psi_0(x) &= \left(\frac{\omega}{\pi\hbar}\right)^{1/4} e^{-\frac{1}{2}\frac{\omega}{\hbar}x^2}, \\
\psi_1(x) &= \left[\frac{4}{\pi}\left(\frac{\omega}{\hbar}\right)^3\right]^{1/4} x e^{-\frac{1}{2}\frac{\omega}{\hbar}x^2}, \\
\psi_2(x) &= \left(\frac{\omega}{4\pi\hbar}\right)^{1/4} \left(2\frac{\omega}{\hbar}x^2 - 1\right) e^{-\frac{1}{2}\frac{\omega}{\hbar}x^2}, \\
&\vdots \\
\psi_n(x) &= \left[\frac{1}{2^n n!}\left(\frac{\hbar}{\omega}\right)^n\right]^{1/2} \left(\frac{\omega}{\pi\hbar}\right)^{1/4} \left(\frac{\omega}{\hbar}x - \frac{\mathrm{d}}{\mathrm{d}x}\right)^n e^{-\frac{1}{2}\frac{\omega}{\hbar}x^2}.
\end{aligned} \tag{1.54}$$

The Heisenberg picture of quantum mechanics is better suited than the Schrödinger picture to a generalization to *quantum field theory.* In the Heisenberg picture, and contrary to the Schrödinger picture, operators are explicitly time dependent. For any operator $\hat{O}$, the solution to the equation of motion[6]

$$\mathrm{i}\hbar \frac{\mathrm{d}\hat{O}(t)}{\mathrm{d}t} = [\hat{O}(t), \hat{H}] \tag{1.55a}$$

that replaces Schrödinger equation is

$$\hat{O}(t) = e^{+\mathrm{i}\frac{\hat{H}t}{\hbar}}\, \hat{O}(t=0)\, e^{-\mathrm{i}\frac{\hat{H}t}{\hbar}}. \tag{1.55b}$$

By definition, the algebra obeyed by operators in the Schrödinger picture holds true in the Heisenberg picture provided operators are taken at *equal time*. For example,

$$\hat{x}(t) := e^{+\mathrm{i}\frac{\hat{H}t}{\hbar}}\, \hat{x}(t=0)\, e^{-\mathrm{i}\frac{\hat{H}t}{\hbar}}, \qquad \hat{p}(t) := e^{+\mathrm{i}\frac{\hat{H}t}{\hbar}}\, \hat{p}(t=0)\, e^{-\mathrm{i}\frac{\hat{H}t}{\hbar}}, \tag{1.56}$$

obey by construction the equal-time commutator

$$[\hat{x}(t), \hat{p}(t)] = \mathrm{i}\hbar, \qquad \forall t \in \mathbb{R}. \tag{1.57}$$

Finding the commutator of $\hat{x}(t)$ and $\hat{p}(t')$ at unequal times $t \neq t'$ requires solving the dynamics of the system, i.e., Eq. (1.55a) with $\hat{O}$ substituted for $\hat{x}$ and $\hat{p}$, respectively,

$$\frac{\mathrm{d}\hat{x}(t)}{\mathrm{d}t} = +\hat{p}(t), \qquad \frac{\mathrm{d}\hat{p}(t)}{\mathrm{d}t} = -\omega^2 \hat{x}(t). \tag{1.58}$$

In other words, the Heisenberg operators $\hat{x}(t)$ and $\hat{p}(t)$ satisfy the same equations of motion as the classical variables they replace,

$$\frac{\mathrm{d}^2\hat{x}(t)}{\mathrm{d}t^2} + \omega^2 \hat{x}(t) = 0, \qquad \hat{p}(t) = \frac{\mathrm{d}\hat{x}(t)}{\mathrm{d}t}. \tag{1.59}$$

[6] The assumption that the system is non-dissipative has been used here in that $\hat{H}$ does not depend explicitly on time, $\partial_t \hat{H} = 0$.

The solution (1.48) can thus be borrowed with the caveat that A and B should be replaced by time-independent operators $\hat{A}$ and $\hat{B}$.

At this stage, it is more productive to depart from following a strategy dictated by the real-valued classical solution (1.48). The key observation is that the quantum Hamiltonian for the harmonic oscillator takes the quadratic form[7]

$$\hat{H} = \hbar\omega \left[\hat{a}^{\dagger}(t)\,\hat{a}(t) + \frac{1}{2}\right], \tag{1.60a}$$

if the pair of canonically conjugate Hermitian operators $\hat{x}(t)$ and $\hat{p}(t)$ is traded for the pair $\hat{a}^{\dagger}(t)$ and $\hat{a}(t)$ of operators defined by

$$\hat{x}(t) =: \sqrt{\frac{\hbar}{2\omega}}\left[\hat{a}(t) + \hat{a}^{\dagger}(t)\right], \qquad \hat{p}(t) =: \sqrt{\frac{\hbar}{2\omega}}\left[-\mathrm{i}\omega\hat{a}(t) + \mathrm{i}\omega\hat{a}^{\dagger}(t)\right]. \tag{1.60b}$$

Once the equal-time commutator $[\hat{a}(t), \hat{a}^{\dagger}(t)]$ is known, the Heisenberg equations of motion are easily derived from

$$\mathrm{i}\hbar\frac{\mathrm{d}\hat{a}^{\dagger}(t)}{\mathrm{d}t} = [\hat{a}^{\dagger}(t), \hat{H}], \qquad \mathrm{i}\hbar\frac{\mathrm{d}\hat{a}(t)}{\mathrm{d}t} = [\hat{a}(t), \hat{H}]. \tag{1.61}$$

With the help of

$$\hat{a}^{\dagger}(t) = \sqrt{\frac{\omega}{2\hbar}}\left[\hat{x}(t) - \mathrm{i}\frac{\hat{p}(t)}{\omega}\right], \qquad \hat{a}(t) = \sqrt{\frac{\omega}{2\hbar}}\left[\hat{x}(t) + \mathrm{i}\frac{\hat{p}(t)}{\omega}\right], \tag{1.62a}$$

one verifies that

$$\begin{aligned} &[\hat{x}(t), \hat{p}(t)] = \mathrm{i}\hbar, \qquad [\hat{x}(t), \hat{x}(t)] = [\hat{p}(t), \hat{p}(t)] = 0 \Longleftrightarrow \\ &\qquad [\hat{a}(t), \hat{a}^{\dagger}(t)] = 1, \qquad [\hat{a}(t), \hat{a}(t)] = [\hat{a}^{\dagger}(t), \hat{a}^{\dagger}(t)] = 0. \end{aligned} \tag{1.62b}$$

The change of Hermitian operator-valued variables to non-Hermitian operator-valued variables is advantageous in that the equations of motion for $\hat{a}^{\dagger}(t)$ and $\hat{a}(t)$ decouple according to

$$\begin{aligned} \frac{\mathrm{d}\hat{a}^{\dagger}(t)}{\mathrm{d}t} &= +\mathrm{i}\omega\,\hat{a}^{\dagger}(t), \qquad \hat{a}^{\dagger}(t) = \hat{a}^{\dagger}(t=0)\; e^{+\mathrm{i}\omega t}, \\ \frac{\mathrm{d}\hat{a}(t)}{\mathrm{d}t} &= -\mathrm{i}\omega\,\hat{a}(t), \qquad \hat{a}(t) = \hat{a}(t=0)\; e^{-\mathrm{i}\omega t}, \end{aligned} \tag{1.63}$$

respectively. Below, we will write $\hat{a}$ for $\hat{a}(t=0)$ and similarly for $\hat{a}^{\dagger}$. The time evolution of $\hat{x}(t)$, $\hat{p}(t)$, and $\hat{H}$ is now explicitly given by

$$\hat{x}(t) = \sqrt{\frac{\hbar}{2\omega}}\left(\hat{a}\,e^{-\mathrm{i}\omega t} + \hat{a}^{\dagger}\,e^{+\mathrm{i}\omega t}\right), \tag{1.64a}$$

$$\hat{p}(t) = -\mathrm{i}\sqrt{\frac{\hbar\omega}{2}}\left(\hat{a}\,e^{-\mathrm{i}\omega t} - \hat{a}^{\dagger}\,e^{+\mathrm{i}\omega t}\right), \tag{1.64b}$$

$$\hat{H} = \hbar\omega\left(\hat{a}^{\dagger}\hat{a} + \frac{1}{2}\right). \tag{1.64c}$$

As must be by the absence of dissipation, $\hat{H}$ is explicitly time independent, $\partial_t\hat{H} = 0$.

[7] We are anticipating that $\hat{H}$ does not depend explicitly on time.

The Hilbert space can now be constructed explicitly with purely algebraic methods. The Hilbert space is defined by all possible linear combinations of the eigenstates

$$|n\rangle := \frac{(\hat{a}^\dagger)^n}{\sqrt{n!}}|0\rangle, \qquad \hat{H}|n\rangle = \varepsilon_n|n\rangle, \qquad n = 0,1,2,\cdots. \tag{1.65}$$

Here, the ground state or vacuum $|0\rangle$ is defined by the condition

$$\hat{a}|0\rangle = 0. \tag{1.66}$$

One verifies that $\psi_0(x)$ in Eq. (1.54) *uniquely (up to a phase)* satisfies Eq. (1.66) by using the position-space representation of the operator $\hat{a}$.

1.3.2 *Discrete limit*

In the spirit of the Heisenberg picture for the harmonic oscillator and guided by the Fourier expansions in Eqs. (1.16) and (1.17), we begin by *defining* the operators

$$\begin{aligned}
\hat{\eta}_n(t) :=& \frac{1}{\sqrt{N}}\sum_{l=1}^{N}\sqrt{\frac{\hbar}{2\omega_l}}\left[\hat{a}_l\, e^{+\mathrm{i}(k_l n-\omega_l t)} + \hat{a}_l^\dagger\, e^{-\mathrm{i}(k_l n-\omega_l t)}\right], \quad n=1,\cdots,N,\\
\hat{\pi}_n(t) :=& -\mathrm{i}\frac{1}{\sqrt{N}}\sum_{l=1}^{N}\sqrt{\frac{\hbar\omega_l}{2}}\left[\hat{a}_l\, e^{+\mathrm{i}(k_l n-\omega_l t)} - \hat{a}_l^\dagger\, e^{-\mathrm{i}(k_l n-\omega_l t)}\right], \quad n=1,\cdots,N,
\end{aligned} \tag{1.67a}$$

where the frequency ω_l and the integer label l are related by Eq. (1.15), i.e., (remember that we have chosen units in which the mass is given by $m=1$)

$$\omega_l = \sqrt{2\kappa\,(1-\cos k_l)+\Omega^2}, \qquad k_l := \frac{2\pi}{N}l, \qquad l=1,\cdots,N, \tag{1.67b}$$

and the operator-valued expansion coefficients $\hat{a}_l^\dagger$ and $\hat{a}_l$ obey the harmonic oscillator algebra

$$[\hat{a}_l,\hat{a}_{l'}^\dagger] = \delta_{l,l'}, \qquad [\hat{a}_l,\hat{a}_{l'}] = [\hat{a}_l^\dagger,\hat{a}_{l'}^\dagger] = 0, \qquad l,l'=1,\cdots,N. \tag{1.67c}$$

The normalization factor $1/\sqrt{N}$ is needed to cancel the factor of N present in the Fourier series

$$\sum_{l=1}^{N} e^{\mathrm{i}k_l(m-n)} = N\delta_{m,n}, \tag{1.68}$$

which shows up when one verifies that the equal-time commutators

$$[\hat{\eta}_m(t),\hat{\pi}_n(t)] = \mathrm{i}\hbar\,\delta_{m,n}, \quad [\hat{\eta}_m(t),\hat{\eta}_n(t)] = [\hat{\pi}_m(t),\hat{\pi}_n(t)] = 0, \quad m,n=1,\cdots,N, \tag{1.69}$$

hold for all times. We are now ready to *define in a consistent way* the Hamiltonian $\hat{H}$ for the quantum one-dimensional harmonic crystal [compare with Eq. (1.18)]

$$\hat{H} := \sum_{n=1}^{N}\frac{1}{2}\left\{[\hat{\pi}_n(t)]^2 + \kappa\left[\hat{\eta}_{n+1}(t)-\hat{\eta}_n(t)\right]^2 + \Omega^2\left[\hat{\eta}_n(t)\right]^2\right\}. \tag{1.70}$$

With the help of the algebra (1.67c), one verifies that $\hat{H}$ is explicitly time independent and given by

$$\hat{H} = \sum_{l=1}^{N} \hbar\omega_l \left(\hat{a}_l^\dagger \, \hat{a}_l + \frac{1}{2} \right). \tag{1.71}$$

The next task is to construct the Hilbert space for the one-dimensional quantum crystal by algebraic methods. Assume that there exists a *unique (up to a phase)* normalized state $|0\rangle$, the ground state or vacuum, defined by

$$\langle 0|0\rangle = 1, \qquad \hat{a}_l|0\rangle = 0, \qquad l = 1, \cdots, N. \tag{1.72}$$

If so, the state

$$|n_1, n_2, \cdots, n_N\rangle := \prod_{l=1}^{N} \frac{1}{\sqrt{n_l!}} \left(\hat{a}_l^\dagger\right)^{n_l} |0\rangle, \qquad n_1, n_2, \cdots, n_N = 0, 1, 2, \cdots, \tag{1.73}$$

is normalized to one and is an eigenstate of $\hat{H}$ with the energy eigenvalue

$$\varepsilon_{n_1, \cdots, n_N} := \sum_{l=1}^{N} \hbar\omega_l \left(n_l + \frac{1}{2} \right). \tag{1.74}$$

The ground-state energy is of order N and given by

$$\varepsilon_{0, \cdots, 0} := \frac{1}{2} \sum_{l=1}^{N} \hbar\omega_l. \tag{1.75}$$

Excited states have at least one $n_l > 0$. They are called *phonons*. The eigenstate $|n_1, n_2, \cdots, n_N\rangle$ is said to have n_1 phonons in the first mode, n_2 phonons in the second mode, and so on. Phonons can be thought of as identical elementary particles since they possess a definite energy and momentum. Because the phonon occupation number

$$n_l = \langle n_1, \cdots, n_l, \cdots, n_N | \hat{a}_l^\dagger \, \hat{a}_l | n_1, \cdots, n_l, \cdots, n_N \rangle \tag{1.76}$$

is an arbitrary positive integer, phonons obey *Bose-Einstein statistics.* Upon switching on a suitable interaction [say by including cubic and quartic terms in the expansion (1.4)], phonons scatter off one other just as other "-ons" (mesons, photons, gluons, and so on) known to physics do. Although we are en route towards constructing the quantum field $\hat{\eta}(x, t)$ out of $\hat{\eta}_n(t)$, we have encountered particles. The duality between quantum fields and particles is the essence of quantum field theory.

The vector space spanned by the states labeled by the phonon occupation numbers $(n_1, \cdots, n_N) \in \{0, 1, 2, \cdots\}^N$ in Eq. (1.73) is the Hilbert space of the one-dimensional quantum crystal. The mathematical structure of this Hilbert space is a symmetric tensor product of N copies of the Hilbert space for the harmonic oscillator. In physics, this symmetric tensor product is called a *Fock space* when the emphasis is to think of the phonon as an "*elementary particle*".

1.3.3 *Thermodynamic limit*

Taking the thermodynamic limit $N \to \infty$ is a direct application to section 1.3.2 of the rules established in the context of the classical description of section 1.2.2. Hence, with the identifications[8]

$$
\begin{aligned}
&\sum_{n=1}^{N} \mathfrak{a} \longrightarrow \int \mathrm{d}x,\\
&\frac{1}{N\mathfrak{a}}\sum_{l=1}^{N} \longrightarrow \int \frac{\mathrm{d}k}{2\pi},\\
&\omega_l \longrightarrow \omega(k) = \sqrt{2\frac{v^2}{\mathfrak{a}^2}[1-\cos(k\,\mathfrak{a})]+\Omega^2} \approx \sqrt{v^2k^2+\Omega^2}, \text{ if } |k\,\mathfrak{a}| \ll 1, \qquad (1.77)\\
&k_l n \longrightarrow kx,\\
&\hat{a}_l \longrightarrow \frac{1}{\sqrt{N\,\mathfrak{a}}}\,\hat{a}(k),\\
&\hat{\eta}_n(t) \longrightarrow \sqrt{\mathfrak{a}}\,\hat{\eta}(x,t),\\
&\hat{\pi}_n(t) \longrightarrow \sqrt{\mathfrak{a}}\,\hat{\pi}(x,t),
\end{aligned}
$$

the canonically conjugate pairs of operators $\hat{\eta}_n(t)$ and $\hat{\pi}_n(t)$ are replaced by the *quantum fields*

$$
\begin{aligned}
\hat{\eta}(x,t) &:= \int \frac{\mathrm{d}k}{2\pi}\sqrt{\frac{\hbar}{2\omega(k)}}\left\{\hat{a}(k)\,e^{+\mathrm{i}[kx-\omega(k)t]} + \hat{a}^\dagger(k)\,e^{-\mathrm{i}[kx-\omega(k)t]}\right\},\\
\hat{\pi}(x,t) &:= -\mathrm{i}\int \frac{\mathrm{d}k}{2\pi}\sqrt{\frac{\hbar\omega(k)}{2}}\left\{\hat{a}(k)\,e^{+\mathrm{i}[kx-\omega(k)t]} - \hat{a}^\dagger(k)\,e^{-\mathrm{i}[kx-\omega(k)t]}\right\},
\end{aligned}
\qquad (1.78)
$$

respectively.[9] Their equal-time commutators follow from the harmonic oscillator algebra

$$
[\hat{a}(k),\hat{a}^\dagger(k')] = 2\pi\delta(k-k'), \qquad [\hat{a}(k),\hat{a}(k')] = [\hat{a}^\dagger(k),\hat{a}^\dagger(k')] = 0. \qquad (1.79)
$$

They are

$$
[\hat{\eta}(x,t),\hat{\pi}(y,t)] = \mathrm{i}\hbar\,\delta(x-y), \qquad [\hat{\eta}(x,t),\hat{\eta}(y,t)] = [\hat{\pi}(x,t),\hat{\pi}(y,t)] = 0. \qquad (1.80)
$$

The Hamiltonian is

$$
\begin{aligned}
\hat{H} &= \int \mathrm{d}x\,\frac{1}{2}\left\{[\hat{\pi}(x,t)]^2 + v^2\,[\partial_x\hat{\eta}(x,t)]^2 + \Omega^2\,[\hat{\eta}(x,t)]^2\right\}\\
&= \int \frac{\mathrm{d}k}{2\pi}\frac{1}{2}\hbar\omega(k)\left[\hat{a}^\dagger(k)\hat{a}(k) + \hat{a}(k)\hat{a}^\dagger(k)\right].
\end{aligned}
\qquad (1.81)
$$

---

[8] Limits of integrations in position and reciprocal spaces are left unspecified at this stage as we want to remain free to choose how the thermodynamic limit $N \to \infty$ is to be taken. For example, we could keep $\mathfrak{a}$ finite in which case the thermodynamic limit $N \to \infty$ implies $L \to \infty$. Alternatively, we could keep L finite in which case the thermodynamic limit $N \to \infty$ implies $\mathfrak{a} \to 0$.

[9] The substitution rules $\hat{a}_l \longrightarrow \frac{1}{\sqrt{N\,\mathfrak{a}}}\hat{a}(k)$, $\hat{\eta}_n(t) \longrightarrow \sqrt{\mathfrak{a}}\,\hat{\eta}(x,t)$, and $\hat{\pi}_n(t) \longrightarrow \sqrt{\mathfrak{a}}\,\hat{\pi}(x,t)$, are needed to cancel the volume factor $N\,\mathfrak{a}$ in $\sum_{l=1}^{N} \longrightarrow N\,\mathfrak{a}\int\frac{\mathrm{d}k}{2\pi}$.

The excitation spectrum is obtained by making use of the commutator between $\hat{a}^\dagger(k)$ and $\hat{a}(k)$. It is given by

$$\hat{H} - E_0 := \int \frac{\mathrm{d}k}{2\pi} \hbar\,\omega(k)\,\hat{a}^\dagger(k)\,\hat{a}(k), \tag{1.82}$$

and is observed to vanish for the vacuum $|0\rangle$. The operation of subtracting from the Hamiltonian the ground state energy E_0 is called *normal ordering.* It amounts to placing all *annihilation* operators $\hat{a}(k)$ to the right of the *creation* operators $\hat{a}^\dagger(k)$. The ground state energy

$$\begin{aligned} E_0 &:= \langle 0|\hat{H}|0\rangle \\ &= \int \frac{\mathrm{d}k}{2\pi}\,\frac{1}{2}\hbar\omega(k) \times 2\pi\delta(k=0) \\ &= (\text{Volume in position space}) \times \int \frac{\mathrm{d}k}{2\pi}\frac{1}{2}\hbar\,\omega(k) \\ &= \sum_{\text{modes}} \frac{1}{2}\hbar\,\omega_{\text{modes}} \end{aligned} \tag{1.83}$$

can be ill-defined for two distinct reasons. First, if $N \to \infty$ with $\mathfrak{a}$ held fixed, there exists an upper cut-off to the integral over reciprocal space at the Brillouin zone boundaries $\pm\pi/\mathfrak{a}$ and E_0 is only *infrared divergent* due to the fact that $2\pi\delta(k=0)$ is the diverging volume $L = N\,\mathfrak{a}$ in position space. Second, even if $L = N\,\mathfrak{a}$ is kept finite while both the infrared $N \to \infty$ and *ultraviolet* $\mathfrak{a} \to 0$ limits are taken, the absence of an upper cut-off in the k integral can cause the *zero-point energy density* E_0/L to diverge as well. Divergences of E_0 or E_0/L are only of practical relevance if one can control experimentally $\omega(k)$ or the density of states $\sum_{\text{modes}}$ and thereby measure *changes* in E_0 or E_0/L. For example, this can be achieved in a resonant cavity whose size is variable. If so, changes of E_0 with the cavity size can be measured. These changes in the zero point energy are known as the *Casimir energy.* Sensitivity to E_0 with measurable consequences also occurs when, upon tuning of some internal parameters entering the microscopic Hamiltonian, the vacuum state $|0\rangle$ becomes unstable, i.e., is not the true ground state anymore. The system then undergoes a *quantum phase transition.* Finally, divergences of E_0/L matter greatly if the energy-momentum tensors of "matter fields" are dynamical variables as is the case in cosmological models.

1.4 Higher-dimensional generalizations

Generalizations to higher dimensions are straightforward. The coordinates $x \in \mathbb{R}^1$ and $k \in \mathbb{R}^1$ in position and reciprocal one-dimensional spaces need only be replaced by the vectors $\boldsymbol{r} \in \mathbb{R}^d$ and $\boldsymbol{k} \in \mathbb{R}^d$, in position and reciprocal d-dimensional spaces, respectively.

1.5 Problems

1.5.1 *Absence of crystalline order in one and two dimensions*

Introduction

We are going to prove the Mermin-Wagner theorem for the case of crystalline order in two (and one) dimensions of position space [23]. The Mermin-Wagner theorem states that classical particles in a box, i.e., particles that are subject to hard-wall boundary conditions, cannot exhibit crystalline order in one and two dimensions, provided that the pair potential $\Phi(\boldsymbol{r})$ through which they interact satisfies certain conditions [see Eq. (1.109)].

Before we start with the derivation, let us set up some notation. Given the pair potential $\Phi(\boldsymbol{r})$, the internal energy of a configuration of N particles with coordinates $\boldsymbol{r}_1, \cdots, \boldsymbol{r}_N$ in d-dimensional position space is given by

$$U(\boldsymbol{r}_1, \cdots, \boldsymbol{r}_N) = \frac{1}{2} \sum_{i \neq j}^{N} \Phi(\boldsymbol{r}_i - \boldsymbol{r}_j). \tag{1.84}$$

Using this, we can define the (classical) ensemble average of a real-valued function f of the coordinates $\boldsymbol{r}_1, \cdots, \boldsymbol{r}_N$ by

$$\langle f \rangle := \frac{1}{Z} \int\limits_{\mathrm{B}} \left(\prod_{i=1}^{N} \mathrm{d}^d \boldsymbol{r}_i \right) e^{-\beta U(\boldsymbol{r}_1, \cdots, \boldsymbol{r}_N)} f(\boldsymbol{r}_1, \cdots, \boldsymbol{r}_N), \tag{1.85a}$$

and

$$Z := \int\limits_{\mathrm{B}} \left(\prod_{i=1}^{N} \mathrm{d}^d \boldsymbol{r}_i \right) e^{-\beta U(\boldsymbol{r}_1, \cdots, \boldsymbol{r}_N)}. \tag{1.85b}$$

Here, β is the inverse temperature after the Boltzmann constant k_{B} has been set to unity and B denotes the box over which the integration is taken.

Step 1: Proof of Bogoliubov inequality

The proof of the Mermin-Wagner theorem will be crucially based on an inequality due to Bogoliubov, which for our purposes can be formulated as

$$\left\langle \left| \sum_{i=1}^{N} \psi(\boldsymbol{r}_i) \right|^2 \right\rangle \geq \frac{\left| \sum\limits_{i=1}^{N} \langle \varphi(\boldsymbol{r}_i) \boldsymbol{\nabla} \psi(\boldsymbol{r}_i) \rangle \right|^2}{\left\langle \frac{\beta}{2} \sum\limits_{i,j=1}^{N} \Delta\Phi(\boldsymbol{r}_i - \boldsymbol{r}_j) \left| \varphi(\boldsymbol{r}_i) - \varphi(\boldsymbol{r}_j) \right|^2 + \sum\limits_{i=1}^{N} \left| \boldsymbol{\nabla} \varphi(\boldsymbol{r}_i) \right|^2 \right\rangle}, \tag{1.86}$$

for a real-valued function φ that is continuous and differentiable and vanishes on the boundary $\partial \mathrm{B}$ of B, while ψ is complex valued and sufficiently smooth. Our first task is to prove Eq. (1.86).

Exercise 1.1: Convince yourself that the bilinear map

$$\langle\cdot,\cdot\rangle : \mathfrak{L}\times\mathfrak{L} \longrightarrow \mathbb{R}, \qquad (\varphi,\psi) \longmapsto \langle\varphi,\psi\rangle := \langle\varphi^*\,\psi\rangle\,, \tag{1.87}$$

for two complex-valued functions φ and ψ belonging to the set $\mathfrak{L}$ of continuous differentiable functions from B to $\mathbb{R}$ with the standard definition of a product of two functions, is a scalar product. We then have the Schwarz inequality

$$\left\langle|f_1|^2\right\rangle\left\langle|f_2|^2\right\rangle \geq |\langle f_1\,f_2\rangle|^2\,, \tag{1.88}$$

for any pair of functions f_1 and f_2 from $\mathfrak{L}$ at our disposal.

Exercise 1.2: Use the Schwarz inequality (1.88) with the choice

$$f_1(\boldsymbol{r}_1,\cdots,\boldsymbol{r}_N) := \sum_{i=1}^{N}\psi(\boldsymbol{r}_i), \tag{1.89a}$$

$$f_2(\boldsymbol{r}_1,\cdots,\boldsymbol{r}_N) := -\frac{1}{\beta}\,e^{\beta\,U(\boldsymbol{r}_1,\cdots,\boldsymbol{r}_N)}\sum_{i=1}^{N}\boldsymbol{\nabla}_i\left[\varphi(\boldsymbol{r}_i)\,e^{-\beta\,U(\boldsymbol{r}_1,\cdots,\boldsymbol{r}_N)}\right], \tag{1.89b}$$

to prove Eq. (1.86). *Hint:* Use partial integration.

Step 2: Densities on the lattice

We now want to use the Bogoliubov inequality (1.86) to probe the tendency towards crystalline order (we specialize to $d = 2$ for simplicity, but without loss of generality). Suppose that the crystalline order has the Bravais lattice vectors $\boldsymbol{a}_1$ and $\boldsymbol{a}_2$ and consists of $N_1 \times N_2$ sites so that

$$\mathrm{B} = \left\{\boldsymbol{r}\in\mathbb{R}^d\,|\,\boldsymbol{r} = x_1\,\boldsymbol{a}_1\,N_1 + x_2\,\boldsymbol{a}_2\,N_2, \qquad 0 \leq x_1, x_2 < 1\right\}. \tag{1.90}$$

The reciprocal lattice vectors $\boldsymbol{K}$ are given by

$$\boldsymbol{K} := n_1\,\boldsymbol{b}_1 + n_2\,\boldsymbol{b}_2, \qquad n_1,n_2\in\mathbb{Z}, \qquad \boldsymbol{b}_i\cdot\boldsymbol{a}_j = 2\pi\,\delta_{ij}, \qquad i,j = 1,2, \tag{1.91}$$

and a general wave vector $\boldsymbol{k}$ is given by

$$\boldsymbol{k} := \frac{n_1}{N_1}\,\boldsymbol{b}_1 + \frac{n_2}{N_2}\,\boldsymbol{b}_2, \qquad n_1,n_2\in\mathbb{Z}. \tag{1.92}$$

To probe whether particles form a crystal, we have to compute their density at the reciprocal lattice vectors. In position space, the density of a configuration of N particles is

$$\rho(\boldsymbol{r}) := \sum_{i=1}^{N}\delta(\boldsymbol{r}-\boldsymbol{r}_i). \tag{1.93}$$

Its Fourier component at momentum $\boldsymbol{k}$ is given by

$$\rho_{\boldsymbol{k}} := \int\limits_{\mathrm{B}} \mathrm{d}^2\boldsymbol{r}\,e^{-\mathrm{i}\boldsymbol{k}\cdot\boldsymbol{r}}\,\rho(\boldsymbol{r}) = \sum_{i=1}^{N} e^{-\mathrm{i}\boldsymbol{k}\cdot\boldsymbol{r}_i}. \tag{1.94}$$

This allows us to sharpen a criterion for crystalline order as follows. A crystal has formed, if

$$\begin{aligned}\lim_{N_1,N_2\to\infty}\frac{1}{N}\langle\rho_{\boldsymbol{k}}\rangle&=0,\quad\text{if } \boldsymbol{k} \text{ is not a reciprocal lattice vector,}\\ \lim_{N_1,N_2\to\infty}\frac{1}{N}\langle\rho_{\boldsymbol{K}}\rangle&\neq 0,\quad\text{for at least one reciprocal lattice vector } \boldsymbol{K}.\end{aligned}\tag{1.95}$$

The thermodynamic limit is taken in such a way that the filling of the system with particles $n := N/(N_1\,N_2)$ is held constant.

Exercise 2.1: Define the following momenta. Let $\boldsymbol{k}$ be an arbitrary wave vector from the first Brillouin zone as given by Eq. (1.92). Its components with respect to the basis $\boldsymbol{b}_1$ and $\boldsymbol{b}_2$ of the reciprocal lattice are

$$\boldsymbol{k}_i = n_i\,\boldsymbol{b}_i/N_i,\tag{1.96}$$

for $i = 1, 2$. Let $\boldsymbol{K}$ be a reciprocal lattice vector, for which Eq. (1.95) is claimed not to vanish.

(a) Show that the functions defined by

$$\psi(\boldsymbol{r}) := e^{-\mathrm{i}(\boldsymbol{k}+\boldsymbol{K})\cdot\boldsymbol{r}},\qquad \varphi(\boldsymbol{r}) := \sin(\boldsymbol{k}_1\cdot\boldsymbol{r})\,\sin(\boldsymbol{k}_2\cdot\boldsymbol{r}),\tag{1.97}$$

are such that φ vanishes on the boundary $\partial\mathrm{B}$ of the box B.

(b) Show that the Bogoliubov inequality (1.86) with these functions yields

$$\left\langle\rho_{+\boldsymbol{k}+\boldsymbol{K}}\,\rho_{-\boldsymbol{k}-\boldsymbol{K}}\right\rangle \geq \frac{N_{\boldsymbol{k},\boldsymbol{K}}}{D_{\boldsymbol{k},\boldsymbol{K}}},\tag{1.98a}$$

where the numerator is

$$N_{\boldsymbol{k},\boldsymbol{K}} := \frac{(\boldsymbol{k}+\boldsymbol{K})^2}{16\,\beta}\left|\left\langle\left(\rho_{\boldsymbol{K}}+\rho_{\boldsymbol{K}+2\boldsymbol{k}}-\rho_{\boldsymbol{K}+2\boldsymbol{k}_1}-\rho_{\boldsymbol{K}+2\boldsymbol{k}_2}\right)\right\rangle\right|^2,\tag{1.98b}$$

while the denominator is (Δ is Laplace operator in two-dimensional position space)

$$\begin{aligned}D_{\boldsymbol{k},\boldsymbol{K}} :=&\ \frac{1}{2}\sum_{i,j=1}^{N}\left\langle\Delta\Phi(\boldsymbol{r}_i-\boldsymbol{r}_j)\left[\sin(\boldsymbol{k}_1\cdot\boldsymbol{r}_i)\,\sin(\boldsymbol{k}_2\cdot\boldsymbol{r}_i)-\sin(\boldsymbol{k}_1\cdot\boldsymbol{r}_j)\,\sin(\boldsymbol{k}_2\cdot\boldsymbol{r}_j)\right]^2\right\rangle\\ &+\frac{1}{\beta}\sum_{i=1}^{N}\left\langle|\boldsymbol{k}_2\,\sin(\boldsymbol{k}_1\cdot\boldsymbol{r}_i)\cos(\boldsymbol{k}_2\cdot\boldsymbol{r}_i)+(1\leftrightarrow 2)|^2\right\rangle.\end{aligned}\tag{1.98c}$$

Exercise 2.2:

(a) Show that there exists an $A>0$ that depends only on $\boldsymbol{b}_1$ and $\boldsymbol{b}_2$ such that the estimates

$$A\,(\boldsymbol{k}_1+\boldsymbol{k}_2)^2\geq\boldsymbol{k}_1^2+\boldsymbol{k}_2^2,\qquad A\,(\boldsymbol{k}_1+\boldsymbol{k}_2)^2\geq(\nu_1\,\boldsymbol{k}_1+\nu_2\,\boldsymbol{k}_2)^2,\tag{1.99}$$

for any pair ν_1 and ν_2 of real numbers of magnitudes less or equal to one, $|\nu_1|\leq 1$, $|\nu_2|\leq 1$, hold. We are now going to make use of these inequalities.

(b) Establish upper bounds on the trigonometric functions in the denominator (1.98c) to infer that

$$\frac{1}{N}\left\langle \rho_{+\boldsymbol{k}+\boldsymbol{K}}\,\rho_{-\boldsymbol{k}-\boldsymbol{K}}\right\rangle \geq \frac{1}{N^2}\,\frac{\beta\,N_{\boldsymbol{k},\boldsymbol{K}}}{A\,\boldsymbol{k}^2\left(1+\frac{\beta}{N}\sum\limits_{i,j=1}^{N}\left\langle\left|\Delta\Phi(\boldsymbol{r}_i-\boldsymbol{r}_j)\right|(\boldsymbol{r}_i-\boldsymbol{r}_j)^2\right\rangle\right)}. \tag{1.100}$$

It will be the quadratic $\boldsymbol{k}$-dependence in the denominator on which the argument crucially relies (it would break down for a $\boldsymbol{k}$-linear or constant term). From here on, the task is to find suitable estimates for the remaining factors.

To refine the estimate of the denominator on the right-hand side of Eq. (1.100), we have to impose conditions on the asymptotic behavior of the pair potential $\Phi(\boldsymbol{r})$ at small and large $\boldsymbol{r}$. To that end, we consider a family of pair potentials labeled by a real number $\lambda > 0$

$$\Phi_\lambda(\boldsymbol{r}) := \Phi(\boldsymbol{r}) - \lambda\,\boldsymbol{r}^2\,|\Delta\Phi(\boldsymbol{r})|. \tag{1.101}$$

We define the free energy F to be the functional from the space of pair potentials to the real-valued numbers that assigns to any pair potential $\tilde{\Phi}$ the value

$$-\beta\,F[\tilde{\Phi}] := \ln\int\limits_{\mathrm{B}}\left(\prod_{i=1}^{N}\mathrm{d}^d\boldsymbol{r}_i\right)\exp\left(-\frac{\beta}{2}\sum_{i\neq j}^{N}\tilde{\Phi}(\boldsymbol{r}_i-\boldsymbol{r}_j)\right). \tag{1.102}$$

Exercise 2.3: Show that

$$\frac{F[\Phi_0]-F[\Phi_\lambda]}{N\lambda} \geq \frac{1}{2N}\sum_{i,j=1}^{N}\left\langle(\boldsymbol{r}_i-\boldsymbol{r}_j)^2|\Delta\Phi(\boldsymbol{r}_i-\boldsymbol{r}_j)|\right\rangle \geq 0. \tag{1.103}$$

Hint: Use the representation

$$F[\Phi_0]-F[\Phi_\lambda] = -\int\limits_0^\lambda \mathrm{d}\lambda'\,\frac{\partial F[\Phi_{\lambda'}]}{\partial\lambda'}, \tag{1.104}$$

and the fact that (prove!)

$$-\frac{\partial F[\Phi_\lambda]}{\partial\lambda} = \langle D\rangle_\lambda, \qquad -\frac{\partial^2 F[\Phi_\lambda]}{\partial\lambda^2} = \beta\left\langle\left(D-\langle D\rangle_\lambda\right)^2\right\rangle_\lambda, \tag{1.105}$$

where $\langle\cdots\rangle_\lambda$ denotes the ensemble average using the potential Φ_λ and

$$D(\boldsymbol{r}_1,\cdots,\boldsymbol{r}_N) := \frac{1}{2}\sum_{i,j=1}^{N}(\boldsymbol{r}_i-\boldsymbol{r}_j)^2|\Delta\Phi(\boldsymbol{r}_i-\boldsymbol{r}_j)|. \tag{1.106}$$

The inequality that we are seeking applies to a restricted class of two-body potentials $\{\Phi\}$. This restriction comes about because we need to insure that the thermodynamic limit $N_1, N_2 \to \infty$ is well defined. More precisely, we need the existence of the free energy per particle

$$f_0 := \lim_{N_1,N_2\to\infty}\frac{F[\Phi_0]}{N} < \infty, \tag{1.107}$$

where we made use of the definition (1.101) for $\Phi_0 \equiv \Phi$. Furthermore, we need the existence of at least one $\lambda > 0$ for which

$$f_\lambda := \lim_{N_1,N_2\to\infty} \frac{F[\Phi_\lambda]}{N} < \infty, \tag{1.108}$$

i.e., f_λ is intensive.[10] For any $\lambda > 0$ that satisfies Eq. (1.108), we can then use the fraction $(f_0 - f_\lambda)/\lambda$ to estimate the right-hand side of Eq. (1.100) by writing

$$\frac{1}{N}\left\langle \rho_{+\boldsymbol{k}+\boldsymbol{K}}\, \rho_{-\boldsymbol{k}-\boldsymbol{K}} \right\rangle \geq \frac{1}{N^2} \frac{\beta\, N_{\boldsymbol{k},\boldsymbol{K}}}{A\, \boldsymbol{k}^2 \left[1 + \frac{2\beta}{\lambda}(f_0 - f_\lambda)\right]}. \tag{1.110}$$

Exercise 2.4: By assumption (1.95), the averages $\langle \rho_{\boldsymbol{K}+2\boldsymbol{k}}\rangle$, $\langle \rho_{\boldsymbol{K}+2\boldsymbol{k}_1}\rangle$, and $\langle \rho_{\boldsymbol{K}+2\boldsymbol{k}_2}\rangle$ vanish in the thermodynamic limit if $2\boldsymbol{k}$, $2\boldsymbol{k}_1$, and $2\boldsymbol{k}_2$ are not reciprocal lattice vectors, respectively. Starting from Eq. (1.110), show that

$$\frac{1}{V N}\sum_{\boldsymbol{q}} g(|\boldsymbol{q}|)\left\langle \rho_{\boldsymbol{q}}\, \rho_{-\boldsymbol{q}} \right\rangle \geq \underbrace{\frac{1}{64\, A} \frac{\boldsymbol{K}_0^2\, g(|\boldsymbol{K}| + |\boldsymbol{K}_0|/2)}{1 + \frac{2\beta}{\lambda}(f_0 - f_\lambda)}}\ \frac{\langle \rho_{\boldsymbol{K}}\rangle^2}{N^2} \frac{1}{V} \sum_{|\boldsymbol{k}|<|\boldsymbol{K}_0|/2} \frac{1}{\boldsymbol{k}^2}, \tag{1.111}$$

where $\boldsymbol{K}_0$ is the reciprocal lattice vector with smallest magnitude and the positive function $g:\mathbb{R}\to\mathbb{R}^+$, $k \to g(k) > 0$ is a Gaussian centered at the origin.

The strategy to complete the proof will now be as follows. By inspection of the right-hand side of Eq. (1.111), we anticipate that the factor that is underbraced is non-vanishing but finite in the thermodynamic limit. In contrast, the sum over $1/\boldsymbol{k}^2$, once turned into an integral, diverges logarithmically near the origin. (What happens if $d = 1$ or $d > 2$?) If the left-hand side of Eq. (1.111) turns out to have a *finite* upper bound in the thermodynamic limit, the logarithmic divergence forces

$$\frac{\langle \rho_{\boldsymbol{K}}\rangle^2}{N^2} \overset{N_1,N_2\to\infty}{\longrightarrow} 0. \tag{1.112}$$

This is not compatible with our criterion for crystalline order.

To make this line of arguments work, it thus remains to show that the left-hand side of Eq. (1.111) is bounded from above in the thermodynamic limit. To that end, we define the function

$$\delta\Phi(\boldsymbol{r}) := \int \frac{\mathrm{d}^2\boldsymbol{q}}{(2\pi)^2}\, g(|\boldsymbol{q}|)\, e^{\mathrm{i}\boldsymbol{q}\cdot\boldsymbol{r}}. \tag{1.113}$$

Exercise 2.5: Show that, in the thermodynamic limit,

$$\begin{aligned} \delta\Phi(0) + 2\left(\frac{F[\Phi] - F[\Phi - \delta\Phi]}{N}\right) &\geq \delta\Phi(0) + \frac{1}{N}\sum_{i\neq j}^{N}\left\langle \delta\Phi(\boldsymbol{r}_i - \boldsymbol{r}_j)\right\rangle \\ &= \frac{1}{V N}\sum_{\boldsymbol{q}} g(|\boldsymbol{q}|)\left\langle \rho_{+\boldsymbol{q}}\, \rho_{-\boldsymbol{q}}\right\rangle, \end{aligned} \tag{1.114}$$

[10] It can be shown (see Ref. [23]) that a sufficient condition on Φ_λ for Eq. (1.108) to hold is that

$$\Phi_\lambda(\boldsymbol{r}) \overset{|\boldsymbol{r}|\to\infty}{\sim} |\boldsymbol{r}|^{-(2+\epsilon)}, \qquad \Phi_\lambda(\boldsymbol{r}) \overset{|\boldsymbol{r}|\to 0}{>} \text{const} \times |\boldsymbol{r}|^{-(2+\epsilon')}, \tag{1.109}$$

where const is a positive number and ϵ, ϵ' are two positive numbers.

where $F[\Phi - \delta\Phi]$ is the free energy for the system with the pair potential $\Phi - \delta\Phi$ as defined in Eq. (1.102).

The left-hand side of Eq. (1.114) contains the difference in free energy per particle for the pair potential Φ and $\Phi - \delta\Phi$. We require (and used above already) that $F[\Phi]/N$ is finite in the thermodynamic limit. If a Gaussian is added to the pair potential, this behavior is unaffected, as the additive contribution $\delta\Phi(\boldsymbol{r})$ is well behaved both for small and large $\boldsymbol{r}$. It follows that also $F[\Phi - \delta\Phi]/N$ is finite in the thermodynamic limit.

The estimate (1.114) thus allows us to recast the inequality in (1.111) for sufficiently large N in the form

$$c > \frac{\langle \rho_{\boldsymbol{K}} \rangle^2}{N^2} \frac{1}{V} \sum_{|\boldsymbol{k}|<|\boldsymbol{K}_0|/2} \frac{1}{\boldsymbol{k}^2}, \tag{1.115}$$

where $c < \infty$ is a constant independent of N. For large but finite N, the $\boldsymbol{k}$-sum diverges as

$$\frac{1}{V} \sum_{|\boldsymbol{k}|<|\boldsymbol{K}_0|/2} \frac{1}{\boldsymbol{k}^2} \sim \ln N. \tag{1.116}$$

Hence, the density at every given reciprocal lattice vector goes to zero as

$$\frac{\langle \rho_{\boldsymbol{K}} \rangle}{N} \sim \frac{1}{\sqrt{\ln N}}. \tag{1.117}$$

In this sense, the divergence that leads to the conclusion that there is no crystalline order in two dimensions is very weak. (What about one- or three-dimensional position space?) In turn, weak violations of the assumptions that lead us to this conclusion may already cause some crystalline ordering phenomenon in two dimensions. One example, where this happens, is graphene. There, a weak "buckling" of the plane in which atoms arrange themselves, i.e., slight deviations from the strictly two-dimensional geometry, suffices to allow for a crystalline ordering.

Chapter 2

Bogoliubov theory of a dilute Bose gas

Outline

Second quantization for bosons is reviewed. Bose-Einstein condensation for non-interacting bosons is interpreted as an example of spontaneous-symmetry breaking. The spectrum of a dilute Bose gas with hardcore repulsion is calculated within Bogoliubov mean-field theory using the operator formalism. It is shown that a Goldstone mode, an acoustic phonon, emerges in association with spontaneous-symmetry breaking. Landau criterion for superfluidity is presented. Bose-Einstein condensation as well as superfluidity at non-vanishing temperatures are treated using the path integral formalism.

2.1 Introduction

This chapter is devoted to the study of a dilute Bose gas with a repulsive contact interaction. We shall see that the phenomenon of superfluidity takes place at sufficiently low temperatures. Superfluidity is an example of the spontaneous breaking of a continuous symmetry. The continuous symmetry is the global $U(1)$ gauge symmetry that is responsible for conservation of total particle number. We shall also carefully distinguish Bose-Einstein condensation from superfluidity. Interactions are necessary for superfluidity to take place. Interactions are not needed for Bose-Einstein condensation.

We begin this chapter with the formalism of second quantization for bosons. We then interpret Bose-Einstein condensation at zero temperature as an example of the spontaneous breaking of a continuous symmetry through an explicit construction of a ground state that breaks the global $U(1)$ gauge symmetry. The emphasis is here on how the global $U(1)$ gauge symmetry organizes the Hilbert space spanned by eigenstates of the Hamiltonian. In this construction, the thermodynamic limit plays an essential role.

Next, we treat a repulsive contact interaction through a mean-field approximation first proposed by Bogoliubov.

We revisit this approximation using path-integral techniques to show that it is nothing but a saddle-point approximation. We also present two effective field theories with different physical contents. The first one deals with single-particle excitations. The second one deals with collective excitations.

2.2 Second quantization for bosons

The terminology "second quantization" is rather unfortunate in that it might be perceived as implying concepts more difficult to grasp than the passage from classical to quantum mechanics. Quite to the contrary the relation between "second" and "first" quantization[1] is nothing but a matter of convenience. Going from first to second quantization is like going from a real-space representation of Schrödinger equation to a momentum-space representation when the Hamiltonian has translation symmetry.

Second quantization is a formalism that aims at describing a system made of identical "particles", bosons or fermions, in which creation and annihilation of particles is easily and naturally accounted for. Hence, the quantum "particle number" need not be sharp in this representation, very much in the same way as position is not a sharp quantum number for a momentum eigenstate. Another analogy for the relationship between first quantization, in which the quantum "particle number" is a sharp quantum number, and second quantization, in which it need not be, is that between the canonical and grand-canonical ensembles of statistical mechanics. In the canonical ensemble, particle number is given. In the grand-canonical ensemble, particle number fluctuates statistically as it has been traded for a fixed chemical potential.

The formalism of second quantization can already be introduced at the level of a single harmonic oscillator, but it is for interacting many-body systems that it becomes very powerful. It is nevertheless instructive to develop the formalism already at the level of a single-particle Hamiltonian since, to a large extent, many-body physics is glorified perturbative physics about some non-interacting limit.

We shall now generalize the construction of a second-quantized formalism in terms of creation and annihilation operators for the one-dimensional harmonic oscillator that we presented in chapter 1. We shall thus consider a finite volume V of d-dimensional space on which the single-particle Hilbert space $\mathcal{H}^{(1)}$ of square-integrable and twice-differentiable functions is defined. In turn, the single-particle Hamiltonian is represented by ($\hbar = 1$ and Δ is Laplace's operator in d-dimensional space)

$$H = -\frac{\Delta}{2m} + U(\boldsymbol{r}), \tag{2.1a}$$

[1] By first quantization is meant Schrödinger equation.

and possesses the complete, orthogonal, and normalized basis of eigenfunctions

$$H\,\varphi_n(\boldsymbol{r}) = \varepsilon_n\,\varphi_n(\boldsymbol{r}), \qquad \int\limits_V \mathrm{d}^d\boldsymbol{r}\,\varphi^*_m(\boldsymbol{r})\,\varphi_n(\boldsymbol{r}) = \delta_{m,n}, \qquad \sum_n \varphi^*_n(\boldsymbol{r})\,\varphi_n(\boldsymbol{r}') = \delta(\boldsymbol{r}-\boldsymbol{r}'). \tag{2.1b}$$

The index n belongs to a countable set after appropriate boundary conditions, say periodic, have been imposed at the boundaries of the finite volume V. We assume that the single-particle potential $U(\boldsymbol{r})$ is bounded from below, i.e., there exists a single-particle and non-degenerate ground-state energy, say ε_0. Hence the energy eigenvalue index runs over the positive integers, $n = 0, 1, 2, \cdots$. The time evolution of any solution of Schrödinger equation

$$\mathrm{i}\partial_t \Psi(\boldsymbol{r},t) = H\,\Psi(\boldsymbol{r},t), \qquad \Psi(\boldsymbol{r},t=0) \text{ given}, \tag{2.2a}$$

can be written as

$$\Psi(\boldsymbol{r},t) = \sum_n A_n\,\varphi_n(\boldsymbol{r})\,e^{-\mathrm{i}\varepsilon_n t}, \qquad A_n = \int\limits_V \mathrm{d}^d\boldsymbol{r}\,\varphi^*_n(\boldsymbol{r})\,\Psi(\boldsymbol{r},t=0). \tag{2.2b}$$

The formalism of second quantization starts with the following two postulates.

(1) There exists a set of pairs of adjoint operators $\hat{a}^\dagger_n$ (creation operator) and $\hat{a}_n$ (annihilation operator) labeled by the energy eigenvalue index n and obeying the *bosonic algebra*[2]

$$[\hat{a}_m, \hat{a}^\dagger_n] = \delta_{m,n}, \qquad [\hat{a}_m, \hat{a}_n] = [\hat{a}^\dagger_m, \hat{a}^\dagger_n] = 0, \qquad m,n = 0,1,2,\cdots. \tag{2.3}$$

(2) There exists a non degenerate *vacuum state* $|0\rangle$ that is annihilated by all annihilation operators,

$$\hat{a}_n|0\rangle = 0, \qquad n = 0,1,2,\cdots. \tag{2.4}$$

With these postulates in hand, we define the Heisenberg representation for the operator-valued field (in short, quantum field),

$$\hat{\varphi}^\dagger(\boldsymbol{r},t) := \sum_n \hat{a}^\dagger_n\,\varphi^*_n(\boldsymbol{r})\,e^{+\mathrm{i}\varepsilon_n t}, \tag{2.5a}$$

together with its adjoint

$$\hat{\varphi}(\boldsymbol{r},t) := \sum_n \hat{a}_n\,\varphi_n(\boldsymbol{r})\,e^{-\mathrm{i}\varepsilon_n t}. \tag{2.5b}$$

The bosonic algebra (2.3) endows the quantum fields $\hat{\varphi}^\dagger(\boldsymbol{r},t)$ and $\hat{\varphi}(\boldsymbol{r},t)$ with the equal-time algebra[3]

$$[\hat{\varphi}(\boldsymbol{r},t), \hat{\varphi}^\dagger(\boldsymbol{r}',t)] = \delta(\boldsymbol{r}-\boldsymbol{r}'), \qquad [\hat{\varphi}(\boldsymbol{r},t), \hat{\varphi}(\boldsymbol{r}',t)] = [\hat{\varphi}^\dagger(\boldsymbol{r},t), \hat{\varphi}^\dagger(\boldsymbol{r}',t)] = 0. \tag{2.9}$$

---

[2] The conventions for the commutator and anticommutator of any two "objects" A and B are $[A,B] := AB - BA$ and $\{A,B\} := AB + BA$, respectively.

[3] Alternatively, if we start from the classical Lagrangian density

$$\mathcal{L} := (\varphi^*\mathrm{i}\partial_t\varphi)(\boldsymbol{r},t) - \frac{1}{2m}|\nabla\varphi|^2(\boldsymbol{r},t) - |\varphi^*|^2(\boldsymbol{r},t)\,U(\boldsymbol{r}), \tag{2.6}$$

The quantum fields $\hat{\varphi}^{\dagger}(\boldsymbol{r},t)$ and $\hat{\varphi}(\boldsymbol{r},t)$ act on the *"big" many-particle space*

$$\mathcal{F} := \bigoplus_{N=0}^{\infty}\left\{\bigotimes_{\text{sym}}^{N}\mathcal{H}^{(1)}\right\}. \tag{2.10a}$$

Here, each $\bigotimes_{\text{sym}}^{N}\mathcal{H}^{(1)}$ is spanned by states of the form

$$|m_0,\cdots,m_{i-1},m_i,m_{i+1},\cdots\rangle := \prod_i \frac{\left(\hat{a}_i^{\dagger}\right)^{m_i}}{\sqrt{m_i!}}|0\rangle, \qquad m_i = 0,1,2,\cdots, \tag{2.10b}$$

with the condition on the non-negative integers m_i that

$$\sum_i m_i = N. \tag{2.10c}$$

The algebra obeyed by the $\hat{a}$'s and their adjoints ensures that $\bigotimes_{\text{sym}}^{N}\mathcal{H}^{(1)}$ is the N-th symmetric power of $\mathcal{H}^{(1)}$, i.e., that the state $|m_0,\cdots,m_{i-1},m_i,m_{i+1},\cdots\rangle$ made of N identical particles of which m_i have energy ε_i is left unchanged by any permutation of the N particles. Hence, the "big" many-particle Hilbert space (2.10a) is the sum over the subspaces spanned by wave functions for N identical particles that are symmetric under any permutation of the particles labels. This "big" many-particle Hilbert space is called the *bosonic Fock space* in physics.

The rule to change the representation of operators from the Schrödinger picture to the second quantized language is best illustrated by the following examples.

Example 1: The second-quantized representation $\hat{H}$ of the single-particle Hamiltonian (2.1a) is

$$\begin{aligned}\hat{H} &:= \int\limits_V \mathrm{d}^d\boldsymbol{r}\,\hat{\varphi}^{\dagger}(\boldsymbol{r},t)\,H\,\hat{\varphi}(\boldsymbol{r},t)\\ &= \sum_n \varepsilon_n \hat{a}_n^{\dagger}\,\hat{a}_n.\end{aligned} \tag{2.11}$$

As it should be, it is explicitly time independent.

Example 2: The second-quantized total particle-number operator $\hat{Q}$ is

$$\begin{aligned}\hat{Q} &:= \int\limits_V \mathrm{d}^d\boldsymbol{r}\,\hat{\varphi}^{\dagger}(\boldsymbol{r},t)\,\mathbb{1}\,\hat{\varphi}(\boldsymbol{r},t)\\ &= \sum_n \hat{a}_n^{\dagger}\,\hat{a}_n.\end{aligned} \tag{2.12}$$

---

we can elevate the field $\varphi(\boldsymbol{r},t)$ and its momentum conjugate

$$\pi(\boldsymbol{r},t) := \frac{\delta\mathcal{L}}{\delta(\partial_t\varphi)(\boldsymbol{r},t)} = \mathrm{i}\varphi^{*}(\boldsymbol{r},t) \tag{2.7}$$

to the status of quantum fields $\hat{\varphi}(\boldsymbol{r},t)$ and $\hat{\pi}(\boldsymbol{r},t) = \mathrm{i}\hat{\varphi}^{\dagger}(\boldsymbol{r},t)$ obeying the equal-time bosonic algebra

$$[\hat{\varphi}(\boldsymbol{r},t),\hat{\pi}(\boldsymbol{r}',t)] = \mathrm{i}\delta(\boldsymbol{r}-\boldsymbol{r}'), \qquad [\hat{\varphi}(\boldsymbol{r},t),\hat{\varphi}(\boldsymbol{r}',t)] = [\hat{\pi}(\boldsymbol{r},t),\hat{\pi}(\boldsymbol{r}',t)] = 0. \tag{2.8}$$

It is explicitly time independent as follows from the continuity equation

$$\begin{aligned}
0 &= (\partial_t \rho)(\boldsymbol{r},t) + (\boldsymbol{\nabla}\cdot\boldsymbol{J})(\boldsymbol{r},t),\\
\rho(\boldsymbol{r},t) &:= |\Psi(\boldsymbol{r},t)|^2,\\
\boldsymbol{J}(\boldsymbol{r},t) &:= \frac{1}{2mi}\left[\Psi^*(\boldsymbol{r},t)\left(\boldsymbol{\nabla}\Psi\right)(\boldsymbol{r},t) - \left(\boldsymbol{\nabla}\Psi^*\right)(\boldsymbol{r},t)\Psi(\boldsymbol{r},t)\right],
\end{aligned} \tag{2.13a}$$

obeyed by Schrödinger equation (2.2a). The number operator $\hat{Q}$ is the infinitesimal generator of global gauge transformations by which all states in the bosonic Fock space are multiplied by the *same* operator-valued phase factor. Thus, for any $q \in \mathbb{R}$, a global gauge transformation on the Fock space is implemented by the operation

$$|m_0,\cdots,m_{i-1},m_i,m_{i+1},\cdots\rangle \to e^{+\mathrm{i}q\hat{Q}}\,|m_0,\cdots,m_{i-1},m_i,m_{i+1},\cdots\rangle \tag{2.14}$$

on states, or, equivalently,[4]

$$\hat{a}_n \to e^{+\mathrm{i}q\hat{Q}}\,\hat{a}_n\,e^{-\mathrm{i}q\hat{Q}} = e^{-\mathrm{i}q}\,\hat{a}_n, \tag{2.16a}$$

and

$$\hat{a}^\dagger_n \to e^{+\mathrm{i}q\hat{Q}}\,\hat{a}^\dagger_n\,e^{-\mathrm{i}q\hat{Q}} = e^{+\mathrm{i}q}\,\hat{a}^\dagger_n, \tag{2.16b}$$

for all pairs of creation and annihilation operators, respectively. Equation (2.16b) teaches us that any creation operator carries the particle number $+1$. Equation (2.16a) teaches us that any annihilation operator carries the particle number -1.

Example 3: The second-quantized local particle-number density operator $\hat{\rho}$ and the particle number current density operator $\hat{\boldsymbol{J}}$ are

$$\hat{\rho}(\boldsymbol{r},t) := \hat{\varphi}^\dagger(\boldsymbol{r},t)\,\mathbb{1}\,\hat{\varphi}(\boldsymbol{r},t), \tag{2.17a}$$

and

$$\hat{\boldsymbol{J}}(\boldsymbol{r},t) := \frac{1}{2mi}\left[\hat{\varphi}^\dagger(\boldsymbol{r},t)\left(\boldsymbol{\nabla}\hat{\varphi}\right)(\boldsymbol{r},t) - \left(\boldsymbol{\nabla}\hat{\varphi}^\dagger\right)(\boldsymbol{r},t)\hat{\varphi}(\boldsymbol{r},t)\right], \tag{2.17b}$$

respectively. The continuity equation

$$0 = (\partial_t\hat{\rho})(\boldsymbol{r},t) + \left(\boldsymbol{\nabla}\cdot\hat{\boldsymbol{J}}\right)(\boldsymbol{r},t), \tag{2.17c}$$

which follows from evaluating the commutator between $\hat{\rho}$ and $\hat{H}$, is obeyed as an operator equation.

The operators $\hat{H}$, $\hat{Q}$, $\hat{\rho}$, and $\hat{\boldsymbol{J}}$ all act on the Fock space $\mathcal{F}$. They are thus distinct from their single-particle counterparts H, Q, ρ, and $\boldsymbol{J}$ whose actions are restricted to the Hilbert space $\mathcal{H}^{(1)}$. By construction, the action of $\hat{H}$, $\hat{Q}$, $\hat{\rho}$ and $\hat{\boldsymbol{J}}$ on the subspace $\bigotimes^1_{\mathrm{sym}}\mathcal{H}^{(1)}$ of $\mathcal{F}$ coincides with the action of H, Q, ρ, and $\boldsymbol{J}$ on $\mathcal{H}^{(1)}$, respectively.

[4] We made use of

$$[\hat{a}^\dagger\hat{a},\hat{a}] = \hat{a}^\dagger\hat{a}\hat{a} - \hat{a}\hat{a}^\dagger\hat{a} = \hat{a}^\dagger\hat{a}\hat{a} - \hat{a}^\dagger\hat{a}\hat{a} + \hat{a}^\dagger\hat{a}\hat{a} - \hat{a}\hat{a}^\dagger\hat{a} = \hat{a}^\dagger[\hat{a},\hat{a}] + [\hat{a}^\dagger,\hat{a}]\hat{a} = -\hat{a}, \tag{2.15a}$$

and, similarly,

$$[\hat{a}^\dagger\hat{a},\hat{a}^\dagger] = +\hat{a}^\dagger. \tag{2.15b}$$

2.3 Bose-Einstein condensation and spontaneous symmetry breaking

Given a many-body system made of identical bosons, say atoms carrying an integer-valued total angular momentum, how does one construct the ground state? The simplest answer to this question occurs when bosons are non-interacting. In this case, the ground state is simply obtained by putting all bosons in the lowest energy single-particle state. If the number of bosons is taken to be N, then the ground state is $|N, 0, \cdots\rangle$ with energy $N\varepsilon_0$. This straightforward observation underlies the phenomenon of *Bose-Einstein condensation*. A non-vanishing fraction of bosons occupies the single-particle energy level ε_0 below the Bose-Einstein transition temperature T_{BE} *in the thermodynamic limit of infinite volume V but non-vanishing particle density.*

From a conceptual point of view, it is more fruitful to associate Bose-Einstein condensation with the phenomenon of the spontaneous breaking of a continuous symmetry than with macroscopic occupation of a single-particle level. The continuous symmetry in question is the freedom in the choice of the global phase of the many-particle wave functions. This symmetry is responsible for total particle-number conservation. In mathematical terms, the vanishing commutator

$$[\hat{H}, \hat{Q}] = 0 \tag{2.18}$$

between the total number operator $\hat{Q}$ and the single-particle Hamiltonian $\hat{H}$ implies a global $U(1)$ gauge symmetry.

The concept of spontaneous symmetry breaking is subtle. For one thing it can never take place when the normalized ground state $|\Phi_0\rangle$ of the many-particle Hamiltonian (possibly interacting) is *non-degenerate*, i.e., unique up to a phase factor. Indeed, the transformation law of the ground state $|\Phi_0\rangle$ under any symmetry of the Hamiltonian must then be multiplication by a phase factor. Correspondingly, the ground state $|\Phi_0\rangle$ must transform according to the *trivial* representation of the symmetry group, i.e., $|\Phi_0\rangle$ transforms as a *singlet.* In this case there is no room for the phenomenon of spontaneous symmetry breaking by which the ground state transforms non-trivially under some symmetry group of the Hamiltonian.

Now, the Perron-Frobenius theorem for finite dimensional matrices with positive entries, see Refs. [24] and [25], or its extension, see Ref. [26], to single-particle Hamiltonians of the form (2.1a) guarantees that the ground state is non-degenerate for a non-interacting N-body Hamiltonian defined on the Hilbert space $\bigotimes_{\mathrm{sym}}^{N} \mathcal{H}^{(1)}$. When the ground state of an interacting Hamiltonian defined on the Hilbert space $\bigotimes_{\mathrm{sym}}^{N} \mathcal{H}^{(1)}$ is non-degenerate, then spontaneous symmetry breaking is ruled out for this interacting Hamiltonian.

Before evading this "no-go theorem" by taking advantage of the thermodynamic limit of infinite volume V but non-vanishing particle density, we want to investigate more closely the consequences of having a non-degenerate ground state. We consider the cases of both non-interacting many-body Hamiltonians such as $\hat{H}$ in Eq. (2.11) and interacting many-body Hamiltonians[5] that commute with $\hat{Q}$. The Hilbert space will be the bosonic Fock space $\mathcal{F}$ in Eq. (2.10a) on which the quantum field operator $\hat{\varphi}(\boldsymbol{r},t)$ in Eq. (2.5b) is defined. We shall see that the expectation value of $\hat{\varphi}(\boldsymbol{r},t)$ in the ground state $|\Phi_0\rangle$ of the many-body system can be used as a signature of the spontaneous breaking of the $U(1)$ symmetry. More generally, we shall interpret the quantum statistical average of $\hat{\varphi}(\boldsymbol{r},t)$ as a temperature dependent *order parameter*.

As follows from Eq. (2.16a), the quantum field $\hat{\varphi}(\boldsymbol{r},t)$ transforms according to

$$e^{+\mathrm{i}q\hat{Q}}\,\hat{\varphi}(\boldsymbol{r},t)\,e^{-\mathrm{i}q\hat{Q}} = e^{-\mathrm{i}q}\,\hat{\varphi}(\boldsymbol{r},t), \qquad \forall\boldsymbol{r},t, \tag{2.19}$$

under any global gauge transformation labeled by the real-valued number q. The quantum field $\hat{\varphi}(\boldsymbol{r},t)$ carries $U(1)$ charge -1 as it lowers the bosonic occupation numbers $\sum_i m_i$ by one on any state (2.10b) of the bosonic Fock space $\mathcal{F}$. By hypothesis, the ground state $|\Phi_0\rangle$ of $\hat{H}$ is non-degenerate. Thus, it transforms like a singlet under $U(1)$,

$$\exists\, Q_0\in\mathbb{R}, \qquad e^{-\mathrm{i}q\hat{Q}}\,|\Phi_0\rangle = e^{-\mathrm{i}q\,Q_0}\,|\Phi_0\rangle, \qquad \langle\Phi_0|\,e^{+\mathrm{i}q\hat{Q}} = \langle\Phi_0|\,e^{+\mathrm{i}q\,Q_0}. \tag{2.20}$$

What then follows for the expectation value $\langle\Phi_0|\hat{\varphi}(\boldsymbol{r},t)|\Phi_0\rangle$?

It must vanish. Indeed,

$$\langle\Phi_0|\left(e^{+\mathrm{i}q\hat{Q}}\,\hat{\varphi}(\boldsymbol{r},t)e^{-\mathrm{i}q\hat{Q}}\right)|\Phi_0\rangle = e^{-\mathrm{i}q}\,\langle\Phi_0|\hat{\varphi}(\boldsymbol{r},t)|\Phi_0\rangle, \qquad \forall\boldsymbol{r},t, \tag{2.21}$$

by Eq. (2.19) and

$$\left(\langle\Phi_0|e^{+\mathrm{i}q\hat{Q}}\right)\hat{\varphi}(\boldsymbol{r},t)\left(e^{-\mathrm{i}q\hat{Q}}|\Phi_0\rangle\right) = \langle\Phi_0|\hat{\varphi}(\boldsymbol{r},t)|\Phi_0\rangle, \qquad \forall\boldsymbol{r},t, \tag{2.22}$$

by Eq. (2.20) hold simultaneously for any $q\in\mathbb{R}$. The vanishing of $\langle\Phi_0|\hat{\varphi}(\boldsymbol{r},t)|\Phi_0\rangle$, in view of the fact that $\hat{\varphi}(\boldsymbol{r},t)$ carries $U(1)$ charge -1 and thus transforms non-trivially under $U(1)$, can be traced to the assumption that the ground state $|\Phi_0\rangle$ is unique, i.e., that $|\Phi_0\rangle$ is an eigenstate of $\hat{Q}$. In more intuitive terms, the action of $\hat{\varphi}(\boldsymbol{r},t)$ on an eigenstate of $\hat{Q}$ such as $|\Phi_0\rangle$ is to lower the total number of particles by one, thereby producing a state orthogonal to $|\Phi_0\rangle$. Conversely, a non-vanishing expectation value of $\hat{\varphi}(\boldsymbol{r},t)$ in some state $|\phi\rangle\in\mathcal{F}$ is only possible if $|\phi\rangle\in\mathcal{F}$ is not an eigenstate of $\hat{Q}$.[6]

Evading the "no-go theorem" for spontaneous symmetry breaking thus requires quantum degeneracy of the ground state with *orthogonal* ground states that are related by the action of the $U(1)$ symmetry group. In turn, this can be achieved by constructing a ground state $|\phi\rangle\in\mathcal{F}$ that is an eigenstate of $\hat{\varphi}(\boldsymbol{r},t)$ and thus cannot be an eigenstate of $\hat{Q}$.

A prerequisite to evade the "no-go theorem" for spontaneous symmetry breaking is that the thermodynamic limit of infinite volume V but non-vanishing particle density be taken. This idealized mathematical limit is often an excellent approximation in condensed-matter physics or in cold-atom physics. When the thermodynamic

[5] Interactions are easily introduced through polynomials in creation and annihilation operators of degree larger than 2.

[6] It is impossible for $\hat{\varphi}(\boldsymbol{r},t)$ to acquire an expectation value on $\bigotimes_{\mathrm{sym}}^{N}\mathcal{H}^{(1)}$.

limit $N, V \to \infty$ with N/V held fixed is well defined, there is no difference between approaching this limit by working *at fixed volume and at fixed particle number* with the Hilbert space $\bigotimes_{\mathrm{sym}}^{N} \mathcal{H}^{(1)}$ or approaching the thermodynamic limit by working *at fixed external pressure and at fixed chemical potential* with the Fock space $\mathcal{F} = \sum_{N=0}^{\infty} \bigotimes_{\mathrm{sym}}^{N} \mathcal{H}^{(1)}$. The first approach to the thermodynamic limit defines the so-called *canonical ensemble* of quantum statistical mechanics. The second approach to the thermodynamic limit defines the so-called *grand-canonical ensemble* of quantum statistical mechanics. The thermodynamic limit is also needed to recover spontaneous symmetry breaking even when the Hilbert space of finitely-many degrees of freedom is endowed with the structure of a Fock space.[7]

To underscore the role played by the thermodynamic limit to evade the "no-go theorem" for spontaneous symmetry breaking, we now restrict ourself to the *many-body and non-interacting* Hamiltonian

$$\hat{H}_{\mu} := \hat{H} - \mu \hat{Q}, \tag{2.23a}$$

with

$$H = -\frac{\Delta}{2m}, \tag{2.23b}$$

in Eq. (2.1a) so that translation invariance holds at the single-particle level. The real-valued parameter μ is called the chemical potential. Since $\hat{H}$ commutes with $\hat{Q}$ by hypothesis, an eigenstate of $\hat{H}$ is also an eigenstate of $\hat{H}_{\mu}$ and conversely. Eigenenergies of $\hat{H}$ and $\hat{H}_{\mu}$ may differ, however. For example, the single-particle eigenfunctions $\varphi_n(\boldsymbol{r})$ of H in Eq. (2.1a) are also single-particle eigenfunctions of $\hat{H}_{\mu}$ on $\mathcal{H}^{(1)}$ but with the rigidly shifted spectrum of energy eigenvalues $\varepsilon_n - \mu$. Furthermore, the dimensionalities of the eigenspaces of $\hat{H}$ can change dramatically by the addition of $-\mu\hat{Q}$. To see this, observe that the choice $\mu = \varepsilon_0$ insures that the single-particle ground-state energy of $\hat{H}_{\mu}$ vanishes and that the corresponding normalized eigenfunction $\varphi(\boldsymbol{r}) = 1/\sqrt{V}$.[8] This choice also guarantees that all states

$$|m_0, 0, \cdots\rangle = \frac{(\hat{a}_0^{\dagger})^{m_0}}{\sqrt{m_0!}}|0\rangle, \qquad m_0 = 0, 1, 2, \cdots, \tag{2.25}$$

are *orthogonal* eigenstates of $\hat{H}_{\mu}$ in $\mathcal{F}$ with the same vanishing energy.[9] The choice $\mu = \varepsilon_0$ guarantees that $\hat{H}_{\mu}$ has countably-many orthogonal ground states provided the volume V is finite.

[7] This occurs when the bosons of the many-body system are *collective excitations*, say phonons in a solid, spin waves in an antiferromagnet, or excitons in a semiconductor, i.e., when the finitely-many degrees of freedom are ions, spins, or band electrons, respectively.

[8] A time-dependent gauge transformation plays the same role as the chemical potential if one chooses to work in the canonical instead of the grand-canonical statistical ensemble. For example, setting ε_0 to 0 in the single-particle Hilbert space $\mathcal{H}^{(1)}$ is achieved with the help of the time-dependent gauge transformation

$$\Psi(\boldsymbol{r}, t) \to e^{\mathrm{i}\varepsilon_0 t}\Psi(\boldsymbol{r}, t) \tag{2.24}$$

on the single-particle Schrödinger equation (2.2a).

[9] The same states are also eigenstates of $\hat{H}$ in $\mathcal{F}$ but with *distinct* energy eigenvalues $m_0\varepsilon_0$.

Any linear combination of states of the form (2.25) is a ground state of $\hat{H}_{\mu}$ with $\mu = \varepsilon_0$. Of all these possible linear combinations, consider the continuous family of normalized[10] ground states labeled by the complex-valued parameter ϕ,

$$
\begin{aligned}
|\phi\rangle_{\mathrm{gs}} &:= e^{-\frac{V}{2}|\phi|^2} \sum_{m_0=0}^{\infty} \frac{\left(\sqrt{V}\phi\right)^{m_0}}{\sqrt{m_0!}} |m_0, 0, \cdots\rangle \\
&= e^{-\frac{V}{2}|\phi|^2}\, e^{\sqrt{V}\phi\, \hat{a}_0^{\dagger}} |0\rangle \\
\overset{\hat{a}_0|0\rangle = 0}{} &= e^{-\frac{V}{2}|\phi|^2}\, e^{+\sqrt{V}\phi\, \hat{a}_0^{\dagger}}\, e^{-\sqrt{V}\phi^* \hat{a}_0} |0\rangle \\
&= e^{\sqrt{V}(\phi\, \hat{a}_0^{\dagger} - \phi^* \hat{a}_0)} |0\rangle \\
&=: \hat{D}(\sqrt{V}\phi, 0, \cdots)|0\rangle .
\end{aligned}
\tag{2.27}
$$

To reach the penultimate line, we made use of $[[A,B],A] = [[A,B],B] = 0 \Longrightarrow e^A\, e^B = e^{[A,B]/2}\, e^{A+B}$. Here, the unitary operator $\hat{D}(\sqrt{V}\phi, 0, \cdots)$ rotates the vacuum into the *bosonic coherent state* (see appendix A)

$$
|\sqrt{V}\phi, 0, \cdots\rangle_{\mathrm{cs}} := e^{\sqrt{V}\phi\, \hat{a}_0^{\dagger}} |0\rangle,
\tag{2.28}
$$

up to the proportionality constant $\exp(-\frac{V}{2}|\phi|^2)$. Bosonic coherent states form an overcomplete set of the Fock space (see appendix A). The overlap between any two coherent states is always non-vanishing (see appendix A),

$$
\begin{aligned}
{}_{\mathrm{cs}}\langle \alpha_0, \alpha_1, \cdots | \beta_0, \beta_1, \cdots \rangle_{\mathrm{cs}} &= \prod_n e^{\alpha_n^* \beta_n}, \qquad \alpha_n, \beta_n \in \mathbb{C}, \\
{}_{\mathrm{cs}}\langle \alpha_0, \alpha_1, \cdots | &:= \langle 0| \prod_n e^{\alpha_n^* \hat{a}_n}, \qquad \alpha_n \in \mathbb{C}, \\
|\beta_0, \beta_1, \cdots \rangle_{\mathrm{cs}} &:= \prod_n e^{\beta_n \hat{a}_n^{\dagger}} |0\rangle, \qquad \beta_n \in \mathbb{C}.
\end{aligned}
\tag{2.29}
$$

The same is true of the overlaps (see appendix A)

$$
\begin{aligned}
{}_{\mathrm{gs}}\langle \phi | 0 \rangle &= e^{-\frac{V}{2}|\phi|^2}, \\
{}_{\mathrm{gs}}\langle \phi | \phi' \rangle_{\mathrm{gs}} &= e^{-V\frac{|\phi-\phi'|^2}{2}} .
\end{aligned}
\tag{2.30}
$$

The rational for having scaled the arguments of the unitary operator $\hat{D}(\sqrt{V}\phi, 0, \cdots)$ by the square root of the volume V of the system in Eq. (2.27) is to guarantee that all the rotated vacua in Eq. (2.27) become orthogonal in the thermodynamic limit. The thermodynamic limit is thus essential in providing an escape to the absence of spontaneous symmetry breaking in systems of finite sizes. In the thermodynamic limit, we need not distinguish $\hat{H}$ defined on $\bigotimes_{\mathrm{sym}}^{N} \mathcal{H}^{(1)}$ from $\hat{H}_{\mu}$ defined on $\mathcal{F}$.

---

[10] Observe that the operator

$$
D(\phi_1, \phi_2, \cdots) := \prod_n e^{\left(\phi_n \hat{a}_n^{\dagger} - \phi_n^* \hat{a}_n\right)}, \qquad \phi_1, \phi_2, \cdots \in \mathbb{C},
\tag{2.26}
$$

is unitary.

It is only in the thermodynamic limit that the ground-state manifold $\cong \mathbb{C}$ of $\hat{H}_{\mu}$, $\mu = \varepsilon_0$, in Eq. (2.27) becomes the ground-state manifold $\cong \mathbb{C}$ of $\hat{H}$. Where does this degeneracy of $\hat{H}$ comes from? When V and N are finite and $\hat{H}$ is restricted to $\bigotimes^{N}_{\text{sym}} \mathcal{H}^{(1)}$ the ground-state energy is $N\varepsilon_0$. The ground-state energy of $\hat{H}$ in $\bigotimes^{N\pm1}_{\text{sym}} \mathcal{H}^{(1)}$ differs from that in $\bigotimes^{N}_{\text{sym}} \mathcal{H}^{(1)}$ by a term of order N^0 namely $\pm\varepsilon_0$. In the Fock space $\mathcal{F}$, the energy difference per particle between $\langle N, 0, \cdots |\hat{H}| N, 0, \cdots\rangle$ and $\langle N \pm \delta N, 0, \cdots |\hat{H}| N \pm \delta N, 0, \cdots\rangle$ scales like $1/N$ as the thermodynamic limit $N \to \infty$, $\delta N/N \to 0$, and N/V non-vanishing is taken. Hence, more and more states have an energy of order N^0 above the ground-state energy $N\varepsilon_0$ as the system size is increased. The surprising result is that it is not a mere countable infinity of states that become degenerate with the ground state in the thermodynamic limit but an *uncountable infinity*.

It remains to verify that each ground state $|\phi\rangle_{\text{gs}}$ in Eq. (2.27) is an eigenstate of the quantum fields $\hat{\varphi}(\boldsymbol{r}, t)$,[11] but is not an eigenstate of $\hat{Q}$,

$$\begin{aligned} \hat{\varphi}(\boldsymbol{r}, t)\, |\phi\rangle_{\text{gs}} &= \phi\, |\phi\rangle_{\text{gs}}, \\ e^{-\mathrm{i}\alpha\hat{Q}}\, |\phi\rangle_{\text{gs}} &= |e^{-\mathrm{i}\alpha}\phi\rangle_{\text{gs}}. \end{aligned} \tag{2.31}$$

The $U(1)$ "multiplet" structure of the manifold of ground states $\cong \mathbb{C}$ in Eq. (2.27) is displayed by Eq. (2.31). Circles in the complex plane $\phi \in \mathbb{C}$ correspond to $U(1)$ "multiplets". Normalization of the single-particle eigenfunction $\varphi_0(\boldsymbol{r}) = 1/\sqrt{V}$ and the property that coherent states are eigenstates of annihilation operators guaranty that the quantum field $\hat{\varphi}(\boldsymbol{r}, t)$ acquires the expectation value $\phi \in \mathbb{C}$ with the particle density $|\phi|^2$ in the ground-state manifold (2.27),

$$\begin{aligned} {}_{\text{gs}}\langle\phi|\hat{\varphi}(\boldsymbol{r}, t)|\phi\rangle_{\text{gs}} &= \phi, \\ {}_{\text{gs}}\langle\phi|\hat{\varphi}^{\dagger}(\boldsymbol{r}, t)\hat{\varphi}(\boldsymbol{r}, t)|\phi\rangle_{\text{gs}} &= |\phi|^2. \end{aligned} \tag{2.32}$$

In an interacting system the non-interacting trick relying on fine tuning of the chemical potential $\mu \to \varepsilon_0$ to construct explicitly the many-body ground state breaks down. The chemical potential is chosen instead by demanding that the particle density,

$$\langle\Phi_0|\hat{\varphi}^{\dagger}(\boldsymbol{r}, t)\hat{\varphi}(\boldsymbol{r}, t)|\Phi_0\rangle = \frac{N}{V}, \tag{2.33}$$

at zero temperature,[12] be held fixed to the value N/V as the thermodynamic limit is taken. At non-vanishing temperature the right-hand side is unchanged whereas the left-hand side becomes a statistical average in the grand-canonical ensemble. A degenerate manifold of ground states satisfying Eqs. (2.32) is not anymore

[11] Remember that the single-particle ground-state wave function $\varphi_0(\mathrm{r})$ is the constant $1/\sqrt{V}$. Make then use of the expansion (2.5b) applied to (2.27) whereby $\frac{1}{\sqrt{V}}\hat{a}_0|\sqrt{V}\phi\rangle_{\text{cs}} = \frac{1}{\sqrt{V}}(\sqrt{V}\phi)|\sqrt{V}\phi\rangle_{\text{cs}}$ must be used.

[12] As before, $|\Phi_0\rangle$ denotes the many-body ground state which, in practice, cannot be constructed exactly when interactions are present. We are implicitly assuming translation invariance. This is the reason why the right-hand side does not depend on $\boldsymbol{r}$.

parametrized by $\phi \in \mathbb{C}$ but by $\arg(\phi) \in [0, 2\pi[$, since the modulus $|\phi|^2 = N/V$ is now given. The $U(1)$ symmetry group parametrized by $\exp(\mathrm{i}\alpha\,\hat{Q})$, $\alpha \in [0, 2\pi[$ is said to act *transitively* on the ground-state manifold. Construction of the ground-state manifold relies on approximate schemes such as *mean-field theory*. These approximations are *non-perturbative* in the sense that they yield variational wave functions that cannot be derived from the non-interacting limit to any finite order of the perturbation theory in the interaction strength.

Spontaneous symmetry breaking is said to occur when the ground state $|\Phi_0\rangle$ of a many-body system is no longer a singlet under the action of a symmetry group of the system. A quantity like $\langle\Phi_0|\hat{\varphi}(\boldsymbol{r},t)|\Phi_0\rangle$ that must vanish when the ground state is a singlet, but becomes non-vanishing in a phase with spontaneous symmetry breaking is called an *order parameter*. An order parameter is a probe to detect spontaneous symmetry breaking. In condensed-matter physics, some order parameters can be directly observed in static measurements. For example, elastic-neutron scattering can show Bragg peaks corresponding to crystalline or magnetic order. An order parameter can also be indirectly observed in a dynamical measurement. For instance, inelastic-neutron scattering can show a gapless branch of excitations, *Goldstone modes*, corresponding to phonons or spin waves. Some consequences of symmetries such as selections rules and degeneracies of the excitation spectrum no longer hold in their simplest forms when the phenomenon of spontaneous symmetry breaking occurs. The mass distributions of mesons, hadrons, photon, W and Z bosons are interpreted as a manifestation of spontaneous symmetry breaking leading to the standard model of strong, weak, and electromagnetic interactions.

How does one go about detecting spontaneous symmetry breaking in the canonical ensemble? This question is of relevance to numerical simulations where the dimensionality of the Hilbert space is necessarily finite. A probe for spontaneous symmetry breaking is *off-diagonal long-range order*. Let $|\Phi_N\rangle$ be the ground state of the many-body system in the Hilbert space $\bigotimes_{\mathrm{sym}}^{N}\mathcal{H}^{(1)}$. We denote with $\hat{\varphi}(\boldsymbol{r})$ the quantum field $\hat{\varphi}(\boldsymbol{r}, t = 0)$ in the Schrödinger picture. Here, the Schrödinger picture can be implemented numerically through exact diagonalization of matrices say. We assume translation invariance, i.e., the single-particle potential $U(\boldsymbol{r}) = 0$ in Eq. (2.1a). Define the one-particle density matrix by

$$R(\boldsymbol{r}',\boldsymbol{r}) := \frac{1}{V}\langle\Phi_N|\hat{\varphi}^{\dagger}(\boldsymbol{r}')\,\hat{\varphi}(\boldsymbol{r})|\Phi_N\rangle. \tag{2.34}$$

By translation invariance

$$R(\boldsymbol{r}',\boldsymbol{r}) = R(\boldsymbol{r}'-\boldsymbol{r}), \tag{2.35}$$

which we use to deduce the dependence on $\boldsymbol{r}'-\boldsymbol{r}$. First, we insert into Eq. (2.34) the Fourier expansions (2.5)

$$R(\boldsymbol{r}',\boldsymbol{r}) = \int\frac{\mathrm{d}^d\boldsymbol{k}}{(2\pi)^d}\int\frac{\mathrm{d}^d\boldsymbol{k}'}{(2\pi)^d}\,e^{\mathrm{i}(\boldsymbol{k}\cdot\boldsymbol{r}-\boldsymbol{k}'\cdot\boldsymbol{r}')}\,\langle\Phi_N|\hat{a}^{\dagger}_{\boldsymbol{k}'}\hat{a}_{\boldsymbol{k}}|\Phi_N\rangle. \tag{2.36}$$

Second, we take advantage of $R(\boldsymbol{r}',\boldsymbol{r}) = R(\boldsymbol{r}'-\boldsymbol{r})$ to do

$$R(\boldsymbol{r}',\boldsymbol{r}) = \frac{1}{V}\int \mathrm{d}^d\boldsymbol{y} R(\boldsymbol{r}'+\boldsymbol{y},\boldsymbol{r}+\boldsymbol{y}). \tag{2.37}$$

Third, we combine Eqs. (2.36) and (2.37) into

$$\begin{aligned}
R(\boldsymbol{r}',\boldsymbol{r}) &= \frac{1}{V}\int \mathrm{d}^d\boldsymbol{y}\int \frac{\mathrm{d}^d\boldsymbol{k}}{(2\pi)^d}\int \frac{\mathrm{d}^d\boldsymbol{k}'}{(2\pi)^d}\, e^{\mathrm{i}[\boldsymbol{k}\cdot(\boldsymbol{r}+\boldsymbol{y})-\boldsymbol{k}'\cdot(\boldsymbol{r}'+\boldsymbol{y})]}\,\langle\Phi_N|\hat{a}^\dagger_{\boldsymbol{k}'}\hat{a}_{\boldsymbol{k}}|\Phi_N\rangle \\
&= \frac{1}{V}\int \frac{\mathrm{d}^d\boldsymbol{k}}{(2\pi)^d}\int \frac{\mathrm{d}^d\boldsymbol{k}'}{(2\pi)^d}\,(2\pi)^d\delta(\boldsymbol{k}-\boldsymbol{k}')e^{\mathrm{i}(\boldsymbol{k}\cdot\boldsymbol{r}-\boldsymbol{k}'\cdot\boldsymbol{r}')}\,\langle\Phi_N|\hat{a}^\dagger_{\boldsymbol{k}'}\hat{a}_{\boldsymbol{k}}|\Phi_N\rangle.
\end{aligned} \tag{2.38}$$

Finally, the integration over the momentum $\boldsymbol{k}'$ yields

$$\begin{aligned}
R(\boldsymbol{r}',\boldsymbol{r}) &= \frac{1}{V}\int \frac{\mathrm{d}^d\boldsymbol{k}}{(2\pi)^d} e^{\mathrm{i}\boldsymbol{k}\cdot(\boldsymbol{r}-\boldsymbol{r}')}\,\langle\Phi_N|\hat{a}^\dagger_{\boldsymbol{k}}\hat{a}_{\boldsymbol{k}}|\Phi_N\rangle \\
&=: \int \frac{\mathrm{d}^d\boldsymbol{k}}{(2\pi)^d} e^{\mathrm{i}\boldsymbol{k}\cdot(\boldsymbol{r}-\boldsymbol{r}')}\, n_{\boldsymbol{k}}.
\end{aligned} \tag{2.39}$$

The ground-state expectation value $n_{\boldsymbol{k}}$ is the number of particles per unit volume with momentum $\boldsymbol{k}$. When $\boldsymbol{r}'-\boldsymbol{r}=0$, the one-particle density matrix $R(\boldsymbol{r}'-\boldsymbol{r})$ is just the total number of particles per unit volume $n_0 = N/V$. Bose-Einstein condensation means that

$$n_{\boldsymbol{k}} = n_0(2\pi)^d\delta(\boldsymbol{k}) + f(\boldsymbol{k}), \tag{2.40a}$$

with $f(\boldsymbol{k})$ some smooth function that satisfies

$$\int \frac{\mathrm{d}^d\boldsymbol{k}}{(2\pi)^d} f(\boldsymbol{k}) = 0. \tag{2.40b}$$

In position space, Bose-Einstein condensation thus amounts to

$$R(\boldsymbol{r}',\boldsymbol{r}) = n_0 + F(\boldsymbol{r}'-\boldsymbol{r}), \qquad F(\boldsymbol{r}) := \int \frac{\mathrm{d}^d\boldsymbol{k}}{(2\pi)^d}\, e^{-\mathrm{i}\boldsymbol{k}\cdot\boldsymbol{r}} f(\boldsymbol{k}), \qquad \lim_{|\boldsymbol{r}|\to\infty} F(\boldsymbol{r}) = 0. \tag{2.41}$$

The non-vanishing of $\lim_{|\boldsymbol{r}|\to\infty} R(\boldsymbol{r},\boldsymbol{0})$ is another signature of spontaneous symmetry breaking associated to Bose-Einstein condensation.

We conclude this section with some field-theoretical terminology. States $|\Theta\rangle$ for which

$$\lim_{|\boldsymbol{r}_1-\boldsymbol{r}_2|\to\infty} \langle\Theta|\hat{O}_1(\boldsymbol{r}_1)\hat{O}_2(\boldsymbol{r}_2)|\Theta\rangle = \langle\Theta|\hat{O}_1(\boldsymbol{r}_1)|\Theta\rangle\langle\Theta|\hat{O}_2(\boldsymbol{r}_2)|\Theta\rangle \tag{2.42}$$

holds for any pair of operators $\hat{O}_1(\boldsymbol{r})$ and $\hat{O}_2(\boldsymbol{r})$ defined on the Fock space $\mathcal{F}$ are said to satisfy the *cluster decomposition property* or to be *clustering*. The ground state $|\Phi_N\rangle$ in Eq. (2.41) does not satisfy the clustering property.[13] The manifold of states $|\phi\in\mathbb{C}\rangle_{\mathrm{gs}}$ in Eq. (2.27) does satisfy the clustering property by Eq. (2.32).

[13] Choose $\hat{O}_1 = \hat{\varphi}^\dagger$ and $\hat{O}_2 = \hat{\varphi}$. The left-hand side of Eq. (2.42) is non-vanishing. On the other hand, since the ground state has a well-defined number N of particle, the right-hand side must vanish.

2.4 Dilute Bose gas: Operator formalism at vanishing temperature

2.4.1 *Operator formalism*

Bogoliubov introduced in 1947 an interacting model for superfluid ^{4}He [27]. This model turns out not to be a very good one for superfluid ^{4}He in that the assumption of pairwise interactions made by Bogoliubov fails. However, this model has been conceptually very important. Moreover, this is a realistic model in the field of cold atoms that came into maturity in 1995 with the experimental realization of Bose-Einstein condensation [28].

The model for weakly interacting bosons proposed by Bogoliubov, a dilute Bose gas in short, is defined by the second-quantized Hamiltonian

$$\hat{H}_{\mu,\lambda} = \int\limits_V \mathrm{d}^d\boldsymbol{r} \left[\hat{\varphi}^\dagger(\boldsymbol{r},t)\left(-\frac{\Delta}{2m}-\mu\right)\hat{\varphi}(\boldsymbol{r},t)+\frac{\lambda}{2}\left(\hat{\varphi}^\dagger\hat{\varphi}\right)^2(\boldsymbol{r},t)\right]. \tag{2.43a}$$

The chemical potential μ determines the number $N(\mu)$ of particles in the interacting ground state $|\Phi_{\mathrm{gs}}\rangle$ from

$$N(\mu) = \left\langle \Phi_{\mathrm{gs}} \left| \int\limits_V \mathrm{d}^d\boldsymbol{r}\,\left(\hat{\varphi}^\dagger\hat{\varphi}\right)(\boldsymbol{r},t) \right| \Phi_{\mathrm{gs}} \right\rangle. \tag{2.43b}$$

Conversely, fixing the total particle number to $\overline{N}$ determines $\mu(\overline{N})$. The interaction is a two-body, short-range, and repulsive density-density interaction. In the limit in which the range of this interaction is much smaller than the average particle separation, this interaction is well approximated by a delta function repulsion (this is the justification for the adjective dilute),

$$\hat{H}_\lambda := \frac{\lambda}{2}\int\limits_V \mathrm{d}^d\boldsymbol{r}\int\limits_V \mathrm{d}^d\boldsymbol{r}'\;\hat{\rho}(\boldsymbol{r},t)\,\delta(\boldsymbol{r}-\boldsymbol{r}')\,\hat{\rho}(\boldsymbol{r}',t), \qquad \hat{\rho}(\boldsymbol{r},t) := \left(\hat{\varphi}^\dagger\hat{\varphi}\right)(\boldsymbol{r},t). \tag{2.43c}$$

The real-valued parameter $\lambda \geq 0$ measures the strength of the repulsive interaction and carries the units of (energy×volume). Bosons are said to have a hardcore.

When periodic boundary conditions are imposed in the volume V, it is natural to expand the pair of canonical conjugate quantum fields $\hat{\varphi}(\boldsymbol{r},t)$ and $\mathrm{i}\hat{\varphi}^\dagger(\boldsymbol{r},t)$ in the basis of plane waves,

$$\hat{\varphi}^\dagger(\boldsymbol{r},t) = \frac{1}{\sqrt{V}}\sum_{\boldsymbol{k}} \hat{a}^\dagger_{\boldsymbol{k}}\, e^{-\mathrm{i}(\boldsymbol{k}\cdot\boldsymbol{r}-\varepsilon_{\boldsymbol{k}}t)}, \qquad \hat{\varphi}(\boldsymbol{r},t) = \frac{1}{\sqrt{V}}\sum_{\boldsymbol{k}} \hat{a}_{\boldsymbol{k}}\, e^{+\mathrm{i}(\boldsymbol{k}\cdot\boldsymbol{r}-\varepsilon_{\boldsymbol{k}}t)}. \tag{2.44a}$$

Here, the summation over reciprocal space is infinite but countable,

$$\boldsymbol{k} = \frac{2\pi}{L}\boldsymbol{l}, \qquad \boldsymbol{l}\in\mathbb{Z}^d, \qquad L^d \equiv V, \tag{2.44b}$$

and we have introduced the single-particle dispersion

$$\varepsilon_{\boldsymbol{k}} = \frac{\boldsymbol{k}^2}{2m}. \tag{2.44c}$$

We observe that the single-particle plane wave with the lowest energy is

$$\varphi_{\mathbf{0}}(\boldsymbol{r}) = \frac{1}{\sqrt{V}}, \qquad \varepsilon_{\mathbf{0}} = 0. \tag{2.45}$$

The representation of the Hamiltonian in terms of creation and annihilation operators $\hat{a}^{\dagger}_{\boldsymbol{k}}$ and $\hat{a}_{\boldsymbol{k}}$, respectively, is

$$\hat{H}_{\mu,\lambda} = \sum_{\boldsymbol{k}} \left(\varepsilon_{\boldsymbol{k}} - \mu + \frac{\lambda}{2}\delta(\boldsymbol{r}=\mathbf{0}) \right) \hat{a}^{\dagger}_{\boldsymbol{k}} \hat{a}_{\boldsymbol{k}} + \frac{\lambda}{2V} \sum_{\boldsymbol{k}_1,\boldsymbol{k}_2,\boldsymbol{k}_3,\boldsymbol{k}_4} \delta_{\boldsymbol{k}_1+\boldsymbol{k}_2,\boldsymbol{k}_3+\boldsymbol{k}_4} \hat{a}^{\dagger}_{\boldsymbol{k}_1} \hat{a}^{\dagger}_{\boldsymbol{k}_2} \hat{a}_{\boldsymbol{k}_3} \hat{a}_{\boldsymbol{k}_4}. \tag{2.46}$$

Normal ordering has resulted in the (divergent) shift in the chemical potential $-\frac{\lambda}{2}\delta(\boldsymbol{r}=\mathbf{0})$.

The strategy that we shall use to study the energy spectrum of the dilute Bose gas is to try a variational Ansatz for the ground state. This variational state is taken to be the ground state in the non-interacting limit. Define the Bose-condensate wave function $|\Phi_0\rangle$ to be state (2.27) with

$$\sqrt{V}\phi = \sqrt{N_0}. \tag{2.47}$$

The variational Ansatz $|\Phi_0\rangle$ is the ground state of Eq. (2.43a) with $\mu = \lambda = 0$. It depends on a single variational parameter, the expectation value N_0 of the number operator $\hat{a}^{\dagger}_{\mathbf{0}}\hat{a}_{\mathbf{0}}$ in the state $|\Phi_0\rangle$. The presence of repulsive interactions results in the possibility that N_0 is smaller than $\overline{N}$, i.e., causes a depletion of the Bose condensate in the non-interacting limit. By construction, N_0/V remains non-vanishing in the thermodynamic limit whereas the expectation value of $\hat{a}^{\dagger}_{\boldsymbol{k}}\hat{a}_{\boldsymbol{k}}$ in the state $|\Phi_0\rangle$ vanishes for all $\boldsymbol{k} \neq \mathbf{0}$.

In view of the very special role played by the reciprocal vector $\boldsymbol{k} = \mathbf{0}$, all contributions to the Hamiltonian that depend on $\boldsymbol{k} = \mathbf{0}$ are singled out,

$$\begin{aligned}
\hat{H}_{\mu,\lambda} =& \left(\frac{\lambda}{2}\delta(\boldsymbol{r}=\mathbf{0}) - \mu \right) \hat{a}^{\dagger}_{\mathbf{0}}\hat{a}_{\mathbf{0}} + \frac{\lambda}{2V} \hat{a}^{\dagger}_{\mathbf{0}}\hat{a}^{\dagger}_{\mathbf{0}}\hat{a}_{\mathbf{0}}\hat{a}_{\mathbf{0}} \\
&+ \sum_{\boldsymbol{k}\neq\mathbf{0}} \left(\varepsilon_{\boldsymbol{k}} - \mu + \frac{\lambda}{2}\delta(\boldsymbol{r}=\mathbf{0}) \right) \hat{a}^{\dagger}_{\boldsymbol{k}}\hat{a}_{\boldsymbol{k}} \\
&+ \frac{\lambda}{2V} \sum_{\boldsymbol{k}\neq\mathbf{0}} \left(4\hat{a}^{\dagger}_{\boldsymbol{k}}\hat{a}^{\dagger}_{\mathbf{0}}\hat{a}_{\boldsymbol{k}}\hat{a}_{\mathbf{0}} + \hat{a}^{\dagger}_{+\boldsymbol{k}}\hat{a}^{\dagger}_{-\boldsymbol{k}}\hat{a}_{\mathbf{0}}\hat{a}_{\mathbf{0}} + \hat{a}^{\dagger}_{\mathbf{0}}\hat{a}^{\dagger}_{\mathbf{0}}\hat{a}_{+\boldsymbol{k}}\hat{a}_{-\boldsymbol{k}} \right) \\
&+ \frac{\lambda}{2V} \sum_{\boldsymbol{k},\boldsymbol{k}'\neq\mathbf{0}} \Big(\hat{a}^{\dagger}_{\mathbf{0}}\hat{a}^{\dagger}_{\boldsymbol{k}+\boldsymbol{k}'}\hat{a}_{\boldsymbol{k}}\hat{a}_{\boldsymbol{k}'} + \hat{a}^{\dagger}_{\boldsymbol{k}+\boldsymbol{k}'}\hat{a}^{\dagger}_{\mathbf{0}}\hat{a}_{\boldsymbol{k}}\hat{a}_{\boldsymbol{k}'} + \hat{a}^{\dagger}_{\boldsymbol{k}}\hat{a}^{\dagger}_{\boldsymbol{k}'}\hat{a}_{\mathbf{0}}\hat{a}_{\boldsymbol{k}+\boldsymbol{k}'} \\
&\qquad\qquad + \hat{a}^{\dagger}_{\boldsymbol{k}}\hat{a}^{\dagger}_{\boldsymbol{k}'}\hat{a}_{\boldsymbol{k}+\boldsymbol{k}'}\hat{a}_{\mathbf{0}} \Big) \\
&+ \frac{\lambda}{2V} \sum_{\boldsymbol{k}_1,\boldsymbol{k}_2,\boldsymbol{k}_3,\boldsymbol{k}_4\neq\mathbf{0}} \delta_{\boldsymbol{k}_1+\boldsymbol{k}_2,\boldsymbol{k}_3+\boldsymbol{k}_4} \hat{a}^{\dagger}_{\boldsymbol{k}_1}\hat{a}^{\dagger}_{\boldsymbol{k}_2}\hat{a}_{\boldsymbol{k}_3}\hat{a}_{\boldsymbol{k}_4}.
\end{aligned} \tag{2.48}$$

Interaction terms have been arranged by decreasing number of $\hat{a}^{\dagger}_{\mathbf{0}}$ or $\hat{a}_{\mathbf{0}}$. Momentum conservation prevents terms linear (cubic) in $\hat{a}^{\dagger}_{\mathbf{0}}$ or $\hat{a}_{\mathbf{0}}$ arising from the kinetic energy (interaction). Only the first line contributes to the expectation value of $\hat{H}$ in the variational state $|\Phi_0\rangle$. The new ground-state energy, to first order in λ/V, is thus

$$\left(\frac{\lambda}{2}\delta(\boldsymbol{r}=\boldsymbol{0})-\mu\right)\left(\sqrt{N_0}\right)^2+\frac{\lambda}{2V}\left(\sqrt{N_0}\right)^4. \tag{2.49}$$

It is permissible to replace any $\hat{a}^{\dagger}_{\boldsymbol{0}}$ or $\hat{a}_{\boldsymbol{0}}$ by $\sqrt{N_0}$ on the subspace spanned by acting with the creation operators $\hat{a}^{\dagger}_{\boldsymbol{k}}$, $\boldsymbol{k}\neq\boldsymbol{0}$, on the variational Ansatz $|\Phi_0\rangle$. Hence, on this subspace,

$$\begin{aligned}
\hat{H}_{\mu,\lambda}\to&\left(\frac{\lambda}{2}\delta(\boldsymbol{r}=\boldsymbol{0})-\mu\right)N_0+\frac{\lambda}{2V}N_0^2\\
&+\sum_{\boldsymbol{k}\neq\boldsymbol{0}}\left(\varepsilon_{\boldsymbol{k}}-\mu+\frac{\lambda}{2}\delta(\boldsymbol{r}=\boldsymbol{0})\right)\hat{a}^{\dagger}_{\boldsymbol{k}}\hat{a}_{\boldsymbol{k}}\\
&+\frac{\lambda}{2V}N_0\sum_{\boldsymbol{k}\neq\boldsymbol{0}}\left(4\hat{a}^{\dagger}_{\boldsymbol{k}}\hat{a}_{\boldsymbol{k}}+\hat{a}^{\dagger}_{+\boldsymbol{k}}\hat{a}^{\dagger}_{-\boldsymbol{k}}+\hat{a}_{+\boldsymbol{k}}\hat{a}_{-\boldsymbol{k}}\right)\\
&+2\frac{\lambda}{2V}\sqrt{N_0}\sum_{\boldsymbol{k},\boldsymbol{k}'\neq\boldsymbol{0}}\left(\hat{a}^{\dagger}_{\boldsymbol{k}+\boldsymbol{k}'}\hat{a}_{\boldsymbol{k}}\hat{a}_{\boldsymbol{k}'}+\hat{a}^{\dagger}_{\boldsymbol{k}}\hat{a}^{\dagger}_{\boldsymbol{k}'}\hat{a}_{\boldsymbol{k}+\boldsymbol{k}'}\right)\\
&+\frac{\lambda}{2V}\sum_{\boldsymbol{k}_1,\boldsymbol{k}_2,\boldsymbol{k}_3,\boldsymbol{k}_4\neq\boldsymbol{0}}\delta_{\boldsymbol{k}_1+\boldsymbol{k}_2,\boldsymbol{k}_3+\boldsymbol{k}_4}\hat{a}^{\dagger}_{\boldsymbol{k}_1}\hat{a}^{\dagger}_{\boldsymbol{k}_2}\hat{a}_{\boldsymbol{k}_3}\hat{a}_{\boldsymbol{k}_4}.
\end{aligned} \tag{2.50}$$

After absorbing the divergent $\mathbb{C}$-number $-\frac{\lambda}{2}\delta(\boldsymbol{r}=\boldsymbol{0})$ into a redefinition

$$\mu_{\text{ren}}:=\mu-\frac{\lambda}{2}\delta(\boldsymbol{r}=0) \tag{2.51}$$

of the chemical potential μ, Eq. (2.50) suggests the approximation by which the right-hand side is truncated to the first two leading terms in powers of $\sqrt{N_0}$, i.e., the first two lines, provided the full Fock space $\mathcal{F}$ is restricted to the subspace spanned by the tower of states obtained from acting on $|\Phi_0\rangle$ with $\hat{a}^{\dagger}_{\boldsymbol{k}\neq\boldsymbol{0}}$. Hence, the task of solving for the spectrum of $\hat{H}$ in the Fock space $\mathcal{F}$ has been replaced by the simpler problem of solving for the spectrum of $\hat{H}_{\text{mf}}$ in the Fock space $\mathcal{F}_{\text{mf}}$,

$$\begin{aligned}
\hat{H}_{\mu,\lambda}\to\hat{H}_{\text{mf}}:=&\sum_{\boldsymbol{k}\neq\boldsymbol{0}}\left(\varepsilon_{\boldsymbol{k}}-\mu_{\text{ren}}\right)\hat{a}^{\dagger}_{\boldsymbol{k}}\hat{a}_{\boldsymbol{k}}-\mu_{\text{ren}}N_0\\
&+\frac{\lambda}{2V}N_0\left(N_0+\sum_{\boldsymbol{k}\neq\boldsymbol{0}}\left(4\hat{a}^{\dagger}_{\boldsymbol{k}}\hat{a}_{\boldsymbol{k}}+\hat{a}^{\dagger}_{+\boldsymbol{k}}\hat{a}^{\dagger}_{-\boldsymbol{k}}+\hat{a}_{+\boldsymbol{k}}\hat{a}_{-\boldsymbol{k}}\right)\right),
\end{aligned} \tag{2.52}$$

$$\mathcal{F}\to\mathcal{F}_{\text{mf}}:=\text{span}\left\{\prod_{\boldsymbol{k}\neq\boldsymbol{0}}\left(\hat{a}^{\dagger}_{\boldsymbol{k}}\right)^{m_{\boldsymbol{k}}}|\Phi_0\rangle,\quad m_{\boldsymbol{k}}=0,1,2,\cdots\right\}.$$

This approximation is called a *mean-field* approximation. It is useful because it can be solved exactly, for $\hat{H}_{\text{mf}}$ is quadratic in creation and annihilation operators. It should be a good approximation if N_0 is very close to $\overline{N}$. The self-consistency of this approximation is verified once the variational parameter N_0 has been expressed in terms of the total number of bosons, or, equivalently, in terms of the chemical potential.

We note the presence of the additive $\mathbb{C}$-number

$$\frac{\lambda}{2V}N_0^2 - \mu_{\rm ren}N_0 \tag{2.53}$$

in the mean-field Hamiltonian (2.52). A first estimate of the variational parameter N_0 follows from minimization of this $\mathbb{C}$-number,

$$\frac{N_0}{V} = \frac{\mu_{\rm ren}}{\lambda}. \tag{2.54}$$

Insertion of $N_0 = \mu_{\rm ren}V/\lambda$ into the mean-field Hamiltonian then yields

$$\hat{H}_{\rm mf} = \sum_{\boldsymbol{k}\neq\boldsymbol{0}} \left(\varepsilon_{\boldsymbol{k}} + \mu_{\rm ren}\right) \hat{a}^\dagger_{\boldsymbol{k}}\, \hat{a}_{\boldsymbol{k}} + \frac{\mu_{\rm ren}}{2} \sum_{\boldsymbol{k}\neq\boldsymbol{0}} \left(\hat{a}^\dagger_{+\boldsymbol{k}}\, \hat{a}^\dagger_{-\boldsymbol{k}} + \hat{a}_{+\boldsymbol{k}}\, \hat{a}_{-\boldsymbol{k}}\right) - \frac{V}{2\lambda}\mu^2_{\rm ren}. \tag{2.55}$$

We will discard the last $\mathbb{C}$-number, since we are only interested in the dependence on $\boldsymbol{k}$ of the excitation spectrum of $\hat{H}_{\rm mf}$ and in the change in the variational wave function $|\Phi_0\rangle$ induced by the interactions within the mean-field approximation.

Diagonalization of Eq. (2.55) on the Fock space $\mathcal{F}_{\rm mf}$ is performed with the help of a *canonical* transformation (also called a *Bogoliubov* transformation in this context)[14]

$$\begin{aligned} \hat{a}^\dagger_{+\boldsymbol{k}} &= \sinh(\theta_{+\boldsymbol{k}})\, \hat{b}_{-\boldsymbol{k}} + \cosh(\theta_{+\boldsymbol{k}})\, \hat{b}^\dagger_{+\boldsymbol{k}}, \\ \hat{a}_{+\boldsymbol{k}} &= \cosh(\theta_{+\boldsymbol{k}})\, \hat{b}_{+\boldsymbol{k}} + \sinh(\theta_{+\boldsymbol{k}})\, \hat{b}^\dagger_{-\boldsymbol{k}}, \end{aligned} \tag{2.57a}$$

where (see chapter 2 of Ref. [9] or section 35 in chapter 10 of Ref. [12])

$$\cosh(2\theta_{\boldsymbol{k}}) = \frac{\varepsilon_{\boldsymbol{k}} + \mu_{\rm ren}}{\sqrt{(\varepsilon_{\boldsymbol{k}} + \mu_{\rm ren})^2 - \mu^2_{\rm ren}}}, \qquad \sinh(2\theta_{\boldsymbol{k}}) = \frac{-\mu_{\rm ren}}{\sqrt{(\varepsilon_{\boldsymbol{k}} + \mu_{\rm ren})^2 - \mu^2_{\rm ren}}}. \tag{2.57b}$$

This transformation preserves the bosonic algebra (hence the terminology canonical),

$$\begin{aligned} [\hat{a}_{\boldsymbol{k}}, \hat{a}^\dagger_{\boldsymbol{k}'}] = \delta_{\boldsymbol{k},\boldsymbol{k}'},\ [\hat{a}_{\boldsymbol{k}}, \hat{a}_{\boldsymbol{k}'}] = [\hat{a}^\dagger_{\boldsymbol{k}}, \hat{a}^\dagger_{\boldsymbol{k}'}] = 0 \Longleftrightarrow& \\ [\hat{b}_{\boldsymbol{k}}, \hat{b}^\dagger_{\boldsymbol{k}'}] = \delta_{\boldsymbol{k},\boldsymbol{k}'},\quad [\hat{b}_{\boldsymbol{k}}, \hat{b}_{\boldsymbol{k}'}] = [\hat{b}^\dagger_{\boldsymbol{k}}, \hat{b}^\dagger_{\boldsymbol{k}'}] = 0.& \end{aligned} \tag{2.58}$$

Correspondingly, there exists a unitary transformation $\hat{U}$ on the mean-field Fock space such that $\hat{b}_{\boldsymbol{k}} = \hat{U}\hat{a}_{\boldsymbol{k}}\hat{U}^{-1}$,

$$\hat{U} = \exp\left(+\sum_{\boldsymbol{k}\neq\boldsymbol{0}} \theta_{\boldsymbol{k}} \left(\hat{a}^\dagger_{\boldsymbol{k}}\hat{a}^\dagger_{-\boldsymbol{k}} - \hat{a}_{-\boldsymbol{k}}\hat{a}_{\boldsymbol{k}}\right)\right). \tag{2.59}$$

[14] In matrix form the Bogoliubov transformation reads

$$\begin{pmatrix} \hat{a}_{+\boldsymbol{k}} \\ \hat{a}^\dagger_{-\boldsymbol{k}} \end{pmatrix} = \begin{pmatrix} \cosh\theta_{\boldsymbol{k}} & \sinh\theta_{\boldsymbol{k}} \\ \sinh\theta_{\boldsymbol{k}} & \cosh\theta_{\boldsymbol{k}} \end{pmatrix} \begin{pmatrix} \hat{b}_{+\boldsymbol{k}} \\ \hat{b}^\dagger_{-\boldsymbol{k}} \end{pmatrix} \Longleftrightarrow \begin{pmatrix} \hat{b}_{+\boldsymbol{k}} \\ \hat{b}^\dagger_{-\boldsymbol{k}} \end{pmatrix} = \begin{pmatrix} \cosh\theta_{\boldsymbol{k}} & -\sinh\theta_{\boldsymbol{k}} \\ -\sinh\theta_{\boldsymbol{k}} & \cosh\theta_{\boldsymbol{k}} \end{pmatrix} \begin{pmatrix} \hat{a}_{+\boldsymbol{k}} \\ \hat{a}^\dagger_{-\boldsymbol{k}} \end{pmatrix} \tag{2.56}$$

where $\theta_{\boldsymbol{k}} = \theta_{-\boldsymbol{k}}$.

Up to the $\mathbb{C}$-number

$$E_0 := -\frac{V}{2\lambda}\mu_{\text{ren}}^2 - \sum_{\boldsymbol{k}\neq\boldsymbol{0}} \frac{1}{2}\left[\left(\varepsilon_{\boldsymbol{k}} + \mu_{\text{ren}}\right) - \xi_{\boldsymbol{k}}\right], \tag{2.60a}$$

the mean-field Hamiltonian has become

$$\hat{H}_{\text{mf}} = \sum_{\boldsymbol{k}\neq\boldsymbol{0}} \xi_{\boldsymbol{k}}\, \hat{b}^{\dagger}_{\boldsymbol{k}}\, \hat{b}_{\boldsymbol{k}}, \qquad \xi_{\boldsymbol{k}} := \sqrt{(\varepsilon_{\boldsymbol{k}} + \mu_{\text{ren}})^2 - \mu_{\text{ren}}^2}. \tag{2.60b}$$

This is the Hamiltonian of a gas of *free* bosons with dispersion $\xi_{\boldsymbol{k}}$. For small $|\boldsymbol{k}|$,

$$\xi_{\boldsymbol{k}} \approx \sqrt{\frac{\mu_{\text{ren}}}{m}}\,|\boldsymbol{k}| = \sqrt{\frac{\lambda\, N_0}{m\, V}}\,|\boldsymbol{k}| \equiv v_0\,|\boldsymbol{k}|. \tag{2.61a}$$

This is the dispersion relation of *sound waves* in a fluid that propagate with the speed

$$v_0 := \sqrt{\frac{\lambda\, N_0}{m\, V}}. \tag{2.61b}$$

For large $|\boldsymbol{k}|$, the dispersion crosses over to the usual free-particle expression

$$\xi_{\boldsymbol{k}} \approx \frac{\boldsymbol{k}^2}{2m}. \tag{2.61c}$$

Having found the mean-field excitation spectrum, we must evaluate the change on the unperturbed ground state $|\Phi_0\rangle$ induced by the Bogoliubov transformation $\hat{U}$. The "rotated" ground state is the one annihilated by all $\hat{b}_{\boldsymbol{k}}$, $\boldsymbol{k}\neq\boldsymbol{0}$. The state annihilated by all $\hat{b}_{\boldsymbol{k}}$ is

$$|\Phi_{\text{mf}}\rangle := \hat{U}\,|\Phi_0\rangle. \tag{2.62}$$

With the mean-field ground state at hand, and recalling that the total number of particle $N(\mu)$ is the expectation value of the total particle number operator $\hat{Q}$ in the ground state, we find the relation

$$\begin{aligned}
N(\mu) &\approx \left\langle \Phi_{\text{mf}} \left| \left(\hat{a}^{\dagger}_{\boldsymbol{0}}\, \hat{a}_{\boldsymbol{0}} + \sum_{\boldsymbol{k}\neq\boldsymbol{0}} \hat{a}^{\dagger}_{\boldsymbol{k}}\, \hat{a}_{\boldsymbol{k}} \right) \right| \Phi_{\text{mf}} \right\rangle \\
&= \left\langle \Phi_{\text{mf}} \left| \left(N_0 + \sum_{\boldsymbol{k}\neq\boldsymbol{0}} \hat{b}_{\boldsymbol{k}}\, \hat{b}^{\dagger}_{\boldsymbol{k}} \sinh^2\theta_{-\boldsymbol{k}} \right) \right| \Phi_{\text{mf}} \right\rangle \\
&= \left\langle \Phi_{\text{mf}} \left| \left(N_0 + \sum_{\boldsymbol{k}\neq\boldsymbol{0}} \left(\hat{b}^{\dagger}_{\boldsymbol{k}}\, \hat{b}_{\boldsymbol{k}} + 1\right) \frac{1}{2} \left(\cosh(2\theta_{\boldsymbol{k}}) - 1\right) \right) \right| \Phi_{\text{mf}} \right\rangle \\
\text{Eq. (2.57b)} \quad &= N_0 + \frac{1}{2} \sum_{\boldsymbol{k}\neq\boldsymbol{0}} \left(\frac{\varepsilon_{\boldsymbol{k}} + \mu_{\text{ren}}}{\sqrt{(\varepsilon_{\boldsymbol{k}} + \mu_{\text{ren}})^2 - \mu_{\text{ren}}^2}} - 1 \right).
\end{aligned} \tag{2.63}$$

For comparison, had we estimated $N(\mu)$ using the variational state $\langle\Phi_0\rangle$, we would have found

$$\begin{aligned} N(\mu) &\approx \left\langle \Phi_0 \left| \left(\hat{a}_{\mathbf{0}}^{\dagger}\, \hat{a}_{\mathbf{0}} + \sum_{\boldsymbol{k}\neq\mathbf{0}} \hat{a}_{\boldsymbol{k}}^{\dagger}\, \hat{a}_{\boldsymbol{k}} \right) \right| \Phi_0 \right\rangle \\ &= N_0. \end{aligned} \tag{2.64}$$

The number $N(\mu)$ of particles present in $|\Phi_{\mathrm{mf}}\rangle$ exceeds the number N_0 present in the single-particle condensate $|\Phi_0\rangle$ by the sum over momenta on the right-hand side of Eq. (2.63). Conversely, had we fixed the number of bosons to be $\overline{N}$ rather than fixing the chemical potential μ, then the number N_0 of weakly interacting bosons that form a Bose-Einstein condensate in the mean-field ground state $|\Phi_{\mathrm{mf}}\rangle$ is smaller than $\overline{N}$ by an amount that depends on the dimensionality of space, the density $\overline{N}/V$, and the coupling strength λ. The mean-field approximation is self-consistent if this amount is small, i.e., if and only if the sum over momenta on the right-hand side of Eq. (2.63) can be shown to be small. It is shown with Eq. (2.114) that this is the case in three-dimensional space for either a dilute hardcore Bose gas or for small λ.

Had we not chosen N_0 by minimization, we would have found that $\xi_{\boldsymbol{k}}^2$ would not vanish anymore in the limit $\boldsymbol{k} \to \mathbf{0}$. Hence, for any other value of N_0 than the one in Eq. (2.54), we could lower the trial energy by either removing or adding particles in the condensate, i.e., varying the parameter N_0 of the trial wave function $|\Phi_0\rangle$.

We close the discussion of this mean-field theory with a word of caution. The main prediction of this mean-field analysis is the existence of a mean-field gapless spectrum. Is this prediction robust? This prediction is predicated on the minimization (2.54). As such, it would be robust if and only if this local minimum is the global one, as shall become clear when we derive the mean-field approximation from the path-integral formalism. In practice, such a proof cannot be achieved and the "validity" of a mean-field approximation rests on two verifications, namely that it is self-consistent and that it agrees with experiments.

2.4.2 *Landau criterion for superfluidity*

We shall assume that the mean-field spectrum that was derived for the dilute Bose gas is exact. We shall also assume that the excitations, phonons, described by the pair $\hat{b}_{\boldsymbol{k}}^{\dagger}$ and $\hat{b}_{\boldsymbol{k}}$ of annihilation and creation operators are the only ones. Although neither assumptions are realistic, the point made by Landau is that they are sufficient to understand the phenomenon of *superfluidity*.

Consider a body of large mass M moving in the dilute Bose gas (the fluid from now on) at velocity $\boldsymbol{V}$. By hypothesis, the only way for the body to experience a retarding force or *drag* is for it to emit some phonons. In doing so, energy

$$\delta\varepsilon := \frac{1}{2}M\boldsymbol{V}^2 - \frac{1}{2}M\,(\boldsymbol{V}-\delta\boldsymbol{V})^2 = M\boldsymbol{V}\cdot\delta\boldsymbol{V} + \mathcal{O}[(\delta\boldsymbol{V})^2], \tag{2.65a}$$

and momentum

$$\delta \boldsymbol{k} := M\boldsymbol{V} - M(\boldsymbol{V} - \delta \boldsymbol{V}) = M\delta \boldsymbol{V}, \tag{2.65b}$$

are lost to the phonons with momenta $\boldsymbol{k}_i$ and energies $\varepsilon_{\boldsymbol{k}_i}$, i.e.,

$$\delta\varepsilon = +\sum_i \varepsilon_{\boldsymbol{k}_i}, \tag{2.65c}$$

$$\delta \boldsymbol{k} = +\sum_i \boldsymbol{k}_i. \tag{2.65d}$$

By hypothesis phonons in the model have a non-vanishing minimum phase velocity

$$v_0 = \inf_{\boldsymbol{k}} \left\{ \frac{\varepsilon_{\boldsymbol{k}}}{|\boldsymbol{k}|} \right\} > 0. \tag{2.66}$$

The chain of inequalities

$$|\delta\varepsilon| = \sum_i \varepsilon_{\boldsymbol{k}_i} \geq v_0 \sum_i |\boldsymbol{k}_i| \geq v_0 \left| \sum_i \boldsymbol{k}_i \right| = v_0 \, |\delta \boldsymbol{k}| \tag{2.67}$$

then follows. To leading order in M^{-1}, we have established that

$$|\delta \boldsymbol{k}||\boldsymbol{V}| \geq |\delta \boldsymbol{k} \cdot \boldsymbol{V}| = |\delta\varepsilon| + \mathcal{O}(M^{-1}) \geq v_0 |\delta \boldsymbol{k}| + \mathcal{O}(M^{-1}) \tag{2.68}$$

for any permitted $\delta \boldsymbol{k}$. Such a $\delta \boldsymbol{k}$ can only exist if

$$|\boldsymbol{V}| \geq v_0, \tag{2.69}$$

i.e., the body must exceed a minimum velocity before experiencing any drag. By moving sufficiently slowly, a heavy body suffers no loss of energy and momentum from the medium. This property of the medium is called *superfluidity*. It originates here from the fact that the mean-field excitation spectrum is bounded from below by a linear dispersion. In turn, this is a consequence of the interactions conspiring together with spontaneous symmetry breaking in the existence of Goldstone modes, acoustic phonons. Interactions are essential to superfluidity. The excitation spectrum remains quadratic in the non-interacting limit and the velocity threshold below which a moving body does not suffer drag is $v_0 = 0$. Bose-Einstein condensation alone (i.e., without Goldstone modes) is not sufficient for superfluidity to occur.

2.5 Dilute Bose gas: Path-integral formalism at any temperature

The partition function for the dilute Bose gas at inverse temperature β and chemical potential μ is

$$\begin{aligned} Z(\beta,\mu) &:= \operatorname{Tr}\left(e^{-\beta \hat{H}_{\mu,\lambda}}\right), \\ \hat{H}_{\mu,\lambda} &= \int\limits_V \mathrm{d}^d \boldsymbol{r} \left[\hat{\varphi}^\dagger(\boldsymbol{r}) \left(-\frac{\Delta}{2m} - \mu \right) \hat{\varphi}(\boldsymbol{r}) + \frac{\lambda}{2} \left(\hat{\varphi}^\dagger \hat{\varphi} \right)^2 (\boldsymbol{r}) \right]. \end{aligned} \tag{2.70a}$$

The total number of bosons $N(\beta,\mu)$ at inverse temperature β and chemical potential μ is obtained from

$$\begin{aligned} N(\beta,\mu) := & \left\langle \int\limits_V \mathrm{d}^d\boldsymbol{r}\, \left(\hat{\varphi}^\dagger\hat{\varphi}\right)(\boldsymbol{r}) \right\rangle_{Z(\beta,\mu)} \\ \equiv & \beta^{-1}\partial_\mu \ln Z(\beta,\mu). \end{aligned} \tag{2.70b}$$

We have seen in appendix A that the path-integral representation

$$\begin{aligned} Z(\beta,\mu) =& \int \mathcal{D}[\varphi^*,\varphi]\, \exp\left(-S_{\mathrm{E}}\right) \\ =& \int \mathcal{D}[\varphi^*,\varphi]\, \exp\left(-\int\limits_0^\beta \mathrm{d}\tau \int\limits_V \mathrm{d}^d\boldsymbol{r} \mathcal{L}_{\mathrm{E}} \right), \end{aligned} \tag{2.71a}$$

where

$$\mathcal{L}_{\mathrm{E}} = \varphi^*(\boldsymbol{r},\tau)\left(\partial_\tau - \frac{\Delta}{2m} - \mu + \frac{\lambda}{2}\delta(\boldsymbol{r}=\boldsymbol{0})\right)\varphi(\boldsymbol{r},\tau) + \frac{\lambda}{2}\left[\varphi^*(\boldsymbol{r},\tau)\right]^2\left[\varphi(\boldsymbol{r},\tau)\right]^2, \tag{2.71b}$$

of this partition function exists. Integration variables are the real and imaginary parts of the complex-valued function $\varphi(\boldsymbol{r},\tau)$ or, equivalently, its complex conjugate $\varphi^*(\boldsymbol{r},\tau)$. They obey periodic boundary conditions in imaginary time τ,

$$\varphi^*(\boldsymbol{r},\tau) = \varphi^*(\boldsymbol{r},\tau+\beta), \qquad \varphi(\boldsymbol{r},\tau) = \varphi(\boldsymbol{r},\tau+\beta). \tag{2.71c}$$

Boundary conditions in space, say periodic ones, are also present. The total number of bosons $N(\beta,\mu)$ is now represented by

$$\begin{aligned} N(\beta,\mu) =& \beta^{-1}\left\langle \int\limits_0^\beta \mathrm{d}\tau \int\limits_V \mathrm{d}^d\boldsymbol{r}\, \left(\varphi^*\varphi\right)(\boldsymbol{r},\tau) \right\rangle_{Z(\beta,\mu)} \\ =& \beta^{-1}\partial_\mu \ln Z(\beta,\mu). \end{aligned} \tag{2.71d}$$

The choice of periodic boundary conditions in space *and* time suggests to change integration variable in the path-integral representation of the partition function by performing the Fourier transforms

$$\begin{aligned} \varphi^*(\boldsymbol{r},\tau) =& \frac{1}{\sqrt{\beta V}} \sum_{\boldsymbol{k}}\sum_{l} a^*_{\boldsymbol{k},\varpi_l}\, e^{-\mathrm{i}(\boldsymbol{k}\boldsymbol{r}-\varpi_l\tau)}, \\ a^*_{\boldsymbol{k},\varpi_l} =& \frac{1}{\sqrt{\beta V}} \int\limits_0^\beta \mathrm{d}\tau \int\limits_V \mathrm{d}^d\boldsymbol{r}\, \varphi^*(\boldsymbol{r},\tau)\, e^{+\mathrm{i}(\boldsymbol{k}\boldsymbol{r}-\varpi_l\tau)}, \end{aligned} \tag{2.72a}$$

on the one hand, and

$$\begin{aligned} \varphi(\boldsymbol{r},\tau) =& \frac{1}{\sqrt{\beta V}} \sum_{\boldsymbol{k}}\sum_{l} a_{\boldsymbol{k},\varpi_l}\, e^{+\mathrm{i}(\boldsymbol{k}\boldsymbol{r}-\varpi_l\tau)}, \\ a_{\boldsymbol{k},\varpi_l} =& \frac{1}{\sqrt{\beta V}} \int\limits_0^\beta \mathrm{d}\tau \int\limits_V \mathrm{d}^d\boldsymbol{r}\, \varphi(\boldsymbol{r},\tau)\, e^{-\mathrm{i}(\boldsymbol{k}\boldsymbol{r}-\varpi_l\tau)}, \end{aligned} \tag{2.72b}$$

on the other hand. Here,

$$\varpi_l = \frac{2\pi}{\beta} l, \qquad l \in \mathbb{Z}, \qquad \boldsymbol{k} = \frac{2\pi}{L} \boldsymbol{l}, \qquad \boldsymbol{l} \in \mathbb{Z}^d. \tag{2.72c}$$

This change of integration variable turns the path-integral representation of the partition function into

$$Z(\beta, \mu) = \int \mathcal{D}[a^*, a] \, \exp\left(-S_{\mathrm{E}}\right) \tag{2.73a}$$

with the Euclidean action

$$\begin{aligned} S_{\mathrm{E}} =& \sum_{l} \sum_{\boldsymbol{k}} a^*_{\boldsymbol{k},\varpi_l} \left(-\mathrm{i}\varpi_l + \frac{\boldsymbol{k}^2}{2m} - \mu + \frac{\lambda}{2} \delta(\boldsymbol{r} = \boldsymbol{0}) \right) a_{\boldsymbol{k},\varpi_l} \\ &+ \frac{\lambda}{2} \frac{1}{\beta V} \sum_{\boldsymbol{k}_1, \boldsymbol{k}_2, \boldsymbol{k}_3, \boldsymbol{k}_4}^{l_1, l_2, l_3, l_4} \delta_{l_1 + l_2, l_3 + l_4} \delta_{\boldsymbol{k}_1 + \boldsymbol{k}_2, \boldsymbol{k}_3 + \boldsymbol{k}_4} \, a^*_{\boldsymbol{k}_1, \varpi_{l_1}} a^*_{\boldsymbol{k}_2, \varpi_{l_2}} a_{\boldsymbol{k}_3, \varpi_{l_3}} a_{\boldsymbol{k}_4, \varpi_{l_4}}. \end{aligned} \tag{2.73b}$$

The total number of bosons $N(\beta, \mu)$ is represented by

$$\begin{aligned} N(\beta, \mu) =& \beta^{-1} \left\langle \sum_{l} \sum_{\boldsymbol{k}} a^*_{\boldsymbol{k},\varpi_l} a_{\boldsymbol{k},\varpi_l} \right\rangle_{Z(\beta,\mu)} \\ =& \beta^{-1} \partial_\mu \ln Z(\beta, \mu). \end{aligned} \tag{2.73c}$$

2.5.1 *Non-interacting limit* $\lambda = 0$

In the non-interacting limit $\lambda = 0$, we need to solve the quadratic problem

$$\begin{aligned} Z(\beta, \mu) =& \int \mathcal{D}[a^*, a] \, \exp\left(-S_{\mathrm{E}}\right), \\ S_{\mathrm{E}} =& \sum_{l} \sum_{\boldsymbol{k}} a^*_{\boldsymbol{k},\varpi_l} \left(-\mathrm{i}\varpi_l + \frac{\boldsymbol{k}^2}{2m} - \mu \right) a_{\boldsymbol{k},\varpi_l}. \end{aligned} \tag{2.74a}$$

The path integral is a multi-dimensional Gaussian integral, one Gaussian integral of the form

$$\begin{aligned} \int \frac{\mathrm{d}z^* \, \mathrm{d}z}{2\pi\mathrm{i}} \, e^{-z^* K z} \equiv& \int \frac{\mathrm{d}(x - \mathrm{i}y) \, \mathrm{d}(x + \mathrm{i}y)}{2\pi\mathrm{i}} \, e^{-(x - \mathrm{i}y) \, K \, (x + \mathrm{i}y)} \\ =& \frac{1}{\pi} \int_{-\infty}^{+\infty} \mathrm{d}x \int_{-\infty}^{+\infty} \mathrm{d}y \, e^{-K \, (x^2 + y^2)} \\ =& \frac{1}{\pi} (2\pi) \int_{0}^{+\infty} \mathrm{d}r \, r \, e^{-K \, r^2} \\ =& \frac{1}{\pi} (2\pi) \frac{1}{2K} \, e^{-K \, r^2} \Big|_{+\infty}^{0} \\ =& \frac{1}{K}, \qquad K \in \mathbb{R}^+, \end{aligned} \tag{2.75}$$

for each pair $(a^*_{\boldsymbol{k},\varpi_l}, a_{\boldsymbol{k},\varpi_l})$, provided the counterpart $-\mathrm{i}\varpi_l + \frac{\boldsymbol{k}^2}{2m} - \mu$ to K has a positive real part, i.e.,

$$\frac{\boldsymbol{k}^2}{2m} > \mu. \tag{2.76}$$

This is symbolically written as an *inverse determinant* (see appendix B.1)

$$\begin{aligned} Z(\beta,\mu) &= \frac{1}{\mathrm{Det}\,\left(\partial_\tau - \frac{\Delta}{2m} - \mu\right)} \\ &\equiv \prod_l \prod_{\boldsymbol{k}} \frac{1}{-\mathrm{i}\varpi_l + \frac{\boldsymbol{k}^2}{2m} - \mu}, \qquad \mu < 0. \end{aligned} \tag{2.77}$$

The total number of bosons $N(\beta,\mu)$ is thus given by the expression

$$\begin{aligned} N(\beta,\mu) &= \beta^{-1}\partial_\mu \ln Z(\beta,\mu) \\ &= \beta^{-1}\sum_l \sum_{\boldsymbol{k}} \frac{1}{-\mathrm{i}\varpi_l + \frac{\boldsymbol{k}^2}{2m} - \mu}, \qquad \mu < 0, \end{aligned} \tag{2.78}$$

in the non-interacting limit. It is shown in appendix B.2 that the imaginary-time summation can be written as a contour (Γ) integral in the complex z-plane for any given $\boldsymbol{k}$,

$$\begin{aligned} \sum_l \frac{1}{-\mathrm{i}\varpi_l + \frac{\boldsymbol{k}^2}{2m} - \mu} &= +\beta \int_\Gamma \frac{\mathrm{d}z}{2\pi\mathrm{i}} \frac{f_{\mathrm{BE}}(z)}{-z + \frac{\boldsymbol{k}^2}{2m} - \mu} \\ &= -\beta \int_\Gamma \frac{\mathrm{d}z}{2\pi\mathrm{i}} \frac{f_{\mathrm{BE}}(z)}{z - \frac{\boldsymbol{k}^2}{2m} + \mu} \\ &\equiv -\beta \int_\Gamma \frac{\mathrm{d}z}{2\pi\mathrm{i}} \frac{f_{\mathrm{BE}}(z)}{z - \varepsilon_{\boldsymbol{k}} + \mu}, \qquad \mu < 0, \end{aligned} \tag{2.79a}$$

where $f_{\mathrm{BE}}(z)$ is the Bose-Einstein distribution function

$$f_{\mathrm{BE}}(z) := \frac{1}{e^{\beta z} - 1}, \tag{2.79b}$$

and $\varepsilon_{\boldsymbol{k}}$ is the single-particle dispersion,

$$\varepsilon_{\boldsymbol{k}} := \frac{\boldsymbol{k}^2}{2m}. \tag{2.79c}$$

A second application of the residue theorem (see appendix B.2) turns the z-integral over the counterclockwise Γ contour into

$$-\beta \int_\Gamma \frac{\mathrm{d}z}{2\pi\mathrm{i}} \frac{f_{\mathrm{BE}}(z)}{z - \varepsilon_{\boldsymbol{k}} + \mu} = (-)^2 \beta f_{\mathrm{BE}}(\varepsilon_{\boldsymbol{k}} - \mu), \qquad \mu < 0. \tag{2.80}$$

We conclude that the total number of bosons is given by

$$\begin{aligned} N(\beta,\mu) &= \sum_{\boldsymbol{k}} f_{\mathrm{BE}}(\varepsilon_{\boldsymbol{k}} - \mu) \\ &= \sum_{\boldsymbol{k}} \left[\exp\left(\beta\left(\frac{\boldsymbol{k}^2}{2m} - \mu\right)\right) - 1\right]^{-1}, \qquad \mu < 0. \end{aligned} \tag{2.81}$$

When β is fixed, $N(\beta,\mu)$ is a monotonically increasing function of μ. When μ is fixed, $N(\beta,\mu)$ is a monotonically decreasing function of β. If the temperature dependence of μ is determined by fixing the left-hand side of Eq. (2.81) to be some constant number, say the average total particle number $\overline{N}$ in the grand-canonical ensemble, then $\mu(\beta)$ is a monotonically increasing function of β. However, $\mu(\beta)$ is necessarily bounded from above by

$$\mu_{\mathrm{c}} := \inf_{\boldsymbol{k}} \varepsilon_{\boldsymbol{k}} = 0, \tag{2.82}$$

for, if it was not, there would be an inverse critical temperature β_{c} above which $\mu(\beta) > \mu_{\mathrm{c}}$ and the integral over the so-called *zero mode*[15] $(a^*_{\boldsymbol{k}=\boldsymbol{0},\varpi_l=0}, a_{\boldsymbol{k}=\boldsymbol{0},\varpi_l=0})$,

$$\int \frac{\mathrm{d}a^*_{\boldsymbol{0},0}\mathrm{d}a_{\boldsymbol{0},0}}{2\pi\mathrm{i}} \exp\left(+|\mu(\beta)|a^*_{\boldsymbol{0},0}a_{\boldsymbol{0},0}\right), \tag{2.85}$$

would diverge in contradiction with the assumption that there exists a well defined vacuum $|0\rangle$. The alternative scenario by which $\mu(\beta)$ is pinned to μ_{c} above β_{c} is actually what transpires from a numerical solution of Eq. (2.81) which now reads

$$\overline{N} = \begin{cases} \sum\limits_{\boldsymbol{k}} f_{\mathrm{BE}}(\varepsilon_{\boldsymbol{k}} - \mu), & \text{if } \beta < \beta_{\mathrm{c}}. \\ \\ \beta^{-1}|a_{\boldsymbol{0},0}|^2 + \sum\limits_{\boldsymbol{k}\neq\boldsymbol{0}} f_{\mathrm{BE}}(\varepsilon_{\boldsymbol{k}} - \mu_{\mathrm{c}}), & \text{if } \beta \geq \beta_{\mathrm{c}}. \end{cases} \tag{2.86}$$

Equation (2.86) determines the chemical potential as a function of the inverse temperature. It also determines the macroscopic number of bosons $\beta^{-1}|a_{\boldsymbol{0},0}|^2(\beta)$ that occupy the lowest single-particle energy $\varepsilon_{\boldsymbol{0}}$ above the critical inverse temperature β_{c}. In the limit of vanishing temperature, $\lim_{\beta\to\infty} f_{\mathrm{BE}}(\varepsilon_{\boldsymbol{k}} - \mu_{\mathrm{c}}) = 0$, $\boldsymbol{k} \neq 0$, and all $\overline{N}$ bosons occupy the single-particle ground-state energy $\varepsilon_{\boldsymbol{0}}$. Before tackling the interacting case, it is important to realize that Bose-Einstein condensation did not alter the single-particle dispersion $\varepsilon_{\boldsymbol{k}}$. According to the Landau criterion, superfluidity cannot take place in the non-interacting limit. We shall see below how the excitation spectrum becomes linear at long wavelengths due to a conspiracy between spontaneous symmetry breaking and interactions, thus enabling superfluidity.

Before leaving the non-interacting limit, we introduce the single-particle Green function

$$G_{\boldsymbol{k},\varpi_l} := -\frac{1}{-\mathrm{i}\varpi_l + \frac{\boldsymbol{k}^2}{2m} - \mu}. \tag{2.87}$$

[15] A zero mode is a configuration $\varphi(\boldsymbol{r},\tau)$ that does not depend on space or time:

$$\varphi(\boldsymbol{r},\tau) = \varphi_0. \tag{2.83}$$

The only non-vanishing Fourier component of φ_0 is

$$a_{\boldsymbol{k}=\boldsymbol{0},\varpi_l=0} = \sqrt{\beta V}\varphi_0. \tag{2.84}$$

The sign is convention. The Green function (2.87) is, up to a sign, the inverse of the Kernel in Eq. (2.74a). Furthermore, because of the identities (2.75) and

$$\begin{aligned}\int\frac{\mathrm{d}z^*\mathrm{d}z}{2\pi\mathrm{i}}(z^*z)\,e^{-z^*Kz}=&\frac{\partial^2}{\partial J^*\partial J}\int\frac{\mathrm{d}z^*\mathrm{d}z}{2\pi\mathrm{i}}\,e^{-z^*Kz+J^*z+Jz^*}\Big|_{J^*=J=0}\\ =&\frac{\partial^2}{\partial J^*\partial J}\int\frac{\mathrm{d}z^*\mathrm{d}z}{2\pi\mathrm{i}}\,e^{-\left(z-\frac{J}{K}\right)^*K\left(z-\frac{J}{K}\right)+J^*\frac{1}{K}J}\Big|_{J^*=J=0}\\ \overset{\text{Eq. (2.75)}}{=}&(1/K)\,\frac{\partial^2e^{+J^*\frac{1}{K}J}}{\partial J^*\partial J}\Big|_{J^*=J=0}\\ =&\frac{(1/K)}{K},\qquad K\in\mathbb{R}^+,\end{aligned}\tag{2.88}$$

the Green function (2.87) is the covariance or two-point function

$$\begin{aligned}G_{\boldsymbol{k},\varpi_l}:=&-\left\langle a^*_{\boldsymbol{k},\varpi_l}a_{\boldsymbol{k},\varpi_l}\right\rangle_Z\\ \equiv&-\frac{\int\mathcal{D}[a^*,a]\,e^{-S_{\mathrm{E}}}\,\left(a^*_{\boldsymbol{k},\varpi_l}a_{\boldsymbol{k},\varpi_l}\right)}{\int\mathcal{D}[a^*,a]\,e^{-S_{\mathrm{E}}}}.\end{aligned}\tag{2.89}$$

2.5.2 *Random-phase approximation*

The first change relative to the analysis of the non-interacting limit that is brought by switching a repulsive contact interaction, $\lambda>0$, is the breakdown of the stability argument that leads to the pinning of the chemical potential. To see this, consider as in Eq. (2.85) the action of the zero mode

$$\varphi(\boldsymbol{r},\tau):=\varphi_0,\ \forall\boldsymbol{r},\tau\Longrightarrow S^{(0)}_{\mathrm{E}}[\varphi^*_0,\varphi_0]:=\beta V\left(-\mu_{\mathrm{ren}}|\varphi_0|^2+\frac{\lambda}{2}|\varphi_0|^4\right),\tag{2.90}$$

where *now* $\lambda>0$ implies that the renormalized chemical potential

$$\mu_{\mathrm{ren}}:=\mu-\frac{\lambda}{2}\delta(\boldsymbol{r}=\boldsymbol{0})\tag{2.91}$$

can become arbitrarily large as a function of inverse temperature without endangering the convergence of the contribution

$$Z_0:=\int\frac{\mathrm{d}\varphi^*_0\mathrm{d}\varphi_0}{2\pi\mathrm{i}}\,e^{-S^{(0)}_{\mathrm{E}}[\varphi^*_0,\varphi_0]}\tag{2.92}$$

from the zero modes to the partition function.

An estimate of Eq. (2.92) that becomes exact in the limit of $\beta,V\to\infty$ is obtained from the *saddle-point approximation.* In the saddle-point approximation, the *modulus* $|\varphi_0(\mu_{\mathrm{ren}})|^2$ is given by the solution [compare with Eq. (2.54)]

$$|\varphi_0(\mu_{\mathrm{ren}})|^2=\begin{cases}0, & \text{if } \mu_{\mathrm{ren}}<0,\\ \\ \frac{\mu_{\mathrm{ren}}}{\lambda}, & \text{if } \mu_{\mathrm{ren}}\geq0,\end{cases}\tag{2.93}$$

to the classical equation of motion

$$0=\frac{\delta S_{\rm E}^{(0)}}{\delta|\varphi_0|^2}, \tag{2.94}$$

and

$$Z_0\approx\begin{cases}1, & \text{if } \mu_{\rm ren}<0.\\ \\ \exp\left(+\beta V\frac{\mu_{\rm ren}^2}{2\lambda}\right), & \text{if } \mu_{\rm ren}\geq 0.\end{cases} \tag{2.95}$$

In turn, the dependence on β of the renormalized chemical potential is determined by demanding that

$$\overline{N}=V|\varphi_0(\mu_{\rm ren})|^2+\beta^{-1}\left\langle\int\limits_0^\beta \mathrm{d}\tau\int\limits_V \mathrm{d}^d\boldsymbol{r}\,\left(\widetilde{\varphi}^*\widetilde{\varphi}\right)(\boldsymbol{r},\tau)\right\rangle_{Z/Z_0}. \tag{2.96}$$

The tilde over $\widetilde{\varphi}^*(\boldsymbol{r},\tau)$ and $\widetilde{\varphi}(\boldsymbol{r},\tau)$ as well as the subscript Z/Z_0 are reminders that zero modes should be removed from the path integral in the second term on the right-hand side [compare with Eq. (2.71d)], as they would be counted twice when $\mu_{\rm ren}\geq 0$ otherwise.

The strategy that we shall pursue to go beyond the zero-mode approximation consists in the following steps.
Step 1: We assume that Eq. (2.93) holds with some $\mu_{\rm ren}>0$.
Step 2: We choose a convenient parametrization of the fluctuations $\widetilde{\varphi}^*(\boldsymbol{r},\tau)$ and $\widetilde{\varphi}(\boldsymbol{r},\tau)$ about the zero modes.
Step 3: We construct an effective theory in $\widetilde{\varphi}^*(\boldsymbol{r},\tau)$ and $\widetilde{\varphi}(\boldsymbol{r},\tau)$ to the desired accuracy.
Step 4: We solve Eq. (2.96) with the effective theory of step 3 and verify the self-consistency of step 1 within the accuracy of the approximation made in step 3.

This approximate scheme is called the random-phase approximation (RPA) when the effective theory in step 3 is non-interacting. It is nothing but an expansion of the action up to quadratic order in the fluctuations about the saddle-point or mean-field solution.

The zero-mode approximation in Eq. (2.93) leaves the choice of the phase of the zero mode φ_0 arbitrary. This is the classical implementation of the spontaneous breaking of the $U(1)$ symmetry associated with total particle-number conservation. Without loss of generality, the (*linear*) parametrization that we choose is

$$\varphi^*(\boldsymbol{r},\tau)=\sqrt{\frac{\mu_{\rm ren}}{\lambda}}+\widetilde{\varphi}^*(\boldsymbol{r},\tau),\qquad \varphi(\boldsymbol{r},\tau)=\sqrt{\frac{\mu_{\rm ren}}{\lambda}}+\widetilde{\varphi}(\boldsymbol{r},\tau). \tag{2.97}$$

Here, we are also assuming that

$$0=\langle\widetilde{\varphi}^*(\boldsymbol{r},\tau)\rangle_{Z/Z_0}=\langle\widetilde{\varphi}(\boldsymbol{r},\tau)\rangle_{Z/Z_0}. \tag{2.98}$$

This parametrization is the natural one if the approximation to the action is meant to linearize equations of motion. Correspondingly, the action is expanded up to second order in the deviations $\widetilde{\varphi}^*(\boldsymbol{r},\tau)$ and $\widetilde{\varphi}(\boldsymbol{r},\tau)$ from the saddle-point or mean-field Ansatz (2.93)

$$S_{\rm E}[\varphi^*,\varphi]\approx S^{(0)}_{\rm E}[\sqrt{\mu_{\rm ren}/\lambda},\sqrt{\mu_{\rm ren}/\lambda}] + \int_0^\beta \mathrm{d}\tau \int_V \mathrm{d}^d\boldsymbol{r}\,\left[\widetilde{\varphi}^*\left(\partial_\tau-\frac{\Delta}{2m}\right)\widetilde{\varphi}+\frac{\mu_{\rm ren}}{2}\left(\widetilde{\varphi}^*+\widetilde{\varphi}\right)^2\right](\boldsymbol{r},\tau) + \cdots . \tag{2.99}$$

If partial integrations are performed and all space or time total derivatives are dropped owing to the periodic boundary conditions in space and time, we find the approximation

$$\begin{aligned} S_{\rm E}[\varphi^*,\varphi]\approx\ & S^{(0)}_{\rm E}[\sqrt{\mu_{\rm ren}/\lambda},\sqrt{\mu_{\rm ren}/\lambda}] \\ &+\int_0^\beta \mathrm{d}\tau \int_V \mathrm{d}^d\boldsymbol{r}\,\left[(\mathrm{Re}\,\widetilde{\varphi})\left(-\frac{\Delta}{2m}+2\mu_{\rm ren}\right)(\mathrm{Re}\,\widetilde{\varphi})\right](\boldsymbol{r},\tau) \\ &+\int_0^\beta \mathrm{d}\tau \int_V \mathrm{d}^d\boldsymbol{r}\,\left[(\mathrm{Im}\,\widetilde{\varphi})\left(-\frac{\Delta}{2m}\right)(\mathrm{Im}\,\widetilde{\varphi})\right](\boldsymbol{r},\tau) \\ &+\int_0^\beta \mathrm{d}\tau \int_V \mathrm{d}^d\boldsymbol{r}\,\left[(\mathrm{Re}\,\widetilde{\varphi})\mathrm{i}\partial_\tau(\mathrm{Im}\,\widetilde{\varphi})-(\mathrm{Im}\,\widetilde{\varphi})\mathrm{i}\partial_\tau(\mathrm{Re}\,\widetilde{\varphi})\right](\boldsymbol{r},\tau) \\ &+\cdots . \end{aligned} \tag{2.100}$$

A quite remarkable phenomenon is displayed in Eq. (2.100). A purely imaginary term has appeared in the Euclidean effective action. Hence, this effective action cannot be interpreted as some classical action. The purely imaginary term is an example of a Berry phase. Upon canonical quantization, the commutator between $\mathrm{Re}\,\widehat{\widetilde{\varphi}}(\boldsymbol{r})$ and $\mathrm{Im}\,\widehat{\widetilde{\varphi}}(\boldsymbol{r})$ is the same as that between the position and momentum operators, respectively, in quantum mechanics (see chapter 1).

If we ignore for one instant the Berry phase term, i.e., ignore quantum mechanics, we can interpret $\mathrm{Re}\,\widetilde{\varphi}(\boldsymbol{r},\tau)$ as a massive mode and $\mathrm{Im}\,\widetilde{\varphi}(\boldsymbol{r},\tau)$ as a massless mode. The massive mode $\mathrm{Re}\,\widetilde{\varphi}(\boldsymbol{r},\tau)$ originates from the radial motion of a classical particle which, at rest, is sitting somewhere at the bottom of the circular potential well

$$U_0[\varphi^*_0,\varphi_0]:=\frac{S^{(0)}_{\rm E}[\varphi^*_0,\varphi_0]}{\beta V}=-\mu_{\rm ren}|\varphi_0|^2+\frac{\lambda}{2}|\varphi_0|^4. \tag{2.101}$$

The massless mode $\mathrm{Im}\,\widetilde{\varphi}(\boldsymbol{r},\tau)$ originates from the angular motion of this particle along the bottom of the circular potential well $U_0[\varphi^*_0,\varphi_0]$. At the classical level, the dispersions above the gap thresholds $2\mu_{\rm ren}$ and 0 of $\mathrm{Re}\,\widetilde{\varphi}(\boldsymbol{r},\tau)$ and $\mathrm{Im}\,\widetilde{\varphi}(\boldsymbol{r},\tau)$, respectively, are both quadratic in the momentum $\boldsymbol{k}$. Including quantum fluctuations through the Berry phase dramatically alters this picture. Indeed, these "two classical modes" are not independent, since they interact through the Berry phase terms.

As we now show, including the Berry phase couplings allows us to interpret $\widetilde{\varphi}$ as a mode with a linear (quadratic) dispersion relation at long (short) wavelengths.

The explicit dispersion can be obtained from Fourier transforming Eq. (2.100) into

$$\begin{aligned}
S_{\mathrm{E}} &\approx S_{\mathrm{E}}^{(0)} + \sum_{l\in\mathbb{Z}} \sum_{\frac{L\boldsymbol{k}}{2\pi}\in\mathbb{Z}^d} \begin{pmatrix} (\mathrm{Re}\,\widetilde{\varphi})_{-\boldsymbol{k},-\varpi_l} \\ (\mathrm{Im}\,\widetilde{\varphi})_{-\boldsymbol{k},-\varpi_l} \end{pmatrix}^{\mathsf{T}} \begin{pmatrix} \frac{\boldsymbol{k}^2}{2m} + 2\mu_{\mathrm{ren}} & +\varpi_l \\ -\varpi_l & \frac{\boldsymbol{k}^2}{2m} \end{pmatrix} \begin{pmatrix} (\mathrm{Re}\,\widetilde{\varphi})_{+\boldsymbol{k},+\varpi_l} \\ (\mathrm{Im}\,\widetilde{\varphi})_{+\boldsymbol{k},+\varpi_l} \end{pmatrix} \\
&= S_{\mathrm{E}}^{(0)} + \sum_{l\in\mathbb{Z}} \sum_{\frac{L\boldsymbol{k}}{2\pi}\in\mathbb{Z}^d} \begin{pmatrix} (\mathrm{Re}\,\widetilde{\varphi})_{\boldsymbol{k},\varpi_l} \\ (\mathrm{Im}\,\widetilde{\varphi})_{\boldsymbol{k},\varpi_l} \end{pmatrix}^{\dagger} \begin{pmatrix} \frac{\boldsymbol{k}^2}{2m} + 2\mu_{\mathrm{ren}} & +\varpi_l \\ -\varpi_l & \frac{\boldsymbol{k}^2}{2m} \end{pmatrix} \begin{pmatrix} (\mathrm{Re}\,\widetilde{\varphi})_{\boldsymbol{k},\varpi_l} \\ (\mathrm{Im}\,\widetilde{\varphi})_{\boldsymbol{k},\varpi_l} \end{pmatrix} \\
&= S_{\mathrm{E}}^{(0)} + \sum_{l\in\mathbb{Z}} \sum_{\frac{L\boldsymbol{k}}{2\pi}\in\mathbb{Z}^d} \begin{pmatrix} (\mathrm{Re}\,\widetilde{\varphi})_{-\boldsymbol{k},-\varpi_l} \\ (\mathrm{Im}\,\widetilde{\varphi})_{-\boldsymbol{k},-\varpi_l} \end{pmatrix}^{\dagger} \begin{pmatrix} \frac{\boldsymbol{k}^2}{2m} + 2\mu_{\mathrm{ren}} & -\varpi_l \\ +\varpi_l & \frac{\boldsymbol{k}^2}{2m} \end{pmatrix} \begin{pmatrix} (\mathrm{Re}\,\widetilde{\varphi})_{-\boldsymbol{k},-\varpi_l} \\ (\mathrm{Im}\,\widetilde{\varphi})_{-\boldsymbol{k},-\varpi_l} \end{pmatrix}.
\end{aligned} \tag{2.102}$$

To reach the second line, we have used the fact that the real and imaginary parts of $\varphi(\boldsymbol{r},\tau)$ are real-valued functions. To reach the third line, we made the relabeling $\boldsymbol{k}\to-\boldsymbol{k}$ and $\varpi_l\to-\varpi_l$, under which the ϖ_l-dependence of the kernel is odd while the $\boldsymbol{k}$-dependence of the kernel is even.

Define the 2×2 matrix-valued Green function by its matrix elements in the $\boldsymbol{k}$-ϖ_l basis,

$$\mathbb{G}_{-\boldsymbol{k},-\varpi_l} := -\left\langle \begin{pmatrix} (\mathrm{Re}\,\widetilde{\varphi})_{-\boldsymbol{k},-\varpi_l} \\ (\mathrm{Im}\,\widetilde{\varphi})_{-\boldsymbol{k},-\varpi_l} \end{pmatrix}^{\dagger} \begin{pmatrix} (\mathrm{Re}\,\widetilde{\varphi})_{-\boldsymbol{k},-\varpi_l} & (\mathrm{Im}\,\widetilde{\varphi})_{-\boldsymbol{k},-\varpi_l} \end{pmatrix} \right\rangle_{Z/Z_0}. \tag{2.103}$$

The "$-$" sign is convention. Evaluation of the Green function (2.103) is now an exercise in Gaussian integration over independent complex-valued integration variables that is summarized in appendix B.1. Hence, for any $\mathbf{0}\neq\frac{L\boldsymbol{k}}{2\pi}\in\mathbb{Z}^d$ and $l\in\mathbb{Z}$,

$$\begin{aligned}
\mathbb{G}_{-\boldsymbol{k},-\varpi_l} &= -\frac{1}{2} \begin{pmatrix} \frac{\boldsymbol{k}^2}{2m} + 2\mu_{\mathrm{ren}} & -\varpi_l \\ +\varpi_l & \frac{\boldsymbol{k}^2}{2m} \end{pmatrix}^{-1} \\
&= -\frac{1}{2} \frac{1}{\frac{\boldsymbol{k}^2}{2m}\left(\frac{\boldsymbol{k}^2}{2m} + 2\mu_{\mathrm{ren}}\right) + (\varpi_l)^2} \begin{pmatrix} \frac{\boldsymbol{k}^2}{2m} & +\varpi_l \\ -\varpi_l & \frac{\boldsymbol{k}^2}{2m} + 2\mu_{\mathrm{ren}} \end{pmatrix}.
\end{aligned} \tag{2.104}$$

The factor $1/2$ comes from the fact that only half of $\mathbf{0}\neq\frac{L\boldsymbol{k}}{2\pi}\in\mathbb{Z}^d$ are to be counted as independent labels, for $(\mathrm{Re}\,\widetilde{\varphi})(\boldsymbol{r},\tau)$ and $(\mathrm{Im}\,\widetilde{\varphi})(\boldsymbol{r},\tau)$ are real valued.

The Green function (2.104) has first-order poles whenever

$$\begin{aligned}
\mathrm{i}\varpi_l &= \pm\sqrt{\frac{\boldsymbol{k}^2}{2m}\left(\frac{\boldsymbol{k}^2}{2m} + 2\mu_{\mathrm{ren}}\right)} \\
&= \pm\frac{|\boldsymbol{k}|}{2m}\sqrt{\boldsymbol{k}^2 + 4\,m\,\mu_{\mathrm{ren}}} \\
&\equiv \pm\frac{|\boldsymbol{k}|}{2m}\sqrt{\boldsymbol{k}^2 + (k_0)^2} \\
&\equiv \pm\,\xi_{\boldsymbol{k}}.
\end{aligned} \tag{2.105}$$

We have recovered with $\xi_{\boldsymbol{k}}$ the dispersion in Eq. (2.60b). For long wavelengths, the dispersion is linear

$$\xi_{\boldsymbol{k}} = \frac{k_0}{2m}|\boldsymbol{k}| + \mathcal{O}\left[\left(\frac{|\boldsymbol{k}|}{k_0}\right)^2\right], \tag{2.106a}$$

with the speed of sound

$$v_0 \equiv \frac{k_0}{2m} := \sqrt{\frac{\mu_{\mathrm{ren}}}{m}}. \tag{2.106b}$$

For short wavelengths, the dispersion relation of a free particle emerges,

$$\xi_{\boldsymbol{k}} = \frac{\boldsymbol{k}^2}{2m} + \mathcal{O}\left[\left(\frac{k_0}{|\boldsymbol{k}|}\right)^2\right]. \tag{2.107}$$

It is time to verify the self-consistency of the assumptions encoded by Eqs. (2.93), (2.97), and (2.100). This we do by solving Eq. (2.96) in the Gaussian approximation. Fourier transform of Eq. (2.96) yields

$$\overline{N} = V(\varphi_0)^2 + \delta N, \tag{2.108a}$$

where

$$\begin{aligned}
\delta N :=& \beta^{-1}\left\langle \int_0^\beta \mathrm{d}\tau \int_V \mathrm{d}^d\boldsymbol{r}\, (\widetilde{\varphi}^*\widetilde{\varphi})\,(\boldsymbol{r},\tau)\right\rangle_{Z/Z_0} \\
=& \beta^{-1}\sum_{l\in\mathbb{Z}}\sum_{\frac{L\boldsymbol{k}}{2\pi}\in\mathbb{Z}^d}\left\langle\left(\mathrm{Re}\,\widetilde{\varphi}_{+\boldsymbol{k},+\varpi_l}\right)\left(\mathrm{Re}\,\widetilde{\varphi}_{-\boldsymbol{k},-\varpi_l}\right) + (\mathrm{Re}\to\mathrm{Im})\right\rangle_{Z/Z_0} \\
&+ \mathrm{i}\beta^{-1}\sum_{l\in\mathbb{Z}}\sum_{\frac{L\boldsymbol{k}}{2\pi}\in\mathbb{Z}^d}\left\langle\left(\mathrm{Re}\,\widetilde{\varphi}_{+\boldsymbol{k},+\varpi_l}\right)\left(\mathrm{Im}\,\widetilde{\varphi}_{-\boldsymbol{k},-\varpi_l}\right) - (\mathrm{Re}\leftrightarrow\mathrm{Im})\right\rangle_{Z/Z_0}.
\end{aligned} \tag{2.108b}$$

The Gaussian approximation gives the estimate

$$\delta N \approx \beta^{-1}\sum_{l\in\mathbb{Z}}\sum_{\frac{L\boldsymbol{k}}{2\pi}\in\mathbb{Z}^d}\frac{e^{\mathrm{i}0^+\varpi_l}}{\varpi_l^2+\xi_{\boldsymbol{k}}^2}\left(\frac{\boldsymbol{k}^2}{2m}+\mu_{\mathrm{ren}}+\mathrm{i}\varpi_l\right), \tag{2.109}$$

as can be read from the Green function (2.104). A convergence factor $\exp(\mathrm{i}0^+\varpi_l)$ was introduced to regulate the poles of the Green function (2.104). One verifies that the summand with $\boldsymbol{k}$ and l fixed, while $\mu = 0$ in Eq. (2.78) is recovered in the non-interacting limit $\mu_{\mathrm{ren}} = 0$.

At zero temperature, the summation over l turns into an integral

$$\sum_{l\in\mathbb{Z}} \to \beta\int_{-\infty}^{+\infty}\frac{\mathrm{d}\varpi}{2\pi}. \tag{2.110}$$

The integrand is nothing but $n_{\boldsymbol{k}}$. As a function of $\varpi\in\mathbb{C}$, it has two first-order poles along the imaginary axis at $\pm\mathrm{i}\xi_{\boldsymbol{k}}$ with residues $\pm\frac{1}{\mathrm{i}2\xi_{\boldsymbol{k}}}\left(\frac{\boldsymbol{k}^2}{2m}+\mu_{\mathrm{ren}}\mp\xi_{\boldsymbol{k}}\right)$. The convergence factor $\exp(\mathrm{i}0^{+}\varpi)$ allows one to close the real-line integral by a very large circle in the upper complex plane $\varpi \in \mathbb{C}$. Application of the residue theorems then yields

$$\begin{aligned} n_{\boldsymbol{k}} &\approx \frac{1}{2\xi_{\boldsymbol{k}}}\left(\frac{\boldsymbol{k}^2}{2m}+\mu_{\mathrm{ren}}-\xi_{\boldsymbol{k}}\right) \\ &= \frac{1}{2}\left(\frac{\boldsymbol{k}^2+(k_0)^2/2}{|\boldsymbol{k}|\sqrt{\boldsymbol{k}^2+(k_0)^2}}-1\right). \end{aligned} \tag{2.111}$$

In the thermodynamic limit,

$$V^{-1}\sum_{\boldsymbol{k}\neq 0} n_{\boldsymbol{k}} \approx \gamma_d (k_0)^d, \qquad \gamma_d := \frac{\Omega_d}{2(2\pi)^d}\int_0^\infty \mathrm{d}x x^{d-2}\left(\frac{x^2+1/2}{\sqrt{x^2+1}}-x\right). \tag{2.112}$$

Here, Ω_d is the area of the unit sphere in d dimensions. Since $k_0 = 2\sqrt{m\lambda}\varphi_0$, we conclude that φ_0 is determined by

$$\begin{aligned} \overline{n} &:= \frac{N}{V} \\ &= (\varphi_0)^2 + 2^d\gamma_d(m\lambda)^{d/2}(\varphi_0)^d \\ &= (\varphi_0)^2\left[1+2^d\gamma_d(m\lambda)^{d/2}(\varphi_0)^{d-2}\right]. \end{aligned} \tag{2.113}$$

The Gaussian approximation is selfconsistent if the quantum correction $2^d\gamma_d(m\lambda)^{d/2}(\varphi_0)^d$ is smaller than the semi-classical result $(\varphi_0)^2$, i.e., if

$$\begin{aligned} \varphi_0 &\sim \sqrt{\overline{n}} \\ &\ll \left[2^d\gamma_d(m\lambda)^{d/2}\right]^{-1/(d-2)}. \end{aligned} \tag{2.114}$$

The constant γ_d is finite if and only if $1 < d < 4$. For $d = 1$, γ_1 has an infrared logarithmic divergence. For $d = 4$, γ_4 has an ultraviolet logarithmic divergence. When Eq. (2.114) is satisfied,

$$(\varphi_0)^2 \approx \overline{n} - 2^d\gamma_d(m\lambda)^{d/2}\overline{n}^{d/2}. \tag{2.115}$$

The RPA (Gaussian approximation) in $d = 3$ is thus appropriate in the dilute limit or when the interacting coupling constant λ is small. For $d = 2$, Eq. (2.115) indicates that quantum corrections to the semi-classical result scale in the *same way* as a function of the particle density $\overline{n}$, but with the opposite sign. This is an indication that fluctuations are very important in two-dimensional space and that the RPA might then break down.

The case of two-dimensional space is indeed very special. To see this one can estimate the size of the fluctuations about the semi-classical value of the order parameter by calculating the root-mean-square deviation

$$\sqrt{\frac{\langle|\varphi_0|^2-\varphi^*(\boldsymbol{r},\tau)\varphi(\boldsymbol{r},\tau)\rangle_Z}{|\varphi_0|^2}}, \tag{2.116}$$

within the RPA. The root-mean-square deviation should be smaller than one in the thermodynamic limit if there is true long-range order, i.e., below the transition temperature. It should diverge upon approaching the transition temperature from below. To evaluate

$$\langle \varphi^*(\boldsymbol{r},\tau)\varphi(\boldsymbol{r},\tau)\rangle_Z \tag{2.117}$$

at inverse temperature β one must perform a Fourier integral over the entries of the Green function (2.104). These integrals are dominated at long wavelengths by the contribution

$$\beta^{-1}\int\frac{\mathrm{d}^d\boldsymbol{k}}{\boldsymbol{k}^2} \tag{2.118}$$

coming from the acoustic mode. This contribution is logarithmically (linearly) divergent in $d=2$ ($d=1$) whenever $\beta<\infty$. Hence, the root-mean-square deviation diverges in the thermodynamic limit and within the RPA, signaling the breakdown of spontaneous symmetry breaking and off-diagonal long-range order at any non-vanishing temperature when $d\leq 2$. It is said that $d=2$ is the *lower-critical dimension* at which and below which the $U(1)$ continuous symmetry cannot be spontaneously broken at any non-vanishing temperature within the RPA. Absence of spontaneous symmetry breaking of the $U(1)$ symmetry in the dilute Bose gas within the RPA is an example of the *Hohenberg-Mermin-Wagner-Coleman theorem.* It can be shown that thermal fluctuations due to acoustic modes downgrade the long-range order of the ground state to quasi-long-range order within the RPA. Quasi-long-range order is the property that the one-particle density matrix in Eq. (2.34) decays algebraically fast with $|\boldsymbol{r}'-\boldsymbol{r}|$ at long separations. Quasi-long-range order cannot maintain itself at arbitrary high temperatures. The mechanism by which quasi-long-range order is traded for exponentially fast decaying spatial correlations is called the *Kosterlitz-Thouless* transition. The *Kosterlitz-Thouless* transition cannot be accounted for within the RPA, since it is intrinsically a non-linear phenomenon. Chapter 4 is devoted to the Kosterlitz-Thouless transition.

2.5.3 *Beyond the random-phase approximation*

We close the discussion of a repulsive dilute Bose gas by sketching how one can go beyond the RPA defined by Eq. (2.100). The key to capturing physics beyond the RPA (2.100) is to choose the parametrization

$$\varphi^*(\boldsymbol{r},\tau)=\sqrt{\rho(\boldsymbol{r},\tau)}\,e^{-\mathrm{i}\theta(\boldsymbol{r},\tau)},\qquad \varphi(\boldsymbol{r},\tau)=\sqrt{\rho(\boldsymbol{r},\tau)}\,e^{+\mathrm{i}\theta(\boldsymbol{r},\tau)}, \tag{2.119}$$

of the fields entering the path integral (2.71). This parametrization is *non-linear.* It reduces to the *linear* parametrization (2.97) if one works to linear order in θ and makes the identifications

$$\rho(\boldsymbol{r},\tau)\to\frac{\mu_{\mathrm{ren}}}{\lambda},\qquad \mathrm{i}\rho(\boldsymbol{r},\tau)\theta(\boldsymbol{r},\tau)\to\widetilde{\varphi}(\boldsymbol{r},\tau). \tag{2.120}$$

If we were able to solve the repulsive dilute Bose gas model exactly, the choice of parametrization of the fields in the path integral (2.71) would not matter. However, performing approximations, say linearization of the equations of motion, can lead to very different physics depending on the initial choice of parametrization of the fields in the path integral. For example, the RPA on the linear parametrization (2.97) breaks down in $d = 1, 2$ due to the dominant role played by infrared fluctuations of the Goldstone mode. On the other hand, we shall argue that the non-linear parametrization (2.119) can account for these strong fluctuations. The physical content of Eq. (2.119) is to parametrize the fields in the path integral of Eq. (2.71) in terms of the density

$$\rho(\boldsymbol{r},\tau) := \varphi^*(\boldsymbol{r},\tau)\,\varphi(\boldsymbol{r},\tau), \tag{2.121}$$

and the currents

$$\begin{aligned} \boldsymbol{J}(\boldsymbol{r},\tau) &:= \frac{1}{2m\mathrm{i}}\left[\varphi^*(\boldsymbol{r},\tau)\,(\boldsymbol{\nabla}\varphi)\,(\boldsymbol{r},\tau) - (\boldsymbol{\nabla}\varphi^*)\,(\boldsymbol{r},\tau)\varphi(\boldsymbol{r},\tau)\right] \\ &= \frac{1}{m}\rho(\boldsymbol{r},\tau)(\boldsymbol{\nabla}\theta)(\boldsymbol{r},\tau), \end{aligned} \tag{2.122}$$

associated to the global $U(1)$ gauge invariance of the theory.

We begin by inserting Eq. (2.119) into the Lagrangian in Eq. (2.71)

$$\begin{aligned} \mathcal{L}_{\mathrm{E}} =& \mathrm{i}\rho\partial_\tau\theta + \frac{1}{2m}\left(\frac{1}{4\rho}(\boldsymbol{\nabla}\rho)^2 + \rho(\boldsymbol{\nabla}\theta)^2\right) - \mu_{\mathrm{ren}}\rho + \frac{\lambda}{2}\rho^2 \\ &+ \frac{1}{2}(\partial_\tau\rho) - \frac{1}{2}(\boldsymbol{\nabla}^2\rho) - \mathrm{i}\boldsymbol{\nabla}\left[\rho(\boldsymbol{\nabla}\theta)\right]. \end{aligned} \tag{2.123}$$

The second line does not contribute to the action, since fields obey periodic boundary conditions in imaginary time and in space.

Next, we expand the first line to quadratic order in powers of $\delta\rho$, to zero-th order in powers of $(\boldsymbol{\nabla}\delta\rho)^2$, and to zero-th order in powers of $(\delta\rho)(\boldsymbol{\nabla}\theta)^2$, where

$$\rho(\boldsymbol{r},\tau) = \rho_0 + \delta\rho(\boldsymbol{r},\tau), \qquad \rho_0 := \frac{\mu_{\mathrm{ren}}}{\lambda}, \qquad \int\limits_0^\beta \mathrm{d}\tau \int\limits_V \mathrm{d}^d\boldsymbol{r}\,\delta\rho(\boldsymbol{r},\tau) = 0. \tag{2.124}$$

This expansion is a good one at low temperatures when the renormalized chemical potential is strictly positive. In particular, note that, to the contrary of the RPA (2.100), we are not assuming that θ is small (only $\boldsymbol{\nabla}\theta$ is taken small). We find the quadratic action

$$\begin{aligned} S_{\mathrm{E}} =& S_{\mathrm{E}}^{(0)} + \int\limits_0^\beta \mathrm{d}\tau \int\limits_V \mathrm{d}^d\boldsymbol{r}\left[\mathrm{i}(\delta\rho)\partial_\tau\theta + \frac{\rho_0}{2m}(\boldsymbol{\nabla}\theta)^2 + \frac{\lambda}{2}(\delta\rho)^2 - (\delta\mu)(\delta\rho)\right] + \cdots \\ =& S_{\mathrm{E}}^{(0)} + \sum_l \sum_{\frac{L\boldsymbol{k}}{2\pi}\in\mathbb{Z}^d} \begin{pmatrix} \delta\rho_{+\boldsymbol{k},+\varpi_l} \\ \theta_{+\boldsymbol{k},+\varpi_l} \end{pmatrix}^\dagger \begin{pmatrix} \frac{\lambda}{2} & +\frac{\varpi_l}{2} \\ -\frac{\varpi_l}{2} & \frac{\rho_0}{2m}\boldsymbol{k}^2 \end{pmatrix} \begin{pmatrix} \delta\rho_{+\boldsymbol{k},+\varpi_l} \\ \theta_{+\boldsymbol{k},+\varpi_l} \end{pmatrix} \\ &- \sum_l \sum_{\frac{L\boldsymbol{k}}{2\pi}\in\mathbb{Z}^d} \delta\mu_{-\boldsymbol{k},-\varpi_l}\delta\rho_{+\boldsymbol{k},+\varpi_l} + \cdots, \end{aligned} \tag{2.125}$$

and the currents

$$\boldsymbol{J}(\boldsymbol{r},\tau)=\frac{\rho_0}{m}(\boldsymbol{\nabla}\theta)(\boldsymbol{r},\tau)+\cdots. \tag{2.126}$$

Here, we have also substituted $\mu_{\rm ren}$ by $\mu_{\rm ren}+\delta\mu(\boldsymbol{r},\tau)$. The source term $\delta\mu(\boldsymbol{r},\tau)$, $\int_0^\beta \mathrm{d}\tau \int_V \mathrm{d}^d\boldsymbol{r}\,(\delta\mu)(\boldsymbol{r},\tau)=0$, is a mathematical device to probe the response to an external "scalar" potential.

The first term on the right-hand side of the first line of Eq. (2.125) is the Berry phase that converts the radial ($\delta\rho$) and angular (θ) semi-classical modes into a quantum harmonic oscillator. It implies that $\delta\rho$ and θ are coupled through the classical equations of motion

$$\begin{aligned}0&=+\mathrm{i}\partial_\tau\theta+\lambda\delta\rho-\delta\mu\equiv+\mathrm{i}\partial_\tau\theta-\delta\mu_{\rm eff},\\ 0&=-\mathrm{i}\partial_\tau\delta\rho-\frac{\rho_0}{m}\Delta\theta\equiv-\mathrm{i}\partial_\tau\delta\rho+\boldsymbol{\nabla}\cdot\boldsymbol{J},\end{aligned} \tag{2.127}$$

in imaginary time. These equations are called the *Josephson equations*. Chapter 8 is devoted to their study in the context of superconductivity and quantum decoherence.

The second term on the right-hand side of the first line of Eq. (2.125) is only present below the $U(1)$ symmetry-breaking transition temperature. It endows the angular degree of freedom θ with a rigidity since, classically, $\frac{\rho_0}{2m}(\boldsymbol{\nabla}\theta)^2$ is the penalty in elastic energy paid by a gradient of the phase. Alternatively, this term corresponds to the kinetic energy of a "point-like particle" of mass m/ρ_0.

The third term on the right-hand side of the first line of Eq. (2.125) represents the potential-energy cost induced by the curvature of the semi-classical potential well (2.101) if the "point-like particle" moves by the amount $\delta\rho$ away from the bottom of the well.

Poles in the counterpart

$$\mathbb{G}^{\rm col}_{\boldsymbol{k},\varpi_l}=-\frac{1/2}{\frac{1}{4}\frac{\lambda\rho_0}{m}\boldsymbol{k}^2+\left(\frac{\varpi_l}{2}\right)^2}\begin{pmatrix}\frac{\rho_0}{2m}\boldsymbol{k}^2 & -\frac{\varpi_l}{2}\\ +\frac{\varpi_l}{2} & \frac{\lambda}{2}\end{pmatrix} \tag{2.128}$$

for $\delta\rho$ and θ to the Green function (2.104) give the linear dispersion

$$\xi^{\rm col}_{\boldsymbol{k}}=\sqrt{\frac{\lambda\rho_0}{m}}\,|\boldsymbol{k}|=\sqrt{\frac{\mu_{\rm ren}}{m}}\,|\boldsymbol{k}|. \tag{2.129}$$

There is a one-to-one correspondence between the existence of a linear dispersion relation and the existence of a rigidity $\propto\rho_0$ in our effective model. If we interpret the rigidity of the phase θ as superfluidity, we have establish the one-to-one correspondence between superfluidity and the existence of a *Goldstone mode* associated with spontaneous symmetry breaking. In turn, this Goldstone mode can only exist when $\lambda>0$, i.e., Bose-Einstein condensation at $\lambda=0$ cannot produce superfluidity.

The Green functions (2.104) and (2.128), although very similar, have a very different physical content. In the former case, the Green function describes single-particle properties. In the latter case, the Green function describes collective excitations ($\delta\rho$ and θ). For example, the equal-time density-density correlation function

$S_{\boldsymbol{k}}$ is given by

$$\begin{aligned}
S_{\boldsymbol{k}} &:= \frac{V}{N}\frac{1}{\beta}\sum_{l}\langle\delta\rho_{+l,+\boldsymbol{k}}\delta\rho_{-l,-\boldsymbol{k}}\rangle \\
&= \frac{V}{N}\frac{1}{\beta}\sum_{l}\frac{\frac{\rho_0}{m}|\boldsymbol{k}|^2}{(\xi^{\mathrm{col}}_{\boldsymbol{k}})^2+(\varpi_l)^2} \\
&= \frac{V}{N}\frac{\frac{\rho_0}{2m}|\boldsymbol{k}|^2}{\xi^{\mathrm{col}}_{\boldsymbol{k}}} + \mathcal{O}(\beta^{-1}) \\
\stackrel{\rho_0\approx\frac{N}{V}}{}&\approx \frac{|\boldsymbol{k}|^2}{2\,m\,\xi^{\mathrm{col}}_{\boldsymbol{k}}} + \mathcal{O}(\beta^{-1}).
\end{aligned} \tag{2.130}$$

The so-called *Feynman relation*

$$\xi^{\mathrm{col}}_{\boldsymbol{k}} \approx \frac{|\boldsymbol{k}|^2}{2mS_{\boldsymbol{k}}}, \tag{2.131}$$

which is valid at zero temperature, implies that the long-range correlation

$$S_{\boldsymbol{k}} \propto |\boldsymbol{k}| \tag{2.132}$$

is equivalent to an acoustic wave dispersion for density fluctuations. Feynman relation can be used to establish the existence of superfluidity in $d = 2$ when the criterion built on the RPA (2.100) fails. However, effective theory (2.125) is still too crude for a description of superfluidity in two-dimensional space.

2.6 Problems

2.6.1 *Magnons in quantum ferromagnets and antiferromagnets as emergent bosons*

Introduction

We are going to study spin-wave excitations (magnons) on top of a ferromagnetically and antiferromagnetically ordered state of quantum spins arranged on a lattice. We will treat this problem using the so-called Holstein-Primakoff transformation [29], which allows to rewrite the spin operators at each lattice site in terms of creation and annihilation operators of a boson. The virtue of this transformation is that the algebra of bosons is much simpler than that of spins, for example perturbation theory simplifies greatly by the availability of Wick theorem for bosons. However, the price to pay is that the bosonic operators are acting on a larger local Hilbert space at every lattice site than the spin operators. In this way, one might end up with solutions of the bosonic problem, that do not map back to physical states in the spin variables. This problem is avoided in the approximation by which spins only deviate slightly from the ferromagnetic or antiferromagnetic orientation, as we shall see below.

Our goal is to derive the dispersion relations of spin waves and to understand their connections to the symmetries of the problem. For small momenta $\boldsymbol{k}$, the dispersions will be quadratic in $\boldsymbol{k}$ in the ferromagnet and linear in $\boldsymbol{k}$ in the antiferromagnet.

We consider a Bravais lattice Λ that is spanned by the orthonormal vectors $\boldsymbol{a}_1$, $\cdots$, $\boldsymbol{a}_d$ with integer-valued coefficients and made of N sites labeled by $\boldsymbol{r}$. Each site $\boldsymbol{r} \in \Lambda$ is assigned a spin-S degree of freedom. The three components $\hat{S}^{\alpha}_{\boldsymbol{r}}$ with $\alpha = x, y, z$ of the spin operator $\hat{\boldsymbol{S}}_{\boldsymbol{r}} = \boldsymbol{e}_x \hat{S}^x_{\boldsymbol{r}} + \boldsymbol{e}_y \hat{S}^y_{\boldsymbol{r}} + \boldsymbol{e}_z \hat{S}^z_{\boldsymbol{r}}$ act on the $(2S+1)$-dimensional local Hilbert space

$$\mathcal{H}^{\mathrm{s}}_{\boldsymbol{r}} = \mathrm{span}\{|-S\rangle_{\boldsymbol{r}}, |-S+1\rangle_{\boldsymbol{r}}, \cdots, |S-1\rangle_{\boldsymbol{r}}, |S\rangle_{\boldsymbol{r}}\}, \tag{2.133a}$$

where we have chosen to represent $\mathcal{H}^{\mathrm{s}}_{\boldsymbol{r}}$ with the eigenbasis of $\hat{S}^z_{\boldsymbol{r}}$ $(\hbar = 1)$,

$$\hat{S}^z_{\boldsymbol{r}} |m\rangle_{\boldsymbol{r}} = m\, |m\rangle_{\boldsymbol{r}}, \qquad m = -S, -S+1, \cdots, S-1, S. \tag{2.133b}$$

The components of the spin-operator at each site $\boldsymbol{r} \in \Lambda$ obey the algebra

$$[\hat{S}^{\alpha}_{\boldsymbol{r}}, \hat{S}^{\beta}_{\boldsymbol{r}}] = \mathrm{i}\,\epsilon^{\alpha\beta\gamma}\, \hat{S}^{\gamma}_{\boldsymbol{r}}, \qquad \alpha, \beta, \gamma = x, y, z, \tag{2.134}$$

and commute between different sites. Equivalently, the operators

$$\hat{S}^{\pm}_{\boldsymbol{r}} := \hat{S}^x_{\boldsymbol{r}} \pm \mathrm{i}\, \hat{S}^y_{\boldsymbol{r}} \tag{2.135a}$$

obey the algebra

$$[\hat{S}^{+}_{\boldsymbol{r}}, \hat{S}^{-}_{\boldsymbol{r}}] = 2\hat{S}^z_{\boldsymbol{r}}, \qquad [\hat{S}^{\pm}_{\boldsymbol{r}}, \hat{S}^z_{\boldsymbol{r}}] = \mp \hat{S}^{\pm}_{\boldsymbol{r}}, \tag{2.135b}$$

at each site $\boldsymbol{r} \in \Lambda$ and commute between different sites.

Ferromagnetic spin waves

The Heisenberg Hamiltonian for a ferromagnet in the uniform magnetic field $B\,\boldsymbol{e}_z$, that only includes the nearest-neighbor Heisenberg exchange coupling $J > 0$ is given by

$$\hat{H}_{\mathrm{F}} := -J \sum_{\langle \boldsymbol{r}, \boldsymbol{r}' \rangle} \hat{\boldsymbol{S}}_{\boldsymbol{r}} \cdot \hat{\boldsymbol{S}}_{\boldsymbol{r}'} - B \sum_{\boldsymbol{r} \in \Lambda} \hat{S}^z_{\boldsymbol{r}}, \tag{2.136}$$

where $\langle \boldsymbol{r}, \boldsymbol{r}' \rangle$ indicates that the sum is only taken over directed nearest-neighbor lattice sites. Periodic boundary conditions are imposed.

If $B > 0$, all the magnetic moments align along the positive $\boldsymbol{e}_z$ direction in internal spin space and the ground state is given by

$$|0\rangle = \bigotimes_{\boldsymbol{r} \in \Lambda} |S\rangle_{\boldsymbol{r}}. \tag{2.137}$$

Consider introducing a bosonic degree of freedom at every lattice site $\boldsymbol{r} \in \Lambda$ with creation and annihilation operators $\hat{a}^{\dagger}_{\boldsymbol{r}}$ and $\hat{a}_{\boldsymbol{r}}$, respectively, that act on the local infinite-dimensional Hilbert space

$$\mathcal{H}^{\mathrm{b}}_{\boldsymbol{r}} = \mathrm{span}\left\{ |n\rangle_{\boldsymbol{r}} := \frac{(\hat{a}^{\dagger}_{\boldsymbol{r}})^n}{\sqrt{n!}} |0\rangle_{\boldsymbol{r}} \,\middle|\, n = 0, 1, 2, \cdots, \qquad \hat{a}_{\boldsymbol{r}} |0\rangle_{\boldsymbol{r}} = 0 \right\}, \tag{2.138}$$

and obey

$$[\hat{a}_{\boldsymbol{r}}, \hat{a}^{\dagger}_{\boldsymbol{r}'}] = \delta_{\boldsymbol{r},\boldsymbol{r}'}, \tag{2.139}$$

with all other commutators vanishing. The Holstein-Primakoff transformation is defined by the following substitutes for the spin operators

$$\hat{\mathcal{S}}^{+}_{\boldsymbol{r}} := \left(\sqrt{2S - \hat{a}^{\dagger}_{\boldsymbol{r}}\,\hat{a}_{\boldsymbol{r}}}\right)\,\hat{a}_{\boldsymbol{r}}, \tag{2.140a}$$

$$\hat{\mathcal{S}}^{-}_{\boldsymbol{r}} := \hat{a}^{\dagger}_{\boldsymbol{r}}\sqrt{2S - \hat{a}^{\dagger}_{\boldsymbol{r}}\,\hat{a}_{\boldsymbol{r}}}, \tag{2.140b}$$

$$\hat{\mathcal{S}}^{z}_{\boldsymbol{r}} := S - \hat{a}^{\dagger}_{\boldsymbol{r}}\,\hat{a}_{\boldsymbol{r}}. \tag{2.140c}$$

Exercise 1.1: Show that the operators $\hat{\mathcal{S}}^{+}_{\boldsymbol{r}}$, $\hat{\mathcal{S}}^{-}_{\boldsymbol{r}}$, and $\hat{\mathcal{S}}^{z}_{\boldsymbol{r}}$ defined in Eq. (2.140) on $\mathcal{H}^{\mathrm{b}}_{\boldsymbol{r}}$ obey the same algebra (2.135b) as the operators $\hat{S}^{+}_{\boldsymbol{r}}$, $\hat{S}^{-}_{\boldsymbol{r}}$ and $\hat{S}^{z}_{\boldsymbol{r}}$ on $\mathcal{H}^{\mathrm{s}}_{\boldsymbol{r}}$.

Exercise 1.2: As we are going to study wave-like excitations, it will be convenient to express the theory in the Fourier components of the bosonic operators, that is

$$\hat{c}^{\dagger}_{\boldsymbol{k}} := \frac{1}{\sqrt{N}}\sum_{\boldsymbol{r}\in\Lambda} e^{-\mathrm{i}\boldsymbol{k}\cdot\boldsymbol{r}}\,\hat{a}^{\dagger}_{\boldsymbol{r}}, \qquad \hat{c}_{\boldsymbol{k}} := \frac{1}{\sqrt{N}}\sum_{\boldsymbol{r}\in\Lambda} e^{+\mathrm{i}\boldsymbol{k}\cdot\boldsymbol{r}}\,\hat{a}_{\boldsymbol{r}}, \tag{2.141}$$

where the wave number $\boldsymbol{k}$ belongs to the first Brillouin zone. Show that $\hat{c}^{\dagger}_{\boldsymbol{k}}$ and $\hat{c}_{\boldsymbol{k}}$ obey the algebra

$$[\hat{c}_{\boldsymbol{k}}, \hat{c}^{\dagger}_{\boldsymbol{k}'}] = \delta_{\boldsymbol{k},\boldsymbol{k}'}, \tag{2.142}$$

with all other commutators vanishing.

Now, the central assumption that will allow us to simplify the theory when written in the bosonic variables is that the fraction of reversed spins above the ferromagnetic ground state is small

$$\frac{\langle \hat{a}^{\dagger}_{\boldsymbol{r}}\,\hat{a}_{\boldsymbol{r}}\rangle}{S} \ll 1 \tag{2.143}$$

for all $\boldsymbol{r}\in\Lambda$. Within the range of validity of this assumption, it is justified to expand the square-roots that enter Eq. (2.140) in $\hat{a}^{\dagger}_{\boldsymbol{r}}\hat{a}_{\boldsymbol{r}}/2S$.

Exercise 1.3: Using the spin-wave representation (2.140) of the spin operators, show that the Hamiltonian (2.136) becomes

$$\hat{H}'_{\mathrm{F}} = -J\,N\,z\,S^{2} - B\,N\,S - 2\,J\,z\,S\sum_{\boldsymbol{k}}\left(\frac{\gamma_{\boldsymbol{k}}+\gamma_{-\boldsymbol{k}}}{2} - 1\right)\hat{c}^{\dagger}_{\boldsymbol{k}}\,\hat{c}_{\boldsymbol{k}} + B\sum_{\boldsymbol{k}}\hat{c}^{\dagger}_{\boldsymbol{k}}\,\hat{c}_{\boldsymbol{k}}, \tag{2.144}$$

when expanded to second order in the spin-wave variables $\hat{c}^{\dagger}_{\boldsymbol{k}}$ and $\hat{c}_{\boldsymbol{k}}$ and all higher-order terms are neglected. Here, z is half the coordination number of the Bravais lattice and the form factor

$$\gamma_{\boldsymbol{k}} := \frac{1}{z}\sum_{\boldsymbol{\delta}} e^{+\mathrm{i}\boldsymbol{k}\cdot\boldsymbol{\delta}} \tag{2.145}$$

includes a sum over the z vectors $\boldsymbol{\delta}$ that are the directed connections to nearest-neighbor sites. (Verify that $\sum_{\boldsymbol{k}}\gamma_{\boldsymbol{k}} = 0$.)

With the help of $\gamma_{\boldsymbol{k}} = \gamma^{*}_{-\boldsymbol{k}}$, it follows that the non-constant part of the Hamiltonian (2.144) reads

$$\hat{H}''_{\mathrm{F}} = \sum_{\boldsymbol{k}} \left[2J\,z\,S(1 - \operatorname{Re}\gamma_{\boldsymbol{k}}) + B\right] \hat{c}^{\dagger}_{\boldsymbol{k}}\, \hat{c}_{\boldsymbol{k}}. \tag{2.146}$$

This allows to directly read off the dispersion relation of ferromagnetic magnons on Bravais lattices

$$\omega_{\boldsymbol{k}} := 2\,J\,z\,S(1 - \operatorname{Re}\gamma_{\boldsymbol{k}}) + B. \tag{2.147}$$

Exercise 1.4: Show that for a hypercubic lattice with lattice constant $\mathfrak{a}$, the magnon dispersion becomes

$$\omega_{\boldsymbol{k}} \approx B + J\,S\,\mathfrak{a}^{2}\,|\boldsymbol{k}|^{2}, \tag{2.148}$$

in the limit $\mathfrak{a}\,|\boldsymbol{k}| \ll 1$. This is equivalent to the dispersion relation of a free massive particle of mass $m^{*} := 1/(2\,J\,S\,\mathfrak{a}^{2})$.

Exercise 1.5: Suppose we had considered Hamiltonian (2.136) with $B < 0$. Then, the naive expectation for the ground state is the one with all the magnetic moments aligned along the *negative* $\boldsymbol{e}_z$-axis in internal spin space,

$$|0\rangle = \bigotimes_{\boldsymbol{r}\in\Lambda} |-S\rangle_{\boldsymbol{r}}. \tag{2.149}$$

What is the counterpart to the operators (2.140) for this situation?

Antiferromagnetic spin waves

To study spin waves above an antiferromagnetic ground state, our starting point has to be modified in three respects. First, we have to assume that the lattice Λ is bipartite, i.e., it can be divided into two interpenetrating sublattices Λ_A and Λ_B such that all nearest neighbors of any lattice site in Λ_A are lattice sites in Λ_B and vice versa. Hence,

$$\Lambda = \Lambda_A \cup \Lambda_B, \qquad \Lambda_A \cap \Lambda_B = \emptyset. \tag{2.150}$$

For example, in two dimensions, the square lattice is bipartite, but the triangular lattice is not (the triangular lattice is tripartite). Second, for an antiferromagnetic coupling, the sign of the interaction parameter J that enters the Hamiltonian (2.136) has to be changed (we will implement this sign change explicitly below, keeping $J > 0$ as before). Third, the ferromagnetic source field $B\,\boldsymbol{e}_z$ must be replaced by a staggered source field $\pm B\,\boldsymbol{e}_z$, where one sign is assigned to sublattice Λ_A and the other sign is assigned to sublattice Λ_B.

We will thus consider the Hamiltonian

$$\hat{H}_{\mathrm{AF}} := J \sum_{\langle \boldsymbol{r},\boldsymbol{r}'\rangle} \hat{\boldsymbol{S}}_{\boldsymbol{r}} \cdot \hat{\boldsymbol{S}}_{\boldsymbol{r}'} - B \sum_{\boldsymbol{r}\in\Lambda_A} \hat{S}^{z}_{\boldsymbol{r}} + B \sum_{\boldsymbol{r}\in\Lambda_B} \hat{S}^{z}_{\boldsymbol{r}}, \tag{2.151}$$

where $J > 0$ and $B > 0$ and the sums in the second-to-last and last term on the right-hand side only run over lattice sites in sublattices Λ_A and Λ_B, respectively. Periodic boundary conditions are imposed.

This time, the Holstein-Primakoff transformation differs on sublattice Λ_A from that on sublattice Λ_B (recall exercise 1.5). It is given by

$$\hat{\mathcal{S}}^{+}_{\boldsymbol{r}} := \left(\sqrt{2S - \hat{a}^{\dagger}_{\boldsymbol{r}}\,\hat{a}_{\boldsymbol{r}}}\right)\hat{a}_{\boldsymbol{r}}, \qquad \hat{\mathcal{S}}^{-}_{\boldsymbol{r}} := \hat{a}^{\dagger}_{\boldsymbol{r}}\sqrt{2S - \hat{a}^{\dagger}_{\boldsymbol{r}}\,\hat{a}_{\boldsymbol{r}}}, \qquad \hat{\mathcal{S}}^{z}_{\boldsymbol{r}} := S - \hat{a}^{\dagger}_{\boldsymbol{r}}\,\hat{a}_{\boldsymbol{r}}, \tag{2.152a}$$

for all sites on sublattice Λ_A and

$$\hat{\mathcal{S}}^{+}_{\boldsymbol{r}} := \hat{b}^{\dagger}_{\boldsymbol{r}}\sqrt{2S - \hat{b}^{\dagger}_{\boldsymbol{r}}\,\hat{b}_{\boldsymbol{r}}}, \qquad \hat{\mathcal{S}}^{-}_{\boldsymbol{r}} := \left(\sqrt{2S - \hat{b}^{\dagger}_{\boldsymbol{r}}\,\hat{b}_{\boldsymbol{r}}}\right)\hat{b}_{\boldsymbol{r}}, \quad \hat{\mathcal{S}}^{z}_{\boldsymbol{r}} := \hat{b}^{\dagger}_{\boldsymbol{r}}\,\hat{b}_{\boldsymbol{r}} - S, \tag{2.152b}$$

for all sites on sublattice Λ_B.

Exercise 2.1: We define the magnon variables on the two sublattices as

$$\hat{c}^{\dagger}_{\boldsymbol{k}} := \frac{1}{\sqrt{N_A}}\sum_{\boldsymbol{r}\in\Lambda_A} e^{-\mathrm{i}\,\boldsymbol{k}\cdot\boldsymbol{r}}\,\hat{a}^{\dagger}_{\boldsymbol{r}}, \qquad \hat{c}_{\boldsymbol{k}} := \frac{1}{\sqrt{N_A}}\sum_{\boldsymbol{r}\in\Lambda_A} e^{+\mathrm{i}\,\boldsymbol{k}\cdot\boldsymbol{r}}\,\hat{a}_{\boldsymbol{r}}, \tag{2.153a}$$

$$\hat{d}^{\dagger}_{\boldsymbol{k}} := \frac{1}{\sqrt{N_B}}\sum_{\boldsymbol{r}\in\Lambda_B} e^{+\mathrm{i}\,\boldsymbol{k}\cdot\boldsymbol{r}}\,\hat{b}^{\dagger}_{\boldsymbol{r}}, \qquad \hat{d}_{\boldsymbol{k}} := \frac{1}{\sqrt{N_B}}\sum_{\boldsymbol{r}\in\Lambda_B} e^{-\mathrm{i}\,\boldsymbol{k}\cdot\boldsymbol{r}}\,\hat{b}_{\boldsymbol{r}}, \tag{2.153b}$$

where $N_A = N_B = N/2$ are the numbers of sites on sublattice Λ_A and Λ_B, which we take to be equal. What is the first Brillouin zone for $\boldsymbol{k}$, how does it compare in size to that of the case of a ferromagnet with the same lattice Λ? Assuming the limit in which the spin-wave approximation (2.143) is valid, expand the operators $\hat{\mathcal{S}}^{+}_{\boldsymbol{r}}$ and $\hat{\mathcal{S}}^{z}_{\boldsymbol{r}}$ from Eq. (2.152) on both sublattices Λ_A and Λ_B to second order in the bosonic variables $\hat{a}^{\dagger}_{\boldsymbol{r}}$, $\hat{a}_{\boldsymbol{r}}$, and $\hat{b}^{\dagger}_{\boldsymbol{r}}$, $\hat{b}_{\boldsymbol{r}}$. Then, express the result in terms of the magnon variables $\hat{c}^{\dagger}_{\boldsymbol{k}}$, $\hat{c}_{\boldsymbol{k}}$, $\hat{d}^{\dagger}_{\boldsymbol{k}}$, and $\hat{d}_{\boldsymbol{k}}$ defined in Eq. (2.153).

Exercise 2.2: Show that the Hamiltonian (2.151), when expanded to quadratic order in the magnon variables, becomes

$$\begin{aligned} H'_{\mathrm{AF}} = &- J\,N\,z\,S^2 - B\,N\,S \\ &+ 2J\,z\,S\sum_{\boldsymbol{k}} \mathrm{Re}\,(\gamma_{\boldsymbol{k}})\left(\hat{c}^{\dagger}_{\boldsymbol{k}}\,\hat{d}^{\dagger}_{\boldsymbol{k}} + \hat{c}_{\boldsymbol{k}}\,\hat{d}_{\boldsymbol{k}}\right) + (B + 2J\,z\,S)\sum_{\boldsymbol{k}}\left(\hat{c}^{\dagger}_{\boldsymbol{k}}\,\hat{c}_{\boldsymbol{k}} + \hat{d}^{\dagger}_{\boldsymbol{k}}\,\hat{d}_{\boldsymbol{k}}\right). \end{aligned} \tag{2.154}$$

The remaining task is to diagonalize the Hamiltonian (2.154), which is already quadratic in the bosonic variables. To that end, we make the Ansatz

$$\hat{\alpha}_{\boldsymbol{k}} := u_{\boldsymbol{k}}\,\hat{c}_{\boldsymbol{k}} - v_{\boldsymbol{k}}\,\hat{d}^{\dagger}_{\boldsymbol{k}}, \qquad \hat{\beta}_{\boldsymbol{k}} := u_{\boldsymbol{k}}\,\hat{d}_{\boldsymbol{k}} - v_{\boldsymbol{k}}\,\hat{c}^{\dagger}_{\boldsymbol{k}}, \tag{2.155a}$$

for some real-valued functions $u_{\boldsymbol{k}}$ and $v_{\boldsymbol{k}}$ with

$$u^2_{\boldsymbol{k}} - v^2_{\boldsymbol{k}} = 1. \tag{2.155b}$$

Condition (2.155b) ensures the bosonic commutation relations

$$[\hat{\alpha}_{\boldsymbol{k}}, \hat{\alpha}^{\dagger}_{\boldsymbol{k}}] = [\hat{\beta}_{\boldsymbol{k}}, \hat{\beta}^{\dagger}_{\boldsymbol{k}}] = 1. \tag{2.156}$$

Exercise 2.3: Show that for an appropriate choice of $u_{\boldsymbol{k}}$ and $v_{\boldsymbol{k}}$, the Hamiltonian becomes

$$H'_{\mathrm{AF}} = -J\,N\,z\,S(S+1) - \frac{1}{2}\,B\,N\,(2S+1) + \sum_{\boldsymbol{k}}\omega_{\boldsymbol{k}}\left(\hat{\alpha}^{\dagger}_{\boldsymbol{k}}\,\hat{\alpha}_{\boldsymbol{k}} + \hat{\beta}^{\dagger}_{\boldsymbol{k}}\,\hat{\beta}_{\boldsymbol{k}} + 1\right), \tag{2.157a}$$

with the magnon dispersion

$$\omega_{\boldsymbol{k}} := +\sqrt{(2\,J\,z\,S+B)^2-(2\,J\,z\,S\,\gamma_{\boldsymbol{k}})^2}. \tag{2.157b}$$

Why did we choose the positive square root and not the negative one?

Exercise 2.4: Show that in the limit of small momenta $\mathfrak{a}\,|\boldsymbol{k}|\ll 1$ and vanishing staggered source field $\pm B\,\boldsymbol{e}_z=0$, the magnon dispersion on a simple cubic lattice with lattice constant $\mathfrak{a}$ is given by the two degenerate solutions

$$\omega_{\boldsymbol{k}} = 2\,\sqrt{3}\,J\,S\,\mathfrak{a}\,|\boldsymbol{k}|. \tag{2.158}$$

When a ferromagnet and an antiferromagnet share the same lattice Λ with the cardinality N very large but finite, is there a difference in the total number of their magnons?

Comparison of antiferromagnetic and ferromagnetic cases

Having worked out the Holstein-Primakoff treatment of spin wave for both ferromagnetic and antiferromagnetic order, we will now comment on the reasons for the differences between the two cases. Both cases can be understood as an instance of spontaneous symmetry breaking, if the magnitude $|B|$ of the source field is taken to zero at the end of the calculation. Then, the operator that multiplies the source field, i.e.,

$$\hat{M}_{\mathrm{F}} := \sum_{\boldsymbol{r}\in\Lambda}\hat{S}^z_{\boldsymbol{r}}, \qquad \hat{M}_{\mathrm{AF}} := \sum_{\boldsymbol{r}\in\Lambda_A}\hat{S}^z_{\boldsymbol{r}} - \sum_{\boldsymbol{r}\in\Lambda_B}\hat{S}^z_{\boldsymbol{r}}, \tag{2.159}$$

is the order parameter for either case.

Exercise 3.1: Convince yourself that

$$[\hat{H}_0,\hat{M}_{\mathrm{F}}]=0 \tag{2.160a}$$

while

$$[\hat{H}_0,\hat{M}_{\mathrm{AF}}]\neq 0, \tag{2.160b}$$

where $\hat{H}_0$ is

$$\hat{H}_0 := -J\sum_{\langle\boldsymbol{r},\boldsymbol{r}'\rangle}\hat{\boldsymbol{S}}_{\boldsymbol{r}}\cdot\hat{\boldsymbol{S}}_{\boldsymbol{r}'}. \tag{2.161}$$

We thus observe the following fundamental difference. The Hamiltonian $\hat{H}_0$ commutes with the symmetry-breaking field for the case of ferromagnetism and it does not commute with the symmetry-breaking field in the case of antiferromagnetism.

We can expand on this observation by considering the symmetries of Hamiltonians (2.136) and (2.151) in the limit $B=0$, i.e., Hamiltonian (2.161) with $J>0$ and $J<0$, respectively, when Λ is bipartite. The Hamiltonian $\hat{H}_0$, if the lattice Λ has the point group $\mathcal{P}_\Lambda$ as symmetry group, has the symmetry group

$$G^\Lambda = SO(3)\times\mathbb{Z}_2\times\mathcal{P}_\Lambda. \tag{2.162}$$

Here, the factor groups are the symmetry group of proper rotations in spin space $SO(3) \cong SU(2)/\mathbb{Z}_2$, the symmetry group $\mathbb{Z}_2$ that represents time reversal, which acts like inversion in spin space, and the point group $\mathcal{P}_\Lambda$. For the antiferromagnet, we anticipate the breaking of the sublattice symmetry by factoring the symmetry group $\mathbb{Z}_2$ that interchanges the two sublattices from the point group $\mathcal{P}_\Lambda$ of the lattice Λ, so that

$$\mathcal{P}_\Lambda = \mathbb{Z}_2 \times \mathcal{P}_{\Lambda_A}. \tag{2.163}$$

Here, the factor group $\mathbb{Z}_2$ is the group generated by the interchange of sublattices Λ_A and Λ_B. The factor group $\mathcal{P}_{\Lambda_A}$ is made of the point-group transformations of sublattice Λ_A. Note that reversal of time is represented by an antiunitary operator, while all other symmetries are unitary.

For the ferromagnet, the order parameter $\hat{M}_{\mathrm{F}}$ is one of the generators of the continuous global symmetry group $SO(3) \cong SU(2)/\mathbb{Z}_2$. As a corollary, $\hat{M}_{\mathrm{F}}$ commutes with $\hat{H}_0$, as we have verified explicitly. Thus, $\hat{M}_{\mathrm{F}}$ and with it the ferromagnetic ground state both break the global symmetry group $SO(3)$ down to the subgroup $SO(2)$. At the same time, $\hat{M}_{\mathrm{F}}$ breaks the time-reversal symmetry, i.e., the inversion symmetry in spin space. The symmetry-breaking pattern of the ferromagnet with the Hamiltonian $\hat{H}_0$ obtained by taking the limit $N \to \infty$ before taking the limit $B \to 0$ in Hamiltonian (2.136) is thus

$$G^\Lambda = SO(3) \times \mathbb{Z}_2 \times \mathcal{P}_\Lambda \longrightarrow H^\Lambda_{\mathrm{F}} = SO(2) \times \mathcal{P}_\Lambda. \tag{2.164}$$

For the antiferromagnet, the order parameter $\hat{M}_{\mathrm{AF}}$ also breaks the rotation group $SO(3)$ down to the subgroup $SO(2)$ and breaks time-reversal symmetry. However, in contrast to the ferromagnet, $\hat{M}_{\mathrm{AF}}$ is unchanged under a composition of time-reversal and exchange of sublattices Λ_A and Λ_B. The symmetry breaking pattern of the antiferromagnet with the Hamiltonian $\hat{H}_0$ obtained by taking the limit $N \to \infty$ before taking the limit $B \to 0$ in Hamiltonian (2.151) is thus

$$G^\Lambda = SO(3) \times \mathbb{Z}_2 \times \mathbb{Z}_2 \times \mathcal{P}_{\Lambda_A} \longrightarrow H^\Lambda_{\mathrm{AF}} = SO(2) \times \mathbb{Z}_2 \times \mathcal{P}_{\Lambda_A}. \tag{2.165}$$

A fundamental difference between the broken symmetry groups H^Λ_{F} and H^Λ_{AF} is that all symmetries in H^Λ_{F} are represented by unitary operators, while the $\mathbb{Z}_2$ symmetry in H^Λ_{AF} is represented by an antiunitary operator, i.e., a composition of sublattice exchange and time-reversal. In that sense, it is seen that the antiferromagnet preserves an effective or emergent time-reversal symmetry.

The fact that $\hat{M}_{\mathrm{F}}$ commutes with $\hat{H}_0$ makes the classical [eigenstate of $\hat{H}$ defined by Eq. (2.136) in the limit $B/J \to \infty$ taken after the thermodynamic limit $N \to \infty$] and the quantum mechanical [eigenstate of $\hat{H}$ defined by Eq. (2.136) in the limit $B/J \to 0$ taken after the thermodynamic limit $N \to \infty$] ground states coincide.

Furthermore, the ferromagnet allows for an *exact* treatment of the one-magnon excitations above the ground state, as we shall now explore. For simplicity, we consider a one-dimensional lattice Λ with a spin-1/2 degree of freedom on every lattice site $r \in \Lambda$. We impose periodic boundary conditions. The state

$$|0\rangle = |\uparrow\uparrow \cdots \uparrow\uparrow\rangle \tag{2.166}$$

is a ground state of Hamiltonian (2.136) for ferromagnetic $J > 0$. We first try the states

$$|r\rangle := \hat{S}_r^- |0\rangle, \qquad r \in \Lambda, \tag{2.167}$$

as candidates for excited states. However, these are not eigenstates of the Hamiltonian (2.136). Instead, consider the superposition

$$|\Psi_\alpha\rangle := \sum_{r\in\Lambda} \alpha_r\, |r\rangle, \tag{2.168}$$

with some coefficients $\alpha_r \in \mathbb{C}$.

Exercise 3.2: Assuming a one-dimensional lattice Λ with the lattice spacing $\mathfrak{a}$, and with the Ansatz $\alpha_r = e^{ik\,r}/\sqrt{N}$, show that the magnon dispersion

$$E_k = E_0 + J\big[1 - \cos(k\,\mathfrak{a})\big] \tag{2.169}$$

follows from $\hat{H}_0$. Here, $E_0 = -J\,N/4$ is the ground state energy. Which values of k are allowed by the periodic boundary conditions?

The exact one-magnon dispersion (2.169) coincides with the dispersion (2.147) when Λ is a linear chain with the lattice spacing $\mathfrak{a}$. However, in the derivation of Eq. (2.147), we had made the approximation of small fractional spin reversal (2.143), not knowing that we would nevertheless obtain an exact result.

For the antiferromagnet, such an exact treatment cannot be carried out. The reason is that, unlike with the ferromagnet, the classical [eigenstate of $\hat{H}$ defined by Eq. (2.151) in the limit $B/J \to \infty$ taken after the thermodynamic limit $N \to \infty$] and the quantum mechanical [eigenstate of $\hat{H}$ defined by Eq. (2.151) in the limit $B/J \to 0$] taken after the thermodynamic limit $N \to \infty$] ground states of the antiferromagnet do not coincide due to the lack of commutativity between $\hat{H}_0$ and the antiferromagnetic order parameter (the staggered magnetization) [see Eq. (2.160)]. While in the ferromagnetic ground state at zero temperature, all spins are fully aligned saturating the magnetization to its maximal value, this is not the case for the staggered sublattice magnetization of the antiferromagnet.

We are going to use the quantities introduced in the Holstein-Primakoff treatment of antiferromagnetic magnons to calculate the deviation ΔM from the fully polarized sublattice magnetization $N\,S/2$ of the antiferromagnet. It is given by

$$\Delta M := \frac{N}{2} S - \left\langle \sum_{\boldsymbol{r}\in\Lambda_A} \hat{S}_{\boldsymbol{r}} \right\rangle. \tag{2.170}$$

Here, $\langle\cdot\rangle$ denotes the ground state expectation value, which is defined by the occupation numbers $\hat{\alpha}^\dagger_{\boldsymbol{k}}\, \hat{\alpha}_{\boldsymbol{k}}$ and $\hat{\beta}^\dagger_{\boldsymbol{k}}\, \hat{\beta}_{\boldsymbol{k}}$ from Eq. (2.155a) being zero for all $\boldsymbol{k}$ from the first Brillouin zone of Λ_A.

Exercise 3.3: Using Eqs. (2.152), (2.153), and the inverse transform of Eq. (2.155a) for the case when $B = 0$, show that

$$\Delta M = \frac{1}{2} \sum_{\boldsymbol{k}} \left(\frac{1}{\sqrt{1 - (\operatorname{Re}\gamma_{\boldsymbol{k}})^2}} - 1 \right). \tag{2.171}$$

Hint: Use that $\langle \hat{\alpha}^\dagger_{\boldsymbol{k}} \hat{\alpha}_{\boldsymbol{k}} \rangle = 0$, $\langle \hat{\beta}^\dagger_{\boldsymbol{k}} \hat{\beta}_{\boldsymbol{k}} \rangle = 0$, while all off-diagonal terms such as $\langle \hat{\alpha}^\dagger_{\boldsymbol{k}} \hat{\beta}_{\boldsymbol{k}} \rangle$ vanish.

Exercise 3.4: Evaluate Eq. (2.171) numerically for a simple cubic lattice with lattice spacing $\mathfrak{a} = 1$ in the thermodynamic limit, i.e., evaluate the integral

$$\begin{aligned} \Delta M &= -\frac{N}{4} + \frac{1}{2(2\pi)^3} \int \mathrm{d}^3\boldsymbol{k} \, \frac{1}{\sqrt{1 - (\mathrm{Re}\,\gamma_{\boldsymbol{k}})^2}} \\ &\approx 0.0784 \, \frac{N}{2}. \end{aligned} \tag{2.172}$$

Chapter 3

Non-Linear Sigma Models

Outline

The $O(N)$ Non-Linear Sigma model (NLσM) is defined as an effective field theory that encodes the pattern of symmetry breaking of an $O(N)$ classical Heisenberg model defined on a square lattice. A geometric and group-theoretical interpretation of the $O(N)$ NLσM is given. The notions of fixed points and the notions of relevant, marginal, and irrelevant perturbations are introduced. The real-valued scalar (free) field theory in two-dimensional Euclidean space is taken as and example of a critical field theory and the Callan-Symanzik equations obeyed by $(m + n)$ point correlation functions are derived for the two-dimensional $O(2)$ NLσM. The Callan-Symanzik equations are generalized to include relevant coupling constants. The Callan-Symanzik equation obeyed by the spin-spin correlator in the two-dimensional $O(N > 2)$ NLσM is derived. The beta function and the wave-function renormalization are computed up to second order in the coupling constant of the two-dimensional $O(N > 2)$ NLσM. It is shown that the free-field fixed point is IR unstable in that the coupling constant flows to strong coupling upon coarse graining and that there exists a finite correlation length that increases exponentially fast upon approaching the free-field unstable fixed point. The beta function for the $O(N)$ NLσM is also computed in more than two-dimensions, in which case it exhibits a non-trivial fixed point at a small but finite value of the coupling constant.

3.1 Introduction

The so-called Non-Linear Sigma Models (NLσMs) will occupy us for this and the following chapters. NLσM were first introduced in high-energy physics in the context of *chiral symmetry breaking.* NLσM also play an essential role in condensed matter physics where they appear naturally as effective field theories describing the low-energy and long-wavelength limit of numerous microscopic models. We begin by defining the $O(3)$ NLσM. We then proceed with NLσM whose target manifolds are Riemannian manifolds, homogeneous spaces, and symmetric spaces.

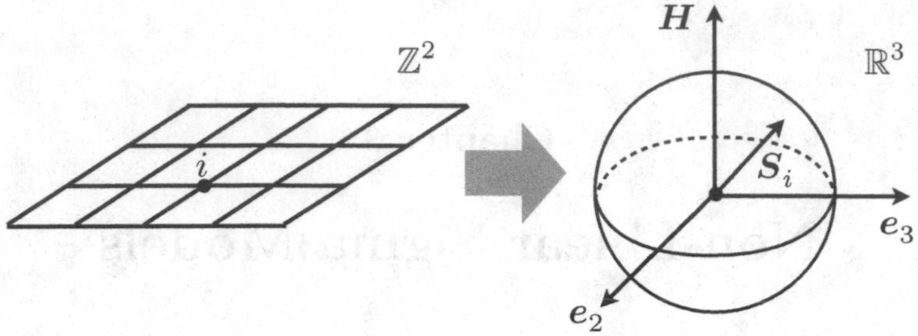

Fig. 3.1 Heisenberg model on a square lattice with $O(3)/O(2)$ symmetry in the presence of a uniform magnetic field $\boldsymbol{H}$. A symmetry breaking magnetic field enforces the pattern of spontaneous-symmetry breaking into the ferromagnetic state. The symmetry group of the Heisenberg exchange interaction is $O(3)$. A uniform magnetic field $\boldsymbol{H}$ breaks the $O(3)$ symmetry down to the subgroup $O(2)$ of rotations about the axis pointing along $\boldsymbol{H}$. There is a one-to-one correspondence between elements of the coset space $O(3)/O(2)$ and points of the two-sphere, i.e., the surface of the unit sphere in $\mathbb{R}^3$.

3.2 Non-Linear Sigma Models (NLσM)

We shall set the Boltzmann constant $k_{\mathrm{B}} = 1.38065 \times 10^{-23}$ J/K to unity, in which case $\beta \equiv 1/(k_{\mathrm{B}}T) = 1/T$ is the inverse temperature throughout this chapter.

3.2.1 *Definition of $O(N)$ NLσM*

An example for an *unfrustrated* classical $O(N)$ Heisenberg magnet at the inverse temperature β and in the presence of an external uniform magnetic field $\boldsymbol{H}$ is given by the classical partition function

$$Z_{\mathcal{N},\beta J,\beta \boldsymbol{H}} := \int \mathcal{D}[\boldsymbol{S}] \exp\left(\beta J \sum_{\langle ij\rangle} \boldsymbol{S}_i \cdot \boldsymbol{S}_j + \beta \sum_i \boldsymbol{H} \cdot \boldsymbol{S}_i\right), \tag{3.1a}$$

where the local degrees of freedom

$$\boldsymbol{S}_i = \boldsymbol{S}_{i+L\boldsymbol{e}_\mu} \in \mathbb{R}^N, \quad \boldsymbol{S}_i^2 = 1, \quad i \in \mathbb{Z}^d, \quad \boldsymbol{H} \in \mathbb{R}^N, \quad N = 1, 2, \cdots, \tag{3.1b}$$

are defined on a lattice spanned by the basis vectors

$$\boldsymbol{e}_\mu^{\mathsf{T}} = (0 \ \cdots \ 0 \ \mathfrak{a} \ 0 \ \cdots 0), \quad \mu = 1, \cdots, d. \tag{3.1c}$$

Accordingly, on each site $i \in \mathbb{Z}^d$ of a d-dimensional hypercubic lattice made of $\mathcal{N} = (L/\mathfrak{a})^d$ sites, a unit vector $\boldsymbol{S}_i$ of $\mathbb{R}^N$ interacts with its $2d$ nearest-neighbors $\boldsymbol{S}_{i\pm\boldsymbol{e}_\mu}$, $\mu = 1, \cdots, d$, $[\langle ij\rangle \equiv \langle i(i+\boldsymbol{e}_\mu)\rangle]$ through the ferromagnetic Heisenberg coupling constant (also called spin stiffness) $J > 0$ as well as with an external uniform magnetic field $\boldsymbol{H}$. Periodic boundary conditions are imposed. The measure $\mathcal{D}[\boldsymbol{S}]$ is the product measure over the infinitesimal surface element of the unit sphere in $\mathbb{R}^N$. Alternatively and in anticipation of taking the continuum limit, the lattice spacing $\mathfrak{a}$ can be made explicit through rescaling,

$$Z_{\mathcal{N},\beta J,\beta \boldsymbol{H}} \propto \left[\prod_i \int_{-\infty}^{+\infty} \mathrm{d}^N \boldsymbol{s}_i\, \delta\left(\frac{1}{g^2} - \boldsymbol{s}_i^2\right)\right] e^{\mathfrak{a}^{d-2} \sum_{\langle ij \rangle} \boldsymbol{s}_i \cdot \boldsymbol{s}_j + \mathfrak{a}^d \sum_i \boldsymbol{h}\cdot\boldsymbol{s}_i}, \tag{3.2a}$$

or

$$Z_{\mathcal{N},\beta J,\beta \boldsymbol{H}} \propto \lim_{\lambda\to\infty} \left(\prod_i \int_{-\infty}^{+\infty} \mathrm{d}^N \boldsymbol{s}_i\right) e^{\mathfrak{a}^{d-2} \sum_{\langle ij \rangle} \boldsymbol{s}_i \cdot \boldsymbol{s}_j + \mathfrak{a}^d \sum_i \left[\boldsymbol{h}\cdot\boldsymbol{s}_i - \lambda\left(\frac{1}{g^2} - \boldsymbol{s}_i^2\right)^2\right]}, \tag{3.2b}$$

or

$$Z_{\mathcal{N},\beta J,\beta \boldsymbol{H}} \propto \left(\prod_i \int_{-\infty}^{+\infty} \mathrm{d}^N \boldsymbol{s}_i \int_{-\infty}^{+\infty} \frac{\mathrm{d}\lambda_i}{2\pi}\right) e^{\mathfrak{a}^{d-2} \sum_{\langle ij \rangle} \boldsymbol{s}_i \cdot \boldsymbol{s}_j + \mathfrak{a}^d \sum_i \left[\boldsymbol{h}\cdot\boldsymbol{s}_i + \mathrm{i}\lambda_i\left(\frac{1}{g^2} - \boldsymbol{s}_i^2\right)\right]}, \tag{3.2c}$$

where

$$g^2 := \mathfrak{a}^{d-2}\frac{1}{\beta J}, \qquad \boldsymbol{h} := \mathfrak{a}^{-d+\frac{d-2}{2}}\sqrt{\frac{\beta}{J}}\,\boldsymbol{H} = \mathfrak{a}^{-d}\beta g \boldsymbol{H}. \tag{3.2d}$$

Both the measure and the Heisenberg interaction are invariant under any *global* rotation $Q \in O(N)$ of the spins $\boldsymbol{S}_i$[1]

$$\boldsymbol{S}_i = Q\,\tilde{\boldsymbol{S}}_i, \qquad \forall i. \tag{3.4}$$

The uniform magnetic field $\boldsymbol{H}$ breaks this global $O(N)$ invariance down to the subgroup $O(N-1)$ of global rotations in the $(N-1)$-dimensional subspace of $\mathbb{R}^N$ orthogonal to $\boldsymbol{H}$ (see Fig. 3.1). A magnetic field should here be thought of as either a formal device to break the $O(N)$ symmetry down to $O(N-1)$ when it is uniform or as a source term inserted for mathematical convenience to compute correlation functions, in which case it can be taken to be non-uniform. In both interpretations, it must be set to zero at the end of the day [see Eq. (3.12)].

Invariance of the partition function under the transformation

$$J \to -J, \qquad \boldsymbol{S}_i \to (-)^{||i/\mathfrak{a}||}\,\boldsymbol{S}_i, \qquad \boldsymbol{H} \to (-)^{||i/\mathfrak{a}||}\,\boldsymbol{H}, \tag{3.5a}$$

where

$$||i/\mathfrak{a}|| \equiv \sum_{\mu=1}^{d} i_\mu, \qquad i \equiv \sum_{\mu=1}^{d} i_\mu\, e_\mu, \qquad i_\mu \in \mathbb{Z}, \tag{3.5b}$$

defines absence of frustration. In general, a lattice is said to be *geometrically frustrated* when it cannot be decomposed into two interpenetrating sublattices. For frustrated lattices, ferromagnetic $(J > 0)$ and antiferromagnetic $(J < 0)$ couplings

[1] The group of orthogonal matrices $O(N)$ is made of all $N \times N$ matrices Q with real-valued matrix elements, non-vanishing determinant, and obeying

$$Q^{\mathsf{T}}\,Q = Q\,Q^{\mathsf{T}} = \mathbb{1}_N. \tag{3.3}$$

Equation (3.3) implies that the determinant of an orthogonal matrix is ± 1. The subgroup $SO(N) \subset O(N)$ is made of all orthogonal matrices with determinant one.

are not equivalent. A next-nearest-neighbor coupling constant on a square lattice is another way by which frustration arises.

The spin configuration with the lowest energy has all $\mathcal{N}$ spins parallel to the external magnetic field $\boldsymbol{H}$, i.e., is fully polarized into the *ferromagnetic ground state.* The *uniform magnetization*

$$\boldsymbol{M} := \frac{1}{\mathcal{N}} \sum_{i} \boldsymbol{S}_i \tag{3.6}$$

is maximal in magnitude in the ferromagnetic ground state,

$$|\boldsymbol{M}| \leq |\boldsymbol{M}_{\text{ferro}}| = 1, \qquad \boldsymbol{M}_{\text{ferro}} := \frac{\boldsymbol{H}}{|\boldsymbol{H}|}. \tag{3.7}$$

The expectation value of the magnetization

$$\lim_{|\boldsymbol{H}|\to 0} \langle \boldsymbol{M} \rangle_{Z_{\mathcal{N},\beta J,\beta \boldsymbol{H}}} := \lim_{|\boldsymbol{H}|\to 0} \frac{1}{\mathcal{N}\beta} \partial_{\boldsymbol{H}} \ln Z_{\mathcal{N},\beta J,\beta \boldsymbol{H}} \tag{3.8}$$

vanishes for any finite $\mathcal{N}$.[2] In the thermodynamic limit, the expectation value of the magnetization (3.8) depends crucially on the order in which the two limits $\mathcal{N} \to \infty$ and $|\boldsymbol{H}| \to 0$ are taken. On the one hand, if the limit $|\boldsymbol{H}| \to 0$ is taken before the thermodynamic limit $\mathcal{N} \to \infty$, then the expectation value of the magnetization vanishes at any temperature. On the other hand, if the limit $|\boldsymbol{H}| \to 0$ is taken after the thermodynamic limit $\mathcal{N} \to \infty$, then the expectation value of the magnetization need not vanish anymore (the answer depends on the dimensionality d of the lattice), since any two configurations $\{\boldsymbol{S}_i\}_{i\in\mathbb{Z}^d}$ and $\{\widetilde{\boldsymbol{S}}_i\}_{i\in\mathbb{Z}^d}$ of the spins that differ by a global or rigid rotation $Q \neq \mathbb{1}_N \in O(N)$ of all spins,

$$\boldsymbol{S}_i = Q\, \widetilde{\boldsymbol{S}}_i, \qquad \forall i \in \mathbb{Z}^d, \tag{3.9}$$

differ in energy by the infinitely high potential barrier

$$\lim_{\mathcal{N}\to\infty} \left[\mathcal{N} \times |\boldsymbol{H} \cdot (\mathbb{1}_N - Q)\, \boldsymbol{M}|\right] = \infty. \tag{3.10}$$

How should one decide if *spontaneous symmetry breaking* at zero temperature as defined by

$$1 = \lim_{|\boldsymbol{H}|\to 0} \lim_{\mathcal{N}\to\infty} \lim_{\beta\to\infty} \langle \boldsymbol{M} \rangle_{Z_{\mathcal{N},\beta J,\beta \boldsymbol{H}}}, \qquad 0 = \lim_{\mathcal{N}\to\infty} \lim_{|\boldsymbol{H}|\to 0} \lim_{\beta\to\infty} \langle \boldsymbol{M} \rangle_{Z_{\mathcal{N},\beta J,\beta \boldsymbol{H}}}, \tag{3.11}$$

extends to finite temperature, i.e.,

$$0 \neq \lim_{|\boldsymbol{H}|\to 0} \lim_{\mathcal{N}\to\infty} \langle \boldsymbol{M} \rangle_{Z_{\mathcal{N},\beta J,\beta \boldsymbol{H}}}, \qquad 0 = \lim_{\mathcal{N}\to\infty} \lim_{|\boldsymbol{H}|\to 0} \langle \boldsymbol{M} \rangle_{Z_{\mathcal{N},\beta J,\beta \boldsymbol{H}}}? \tag{3.12}$$

Since the thermodynamic limit must matter for spontaneous-symmetry breaking to take place, we can limit ourselves to very long wavelengths. Since we want to

[2] This is so because only the magnitude of the magnetization is fixed in the ferromagnetic ground state when the external magnetic field has been switched off. The direction in which the magnetization points is arbitrary. Hence, the path integral over all spin configurations can be restricted to a path integral over all spins pointing to the northern hemisphere of the N-dimensional unit sphere in some arbitrarily chosen spherical coordinate system provided $-\boldsymbol{M}$ is added to $+\boldsymbol{M}$ between the brackets on the left-hand side of Eq. (3.8).

know whether or not zero-temperature spontaneous-symmetry breaking is destroyed by thermal fluctuations for arbitrarily small temperatures, we can limit ourselves to low energies. If we are after some sort of perturbation theory, the dimensionless bare coupling constant

$$\mathfrak{a}^{-(d-2)}\,g^2 = \frac{1}{\beta J} \tag{3.13}$$

might be a good candidate at very low temperatures and very large spin stiffness J.

As the simplest possible effective field theory sharing the global $O(N)$ symmetry of Eqs. (3.1) or (3.2) in the absence of a symmetry breaking external magnetic field, we might try the Euclidean field theory[3]

$$Z_{\beta J,\beta \boldsymbol{H}} := \int \mathcal{D}[\boldsymbol{n}]\,\delta\left(1-\boldsymbol{n}^2\right)\exp\left(-\frac{1}{2g^2}\int\limits_{\mathbb{R}^d}\mathrm{d}^d\boldsymbol{x}\left[\left(\partial_\mu \boldsymbol{n}\right)^2 - 2\mathfrak{a}^{-d}\beta g^2 \boldsymbol{H}\cdot\boldsymbol{n}\right]\right), \tag{3.14a}$$

which defines the $O(N)$ NLσM when $\boldsymbol{H}=0$. The partition function (3.14a) is proportional to the partition function

$$Z_{\beta J,\beta \boldsymbol{H}} \propto \int \mathcal{D}[\boldsymbol{m}]\,\delta\left(\frac{1}{g^2}-\boldsymbol{m}^2\right)\exp\left(-\frac{1}{2}\int\limits_{\mathbb{R}^d}\mathrm{d}^d\boldsymbol{x}\left[\left(\partial_\mu \boldsymbol{m}\right)^2 - 2\boldsymbol{h}\cdot\boldsymbol{m}\right]\right), \tag{3.14b}$$

which is obtained by rescaling the dimensionless field $\boldsymbol{n}$ through a division by the positive square root $g=+\sqrt{g^2}$ of the coupling constant g^2. In turn, we can exponentiate the constraint on the length of the vector field $\boldsymbol{m}$ in two different ways. First,

$$Z_{\beta J,\beta \boldsymbol{H}} \propto \lim_{\lambda\to\infty}\left[\prod_{\boldsymbol{x}\in\mathbb{R}^d}\int\limits_{-\infty}^{+\infty}\mathrm{d}^N\boldsymbol{m}(\boldsymbol{x})\right] e^{-\frac{1}{2}\int\limits_{\mathbb{R}^d}\mathrm{d}^d\boldsymbol{x}\left\{\left(\partial_\mu\boldsymbol{m}\right)^2-2\left[\boldsymbol{h}\cdot\boldsymbol{m}-\lambda\left(\frac{1}{g^2}-\boldsymbol{m}^2\right)^2\right]\right\}}. \tag{3.14c}$$

Second,

$$Z_{\beta J,\beta \boldsymbol{H}} \propto \left[\prod_{\boldsymbol{x}\in\mathbb{R}^d}\int\limits_{-\infty}^{+\infty}\mathrm{d}^N\boldsymbol{m}(\boldsymbol{x})\int\limits_{-\infty}^{+\infty}\frac{\mathrm{d}\lambda(\boldsymbol{x})}{2\pi}\right] e^{-\frac{1}{2}\int\limits_{\mathbb{R}^d}\mathrm{d}^d\boldsymbol{x}\left\{\left(\partial_\mu\boldsymbol{m}\right)^2-2\left[\boldsymbol{h}\cdot\boldsymbol{m}-\mathrm{i}\lambda\left(\frac{1}{g^2}-\boldsymbol{m}^2\right)\right]\right\}}. \tag{3.14d}$$

3.2.2 *O(N) NLσM as a field theory on a Riemannian manifold*

To better understand the relationship between the theory (3.14) in the continuum and the theory (3.2) on the lattice, choose a coordinate system of $\mathbb{R}^N$ in which the

[3] For any pair of directed nearest-neighbor sites $\langle ij\rangle$, we write $\boldsymbol{S}_i\cdot\boldsymbol{S}_j = -\frac{1}{2}\left(\boldsymbol{S}_i-\boldsymbol{S}_j\right)^2+\frac{1}{2}\boldsymbol{S}_i^2+\frac{1}{2}\boldsymbol{S}_j^2$ in Eq. (3.1). We may then replace finite differences by derivatives in the spirit of a naive continuum limit.

symmetry breaking magnetic field $\boldsymbol{h}$ is aligned along the direction $\boldsymbol{e}_1$ (see Fig. 3.1),

$$\boldsymbol{h}\cdot\boldsymbol{m} = |\boldsymbol{h}|m_1. \tag{3.15}$$

We first observe that[4]

$$\begin{aligned}\int\mathcal{D}[\boldsymbol{m}]\,\delta\left(\frac{1}{g^2}-\boldsymbol{m}^2\right)(\cdots) =& \left\{\prod_{\boldsymbol{x}\in\mathbb{R}^d}\int_{-\infty}^{+\infty}\mathrm{d}[m_2(\boldsymbol{x})]\cdots\int_{-\infty}^{+\infty}\mathrm{d}[m_N(\boldsymbol{x})]\right\}\\ &\times\Theta\left(\frac{1}{g^2}-\sum_{j=2}^{N}m_j^2(\boldsymbol{x})\right)\left[\left.\frac{(\cdots)}{2\,m_1(\boldsymbol{x})}\right|_{m_1(\boldsymbol{x})=+\sigma(\boldsymbol{x})}+\left.\frac{(\cdots)}{2|m_1(\boldsymbol{x})|}\right|_{m_1(\boldsymbol{x})=-\sigma(\boldsymbol{x})}\right]\end{aligned} \tag{3.16a}$$

where $\Theta(x)$ is the Heaviside step function and we have introduced

$$\sigma(\boldsymbol{x}) := \sqrt{\frac{1}{g^2}-\sum_{j=2}^{N}m_j^2(\boldsymbol{x})}. \tag{3.16b}$$

Motivated by Eq. (3.16), we shall restrict all the local configurations $\boldsymbol{m}(\boldsymbol{x})$ entering in the path integral (3.14) to configurations called *spin waves* which are defined by the conditions that:

(1) Local longitudinal fluctuations $\sigma(\boldsymbol{x})$ about the ferromagnetic state

$$\boldsymbol{m}(\boldsymbol{x}) = \frac{1}{g}\,\boldsymbol{e}_1, \qquad \forall\boldsymbol{x}, \qquad \boldsymbol{e}_1^{\mathsf{T}} = (1\,0\cdots 0), \tag{3.17a}$$

are strictly positive

$$0 < \sigma(\boldsymbol{x}) \equiv m_1(\boldsymbol{x}) \leq 1/g, \qquad \forall\boldsymbol{x}. \tag{3.17b}$$

(2) Local transverse fluctuations $\boldsymbol{\pi}$,

$$\pi_i(\boldsymbol{x}) \equiv m_{i+1}(\boldsymbol{x}), \qquad i = 1,\cdots,N-1, \tag{3.17c}$$

about the ferromagnetic state are smaller in magnitude than $1/g$,

$$0 < \sigma(\boldsymbol{x}) = +\sqrt{\frac{1}{g^2}-\boldsymbol{\pi}^2(\boldsymbol{x})} \leq \frac{1}{g}. \tag{3.17d}$$

(3) Transverse fluctuations $\boldsymbol{\pi}$ are smooth, i.e., the Taylor expansion

$$\pi_i(\boldsymbol{x}+y\boldsymbol{e}_\mu) = \sum_{l=0}^{\infty}\frac{1}{l!}\left(\partial_\mu^l\pi_i\right)(\boldsymbol{x})y^l, \qquad i=1,\cdots,N-1, \qquad \mu=1,\cdots,d, \tag{3.17e}$$

converges very rapidly when $|y|$ is of the order of the lattice spacing $\mathfrak{a}$.

[4] Use $\delta(x^2-\mathfrak{a}^2) = \delta[(x-\mathfrak{a})(x+\mathfrak{a})] = \sum_{\pm}\frac{1}{2|\pm\mathfrak{a}|}\delta[x-(\pm\mathfrak{a})]$.

Only the northern-half hemisphere of the surface of the sphere with radius $1/g$ in $\mathbb{R}^N$ is thus parametrized in the *spin-wave approximation*. Accessing configurations of spins in which the field $\boldsymbol{m}$ points locally, say at $\boldsymbol{x}$, towards the southern hemisphere, $m_1(\boldsymbol{x}) < 0$, is impossible within the *spin-wave parametrization* (3.17d). In the *spin-wave approximation*, the second additive term on the right-hand side of Eq. (3.16a) is neglected. This approximation is good energetically since configurations of spins in which $m_1(\boldsymbol{x})$ is negative over some large but bounded region Ω of $\mathbb{R}^d$ is suppressed by the exponential factor of order $\exp\left(+|\boldsymbol{h}|\int\limits_{\Omega} \mathrm{d}^d\boldsymbol{x}\, m_1(\boldsymbol{x})\right)$ in $(\cdots)$ of Eq. (3.16a). However, this argument fails to account for the entropy of the excursions of m_1 into the southern hemisphere, i.e., the multiplicity of spin configurations that are suppressed by an exponentially small penalty in energy for pointing antiparallel to $\boldsymbol{h}$ in some region of $\mathbb{R}^d$. The spin-wave approximation breaks down whenever the entropy of *defects* by which m_1 is antiparallel to $\boldsymbol{h}$ overcomes the loss of energy incurred by this excursion of m_1 into the southern hemisphere.

In the spin-wave approximation, the Euclidean action of the NLσM becomes

$$\begin{aligned} S_{\mathrm{sw}\,\beta,\boldsymbol{h}} &:= \frac{1}{2}\int\limits_{\mathbb{R}^d} \mathrm{d}^d\boldsymbol{x} \left[(\partial_\mu\sigma)^2 + (\partial_\mu\boldsymbol{\pi})^2 - 2|\boldsymbol{h}|\sigma\right] \\ &= \frac{1}{2}\int\limits_{\mathbb{R}^d} \mathrm{d}^d\boldsymbol{x} \left[\frac{(\pi_i\partial_\mu\pi_i)(\pi_j\partial_\mu\pi_j)}{\frac{1}{g^2}-\boldsymbol{\pi}^2} + (\partial_\mu\boldsymbol{\pi})^2 - 2|\boldsymbol{h}|\sqrt{\frac{1}{g^2}-\boldsymbol{\pi}^2}\right] \\ &= \frac{1}{2}\int\limits_{\mathbb{R}^d} \mathrm{d}^d\boldsymbol{x} \left[(\partial_\mu\pi_i)\,\mathfrak{g}_{ij}\,(\partial_\mu\pi_j) - 2|\boldsymbol{h}|\sqrt{\frac{1}{g^2}-\boldsymbol{\pi}^2}\right], \end{aligned} \tag{3.18a}$$

where the symmetric (metric) tensor

$$\mathfrak{g}_{ij}(\boldsymbol{x}) := \frac{g^2(\pi_i\pi_j)(\boldsymbol{x})}{1-g^2\boldsymbol{\pi}^2(\boldsymbol{x})} + \delta_{ij}, \qquad i,j = 1,\cdots,N-1, \tag{3.18b}$$

has been introduced and summation over repeated indices is understood.

The metric tensor transforms according to

$$(\partial_\mu\pi_i)\,\mathfrak{g}_{ij}\,(\partial_\mu\pi_j) = (\partial_\mu\widetilde{\pi}_k)\,\widetilde{\mathfrak{g}}_{kl}\,(\partial_\mu\widetilde{\pi}_l), \tag{3.19a}$$

where

$$\begin{aligned} \widetilde{\mathfrak{g}}_{kl} &= R_{ik}\,\mathfrak{g}_{ij}\,R_{jl} \\ &= \left(R^{\mathsf{T}}\right)_{ki}\,\mathfrak{g}_{ij}\,R_{jl}, \qquad k,l = 1,\cdots,N-1, \end{aligned} \tag{3.19b}$$

under the global rotation $R \in O(N-1)$ of the transverse modes $\boldsymbol{\pi}$ under which

$$\boldsymbol{\pi}(\boldsymbol{x}) = R\,\widetilde{\boldsymbol{\pi}}(\boldsymbol{x}). \tag{3.19c}$$

In matrix form, Eq. (3.19) reads

$$\widetilde{\mathfrak{g}}(\boldsymbol{x}) = R^{\mathsf{T}}\,\mathfrak{g}(\boldsymbol{x})\,R, \quad \forall R \in O(N-1) \Longleftrightarrow \mathfrak{g}(\boldsymbol{x}) = R\,\widetilde{\mathfrak{g}}(\boldsymbol{x})\,R^{\mathsf{T}}, \quad \forall R \in O(N-1). \tag{3.20}$$

A useful invariant under global $O(N-1)$ rotations of the transverse modes $\boldsymbol{\pi}$ is the determinant of the metric tensor (3.18b),

$$\begin{aligned}\det[\mathfrak{g}(\boldsymbol{x})] &= \det\left[R\,\widetilde{\mathfrak{g}}(\boldsymbol{x})\,R^{\mathsf{T}}\right] \\ &= \det(R)\,\det[\widetilde{\mathfrak{g}}(\boldsymbol{x})]\,\det\left(R^{\mathsf{T}}\right) \\ &= [\det(R)]^2\,\det[\widetilde{\mathfrak{g}}(\boldsymbol{x})] \\ &= \det[\widetilde{\mathfrak{g}}(\boldsymbol{x})]\,, \quad \forall R\in O(N-1).\end{aligned} \tag{3.21}$$

Equation (3.21) also extends to the situation when the matrix $R \in O(N-1)$ is allowed to vary in space, although it should then be remembered that Eq. (3.19) does not hold anymore. This observation is useful in that it allows to compute $\det[\mathfrak{g}(\boldsymbol{x})]$ by choosing the local rotation $R(\boldsymbol{x}) \in O(N-1)$ that rotates $\boldsymbol{\pi}(\boldsymbol{x})$ along $\boldsymbol{e}_2$, say, in which case $\mathfrak{g}(\boldsymbol{x})$ is purely diagonal with the eigenvalue 1 $(N-2)$-fold degenerate and the eigenvalue $\frac{g^2\boldsymbol{\pi}^2}{1-g^2\boldsymbol{\pi}^2}+1$. Thus, we infer that

$$\det[\mathfrak{g}(\boldsymbol{x})] = \frac{1}{1-g^2\boldsymbol{\pi}^2(\boldsymbol{x})}, \qquad \forall \boldsymbol{x}\in\mathbb{R}^d. \tag{3.22}$$

We are now ready to write in a compact manner the spin-wave approximation

$$\begin{aligned} Z_{\mathrm{sw}\,\beta,\boldsymbol{h}} :=& \left(\prod_{\boldsymbol{x}\in\mathbb{R}^d}\int\limits_{\mathbb{R}^{N-1}} \frac{\mathrm{d}^{N-1}\boldsymbol{\pi}(\boldsymbol{x})}{2}\sqrt{g^2\det\mathfrak{g}(\boldsymbol{x})}\;\Theta\left(\left[g^2\det\mathfrak{g}(\boldsymbol{x})\right]^{-1}\right)\right) \\ &\times\exp\left(-\frac{1}{2}\int\limits_{\mathbb{R}^d}\mathrm{d}^d\boldsymbol{x}\left[(\partial_\mu\pi_i)\mathfrak{g}_{ij}(\partial_\mu\pi_j) - 2|\boldsymbol{h}|\left(g^2\det\mathfrak{g}\right)^{-1/2}\right]\right)\end{aligned} \tag{3.23a}$$

to the partition function of the $O(N)$ NLσM,

$$\begin{aligned} Z_{\beta,\boldsymbol{h}} :=& \left(\prod_{\boldsymbol{x}\in\mathbb{R}^d}\int\limits_{\mathbb{R}^{N-1}} \frac{\mathrm{d}^{N-1}\boldsymbol{\pi}(\boldsymbol{x})}{2}\sqrt{g^2\det\mathfrak{g}(\boldsymbol{x})}\;\Theta\left(\left[g^2\det\mathfrak{g}(\boldsymbol{x})\right]^{-1}\right)\right) \\ &\times\exp\left(-\frac{1}{2}\int\limits_{\mathbb{R}^d}\mathrm{d}^d\boldsymbol{x}\left[(\partial_\mu\pi_i)\mathfrak{g}_{ij}(\partial_\mu\pi_j) - 2|\boldsymbol{h}|\left(g^2\det\mathfrak{g}\right)^{-1/2}\right]\right) \\ &+\left(\prod_{\boldsymbol{x}\in\mathbb{R}^d}\int\limits_{\mathbb{R}^{N-1}} \frac{\mathrm{d}^{N-1}\boldsymbol{\pi}(\boldsymbol{x})}{2}\sqrt{g^2\det\mathfrak{g}(\boldsymbol{x})}\;\Theta\left(\left[g^2\det\mathfrak{g}(\boldsymbol{x})\right]^{-1}\right)\right) \\ &\times\exp\left(-\frac{1}{2}\int\limits_{\mathbb{R}^d}\mathrm{d}^d\boldsymbol{x}\left[(\partial_\mu\pi_i)\mathfrak{g}_{ij}(\partial_\mu\pi_j) + 2|\boldsymbol{h}|\left(g^2\det\mathfrak{g}\right)^{-1/2}\right]\right).\end{aligned} \tag{3.23b}$$

We made two approximations to reach Eq. (3.23a). First, we performed the naive continuum limit (3.14) consisting in expanding the lattice action to Gaussian order in the derivative of the spin field. Second, we ignored the field configurations of the spins that have, locally, any antiparallel component with respect to the external magnetic field in Eq. (3.23b). What is left out from the naive continuum limit is the possibility that "singular" lattice configurations of the spins matter in the thermodynamic limit.[5] These "singular" configurations of the spins on the lattice correspond in the continuum approximation to spin fields whose orientations along the external magnetic field can change from parallel to antiparallel as a function of space. It turns out that singular lattice configurations of the spins are essential to the understanding of the phase diagram of the $O(2)$ NLσM in $d = 2$ as we shall see in the chapter devoted to the Kosterlitz-Thouless transition. The usefulness of the spin-wave approximation is that it allows for an answer to the question of whether thermal fluctuations in the form of spin waves are sufficient to destroy spontaneous-symmetry breaking at zero temperature.

Representation (3.23b) of the $O(N)$ NLσM is geometric in nature. The $(N-1)$-sphere is an example of a Riemannian manifold on which a special choice of coordinate system, encoded by the metric (3.18b), has been made. The action in representation (3.23b) is covariant under a change of coordinate system of the $(N-1)$-sphere. The determinant of the metric in the functional measure of integration over the fields $\boldsymbol{\pi}(\boldsymbol{x})$ guarantees that the functional measure is a geometrical invariant under $O(N)$ induced transformations. If Q denotes an element of $O(N)$, one can always define the matrix-valued function $Q(\boldsymbol{x})$ that relates $\mathbf{m}(\boldsymbol{x})$ to $\boldsymbol{e}_1$ through

$$g\,\mathbf{m}(\boldsymbol{x}) =: Q(\boldsymbol{x})\,\boldsymbol{e}_1, \qquad g = +\sqrt{g^2} \geq 0, \tag{3.24a}$$

The subgroup of $O(N)$ that leaves $\boldsymbol{e}_1$ invariant is called the *little group* (or *stabilizer*) of $\boldsymbol{e}_1$. Here, it is the subgroup $O(N-1)$ of $O(N)$. If $R(\boldsymbol{x})$ takes values in the little group of $\boldsymbol{e}_1$,

$$Q(\boldsymbol{x})\,R(\boldsymbol{x})\,\boldsymbol{e}_1 = Q(\boldsymbol{x})\,\boldsymbol{e}_1 = g\,\mathbf{m}(\boldsymbol{x}). \tag{3.24b}$$

Relations (3.24a) and (3.24b) exhibit the isomorphism between the coset (homogeneous) space $O(N)/O(N-1)$ and the $(N-1)$-sphere S_{N-1}. More generally, Eq. (3.23b) can be taken as the definition of a NLσM on a $(N-1)$-dimensional Riemannian manifold with local metric $\mathfrak{g}_{ij}$.[6] This definition of a NLσM is more general than that of the $O(N)$ NLσM (3.14) as it is not always possible to establish an isomorphism between any given Riemannian manifold and some coset (homogeneous) space. Appendix C is devoted to a detailed study of a NLσM on a generic Riemannian manifold.

---

[5] A configuration $\{\boldsymbol{s}_i\}$ of spins on the lattice is said to be singular if its naive continuum limit counterpart $\boldsymbol{m}$ is not smooth everywhere in $\mathbb{R}^d$, i.e., is singular at isolated points. On the lattice there is no notion of smoothness.

[6] A Riemannian manifold is a smooth manifold on which a continuous 2-covariant symmetric and non-degenerate tensor field called the metric tensor can be defined, i.e., for any point $\mathfrak{p}$ on the manifold there exists a symmetric and non-degenerate bilinear form $g_{\mathfrak{p}}$ from the tangent vector space at x to the real numbers.

3.2.3 $O(N)$ NLσM as a field theory on a symmetric space

Representation (3.23) puts the emphasis on the geometrical structure behind NLσM. The initial question on spontaneous-symmetry breaking is cast in the language of group theory. Is there a representation of the $O(N)$ NLσM that puts the emphasis on the underlying group theoretical structure, i.e., renders the pattern of symmetry breaking explicit?

On the one hand, Eq. (3.23) say that, for any given $\boldsymbol{x} \in \mathbb{R}^d$, $(N-1)$ real parameters $\pi_1(\boldsymbol{x}), \cdots, \pi_{N-1}(\boldsymbol{x})$, are needed to parametrize the $(N-1)$-sphere.[7] On the other hand, the number of independent generators of the coset space $O(N)/O(N-1)$ is also $N-1$.[8]

This agreement is not coincidental as we saw in Eqs. (3.24a) and (3.24b). Indeed, we recall that for any point $g\,\boldsymbol{m}(\boldsymbol{x})$ on the unit sphere S_{N-1} with the coordinates $g\,\boldsymbol{\pi}(\boldsymbol{x})$, there exists the $N \times N$ orthogonal matrix $Q(\boldsymbol{x}) \in O(N)$ such that

$$g\,\boldsymbol{\pi}(\boldsymbol{x}) \sim g\,\boldsymbol{m}(\boldsymbol{x}) =: Q(\boldsymbol{x})\,\boldsymbol{e}_1. \tag{3.29}$$

Evidently, the relation between the point $g\,\boldsymbol{m}(\boldsymbol{x})$ of the unit sphere S_{N-1} and the $N \times N$ rotation matrix $Q(\boldsymbol{x})$ is one to many, since right multiplication of $Q(\boldsymbol{x})$ by any element $R(\boldsymbol{x})$ from the little group $O(N-1)$ that leaves the north pole $\boldsymbol{e}_1$ unchanged yields

$$Q(\boldsymbol{x})\,\boldsymbol{e}_1 = Q(\boldsymbol{x})\,R(\boldsymbol{x})\,\boldsymbol{e}_1, \quad \forall R(\boldsymbol{x}) \in O(N-1). \tag{3.30}$$

The one-to-one relationship that we are seeking is between the unit sphere S_{N-1} and the quotient space of $N \times N$ matrices

$$O(N)/O(N-1) := \{[Q] \,|\, Q \in O(N)\}, \tag{3.31a}$$

[7] The $(N-1)$-sphere is the $(N-1)$-dimensional surface embedded in $\mathbb{R}^N$ and defined by

$$g^2\left(\sigma^2 + \sum_{j=1}^{N-1} \pi_j^2\right) = 1, \qquad \forall \begin{pmatrix} \sigma \\ \boldsymbol{\pi} \end{pmatrix} \in \mathbb{R}^N. \tag{3.25}$$

The $(N-1)$-sphere is often denoted $S_{N-1} \subset \mathbb{R}^N$.

[8] For any $Q \in O(N)$ $\det Q = \pm 1$. If $\det Q = 1$, i.e., $Q \in SO(N)$, it is always possible to write $Q = \exp(A)$ and $Q^{\mathsf{T}} = \exp(A^{\mathsf{T}})$ where A is also a $N \times N$ matrix with real-valued matrix elements. Equation (3.3) in foootnote 1 implies that A and A^{T} obey

$$A + A^{\mathsf{T}} = 0, \tag{3.26}$$

i.e., that A is a $N \times N$ real-valued antisymmetric matrix. The number of independent real-valued matrix elements in A equals the number of entries above the diagonal, i.e.,

$$\frac{1}{2}\left(N^2 - N\right) = \frac{1}{2}N(N-1). \tag{3.27}$$

As real vector spaces, the dimensionality of $O(N)$ is thus $\frac{1}{2}N(N-1)$ and the dimensionality of $O(N-1)$ is $\frac{1}{2}(N-1)(N-2)$. The dimensionality of the coset space $O(N)/O(N-1)$ is, by definition, the difference between the dimensionality of $O(N)$ and $O(N-1)$,

$$\dim O(N)/O(N-1) := \dim O(N) - \dim O(N-1) = \frac{1}{2}(N-1)\,[N-(N-2)] = N-1. \tag{3.28}$$

where the equivalence class

$$[Q] := \{Q' \in O(N) \,|\, Q \sim Q'\} \tag{3.31b}$$

is defined through the equivalence relation $Q \sim Q'$ if and only if there exists $R \in O(N)$ such that $Q = Q'R$ and $R\,\boldsymbol{e}_1 = \boldsymbol{e}_1$. In other words, the coordinate $g\,\boldsymbol{\pi}(\boldsymbol{x})$ of the point $g\,\boldsymbol{m}(\boldsymbol{x})$ on the unit sphere is identified with the set

$$\{Q(\boldsymbol{x})R(\boldsymbol{x}) | g\,\boldsymbol{m}(\boldsymbol{x}) = Q(\boldsymbol{x})\,\boldsymbol{e}_1 \text{ and } R(\boldsymbol{x})\,\boldsymbol{e}_1 = \boldsymbol{e}_1\}. \tag{3.31c}$$

We are now going to represent the $O(N)$ NLσM in terms of the elements of $O(N)$. To this end, observe that any real-valued $N \times N$ antisymmetric matrix $A(\boldsymbol{x})$ can be written as[9]

$$A(\boldsymbol{x}) = \frac{g}{2} \sum_{1\leq i<j\leq N} \alpha_{ji}(\boldsymbol{x})\, T_{ij}, \qquad T_{ij} := E_{ij} - E_{ji}, \tag{3.34}$$

where the $N \times N$ matrices E_{ij} has one single non-vanishing matrix element equal to 1 for line i and column j, $\alpha_{ji}(\boldsymbol{x})$ are real-valued numbers, and α_{ji} are smooth functions $\mathbb{R}^d \to \mathbb{R}$. The factor of 1/2 is convention [see Eq. (3.33) in footnote 9] and we have endowed $\alpha_{ij}(\boldsymbol{x})$ with the dimensions of g^{-1}. We shall assume that g is infinitesimal, in which case A is also infinitesimal. According to Eqs. (3.26) and (3.33) in footnotes 8 and 9, respectively, for any infinitesimal $A = -A^{\mathsf{T}}$ we deduce the following chain of equalities

$$Q(\boldsymbol{x}) = e^{A(\boldsymbol{x})} \approx \mathbb{1}_N + A(\boldsymbol{x}) \in SO(N), \tag{3.35a}$$

$$\left(Q^{\mathsf{T}}\partial_\mu Q\right)(\boldsymbol{x}) \approx \frac{g}{2} \sum_{1\leq i<j\leq N} \left(\partial_\mu \alpha_{ji}\right)(\boldsymbol{x}) T_{ij}, \tag{3.35b}$$

$$\left(Q^{\mathsf{T}}\partial_\mu Q\right)^{\mathsf{T}}(\boldsymbol{x}) = -\left(Q^{\mathsf{T}}\partial_\mu Q\right)(\boldsymbol{x}) \approx -\frac{g}{2} \sum_{1\leq i<j\leq N} \left(\partial_\mu \alpha_{ji}\right)(\boldsymbol{x}) T_{ij}, \tag{3.35c}$$

and

$$\operatorname{tr}\left[\left(Q^{\mathsf{T}}\partial_\mu Q\right)^{\mathsf{T}}(\boldsymbol{x})\left(Q^{\mathsf{T}}\partial_\mu Q\right)(\boldsymbol{x})\right] \approx \frac{g^2}{2} \sum_{1\leq i<j\leq N} \left(\partial_\mu \alpha_{ji}\right)^2(\boldsymbol{x}). \tag{3.35d}$$

---

[9] The algebra

$$[T_{ij}, T_{kl}] = \delta_{ik} T_{lj} + \delta_{jl} T_{ki} + \delta_{il} T_{jk} + \delta_{jk} T_{il}, \qquad 1 \leq i < j \leq N, \qquad 1 \leq k < l \leq N, \tag{3.32}$$

defines the Lie algebra of the Lie group $SO(N)$. Since T_{ij} is antisymmetric with only two non-vanishing entries, +1 for line i and column j and −1 for line j and column i,

$$\begin{aligned}\operatorname{tr}\left(T_{ij} T_{kl}\right) &= \sum_{m,n=1}^{N} \left(T_{ij}\right)_{mn} \left(T_{kl}\right)_{nm} \\ &= \sum_{m,n=1}^{N} \left(\delta_{im}\delta_{jn} - \delta_{in}\delta_{jm}\right)\left(\delta_{kn}\delta_{lm} - \delta_{km}\delta_{ln}\right) \\ &= 2\left(\delta_{il}\delta_{jk} - \delta_{ik}\delta_{jl}\right), \qquad i,j,k,l = 1,\cdots,N.\end{aligned} \tag{3.33}$$

The scaling factor of 1/2 in Eq. (3.34) insures that the trace of $T_{ij}/2$ with itself gives −1/2.

The right-hand side of Eq. (3.35d) is positive and can thus be used to construct a Boltzmann weight.

Next, we *define* the partition function

$$Z_{g^2,H_0} := \left(\prod_{\boldsymbol{x}\in\mathbb{R}^d}\int\limits_{O(N)} \mathrm{d}Q(\boldsymbol{x})\right) e^{-\int\limits_{\mathbb{R}^d}\mathrm{d}^d\boldsymbol{x}\,\frac{1}{g^2}\mathrm{tr}\left[(Q^\mathsf{T}D_\mu Q)^\mathsf{T}(Q^\mathsf{T}D_\mu Q)-H_0\,I_{1,N-1}\,Q\right]}, \tag{3.36a}$$

where we are making use of the covariant derivative

$$D_\mu Q := \partial_\mu Q - Q\,A_\mu, \tag{3.36b}$$

with the gauge field defined by

$$A_\mu(\boldsymbol{x}) := \text{Projection of } \left(Q^\mathsf{T}\partial_\mu Q\right)(\boldsymbol{x}) \text{ onto the little group.} \tag{3.36c}$$

[The little group was defined in Eq. (3.24).] The (Haar) measure of $O(N)$ accounts for the fact that $O(N)$ is not simply connected, for it is impossible to smoothly change the sign of the determinant of an orthogonal matrix.[10] In other words, one must sum separately over the two connected components of $O(N)$, i.e., over those pure rotations with determinant $+1$, and those rotations that have been composed with the inversion of one and only one coordinate (the combined operation thus has determinant -1). This observation is nothing but a manifestation of the additive decomposition (3.16a). In turn, the measure on either of the two connected components of $O(N)$ is the Haar measure of the group $SO(N)$, an example of a simple and compact Lie group. In this book, it will be sufficient to know that a Haar measure can be constructed for any compact Lie group and that this measure is invariant under left or right group multiplication. The so-called covariant derivative D_μ expresses the fact that, out of the $N(N-1)/2$ degrees of freedom associated to the connected subgroup $SO(N)$, $(N-1)(N-2)/2$ of them are redundant. Indeed, the "magnetic field" $H_0\,I_{1,N-1}$, here represented by the real number H_0 multiplying the diagonal matrix

$$I_{1,N-1} := \begin{pmatrix} +1 & 0 & 0 & \cdots & 0 \\ 0 & -1 & 0 & \cdots & 0 \\ \vdots & \vdots & \vdots & \vdots & \vdots \\ 0 & 0 & \cdots & 0 & -1 \end{pmatrix} \tag{3.37}$$

in $O(N)$, defines the subgroup $O(N-1) \subset O(N)$ made of all matrices from $O(N)$ that commute with $I_{1,N-1}$. This is the little group. Any two *local* elements $Q_1(\boldsymbol{x}) \in SO(N)$ and $Q_2(\boldsymbol{x}) \in SO(N)$ from the target manifold that differ by the *right* multiplication with the *local* matrix $R(\boldsymbol{x})$ from $SO(N-1)$,

$$Q_1(\boldsymbol{x}) = Q_2(\boldsymbol{x})\,R(\boldsymbol{x}), \tag{3.38}$$

[10] A gentle introduction to the mathematics of Haar measures can be found in chapter 15 of Ref. [30].

are physically equivalent and should only be counted once in the path integral. To put it differently, the covariant derivative insures that the path integral over $SO(N)$ reduces to a path integral over all equivalence classes in the coset space $SO(N)/SO(N-1)$. This redundancy under local right multiplication is an example of a local gauge symmetry. The transformation laws of the non-Abelian gauge field and covariant derivative under the right multiplication

$$Q(\boldsymbol{x}) \to Q(\boldsymbol{x})\, R(\boldsymbol{x}), \qquad R(\boldsymbol{x}) \in SO(N-1), \tag{3.39}$$

are

$$A_\mu \to R^{\mathsf{T}} A_\mu R + R^{\mathsf{T}} \partial_\mu R, \tag{3.40}$$

and

$$D_\mu Q \to \left(D_\mu Q\right) R, \tag{3.41}$$

respectively. Hence, when $H_0 = 0$, the action in Eq. (3.36a) is locally gauge invariant due to the cyclicity of the trace. Evidently, when $H_0 = 0$, the action in Eq. (3.36a) is also invariant under the $O(N)$ global left multiplication

$$Q(\boldsymbol{x}) \to L_0\, Q(\boldsymbol{x}), \tag{3.42}$$

since

$$\left(Q^{\mathsf{T}} \partial_\mu Q\right)(\boldsymbol{x}) \to \left(Q^{\mathsf{T}} \partial_\mu Q\right)(\boldsymbol{x}), \tag{3.43}$$

while

$$\left(D_\mu Q\right)(\boldsymbol{x}) \to L_0 \left(D_\mu Q\right)(\boldsymbol{x}). \tag{3.44}$$

When $H_0 \neq 0$, by the cyclicity of the trace, the global symmetry group is the transformation

$$Q(\boldsymbol{x}) \to L_0^{\mathsf{T}}\, Q(\boldsymbol{x})\, L_0, \tag{3.45}$$

where L_0 is any $N \times N$ matrix from the subgroup $O(N-1)$ of matrices in $O(N)$ that commute with $I_{1,N-1}$.

The partition function (3.36) shares the same global symmetries as the partition function (3.23b). Deriving the partition function (3.23b) from the partition function (3.36) requires an explicit parametrization of the $N \times N$ orthogonal matrices Q. One possible choice can be found in chapter 6 from Ref. [31]. However, the message of section 3.2.3 is that a classical partition function can be interpreted as a gauge theory if a redundant description of the degrees of freedom is chosen.

We close this discussion with a brief description of some mathematical background.

An N-dimensional Riemannian manifold can be pictured as a smooth N-dimensional surface embedded in some Euclidean (flat) space through the imposition of a constraint (see appendix C). For example, the unit sphere S_{N-1} is the set of all N-dimensional real-valued vectors with unit length.

Riemannian manifolds are endowed with a metric, i.e., a notion of distance (see appendix C). For the case of the unit sphere S_{N-1} with the coordinates $\pi_1, \cdots, \pi_{N-1}$, and the metric (3.18b), the distance between any two points follows from minimizing the length

$$L[c] := g \int_0^1 \mathrm{d}t \sqrt{\mathfrak{g}_{ij} \frac{\mathrm{d}\pi_i}{\mathrm{d}t} \frac{\mathrm{d}\pi_j}{\mathrm{d}t}}, \tag{3.46}$$

with respect to the choice made for the curve $c(t)$ parametrized by $0 \leq t \leq 1$ that connects the two points on the sphere. The minimal curve is called a *geodesic*.

The unit sphere S_{N-1} has, however, more than a metric. Any rotation of the Cartesian coordinate system in the embedding Euclidean space $\mathbb{R}^N$ leaves the distance between any two points from the unit sphere unchanged. As a corollary, Eq. (3.29) holds.

This property of the unit sphere S_{N-1} can be generalized as follows. A Riemannian manifold $\mathfrak{M}$ is said to be *homogeneous* if it can be associated to a Lie group G in such a way that for any two point $\mathfrak{x}$ and $\mathfrak{y}$ in $\mathfrak{M}$ (i) there exists an element from $g \in G$ with $g\,\mathfrak{x} = \mathfrak{y}$ (*transitivity*) and (ii) the distance between $\mathfrak{x}$ and $\mathfrak{y}$ is the same as the distance between $g\,\mathfrak{x}$ and $g\,\mathfrak{y}$ (*isometry*).

For example, the unit sphere S_{N-1} is an homogeneous Riemannian manifold with the transitive isometric group $O(N)$.

An homogeneous Riemannian manifold $\mathfrak{M}$ is characterized by the coset G/H where H is the subgroup of G that leaves an arbitrary point $\mathfrak{x}$ of $\mathfrak{M}$ invariant,

$$\mathfrak{x} = h\,\mathfrak{x} \tag{3.47}$$

for any $h \in H$. An homogeneous Riemannian manifold $\mathfrak{M}$ is said to be *symmetric* if its symmetry group G (a semi-simple compact Lie group) is also characterized by a mapping on itself that preserves the group structure (an *automorphism*) and is *involutive* (it becomes the identity mapping if composed with itself). All elements of the Lie algebra of G are then either odd or even under this involution. The little group H is then generated by all the even generators from the Lie algebra of G under the involutive automorphism.

For example, a family of involutive automorphisms on $O(N)$ are the mappings

$$Q \to I^{-1}_{p,q}\, Q\, I_{p,q} \tag{3.48a}$$

where the $N \times N$ diagonal matrices $I_{p,q}$ are

$$I_{p,q} = \mathrm{diag}(\underbrace{+1, \cdots, +1}_{p\text{-times}}, \underbrace{-1, \cdots, -1}_{q\text{-times}}), \qquad N = p + q. \tag{3.48b}$$

The family of subgroups of $O(N)$ left invariant by these automorphisms is

$$\{O(p) \times O(N-p)\,|\,p = 1, \cdots, N-1\}. \tag{3.49}$$

The corresponding family of coset spaces

$$\{\mathcal{G}_p \equiv O(N)/O(p) \times O(N-p)\,|\,p = 1, \cdots, N-1\}, \tag{3.50}$$

are called Grassmannian manifolds. The case $p = 1$ corresponds to the choice (3.37) that we made for the symmetry-breaking term in the partition function (3.36a). Thus, the $O(N)$ NLσM is the special case when the target space is the $p = 1$ Grassmannian manifold $\mathcal{G}_1$. A one-to-one realization of $\mathcal{G}_p$ in $O(N)$ is given by

$$\boldsymbol{x} \to T(\boldsymbol{x})\, I_{p,N-p}\, T^{-1}(\boldsymbol{x}), \qquad T(\boldsymbol{x}) \in O(N), \tag{3.51}$$

since right multiplication

$$T(\boldsymbol{x}) \to T(\boldsymbol{x})\, R(\boldsymbol{x}), \qquad R(\boldsymbol{x}) \in O(p) \times O(N-p) \subset O(N), \tag{3.52}$$

leaves $T(\boldsymbol{x})\, I_{p,N-p}\, T^{-1}(\boldsymbol{x})$ unchanged. The action

$$S_{g^2,H_0}[Q] := \int\limits_{\mathbb{R}^d} \mathrm{d}^d\boldsymbol{x}\, \frac{1}{g^2} \mathrm{tr}\left[\left(Q^{\mathsf{T}}\partial_\mu Q\right)^{\mathsf{T}} \left(Q^{\mathsf{T}}\partial_\mu Q\right) - H_0\, I_{p,N-p}\, Q\right], \tag{3.53}$$

where $Q(\boldsymbol{x}) \in O(N)$ is parametrized according to Eq. (3.51), delivers the Riemannian metric of $\mathcal{G}_p$ once the trace has been evaluated (see Ref. [25]).

3.2.4 *Other examples of NLσM*

(1) Classical ferromagnetism with the group $O(3)$.
(2) Liquid crystals.
(3) Quantum antiferromagnets on a square lattice with the group $O(3)$.
(4) Spin-1/2 quantum spin chains with the group $SU(2)$ which is locally isomorphic to $O(3)$.
(5) Anderson localization, polymers, and other disordered systems.
(6) Strongly correlated systems with the groups $SO(5)$ and $\mathbb{C}P_{N-1}$, the latter being locally isomorphic to $SU(N)/U(N-1)$.

3.3 Fixed point theories, engineering and scaling dimensions, irrelevant, marginal, and relevant interactions

For notational simplicity, we shall consider the case $N = 1$ in this section. No fluctuations about the ferromagnetic state is allowed, irrespective of temperature, in the $O(1)$ NLσM,

$$\int \mathcal{D}[\varphi]\delta\left(\frac{1}{g^2} - \varphi^2\right) \exp\left(-\frac{1}{2}\int\limits_{\mathbb{R}^d} \mathrm{d}^d\boldsymbol{x}\left[\left(\partial_\mu\varphi\right)^2 - 2h\varphi\right]\right) = e^{+\frac{1}{|g|}\int_{\mathbb{R}^d} \mathrm{d}^d\boldsymbol{x}\, h}. \tag{3.54}$$

Rather than the $O(1)$ NLσM, we consider the interacting field theory for the real-valued scalar field φ defined by the (Euclidean) partition function

$$Z = \int \mathcal{D}[\varphi] e^{-(S_0+S_1)}, \tag{3.55a}$$

with the non-interacting action (Lagrangian)

$$S_0 = \int\limits_{\mathbb{R}^d} \mathrm{d}^d\boldsymbol{x}\, \mathcal{L}_0, \qquad \mathcal{L}_0 = \frac{1}{2}\left(\partial_\mu\varphi\right)^2, \tag{3.55b}$$

and the interacting action (Lagrangian)

$$S_1 = \int\limits_{\mathbb{R}^d} \mathrm{d}^d\boldsymbol{x}\, \mathcal{L}_1, \qquad \mathcal{L}_1 = V(\varphi). \tag{3.55c}$$

As usual units are chosen such that the fundamental constants $\hbar = c = 1$. If so, the action

$$S = S_0 + S_1 \tag{3.56}$$

must be dimensionless, sitting as it is in the argument of an exponential. Consequently, the Lagrangian density

$$\mathcal{L} = \mathcal{L}_0 + \mathcal{L}_1 \tag{3.57}$$

has dimension

$$[\mathcal{L}] = (\text{length})^{-d}. \tag{3.58}$$

By convention (stemming from high-energy physics whereby dimensions are counted in inverse powers of length, i.e., in powers of momentum) the *engineering dimension* of $\mathcal{L}$ is d. The engineering dimension of the scalar field can be read from the kinetic energy $\mathcal{L}_0$,

$$[\varphi] = (\text{length})^{-(d-2)/2}. \tag{3.59}$$

Thus, φ has engineering dimension $(d-2)/2$. In particular, the engineering dimension of the scalar field is: (i) $-1/2$ if $d=1$, (ii) 0 if $d=2$, (iii) $+1/2$ if $d=3$, and (iv) 1 if $d=4$.

When the interaction potential vanishes,

$$V(\varphi) = 0, \tag{3.60}$$

any rescaling of space

$$\boldsymbol{x} = \kappa\widetilde{\boldsymbol{x}} \Longleftrightarrow \mathfrak{a} = \kappa\,\tilde{\mathfrak{a}}, \qquad 0 < \kappa < \infty, \tag{3.61}$$

($\mathfrak{a}$ is the initial microscopic length scale, say the lattice spacing, $\tilde{\mathfrak{a}}$ is the rescaled microscopic length scale) can be compensated by the rescaling

$$\varphi = \kappa^{-(d-2)/2}\widetilde{\varphi} \tag{3.62}$$

of the scalar field φ so as to insure the invariance of the action $S = S_0$ under this rescaling,

$$S = \int\limits_{\mathbb{R}^d} \mathrm{d}^d\boldsymbol{x}\, \frac{1}{2}\left(\partial_\mu\varphi\right)^2 = \int\limits_{\mathbb{R}^d} \mathrm{d}^d\widetilde{\boldsymbol{x}}\, \frac{1}{2}\left(\widetilde{\partial}_\mu\widetilde{\varphi}\right)^2 = \widetilde{S}. \tag{3.63}$$

Equation (3.63) encodes the property of *scale invariance.* The action of the free scalar field theory is scale invariant. The partition function $Z \equiv Z_0$ of the free scalar field theory is not scale invariant since Eq. (3.62) changes the partition function by an infinite multiplicative factor (the factor $\kappa^{-(d-2)/2}$ for each $\boldsymbol{x}$). However, this infinite multiplicative factor drops out of all correlation functions

$$\begin{aligned}\langle\varphi(\boldsymbol{x}_1)\cdots\varphi(\boldsymbol{x}_m)\varphi(\boldsymbol{y}_1)\cdots\varphi(\boldsymbol{y}_n)\rangle_{Z_0} &:= \frac{\int\mathcal{D}[\varphi]\;\varphi(\boldsymbol{x}_1)\cdots\varphi(\boldsymbol{x}_m)\varphi(\boldsymbol{y}_1)\cdots\varphi(\boldsymbol{y}_n)\;e^{-S_0}}{\int\mathcal{D}[\varphi]\;e^{-S_0}}\\ &= (-)^{m+n}\,\frac{1}{Z_0}\,\frac{\partial^{m+n}Z_{0\,J}}{\partial J(\boldsymbol{x}_1)\cdots\partial J(\boldsymbol{y}_n)}\bigg|_{J=0},\end{aligned} \tag{3.64a}$$

where

$$Z_{0\,J} := \int\mathcal{D}[\varphi]\exp\left(-S_0 - \int\limits_{\mathbb{R}^d}\mathrm{d}^d\boldsymbol{x}\;J\varphi\right). \tag{3.64b}$$

Scale invariance of the action S_0 fixes the engineering dimension of $(m+n)$-point correlation functions of the free scalar field,

$$\left[\langle\varphi(\boldsymbol{x}_1)\cdots\varphi(\boldsymbol{x}_m)\varphi(\boldsymbol{y}_1)\cdots\varphi(\boldsymbol{y}_n)\rangle_{Z_0}\right] = (\text{length})^{-(d-2)(m+n)/2}. \tag{3.65}$$

Equation (3.65) suggests the guess that, up to some dimensionless multiplicative prefactor,

$$\langle\varphi(\boldsymbol{x})\varphi(\boldsymbol{y})\rangle_{Z_0} \propto \left(\frac{1}{|\boldsymbol{x}-\boldsymbol{y}|^2}\right)^{(d-2)/2}, \qquad d = 1,3,4,\cdots. \tag{3.66}$$

This guess is confirmed by direct computation of the Fourier transform of the free scalar field propagator $1/\boldsymbol{k}^2$ in momentum space,

$$\begin{aligned}D(\boldsymbol{x}) &:= \int\frac{\mathrm{d}^d\boldsymbol{k}}{(2\pi)^d}\,\frac{1}{\boldsymbol{k}^2}\,e^{\mathrm{i}\boldsymbol{k}\cdot\boldsymbol{x}}\\ &= \frac{\Gamma\left(\frac{d-2}{2}\right)}{4\pi^{d/2}}\left(\frac{1}{|\boldsymbol{x}|^2}\right)^{(d-2)/2}, \qquad d = 1,3,4,\cdots.\end{aligned} \tag{3.67}$$

The case of two space-time dimension, $d = 2$, is very special in that φ is itself scale invariant. This is reflected by the singularity of the gamma function at the origin.[11] We will devote a subsection to this case below.

Predictions from scale invariance of S_0 can be circumvented by other symmetries of S_0. For example, S_0 is also invariant under the discrete symmetry

$$\varphi = -\widetilde{\varphi}. \tag{3.69}$$

[11] The gamma function has the integral representation

$$\Gamma(z) := \int\limits_0^\infty \mathrm{d}t\,t^{z-1}\,e^{-t}, \qquad z\in\mathbb{C}. \tag{3.68}$$

The gamma function is single valued and analytic over the entire complex plane, save for the points $z = 0, -1, -2, -3, \cdots$ where it possesses single poles with residues $(-1)^n/n!$.

This symmetry implies that the $(m+n)$-point correlation function (3.64a) vanishes whenever $m+n$ is odd.

Scale invariance of S_0 has a limited predictive power for the spatial dependence of $2m$-point correlation function, $m > 1$. One must rely on a direct calculation to show that the $2m$-point correlation function reduces to a sum over m products of two-point functions. This result is known as the Wick theorem in an operator representation of quantum field theory. With path integral techniques this result simply follows from application of the product rule for differentiation.[12]

3.3.1 *Fixed-point theories*

Consider the family of partition functions $\{Z\}_V$ labeled by the interaction potential $V(\varphi)$ in Eq. (3.55). A *fixed-point theory* $Z^\star$ from this family of theories has an action $S^\star = S_0 + S_1^\star$ that is scale invariant under simultaneous rescaling of space-time $\boldsymbol{x}$ and φ. We have already encountered one fixed-point theory, the *free-field-fixed-point theory* when the interaction potential $V(\varphi)$ vanishes. One could imagine that there are other potentials for which scale invariance is realized. At a fixed point, scale invariance dictates that $(m+n)$-point correlation functions [Eq. (3.64) with $S_0 \to S_0 + S_1^\star$] are algebraic functions in any dimensions other than $d = 2$. The 2-point function can then be used to define the scaling dimension δ_φ of the scalar field at a fixed point,

$$\langle \varphi(\boldsymbol{x})\varphi(\boldsymbol{y})\rangle_{Z^\star} \propto \mathfrak{a}^{-(d-2)} \left(\frac{\mathfrak{a}^2}{|\boldsymbol{x}-\boldsymbol{y}|^2}\right)^{\delta_\varphi}, \qquad d = 1,3,4,\cdots . \tag{3.70}$$

The engineering dimension of the correlation function is made explicit by the introduction of the microscopic length scale $\mathfrak{a}$, say the lattice spacing. The proportionality constant is some dimensionless numerical factor. The free-field-fixed-point theory is characterized by the fact that engineering and scaling dimensions coincide. This need not be true anymore at some putative interacting fixed-point theory where $V^\star(\varphi) \neq 0$.

The physical significance of a fixed-point theory depends on the way any perturbation to the fixed-point theory behaves under rescaling. Consider for example the perturbation

$$V_m(\varphi) := \frac{1}{2m}\lambda_m \varphi^{2m}, \qquad m = 1,2,\cdots, \qquad 0 < \lambda_m \in \mathbb{R}, \tag{3.71}$$

to the free-field fixed point theory $Z^\star = Z_0$. At the free-field fixed point, we need not distinguish engineering from scaling dimensions. The dimension of the coupling

[12] Correlation functions in a local field theory defined out of the local field φ are obtained from the partition function Z_j in the presence of a source field j that couples linearly to the local field φ. For a free-field theory, the generating function Z_j is proportional to $\exp(+j\,G\,j/2)$ where G is the free-field Green function, since the path integral is Gaussian. Repeated differentiation with respect to the source field of the generating function yields all correlation functions once the limit $j \to 0$ is taken. Wick theorem is just an application of the product rule to the n-th order differentiation of $\exp(+j\,G\,j/2)$ with respect to j.

constant λ_m is

$$[\lambda_m] = (\text{length})^{-d+(d-2)m}\,, \tag{3.72}$$

since

$$S_1 = \int\limits_{\mathbb{R}^d} \mathrm{d}^d \boldsymbol{x}\, \frac{1}{2m} \lambda_m \varphi^{2m} \tag{3.73}$$

is dimensionless and the *scaling* dimension of φ is fixed by Eq. (3.59). Thus, under length rescaling (3.61),

$$\lambda_m = \kappa^{-d+(d-2)m} \widetilde{\lambda_m}. \tag{3.74}$$

Choose the rescaling factor $0 < \kappa < 1$. The rescaled coupling constant

$$\widetilde{\lambda_m} = \kappa^{d-(d-2)m} \lambda_m, \qquad 0 < \kappa < 1, \tag{3.75}$$

- is smaller than the original one if

$$(d-2)m < d, \tag{3.76}$$

- is unchanged if

$$(d-2)m = d, \tag{3.77}$$

- is larger than the original one if

$$(d-2)m > d. \tag{3.78}$$

Correspondingly, the interaction V_m is said to be

- UV irrelevant if

$$(d-2)m < d, \tag{3.79}$$

- marginal if

$$(d-2)m = d, \tag{3.80}$$

- UV relevant if

$$(d-2)m > d. \tag{3.81}$$

The terminology of irrelevance and relevance depends on the choice between $0 < \kappa < 1$ or $1 < \kappa < \infty$. Irrelevant interactions when $0 < \kappa < 1$ become relevant interactions when $1 < \kappa < \infty$ and vice versa. The choice $0 < \kappa < 1$ consists in zooming into the microscopic scale in that the new microscopic length scale $\tilde{\mathfrak{a}} = \mathfrak{a}/\kappa$ by which all lengths are measured *now appears larger* than the original one $\mathfrak{a}$. The choice $0 < \kappa < 1$ is made if one is interested in the asymptotic behavior of correlation functions at short distances or short wavelengths. This is the ultraviolet (UV) limit of primary interest in high-energy physics. The choice $1 < \kappa < \infty$ consists in zooming away from the microscopic scale in that the new microscopic length scale $\tilde{\mathfrak{a}} = \mathfrak{a}/\kappa$ by which all lengths are measured *now appears smaller* than the original one $\mathfrak{a}$. The choice $1 < \kappa < \infty$ is made if one is interested in the asymptotic behavior of correlation functions at long distances or long wavelengths. This is the infrared (IR) limit of primary interest in condensed matter physics.[13] The rescaled coupling constant

$$\widetilde{\lambda_m} = \kappa^{d-(d-2)m}\lambda_m, \qquad 1 < \kappa < \infty, \tag{3.85}$$

- is larger than the original one if

$$(d-2)m < d, \tag{3.86}$$

- is unchanged if

$$(d-2)m = d, \tag{3.87}$$

- is smaller than the original one if

$$(d-2)m > d. \tag{3.88}$$

Correspondingly, the interaction V_m is said to be

- IR relevant if

$$(d-2)m < d, \tag{3.89}$$

- marginal if

$$(d-2)m = d, \tag{3.90}$$

- IR irrelevant if

$$(d-2)m > d. \tag{3.91}$$

[13] Assume that the short-distance cut-off is $\mathfrak{a} - \mathrm{d}\mathfrak{a}$ to begin with. Imagine that one integrates over all length scales between $\mathfrak{a} - \mathrm{d}\mathfrak{a}$ and $\mathfrak{a}$, say by breaking up integrals into

$$\int\limits_{\mathfrak{a}-\mathrm{d}\mathfrak{a}}^{\infty} \mathrm{d}r \cdots = \int\limits_{\mathfrak{a}-\mathrm{d}\mathfrak{a}}^{\mathfrak{a}} \mathrm{d}r \cdots + \int\limits_{\mathfrak{a}}^{\infty} \mathrm{d}r \cdots . \tag{3.82}$$

Integration over the interval $[\mathfrak{a}-\mathrm{d}\mathfrak{a}, \mathfrak{a}]$ can sometimes be absorbed into a redefinition of the coupling constants of the theory. If so, one is left with

$$\int\limits_{\mathfrak{a}}^{\infty} \mathrm{d}r \cdots = \int\limits_{\mathfrak{a}-\mathrm{d}\mathfrak{a}}^{\infty} \mathrm{d}\tilde{r} \cdots . \tag{3.83}$$

Here, form invariance has been restored on the right-hand side with the help of the rescaling [compare with Eq. (3.61)]

$$r = \frac{\mathfrak{a}}{\mathfrak{a} - \mathrm{d}\mathfrak{a}}\tilde{r}. \tag{3.84}$$

Hence, $\kappa = \mathfrak{a}/(\mathfrak{a} - \mathrm{d}\mathfrak{a})$ is indeed larger than unity.

Observe that a mass term is relevant in any dimensions in the IR limit [Eq. (3.89) with $m = 1$].

At a generic IR fixed-point, it is the scaling dimension δ_O of a field O, not the engineering dimension $-\log[O]$ in units of length, that decides of the relevance, marginality, or irrelevance of the "small perturbation" O, whereby it is imagined that

$$S_O = \int\limits_{\mathbb{R}^d} \mathrm{d}^d \boldsymbol{x}\, \lambda_O\, O, \qquad 0 \leq \mathfrak{a}^{d+\log[O]} \lambda_O \ll 1, \tag{3.92}$$

has been added to the fixed-point action $S^{\star}$. Under rescaling $\mathfrak{a} = \kappa \tilde{\mathfrak{a}}$,

$$S_O = \int\limits_{\mathbb{R}^d} \mathrm{d}^d \widetilde{\boldsymbol{x}}\, \kappa^{d-\delta_O} \lambda_O\, \widetilde{O}, \tag{3.93}$$

and the perturbation O is said to be

- IR relevant if

$$\delta_O < d, \tag{3.94}$$

- marginal if

$$\delta_O = d, \tag{3.95}$$

- IR irrelevant if

$$\delta_O > d. \tag{3.96}$$

3.3.2 *Two-dimensional $O(2)$ NLσM in the spin-wave approximation*

To illustrate the peculiarities of two-dimensional space-time, consider the $O(2)$ NLσM in $d = 2$ with the partition function

$$Z_{\beta J, \beta \boldsymbol{H}} := \int \mathcal{D}[\boldsymbol{n}] \delta\left(1 - \boldsymbol{n}^2\right) \exp\left(-\frac{1}{2g^2} \int\limits_{\mathbb{R}^2} \mathrm{d}^2 \boldsymbol{x} \left[\left(\partial_\mu \boldsymbol{n}\right)^2 - 2\mathfrak{a}^{-2} \beta g^2 \boldsymbol{H} \cdot \boldsymbol{n} \right] \right). \tag{3.97}$$

The surface of the unit sphere in $\mathbb{R}^2$, the one-sphere, is simply the unit circle. Planar spins of unit length can be represented as complex numbers of unit length, i.e., phases,

$$\boldsymbol{n} = \begin{pmatrix} n_1 \\ n_2 \end{pmatrix}, \qquad n_1 = \cos\phi = \operatorname{Re}\varphi, \qquad n_2 = \sin\phi = \operatorname{Im}\varphi, \qquad \varphi = e^{\mathrm{i}\phi}. \tag{3.98}$$

Hence,[14]

$$Z_{\beta J,\beta \boldsymbol{H}} \propto \int \mathcal{D}[\varphi^*,\varphi]\delta\left(1-|\varphi|^2\right)\, e^{-\frac{1}{2g^2}\int\limits_{\mathbb{R}^2} \mathrm{d}^2\boldsymbol{x}\left[\left|\partial_\mu\varphi\right|^2-2\mathfrak{a}^{-2}\beta g^2\left(H_1\frac{\varphi+\varphi^*}{2}+H_2\frac{\varphi-\varphi^*}{2\mathrm{i}}\right)\right]}. \tag{3.100}$$

Without loss of generality, spontaneous-symmetry breaking into the ferromagnetic ground state is enforced by taking the external magnetic field to be $\boldsymbol{H} = |\boldsymbol{H}|\boldsymbol{e}_1$ and letting $|\boldsymbol{H}| \to 0$ at the end of the day. Hence, the ferromagnetic state is

$$\varphi(\boldsymbol{x}) = 1, \qquad \forall \boldsymbol{x} \in \mathbb{R}^2. \tag{3.101}$$

At non-vanishing temperature, the path-integral representation of the partition function will be restricted to small deviations $\varphi(\boldsymbol{x})$ about the ferromagnetic state (3.101). To be more precise, the spin-wave approximation by which the angular field $\phi(\boldsymbol{x}) = \arg[\varphi(\boldsymbol{x})]$ is rotation free,

$$0 = \oint\limits_{\boldsymbol{x}} \mathrm{d}x_\mu \epsilon_{\mu\nu}\partial_\nu\phi, \qquad \forall \boldsymbol{x} \in \mathbb{R}^2, \tag{3.102a}$$

is made. Here, $\oint\limits_{\boldsymbol{x}}$ denotes any closed line integral that encloses $\boldsymbol{x}$ and

$$\epsilon_{12} = -\epsilon_{21} = 1, \qquad \epsilon_{11} = \epsilon_{22} = 0, \tag{3.102b}$$

i.e., ϕ is smooth and single valued everywhere. Inclusion in the path integral of multi-valued configurations of ϕ leads to the so-called Kosterlitz-Thouless transition. However, We will ignore this important aspect of the problem as our goal is to illustrate how a perturbative RG procedure can be performed on the $O(N)$ NLσM whereas the physics of the Kosterlitz-Thouless transition is non-perturbative (with respect to g^2) by nature.

We want to compute the $(m+n)$-point correlation function

$$\begin{aligned} G^{(m,n)}_{\mathrm{sw}\, g^2,\boldsymbol{H}=\boldsymbol{0}}&(\boldsymbol{x}_1,\ldots,\boldsymbol{x}_m,\boldsymbol{y}_1,\ldots,\boldsymbol{y}_n) := \\ &\langle\varphi(\boldsymbol{x}_1)\cdots\varphi(\boldsymbol{x}_m)\varphi^*(\boldsymbol{y}_1)\cdots\varphi^*(\boldsymbol{y}_n)\rangle_{\mathrm{sw}\, g^2,\boldsymbol{H}=\boldsymbol{0}} \end{aligned} \tag{3.103a}$$

within the spin-wave approximation of the $O(2)$ NLσM in two-dimensional space, i.e., the angular brackets denotes averaging with the partition function (3.100) whereby the periodic nature of argφ is neglected or, equivalently, the $(m+n)$-point correlation function

$$\begin{aligned} G^{(m,n)}_{\mathrm{sw}\, g^2,\boldsymbol{H}=\boldsymbol{0}}&(\boldsymbol{x}_1,\ldots,\boldsymbol{x}_m,\boldsymbol{y}_1,\ldots,\boldsymbol{y}_n) = \\ &\left\langle e^{+\mathrm{i}\phi(\boldsymbol{x}_1)}\cdots e^{+\mathrm{i}\phi(\boldsymbol{x}_m)}e^{-\mathrm{i}\phi(\boldsymbol{y}_1)}\cdots e^{-\mathrm{i}\phi(\boldsymbol{y}_n)}\right\rangle_{\mathrm{sw}\, g^2,\boldsymbol{H}=\boldsymbol{0}} \end{aligned} \tag{3.103b}$$

[14] The proportionality constant results from the change of the normalization of the measure in the path integral,

$$\mathrm{d}n_1(\boldsymbol{x})\,\mathrm{d}n_2(\boldsymbol{x}) \to \frac{\mathrm{d}n_1(\boldsymbol{x})\,\mathrm{d}n_2(\boldsymbol{x})}{\pi} \equiv \frac{\mathrm{d}\varphi^*(\boldsymbol{x})\,\mathrm{d}\varphi(\boldsymbol{x})}{2\pi\mathrm{i}}. \tag{3.99}$$

In this way a Gaussian integral is normalized to the inverse of the determinant of the kernel.

within the spin-wave approximation of the $O(2)$ NLσM in two-dimensional space, i.e., the angular brackets denotes averaging with the partition function

$$Z_{\mathrm{sw}\,\beta J,\beta \boldsymbol{H}} \propto \int \mathcal{D}[\phi] e^{-\frac{1}{2g^2}\int_{\mathbb{R}^2} \mathrm{d}^2\boldsymbol{x}\left[(\partial_\mu\phi)^2 - 2\mathfrak{a}^{-2}\beta g^2 (H_1\cos\phi + H_2\sin\phi)\right]}, \tag{3.103c}$$

whereby the periodic nature of ϕ is ignored.

Observe that the partition function (3.103c) is unchanged under

$$\phi = \widetilde{\phi} + \mathrm{const}, \tag{3.104}$$

in the thermodynamic limit and when $\boldsymbol{H}$ vanishes. This immediately implies that the correlation function (3.103a) vanishes unless

$$m = n. \tag{3.105}$$

With a vanishing external magnetic field, the identity

$$\left\langle e^{+\mathrm{i}\int \mathrm{d}^2\boldsymbol{x}\, J(\boldsymbol{x})\,\phi(\boldsymbol{x})} \right\rangle_{\mathrm{sw}\, g^2, \boldsymbol{H}=0} = e^{-\frac{1}{2}\int \mathrm{d}^2\boldsymbol{x} \int \mathrm{d}^2\boldsymbol{y} J(\boldsymbol{x})\, G(\boldsymbol{x},\boldsymbol{y})\, J(\boldsymbol{y})} \tag{3.106a}$$

holds for any source $J(\boldsymbol{x})$. Here, $G(\boldsymbol{x},\boldsymbol{y})$ is the Green function defined by

$$\left((-)\frac{1}{g^2}\partial_\mu^2\right) G(\boldsymbol{x},\boldsymbol{y}) = \delta(\boldsymbol{x}-\boldsymbol{y}). \tag{3.106b}$$

In other words,

$$\begin{aligned} G(\boldsymbol{x},\boldsymbol{y}) &= \lim_{M^2\to 0} g^2 \int \frac{\mathrm{d}^2\boldsymbol{k}}{(2\pi)^2} \frac{e^{+\mathrm{i}\boldsymbol{k}\cdot(\boldsymbol{x}-\boldsymbol{y})}}{\boldsymbol{k}^2+M^2} \\ &= \lim_{M^2\to 0} g^2 \left[-\frac{1}{2\pi}\ln\left(M|\boldsymbol{x}-\boldsymbol{y}|\right) + \mathrm{const} + \mathcal{O}(M|\boldsymbol{x}-\boldsymbol{y}|)\right]. \end{aligned} \tag{3.106c}$$

Strictly speaking, the Green function is ill-defined because of the logarithmic singularities in the infrared limit $M \to 0$ and in the ultraviolet limit $|\boldsymbol{x}-\boldsymbol{y}| \to 0$. The ultraviolet singularity can be removed by introducing a high-energy cut-off, say the inverse lattice spacing $1/\mathfrak{a}$. The infrared cut-off M then drops out of the difference

$$G(\boldsymbol{x},\boldsymbol{y}) - G(\boldsymbol{x},\boldsymbol{x}+\mathfrak{a}\,\widehat{\boldsymbol{r}}) = -\frac{g^2}{2\pi}\ln\frac{|\boldsymbol{x}-\boldsymbol{y}|}{\mathfrak{a}} + \mathcal{O}(M|\boldsymbol{x}-\boldsymbol{y}|) \tag{3.106d}$$

($\widehat{\boldsymbol{r}}$ is some unit length vector) to leading order in $|\boldsymbol{x}-\boldsymbol{y}|/\mathfrak{a}$.

To compute the correlation function (3.103a) it suffices to choose the source

$$J(\boldsymbol{x}) := \sum_{i=1}^{m}\delta(\boldsymbol{x}-\boldsymbol{x}_i) - \sum_{j=1}^{n}\delta(\boldsymbol{x}-\boldsymbol{y}_j) \tag{3.107}$$

in Eq. (3.106a). This gives

$$
\begin{aligned}
&G^{(m,n)}_{\mathrm{sw}\,g^2,\boldsymbol{H}=\boldsymbol{0}}(\boldsymbol{x}_1,\ldots,\boldsymbol{x}_m,\boldsymbol{y}_1,\ldots,\boldsymbol{y}_n)\\
&=e^{-\frac{1}{2}\sum\limits_{i,j=1}^{m}G(\boldsymbol{x}_i-\boldsymbol{x}_j)}\times e^{-\frac{1}{2}\sum\limits_{k,l=1}^{n}G(\boldsymbol{y}_k-\boldsymbol{y}_l)}\times e^{+2\times\frac{1}{2}\sum\limits_{i=1}^{m}\sum\limits_{l=1}^{n}G(\boldsymbol{x}_i-\boldsymbol{y}_l)}\\
&=(M\mathfrak{a})^{+\frac{g^2}{4\pi}(m+n)}\prod_{1\le i\neq j\le m}^{1\le k\neq l\le n}\left[(M|\boldsymbol{x}_i-\boldsymbol{x}_j|)\,(M|\boldsymbol{y}_k-\boldsymbol{y}_l|)\right]^{+\frac{g^2}{4\pi}}\prod_{i=1}^{m}\prod_{l=1}^{n}(M|\boldsymbol{x}_i-\boldsymbol{y}_l|)^{-\frac{g^2}{2\pi}}\\
&=M^{+\frac{g^2}{4\pi}(m-n)^2}\mathfrak{a}^{+\frac{g^2}{4\pi}(m+n)}\left[\frac{\left(\prod\limits_{1\le i<j\le m}|\boldsymbol{x}_i-\boldsymbol{x}_j|\right)\left(\prod\limits_{1\le k<l\le n}|\boldsymbol{y}_k-\boldsymbol{y}_l|\right)}{\left(\prod\limits_{i=1}^{m}\prod\limits_{l=1}^{n}|\boldsymbol{x}_i-\boldsymbol{y}_l|\right)}\right]^{+\frac{g^2}{2\pi}}.
\end{aligned}
\tag{3.108}
$$

This expression remains well defined when the infrared cut-off is removed, $M \to 0$, as long as the short distance cut-off $\mathfrak{a}$ is kept non-vanishing, in which case

$$
\begin{aligned}
&G^{(m,n)}_{\mathrm{sw}\,g^2,\boldsymbol{H}=\boldsymbol{0}}(\boldsymbol{x}_1,\ldots,\boldsymbol{x}_m,\boldsymbol{y}_1,\ldots,\boldsymbol{y}_n)\\
&\qquad=\begin{cases}0, & \text{if } m\neq n,\\[1ex]
\mathfrak{a}^{+\frac{g^2}{2\pi}m}\left[\dfrac{\prod\limits_{1\le i<j\le m}^{1\le k<l\le m}(|\boldsymbol{x}_i-\boldsymbol{x}_j|)(|\boldsymbol{y}_k-\boldsymbol{y}_l|)}{\left(\prod\limits_{i,j=1}^{m}|\boldsymbol{x}_i-\boldsymbol{y}_j|\right)}\right]^{+\frac{g^2}{2\pi}}, & \text{if } m=n.\end{cases}
\end{aligned}
\tag{3.109}
$$

The case $m = n = 1$ gives the two-point function

$$
G^{(1,1)}_{\mathrm{sw}\,g^2,\boldsymbol{H}=\boldsymbol{0}}(\boldsymbol{x},\boldsymbol{y})=\left(\frac{\mathfrak{a}}{|\boldsymbol{x}-\boldsymbol{y}|}\right)^{+\frac{g^2}{2\pi}}. \tag{3.110}
$$

All correlation functions of the form (3.103a) are thus algebraic functions for any given non-vanishing value of g^2. At zero temperature, i.e., when $g^2 = 0$, all correlation functions are constant as it should be if the ground state supports ferromagnetic long-range order (LRO). Within the spin-wave approximation, LRO at zero temperature ($g^2 = 0$) is downgraded to *algebraic* order or *quasi-long-range* order (QLRO) at any non-vanishing temperature. Equation (3.108) defines a *critical* phase of matter for any given g^2. At criticality scale invariance manifests itself by algebraic decaying correlation functions. Here, the critical phase of matter is called the *spin-wave phase.* Direct inspection of the two-point function (3.110) allows us to infer that the scaling dimension δ_φ of the field φ is given by

$$
\delta_\varphi=\frac{g^2}{4\pi}. \tag{3.111}
$$

This scaling dimension is a smooth function of g^2 and is different from the engineering dimension of φ which is zero.

Correlation functions in the spin-wave phase are ambiguous in the limit $\mathfrak{a} \to 0$ in which the ultraviolet cut-off is removed. This ambiguity can be interpreted as follows. The accuracy of the spin-wave approximation improves at low energies and long distances, i.e., scaling exponents controlling the algebraic decay of correlation functions can be thought of as being exact or, more precisely, *universal* in that they do not depend on the prescription used to regularize the theory at short distances. Short-distance regularizations in condensed matter physics are much more than a mathematical artifact as they refer to a specific lattice or microscopic model. The mathematical ambiguity in the choice of an ultraviolet cut-off reflects the property that lattice models that differ on the microscopic scale might nevertheless share the same properties at low energies and long distances. From the point of view of physics this is a very important property called *universality* without which the task of classifying and predicting phases of condensed matter would otherwise be hopeless.

The mathematical ambiguity in the choice of an ultraviolet cut-off can be encoded in a differential equation obeyed by correlation functions. This differential equation is called the *Callan-Symanzik* equation. The construction of the Callan-Symanzik equation in the spin-wave phase proceeds as follows. The ambiguity in the choice of the ultraviolet cut-off $1/\mathfrak{a}$ can be quantified by introducing a *renormalization point* or *renormalization mass* μ through

$$\begin{aligned}
&\mathfrak{a}^{+\frac{g^2}{2\pi}m} \left[\frac{\prod_{1\leq i<j\leq m}^{1\leq k<l\leq m} |\boldsymbol{x}_i - \boldsymbol{x}_j||\boldsymbol{y}_k - \boldsymbol{y}_l|}{\prod_{i,j=1}^{m} |\boldsymbol{x}_i - \boldsymbol{y}_j|} \right]^{+\frac{g^2}{2\pi}} \\
&\qquad = (\mathfrak{a}\mu)^{+\frac{g^2}{2\pi}m} \left[\frac{\prod_{1\leq i<j\leq m}^{1\leq k<l\leq m} (\mu|\boldsymbol{x}_i - \boldsymbol{x}_j|)\,(\mu|\boldsymbol{y}_k - \boldsymbol{y}_l|)}{\prod_{i,j=1}^{m} \mu|\boldsymbol{x}_i - \boldsymbol{y}_l|} \right]^{+\frac{g^2}{2\pi}}
\end{aligned} \tag{3.112}$$

for the $(2m)$-point function (3.109) and

$$\left(\frac{\mathfrak{a}}{|\boldsymbol{x}-\boldsymbol{y}|}\right)^{+\frac{g^2}{2\pi}} = (\mathfrak{a}\mu)^{+\frac{g^2}{2\pi}} \left(\frac{1}{\mu|\boldsymbol{x}-\boldsymbol{y}|}\right)^{+\frac{g^2}{2\pi}} \tag{3.113}$$

for the 2-point function (3.110). Define the *renormalized field*

$$\varphi^{(\mathrm{R})} := \frac{1}{\sqrt{Z}}\varphi, \tag{3.114}$$

whereby

$$\sqrt{Z} := (\mathfrak{a}\mu)^{+\frac{g^2}{4\pi}} = e^{+\frac{g^2}{4\pi}\ln(\mathfrak{a}\mu)}. \tag{3.115}$$

The original field φ is called the *bare* or *unrenormalized* field. The dimensionless number Z is called the *wave-function renormalization factor*. Correlation functions

(3.108) and (3.110) can be expressed in terms of the renormalized fields as

$$\begin{aligned}
G^{(m,m)(\mathrm{R})}_{\mathrm{sw}\,g^2,\boldsymbol{H}=\boldsymbol{0}}&(\boldsymbol{x}_1,\ldots,\boldsymbol{x}_m,\boldsymbol{y}_1,\ldots,\boldsymbol{y}_m)\\
&:=\left\langle\varphi^{(\mathrm{R})}(\boldsymbol{x}_1)\cdots\varphi^{(\mathrm{R})}(\boldsymbol{x}_m)\varphi^{(\mathrm{R})*}(\boldsymbol{y}_1)\cdots\varphi^{(\mathrm{R})*}(\boldsymbol{y}_m)\right\rangle_{\mathrm{sw}\,g^2,\boldsymbol{H}=\boldsymbol{0}}\\
&=\left[\frac{\left(\prod\limits_{1\le i<j\le m}\mu|\boldsymbol{x}_i-\boldsymbol{x}_j|\right)\left(\prod\limits_{1\le k<l\le m}\mu|\boldsymbol{y}_k-\boldsymbol{y}_l|\right)}{\left(\prod\limits_{i,j=1}^{m}\mu|\boldsymbol{x}_i-\boldsymbol{y}_j|\right)}\right]^{+\frac{g^2}{2\pi}}
\end{aligned}\tag{3.116}$$

and

$$\left\langle\varphi^{(\mathrm{R})}(\boldsymbol{x})\varphi^{(\mathrm{R})*}(\boldsymbol{y})\right\rangle_{\mathrm{sw}\,g^2,\boldsymbol{H}=\boldsymbol{0}}=\left(\frac{1}{\mu|\boldsymbol{x}-\boldsymbol{y}|}\right)^{+\frac{g^2}{2\pi}},\tag{3.117}$$

respectively. We have thus traded the ultraviolet cut-off $1/\mathfrak{a}$ for μ. The Callan-Symanzik equation obeyed by the correlation function (3.109) follows from the observation that Eq. (3.109) does not depend on μ,

$$\begin{aligned}
0&=\mu\frac{\mathrm{d}}{\mathrm{d}\mu}G^{(m,m)}_{\mathrm{sw}\,g^2,\boldsymbol{H}=\boldsymbol{0}}(\boldsymbol{x}_1,\ldots,\boldsymbol{x}_m,\boldsymbol{y}_1,\ldots,\boldsymbol{y}_m)\\
&=\mu\frac{\mathrm{d}}{\mathrm{d}\mu}\left[Z^m G^{(m,m)(\mathrm{R})}_{\mathrm{sw}\,g^2,\boldsymbol{H}=\boldsymbol{0}}(\boldsymbol{x}_1,\ldots,\boldsymbol{x}_m,\boldsymbol{y}_1,\ldots,\boldsymbol{y}_m)\right]\quad\text{by Eqs. (3.116) and (3.114)}\\
&=Z^m\left(\mu\frac{\partial}{\partial\mu}+2m\,\underbrace{\mu\frac{\partial\ln\sqrt{Z}}{\partial\mu}}\right)G^{(m,m)(\mathrm{R})}_{\mathrm{sw}\,g^2,\boldsymbol{H}=\boldsymbol{0}}(\boldsymbol{x}_1,\ldots,\boldsymbol{x}_m,\boldsymbol{y}_1,\ldots,\boldsymbol{y}_m)\\
\text{Eq. (3.115)}\quad&=Z^m\left(\mu\frac{\partial}{\partial\mu}+2m\,\frac{g^2}{4\pi}\right)G^{(m,m)(\mathrm{R})}_{\mathrm{sw}\,g^2,\boldsymbol{H}=\boldsymbol{0}}(\boldsymbol{x}_1,\ldots,\boldsymbol{x}_m,\boldsymbol{y}_1,\ldots,\boldsymbol{y}_m)\\
&\equiv Z^m\left[\mu\frac{\partial}{\partial\mu}+2m\,\gamma(g^2)\right]G^{(m,m)(\mathrm{R})}_{\mathrm{sw}\,g^2,\boldsymbol{H}=\boldsymbol{0}}(\boldsymbol{x}_1,\ldots,\boldsymbol{x}_m,\boldsymbol{y}_1,\ldots,\boldsymbol{y}_m).
\end{aligned}\tag{3.118}$$

The *anomalous scaling dimension*

$$\begin{aligned}
\gamma(g^2)&:=\mu\frac{\partial\ln\sqrt{Z}}{\partial\mu}\\
&=\frac{g^2}{4\pi}
\end{aligned}\tag{3.119}$$

has been introduced. It is the difference between the scaling and the engineering dimension of the field φ.

The lessons learned from the example of the two-dimensional $O(2)$ NLσM are:

(1) The vanishing of the $m \neq n$ correlator (3.103a) as the infrared cut-off $M \to 0$ guarantees that the $O(2)$ symmetry is not spontaneously broken at any non-vanishing temperature. This is an example of the Hohenberg-Mermin-Wagner-Coleman theorem that asserts that no continuous global symmetry can be spontaneously broken in $d \leq 2$.
(2) Correlator (3.103a) depends on an ultraviolet cut-off. This dependence can be quantified by the Callan-Symanzik equation obeyed by renormalized fields.
(3) Anomalous scaling dimensions that appear in the Callan-Symanzik equation are universal in that they are independent of the choice of the ultraviolet cut-off.
(4) The spin-wave phase is a critical or QLRO phase in which correlator (3.103a) is an algebraic decaying function at any non-vanishing temperature. Anomalous scaling dimensions are continuous functions of the temperature.

3.4 General method of renormalization

In this section, we are going to set up the Callan-Symanzik equation obeyed by correlation functions in all generality. Consider some bare correlation function

$$G_{\mathrm{B}}^{(m,n)}(\boldsymbol{z}; g_{\mathrm{B}}, \Lambda) \tag{3.120a}$$

between $(m+n)$ local fields. Here, $\boldsymbol{z}$ denotes collectively the $(m+n)$ space arguments of the local fields,

$$\boldsymbol{z} = \{\boldsymbol{x}_1, \ldots, \boldsymbol{x}_m, \boldsymbol{y}_1, \ldots, \boldsymbol{y}_n\}, \tag{3.120b}$$

and g_{B} denotes collectively all (IR) relevant coupling constants at the free-field fixed point,

$$g_{\mathrm{B}} = \left\{ g_{\mathrm{B}}^{(1)}, g_{\mathrm{B}}^{(2)}, \ldots \right\}. \tag{3.120c}$$

It is commonly assumed that the number of relevant coupling constants is finite but this need not be so, for example when dealing with disordered systems. The inverse of the lattice spacing

$$\Lambda = \frac{1}{\mathfrak{a}} \tag{3.121}$$

is taken as the UV cut-off. We now assume that it is possible to express the correlator (3.120a) in terms of a wave-function renormalization factor Z, renormalized coupling constants g_{R}, and a new renormalization point μ according to

$$G_{\mathrm{B}}^{(m,n)}(\boldsymbol{z}; g_{\mathrm{B}}, \Lambda) = [Z(g_{\mathrm{R}}, \mu/\Lambda)]^{+\frac{m+n}{2}} \times G_{\mathrm{R}}^{(m,n)}(\boldsymbol{z}; g_{\mathrm{R}}, \mu). \tag{3.122}$$

Equation (3.122) is certainly not correct when a pair of spatial arguments of the correlator is within a distance of the order of the lattice spacing $\mathfrak{a}$ as the renormalized correlator must then also depend on Λ. However, Eq. (3.122) becomes plausible when all spatial arguments are separated pairwise by an amount much larger than the lattice spacing $\mathfrak{a}$. In any case Eq. (3.122) is to be verified by explicit computation as we did for the spin-wave phase of the two-dimensional $O(2)$ NLσM. Assumption (3.122) implies the Callan-Symanzik equation

$$\begin{aligned} 0 &= \left[\mu\partial_\mu + \beta(g_{\mathrm{R}})\partial_{g_{\mathrm{R}}} + (m+n)\gamma(g_{\mathrm{R}})\right] G^{(m,n)}_{\mathrm{R}}(\boldsymbol{z}; g_{\mathrm{R}}, \mu), \\ \beta(g_{\mathrm{R}}) &:= \mu\partial_\mu g_{\mathrm{R}} \qquad \text{at fixed } g_{\mathrm{B}} \text{ and } \Lambda, \\ \gamma(g_{\mathrm{R}}) &:= \mu\partial_\mu \ln\sqrt{Z} \qquad \text{at fixed } g_{\mathrm{B}} \text{ and } \Lambda. \end{aligned} \tag{3.123}$$

Observe that we could have equally well written

$$G^{(m,n)}_{\mathrm{R}}(\boldsymbol{z}; g_{\mathrm{R}}, \mu) = Z^{-\frac{m+n}{2}}(g_{\mathrm{B}}, \mu/\Lambda) \times G^{(m,n)}_{\mathrm{B}}(\boldsymbol{z}; g_{\mathrm{B}}, \Lambda), \tag{3.124}$$

rather than Eq. (3.122) to derive the Callan-Symanzik equation

$$\begin{aligned} 0 &= \left[\Lambda\partial_\Lambda + \widetilde{\beta}(g_{\mathrm{B}})\partial_{g_{\mathrm{B}}} - (m+n)\widetilde{\gamma}(g_{\mathrm{B}})\right] G^{(m,n)}_{\mathrm{B}}(\boldsymbol{z}; g_{\mathrm{B}}, \Lambda), \\ \widetilde{\beta}(g_{\mathrm{B}}) &:= \Lambda\partial_\Lambda g_{\mathrm{B}} \qquad \text{at fixed } g_{\mathrm{R}} \text{ and } \mu, \\ \widetilde{\gamma}(g_{\mathrm{B}}) &:= \Lambda\partial_\Lambda \ln\sqrt{Z} \qquad \text{at fixed } g_{\mathrm{R}} \text{ and } \mu. \end{aligned} \tag{3.125}$$

The function $\beta(g_{\mathrm{R}})$ $[\widetilde{\beta}(g_{\mathrm{B}})]$ quantifies the rate of change of the renormalized (bare) coupling constants as the renormalization point (lattice spacing) is varied. The flow of the coupling constants under an infinitesimal change in the renormalization point (lattice spacing) is thus controlled by the so-called beta function. For the two-dimensional $O(2)$ NLσM in the spin-wave approximation there is only one coupling constant g^2 that does not flow, i.e., the beta function of g^2 vanishes identically as it should be at a critical point.

3.5 Perturbative expansion of the two-point correlation function up to one loop for the two-dimensional $O(N)$ NLσM

We are after the expansion of

$$\begin{aligned} G^{(1,1)}_{\mathrm{sw}\,g^2,\boldsymbol{H}=0}(\boldsymbol{x},\boldsymbol{y};\mathfrak{a}) :=& \left\langle \boldsymbol{n}(\boldsymbol{x})\cdot\boldsymbol{n}(\boldsymbol{y})\right\rangle_{\mathrm{sw};\mathfrak{a}} \\ =& g^2\left\langle \boldsymbol{m}(\boldsymbol{x})\cdot\boldsymbol{m}(\boldsymbol{y})\right\rangle_{\mathrm{sw};\mathfrak{a}} \\ =& g^2\left\langle\sqrt{\frac{1}{g^2}-\boldsymbol{\pi}^2(\boldsymbol{x})}\sqrt{\frac{1}{g^2}-\boldsymbol{\pi}^2(\boldsymbol{y})}\right\rangle_{\mathrm{sw};\mathfrak{a}} + g^2\left\langle\boldsymbol{\pi}(\boldsymbol{x})\cdot\boldsymbol{\pi}(\boldsymbol{y})\right\rangle_{\mathrm{sw};\mathfrak{a}} \end{aligned} \tag{3.126}$$

up to order g^4, where the expectation value $\langle(\cdots)\rangle_{\mathrm{sw};\mathfrak{a}}$ is defined by

$$\langle(\cdots)\rangle_{\mathrm{sw};\mathfrak{a}} := \frac{\int\mathrm{d}[\boldsymbol{\pi}]\Theta(1-g^2\boldsymbol{\pi}^2)e^{-\frac{1}{2}\int\limits_{\mathbb{R}^2}\frac{\mathrm{d}^2\boldsymbol{x}}{\mathfrak{a}^2}\ln\left(1-g^2\boldsymbol{\pi}^2\right)-\frac{1}{2}\int\limits_{\mathbb{R}^2}\mathrm{d}^2\boldsymbol{x}(\partial_\mu\pi_i)\left[\frac{g^2\pi_i\pi_j}{1-g^2\boldsymbol{\pi}^2}+\delta_{ij}\right](\partial_\mu\pi_j)}(\cdots)}{\int\mathrm{d}[\boldsymbol{\pi}]\Theta(1-g^2\boldsymbol{\pi}^2)e^{-\frac{1}{2}\int\limits_{\mathbb{R}^2}\frac{\mathrm{d}^2\boldsymbol{x}}{\mathfrak{a}^2}\ln\left(1-g^2\boldsymbol{\pi}^2\right)-\frac{1}{2}\int\limits_{\mathbb{R}^2}\mathrm{d}^2\boldsymbol{x}(\partial_\mu\pi_i)\left[\frac{g^2\pi_i\pi_j}{1-g^2\boldsymbol{\pi}^2}+\delta_{ij}\right](\partial_\mu\pi_j)}}. \tag{3.127}$$

Observe that the Jacobian

$$\begin{aligned}\prod_{\boldsymbol{x}\in\mathbb{R}^2} \det\sqrt{\mathfrak{g}_{ij}(\boldsymbol{x})} :=& \prod_{\boldsymbol{x}\in\mathbb{R}^2} \frac{1}{\sqrt{1-g^2\boldsymbol{\pi}^2}}\\
=&\exp\Bigg(-\frac{1}{2}\sum_{\boldsymbol{x}\in\mathbb{R}^2}\ln\left(1-g^2\boldsymbol{\pi}^2\right)\Bigg)\\
=&\exp\Bigg(-\frac{1}{2}\int\limits_{\mathbb{R}^2}\frac{\mathrm{d}^2\boldsymbol{x}}{\mathfrak{a}^2}\ln\left(1-g^2\boldsymbol{\pi}^2\right)\Bigg)\\
=&\exp\Bigg(-\delta(\boldsymbol{r}=\boldsymbol{0})\frac{1}{2}\int\limits_{\mathbb{R}^2}\mathrm{d}^2\boldsymbol{x}\ln\left(1-g^2\boldsymbol{\pi}^2\right)\Bigg)\end{aligned} \tag{3.128a}$$

depends explicitly on the short distance cut-off $\mathfrak{a}$ that regularizes the delta function in position space,

$$\delta(\boldsymbol{r}) = \begin{cases} 0, & \text{if } \boldsymbol{r}\neq\boldsymbol{0}.\\ \frac{1}{\mathfrak{a}^2}, & \text{if } \boldsymbol{r}=\boldsymbol{0}.\end{cases} \tag{3.128b}$$

In the sequel, we can forget the Heaviside step function in the measure for the spin waves as it plays no role in perturbation theory in powers of g^2.

To organize the perturbative expansion, note that we need to expand

- the argument of the expectation value in powers of g^2, i.e., we need

$$\sqrt{1-x} = 1 - \frac{1}{2}x - \frac{1}{2}\frac{1}{4}x^2 + \cdots . \tag{3.129}$$

- the action in powers of g^2, i.e., we need

$$\ln(1-x) = -x - \frac{1}{2}x^2 + \cdots . \tag{3.130}$$

- the Boltzmann weight in powers of g^2, i.e., we need

$$e^{-x} = 1 - x + \frac{1}{2}x^2 + \cdots . \tag{3.131}$$

- the inverse of the partition function in powers of g^2, i.e., we need

$$\frac{1}{1-x} = 1 + x + x^2 + \cdots . \tag{3.132}$$

Expansion in powers of g^2 of the argument in the expectation value (3.126) gives

$$\begin{aligned}G^{(1,1)}_{\mathrm{sw}\,g^2,\boldsymbol{H}=\boldsymbol{0}}(\boldsymbol{x},\boldsymbol{y};\mathfrak{a}) =&\ 1 + g^2\left\langle -\frac{1}{2}\boldsymbol{\pi}(\boldsymbol{x})\cdot\boldsymbol{\pi}(\boldsymbol{x}) - \frac{1}{2}\boldsymbol{\pi}(\boldsymbol{y})\cdot\boldsymbol{\pi}(\boldsymbol{y}) + \boldsymbol{\pi}(\boldsymbol{x})\cdot\boldsymbol{\pi}(\boldsymbol{y})\right\rangle_{\mathrm{sw};\mathfrak{a}}\\
&+g^4\left\langle +\frac{1}{4}\boldsymbol{\pi}^2(\boldsymbol{x})\boldsymbol{\pi}^2(\boldsymbol{y}) - \frac{1}{8}\boldsymbol{\pi}^2(\boldsymbol{x})\boldsymbol{\pi}^2(\boldsymbol{x}) - \frac{1}{8}\boldsymbol{\pi}^2(\boldsymbol{y})\boldsymbol{\pi}^2(\boldsymbol{y})\right\rangle_{\mathrm{sw};\mathfrak{a}}\\
&+\mathcal{O}(g^6).\end{aligned} \tag{3.133}$$

Before proceeding with the expansion, we introduce the notation

$$\begin{aligned}\mathcal{L}_{\mathrm{sw};\mathfrak{a}} :=& \underbrace{\delta(\boldsymbol{r}=\boldsymbol{0})}_{=1/\mathfrak{a}^2}\frac{1}{2}\ln\left(1-g^2\boldsymbol{\pi}^2\right)+\frac{1}{2}(\partial_\mu\pi_i)\left(\frac{g^2\pi_i\pi_j}{1-g^2\boldsymbol{\pi}^2}+\delta_{ij}\right)(\partial_\mu\pi_j) \\ \equiv& \mathcal{L}_0+g^2\mathcal{L}_{1,1}+g^2\mathcal{L}_{1,2;\mathfrak{a}}+\mathcal{O}(g^4),\end{aligned} \tag{3.134a}$$

where

$$\begin{aligned}\mathcal{L}_0 :=& \frac{1}{2}(\partial_\mu\boldsymbol{\pi})\cdot(\partial_\mu\boldsymbol{\pi}), \\ \mathcal{L}_{1,1} :=& \frac{1}{2}\left(\boldsymbol{\pi}\cdot\partial_\mu\boldsymbol{\pi}\right)\left(\boldsymbol{\pi}\cdot\partial_\mu\boldsymbol{\pi}\right), \\ \mathcal{L}_{1,2;\mathfrak{a}} :=& -\underbrace{\delta(\boldsymbol{r}=\boldsymbol{0})}_{=1/\mathfrak{a}^2}\frac{1}{2}\boldsymbol{\pi}^2.\end{aligned} \tag{3.134b}$$

The four actions obtained from the four Lagrangians $\mathcal{L}_{\mathrm{sw};\mathfrak{a}}$, $\mathcal{L}_0$, $\mathcal{L}_{1,1}$, and $\mathcal{L}_{1,2;\mathfrak{a}}$, are denoted $S_{\mathrm{sw};\mathfrak{a}}$, S_0, $S_{1,1}$, and $S_{1,2;\mathfrak{a}}$, respectively. We will also need the expansion

$$\begin{aligned}g^2\left\langle AB\right\rangle :=& g^2\frac{\int\mathrm{d}[\theta]\,e^{-S_0-g^2S_1}\,AB}{\int\mathrm{d}[\theta]\,e^{-S_0-g^2S_1}} \\ =& g^2\frac{\int\mathrm{d}[\theta]\,e^{-S_0}\,AB\,(1-g^2S_1+\cdots)}{\int\mathrm{d}[\theta]\,e^{-S_0}\quad(1-g^2S_1+\cdots)} \\ =& g^2\left\langle AB\right\rangle_0 \\ &+g^4\left[-\left\langle ABS_1\right\rangle_0+\left\langle AB\right\rangle_0\left\langle S_1\right\rangle_0\right] \\ &+\mathcal{O}(g^6),\end{aligned} \tag{3.135a}$$

where $S_1=S_{1,1}+S_{1,2;\mathfrak{a}}$ and

$$\left\langle(\cdots)\right\rangle_0 := \frac{\int\mathrm{d}[\theta]\,e^{-S_0}\,(\cdots)}{\int\mathrm{d}[\theta]\,e^{-S_0}}. \tag{3.135b}$$

Altogether, the final expansion for the spin-spin correlator reads

$$\begin{aligned}G^{(1,1)}_{\mathrm{sw}\,g^2,\boldsymbol{H}=\boldsymbol{0}}(\boldsymbol{x},\boldsymbol{y};\mathfrak{a}) =& 1+g^2\left\langle-\frac{1}{2}\boldsymbol{\pi}(\boldsymbol{x})\cdot\boldsymbol{\pi}(\boldsymbol{x})-\frac{1}{2}\boldsymbol{\pi}(\boldsymbol{y})\cdot\boldsymbol{\pi}(\boldsymbol{y})+\boldsymbol{\pi}(\boldsymbol{x})\cdot\boldsymbol{\pi}(\boldsymbol{y})\right\rangle_{S_0} \\ &+g^4\left\langle+\frac{1}{4}\boldsymbol{\pi}^2(\boldsymbol{x})\boldsymbol{\pi}^2(\boldsymbol{y})-\frac{1}{8}\boldsymbol{\pi}^2(\boldsymbol{x})\boldsymbol{\pi}^2(\boldsymbol{x})-\frac{1}{8}\boldsymbol{\pi}^2(\boldsymbol{y})\boldsymbol{\pi}^2(\boldsymbol{y})\right\rangle_{S_0} \\ &+g^4\left\langle\left[+\frac{1}{2}\boldsymbol{\pi}^2(\boldsymbol{x})+\frac{1}{2}\boldsymbol{\pi}^2(\boldsymbol{y})-\boldsymbol{\pi}(\boldsymbol{x})\cdot\boldsymbol{\pi}(\boldsymbol{y})\right](S_{1,1}+S_{1,2;\mathfrak{a}})\right\rangle_{S_0} \\ &+g^4\left\langle\left[-\frac{1}{2}\boldsymbol{\pi}^2(\boldsymbol{x})-\frac{1}{2}\boldsymbol{\pi}^2(\boldsymbol{y})+\boldsymbol{\pi}(\boldsymbol{x})\cdot\boldsymbol{\pi}(\boldsymbol{y})\right]\right\rangle_{S_0}\left\langle(S_{1,1}+S_{1,2;\mathfrak{a}})\right\rangle_{S_0} \\ &+\mathcal{O}(g^6).\end{aligned} \tag{3.136}$$

This is the expansion of the two-point function in the $O(N)$ NLσM up to order g^4 in the coupling constant.

If we recall that the limit $g^2 \to 0$ corresponds to zero temperature, we infer that the two-point function is constant to zero-th order in g^2. This is the signature of spontaneous-symmetry breaking through ferromagnetic LRO.

Spin waves disturb the ferromagnetic LRO at any non-vanishing temperature. Noting that

$$\begin{aligned}\left\langle\pi_i(\boldsymbol{x})\pi_j(\boldsymbol{y})\right\rangle_{S_0} \to& \delta_{ij}\int\frac{\mathrm{d}^2\boldsymbol{k}}{(2\pi)^2}\frac{e^{\mathrm{i}\boldsymbol{k}\cdot(\boldsymbol{x}-\boldsymbol{y})}}{\boldsymbol{k}^2+M^2}\\ =& -\delta_{ij}\frac{1}{2\pi}\ln\left(M|\boldsymbol{x}-\boldsymbol{y}|\right)+\cdots\\ \equiv& \delta_{ij}G(\boldsymbol{x},\boldsymbol{y})+\cdots, \qquad i,j=1,\cdots,N-1,\end{aligned} \tag{3.137}$$

we see that the deviations from ferromagnetic LRO induced by spin waves are logarithmically large to order g^2,

$$\begin{aligned}G^{(1,1)}_{\mathrm{sw}\,g^2,\boldsymbol{H}=\boldsymbol{0}}(\boldsymbol{x},\boldsymbol{y};\mathfrak{a}) =& 1+g^2\left\langle-\frac{1}{2}\boldsymbol{\pi}(\boldsymbol{x})\cdot\boldsymbol{\pi}(\boldsymbol{x})-\frac{1}{2}\boldsymbol{\pi}(\boldsymbol{y})\cdot\boldsymbol{\pi}(\boldsymbol{y})+\boldsymbol{\pi}(\boldsymbol{x})\cdot\boldsymbol{\pi}(\boldsymbol{y})\right\rangle_{S_0}+\mathcal{O}(g^4)\\ =& 1+g^2\left[-\frac{1}{2}(N-1)G(\boldsymbol{x},\boldsymbol{x})-\frac{1}{2}(N-1)G(\boldsymbol{y},\boldsymbol{y})+(N-1)G(\boldsymbol{x},\boldsymbol{y})\right]+\mathcal{O}(g^4)\\ =& 1+g^2(N-1)\left[G(\boldsymbol{x},\boldsymbol{y})-G(\boldsymbol{0},\boldsymbol{0})\right]+\mathcal{O}(g^4),\end{aligned} \tag{3.138}$$

where it is understood that the IR cut-off M drops out from

$$\begin{aligned}G(\boldsymbol{x},\boldsymbol{y})-G(\boldsymbol{0},\boldsymbol{0}) =& -\frac{1}{2\pi}\left[\ln(M|\boldsymbol{x}-\boldsymbol{y}|)-\ln(M\mathfrak{a})\right]+\cdots\\ =& -\frac{1}{2\pi}\ln\frac{|\boldsymbol{x}-\boldsymbol{y}|}{\mathfrak{a}}+\cdots.\end{aligned} \tag{3.139}$$

Perturbation theory in powers of g^2 thus appear to be hopeless except for the possibility that the contribution of order g^4 to the expansion be proportional to

$$\left[G(\boldsymbol{x},\boldsymbol{y})-G(\boldsymbol{0},\boldsymbol{0})\right]^2. \tag{3.140}$$

Indeed, this possibility could signal that the inclusion of spin waves renders the anomalous scaling dimensions of $\boldsymbol{\pi}$ non-vanishing, as was the case for the $O(2)$ NLσM[15], and that an expansion in powers of g^2 could be reinterpreted in a sensible way through a RG analysis based on a Callan-Symanzik equation.

There are several contributions to account for to order g^4. The second line of Eq. (3.136) gives, with the application

$$\langle ABCD\rangle_0=\langle AB\rangle_0\langle CD\rangle_0+\langle AC\rangle_0\langle BD\rangle_0+\langle AD\rangle_0\langle BC\rangle_0 \tag{3.141}$$

[15]This can be seen by expanding the right-hand side of Eq. (3.110) in powers of g^2.

of Wick's theorem and with the help of translation invariance,

$$
\begin{aligned}
&+g^4\left\langle+\frac{1}{4}\boldsymbol{\pi}^2(\boldsymbol{x})\boldsymbol{\pi}^2(\boldsymbol{y})-\frac{1}{8}\boldsymbol{\pi}^2(\boldsymbol{x})\boldsymbol{\pi}^2(\boldsymbol{x})-\frac{1}{8}\boldsymbol{\pi}^2(\boldsymbol{y})\boldsymbol{\pi}^2(\boldsymbol{y})\right\rangle_{S_0}\\
&=+g^4\left[+\frac{1}{4}\left\langle\boldsymbol{\pi}^2(\boldsymbol{x})\right\rangle_{S_0}\left\langle\boldsymbol{\pi}^2(\boldsymbol{y})\right\rangle_{S_0}+\frac{1}{2}\left\langle\pi_i(\boldsymbol{x})\pi_j(\boldsymbol{y})\right\rangle_{S_0}\left\langle\pi_i(\boldsymbol{x})\pi_j(\boldsymbol{y})\right\rangle_{S_0}\right]\\
&\quad+g^4\left[-\frac{1}{8}\left\langle\boldsymbol{\pi}^2(\boldsymbol{x})\right\rangle_{S_0}\left\langle\boldsymbol{\pi}^2(\boldsymbol{x})\right\rangle_{S_0}-\frac{1}{4}\left\langle\pi_i(\boldsymbol{x})\pi_j(\boldsymbol{x})\right\rangle_{S_0}\left\langle\pi_i(\boldsymbol{x})\pi_j(\boldsymbol{x})\right\rangle_{S_0}\right]\\
&\quad+g^4\left[-\frac{1}{8}\left\langle\boldsymbol{\pi}^2(\boldsymbol{y})\right\rangle_{S_0}\left\langle\boldsymbol{\pi}^2(\boldsymbol{y})\right\rangle_{S_0}-\frac{1}{4}\left\langle\pi_i(\boldsymbol{y})\pi_j(\boldsymbol{y})\right\rangle_{S_0}\left\langle\pi_i(\boldsymbol{y})\pi_j(\boldsymbol{y})\right\rangle_{S_0}\right]\\
&=+g^4\left[+\frac{1}{2}\left\langle\pi_i(\boldsymbol{x})\pi_j(\boldsymbol{y})\right\rangle_{S_0}\left\langle\pi_i(\boldsymbol{x})\pi_j(\boldsymbol{y})\right\rangle_{S_0}-\frac{1}{2}\left\langle\pi_i(\boldsymbol{0})\pi_j(\boldsymbol{0})\right\rangle_{S_0}\left\langle\pi_i(\boldsymbol{0})\pi_j(\boldsymbol{0})\right\rangle_{S_0}\right]\\
&=+g^4\frac{1}{2}(N-1)\left[G^2(\boldsymbol{x},\boldsymbol{y})-G^2(\boldsymbol{0},\boldsymbol{0})\right].
\end{aligned}
\tag{3.142}
$$

The third line of Eq. (3.136) demands the evaluation of

$$
\begin{aligned}
&+\frac{g^4}{4}\int\limits_{\mathbb{R}^2}\mathrm{d}^2\boldsymbol{r}\,\left\langle\pi_i(\boldsymbol{x})\pi_i(\boldsymbol{x})[\pi_j(\boldsymbol{r})\partial_\mu\pi_j(\boldsymbol{r})][\pi_k(\boldsymbol{r})\partial_\mu\pi_k(\boldsymbol{r})]\right\rangle_{S_0}\\
&+\frac{g^4}{4}\int\limits_{\mathbb{R}^2}\mathrm{d}^2\boldsymbol{r}\,\left\langle\pi_i(\boldsymbol{y})\pi_i(\boldsymbol{y})[\pi_j(\boldsymbol{r})\partial_\mu\pi_j(\boldsymbol{r})][\pi_k(\boldsymbol{r})\partial_\mu\pi_k(\boldsymbol{r})]\right\rangle_{S_0}\\
&-\frac{g^4}{2}\int\limits_{\mathbb{R}^2}\mathrm{d}^2\boldsymbol{r}\,\left\langle\pi_i(\boldsymbol{x})\pi_i(\boldsymbol{y})[\pi_j(\boldsymbol{r})\partial_\mu\pi_j(\boldsymbol{r})][\pi_k(\boldsymbol{r})\partial_\mu\pi_k(\boldsymbol{r})]\right\rangle_{S_0}\\
&+\frac{g^4}{4}\left(-\frac{1}{\mathfrak{a}^2}\right)\int\limits_{\mathbb{R}^2}\mathrm{d}^2\boldsymbol{r}\,\left\langle\pi_i(\boldsymbol{x})\pi_i(\boldsymbol{x})\pi_j(\boldsymbol{r})\pi_j(\boldsymbol{r})\right\rangle_{S_0}\\
&+\frac{g^4}{4}\left(-\frac{1}{\mathfrak{a}^2}\right)\int\limits_{\mathbb{R}^2}\mathrm{d}^2\boldsymbol{r}\,\left\langle\pi_i(\boldsymbol{y})\pi_i(\boldsymbol{y})\pi_j(\boldsymbol{r})\pi_j(\boldsymbol{r})\right\rangle_{S_0}\\
&-\frac{g^4}{2}\left(-\frac{1}{\mathfrak{a}^2}\right)\int\limits_{\mathbb{R}^2}\mathrm{d}^2\boldsymbol{r}\,\left\langle\pi_i(\boldsymbol{x})\pi_i(\boldsymbol{y})\pi_j(\boldsymbol{r})\pi_j(\boldsymbol{r})\right\rangle_{S_0}.
\end{aligned}
\tag{3.143}
$$

The fourth line of Eq. (3.136) subtracts

$$
\begin{aligned}
&+\frac{g^4}{4}\int\limits_{\mathbb{R}^2} \mathrm{d}^2\boldsymbol{r}\,\langle\pi_i(\boldsymbol{x})\pi_i(\boldsymbol{x})\rangle_{S_0}\left\langle[\pi_j(\boldsymbol{r})\partial_\mu\pi_j(\boldsymbol{r})][\pi_k(\boldsymbol{r})\partial_\mu\pi_k(\boldsymbol{r})]\right\rangle_{S_0}\\
&+\frac{g^4}{4}\int\limits_{\mathbb{R}^2} \mathrm{d}^2\boldsymbol{r}\,\langle\pi_i(\boldsymbol{y})\pi_i(\boldsymbol{y})\rangle_{S_0}\left\langle[\pi_j(\boldsymbol{r})\partial_\mu\pi_j(\boldsymbol{r})][\pi_k(\boldsymbol{r})\partial_\mu\pi_k(\boldsymbol{r})]\right\rangle_{S_0}\\
&-\frac{g^4}{2}\int\limits_{\mathbb{R}^2} \mathrm{d}^2\boldsymbol{r}\,\langle\pi_i(\boldsymbol{x})\pi_i(\boldsymbol{y})\rangle_{S_0}\left\langle[\pi_j(\boldsymbol{r})\partial_\mu\pi_j(\boldsymbol{r})][\pi_k(\boldsymbol{r})\partial_\mu\pi_k(\boldsymbol{r})]\right\rangle_{S_0}\\
&+\frac{g^4}{4}\left(-\frac{1}{\mathfrak{a}^2}\right)\int\limits_{\mathbb{R}^2} \mathrm{d}^2\boldsymbol{r}\,\langle\pi_i(\boldsymbol{x})\pi_i(\boldsymbol{x})\rangle_{S_0}\left\langle\pi_j(\boldsymbol{r})\pi_j(\boldsymbol{r})\right\rangle_{S_0}\\
&+\frac{g^4}{4}\left(-\frac{1}{\mathfrak{a}^2}\right)\int\limits_{\mathbb{R}^2} \mathrm{d}^2\boldsymbol{r}\,\langle\pi_i(\boldsymbol{y})\pi_i(\boldsymbol{y})\rangle_{S_0}\left\langle\pi_j(\boldsymbol{r})\pi_j(\boldsymbol{r})\right\rangle_{S_0}\\
&-\frac{g^4}{2}\left(-\frac{1}{\mathfrak{a}^2}\right)\int\limits_{\mathbb{R}^2} \mathrm{d}^2\boldsymbol{r}\,\langle\pi_i(\boldsymbol{x})\pi_i(\boldsymbol{y})\rangle_{S_0}\left\langle\pi_j(\boldsymbol{r})\pi_j(\boldsymbol{r})\right\rangle_{S_0}
\end{aligned}
\tag{3.144}
$$

from the third line of Eq. (3.136), i.e., it is sufficient to evaluate Eq. (3.143) with the help of Wick's theorem with the additional rule that no Wick contraction between the two points $\boldsymbol{x}$ and $\boldsymbol{x}$ or $\boldsymbol{y}$ and $\boldsymbol{y}$ or $\boldsymbol{x}$ and $\boldsymbol{y}$ can occur.[16]

Because of translation invariance, Eqs. (3.143) and (3.144) simplify to

$$
\begin{aligned}
&+\frac{g^4}{2}\int\limits_{\mathbb{R}^2} \mathrm{d}^2\boldsymbol{r}\left\langle\pi_i(\boldsymbol{0})\pi_i(\boldsymbol{0})[\pi_j(\boldsymbol{r})\partial_\mu\pi_j(\boldsymbol{r})][\pi_k(\boldsymbol{r})\partial_\mu\pi_k(\boldsymbol{r})]\right\rangle_{S_0}\\
&-\frac{g^4}{2}\int\limits_{\mathbb{R}^2} \mathrm{d}^2\boldsymbol{r}\left\langle\pi_i(\boldsymbol{x})\pi_i(\boldsymbol{y})[\pi_j(\boldsymbol{r})\partial_\mu\pi_j(\boldsymbol{r})][\pi_k(\boldsymbol{r})\partial_\mu\pi_k(\boldsymbol{r})]\right\rangle_{S_0}\\
&+\frac{g^4}{2}\left(-\frac{1}{\mathfrak{a}^2}\right)\int\limits_{\mathbb{R}^2} \mathrm{d}^2\boldsymbol{r}\left\langle\pi_i(\boldsymbol{0})\pi_i(\boldsymbol{0})\pi_j(\boldsymbol{r})\pi_j(\boldsymbol{r})\right\rangle_{S_0}\\
&-\frac{g^4}{2}\left(-\frac{1}{\mathfrak{a}^2}\right)\int\limits_{\mathbb{R}^2} \mathrm{d}^2\boldsymbol{r}\left\langle\pi_i(\boldsymbol{x})\pi_i(\boldsymbol{y})\pi_j(\boldsymbol{r})\pi_j(\boldsymbol{r})\right\rangle_{S_0}
\end{aligned}
\tag{3.147}
$$

[16] Wick's theorem reduces a Gaussian expectation value $\langle(\cdots)\rangle_0$ of $2m$ variables to the sum over all possible products of two-point functions, say for $m=3$,

$$
\begin{aligned}
\langle ABCDEF\rangle_0 =&\ \langle AB\rangle_0\langle CDEF\rangle_0+\langle AC\rangle_0\langle BDEF\rangle_0+\langle AD\rangle_0\langle BCEF\rangle_0\\
&+\langle AE\rangle_0\langle BCDF\rangle_0+\langle AF\rangle_0\langle BCDE\rangle_0,
\end{aligned}
\tag{3.145}
$$

where

$$
\langle ABCD\rangle_0=\langle AB\rangle_0\langle CD\rangle_0+\langle AC\rangle_0\langle BD\rangle_0+\langle AD\rangle_0\langle BC\rangle_0. \tag{3.146}
$$

There are thus $5\times 3=15$ contributions when $m=3$. Subtraction from $\langle ABCDEF\rangle_0$ of $\langle AB\rangle_0\langle CDEF\rangle_0$ gives 12 contributions.

and

$$
\begin{aligned}
&+\frac{g^4}{2}\int\limits_{\mathbb{R}^2} \mathrm{d}^2\boldsymbol{r}\,\langle\pi_i(\boldsymbol{0})\pi_i(\boldsymbol{0})\rangle_{S_0}\left\langle[\pi_j(\boldsymbol{r})\partial_\mu\pi_j(\boldsymbol{r})][\pi_k(\boldsymbol{r})\partial_\mu\pi_k(\boldsymbol{r})]\right\rangle_{S_0} \\
&-\frac{g^4}{2}\int\limits_{\mathbb{R}^2} \mathrm{d}^2\boldsymbol{r}\,\langle\pi_i(\boldsymbol{x})\pi_i(\boldsymbol{y})\rangle_{S_0}\left\langle[\pi_j(\boldsymbol{r})\partial_\mu\pi_j(\boldsymbol{r})][\pi_k(\boldsymbol{r})\partial_\mu\pi_k(\boldsymbol{r})]\right\rangle_{S_0} \\
&+\frac{g^4}{2}\left(-\frac{1}{\mathfrak{a}^2}\right)\int\limits_{\mathbb{R}^2} \mathrm{d}^2\boldsymbol{r}\,\langle\pi_i(\boldsymbol{0})\pi_i(\boldsymbol{0})\rangle_{S_0}\left\langle\pi_j(\boldsymbol{r})\pi_j(\boldsymbol{r})\right\rangle_{S_0} \\
&-\frac{g^4}{2}\left(-\frac{1}{\mathfrak{a}^2}\right)\int\limits_{\mathbb{R}^2} \mathrm{d}^2\boldsymbol{r}\,\langle\pi_i(\boldsymbol{x})\pi_i(\boldsymbol{y})\rangle_{S_0}\left\langle\pi_j(\boldsymbol{r})\pi_j(\boldsymbol{r})\right\rangle_{S_0},
\end{aligned}
\tag{3.148}
$$

respectively. It is then sufficient to evaluate

$$
\begin{aligned}
-\frac{g^4}{2}\int\limits_{\mathbb{R}^2} \mathrm{d}^2\boldsymbol{r}\,\Big[&\left\langle\pi_i(\boldsymbol{x})\pi_i(\boldsymbol{y})[\pi_j(\boldsymbol{r})\partial_\mu\pi_j(\boldsymbol{r})][\pi_k(\boldsymbol{r})\partial_\mu\pi_k(\boldsymbol{r})]\right\rangle_{S_0} \\
&-\langle\pi_i(\boldsymbol{x})\pi_i(\boldsymbol{y})\rangle_{S_0}\left\langle[\pi_j(\boldsymbol{r})\partial_\mu\pi_j(\boldsymbol{r})][\pi_k(\boldsymbol{r})\partial_\mu\pi_k(\boldsymbol{r})]\right\rangle_{S_0}\Big]
\end{aligned}
\tag{3.149}
$$

and

$$
-\frac{g^4}{2}\left(-\frac{1}{\mathfrak{a}^2}\right)\int\limits_{\mathbb{R}^2} \mathrm{d}^2\boldsymbol{r}\,\Big[\left\langle\pi_i(\boldsymbol{x})\pi_i(\boldsymbol{y})\pi_j(\boldsymbol{r})\pi_j(\boldsymbol{r})\right\rangle_{S_0}-\langle\pi_i(\boldsymbol{x})\pi_i(\boldsymbol{y})\rangle_{S_0}\left\langle\pi_j(\boldsymbol{r})\pi_j(\boldsymbol{r})\right\rangle_{S_0}\Big]
\tag{3.150}
$$

since one can always choose $\boldsymbol{x}=\boldsymbol{y}$. Remarkably, contribution (3.150) is contained in contribution (3.149) but with the opposite sign and thus cancels out of the spin-spin correlator. To see this, make use of translation invariance and of Eqs. (3.145) and (3.146) to write the Wick decomposition

$$
\begin{aligned}
-\frac{g^4}{2}\int\limits_{\mathbb{R}^2} \mathrm{d}^2\boldsymbol{r}\,\Big[&\left\langle\pi_i(\boldsymbol{x})\pi_i(\boldsymbol{y})[\pi_j(\boldsymbol{r})\partial_\mu\pi_j(\boldsymbol{r})][\pi_k(\boldsymbol{r})\partial_\mu\pi_k(\boldsymbol{r})]\right\rangle_{S_0} \\
&-\langle\pi_i(\boldsymbol{x})\pi_i(\boldsymbol{y})\rangle_{S_0}\left\langle[\pi_j(\boldsymbol{r})\partial_\mu\pi_j(\boldsymbol{r})][\pi_k(\boldsymbol{r})\partial_\mu\pi_k(\boldsymbol{r})]\right\rangle_{S_0}\Big] \\
=-1\times 2\times\frac{g^4}{2}&\int\limits_{\mathbb{R}^2} \mathrm{d}^2\boldsymbol{r}\,\left\langle\pi_i(\boldsymbol{x})\pi_j(\boldsymbol{r})\right\rangle_{S_0}\left\langle[\pi_k(\boldsymbol{r})\partial_\mu\pi_k(\boldsymbol{r})]\right\rangle_{S_0}\left\langle[\partial_\mu\pi_j(\boldsymbol{r})]\pi_i(\boldsymbol{y})\right\rangle_{S_0} \\
-1\times 2\times\frac{g^4}{2}&\int\limits_{\mathbb{R}^2} \mathrm{d}^2\boldsymbol{r}\,\left\langle\pi_i(\boldsymbol{x})\pi_j(\boldsymbol{r})\right\rangle_{S_0}\left\langle[\pi_k(\boldsymbol{r})\partial_\mu\pi_j(\boldsymbol{r})]\right\rangle_{S_0}\left\langle[\partial_\mu\pi_k(\boldsymbol{r})]\pi_i(\boldsymbol{y})\right\rangle_{S_0}
\end{aligned}
$$

$$
\begin{aligned}
&-1\times 2\times\frac{g^4}{2}\int\limits_{\mathbb{R}^2}\mathrm{d}^2\boldsymbol{r}\,\langle\pi_i(\boldsymbol{y})\pi_j(\boldsymbol{r})\rangle_{S_0}\,\langle[\pi_k(\boldsymbol{r})\partial_\mu\pi_k(\boldsymbol{r})]\rangle_{S_0}\,\langle[\partial_\mu\pi_j(\boldsymbol{r})]\pi_i(\boldsymbol{x})\rangle_{S_0}\\
&-1\times 2\times\frac{g^4}{2}\int\limits_{\mathbb{R}^2}\mathrm{d}^2\boldsymbol{r}\,\langle\pi_i(\boldsymbol{y})\pi_j(\boldsymbol{r})\rangle_{S_0}\,\langle[\pi_k(\boldsymbol{r})\partial_\mu\pi_j(\boldsymbol{r})]\rangle_{S_0}\,\langle[\partial_\mu\pi_k(\boldsymbol{r})]\pi_i(\boldsymbol{x})\rangle_{S_0}\\
&-1\times 2\times\frac{g^4}{2}\int\limits_{\mathbb{R}^2}\mathrm{d}^2\boldsymbol{r}\,\langle\pi_i(\boldsymbol{x})[\partial_\mu\pi_j(\boldsymbol{r})]\rangle_{S_0}\,\langle\pi_j(\boldsymbol{r})\pi_k(\boldsymbol{r})\rangle_{S_0}\,\langle[\partial_\mu\pi_k(\boldsymbol{r})]\pi_i(\boldsymbol{y})\rangle_{S_0}\\
&-1\times 2\times\frac{g^4}{2}\int\limits_{\mathbb{R}^2}\mathrm{d}^2\boldsymbol{r}\,\langle\pi_i(\boldsymbol{x})\pi_j(\boldsymbol{r})\rangle_{S_0}\,\langle[\partial_\mu\pi_j(\boldsymbol{r})][\partial_\mu\pi_k(\boldsymbol{r})]\rangle_{S_0}\,\langle\pi_k(\boldsymbol{r})\pi_i(\boldsymbol{y})\rangle_{S_0}\,.
\end{aligned}
\tag{3.151}
$$

Insertion of the unperturbed Green function (3.137) turns Eq. (3.151) into

$$
\begin{aligned}
&-\frac{g^4}{2}\int\limits_{\mathbb{R}^2}\mathrm{d}^2\boldsymbol{r}\,\Big[\langle\pi_i(\boldsymbol{x})\pi_i(\boldsymbol{y})[\pi_j(\boldsymbol{r})\partial_\mu\pi_j(\boldsymbol{r})][\pi_k(\boldsymbol{r})\partial_\mu\pi_k(\boldsymbol{r})]\rangle_{S_0}\\
&\qquad\qquad-\langle\pi_i(\boldsymbol{x})\pi_i(\boldsymbol{y})\rangle_{S_0}\,\langle[\pi_j(\boldsymbol{r})\partial_\mu\pi_j(\boldsymbol{r})][\pi_k(\boldsymbol{r})\partial_\mu\pi_k(\boldsymbol{r})]\rangle_{S_0}\Big]\\
&=-1\times 2\times\frac{g^4}{2}\int\limits_{\mathbb{R}^2}\mathrm{d}^2\boldsymbol{r}\delta_{ij}G(\boldsymbol{x},\boldsymbol{r})\delta_{kk}\lim_{\widetilde{\boldsymbol{r}}\to\boldsymbol{r}}\left(\partial_{\widetilde{\boldsymbol{r}}_\mu}G\right)(\boldsymbol{r},\widetilde{\boldsymbol{r}})\delta_{ji}\left(\partial_{\boldsymbol{r}_\mu}G\right)(\boldsymbol{r},\boldsymbol{y})\\
&-1\times 2\times\frac{g^4}{2}\int\limits_{\mathbb{R}^2}\mathrm{d}^2\boldsymbol{r}\delta_{ij}G(\boldsymbol{x},\boldsymbol{r})\delta_{kj}\lim_{\widetilde{\boldsymbol{r}}\to\boldsymbol{r}}\left(\partial_{\widetilde{\boldsymbol{r}}_\mu}G\right)(\boldsymbol{r},\widetilde{\boldsymbol{r}})\delta_{ki}\left(\partial_{\boldsymbol{r}_\mu}G\right)(\boldsymbol{r},\boldsymbol{y})\\
&-1\times 2\times\frac{g^4}{2}\int\limits_{\mathbb{R}^2}\mathrm{d}^2\boldsymbol{r}\delta_{ij}G(\boldsymbol{y},\boldsymbol{r})\delta_{kk}\lim_{\widetilde{\boldsymbol{r}}\to\boldsymbol{r}}\left(\partial_{\widetilde{\boldsymbol{r}}_\mu}G\right)(\boldsymbol{r},\widetilde{\boldsymbol{r}})\delta_{ji}\left(\partial_{\boldsymbol{r}_\mu}G\right)(\boldsymbol{r},\boldsymbol{x})\\
&-1\times 2\times\frac{g^4}{2}\int\limits_{\mathbb{R}^2}\mathrm{d}^2\boldsymbol{r}\delta_{ij}G(\boldsymbol{y},\boldsymbol{r})\delta_{kj}\lim_{\widetilde{\boldsymbol{r}}\to\boldsymbol{r}}\left(\partial_{\widetilde{\boldsymbol{r}}_\mu}G\right)(\boldsymbol{r},\widetilde{\boldsymbol{r}})\delta_{ki}\left(\partial_{\boldsymbol{r}_\mu}G\right)(\boldsymbol{r},\boldsymbol{x})\\
&-1\times 2\times\frac{g^4}{2}\int\limits_{\mathbb{R}^2}\mathrm{d}^2\boldsymbol{r}\delta_{ij}\left(\partial_{\boldsymbol{r}_\mu}G\right)(\boldsymbol{x},\boldsymbol{r})\delta_{jk}G(\boldsymbol{0},\boldsymbol{0})\delta_{ki}\left(\partial_{\boldsymbol{r}_\mu}G\right)(\boldsymbol{r},\boldsymbol{y})\\
&-1\times 2\times\frac{g^4}{2}\int\limits_{\mathbb{R}^2}\mathrm{d}^2\boldsymbol{r}\delta_{ij}G(\boldsymbol{x},\boldsymbol{r})\delta_{jk}\lim_{\widetilde{\boldsymbol{r}}\to\boldsymbol{r}}\left(\partial_{\boldsymbol{r}_\mu}\partial_{\widetilde{\boldsymbol{r}}_\mu}G\right)(\boldsymbol{r},\widetilde{\boldsymbol{r}})\delta_{ki}G(\boldsymbol{r},\boldsymbol{y}).
\end{aligned}
\tag{3.152}
$$

The first four lines on the right-hand side of Eq. (3.152) vanish since

$$
\lim_{\widetilde{\boldsymbol{r}}\to\boldsymbol{r}}\left(\partial_{\widetilde{\boldsymbol{r}}_\mu}G\right)(\boldsymbol{r},\widetilde{\boldsymbol{r}})\sim\int\limits_{\mathbb{R}^2}\frac{\mathrm{d}^2\boldsymbol{q}}{(2\pi)^2}\frac{q_\mu}{\boldsymbol{q}^2}=0.
\tag{3.153}
$$

The fifth line gives (with the help of $-\Delta G(\boldsymbol{r},\boldsymbol{y})=\delta(\boldsymbol{r}-\boldsymbol{y}) \Leftrightarrow \boldsymbol{q}^2 G_{\boldsymbol{q}}=1$)

$$
\begin{aligned}
&-1\times 2\times\frac{g^4}{2}\int\limits_{\mathbb{R}^2}\mathrm{d}^2\boldsymbol{r}\delta_{ij}\left(\partial_{r_\mu}G\right)(\boldsymbol{x},\boldsymbol{r})\delta_{jk}G(\boldsymbol{0},\boldsymbol{0})\delta_{ki}\left(\partial_{r_\mu}G\right)(\boldsymbol{r},\boldsymbol{y})\\
=&-1\times 2\times\frac{g^4}{2}(N-1)G(\boldsymbol{0},\boldsymbol{0})\int\limits_{\mathbb{R}^2}\mathrm{d}^2\boldsymbol{r}\left(\partial_{r_\mu}G\right)(\boldsymbol{x},\boldsymbol{r})\left(\partial_{r_\mu}G\right)(\boldsymbol{r},\boldsymbol{y})\\
=&-1\times 2\times\frac{g^4}{2}(N-1)G(\boldsymbol{0},\boldsymbol{0})\int\limits_{\mathbb{R}^2}\mathrm{d}^2\boldsymbol{r}\,G(\boldsymbol{x},\boldsymbol{r})\left[(-)\partial_{r_\mu}\partial_{r_\mu}G\right](\boldsymbol{r},\boldsymbol{y})\\
=&-1\times 2\times\frac{g^4}{2}(N-1)G(\boldsymbol{0},\boldsymbol{0})\int\limits_{\mathbb{R}^2}\mathrm{d}^2\boldsymbol{r}\,G(\boldsymbol{x},\boldsymbol{r})\delta(\boldsymbol{r}-\boldsymbol{y})\\
=&-g^4(N-1)G(\boldsymbol{0},\boldsymbol{0})G(\boldsymbol{x},\boldsymbol{y}).
\end{aligned}
\tag{3.154}
$$

The last line gives

$$
\begin{aligned}
&-1\times 2\times\frac{g^4}{2}\int\limits_{\mathbb{R}^2}\mathrm{d}^2\boldsymbol{r}\delta_{ij}G(\boldsymbol{x},\boldsymbol{r})\delta_{jk}\lim_{\widetilde{\boldsymbol{r}}\to\boldsymbol{r}}\left(\partial_{r_\mu}\partial_{\widetilde{r}_\mu}G\right)(\boldsymbol{r},\widetilde{\boldsymbol{r}})\delta_{ki}G(\boldsymbol{r},\boldsymbol{y})\\
=&-g^4(N-1)\int\limits_{\mathbb{R}^2}\mathrm{d}^2\boldsymbol{r}\,G(\boldsymbol{x},\boldsymbol{r})\lim_{\widetilde{\boldsymbol{r}}\to\boldsymbol{r}}\delta(\boldsymbol{r}-\widetilde{\boldsymbol{r}})G(\boldsymbol{r},\boldsymbol{y})\\
=&+g^4(N-1)\left(-\frac{1}{\mathfrak{a}^2}\right)\int\limits_{\mathbb{R}^2}\mathrm{d}^2\boldsymbol{r}\,G(\boldsymbol{x},\boldsymbol{r})G(\boldsymbol{r},\boldsymbol{y}).
\end{aligned}
\tag{3.155}
$$

This is nothing but the same as contribution (3.150) up to an overall sign. As promised contribution (3.150) cancels out. Adding up all non-vanishing contributions of order g^4 to the spin-spin correlator gives

$$
\begin{aligned}
+g^4\frac{1}{2}(N-1)\left[G(\boldsymbol{x},\boldsymbol{y})-G(\boldsymbol{0},\boldsymbol{0})\right]^2=&+g^4\frac{1}{2}(N-1)\left[G^2(\boldsymbol{x},\boldsymbol{y})-G^2(\boldsymbol{0},\boldsymbol{0})\right]\\
&-g^4\ (N-1)G(\boldsymbol{0},\boldsymbol{0})G(\boldsymbol{x},\boldsymbol{y})\\
&+g^4\ (N-1)G(\boldsymbol{0},\boldsymbol{0})G(\boldsymbol{0},\boldsymbol{0}).
\end{aligned}
\tag{3.156}
$$

In summary, the expansion of the spin-spin correlator up to order g^4 is

$$
\begin{aligned}
G^{(1,1)}_{\mathrm{sw}\,g^2,\boldsymbol{H}=\boldsymbol{0}}(\boldsymbol{x},\boldsymbol{y};\mathfrak{a})=1&\\
&+g^2\ (N-1)\left[G(\boldsymbol{x},\boldsymbol{y})-G(\boldsymbol{0},\boldsymbol{0})\right]\\
&+g^4\frac{1}{2}(N-1)\left[G(\boldsymbol{x},\boldsymbol{y})-G(\boldsymbol{0},\boldsymbol{0})\right]^2\\
&+\mathcal{O}(g^6).
\end{aligned}
\tag{3.157}
$$

As a check, we recognize the first two terms in the expansion in powers of g^2 of

$$
\left(\frac{\mathfrak{a}}{|\boldsymbol{x}-\boldsymbol{y}|}\right)^{+\frac{g^2}{2\pi}}=\exp\left(+g^2\left[G(\boldsymbol{x},\boldsymbol{y})-G(\boldsymbol{0},\boldsymbol{0})\right]\right),
\tag{3.158}
$$

if we set $N = 2$.

The origin of the divergent logarithms occurring in the expansion in powers of g^2 is the existence in two dimensions of very strong fluctuations. Spin waves destroy ferromagnetic LRO. In mathematical terms, the engineering dimension of the spin degrees of freedom differs from the scaling dimension. Correspondingly, LRO is downgraded to QLRO.

The factor $N - 1$ counts all the "Goldstone modes", i.e., those independent degrees of freedom that parametrize small fluctuations orthogonal to the ferromagnetic magnetization axis of the ferromagnetic ground state. The lattice spacing $\mathfrak{a}$ ($1/\mathfrak{a}$) plays the role of a short distance (ultraviolet) cut-off. Perturbative expansion (3.157) suggests that the expansion parameter is not simply g^2 but $g^2 \ln |\boldsymbol{x} - \boldsymbol{y}|/\mathfrak{a}$. This hypothesis is verified when $N = 2$. Correspondingly, expansion (3.157) of the spin-spin correlator is not uniformly convergent as a function of $|\boldsymbol{x} - \boldsymbol{y}|/\mathfrak{a}$. The most likely interpretation of this mathematical difficulty is that spin-wave fluctuations destroy the ferromagnetic long-range order (LRO) of the ground state at any finite temperature as it does when $N = 2$. However, the destruction of ferromagnetic LRO by spin waves when $N > 2$ is qualitatively different from the $N = 2$ case as we shall argue that ferromagnetic LRO at zero temperature ($g^2 = 0$) is replaced by paramagnetism at any finite temperature ($g^2 > 0$) using renormalization-group (RG) methods when $N > 2$. More precisely, we will derive the RG flow obeyed by g^2 and show that the coupling constant g^2 is infrared (IR) relevant relative to the ferromagnetic fixed point $g^2 = 0$ up to order g^4 when $N > 2$. Assuming that this relevance holds for all g^2, this implies that the infinite temperature fixed point $g^2 = \infty$ is stable. But the attractive fixed point $g^2 = \infty$ is, on physical grounds, nothing but the paramagnetic phase with a correlation length of the order of the lattice spacing. To put it in more quantitative terms, there must exist a correlation length ξ which is a function of g^2 that diverges as $g^2 \to 0$ and is of the order of the lattice spacing $\mathfrak{a}$ as $g^2 \to \infty$ such that

$$\langle \boldsymbol{n}(\boldsymbol{x}) \cdot \boldsymbol{n}(\boldsymbol{y}) \rangle_{\mathrm{sw}\, g^2, \boldsymbol{H}=\boldsymbol{0}} = c\, e^{-\frac{|\boldsymbol{x}-\boldsymbol{y}|}{\xi}} \tag{3.159}$$

when $N > 2$. The proportionality constant c is a dimensionless number that depends on g^2 among others. We shall show how one can compute the correlation length ξ as an expansion in powers of g^2 up to order g^4, thereby obtaining a non-perturbative result with respect to the expansion (3.157). The RG approach that we will follow is based on the observation that the short-distance cut-off $\mathfrak{a}$ in the spin-spin correlator (3.157) can be arbitrarily chosen for very large separations $|\boldsymbol{x} - \boldsymbol{y}|/\mathfrak{a} \gg 1$.

3.6 Callan-Symanzik equation obeyed by the spin-spin correlator for the two-dimensional $O(N)$ NLσM with $N > 2$

We shall derive the Callan-Symanzik equation obeyed by the spin-spin correlator in the two-dimensional $O(N > 2)$ NLσM. We will follow the conventions used in high-energy physics, i.e., we will work with the UV momentum cut-off

$$\Lambda \equiv \frac{1}{a}, \tag{3.160a}$$

and the *bare* spin-spin correlator

$$\begin{aligned} G_{\mathrm{B}}(\boldsymbol{x}) &:= \langle \boldsymbol{n}(\boldsymbol{x}) \cdot \boldsymbol{n}(\boldsymbol{0}) \rangle_{\mathrm{sw}\, g^2, \boldsymbol{H}=\boldsymbol{0}; \Lambda} \\ &= 1 - (N-1)\frac{g_{\mathrm{B}}^2}{2\pi} \ln(\Lambda|\boldsymbol{x}|) + \frac{1}{2}(N-1)\frac{g_{\mathrm{B}}^4}{(2\pi)^2} \ln^2(\Lambda|\boldsymbol{x}|) + \mathcal{O}(g_{\mathrm{B}}^6). \end{aligned} \tag{3.160b}$$

As a warm up, we multiply and divide the UV cut-off Λ by the new UV cut-off μ so as to trade the bare spin-spin correlator $G_{\mathrm{B}}(\boldsymbol{x})$ that depends on the original UV regulator Λ for a renormalized spin-spin correlator $G_{\mathrm{R}}(\boldsymbol{x})$ that depends on the new UV regulator μ. This is done up to order g_{B}^2 for which

$$\begin{aligned} G_{\mathrm{B}}(\boldsymbol{x}) &= \langle \boldsymbol{n}(\boldsymbol{x}) \cdot \boldsymbol{n}(\boldsymbol{0}) \rangle_{\mathrm{sw}\, g^2, \boldsymbol{H}=\boldsymbol{0}; \Lambda} \\ &= 1 - (N-1)\frac{g_{\mathrm{B}}^2}{2\pi} \ln(\Lambda|\boldsymbol{x}|) + \mathcal{O}(g_{\mathrm{B}}^4) \\ &= 1 - (N-1)\frac{g_{\mathrm{B}}^2}{2\pi} \ln\left(\frac{\Lambda}{\mu}\right) - (N-1)\frac{g_{\mathrm{B}}^2}{2\pi} \ln(\mu|\boldsymbol{x}|) + \mathcal{O}(g_{\mathrm{B}}^4) \\ &= \left[1 - (N-1)\frac{g_{\mathrm{R}}^2}{2\pi} \ln\left(\frac{\Lambda}{\mu}\right) + \mathcal{O}(g_{\mathrm{B}}^4)\right]\left[1 - (N-1)\frac{g_{\mathrm{R}}^2}{2\pi} \ln(\mu|\boldsymbol{x}|) + \mathcal{O}(g_{\mathrm{B}}^4)\right] \\ &\equiv Z(g_{\mathrm{R}}^2)\, G_{\mathrm{R}}(\boldsymbol{x}), \end{aligned} \tag{3.161a}$$

where

$$\begin{aligned} g_{\mathrm{R}}^2 &:= g_{\mathrm{B}}^2 + \mathcal{O}(g_{\mathrm{B}}^4), \\ Z(g_{\mathrm{R}}^2) &:= 1 - (N-1)\frac{g_{\mathrm{R}}^2}{2\pi} \ln\left(\frac{\Lambda}{\mu}\right) + \mathcal{O}(g_{\mathrm{B}}^4), \\ G_{\mathrm{R}}(\boldsymbol{x}) &:= 1 - (N-1)\frac{g_{\mathrm{R}}^2}{2\pi} \ln(\mu|\boldsymbol{x}|) + \mathcal{O}(g_{\mathrm{B}}^4). \end{aligned} \tag{3.161b}$$

Observe that

- The wave-function renormalization is given by

$$Z(g_{\mathrm{R}}^2) = \lim_{|\boldsymbol{x}| \to 1/\mu} G_{\mathrm{B}}(\boldsymbol{x}) \tag{3.162}$$

  up to order g_{R}^2. Equivalently,

$$\lim_{|\boldsymbol{x}| \to 1/\mu} G_{\mathrm{R}}(\boldsymbol{x}) = 1 \tag{3.163}$$

  up to order g_{R}^2.
- The renormalized coupling constant is given by

$$\lim_{|\boldsymbol{x}| \to 1/\mu} \frac{\partial G_{\mathrm{R}}(\boldsymbol{x})}{\partial \ln |\boldsymbol{x}|} = -(N-1)\frac{g_{\mathrm{R}}^2}{2\pi} \tag{3.164}$$

  up to order g_{R}^2.

We shall first extend the expansion of the renormalized coupling constant g^2_{R}, the wave-function renormalization $Z(g^2_{\mathrm{R}})$, and the renormalized spin-spin correlator $G_{\mathrm{R}}(\boldsymbol{x})$ up to order g^4_{R}. We shall then compute the Callan-Symanzik equation obeyed by the spin-spin correlator. However, we need to define the renormalized coupling constant g^2_{R} and the wave-function renormalization $Z(g^2_{\mathrm{R}})$ *non-perturbatively* to begin with.

3.6.1 *Non-perturbative definitions of the renormalized coupling constant and the wave-function renormalization*

The wave-function renormalization $Z(g^2_{\mathrm{B}})$ is defined, to all orders in g^2_{B}, by demanding that

$$Z(g^2_{\mathrm{B}}) := G_{\mathrm{B}}(\boldsymbol{x}) \quad \text{when } |\boldsymbol{x}| = \tfrac{1}{\mu}. \tag{3.165}$$

This definition is equivalent to demanding that

$$G_{\mathrm{R}}(\boldsymbol{x}) = 1 \quad \text{when } |\boldsymbol{x}| = \tfrac{1}{\mu} \tag{3.166}$$

since

$$G_{\mathrm{R}}(\boldsymbol{x}) = \frac{1}{Z} G_{\mathrm{B}}(\boldsymbol{x}). \tag{3.167}$$

Observe that this definition is consistent with Eq. (3.161). For given μ, the condition

$$|\boldsymbol{x}| = 1/\mu \tag{3.168}$$

defines the renormalization point. The definition of the renormalized coupling g^2_{R} is also motivated by Eq. (3.161) as it is given by the condition

$$\lim_{|\boldsymbol{x}|\to 1/\mu} \frac{\partial G_{\mathrm{R}}(\boldsymbol{x})}{\partial \ln |\boldsymbol{x}|} = -(N-1)\frac{g^2_{\mathrm{R}}}{2\pi}, \tag{3.169}$$

which must now hold non-perturbatively in powers of g^2_{B}.

3.6.2 *Expansion of the renormalized coupling constant, the wave-function renormalization, and the renormalized spin-spin correlator up to order g^4_{B}*

Inputs are

$$G_{\mathrm{B}}(\boldsymbol{x}) = 1 - (N-1)\frac{g^2_{\mathrm{B}}}{2\pi}\ln(\Lambda|\boldsymbol{x}|) + \frac{1}{2}(N-1)\frac{g^4_{\mathrm{B}}}{(2\pi)^2}\ln^2(\Lambda|\boldsymbol{x}|) + \mathcal{O}(g^6_{\mathrm{B}}), \tag{3.170a}$$

$$\begin{aligned} g^2_{\mathrm{R}}(g^2_{\mathrm{B}}) &:= -\frac{2\pi}{N-1}\left.\frac{\partial G_{\mathrm{R}}}{\partial \ln |\boldsymbol{x}|}\right|_{|\boldsymbol{x}|=1/\mu} \\ &= -\frac{2\pi}{N-1}\left.\frac{\partial\left(Z^{-1}G_{\mathrm{B}}\right)}{\partial \ln |\boldsymbol{x}|}\right|_{|\boldsymbol{x}|=1/\mu} \\ &= -\frac{2\pi}{N-1}\frac{1}{Z(g^2_{\mathrm{B}})}\left[-(N-1)\frac{g^2_{\mathrm{B}}}{2\pi} + (N-1)\frac{g^4_{\mathrm{B}}}{(2\pi)^2}\ln\left(\frac{\Lambda}{\mu}\right) + \mathcal{O}(g^6_{\mathrm{B}})\right], \end{aligned} \tag{3.170b}$$

$$
\begin{aligned}
Z(g_{\mathrm{B}}^2) &:= G_{\mathrm{B}}(\boldsymbol{x})\big|_{|\boldsymbol{x}|=1/\mu} \\
&= 1-(N-1)\frac{g_{\mathrm{B}}^2}{2\pi}\ln\left(\frac{\Lambda}{\mu}\right)+\frac{1}{2}(N-1)\frac{g_{\mathrm{B}}^4}{(2\pi)^2}\ln^2\left(\frac{\Lambda}{\mu}\right)+\mathcal{O}(g_{\mathrm{B}}^6),
\end{aligned}
\tag{3.170c}
$$

and

$$
\frac{1}{Z(g_{\mathrm{B}}^2)}=1+(N-1)\frac{g_{\mathrm{B}}^2}{2\pi}\ln\left(\frac{\Lambda}{\mu}\right)+(N-1)(N-3/2)\frac{g_{\mathrm{B}}^4}{(2\pi)^2}\ln^2\left(\frac{\Lambda}{\mu}\right)+\mathcal{O}(g_{\mathrm{B}}^6), \tag{3.170d}
$$

from which follows that

$$
\begin{aligned}
g_{\mathrm{R}}^2(g_{\mathrm{B}}^2) &= -\frac{2\pi}{N-1}\,\frac{-(N-1)\frac{g_{\mathrm{B}}^2}{2\pi}+(N-1)\frac{g_{\mathrm{B}}^4}{(2\pi)^2}\ln\left(\frac{\Lambda}{\mu}\right)+\mathcal{O}(g_{\mathrm{B}}^6)}{1-(N-1)\frac{g_{\mathrm{B}}^2}{2\pi}\ln\left(\frac{\Lambda}{\mu}\right)+\mathcal{O}(g_{\mathrm{B}}^4)} \\
&= \frac{+g_{\mathrm{B}}^2-\frac{g_{\mathrm{B}}^4}{2\pi}\ln\left(\frac{\Lambda}{\mu}\right)+\mathcal{O}(g_{\mathrm{B}}^6)}{1-(N-1)\frac{g_{\mathrm{B}}^2}{2\pi}\ln\left(\frac{\Lambda}{\mu}\right)+\mathcal{O}(g_{\mathrm{B}}^4)} \\
&= g_{\mathrm{B}}^2+(N-2)\frac{g_{\mathrm{B}}^4}{2\pi}\ln\left(\frac{\Lambda}{\mu}\right)+\mathcal{O}(g_{\mathrm{B}}^6),
\end{aligned}
\tag{3.171}
$$

on the one hand, and

$$
\begin{aligned}
G_{\mathrm{R}}(\boldsymbol{x}) :&= Z^{-1}(g_{\mathrm{B}}^2)\,G_{\mathrm{B}}(\boldsymbol{x}) \\
&= 1 \\
&\quad -(N-1)\frac{g_{\mathrm{B}}^2}{2\pi}\ln(\Lambda|\boldsymbol{x}|)+(N-1)\frac{g_{\mathrm{B}}^2}{2\pi}\ln\left(\frac{\Lambda}{\mu}\right) \\
&\quad +\frac{1}{2}(N-1)\frac{g_{\mathrm{B}}^4}{(2\pi)^2}\ln^2(\Lambda|\boldsymbol{x}|)+(N-1)(N-3/2)\frac{g_{\mathrm{B}}^4}{(2\pi)^2}\ln^2\left(\frac{\Lambda}{\mu}\right) \\
&\quad -(N-1)^2\frac{g_{\mathrm{B}}^4}{(2\pi)^2}\ln(\Lambda|\boldsymbol{x}|)\ln\left(\frac{\Lambda}{\mu}\right)+\mathcal{O}(g_{\mathrm{B}}^6),
\end{aligned}
\tag{3.172}
$$

on the other hand.

3.6.3 *Expansion of the bare coupling constant, the wave-function renormalization, and the renormalized spin-spin correlator up to order g_{R}^4*

Inverting

$$
g_{\mathrm{R}}^2(g_{\mathrm{B}}^2)=g_{\mathrm{B}}^2+(N-2)\frac{g_{\mathrm{B}}^4}{2\pi}\ln\left(\frac{\Lambda}{\mu}\right)+\mathcal{O}(g_{\mathrm{B}}^6), \tag{3.173}
$$

gives

$$
g_{\mathrm{B}}^2(g_{\mathrm{R}}^2)=g_{\mathrm{R}}^2-(N-2)\frac{g_{\mathrm{R}}^4}{2\pi}\ln\left(\frac{\Lambda}{\mu}\right)+\mathcal{O}(g_{\mathrm{R}}^6). \tag{3.174}
$$

Insertion of Eq. (3.174) into the right-hand sides of Eqs. (3.170a), (3.170c), and (3.170d) gives

$$\begin{aligned}
G_{\rm B}(\boldsymbol{x}) =&\ 1-(N-1)\frac{1}{2\pi}\left[g_{\rm R}^2-(N-2)\frac{g_{\rm R}^4}{2\pi}\ln\left(\frac{\Lambda}{\mu}\right)+\mathcal{O}(g_{\rm R}^6)\right]\ln(\Lambda|\boldsymbol{x}|)\\
&+\frac{1}{2}(N-1)\frac{g_{\rm R}^4}{(2\pi)^2}\ln^2(\Lambda|\boldsymbol{x}|)+\mathcal{O}(g_{\rm R}^6)\\
=&\ 1-(N-1)\frac{g_{\rm R}^2}{2\pi}\ln(\Lambda|\boldsymbol{x}|)+\frac{1}{2}(N-1)\frac{g_{\rm R}^4}{(2\pi)^2}\ln^2(\Lambda|\boldsymbol{x}|)\\
&+(N-1)(N-2)\frac{g_{\rm R}^4}{(2\pi)^2}\ln(\Lambda|\boldsymbol{x}|)\ln\left(\frac{\Lambda}{\mu}\right)+\mathcal{O}(g_{\rm R}^6)\\
=&\ 1-(N-1)\frac{g_{\rm R}^2}{2\pi}\ln(\Lambda|\boldsymbol{x}|)+\frac{1}{2}(N-1)\frac{g_{\rm R}^4}{(2\pi)^2}\ln^2(\Lambda|\boldsymbol{x}|)\\
&+\left[(N-1)^2-(N-1)\right]\frac{g_{\rm R}^4}{(2\pi)^2}\ln(\Lambda|\boldsymbol{x}|)\ln\left(\frac{\Lambda}{\mu}\right)+\mathcal{O}(g_{\rm R}^6),
\end{aligned} \tag{3.175a}$$

$$\begin{aligned}
Z(g_{\rm R}^2) =&\ 1-(N-1)\frac{1}{2\pi}\left[g_{\rm R}^2-(N-2)\frac{g_{\rm R}^4}{2\pi}\ln\left(\frac{\Lambda}{\mu}\right)+\mathcal{O}(g_{\rm R}^6)\right]\ln\left(\frac{\Lambda}{\mu}\right)\\
&+\frac{1}{2}(N-1)\frac{g_{\rm R}^4}{(2\pi)^2}\ln^2\left(\frac{\Lambda}{\mu}\right)+\mathcal{O}(g_{\rm R}^6)\\
=&\ 1-(N-1)\frac{g_{\rm R}^2}{2\pi}\ln\left(\frac{\Lambda}{\mu}\right)\\
&+(N-1)(N-3/2)\frac{g_{\rm R}^4}{(2\pi)^2}\ln^2\left(\frac{\Lambda}{\mu}\right)+\mathcal{O}(g_{\rm R}^6),
\end{aligned} \tag{3.175b}$$

and

$$\begin{aligned}
\frac{1}{Z(g_{\rm R}^2)} =&\ \frac{1}{1-(N-1)\frac{g_{\rm R}^2}{2\pi}\ln\left(\frac{\Lambda}{\mu}\right)+(N-1)(N-3/2)\frac{g_{\rm R}^4}{(2\pi)^2}\ln^2\left(\frac{\Lambda}{\mu}\right)+\mathcal{O}(g_{\rm R}^6)}\\
=&\ 1+(N-1)\frac{g_{\rm R}^2}{2\pi}\ln\left(\frac{\Lambda}{\mu}\right)\\
&+\left[(N-1)^2-(N-1)(N-3/2)\right]\frac{g_{\rm R}^4}{(2\pi)^2}\ln^2\left(\frac{\Lambda}{\mu}\right)+\mathcal{O}(g_{\rm R}^6)\\
=&\ 1+(N-1)\frac{g_{\rm R}^2}{2\pi}\ln\left(\frac{\Lambda}{\mu}\right)\\
&+(N-1)\left(N-1-N+3/2\right)\frac{g_{\rm R}^4}{(2\pi)^2}\ln^2\left(\frac{\Lambda}{\mu}\right)+\mathcal{O}(g_{\rm R}^6)\\
=&\ 1+(N-1)\frac{g_{\rm R}^2}{2\pi}\ln\left(\frac{\Lambda}{\mu}\right)+\frac{1}{2}(N-1)\frac{g_{\rm R}^4}{(2\pi)^2}\ln^2\left(\frac{\Lambda}{\mu}\right)+\mathcal{O}(g_{\rm R}^6),
\end{aligned} \tag{3.175c}$$

respectively. Multiplication of Eq. (3.175a) by Eq. (3.175c) gives the desired expansion of the renormalized spin-spin correlator

$$G_{\rm R}(\boldsymbol{x}) = 1-(N-1)\frac{g_{\rm R}^2}{2\pi}\ln(\mu|\boldsymbol{x}|)+\frac{1}{2}(N-1)\frac{g_{\rm R}^4}{(2\pi)^2}\ln^2(\mu|\boldsymbol{x}|)+\mathcal{O}(g_{\rm R}^6), \tag{3.176}$$

since the cross term of order g_{R}^4 cancels the term underlined in Eq. (3.175a). The fact that G_{R} is obtained from G_{B} with the substitution

$$g_{\mathrm{B}}^2 \longleftrightarrow g_{\mathrm{R}}^2, \qquad \Lambda \longleftrightarrow \mu \tag{3.177}$$

is an artifact of the expansion to order g_{R}^4 coupled with the choice of the renormalization point made in section 3.6.1.

3.6.4 *Callan-Symanzik equation obeyed by the spin-spin correlator*

The Callan-Symanzik equation obeyed by the renormalized spin-spin correlator is

$$\begin{aligned} 0 &= \left[\mu\frac{\partial}{\partial\mu} + \beta(g_{\mathrm{R}}^2)\frac{\partial}{\partial g_{\mathrm{R}}^2} + 2\gamma(g_{\mathrm{R}}^2)\right] G_{\mathrm{R}}(\boldsymbol{x}), \\ 1 &= \lim_{|\boldsymbol{x}|\to 1/\mu} G_{\mathrm{R}}(\boldsymbol{x}). \end{aligned} \tag{3.178}$$

We try the Ansatz

$$\begin{aligned} \beta(g_{\mathrm{R}}^2) &:= b_2 g_{\mathrm{R}}^4 + \mathcal{O}(g_{\mathrm{R}}^6), \\ \gamma(g_{\mathrm{R}}^2) &:= a_1 g_{\mathrm{R}}^2 + a_2 g_{\mathrm{R}}^4 + \mathcal{O}(g_{\mathrm{R}}^6). \end{aligned} \tag{3.179}$$

With the help of

$$\begin{aligned} \mu\frac{\partial}{\partial\mu}G_{\mathrm{R}}(\boldsymbol{x}) &= \underline{-(N-1)\frac{g_{\mathrm{R}}^2}{2\pi}} + \underline{\underline{\underline{(N-1)\frac{g_{\mathrm{R}}^4}{(2\pi)^2}\ln(\mu|\boldsymbol{x}|)}}} + \mathcal{O}(g_{\mathrm{R}}^6), \\ \beta(g_{\mathrm{R}}^2)\frac{\partial}{\partial g_{\mathrm{R}}^2}G_{\mathrm{R}}(\boldsymbol{x}) &= \left[b_2 g_{\mathrm{R}}^4 + \mathcal{O}(g_{\mathrm{R}}^6)\right] \\ &\quad\times\left[-(N-1)\frac{1}{2\pi}\ln(\mu|\boldsymbol{x}|) + (N-1)\frac{g_{\mathrm{R}}^2}{(2\pi)^2}\ln^2(\mu|\boldsymbol{x}|) + \mathcal{O}(g_{\mathrm{R}}^6)\right] \\ &= \underline{\underline{\underline{-b_2(N-1)\frac{g_{\mathrm{R}}^4}{2\pi}\ln(\mu|\boldsymbol{x}|)}}} + \mathcal{O}(g_{\mathrm{R}}^6), \\ 2\gamma(g_{\mathrm{R}}^2)G_{\mathrm{R}}(\boldsymbol{x}) &= 2\left[a_1 g_{\mathrm{R}}^2 + a_2 g_{\mathrm{R}}^4 + \mathcal{O}(g_{\mathrm{R}}^6)\right] \\ &\quad\times\left[1-(N-1)\frac{g_{\mathrm{R}}^2}{2\pi}\ln(\mu|\boldsymbol{x}|) + \frac{1}{2}(N-1)\frac{g_{\mathrm{R}}^4}{(2\pi)^2}\ln^2(\mu|\boldsymbol{x}|) + \mathcal{O}(g_{\mathrm{R}}^6)\right] \\ &= \underline{2a_1 g_{\mathrm{R}}^2} + \underline{\underline{2a_2 g_{\mathrm{R}}^4}} \underline{\underline{\underline{- 2a_1(N-1)\frac{g_{\mathrm{R}}^4}{2\pi}\ln(\mu|\boldsymbol{x}|)}}} + \mathcal{O}(g_{\mathrm{R}}^6), \end{aligned} \tag{3.180}$$

one needs to solve the equations

$$\begin{aligned} 0 &= \underline{-(N-1)\frac{g_{\mathrm{R}}^2}{2\pi}} + \underline{2a_1 g_{\mathrm{R}}^2}, \\ 0 &= \underline{\underline{2a_2 g_{\mathrm{R}}^4}}, \\ 0 &= \underline{\underline{\underline{(N-1)\frac{g_{\mathrm{R}}^4}{(2\pi)^2}\ln(\mu|\boldsymbol{x}|)}}} \underline{\underline{\underline{- b_2(N-1)\frac{g_{\mathrm{R}}^4}{2\pi}\ln(\mu|\boldsymbol{x}|)}}} \underline{\underline{\underline{- 2a_1(N-1)\frac{g_{\mathrm{R}}^4}{2\pi}\ln(\mu|\boldsymbol{x}|)}}}, \end{aligned} \tag{3.181}$$

i.e.,

$$\begin{aligned} 2a_1 &= \frac{N-1}{2\pi}, \\ 2a_2 &= 0, \\ b_2 &= -\frac{N-2}{2\pi}. \end{aligned} \tag{3.182}$$

We conclude that the Callan-Symanzik equation obeyed by the renormalized spin-spin correlator is given by

$$0 = \left[\mu\frac{\partial}{\partial\mu} + \beta(g_{\mathrm{R}}^2)\frac{\partial}{\partial g_{\mathrm{R}}^2} + 2\gamma(g_{\mathrm{R}}^2)\right] G_{\mathrm{R}}(\boldsymbol{x}), \tag{3.183a}$$

$$\beta(g_{\mathrm{R}}^2) \equiv \mu\frac{\partial g_{\mathrm{R}}^2}{\partial\mu} = -(N-2)\frac{g_{\mathrm{R}}^4}{2\pi} + \mathcal{O}(g_{\mathrm{R}}^6), \tag{3.183b}$$

$$\gamma(g_{\mathrm{R}}^2) \equiv \mu\frac{\partial\sqrt{\ln Z}}{\partial\mu} = +\frac{N-1}{2}\frac{g_{\mathrm{R}}^2}{2\pi} + \mathcal{O}(g_{\mathrm{R}}^6), \tag{3.183c}$$

with the non-perturbative condition

$$\lim_{|\boldsymbol{x}|\to 1/\mu} G_{\mathrm{R}}(\boldsymbol{x}) = 1. \tag{3.183d}$$

3.6.5 *Physical interpretation of the Callan-Symanzik equation*

The Callan-Symanzik equation (3.183) is a set of three first-order differential equations obeyed by the spin-spin correlator, the coupling constant, and the wave-function renormalization in the two-dimensional $O(N>2)$ NLσM. As such it has a unique solution if and only if the value of the spin-spin correlator is specified at "one point" $(\mu\boldsymbol{x}, g_{\mathrm{R}}^2)$, the so-called renormalization point. The renormalization point that we chose is

$$\mu|\boldsymbol{x}| = 1, \qquad g_{\mathrm{R}}^2 = -\frac{2\pi}{N-1}\lim_{|\boldsymbol{x}|\to 1/\mu}\frac{\partial G_{\mathrm{R}}}{\partial \ln|\boldsymbol{x}|}, \tag{3.184}$$

(by translation invariance, the spin-spin correlator is a function of $|\boldsymbol{x}|$ only) at which we took the spin-spin correlator to be unity. The numerical values taken by the expansion coefficients of the beta function $\beta(g_{\mathrm{R}}^2)$ and the anomalous scaling dimension $\gamma(g_{\mathrm{R}}^2)$ depend on the choice of the renormalization point [the point $(\mu\boldsymbol{x}, g_{\mathrm{R}}^2)$ at which the renormalized spin-spin correlator is unity, say]. The signs of $\beta(g_{\mathrm{R}}^2)$ and $\gamma(g_{\mathrm{R}}^2)$ in the vicinity of the free-field fixed point $g_{\mathrm{R}}^2 = 0$ are independent of the renormalization point.

The Callan-Symanzik equation (3.183) can be solved by the method of characteristics by which Eq. (3.183) is recast into

$$\begin{aligned} 0 &= \left[\frac{\mathrm{d}}{\mathrm{d}t} + 2\gamma\big(g_{\mathrm{R}}^2(t)\big)\right] \widetilde{G}_{\mathrm{R}}\big(\mu(t)\boldsymbol{x}\big), \\ \mu(t) &:= \frac{\mathrm{d}\mu(t)}{\mathrm{d}t}, \\ \beta\big(g_{\mathrm{R}}^2(t)\big) &:= \frac{\mathrm{d}g_{\mathrm{R}}^2(t)}{\mathrm{d}t}, \end{aligned} \tag{3.185a}$$

with some initial data at "time" t_0, say,

$$\mu(t_0) \equiv \Lambda, \qquad g^2_{\mathrm{R}}(t_0) \equiv g^2_{\mathrm{B}}, \qquad \widetilde{G}_{\mathrm{R}}\big(\mu(t_0)\boldsymbol{x}\big) \equiv \widetilde{G}_{\mathrm{B}}(\Lambda\boldsymbol{x}). \tag{3.185b}$$

The curve parametrized by t and defined by the set of points $\big(\mu(t)|\boldsymbol{x}|, g^2_{\mathrm{R}}(t)\big) \in \mathbb{R}^2$ is called a characteristic of the Callan-Symanzik equation (3.185). In this incarnation, the Callan-Symanzik equation encodes the notion of scaling in that the spatial argument of the spin-spin correlator only depends on the dimensionless ratio of length scales $\mu(t)\boldsymbol{x}$,

$$\widetilde{G}_{\mathrm{R}}\big(\mu(t)\boldsymbol{x}\big) := G_{\mathrm{R}}(\boldsymbol{x}), \qquad \widetilde{G}_{\mathrm{B}}(\Lambda\boldsymbol{x}) := G_{\mathrm{B}}(\boldsymbol{x}), \tag{3.186}$$

as we have verified explicitly up to second order in perturbation theory with Eqs. (3.176) and (3.160b), respectively. The coupling constant $g^2_{\mathrm{R}}(t)$ is reinterpreted as a "running" coupling constant, i.e., as a scale dependent coupling constant.

In the representation (3.185), the Callan-Symanzik equation can be integrated to, say,

$$\begin{aligned}
\widetilde{G}_{\mathrm{R}}\big(\mu(t)\boldsymbol{x}\big) &= \widetilde{G}_{\mathrm{B}}(\Lambda\boldsymbol{x}) \times \exp\left(-2\int\limits_{t_0}^{t} \mathrm{d}t'\gamma\big(g^2_{\mathrm{R}}(t')\big)\right),\\
\mu(t) &= \Lambda\, e^{t-t_0},\\
t - t_0 &= \int\limits_{g^2_{\mathrm{B}}}^{g^2_{\mathrm{R}}} \frac{\mathrm{d}g^2}{\beta(g^2)}.
\end{aligned} \tag{3.187}$$

By choosing the initial time t_0 so that $\Lambda|\boldsymbol{x}|$ is at the renormalization point

$$\Lambda|\boldsymbol{x}| = 1, \qquad g^2_{\mathrm{B}}, \tag{3.188}$$

Eq. (3.187) becomes

$$\begin{aligned}
\widetilde{G}_{\mathrm{R}}\left(e^{+(t-t_0)}\right) &= \exp\left(-2\int\limits_{t_0}^{t} \mathrm{d}t'\gamma\big(g^2_{\mathrm{R}}(t')\big)\right),\\
t - t_0 &= \int\limits_{g^2_{\mathrm{B}}}^{g^2_{\mathrm{R}}} \frac{\mathrm{d}g^2}{\beta(g^2)}.
\end{aligned} \tag{3.189}$$

By choosing the final time t so that $\mu(t)|\boldsymbol{x}|$ is at the renormalization point

$$\mu(t)|\boldsymbol{x}| = 1, \qquad g^2_{\mathrm{R}}, \tag{3.190}$$

Eq. (3.187) becomes

$$\begin{aligned}
\widetilde{G}_{\mathrm{B}}\left(e^{-(t-t_0)}\right) &= \exp\left(+2\int\limits_{t_0}^{t} \mathrm{d}t'\gamma\big(g^2_{\mathrm{R}}(t')\big)\right),\\
t - t_0 &= \int\limits_{g^2_{\mathrm{B}}}^{g^2_{\mathrm{R}}} \frac{\mathrm{d}g^2}{\beta(g^2)}.
\end{aligned} \tag{3.191}$$

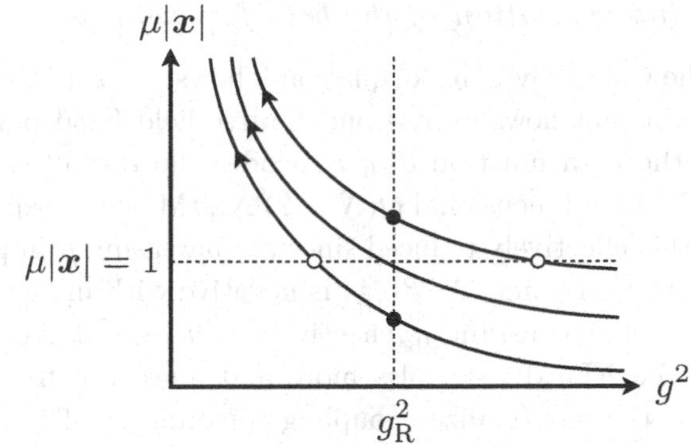

Fig. 3.2 RG characteristics for the Callan-Symanzik equation in an asymptotically free theory at short distances. Initial data are depicted as open circles along the constant line $\mu|\boldsymbol{x}| = 1$ in the $\mu|\boldsymbol{x}|$-g^2 plane. Final data are depicted as filled circles along the characteristics emanating from the initial data.

Qualitative RG characteristics for the spin-spin correlators are displayed in Fig. 3.2, whereby it is assumed that $\beta(g^2) < 0$ for all $g^2 > 0$. Figure 3.2 gives a pictorial answer to the question of what range of g^2 is needed to integrate $\widetilde{G}_{\mathrm{R}}(\mu(t')\boldsymbol{x})$ from its initial reference value $1 = \widetilde{G}_{\mathrm{R}}(\mu(t_0)|\boldsymbol{x}| = 1)$ (an open circle in Fig. 3.2 along the horizontal line $\mu|\boldsymbol{x}| = 1$ at which $g^2 \equiv g_{\mathrm{B}}^2$) to its final value $\widetilde{G}_{\mathrm{R}}(\mu(t)\boldsymbol{x})$ at some *given* $g_{\mathrm{R}}^2(t)$ [a closed circle in Fig. 3.2 along the vertical line $g^2 = g_{\mathrm{R}}^2(t)$]. We can distinguish two families of characteristics.

There are those characteristics that intercept the fixed vertical line $g^2 = g_{\mathrm{R}}^2(t)$ at a value of $\mu(t)|\boldsymbol{x}| < 1$. The range of g^2 interpolating between the initial g_B^2 at t_0 (open circle) and the final $g_{\mathrm{R}}^2(t)$ is then the finite segment $[0, g_{\mathrm{R}}^2(t)]$ as $\mu(t)|\boldsymbol{x}| \to 0$.

There are those characteristics that intercept the fixed vertical line $g^2 = g_{\mathrm{R}}^2(t)$ at a value of $\mu(t)|\boldsymbol{x}| > 1$. The range of g^2 interpolating between the initial g_B^2 at t_0 (open circle) and final $g_{\mathrm{R}}^2(t)$ is then the semi-infinite segment $[g_{\mathrm{R}}^2(t), \infty[$ as $\mu(t)|\boldsymbol{x}| \to \infty$.

If we seek values of $\widetilde{G}_{\mathrm{R}}(\mu(t)\boldsymbol{x})$ with $\mu(t)|\boldsymbol{x}| \gg 1$ given $\widetilde{G}_{\mathrm{R}}(\mu(t_0)\boldsymbol{x})$ perturbation theory is thus condemned to failure. On the other hand, perturbation theory can be accurate if we seek values of $\widetilde{G}_{\mathrm{R}}(\mu(t)\boldsymbol{x})$ given $\widetilde{G}_{\mathrm{R}}(\mu(t_0)\boldsymbol{x})$ with $\mu(t)|\boldsymbol{x}| \ll 1$.

As scale invariance implies that we can equally well regard variations of $\mu\boldsymbol{x}$ as being variations of $\boldsymbol{x}$ at fixed μ or conversely, we infer that the accuracy of perturbation theory improves as the Callan-Symanzik equation is integrated to probe the spin-spin correlation function at arbitrary small $|\boldsymbol{x}|$, a property called *asymptotic freedom at short distances*. Conversely, the accuracy of perturbation theory diminishes (breaks down) as the Callan-Symanzik equation is integrated to probe the spin-spin correlation function at arbitrary large $|\boldsymbol{x}|$.

3.6.6 *Physical interpretation of the beta function*

This property of the Callan-Symanzik equation follows from the fact that the renormalized coupling constant flows away from the free-field fixed point $g^2 = 0$ in the IR limit. Indeed, the beta function $\beta(g_{\mathrm{R}}^2)$ encodes the rate of change of the coupling constant of the two-dimensional $O(N > 2)$ NLσM as the separation $|\boldsymbol{x}|$ in the spin-spin correlator is effectively reduced since an increasing μ implies a decreasing $|\boldsymbol{x}|$ at the renormalization point. As $\beta(g_{\mathrm{R}}^2)$ is negative with increasing μ for $N > 2$, the renormalized coupling constant g_{R}^2 effectively *decreases* at shorter distances. At shorter distances, the NLσM resembles more and more the free-field fixed point $g_{\mathrm{R}}^2 = 0$. Conversely, the renormalized coupling constant g_{R}^2 effectively *increases* at longer distances. Within the RG terminology, the coupling constant of the two-dimensional $O(N > 2)$ NLσM is UV irrelevant, or, equivalently, IR relevant at the free-field fixed point. The free-field fixed point $g_{\mathrm{R}}^2 = 0$ is UV stable, or, equivalently, IR unstable when $N > 2$. Our perturbative RG analysis can thus only be trusted in the close vicinity of the UV limit $\lim_{t\to\infty}\mu(t) = \infty$. As perturbation theory breaks down in the IR limit, $\lim_{t\to\infty}\mu(t) = 0$, one must rely on alternative methods [Bethe Ansatz, numerical simulations on the underlying lattice model, high temperature $(g^2 \gg 1)$ expansions] to probe the physics of the two-dimensional $O(N > 2)$ NLσM at long distances.

As the RG trajectories flow out of the regime of applicability of perturbation theory in g^2, we cannot infer from our calculation the behavior of the spin-spin correlator for very large separations. The most economical hypothesis is to imagine that the flow is to an IR stable fixed point describing a paramagnetic phase as $g^2 \to \infty$. In the paramagnetic phase, the exponential decay

$$\exp\left(-\frac{|\boldsymbol{x}|}{\xi}\right) \tag{3.192}$$

with large separation $|\boldsymbol{x}|$ of the spin-spin correlator allows the identification of the length scale ξ, the so-called paramagnetic correlation length, which is of the order of the lattice spacing. Although our RG analysis cannot alone establish the existence of the IR stable paramagnetic phase and of the concomitant finite correlation length of order of the lattice spacing, it can predict the small g^2 dependence of a finite (though large) correlation length in the close vicinity of the ferromagnetic IR unstable fixed point.

By dimensional analysis, the rescaling

$$\mathfrak{a} \to \frac{\mathfrak{a}}{b} \equiv \mathfrak{a} + \mathrm{d}\mathfrak{a}, \qquad \frac{\mathrm{d}\mathfrak{a}}{\mathfrak{a}} - \mathcal{O}\big((\mathrm{d}\mathfrak{a}/\mathfrak{a})^2\big) = \ln(1/b), \tag{3.193}$$

implies that the correlation length $\xi(g_{\mathfrak{a}}^2)$ calculated with the lattice spacing $\mathfrak{a}$ is related to the correlation length $\xi(g_{\mathfrak{a}/b}^2)$ calculated with the lattice spacing $\mathfrak{a}/b$ by

$$\xi(g_{\mathfrak{a}}^2) = \frac{1}{b}\xi(g_{\mathfrak{a}/b}^2). \tag{3.194}$$

To proceed, integrate (note the sign difference relative to a variation with respect to the momentum cut-off μ)

$$\mathfrak{a}\frac{\partial g^2}{\partial \mathfrak{a}} = +(N-2)\frac{g^4}{2\pi} + \mathcal{O}(g^6) \tag{3.195}$$

to find

$$\begin{aligned} 0 &= \int_{g^2_{\mathfrak{a}}}^{g^2_{\mathfrak{a}/b}} \frac{dg^2}{g^4} - \frac{N-2}{2\pi}\int_{\mathfrak{a}}^{\mathfrak{a}/b} \frac{d\mathfrak{a}}{\mathfrak{a}} \\ &= \left(\frac{1}{g^2_{\mathfrak{a}}} - \frac{1}{g^2_{\mathfrak{a}/b}}\right) - \frac{N-2}{2\pi}\left[\ln(\mathfrak{a}/b) - \ln \mathfrak{a}\right] \\ &= \left(\frac{1}{g^2_{\mathfrak{a}}} - \frac{1}{g^2_{\mathfrak{a}/b}}\right) - \frac{N-2}{2\pi}\ln(1/b). \end{aligned} \tag{3.196}$$

Assume now that $\mathfrak{a}/b$ is chosen so that $\xi(g^2_{\mathfrak{a}/b})$ in Eq. (3.194) is of order of the lattice spacing so that, when combined with Eq. (3.196), it is found that

$$\begin{aligned} \xi(g^2_{\mathfrak{a}}) &= \frac{1}{b}\xi(g^2_{\mathfrak{a}/b}) \\ &\approx \exp\left(\ln(1/b)\right) \times \mathfrak{a} \\ &\approx \exp\left(+\frac{2\pi}{N-2}\frac{1}{g^2_{\mathfrak{a}}}\left[1 + \mathcal{O}\left(\frac{g^2_{\mathfrak{a}}}{g^2_{\mathfrak{a}/b}}\right)\right]\right) \times \mathfrak{a}. \end{aligned} \tag{3.197}$$

The very rapid divergence of the correlation length as $g^2 \to 0$ corresponds to a weak singularity of the free energy (the logarithm of the partition function). We have uncovered a second important property of the two-dimensional $O(N > 2)$ NLσM aside from UV asymptotic freedom, namely that of *dimensional transmutation*, whereby a macroscopic length scale in the form of a correlation length is generated out of a field theory depending on one dimensionless coupling constant and one microscopic UV cut-off.

Before closing this section observe that the prefactor to the exponential dependence on g^2 in the correlation length (3.197) can also be g^2 dependent. To see this it suffices to include the first non-vanishing contribution of higher order than g^4 to the beta function in the two-dimensional $O(N > 2)$ NLσM, say the term $\beta_3 g^6$, in the expansion

$$\beta(g^2) = \sum_{n=0}^{\infty} \beta_n g^{2n}, \qquad \beta_0 = 0, \qquad \beta_1 = 0, \qquad \beta_2 = -\frac{N-2}{2\pi}, \tag{3.198}$$

where β_3 is yet to be calculated. Assuming that β_3 is non-vanishing, one finds

$$\xi(g^2_{\mathfrak{a}}) \approx \left(\frac{g^2_{\mathfrak{a}}}{g^2_{\mathfrak{a}/b}}\right)^{(2\pi)^2\beta_3/(N-2)^2} \exp\left(+\frac{2\pi}{N-2}\frac{1}{g^2_{\mathfrak{a}}}\left[1 + \mathcal{O}\left(\frac{g^2_{\mathfrak{a}}}{g^2_{\mathfrak{a}/b}}\right)\right]\right) \times \mathfrak{a}. \tag{3.199}$$

3.7 Beta function for the d-dimensional $O(N)$ NLσM with $d > 2$ and $N > 2$

The derivation of the Callan-Symanzik equation obeyed by the spin-spin correlator in section 3.6 was done in the spirit of RG approach used in high-energy physics in the 50's and early 70's. High-energy physics in the 50's and in the 70's relied heavily on quantum field theory to describe the electromagnetic, weak, and strong interactions. Locality, causality, and relativistic invariance were elevated to the status of fundamental principles of nature. The mathematical starting point was an action for local fields describing elementary (i.e., point-like) relativistic particles interacting through gauge fields.

The price to be paid in this approach is the occurrence of divergences caused by the point-like nature of the quantum fields, i.e., the absence of a high-energy (UV) cut-off. The severity of the UV divergences plaguing quantum field theories is measured by the notion of whether or not a theory can be renormalized. The idea behind the program of renormalization of quantum field theories is to demand that the scattering cross sections be finite so as to allow a comparison with measured cross sections in colliders.

This selection criterion for quantum field theories describing the fundamental interactions of nature led in the 50's and 70's to the realization that *all* the divergences associated to the point-like nature of *renormalizable local quantum field theories* can be consistently absorbed into a redefinition of a *finite* number of bare coupling constants in the Lagrangian, while leaving all measurable cross sections finite. Absorbing all UV divergences of a renormalizable local quantum field theory into a redefinition of the coupling constants means that the coupling constants depend on the UV cut-off whereas physical quantities are cut-off independent. Fundamental (physical) objects are, typically, gauge-invariant correlation functions made up of the local fields entering the theory. The independence on the UV cut-off of physical correlation functions implies that they obey a Callan-Symanzik equation through the implicit dependence of the coupling constants on the UV cut-off. From this point of view, the Lagrangian, action, and partition function are not considered to be as fundamental as correlation functions that can be measured in a collider.

In statistical physics the partition function (intensive free energy) plays a much more fundamental role than in quantum field theory. It can be considered as a fundamental physical quantity as it is well defined in the thermodynamic limit due to the presence of UV cut-off such as the lattice spacing. Statistical models that correspond to unrenormalizable field theories if the UV cut-off were to be removed are not a priori ruled out. Correspondingly, it is desirable to compute the partition function (intensive free energy) and to decide on a case by case basis if and how some correlation functions become independent of the UV cut-off as the thermodynamic limit is taken. With the advent of powerful computers it is possible to compute the partition function (intensive free energy) for very large system sizes. It is thus not surprising that RG approaches were developed by the condensed matter and statistical physics communities to evaluate directly the partition function (intensive free energy). One popular method is to integrate high-energy degrees of freedom through a momentum-shell integration. The momentum-shell integration can be easily implemented on the $O(N>2)$ NLσM if one is only after the beta function up to order g^4 [32]. This is the method that we will use to derive the IR RG equation obeyed by g^2 in the $O(N>2)$ NLσM in dimensions larger than 2. Another method consists in performing the RG analysis in position space, as will be illustrated in the chapter on the Kosterlitz-Thouless transition.

The RG analysis of the $O(N>2)$ NLσM in dimensions d larger than 2 can be performed on the partition function

$$\begin{aligned} Z &:= \int \mathrm{d}[\boldsymbol{n}]\, \delta\left(\boldsymbol{n}^2-1\right) e^{-\frac{1}{2\mathfrak{a}^{d-2}g^2}\int\limits_{\mathbb{R}^d}\mathrm{d}^d\boldsymbol{x}\left(\partial_\mu \boldsymbol{n}\right)^2} \\ &\propto \int \mathrm{d}[\boldsymbol{m}]\, \delta\left(\boldsymbol{m}^2-\frac{1}{\mathfrak{a}^{d-2}g^2}\right) e^{-\frac{1}{2}\int\limits_{\mathbb{R}^d}\mathrm{d}^d\boldsymbol{x}\left(\partial_\mu \boldsymbol{m}\right)^2}. \end{aligned} \tag{3.200}$$

Observe that the partition function depends explicitly on the lattice spacing $\mathfrak{a}$ that plays the role of the UV cut-off as we have chosen to keep g^2 dimensionless when $d\neq 2$. Choose the parametrization

$$\begin{aligned} m_1 &:= \sqrt{\frac{1-\mathfrak{a}^{d-2}g^2\boldsymbol{\pi}^2}{\mathfrak{a}^{d-2}g^2}}\cos\theta, \\ m_2 &:= \sqrt{\frac{1-\mathfrak{a}^{d-2}g^2\boldsymbol{\pi}^2}{\mathfrak{a}^{d-2}g^2}}\sin\theta, \\ m_3 &:= \pi_1, \\ &\vdots \\ m_N &:= \pi_{N-2}, \end{aligned} \tag{3.201}$$

motivated as we are by the $O(2)$ NLσM, under which the Lagrangian

$$\begin{aligned} \mathcal{L} &= \frac{1}{2}\left(\partial_\mu \boldsymbol{m}\right)^2 \\ &= \frac{1}{2}\left(\left(\partial_\mu m_1\right)^2+\left(\partial_\mu m_2\right)^2+\sum_{j=3}^{N}\left(\partial_\mu m_j\right)^2\right) \end{aligned} \tag{3.202}$$

becomes

$$\mathcal{L}=\frac{1}{2}\left(\frac{1-\mathfrak{a}^{d-2}g^2\boldsymbol{\pi}^2}{\mathfrak{a}^{d-2}g^2}\left(\partial_\mu\theta\right)^2+\frac{\mathfrak{a}^{d-2}g^2}{1-\mathfrak{a}^{d-2}g^2\boldsymbol{\pi}^2}\left(\boldsymbol{\pi}\cdot\partial_\mu\boldsymbol{\pi}\right)^2+\left(\partial_\mu\boldsymbol{\pi}\right)^2\right)-\ln|\mathcal{J}(\boldsymbol{\pi})|. \tag{3.203}$$

Here, $\mathcal{J}(\boldsymbol{\pi})$ is the Jacobian of the transformation

$$\boldsymbol{m} \text{ whereby } \boldsymbol{m}^2=\frac{1}{\mathfrak{a}^{d-2}g^2}\longrightarrow\begin{pmatrix}\theta\\ \boldsymbol{\pi}\end{pmatrix}. \tag{3.204}$$

As $\mathcal{J}(\boldsymbol{\pi})$ does not depend on $\boldsymbol{\pi}$, it will be dropped.[17] Thus, we can make the replacement

$$\begin{aligned}
\mathcal{L} &\longrightarrow \mathcal{L}_0 + \mathcal{L}_1 + \mathcal{L}_2,\\
\mathcal{L}_0 &= \frac{1}{2}\frac{1}{\mathfrak{a}^{d-2}g^2}\left(\partial_\mu\theta\right)^2,\\
\mathcal{L}_1 &= \frac{1}{2}\left(\partial_\mu\boldsymbol{\pi}\right)^2 - \frac{1}{2}\boldsymbol{\pi}^2\left(\partial_\mu\theta\right)^2,\\
\mathcal{L}_2 &= \mathfrak{a}^{d-2}g^2\left(\boldsymbol{\pi}\cdot\partial_\mu\boldsymbol{\pi}\right)^2\sum_{n=0}^{\infty}\left(\mathfrak{a}^{d-2}g^2\boldsymbol{\pi}^2\right)^n.
\end{aligned} \tag{3.206a}$$

Finally, we can neglect $\mathcal{L}_2$ if we are only after RG equations up to order g^4, in which case we need to perform an RG analysis of the partition function

$$Z_{\mathrm{sw}} := \int \mathrm{d}[\theta,\boldsymbol{\pi}]\, e^{-\int_{\mathbb{R}^d} \mathrm{d}^d\boldsymbol{x}\left(\mathcal{L}_0+\mathcal{L}_1+\mathcal{O}(g^2)\right)}. \tag{3.206b}$$

Observe that the field θ is much more rigid or stiff than the fields $\boldsymbol{\pi}$ in the limit of very low temperatures $g^2 \ll 1$. In other words, θ varies appreciably on much longer length scales than $\boldsymbol{\pi}$ does.

We would like to integrate over the fast modes in the partition function. To this end, we choose the asymmetric Fourier convention

$$f(\boldsymbol{x}) := \int_{\mathbb{R}^d} \frac{\mathrm{d}^d k}{(2\pi)^d} e^{+\mathrm{i}\boldsymbol{k}\cdot\boldsymbol{x}} f(\boldsymbol{k}), \qquad f(\boldsymbol{k}) := \int_{\mathbb{R}^d} \mathrm{d}^d x\, e^{-\mathrm{i}\boldsymbol{k}\cdot\boldsymbol{x}} f(\boldsymbol{x}), \tag{3.207}$$

for some complex-valued function f. Without an UV cut-off in momentum space, the momentum-space representation of

$$S_0 + S_1 \equiv \int_{\mathbb{R}^d} \mathrm{d}^d\boldsymbol{x}\,\left(\mathcal{L}_0 + \mathcal{L}_1\right) \tag{3.208a}$$

is then

$$S_0 + S_1 = \int_{\mathbb{R}^d} \frac{\mathrm{d}^d\boldsymbol{k}}{(2\pi)^d}\frac{1}{2}\left(\frac{1}{\mathfrak{a}^{d-2}g^2}\boldsymbol{k}^2\theta(+\boldsymbol{k})\theta(-\boldsymbol{k}) + \boldsymbol{k}^2\boldsymbol{\pi}(+\boldsymbol{k})\cdot\boldsymbol{\pi}(-\boldsymbol{k}) - f(+\boldsymbol{k})g(-\boldsymbol{k})\right), \tag{3.208b}$$

[17] The Jacobian $\mathcal{J}(\boldsymbol{\pi})$ can be read from

$$\begin{aligned}
\prod_{\boldsymbol{x}\in\mathbb{R}^d}\pi\int_{\mathbb{R}^{N-2}}\mathrm{d}^{N-2}\boldsymbol{\pi} &= \prod_{\boldsymbol{x}\in\mathbb{R}^d}(2\pi)\int_{\mathbb{R}^{N-2}}\mathrm{d}^{N-2}\boldsymbol{\pi}\int_0^{+\infty}\mathrm{d}r\,\frac{r}{|2r|}\delta\left(r-\sqrt{\frac{1}{\mathfrak{a}^{d-2}g^2}-\boldsymbol{\pi}^2}\right)\\
&= \prod_{\boldsymbol{x}\in\mathbb{R}^d}\int_0^{+\infty}\mathrm{d}r\,r\int_0^{2\pi}\mathrm{d}\theta\int_{\mathbb{R}^{N-2}}\mathrm{d}^{N-2}\boldsymbol{\pi}\,\delta\left(r^2-\frac{1}{\mathfrak{a}^{d-2}g^2}+\boldsymbol{\pi}^2\right)\\
&= \prod_{\boldsymbol{x}\in\mathbb{R}^d}\int_{-\infty}^{+\infty}\mathrm{d}m_1\int_{-\infty}^{+\infty}\mathrm{d}m_2\cdots\int_{-\infty}^{+\infty}\mathrm{d}m_N\,\delta\left(\boldsymbol{m}^2-\frac{1}{\mathfrak{a}^{d-2}g^2}\right).
\end{aligned} \tag{3.205}$$

where

$$f(+\boldsymbol{k}) := \int\limits_{\mathbb{R}^d} \frac{d^d \boldsymbol{q}}{(2\pi)^d} \boldsymbol{\pi}(\boldsymbol{k}+\boldsymbol{q}) \cdot \boldsymbol{\pi}(-\boldsymbol{q}), \tag{3.208c}$$

and

$$g(+\boldsymbol{k}) := \int\limits_{\mathbb{R}^d} \frac{d^d \boldsymbol{q}}{(2\pi)^d} \left[-i(\boldsymbol{k}+\boldsymbol{q})\right] \cdot \left[-i(-\boldsymbol{q})\right] \theta(\boldsymbol{k}+\boldsymbol{q})\theta(-\boldsymbol{q}) \tag{3.208d}$$

are the Fourier transforms of

$$f(\boldsymbol{x}) \equiv \boldsymbol{\pi}^2(\boldsymbol{x}), \tag{3.208e}$$

and

$$g(\boldsymbol{x}) \equiv (\partial_\mu \theta)^2(\boldsymbol{x}), \tag{3.208f}$$

respectively.

When $g \ll 1$, the same variation of θ and $\boldsymbol{\pi}$ takes place on vastly different characteristic length scales. In momentum space this means that θ is much more strongly peaked about $\boldsymbol{k} = \boldsymbol{0}$ than $\boldsymbol{\pi}$ is. This fact suggests the introduction of a thin momentum shell

$$b\Lambda < |\boldsymbol{k}| < \Lambda, \qquad b = 1 - \epsilon, \qquad \epsilon \text{ a positive infinitesimal number}, \tag{3.209}$$

below the UV momentum cut-off Λ and to perform the approximation by which $\boldsymbol{k}^2\theta(+\boldsymbol{k})\theta(-\boldsymbol{k})$ is negligible relative to $\boldsymbol{k}^2\boldsymbol{\pi}(+\boldsymbol{k})\boldsymbol{\pi}(-\boldsymbol{k})$ and $g(-\boldsymbol{k}) \approx (2\pi)^d\, g(\boldsymbol{0})\, \delta(\boldsymbol{k})$ in the momentum shell (3.209), i.e.,

$$\begin{aligned} S_0 + S_1 \approx & \int\limits_{|\boldsymbol{k}|<b\Lambda} \frac{d^d \boldsymbol{k}}{(2\pi)^d} \frac{1}{2} \Bigg(\frac{b^{d-2}}{(ba)^{d-2} g^2} \boldsymbol{k}^2 \theta(+\boldsymbol{k})\theta(-\boldsymbol{k}) \\ & \qquad + \boldsymbol{k}^2 \boldsymbol{\pi}(+\boldsymbol{k}) \cdot \boldsymbol{\pi}(-\boldsymbol{k}) - f_{b\Lambda}(+\boldsymbol{k}) g_{b\Lambda}(-\boldsymbol{k}) \Bigg) \\ & + \int\limits_{b\Lambda<|\boldsymbol{k}|<\Lambda} \frac{d^d \boldsymbol{k}}{(2\pi)^d} \frac{1}{2} \left(\boldsymbol{k}^2 \boldsymbol{\pi}(+\boldsymbol{k}) \cdot \boldsymbol{\pi}(-\boldsymbol{k}) - \boldsymbol{\pi}(+\boldsymbol{k}) \cdot \boldsymbol{\pi}(-\boldsymbol{k}) g_{b\Lambda}(\boldsymbol{q} = \boldsymbol{0}) \right). \end{aligned} \tag{3.210}$$

The Fourier transforms $f_{b\Lambda}(+\boldsymbol{k})$ and $g_{b\Lambda}(-\boldsymbol{k})$ are defined as in Eqs. (3.208c) and (3.208d) except for the sharp UV momentum cut-off $b\Lambda$, i.e., it is understood that the replacements

$$\theta(\boldsymbol{k}) \longrightarrow \theta(\boldsymbol{k})\Theta(b\Lambda - |\boldsymbol{k}|) \tag{3.211}$$

and

$$\boldsymbol{\pi}(\boldsymbol{k}) \longrightarrow \boldsymbol{\pi}(\boldsymbol{k})\Theta(b\Lambda - |\boldsymbol{k}|) \tag{3.212}$$

have been made [$\Theta(x)$ is the Heaviside step function].

Integration over the partial measure

$$\prod_{b\Lambda<|\boldsymbol{k}|<\Lambda} \mathrm{d}\boldsymbol{\pi}(+\boldsymbol{k}) = \prod_{b\Lambda<|\boldsymbol{k}|<\Lambda}^{k_1>0} \mathrm{d}\boldsymbol{\pi}(-\boldsymbol{k})\mathrm{d}\boldsymbol{\pi}(+\boldsymbol{k}) = \prod_{b\Lambda<|\boldsymbol{k}|<\Lambda}^{k_1>0} \mathrm{d}\boldsymbol{\pi}^*(+\boldsymbol{k})\mathrm{d}\boldsymbol{\pi}(+\boldsymbol{k}) \quad (3.213)$$

of the partition function with action (3.210) is Gaussian and given by

$$\begin{aligned}
Z_{\mathrm{sw}} &\propto \int\limits_{|\boldsymbol{k}|<b\Lambda} \mathrm{d}[\theta,\boldsymbol{\pi}] e^{-S_0'-S_1'-\delta S},\\
S_0' &= \int\limits_{|\boldsymbol{k}|<b\Lambda} \frac{\mathrm{d}^d\boldsymbol{k}}{(2\pi)^d}\frac{1}{2}\frac{1}{b^{d-2}}\frac{1}{(\mathfrak{a}/b)^{d-2}g^2}\boldsymbol{k}^2\theta(+\boldsymbol{k})\theta(-\boldsymbol{k}),\\
S_1' &= \int\limits_{|\boldsymbol{k}|<b\Lambda} \frac{\mathrm{d}^d\boldsymbol{k}}{(2\pi)^d}\frac{1}{2}\left[\boldsymbol{k}^2\boldsymbol{\pi}(+\boldsymbol{k})\cdot\boldsymbol{\pi}(-\boldsymbol{k}) - f_{b\Lambda}(+\boldsymbol{k})g_{b\Lambda}(-\boldsymbol{k})\right],\\
\delta S &= -\int\limits_{b\Lambda<|\boldsymbol{k}|<\Lambda} \frac{\mathrm{d}^d\boldsymbol{k}}{(2\pi)^d}\frac{N-2}{2}\ln\left(\frac{1}{\boldsymbol{k}^2 - g_{b\Lambda}(\boldsymbol{q}=\boldsymbol{0})}\right).
\end{aligned} \quad (3.214)$$

With the estimate

$$\begin{aligned}
\delta S &= -\int\limits_{b\Lambda<|\boldsymbol{k}|<\Lambda} \frac{\mathrm{d}^d\boldsymbol{k}}{(2\pi)^d}\frac{N-2}{2}\left[\ln\frac{1}{\boldsymbol{k}^2} + \ln\left(\frac{1}{1-\frac{g_{b\Lambda}(\boldsymbol{q}=\boldsymbol{0})}{\boldsymbol{k}^2}}\right)\right]\\
&= +\int\limits_{b\Lambda<|\boldsymbol{k}|<\Lambda} \frac{\mathrm{d}^d\boldsymbol{k}}{(2\pi)^d}(N-2)\ln|\boldsymbol{k}|\\
&\quad -\int\limits_{b\Lambda<|\boldsymbol{k}|<\Lambda} \frac{\mathrm{d}^d\boldsymbol{k}}{(2\pi)^d}\frac{N-2}{2}\sum_{n=1}^{\infty}\frac{1}{n}\left(\frac{g_{b\Lambda}(\boldsymbol{q}=\boldsymbol{0})}{\boldsymbol{k}^2}\right)^n\\
&= +\int\limits_{b\Lambda<|\boldsymbol{k}|<\Lambda} \frac{\mathrm{d}^d\boldsymbol{k}}{(2\pi)^d}(N-2)\ln|\boldsymbol{k}|\\
&\quad -\frac{\Omega(d)}{(2\pi)^d}\frac{N-2}{2}\sum_{n=1}^{\infty}\frac{1}{n}\left[g_{b\Lambda}(\boldsymbol{q}=\boldsymbol{0})\right]^n\frac{\Lambda^{d-2n}-(b\Lambda)^{d-2n}}{d-2n},
\end{aligned} \quad (3.215)$$

where it is understood that $\Omega(d)$ is the area of the d-dimensional unit sphere and that integration over the momentum shell gives a logarithm and not a power law when $d = 2n$, the original action with the sharp UV cut-off Λ is modified in three ways:

- The new sharp UV cut-off is $b\Lambda$, i.e.,

$$\Lambda \longrightarrow b\Lambda = e^{\ln b}\Lambda = \left[1 + \ln b + \mathcal{O}(\ln^2 b)\right]\Lambda. \quad (3.216)$$

- The action has changed by an additive constant

$$+\int\limits_{b\Lambda<|\boldsymbol{k}|<\Lambda} \frac{\mathrm{d}^d\boldsymbol{k}}{(2\pi)^d}(N-2)\ln|\boldsymbol{k}|. \quad (3.217)$$

- To leading order in the UV momentum cut-off $\Lambda = 1/\mathfrak{a}$ the coupling constant $\mathfrak{a}^{d-2}g^2$ has changed by

$$\frac{1}{\mathfrak{a}^{d-2}g^2} \longrightarrow \frac{1}{\mathfrak{a}^{d-2}g^2} - \frac{\Omega(d)}{(2\pi)^d}\frac{N-2}{d-2}\left(1-b^{d-2}\right)\left(\frac{1}{\mathfrak{a}}\right)^{d-2}. \tag{3.218}$$

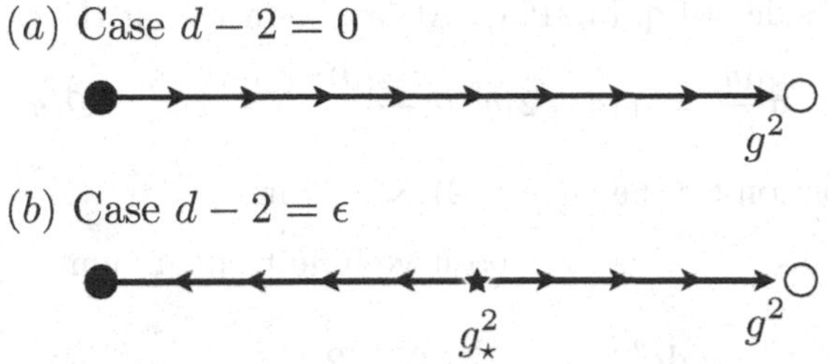

Fig. 3.3 Qualitative IR flow of the coupling constant g^2 in the $O(N>2)$ NLσM as a function of dimensionality $d-2 \geq 0$. The filled circle depicts the ferromagnetic fixed point at zero temperature, i.e., at $g^2=0$. The empty circle depicts the infinite temperature paramagnetic fixed point. The star depicts a finite temperature $g^2_\star := \frac{2\pi}{N-2}\epsilon$ critical point below which the system develops ferromagnetic LRO and above which the system is paramagnetic.

The beta function for the coupling constant g^2 is obtained from

$$\frac{1}{(\mathfrak{a}')^{d-2}(g^2)'} := \frac{1}{\mathfrak{a}^{d-2}g^2} - \frac{\Omega(d)}{(2\pi)^d}\frac{N-2}{d-2}\left(1-b^{d-2}\right)\left(\frac{1}{\mathfrak{a}}\right)^{d-2} \tag{3.219a}$$

whereby

$$\begin{aligned}\mathfrak{a}' &= \mathfrak{a} + \mathrm{d}\mathfrak{a} + \mathcal{O}[(\mathrm{d}\mathfrak{a})^2]\\ &:= \frac{\mathfrak{a}}{b}\\ &= \left[1 - \ln b + \mathcal{O}(\ln^2 b)\right]\mathfrak{a}\end{aligned} \tag{3.219b}$$

and

$$(g^2)' \equiv g^2 + \mathrm{d}(g^2) + \mathcal{O}\left\{\left[\mathrm{d}(g^2)\right]^2\right\}. \tag{3.219c}$$

On the one hand, we have the expansion

$$\begin{aligned}\frac{1}{(\mathfrak{a}')^{d-2}(g^2)'} &= \frac{1}{\mathfrak{a}^{d-2}\left[1-(d-2)\ln b + \mathcal{O}(\ln^2 b)\right]\left(g^2 + \mathrm{d}(g^2) + \mathcal{O}\left\{\left[\mathrm{d}(g^2)\right]^2\right\}\right)}\\ &= \frac{1}{\mathfrak{a}^{d-2}g^2\left(1-(d-2)\ln b + \frac{\mathrm{d}g^2}{g^2} + \mathcal{O}(\ln^2 b) + \mathcal{O}\left\{\left[\mathrm{d}(g^2)\right]^2\right\}\right)}\\ &= \frac{1}{\mathfrak{a}^{d-2}g^2}\left(1+(d-2)\ln b - \frac{\mathrm{d}g^2}{g^2} + \mathcal{O}(\ln^2 b) + \mathcal{O}\left\{\left[\mathrm{d}(g^2)\right]^2\right\}\right)\\ &= \frac{1}{\mathfrak{a}^{d-2}g^2}\left(1-(d-2)\frac{\mathrm{d}\mathfrak{a}}{\mathfrak{a}} - \frac{\mathrm{d}g^2}{g^2} + \mathcal{O}\left[(\mathrm{d}\mathfrak{a}/\mathfrak{a})^2\right] + \mathcal{O}\left\{\left[\mathrm{d}(g^2)\right]^2\right\}\right)\end{aligned} \tag{3.220}$$

of the left-hand side of Eq. (3.219a). On the other hand, we have the expansion

$$\begin{aligned}\frac{1}{(\mathfrak{a}')^{d-2}(g^2)'} &= \frac{1}{\mathfrak{a}^{d-2}g^2}\left\{1+\frac{\Omega(d)}{(2\pi)^d}\frac{N-2}{d-2}g^2\left[(d-2)\ln b+\mathcal{O}(\ln^2 b)\right]\right\}\\ &= \frac{1}{\mathfrak{a}^{d-2}g^2}\left(1-\frac{\Omega(d)}{(2\pi)^d}\frac{N-2}{d-2}g^2\left\{(d-2)\frac{\mathrm{d}\mathfrak{a}}{\mathfrak{a}}+\mathcal{O}\left[(\mathrm{d}\mathfrak{a}/\mathfrak{a})^2\right]\right\}\right)\end{aligned} \tag{3.221}$$

of the right-hand side of Eq. (3.219a). At last we arrive at the beta function

$$-\mathfrak{a}\frac{\mathrm{d}g^2}{\mathrm{d}\mathfrak{a}} = +(d-2)g^2 - \frac{\Omega(d)}{(2\pi)^d}(N-2)g^4 + \mathcal{O}(g^6). \tag{3.222}$$

The UV beta function for the $O(N>2)$ NLσM in

$$d = 2+\epsilon, \qquad \epsilon \text{ a positive infinitesimal number,} \tag{3.223}$$

reduces to

$$-\mathfrak{a}\frac{\mathrm{d}g^2}{\mathrm{d}\mathfrak{a}} = +\epsilon\, g^2 - \frac{N-2}{2\pi}g^4 + \mathcal{O}(\epsilon g^4, g^6) \tag{3.224}$$

as

$$\Omega(2+\epsilon) = 2\pi + \mathcal{O}(\epsilon), \qquad (2\pi)^{2+\epsilon} = (2\pi)^2 + \mathcal{O}(\epsilon). \tag{3.225}$$

The UV beta function

$$\beta(g^2) := +\epsilon\, g^2 - \frac{N-2}{2\pi}g^4 \tag{3.226}$$

vanishes when

$$g^2 = g^2_\star := \frac{2\pi}{N-2}\epsilon. \tag{3.227}$$

For $0 < g^2 < g^2_\star$, the UV beta function is positive, i.e., g^2 is UV relevant (IR irrelevant). For $g^2 = g^2_\star$, the theory is critical as the beta function vanishes. For $g^2 > g^2_\star$, the UV beta function is negative, i.e., g^2 is UV irrelevant (IR relevant). The critical point $g^2 = g^2_\star$ is an IR unstable fixed point. To the right of this fixed point the system flows in the IR limit to the infinite temperature paramagnet fixed point. To the left of this fixed point the system flows in the IR limit to the zero temperature ferromagnetic fixed point. Within the spin-wave approximation, the ferromagnetic LRO at zero temperature thus extends to a finite critical temperature which is proportional to $\epsilon = d-2$. Figure 3.3 depicts the qualitative behavior of the beta function in the $O(N>2)$ NLσM as a function of dimensionality.

A final comment is of order. We analytically continued dimensionality $d = 1,2,\cdots$ of space to real values $d = 2+\epsilon$. If so, one might also wonder if it makes sense to analytically continue $N = 2,3,\cdots$ to real values below 2 and, in particular, to the limit $N = 0$. This limit is interesting as it changes the sign of the beta function in $d = 2$. It turns out that some problems in statistical physics such as polymers or the problem of Anderson localization demand analytical continuations of the type $N \to 0$. We would like to give a geometric interpretation to this change in sign.

Let

$$S[\phi] := \int \frac{\mathrm{d}^d x}{\mathfrak{a}^{d-2}}\frac{1}{2}\sum_{a,b=1}^{N}\sum_{\mu=1}^{d}\mathfrak{g}_{ab}(\phi)\,\partial_\mu\phi^a\,\partial_\mu\phi^b \tag{3.228a}$$

denotes the action of a NLσM on the Riemannian manifold $\mathfrak{M}$. It is shown in appendix C that it is the Ricci curvature tensor of the target space that controls the RG flow of the beta function up to first order in the loop expansion,

$$\mathfrak{a}\frac{\partial}{\partial\mathfrak{a}}\mathfrak{g}_{ab}=\epsilon\,\mathfrak{g}_{ab}-\frac{1}{2\pi}R_{ab},\qquad \epsilon \text{ a positive infinitesimal number.} \tag{3.228b}$$

Imagine that it is permissible to analytically continue the coordinates on the target space $\mathfrak{M}$ from real values to imaginary values according to

$$\phi^{a}=\mathrm{i}\phi^{\star\,a},\qquad a=1,\cdots,N, \tag{3.229a}$$

so as to obtain a NLσM on the Riemannian manifold $\mathfrak{M}^{\star}$ *defined* by the metric tensor[18]

$$\mathfrak{g}^{\star}_{ab}(\phi^{\star}):=-\mathfrak{g}_{ab}(\phi^{\star}) \tag{3.229b}$$

and the action

$$S^{\star}[\phi^{\star}]:=\int\frac{\mathrm{d}^{d}x}{\mathfrak{a}^{d-2}}\frac{1}{2}\sum_{a,b=1}^{N}\sum_{\mu=1}^{d}\mathfrak{g}^{\star}_{ab}(\phi^{\star})\partial_{\mu}\phi^{\star\,a}\,\partial_{\mu}\phi^{\star\,b}. \tag{3.229c}$$

Then, there follows the one-loop RG flow

$$\mathfrak{a}\frac{\partial}{\partial\mathfrak{a}}\mathfrak{g}^{\star}_{ab}=\epsilon\,\mathfrak{g}^{\star}_{ab}-\frac{1}{2\pi}R^{\star}_{ab},\qquad \epsilon \text{ a positive infinitesimal number.} \tag{3.229d}$$

We ignore the important question of the convergence of the path integral upon this analytic continuation.

We need to answer two questions. When does the one-loop RG flow (3.228b) reduce to a one-loop RG flow of the form (3.224)? What lessons do we learn from the analytical continuation (3.229) in this case?

For a generic N-dimensional Riemannian manifold, the one-loop RG flow (3.228b) involves at most $N+N(N-1)/2$ independent running couplings, for the metric tensor is a symmetric matrix. Symmetry properties of a N-dimensional Riemannian manifold can reduce the number of independent running couplings. For symmetric spaces, this reduction in the number of independent running couplings is the most dramatic. For any compact symmetric space $\mathfrak{M}$, it is possible to rewrite the UV one-loop RG flow (3.228b) as

$$-\mathfrak{a}\frac{\partial}{\partial\mathfrak{a}}g^{2}=\epsilon\,g^{2}-\frac{c_{v}}{4\pi}g^{4},\qquad \epsilon \text{ a positive infinitesimal number.} \tag{3.230}$$

Here, the positive number c_{v} is the quadratic Casimir invariant of the global symmetry group associated to the symmetric space $\mathfrak{M}$. Upon the analytic continuation (3.229), the UV one-loop RG flow (3.230) turns into [see Eq. (C.64)]

$$-\mathfrak{a}\frac{\partial}{\partial\mathfrak{a}}g^{2}=\epsilon\,g^{2}+\frac{c_{v}}{4\pi}g^{4},\qquad \epsilon \text{ a positive infinitesimal number.} \tag{3.231}$$

18 Observe that the step (3.229b) is not equivalent to the transformation law $\mathfrak{g}^{\star}_{ab}(\phi^{\star}):=-\mathfrak{g}_{ab}(\mathrm{i}\phi^{\star})$ under the reparametrization $\phi^{a}=\mathrm{i}\phi^{\star\,a}$. The metric $\mathfrak{g}^{\star}_{ab}(\phi^{\star}):=-\mathfrak{g}_{ab}(\mathrm{i}\phi^{\star})$ with the action (3.229c) deliver the RG equations (3.228b).

The analytic continuation (3.229) has induced a change of the sign by which the contribution arising from the Ricci curvature tensor enters in the one-loop RG flow (3.231). This sign change is interpreted as the fact that the symmetric space $\mathfrak{M}^\star$ defined by the analytic continuation (3.229) on the symmetric space $\mathfrak{M}$ is non-compact. For example, the analytic continuation (3.229) on the $O(N)$ NLσM delivers the $O(1, N-1)$ NLσM. Hereto, we would like to interpret the change of sign of the term proportional to g^4 in the beta function of the $O(N)$ NLσM when $N \to 0$ as the fact that this limit "defines" a non-compact target manifold.

Symmetric spaces have been classified by Cartan into families of triplets [33]. Two of the three symmetric spaces making up a triplet have *sectional curvatures* of opposite sign. The third member of the triplet has vanishing sectional curvature. The $O(2)$ NLσM is an example of a target Riemannian manifold with vanishing sectional curvature. In random matrix theory, the statistical properties of diverse quantities are controlled by statistical ensemble of matrices closely related to symmetric spaces. Statistical correlations of random energy eigenvalues follow from choosing symmetric spaces with vanishing sectional curvature. Statistical correlations of the random eigenvalues of unitary matrices follow from choosing symmetric spaces with positive sectional curvature. Statistical correlations of the eigenvalues of pseudo-unitary matrices (the so-called Lyapunov exponents) follow from choosing symmetric spaces with negative sectional curvature. In turn, the global symmetry group characterizing the symmetric space is dictated by the intrinsic symmetries respected by the ensemble of statistical matrices. These intrinsic symmetries are the presence or absence of time-reversal symmetry, of spin-rotation symmetry, of particle-hole symmetry in both its unitary and antiunitary incarnations as defined by the application on spinors (fermions). More generally, the NLσM's that describe the physics of Anderson localization when the dimensionality of base space is $d = 0, 1, 2, \cdots$ have (supersymmetric) target spaces with both a compact and non-compact component.

As will be illustrated in the chapter on the Kosterlitz-Thouless phase transition, the global structure of the target manifold plays no role in the perturbative analysis (spin-wave approximation) that we have performed so far. For example, the $O(2)$ NLσM has the circle as a target manifold. Locally the real line and the circle cannot be distinguished. The spin-wave approximation on the two-dimensional $O(2)$ NLσM neglects the fact that the circle is a compact manifold, i.e., replaces the circle by the line as a target manifold. Any perturbative treatment in powers of the coupling g^2 on a symmetric space, even if it leads to non-perturbative results such as the essential singularity in the dependence of the correlation length on g^2, is bound to ignore the compact nature of the symmetric space and to fail if this property is essential. A classification by mathematicians of the global structures of Riemannian manifolds has been undertaken and applied to physics. When the Riemannian manifold has a non-trivial global structure (non-trivial topology) such as is the case for a circle in opposition to the real line, it is possible, on a case by case basis, to supplement the action of the NLσM by a new term or by new degrees of freedom that account for the superseding global structure. The inclusion of these "global" degrees of freedom leads to a finite temperature transition from the spin-wave phase to a paramagnetic phase called the Kosterlitz-Thouless phase transition for the case of the two-dimensional $O(2)$ NLσM. For Heisenberg quantum spin chains with nearest-neighbor antiferromagnetic exchange interactions, the addition of a topological term (i.e., a contribution to the action that is necessarily quantized) to the $O(3)$ NLσM induces critical behavior when the spin degrees of freedom carry half-integer representations of $SU(2)$.

3.8 Problems

3.8.1 *The Mermin-Wagner theorem for quantum spin Hamiltonians*

Introduction

We are going to prove the Mermin-Wagner theorem as it is stated in the original paper by Mermin and Wagner [34]. It says that the quantum Heisenberg Hamiltonian

$$\hat{H} := -\sum_{\boldsymbol{r},\boldsymbol{r}'\in\Lambda} J_{\boldsymbol{r}-\boldsymbol{r}'}\,\hat{\boldsymbol{S}}_{\boldsymbol{r}}\cdot\hat{\boldsymbol{S}}_{\boldsymbol{r}'} - h\sum_{\boldsymbol{r}\subset\Lambda}\hat{S}^{z}_{\boldsymbol{r}}\,e^{-\mathrm{i}\boldsymbol{K}\cdot\boldsymbol{r}} \tag{3.232}$$

has no long-range ferromagnetic or antiferromagnetic order in $d=1$ and $d=2$ dimensions of position space at non-vanishing temperature $T>0$, if the field h is taken to zero and the interaction $J_{\boldsymbol{r}}$ is short-ranged, that is, at long distances it decays faster than $|\boldsymbol{r}|^{-d-2}$. Here, the sum over $\boldsymbol{r}$ runs over the sites of a lattice Λ with periodic boundary conditions in place and the operators $\hat{\boldsymbol{S}}_{\boldsymbol{r}} = (\hat{S}^x_{\boldsymbol{r}}, \hat{S}^y_{\boldsymbol{r}}, \hat{S}^z_{\boldsymbol{r}})^{\mathsf{T}}$ describe a quantum spin-S, thus obeying the $SU(2)$ algebra (2.134) at every lattice site $\boldsymbol{r}\in\Lambda$. Translation invariance of the Hamiltonian is manifest in the fact that $J_{\boldsymbol{r}-\boldsymbol{r}'}$ depends only on the difference of lattice sites. We shall further assume

$$J_{\boldsymbol{r}} = J_{-\boldsymbol{r}}, \qquad J_{0} = 0. \tag{3.233}$$

To probe the tendency toward ferromagnetic order, we choose $\boldsymbol{K}=0$ to describe a homogeneous field. To probe the tendency toward antiferromagnetic order, we assume that Λ can be bipartitioned into two sublattices and we choose $\boldsymbol{K}$ such that $e^{-\mathrm{i}\boldsymbol{K}\cdot\boldsymbol{r}} = -1$ if $\boldsymbol{r}$ connects sites on different sublattices, while $e^{-\mathrm{i}\boldsymbol{K}\cdot\boldsymbol{r}} = +1$ if $\boldsymbol{r}$ connects sites on the same sublattice. The amplitude h is thus that of a source field that selects a collinear order parameter characterized by the wave vector $\boldsymbol{K}$.

Magnetic collinear long-range order is signaled by a non-vanishing value of the order parameter

$$s^{z}(\beta,h) := \frac{1}{N}\sum_{\boldsymbol{r}\in\Lambda}\left\langle \hat{S}^{z}_{\boldsymbol{r}}\,e^{-\mathrm{i}\boldsymbol{K}\cdot\boldsymbol{r}}\right\rangle, \tag{3.234a}$$

where N is the number of lattice sites and the expectation value of any operator $\hat{A}$ is given by

$$\langle A \rangle := \frac{\mathrm{Tr}\left(e^{-\beta\hat{H}}\,\hat{A}\right)}{\mathrm{Tr}\left(e^{-\beta\hat{H}}\,\right)}, \tag{3.234b}$$

with $\beta := 1/(k_{\mathrm{B}}T)$ and Tr the trace over the Hilbert space. Our task is thus to compute the function $s^z(\beta,\cdot)$ and consider it in the limit $h \to 0$. If $s^z(\beta,\cdot)$ remains non-vanishing in this limit, spontaneous collinear long-range order takes place. Otherwise we can rule out spontaneous symmetry breaking. It turns out that $s^z(\beta,h)$ cannot be computed exactly. Instead, we will be able to find an upper bound to $s^z(\beta,h)$ and show that this upper bound vanishes in the limit $h \to 0$ at any non-vanishing temperature.

Observe that the Hamiltonian has $SU(2)$ spin-rotation invariance in the limit $h \to 0$. Even though we will not explicitly make use of this symmetry in our calculation, its presence is of crucial importance for the result to hold. The use of the terminology Mermin-Wagner theorem often refers to the following generalization of the statement above. There exists no spontaneous breaking of a *continuous* symmetry group in $(d=1)$ and $(d=2)$ dimensions of space at at non-vanishing temperature T, if the Hamiltonian has only short-ranged interactions. In contrast, *discrete* symmetries can very well be broken spontaneously in $(d=2)$ dimensions for a non-vanishing temperature T by short-range interactions. A prominent example is the Ising model.

Proof of Bogoliubov inequality

Before we turn to the proof of the Mermin-Wagner theorem itself, we want to establish an inequality due to Bogoliubov on which the proof relies. (This inequality will allow us to establish an upper bound on the order parameter s^z.) The (Bogoliubov) inequality states that

$$\mathrm{Tr}\left(\left\{\hat{A},\hat{A}^{\dagger}\right\}e^{\hat{D}}\right)\,\mathrm{Tr}\left(\left[[\hat{D},\hat{C}],\hat{C}^{\dagger}\right]e^{\hat{D}}\right) \geq 2\left|\mathrm{Tr}\left(\left[\hat{C},\hat{A}\right]e^{\hat{D}}\right)\right|^2, \tag{3.235}$$

where $\hat{A}$, $\hat{C}$, and $\hat{D}$ are bounded linear operators on a Hilbert space $\mathfrak{H}$, $\hat{D}=\hat{D}^{\dagger}$ is Hermitian, the brackets $[\cdot,\cdot]$ and $\{\cdot,\cdot\}$ denote the commutator and the anticommutator, respectively, and the trace Tr is taken over $\mathfrak{H}$.

To prove the inequality (3.235), we use a basis of $\mathfrak{H}$ in which $\hat{D}$ is diagonal. We denote the orthonormal eigenvectors of $\hat{D}$ with $|i\rangle$, $i=1,2,\cdots$, and the corresponding eigenvalues with d_i, $i=1,2,\cdots$. The same basis also diagonalizes the operator $e^{\hat{D}}$ and its eigenvalues are given by $w_i = e^{d_i}$, $i=1,2,\cdots$. In summary, we have

$$\hat{D}\,|i\rangle = d_i|i\rangle, \qquad e^{\hat{D}}\,|i\rangle = w_i\,|i\rangle, \qquad i=1,2,\cdots. \tag{3.236}$$

We can now define an inner product $(\cdot,\cdot)$ between any two bounded linear operators $\hat{A}$ and $\hat{B}$ on $\mathfrak{H}$ as

$$(\hat{A},\hat{B}) := \sum_{\substack{i,j\\ d_i\neq d_j}} \left\langle j\left|\hat{A}^{\dagger}\right|i\right\rangle\left\langle i\left|\hat{B}\right|j\right\rangle \frac{w_i - w_j}{d_i - d_j}, \tag{3.237}$$

where it is understood that the sum runs over all pairs i,j, except for those which have degenerate eigenvalues $d_i = d_j$.

Exercise 1.1: Show that the definition Eq. (3.237) has the properties of an inner product, that is

$$(\hat{A},\hat{B}) = (\hat{B},\hat{A})^{*} \qquad \text{conjugate symmetry,} \tag{3.238a}$$

$$(\hat{A},\beta_1\,\hat{B}_1+\beta_2\,\hat{B}_2) = \beta_1\,(\hat{A},\hat{B}_1)+\beta_2\,(\hat{A},\hat{B}_2) \qquad \text{linearity,} \tag{3.238b}$$

$$(\hat{A},\hat{A}) \geq 0 \qquad \text{positive definiteness,} \tag{3.238c}$$

for any pair β_1 and β_2 of complex numbers.

With Eq. (3.237) defining an inner product, we conclude that the Schwarz inequality

$$(\hat{A},\hat{A})\,(\hat{B},\hat{B}) \geq \left|(\hat{A},\hat{B})\right|^{2} \tag{3.239}$$

holds.

Exercise 1.2: Show that the inequality

$$(\hat{A},\hat{A}) \leq \frac{1}{2}\,\mathrm{Tr}\left(\left\{\hat{A},\hat{A}^{\dagger}\right\}e^{\hat{D}}\right) \tag{3.240}$$

holds. *Hint:* Start from the definition (3.237) of the inner product $(\hat{A},\hat{A})$ and show that

$$\frac{w_i - w_j}{d_i - d_j} \leq \frac{w_i + w_j}{2}, \qquad d_i \neq d_j. \tag{3.241}$$

Exercise 1.3: Show that the two equalities

$$(\hat{A},[\hat{C}^{\dagger},\hat{D}]) = \mathrm{Tr}\left([\hat{A}^{\dagger},\hat{C}^{\dagger}]\,e^{\hat{D}}\right) \tag{3.242a}$$

and

$$([\hat{C}^{\dagger},\hat{D}],[\hat{C}^{\dagger},\hat{D}]) = \mathrm{Tr}\left([[\hat{D},\hat{C}],\hat{C}^{\dagger}]\,e^{\hat{D}}\right) \tag{3.242b}$$

hold. Using them, as well as the inequality (3.240), and the Schwarz inequality (3.239), prove the Bogoliubov inequality (3.235).

Application of Bogoliubov inequality to the quantum Heisenberg Hamiltonian

Define the operators

$$\hat{S}^{+}_{\boldsymbol{r}} := \hat{S}^{x}_{\boldsymbol{r}} + \mathrm{i}\hat{S}^{y}_{\boldsymbol{r}}, \qquad \hat{S}^{-}_{\boldsymbol{r}} := \hat{S}^{x}_{\boldsymbol{r}} - \mathrm{i}\hat{S}^{y}_{\boldsymbol{r}}, \qquad \boldsymbol{r}\in\Lambda, \tag{3.243}$$

and the Fourier transform $f_{\boldsymbol{k}}$ of any operator or function $f_{\boldsymbol{r}}$ that is defined on the lattice

$$f_{\boldsymbol{k}} := \sum_{\boldsymbol{r}\in\Lambda} e^{-\mathrm{i}\boldsymbol{k}\cdot\boldsymbol{r}} f_{\boldsymbol{r}} \tag{3.244}$$

such that

$$f_{\boldsymbol{r}} = \frac{1}{N} \sum_{\boldsymbol{k}\in\mathrm{BZ}} e^{+\mathrm{i}\boldsymbol{k}\cdot\boldsymbol{r}} f_{\boldsymbol{k}}, \tag{3.245}$$

where $\boldsymbol{k}$ takes values in the first Brillouin zone (BZ).

Exercise 2.1: Rewrite the Hamiltonian (3.232) in terms of the Fourier transformed operators $\hat{S}^{+}_{\boldsymbol{k}}$, $\hat{S}^{-}_{\boldsymbol{k}}$, $\hat{S}^{z}_{\boldsymbol{k}}$ and the function $J_{\boldsymbol{k}}$.

Exercise 2.2: Familiarize yourself with the algebra obeyed by the operators $\hat{S}^{+}_{\boldsymbol{k}}$, $\hat{S}^{-}_{\boldsymbol{k}}$, and $\hat{S}^{z}_{\boldsymbol{k}}$ by computing the commutators

$$[\hat{S}^{+}_{\boldsymbol{k}}, \hat{S}^{-}_{\boldsymbol{k}'}], \qquad [\hat{S}^{+}_{\boldsymbol{k}}, \hat{S}^{z}_{\boldsymbol{k}'}], \qquad [\hat{S}^{-}_{\boldsymbol{k}}, \hat{S}^{z}_{\boldsymbol{k}'}]. \tag{3.246}$$

Then, use this algebra and the momentum-space representation of Hamiltonian (3.232) that was obtained in exercise 2.1 to verify that

$$\begin{aligned} g_{\boldsymbol{k}} :&= \left\langle [[\hat{S}^{+}_{\boldsymbol{k}}, \hat{H}], \hat{S}^{-}_{-\boldsymbol{k}}] \right\rangle \\ &= \frac{1}{N} \sum_{\boldsymbol{q}} \left(J_{\boldsymbol{q}} - J_{\boldsymbol{q}+\boldsymbol{k}}\right) \left\langle \hat{S}^{+}_{\boldsymbol{q}}\, \hat{S}^{-}_{-\boldsymbol{q}} + \hat{S}^{-}_{\boldsymbol{q}}\, \hat{S}^{+}_{-\boldsymbol{q}} + 4\, \hat{S}^{z}_{\boldsymbol{q}}\, \hat{S}^{z}_{-\boldsymbol{q}} \right\rangle + 2\, h\, N\, s^{z}. \end{aligned} \tag{3.247}$$

Exercise 2.3: Use the Bogoliubov inequality (3.235) with the following choice for the operators

$$\hat{C} = \hat{S}^{+}_{\boldsymbol{k}}, \qquad \hat{A} = \hat{S}^{-}_{-\boldsymbol{k}-\boldsymbol{K}}, \qquad \hat{D} = -\beta\, \hat{H}, \tag{3.248}$$

to show that

$$\frac{1}{2} \left\langle \left\{ \hat{S}^{-}_{-\boldsymbol{k}-\boldsymbol{K}}, \hat{S}^{+}_{\boldsymbol{k}+\boldsymbol{K}} \right\} \right\rangle \geq \frac{4N^2}{\beta} \frac{|s^z|^2}{g_{\boldsymbol{k}}}. \tag{3.249}$$

We can already anticipate that the inequality (3.249) might allow us to establish an upper bound on the order parameter s^z. The idea is to sum both sides over $\boldsymbol{k} \in \mathrm{BZ}$ and use the identity

$$\begin{aligned} \sum_{\boldsymbol{k}\in\mathrm{BZ}} \hat{\boldsymbol{S}}_{\boldsymbol{k}} \cdot \hat{\boldsymbol{S}}_{-\boldsymbol{k}} &= \sum_{\boldsymbol{k}\in\mathrm{BZ}} \sum_{\boldsymbol{r}\in\Lambda} \sum_{\boldsymbol{r}'\in\Lambda} e^{-\mathrm{i}\boldsymbol{k}\cdot\boldsymbol{r}}\, e^{\mathrm{i}\boldsymbol{k}\cdot\boldsymbol{r}'}\, \hat{\boldsymbol{S}}_{\boldsymbol{r}} \cdot \hat{\boldsymbol{S}}_{\boldsymbol{r}'} \\ &= N \sum_{\boldsymbol{r}} \hat{\boldsymbol{S}}_{\boldsymbol{r}} \cdot \hat{\boldsymbol{S}}_{\boldsymbol{r}} \\ &= N^2\, S\, (S+1), \end{aligned} \tag{3.250}$$

to establish an upper bound for the left-hand side of inequality (3.249)

$$N^2\, S\, (S+1) \geq \sum_{\boldsymbol{k}\in\mathrm{BZ}} \frac{1}{2} \left\langle \left\{ \hat{S}^{-}_{-\boldsymbol{k}-\boldsymbol{K}}, \hat{S}^{+}_{\boldsymbol{k}+\boldsymbol{K}} \right\} \right\rangle \geq \frac{4N^2}{\beta}\, |s^z|^2 \sum_{\boldsymbol{k}\in\mathrm{BZ}} \frac{1}{g_{\boldsymbol{k}}}. \tag{3.251}$$

To evaluate the right-hand side, that is, $\sum_{\boldsymbol{k}\in\mathrm{BZ}} g_{\boldsymbol{k}}^{-1}$, is essentially intractable. Rather, we will establish an appropriate lower bound to this quantity by finding an upper bound to $g_{\boldsymbol{k}}$.

Exercise 2.4: Show that

$$g_{\boldsymbol{k}} \leq 4N\, S\,(S+1)\, \boldsymbol{k}^2 \sum_{\boldsymbol{r}\in\Lambda} \boldsymbol{r}^2\, |J_{\boldsymbol{r}}| + 2N|hs^z| \tag{3.252}$$

and use this result to rewrite the inequality (3.251) as

$$|s^z|^2 \leq \frac{S\,(S+1)\,\beta}{2\,\Theta}, \tag{3.253a}$$

where

$$\Theta := \frac{1}{N} \sum_{\boldsymbol{k}\in\mathrm{BZ}} \frac{1}{2\,\mathcal{J}\,\boldsymbol{k}^2 + |h\,s^z|}, \tag{3.253b}$$

and

$$\mathcal{J} := S\,(S+1) \sum_{\boldsymbol{r}\in\Lambda} \boldsymbol{r}^2\, |J_{\boldsymbol{r}}|. \tag{3.253c}$$

In the final step, we are going to the thermodynamic limit, in which we can replace the summation

$$\frac{1}{N} \sum_{\boldsymbol{k}\in\mathrm{BZ}} f_{\boldsymbol{k}} \to \int \frac{\mathrm{d}^d \boldsymbol{k}}{\Omega}\, f(\boldsymbol{k}), \tag{3.254}$$

where Ω is the volume of the BZ. As we are only after an upper bound for Eq. (3.253), and the integrand of Θ is positive definite, we can restrict the integration to a ball of radius $k_0 > 0$ that entirely fits in the BZ. Physically, this is a valid approximation, as the tendency to long-range order is determined by the contributions at small momenta only.

Exercise 2.5: Show that within this approximation, one obtains the following leading expansions for small fields h

$$|s^z| < \begin{cases} \beta^{2/3}\,|h|^{1/3} \times \text{const}, & d = 1, \\[2ex] \dfrac{\beta^{1/2}}{\sqrt{|\ln|h||}} \times \text{const}, & d = 2. \end{cases} \tag{3.255}$$

Equation (3.255) shows that the order parameter s^z vanishes in the limit $h \to 0$ in one and two dimensions. Note that we have implicitly used the fact that only short-range interactions are permitted, by assuming that the constant $\mathcal{J}$ remains finite when the thermodynamic limit is taken. This is indeed the case if $J_{\boldsymbol{r}}$ decays faster than $|\boldsymbol{r}|^{-d-2}$ as the distance $|\boldsymbol{r}|$ tends to infinity.

3.8.2 Quantum spin coherent states and the $O(3)$ QNLσM

Introduction

We are after a path-integral representation of a quantum spin Hamiltonian in terms of the coherent-state representation of the irreducible representations of the group $SU(2)$. The lack of a version of Wick theorem for quantum spin degrees of freedom complicates perturbation theory enormously. One way out is to represent the spin algebra in terms of fermions or bosons. However, the price paid is the enlargement of the Hilbert space, i.e., the introduction of gauge degrees of freedom. Another way out is to work with a basis of the Hilbert space that mimics the classical limit of the quantum spin system as closely as possible. Our first goal is to show how the latter approach can be achieved. Our second goal is to derive the $O(3)$ Quantum Non-Linear Sigma Model (QNLσM) representation of a quantum antiferromagnet whose classical ground state supports collinear antiferromagnetic long-range order.

The $O(3)$ QNLσM is a long-wavelength and low-energy effective field theory that is believed to capture qualitatively the properties of a quantum antiferromagnet at very low temperatures, provided the quantum ground state supports collinear antiferromagnetic correlations on the scale of few lattice spacings.

Although quantum antiferromagnets have a long and illustrious history dating back to the Bethe solution to the quantum spin-1/2 Heisenberg chain in the early days of quantum mechanics, it is only through the work of Haldane in the early 80's that the connection between the QNLσM and quantum spin Hamiltonians on bipartite lattice was established in Ref. [35]. The insights brought by this connection were revolutionary.

It had been believed for one generation, based on the Bethe Ansatz solution to the spin-1/2 antiferromagnetic Heisenberg chain and numerical simulations thereof, that all quantum spin $S = 1/2, 1, 3/2, \cdots$ antiferromagnetic Heisenberg chains were characterized by quasi-long-range order in their ground states and that the excitation spectrum above these ground states were gapless.

Haldane deduced from his mapping of the quantum spin-S antiferromagnetic Heisenberg chain to the $O(3)$ QNLσM that the case of integer spin chains differs qualitatively from the case of half-odd-integer spin chains. To the contrary of the half-odd-integer case, the integer case was conjectured by Haldane to display a ground state without quasi-long-range order for the spin degrees of freedom and supporting a gap to all spin excitations.

The prediction of Haldane was initially controversial as it relied on a mapping to the $O(3)$ QNLσM that is approximate with an error of order $1/S$. As we shall see, this is a semi-classical approximation, one reason for which it is surprising that this approximation captures a quantum manifestation as dramatic as the distinction between integer and half-odd-integer spins. Exactly soluble models, numerical simulations, and the discovery of quasi-one-dimensional quantum antiferromagnets in "real life" have vindicated Haldane since then.

We shall consider the quantum lattice model

$$\hat{H}_{S,\boldsymbol{H}}[\hat{\boldsymbol{S}}] := -\frac{1}{2} \sum_{i,j\in\Lambda} J_{ij}\, \hat{\boldsymbol{S}}_i \cdot \hat{\boldsymbol{S}}_j - \sum_{i\in\Lambda} \boldsymbol{H}_i \cdot \hat{\boldsymbol{S}}_i. \tag{3.256a}$$

Here, the sites i and j belong to a lattice Λ. There is a classical local magnetic field $\boldsymbol{H}_i$ that couples to the local spin operator $\hat{\boldsymbol{S}}_i$ through the Zeeman term. The three components $\hat{S}^a_i$ with $a = 1, 2, 3 \equiv x, y, z$ of the local spin operator $\hat{\boldsymbol{S}}_i$ satisfy the commutation relations

$$\left[\hat{S}^a_i, \hat{S}^b_j\right] = \mathrm{i}\delta_{ij}\, \epsilon^{abc}\, \hat{S}^c_j, \qquad a, b, c = 1, 2, 3, \qquad i, j \in \Lambda. \tag{3.256b}$$

We fix the irreducible representation of this algebra defined by the Casimir operator taking the value

$$\hat{\boldsymbol{S}}^2_i = S(S+1), \qquad i \in \Lambda. \tag{3.256c}$$

Here, S is either a positive half odd integer or a positive integer. The Heisenberg exchange couplings obey

$$J_{ij} = J_{ji}, \tag{3.256d}$$

for any pair of sites $i, j \in \Lambda$. The Heisenberg exchange interaction $J_{ij}\, \hat{\boldsymbol{S}}_i \cdot \hat{\boldsymbol{S}}_j$ is the simplest interaction between two quantum spins that is invariant under a global $SU(2)$ rotation of all the quantum spins. The Zeeman term breaks the local $SU(2)$ symmetry in spin space down to the subgroup $U(1)$ of local rotations around the direction in spin space corresponding to $\boldsymbol{H}_i$.

We shall limit ourselves to the case when the lattice Λ is assumed to be bipartite and made of $N \gg 1$ sites. More precisely, the lattice will be taken to be a macroscopically large subset of the hypercubic lattice $\mathbb{Z}^d$ with lattice spacing $\mathfrak{a}$. We shall also assume that the couplings J_{ij} are only non-vanishing if i belongs to one sublattice, while j belongs to the other sublattice, in which case they are taken negative $J_{ij} < 0$. The latter condition ($J_{ij} \leq 0$) defines a quantum spin-S Heisenberg antiferromagnet, while the former condition insures the absence of geometric frustration. The interaction $|J_{ij}|\, \hat{\boldsymbol{S}}_i \cdot \hat{\boldsymbol{S}}_j$ favors the singlet state for the two-site problem. If the degrees of freedom $\hat{\boldsymbol{S}}_i$ and $\hat{\boldsymbol{S}}_j$ were not operator-valued vectors but classical vectors in $\mathbb{R}^3$ of a fixed magnitude, the classical interaction $|J_{ij}|\, \hat{\boldsymbol{S}}_i \cdot \hat{\boldsymbol{S}}_j$ would favor an antiparallel alignment of these classical vectors. If so, the classical configuration that minimizes the classical energy when $\boldsymbol{H}_i = 0$ for all $i \in \Lambda$ of the classical counterpart to Eq. (3.256a) has all spins pointing along one direction on one sublattice and all spins pointing in the opposite direction on the other sublattice for any $d \geq 1$. This is the so-called Néel collinear antiferromagnetic state. The fundamental question to be addressed at the quantum level when $\boldsymbol{H}_i = 0$ for all $i \in \Lambda$ is what is the fate of the classical long-range order in the quantum ground state as a result of quantum fluctuations.

The two-site problem

Exercise 1.1:

(a) Compute exactly the partition function

$$Z_{\beta,S,\boldsymbol{H}} := \operatorname{tr}\left(e^{-\beta \hat{H}_{S,\boldsymbol{H}}[\hat{\boldsymbol{S}}]}\right), \tag{3.257}$$

with $\hat{H}_{S,\boldsymbol{H}}[\hat{\boldsymbol{S}}]$ given by Eq. (3.256a), when the lattice is made of two sites and $J_{ij} = J$.

(b) Comment on the difference when $J > 0$ and $J < 0$.

Semi-classical limit

Exercise 2.1:

(a) Perform the rescaling

$$\hat{\boldsymbol{S}}_i =: S\,\hat{\boldsymbol{s}}_i \tag{3.258}$$

and deduce the algebra obeyed by the operators $\hat{\boldsymbol{s}}_i$. From the algebra obeyed by the $\hat{\boldsymbol{s}}_i$ justify why the limit $S \to \infty$ can be interpreted as the semi-classical limit.

(b) We now consider the classical counterpart to the Hamiltonian (3.256a) obtained by replacing the operator-valued $\hat{\boldsymbol{S}}_i$ by classical unit vectors $\boldsymbol{N}_i$ from $\mathbb{R}^3$.
   - Assume that $\boldsymbol{H}_i = 0$ for all $i \in \Lambda$, that the lattice is the square lattice, and that J_{ij} is non-vanishing and negative on nearest-neighboring sites only. Construct the classical manifold of configurations that minimizes the classical energy (3.256a).
   - Assume that $\boldsymbol{H}_i = 0$ for all $i \in \Lambda$, that the lattice is the triangular lattice, and that J_{ij} is non-vanishing and negative on nearest-neighboring sites only. Construct the classical manifold of configurations that minimizes the classical energy (3.256a).

Single quantum spin coherent states

In this warm-up we begin with a single quantum spin $\hat{\boldsymbol{S}}$ and will therefore drop the site index. We shall denote the quantum Hamiltonian of the single quantum spin $\hat{\boldsymbol{S}}$ by $\hat{H}[\hat{\boldsymbol{S}}]$. Here, if the operator-valued argument $\hat{\boldsymbol{S}}$ is replaced by a classical vector in $\mathbb{R}^3$ and if we then drop the hat over $\hat{H}$, then H should be thought of as some smooth scalar-valued function. The Hilbert space is spanned by the $(2S+1)$ orthonormal states of the quantum spin-S irreducible representation of the group $SU(2)$

$$|S,m\rangle\,, \qquad m = -S, -S+1, \cdots, S-1, S, \tag{3.259a}$$

where S takes integer or half-odd-integer values, and

$$\hat{S}^z\,|S,m\rangle = m\,|S,m\rangle\,, \qquad \hat{\boldsymbol{S}}^2\,|S,m\rangle = S(S+1)\,|S,m\rangle\,. \tag{3.259b}$$

The $(2S+1)$ states in Eq. (3.259a) can be constructed with the help of the ladder operators

$$\hat{S}^{+} := \hat{S}^{x} + \mathrm{i}\hat{S}^{y}, \qquad \hat{S}^{-} = \hat{S}^{x} - \mathrm{i}\hat{S}^{y}. \tag{3.259c}$$

We call the state $|S,S\rangle$ the highest weight state. Observe that

$$\hat{S}^{+}|S,S\rangle = 0, \tag{3.259d}$$

so that we can interpret the highest weight state as the counterpart to the vacuum state for the boson annihilation operators in Eq. (2.10b). Successive action of $\hat{S}^{-}$ on the highest weight state $|S,S\rangle$ yields all the states in Eq. (3.259a), very much in the same way as application of all powers of the boson creation operators on the bosonic vacuum generates a basis of the bosonic Fock space (2.10a). Notice that $\hat{S}^{-}|S,-S\rangle = 0$ implies that we could equally have chosen $|S,-S\rangle = 0$ as the highest weight state. In this basis, the resolution of the identity is the representation

$$\mathbb{1} = \sum_{m=-S,-S+1,\cdots,S-1,S} |S,m\rangle\langle S,m| \tag{3.259e}$$

of the unit $(2S+1)\times(2S+1)$ matrix $\mathbb{1}$.

Exercise 3.1: Using the commutation relations Eq. (3.256b), compute the commutator of $\hat{S}^{+}$ with $\hat{S}^{-}$.

For the derivation of the path integral, we seek a set of states for which the matrix elements of $\hat{\boldsymbol{S}}$ are "as classical as may be". For this purpose, the states in Eq. (3.259a) are not convenient. Instead, we shall use the so-called spin coherent states. These are an infinite and over-complete set of states $|\boldsymbol{N}\rangle$, labeled by the points $\boldsymbol{N}$ on the surface of the unit sphere in $\mathbb{R}^3$,

$$\boldsymbol{N}^{\mathsf{T}} := (\sin\theta\,\cos\phi, \sin\theta\,\sin\phi, \cos\theta), \tag{3.260a}$$

that obey the following properties,

$$\langle \boldsymbol{N}_1|\,\boldsymbol{N}_2\rangle = e^{\mathrm{i}S\,\Phi(\boldsymbol{N}_0,\boldsymbol{N}_1,\boldsymbol{N}_2)}\left(\frac{1+\boldsymbol{N}_1\cdot\boldsymbol{N}_2}{2}\right)^{S}, \tag{3.260b}$$

$$\langle \boldsymbol{N}|\,\hat{\boldsymbol{S}}\,|\boldsymbol{N}\rangle = S\boldsymbol{N}, \tag{3.260c}$$

$$\mathbb{1} = \frac{2S+1}{4\pi}\int \mathrm{d}\mu(\boldsymbol{N})\,|\boldsymbol{N}\rangle\langle\boldsymbol{N}|. \tag{3.260d}$$

Equation (3.260b) implies that the states $\{|\boldsymbol{N}\rangle\}$ are not orthogonal for any finite S. The phase of the overlap between states $|\boldsymbol{N}_1\rangle$ and $|\boldsymbol{N}_2\rangle$ has a geometrical origin as $\Phi(\boldsymbol{N}_0,\boldsymbol{N}_1,\boldsymbol{N}_2)$ is the oriented area of the spherical triangle with vertices $\boldsymbol{N}_1$, $\boldsymbol{N}_2$, and some arbitrarily chosen reference unit vector $\boldsymbol{N}_0$ (see Fig. 3.4).

Exercise 3.2:

(a) Explain why the ambiguity in defining $\Phi(\boldsymbol{N}_0,\boldsymbol{N}_1,\boldsymbol{N}_2)$ does not matter in Eq. (3.260b).
(b) What is the value of the overlap (3.260b) in the limit $S\to\infty$?

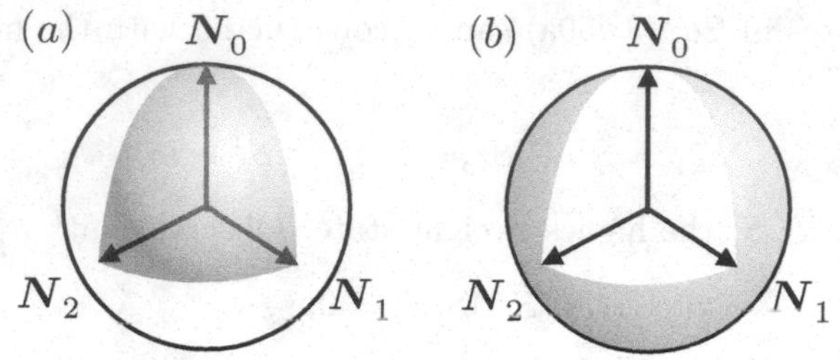

Fig. 3.4 A spherical triangle with vertices $\boldsymbol{N}_0$, $\boldsymbol{N}_1$, and $\boldsymbol{N}_2$. The definition of the area of this spherical triangle is ambiguous. It can be interpreted either as the inner area (a) or as the outer area (b).

Equation (3.260c) defines "as classical as may be". Equation (3.260d) is the resolution of unity, where it is understood that the integral is over the unit sphere,

$$\int \mathrm{d}\mu(\boldsymbol{N}) \equiv \int\limits_{\mathbb{R}^3} \mathrm{d}^3\boldsymbol{N}\, \delta(\boldsymbol{N}^2 - 1). \tag{3.261}$$

Equations (3.260b), (3.260c), and (3.260d) are the only ingredients that we will need to derive the path integral representation of a quantum spin-S Hamiltonian. We shall now construct explicitly coherent states satisfying Eqs. (3.260).

Given the unit vector (3.260a), we define the state $|\boldsymbol{N}\rangle$ by

$$|\boldsymbol{N}\rangle := e^{\zeta\, \hat{S}^{+} - \zeta^{*}\, \hat{S}^{-}}\; |S, S\rangle, \tag{3.262a}$$

where

$$\zeta := -\frac{\theta}{2}\, e^{-\mathrm{i}\phi}. \tag{3.262b}$$

In this representation the highest weight state $|S, S\rangle$ corresponds to the north pole of the unit sphere $\boldsymbol{N}^{\mathrm{T}} = (0, 0, 1)$ at which $\theta = 0$. We now specialize to the case of a spin-1/2 for which we shall prove explicitly Eqs. (3.260). We thus choose the representation, in units for which $\hbar = 1$,

$$\hat{\boldsymbol{S}} = \frac{1}{2}\boldsymbol{\sigma} \tag{3.263}$$

of the spin operators in terms of the 2×2 Pauli matrices.

Exercise 3.3:

(a) Show that for $S = 1/2$ the spin coherent states Eq. (3.262) can be written as

$$|\boldsymbol{N}\rangle = \cos\frac{\theta}{2}\; |1/2, 1/2\rangle + e^{\mathrm{i}\phi}\, \sin\frac{\theta}{2}\; |1/2, -1/2\rangle\,, \tag{3.264}$$

by making use of the identity (σ_0 is the 2×2 unit matrix)

$$e^{-\frac{\mathrm{i}}{2}\,\theta\,\boldsymbol{n}\cdot\boldsymbol{\sigma}} = \sigma_0\, \cos\frac{\theta}{2} - \mathrm{i}\boldsymbol{n}\cdot\boldsymbol{\sigma}\, \sin\frac{\theta}{2}, \tag{3.265}$$

where $\boldsymbol{n} \in \mathbb{R}^3$ is a unit vector.

(b) With the help of Eq. (3.264), show that the spin-1/2 coherent states satisfy the completeness relation

$$\begin{aligned}\sigma_0 &= \sum_{m=-\frac{1}{2},\frac{1}{2}} |1/2, m\rangle \langle 1/2, m| \\ &= \frac{1}{2\pi} \int \mathrm{d}\mu(\boldsymbol{N})\, |\boldsymbol{N}\rangle \langle \boldsymbol{N}| \\ &= \left[\frac{2S+1}{4\pi} \int \mathrm{d}\mu(\boldsymbol{N})\, |\boldsymbol{N}\rangle \langle \boldsymbol{N}|\right]_{S=1/2}. \end{aligned} \tag{3.266}$$

(c) With the help of Eq. (3.264) and Eq. (3.263), show that

$$\langle \boldsymbol{N}| \,\hat{\boldsymbol{S}}\, |\boldsymbol{N}\rangle = S\boldsymbol{N} = \frac{1}{2}\boldsymbol{N}. \tag{3.267}$$

(d) With the help of Eq. (3.264), show that

$$\langle \boldsymbol{N}_1 |\, \boldsymbol{N}_2\rangle = e^{(\mathrm{i}/2)\,\Phi(\boldsymbol{N}_0,\boldsymbol{N}_1,\boldsymbol{N}_2)} \left(\frac{1+\boldsymbol{N}_1\cdot\boldsymbol{N}_2}{2}\right)^{1/2}. \tag{3.268}$$

Here, the three orthonormal vectors $\boldsymbol{N}_2$, $\boldsymbol{N}_1$, and $\boldsymbol{N}_0$ define a Cartesian basis of internal spin-1/2 space $\mathbb{R}^3$ with the equatorial plane of the two-sphere S^2 depicted in Fig. 3.4 spanned by $\boldsymbol{N}_2$ and $\boldsymbol{N}_1$.

Coherent-state path integral for a single quantum spin

Having defined the spin-coherent states for a single quantum spin, we are now ready to derive the coherent state path integral for the partition function

$$Z_\beta := \mathrm{Tr}\left(e^{-\beta\,\hat{H}[\hat{\boldsymbol{S}}]}\right) = \frac{2S+1}{4\pi} \int \mathrm{d}\mu(\boldsymbol{N})\, \langle \boldsymbol{N}|\, e^{-\beta\,\hat{H}[\hat{\boldsymbol{S}}]}\, |\boldsymbol{N}\rangle\,, \tag{3.269}$$

where β is the inverse temperature in units for which the Boltzmann constant is unity and $\hat{H}$ is a linear function of the spin operator $\hat{\boldsymbol{S}}$. Although we are restricting ourselves to a single quantum spin, the generalization to many quantum spins is straightforward. As usual, we break the above exponential into a product of exponentials of infinitesimal time evolution operators

$$Z_\beta = \lim_{M\to\infty} \prod_{i=1}^{M} e^{-\Delta\tau_i\,\hat{H}[\hat{\boldsymbol{S}}]}, \qquad \Delta\tau_i := \beta/M, \tag{3.270}$$

and insert the resolution of the identity Eq. (3.260d) between each exponential.

Exercise 4.1:

(a) With the help of Eq. (3.260c), evaluate the time evolution during the infinitesimally small "time" $\Delta\tau$,

$$\left\langle \boldsymbol{N}(\tau) \,\middle|\, e^{-\Delta\tau\,\hat{H}[\hat{\boldsymbol{S}}]} \,\middle|\, \boldsymbol{N}(\tau-\Delta\tau)\right\rangle, \tag{3.271}$$

where one neglects terms of order $(\Delta\tau)^2$ and higher.

(b) Insert this result into Eq. (3.270) and show that the functional integral for the partition function (3.269) is given by

$$Z_\beta = \int \mathcal{D}\boldsymbol{N}(\tau)\, e^{-S_{\mathrm{B}} - \int_0^\beta \mathrm{d}\tau\, H[S\,\boldsymbol{N}(\tau)]}, \tag{3.272a}$$

where

$$S_{\mathrm{B}} = \int_0^\beta \mathrm{d}\tau \left\langle \boldsymbol{N}(\tau) \middle| \left(\frac{\mathrm{d}\boldsymbol{N}}{\mathrm{d}\tau}\right)(\tau) \right\rangle, \tag{3.272b}$$

and periodic boundary conditions

$$|\boldsymbol{N}(0)\rangle = |\boldsymbol{N}(\beta)\rangle \tag{3.272c}$$

are used. The real-valued $H(S\,\boldsymbol{N})$ is obtained by replacing every occurrence of $\hat{\boldsymbol{S}}$ in the quantum Hamiltonian $\hat{H}[\hat{\boldsymbol{S}}]$ by $S\boldsymbol{N}$ and removing the hat above the functional for the quantum Hamiltonian.

(c) Show that the first term in the argument of the exponential in Eq. (3.272a), i.e., S_{B} given by Eq. (3.272b), leads to a phase factor by verifying that S_{B} is pure imaginary.

The term S_{B} is called the Berry phase [36]. It represents the overlap of the coherent states at infinitesimally separated imaginary times. In differential geometry, it is interpreted as a "gauge connection". In the physics literature, it is interpreted as a gauge field of geometrical origin, with the geometry being that of the set made of an overcomplete basis of the Hilbert space that varies adiabatically as a function of a continuous parameter, here imaginary time. In Eq. (3.272b) the Berry phase is given in terms of an integral along the closed curve $\boldsymbol{N}(\tau)$ on the unit sphere. In order to bring the Berry phase into a geometrically more transparent form, we take advantage of the properties of the spin coherent states and transform the line integral into a surface integral . Thereto, we make use of the identity[19]

$$\frac{\mathrm{d}}{\mathrm{d}\tau} e^{\hat{O}} = \int_0^1 \mathrm{d}u\, e^{\hat{O}\,(1-u)} \frac{\mathrm{d}\hat{O}}{\mathrm{d}\tau} e^{\hat{O}\,u}, \tag{3.273}$$

which is the generalization of $\frac{\mathrm{d}}{\mathrm{d}\tau} e^{x} = \left(\frac{\mathrm{d}x}{\mathrm{d}\tau}\right) e^{x}$ to an operator $\hat{O}$ that does not commute with its derivative $\mathrm{d}\hat{O}/\mathrm{d}\tau$.

Exercise 4.2:

(a) With the help of Eqs. (3.260b), (3.260c), (3.262b), and (3.273), show that

$$S_{\mathrm{B}} = -S \int_0^\beta \mathrm{d}\tau \int_0^1 \mathrm{d}u \left[\zeta\,(\partial_\tau N^{+})(\tau,u) - \zeta^*\,(\partial_\tau N^{-})(\tau,u)\right], \tag{3.274a}$$

[19] Alternatively, one could use Stoke theorem (see Ref. [36]).

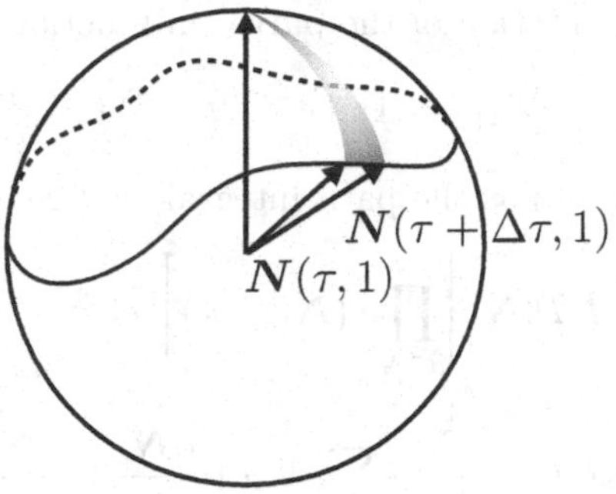

Fig. 3.5 $N(\tau,u)$ parametrizes the area on the unit sphere bounded by $N(\tau,1)$.

where

$$N^{\mathsf{T}}(\tau,u) = \Big(\sin\left(u\,\theta(\tau)\right)\,\cos\phi(\tau), \sin\left(u\,\theta(\tau)\right)\,\sin\phi(\tau), \cos\left(u\,\theta(\tau)\right)\Big), \tag{3.274b}$$

and

$$N^{+} := N^{x} + \mathrm{i}N^{y}, \qquad N^{-} := N^{x} - \mathrm{i}N^{y}. \tag{3.274c}$$

(b) Show that

$$N^{z}\partial_{u}N^{-} - N^{-}\partial_{u}N^{z} = -2\zeta, \qquad N^{z}\partial_{u}N^{+} - N^{+}\partial_{u}N^{z} = -2\zeta^{*}. \tag{3.275}$$

(c) Conclude that the Berry phase can be written as

$$S_{\mathrm{B}} = \mathrm{i}S\int\limits_{0}^{\beta}\mathrm{d}\tau\int\limits_{0}^{1}\mathrm{d}u\,N\cdot\left(\frac{\partial N}{\partial u}\wedge\frac{\partial N}{\partial \tau}\right). \tag{3.276}$$

From Eq. (3.274b) we infer that $N(\tau,u)$ moves with u along the great circle between the north pole and the physical value $N(\tau,1)$ (see Fig. 3.5). Hence the integral in Eq. (3.276) is simply the oriented area on the unit sphere bounded by $N(\tau)$. The value of this area depends on the fact that $N(\tau,0)$ corresponds to the north pole. This was a gauge choice. By making a different choice of phase for the coherent states, the point $N(\tau,0)$ can be chosen anywhere on the sphere. However, $e^{S_{\mathrm{B}}}$ in the coherent state path integral (3.272) is independent on the location of $N(\tau,0)$ up to a factor $e^{\mathrm{i}4\pi\,S}$. Since $2S$ is an integer, this factor leaves the Boltzmann weight entering the path integral (3.272a) unchanged.

Quantum antiferromagnets

We are going to generalize the path-integral representation for a single-spin Hamiltonian to derive the $O(3)$ Quantum Non-Linear Sigma Model (QNLσM) for a quantum antiferromagnet.

We assume the nearest-neighbor Heisenberg exchange couplings

$$J_{ij} = \begin{cases} -J < 0, & \text{if } i \text{ and } j \text{ are nearest-neighbor sites of } \Lambda, \\ 0, & \text{otherwise.} \end{cases} \tag{3.277}$$

We start from the representation of the partition function

$$Z_{\mathrm{AF}} := \operatorname{Tr}\left(e^{-\beta\,\hat{H}_{S,\boldsymbol{H}}[\hat{\boldsymbol{S}}]}\right), \tag{3.278}$$

for the Hamiltonian (3.256a) as the path integral over $SU(2)$-coherent states

$$Z_{\mathrm{AF}} = \int \mathcal{D}[\boldsymbol{N}] \left[\prod_{i\in\Lambda} \delta\left(\boldsymbol{N}_i^2 - 1\right)\right] e^{-S_B - S_{\mathrm{AF}}}, \tag{3.279a}$$

$$S_B := \mathrm{i}S \int_0^\beta \mathrm{d}\tau \int_0^1 \mathrm{d}u \sum_{i\in\Lambda} \boldsymbol{N}_i \cdot \left(\frac{\partial \boldsymbol{N}_i}{\partial u} \wedge \frac{\partial \boldsymbol{N}_i}{\partial \tau}\right), \tag{3.279b}$$

$$S_{\mathrm{AF}} := \int_0^\beta \mathrm{d}\tau \left(S^2 J \sum_{\langle ij\rangle} \boldsymbol{N}_i \cdot \boldsymbol{N}_j - S \sum_{i\in\Lambda} \boldsymbol{H}_i \cdot \boldsymbol{N}_i\right), \tag{3.279c}$$

which is a straightforward generalization of Eq. (3.272) and Eq. (3.276) to many spins. The local constraint $\boldsymbol{N}_i^2 = 1$ for all $i \in \Lambda$ has been enforced by a local delta function in order to trade the local integrals over the unit sphere S^2 for local integrals over $\mathbb{R}^3$. The first sum in (3.279c) is over all directed nearest-neighbor pairs $\langle ij\rangle$ on the bipartite lattice Λ. Periodic boundary conditions are imposed across the lattice.

We now assume that the lattice Λ is bipartite as was the case in Eq. (2.150), from which we borrow the convention for the notation of the two sublattices. It is known that, at zero temperature and in the classical limit $S \to \infty$, the ground state of a nearest-neighbor Heisenberg antiferromagnet on a bipartite lattice has spins oriented in opposite directions on the two sublattices of the bipartite lattice. This classical ground state is called the Néel ordered state. We assume that the partition function (3.279) is close to an antiferromagnetic fixed point. At this fixed point, the ground state breaks the translation symmetry of the lattice Λ down to the translation symmetry of any one of its sublattices. We aim at deriving an effective action for the Euclidean action $S \equiv S_{\mathrm{B}} + S_{\mathrm{AF}}$, which is valid in the long-wavelength and low-energy limit.

From now on, we choose the bipartite lattice Λ to be hypercubic and spanned by the orthonormal (Cartesian) unit vectors $\boldsymbol{e}_1, \cdots, \boldsymbol{e}_d$. The volume of Λ is $L^d = N\,\mathfrak{a}^d$ where L is the length of an edge of the hypercube measured in units of the lattice spacing $\mathfrak{a}$ and N is the number of sites in Λ. We are going to work with a unit cell of the hypercubic lattice Λ with two non-equivalent sites per unit cell.

Exercise 5.1:

(a) Choose a site $\boldsymbol{r}_i$ on sublattice Λ_A. The site $\boldsymbol{r}_i$ has $2\times d$ nearest-neighbors $\boldsymbol{r}_i \pm \mathfrak{a}\,\boldsymbol{e}_\mu$ with $\mu = 1, 2, 3$ sitting on sublattice Λ_B. The number $2 \times d$ is the coordination number of the hypercubic lattice. How many next-nearest-neighbor sites has $\boldsymbol{r}_i$ in $d = 1$, $d = 2$, and $d = 3$ and what are their Cartesian coordinates relative to $\boldsymbol{r}_i$?

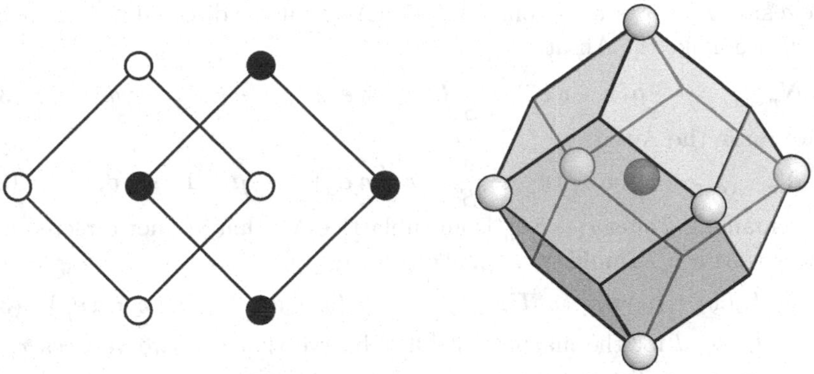

Fig. 3.6 Two-site unit cell centered about a site of sublattice Λ_A for $d = 2$ and $d = 3$.

(b) What type of lattice is Λ_A in $d = 1$, $d = 2$, and $d = 3$?

(c) Define a unit cell to be hypercubic with the edge-length $2\mathfrak{a}$ and volume $(2\mathfrak{a})^d$. Place the corners of this unit cell on sublattice Λ_B, say. This unit cell fills $\mathbb{R}^d$ by translations built from linear superpositions with integer-valued coefficients of the Cartesian unit vectors $2\mathfrak{a}\,\boldsymbol{e}_\mu$ with $\mu = 1, \cdots, d$. How many sites from sublattice Λ_B sit in the interior and on the vertices, edges, and faces of this unit cell in $d = 1$, $d = 2$, and $d = 3$? How many sites from sublattice Λ_A sit in the interior and on the vertices, edges, and faces of this unit cell in $d = 1$, $d = 2$, and $d = 3$? Deduce from these numbers that the unit cell contains two, four, and eight non-equivalent sites in $d = 1$, $d = 2$, and $d = 3$, respectively. The same conclusion follows from $(2\mathfrak{a})^d/2^d = \mathfrak{a}^d$.

(d) Define the following unit cells in $d = 1, 2, 3, \cdots$. Draw lines connecting $\boldsymbol{r}_i \in \Lambda_A$ to all its next-nearest-neighbor sites in Λ. For any of these connecting lines, draw the hypersurface of dimension $d-1$ normal to this connecting line in such a way that this hypersurface intersects the line at its mid-point. The resulting volume bounded by these $(d-1)$-dimensional hypersurface is the unit cell. It is an example of a geometrical object called a polytope. How many vertices, edges, and faces characterizes this example of a polytope in $d = 1$, $d = 2$, and $d = 3$? Where are the nearest-neighbor sites to $\boldsymbol{r}_i \in \Lambda_A$ about which this polytope is centered in $d = 1$, $d = 2$, and $d = 3$? Show that this unit cell contains two non-equivalent sites for any $d = 1, 2, 3, \cdots$. Show that the volume of this unit cell is $2 \times \mathfrak{a}^d$ for any $d = 1, 2, 3, \cdots$. Filling $\mathbb{R}^d$ with this unit cell requires both translations and rotations. This will be the unit cell with two non-equivalent sites, shown in Fig. 3.6 for $d = 2$ and $d = 3$, that we are going to use to construct the continuum limit.

For any site $\boldsymbol{r}_i \in \mathbb{R}^d$ from sublattice Λ_A, we make the Ansatz

$$\boldsymbol{r}_i \sim \boldsymbol{x} \in \mathbb{R}^d, \qquad \boldsymbol{N}_{\boldsymbol{r}_i} \sim +\boldsymbol{n}(\boldsymbol{x}) + \frac{\mathfrak{a}}{S}\,\boldsymbol{L}(\boldsymbol{x}). \tag{3.280a}$$

For the d sites $\boldsymbol{r}_j \equiv \boldsymbol{r}_i + \mathfrak{a}\,\boldsymbol{e}_\mu$ from sublattice Λ_B that are directed nearest-neighbors of site $\boldsymbol{r}_i$, we make the Ansatz

$$\boldsymbol{N}_{\boldsymbol{r}_i+\mathfrak{a}\,\boldsymbol{e}_\mu} \sim -\boldsymbol{n}(\boldsymbol{x}+\mathfrak{a}\,\boldsymbol{e}_\mu) + \frac{\mathfrak{a}}{S}\,\boldsymbol{L}(\boldsymbol{x}+\mathfrak{a}\,\boldsymbol{e}_\mu), \qquad \mu = 1,\cdots,d, \tag{3.280b}$$

while we make the Ansatz

$$\boldsymbol{N}_{\boldsymbol{r}_i-\mathfrak{a}\,\boldsymbol{e}_\mu} \sim -\boldsymbol{n}(\boldsymbol{x}-\mathfrak{a}\,\boldsymbol{e}_\mu) + \frac{\mathfrak{a}}{S}\,\boldsymbol{L}(\boldsymbol{x}-\mathfrak{a}\,\boldsymbol{e}_\mu), \qquad \mu = 1,\cdots,d, \tag{3.280c}$$

for the remaining d sites $\boldsymbol{r}_i - \mathfrak{a}\,\boldsymbol{e}_\mu$ from sublattice Λ_B that are not directed nearest-neighbors to site $\boldsymbol{r}_i$. Similarly, we make the Ansatz,

$$\boldsymbol{H}_{\boldsymbol{r}_i} \sim +\boldsymbol{h}_{\mathrm{s}}(\boldsymbol{x}) + \boldsymbol{h}_{\mathrm{u}}(\boldsymbol{x}), \qquad \boldsymbol{H}_{\boldsymbol{r}_i\pm\mathfrak{a}\,\boldsymbol{e}_\mu} \sim -\boldsymbol{h}_{\mathrm{s}}(\boldsymbol{x}\pm\mathfrak{a}\,\boldsymbol{e}_\mu) + \boldsymbol{h}_{\mathrm{u}}(\boldsymbol{x}\pm\mathfrak{a}\,\boldsymbol{e}_\mu), \tag{3.280d}$$

with $\mu = 1,\cdots,d$ for the magnetic field. Observe that the two vertices $\boldsymbol{r}_i + \mathfrak{a}\,\boldsymbol{e}_\mu$ and $\boldsymbol{r}_i - \mathfrak{a}\,\boldsymbol{e}_\mu$ are a distance $2\times\mathfrak{a}$ along the Cartesian axis $\boldsymbol{e}_\mu$ apart. The smooth vector fields $\boldsymbol{n}$, $\boldsymbol{L}$, $\boldsymbol{h}_{\mathrm{s}}$, and $\boldsymbol{h}_{\mathrm{u}}$, that assign to the continuous position $\boldsymbol{x}\in\mathbb{R}^d$ the vectors $\boldsymbol{n}(\boldsymbol{x})$, $\boldsymbol{L}(\boldsymbol{x})$, $\boldsymbol{h}_{\mathrm{s}}(\boldsymbol{x})$, and $\boldsymbol{h}_{\mathrm{u}}(\boldsymbol{x})$ from the target space $\mathbb{R}^3$, are assumed to vary slowly on the scale of the lattice spacing $\mathfrak{a}$. We impose the constraints

$$|\boldsymbol{n}(\boldsymbol{x})|^2 = 1, \qquad \boldsymbol{n}(\boldsymbol{x})\cdot\boldsymbol{L}(\boldsymbol{x}) = 0, \qquad \forall\boldsymbol{x}\in\mathbb{R}^d, \tag{3.280e}$$

Constraints (3.280e) imply that $[\boldsymbol{n}(\boldsymbol{x})\pm\mathfrak{a}\,\boldsymbol{L}(\boldsymbol{x})/S]^2 = 1$ hold up to first order in $(\mathfrak{a}/S)\,|\boldsymbol{L}(\boldsymbol{x}_i)|$, an approximation that becomes exact in the classical limit $S\to\infty$.

Exercise 5.2:

(a) Show that

$$\begin{aligned} H_S[S\,\boldsymbol{N}] =& +S^2\,J\sum_{i\in\Lambda_A}\sum_{\mu=1}^{d}\boldsymbol{N}_{\boldsymbol{r}_i}\cdot\left(\boldsymbol{N}_{\boldsymbol{r}_i+\mathfrak{a}\,\boldsymbol{e}_\mu}+\boldsymbol{N}_{\boldsymbol{r}_i-\mathfrak{a}\,\boldsymbol{e}_\mu}\right)\\ &-S\sum_{i\in\Lambda_A}\left(\boldsymbol{H}_{\boldsymbol{r}_i}\cdot\boldsymbol{N}_{\boldsymbol{r}_i}+\boldsymbol{H}_{\boldsymbol{r}_i+\mathfrak{a}\,\boldsymbol{e}_1}\cdot\boldsymbol{N}_{\boldsymbol{r}_i+\mathfrak{a}\,\boldsymbol{e}_1}\right). \end{aligned} \tag{3.281}$$

(b) Show that the continuum limit of Eq. (3.281) with the Ansatz (3.280) is (the summation convention over the repeated index μ is assumed)

$$\begin{aligned} H_S[S\,\boldsymbol{N}] \sim& -S^2\,J\,d\,\frac{L^d}{\mathfrak{a}^d}\\ &+\int\frac{\mathrm{d}^d\boldsymbol{x}}{2\mathfrak{a}^d}\left[S^2\,J\,\mathfrak{a}^2\,(\partial_\mu\boldsymbol{n})^2+2d\,J\,\mathfrak{a}^2\,\boldsymbol{L}^2\right](\boldsymbol{x})\\ &-2\times\int\frac{\mathrm{d}^d\boldsymbol{x}}{2\mathfrak{a}^d}\,(S\,\boldsymbol{h}_{\mathrm{s}}\cdot\boldsymbol{n}+\mathfrak{a}\,\boldsymbol{h}_{\mathrm{u}}\cdot\boldsymbol{L})(\boldsymbol{x}), \end{aligned} \tag{3.282}$$

to leading order in a gradient expansion.

The constant term in Eq. (3.282) is the exchange energy $S^2\,J$ per directed nearest-neighbor bond times the number of directed nearest-neighbor bonds $d\times L^d/\mathfrak{a}^d$ on a lattice Λ made of $L^d/\mathfrak{a}^d$ sites. It only leads to a change of the normalization of the partition function (3.279a). Therefore, it can be dropped.

To complete the expression for the coherent-state path integral of the quantum antiferromagnet in the continuum limit, we need to express the Berry phase S_{B} in terms of the staggered and uniform fields.

Exercise 5.3:

(a) The Ansatz (3.280) applied to Eq. (3.279b) yields the expansion

$$S_B \sim S'_B + S''_B + S'''_B + S''''_B, \tag{3.283}$$

where S'_B is of first order in S, S''_B is of zeroth order in S, S'''_B is of order S^{-1}, and S''''_B is of order S^{-2}. Show that S''_B is

$$\begin{aligned} S''_{\rm B} \sim +\mathrm{i}\,2\times\mathfrak{a}\int\frac{\mathrm{d}^d\boldsymbol{x}}{2\mathfrak{a}^d}\int\limits_0^\beta \mathrm{d}\tau\int\limits_0^1 \mathrm{d}u\,\Bigg[&\boldsymbol{n}\cdot\left(\frac{\partial\boldsymbol{n}}{\partial u}\wedge\frac{\partial\boldsymbol{L}}{\partial\tau}\right)\\ &+\boldsymbol{n}\cdot\left(\frac{\partial\boldsymbol{L}}{\partial u}\wedge\frac{\partial\boldsymbol{n}}{\partial\tau}\right)+\boldsymbol{L}\cdot\left(\frac{\partial\boldsymbol{n}}{\partial u}\wedge\frac{\partial\boldsymbol{n}}{\partial\tau}\right)\Bigg]. \end{aligned} \tag{3.284}$$

(b) Using the fact that the vector $\left(\frac{\partial \boldsymbol{n}}{\partial u}\wedge\frac{\partial \boldsymbol{n}}{\partial \tau}\right)\in\mathbb{R}^3$ is directed along $\boldsymbol{n}$ and with the help of Eq. (3.280e), show that Eq. (3.284) can be expressed as the difference of two total derivatives,

$$\begin{aligned} S''_{\rm B} \sim +\mathrm{i}\,2\times\mathfrak{a}\int\frac{\mathrm{d}^d\boldsymbol{x}}{2\mathfrak{a}^d}\int\limits_0^\beta \mathrm{d}\tau\int\limits_0^1 \mathrm{d}u\,\Bigg\{&\frac{\partial}{\partial\tau}\left[\boldsymbol{n}\cdot\left(\frac{\partial\boldsymbol{n}}{\partial u}\wedge\boldsymbol{L}\right)\right]\\ &-\frac{\partial}{\partial u}\left[\boldsymbol{n}\cdot\left(\frac{\partial\boldsymbol{n}}{\partial \tau}\wedge\boldsymbol{L}\right)\right]\Bigg\}. \end{aligned} \tag{3.285}$$

(c) Using the periodicity of the fields $\boldsymbol{n}$ and $\boldsymbol{L}$ in τ and the fact that $\boldsymbol{L}(\tau,0)=0$ for all τ, show that Eq. (3.285) simplifies to

$$S''_{\rm B} \sim -\mathrm{i}\,2\times\mathfrak{a}\int\frac{\mathrm{d}^d\boldsymbol{x}}{2\mathfrak{a}^d}\int\limits_0^\beta \mathrm{d}\tau\,\boldsymbol{L}\cdot\left(\boldsymbol{n}\wedge\frac{\partial\boldsymbol{n}}{\partial\tau}\right). \tag{3.286}$$

We shall ignore $S'''_{\rm B}$ and $S''''_{\rm B}$ since they are subleading in the expansion in powers of $1/S$.

After combining Eqs. (3.282), (3.286), and (3.279), we obtain the path integral for the partition function of the quantum spin-S antiferromagnet

$$\begin{aligned} Z_{\rm AF} &\sim \int\mathcal{D}[\boldsymbol{n},\boldsymbol{L}]\,\delta\left(\boldsymbol{n}^2-1\right)\,\delta\left(\boldsymbol{n}\cdot\boldsymbol{L}\right)\,e^{-S'_B-S'_{\rm AF}},\\ S'_{\rm AF} &= \frac{1}{2}\int\limits_0^\beta \mathrm{d}\tau\int\mathrm{d}^d\boldsymbol{x}\,\left[S^2\,J\,\mathfrak{a}^{2-d}\left(\partial_\mu\boldsymbol{n}\right)^2+2\,d\,J\,\mathfrak{a}^{2-d}\,\boldsymbol{L}^2\right]\\ &\quad-\int\limits_0^\beta\mathrm{d}\tau\int\mathrm{d}^d\boldsymbol{x}\,\left[S\,\mathfrak{a}^{-d}\,\boldsymbol{n}\cdot\boldsymbol{h}_{\rm s}+\mathrm{i}\,\mathfrak{a}^{1-d}\,\boldsymbol{L}\cdot\left(\boldsymbol{n}\wedge\frac{\partial\boldsymbol{n}}{\partial\tau}-\mathrm{i}\,\boldsymbol{h}_{\rm u}\right)\right]. \end{aligned} \tag{3.287}$$

Exercise 5.4:

(a) Compute the integration over $\boldsymbol{L}$ by completing the square and show that Z_{AF} simplifies to

$$
\begin{aligned}
Z_{\mathrm{AF}} &\sim \int \mathcal{D}[\boldsymbol{n}]\,\delta\left(\boldsymbol{n}^2-1\right)\,e^{-S'_B-S''_{\mathrm{AF}}},\\
S''_{\mathrm{AF}} &\sim \frac{1}{2}\int_0^\beta \mathrm{d}\tau \int \mathrm{d}^d\boldsymbol{x}\left[S^2\,J\,\mathfrak{a}^{2-d}\,\left(\partial_\mu \boldsymbol{n}\right)^2+\frac{\mathfrak{a}^{-d}}{2\,d\,J}\left(\partial_\tau \boldsymbol{n}-\mathrm{i}\,\boldsymbol{h}_{\mathrm{u}}\wedge\boldsymbol{n}\right)^2\right]\\
&\quad -\int_0^\beta \mathrm{d}\tau \int \mathrm{d}^d\boldsymbol{x}\,S\,\mathfrak{a}^{-d}\boldsymbol{n}\cdot\boldsymbol{h}_{\mathrm{s}}.
\end{aligned}
\tag{3.288}
$$

(b) The Euclidean action S'_{B} in Eq. (3.288) arises from (i) inserting the Ansatz (3.280) into the Berry phase (3.279b), (ii) selecting the term of order S, (iii) and performing a gradient expansion. Carrying this program is subtle, because step (ii) contains a term that oscillates in sign between the two sublattices. Before evaluating S'_{B} in one-dimensional position space, show that step (ii) gives

$$
\begin{aligned}
S'_{\mathrm{B}} \sim +\mathrm{i}\,S \int \frac{\mathrm{d}^d\boldsymbol{x}}{2\mathfrak{a}^d}\int_0^\beta \mathrm{d}\tau \int_0^1 \mathrm{d}u\,\Bigg[&\boldsymbol{n}(\boldsymbol{x},\tau,u)\cdot\left(\frac{\partial \boldsymbol{n}}{\partial u}\wedge\frac{\partial \boldsymbol{n}}{\partial \tau}\right)(\boldsymbol{x},\tau,u)\\
&-\boldsymbol{n}(\boldsymbol{x}+\mathfrak{a}\,\boldsymbol{e}_1,\tau,u)\cdot\left(\frac{\partial \boldsymbol{n}}{\partial u}\wedge\frac{\partial \boldsymbol{n}}{\partial \tau}\right)(\boldsymbol{x}+\mathfrak{a}\,\boldsymbol{e}_1,\tau,u)\Bigg].
\end{aligned}
\tag{3.289}
$$

Evaluation of S'_{B} in one dimension

Exercise 6.1:

(a) Take advantage of the periodic boundary conditions and assume an even number of sites in one dimension to show that the expansion of S'_{B} to zeroth-order in $(2\mathfrak{a})\,\boldsymbol{L}/S$ is approximately given by

$$
S'_{\mathrm{B}} \sim -\mathrm{i}\frac{S}{2}\int \mathrm{d}x \int_0^\beta \mathrm{d}\tau\,\boldsymbol{n}\cdot\left(\frac{\partial \boldsymbol{n}}{\partial x}\wedge\frac{\partial \boldsymbol{n}}{\partial \tau}\right).
\tag{3.290}
$$

Hint: Draw the two areas on the unit sphere of Fig. 3.5, one associated to the Berry phase arising from $\boldsymbol{n}(\boldsymbol{x},\tau,u)$ and one associated to the Berry phase arising from $\boldsymbol{n}(\boldsymbol{x}+\mathfrak{a}\,\boldsymbol{e}_1,\tau,u)$, assuming that their boundaries $\boldsymbol{n}(\boldsymbol{x},\tau,1)$ and $\boldsymbol{n}(\boldsymbol{x}+\mathfrak{a}\,\boldsymbol{e}_1,\tau,1)$, respectively, are infinitesimally far apart.

(b) Show that the contribution of the term S'_{B} to the partition function Z_{AF} is given by

$$
e^{+\mathrm{i}2\pi\,S\,Q},
\tag{3.291}
$$

where

$$Q := \frac{1}{4\pi}\int \mathrm{d}x \int_0^\beta \mathrm{d}\tau\, \boldsymbol{n}\cdot\left(\frac{\partial \boldsymbol{n}}{\partial x}\wedge\frac{\partial \boldsymbol{n}}{\partial \tau}\right), \tag{3.292}$$

and discuss the cases S integer and S half-odd integer. Convince yourself that Q is an integer. The integral Q is an example of a topological invariant from algebraic topology.

The fact that the long-wavelength and low-energy properties of the one-dimensional quantum spin chain depend, through the Berry phase, in a dramatic fashion on S being an integer or a half-odd integer was conjectured by Haldane [35]. The path integral (3.288) can be interpreted as an integration over all possible spin fluctuations. Those fluctuations that do not depend on τ are the classical fluctuations. Those fluctuations that depend on τ are the quantum fluctuations. For the case of S integer, the effective action $S_{\text{eff}} = S'_{\text{B}} + S''_{\text{AF}}$ is such that $e^{-S_{\text{eff}}}$ is always positive as a result of the Berry phase being inoperative. Hence, all the quantum fluctuations contribute with the same sign. However, for the case of S a half-odd integer, the Berry phase is operative and quantum fluctuations are suppressed owing to the destructive interference caused by the alternating sign of (3.291). For S integer, there is a finite gap in the spin excitation spectrum (Haldane gap). For S half-odd integer, the ground state is quasi-long-range ordered (the best the system can do short of long-range order in view of the Mermin-Wagner theorem) and excitations are gapless.

Conjecture for two-dimensional quantum antiferromagnets

Exercise 7.1: Consider two decoupled spin-1/2 antiferromagnetic chains, each of which is described by the QNLσM in one-dimensional position space with the topological term (3.290). Assume that the two chains are weakly coupled by an antiferromagnetic Heisenberg exchange coupling along the "rungs" of a "ladder" in such a way that the long-wavelength and low-energy effective action for the ladder is, to zeroth order in the rung coupling, two copies of the QNLσM in one-dimensional position space and at zero temperature. Argue how should one choose the relative sign of the topological terms along each chain for an infinitesimal antiferromagnetic Heisenberg exchange coupling along the rungs. Decide from this thought experiment whether n weakly coupled spin-1/2 antiferromagnetic chains have or do not have a topological term in their long-wavelength and low-energy effective action.

Exercise 7.2: Assume that the QNLσM that captures the physics of n weakly coupled spin-1/2 antiferromagnetic ladders at zero temperature can be brought to the form of the classical two-dimensional $O(3)$ NLσM studied in chapter 3 with the effective spin $2n \times S = n$. What is the dependence on n of the correlation length derived in chapter 3? If the limit $n \to \infty$ was taken to define a two-dimensional spin-1/2 antiferromagnetic Heisenberg model on the square lattice with anisotropic exchange couplings, would the ground state be separated from the excitations by a gap or would it be gapless? Is your conjectured excitation spectrum consistent with taking the two-dimensional limit by using $(2n+1)$ weakly coupled spin-1/2 antiferromagnetic chains instead of n ladders?

3.8.3 Classical $O(N)$ NLσM with $N > 2$: One-loop RG using the Berezinskii-Blank parametrization of spin waves

Introduction

Our goal is to perform a one-loop RG analysis on the QNLσM defined by Eq. (3.288) in the absence of the topological term S'_{B}. The RG technique that we are going to use relies on a parametrization of spin waves introduced by Berezinskii and Blank in Ref. [37].

As a warm up, we are going to perform the one-loop RG analysis of the classical limit of the $O(N>2)$ QNLσM using the Berezinskii-Blank parametrization of spin waves as was done by Polyakov in Ref. [32].

Definitions

The classical $O(N)$ NLσM is defined by the partition function

$$Z := \int\limits_{\mathbb{R}^N} \mathcal{D}[\boldsymbol{n}]\, \delta(\boldsymbol{n}^2 - 1)\, e^{-\int \mathrm{d}^d\boldsymbol{r}\, \mathcal{L}}, \tag{3.293a}$$

$$\mathcal{L} := \frac{1}{2g} \left(\partial_\mu \boldsymbol{n}\right)^2. \tag{3.293b}$$

Summation over repeated Greek indices ($\mu = 1, \cdots, d$) is assumed throughout. The coupling constant g has dimension

$$[g] = \text{length}^{d-2}. \tag{3.293c}$$

Observe that g is dimensionless if and only if $d = 2$. The measure of the NLσM is defined to be $\mathcal{D}[\boldsymbol{n}]\,\delta(\boldsymbol{n}^2 - 1)$ where $\boldsymbol{n} : \mathbb{R}^d \to \mathbb{R}^N$, $\boldsymbol{r} \to \boldsymbol{n}(\boldsymbol{r})$, is a real-valued dimensionless vector field. What makes the NLσM non-trivial is the constraint on $\boldsymbol{n}$ that is implemented by $\delta(\boldsymbol{n}^2 - 1)$.

In order to exploit the techniques of renormalization group (RG), e.g., for the computation of the beta function, we shall use Berezinskii and Blank's parametrization of the vector field $\boldsymbol{n}(\boldsymbol{r})$ (see Refs. [37] and [32]),

$$\boldsymbol{n}(\boldsymbol{r}) := \sqrt{1 - \boldsymbol{\phi}^2(\boldsymbol{r})}\; \boldsymbol{n}^0(\boldsymbol{r}) + \sum_{a=1}^{N-1} \phi_a(\boldsymbol{r})\, \boldsymbol{e}^a(\boldsymbol{r}), \tag{3.294}$$

where $\{\boldsymbol{n}^0(\boldsymbol{r}), \boldsymbol{e}^1(\boldsymbol{r}), \cdots, \boldsymbol{e}^{N-1}(\boldsymbol{r})\}$ is an orthonormal basis of $\mathbb{R}^N$ for any given $\boldsymbol{r} \in \mathbb{R}^d$. It is assumed that $\boldsymbol{n}^0(\boldsymbol{r})$ deviates only slightly from a given fixed coordinate axis, $\boldsymbol{e}^0$ say, for all $\boldsymbol{r} \in \mathbb{R}^d$. That is, we want to describe the effect of spin fluctuations at finite g in an antiferromagnetically ordered system when $g = 0$.

Therefore, we can regard $\boldsymbol{n}^0(\boldsymbol{r})$ as a slowly varying vector, with Fourier wave vectors in the range $|\boldsymbol{p}| < \widetilde{\Lambda}$, say. The "fast" degrees of freedom are contained in $\boldsymbol{\phi}(\boldsymbol{r}) = (\phi_1(\boldsymbol{r}), \cdots, \phi_{N-1}(\boldsymbol{r}))$, which have wave vectors in the range $\widetilde{\Lambda} < |\boldsymbol{p}| < \Lambda$. Note that there is an arbitrariness in choosing the vectors $\boldsymbol{e}^1(\boldsymbol{r}), \cdots, \boldsymbol{e}^{N-1}(\boldsymbol{r})$. At each point $\boldsymbol{r} \in \mathbb{R}^d$ the orthonormal basis $\boldsymbol{e}^1(\boldsymbol{r}), \cdots, \boldsymbol{e}^{N-1}(\boldsymbol{r})$ is only defined up to a $O(N-1)$ rotation [$O(N-1)$ gauge symmetry]. One way to eliminate this arbitrariness is by choosing the "Coulomb gauge"

$$\partial_\mu \left[\boldsymbol{e}^a(\boldsymbol{r}) \cdot \partial_\mu \boldsymbol{n}^0(\boldsymbol{r})\right] = 0, \qquad a = 1, \ldots, N-1, \tag{3.295}$$

which is a first-order differential equation in the $\boldsymbol{e}^a$'s.

Exercise 1.1:

(a) What is problematic, if the unit length vector $\boldsymbol{n}^0$ only has non-zero wave vectors in the restricted range $|\boldsymbol{p}| < \widetilde{\Lambda}$? We will ignore this issue in the sequel.

(b) Show that $\left(\partial_\mu \boldsymbol{n}^0\right)(\boldsymbol{r})$ is orthogonal to $\boldsymbol{n}^0(\boldsymbol{r})$ for all $\boldsymbol{r} \in \mathbb{R}^d$. Conclude that there exist $N-1$ expansion coefficients $A^0_{1\mu}(\boldsymbol{r}), A^0_{2\mu}(\boldsymbol{r}), \cdots, A^0_{(N-1)\mu}(\boldsymbol{r})$, such that

$$\left(\partial_\mu \boldsymbol{n}^0\right)(\boldsymbol{r}) = \sum_{b=1}^{N-1} A^0_{b\mu}(\boldsymbol{r})\, \boldsymbol{e}^b(\boldsymbol{r}). \tag{3.296}$$

(c) Show that $\left(\partial_\mu \boldsymbol{e}^a\right)(\boldsymbol{r})$ is orthogonal to $\boldsymbol{e}^a(\boldsymbol{r})$ for $a = 1, \cdots, N-1$ and all $\boldsymbol{r} \in \mathbb{R}^d$. Infer that

$$\left(\partial_\mu \boldsymbol{e}^a\right)(\boldsymbol{r}) = \sum_{b \neq a} A^a_{b\mu}(\boldsymbol{r})\, \boldsymbol{e}^b(\boldsymbol{r}) - A^0_{a\mu}(\boldsymbol{r})\, \boldsymbol{n}^0(\boldsymbol{r}), \qquad a = 1, \cdots, N-1, \tag{3.297}$$

with the expansion coefficients $A^a_{1\mu}(\boldsymbol{r}), \cdots, A^a_{(N-1)\mu}(\boldsymbol{r})$, where $A^a_{a\mu}(\boldsymbol{r}) = 0$. In particular, verify that $A^0_{a\mu}(\boldsymbol{r})$ in Eq. (3.297) is indeed the same field as the expansion coefficient $A^0_{a\mu}(\boldsymbol{r})$ in Eq. (3.296).

(d) Show that $A^a_{b\mu}$ is antisymmetric, i.e.,

$$A^a_{b\mu} = -A^b_{a\mu}, \qquad a, b = 1, \cdots, N-1. \tag{3.298}$$

Verify that

$$A^0_{a\mu} = \boldsymbol{e}^a \cdot \left(\partial_\mu \boldsymbol{n}^0\right), \qquad A^a_{b\mu} = \boldsymbol{e}^b \cdot \left(\partial_\mu \boldsymbol{e}^a\right). \tag{3.299}$$

(e) With the help of Eqs. (3.296) and (3.297), show that substituting the Berezinskii-Blank parametrization (3.294) into the Lagrangian (3.293b) of the NLσM gives

$$\mathcal{L} = \frac{1}{2g}\left\{\left[\partial_\mu \left(1-\boldsymbol{\phi}^2\right)^{1/2} - \phi_a\, A^0_{a\mu}\right]^2 + \left[\left(1-\boldsymbol{\phi}^2\right)^{1/2} A^0_{a\mu} + \partial_\mu \phi_a + \phi_b\, A^b_{a\mu}\right]^2\right\}, \tag{3.300}$$

where repeated indices are to be summed over. In order to compute the beta function in "a quick and dirty way" we shall only retain fast fluctuations up to second order, that is we need to isolate the term that is quadratic in the ϕ_a's.

(f) Isolate in Eq. (3.300) the term that is quadratic in the ϕ_a's. Show that this yields[20]

$$\mathcal{L} = \frac{1}{2g}\Big[\left(\partial_\mu \phi_a + A^b_{a\mu}\,\phi_b\right)^2 + A^0_{a\mu}\,A^0_{b\mu}\left(\phi_a\,\phi_b - \boldsymbol{\phi}^2\,\delta_{ab}\right) + A^0_{a\mu}\,A^0_{a\mu} \\ + 2A^0_{a\mu}\left(\partial_\mu \phi_a + \phi_b\,A^b_{a\mu}\right)\Big]. \tag{3.301}$$

Define the Lagrangian densities

$$\begin{aligned} \mathcal{L}_0 &:= \mathcal{L}_0^{\mathrm{slow}} + \mathcal{L}_0^{\mathrm{fast}} + \mathcal{L}_0^{\mathrm{int}}, \\ \mathcal{L}_0^{\mathrm{slow}} &:= \frac{1}{2g}\,A^0_{a\mu}\,A^0_{a\mu}, \\ \mathcal{L}_0^{\mathrm{fast}} &:= \frac{1}{2g}\,\left(\partial_\mu \phi_a\right)^2, \\ \mathcal{L}_0^{\mathrm{int}} &:= \frac{1}{2g}\,A^0_{a\mu}\,A^0_{b\mu}\left(\phi_a\,\phi_b - \boldsymbol{\phi}^2\,\delta_{ab}\right). \end{aligned} \tag{3.302}$$

(g) With the help of Eq. (3.299), show that the last term in Eq. (3.301) vanishes if we work in the Coulomb gauge given by Eq. (3.295).

(h) Show that the gauge choice Eq. (3.295) is consistent with orthonormality of the $\boldsymbol{e}^a$'s.

We shall assume that of all the terms generated by a momentum shell integration of $\boldsymbol{\phi}$ in the shell $\widetilde{\Lambda} < |\boldsymbol{p}| < \Lambda$ the single most relevant one can be absorbed by renormalization of the coupling constant g, i.e., the renormalized Lagrangian takes the form

$$\begin{aligned} \widetilde{\mathcal{L}} &= \frac{1}{2\widetilde{g}}\,\left(\partial_\mu \boldsymbol{n}^0\right)^2 + \cdots \\ &= \frac{1}{2\widetilde{g}}\,\left(A^0_{a\mu}\right)^2 + \cdots. \end{aligned} \tag{3.303}$$

Needed is the change of g induced by the momentum shell integration as was done by Polyakov in 1975 in his pioneering work [32]. In doing so we shall ignore the renormalization associated to the gauge fields as the gauge invariance of the theory dictates that these effects are less important than the renormalization of g. We shall assume that the renormalization of g comes solely from the second term in Eq. (3.301).

Exercise 1.2:

(a) A kinetic energy term for the non-Abelian gauge fields $A^a_{b\mu}$, $a,b = 1,\cdots,N-1$, will be included in the $\cdots$ of Eq. (3.303)

$$\frac{1}{\widetilde{e}}\,F_{\mu\nu}\,F_{\mu\nu}. \tag{3.304}$$

On the basis of symmetry alone, write down the explicit form of $F_{\mu\nu}$. What is the naive scaling dimension of the effective "charge" $\widetilde{e}$?

[20] Terms of order $\phi_a\,\partial_\mu\phi_b\,\phi_c$ can be dropped.

(b) Substitute in the partition function (3.293) the original Lagrangian by the one defined in Eq. (3.302). Expand the exponential of $\mathcal{L}_0^{\rm int}$ and integrate out the "fast" fields. If so, show that

$$\begin{aligned} Z \approx & \int\limits_{\boldsymbol{n}_0^2=1} \mathcal{D}[\boldsymbol{n}_0]\, e^{-\frac{1}{2g}\int^{\widetilde{\Lambda}} \mathrm{d}^d\boldsymbol{p}\, A^0_{a\mu}(+\boldsymbol{p})\cdot A^0_{a\mu}(-\boldsymbol{p})} \\ & \times \left[1-\frac{f_{ab}(\Lambda,\widetilde{\Lambda})}{2g}\int^{\widetilde{\Lambda}} \mathrm{d}^d\boldsymbol{k}\, A^0_{a\mu}(+\boldsymbol{k})\, A^0_{b\mu}(-\boldsymbol{k})\right], \end{aligned} \tag{3.305a}$$

where, for $d=2$,

$$f_{ab}(\Lambda,\widetilde{\Lambda}) = -\frac{N-2}{2\pi}\, g\, \ln\frac{\Lambda}{\widetilde{\Lambda}}\, \delta_{ab}. \tag{3.305b}$$

Conclude that, in $d=2$, the integration over the fast modes with Fourier components restricted to the momentum shell $\widetilde{\Lambda}<|\boldsymbol{p}|<\Lambda$ in Eq. (3.305a) yields the renormalization of the coupling constant

$$\frac{1}{\widetilde{g}} \approx \frac{1}{g} + \frac{(2-N)}{2\pi}\log\frac{\Lambda}{\widetilde{\Lambda}}. \tag{3.306}$$

(c) Generalize Eqs. (3.305b) and (3.306) to the case of $d>2$. *Hint:* Account for the fact that g is dimensiontul when $d>2$.

(d) Show for $d=2$ that after repeating the momentum shell integration a sufficient number of times the coupling constant for momentum cut-off q is given by

$$\widetilde{g}(q) = \frac{g}{1-g\,(N-2)\,(1/2\pi)\ln(\Lambda/q)}, \tag{3.307}$$

and compute the beta function

$$\beta(\widetilde{g}) = \frac{\mathrm{d}\widetilde{g}(q)}{\mathrm{d}\ln(\Lambda/q)}. \tag{3.308}$$

How does this result compare with Eq. (3.183b)?

(e) Using the parametrization (3.294) and for $d=2$, show that the correlator

$$\mathcal{D}(r;\Lambda) = \langle \boldsymbol{n}(\boldsymbol{0})\cdot\boldsymbol{n}(\boldsymbol{r})\rangle, \qquad r\equiv|\boldsymbol{r}|, \tag{3.309}$$

at the renormalization point Λ, i.e., defined by imposing the upper momentum cut-off Λ, and the correlator $\mathcal{D}$ at the renormalization point $\widetilde{\Lambda}$ are approximately related by

$$\begin{aligned} \mathcal{D}(\boldsymbol{r},\Lambda) &\approx \left(1-\frac{N-1}{2\pi}\,g(\Lambda)\,\ln\frac{\Lambda}{\widetilde{\Lambda}}\right)\mathcal{D}(\boldsymbol{r},\widetilde{\Lambda}) \\ &\equiv Z\,\mathcal{D}(\boldsymbol{r},\widetilde{\Lambda}). \end{aligned} \tag{3.310}$$

The function Z is called the wave-function renormalization. Compute the anomalous dimension $\gamma(\widetilde{g})$

$$\gamma(\widetilde{g}) := \frac{1}{2}\,\frac{\mathrm{d}Z}{\mathrm{d}\ln(\Lambda/\widetilde{\Lambda})}. \tag{3.311}$$

How does this result compare with Eq. (3.183c)?

3.8.4 $O(N)$ QNLσM: Large-N expansion

Introduction

It is time to study the $O(N)$ QNLσM. We do this in the limit $N \to \infty$ to begin with [38]. The one-loop RG analysis is done in section 3.8.5.

Definitions

The $O(N)$ QNLσM is defined by the partition function

$$Z_{\Omega,g}[\boldsymbol{J}] := \int\limits_{\mathbb{R}^N} \mathcal{D}[\boldsymbol{n}]\,\delta(\boldsymbol{n}^2-1)\,e^{-\frac{1}{2\,c\,g}\int\limits_0^\beta \mathrm{d}\tau \int\limits_0^L \mathrm{d}^d\boldsymbol{r}\left[(\partial_\tau \boldsymbol{n})^2 + c^2(\boldsymbol{\nabla}\boldsymbol{n})^2 - 2\,c^2\,g\,\boldsymbol{n}\cdot\boldsymbol{J}\right]}. \tag{3.312}$$

Use the notation $x_0 \equiv c\tau$ and $\boldsymbol{x} \equiv \boldsymbol{r}$,

$$Z_{\Omega,g}[\boldsymbol{J}] = \int\limits_{\mathbb{R}^N} \mathcal{D}[\boldsymbol{n}]\,\delta(\boldsymbol{n}^2-1)\,e^{-\frac{1}{2\,g}\int\limits_0^{c\beta} \mathrm{d}x_0 \int\limits_0^L \mathrm{d}^d\boldsymbol{x}\left[(\partial_0 \boldsymbol{n})^2 + (\boldsymbol{\partial}\boldsymbol{n})^2 - 2\,g\,\boldsymbol{n}\cdot\boldsymbol{J}\right]}. \tag{3.313}$$

We set $c=1$ from now on,[21] and use the short-hand notation

$$Z_{\Omega,g}[\boldsymbol{J}] \equiv \int\limits_{\mathbb{R}^N} \mathcal{D}[\boldsymbol{n}]\,\delta(\boldsymbol{n}^2-1)\,e^{-\frac{1}{2\,g}\int\limits_{\Omega}\left[(\partial_0 \boldsymbol{n})^2 + (\boldsymbol{\partial}\boldsymbol{n})^2 - 2\,g\,\boldsymbol{n}\cdot\boldsymbol{J}\right]}. \tag{3.314}$$

The base space that defines the domain of definition of the field $\boldsymbol{n}$ is the volume Ω of $(d+1)$-dimensional space-time. The field $\boldsymbol{n}$ takes values in $\mathbb{R}^N$ subject to the constraint that it has unit length, i.e., the allowed contributions to the path integral belong to the unit sphere

$$\begin{aligned} S^{N-1} &:= \left\{\boldsymbol{n} \in \mathbb{R}^N \middle| \boldsymbol{n}^2 = 1\right\} \\ &= O(N)/O(N-1), \end{aligned} \tag{3.315}$$

that defines the target space of the QNLσM.

Periodic boundary conditions are imposed on the field $\boldsymbol{n}$ at the boundary of Ω. The field $\boldsymbol{J}$ is a source field. We seek to define the limit $N \to \infty$ properly and to evaluate the partition function of the $O(N)$ QNLσM as well as the relevant correlation functions in this limit.

The scaling limit $N \to \infty$

We insert the representation

$$\delta(\boldsymbol{n}^2-1) = \int\limits_{\mathrm{i}\times\mathbb{R}} \mathcal{D}[\lambda]\,e^{-\int\limits_\Omega \lambda\,(\boldsymbol{n}^2-1)} \tag{3.316}$$

[21] The coordinate x_0 has the same units as $\boldsymbol{x}$, namely those of the lattice spacing $\mathfrak{a}$. The $(d+1)$-dimensional volume Ω has units of $\mathfrak{a}^{d+1}$. The coupling g has units of $\mathfrak{a}^{d-1}$. The field $\boldsymbol{J}$ has units of $1/\Omega$.

into Eq. (3.314). The field λ has the units of $1/\Omega$. We integrate over the field $\boldsymbol{n}$,

$$\begin{aligned} Z_{\Omega,g}[\boldsymbol{J}] :=& \int\limits_{\mathrm{i}\times\mathbb{R}^{N+1}} \mathcal{D}[\boldsymbol{n},\lambda]\, e^{-\frac{1}{2g}\int\limits_{\Omega}\left[(\partial_\mu \boldsymbol{n})^2+2g\lambda(\boldsymbol{n}^2-1)-2g\boldsymbol{J}\cdot\boldsymbol{n}\right]} \\ =& \int\limits_{\mathrm{i}\times\mathbb{R}} \mathcal{D}[\lambda]\, e^{\int\limits_{\Omega}\lambda} \int\limits_{\mathbb{R}^N} \mathcal{D}[\boldsymbol{n}]\, e^{-\frac{1}{2g}\int\limits_{\Omega}\left[\boldsymbol{n}\cdot(-\partial_\mu^2+2g\lambda)\boldsymbol{n}-2g\boldsymbol{J}\cdot\boldsymbol{n}\right]}. \end{aligned} \tag{3.317}$$

To reach the second line, we have assumed that the path integrals over $\boldsymbol{n}$ and λ can be interchanged.

Exercise 1.1:

(a) Show that integration over $\boldsymbol{n}$ gives

$$Z_{\Omega,g}[\boldsymbol{J}] = \int\limits_{\mathrm{i}\times\mathbb{R}} \mathcal{D}[\lambda]\, e^{N\left[\int\limits_{\Omega}\frac{1}{N}\lambda+\frac{g}{2N}\int\limits_{\Omega}\boldsymbol{J}\cdot\left(\frac{1}{-\partial_\mu^2+2g\lambda}\boldsymbol{J}\right)-\frac{1}{2}\operatorname{Tr}\left(\ln\frac{-\partial_\mu^2+2g\lambda}{g}\right)\right]}. \tag{3.318}$$

(b) Find the transformation law relating λ, $\boldsymbol{J}$, and g to λ', $\boldsymbol{J}'$, and g' and in terms of which

$$Z_{\Omega,g}[\boldsymbol{J}] = \int\limits_{\mathrm{i}\times\mathbb{R}} \mathcal{D}[\lambda']\, e^{-F_{\Omega,g'}(\lambda',\boldsymbol{J}')+\frac{N}{2}\operatorname{Tr}(\ln g)}, \tag{3.319a}$$

where

$$\begin{aligned} F_{\Omega,g'}[\lambda',\boldsymbol{J}'] :=& -N\Bigg\{\int\limits_{\Omega}\lambda' + \frac{1}{2}\int\limits_{\Omega}\boldsymbol{J}'\cdot\left(\frac{1}{\partial_\mu^2+2g'\lambda'}\boldsymbol{J}'\right) \\ & -\frac{1}{2}\operatorname{Tr}\left[\ln\left(-\partial_\mu^2+2g'\lambda'\right)\right]\Bigg\}. \end{aligned} \tag{3.319b}$$

(c) What are the units of λ', $\boldsymbol{J}'$, and g'?

The partition function of the $O(N)$ QNLσM in the large-N limit, defined by the limit $N\to\infty$ with g' held fixed, can now be evaluated by the method of steepest descent.

Saddle-point equations

We try the Ansatz

$$0<\lambda'(x)=\frac{m^2}{2g'},\qquad \boldsymbol{J}'(x)=\boldsymbol{H},\qquad x\in\mathbb{R}^{d+1}. \tag{3.320}$$

We then define the free energy per unit volume $|\Omega|$ and per channel (free energy divided by N)

$$\begin{aligned} \mathcal{F}_{\Omega,g'}(m^2,\boldsymbol{H}) :=& -\frac{m^2}{2g'}+\frac{1}{2|\Omega|}\operatorname{Tr}\left[\ln\left(-\partial_\mu^2+m^2\right)\right]-\frac{\boldsymbol{H}^2}{2|\Omega|}\int\limits_{\Omega}\frac{1}{0+m^2} \\ =& -\frac{m^2}{2g'}+\frac{1}{2|\Omega|}\operatorname{Tr}\left[\ln\left(-\partial_\mu^2+m^2\right)\right]-\frac{\boldsymbol{H}^2}{2m^2} \end{aligned} \tag{3.321}$$

or, equivalently, its Legendre transform with respect to the space-time constant staggered magnetic field $\boldsymbol{H}$

$$\mathcal{V}_{\Omega,g'}(m^2,\boldsymbol{M}) := \mathcal{F}_{\Omega,g'}(m^2,\boldsymbol{M}) - \boldsymbol{H}\cdot\left(\frac{\partial\mathcal{F}_{\Omega,g'}}{\partial\boldsymbol{H}}\right)(m^2,\boldsymbol{M}), \tag{3.322a}$$

$$\boldsymbol{M}(m^2,\boldsymbol{H}) := -\left(\frac{\partial\mathcal{F}_{\Omega,g'}}{\partial\boldsymbol{H}}\right)(m^2,\boldsymbol{H}). \tag{3.322b}$$

Exercise 2.1:

(a) Show that $(0 < m^2)$

$$\mathcal{V}_{\Omega,g'}(m^2,\boldsymbol{M}) = -\frac{m^2}{2\,g'} + \frac{1}{2\,|\Omega|}\mathrm{Tr}\,\left[\ln\left(-\partial_\mu^2 + m^2\right)\right] + \frac{1}{2}\,m^2\,\boldsymbol{M}^2. \tag{3.323}$$

The saddle-point equations are then

$$0 = 2\,g'\left(\frac{\partial\mathcal{V}_{\Omega,g'}}{\partial m^2}\right)(m^2,\boldsymbol{M}), \qquad 0 < m^2, \tag{3.324a}$$

$$0 = \left(\frac{\partial\mathcal{V}_{\Omega,g'}}{\partial\boldsymbol{M}}\right)(m^2,\boldsymbol{M}), \qquad 0 < m^2, \tag{3.324b}$$

i.e.,

$$1 = \frac{1}{|\Omega|}\,\mathrm{Tr}\left(\frac{g'}{-\partial_\mu^2 + m^2}\right) + g'\,\boldsymbol{M}^2, \qquad 0 < m^2, \tag{3.325a}$$

$$0 = m^2\,\boldsymbol{M}, \qquad 0 < m^2. \tag{3.325b}$$

(b) What does a non-vanishing solution for m^2 imply for the uniform and static staggered magnetization $\boldsymbol{M}$?

(c) What does a vanishing solution for m^2 imply for the uniform and static staggered magnetization $\boldsymbol{M}$?

We are going to treat the limit $\beta \to \infty$ of zero temperature first and then that of finite temperature, while the infinite volume limit $L \to \infty$ is here always understood.

Saddle-point equation with $m^2 > 0$ at vanishing temperature

Exercise 3.1:

(a) Assume that $m := +\sqrt{m^2} > 0$, $\beta = \infty$, and that $\boldsymbol{H} = 0$. Show that the saddle-point equations reduce then to

$$1 = \int\limits_{\mathbb{R}^{d+1}} \frac{\mathrm{d}^{d+1}k}{(2\pi)^{d+1}}\,\frac{g'}{k^2 + m^2}, \qquad m := +\sqrt{m^2} > 0. \tag{3.326}$$

Define the $(d+1)$-dimensional integral

$$I(d,g',m) := \int\limits_{\mathbb{R}^{d+1}} \frac{\mathrm{d}^{d+1}k}{(2\pi)^{d+1}}\,\frac{g'}{k^2 + m^2}, \qquad m := \sqrt{m^2} > 0. \tag{3.327}$$

(b) Under what conditions on $d = 0, 1, 2, \cdots$, and $m \geq 0$ is this integral well defined?
(c) Give a prescription to tame the UV divergences that preserves the formal $O(d+1)$ symmetry of the integrand. *Hint:* Use a momentum cut-off $\Lambda = \pi/\mathfrak{a}$ where $\mathfrak{a}$ is some underlying lattice spacing.
(d) Give a prescription to tame the UV divergences that break the formal $O(d+1)$ symmetry of the integrand down to the subgroup $O(d)$. *Hint:* use a momentum cut-off $\Lambda = \pi/\mathfrak{a}$ where $\mathfrak{a}$ is some underlying lattice spacing.
(e) In what way do the cases of $d = 0$ and $d = 1$ differ from the cases of $d \geq 2$?
(f) What dimension d is plagued with both IR and UV divergences? This dimension is called the lower critical dimension.

Solutions with $m^2 > 0$ and isotropic UV cut-off for frequencies and wave vectors

To preserve the $O(d+1)$ invariance of the integrand in the saddle-point equation, define the UV regulated $(d+1)$-dimensional integral

$$
\begin{aligned}
I(d, g', m, \pi/\mathfrak{a}) &:= \int\limits_{|k|<\pi/\mathfrak{a}} \frac{\mathrm{d}^{d+1}k}{(2\pi)^{d+1}} \frac{g'}{k^2+m^2} \\
&= C(d) \int\limits_0^{\pi/\mathfrak{a}} \mathrm{d}k k^d \frac{g'}{k^2+m^2} \\
&= C(d) \int\limits_0^{\pi/\mathfrak{a}} \mathrm{d}k k^d \left(\frac{m}{m}\right)^{d+1} \frac{g'}{k^2+m^2} \\
\underset{k=:my}{=} & C(d)\, m^{d-1} g' \int\limits_0^{\pi/(\mathfrak{a}\, m)} \mathrm{d}y \frac{y^d}{y^2+1}, \qquad m > 0. \qquad (3.328)
\end{aligned}
$$

The d-dependent numerical constant $C(d)$ is the area of the unit sphere $\left\{\boldsymbol{x} \in \mathbb{R}^{d+1} \middle| \boldsymbol{x}^2 = 1\right\}$ in $(d+1)$-dimensions divided by $(2\pi)^{d+1}$.

Exercise 4.1: Verify that

$$C(d) := \frac{1}{(2\pi)^{d+1}} \times \frac{2\pi^{(d+1)/2}}{\Gamma\big((d+1)/2\big)} = \frac{1}{2^d\, \pi^{(d+1)/2}\, \Gamma\big((d+1)/2\big)}, \qquad (3.329\text{a})$$

$$\Gamma(z) := \int\limits_0^\infty \mathrm{d}t\, e^{-t}\, t^{z-1}, \qquad \operatorname{Re} z > 0, \qquad (3.329\text{b})$$

$$\Gamma(1/2) = \sqrt{\pi}, \quad \Gamma(n) = (n-1)!, \ n = 0, 1, 2, \cdots, \quad \Gamma(x+1) = x\,\Gamma(x), \ x > 0. \qquad (3.329\text{c})$$

Verify that

$$d=0: \qquad \frac{2\pi^{(d+1)/2}}{\Gamma\big((d+1)/2\big)}=2, \qquad C(0)=\frac{1}{\pi}, \tag{3.330a}$$

$$d=1: \qquad \frac{2\pi^{(d+1)/2}}{\Gamma\big((d+1)/2\big)}=2\pi, \qquad C(1)=\frac{1}{2\pi}, \tag{3.330b}$$

$$d=2: \qquad \frac{2\pi^{(d+1)/2}}{\Gamma\big((d+1)/2\big)}=4\pi, \qquad C(2)=\frac{1}{2\pi^2}, \tag{3.330c}$$

$$d=3: \qquad \frac{2\pi^{(d+1)/2}}{\Gamma\big((d+1)/2\big)}=2\pi^2, \qquad C(3)=\frac{1}{8\pi^2}, \tag{3.330d}$$

$$d=4: \qquad \frac{2\pi^{(d+1)/2}}{\Gamma\big((d+1)/2\big)}=\frac{8\pi^2}{3}, \qquad C(4)=\frac{1}{12\pi^3}. \tag{3.330e}$$

Exercise 4.2:

(a) Is $I(d,g',m,\pi/\mathfrak{a})$ an increasing or decreasing function of m when holding all other variables fixed?
Verify the recursion relation

$$\int \mathrm{d}y\,\frac{y^d}{y^2+1}=\frac{y^{d-1}}{d-1}-\int \mathrm{d}y\,\frac{y^{d-2}}{y^2+1}, \qquad d=2,3,4,\cdots. \tag{3.331}$$

Verify that recursion relation (3.331) with the seeds

$$d=0: \qquad \int \mathrm{d}y\,\frac{1}{y^2+1}=\arctan y, \tag{3.332a}$$

$$d=1: \qquad \int \mathrm{d}y\,\frac{y}{y^2+1}=\frac{1}{2}\ln(1+y^2), \tag{3.332b}$$

yields

$$d=2: \qquad \int \mathrm{d}y\,\frac{y^2}{y^2+1}=y-\arctan y, \tag{3.332c}$$

$$d=3: \qquad \int \mathrm{d}y\,\frac{y^3}{y^2+1}=\frac{y^2}{2}-\frac{1}{2}\ln(1+y^2), \tag{3.332d}$$

$$d=4: \qquad \int \mathrm{d}y\,\frac{y^4}{y^2+1}=\frac{y^3}{3}-y+\arctan y. \tag{3.332e}$$

Verify that, for any $m>0$, this gives

$$d=0: \quad I(d,g',m,\pi/\mathfrak{a})=\frac{g'}{\pi m}\quad \arctan\left(\frac{\pi}{\mathfrak{a}m}\right), \tag{3.333a}$$

$$d=1: \quad I(d,g',m,\pi/\mathfrak{a})=\frac{g'}{4\pi}\quad \ln\left(1+\left(\frac{\pi}{\mathfrak{a}m}\right)^2\right), \tag{3.333b}$$

$$d=2: \quad I(d,g',m,\pi/\mathfrak{a})=\frac{mg'}{2\pi^2}\left[\left(\frac{\pi}{\mathfrak{a}m}\right)-\arctan\left(\frac{\pi}{\mathfrak{a}m}\right)\right], \tag{3.333c}$$

$$d=3: \quad I(d,g',m,\pi/\mathfrak{a})=\frac{m^2g'}{16\pi^2}\left[\left(\frac{\pi}{\mathfrak{a}m}\right)^2-\ln\left(1+\left(\frac{\pi}{\mathfrak{a}m}\right)^2\right)\right], \tag{3.333d}$$

$$d=4: \quad I(d,g',m,\pi/\mathfrak{a}) = \frac{m^3 g'}{36\pi^3}\left[\left(\frac{\pi}{\mathfrak{a}\,m}\right)^3 - 3\left(\frac{\pi}{\mathfrak{a}\,m}\right) + 3\arctan\left(\frac{\pi}{\mathfrak{a}\,m}\right)\right]. \tag{3.333e}$$

(b) What are the singularities, if any, of

- Eq. (3.333a) in the IR limit $m \downarrow 0$ and UV limit $\mathfrak{a} \downarrow 0$?
- Eq. (3.333b) in the IR limit $m \downarrow 0$ and UV limit $\mathfrak{a} \downarrow 0$?
- Eq. (3.333c) in the IR limit $m \downarrow 0$ and UV limit $\mathfrak{a} \downarrow 0$?
- Eq. (3.333d) in the IR limit $m \downarrow 0$ and UV limit $\mathfrak{a} \downarrow 0$?
- Eq. (3.333e) in the IR limit $m \downarrow 0$ and UV limit $\mathfrak{a} \downarrow 0$?

(c) For what dimensions can the saddle-point equation $1 = I(d,g',m,\pi/\mathfrak{a})$ admit a solution at $m=0$ for some critical value of g'?

(d) Show that, for sufficiently small values of m, the saddle-point equation $1 = I(d,g',m,\pi/\mathfrak{a})$ has the approximate solutions

$$d=0: \qquad m \approx \frac{g'}{2}, \tag{3.334a}$$

$$d=1: \qquad m \approx e^{-2\pi/g'}\left(\frac{\pi}{\mathfrak{a}}\right), \tag{3.334b}$$

when $d=0$ and $d=1$, respectively.

(e) Show that, in the IR limit $m \downarrow 0$, the saddle-point equation $1 = I(d,g',m=0,\pi/\mathfrak{a})$ implies that $g' = g'_{\mathrm{cr}}$ where

$$d=2: \qquad g'_{\mathrm{cr}} = 2\pi^2\left(\frac{\mathfrak{a}}{\pi}\right), \tag{3.335a}$$

$$d=3: \qquad g'_{\mathrm{cr}} = 16\pi^2\left(\frac{\mathfrak{a}}{\pi}\right)^2, \tag{3.335b}$$

$$d=4: \qquad g'_{\mathrm{cr}} = 36\pi^3\left(\frac{\mathfrak{a}}{\pi}\right)^3, \tag{3.335c}$$

when $d=2,3,4$, respectively.

(f) Explain why real-valued roots $m>0$ to the saddle-point equation $1 = I(d,g',m>0,\pi/\mathfrak{a})$ are only possible for $g' > g'_{\mathrm{cr}}$ when $d=2,3,4$.
To solve for these roots we express Eq. (3.333c), Eq. (3.333d), and Eq. (3.333e) in terms of Eq. (3.335a), Eq. (3.335b), and Eq. (3.335c), respectively,

$$d=2: \quad I(d,g',m,\pi/\mathfrak{a}) = \frac{g'}{g'_{\mathrm{cr}}}\left[1-\left(\frac{\mathfrak{a}\,m}{\pi}\right)\arctan\left(\frac{\pi}{\mathfrak{a}\,m}\right)\right], \tag{3.336a}$$

$$d=3: \quad I(d,g',m,\pi/\mathfrak{a}) = \frac{g'}{g'_{\mathrm{cr}}}\left[1-\left(\frac{\mathfrak{a}\,m}{\pi}\right)^2\ln\left(1+\left(\frac{\pi}{\mathfrak{a}\,m}\right)^2\right)\right], \tag{3.336b}$$

$$d=4: \quad I(d,g',m,\pi/\mathfrak{a}) = \frac{g'}{g'_{\mathrm{cr}}}\left[1-3\left(\frac{\mathfrak{a}\,m}{\pi}\right)^2+3\left(\frac{\mathfrak{a}\,m}{\pi}\right)^3\arctan\left(\frac{\pi}{\mathfrak{a}\,m}\right)\right]. \tag{3.336c}$$

(g) Show that solutions to the saddle-point equations

$$d=2: \qquad 1=\frac{g'}{g'_{\rm cr}}\left[1-\left(\frac{\mathfrak{a}\,m}{\pi}\right)\arctan\left(\frac{\pi}{\mathfrak{a}\,m}\right)\right], \tag{3.337a}$$

$$d=3: \qquad 1=\frac{g'}{g'_{\rm cr}}\left[1-\left(\frac{\mathfrak{a}\,m}{\pi}\right)^2\ln\left(1+\left(\frac{\pi}{\mathfrak{a}\,m}\right)^2\right)\right], \tag{3.337b}$$

$$d=4: \qquad 1=\frac{g'}{g'_{\rm cr}}\left\{1-3\left(\frac{\mathfrak{a}\,m}{\pi}\right)^2\left[1-\left(\frac{\mathfrak{a}\,m}{\pi}\right)\arctan\left(\frac{\pi}{\mathfrak{a}\,m}\right)\right]\right\}, \tag{3.337c}$$

in the limit

$$\frac{\pi}{\mathfrak{a}}\gg m \tag{3.338}$$

are

$$d=2: \qquad m\approx\frac{2}{\pi}\left(\frac{g'-g'_{\rm cr}}{g'}\right)\left(\frac{\pi}{\mathfrak{a}}\right), \tag{3.339a}$$

$$d=3: \qquad m\approx\left[\left(\frac{g'-g'_{\rm cr}}{g'}\right)\right]^{1/2}\left(\frac{\pi}{\mathfrak{a}}\right), \tag{3.339b}$$

$$d=4: \qquad m\approx\left[\frac{1}{3}\left(\frac{g'-g'_{\rm cr}}{g'}\right)\right]^{1/2}\left(\frac{\pi}{\mathfrak{a}}\right), \tag{3.339c}$$

provided

$$g'>g'_{\rm cr}. \tag{3.340}$$

The lessons that we learn are:

(i) The nature of the root m of the saddle-point equation is very different depending on whether the UV limit $\mathfrak{a}\downarrow 0$ is finite ($d=0$) or diverges in a power law fashion ($d\geq 2$) with the case $d=1$ being marginal.

(ii) There is a qualitative difference in the algebraic dependence of m on $(g'-g'_{\rm cr})/g'$ when $d=2$ compared with when $d\geq 4$, with $d=3$ being the marginal case. For $d=2$ the critical exponent ν defined by

$$m\sim\left(\frac{g'-g'_{\rm cr}}{g'_{\rm cr}}\right)^{\nu} \tag{3.341}$$

is $\nu=1$ while it is always $\nu=1/2$ for $d\geq 4$.

(h) Why is "$\sim$" used in Eq. (3.341) instead of "$\approx$" as in Eqs. (3.339)?

(i) Give one explanation for the following observation: "The case $d=3$ is marginal in that regard as the algebraic law with $\nu=1/2$ holds up to logarithmic corrections."

This observation has a simple explanation that follows from isolating the UV divergent contribution from the finite one in the saddle-point equation (3.326) in combination with dimensional analysis. To this end

$$\begin{aligned}1 &= g' \int \frac{\mathrm{d}^{d+1}k}{(2\pi)^{d+1}} \frac{1}{k^2+m^2} \\ &= g' \int \frac{\mathrm{d}^{d+1}k}{(2\pi)^{d+1}} \frac{1}{k^2} - g' \int \frac{\mathrm{d}^{d+1}k}{(2\pi)^{d+1}} \left(\frac{1}{k^2} - \frac{1}{k^2+m^2} \right) \\ &\equiv \frac{g'}{g'_{\mathrm{cr}}} - g' \int \frac{\mathrm{d}^{d+1}k}{(2\pi)^{d+1}} \frac{m^2}{k^2(k^2+m^2)},\end{aligned} \tag{3.342}$$

implies the relation (remember that g' has the dimension $\mathfrak{a}^{d-1}$)

$$1 = \frac{g'}{g'_{\mathrm{cr}}} - (\text{UV finite constant}) \times g' m^{d-1} \Leftrightarrow m \sim \left(\frac{g' - g'_{\mathrm{cr}}}{g'_{\mathrm{cr}}} \right)^{1/(d-1)}, \tag{3.343}$$

for $1 < d < 3$. When the integral on the right-hand side of Eq. (3.342) is no longer UV finite, i.e., when $d \geq 3$, application of the recursion relation (3.331) gives the mean field exponent $\nu = 1/2$.

(j) The naive continuum limit is obtained as $\mathfrak{a} \downarrow 0$. Deduce from Eqs. (3.339) how the continuum limit should really be understood when $d = 2, 3, 4$. *Hint:* Ask yourself how the limit $\mathfrak{a} \downarrow 0$ can be taken so that Eqs. (3.339) make sense.

Solutions with $m^2 > 0$ and with isotropic UV cut-off for the wave vectors only

We now give up the $O(d+1)$ invariance of the integrand. We define the UV regulated $(d+1)$-dimensional integral

$$I_\infty(d, g', m, \pi/\mathfrak{a}) := \int\limits_{\mathbb{R}} \frac{\mathrm{d}\varpi}{2\pi} \int\limits_{|\boldsymbol{k}|<\pi/\mathfrak{a}} \frac{\mathrm{d}^d \boldsymbol{k}}{(2\pi)^d} \frac{g'}{\varpi^2 + \boldsymbol{k}^2 + m^2}, \tag{3.344}$$

for $d = 1, 2, \cdots$ and $m > 0$.

Exercise 5.1:

(a) Why did we forbid the case $d = 0$?

(b) Introduce the notation

$$\omega(\boldsymbol{k}) := \sqrt{\boldsymbol{k}^2 + m^2}, \tag{3.345}$$

and show that

$$I_\infty(d, g', m, \pi/\mathfrak{a}) = \frac{g'}{2} \int\limits_{|\boldsymbol{k}|<\pi/\mathfrak{a}} \frac{\mathrm{d}^d \boldsymbol{k}}{(2\pi)^d} \frac{1}{\omega(\boldsymbol{k})} \tag{3.346a}$$

$$= C(d-1) m^{d-1} \frac{g'}{2} \int\limits_0^{\pi/(\mathfrak{a}\, m)} \mathrm{d}y\, y^{d-1} \frac{1}{\sqrt{y^2+1}}, \tag{3.346b}$$

with $d = 1, 2, \cdots$, and $m > 0$.

Assume the recursion relation

$$\int \mathrm{d}y \, \frac{y^{d-1}}{\sqrt{y^2+1}} = \frac{1}{d-1} y^{d-2}\sqrt{y^2+1} - \frac{d-2}{d-1}\int \mathrm{d}y \frac{y^{d-3}}{\sqrt{y^2+1}}, \qquad d=3,4,\cdots, \tag{3.347}$$

with the seeds

$$d=1: \qquad \int \mathrm{d}y \, \frac{1}{\sqrt{y^2+1}} = \ln\left(y+\sqrt{y^2+1}\right), \tag{3.348a}$$

$$d=2: \qquad \int \mathrm{d}y \, \frac{y}{\sqrt{y^2+1}} = \sqrt{y^2+1}. \tag{3.348b}$$

(c) Show that, under the assumption that m is small,

$$d=1: \qquad m \approx \left(\frac{\pi}{\mathfrak{a}}\right) e^{-2\pi/g'}, \tag{3.349a}$$

as in Eq. (3.334b).

(d) Show that the roots

$$d=2: \qquad g'_{\mathrm{cr}} = \ 4\pi\left(\frac{\mathfrak{a}}{\pi}\right), \tag{3.349b}$$

$$d=3: \qquad g'_{\mathrm{cr}} = \ 8\pi^2\left(\frac{\mathfrak{a}}{\pi}\right)^2, \tag{3.349c}$$

$$d=4: \qquad g'_{\mathrm{cr}} = 48\pi^2\left(\frac{\mathfrak{a}}{\pi}\right)^3, \tag{3.349d}$$

of the saddle-point equation $1 = I_\infty(d, g', m, \pi/\mathfrak{a})$ in the IR limit $m \downarrow 0$ are smaller than their counterparts Eq. (3.335a), Eq. (3.335b), and Eq. (3.335c) by the geometrical ratio

$$\frac{C(d)}{C(d-1)/2}, \qquad d=2,3,4, \tag{3.350}$$

respectively.

(e) Show that the roots of these saddle-point equations are, in the limit

$$\frac{\pi}{\mathfrak{a}} \gg m, \tag{3.351}$$

given by the elementary functions (the case $d=3$ is again special in view of the logarithmic correction)

$$d=2: \qquad m \approx \left(\frac{g'-g'_{\mathrm{cr}}}{g'}\right)\left(\frac{\pi}{\mathfrak{a}}\right), \tag{3.352a}$$

$$d=3: \qquad m \approx \left[\left(\frac{g'-g'_{\mathrm{cr}}}{g'}\right)\right]^{1/2}\left(\frac{\pi}{\mathfrak{a}}\right), \tag{3.352b}$$

$$d=4: \qquad m \approx \left[\frac{1}{2}\left(\frac{g'-g'_{\mathrm{cr}}}{g'}\right)\right]^{1/2}\left(\frac{\pi}{\mathfrak{a}}\right), \tag{3.352c}$$

provided

$$g' > g'_{\mathrm{cr}}. \tag{3.353}$$

(f) Equations (3.352) should be compared to Eqs. (3.339). In what ways do they differ and agree?

Saddle-point equation with $m^2 > 0$ at non-vanishing temperature

The case of non-vanishing temperature differs from the vanishing one through the nature of the trace in the saddle-point equation

$$\begin{aligned} 1 &= \frac{1}{|\Omega|}\mathrm{Tr}\left(\frac{g'}{-\partial_\mu^2 + m^2}\right) \\ &= \frac{1}{\beta}\sum_{\varpi_n = 2\pi n/\beta}^{n\in\mathbb{Z}} \int\limits_{|\boldsymbol{k}|<\pi/\mathfrak{a}} \frac{\mathrm{d}^d\boldsymbol{k}}{(2\pi)^d}\frac{g'}{\varpi_n^2 + \boldsymbol{k}^2 + m^2} \\ \text{Eq. (3.328)} &= \frac{1}{\beta}\sum_{\varpi_n = 2\pi n/\beta}^{n\in\mathbb{Z}} I\left(d-1, g', \sqrt{\varpi_n^2 + m^2}, \pi/\mathfrak{a}\right), \qquad m > 0. \end{aligned} \tag{3.354}$$

Exercise 6.1:

(a) Write

$$\begin{aligned} 1 = &\frac{1}{\beta}\int\limits_{|\boldsymbol{k}|<\pi/\mathfrak{a}} \frac{\mathrm{d}^d\boldsymbol{k}}{(2\pi)^d}\frac{g'}{\boldsymbol{k}^2 + m^2} \\ &+\frac{1}{\beta}\sum_{\varpi_n = 2\pi n/\beta}^{n\in\mathbb{Z}\setminus\{0\}} \int\limits_{|\boldsymbol{k}|<\pi/\mathfrak{a}} \frac{\mathrm{d}^d\boldsymbol{k}}{(2\pi)^d}\frac{g'}{\varpi_n^2 + \boldsymbol{k}^2 + m^2}, \qquad m > 0. \end{aligned} \tag{3.355}$$

With the help of Eq. (3.328),

$$\begin{aligned} 1 = &\frac{1}{\beta}I(d-1, g', m, \pi/\mathfrak{a}) \\ &+\frac{1}{\beta}\sum_{\varpi_n = 2\pi n/\beta}^{n\in\mathbb{Z}\setminus\{0\}} \int\limits_{|\boldsymbol{k}|<\pi/\mathfrak{a}} \frac{\mathrm{d}^d\boldsymbol{k}}{(2\pi)^d}\frac{g'}{\varpi_n^2 + \boldsymbol{k}^2 + m^2}, \qquad m > 0. \end{aligned} \tag{3.356}$$

Depending on d, when is the right-hand side of the saddle-point equation IR or UV singular and determine the singularities (pole-like or branch cut)?

(b) For what d in the large N-limit can there be long-range order at any finite temperature?

(c) Show that the summation over the Matsubara frequencies $\{\varpi_n = 2\pi n/\beta,\ n \in \mathbb{Z}\}$ in the saddle-point equation (3.354) can be performed exactly to yield the saddle-point equation

$$\begin{aligned} 1 = &\frac{g'}{2}\int\limits_{|\boldsymbol{k}|<\pi/\mathfrak{a}} \frac{\mathrm{d}^d\boldsymbol{k}}{(2\pi)^d}\frac{1}{\omega(\boldsymbol{k})}\coth\left(\beta\,\omega(\boldsymbol{k})/2\right) \\ \equiv &\, I(d, g', m, \pi/\mathfrak{a}, \beta), \qquad m > 0. \end{aligned} \tag{3.357}$$

(d) In what two limits can the integrand of $I(d, g', m, \pi/\mathfrak{a}, \beta)$ reduce to the integrand of $\frac{1}{\beta}I(d-1, g', m, \pi/\mathfrak{a})$ on the right-hand side of Eq. (3.356)?

(e) In what limit does one recover Eq. (3.346a) and its subsequent analysis.

(f) Explain how the existence of a root at $m = 0$ to the saddle-point equation (3.357) implies the existence of a positive-valued function $\beta_{\mathrm{cr}}(g')$ that is a monotonously increasing function of g'.

Exercise 6.2: We now consider the cases $d = 1$ and $d = 2$ of the saddle-point equation (3.357).

(a) Explain why the integral over wave vectors on the right-hand side of the saddle-point equation (3.357) can be dominated by the IR limit $\boldsymbol{k} \to 0$ when $d = 2$ but not in $d = 1$.
(b) Assume that for $d = 2$,

$$d = 2: \qquad \lim_{\beta\to\infty} \lim_{|\boldsymbol{k}|\to 0} \beta\,\omega(\boldsymbol{k}) \ll 1 \tag{3.358}$$

is self-consistent. Show under assumption (3.358) that the estimate

$$d = 2: \qquad 1 = \frac{1}{\beta} I(d-1, g', m, \pi/\mathfrak{a}) + \cdots, \tag{3.359}$$

to the saddle-point equation (3.357) implies

$$d = 2: \qquad m(g', \beta) \approx \left(\frac{\pi}{\mathfrak{a}}\right) e^{-2\pi\beta/g'}. \tag{3.360}$$

(c) Check the self-consistency of assumption (3.358) implied by Eq. (3.360).

Exercise 6.3: The analysis when $d = 2$ can be refined as the saddle-point equation (3.354) or, equivalently, (3.357) can be solved in *closed form* if it is modified so as to remain finite in the UV limit $\mathfrak{a} \downarrow 0$. To this end, we define the Pauli-Villars regularization ($m > 0$)

$$d = 2: \qquad 1 = \lim_{\mathfrak{a}\downarrow 0} \Big[I(d, g', m, \pi/\mathfrak{a}, \beta) - I(d, g', M, \pi/\mathfrak{a}, \beta) \Big], \tag{3.361}$$

to the saddle-point equation (3.354). The only condition made on the choice of the momentum scale M is that it is much larger than the temperature,

$$1 \ll \beta M. \tag{3.362}$$

(a) Show that the integral $I(d, g', m, \pi/\mathfrak{a}, \beta)$, Eq. (3.357), when $d = 2$ can be evaluated in closed form

$$\begin{aligned} I(d, g', m, \pi/\mathfrak{a}, \beta) &= I_\infty(d, g', m, \pi/\mathfrak{a}) + g' \int\limits_{|\boldsymbol{k}|<\pi/\mathfrak{a}} \frac{\mathrm{d}^d \boldsymbol{k}}{(2\pi)^d} \frac{1}{\omega(\boldsymbol{k})} \frac{1}{e^{\beta\,\omega(\boldsymbol{k})} - 1} \\ &= \frac{g'}{4\mathfrak{a}} \sqrt{1 + \left(\frac{\mathfrak{a}\, m}{\pi}\right)^2} \qquad \text{first order pole in } \mathfrak{a} \text{ with residue } g'/4 \text{ as } \mathfrak{a} \downarrow 0 \\ &\quad - \frac{g'}{2\pi\beta} \ln\Big(2\sinh(\beta m/2)\Big) \qquad \text{remains finite as } \mathfrak{a} \downarrow 0 \\ &\quad + \frac{g'}{2\pi\beta} \ln\left(1 - e^{-\beta\sqrt{(\pi/\mathfrak{a})^2 + m^2}}\right), \qquad \text{vanishes as } \mathfrak{a} \downarrow 0 \end{aligned} \tag{3.363}$$

and that the Pauli-Villars regularization of the saddle-point equation reduces to

$$
\begin{aligned}
1=\frac{g'}{2\pi\beta}\ln\left(\frac{\sinh(\beta M/2)}{\sinh(\beta m/2)}\right)\Longleftrightarrow m &= \frac{2}{\beta}\operatorname{arcsinh}\left(e^{-\frac{2\pi\beta}{g'}}\sinh\left(\beta M/2\right)\right)\\
&= \frac{2}{\beta}\operatorname{arcsinh}\frac{e^{-2\pi\beta\left(\frac{1}{g'}-\frac{1}{2\pi\beta}\ln\left(2\sinh(\beta M/2)\right)\right)}}{2}\\
&\equiv \frac{2}{\beta}\operatorname{arcsinh}\frac{e^{-2\pi\beta\left(\frac{1}{g'}-\frac{1}{g'_{\rm cr}}\right)}}{2}\\
{}_{\beta M\gg 1}&\approx \frac{2}{\beta}\operatorname{arcsinh}\frac{e^{-2\pi\beta\left(\frac{1}{g'}-\frac{M}{4\pi}\right)}}{2}.
\end{aligned}
\tag{3.364}
$$

(b) When $d=2$, show that the "critical coupling"

$$
g'_{\rm cr}:=\frac{2\pi\beta}{\ln\left(2\sinh\left(\beta M/2\right)\right)}
\tag{3.365a}
$$

only becomes truly independent of β when

$$
\beta M\gg 1,
\tag{3.365b}
$$

in which case it is given by

$$
g'_{\rm cr}=\frac{4\pi}{M}+\mathcal{O}\left(\beta^{-1}e^{-\beta M}\right).
\tag{3.365c}
$$

Equation (3.365c) should be compared with the one derived in Eq. (3.349b).

(c) When $d=2$, show that the asymptotic dependence of m on β depends on whether g' is equal, larger or smaller than $g'_{\rm cr}$,[22]

$$
m(g',\beta)=
\begin{cases}
\frac{2\ln(2^{-1}+\sqrt{2^{-2}+1})}{\beta}, & \text{if either } \beta\downarrow 0 \text{ or } g'=g'_{\rm cr},\\[2ex]
\frac{1}{\beta}e^{-2\pi\beta\left|\frac{1}{g'}-\frac{1}{g'_{\rm cr}}\right|}, & \text{if } \beta\to\infty \text{ and } g'<g'_{\rm cr},\\[2ex]
4\pi\left|\frac{1}{g'}-\frac{1}{g'_{\rm cr}}\right|, & \text{if } \beta\to\infty \text{ and } g'>g'_{\rm cr}.
\end{cases}
\tag{3.367}
$$

For any $d=1,2,\cdots$, the Laurent expansion of the coth in the integrand on the right-hand side of the saddle-point equation (3.357) is of course valid for all wave vectors in the high-temperature limit $\beta\downarrow 0$ and can be used to estimate the correlation length $1/m$ in the high-temperature limit.

[22] We here make use of

$$
\operatorname{arcsinh}x=\ln\left(x+\sqrt{1+x^2}\right),\qquad x\geq 0,
\tag{3.366a}
$$

$$
\operatorname{arcsinh}x=\ln x,\qquad x\gg 1,
\tag{3.366b}
$$

$$
\operatorname{arcsinh}x=x-\frac{1}{2\times 3}x^3\pm\cdots,\qquad 0<x\ll 1.
\tag{3.366c}
$$

Staggered spin susceptibility when $m^2 > 0$

Exercise 7.1: Explain why a finite root m to the saddle-point equation can be thought of as the inverse correlation length at the AF wave vector.

Antiferromagnetic transition (Néel) temperature when $d = 3, 4$

We have shown that the saddle-point equation (3.357) can be written as

$$\begin{aligned} 1 &= I(d, g', m, \pi/\mathfrak{a}, \beta) \\ &= I_{\infty}(d, g', m, \pi/\mathfrak{a}) + g' \int\limits_{|\boldsymbol{k}|<\pi/\mathfrak{a}} \frac{\mathrm{d}^d\boldsymbol{k}}{(2\pi)^d} \frac{1}{\omega(\boldsymbol{k})} \frac{1}{e^{\beta\,\omega(\boldsymbol{k})} - 1}. \end{aligned} \tag{3.368}$$

Exercise 8.1:

(a) What are the possible singularities as a function of d of the contribution

$$g' \int\limits_{|\boldsymbol{k}|<\pi/\mathfrak{a}} \frac{\mathrm{d}^d\boldsymbol{k}}{(2\pi)^d} \frac{1}{\omega(\boldsymbol{k})} \frac{1}{e^{\beta\,\omega(\boldsymbol{k})} - 1} \tag{3.369}$$

to the saddle-point equation that encodes the temperature dependence?

(b) Define the AF temperature that signals long-range order in the large N-limit of the $O(N)$ NLσM by writing down the proper limit of the saddle-point equation (3.368).

(c) Assuming that

$$\frac{\pi}{\mathfrak{a}} \gg m \tag{3.370}$$

show that

$$1 \approx \frac{g'}{g'_{\mathrm{cr}}} + \mathrm{const} \times \beta^{1-d}\, g', \qquad d = 3, 4, \cdots. \tag{3.371}$$

(d) Show here that the combination $\beta^{1-d} g'$ follows from dimensional analysis alone under the condition that the integral

$$\int\limits_{|\boldsymbol{k}|<\pi/\mathfrak{a}} \frac{\mathrm{d}^d\boldsymbol{k}}{(2\pi)^d} \frac{1}{\omega(\boldsymbol{k})} \frac{1}{e^{\beta\,\omega(\boldsymbol{k})} - 1} \tag{3.372}$$

is finite, i.e., $d = 3, 4, \cdots$.

(e) Use Eq. (3.371) to solve for the onset temperature for AF long-range order.

(f) Can there be AF long-range order when

$$g' > g'_{\mathrm{cr}}? \tag{3.373}$$

Elaborate for the case

$$g' < g'_{\mathrm{cr}}. \tag{3.374}$$

(g) Does the AF (Néel) temperature β_{AF}^{-1} increase or decrease as g' is decreased away from g'_{cr}?

3.8.5 *O(N) QNLσM with N > 2: One-loop RG using the Berezinskii-Blank parametrization of spin waves*

Introduction

We are ready to perform a one-loop RG analysis of the $O(N > 2)$ quantum non-linear σ model. We shall reproduce the results obtained in Ref. [39].

Definitions

The $O(N > 2)$ QNLσM is defined by the partition function

$$Z[\boldsymbol{h}] := \int\limits_{\mathbb{R}^N} \mathcal{D}[\boldsymbol{n}]\, \delta(\boldsymbol{n}^2 - 1)\, e^{-\left(S^{(1)}+S^{(2)}\right)}, \tag{3.375a}$$

where

$$S^{(1)} := \int \mathcal{L}^{(1)} \equiv \int\limits_0^\beta \mathrm{d}\tau \int\limits_{\mathfrak{a}}^{L} \mathrm{d}^d\boldsymbol{r}\, \frac{c}{2\,\mathfrak{a}^{d-1}\,g} \left(\partial_\mu \boldsymbol{n}\right)^2, \tag{3.375b}$$

and

$$S^{(2)} := \int \mathcal{L}^{(2)} \equiv -\int\limits_0^\beta \mathrm{d}\tau \int\limits_{\mathfrak{a}}^{L} \mathrm{d}^d\boldsymbol{r}\, \frac{c}{\mathfrak{a}^{d+1}}\, Z_h\, \boldsymbol{h}\cdot\boldsymbol{n}. \tag{3.375c}$$

Here, the lattice spacing $\mathfrak{a}$ plays the role of the microscopic ultraviolet (UV) cut-off, i.e., $\Lambda \sim 1/\mathfrak{a}$ that of an upper cut-off on momenta. The linear size L is the largest length scale of the problem. The derivative $\partial_\mu = (\partial_{c\tau}, \boldsymbol{\nabla})$ depends on the spin wave velocity c in the plane, $c = \mathfrak{a}\, g\, J/\hbar$, where J is a characteristic energy scale such as a Heisenberg exchange coupling. The dimensionless coupling constant g depends on the microscopic details of the intraplane interactions. The dimensionless background field $\boldsymbol{h}$, where $h = |\boldsymbol{h}|$, is the external source for a static staggered magnetic field conjugate to the planar antiferromagnetic order parameter of the underlying lattice model. It breaks the $O(N > 2)$ symmetry of Lagrangian (3.375b) down to $O(N-1)$ and as such acts as an infrared (IR) regulator. The dimensionless coupling Z_h is the field renormalization constant associated to $\boldsymbol{n}$. The use of the continuum limit is justified if we are after the physics on length scales much longer than $\mathfrak{a}$.

We assume that the $O(N > 2)$ QNLσM (3.375) with $\boldsymbol{h} = 0$ is renormalizable in that all the effects induced by the integration over the fast modes with momenta belonging to the infinitesimal momentum shell

$$\widetilde{\Lambda} \le |\boldsymbol{k}| \le \Lambda \tag{3.376}$$

can be absorbed in a redefinition of the two dimensionless coupling constants

$$g \qquad \text{and} \qquad t \equiv 1/(J\,\beta), \tag{3.377}$$

to leading order in g/c. We use the Berezinskii-Blank parametrization (3.294).

Exercise 1.1:

(a) Explain why g/t can be interpreted as the dimensionless slab thickness in the imaginary-time direction.

(b) With the same level of rigor as before show that the renormalized values $\widetilde{g}$ and $\widetilde{t}$ at the scale $\widetilde{\Lambda}$ induced by averaging over the fast modes ϕ with momenta in the infinitesimal momentum shell (3.376) are given by

$$\frac{\widetilde{g}}{\widetilde{t}} = \left(\frac{\Lambda}{\widetilde{\Lambda}}\right)^{-1} \frac{g}{t}, \tag{3.378}$$

$$\frac{1}{\widetilde{g}} = \frac{1}{g} \left(\frac{\Lambda}{\widetilde{\Lambda}}\right)^{d-1} \left(1 + \langle \phi^a \phi^b - \phi^2 \delta^{ab} \rangle \right), \tag{3.379}$$

or, equivalently, by

$$\frac{1}{\widetilde{g}} = \frac{1}{g} \left(\frac{\Lambda}{\widetilde{\Lambda}}\right)^{d-1} \left(1 + \langle \phi^a \phi^b - \phi^2 \delta^{ab} \rangle \right), \tag{3.380a}$$

$$\frac{1}{\widetilde{t}} = \frac{1}{t} \left(\frac{\Lambda}{\widetilde{\Lambda}}\right)^{d-2} \left(1 + \langle \phi^a \phi^b - \phi^2 \delta^{ab} \rangle \right). \tag{3.380b}$$

Here, when $d = 2$,

$$\langle \phi^a \phi^b \rangle = \delta^{ab} \ln \frac{\Lambda}{\widetilde{\Lambda}} \frac{g}{4\pi} \coth \frac{g}{2t}. \tag{3.381}$$

These RG equations were first derived by Chakravarty, Halperin, and Nelson in Ref. [39].

(c) Explain the sentence: *Equation (3.381) is the fluctuation bubble accounting for the quantum fluctuations induced by the fast modes with momenta in an infinitesimal momentum shell.* In what ways does it differ from its counterpart in Eq. (3.305b).

(d) With the same level of rigor as before, show that the wave-function renormalization is

$$\frac{\widetilde{Z}_h}{Z_h} = \left(1 - \frac{1}{2} \langle \phi^2 \rangle + \cdots \right). \tag{3.382}$$

(e) Use the notation

$$\ell := \ln \frac{\Lambda}{\widetilde{\Lambda}} \tag{3.383}$$

and turn Eqs. (3.380) into two coupled differential equations for the rate of change of g and t with ℓ.

(f) Consider the limits $g/2t \to 0$ and $g/2t \to \infty$ of the coupled equations

$$\frac{\mathrm{d}\, g}{\mathrm{d}\, \ell} = \beta_1(g, t), \qquad \frac{\mathrm{d}\, t}{\mathrm{d}\, \ell} = \beta_2(g, t), \tag{3.384}$$

that you have derived. One of these limits is the low-temperature limit, the other the high-temperature limit. To which RG equations, that you have already encountered, reduce the flows (3.384) in these two limits?

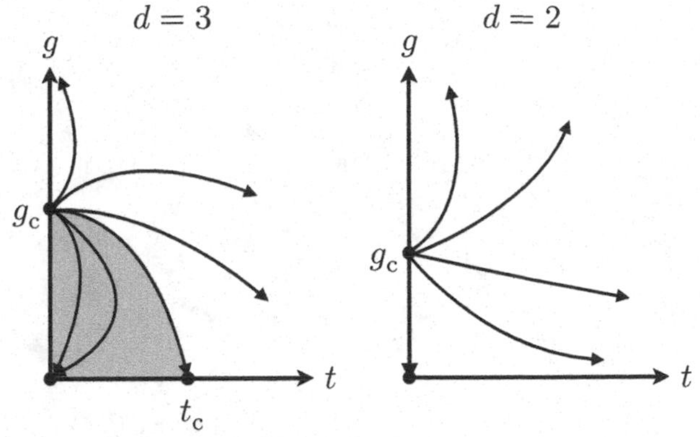

Fig. 3.7 Phase diagram of quantum antiferromagnets in three- and two-dimensional position space.

(g) The phase diagram that follows from the RG flows (3.384) is shown in Fig. 3.7

(i) What is the meaning of the black circles?
(ii) Why is g_c represented smaller for $d = 2$ than for $d = 3$?
(iii) What is the meaning of the shaded region when $d = 3$?
(iv) What is the meaning of the complementary region to the shaded one when $d = 3$?
(v) What is the meaning of the line that joins g_c to t_c when $d = 3$?
(vi) Why are there arrows on the RG trajectories and how can they be deduced from the RG equations?

Chapter 4

Kosterlitz-Thouless transition

Outline

The classical two-dimensional XY ($2d$–XY) model is defined. The quasi-long-range ordered phase in the continuum spin-wave approximation derived in section 3.3.2 is shown to be unstable to the deconfining transition of topological defects, i.e., vortices, that drives the model into a high-temperature paramagnetic phase [40–42]. This transition, which we shall call the *Kosterlitz-Thouless (KT) transition*, although the terminology *Berezinskii-Kosterlitz-Thouless (BKT) transition* is also used in the literature, is studied within a perturbative renormalization-group analysis. To set up the renormalization-group analysis, the vortex sector of the $2d-XY$ is first shown to be equivalent to a $2d$–Coulomb gas [40]. In turn the $2d$–Coulomb gas is shown to be equivalent to the $2d$–Sine-Gordon model. The renormalization-group analysis is made within the $2d$–Sine-Gordon model.

4.1 Introduction

The classical two-dimensional XY ($2d - XY$) model is extremely important, both conceptually and for its applicability to materials. Conceptually, it is perhaps the simplest example in which a phase transition of a topological nature takes place. From a practical point of view, the classical $2d - XY$ model is relevant to two-dimensional superconductivity, two-dimensional superfluidity, two-dimensional arrays of Josephson junctions, the melting of crystalline thin films, roughening transition of crystalline surfaces, one-dimensional metals, and quantum magnetism in one-space dimension. This list is not exhaustive.

4.2 Classical two-dimensional XY model

Consider a *square* lattice with the lattice spacing a embedded in two-dimensional Euclidean space. Assign to each site $i \in \mathbb{Z}^2$ an *angle* $\phi_i \in S^1$, where S^1 is the one-sphere, i.e., the circle in the complex plane (see Fig. 4.1). The partition function of

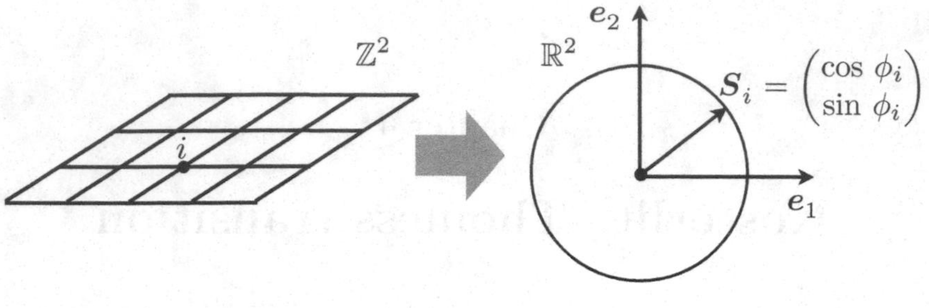

Fig. 4.1 Heisenberg model on a square lattice with the lattice spacing $\mathfrak{a}$ and with the $O(2) \cong U(1)$ internal symmetry.

the classical XY model is defined by

$$Z_{XY} := \int \mathcal{D}[\phi] \, \exp\left(-S_{XY}\right), \tag{4.1a}$$

with the action

$$S_{XY} := K \sum_{\langle ij \rangle} \left[1 - \cos\left(\phi_i - \phi_j\right)\right], \tag{4.1b}$$

and the dimensionless coupling constant K

$$K := \beta J, \qquad \beta = 1/(k_{\mathrm{B}} T). \tag{4.1c}$$

We shall set the Boltzmann constant $k_{\mathrm{B}} = 1.38065 \times 10^{-23}$ J/K to unity, in which case β is the inverse temperature. The dimensionless coupling constant K is the reduced ferromagnetic exchange coupling between nearest-neighbor planar spins

$$\begin{pmatrix} \cos\phi_i \\ \sin\phi_i \end{pmatrix} \quad \text{and} \quad \begin{pmatrix} \cos\phi_j \\ \sin\phi_j \end{pmatrix}, \tag{4.2}$$

i.e., for any directed pair of nearest-neighbor sites $\langle ij \rangle$ one assigns the link energy

$$J \left[1 - \begin{pmatrix} \cos\phi_i & \sin\phi_i \end{pmatrix} \begin{pmatrix} \cos\phi_j \\ \sin\phi_j \end{pmatrix}\right] = J \left[1 - \cos\left(\phi_i - \phi_j\right)\right], \tag{4.3}$$

and the Boltzmann weight

$$\exp\Big(- \beta J \left[1 - \cos\left(\phi_i - \phi_j\right)\right]\Big), \tag{4.4}$$

respectively. The reduced ferromagnetic exchange coupling K can be varied by changing the ferromagnetic spin stiffness $J > 0$ or by changing the inverse temperature β. Here, we will use the inverse temperature to vary K. The classical $2d$–XY model (4.1) thus describes a classical isotropic ferromagnet for planar spins located on a square lattice with the lattice spacing $\mathfrak{a}$.

The link energy (4.3) is minimized by choosing $\phi_i = \phi_j$ (see Fig. 4.2). Moreover, minimization of the sum over the four link energies, whereby the four links define an elementary square unit cell (plaquette) of the lattice, is also achieved by choosing all angles at the four corners of the plaquette equal. The system is thus not frustrated

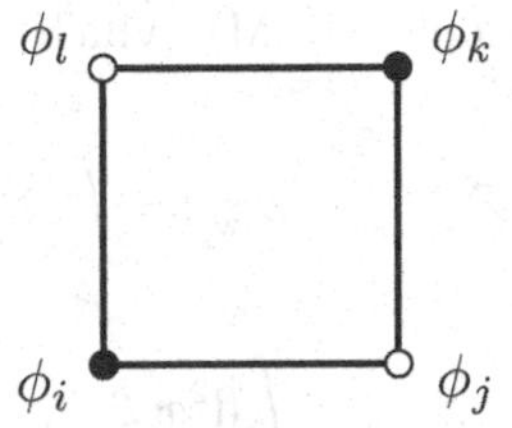

Fig. 4.2 An elementary plaquette of the square lattice. The sublattice structure is made explicit by drawing lattice sites with white and black circles. The four angles ϕ_i, ϕ_j, ϕ_k, ϕ_l are all equal in the ferromagnetic state.

and the configuration of angles with the lowest energy has all angles equal. This is the *ferromagnetic state*

$$\phi_i = \phi_{\text{ferro}}, \qquad \forall i \in \mathbb{Z}^2. \tag{4.5}$$

Since ϕ_{ferro} can be any real number between 0 and 2π, the ground state of the XY model spontaneously breaks the global invariance of the XY model under the continuous $U(1)$ transformation

$$\phi_i = \tilde{\phi}_i + \alpha, \qquad \forall i \in \mathbb{Z}^2, \qquad 0 \leq \alpha < 2\pi. \tag{4.6}$$

Thermal fluctuations disturb the long-range order of the ground state (4.5). In fact, it is reasonable to anticipate that the long-range ferromagnetic order disappears in the limit of infinite temperature $\beta \to 0$, i.e., $K \to 0$, as all configurations of angles $\{\phi_i\}_{i\in\mathbb{Z}^2}$ are equally weighted in the partition function in this limit. Since there are "many more" configurations of angles that deviate strongly from the ferromagnetic state than those that do not, one is tempted to identify the limit $K \to 0$ with a paramagnetic phase. High-temperature expansions [expansions of the Boltzmann weight (4.4) in powers of K] are in fact consistent with the existence of a *paramagnetic phase* for a finite temperature range in the vicinity of $\beta = 0$. Hence, there must be some type of phase transition at some critical value K_{c} of the dimensionless coupling K that separates a low-temperature phase related to the ferromagnetic ground state from the paramagnetic high-temperature phase. The questions to be answered are:

(1) Is $K_{\text{c}} < \infty$?
(2) What is the nature of the low-temperature phase, how does it connect to the ferromagnetic long-range order at zero temperature?
(3) What is the nature of the phase transition between the low- and high-temperature phases?

A first attempt to provide an answer to these questions is to study the *stability* of the ferromagnetic long-range order at zero temperature in the presence of spin-wave fluctuations. In the continuum spin-wave approximation, the partition function Z_{XY}, which is defined on the lattice, is replaced by the partition function for the

2*d* $O(2)$ Non-Linear Sigma Model (NLσM), which is defined in the continuum by the partition function

$$Z_{XY} \to Z_{\mathrm{sw}}, \qquad Z_{\mathrm{sw}} := \int \mathcal{D}[\phi]\, e^{-S_{\mathrm{sw}}}, \tag{4.7a}$$

with the action

$$S_{\mathrm{sw}} := \int \mathrm{d}^2\boldsymbol{x}\, \mathcal{L}_{\mathrm{sw}}, \tag{4.7b}$$

and the Lagrangian density

$$\mathcal{L}_{\mathrm{sw}} := \frac{K}{2} \left(\partial_\mu \phi\right)^2. \tag{4.7c}$$

We have performed the *continuum-spin-wave approximation* that is nothing but the naive continuum limit

$$\begin{aligned} S_{XY} &= K \sum_{\langle ij \rangle} \left[1 - \cos\left(\phi_i - \phi_j\right)\right] \\ &= K \sum_{\langle ij \rangle} \mathfrak{a}^2 \left[\frac{1}{2}\left(\frac{\phi_i - \phi_j}{\mathfrak{a}}\right)^2 - \frac{\mathfrak{a}^2}{4!}\left(\frac{\phi_i - \phi_j}{\mathfrak{a}}\right)^4 + \cdots\right] \\ &= K \sum_{\langle ij \rangle} \mathfrak{a}^2 \left\{\frac{1}{2}\left(\frac{\phi_i - \phi_j}{\mathfrak{a}}\right)^2 + \mathcal{O}\left[\mathfrak{a}^2 \left(\frac{\phi_i - \phi_j}{\mathfrak{a}}\right)^4\right]\right\} \\ &\to K \int \mathrm{d}^2\boldsymbol{x} \left\{\frac{1}{2}\left(\partial_\mu\phi\right)^2 + \mathcal{O}\left[\left(\partial_\mu \phi\right)^4\right]\right\}. \end{aligned} \tag{4.8}$$

The second line is called a gradient expansion, i.e., it is an expansion in inverse powers of the lattice spacing $\mathfrak{a}$. The *lattice-spin-wave approximation* truncates the gradient expansion to the Gaussian order as is done on the third line. Before performing the gradient expansion, the integration measure

$$\mathcal{D}[\phi] := \prod_{i \in \mathbb{Z}^2} \mathrm{d}\phi_i, \qquad 0 \leq \phi_i < 2\pi, \tag{4.9}$$

encodes the compact nature of the angles ϕ_i with $i \in \mathbb{Z}^2$, i.e., the fact that the link interaction energy is, through $\cos(\phi_i - \phi_j)$, a periodic function of ϕ_i (ϕ_j) with periodicity 2π. Any truncation of the gradient expansion destroys this periodicity. To make sense of an approximation by which the gradient expansion is truncated to finite order, the integration measure must be modified accordingly,

$$\mathcal{D}[\phi] \to \prod_{i \in \mathbb{Z}^2} \mathrm{d}\phi_i, \qquad \phi_i \in \mathbb{R}. \tag{4.10}$$

Often, the spin-wave approximation is understood to be the continuum limit on the last line of Eq. (4.8). The gradient expansion assumes smoothness of the spin configuration $\{\phi_i\}_{i \in \mathbb{Z}^2}$ in that $\phi_i - \phi_j$ is small on the scale of the lattice spacing $\mathfrak{a}$, i.e., one can replace $(\phi_i - \phi_j)/\mathfrak{a}$ by the function $\partial_\mu \phi$. The corresponding integration measure

$$\mathcal{D}[\phi] := \prod_{\boldsymbol{x} \in \mathbb{R}^2} \mathrm{d}\phi(\boldsymbol{x}), \qquad \phi(\boldsymbol{x}) \in \mathbb{R}, \tag{4.11}$$

is then restricted to smooth, i.e., differentiable, *single-valued* functions

$$\phi : \mathbb{R}^2 \to \mathbb{R} \tag{4.12}$$

that vanish at infinity. The main assumption made in the spin-wave approximation, be it before or after the continuum limit, is, as it turns out, not so much the replacement of a non-linear theory by a Gaussian theory than the neglect of the compactness of $\phi_i \in S^1$ in the $2d$–XY model.

We have already studied the $2d$ $O(2)$ NLσM in section 3.3.2. There, we saw that the $(2n)$-point function

$$\left\langle e^{+\mathrm{i}\phi(\boldsymbol{x}_1)+\cdots+\mathrm{i}\phi(\boldsymbol{x}_n)-\mathrm{i}\phi(\boldsymbol{y}_1)-\cdots-\mathrm{i}\phi(\boldsymbol{y}_n)}\right\rangle_{Z_{\mathrm{sw}}} = \mathfrak{a}^{\frac{2n}{4\pi K}} \left(\frac{\prod\limits_{1\leq i<j\leq n} |\boldsymbol{x}_i - \boldsymbol{x}_j||\boldsymbol{y}_i - \boldsymbol{y}_j|}{\prod\limits_{i=1}^{n}\prod\limits_{j=1}^{n} |\boldsymbol{x}_i - \boldsymbol{y}_j|} \right)^{\frac{1}{2\pi K}} \tag{4.13}$$

is algebraic at *any finite temperature*. Within the spin-wave approximation, ferromagnetic long-range order is downgraded to quasi-long-range order (algebraic order) at any finite temperature, i.e., all spin-spin correlation functions decay algebraically fast for large separations. The spin-wave approximation thus captures an instability of the ferromagnetic long-range order at vanishing temperature. This instability is an example of the *Mermin-Wagner theorem*. However, the spin-wave approximation is deficient in that it fails to predict a phase other than one with quasi-long-range order.[1]

The failure at high temperatures of the spin-wave approximation is rooted in that it only allows for small and smooth deviations (gradient expansion) about the ferromagnetic ordered state. In particular, it is ruled out that the field ϕ be singular at some isolated point in the sense that it is everywhere *single valued*,

$$\int\limits_{\gamma_{\boldsymbol{x}}} \mathrm{d}\phi = 0 \tag{4.14}$$

for any closed path that encloses $\boldsymbol{x}$ with $\boldsymbol{x}$ arbitrarily chosen in $\mathbb{R}^2$. In the $2d$–XY model, only the spin

$$\begin{pmatrix} \cos\phi_i \\ \sin\phi_i \end{pmatrix} \sim e^{\mathrm{i}\phi_i} \tag{4.15}$$

must be single valued. Hence, ϕ_i and $\phi_i + 2\pi$ are physically indistinguishable. There is thus no a priori reason to demand that, after the continuum limit has been taken, the field

$$\phi : \mathbb{R}^2 \to \mathbb{R} \tag{4.16}$$

is single valued.

[1] Within the spin-wave approximation, we are free to absorb K into a redefinition of the non-compact lattice degree of freedom ϕ_i on the lattice (or $\phi(\boldsymbol{x})$ in the continuum). This rescaling shows up when calculating correlation functions. This rescaling also tells us that if we know how to solve the theory for, say, $K = 1$, we know how to solve the theory for all K's, as is expressed by Eq. (4.13). On the other hand, the infinitesimal rescaling of $\phi_i = 2\pi - \epsilon$ into $\phi_i = 2\pi + \epsilon$ is dramatic since it turns a large angle, $2\pi - \epsilon$, into a small one, ϵ, in the $2d$–XY model. The $2d$–XY model is thus certainly not scale invariant.

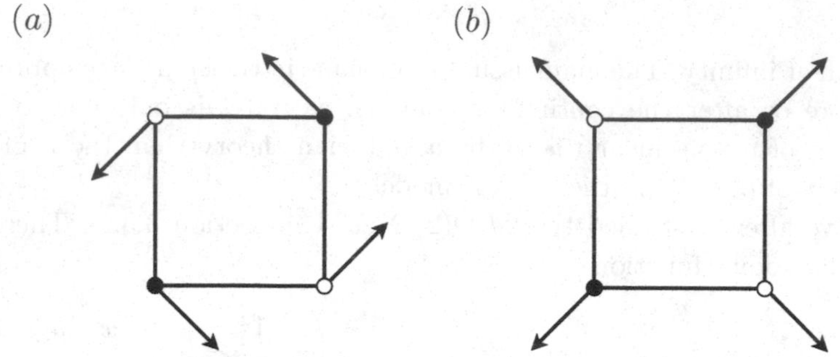

Fig. 4.3 (a) An elementary plaquette supporting a configuration of four planar spins of unit length that would match the magnetic field induced by a thin current-carrying wire threading the center of the plaquette in three-dimensional electrostatics if two-dimensional space is embedded in three-dimensional space. (b) An elementary plaquette supporting a configuration of four planar spins of unit length that would match the electric field induced by a thin charged wire threading the center of the plaquette in three-dimensional electrostatics if two-dimensional space is embedded in three-dimensional space. The representation of vortices in the continuum given in the text corresponds to configurations of spins that vary very little on the lattice scale but that behave like the spin configuration in (a) along closed path that are much longer than the perimeter of an elementary plaquette.

An example of a multivalued field is the function

$$\begin{aligned}&\Theta:\mathbb{R}^2\setminus\{\boldsymbol{x}_1,\cdots,\boldsymbol{x}_M\}\to\mathbb{R},\\ &\boldsymbol{x}\to\Theta(\boldsymbol{x})=\sum_{i=1}^{M}m_i\arctan\left(\frac{(\boldsymbol{x}-\boldsymbol{x}_i)_2}{(\boldsymbol{x}-\boldsymbol{x}_i)_1}\right).\end{aligned}\tag{4.17}$$

The condition that $\exp\left(\mathrm{i}\Theta(\boldsymbol{x})\right)$ be single valued demands that

$$m_i\in\mathbb{Z},\qquad i=1,\cdots,M.\tag{4.18}$$

The field Θ is smooth everywhere except at the isolated points $\boldsymbol{x}_1,\cdots,\boldsymbol{x}_M$. Around any $\boldsymbol{x}_i$, the open disc U_i can be chosen sufficiently small so that counterclockwise integration about its boundary ∂U_i yields

$$\oint\limits_{\partial U_i}\mathrm{d}\Theta=2\pi m_i.\tag{4.19}$$

Correspondingly, each singularity $\boldsymbol{x}_i$ is said to carry *vorticity or charge* m_i. The vorticity m_i counts how many times Θ *winds* about $\boldsymbol{x}_i$. Figure 4.3(a) depicts the configuration

$$\begin{pmatrix}\cos\left(\Theta+\frac{\pi}{2}\right)\\ \sin\left(\Theta+\frac{\pi}{2}\right)\end{pmatrix}\sim e^{\mathrm{i}\left(\Theta+\frac{\pi}{2}\right)}.\tag{4.20a}$$

Figure 4.3(b) depicts the configuration

$$\begin{pmatrix}\cos\Theta\\ \sin\Theta\end{pmatrix}\sim e^{\mathrm{i}\Theta}.\tag{4.20b}$$

The dimensionless energy stored by Θ in a finite region $\Omega \subset \mathbb{R}^2$ of linear size L that contains all the singularities of Θ is

$$S_{\mathrm{Cb}}[\Theta, K, L] := \frac{K}{2} \int_{\Omega} \mathrm{d}^2\boldsymbol{x}\ (\partial_\mu \Theta)^2, \tag{4.21}$$

where the label Cb for the action stands for Coulomb. Observe that this dimensionless energy is form invariant under the rescaling

$$\boldsymbol{x} = \kappa \boldsymbol{x}', \qquad L = \kappa L', \qquad 0 < \kappa \in \mathbb{R}, \tag{4.22}$$

i.e.,

$$S_{\mathrm{Cb}}[\Theta, K, L] = S_{\mathrm{Cb}}[\Theta, K, L']. \tag{4.23}$$

To proceed with the evaluation of the dimensionless energy stored by Θ, observe that the field

$$\begin{aligned} &\widetilde{\Theta} : \mathbb{R}^2 \setminus \{\boldsymbol{x}_1, \cdots, \boldsymbol{x}_M\} \to \mathbb{R}, \\ &\boldsymbol{x} \to \widetilde{\Theta}(\boldsymbol{x}) = - \sum_{i=1}^{M} m_i \ln \left| \frac{\boldsymbol{x} - \boldsymbol{x}_i}{\ell} \right|, \end{aligned} \tag{4.24}$$

where ℓ is some arbitrarily chosen length scale, is related to Θ by the Cauchy-Riemann relation[2]

$$\partial_1 \Theta = +\partial_2 \widetilde{\Theta}, \qquad \partial_2 \Theta = -\partial_1 \widetilde{\Theta}. \tag{4.32}$$

The field $\widetilde{\Theta}$ is called the dual field to Θ. With the notation

$$\widetilde{\partial}_\mu := \epsilon_{\mu\nu} \partial_\nu, \qquad \epsilon_{\mu\nu} = -\epsilon_{\nu\mu}, \qquad \epsilon_{12} = 1, \qquad \partial_\mu \Theta = +\widetilde{\partial}_\mu \widetilde{\Theta}, \qquad \partial_\mu \widetilde{\Theta} = -\widetilde{\partial}_\mu \Theta, \tag{4.33}$$

[2] Let

$$z = x + \mathrm{i}y = |z| e^{\mathrm{i}\varphi} \in \mathbb{C}, \qquad |z| := \sqrt{x^2 + y^2} \equiv r, \qquad \arg(z) := \arctan\left(\frac{y}{x}\right) \equiv \varphi. \tag{4.25}$$

If $\arg(z)$ and $\ln|z|$ are understood as real-valued functions on $\mathbb{C} \setminus \{0\}$, then they obey the Cauchy-Riemann conditions

$$\begin{aligned} (\partial_x \arg)(z) &= \frac{1}{1 + \left(\frac{y}{x}\right)^2} (-) \frac{y}{x^2} = -\frac{y}{r^2} = -(\partial_y \ln|z|), \\ (\partial_y \arg)(z) &= \frac{1}{1 + \left(\frac{y}{x}\right)^2} (+) \frac{1}{x} = +\frac{x}{r^2} = +(\partial_x \ln|z|). \end{aligned} \tag{4.26}$$

The complex-valued function $\log(z) := \ln|z| + \mathrm{i}\arg(z)$ is thus analytic, for $\ln|z|$ and $\arg(z)$ are a pair of conjugate single-valued harmonic functions

$$0 = (\partial_x^2 + \partial_y^2) \ln|z| = (\partial_x^2 + \partial_y^2) \arg(z), \tag{4.27}$$

on the Riemann sheet $0 \leq \arg(z) < 2\pi$ of $\mathbb{C} \setminus \{0\}$. By definition

$$\oint \mathrm{d}\arg(z) = 2\pi \tag{4.28}$$

the property

$$\widetilde{\partial}^2_\mu = \epsilon_{\mu\nu}\epsilon_{\mu\lambda}\partial_\nu\partial_\lambda = \partial^2_\mu, \tag{4.34}$$

and the two-dimensional Green function

$$-\left(\partial^2_\mu\right)^{-1}(\boldsymbol{x},\boldsymbol{y}) = -\frac{1}{2\pi}\ln\left|\frac{\boldsymbol{x}-\boldsymbol{y}}{\ell}\right| \tag{4.35}$$

of Laplace operator $-\partial^2_\mu \equiv -\Delta$, the dimensionless energy stored by Θ becomes

$$\begin{aligned} S_{\mathrm{Cb}}[\Theta,K,L] &= \frac{K}{2}\int_\Omega \mathrm{d}^2\boldsymbol{x}\,\left(\partial_\mu\Theta\right)^2 \\ &= \frac{K}{2}\int_\Omega \mathrm{d}^2\boldsymbol{x}\,\left(\partial_\mu\widetilde{\Theta}\right)^2. \end{aligned} \tag{4.36}$$

By orienting the infinitesimal surface element $\mathrm{d}^2\boldsymbol{x}$ according to $\mathrm{d}^2\boldsymbol{x} \to \mathrm{d}x_1 \wedge \mathrm{d}x_2 = -\mathrm{d}x_2 \wedge \mathrm{d}x_1$ and in combination with partial integration,

$$S_{\mathrm{Cb}}[\Theta,K,L] = \frac{K}{2}\int_\Omega \mathrm{d}x_1\wedge\mathrm{d}x_2\,\partial_\mu\left[\widetilde{\Theta}\left(\partial_\mu\widetilde{\Theta}\right)\right] + \frac{K}{2}\int_\Omega \mathrm{d}^2\boldsymbol{x}\,\widetilde{\Theta}\left(-\partial^2_\mu\widetilde{\Theta}\right). \tag{4.37}$$

With the help of Eq. (4.33),

$$S_{\mathrm{Cb}}[\Theta,K,L] = \frac{K}{2}\int_\Omega \mathrm{d}x_1\wedge\mathrm{d}x_2\,\partial_\mu\left[\widetilde{\Theta}\left(-\widetilde{\partial}_\mu\Theta\right)\right] + \frac{K}{2}\int_\Omega \mathrm{d}^2\boldsymbol{x}\,\widetilde{\Theta}\left(-\partial^2_\mu\widetilde{\Theta}\right). \tag{4.38}$$

Application of Stokes theorem then gives,

$$S_{\mathrm{Cb}}[\Theta,K,L] = \frac{K}{2}\oint_{\partial\Omega} \mathrm{d}x_\mu\,\widetilde{\Theta}\left(-\partial_\mu\Theta\right) + \frac{K}{2}\int_\Omega \mathrm{d}^2\boldsymbol{x}\,\widetilde{\Theta}(\boldsymbol{x})\left[(-)^2 2\pi\sum_{l=1}^{M} m_l\delta(\boldsymbol{x}-\boldsymbol{x}_l)\right]. \tag{4.39}$$

Finally, theorem 20 from section 4.6.2 of Ref. [43] delivers

$$\begin{aligned} S_{\mathrm{Cb}}[\Theta,K,L] =&\ (2\pi)^2\frac{K}{2}\frac{\left(\sum_{i=1}^{M} m_i\right)^2}{2\pi}\ln\left(\frac{L}{\ell}\right) \\ &+(2\pi)^2\frac{K}{2}\sum_{k,l=1}^{M}\left(-\frac{m_k m_l}{2\pi}\ln\left|\frac{\boldsymbol{x}_k-\boldsymbol{x}_l}{\ell}\right|\right). \end{aligned} \tag{4.40}$$

---

for any closed curve winding once counterclockwise around the origin. Observe that, whereas

$$\frac{1}{2\pi}\left(\partial^2_x + \partial^2_y\right)\ln|z| = \delta(|z|), \tag{4.29}$$

the singularity of $\arg(z)$ at the origin is not manifest in

$$\frac{1}{2\pi}\left(\partial^2_x + \partial^2_y\right)\arg(z) = 0, \tag{4.30}$$

it is manifest in

$$\frac{1}{2\pi}\left(\partial_x\partial_y - \partial_y\partial_x\right)\arg(z) = \delta(|z|). \tag{4.31}$$

Correspondingly, the vector field $\boldsymbol{\partial}\arg(z)$ can be interpreted as the magnetic field depicted in Fig. 4.3(a), while the vector field $\boldsymbol{\partial}\ln|z|$ can be interpreted as the electric field depicted in Fig. 4.3(b).

As expected from the form invariance (4.23) of the dimensionless energy stored by Θ under length rescaling, the dependence on ℓ of the boundary contribution cancels the dependence on ℓ of the bulk contribution on the last line of Eq. (4.40). The boundary term is the dimensionless self-energy of a single charge $\sum_i m_i$ concentrated in a radius ℓ about the origin and guarantees that the dimensionless energy stored by Θ is strictly positive. The short-distance behavior of the Green function is regulated by

$$\ln\left|\frac{\boldsymbol{x}-\boldsymbol{y}}{\ell}\right| = \ln\left(\frac{\mathfrak{a}}{\ell}\right), \tag{4.41}$$

when $\boldsymbol{x}$ is within a lattice spacing away from $\boldsymbol{y}$. As explained in sections 3.3.2 and 3.4, the choice for the ultraviolet regularization of the Green function is arbitrary from the point of view of field theory. Within the $2d$–XY model, the Green function is unambiguously defined all the way to the lattice spacing, the characteristic length at which scale invariance is lost and the Green function is not well approximated by the logarithm anymore. The same is also true of the spin-wave approximation on the lattice, i.e., the bilinear action on the penultimate line of Eq. (4.8).

The dimensionless energy stored in Θ is finite in the thermodynamic (infrared) limit $L \to \infty$ *if and only if charge neutrality holds,*

$$\sum_{i=1}^{M} m_i = 0. \tag{4.42}$$

For example, a single-vortex configuration carrying charge m stores the energy

$$\pi m^2 J \ln\left(\frac{L}{\ell}\right) - \pi m^2 J \ln\left(\frac{\mathfrak{a}}{\ell}\right) = \pi m^2 J \ln\left(\frac{L}{\mathfrak{a}}\right), \tag{4.43}$$

which seems insurmountable in the thermodynamic limit. In two dimensions, the fact that the two-dimensional Green function for Laplace operator $-\partial_\mu^2 \equiv -\Delta$ grows logarithmically at long distances as a result of scale invariance means that one must be careful when balancing energy with entropy. The entropy of a single vortex is proportional to the logarithm of all the distinct ways of placing it in the underlying square lattice, the proportionality constant being the Boltzmann constant which was set to unity, i.e.,

$$\ln\left(\frac{L}{\mathfrak{a}}\right)^2. \tag{4.44}$$

Thus, the reduced free energy βF of a single vortex becomes

$$\beta F = \left(\pi m^2 K - 2\right) \ln\left(\frac{L}{\mathfrak{a}}\right) \begin{cases} > 0 \text{ if } K > \frac{2}{\pi m^2}, \\ \\ \leq 0 \text{ if } K \leq \frac{2}{\pi m^2}. \end{cases} \tag{4.45}$$

At sufficiently high temperatures, the free energy for a single vortex is dominated by the entropy gain. At sufficiently low temperatures (the lower $|m|$, the lower one must go down in temperature), the free energy for a single vortex is dominated by the energy cost. By this argument, the *Kosterlitz-Thouless criterion*, vortices become important when $K \sim 2/\pi$ [41].

Equation (4.45) suggests the following scenario that would reconcile the prediction of the spin-wave approximation with the prediction of the high-temperature expansion on the $2d$–XY model.

At zero temperature, no vortices are allowed for energetic reasons. At any finite temperature, the condition of charge neutrality must apply to vortices in the thermodynamic limit, i.e., vortices come in pairs of opposite charges. A quasi-long-range ordered phase driven by spin waves for sufficiently small temperatures can only survive the presence of vortices if vortices form bound states due to their strong logarithmic interaction at large distances, the simplest bound state being a dipole. The lower the temperature the fewer and tighter the bound states. As the size of a bound state becomes of the order of the lattice spacing, the vortices making up the bound state annihilate. Far away from bound states of vortices, the local disturbance to the spin-wave texture induced by vortices, as is depicted in Fig. 4.3, is negligible.

In the opposite limit of high temperatures, there is room for an entropy driven transition by which single-vortices behave like a weakly interacting gas for sufficiently high densities. Here, increasing the temperature increases the density of bound states until a critical density is reached above which screening of the bare logarithmic interaction takes place at long distances and vortex bound states unbind. Vortices strongly disrupt ferromagnetic or quasi-long-range order on a microscopic scale as is depicted in Fig. 4.3 and, if they are driven by entropy to unbind from tight-dipole pairs, they also strongly disrupt ferromagnetic or quasi-long-range order on macroscopic scales. By this entropy-driven mechanism, the quasi-long-range ordered phase would be washed out and turned into a paramagnetic phase.

There are caveats to this scenario which can be simply stated by the following interpretation of Eq. (4.45). There is only one dimensionless parameter available, the ratio K of the Heisenberg exchange interaction and the temperature, when describing the $2d$–XY model or the $2d$ $O(2)$ NLσM augmented by vortices. The density of vortices cannot be tuned independently of K, the density of vortices is a function of K. The concept of screening of the bare logarithmic interaction between vortices is nothing but interpreting K as a scale-dependent coupling constant. A finite-temperature and entropy-driven phase transition between a low-temperature phase — in which vortices are *confined* into bound states — and a high-temperature phase — in which vortices are *deconfined* — demands that the renormalized coupling constant K_{ren} renormalizes in the infrared limit to:

(1) infinity if the initial (bare) value K_{bare} corresponds to a temperature below the critical temperature.
(2) zero if the initial (bare) value K_{bare} corresponds to a temperature above the critical temperature.

If K_{ren} renormalizes to zero whatever the value taken by K_{bare}, entropy always dominates, ferromagnetic long-range order is destroyed at any finite temperature, and the paramagnetic phase extends to all temperatures. If K_{ren} renormalizes to infinity whatever the value taken by K_{bare}, energy always dominates, quasi-long-range order extends to all finite temperatures, and the paramagnetic phase only exist at $T = \infty$. In the latter case, the $2d$ $O(2)$ NLσM augmented by vortices does not capture the physics of the $2d$–XY model.

Needed is thus a field theory for the $2d$ $O(2)$ NLσM augmented by vortices on which a renormalization-group analysis for the dependence of the interaction between vortices on length rescaling can be performed.

A welcome simplifying feature is that spin waves and vortices do not interact with each other at the Gaussian order of the gradient expansion (4.8). The dimensionless energy of a spin wave ϕ superposed to a vortex configuration Θ is additive,

$$\frac{K}{2}\int \mathrm{d}^2\boldsymbol{x}\,(\partial_\mu\phi+\partial_\mu\Theta)^2 = \frac{K}{2}\int \mathrm{d}^2\boldsymbol{x}\,(\partial_\mu\phi)^2 + \frac{K}{2}\int \mathrm{d}^2\boldsymbol{x}\,(\partial_\mu\Theta)^2, \tag{4.46}$$

as follows after partial integration on the cross-term

$$\int \mathrm{d}^2\boldsymbol{x}(\partial_\mu\phi)(\partial_\mu\Theta) = \int \mathrm{d}^2\boldsymbol{x}(\partial_\mu\phi)(\widetilde{\partial}_\mu\widetilde{\Theta}), \tag{4.47}$$

keeping in mind that the two-dimensional vector field $\partial_\mu\Theta = \widetilde{\partial}_\mu\widetilde{\Theta}$ is divergence free and that spin waves vanish at infinity. It is therefore sufficient to study the vortices alone.[3]

Vortices form the so-called *two-dimensional Coulomb gas* ($2d$–Cb–gas). From now on the vorticities will be restricted to

$$m_i = \pm 1. \tag{4.48}$$

As the energy of a single vortex of vorticity $m \in \mathbb{Z}$ is proportional to m^2, this simplification is of no consequence with regard to the existence of a critical value K_{c} at which quasi-long-range order could be destroyed by the entropy-driven deconfinement of vortices.

The simplest "physical" probe as a diagnostic of a putative transition from the spin-wave to the paramagnetic phase is the two-point spin-spin correlation function

$$\begin{aligned}\left\langle e^{+\mathrm{i}[\phi(\boldsymbol{x})+\Theta(\boldsymbol{x})]}e^{-\mathrm{i}[\phi(\boldsymbol{y})+\Theta(\boldsymbol{y})]}\right\rangle_{Z_{\mathrm{sw}}\times Z_{\mathrm{Cb}}} &= \left\langle e^{+\mathrm{i}[\phi(\boldsymbol{x})-\phi(\boldsymbol{y})]}\right\rangle_{Z_{\mathrm{sw}}} \times \left\langle e^{+\mathrm{i}[\Theta(\boldsymbol{x})-\Theta(\boldsymbol{y})]}\right\rangle_{Z_{\mathrm{Cb}}} \\ &= \left|\frac{\mathfrak{a}}{\boldsymbol{x}-\boldsymbol{y}}\right|^{\frac{1}{2\pi K}} \times \left\langle e^{+\mathrm{i}[\Theta(\boldsymbol{x})-\Theta(\boldsymbol{y})]}\right\rangle_{Z_{\mathrm{Cb}}}.\end{aligned} \tag{4.49}$$

[3] One might wonder if this decoupling of spin waves and vortices is an artifact of the Gaussian approximation to the gradient expansion. Villain has shown in Ref. [44] that this decoupling is not an artifact in that he constructed a lattice model, now called the *Villain model*, that shares with the XY model the compact nature of angular degrees of freedom but is nevertheless Gaussian. The decoupling between spin waves and vortices is a rigorous property of the Villain model, as is shown in appendix D. The phase diagram of the Villain model is believed to be identical to the one of the XY model in that the same phases of matter are separated by the same phase transitions. Transition temperatures are not equal, though, as they reflect different microscopic details [45].

Needed is the partition function for the $2d$–Cb–gas and an evaluation of $\left\langle e^{+\mathrm{i}[\Theta(\boldsymbol{x})-\Theta(\boldsymbol{y})]}\right\rangle_{Z_{\mathrm{Cb}}}$. The correlation function $\left\langle e^{+\mathrm{i}[\Theta(\boldsymbol{x})-\Theta(\boldsymbol{y})]}\right\rangle_{Z_{\mathrm{Cb}}}$ must decay algebraically at sufficiently low temperatures and exponentially fast at sufficiently high temperatures if the vortex sector can account for the paramagnetic phase of the $2d$–XY model.

4.3 The Coulomb-gas representation of the classical two-dimensional XY model

We cannot rely solely on the naive continuum limit of the $2d$–XY model to define the $2d$–Cb–gas. The continuum limit is scale invariant and vortices of opposite charges are not prevented from collapsing towards each other under the attractive force

$$\boldsymbol{F}(\boldsymbol{x}_1-\boldsymbol{x}_2)=+\frac{m_1 m_2}{2\pi}\frac{\boldsymbol{x}_1-\boldsymbol{x}_2}{|\boldsymbol{x}_1-\boldsymbol{x}_2|^2}, \tag{4.50a}$$

induced by the two-body potential

$$V_{\mathrm{Cb};\mathfrak{a}}(\boldsymbol{x})=-\frac{1}{2\pi}\ln\left|\frac{\boldsymbol{x}}{\mathfrak{a}}\right|, \tag{4.50b}$$

which solves Poisson equation

$$(-\partial_\mu^2 V_{\mathrm{Cb};\mathfrak{a}})(\boldsymbol{x})=\delta(\boldsymbol{x}), \tag{4.50c}$$

with the boundary condition

$$V_{\mathrm{Cb};\mathfrak{a}}(\mathfrak{a})=0. \tag{4.50d}$$

A hardcore two-body repulsive potential between all vortices that vanishes for separations larger than the lattice spacing must be introduced by hand. In the presence of this hardcore repulsive potential, vortices cannot occupy the same volume $\mathfrak{a}^2$. Another effect of the hardcore repulsive potential is to change the so-called *vortex core energy* $V_{\mathrm{Cb};\mathfrak{a}}(\mathfrak{a})=0$ to the finite positive value $E_{\mathrm{core}}(\mathfrak{a})$. The vortex core energy is unambiguously defined in the $2d$–XY model as the value of the two-point function $\langle\cos(\phi_i-\phi_j)\rangle_{Z_{XY}}$ when $i=j$. The vortex core energy could also be taken as the limit $i=j$ of the two-point function $\langle\cos(\phi_i-\phi_j)\rangle_{Z_{\mathrm{SW}}}$ where Z_{SW} is the lattice partition function defined with the Gaussian link energy on the penultimate line of Eq. (4.8). From the point of view of field theory, $E_{\mathrm{core}}(\mathfrak{a})$ is a high-energy cut-off that requires a lattice regularization for its determination and thus depends on detailed knowledge of the microscopic theory.

The grand-canonical partition function for the $2d$–Cb–gas in a finite volume $\Omega\subset\mathbb{R}^2$ of linear dimension L is

$$\begin{aligned}Z_{\mathrm{Cb}}:=&\sum_{M_+=0}^{\infty}\sum_{M_-=0}^{\infty}\frac{Y_+^{M_+}}{M_+!}\frac{Y_-^{M_-}}{M_-!}\exp\left(+\frac{\beta_{\mathrm{Cb}}}{2}\left(M_+-M_-\right)^2 V_{\mathrm{Cb};\mathfrak{a}}(L)\right)\\&\times\int\limits_{\Omega}\mathcal{D}_M[x]\exp\left(-\beta_{\mathrm{Cb}}\sum_{1\le k<l}^{M\equiv M_++M_-}m_k m_l V_{\mathrm{Cb};\mathfrak{a}}(\boldsymbol{x}_k-\boldsymbol{x}_l)\right).\end{aligned} \tag{4.51a}$$

Here,

$$Y \equiv Y_+ = Y_- = e^{\beta_{\text{Cb}}\mu_{\text{Cb}}}, \qquad \mu_{\text{Cb}} = -\frac{1}{2}E_{\text{core;Cb}}(\mathfrak{a}) \tag{4.51b}$$

is the fugacity of unit-charge vortices, β_{Cb} is the $2d$–Cb–gas inverse temperature, and

$$\begin{aligned}\mathcal{D}_M[x] &:= \frac{\mathrm{d}^2\boldsymbol{x}_1}{\mathfrak{a}^2}\cdots\frac{\mathrm{d}^2\boldsymbol{x}_{M_+}}{\mathfrak{a}^2}\frac{\mathrm{d}^2\boldsymbol{x}_{M_++1}}{\mathfrak{a}^2}\cdots\frac{\mathrm{d}^2\boldsymbol{x}_{M_++M_-}}{\mathfrak{a}^2}f(\boldsymbol{x}_1,\cdots,\boldsymbol{x}_M)\\ &\equiv \underbrace{\frac{\mathrm{d}^2\boldsymbol{x}_1}{\mathfrak{a}^2}\cdots\frac{\mathrm{d}^2\boldsymbol{x}_{M_+}}{\mathfrak{a}^2}\frac{\mathrm{d}^2\boldsymbol{x}_{M_++1}}{\mathfrak{a}^2}\cdots\frac{\mathrm{d}^2\boldsymbol{x}_{M_++M_-}}{\mathfrak{a}^2}}_{\neq},\end{aligned} \tag{4.51c}$$

is the infinitesimal volume element of phase space. The function f vanishes whenever two of its arguments are within a lattice spacing. Otherwise f is unity. The grand-canonical partition of the $2d$–Cb–gas is related to the vortex sector of the $2d$–XY model once the identification [see Eq. (4.40)]

$$(2\pi)^2 K \to \beta_{\text{Cb}}. \tag{4.52}$$

is made.

4.4 Equivalence between the Coulomb gas and Sine-Gordon model

4.4.1 *Definitions and statement of results*

An equivalence between the grand-canonical partition function of the $2d$–Cb–gas (4.51) and the canonical partition function of the two-dimensional Sine-Gordon ($2d$–SG) model will be derived. This equivalence allows to make contact between the $2d$–XY model and quantum systems in one-space and one-time dimensions.

This equivalence can be proved at different levels. The strongest equivalence consists in establishing a one-to-one correspondence between all correlation functions in the $2d$–Cb–gas with all correlation functions in the $2d$–SG model. A weaker equivalence is the proof that the partition function of the $2d$–SG model can be rewritten as the grand-canonical partition function of the $2d$–Cb–gas.

The definition of the $2d$–SG model is that of a real-valued scalar field in two-dimensional Euclidean space that is self-interacting with a cosine potential. Thus, the generating function for the $2d$–SG model is

$$Z_{\text{SG}}[J] := \int \mathcal{D}[\theta]\, \exp\left(-\int\limits_{\Omega} \mathrm{d}^2\boldsymbol{x}\left(\frac{1}{2t}(\partial_\mu\theta)^2 - \frac{h}{t}\cos\theta + J\theta\right)\right). \tag{4.53a}$$

The real and positive parameter t plays the role of a dimensionless temperature. The real parameter h plays the role of a magnetic field. It carries the dimension of inverse area. Differentiation with respect to the source J with dimensions of inverse area generates all correlation functions for the scalar field θ. The model is defined on a large domain Ω of linear extent L in the Euclidean plane. We will always assume the boundary condition [recall Eq. (4.39)]

$$\int_{\Omega} \mathrm{d}^2\boldsymbol{x}\, \partial_\mu \left[\theta(\partial_\mu \theta)\right] = 0. \tag{4.53b}$$

The dependence on the lattice spacing $\mathfrak{a}$ indicates the need to introduce a short-distance cut-off to regularize the theory at short distances.

We show in section 4.4.2 how to obtain the grand-canonical partition function of the $2d$–Cb–gas model through a formal power expansion in the reduced magnetic field of the canonical partition function of the $2d$–SG model. More precisely, we show that, with the inverse

$$\begin{aligned} V_{\mathrm{Cb};L}(\boldsymbol{x}) :=& -\frac{1}{2\pi}\ln\left|\frac{\boldsymbol{x}}{L}\right| \\ =& -\frac{1}{2\pi}\ln\left|\frac{\boldsymbol{x}}{\mathfrak{a}}\right| + \frac{1}{2\pi}\ln\left|\frac{L}{\mathfrak{a}}\right| \\ \text{By Eq. (4.50b)} \equiv& V_{\mathrm{Cb};\mathfrak{a}}(\boldsymbol{x}) - V_{\mathrm{Cb};\mathfrak{a}}(L) \end{aligned} \tag{4.54}$$

of the $2d$–Laplace operator $-\partial_\mu^2 \equiv -\Delta$, i.e., the $2d$–Cb potential for a point charge in two dimensions that implements boundary condition (4.53b), the $2d$–SG partition function (4.53a) becomes

$$\begin{aligned} Z_{\mathrm{SG}}[J=0] =& \int \mathcal{D}[\theta]\, \exp\left(-\int_{\Omega} \mathrm{d}^2\boldsymbol{x} \left(\frac{1}{2t}(\partial_\mu\theta)^2 - \frac{h}{t}\cos\theta\right)\right) \\ \propto& \sum_{M_+=0}^{\infty}\sum_{M_-=0}^{\infty} \frac{z_+^{M_+}\, z_-^{M_-}}{M_+!\, M_-!} e^{+\frac{\beta_{\mathrm{SG}}}{2}(M_+-M_-)^2 V_{\mathrm{Cb};\mathfrak{a}}(L)} \\ &\times \int_{\Omega} \mathcal{D}_M[x]\, \exp\left(-\beta_{\mathrm{SG}} \sum_{1\leq k<l}^{M=M_++M_-} m_k m_l V_{\mathrm{Cb};\mathfrak{a}}(\boldsymbol{x}_k - \boldsymbol{x}_l)\right). \end{aligned} \tag{4.55a}$$

Here, the effective fugacities $z_\pm$, inverse temperature β_{SG}, and phase space measure $\mathcal{D}_M[x]$ are

$$\begin{aligned} & z_\pm := \left(\frac{h}{2t}\right) \times e^{\beta_{\mathrm{SG}}\mu_{\mathrm{SG}}}, \qquad \mu_{\mathrm{SG}} := -\frac{1}{2}E_{\mathrm{core;SG}}(\mathfrak{a}), \qquad \beta_{\mathrm{SG}} := t, \\ & \mathcal{D}_M[x] := \frac{\mathrm{d}^2\boldsymbol{x}_1}{\mathfrak{a}^2}\cdots\frac{\mathrm{d}^2\boldsymbol{x}_M}{\mathfrak{a}^2} f(\boldsymbol{x}_1,\cdots,\boldsymbol{x}_M), \qquad M = M_+ + M_-, \\ & f(\underbrace{\boldsymbol{x}_1,\cdots,\boldsymbol{x}_M}_{\in\Omega\times\cdots\times\Omega}) := \begin{cases} 0, \text{ if } \exists k,l, \text{ such that } 1\leq k<l\leq M, \text{ and } |\boldsymbol{x}_k - \boldsymbol{x}_l| \leq \mathfrak{a}, \\ 1, \text{ otherwise}, \end{cases} \end{aligned} \tag{4.55b}$$

respectively.

In other words, vortices have a hardcore radius $\mathfrak{a} \ll L$ and hardcore energy $E_{\mathrm{core;SG}}(\mathfrak{a})$, respectively. The hardcore energy $E_{\mathrm{core;SG}}(\mathfrak{a})$ must be determined microscopically, i.e., it cannot be derived within the $2d$–SG field theory. Rather, it plays here the role of a (phenomenological) high-energy cut-off. By comparison with Eq. (4.51), we infer that the $2d$–XY, Cb–gas, and SG models are equivalent once

$$(2\pi)^2 K \leftrightarrow \beta_{\mathrm{Cb}} \leftrightarrow t, \qquad e^{-\pi K\, D_{\mathrm{dia}}} \leftrightarrow Y \leftrightarrow \left(\frac{h}{2t}\right) \times e^{-\frac{t}{2} E_{\mathrm{core;SG}}(\mathfrak{a})}, \tag{4.56}$$

have been identified. Here, D_{dia} is defined in appendix D to be the diagonal matrix element of the Green function for the Villain model on the square lattice, see Eq. (D.24a).

4.4.2 *Formal expansion in powers of the reduced magnetic field*

We perform the formal expansion

$$\begin{aligned}
&e^{+\frac{h}{t}\int\limits_{\Omega} \mathrm{d}^2\boldsymbol{x} \cos\theta} \\
&= \sum_{M=0}^{\infty} \frac{1}{M!}\left(\frac{h}{t}\right)^M \int\limits_{\Omega} \mathrm{d}^2\boldsymbol{x}_1 \cdots \int\limits_{\Omega} \mathrm{d}^2\boldsymbol{x}_M (\cos\theta)(\boldsymbol{x}_1)\cdots(\cos\theta)(\boldsymbol{x}_M) \\
&= \sum_{M=0}^{\infty} \frac{1}{M!}\left(\frac{h}{2t}\right)^M \int\limits_{\Omega} \mathrm{d}^2\boldsymbol{x}_1 \cdots \int\limits_{\Omega} \mathrm{d}^2\boldsymbol{x}_M \left[e^{+\mathrm{i}\theta(\boldsymbol{x}_1)} + e^{-\mathrm{i}\theta(\boldsymbol{x}_1)}\right]\cdots\left[e^{+\mathrm{i}\theta(\boldsymbol{x}_M)} + e^{-\mathrm{i}\theta(\boldsymbol{x}_M)}\right].
\end{aligned} \tag{4.57}$$

Insertion of this expansion in the generating function (4.53a) yields, owing to the fact that one can freely rename integration variables,

$$\begin{aligned}
Z_{\mathrm{SG}}[J] = &\sum_{M=0}^{\infty} \frac{1}{M!}\left(\frac{h}{2t}\right)^M \sum_{m=0}^{M} \binom{M}{m} \int\limits_{\Omega} \mathrm{d}^2\boldsymbol{x}_1 \cdots \int\limits_{\Omega} \mathrm{d}^2\boldsymbol{x}_{M-m} \\
&\times \int\limits_{\Omega} \mathrm{d}^2\boldsymbol{y}_1 \cdots \int\limits_{\Omega} \mathrm{d}^2\boldsymbol{y}_m \left\langle e^{\mathrm{i}\sum\limits_{k=1}^{M-m}\theta(\boldsymbol{x}_k) - \mathrm{i}\sum\limits_{l=1}^{m}\theta(\boldsymbol{y}_l)} \right\rangle_J^{\mathrm{unnor}},
\end{aligned} \tag{4.58a}$$

where

$$\langle\, (\cdots)\,\rangle_J^{\mathrm{unnor}} := \int \mathcal{D}[\theta]\; e^{-\int\limits_{\Omega}\mathrm{d}^2\boldsymbol{x}\left(\frac{1}{2t}(\partial_\mu\theta)^2 + J\theta\right)}\;(\cdots) \tag{4.58b}$$

denotes an unnormalized average. Consider the source term with dimension of inverse area given by

$$\mathcal{J}(\boldsymbol{x}) := J(\boldsymbol{x}) - \sum_{k=1}^{M-m} \mathrm{i}\delta(\boldsymbol{x}-\boldsymbol{x}_k) + \sum_{l=1}^{m} \mathrm{i}\delta(\boldsymbol{x}-\boldsymbol{y}_l). \tag{4.59}$$

Needed is the term by term evaluation of

$$
\begin{aligned}
Z_{\mathrm{SG}}[J] = \sum_{M=0}^{\infty} \frac{1}{M!} \left(\frac{h}{2t}\right)^{M} \sum_{m=0}^{M} \binom{M}{m} \int_{\Omega} \mathrm{d}^2 \boldsymbol{x}_1 \cdots \int_{\Omega} \mathrm{d}^2 \boldsymbol{x}_{M-m} \\
\times \int_{\Omega} \mathrm{d}^2 \boldsymbol{y}_1 \cdots \int_{\Omega} \mathrm{d}^2 \boldsymbol{y}_m \left\langle e^{-\int_{\Omega} \mathrm{d}^2 \boldsymbol{x} \mathcal{J} \theta} \right\rangle_{J=0}^{\mathrm{unnor}} .
\end{aligned}
\tag{4.60}
$$

Completion of the square gives

$$
\left\langle e^{-\int_{\Omega} \mathrm{d}^2 \boldsymbol{x} \mathcal{J} \theta} \right\rangle_{J=0}^{\mathrm{unnor}} = \left[\mathrm{Det}\left(-\frac{1}{t}\partial_\mu^2\right)\right]^{-\frac{1}{2}} e^{+\frac{1}{2}\int_{\Omega} \mathrm{d}^2 \boldsymbol{x} \int_{\Omega} \mathrm{d}^2 \boldsymbol{y} \mathcal{J}(\boldsymbol{x}) \left(-\frac{1}{t}\partial_\mu^2\right)^{-1} (\boldsymbol{x}-\boldsymbol{y}) \mathcal{J}(\boldsymbol{y})} .
\tag{4.61}
$$

Equation (4.61) suggests defining

$$
S_{\mathrm{eff}}^{(M,m)}[J] := -\frac{t}{2} \int_{\Omega} \mathrm{d}^2 \boldsymbol{x} \int_{\Omega} \mathrm{d}^2 \boldsymbol{y} \mathcal{J}(\boldsymbol{x})\, V_{\mathrm{Cb};L}(\boldsymbol{x}-\boldsymbol{y})\, \mathcal{J}(\boldsymbol{y}).
\tag{4.62}
$$

Two comments are of order here. First, if $S_{\mathrm{eff}}^{(M,m)}[J]$ is to be interpreted as a dimensionless action, it has to be positive definite. Second, Eq. (4.62) is form invariant under rescaling of the coordinates. The latter observation allows to extract the explicit dependence of Eq. (4.62) on L from

$$
\begin{aligned}
S_{\mathrm{eff}}^{(M,m)}[J] = & -\frac{t}{2} \int_{\Omega} \mathrm{d}^2 \boldsymbol{x} \int_{\Omega} \mathrm{d}^2 \boldsymbol{y}\, J(\boldsymbol{x})\, V_{\mathrm{Cb};L}(\boldsymbol{x}-\boldsymbol{y})\, J(\boldsymbol{y}) \\
& + \mathrm{i}t \sum_{k=1}^{M} m_k \int_{\Omega} \mathrm{d}^2 \boldsymbol{x}\, J(\boldsymbol{x})\, V_{\mathrm{Cb};L}(\boldsymbol{x}-\boldsymbol{z}_k) \\
& + \frac{t}{2} \sum_{k,l=1}^{M} m_k m_l\, V_{\mathrm{Cb};L}(\boldsymbol{z}_k-\boldsymbol{z}_l),
\end{aligned}
\tag{4.63a}
$$

where $\boldsymbol{z}_k = \boldsymbol{x}_k$ ($\boldsymbol{z}_k = \boldsymbol{y}_k$) for $k = 1, \cdots, M-m$ ($k = M-m+1, \cdots, M$) and $m_k = +1$ ($m_k = -1$) for $k = 1, \cdots, M-m$ ($k = M-m+1, \cdots, M$), by trading $V_{\mathrm{Cb};L}$ for $V_{\mathrm{Cb};\mathfrak{a}}$, as is done on the last line of Eq. (4.54). Hence,

$$
\begin{aligned}
S_{\mathrm{eff}}^{(M,m)}[J] = & -\frac{t}{2} \int_{\Omega} \mathrm{d}^2 \boldsymbol{x} \int_{\Omega} \mathrm{d}^2 \boldsymbol{y}\, J(\boldsymbol{x})\, V_{\mathrm{Cb};L}(\boldsymbol{x}-\boldsymbol{y})\, J(\boldsymbol{y}) \\
& - \mathrm{i}t(M-2m) V_{\mathrm{Cb};\mathfrak{a}}(L) \int_{\Omega} \mathrm{d}^2 \boldsymbol{x}\, J(\boldsymbol{x}) + \mathrm{i}t \sum_{k=1}^{M} m_k \int_{\Omega} \mathrm{d}^2 \boldsymbol{x}\, J(\boldsymbol{x}) V_{\mathrm{Cb};\mathfrak{a}}(\boldsymbol{x}-\boldsymbol{z}_k) \\
& - \frac{t}{2}(M-2m)^2 V_{\mathrm{Cb};\mathfrak{a}}(L) + \frac{t}{2} \sum_{k,l=1}^{M} m_k m_l V_{\mathrm{Cb};\mathfrak{a}}(\boldsymbol{z}_k-\boldsymbol{z}_l).
\end{aligned}
\tag{4.63b}
$$

Equation (4.63) shows that the relationships between correlation functions in the $2d$–SG model and the $2d$–Cb–gas are *non-local*. The SG field θ is not related to the spin-wave field ϕ in a local way.

If the source J for the correlation functions in the $2d$–SG model is set to zero,

$$Z_{\mathrm{SG}}[J=0] \propto \sum_{M=0}^{\infty}\sum_{m=0}^{M} \frac{\left(\frac{h}{2t}\right)^{M-m}}{(M-m)!}\frac{\left(\frac{h}{2t}\right)^{m}}{m!} \times e^{+\frac{t}{2}(M-2m)^2 V_{\mathrm{Cb};\mathfrak{a}}(L)} \int\limits_{\Omega} \mathrm{d}^2\boldsymbol{z}_1 \cdots \int\limits_{\Omega} \mathrm{d}^2\boldsymbol{z}_M e^{-\frac{t}{2}\sum\limits_{k,l=1}^{M} m_k m_l V_{\mathrm{Cb};\mathfrak{a}}(\boldsymbol{z}_k-\boldsymbol{z}_l)}. \tag{4.64}$$

Finally, by attaching the contribution to the Boltzmann weight from the self-energy of the Coulomb gas to the dimensionless ratio $h/(2t)$, we obtain

$$Z_{\mathrm{SG}}[J=0] \propto \sum_{M=0}^{\infty}\sum_{m=0}^{M} \frac{\left[\frac{h}{2t}\times e^{-\frac{t}{2}V_{\mathrm{Cb};\mathfrak{a}}(\mathfrak{a})}\right]^{M-m}}{(M-m)!}\frac{\left[\frac{h}{2t}\times e^{-\frac{t}{2}V_{\mathrm{Cb};\mathfrak{a}}(\mathfrak{a})}\right]^{m}}{m!} \times e^{+\frac{t}{2}(M-2m)^2 V_{\mathrm{Cb};\mathfrak{a}}(L)} \int\limits_{\Omega} \mathrm{d}^2\boldsymbol{z}_1 \cdots \int\limits_{\Omega} \mathrm{d}^2\boldsymbol{z}_M\, e^{-t\sum\limits_{1\leq k<l}^{M} m_k m_l V_{\mathrm{Cb};\mathfrak{a}}(\boldsymbol{z}_k-\boldsymbol{z}_l)}. \tag{4.65}$$

Comparison with the partition function (4.51) of the $2d$–Cb–gas yields the $2d$–Cb gas representation of the $2d$–SG-model, Eqs. (4.55a) and (4.56), after a core energy has been added by hand.

4.4.3 *Sine-Gordon representation of the spin-spin correlation function in the two-dimensional XY model*

The two-point spin-spin correlation function in the $2d$–XY model is approximated by Eq. (4.49) in the $2d$ $O(2)$ NLσM augmented by vortices. How should one represent the $2d$–Cb gas correlation function

$$\left\langle e^{+\mathrm{i}[\Theta(\boldsymbol{x})-\Theta(\boldsymbol{y})]}\right\rangle_{Z_{\mathrm{Cb}}} \tag{4.66}$$

within the $2d$–SG field theory?

On the one hand, we perform the following manipulations. Let $\Gamma_{\boldsymbol{x},\boldsymbol{y}}$ be a smooth path connecting $\boldsymbol{x}$ to $\boldsymbol{y}$ within the region Ω defined before Eq. (4.21). The argument of the two-point correlation function (4.66) can then be written as the following line integral

$$\begin{aligned}
\mathrm{i}\left[\Theta(\boldsymbol{y})-\Theta(\boldsymbol{x})\right] &= \mathrm{i}\int\limits_{\Gamma_{\boldsymbol{x},\boldsymbol{y}}} \mathrm{d}s_\mu\,\left(\partial_\mu\Theta\right)(\boldsymbol{s}) \\
\text{By Eq. (4.33)} &= \mathrm{i}\int\limits_{\Gamma_{\boldsymbol{x},\boldsymbol{y}}} \mathrm{d}s_\mu\,\left(\widetilde{\partial}_\mu\widetilde{\Theta}\right)(\boldsymbol{s}) \\
\text{By Eq. (4.24)} &= -\mathrm{i}\sum_i m_i \int\limits_{\Gamma_{\boldsymbol{x},\boldsymbol{y}}} \mathrm{d}s_\mu\,\epsilon_{\mu\nu}\frac{\partial}{\partial s_\nu}\ln\left|\frac{\boldsymbol{s}-\boldsymbol{x}_i}{\ell}\right| \\
\partial_\nu \ln(L/\ell)=0 &= +2\pi\mathrm{i}\sum_i m_i \int\limits_{\Gamma_{\boldsymbol{x},\boldsymbol{y}}} \mathrm{d}s_\mu\,\epsilon_{\mu\nu}\frac{\partial}{\partial s_\nu}V_{\mathrm{Cb};L}(\boldsymbol{s}-\boldsymbol{x}_i).
\end{aligned} \tag{4.67}$$

On the other hand, Eqs. (4.67) and (4.59) suggest choosing the source

$$\begin{aligned}
\mathcal{J}_{\Gamma_{\boldsymbol{x},\boldsymbol{y}}}(\boldsymbol{s}) = &\underbrace{\qquad J_{\Gamma_{\boldsymbol{x},\boldsymbol{y}}}(\boldsymbol{s})\qquad} -\mathrm{i}\left(\sum_{k=1}^{M-m}\delta(\boldsymbol{s}-\boldsymbol{x}_k)-\sum_{l=1}^{m}\delta(\boldsymbol{s}-\boldsymbol{y}_l)\right) \\
:= &\,\alpha\int\limits_{\Gamma_{\boldsymbol{x},\boldsymbol{y}}} \mathrm{d}u_\mu\,\epsilon_{\mu\nu}\frac{\partial}{\partial u_\nu}\delta(\boldsymbol{s}-\boldsymbol{u}) - \mathrm{i}\left(\sum_{k=1}^{M-m}\delta(\boldsymbol{s}-\boldsymbol{x}_k)-\sum_{l=1}^{m}\delta(\boldsymbol{s}-\boldsymbol{y}_l)\right)
\end{aligned} \tag{4.68}$$

in the $2d$–Cb–gas expansion of the SG model (4.60). Indeed,

$$\int \mathrm{d}^2\boldsymbol{s}\,\mathcal{J}_{\Gamma_{\boldsymbol{x},\boldsymbol{y}}}(\boldsymbol{s})\,\theta(\boldsymbol{s}) = \alpha\int\limits_{\Gamma_{\boldsymbol{x},\boldsymbol{y}}} \mathrm{d}u_\mu\,\epsilon_{\mu\nu}\left(\frac{\partial}{\partial u_\nu}\theta\right)(\boldsymbol{u}) - \mathrm{i}\sum_{k=1}^{M} m_k\theta(\boldsymbol{z}_k), \tag{4.69}$$

where the constant α will be fixed shortly and $\boldsymbol{z}_k=\boldsymbol{x}_k$ ($\boldsymbol{z}_k=\boldsymbol{y}_k$) for $k=1,\cdots,M-m$ ($k=M-m+1,\cdots,M$) and $m_k=+1$ ($m_k=-1$) for $k=1,\cdots,M-m$ ($k=M-m+1,\cdots,M$). To fix the constant α for the Ansatz

$$J_{\Gamma_{\boldsymbol{x},\boldsymbol{y}}}(\boldsymbol{s}) := \alpha\int\limits_{\Gamma_{\boldsymbol{x},\boldsymbol{y}}} \mathrm{d}u_\mu\,\epsilon_{\mu\nu}\frac{\partial}{\partial u_\nu}\delta(\boldsymbol{s}-\boldsymbol{u}) \tag{4.70}$$

that we made in Eq. (4.68), observe that Eq. (4.63a) with the Ansatz (4.68) as argument becomes

$$
\begin{aligned}
S_{\rm eff}^{(M,m)}[J_{\Gamma_{\boldsymbol{x},\boldsymbol{y}}}] =& -\frac{t}{2}\int\limits_{\Omega} \mathrm{d}^2\boldsymbol{s} \int\limits_{\Omega} \mathrm{d}^2\boldsymbol{t}\; J_{\Gamma_{\boldsymbol{x},\boldsymbol{y}}}(\boldsymbol{s}) V_{\mathrm{Cb};L}(\boldsymbol{s}-\boldsymbol{t}) J_{\Gamma_{\boldsymbol{x},\boldsymbol{y}}}(\boldsymbol{t}) \\
&+ \mathrm{i}t \sum_{k=1}^{M} m_k \int\limits_{\Omega} \mathrm{d}^2\boldsymbol{s}\; J_{\Gamma_{\boldsymbol{x},\boldsymbol{y}}}(\boldsymbol{s}) V_{\mathrm{Cb};L}(\boldsymbol{s}-\boldsymbol{z}_k) \\
&+ \frac{t}{2} \sum_{k,l=1}^{M} m_k m_l V_{\mathrm{Cb};L}(\boldsymbol{z}_k-\boldsymbol{z}_l) \\
=& -\frac{t}{2}\alpha^2 \int\limits_{\Gamma_{\boldsymbol{x},\boldsymbol{y}}} \mathrm{d}u_\mu \int\limits_{\Gamma_{\boldsymbol{x},\boldsymbol{y}}} \mathrm{d}\bar{u}_{\bar{\mu}}\; \epsilon_{\mu\nu}\epsilon_{\bar{\mu}\bar{\nu}} \left[\partial_\nu \partial_{\bar{\nu}} V_{\mathrm{Cb};L}(\boldsymbol{u}-\bar{\boldsymbol{u}})\right] \\
&+ \underline{\mathrm{i}t\alpha \sum_{k=1}^{M} m_k \int\limits_{\Gamma_{\boldsymbol{x},\boldsymbol{y}}} \mathrm{d}u_\mu\, \epsilon_{\mu\nu}\partial_\nu V_{\mathrm{Cb};L}(\boldsymbol{u}-\boldsymbol{z}_k)} \\
&+ \frac{t}{2} \sum_{k,l=1}^{M} m_k m_l V_{\mathrm{Cb};L}(\boldsymbol{z}_k-\boldsymbol{z}_l).
\end{aligned}
\tag{4.71}
$$

The constant α can now be chosen by demanding that the penultimate (underlined) line reduces to the right-hand side of Eq. (4.67), i.e.,

$$
\alpha = \frac{2\pi}{t} = \frac{1}{2\pi K}.
\tag{4.72}
$$

It is time to turn our attention to the term quadratic in $J_{\Gamma_{\boldsymbol{x},\boldsymbol{y}}}$. It can be simplified with the help of

$$
\epsilon_{\mu\nu}\epsilon_{\bar{\mu}\bar{\nu}} = \delta_{\mu\bar{\mu}}\delta_{\nu\bar{\nu}} - \delta_{\mu\bar{\nu}}\delta_{\nu\bar{\mu}}.
\tag{4.73}
$$

It becomes

$$
\begin{aligned}
D(\boldsymbol{x}-\boldsymbol{y}) :=& -\frac{t}{2}\left(\frac{2\pi}{t}\right)^2 \int\limits_{\Gamma_{\boldsymbol{x},\boldsymbol{y}}} \mathrm{d}u_\mu \int\limits_{\Gamma_{\boldsymbol{x},\boldsymbol{y}}} \mathrm{d}\bar{u}_{\bar{\mu}}\; \epsilon_{\mu\nu}\epsilon_{\bar{\mu}\bar{\nu}} \left[\partial_\nu \partial_{\bar{\nu}} V_{\mathrm{Cb};L}(\boldsymbol{u}-\bar{\boldsymbol{u}})\right] \\
=& -\tfrac{t}{2}\left(\tfrac{2\pi}{t}\right)^2 \int\limits_{\Gamma_{\boldsymbol{x},\boldsymbol{y}}} \mathrm{d}u_\mu \int\limits_{\Gamma_{\boldsymbol{x},\boldsymbol{y}}} \mathrm{d}\bar{u}_{\bar{\mu}} \left[-\delta_{\mu\bar{\mu}}\delta(\boldsymbol{u}-\bar{\boldsymbol{u}}) - \left(-\tfrac{1}{2\pi}\partial_\mu\partial_{\bar{\mu}} \ln\left|\tfrac{\boldsymbol{u}-\bar{\boldsymbol{u}}}{L}\right|\right)\right] \\
=& +\tfrac{t}{2}\left(\tfrac{2\pi}{t}\right)^2 \int\limits_{\Gamma_{\boldsymbol{x},\boldsymbol{y}}} \mathrm{d}u_\mu \int\limits_{\Gamma_{\boldsymbol{x},\boldsymbol{y}}} \mathrm{d}\bar{u}_\mu\; \delta(\boldsymbol{u}-\bar{\boldsymbol{u}}) - \tfrac{2\pi}{t}\left(\ln\left|\tfrac{\mathfrak{a}}{L}\right| - \ln\left|\tfrac{\boldsymbol{y}-\boldsymbol{x}}{L}\right|\right) \\
=& \text{ constant} + \frac{2\pi}{t}\ln\left|\frac{\boldsymbol{x}-\boldsymbol{y}}{\mathfrak{a}}\right| \\
\stackrel{\text{Eq. (4.56)}}{=}& \text{ constant} + \frac{1}{2\pi K}\ln\left|\frac{\boldsymbol{x}-\boldsymbol{y}}{\mathfrak{a}}\right|.
\end{aligned}
\tag{4.74}
$$

To recapitulate, we have found that the effective action (4.62) becomes

$$
S_{\rm eff}^{(M,m)}[J_{\Gamma_{\boldsymbol{x},\boldsymbol{y}}}] = D(\boldsymbol{x}-\boldsymbol{y}) - \mathrm{i}\left[\Theta(\boldsymbol{x}) - \Theta(\boldsymbol{y})\right] + \frac{t}{2}\sum_{k,l=1}^{M} m_k m_l V_{\mathrm{Cb};L}(\boldsymbol{z}_k-\boldsymbol{z}_l),
\tag{4.75}
$$

when the source $J_{\Gamma_{\boldsymbol{x},\boldsymbol{y}}}$ is chosen as in Eqs. (4.70) and (4.72), whereby:

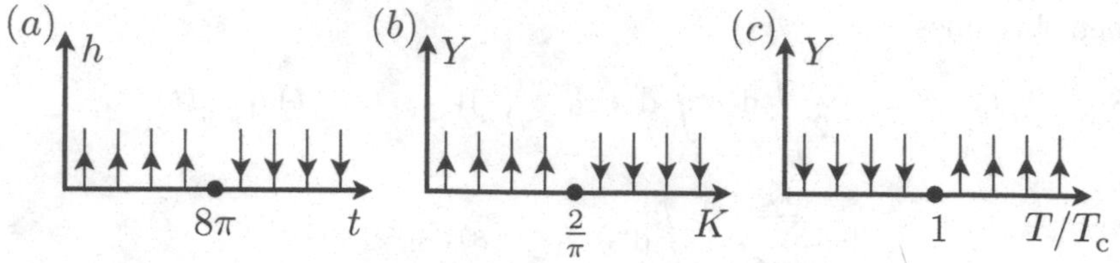

Fig. 4.4 (a) Stability analysis of the critical line $h = 0$ in the $2d$–SG model. Vertical arrows indicate whether the magnetic field decreases or increases under coarse graining. (b) Stability analysis of the critical line $Y = 0$ in the $2d$–Cb–gas model. Vertical arrows indicate whether the fugacity decreases or increases under coarse graining. (c) Same as in (b) but with the horizontal axis K replaced by $T/T_{\mathrm{KT}} = K_{\mathrm{KT}}/K$, $K_{\mathrm{KT}} = 2/\pi$.

- $D(\boldsymbol{x}-\boldsymbol{y})$ is independent of the $M - m$ positive vortices located at $\boldsymbol{x}_k$ and the m negative vortices located at $\boldsymbol{y}_l$.
- Up to a constant, $\exp\big(-D(\boldsymbol{x}-\boldsymbol{y})\big)$ is the spin-wave two-point function,

$$e^{-D(\boldsymbol{x}-\boldsymbol{y})} = e^{\text{constant}} \times \left|\frac{\mathfrak{a}}{\boldsymbol{x}-\boldsymbol{y}}\right|^{+\frac{1}{2\pi K}} = e^{\text{constant}} \times \left\langle e^{+\mathrm{i}[\phi(\boldsymbol{x})-\phi(\boldsymbol{y})]}\right\rangle_{\mathrm{sw}}. \tag{4.76}$$

In other words, we have proved that the spin-spin two-point correlation function (4.49) in the $2d$–XY model within the continuum spin-wave approximation augmented by vortices has the $2d$–SG representation

$$\left\langle e^{+\mathrm{i}[\phi(\boldsymbol{x})+\Theta(\boldsymbol{x})]-\mathrm{i}[\phi(\boldsymbol{y})+\Theta(\boldsymbol{y})]}\right\rangle_{Z_{\mathrm{sw}}\times Z_{\mathrm{Cb}}} = \left\langle e^{-\frac{2\pi}{t}\int\limits_{\Gamma_{\boldsymbol{x},\boldsymbol{y}}} \mathrm{d}u_\mu\, \epsilon_{\mu\nu}\left(\frac{\partial}{\partial u_\nu}\theta\right)(\boldsymbol{u})}\right\rangle_{Z_{\mathrm{SG}}}. \tag{4.77}$$

As a corollary, the $2d$–SG representation of the $2d$–Cb–gas correlation function (4.66) is

$$\left\langle e^{+\mathrm{i}[\Theta(\boldsymbol{x})-\Theta(\boldsymbol{y})]}\right\rangle_{Z_{\mathrm{Cb}}} = \left|\frac{\mathfrak{a}}{\boldsymbol{x}-\boldsymbol{y}}\right|^{-\frac{1}{2\pi K}} \left\langle e^{-\frac{2\pi}{t}\int\limits_{\Gamma_{\boldsymbol{x},\boldsymbol{y}}} \mathrm{d}u_\mu\, \epsilon_{\mu\nu}\left(\frac{\partial}{\partial u_\nu}\theta\right)(\boldsymbol{u})}\right\rangle_{Z_{\mathrm{SG}}}. \tag{4.78}$$

4.4.4 *Stability analysis of the line of fixed points in the two-dimensional Sine-Gordon model*

Now that we have identified the correlation function in the $2d$–SG model that approximates the spin-spin two-point correlation function in the $2d$–XY model, we can deduce from the $2d$–SG model the stability of the spin-wave phase. The spin-wave phase is obtained by switching off vortices. In the $2d$–SG model, turning off the magnetic field h amounts to removing all vortices from the Cb gas sector of the XY model by tuning the vortex core energy to infinity. With the magnetic field h turned off, the $2d$–SG model reduces to the free $2d$ scalar field theory. All correlation functions for the exponentiated SG scalar field $\exp[\mathrm{i}\theta(\boldsymbol{x})]$ are algebraic when the reduced temperature t is finite. The horizontal line $h = 0$ in the t–h plane is a line of critical points. The magnetic field, as a perturbation to this line of fixed points, has engineering dimensions 2 and scaling dimension

$$\Delta_h = \frac{t}{4\pi} = \pi K, \tag{4.79a}$$

since

$$\begin{aligned}
\lim_{h\to 0} \langle \cos\theta(\boldsymbol{x}) \cos\theta(\boldsymbol{y}) \rangle_{Z_{\text{SG}}} &\sim \lim_{h\to 0} \left\langle e^{+\mathrm{i}[\theta(\boldsymbol{x})-\theta(\boldsymbol{y})]} \right\rangle_{Z_{\text{SG}}} \\
\overset{\text{section 3.3.2}}{} &= e^{tV_{\text{Cb};\mathfrak{a}}(\boldsymbol{x}-\boldsymbol{y})} \\
&= \left| \frac{\mathfrak{a}}{\boldsymbol{x}-\boldsymbol{y}} \right|^{\frac{t}{2\pi}} \\
&= \left| \frac{\mathfrak{a}}{\boldsymbol{x}-\boldsymbol{y}} \right|^{2\pi K}.
\end{aligned} \tag{4.79b}$$

In section 3.3.1, we saw that a small perturbation to a critical fixed point in d-dimensions is infrared irrelevant, marginal, or relevant if its scaling dimension is larger, equal, or smaller than d, respectively. Hence, the magnetic field in the $2d$–SG model (or, equivalently, the fugacity in the $2d$–Cb–gas) is infrared:

- irrelevant when

$$t > 8\pi \Longleftrightarrow K > \frac{2}{\pi} \equiv K_{\text{KT}}. \tag{4.80}$$

- marginal when

$$t = 8\pi \Longleftrightarrow K = \frac{2}{\pi} \equiv K_{\text{KT}}. \tag{4.81}$$

- relevant when

$$t < 8\pi \Longleftrightarrow K < \frac{2}{\pi} \equiv K_{\text{KT}}. \tag{4.82}$$

Infrared irrelevance of the cosine potential means that the algebraic (spin-wave) phase is stable to a weak perturbing magnetic field (vortices with large core energy). infrared relevance of the cosine potential means that the algebraic (spin-wave) phase is unstable to a weak perturbing magnetic field (vortices with large core energy). The criterion (4.82) for the instability of the algebraic phase agrees with the Kosterlitz-Thouless criterion (4.45).[4] This stability analysis of the line of fixed point $h = Y = 0$ is summarized by Fig. 4.4. Needed is a better grasp of the flow obeyed by the reduced temperature (reduced spin stiffness) and by the magnetic field (fugacity).

[4] Vortices with charge $m \in \mathbb{Z}$ are induced by the cosine potential $\cos(m\theta)$ in the $2d$–SG model.

4.5 Fugacity expansion of n-point functions in the Sine-Gordon model

We start from the $2d$–SG Lagrangian

$$\mathcal{L}_{\mathrm{SG}}[\theta] := \frac{1}{2t}(\partial_\mu \theta)^2 - \frac{h}{t}\cos\theta. \tag{4.83}$$

Here, we remember the bookkeeping

$$K := \frac{t}{4\pi^2}, \quad Y \sim \frac{h}{2t}, \tag{4.84}$$

to make contact with the $2d$–Cb–gas representation of the $2d$–XY model with reduced spin stiffness K. We shall denote thermal averaging with angular brackets,

$$\begin{aligned}\langle(\cdots)\rangle &:= \frac{1}{Z_{\mathrm{SG}}}\int \mathcal{D}[\theta]\, e^{-S_{\mathrm{SG}}[\theta]}\,(\cdots),\\ Z_{\mathrm{SG}} &:= \int \mathcal{D}[\theta]\, e^{-S_{\mathrm{SG}}[\theta]},\end{aligned} \tag{4.85}$$

where $S_{\mathrm{SG}}[\theta] := \int \mathrm{d}^2\boldsymbol{x}\, \mathcal{L}_{\mathrm{SG}}$ is the SG action obtained from Eq. (4.83).

Let n be a positive integer, choose n points $\boldsymbol{x}_1, \cdots, \boldsymbol{x}_n$ on the Euclidean plane, and define

$$F_{\boldsymbol{x}_1,\cdots,\boldsymbol{x}_n} := e^{\mathrm{i}\epsilon_1\theta(\boldsymbol{x}_1)}\cdots e^{\mathrm{i}\epsilon_n\theta(\boldsymbol{x}_n)}, \qquad \epsilon_1,\cdots,\epsilon_n = \pm 1, \tag{4.86a}$$

$$\langle F_{\boldsymbol{x}_1,\cdots,\boldsymbol{x}_n}\rangle^{\mathrm{unnor}} := \int \mathcal{D}[\theta]\, e^{-S_{\mathrm{SG}}[\theta]}\, F_{\boldsymbol{x}_1,\cdots,\boldsymbol{x}_n}. \tag{4.86b}$$

Thermal averaging of Eq. (4.86a) is obtained by dividing the unnormalized average in Eq. (4.86b) by the SG partition function,

$$\langle F_{\boldsymbol{x}_1,\cdots,\boldsymbol{x}_n}\rangle = \frac{\langle F_{\boldsymbol{x}_1,\cdots,\boldsymbol{x}_n}\rangle^{\mathrm{unnor}}}{Z_{\mathrm{SG}}}. \tag{4.87}$$

We are going to compute both $\langle F_{\boldsymbol{x}_1,\cdots,\boldsymbol{x}_n}\rangle$ and $\langle F_{\boldsymbol{x}_1,\cdots,\boldsymbol{x}_n}\rangle^{\mathrm{unnor}}$ through a formal power expansion in h.

To this end, the key identity that is needed is [recall the expansion done in Eq. (4.58a)]

$$\begin{aligned}\langle F_{\boldsymbol{x}_1,\cdots,\boldsymbol{x}_n}\rangle^{\mathrm{unnor}} = \sum_{m=0}^{\infty}\frac{1}{m!}\left(\frac{h}{2t}\right)^m \sum_{p=0}^{m}\binom{m}{p}\int \underbrace{\mathrm{d}^2\boldsymbol{y}_1\cdots\mathrm{d}^2\boldsymbol{y}_m}_{\neq}\\ \times\left\langle e^{\mathrm{i}\theta(\boldsymbol{y}_1)}\cdots e^{\mathrm{i}\theta(\boldsymbol{y}_p)}e^{-\mathrm{i}\theta(\boldsymbol{y}_{p+1})}\cdots e^{-\mathrm{i}\theta(\boldsymbol{y}_m)}F_{\boldsymbol{x}_1,\cdots,\boldsymbol{x}_n}\right\rangle^{\mathrm{unnor}}_{h=0}.\end{aligned} \tag{4.88}$$

Integrations over coordinates are done with the hardcore constraint that no two points ever coincide as indicated by the underbrace. Implementing the hardcore constraint is equivalent to a renormalization of the magnetic field h by the core energy of the vortices in the $2d$–Cb–gas interpretation. The renormalized magnetic field is essentially the $2d$–Cb–fugacity. Thermal averaging on the last line must be performed with $h = 0$, in which case averaging over θ is Gaussian.

4.5.1 *Fugacity expansion of the two-point function*

We treat the case of the two-point function

$$F_{\boldsymbol{x}_1,\boldsymbol{x}_2} := e^{\mathrm{i}\theta(\boldsymbol{x}_1)-\mathrm{i}\theta(\boldsymbol{x}_2)} \equiv F_{12}. \tag{4.89}$$

The expansion in powers of $h/2t$ of its thermal average is

$$\begin{aligned}\langle F_{12}\rangle &\equiv \sum_{n=0}^{\infty} F_{12}^{(n)}\,(h/2t)^n\\ &=: \frac{\sum\limits_{m=0}^{\infty} f_{12}^{(m)}\,(h/2t)^m}{1+\sum\limits_{n=1}^{\infty} Z^{(n)}\,(h/2t)^n},\end{aligned} \tag{4.90a}$$

where $F_{12}^{(2n+1)} = f_{12}^{(2n+1)} = Z^{(2n+1)} = 0$,

$$F_{12}^{(0)} = f_{12}^{(0)}, \tag{4.90b}$$

$$F_{12}^{(2)} = f_{12}^{(2)} - f_{12}^{(0)} Z^{(2)}, \tag{4.90c}$$

$$F_{12}^{(4)} = f_{12}^{(4)} - \left(f_{12}^{(2)} Z^{(2)} + f_{12}^{(0)} Z^{(4)}\right) + f_{12}^{(0)} Z^{(2)} \times Z^{(2)}, \tag{4.90d}$$

and the generic term

$$\begin{aligned}F_{12}^{(2n)} &= f_{12}^{(2n)}\\ &-\left(f_{12}^{(2n-2)} Z^{(2)} + \cdots + f_{12}^{(0)} Z^{(2n)}\right)\\ &+\left(f_{12}^{(2n-4)}\underbrace{Z^{(2)}\times Z^{(2)}}_{2-\text{times}} + 2 f_{12}^{(2n-6)}\underbrace{Z^{(2)}\times Z^{(4)}}_{2-\text{times}} + \cdots\right) - \cdots\\ &+(-)^n f_{12}^{(0)}\underbrace{Z^{(2)}\times\cdots\times Z^{(2)}}_{n-\text{times}}.\end{aligned} \tag{4.90e}$$

The coefficients in the power expansions in $h/2t$ of the numerator and denominator are

$$\begin{aligned}f_{12}^{(2n)} &= \frac{1}{(n!)^2}\int \underbrace{\mathrm{d}^2\boldsymbol{y}_1\cdots\mathrm{d}^2\boldsymbol{y}_{2n}}_{\neq}\Big\langle e^{\mathrm{i}\theta(\boldsymbol{y}_1)+\cdots+\mathrm{i}\theta(\boldsymbol{y}_n)-\mathrm{i}\theta(\boldsymbol{y}_{n+1})-\cdots-\mathrm{i}\theta(\boldsymbol{y}_{2n})}\\ &\quad\times e^{+\mathrm{i}\theta(\boldsymbol{x}_1)-\mathrm{i}\theta(\boldsymbol{x}_2)}\Big\rangle_{h=0}^{\text{unnor}}\end{aligned} \tag{4.90f}$$

and

$$Z^{(2n)} = \frac{1}{(n!)^2}\int \underbrace{\mathrm{d}^2\boldsymbol{y}_1\cdots\mathrm{d}^2\boldsymbol{y}_{2n}}_{\neq}\Big\langle e^{\mathrm{i}\theta(\boldsymbol{y}_1)+\cdots+\mathrm{i}\theta(\boldsymbol{y}_n)-\mathrm{i}\theta(\boldsymbol{y}_{n+1})-\cdots-\mathrm{i}\theta(\boldsymbol{y}_{2n})}\Big\rangle_{h=0}^{\text{unnor}}, \tag{4.90g}$$

respectively.

4.5.2 Two-point function to lowest order in $h/2t$

We need (the short distance cut-off is set to one: $\mathfrak{a}=1$)

$$F^{(0)}_{12} = f^{(0)}_{12}, \tag{4.91a}$$

$$f^{(0)}_{12} = \left\langle e^{\mathrm{i}\theta(\boldsymbol{x}_1)-\mathrm{i}\theta(\boldsymbol{x}_2)}\right\rangle^{\mathrm{unnor}}_{h=0} = \frac{1}{|\boldsymbol{x}_1-\boldsymbol{x}_2|^{2\pi K}}, \tag{4.91b}$$

where we have used [recall Eq. (3.106d)]

$$-(\partial^2_\mu)^{-1}(\boldsymbol{x}_1,\boldsymbol{x}_2) = -\frac{1}{2\pi}\ln|\boldsymbol{x}_1-\boldsymbol{x}_2|, \quad 2\pi K = \frac{t}{2\pi}. \tag{4.91c}$$

Thus,

$$F^{(0)}_{12} = |\boldsymbol{x}_1-\boldsymbol{x}_2|^{-2\pi K}. \tag{4.92}$$

4.5.3 Two-point function to second order in $h/2t$

We need (the short distance cut-off is set to one: $\mathfrak{a}=1$)

$$F^{(2)}_{12} = f^{(2)}_{12} - f^{(0)}_{12} Z^{(2)}, \tag{4.93a}$$

$$\begin{aligned} f^{(2)}_{12} &= \int \underbrace{\mathrm{d}^2\boldsymbol{y}_1\mathrm{d}^2\boldsymbol{y}_2}_{\neq} \left\langle e^{\mathrm{i}\theta(\boldsymbol{y}_1)-\mathrm{i}\theta(\boldsymbol{y}_2)+\mathrm{i}\theta(\boldsymbol{x}_1)-\mathrm{i}\theta(\boldsymbol{x}_2)}\right\rangle^{\mathrm{unnor}}_{h=0} \\ &= \int \underbrace{\mathrm{d}^2\boldsymbol{y}_1\mathrm{d}^2\boldsymbol{y}_2}_{\neq} \frac{1}{|\boldsymbol{x}_1-\boldsymbol{x}_2|^{2\pi K}|\boldsymbol{y}_1-\boldsymbol{y}_2|^{2\pi K}} \left(\frac{|\boldsymbol{y}_1-\boldsymbol{x}_1||\boldsymbol{y}_2-\boldsymbol{x}_2|}{|\boldsymbol{y}_1-\boldsymbol{x}_2||\boldsymbol{y}_2-\boldsymbol{x}_1|}\right)^{2\pi K}, \end{aligned} \tag{4.93b}$$

$$\begin{aligned} Z^{(2)} &= \int \underbrace{\mathrm{d}^2\boldsymbol{y}_1\mathrm{d}^2\boldsymbol{y}_2}_{\neq} \left\langle e^{\mathrm{i}\theta(\boldsymbol{y}_1)-\mathrm{i}\theta(\boldsymbol{y}_2)}\right\rangle^{\mathrm{unnor}}_{h=0} \\ &= \int \underbrace{\mathrm{d}^2\boldsymbol{y}_1\mathrm{d}^2\boldsymbol{y}_2}_{\neq} \frac{1}{|\boldsymbol{y}_1-\boldsymbol{y}_2|^{2\pi K}}, \end{aligned} \tag{4.93c}$$

$$f^{(0)}_{12} Z^{(2)} = \frac{1}{|\boldsymbol{x}_1-\boldsymbol{x}_2|^{2\pi K}} \int \underbrace{\mathrm{d}^2\boldsymbol{y}_1\mathrm{d}^2\boldsymbol{y}_2}_{\neq} \frac{1}{|\boldsymbol{y}_1-\boldsymbol{y}_2|^{2\pi K}}, \tag{4.93d}$$

where we have used [recall Eq. (3.106d)]

$$-(\partial^2_\mu)^{-1}(\boldsymbol{x}_1,\boldsymbol{x}_2) = -\frac{1}{2\pi}\ln|\boldsymbol{x}_1-\boldsymbol{x}_2|, \quad 2\pi K = \frac{t}{2\pi}. \tag{4.93e}$$

Thus,

$$F^{(2)}_{12} = \frac{1}{|\boldsymbol{x}_1-\boldsymbol{x}_2|^{2\pi K}} \int \underbrace{\mathrm{d}^2\boldsymbol{y}_1\mathrm{d}^2\boldsymbol{y}_2}_{\neq} \frac{\mathcal{K}_{\boldsymbol{x}_1\boldsymbol{x}_2}(\boldsymbol{y}_1,\boldsymbol{y}_2;2\pi K) - \mathcal{K}_{\boldsymbol{x}_1\boldsymbol{x}_2}(\boldsymbol{y}_1,\boldsymbol{y}_2;0)}{|\boldsymbol{y}_1-\boldsymbol{y}_2|^{2\pi K}}, \tag{4.94a}$$

where

$$\mathcal{K}_{\boldsymbol{x}_1\boldsymbol{x}_2}(\boldsymbol{y}_1,\boldsymbol{y}_2;\kappa) := \left(\frac{|\boldsymbol{y}_1-\boldsymbol{x}_1||\boldsymbol{y}_2-\boldsymbol{x}_2|}{|\boldsymbol{y}_1-\boldsymbol{x}_2||\boldsymbol{y}_2-\boldsymbol{x}_1|}\right)^{\kappa}. \tag{4.94b}$$

We now estimate the integrals over $\boldsymbol{y}_1$ and $\boldsymbol{y}_2$. To this end, we introduce the center of mass and relative coordinates

$$\boldsymbol{Y} := \frac{1}{2}(\boldsymbol{y}_1 + \boldsymbol{y}_2), \quad \boldsymbol{y}_{12} := \boldsymbol{y}_1 - \boldsymbol{y}_2 \Longleftrightarrow \boldsymbol{y}_1 = \boldsymbol{Y} + \frac{1}{2}\boldsymbol{y}_{12}, \quad \boldsymbol{y}_2 = \boldsymbol{Y} - \frac{1}{2}\boldsymbol{y}_{12}. \tag{4.95}$$

The first step is to approximate the difference of two logarithms to third order in the relative coordinates,

$$\begin{aligned}
\ln|\boldsymbol{y}_1 - \boldsymbol{x}_1| - \ln|\boldsymbol{y}_2 - \boldsymbol{x}_1| &= \ln|\boldsymbol{Y} - \boldsymbol{x}_1 + \frac{1}{2}\boldsymbol{y}_{12}| - \ln|\boldsymbol{Y} - \boldsymbol{x}_1 - \frac{1}{2}\boldsymbol{y}_{12}| \\
&= +\frac{\partial \ln|\boldsymbol{Y} - \boldsymbol{x}_1|}{\partial \boldsymbol{Y}} \cdot \boldsymbol{y}_{12} + \mathcal{O}(\boldsymbol{y}_{12}^3), \qquad (4.96\text{a}) \\
\ln|\boldsymbol{y}_2 - \boldsymbol{x}_2| - \ln|\boldsymbol{y}_1 - \boldsymbol{x}_2| &= \ln|\boldsymbol{Y} - \boldsymbol{x}_2 - \frac{1}{2}\boldsymbol{y}_{12}| - \ln|\boldsymbol{Y} - \boldsymbol{x}_2 + \frac{1}{2}\boldsymbol{y}_{12}| \\
&= -\frac{\partial \ln|\boldsymbol{Y} - \boldsymbol{x}_2|}{\partial \boldsymbol{Y}} \cdot \boldsymbol{y}_{12} + \mathcal{O}(\boldsymbol{y}_{12}^3). \qquad (4.96\text{b})
\end{aligned}$$

The second step is to expand the exponential of the difference of logarithms in powers of the relative coordinates:

$$\begin{aligned}
\mathcal{K}_{\boldsymbol{x}_1\boldsymbol{x}_2}(\boldsymbol{y}_1, \boldsymbol{y}_2; \kappa) &:= \left(\frac{|\boldsymbol{y}_1 - \boldsymbol{x}_1||\boldsymbol{y}_2 - \boldsymbol{x}_2|}{|\boldsymbol{y}_1 - \boldsymbol{x}_2||\boldsymbol{y}_2 - \boldsymbol{x}_1|}\right)^{\kappa} \\
&= e^{\kappa\left(\frac{\partial(\ln|\boldsymbol{Y}-\boldsymbol{x}_1| - \ln|\boldsymbol{Y}-\boldsymbol{x}_2|)}{\partial \boldsymbol{Y}} \cdot \boldsymbol{y}_{12} + \mathcal{O}(\boldsymbol{y}_{12}^3)\right)} \\
&= 1 \\
&+ \kappa \frac{\partial(\ln|\boldsymbol{Y} - \boldsymbol{x}_1| - \ln|\boldsymbol{Y} - \boldsymbol{x}_2|)}{\partial \boldsymbol{Y}} \cdot \boldsymbol{y}_{12} \\
&+ \frac{1}{2}\kappa^2 \left(\frac{\partial(\ln|\boldsymbol{Y} - \boldsymbol{x}_1| - \ln|\boldsymbol{Y} - \boldsymbol{x}_2|)}{\partial \boldsymbol{Y}} \cdot \boldsymbol{y}_{12}\right)^2 \\
&+ \mathcal{O}(\boldsymbol{y}_{12}^3). \qquad (4.97)
\end{aligned}$$

The third step is to perform the integration over the measure $\mathrm{d}^2\boldsymbol{y}_1\mathrm{d}^2\boldsymbol{y}_2 = \mathrm{d}^2\boldsymbol{Y}\mathrm{d}^2\boldsymbol{y}_{12}$, where the hardcore constraint is implemented by $|\boldsymbol{y}_{12}| > \mathfrak{a}$, with the approximation (4.97) of the function $\mathcal{K}_{\boldsymbol{x}_1\boldsymbol{x}_2}(\boldsymbol{y}_1, \boldsymbol{y}_2; \kappa)$. We will assume that the domain of integration is invariant under $\boldsymbol{y}_{12} \to -\boldsymbol{y}_{12}$. The two terms independent of $\boldsymbol{y}_{12}$ in the numerator of the integrand cancel out. By assumption integration over the terms linear in $\boldsymbol{y}_{12}$ in the numerator of the integrand vanish. Finally, we are left with the contribution from the term quadratic in $\boldsymbol{y}_{12}$ in the numerator of the integrand. We thus need, for any function $f(|\boldsymbol{z}|)$,

$$\begin{aligned}\int \mathrm{d}^2\boldsymbol{z}(\boldsymbol{x}\cdot\boldsymbol{z})^2\, f(|\boldsymbol{z}|) &= \int\limits_0^\infty \mathrm{d}r r \int\limits_0^{2\pi} \mathrm{d}\varphi (x_1 r\cos\varphi + x_2 r\sin\varphi)^2\, f(r) \\ &= \int\limits_0^\infty \mathrm{d}r r \int\limits_0^{2\pi} \mathrm{d}\varphi (x_1^2 r^2 \cos^2\varphi \\ &\quad + x_2^2 r^2 \sin^2\varphi + 2x_1 x_2 r^2 \cos\varphi\sin\varphi)\, f(r) \\ &= \int\limits_0^\infty \mathrm{d}r r \pi \left(x_1^2 r^2 + x_2^2 r^2\right)\, f(r) \\ &= \pi|\boldsymbol{x}|^2 \int\limits_0^\infty \mathrm{d}r r^3\, f(r). \end{aligned} \tag{4.98}$$

Notice that this intermediary result can also be understood as follows. The measure and the integrand are invariant under rotations of the domain of integration. Hence,

$$\int \mathrm{d}^2\boldsymbol{z}(\boldsymbol{x}\cdot\boldsymbol{z})^2\, f(|\boldsymbol{z}|) = |\boldsymbol{x}|^2 \int \mathrm{d}^2\boldsymbol{z}(\hat{\boldsymbol{n}}\cdot\boldsymbol{z})^2\, f(|\boldsymbol{z}|) = |\boldsymbol{x}|^2 \int\limits_0^\infty \mathrm{d}r\, r^3\, f(r) \int\limits_0^{2\pi} \mathrm{d}\varphi\, \cos^2\varphi \tag{4.99}$$

for any $\hat{\boldsymbol{n}}$, $\hat{\boldsymbol{n}}\cdot\hat{\boldsymbol{n}} = 1$. We conclude that (after reinserting the short distance cut-off $\mathfrak{a}$)

$$\begin{aligned}&\int \mathrm{d}^2\boldsymbol{y}_1 \int \mathrm{d}^2\boldsymbol{y}_2 \left[\mathcal{K}_{\boldsymbol{x}_1\boldsymbol{x}_2}(\boldsymbol{y}_1,\boldsymbol{y}_2;\kappa_a) - \mathcal{K}_{\boldsymbol{x}_1\boldsymbol{x}_2}(\boldsymbol{y}_1,\boldsymbol{y}_2;\kappa_b)\right] f(|\boldsymbol{y}_1-\boldsymbol{y}_2|) \\ &\approx \frac{\pi}{2}\left(\kappa_a^2-\kappa_b^2\right) \int\limits_{\mathfrak{a}}^{\mathrm{y}_{12}^{\max}} \mathrm{d}|\boldsymbol{y}_{12}||\boldsymbol{y}_{12}|^3 f(|\boldsymbol{y}_{12}|) \int\limits_0^{\mathrm{Y}^{\max}} \mathrm{d}^2\boldsymbol{Y} \left|\frac{\partial\left(\ln|\boldsymbol{Y}-\boldsymbol{x}_1| - \ln|\boldsymbol{Y}-\boldsymbol{x}_2|\right)}{\partial\boldsymbol{Y}}\right|^2 .\end{aligned} \tag{4.100}$$

With the help of partial integration with respect to the center-of-mass coordinate $\boldsymbol{Y}$ and the Green function (4.93e),

$$\begin{aligned}&\int\limits_0^{\mathrm{Y}^{\max}} \mathrm{d}^2\boldsymbol{Y} \left|\frac{\partial\left(\ln|\boldsymbol{Y}-\boldsymbol{x}_1| - \ln|\boldsymbol{Y}-\boldsymbol{x}_2|\right)}{\partial\boldsymbol{Y}}\right|^2 \\ &= \int\limits_0^{\mathrm{Y}^{\max}} \mathrm{d}^2\boldsymbol{Y}\left(\ln|\boldsymbol{Y}-\boldsymbol{x}_1| - \ln|\boldsymbol{Y}-\boldsymbol{x}_2|\right)(-2\pi)\left[\delta\left(\boldsymbol{Y}-\boldsymbol{x}_1\right) - \delta\left(\boldsymbol{Y}-\boldsymbol{x}_2\right)\right].\end{aligned} \tag{4.101}$$

Insertion of Eq. (4.101) into Eq. (4.100) gives

$$\begin{aligned}&\int \mathrm{d}^2\boldsymbol{y}_1 \int \mathrm{d}^2\boldsymbol{y}_2 \left[\mathcal{K}_{\boldsymbol{x}_1\boldsymbol{x}_2}(\boldsymbol{y}_1,\boldsymbol{y}_2;\kappa_a) - \mathcal{K}_{\boldsymbol{x}_1\boldsymbol{x}_2}(\boldsymbol{y}_1,\boldsymbol{y}_2;\kappa_b)\right] f(|\boldsymbol{y}_1-\boldsymbol{y}_2|) \\ &\approx \frac{4\pi^2}{2}\left(\kappa_a^2-\kappa_b^2\right) \ln\left|\frac{\boldsymbol{x}_1-\boldsymbol{x}_2}{\mathfrak{a}}\right| \int\limits_{\mathfrak{a}}^{\mathrm{y}_{12}^{\max}} \mathrm{d}|\boldsymbol{y}_{12}||\boldsymbol{y}_{12}|^3 f(|\boldsymbol{y}_{12}|),\end{aligned} \tag{4.102}$$

which implies

$$F_{12}^{(2)} \approx \left|\frac{\boldsymbol{x}_{12}}{\mathfrak{a}}\right|^{-2\pi K} 8\pi^4 K^2 \left(\int_{\mathfrak{a}}^{\mathrm{y}_{12}^{\max}} \mathrm{d}|\boldsymbol{y}_{12}|\, |\boldsymbol{y}_{12}|^3 \left|\frac{\boldsymbol{y}_{12}}{\mathfrak{a}}\right|^{-2\pi K} \right) \ln\left|\frac{\boldsymbol{x}_{12}}{\mathfrak{a}}\right|. \tag{4.103}$$

By collecting all terms up to and including second order in the fugacity, one can write

$$\langle F_{12}\rangle \approx \left|\frac{\boldsymbol{x}_{12}}{\mathfrak{a}}\right|^{-2\pi x(1)}, \tag{4.104a}$$

$$x(1) := K - 4\pi^3 K^2 Y^2 \int_1^\infty \mathrm{d}y\, y^{3-2\pi K}. \tag{4.104b}$$

Here, we have introduced the squared dimensionless fugacity

$$Y^2 := \left(\frac{\mathfrak{a}^2 h}{2t}\right)^2. \tag{4.104c}$$

Before proceeding to an RG interpretation of this result, we observe that when K is larger than $2/\pi$, then the y integral is convergent and the scaling exponent $x(1)$ is a well-defined number. The assumption that the density of vortices is small is then consistent. On the other hand, when K is smaller or equal to $2/\pi$, then the y integral is divergent so that the scaling exponent $x(1)$ is ill-defined. The assumption that the density of vortices is small is not consistent. The bare logarithmic interaction between vortices is screened so as to change the functional form of the decay of the spin-spin correlation function.

Notice that if we rewrite

$$\begin{aligned}
\int_1^\infty \mathrm{d}y\, y^{3-2\pi K} &= \int_1^{e^l} \mathrm{d}y\, y^{3-2\pi K} + \int_{e^l}^\infty \mathrm{d}y\, y^{3-2\pi K} \\
&= \frac{e^{l(4-2\pi K)}-1}{4-2\pi K} + e^{l(4-2\pi K)} \int_1^\infty \mathrm{d}y\, y^{3-2\pi K},
\end{aligned} \tag{4.105}$$

then Eq. (4.104b) is form invariant, i.e.,

$$x(1) = K' - 4\pi^3 K^2 (Y')^2 \int_1^\infty \mathrm{d}y\, y^{3-2\pi K}, \tag{4.106a}$$

where

$$\begin{aligned}
K' &:= K - 4\pi^3 K^2 Y^2 \frac{e^{l(4-2\pi K)}-1}{4-2\pi K} \\
&= K - 4\pi^3 K^2 Y^2 l + \mathcal{O}(l^2),
\end{aligned} \tag{4.106b}$$

$$\begin{aligned}
Y' &:= \frac{\mathfrak{a}^2 h'}{2t} \\
&= Y e^{l(2-\pi K)} \\
&= Y[1 + l(2-\pi K) + \mathcal{O}(l^2)].
\end{aligned} \tag{4.106c}$$

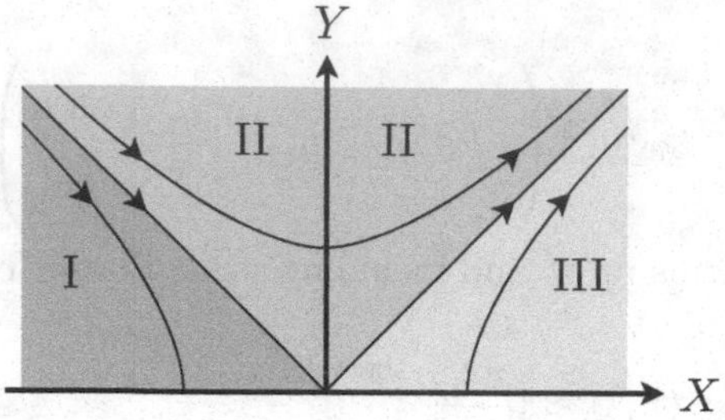

Fig. 4.5 Kosterlitz RG flow in the vicinity of fixed point $X = Y = 0$ in the half-plane $Y > 0$. The flow in the half-plane $Y < 0$ is obtained by reflection symmetry about the line $Y = 0$.

If l is chosen to be arbitrary small, Eqs. (4.106b) and (4.106c) can be rewritten as differential equations,

$$\frac{\mathrm{d}K}{\mathrm{d}l} = -4\pi^3 K^2 Y^2, \tag{4.107a}$$

$$\frac{\mathrm{d}Y}{\mathrm{d}l} = (2 - \pi K)Y. \tag{4.107b}$$

These equations were first derived by Kosterlitz in Ref. [42]. They are often called the Kosterlitz-Thouless RG equations in the literature. Kosterlitz derived these equations by using the Coulomb gas representation of the $2d$–XY model. Here, we used the Sine-Gordon representation of the $2d$–XY model to derive the KT RG equations [46]. Equation (4.107b) reproduces the Kosterlitz-Thouless criterion (4.45) and the scaling analysis (4.82). However, there is more information to be gained from Eq. (4.107a) that encodes the screening of the logarithmic interaction between a pair of vortices far apart as anticipated by Berezinskii in Ref. [40]. Kosterlitz RG equations are left invariant by

$$Y \to -Y. \tag{4.108}$$

This invariance reflects the condition of charge neutrality; vortices occur in pairs of opposite charges in the thermodynamic limit.

4.6 Kosterlitz renormalization-group equations

4.6.1 *Kosterlitz RG equations in the vicinity of* $X = Y = 0$

In this section we want to analyze in details the Kosterlitz RG equations (4.107a) and (4.107b) in the vicinity of $K = 2/\pi$ and $Y = 0$. Define the scaling variable

$$X := 2 - \pi K \Longleftrightarrow K = \frac{(2 - X)}{\pi}. \tag{4.109}$$

The physical interpretation of X is that it is proportional to the deviation in the temperature T of the $2d$–XY model away from the Kosterlitz-Thouless critical temperature (4.82), (recall that the Boltzmann constant is set to unity)

$$X \equiv \pi\left(K_{\mathrm{KT}}-K\right) \equiv \pi J\left(\frac{1}{T_{\mathrm{KT}}}-\frac{1}{T}\right)=\frac{\pi J}{T_{\mathrm{KT}}}\left\{\frac{T-T_{\mathrm{KT}}}{T_{\mathrm{KT}}}+\mathcal{O}\left[\left(\frac{T-T_{\mathrm{KT}}}{T_{\mathrm{KT}}}\right)^{2}\right]\right\}. \tag{4.110}$$

Equations (4.107a) and (4.107b) are rewritten

$$\frac{\mathrm{d}X}{\mathrm{d}l}=+4\pi^{2}(2-X)^{2}Y^{2}, \tag{4.111a}$$

$$\frac{\mathrm{d}Y}{\mathrm{d}l}=XY, \tag{4.111b}$$

in the X–Y coupling plane.

Next, we expand Eqs. (4.111a) and (4.111b) in the vicinity of the fixed point

$$0=X, \qquad 0=Y, \tag{4.112}$$

to the first non-trivial order. We thus find

$$\frac{\mathrm{d}X}{\mathrm{d}l}=+(4\pi)^{2}\,Y^{2}+\mathcal{O}(XY^{2}), \tag{4.113a}$$

$$\frac{\mathrm{d}Y}{\mathrm{d}l}=XY. \tag{4.113b}$$

These two equations are brought to the more symmetric form

$$\frac{\mathrm{d}X}{\mathrm{d}l}=+2Y^{2}\equiv\beta_{x}, \tag{4.114a}$$

$$\frac{\mathrm{d}Y}{\mathrm{d}l}=2XY\equiv\beta_{y}, \tag{4.114b}$$

by another redefinition of the running coupling constants X and Y and of the rescaling parameter l,

$$X\rightarrow 4X, \qquad Y\rightarrow\frac{1}{\pi}Y, \qquad l\rightarrow l/2. \tag{4.114c}$$

It is possible to find curves in the X–Y coupling plane that are invariant under the Kosterlitz RG equations (4.114a) and (4.114b). Define the family of hyperbolas parametrized by $\alpha\in\mathbb{R}$

$$\begin{aligned}\Gamma_{\alpha}:\mathbb{R}&\longrightarrow\mathbb{R}^{2},\\ l&\longmapsto\begin{pmatrix}X(l)\\ Y(l)\end{pmatrix}, \qquad X^{2}(l)-Y^{2}(l)=\alpha.\end{aligned} \tag{4.115}$$

Under the Kosterlitz RG equations (4.114a) and (4.114b)

$$\begin{aligned}\frac{\mathrm{d}}{\mathrm{d}l}\left[X^{2}(l)-Y^{2}(l)\right]&=2\left[X(l)\beta_{x}-Y(l)\beta_{y}\right]\\ &=4\left[(XY^{2})(l)-(YXY)(l)\right]\\ &=0, \qquad \forall l\in\mathbb{R}.\end{aligned} \tag{4.116}$$

In view of the invariance of the Kosterlitz RG equations (4.111a) and (4.111b) under $Y\rightarrow -Y$, we need to distinguish three cases.

(1) When $\alpha > 0$, the hyperbola (4.115) is parametrized by (without loss of generality, $Y \geq 0$)

$$X = (\pm)\sqrt{\alpha}\frac{1+s^2}{1-s^2}, \qquad Y = \sqrt{\alpha}\frac{2s}{1-s^2}, \qquad 0 \leq s < 1. \tag{4.117}$$

Equation (4.114a) reads

$$\begin{aligned} 0 &= \frac{\mathrm{d}X}{\mathrm{d}l} - 2Y^2 \\ &= (\pm)\sqrt{\alpha}\frac{2s(1-s^2)+(1+s^2)2s}{(1-s^2)^2}\left(\frac{\mathrm{d}s}{\mathrm{d}l}\right) - 2\alpha\frac{4s^2}{(1-s^2)^2} \\ &= (\pm)\sqrt{\alpha}\frac{4s}{(1-s^2)^2}\left[\left(\frac{\mathrm{d}s}{\mathrm{d}l}\right) - (\pm)2\sqrt{\alpha}\,s\right]. \end{aligned} \tag{4.118}$$

The RG equation for the parameter s is

$$\frac{\mathrm{d}s}{\mathrm{d}l} = (\pm)2\sqrt{\alpha}\,s, \tag{4.119a}$$

with the solution

$$s(l) = s(l_0)\exp\Big((\pm)2\sqrt{\alpha}\,l\Big). \tag{4.119b}$$

(a) When $\alpha > 0$ and $X(l) > 0$ (high-temperature phase), one must choose the positive root $+\sqrt{\alpha}$ in Eq. (4.117) and the solution to Eq. (4.119a) is

$$s(l) = s(l_0)\exp\left(+2\sqrt{\alpha}\,l\right). \tag{4.120}$$

This is the solution corresponding to an initial inverse reduced temperature $K(l_0)$ *below* $K_{\mathrm{c}} \lesssim 2/\pi$. The fugacity, initially very small, grows exponentially fast until the self-consistency of the perturbative expansion is lost. This behavior is the one expected when vortices are relevant perturbations to the spin-wave phase. From the high-temperature expansion, this phase is believed to be the paramagnetic phase.

(b) When $\alpha > 0$ and $X(l) < 0$ (low temperature phase), one must choose the negative root $-\sqrt{\alpha}$ in Eq. (4.117) and the solution to Eq. (4.119a) is

$$s(l) = s(l_0)\exp\left(-2\sqrt{\alpha}\,l\right). \tag{4.121}$$

This is the solution corresponding to an initial inverse reduced temperature $K(l_0)$ *above* $K_{\mathrm{c}} \gtrsim 2/\pi$. The fugacity, initially very small, decreases exponentially fast. The self-consistency of the perturbative expansion improves under the RG group flow. This behavior is the one expected when vortices are irrelevant perturbations to the spin-wave phase.

(2) When $\alpha < 0$, the hyperbola (4.115) is parametrized by (without loss of generality, $Y \geq 0$)

$$X = \sqrt{|\alpha|}\frac{2s}{1-s^2}, \qquad Y = \sqrt{|\alpha|}\frac{1+s^2}{1-s^2}, \qquad -1 < s < 1. \tag{4.122}$$

Equation (4.114a) reads

$$\begin{aligned}0 &= \frac{\mathrm{d}X}{\mathrm{d}l} - 2Y^2 \\ &= \sqrt{|\alpha|}\frac{2(1-s^2)+4s^2}{(1-s^2)^2}\left(\frac{\mathrm{d}s}{\mathrm{d}l}\right) - 2|\alpha|\frac{(1+s^2)^2}{(1-s^2)^2} \\ &= 2\sqrt{|\alpha|}\frac{(1+s^2)}{(1-s^2)^2}\left[\left(\frac{\mathrm{d}s}{\mathrm{d}l}\right) - \sqrt{|\alpha|}(1+s^2)\right]. \end{aligned} \tag{4.123}$$

The RG equation for the parameter s is

$$\frac{\mathrm{d}s}{\mathrm{d}l} = \sqrt{|\alpha|}\,(1+s^2), \tag{4.124a}$$

with the solution

$$\arctan[s(l)] - \arctan[s(l_0)] = \sqrt{|\alpha|}\,(l-l_0). \tag{4.124b}$$

For any $\alpha < 0$, $s(l)$ increases with l. The fugacity, initially very small, increases under the RG group flow until the self-consistency of the perturbative expansion is lost. This behavior is the one expected when vortices are relevant perturbations to the spin-wave phase.

According to this analysis, the half-plane $Y > 0$ can be divided into three regions separated by the half-lines

$$Y = +X, \quad X > 0, \qquad \text{and} \qquad Y = -X, \quad X < 0, \tag{4.125}$$

respectively (see Fig. 4.5). Region I is defined by $X < 0$ and $|X| > Y > 0$. This is the regime in which the spin-wave phase is stable to the thermal nucleation of vortices with a large core energy. In this regime spin-spin correlation functions decay algebraically fast and the interaction between vortices grows logarithmically for large separations. Vortices can only appear in tight bound states at low temperatures. Quasi-long-range order is associated to an infinitely large correlation length. Region II is defined by $Y > |X| > 0$. Region III is defined by $X > Y > 0$. In both regimes II and III, the spin-wave phase is unstable to the thermal nucleation of vortices. Spin-spin correlation functions decay exponentially fast and the interaction between vortices is screened at long distances. Vortices are deconfined at long distances. The difference between region II and region III is that vortices are also ultraviolet relevant in region II whereas they are ultraviolet irrelevant in region III. In region II, the field theory never reduces to a free scalar field theory obtained by ignoring the cosine potential, be it at long or short distances. In region III, the field theory is *asymptotically free*. The field theory reduces to the free scalar field theory obtained by ignoring the cosine potential at short distances. The property of asymptotic freedom is of little use to the understanding of the $2d$–XY model however, since there is no justification for approximating the $2d$–XY model by a field theory on length scales of the order of the lattice spacing. The separatrix $Y = |X|$ is a line of phase transitions. These transitions are continuous but very weak as we demonstrate by estimating how the correlation length diverges upon approaching $X = Y = 0$ from the high-temperature regime $X > 0$.

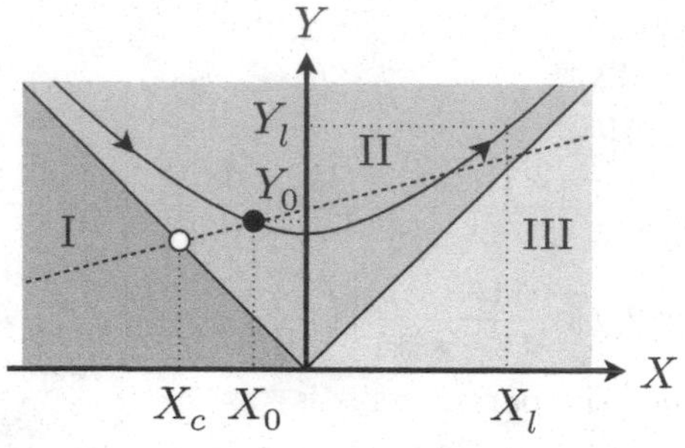

Fig. 4.6 Blow up of Fig. 4.5 in the vicinity of the fixed point $X = Y = 0$ in the half-plane $Y > 0$. The dotted line represents the initial value $Y_0 \equiv Y(l_0)$ of the fugacity as a function of the initial value $X_0 \equiv X(l_0)$. The case considered here is when (X_0, Y_0), depicted by a black dot, is in region II and is very close to the separatrix $Y = -X$. The intercept (white dot) between the separatrix $Y = -X$ and the dotted line defines the critical temperature and fugacity, i.e., X_c and Y_c, respectively.

4.6.2 *Correlation length near* $X = Y = 0$

The initial value of the fugacity $Y_0 \equiv Y(l_0)$ at $X_0 \equiv X(l_0)$ is inferred from Eq. (4.56). We assume that it belongs to region II. There is one hyperbola (4.115), that goes through the coordinate (X_0, Y_0) in region II of the X–Y coupling plane as is depicted in Fig. 4.6 when $Y_0 > X_0$. This hyperbola is labeled by the value

$$\alpha_0 = X_0^2 - Y_0^2 < 0. \tag{4.126}$$

This hyperbola intersects the fugacity axis at the value $Y_{\text{intersection}} = \sqrt{|\alpha_0|}$. The Kosterlitz RG flow (4.114a) and (4.114b) takes (X_0, Y_0) to the point $(X(l), Y(l))$ of region II as is depicted in Fig. 4.6. When $l \gg l_0$, we should expect to be deep in the paramagnetic phase. The correlation length $\xi(l)$ defined by the asymptotic exponential decay length of the spin-spin two-point function should be very small for $(X(l), Y(l)) \sim (1, 1)$, say of the order of the lattice spacing. The question we want to answer is what is the value of the initial correlation length ξ_0 for (X_0, Y_0) very close to $X = Y = 0$?

The answer to this question requires two steps. First, we observe that the transformation law obeyed by the correlation length under the RG flow is

$$\xi(l) = \xi(l_0)\, e^{-(l-l_0)/2}. \tag{4.127}$$

The argument $l/2$ comes from the redefinition of l in Eq. (4.114c). By assumption,

$$\xi(l) \sim \mathfrak{a}. \tag{4.128}$$

The second step consists in expressing l in terms of X_0 and Y_0 given that X_0 and Y_0 are very close to the origin $X = Y = 0$. To this end, one divides Eq. (4.114a) by Eq. (4.114b),

$$\begin{aligned} \frac{\mathrm{d}X}{\mathrm{d}Y} = \frac{Y}{X} &\iff X\mathrm{d}X = Y\mathrm{d}Y \\ &\Longrightarrow X^2(l) - Y^2(l) = \alpha_0. \end{aligned} \tag{4.129}$$

By Eq. (4.114a),

$$
\begin{aligned}
l - l_0 &= \int_{l_0}^{l} \mathrm{d}l' \\
&= \int_{X_0}^{X(l)} \frac{\mathrm{d}X}{2Y^2}.
\end{aligned}
\tag{4.130}
$$

By Eq. (4.129),

$$
\begin{aligned}
l - l_0 &= \frac{1}{2} \int_{X_0}^{X(l)} \frac{\mathrm{d}X}{X^2 - \alpha_0} \\
\text{Region II has } \alpha_0 < 0 &= \frac{1}{2} \int_{X_0}^{X(l)} \frac{\mathrm{d}X}{X^2 + |\alpha_0|} \\
&= \frac{1}{2\sqrt{|\alpha_0|}} \left[\arctan\left(\frac{X(l)}{\sqrt{|\alpha_0|}} \right) - \arctan\left(\frac{X_0}{\sqrt{|\alpha_0|}} \right) \right].
\end{aligned}
\tag{4.131}
$$

By assumption, $|\alpha_0|$ is very small. More precisely, note by inspection of Fig. 4.6 that (X_0, Y_0) is very close to the separatrix $Y = -X$. We may do the linearization

$$
X_0 = -Y_0 + c^2\, Y_0 \left(\frac{T - T_c}{T_c} \right) + \mathcal{O}\left[\left(\frac{T - T_c}{T_c} \right)^2 \right],
\tag{4.132}
$$

where c^2 is some positive number that depends on details at the microscopic level. Hence,

$$
\begin{aligned}
|\alpha_0| &= Y_0^2 - X_0^2 \\
&\approx Y_0^2 - \left[-Y_0 + c^2\, Y_0 \left(\frac{T - T_c}{T_c} \right) \right]^2 \\
&\approx 2c^2 Y_0^2 \left(\frac{T - T_c}{T_c} \right).
\end{aligned}
\tag{4.133}
$$

On the other hand, we have also assumed that $X(l) > 0$ is of order 1. Thus, to a first approximation, we may replace $X(l)/\sqrt{\alpha_0}$ by $+\infty$ and $X_0/\sqrt{\alpha_0}$ by $-\infty$. In this way,

$$
l - l_0 \approx \frac{\pi}{2\sqrt{|\alpha_0|}} \approx \frac{\pi}{2\sqrt{2}\, c\, Y_0} \left(\frac{T - T_c}{T_c} \right)^{-1/2}.
\tag{4.134}
$$

This is the desired relationship for l in terms of the initial conditions. At last we obtain the correlation length [recall the rescaling done in Eq. (4.114c)]

$$
\xi(l_0) \equiv \xi_0 \approx \mathfrak{a} \times \exp\left(\frac{\pi}{4\sqrt{2}\, c\, Y_0} t^{-1/2} \right), \qquad t := \frac{T - T_c}{T_c}.
\tag{4.135}
$$

The correlation length in the paramagnetic phase diverges faster than any power of t upon approaching the Kosterlitz-Thouless transition. It can be shown that the regime of validity of Eq. (4.135) demands that $t < 10^{-2}$ for which values ξ_0 is at least 10^8 lattice spacings. Testing Eq. (4.135) experimentally demands very clean samples and is hopeless numerically. It is known from the theory of critical phenomena that the free energy per unit volume can be decomposed into a regular and a singular contribution at a critical point and that the singular contribution is roughly given by ξ_0^{-2} upon approaching the critical point from the side of the disordered phase. It follows that the free energy per unit volume is of the form $\exp(-|\mathrm{const}| \times t^{-1/2})$ and thus has an essential singularity at $t = 0$. All derivatives of the free energy per unit volume are continuous functions of t through the transition. Consequently, the Kosterlitz-Thouless phase transition lies outside of the 19th century classification of phase transitions in terms of the order at which the derivative of the free energy is discontinuous. The Kosterlitz-Thouless transition suggests that a perhaps better classification of phase transitions should simply be one distinguishing continuous from non-continuous (first order) phase transitions.

A more elegant way of expressing l and thus the correlation length $\xi(l_0)$ in terms of X_0 and Y_0 is to choose once more a new set of running coupling constants. Let

$$\begin{aligned} X_+ &:= Y + X, \\ X_- &:= Y - X, \end{aligned} \iff \begin{aligned} Y &:= \tfrac{1}{2}(X_+ + X_-), \\ X &:= \tfrac{1}{2}(X_+ - X_-). \end{aligned} \tag{4.136}$$

This choice is nothing but a rotation by $\pi/4$ of Figs. 4.5 and 4.6. The separatrix (4.125) are

$$X_- = 0, \qquad X_+ = 0. \tag{4.137}$$

The hyperbolas (4.115) are

$$X_+ X_- = -\alpha. \tag{4.138}$$

The Kosterlitz RG equations (4.114a) and (4.114b) are

$$\begin{aligned} \frac{\mathrm{d}X_+}{\mathrm{d}l} &= 2\left(XY + Y^2\right) \\ &= \frac{1}{2}\left(X_+^2 - X_-^2 + X_+^2 + X_-^2 + 2X_+X_-\right) \\ &= X_+^2 + X_+X_- \\ &= X_+^2 - \alpha, \end{aligned} \tag{4.139a}$$

$$\begin{aligned} \frac{\mathrm{d}X_-}{\mathrm{d}l} &= 2\left(XY - Y^2\right) \\ &= \frac{1}{2}\left(X_+^2 - X_-^2 - X_+^2 - X_-^2 - 2X_+X_-\right) \\ &= -X_-^2 - X_+X_- \\ &= -X_-^2 + \alpha. \end{aligned} \tag{4.139b}$$

These can be rewritten as

$$\frac{\mathrm{d}X_+}{\alpha - X_+^2} = -\mathrm{d}l, \tag{4.140a}$$

$$\frac{\mathrm{d}X_-}{\alpha - X_-^2} = +\mathrm{d}l, \tag{4.140b}$$

which can be immediately integrated. For $X_+(l)$, it is found that

$$\sqrt{|\alpha|}(l-l_0) = \begin{cases} +\arctan\left[\frac{X_+(l)}{\sqrt{|\alpha|}}\right] - \arctan\left[\frac{X_+(l_0)}{\sqrt{|\alpha|}}\right], & \text{if } \alpha < 0, \\ -\frac{1}{2}\left\{\ln\left[\frac{\sqrt{\alpha}+X_+(l)}{\sqrt{\alpha}-X_+(l)}\right] - \ln\left[\frac{\sqrt{\alpha}+X_+(l_0)}{\sqrt{\alpha}-X_+(l_0)}\right]\right\}, & \text{if } \alpha > 0 \text{ and } \alpha > X_+^2, \\ -\frac{1}{2}\left\{\ln\left[\frac{X_+(l)+\sqrt{\alpha}}{X_+(l)-\sqrt{\alpha}}\right] - \ln\left[\frac{X_+(l_0)+\sqrt{\alpha}}{X_+(l_0)-\sqrt{\alpha}}\right]\right\}, & \text{if } \alpha > 0 \text{ and } \alpha < X_+^2, \end{cases} \tag{4.141}$$

whereas

$$\sqrt{|\alpha|}(l-l_0) = \begin{cases} -\arctan\left[\frac{X_-(l)}{\sqrt{|\alpha|}}\right] + \arctan\left[\frac{X_-(l_0)}{\sqrt{|\alpha|}}\right], & \text{if } \alpha < 0, \\ +\frac{1}{2}\left\{\ln\left[\frac{\sqrt{\alpha}+X_-(l)}{\sqrt{\alpha}-X_-(l)}\right] - \ln\left[\frac{\sqrt{\alpha}+X_-(l_0)}{\sqrt{\alpha}-X_-(l_0)}\right]\right\}, & \text{if } \alpha > 0 \text{ and } \alpha > X_-^2, \\ +\frac{1}{2}\left\{\ln\left[\frac{X_-(l)+\sqrt{\alpha}}{X_-(l)-\sqrt{\alpha}}\right] - \ln\left[\frac{X_-(l_0)+\sqrt{\alpha}}{X_-(l_0)-\sqrt{\alpha}}\right]\right\}, & \text{if } \alpha > 0 \text{ and } \alpha < X_-^2, \end{cases} \tag{4.142}$$

for $X_-(l)$. Since we are interested in region II, we use the solution

$$\sqrt{|\alpha_0|}(l-l_0) = +\arctan\left[\frac{X_+(l)}{\sqrt{|\alpha_0|}}\right] - \arctan\left[\frac{X_+(l_0)}{\sqrt{|\alpha_0|}}\right] \tag{4.143}$$

for $X_+(l)$, which, by assumption, is very close to the separatrix $Y = X$, i.e.,

$$X_+(l) \sim 2Y(l) \sim 2X(l) \sim 1 \gg X_+(l_0) > 0. \tag{4.144}$$

The limit in which

$$\frac{X_+(l_0)}{\sqrt{|\alpha_0|}} = \sqrt{\left|\frac{X_+(l_0)}{X_-(l_0)}\right|} \tag{4.145}$$

is small gives

$$\sqrt{|\alpha_0|}(l-l_0) \sim \frac{\pi}{2}, \qquad l - l_0 \sim \frac{\pi}{2\sqrt{|\alpha_0|}}, \qquad \frac{l-l_0}{2} \sim \frac{\pi}{4\sqrt{|\alpha_0|}}, \tag{4.146}$$

so that Eqs. (4.127) and (4.128) become

$$\xi(l_0) \sim \mathfrak{a}\, e^{(l-l_0)/2} \sim \mathfrak{a} \exp\left(\frac{\pi}{4\sqrt{|\alpha_0|}}\right). \tag{4.147}$$

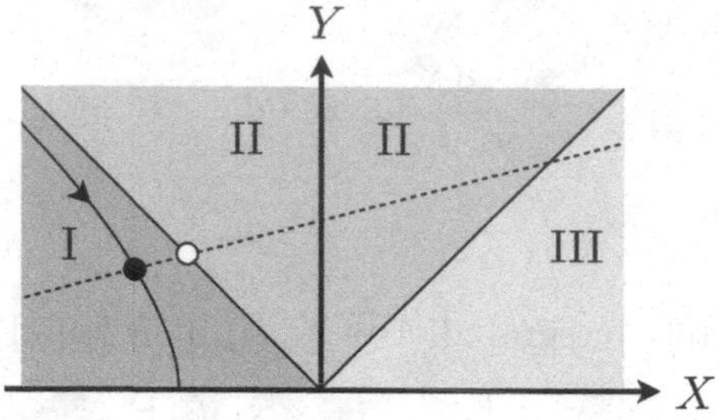

Fig. 4.7 Kosterlitz RG flow in the vicinity of fixed point $X = Y = 0$ in region I. Initial conditions (black dot) correspond to a temperature below T_c (white dot).

4.6.3 *Universal jump of the spin stiffness*

Below the critical temperature T_c (white dot in Fig. 4.7), the RG trajectory takes the initial coupling constants (black dot in Fig. 4.7) to the Gaussian fixed line $Y = 0$ at some particular value of $X(l = \infty) > X_0$ or, equivalently, $K(l = \infty) < K_0$. A finite value of $K(l = \infty)$ means that screening effects of the bare logarithmic vortex interaction are only partially effective in making spin-spin correlation functions decay faster with large separations. The exponent

$$\eta(K_0) = \lim_{l\to\infty} \frac{1}{2\pi K(l)}, \tag{4.148}$$

which characterizes the algebraic decay of the spin-spin two-point function in region I, is non-universal and monotonically decreasing (increasing) function of K (i.e., T). When $K_0 = K_c$ (i.e., $T = T_c$), we may use the fact that $\lim_{l\to\infty} K(l) = K_{\mathrm{KT}} = (2/\pi)$ so that the value

$$\eta(K_c) = \lim_{l\to\infty} \frac{1}{2\pi \times (2/\pi)} = \frac{1}{4} \tag{4.149}$$

is, however, universal.

The quantity $K(l = \infty)$ is called the *spin stiffness* as it measures the sensitivity to changes in the boundary conditions. To illustrate this, choose a rectangular geometry with linear dimensions $L \times L'$ in the $\hat{\boldsymbol{x}}$ and $\hat{\boldsymbol{y}}$ directions. Periodic boundary conditions are imposed in the $\hat{\boldsymbol{y}}$ direction. Twisted boundary conditions are imposed in the $\hat{\boldsymbol{x}}$ direction, i.e., the spin-wave field ϕ obeys

$$\phi(L, y) = \phi(0, y) + \alpha \qquad \phi(x, L') = \phi(x, 0), \qquad \alpha \in \mathbb{R}. \tag{4.150}$$

The change in the dimensionless energy (4.7) induced by changing the boundary condition from periodic to α-twisted is

$$\Delta S_{\mathrm{sw}} = \frac{1}{2} K \left(\frac{\alpha}{L}\right)^2 L L'. \tag{4.151}$$

To see this, observe that if ϕ obeys α-twisted boundary conditions, then $\widetilde{\phi}$, which is defined by

$$\phi(x, y) = \widetilde{\phi}(x, y) + \alpha \frac{x}{L}, \tag{4.152}$$

must obey periodic boundary conditions. The *spin-wave stiffness* Υ_{sw} is defined by

$$\Upsilon_{\mathrm{sw}} := \left(\frac{L}{\mathfrak{a}}\right)^2 \times \frac{\Delta S_{\mathrm{sw}}}{LL'} = \frac{K}{2}. \tag{4.153}$$

It can be shown that in the presence of vortices

$$\Upsilon_{\mathrm{sw}} \to \Upsilon_{\mathrm{sw+Cb}} = \begin{cases} \frac{1}{2}K(l=\infty), & \text{if } K_0 > K_{\mathrm{c}}, \\ \frac{1}{2}K_{\mathrm{KT}}, & \text{if } K_0 = K_{\mathrm{c}}, \\ 0, & \text{if } K_0 < K_{\mathrm{c}}. \end{cases} \tag{4.154}$$

The stiffness $\Upsilon_{\mathrm{sw+Cb}}$ thus exhibits a *universal jump* coming from the paramagnetic phase. As a model of $2d$ superfluidity for thin films of He^4, the $2d$–XY model predicts a jump of the superfluidity density ρ_{s} at the superfluid transition,

$$\frac{\hbar^2 \rho_{\mathrm{s}}(T_{\mathrm{c}})}{m^2 k_{\mathrm{B}} T_{\mathrm{c}}} = \frac{2}{\pi}, \tag{4.155}$$

where m is the mass of the helium atom [47].

4.7 Problems

4.7.1 *The classical two-dimensional random phase XY model*

Introduction

We define the two-dimensional random phase XY model on the square lattice Λ by the partition function

$$Z_{XY}[A] := \left(\prod_{i\in\Lambda}\int_0^{2\pi} \mathrm{d}\phi_i\right) e^{-S_{XY}[A]}, \tag{4.156a}$$

with the classical action

$$S_{XY}[A] := K \sum_{\langle ij\rangle} \left[1 - \cos\left(\phi_i - \phi_j - A_{ij}\right)\right]. \tag{4.156b}$$

The product of the inverse temperature β (the Boltzmann constant $k_{\mathrm{B}} = 1$) with the ferromagnetic exchange coupling $J > 0$ is $K \equiv \beta J \geq 0$, $\langle ij\rangle$ is any directed nearest-neighbor pair of lattice sites from Λ, $0 \leq \phi_i < 2\pi$ is an angle, and the real-valued random numbers $\{A_{ij} \equiv -A_{ji}\}$ obey the distribution law (probability distribution)

$$P[A] := \prod_{\langle ij\rangle} \frac{e^{-A_{ij}^2/(2\,g_A)}}{\sqrt{2\pi\, g_A}}. \tag{4.156c}$$

The choice for this probability distribution is motivated by simplicity, for it only depends on one (dimensionless) coupling g_A and is amenable to (Gaussian) integrations. In the limit $g_A = 0$, the statistical ensemble consist of one element $\{A_{ij} = 0\}$.

This is the clean limit. In the opposite limit $g_A = \infty$, all choices from $\{A_{ij} \in \mathbb{R}\}$ are equally likely. This is the gauge glass limit.

The sample $Z_{XY}[A=0]$ from the statistical ensemble $\{Z_{XY}[A]\}$ of partition functions supports a fully saturated ferromagnetic ordered state at vanishing temperature $1/K = 0$ and a quasi-long-range-ordered phase for $0 < 1/K < \pi/2$ up to the KT transition temperature $1/K = \pi/2$ above which thermal fluctuations select a paramagnetic phase.

Consider any sample $Z_{XY}[A]$ from the statistical ensemble $\{Z_{XY}[A]\}$ of partition functions such that the flux

$$\Phi_{\square} := A_{ij} + A_{jk} + A_{kl} + A_{li} \tag{4.157}$$

through any elementary square plaquette $\square$ from Λ, here labeled counterclockwise by the vertices i, j, k, and l as shown in Fig. 4.2, equals π mod 2π. Such a partition function $Z_{XY}[A]$ realizes a classical two-dimensional frustrated XY magnet whose ground state at vanishing temperature $1/K = 0$ is not the fully saturated ferromagnetic ordered state. (The number of plaquettes in Λ is taken to be even so that $\sum_{\square} \Phi_{\square} = 0 \mod 2\pi$.)

The question to be addressed is which of these two cases is typical of the statistical ensemble $\{Z_{XY}[A]\}$ as a function of K and g_A?

The answer to this question is encoded by the probability distribution

$$P_{XY}[Z] := \int \left(\prod_{\langle ij \rangle} \mathrm{d}A_{ij} \, \frac{e^{-A_{ij}^2/(2 g_A)}}{\sqrt{2\pi \, g_A}} \right) \delta(Z - Z_{XY}[A]). \tag{4.158}$$

However, the probability distribution (4.158) has no more than a symbolic value given that the functional dependence of the partition function $Z_{XY}[A]$ on $\{A_{ij}\}$ is not known in closed form.

The effect of a random phase (disorder) on the phase diagram from Fig. 4.4 and RG flows from Fig. 4.5 of the classical two-dimensional XY model must be studied with the help of less ambitious means than by computing the probability distribution (4.158). Our strategy is going to be to compute the disorder average of some two-point correlation function to the first non-trivial order in the fugacity expansion of section 4.5, where we are now attaching the magnitude of the vorticity whose fugacity is included in the RG analysis. Hence, Y_1 is the (bare) fugacity of vortices carrying the vorticities ± 1. In this way, the stability analysis of the quasi-long-range-ordered phase captured by Fig. 4.4(c) will become the shaded area bounded by the parabola and the horizontal axis in Fig. 4.8(a). According to this calculation, there is a re-entrant phase transition from a quasi-long-range-ordered phase to a paramagnetic phase for any fixed but not too strong disorder strength g_A upon lowering the reduced temperature $1/K$ [48]. What we will not do here is to show that this perturbative stability analysis is misleading [49–51]. Indeed, it can be shown that the regime of stability to the second order in the fugacity expansion shrinks as is shown in Fig. 4.8(b) [52]. The breakdown of this perturbative approach

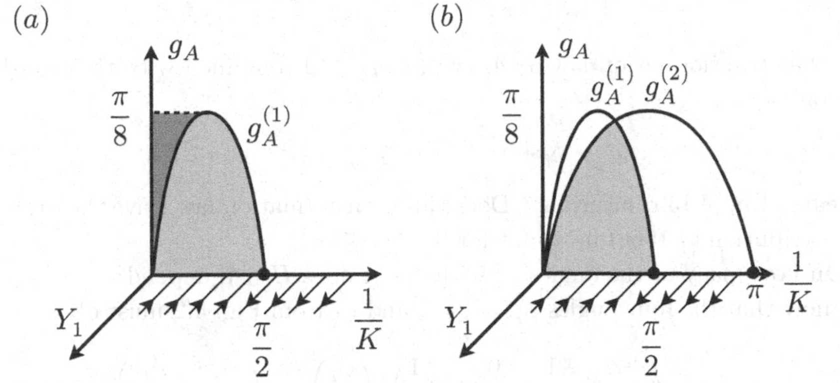

Fig. 4.8 (a) Stability analysis of the critical plane $Y_1 = 0$ in the classical two-dimensional random phase XY model to first-order in perturbation theory in powers of the bare charge-one fugacity Y_1. The coupling constant $1/K$ is the reduced temperature, i.e., the temperature in units of the spin exchange coupling. The coupling constant g_A is the variance of the random phases. It is dimensionless and measures the strength of the disorder. The parabolic boundary $g_A^{(1)}(1/K)$ is the value of the disorder strength beyond which Y_1 becomes relevant in a one-loop RG analysis. The additional dark-shaded area below the horizontal dashed line results from a stability analysis that is non-perturbative in Y_1. (b) If the stability analysis of (a) is extended to second order in perturbation theory in powers of the bare charge-one fugacity, then the regime of stability of the quasi-long-range-ordered phase with the charge-one and charge-two fugacities renormalizing to zero is the shaded intersection between the area below the parabolic boundary $g_A^{(1)}(1/K)$ and the area below the parabolic boundary $g_A^{(2)}(1/K)$.

has two possible interpretations [52]. It could either signal that the quasi-long-range-ordered phase is unstable to any g_A, or that this phase remains critical for sufficiently small g_A, but with critical exponents showing a non-analytic dependence on the disorder strength $g_A > 0$. It is believed that the latter scenario holds and that the quasi-long-range-ordered phase at $Y_1 = 0$ is stable in a region of the plane $Y_1 = 0$ approximately delimited by the segment parallel to the $1/K$ axis that joins $(1/K, g_A) = (0, \pi/8)$ to the maximum of the parabola at $(1/K, g_A) = \left(\pi/4, g_A^{(1)}(\pi/4)\right)$ [the dashed segment in Fig. 4.8(a)] and continues with the branch $g_A^{(1)}(1/K)$ for $\pi/4 < 1/K < \pi/2$ of the parabola [53, 54].

The problem of a single plaquette

Exercise 1.1: Define the plaquette Hamiltonian

$$H_\square[A] := -J \sum_{\langle ij\rangle \in \square} \cos\left(\phi_i - \phi_j - A_{ij}\right), \qquad J > 0, \tag{4.159a}$$

where $\square$ denotes the plaquette shown in in Fig. 4.2, with $\langle ij\rangle$ any directed nearest-neighbor pair of sites from $\square$, and the flux

$$\Phi_\square = \sum_{\langle ij\rangle \in \square} A_{ij} \tag{4.159b}$$

is given.

(a) What transformation law of A_{ij} with $\langle ij\rangle \in \square$ combined with the transformation law

$$\phi_i \to \phi_i + \chi_i, \qquad 0 \leq \chi_i < 2\pi, \qquad i \in \square, \tag{4.160}$$

leaves Eq. (4.159) invariant? Does this transformation law leave the probability distribution (4.156c) invariant for $0 < g_A \leq \infty$?

(b) On how many of the angles ϕ_i with $i \in \square$ does $H_\square[A]$ depend?

(c) Show that the four angles ϕ_i, ϕ_j, ϕ_k, and ϕ_l from Fig. 4.2 must obey

$$\begin{pmatrix} +2 & -1 & 0 & -1 \\ -1 & +2 & -1 & 0 \\ 0 & -1 & +2 & -1 \\ -1 & 0 & -1 & +2 \end{pmatrix} \begin{pmatrix} \phi_i \\ \phi_j \\ \phi_k \\ \phi_l \end{pmatrix} = \begin{pmatrix} A_{ij} - A_{li} \\ A_{jk} - A_{ij} \\ A_{kl} - A_{jk} \\ A_{li} - A_{kl} \end{pmatrix} \tag{4.161}$$

if they are to minimize $H_\square[A]$.

(d) Solve Eq. (4.161) when the flux (4.159b) vanishes or equals π and comment on the nature of the ground states.

Factorization into the spin wave and vortex sectors

We assume the naive continuum limit for the classical two-dimensional random phase XY model (4.156) by which

$$S_{XY}[A] \to K \int \mathrm{d}^2\boldsymbol{x}\, \frac{1}{2}\left(\partial_\mu\phi + \partial_\mu\Theta + A_\mu\right)^2 \tag{4.162a}$$

and

$$P[A] \to e^{-\frac{1}{2 g_A}\int \mathrm{d}^2\boldsymbol{x}\, A_\mu^2}. \tag{4.162b}$$

The summation convention over the repeated index $\mu = 1, 2$ is implied. We also adopt the notation of Eq. (4.46) by which ϕ is vortex free while Θ is not.

Exercise 2.1:

(a) It is shown in appendix D that the factorization into a spin-wave and vortex sector holds for the Villain model in the presence of the random phases $A_{ij} \in \mathbb{R}$. Show that the same holds for the continuum limit (4.162) by making use of the decomposition

$$A_\mu =: \tilde{\partial}_\mu\theta + \partial_\mu\eta, \qquad \tilde{\partial}_\mu := \epsilon_{\mu\nu}\,\partial_\nu, \tag{4.163}$$

both on the action (4.162a) and the probability distribution (4.162b).

(b) Express the random magnetic field

$$b := \epsilon_{\mu\nu}\,\partial_\mu A_\nu \tag{4.164}$$

in terms of θ and η defined by Eq. (4.163).

(c) Compute the disorder averages

$$\overline{A_\mu(\boldsymbol{x})\,A_\nu(\boldsymbol{y})} := \frac{\int \mathcal{D}[A]\,P[A]\,A_\mu(\boldsymbol{x})\,A_\nu(\boldsymbol{y})}{\int \mathcal{D}[A]\,P[A]}, \qquad \mu,\nu = 1,2, \tag{4.165}$$

$$\overline{\theta(\boldsymbol{x})\,\theta(\boldsymbol{y})} := \frac{\int \mathcal{D}[A]\,P[A]\,\theta(\boldsymbol{x})\,\theta(\boldsymbol{y})}{\int \mathcal{D}[A]\,P[A]}, \tag{4.166}$$

$$\overline{\eta(\boldsymbol{x})\,\eta(\boldsymbol{y})} := \frac{\int \mathcal{D}[A]\,P[A]\,\eta(\boldsymbol{x})\,\eta(\boldsymbol{y})}{\int \mathcal{D}[A]\,P[A]}, \tag{4.167}$$

$$\overline{b(\boldsymbol{x})\,b(\boldsymbol{y})} := \frac{\int \mathcal{D}[A]\,P[A]\,b(\boldsymbol{x})\,b(\boldsymbol{y})}{\int \mathcal{D}[A]\,P[A]}. \tag{4.168}$$

(d) Show that the spin-wave sector can be represented by the (random) partition function with the (random) Gaussian action

$$S_{\rm sw}[\phi;\eta] := \int \mathrm{d}^2\boldsymbol{x}\,\frac{K}{2}\,\left(\partial_\mu\phi - \partial_\mu\eta\right)^2. \tag{4.169}$$

The gauge transformation $\phi \to \phi + \eta$ decouples the spin wave ϕ from the longitudinal disorder η.

(e) Show that the vortex sector can be represented by the Coulomb (Cb) gas with the action

$$S_{\rm Cb}[\Theta,\theta] := E_{\rm c}\sum_k (m_k - n_k)^2 - \pi K \sum_{k\neq l}(m_k - n_k)\,(m_l - n_l)\,\ln\left|\frac{\boldsymbol{x}_k - \boldsymbol{x}_l}{\ell}\right| \tag{4.170}$$

where $E_{\rm c}$ is the core energy of vortices, $m_k \in \mathbb{Z}$ is the vorticity of a "thermal" vortex at $\boldsymbol{x}_k$ present in Θ, while $n_k \in \mathbb{R}$ is the vorticity of a "quenched" vortex at $\boldsymbol{x}_k$ present in θ.

(f) Explain why the vorticity of the random magnetic field is not quantized while that of Θ is. The transverse disorder θ cannot be gauged away.

(g) Show that the Cb gas with the action (4.170) can be represented by the SG theory with the action

$$S_{\rm SG}[\chi;\theta] := \int \mathrm{d}^2\boldsymbol{x}\left[\frac{1}{2t}(\partial_\mu\chi)^2 - \frac{h_1}{t}\cos\chi + \frac{\mathrm{i}}{2\pi}\,\chi\,(\partial_\mu^2\theta)\right] \tag{4.171a}$$

and the identifications

$$K = \frac{t}{4\pi^2}, \qquad Y_1 \sim \frac{h_1}{2t}. \tag{4.171b}$$

The index 1 of h_1 or Y_1 refers to the vorticity. The bare fugacity h_m or Y_m of thermal vortices with vorticity $\pm m \in \mathbb{Z}$ larger in magnitude than unity is set to zero. Nevertheless, they are always generated under RG. Their irrelevances in the spin-wave phase and at the KT transition justify safely ignoring them when discussing the clean case at $g_A = 0$. They cannot be neglected as soon as $g_A > 0$.

Moments of the two-point function in the spin-wave phase

Exercise 3.1: A prerequisite for the phase diagram shown in Fig. 4.8 is that the spin-wave phase defined by the condition $Y_1 = 0$ remains critical for any g_A. Show that, in the spin-wave sector,

$$\overline{\left\langle e^{+\mathrm{i}\phi(\boldsymbol{x}_1)}\, e^{-\mathrm{i}\phi(\boldsymbol{x}_2)}\right\rangle^{q}} \propto \left|\frac{\mathfrak{a}}{\boldsymbol{x}_1-\boldsymbol{x}_2}\right|^{\frac{q}{2\pi K}} \overline{e^{+\mathrm{i}q\,\eta(\boldsymbol{x}_1)}\, e^{-\mathrm{i}q\,\eta(\boldsymbol{x}_2)}}$$
$$= \left|\frac{\mathfrak{a}}{\boldsymbol{x}_1-\boldsymbol{x}_2}\right|^{\frac{q+g_A\,K\,q^2}{2\pi K}}, \tag{4.172}$$

for any positive q and with $\mathfrak{a}$ the UV cut-off. Angular brackets denote thermal averaging with the normalized measure $\propto \mathcal{D}[\phi]\, e^{-(K/2)\int \mathrm{d}^2\boldsymbol{y}\,(\partial_\mu\phi-\partial_\mu\eta)^2}$. The overline denotes disorder averaging with the normalized measure $\propto \mathcal{D}[\eta]\, e^{-(1/2g_A)\int \mathrm{d}^2\boldsymbol{y}\,(\partial_\mu\eta)^2}$. The effect of disorder is thus, on average, to increase the value of the critical exponent that characterizes the algebraic decay of the q-th moment of the two-point function in the spin-wave phase.

Moments of the two-point function in the SG theory

Define the thermal two-point function

$$\left\langle F_{\boldsymbol{x}_1,\boldsymbol{x}_2}\right\rangle[\theta] := \frac{\int \mathcal{D}[\chi]\, e^{-S_{\mathrm{SG}}[\chi,\theta]}\, e^{\mathrm{i}\chi(\boldsymbol{x}_1)-\mathrm{i}\chi(\boldsymbol{x}_2)}}{Z_{\mathrm{SG}}[\theta]}. \tag{4.173}$$

We have seen in section 4.5 that by expanding $\left\langle F_{\boldsymbol{x}_1,\boldsymbol{x}_2}\right\rangle[\theta=0]$ in powers of a very small fugacity $h_1/2t$, all coefficients of the expansion in the fugacity are ill-defined unless a short distance cut-off $\mathfrak{a}$ is imposed. The arbitrariness in the choice of the short distance cut-off was used to derive RG equations obeyed by the fugacity and the reduced temperature. The RG equations were integrated to determine whether the initial assumption of a very small fugacity is consistent. The irrelevance, marginality, and relevance of the fugacity then determines the spin-wave phase, KT transition, and disordered phase of the XY model, respectively, i.e., Figs. 4.4 and 4.5.

The transformation

$$\chi \to \chi + \frac{\mathrm{i}t}{2\pi}\,\theta \tag{4.174}$$

plays an essential role in what follows.

Exercise 4.1: Show that the fugacity expansion of $\overline{\left\langle F_{\boldsymbol{x}_1,\boldsymbol{x}_2}\right\rangle[\theta]}$ depends on correlation functions calculated for vanishing "magnetic field" h_1 (fugacity $h_1/2t$) such as

$$\int \mathrm{d}^2\boldsymbol{y}_1\cdots \mathrm{d}^2\boldsymbol{y}_{2n}\, \overline{\langle e^{\mathrm{i}[\chi(\boldsymbol{x}_1)-\chi(\boldsymbol{x}_2)+\chi(\boldsymbol{y}_1)+\cdots-\chi(\boldsymbol{y}_{2n})]}\rangle_{h_1=0}}, \tag{4.175a}$$

on the one hand, but also such as

$$\overline{\langle e^{\mathrm{i}[\chi(\boldsymbol{x}_1)-\chi(\boldsymbol{x}_2)]}\rangle_{h_1=0}\left\langle\int \mathrm{d}^2\boldsymbol{y}_1\mathrm{d}^2\boldsymbol{y}_2\, e^{\mathrm{i}[\chi(\boldsymbol{y}_1)-\chi(\boldsymbol{y}_2)]}\right\rangle^{n}_{h_1=0}}, \tag{4.175b}$$

on the other hand. Here, the overlines denote disorder averaging with the normalized measure $\propto \mathcal{D}[\theta]\, e^{-(1/2g_A)\int \mathrm{d}^2\boldsymbol{y}\,(\partial_\mu\theta)^2}$.

Exercise 4.2: Show that

$$\begin{aligned}\overline{\left\langle e^{+\mathrm{i}\chi(\boldsymbol{y}_1)}\, e^{-\mathrm{i}\chi(\boldsymbol{y}_2)}\right\rangle^{q}_{h_1=0}} &\propto \left|\frac{\mathfrak{a}}{\boldsymbol{y}_1-\boldsymbol{y}_2}\right|^{\frac{q\,t}{2\pi}} \overline{e^{-\frac{q\,t}{2\pi}\theta(\boldsymbol{y}_1)}\, e^{+\frac{q\,t}{2\pi}\theta(\boldsymbol{y}_2)}} \\ &= \left|\frac{\mathfrak{a}}{\boldsymbol{y}_1-\boldsymbol{y}_2}\right|^{2\pi\,K\,q(1-g_A\,K\,q)},\end{aligned} \tag{4.176}$$

for any positive q and with $\mathfrak{a}$ the UV cut-off. Compare this result with Eq. (4.172). Contrast the cases $g_A=0$ and $g_A>0$ and give an interpretation.

Exercise 4.3: The parabola

$$g^{(1)}_A\left(\frac{1}{K}\right) := \frac{1}{K}\left(1-\frac{2}{\pi}\frac{1}{K}\right) \tag{4.177}$$

is obtained by requiring that the scaling exponent on the right-hand side of Eq. (4.176) for the first positive integer moment $q=1$ be "marginal", i.e., equals 4. What happens if the scaling exponent is smaller than 4? *Hint:* See the discussion below Eq. (4.104).

Exercise 4.4: The parabola

$$g^{(q)}_A\left(\frac{1}{K}\right) := \frac{1}{K\,q}\left(1-\frac{2}{\pi}\frac{1}{K\,q}\right) \tag{4.178}$$

for any $q>0$ is obtained by requiring that the scaling exponent on the right-hand side of Eq. (4.176) be "marginal", i.e., equals 4.

Screening of quenched vortices by thermal vortices

The ground state of the Cb gas defined in Eq. (4.170) is the state that nucleates integer-valued vortices out of the thermal field Θ in order to minimize the Cb energy contained in the quenched vortices out of the random field θ. Screening is not perfect because the vorticity of the quenched vortices is not quantized.

Exercise 5.1: Show that screening by thermal vortices m_k of quenched vortices n_k is least efficient if $n_k=\pm 1/2$.

Exercise 5.2: Show that the probability distribution for the field θ with the restriction that its vortices are quantized in units of $1/2$ can be interpreted as the classical two-dimensional XY model on a square lattice with the reduced temperature g_A.

Exercise 5.3: Show that $g_A=\pi/8$ is the KT "transition temperature" if the quenched vortices are quantized in units of $1/2$.

Exercise 5.4: Argue that the segment $(Y_1, 1/K, g_A)=(0,0,g_A)$ in Fig. 4.8 with $0\le g_A\le \pi/8$ must be critical.

PART 2
Fermions

Chapter 5

Non-interacting fermions

Outline

The physics of non-interacting fermions is reviewed.

5.1 Introduction

This chapter is a review devoted to the second quantization of fermions and to the thermodynamic and transport properties of the non-interacting electron gas.

When the dispersion of the non-interacting electron gas is assumed to be the non-relativistic parabolic spectrum of electrons in vacuum, we say that we are dealing with the non-interacting jellium model.

The non-interacting jellium model treats electrons in a metal as if they were in vacuum, except for a homogeneous, inert, and positive background charge that restores charge neutrality and represents the crudest approximation to the ions of a metal. This background charge plays no role in this chapter and will thus be omitted entirely. However, this background charge plays an important role when the Coulomb interaction between electrons in the jellium model is accounted for, as we shall see in the next chapter.

After a quick summary of second quantization for fermions (section 5.2), the notions of the Fermi sea and the Fermi surface will be reviewed (section 5.3). We shall see that thermodynamic properties are controlled by the Fermi surface at sufficiently low temperatures. The same is also true of transport properties.

The sections on the time-ordered Green functions for the non-interacting jellium model (section 5.4), the Grassmann coherent states (appendix E.1), fermionic path integrals (appendix E.2), Jordan-Wigner fermions (appendix E.3), the electronic correlation energy (appendix E.4), and the fluctuation-dissipation theorem (appendix E.5) are included for completeness.

5.2 Second quantization for fermions

The Hilbert space for a many-electron system is constructed by taking the direct sum of all antisymmetric (exterior) tensor products of a single-electron Hilbert space. This construction is called second quantization for electrons and is the natural quantum counterpart of the grand-canonical ensemble in classical statistical mechanics. We will present the formalism of second quantization for fermions by taking the fermions to be spinless in order to simplify notation. This economy also makes sense whenever the electronic spin is a mere bystander that plays no consequential role. Furthermore, there are collective excitations in condensed matter systems that, to a good approximation, behave like spinless electrons (see appendix E.3).

Assume that the single-particle Hamiltonian (from now on, $\hbar = 1$ unless specified)

$$H = -\frac{\Delta}{2m} + U(\boldsymbol{r}), \tag{5.1a}$$

with appropriate boundary conditions, has the countable basis of eigenfunctions

$$H\psi_n(\boldsymbol{r}) = \varepsilon_n \psi_n(\boldsymbol{r}), \qquad \int\limits_V \mathrm{d}^d\boldsymbol{r}\,\psi^*_m(\boldsymbol{r})\psi_n(\boldsymbol{r}) = \delta_{m,n}, \qquad \sum_n \psi^*_n(\boldsymbol{r})\psi_n(\boldsymbol{r}') = \delta(\boldsymbol{r}-\boldsymbol{r}'), \tag{5.1b}$$

in the single-particle Hilbert space $\mathcal{H}^{(1)}$ of square integrable and twice differentiable functions on $\mathbb{R}^d$. We also assume that the single-particle potential $U(\boldsymbol{r})$ is bounded from below, i.e., there exists a single-particle and non-degenerate[1] ground-state energy, say ε_0. Hence, the energy eigenvalue index n can be chosen to run over the non-negative integers, $n = 0, 1, 2, \cdots$. The evolution in time of any solution of Schrödinger equation

$$\mathrm{i}\partial_t \Psi(\boldsymbol{r},t) = H\Psi(\boldsymbol{r},t), \qquad \Psi(\boldsymbol{r},t=0) \text{ given}, \tag{5.2a}$$

can be written as

$$\Psi(\boldsymbol{r},t) = \sum_n C_n\,\psi_n(\boldsymbol{r})\,e^{-\mathrm{i}\varepsilon_n t}, \qquad C_n = \int\limits_V \mathrm{d}^d\boldsymbol{r}\,\psi^*_n(\boldsymbol{r})\Psi(\boldsymbol{r},t=0). \tag{5.2b}$$

The formalism of second quantization starts with the following two postulates.

(1) There exists a set of pairs of adjoint operators $\hat{c}^\dagger_n$ (creation operator) and $\hat{c}_n$ (annihilation operator) labeled by the energy eigenvalue index n and obeying the *fermionic algebra*[2]

$$\{\hat{c}_m, \hat{c}^\dagger_n\} = \delta_{m,n}, \qquad \{\hat{c}_m, \hat{c}_n\} = \{\hat{c}^\dagger_m, \hat{c}^\dagger_n\} = 0, \qquad m,n = 0,1,2,\cdots. \tag{5.3}$$

[1] By hypothesis fermions are spinless and there is no Kramer degeneracy associated to the spin-1/2 degrees of freedom of real electrons.

[2] The conventions for the commutator and anticommutator of any two "objects" A and B are $[A,B] := AB - BA$ and $\{A,B\} := AB + BA$, respectively.

(2) There exists a non-degenerate *vacuum state* $|0\rangle$ that is annihilated by all annihilation operators,

$$\hat{c}_n|0\rangle = 0, \qquad n = 0, 1, 2, \cdots . \tag{5.4}$$

With these postulates in hand, we define the Heisenberg representation for the operator-valued field (in short, quantum field),

$$\hat{\psi}^\dagger(\boldsymbol{r},t) := \sum_n \hat{c}^\dagger_n\, \psi^*_n(\boldsymbol{r})\, e^{+\mathrm{i}\varepsilon_n t}, \tag{5.5a}$$

together with its adjoint

$$\hat{\psi}(\boldsymbol{r},t) := \sum_n \hat{c}_n\, \psi_n(\boldsymbol{r})\, e^{-\mathrm{i}\varepsilon_n t}. \tag{5.5b}$$

The fermionic algebra (5.3) endows the quantum fields $\hat{\psi}^\dagger(\boldsymbol{r},t)$ and $\hat{\psi}(\boldsymbol{r},t)$ with the equal-time algebra[3]

$$\{\hat{\psi}(\boldsymbol{r},t), \hat{\psi}^\dagger(\boldsymbol{r}',t)\} = \delta(\boldsymbol{r}-\boldsymbol{r}'), \qquad \{\hat{\psi}(\boldsymbol{r},t), \hat{\psi}(\boldsymbol{r}',t)\} = \{\hat{\psi}^\dagger(\boldsymbol{r},t), \hat{\psi}^\dagger(\boldsymbol{r}',t)\} = 0. \tag{5.9}$$

The quantum fields $\hat{\psi}^\dagger(\boldsymbol{r},t)$ and $\hat{\psi}(\boldsymbol{r},t)$ act on the *"big" many-particle space*

$$\mathcal{F} := \bigoplus_{N=0}^{\infty} \left\{ \bigwedge^N \mathcal{H}^{(1)} \right\}. \tag{5.10a}$$

Here, each $\bigwedge^N \mathcal{H}^{(1)}$ is spanned by states of the form

$$|m_0, \cdots, m_{i-1}, m_i, m_{i+1}, \cdots\rangle := \prod_i \left(\hat{c}^\dagger_i\right)^{m_i} |0\rangle, \qquad m_i = 0, 1, \tag{5.10b}$$

with the condition on $m_i = 0, 1$ that

$$\sum_i m_i = N. \tag{5.10c}$$

The algebra obeyed by the $\hat{c}$'s and their adjoints ensures that $\bigwedge^N \mathcal{H}^{(1)}$ is the N-th antisymmetric power of $\mathcal{H}^{(1)}$, i.e., that the state $|m_0, \cdots, m_{i-1}, m_i, m_{i+1}, \cdots\rangle$ made of N identical particles of which m_i have energy ε_i changes by a sign under exchange of any two of the N particles. Hence, the "big" many-particle Hilbert space

[3] Alternatively, if we start from the classical Lagrangian density

$$\mathcal{L} := (\psi^* \mathrm{i}\partial_t \psi)(\boldsymbol{r},t) - \frac{|\nabla\psi|^2(\boldsymbol{r},t)}{2m} - |\psi|^2(\boldsymbol{r},t)U(\boldsymbol{r}), \tag{5.6}$$

we can elevate the field $\psi(\boldsymbol{r},t)$ and its momentum conjugate

$$\pi(\boldsymbol{r},t) := \frac{\delta\mathcal{L}}{\delta(\partial_t\psi)(\boldsymbol{r},t)} = \mathrm{i}\psi^*(\boldsymbol{r},t) \tag{5.7}$$

to the status of quantum fields $\hat{\psi}(\boldsymbol{r},t)$ and $\hat{\pi}(\boldsymbol{r},t) = \mathrm{i}\hat{\psi}^\dagger(\boldsymbol{r},t)$ obeying the equal-time fermionic algebra

$$\{\hat{\psi}(\boldsymbol{r},t), \hat{\pi}(\boldsymbol{r}',t)\} = \mathrm{i}\delta(\boldsymbol{r}-\boldsymbol{r}'), \qquad \{\hat{\psi}(\boldsymbol{r},t), \hat{\psi}(\boldsymbol{r}',t)\} = \{\hat{\pi}(\boldsymbol{r},t), \hat{\pi}(\boldsymbol{r}',t)\} = 0. \tag{5.8}$$

(5.10a) is the sum over the subspaces of wave functions for N identical particles that are antisymmetric under any odd permutation of the particles labels.[4] This "big" many-particle Hilbert space is called the *fermion Fock space* in physics.

The rule to change the representation of operators from the Schrödinger picture to the second quantized language is best illustrated by the following examples.

Example 1: The second-quantized representation $\hat{H}$ of the single-particle Hamiltonian (5.1a) is

$$\begin{aligned}\hat{H} :=& \int\limits_V \mathrm{d}^d\boldsymbol{r}\, \hat{\psi}^\dagger(\boldsymbol{r},t)\, H\, \hat{\psi}(\boldsymbol{r},t) \\ =& \sum_n \varepsilon_n \hat{c}^\dagger_n \hat{c}_n.\end{aligned} \tag{5.11}$$

As it should be it is explicitly time independent.

Example 2: The second-quantized total particle-number operator $\hat{Q}$ is

$$\begin{aligned}\hat{Q} :=& \int\limits_V \mathrm{d}^d\boldsymbol{r}\, \hat{\psi}^\dagger(\boldsymbol{r},t)\, \mathbb{1}\, \hat{\psi}(\boldsymbol{r},t) \\ =& \sum_n \hat{c}^\dagger_n \hat{c}_n.\end{aligned} \tag{5.12}$$

It is explicitly time independent, as follows from the continuity equation

$$\begin{aligned}&0 = (\partial_t\rho)(\boldsymbol{r},t) + (\boldsymbol{\nabla}\cdot\boldsymbol{J})(\boldsymbol{r},t),\\ &\rho(\boldsymbol{r},t) := |\Psi(\boldsymbol{r},t)|^2,\\ &\boldsymbol{J}(\boldsymbol{r},t) := \frac{1}{2mi}\left[\Psi^*(\boldsymbol{r},t)\left(\boldsymbol{\nabla}\Psi\right)(\boldsymbol{r},t) - \left(\boldsymbol{\nabla}\Psi^*\right)(\boldsymbol{r},t)\Psi(\boldsymbol{r},t)\right],\end{aligned} \tag{5.13}$$

obeyed by Schrödinger equation (5.2a). The number operator $\hat{Q}$ is the infinitesimal generator of global gauge transformations by which all N-particle states in the fermion Fock space are multiplied by the *same* phase factor. Thus, for any $q \in \mathbb{R}$, a global gauge transformation on the Fock space is implemented by the operation

$$|m_0,\cdots,m_{i-1},m_i,m_{i+1},\cdots\rangle \to e^{+\mathrm{i}q\,\hat{Q}}\,|m_0,\cdots,m_{i-1},m_i,m_{i+1},\cdots\rangle \tag{5.14}$$

on states, or, equivalently,[5]

$$\hat{c}^\dagger_n \to e^{+\mathrm{i}q\hat{Q}}\,\hat{c}^\dagger_n\, e^{-\mathrm{i}q\hat{Q}} = e^{+\mathrm{i}q}\,\hat{c}^\dagger_n, \tag{5.17}$$

and

$$\hat{c}_n \to e^{+\mathrm{i}q\hat{Q}}\,\hat{c}_n\, e^{-\mathrm{i}q\hat{Q}} = e^{-\mathrm{i}q}\,\hat{c}_n, \tag{5.18}$$

[4] An odd permutation is made of an odd product of pairwise exchanges.

[5] We made use of

$$[\hat{c}^\dagger\hat{c},\hat{c}] = \hat{c}^\dagger\hat{c}\hat{c} - \hat{c}\hat{c}^\dagger\hat{c} = \hat{c}^\dagger\hat{c}\hat{c} + \hat{c}^\dagger\hat{c}\hat{c} - \hat{c}^\dagger\hat{c}\hat{c} - \hat{c}\hat{c}^\dagger\hat{c} = \hat{c}^\dagger\{\hat{c},\hat{c}\} - \{\hat{c}^\dagger,\hat{c}\}\hat{c} = -\hat{c}, \tag{5.15}$$

and, similarly,

$$[\hat{c}^\dagger\hat{c},\hat{c}^\dagger] = +\hat{c}^\dagger. \tag{5.16}$$

for all pairs of creation and annihilation operators, respectively. Equation (5.17) teaches us that any creation operator carries the particle number +1. Equation (5.18) teaches us that any annihilation operator carries the particle number −1.

Example 3: The second-quantized local particle-number density operator $\hat{\rho}$ and the particle-number current density operator $\hat{\boldsymbol{J}}$ are

$$\hat{\rho}(\boldsymbol{r},t) = \hat{\psi}^{\dagger}(\boldsymbol{r},t)\mathbb{1}\hat{\psi}(\boldsymbol{r},t), \tag{5.19a}$$

and

$$\hat{\boldsymbol{J}}(\boldsymbol{r},t) := \frac{1}{2mi}\left[\hat{\psi}^{\dagger}(\boldsymbol{r},t)\left(\boldsymbol{\nabla}\hat{\psi}\right)(\boldsymbol{r},t) - \left(\boldsymbol{\nabla}\hat{\psi}^{\dagger}\right)(\boldsymbol{r},t)\hat{\psi}(\boldsymbol{r},t)\right], \tag{5.19b}$$

respectively. The continuity equation

$$0 = (\partial_t\hat{\rho})(\boldsymbol{r},t) + (\boldsymbol{\nabla}\cdot\hat{\boldsymbol{J}})(\boldsymbol{r},t), \tag{5.19c}$$

which follows from evaluating the commutator between $\hat{\rho}$ and $\hat{H}$, is obeyed as an operator equation.

The operators $\hat{H}$, $\hat{Q}$, $\hat{\rho}$, and $\hat{\boldsymbol{J}}$ all act on the Fock space $\mathcal{F}$. They are thus distinct from their single-particle counterparts H, Q, ρ, and $\boldsymbol{J}$ whose actions are restricted to the Hilbert space $\mathcal{H}^{(1)}$. By construction, the action of $\hat{H}$, $\hat{Q}$, $\hat{\rho}$ and $\hat{\boldsymbol{J}}$ on the subspace $\bigwedge^1\mathcal{H}^{(1)}$ of $\mathcal{F}$, say, coincide with the action of H, Q, ρ, and $\boldsymbol{J}$ on $\mathcal{H}^{(1)}$.

Example 1: A single-particle wave function is recovered by defining the single-particle state

$$|m\rangle := \hat{c}^{\dagger}_{m}|0\rangle \tag{5.20}$$

and calculating the overlap

$$\langle 0|\hat{\psi}(\boldsymbol{r},t)|m\rangle = \psi_m(\boldsymbol{r})\,e^{\;\mathrm{i}\varepsilon_m t}. \tag{5.21}$$

Example 2: Let $|\Phi_0\rangle$ be the state defined by filling the N lowest energy eigenstates of H,

$$|\Phi_0\rangle := \prod_{j=1}^{N}\hat{c}^{\dagger}_{j}|0\rangle. \tag{5.22}$$

This state is called the *Fermi sea*. The overlap

$$\begin{aligned}\left\langle 0\left|\prod_{j=1}^{N}\hat{\psi}(\boldsymbol{r}_j,t)\right|\Phi_0\right\rangle &= \begin{vmatrix}\psi_1(\boldsymbol{r}_1)\,e^{-\mathrm{i}\varepsilon_1 t} & \psi_2(\boldsymbol{r}_1)\,e^{-\mathrm{i}\varepsilon_2 t} & \cdots & \psi_N(\boldsymbol{r}_1)\,e^{-\mathrm{i}\varepsilon_N t}\\ \psi_1(\boldsymbol{r}_2)\,e^{-\mathrm{i}\varepsilon_1 t} & \psi_2(\boldsymbol{r}_2)\,e^{-\mathrm{i}\varepsilon_2 t} & \cdots & \psi_N(\boldsymbol{r}_2)\,e^{-\mathrm{i}\varepsilon_N t}\\ \vdots & \vdots & \cdots & \vdots\\ \psi_1(\boldsymbol{r}_N)\,e^{-\mathrm{i}\varepsilon_1 t} & \psi_2(\boldsymbol{r}_N)\,e^{-\mathrm{i}\varepsilon_2 t} & \cdots & \psi_N(\boldsymbol{r}_N)\,e^{-\mathrm{i}\varepsilon_N t}\end{vmatrix}\\ &= \exp\left(-\mathrm{i}\sum_{j=1}^{N}\varepsilon_j t\right)\begin{vmatrix}\psi_1(\boldsymbol{r}_1) & \psi_2(\boldsymbol{r}_1) & \cdots & \psi_N(\boldsymbol{r}_1)\\ \psi_1(\boldsymbol{r}_2) & \psi_2(\boldsymbol{r}_2) & \cdots & \psi_N(\boldsymbol{r}_2)\\ \vdots & \vdots & \cdots & \vdots\\ \psi_1(\boldsymbol{r}_N) & \psi_2(\boldsymbol{r}_N) & \cdots & \psi_N(\boldsymbol{r}_N)\end{vmatrix},\end{aligned} \tag{5.23}$$

is the *Slater determinant* representation of the Fermi sea. N-particle states that can be expressed by a *single* $N\times N$ Slater determinant are said to be *decomposable*. Decomposable states form a very small subset of the totality of N-particle states. The so-called *Hartree-Fock approximation* to the quantum many-body problem seeks the best trial function among decomposable states.

5.3 The non-interacting jellium model

The non-interacting jellium model describes non-interacting electrons with the mass m and the electrical charge $-e$ (the electric charge e is chosen positive by convention) moving freely in a box of linear size L. Mathematically, the non-interacting jellium model in the volume $V = L^3$, at temperature $T = (k_{\mathrm{B}}\beta)^{-1}$, and chemical potential μ is defined by the grand-canonical partition function

$$Z(L^3, \beta, \mu) := \mathrm{Tr}_{\mathcal{F}} \left[e^{-\beta(\hat{H} - \mu \hat{N})} \right], \tag{5.24a}$$

with the Hamiltonian and number operators

$$\hat{H} := \sum_{\sigma} \sum_{\boldsymbol{k}} \varepsilon_{\sigma,\boldsymbol{k}} \, \hat{c}^{\dagger}_{\sigma,\boldsymbol{k}} \hat{c}_{\sigma,\boldsymbol{k}}, \qquad \hat{N} := \sum_{\sigma} \sum_{\boldsymbol{k}} \hat{c}^{\dagger}_{\sigma,\boldsymbol{k}} \hat{c}_{\sigma,\boldsymbol{k}}, \qquad \varepsilon_{\sigma,\boldsymbol{k}} := \frac{\hbar^2 \boldsymbol{k}^2}{2m}, \tag{5.24b}$$

acting on the Fock space

$$\mathcal{F} := \mathrm{span} \left\{ \prod_{\iota \equiv (\sigma, \boldsymbol{k})} \left(\hat{c}^{\dagger}_{\iota} \right)^{m_{\iota}} |0\rangle \middle| \sigma = \uparrow, \downarrow, \quad \frac{L}{2\pi} \boldsymbol{k} \in \mathbb{Z}^3, \quad m_{\iota} = 0, 1, \right.$$
$$\left. \hat{c}_{\iota} |0\rangle = 0, \ \{ \hat{c}_{\iota}, \hat{c}^{\dagger}_{\iota'} \} = \delta_{\iota,\iota'}, \ \{ \hat{c}^{\dagger}_{\iota}, \hat{c}^{\dagger}_{\iota'} \} = \{ \hat{c}_{\iota}, \hat{c}_{\iota'} \} = 0 \right\}. \tag{5.24c}$$

The choice of periodic boundary conditions does not affect bulk properties in the thermodynamic limit $L \to \infty$.

What distinguishes the non-interacting jellium model from other non-interacting electron models is the non-relativistic parabolic dispersion and the unboundness of the allowed momenta. In the presence of a weak single-particle periodic perturbation of the jellium model, momenta can be restricted to the first Brillouin zone, i.e., momenta are bounded from above and below in magnitude, although the dispersion remains unbounded from above. In contrast to the jellium model, tight-binding electronic models have a kinetic energy that is bounded from below and from above. Correspondingly, the tight-binding single-particle dispersion is periodic in the extended zone scheme with the periodicity set by the first Brillouin zone.

In this section, we are going to derive the thermodynamic properties of the non-interacting jellium model in the absence of a magnetic field. We will then review the Sommerfeld semi-classical theory of transport for non-interacting electrons. We close with the effects of a magnetic field in the form of Pauli paramagnetism and of Landau diamagnetism.

5.3.1 *Thermodynamics without magnetic field*

Evaluation of the grand-canonical partition function (5.24) is performed in two steps when interactions are absent. First,

$$\begin{aligned} Z(L^3,\beta,\mu) &= \mathrm{Tr}_{\mathcal{F}}\left[e^{-\beta\sum\limits_{\iota=(\sigma,\boldsymbol{k})}(\varepsilon_\iota-\mu)\hat{c}^\dagger_\iota\hat{c}_\iota}\right] \\ &= \mathrm{Tr}_{\mathcal{F}}\left[\prod_{\iota=(\sigma,\boldsymbol{k})} e^{-\beta(\varepsilon_\iota-\mu)\hat{c}^\dagger_\iota\hat{c}_\iota}\right] \\ &= \prod_{\iota=(\sigma,\boldsymbol{k})} \mathrm{Tr}_{\mathcal{F}_\iota}\left[e^{-\beta(\varepsilon_\iota-\mu)\hat{c}^\dagger_\iota\hat{c}_\iota}\right]. \end{aligned} \tag{5.25a}$$

Here, owing to the lack of interactions, we have interchanged the trace and the product whereby the two-dimensional single-particle Fock space

$$\mathcal{F}_\iota := \mathrm{span}\left\{\left(\hat{c}^\dagger_\iota\right)^{m_\iota}|0\rangle\middle| m_\iota=0,1\right\} \tag{5.25b}$$

is introduced. Second, we can now perform the trace over each Fock space $\mathcal{F}_\iota$ labeled by the single-particle quantum number ι independently,

$$\begin{aligned} Z(L^3,\beta,\mu) &= \prod_{\iota=(\sigma,\boldsymbol{k})}\sum_{m_\iota=0,1} e^{-\beta(\varepsilon_\iota-\mu)m_\iota} \\ &= \prod_{\iota=(\sigma,\boldsymbol{k})}\left(1+e^{-\beta(\varepsilon_\iota-\mu)}\right), \end{aligned} \tag{5.26}$$

where $\sigma=\uparrow,\downarrow$, $\boldsymbol{k}=\frac{2\pi}{L}\boldsymbol{n}$, and $\boldsymbol{n}\in\mathbb{Z}^3$. In terms of the Fermi-Dirac distribution

$$f_{\mathrm{FD}}(\varepsilon_\iota) := \frac{1}{e^{\beta(\varepsilon_\iota-\mu)}+1} \Longleftrightarrow 1-f_{\mathrm{FD}}(\varepsilon_\iota) := \frac{e^{\beta(\varepsilon_\iota-\mu)}}{e^{\beta(\varepsilon_\iota-\mu)}+1}, \tag{5.27a}$$

Eq. (5.26) becomes

$$Z(L^3,\beta,\mu) = \prod_{\iota=(\sigma,\boldsymbol{k})}\frac{1}{1-f_{\mathrm{FD}}(\varepsilon_\iota)}. \tag{5.27b}$$

The internal energy U of the non-interacting jellium model is

$$\begin{aligned} U(L^3,\beta,\mu) &:= \frac{\mathrm{Tr}_{\mathcal{F}}\left[e^{-\beta\left(\hat{H}-\mu\hat{N}\right)}\hat{H}\right]}{\mathrm{Tr}_{\mathcal{F}}\left[e^{-\beta\left(\hat{H}-\mu\hat{N}\right)}\right]} \\ &= -\frac{\partial\ln Z(L^3,\beta,\mu)}{\partial\beta}+\beta^{-1}\mu\frac{\partial\ln Z(L^3,\beta,\mu)}{\partial\mu} \\ &= \sum_{\iota=(\sigma,\boldsymbol{k})}\frac{e^{-\beta(\varepsilon_\iota-\mu)}\varepsilon_\iota}{1+e^{-\beta(\varepsilon_\iota-\mu)}} \\ &= \sum_{\iota=(\sigma,\boldsymbol{k})} f_{\mathrm{FD}}(\varepsilon_\iota)\,\varepsilon_\iota. \end{aligned} \tag{5.28a}$$

The grand-canonical potential F of the non-interacting jellium model is

$$\begin{aligned} F(L^3,\beta,\mu) &:= -\beta^{-1}\ln Z(L^3,\beta,\mu) \\ &= -\beta^{-1}\sum_{\iota=(\sigma,\boldsymbol{k})}\ln\left(1+e^{-\beta(\varepsilon_\iota-\mu)}\right) \\ &= +\beta^{-1}\sum_{\iota=(\sigma,\boldsymbol{k})}\ln\Big(1-f_{\mathrm{FD}}(\varepsilon_\iota)\Big). \end{aligned} \tag{5.28b}$$

The entropy S of the non-interacting jellium model is

$$\begin{aligned} S(L^3,\beta,\mu) &:= -\frac{\partial F(L^3,\beta,\mu)}{\partial T} \\ &= -k_{\mathrm{B}}\frac{\partial F(L^3,\beta,\mu)}{\partial\beta^{-1}} \\ &= -k_{\mathrm{B}}\sum_{\iota=(\sigma,\boldsymbol{k})}\Big[f_{\mathrm{FD}}(\varepsilon_\iota)\ln f_{\mathrm{FD}}(\varepsilon_\iota)+\Big(1-f_{\mathrm{FD}}(\varepsilon_\iota)\Big)\ln\Big(1-f_{\mathrm{FD}}(\varepsilon_\iota)\Big)\Big]. \end{aligned} \tag{5.28c}$$

The pressure P of the non-interacting jellium model is obtained in two steps. First, differentiation yields

$$\begin{aligned} P(L^3,\beta,\mu) &:= -\frac{\partial F(L^3,\beta,\mu)}{\partial L^3} \\ &= +\beta^{-1}\frac{\partial}{\partial L^3}\sum_{\iota=(\sigma,\boldsymbol{k})}\ln\left(1+e^{-\beta(\varepsilon_\iota-\mu)}\right) \\ &= +\beta^{-1}\sum_{\iota=(\sigma,\boldsymbol{k})}\frac{e^{-\beta(\varepsilon_\iota-\mu)}}{1+e^{-\beta(\varepsilon_\iota-\mu)}}(-)\beta\frac{\partial\varepsilon_\iota}{\partial L^3}. \end{aligned} \tag{5.28d}$$

Second, for a quadratic dispersion, $\frac{\partial\varepsilon_\iota}{\partial L^3}\propto\frac{\partial(L^3)^{-2/3}}{\partial L^3}=-(2/3)(L^3)^{-2/3-1}$, so that

$$\begin{aligned} P(L^3,\beta,\mu) &= +\frac{2}{3}L^{-3}\sum_{\iota=(\sigma,\boldsymbol{k})}\frac{e^{-\beta(\varepsilon_\iota-\mu)}\varepsilon_\iota}{1+e^{-\beta(\varepsilon_\iota-\mu)}} \\ &= +\frac{2}{3}L^{-3}\sum_{\iota=(\sigma,\boldsymbol{k})}f_{\mathrm{FD}}(\varepsilon_\iota)\,\varepsilon_\iota \\ &\overset{\text{Eq. (5.28a)}}{=} \frac{2}{3}L^{-3}\times U(L^3,\beta,\mu). \end{aligned} \tag{5.28e}$$

The average number of electrons is

$$\begin{aligned} N_{\mathrm{e}}(L^3,\beta,\mu) &:= \frac{\mathrm{Tr}_{\mathcal{F}}\left[e^{-\beta(\hat{H}-\mu\hat{N})}\hat{N}\right]}{\mathrm{Tr}_{\mathcal{F}}\left[e^{-\beta(\hat{H}-\mu\hat{N})}\right]} \\ &= \beta^{-1}\frac{\partial\ln Z(L^3,\beta,\mu)}{\partial\mu} \\ &= \sum_{\iota=(\sigma,\boldsymbol{k})}\frac{e^{-\beta(\varepsilon_\iota-\mu)}}{1+e^{-\beta(\varepsilon_\iota-\mu)}} \\ &= \sum_{\iota=(\sigma,\boldsymbol{k})}f_{\mathrm{FD}}(\varepsilon_\iota), \end{aligned} \tag{5.28f}$$

while the average occupation number of the single-particle level $\iota = (\sigma, \boldsymbol{k})$ is

$$\begin{aligned}\langle \hat{c}^\dagger_\iota \hat{c}_\iota \rangle_{L^3,\beta,\mu} :=& \frac{\mathrm{Tr}_{\mathcal{F}} \left[e^{-\beta(\hat{H}-\mu\hat{N})} \hat{c}^\dagger_\iota \hat{c}_\iota \right]}{\mathrm{Tr}_{\mathcal{F}} \left[e^{-\beta(\hat{H}-\mu\hat{N})} \right]} \\ =& f_{\mathrm{FD}}(\varepsilon_\iota). \end{aligned} \tag{5.29}$$

We now take the thermodynamic limit $L \to \infty$. In this limit, the single-particle spectrum becomes continuous. Correspondingly, the density of states per unit energy and per unit volume (a distribution)

$$\nu(\varepsilon, L^3) := L^{-3} \sum_{\iota=(\sigma,\boldsymbol{k})} \delta\left(\varepsilon - \varepsilon_\iota\right) \tag{5.30}$$

becomes the continuous function of the single-particle energy ε,[6]

$$\begin{aligned}\nu(\varepsilon) =& \sum_{\sigma=\uparrow,\downarrow} \int \frac{\mathrm{d}^3\boldsymbol{k}}{(2\pi)^3} \, \delta\left(\varepsilon - \frac{\hbar^2 \boldsymbol{k}^2}{2m}\right) \\ =& 2 \times 4\pi \int_0^{+\infty} \frac{\mathrm{d}k\, k^2}{8\pi^3} \, \delta\left(\varepsilon - \frac{\hbar^2 k^2}{2m}\right), \end{aligned} \tag{5.32}$$

as 4π is the area of the unit sphere. With the help of $\omega := \frac{\hbar^2 k^2}{2m}$, $k = \frac{\sqrt{2m\omega}}{\hbar}$, $\mathrm{d}k = \mathrm{d}\omega \sqrt{\frac{m}{2\hbar^2\omega}}$ there then follows that

$$\begin{aligned}\nu(\varepsilon) =& 2 \times \frac{1}{2\pi^2} \frac{m}{\hbar^2} \sqrt{\frac{2m}{\hbar^2}} \int_0^{+\infty} \mathrm{d}\omega \, \omega^{1/2} \, \delta\left(\varepsilon - \omega\right) \\ =& 2 \times \frac{1}{2\pi^2} \frac{m}{\hbar^2} \sqrt{\frac{2\, m\, \varepsilon}{\hbar^2}} \, \Theta(\varepsilon). \end{aligned} \tag{5.33}$$

Here, we have introduced the Heaviside step function

$$\Theta(x) := \begin{cases} 1, & \text{if } x > 0, \\ 0, & \text{if } x < 0. \end{cases} \tag{5.34}$$

In the thermodynamic limit $L \to \infty$, the internal energy per unit volume u, the grand-canonical potential per unit volume f, the entropy per unit volume s, the

---

[6] Dimensional analysis gives the estimate

$$\nu(\varepsilon) \propto |\boldsymbol{k}(\varepsilon)|^d \times \varepsilon^{-1} \propto \varepsilon^{(d/n)-1}, \tag{5.31}$$

in d dimensions and with the dispersion $\varepsilon(\boldsymbol{k}) \propto |\boldsymbol{k}|^n$.

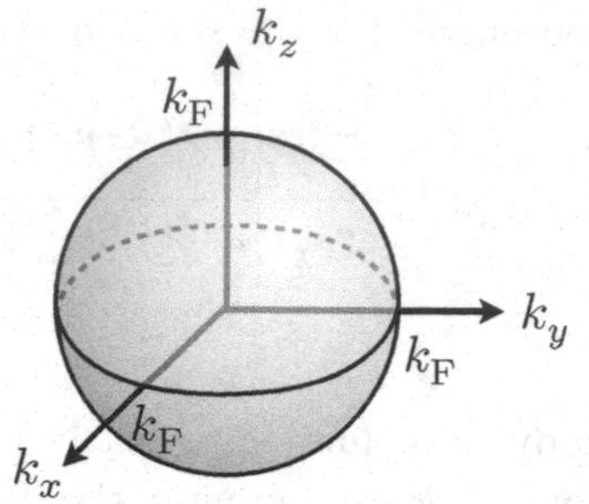

Fig. 5.1 The Fermi sea of the non-interacting jellium model is a sphere in momentum or wave number space. The Fermi surface is the surface of the sphere.

pressure p, and the average number of electrons per unit volume n_{e} are given by

$$\begin{aligned}
u(\beta,\mu) &:= \lim_{L\to\infty} L^{-3}\, U(L^3,\beta,\mu) \\
&= \int_{\mathbb{R}} \mathrm{d}\varepsilon\, \nu(\varepsilon)\, f_{\mathrm{FD}}(\varepsilon)\, \varepsilon, && (5.35\text{a}) \\
f(\beta,\mu) &:= \lim_{L\to\infty} L^{-3}\, F(L^3,\beta,\mu) \\
&= \beta^{-1} \int_{\mathbb{R}} \mathrm{d}\varepsilon\, \nu(\varepsilon)\, \ln\Big(1 - f_{\mathrm{FD}}(\varepsilon)\Big), && (5.35\text{b}) \\
s(\beta,\mu) &:= \lim_{L\to\infty} L^{-3}\, S(L^3,\beta,\mu) && (5.35\text{c}) \\
&= -k_{\mathrm{B}} \int_{\mathbb{R}} \mathrm{d}\varepsilon\, \nu(\varepsilon) \Big[f_{\mathrm{FD}}(\varepsilon) \ln f_{\mathrm{FD}}(\varepsilon) + \Big(1 - f_{\mathrm{FD}}(\varepsilon)\Big) \ln \Big(1 - f_{\mathrm{FD}}(\varepsilon)\Big)\Big], \\
p(\beta,\mu) &:= \lim_{L\to\infty} P(L^3,\beta,\mu) \\
&= \frac{2}{3}\, u(\beta,\mu), && (5.35\text{d}) \\
n_{\mathrm{e}}(\beta,\mu) &:= \lim_{L\to\infty} L^{-3}\, N_{\mathrm{e}}(L^3,\beta,\mu) \\
&= \int_{\mathbb{R}} \mathrm{d}\varepsilon\, \nu(\varepsilon)\, f_{\mathrm{FD}}(\varepsilon), && (5.35\text{e})
\end{aligned}$$

respectively. Equations (5.35a–5.35e) hold for any non-interacting Fermi system once the thermodynamic limit of the single-particle density of states is known.

In the thermodynamic limit $L \to \infty$, the ground state of the non-interacting jellium model is the Fermi sea with the Fermi wave vector k_{F}. The Fermi sea shown in Fig. 5.1 is the sphere with the radius

$$k_{\mathrm{F}} \tag{5.36a}$$

and the volume

$$4\pi k_{\mathrm{F}}^3/3 \tag{5.36b}$$

obtained by filling all single-particle levels ε_ι where $\iota = (\sigma, \boldsymbol{k})$ with the wave vectors satisfying

$$0 \le |\boldsymbol{k}| \le k_{\rm F}. \tag{5.36c}$$

Since there is a total of

$$\frac{4\pi k_{\rm F}^3/3}{(2\pi)^3} = \frac{k_{\rm F}^3}{6\pi^2} \tag{5.37}$$

single-particle wave vectors available per unit volume, the Fermi wave vector is given by

$$n_{\rm e} = 2 \times \frac{k_{\rm F}^3}{6\pi^2} \Longleftrightarrow k_{\rm F} = \left(3\pi^2 n_{\rm e}\right)^{1/3} \tag{5.38}$$

when the number of electrons per unit volume is given by $n_{\rm e}$. The Fermi energy $\varepsilon_{\rm F}$ is the largest single-particle energy that is occupied in the Fermi sea,

$$\varepsilon_{\rm F} = \varepsilon_\iota, \qquad \iota = (\sigma, k_{\rm F}) \Longleftrightarrow \varepsilon_{\rm F} = \varepsilon_\iota = \frac{\hbar^2 k_{\rm F}^2}{2m}. \tag{5.39}$$

The Fermi wave vector (or the Fermi energy) defines the Fermi surface. Single-particle states above the Fermi surface are unoccupied, while they are occupied below it in the ground state of the non-interacting jellium model. The Fermi energy of the non-interacting jellium model takes the form

$$\varepsilon_{\rm F} = \frac{\hbar^2 k_{\rm F}^2}{2m} = \frac{e^2 a_{\rm B} k_{\rm F}^2}{2} = \frac{e^2}{2a_{\rm B}} \left(k_{\rm F} a_{\rm B}\right)^2 = {\rm Ry} \times \left(k_{\rm F} a_{\rm B}\right)^2 \tag{5.40}$$

when expressed in term of the Bohr radius

$$a_{\rm B} := \frac{\hbar^2}{me^2} \tag{5.41}$$

and the ground-state binding energy

$${\rm Ry} := \frac{e^2}{2a_{\rm B}} \tag{5.42}$$

of the hydrogen atom, i.e., 13.6 eV. As good metals have

$$k_{\rm F}\, a_{\rm B} \approx 1 \tag{5.43}$$

of the order unity, their Fermi energy have the magnitude of a typical atomic binding energy. The Fermi wave vector also defines the Fermi velocity

$$v_{\rm F} := \frac{\hbar k_{\rm F}}{m}, \tag{5.44}$$

which is three orders of magnitude smaller than the velocity of light for good metals. Neglecting relativistic effects to describe electrons in good metals is therefore justified to a first approximation. For copper,

$$\begin{aligned} [k_{\rm F}] &= 13.6\, nm^{-1}, & [\varepsilon_{\rm F}] &= 7.03\, eV, \\ [\lambda_{\rm F}] &= 0.46\, nm, & [v_{\rm F}] &= 0.005\, c. \end{aligned} \tag{5.45}$$

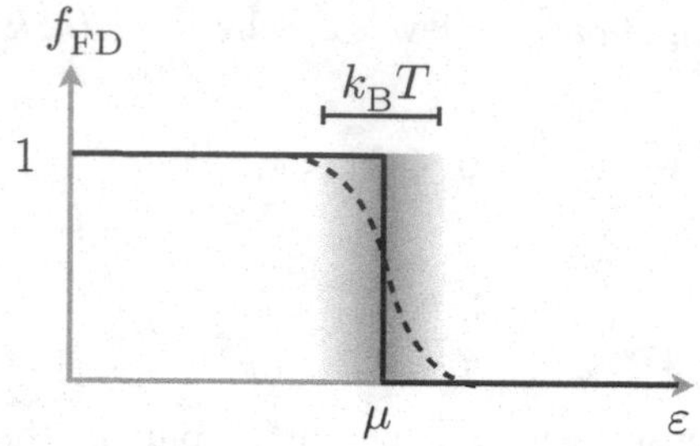

Fig. 5.2 The Fermi-Dirac function $f_{\rm FD}(\varepsilon) := \left(e^{\beta(\varepsilon-\mu)}+1\right)^{-1}$ is an analytic function of the energy ε at any finite temperature. At zero temperature, it is discontinuous at the chemical potential μ. The unit step at $\beta=\infty$ when $\varepsilon=\mu$ turns into a continuous and monotonic decrease over the energy range $\beta^{-1}=k_{\rm B}T$ at any finite temperature. The sharp Fermi surface at zero temperature is smeared over the temperature range $\beta^{-1}=k_{\rm B}T$ as depicted by the shaded box.

At zero temperature, the Fermi-Dirac distribution (5.27a) shown in Fig. 5.2 is the step function

$$\lim_{\beta\to\infty} f_{\rm FD}(\varepsilon)=\Theta\left(\mu-\varepsilon\right), \tag{5.46}$$

whose derivative with respect to energy is the delta function

$$\lim_{\beta\to\infty}\left(\frac{\mathrm{d}f_{\rm FD}}{\mathrm{d}\varepsilon}\right)(\varepsilon)=-\delta\left(\mu-\varepsilon\right). \tag{5.47}$$

This suggests a Taylor expansion about the chemical potential μ of the function

$$h(\varepsilon):=\int\limits_{-\infty}^{\varepsilon}\mathrm{d}\varepsilon'\,g(\varepsilon')\Longleftrightarrow\left(\frac{\mathrm{d}h}{\mathrm{d}\varepsilon}\right)(\varepsilon):=g(\varepsilon), \tag{5.48}$$

which appears in the integral

$$\int\limits_{\mathbb{R}}\mathrm{d}\varepsilon\,g(\varepsilon)\,f_{\rm FD}(\varepsilon)=\int\limits_{\mathbb{R}}\mathrm{d}\varepsilon\,h(\varepsilon)\left(-\frac{\mathrm{d}f_{\rm FD}}{\mathrm{d}\varepsilon}\right)(\varepsilon), \tag{5.49}$$

provided g vanishes as $\epsilon\to-\infty$ and diverges no faster than polynomially for $\epsilon\to+\infty$. The so-called Sommerfeld expansion

$$\begin{aligned}
\int\limits_{\mathbb{R}}\mathrm{d}\varepsilon\,g(\varepsilon)\,f_{\rm FD}(\varepsilon)&=\int\limits_{\mathbb{R}}\mathrm{d}\varepsilon\,h(\varepsilon)\left(-\frac{\mathrm{d}f_{\rm FD}}{\mathrm{d}\varepsilon}\right)(\varepsilon)\\
&=\int\limits_{\mathbb{R}}\mathrm{d}\varepsilon\left[h(\mu)+\sum_{m=1}^{\infty}\frac{(\varepsilon-\mu)^m}{m!}\left(\frac{\mathrm{d}^m h}{\mathrm{d}\varepsilon^m}\right)(\mu)\right]\left(-\frac{\mathrm{d}f_{\rm FD}}{\mathrm{d}\varepsilon}\right)(\varepsilon)\\
&=h(\mu)+\sum_{m=1}^{\infty}\left(\frac{\mathrm{d}^{2m}h}{\mathrm{d}\varepsilon^{2m}}\right)(\mu)\int\limits_{\mathbb{R}}\mathrm{d}\varepsilon\,\frac{(\varepsilon-\mu)^{2m}}{(2m)!}\left(-\frac{\mathrm{d}f_{\rm FD}}{\mathrm{d}\varepsilon}\right)(\varepsilon)\\
&=\int\limits_{-\infty}^{\mu}\mathrm{d}\varepsilon\,g(\varepsilon)+\sum_{m=1}^{\infty}a_m\left(\frac{\mathrm{d}^{2m-1}g}{\mathrm{d}\varepsilon^{2m-1}}\right)(\mu)\,(k_{\rm B}T)^{2m}
\end{aligned} \tag{5.50}$$

follows. To reach the third equality, we used the fact that

$$\left(-\frac{\mathrm{d}f_{\mathrm{FD}}}{\mathrm{d}\varepsilon}\right)(\varepsilon)=\frac{\beta/4}{\cosh^2\left(\frac{\beta}{2}\left(\varepsilon-\mu\right)\right)} \tag{5.51}$$

is an even function of $\varepsilon-\mu$ at any temperature. To reach the last equality, we re-expressed h in terms of g and used the dimensionless integration variable

$$x:=\beta(\varepsilon-\mu) \tag{5.52}$$

to write

$$\begin{aligned}\int_{\mathbb{R}}\mathrm{d}\varepsilon\,\frac{(\varepsilon-\mu)^{2m}}{(2m)!}\left(-\frac{\mathrm{d}f_{\mathrm{FD}}}{\mathrm{d}\varepsilon}\right)(\varepsilon)&=(k_{\mathrm{B}}T)^{2m}\times\frac{1}{(2m)!}\int_{\mathbb{R}}\mathrm{d}x\,x^{2m}\left(-\frac{\mathrm{d}}{\mathrm{d}x}\frac{1}{e^x+1}\right)\\&=(k_{\mathrm{B}}T)^{2m}\times\underbrace{2\sum_{i=1}^{\infty}\frac{(-1)^{i+1}}{i^{2m}}}_{\equiv a_m}\\&=a_m\,(k_{\mathrm{B}}T)^{2m},\qquad m=1,2,\cdots.\end{aligned} \tag{5.53}$$

The coefficients a_m introduced in the last equality are, up to a factor of 2, the values taken by the Dirichlet eta function $\eta(s):=\sum_{n=1}^{\infty}(-1)^{n-1}n^{-s}$. If g varies significantly on the energy scale of μ, i.e.,

$$\left(\frac{\mathrm{d}^{2m-1}g}{\mathrm{d}\varepsilon^{2m-1}}\right)(\mu)\approx\frac{g(\mu)}{\mu^{2m-1}},\qquad m=1,2,\cdots, \tag{5.54}$$

then the ratio of two successive terms in the Sommerfeld expansion is of the order $\mathcal{O}\left((k_{\mathrm{B}}T/\mu)^2\right)$ so that, up to order four in the Sommerfeld expansion,

$$\int_{\mathbb{R}}\mathrm{d}\varepsilon\,g(\varepsilon)\,f_{\mathrm{FD}}(\varepsilon)=\int_{-\infty}^{\mu}\mathrm{d}\varepsilon\,g(\varepsilon)+\frac{\pi^2}{6}\,g'(\mu)\,(k_{\mathrm{B}}T)^2+\frac{7\pi^4}{360}\,g'''(\mu)\,(k_{\mathrm{B}}T)^4. \tag{5.55}$$

The Sommerfeld expansion (5.55) applied to the internal energy density (5.35a) and the average occupation number density (5.35e) yields

$$\begin{aligned}u(T,\mu)&=\int_{-\infty}^{\mu}\mathrm{d}\varepsilon\,\nu(\varepsilon)\,\varepsilon+\frac{\pi^2}{6}\,(k_{\mathrm{B}}T)^2\,[\mu\nu'(\mu)+\nu(\mu)]+\cdots,\\n_{\mathrm{e}}(T,\mu)&=\int_{-\infty}^{\mu}\mathrm{d}\varepsilon\,\nu(\varepsilon)+\frac{\pi^2}{6}\,(k_{\mathrm{B}}T)^2\,\nu'(\mu)+\cdots,\end{aligned} \tag{5.56}$$

respectively.

We now assume that

$$\mu=\varepsilon_{\mathrm{F}}+\mathcal{O}\left((k_{\mathrm{B}}T)^2\right), \tag{5.57}$$

an assumption whose consistency we shall shortly verify. Under this assumption and owing to the vanishing of the density of states for negative energies,

$$\begin{aligned}
u(T,\mu) &= \int_0^{\varepsilon_\mathrm{F}} \mathrm{d}\varepsilon\, \nu(\varepsilon)\, \varepsilon + \varepsilon_\mathrm{F}\,(\mu - \varepsilon_\mathrm{F})\,\nu(\varepsilon_\mathrm{F}) + \frac{\pi^2}{6}\left[\varepsilon_\mathrm{F}\nu'(\varepsilon_\mathrm{F}) + \nu(\varepsilon_\mathrm{F})\right](k_\mathrm{B}T)^2 + \cdots, \\
n_\mathrm{e}(T,\mu) &= \int_0^{\varepsilon_\mathrm{F}} \mathrm{d}\varepsilon\, \nu(\varepsilon) + (\mu - \varepsilon_\mathrm{F})\,\nu(\varepsilon_\mathrm{F}) + \frac{\pi^2}{6}\nu'(\varepsilon_\mathrm{F})\,(k_\mathrm{B}T)^2 + \cdots .
\end{aligned} \tag{5.58}$$

We also assume that the electronic density is temperature independent

$$n_\mathrm{e}(T,\mu) = n_\mathrm{e}(T=0,\mu) = \int_0^{\varepsilon_\mathrm{F}} \mathrm{d}\varepsilon\, \nu(\varepsilon) \equiv n_\mathrm{e}, \qquad \forall T,\mu. \tag{5.59}$$

This implies that the chemical potential μ is a function of temperature and of the electronic density n_e given by

$$\mu = \varepsilon_\mathrm{F} - \frac{\pi^2}{6}\frac{\nu'(\varepsilon_\mathrm{F})}{\nu(\varepsilon_\mathrm{F})}\,(k_\mathrm{B}T)^2 + \cdots , \tag{5.60}$$

while the internal energy density reduces to

$$u(T) = \int_0^{\varepsilon_\mathrm{F}} \mathrm{d}\varepsilon\, \nu(\varepsilon)\, \varepsilon + \frac{\pi^2}{6}\nu(\varepsilon_\mathrm{F})\,(k_\mathrm{B}T)^2 + \cdots . \tag{5.61}$$

This result could have been guessed from the following argument. The difference between the internal energy density at finite and at zero temperature is the product of three factors. First, there is the support $k_\mathrm{B}T$ of the Fermi-Dirac distribution over which it varies significantly at the non-vanishing temperature T. Second, there is the non-vanishing density of states at the Fermi energy $\nu(\varepsilon_\mathrm{F})$. Finally, there is the characteristic excitation energy $k_\mathrm{B}T$ measured relative to the Fermi energy, i.e.,

$$u(T) - \int_0^{\varepsilon_\mathrm{F}} \mathrm{d}\varepsilon\, \nu(\varepsilon)\, \varepsilon \propto \nu(\varepsilon_\mathrm{F})\,(k_\mathrm{B}\,T)^2 . \tag{5.62}$$

At last, the specific heat of the non-interacting jellium model at fixed electronic concentration is

$$\begin{aligned}
C_v(T) &:= \left.\frac{\partial u(T)}{\partial T}\right|_{n_\mathrm{e}} \\
&= \frac{\pi^2}{3}\nu(\varepsilon_\mathrm{F}) k_\mathrm{B}^2 T + \cdots .
\end{aligned} \tag{5.63}$$

This result holds for any non-vanishing single-particle density of states at the Fermi energy. For the non-interacting jellium model

$$C_v(T) = \frac{\pi^2 k_\mathrm{B} n_\mathrm{e}}{2}\left(\frac{k_\mathrm{B}T}{\varepsilon_\mathrm{F}}\right) + \cdots . \tag{5.64}$$

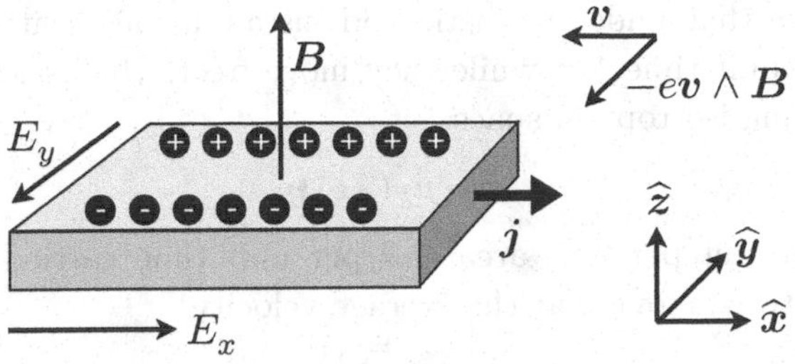

Fig. 5.3 The set up for Hall's experiment is the following. The electric charge e is chosen positive by convention. A *dc* electric field $E_x\,\widehat{\boldsymbol{x}}$ pointing along the positive x-Cartesian axis is applied on a metallic wire. It induces an electronic steady-state current $j_x\widehat{\boldsymbol{x}}$ along the positive x-Cartesian axis. A *dc* magnetic field of magnitude B is pointing along the positive z-Cartesian axis. The Lorentz force $\frac{(-e)\,(-|v_x|)\,B}{c}\widehat{\boldsymbol{x}}\wedge\widehat{\boldsymbol{z}} = -\frac{e\,|v_x|\,B}{c}\widehat{\boldsymbol{y}}$ is balanced by the force induced by the electric field $E_y\,\widehat{\boldsymbol{y}}$ that points along the negative y-Cartesian axis. The latter force is induced by the electric charge that have accumulated on the boundaries along the y-Cartesian axis. Here, we are assuming overall charge neutrality and a steady state. For positive charge carriers, $\boldsymbol{v}$ points along the positive x-Cartesian axis and thus induces an electrical field $E_y\,\widehat{\boldsymbol{y}}$ pointing along the positive y-Cartesian axis. Changing the sign of the charge carrier leaves $j_x = n_{\mathrm{e}}(\mp e)(\mp|v_x|)$ unchanged but reverses the sign of E_y and thus of the the Hall coefficient $R_{\mathrm{H}} := \frac{E_y}{j_x\,B}$.

For comparison, a classical ideal gas has the constant volume specific heat

$$\frac{3k_{\mathrm{B}}n_{\mathrm{e}}}{2}. \tag{5.65}$$

The Fermi-Dirac statistics suppresses the classical result by the multiplicative factor

$$\frac{\pi^2}{3}\frac{k_{\mathrm{B}}T}{\varepsilon_{\mathrm{F}}}. \tag{5.66}$$

The prediction of a linear specific heat for a non-interacting Fermi gas is a simple test of how important electronic interactions are in a metal. It is customary to call the linear coefficient of the temperature dependence of the specific heat the γ coefficient and to plot

$$\left(\frac{C_v}{T}\right) = \gamma + AT^2 \tag{5.67}$$

linearly, i.e., as a function of T^2. For good metals, the linear dependence on temperature of the specific heat becomes comparable to the cubic dependence at a few degrees Kelvin.

5.3.2 *Sommerfeld semi-classical theory of transport*

The semi-classical theory of transport in metals by Sommerfeld is a quantum extension of the classical kinetic theory of transport by Drude. We thus review first the classical theory of transport in metals by Drude.

The Drude model of electrical transport in metals assumes that electricity is carried by small (point-like) hard spheres quantized in the units of e with $e > 0$

the electric charge that undergo elastic and instantaneous scattering events with a probability per unit time $1/\tau$ while they move freely (ballistically) between the collisions. Assuming isotropy in space, let

$$\boldsymbol{j} = n_{\mathrm{e}}\,(-e)\,\boldsymbol{v} \tag{5.68}$$

be the electric current per unit area and per unit time carried by an electronic density n_{e} of electrons moving at the average velocity

$$\boldsymbol{v} = \frac{(-e)\,\boldsymbol{E}\,\tau}{m}, \tag{5.69}$$

induced by a *dc* (static) electric field $\boldsymbol{E}$ between the collisions with probability per unit time $1/\tau$. The linear relation

$$\boldsymbol{j} = \sigma_{\mathrm{D}}\,\boldsymbol{E}, \qquad \sigma_{\mathrm{D}} = \frac{n_{\mathrm{e}}\,e^2\,\tau}{m}, \tag{5.70}$$

defines the Drude conductivity σ_{D} and the Drude resistivity

$$\boldsymbol{E} = \rho_{\mathrm{D}}\,\boldsymbol{j}, \qquad \rho_{\mathrm{D}} = \left(\frac{n_{\mathrm{e}}\,e^2\,\tau}{m}\right)^{-1}. \tag{5.71}$$

Isotropy of space is broken by a *dc* (static) magnetic field $\boldsymbol{B} = B\,\widehat{\boldsymbol{z}}$ perpendicular to a rectangular metallic sample as is shown in Fig. 5.3. One defines the magnetoresistance

$$\rho(B) := \frac{E_x}{j_x} \tag{5.72a}$$

and the Hall coefficient

$$R_{\mathrm{H}} := \frac{E_y}{j_x B} \tag{5.72b}$$

induced by solving the steady-state equation

$$0 = -\left(e\,\boldsymbol{E} + \frac{e\,B}{m\,c}\,(m\,\boldsymbol{v}) \wedge \widehat{\boldsymbol{z}}\right) - \frac{(m\,\boldsymbol{v})}{\tau} \tag{5.72c}$$

with

$$\boldsymbol{E} := E_x\,\widehat{\boldsymbol{x}} + E_y\,\widehat{\boldsymbol{y}}, \qquad \boldsymbol{v} := v_x\,\widehat{\boldsymbol{x}}. \tag{5.72d}$$

Multiplication by

$$-\frac{\sigma_{\mathrm{D}}}{e} = -\frac{n_{\mathrm{e}}\,e\,\tau}{m} \tag{5.73}$$

of the steady-state equation and the introduction of the cyclotron frequency

$$\omega_{\mathrm{c}} := \frac{e\,B}{m\,c} \tag{5.74}$$

yields

$$\sigma_{\mathrm{D}} \begin{pmatrix} E_x \\ E_y \end{pmatrix} = \begin{pmatrix} 1 & +\omega_{\mathrm{c}}\tau \\ -\omega_{\mathrm{c}}\tau & 1 \end{pmatrix} \begin{pmatrix} j_x \\ 0 \end{pmatrix} \Longleftrightarrow \rho(B) = \rho_{\mathrm{D}}, \qquad R_{\mathrm{H}} = -\frac{\omega_{\mathrm{c}}\,\tau}{\sigma_{\mathrm{D}}\,B} = -\frac{1}{n_{\mathrm{e}}\,e\,c}. \tag{5.75}$$

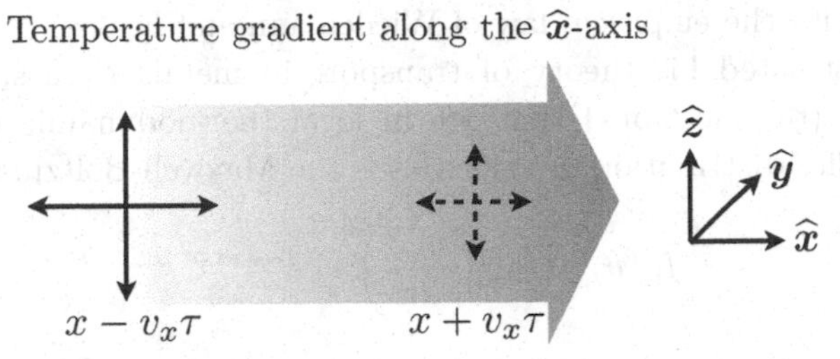

Fig. 5.4 The Drude theory for the thermal current assumes a directionally isotropic distribution of velocities after an elastic and instantaneous scattering event. Two such collisions are depicted by a star of arrows representing the distribution of velocities after scattering. We assume that the direction of the temperature gradient (black arrow) from high to low temperatures is from left to right. This is depicted with the vectors emerging from a scattering event at $x - v_x \tau$ longer than the vectors emerging from a scattering event at $x + v_x \tau$. At a midpoint between the left and right scattering events, the electrons moving from left to right are more energetic than the electrons moving from right to left. This yields a net thermal current to the right that can be modeled by $j_x = \frac{n_e}{2} \times v_x \times E(T(x - v_x \tau)) - \frac{n_e}{2} \times v_x \times E(T(x + v_x \tau)) \approx n_e \times v_x^2 \tau \times \left(\frac{\partial E}{\partial T}\right) \left(-\frac{dT}{dx}\right)$ where n_e is the electronic density, v_x is the velocity at x, $E(T(x \mp v_x \tau))$ is the thermal energy at the last scattering event. Equation (5.77) follows with the identifications $v_x^2 \to \frac{1}{3}\boldsymbol{v}^2$, $n_e \times \left(\frac{\partial E}{\partial T}\right) \to C_v$, and $\left(-\frac{dT}{dx}\right) \to (-\boldsymbol{\nabla} T)$.

The Drude magnetoresistance is independent of the applied magnetic field. The Drude Hall coefficient depends only on the electronic density and on the sign of the charge carrier. Measurements of the Drude conductivity and of the Hall coefficient allow to extract τ and n_e. For good metals n_e is of the order 10^{22} per cubic centimeter and τ is of order 10^{-14} seconds at room temperature (although strongly temperature dependent). Drude's mean free path

$$\ell_D := v_{eqp}\, \tau \tag{5.76a}$$

with the characteristic (equipartition) velocity

$$\frac{1}{2} m\, v_{eqp}^2 := \frac{3}{2} k_B T \tag{5.76b}$$

yields a mean free path of the order of the Ångström at room temperature that is one order of magnitude too small.

The Drude model of thermal transport assumes that the thermal current per unit area and per unit time $\boldsymbol{j}_D$ is given by (see Fig. 5.4)

$$\boldsymbol{j}_D := -\kappa_D\, \boldsymbol{\nabla} T \tag{5.77a}$$

with

$$\kappa_D = \frac{1}{3} v^2\, \tau\, C_v, \qquad \frac{1}{2} m\, v^2 = \frac{3}{2} k_B\, T, \qquad C_v = \frac{3}{2} n_e\, k_B. \tag{5.77b}$$

Drude thus predicts the universal ratio

$$\frac{\kappa_D}{\sigma_D T} = \frac{3}{2} \left(\frac{k_B}{e}\right)^2, \tag{5.78}$$

in agreement with the empirical law of Wiedemann and Franz.

Drude constructed his theory of transport in metals by assuming point-like charge carriers (the electrons) that are in local thermodynamic equilibrium and whose probability distribution of velocities is the Maxwell-Boltzmann distribution

$$f_{\mathrm{B}}(\boldsymbol{v}) := n_{\mathrm{e}} \left(\frac{m\,\beta}{2\pi}\right)^{3/2} e^{-\frac{1}{2}m\,\boldsymbol{v}^2\,\beta}. \tag{5.79}$$

Thus

$$f_{\mathrm{B}}(\boldsymbol{v})\,\mathrm{d}^3\boldsymbol{v} \tag{5.80}$$

is the number of electrons per unit volume with velocities in the range $\mathrm{d}^3\boldsymbol{v}$ about $\boldsymbol{v}$. Sommerfeld's theory of transport in metals simply replaces the Maxwell-Boltzmann distribution (5.79) by the Fermi-Dirac distribution

$$f_{\mathrm{FD}}(\boldsymbol{v}) := 2 \times \frac{1}{(2\pi)^3}\left(\frac{m}{\hbar}\right)^3 \frac{1}{e^{\left(\frac{1}{2}m\,\boldsymbol{v}^2-\mu\right)\beta}+1}. \tag{5.81}$$

This approximation is justified if positions and momenta of electrons can be specified as accurately as necessary without violating the uncertainty principle. Since the typical momentum of an electron in a metal is

$$\hbar\,k_{\mathrm{F}}, \tag{5.82}$$

we must demand that the momentum uncertainty Δp satisfies

$$\Delta p \ll \hbar\,k_{\mathrm{F}}. \tag{5.83}$$

As the uncertainty in the electronic position Δx is given by

$$\Delta x \sim \frac{\hbar}{\Delta p}, \tag{5.84}$$

it follows that

$$\Delta x \gg \frac{1}{k_{\mathrm{F}}}. \tag{5.85}$$

However, for a good metal

$$k_{\mathrm{F}} \sim \frac{1}{a_{\mathrm{B}}}, \tag{5.86}$$

so that

$$\Delta x \gg a_{\mathrm{B}}. \tag{5.87}$$

We conclude that a classical description of electrons requires that the uncertainty in their position be much larger than the Bohr radius. A classical description of transport in metals is prohibited if electrons are localized in space within atomic distances. Two characteristic length scales enter the Sommerfeld's or Drude's theory of transport in metals. First, there is the characteristic range λ of variations in space of the external probes applied to a metal in order to induce transport, say an electromagnetic field or a temperature gradient. One must demand that

$$\lambda \gg k_{\mathrm{F}}^{-1}, \tag{5.88}$$

for a semi-classical treatment à la Sommerfeld to hold. Second, there is the mean free path ℓ_{S} which must therefore also satisfy

$$\ell_{\mathrm{S}} \gg k_{\mathrm{F}}^{-1}, \tag{5.89}$$

for a semi-classical treatment à la Sommerfeld to hold.

The replacement by the Fermi-Dirac distribution (5.81) of the Maxwell-Boltzmann distribution (5.79) only affects transport coefficients that depend on the equilibrium velocity distribution. If one assumes that the rate $1/\tau$ at which elastic scattering occurs between electrons is independent of the electron energy, then the *dc* conductivity, magnetoresistance, and Hall coefficient agree in the Sommerfeld and Drude models. On the other hand, the Drude mean free path (5.76) is changed to

$$\ell_{\mathrm{S}} = v_{\mathrm{F}}\tau, \tag{5.90}$$

which can be larger than ℓ_{D} by two orders of magnitude at room temperature. Similarly, the thermal velocity

$$\sim \sqrt{k_{\mathrm{B}}T/m} = \sqrt{\left((k_{\mathrm{B}}T)/\varepsilon_{\mathrm{F}}\right) \times (\varepsilon_{\mathrm{F}}/m)} \tag{5.91}$$

in the thermal conductivity (5.77) must be replaced by the Fermi velocity

$$\sim \sqrt{\varepsilon_{\mathrm{F}}/m}, \tag{5.92}$$

while the Drude specific heat

$$\sim n_{\mathrm{e}}k_{\mathrm{B}} \tag{5.93}$$

must be replaced by the smaller specific heat

$$\sim \left(\frac{k_{\mathrm{B}}T}{\varepsilon_{\mathrm{F}}}\right) \times n_{\mathrm{e}}k_{\mathrm{B}}. \tag{5.94}$$

The enhancement factor $\varepsilon_{\mathrm{F}}/\left(k_{\mathrm{B}}T\right)$ induced by the use of the Fermi velocity cancels the reduction factor $(k_{\mathrm{B}}\,T)/\varepsilon_{\mathrm{F}}$ induced by the use of the Fermi gas specific heat. The empirical law of Wiedemann and Franz (5.78) is thus also satisfied in the model of Sommerfeld albeit with the universal coefficient

$$\frac{\kappa_{\mathrm{S}}}{\sigma_{\mathrm{D}}T} = \frac{\pi^2}{3}\left(\frac{k_{\mathrm{B}}}{e}\right)^2. \tag{5.95}$$

5.3.3 *Pauli paramagnetism*

So far we have assumed that the single-particle dispersion ε_ι does not depend on the electronic spin. We are now going to treat a simple model in which the single-particle energy dispersion becomes spin dependent by accounting for a Zeeman term, but neglecting the orbital response to the presence of an external magnetic field $\boldsymbol{B}$ with the magnitude $B = |\boldsymbol{B}|$.

To this end, we recall that the Zeeman energy for a magnetic moment $\boldsymbol{\mu}$ (not to be confused with the chemical potential) in the presence of a uniform magnetic field $\boldsymbol{B}$ is

$$-\boldsymbol{\mu}\cdot\boldsymbol{B}. \tag{5.96}$$

The magnetic moment of an electron with the spin operator $\boldsymbol{S}$ is

$$\boldsymbol{\mu} = -\frac{g\,\mu_{\mathrm{B}}}{\hbar}\,\boldsymbol{S} \approx -\mu_{\mathrm{B}}\,\boldsymbol{\sigma}, \tag{5.97}$$

owing to the negative charge of the electron and the electron g-factor being approximately 2. Hence, we work with the grand-canonical partition function

$$\begin{aligned}
&Z(L^3,\beta,\mu,B) := \mathrm{Tr}_{\mathcal{F}}\left[e^{-\beta(\hat{H}-\mu\hat{N})}\right],\\
&\hat{H} := \sum_{\sigma=\pm1}\sum_{\boldsymbol{k}} \varepsilon_{\sigma,\boldsymbol{k}}\,\hat{c}^{\dagger}_{\sigma,\boldsymbol{k}}\hat{c}_{\sigma,\boldsymbol{k}}, \qquad \hat{N} := \sum_{\sigma=\pm1}\sum_{\boldsymbol{k}} \hat{c}^{\dagger}_{\sigma,\boldsymbol{k}}\hat{c}_{\sigma,\boldsymbol{k}}, \qquad \varepsilon_{\sigma,\boldsymbol{k}} := \frac{\hbar^2\boldsymbol{k}^2}{2m} + \sigma\,\mu_{\mathrm{B}}\,B,\\
&\mathcal{F} := \mathrm{span}\left\{ \prod_{\iota\equiv(\sigma,\boldsymbol{k})} \left(\hat{c}^{\dagger}_{\iota}\right)^{m_\iota}|0\rangle \,\middle|\, \sigma=-1,+1, \quad \frac{L}{2\pi}\boldsymbol{k}\in\mathbb{Z}^3, \quad m_\iota = 0,1, \right.\\
&\qquad\qquad \left. \hat{c}_\iota|0\rangle = 0,\ \{\hat{c}_\iota,\hat{c}^{\dagger}_{\iota'}\} = \delta_{\iota,\iota'},\ \{\hat{c}^{\dagger}_{\iota},\hat{c}^{\dagger}_{\iota'}\} = \{\hat{c}_\iota,\hat{c}_{\iota'}\} = 0 \right\},
\end{aligned} \tag{5.98a}$$

where we have introduced the Bohr magneton (the electric charge e is chosen positive by convention)

$$\mu_{\mathrm{B}} := \frac{e\,\hbar}{2\,m\,c}. \tag{5.98b}$$

The Bohr magneton has the units of a magnetic moment.

We want to compute the statistical average

$$\boldsymbol{M}_{\mathrm{P}}(L^3,\beta,\mu,B) := L^{-3}\frac{\mathrm{Tr}_{\mathcal{F}}\left[e^{-\beta(\hat{H}-\mu\hat{N})}\left(-\frac{g\,\mu_{\mathrm{B}}}{\hbar}\,\boldsymbol{S}\right)\right]}{\mathrm{Tr}_{\mathcal{F}}\left[e^{-\beta(\hat{H}-\mu\hat{N})}\right]}. \tag{5.99}$$

Because the uniform magnetic field $\boldsymbol{B}$ only breaks the $SU(2)$ spin-rotation symmetry down to the subgroup $U(1)$ of rotations about the quantization axis in spin space, only the component $M_{\mathrm{P}}(L^3,\beta,\mu,B)$ of the magnetization per unit volume (5.99) along the quantization axis that is selected by the applied magnetic field $\boldsymbol{B}$ is non-vanishing. Hence, we are after

$$M_{\mathrm{P}}(L^3,\beta,\mu,B) \equiv:= +L^{-3}\beta^{-1}\frac{\partial \ln Z(L^3,\beta,\mu,B)}{\partial B}, \tag{5.100a}$$

and the corresponding spin susceptibility

$$\chi_{\mathrm{P}}(L^3,\beta,\mu,B) := \frac{\partial M_{\mathrm{P}}(L^3,\beta,\mu,B)}{\partial B} \tag{5.100b}$$

in the thermodynamic limit $L \to \infty$ holding the electronic density n_{e} fixed.

Each electron with spin parallel to $\boldsymbol{B}$ contributes

$$-L^{-3} \times \mu_{\mathrm{B}} \tag{5.101}$$

to the magnetization density. Each electron with spin antiparallel to $\boldsymbol{B}$ contributes

$$+L^{-3} \times \mu_{\mathrm{B}} \tag{5.102}$$

to the magnetization density. If

$$n_{\mathrm{e}\pm}(\beta,\mu,B) \tag{5.103}$$

denotes the density of electrons with spin parallel (+) and antiparallel (−) to $\boldsymbol{B}$ in the thermodynamic limit, then the magnetization density is

$$M_{\mathrm{P}}(\beta,\mu,B) = -\mu_{\mathrm{B}}\left[n_{\mathrm{e}+}(\beta,\mu,B) - n_{\mathrm{e}-}(\beta,\mu,B)\right], \tag{5.104}$$

in the thermodynamic limit. Of course, the constraint

$$n_{\mathrm{e}} = n_{\mathrm{e}+}(\beta,\mu,B) + n_{\mathrm{e}-}(\beta,\mu,B) \tag{5.105}$$

must hold for all β, μ and B. This constraint fixes the dependence of the chemical potential on β and B. For ease of notation, we drop the arguments of $n_{\mathrm{e}\pm}$, M, and χ from now on.

When $B = 0$, the density of states per unit energy, per unit volume, and per spin $\nu_{\pm}(\varepsilon)$ obeys

$$\nu_{\pm}(\varepsilon) = \frac{1}{2}\nu(\varepsilon), \tag{5.106}$$

with $\nu(\varepsilon)$ defined in Eq. (5.33). When $B \neq 0$,

$$\nu_{\pm}(\varepsilon) = \frac{1}{2}\nu(\varepsilon \mp \mu_{\mathrm{B}}\, B), \tag{5.107}$$

for an electron with spin down relative to the quantization axis $\boldsymbol{B}/B$ in spin space lowers its energy by $\mu_{\mathrm{B}}\, B$. Hence,

$$n_{\mathrm{e}\pm} = \int\limits_{\mathbb{R}} \mathrm{d}\varepsilon\, \nu_{\pm}(\varepsilon)\, f_{\mathrm{FD}}(\varepsilon). \tag{5.108}$$

We shall assume that

$$\mu_{\mathrm{B}}\, B \ll \varepsilon_{\mathrm{F}}. \tag{5.109}$$

This is a reasonable assumption since a B of 10^4 Gauss gives $\mu_{\mathrm{B}}B$ of order $10^{-4}\times\varepsilon_{\mathrm{F}}$. We then do the Taylor expansions

$$\begin{aligned}
\nu_{\pm}(\varepsilon) &= \frac{1}{2}\nu(\varepsilon\mp\mu_{\mathrm{B}}B) \\
&= \frac{1}{2}\nu(\varepsilon)\mp\frac{1}{2}\mu_{\mathrm{B}}B\nu'(\varepsilon)+\cdots, \\
n_{\mathrm{e}\pm} &= \int_{\mathbb{R}} \mathrm{d}\varepsilon\,\nu_{\pm}(\varepsilon)\,f_{\mathrm{FD}}(\varepsilon) \\
&= \frac{1}{2}\int_{\mathbb{R}} \mathrm{d}\varepsilon\,\nu(\varepsilon)f_{\mathrm{FD}}(\varepsilon)\mp\frac{1}{2}\mu_{\mathrm{B}}B\int_{\mathbb{R}} \mathrm{d}\varepsilon\,\nu'(\varepsilon)\,f_{\mathrm{FD}}(\varepsilon)+\cdots, \\
M_{\mathrm{P}} &= -\,\mu_{\mathrm{B}}\left(n_{\mathrm{e}+}-n_{\mathrm{e}-}\right) \\
&= +\,\mu_{\mathrm{B}}^2 B\int_{\mathbb{R}} \mathrm{d}\varepsilon\,\nu'(\varepsilon)\,f_{\mathrm{FD}}(\varepsilon)+\cdots \\
&= +\,\mu_{\mathrm{B}}^2 B\int_{\mathbb{R}} \mathrm{d}\varepsilon\,\nu(\varepsilon)\left(-\frac{\mathrm{d}f'_{\mathrm{FD}}(\varepsilon)}{\mathrm{d}\varepsilon}\right)+\cdots,
\end{aligned} \tag{5.110}$$

subject to the constraint that

$$n_{\mathrm{e}} = \int_{\mathbb{R}} \mathrm{d}\varepsilon\,\nu(\varepsilon)\,f_{\mathrm{FD}}(\varepsilon)+\cdots. \tag{5.111}$$

We can then use Eq. (5.60) to solve for the chemical potential

$$\mu = \varepsilon_{\mathrm{F}}+\cdots. \tag{5.112}$$

At zero temperature

$$M_{\mathrm{P}} = \mu_{\mathrm{B}}^2\,\nu(\varepsilon_{\mathrm{F}})\,B, \qquad \chi_{\mathrm{P}} = \mu_{\mathrm{B}}^2\,\nu(\varepsilon_{\mathrm{F}}), \tag{5.113}$$

with corrections of the order $(k_{\mathrm{B}}T/\varepsilon_{\mathrm{F}})^2$ at finite temperature. This result, known as the Pauli paramagnetism, is a dramatic manifestation of the Pauli principle. It should be contrasted to Curie's law

$$\chi_{\mathrm{P}} = n_{\mathrm{i}}\frac{(g_{\mathrm{L}}\mu_{\mathrm{B}})^2}{3}\frac{J(J+1)}{k_{\mathrm{B}}T}+\mathcal{O}\left(\frac{g_{\mathrm{L}}\mu_{\mathrm{B}}\,B}{k_{\mathrm{B}}\,T}\right), \tag{5.114}$$

for non-interacting ions with density n_{i}, total angular momentum quantum number J, and Landé factor g_{L}.

5.3.4 *Landau levels in a magnetic field*

We take the jellium model in the presence of the magnetic field

$$\boldsymbol{B} = \begin{pmatrix} 0 \\ 0 \\ B \end{pmatrix} = \boldsymbol{\nabla}\wedge\begin{pmatrix} 0 \\ Bx \\ 0 \end{pmatrix} \equiv \boldsymbol{\nabla}\wedge\boldsymbol{A}. \tag{5.115a}$$

The relevant single-particle Hamiltonian is the Pauli Hamiltonian for an electron carrying the negative charge $-e$. It is

$$H = \frac{1}{2m}\left(\frac{\hbar}{\mathrm{i}}\boldsymbol{\nabla} - \frac{(-e)}{c}\boldsymbol{A}\right)^2 \sigma_0 - (-\mu_{\mathrm{B}})\,\sigma_3 B$$
$$= \frac{1}{2m}\left[-\hbar^2\partial_x^2 + \left(\frac{\hbar}{\mathrm{i}}\partial_y + \frac{e}{c}B\,x\right)^2 - \hbar^2\partial_z^2\right]\sigma_0 + \mu_{\mathrm{B}}\,\sigma_3\,B. \qquad (5.115\text{b})$$

The eigenvalue problem

$$H\Psi(\boldsymbol{r}) = \varepsilon\Psi(\boldsymbol{r}), \qquad \Psi(\boldsymbol{r}) := \frac{e^{\mathrm{i}k_z z}}{\sqrt{L}} \times \frac{e^{\mathrm{i}k_y y}}{\sqrt{L}} \times \phi(x) \times \xi_\sigma, \qquad 0 \le x, y, z \le L, \qquad (5.116)$$

with $\xi_\sigma \in \mathbb{C}^2$ a two-component spinor and $\frac{L}{2\pi}k_z = m_z \in \mathbb{Z}$, $\frac{L}{2\pi}k_y = m_y \in \mathbb{Z}$, reduces, for any given $0 \le y \le L$, to solving the one-dimensional harmonic oscillator for the wave function ϕ. The corresponding orthonormal eigenfunctions and energy eigenvalues are

$$\phi_{n,k_y}(x) = \left(2^n n!\sqrt{\pi}\ell_{\mathrm{c}}\right)^{-1/2} \times H_n\left(\frac{x + k_y\ell_{\mathrm{c}}^2}{\ell}\right) \times e^{-\left(x+k_y\ell_{\mathrm{c}}^2\right)^2/(2\ell_{\mathrm{c}}^2)}, \qquad 0 \le x \le L, \qquad (5.117\text{a})$$

and

$$\varepsilon_{n,k_z,\sigma} = \frac{\hbar^2 k_z^2}{2m} + \hbar\omega_c\left(n + \frac{1}{2}\right) + \mu_{\mathrm{B}}\,B\,\sigma, \qquad (5.117\text{b})$$

with

$$n = 0, 1, 2, \cdots, \qquad \sigma = \pm, \qquad \omega_{\mathrm{c}} := \frac{|e\,B|}{m\,c}, \qquad \ell_{\mathrm{c}} := \sqrt{\frac{\hbar c}{|e\,B|}}, \qquad (5.117\text{c})$$

respectively. (The functions H_n are the Hermite polynomials.) Energy eigenvalues do not depend on $k_y = 2\pi m_y/L$. Energy levels are thus degenerate. The degeneracy of the energy level with quantum numbers n, k_z, and σ is

$$\frac{L^2}{2\pi\ell_{\mathrm{c}}^2} \qquad (5.118)$$

as follows from the constraint on $k_y\,\ell_{\mathrm{c}}^2$,

$$0 \le \frac{2\pi m_y}{L}\ell_{\mathrm{c}}^2 \le L, \qquad m_y \in \mathbb{Z} \qquad \Longleftrightarrow \qquad 0 \le m_y \le \frac{L^2}{2\pi\ell_{\mathrm{c}}^2}, \qquad m_y \in \mathbb{Z}. \qquad (5.119)$$

In the thermodynamic limit $L \to \infty$, the density of states per unit energy, per unit volume, per spin, and in the n-th Landau level is

$$\nu(\varepsilon, \sigma, n) := \lim_{L\to\infty} L^{-3} \times \frac{L^2}{2\pi\ell_{\mathrm{c}}^2} \times \sum_{k_z} \delta\left(\varepsilon - \varepsilon_{n,k_z,\sigma}\right)$$
$$= \frac{1}{2\pi\ell_{\mathrm{c}}^2} \times \int_{\mathbb{R}} \frac{\mathrm{d}k_z}{2\pi}\delta\left(\varepsilon - \frac{\hbar^2 k_z^2}{2m} - \hbar\omega_c\left(n + \frac{1}{2}\right) - \mu_{\mathrm{B}}\,B\,\sigma\right)$$
$$= \frac{(2m)^{3/2}\,\omega_{\mathrm{c}}}{8\,\pi^2\,\hbar^2}\frac{\Theta\left(\varepsilon - \hbar\omega_c\left(n + \frac{1}{2}\right) - \mu_{\mathrm{B}}B\sigma\right)}{\sqrt{\varepsilon - \hbar\omega_c\left(n + \frac{1}{2}\right) - \mu_{\mathrm{B}}\,B\,\sigma}}. \qquad (5.120)$$

For a fixed $n = 0, 1, 2, \cdots$ and a fixed $\sigma = \pm$, this density of states has a square root singularity that is typical of a free one-dimensional electron gas. The smooth density of state (5.33) is strongly affected by a magnetic field through the square root singularities. The positions of these singularities depend on the magnetic field.

The grand-canonical partition function of the jellium model perturbed by a static and spatially uniform magnetic field pointing along the z Cartesian axis is given by Eq. (5.26) with the identifications ($n = 0, 1, 2, \cdots$)

$$\iota \to (n, k_z, \sigma), \ \varepsilon_\iota \to \frac{\hbar^2 k_z^2}{2m} + \hbar\omega_c \left(n + \frac{1}{2}\right) + \mu_{\mathrm{B}}\, B\, \sigma, \quad \frac{2\pi}{L} k_z \in \mathbb{Z}, \quad \sigma = \pm. \tag{5.121}$$

The magnetization per unit volume M can be calculated in closed form with the help of the Poisson formula. It is [55]

$$M = \chi_{\mathrm{P}} B \left(1 - \frac{1}{3} + \frac{\pi k_{\mathrm{B}} T}{\mu_{\mathrm{B}} B} \sqrt{\frac{\varepsilon_{\mathrm{F}}}{\mu_{\mathrm{B}} B}} \sum_{m=1}^{\infty} \frac{1}{\sqrt{m}} \frac{\sin\left(\frac{\pi}{4} - \frac{\pi \varepsilon_{\mathrm{F}} m}{\mu_{\mathrm{B}} B}\right)}{\sinh\left(\frac{\pi^2 k_{\mathrm{B}} T m}{\mu_{\mathrm{B}} B}\right)}\right). \tag{5.122}$$

The susceptibility

$$\chi := \frac{\partial M}{\partial B} \tag{5.123}$$

reduces to the sum of the Pauli (paramagnetic) susceptibility

$$\chi_{\mathrm{P}} = \mu_{\mathrm{B}}^2\, \nu(\varepsilon_{\mathrm{F}}) \tag{5.124}$$

and the Landau (diamagnetic) susceptibility

$$\chi_{\mathrm{L}} = -\frac{1}{3} \chi_{\mathrm{P}} \tag{5.125}$$

in the limit

$$\frac{k_{\mathrm{B}} T}{\mu_{\mathrm{B}} B} \gg 1. \tag{5.126}$$

In the opposite limit

$$\frac{k_{\mathrm{B}} T}{\mu_{\mathrm{B}} B} \ll 1 \tag{5.127}$$

of very low temperatures, the dependence of χ on $1/B$ oscillates with the dominant period $\Delta(1/B)$ given by

$$\frac{\pi \varepsilon_{\mathrm{F}}}{\mu_{\mathrm{B}}} \Delta(1/B) = 2\pi, \tag{5.128}$$

i.e.,

$$\begin{aligned}
\Delta(1/B) &= \frac{2\mu_{\mathrm{B}}}{\varepsilon_{\mathrm{F}}} \\
&= 2 \times \frac{e\hbar}{2\, m\, c} \times \frac{2m}{\hbar^2 k_{\mathrm{F}}^2} \\
&= \frac{2\pi e}{\hbar c} \frac{1}{A(k_{\mathrm{F}})},
\end{aligned} \tag{5.129}$$

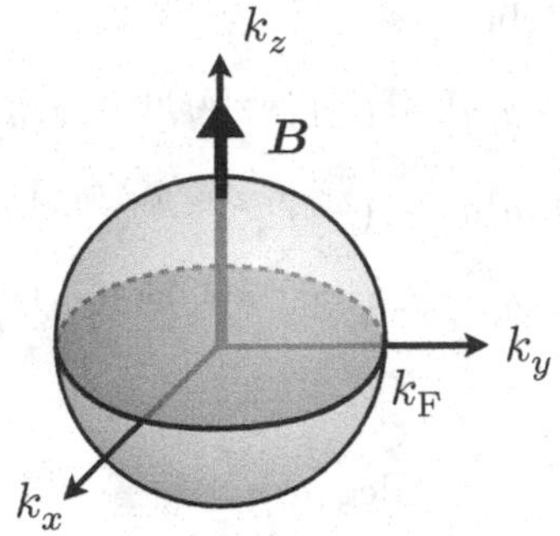

Fig. 5.5 The extremal area among all the discs obtained by intersecting the Fermi sea with planes perpendicular to the applied magnetic field is that of the equatorial plane.

where

$$A(k_{\mathrm{F}}) = \pi k_{\mathrm{F}}^2 \tag{5.130}$$

is the extremal area among all the discs obtained by intersecting the Fermi sea with planes perpendicular to the magnetic field. This oscillatory behavior of the uniform and static magnetic susceptibility for the jellium model was explained by Landau in 1930 within the non-interacting jellium model shortly after qualitatively similar oscillations were measured in metals by de Haas and van Alphen the same year. This is the so-called de-Haas-van-Alphen effect. Onsager showed in 1952 how to generalize Landau's analysis to the nearly free electron model.

5.4 Time-ordered Green functions

5.4.1 *Definitions*

Before specializing to the case of the non-interacting jellium model, we consider the generic case of a conserved many-body Hamiltonian

$$\hat{H}_\mu \equiv \hat{H} - \mu \hat{N} \tag{5.131}$$

acting on a $\mathbb{Z}_2$-graded Fock space $\mathcal{F}$. The Fock space

$$\mathcal{F} := \mathrm{span}\left\{ \prod_\iota \left(\hat{a}_\iota^\dagger\right)^{n_\iota} |0\rangle \,\middle|\, \left[\!\left[\hat{a}_\iota, \hat{a}_{\iota'}^\dagger\right]\!\right] = \delta_{\iota,\iota'},\ \left[\!\left[\hat{a}_\iota, \hat{a}_{\iota'}\right]\!\right] = \left[\!\left[\hat{a}_\iota^\dagger, \hat{a}_{\iota'}^\dagger\right]\!\right] = 0,\ \hat{a}_\iota|0\rangle = 0, \right.$$
$$\left. \deg\left(\hat{a}_\iota\right) = 0 \Rightarrow n_\iota = 0,1,2,\cdots, \qquad \deg\left(\hat{a}_\iota\right) = 1 \Rightarrow n_\iota = 0,1 \right\} \tag{5.132a}$$

is $\mathbb{Z}_2$-graded because any pair $\hat{a}_\iota, \hat{a}_\iota^\dagger$ carries, through its degree

$$\deg\left(\hat{a}_\iota^\dagger\right) \equiv \deg\left(\hat{a}_\iota\right) = 0, 1, \tag{5.132b}$$

the bosonic or fermionic algebra

$$\begin{aligned}
\left[\!\left[\hat{a}_{\iota},\hat{a}^{\dagger}_{\iota'}\right]\!\right] &:= \hat{a}_{\iota}\hat{a}^{\dagger}_{\iota'} - (-1)^{\deg(\hat{a}_{\iota})\deg(\hat{a}_{\iota'})}\hat{a}^{\dagger}_{\iota'}\hat{a}_{\iota} = \delta_{\iota,\iota'},\\
\left[\!\left[\hat{a}^{\dagger}_{\iota},\hat{a}^{\dagger}_{\iota'}\right]\!\right] &:= \hat{a}^{\dagger}_{\iota}\hat{a}^{\dagger}_{\iota'} - (-1)^{\deg(\hat{a}_{\iota})\deg(\hat{a}_{\iota'})}\hat{a}^{\dagger}_{\iota'}\hat{a}^{\dagger}_{\iota} = 0,\\
\left[\!\left[\hat{a}_{\iota},\hat{a}_{\iota'}\right]\!\right] &:= \hat{a}_{\iota}\hat{a}_{\iota'} - (-1)^{\deg(\hat{a}_{\iota})\deg(\hat{a}_{\iota'})}\hat{a}_{\iota'}\hat{a}_{\iota} = 0,
\end{aligned} \tag{5.132c}$$

whenever

$$\deg\left(\hat{a}_{\iota}\right) = 0, \tag{5.132d}$$

or

$$\deg\left(\hat{a}_{\iota}\right) = 1, \tag{5.132e}$$

respectively. We shall assume that the many-body Hamiltonian $\hat{H}_{\mu}$ has a Taylor expansion in powers of the operators $\hat{a}$'s generating the $\mathbb{Z}_2$-graded Fock space (5.132) in such a way that it can be decomposed into the sum of two non-commuting and conserved Hermitian operators $\hat{H}_{0,\mu}$ and $\hat{H}_1$,

$$\hat{H}_{\mu} = \hat{H}_{0,\mu} + \hat{H}_1, \qquad \left[\hat{H}_{0,\mu},\hat{H}_1\right] \neq 0, \tag{5.133}$$

whereby $\hat{H}_{0,\mu}$ is the quadratic form

$$\hat{H}_{0,\mu} = \sum_{\iota} \left(\varepsilon_{\iota} - \mu\right)\hat{a}^{\dagger}_{\iota}\hat{a}_{\iota}, \tag{5.134}$$

while $\hat{H}_1$ is of higher order in the $\hat{a}$'s. We work in the grand-canonical ensemble with the grand-canonical partition function

$$Z(\beta,\mu) := \mathrm{Tr}_{\mathcal{F}}\left(e^{-\beta\hat{H}_{\mu}}\right). \tag{5.135}$$

Let $0 \le \lambda \le 1$ be a dimensionless coupling that allows us to treat the interaction $\hat{H}_1$ adiabatically, i.e., we define

$$\hat{H}_{\mu}(\lambda) := \hat{H}_{0,\mu} + \lambda\hat{H}_1. \tag{5.136}$$

We shall use the notation

$$\hat{H}_{\mu} \equiv \hat{H}_{\mu}(\lambda = 1), \tag{5.137}$$

so that $\hat{H}_{\mu}(\lambda)$ interpolates between $\hat{H}_{0,\mu}$ and $\hat{H}_{\mu}$ as λ varies between 0 and 1.

The grand-canonical potential in the grand-canonical ensemble is defined by

$$\begin{aligned}
F(\beta,\mu;\lambda) :=& U(\beta,\mu;\lambda) - TS(\beta,\mu;\lambda)\\
\equiv& -\frac{1}{\beta}\ln\mathrm{Tr}_{\mathcal{F}}\left[e^{-\beta\hat{H}_{\mu}(\lambda)}\right]\\
\equiv& -\frac{1}{\beta}\ln Z(\beta,\mu;\lambda).
\end{aligned} \tag{5.138}$$

The thermal expectation value of the interaction is

$$\frac{\partial F(\beta,\mu;\lambda)}{\partial\lambda}=\frac{\mathrm{Tr}_{\mathcal{F}}\left[e^{-\beta\hat{H}_{\mu}(\lambda)}\,\hat{H}_1\right]}{\mathrm{Tr}_{\mathcal{F}}\left[e^{-\beta\hat{H}_{\mu}(\lambda)}\right]}\equiv\left\langle\hat{H}_1\right\rangle_{\beta,\mu;\lambda}. \tag{5.139}$$

The change in the grand-canonical potential induced by switching on the interaction adiabatically is

$$F(\beta,\mu;1)-F(\beta,\mu;0)=\int\limits_0^1\mathrm{d}\lambda\,\frac{\partial F(\beta,\mu;\lambda)}{\partial\lambda}=\int\limits_0^1\frac{\mathrm{d}\lambda}{\lambda}\left\langle\lambda\hat{H}_1\right\rangle_{\beta,\mu;\lambda}. \tag{5.140}$$

It turns out (see appendix E.4) that the grand-canonical expectation value

$$\left\langle\lambda\hat{H}_1\right\rangle_{\beta,\mu;\lambda} \tag{5.141}$$

can be related to the so-called time-ordered single-particle Green function. One important physical meaning of the time-ordered single-particle Green function is thus that it encodes the correlation energy (5.140). With this motivation in mind, we define time-ordered Green functions in the grand-canonical ensemble as a conclusion to this chapter.

5.4.2 *Time-ordered Green functions in imaginary time*

Let $\hat{A}$ be any operator acting on the Fock space (5.132) on which the grand-canonical partition function (5.135), is defined. Examples of fermionic operators for the non-interacting jellium model are

$$\begin{aligned}\hat{c}^{\dagger}_{\sigma,\boldsymbol{k}}&=\frac{1}{\sqrt{V}}\int\limits_V\mathrm{d}^3\boldsymbol{r}\,e^{+\mathrm{i}\boldsymbol{k}\cdot\boldsymbol{r}}\hat{\psi}^{\dagger}_{\sigma}(\boldsymbol{r}), &\qquad \hat{c}_{\sigma,\boldsymbol{k}}&=\frac{1}{\sqrt{V}}\int\limits_V\mathrm{d}^3\boldsymbol{r}\,e^{-\mathrm{i}\boldsymbol{k}\cdot\boldsymbol{r}}\hat{\psi}_{\sigma}(\boldsymbol{r}),\\ \hat{\psi}^{\dagger}_{\sigma}(\boldsymbol{r})&=\frac{1}{\sqrt{V}}\sum_{\boldsymbol{k}}e^{-\mathrm{i}\boldsymbol{k}\cdot\boldsymbol{r}}\hat{c}^{\dagger}_{\sigma,\boldsymbol{k}}, &\qquad \hat{\psi}_{\sigma}(\boldsymbol{r})&=\frac{1}{\sqrt{V}}\sum_{\boldsymbol{k}}e^{+\mathrm{i}\boldsymbol{k}\cdot\boldsymbol{r}}\hat{c}_{\sigma,\boldsymbol{k}}.\end{aligned} \tag{5.142}$$

The symmetric convention for the normalization by the volume $V=L^3$ is here chosen so that the $\hat{c}$'s are dimensionless while the $\hat{\psi}$'s have the dimensions of $1/\sqrt{V}$. Examples of bosonic operators for the non-interacting jellium model are

$$\begin{aligned}\hat{\rho}_{\boldsymbol{q}}&\equiv\sum_{\sigma=\uparrow,\downarrow}\sum_{\boldsymbol{k}}\hat{c}^{\dagger}_{\boldsymbol{k},\sigma}\hat{c}_{\boldsymbol{k}+\boldsymbol{q},\sigma}=\int\limits_V\mathrm{d}^3\boldsymbol{r}\,e^{-\mathrm{i}\boldsymbol{q}\cdot\boldsymbol{r}}\,\hat{\rho}(\boldsymbol{r}),\\ \hat{\rho}(\boldsymbol{r})&\equiv\sum_{\sigma=\uparrow,\downarrow}\hat{\psi}^{\dagger}_{\sigma}(\boldsymbol{r})\hat{\psi}_{\sigma}(\boldsymbol{r})=\frac{1}{V}\sum_{\boldsymbol{q}}e^{+\mathrm{i}\boldsymbol{q}\cdot\boldsymbol{r}}\,\hat{\rho}_{\boldsymbol{q}}.\end{aligned} \tag{5.143}$$

The asymmetric convention for the normalization by the volume is here chosen so that the $\hat{\rho}$'s are dimensionless in momentum space while they have the dimensions of $1/V$ in position space.

Operator $\hat{A}$, as any operator acting on the Fock space including the kinetic energy operator $\hat{H}_{0,\mu}$ or the interaction $\hat{H}_1$, is explicitly time independent. Let

$\tau \in \mathbb{R}$ be a real parameter with dimension of time that we call imaginary time. We endow the operator $\hat{A}$ with the explicit dependence on imaginary time

$$\hat{A}_{\mathrm{H}}(\tau, \tau_0) := e^{+(\tau-\tau_0)\hat{H}_\mu}\,\hat{A}(\tau_0)\,e^{-(\tau-\tau_0)\hat{H}_\mu}, \qquad \hat{A}(\tau_0) \equiv \hat{A}. \tag{5.144}$$

The index H stands for the *Heisenberg picture.* Alternatively, we endow the operator $\hat{A}$ with the explicit dependence on imaginary time

$$\hat{A}_{\mathrm{I}}(\tau, \tau_0) := e^{+(\tau-\tau_0)\hat{H}_{0,\mu}}\,\hat{A}(\tau_0)\,e^{-(\tau-\tau_0)\hat{H}_{0,\mu}}, \qquad \hat{A}(\tau_0) \equiv \hat{A}. \tag{5.145}$$

The index I stands for the *interacting picture.*

The equations of motion obeyed by $\hat{A}$ in the Heisenberg and interacting pictures follow from taking imaginary time $\tau - \tau_0$ to be infinitesimal. They are

$$\left(\partial_\tau \hat{A}_{\mathrm{H}}\right)(\tau, \tau_0) = \left[\hat{H}_\mu, \hat{A}_{\mathrm{H}}(\tau, \tau_0)\right], \qquad \hat{A}_{\mathrm{H}}(\tau_0) = \hat{A}, \tag{5.146}$$

in the Heisenberg picture and

$$\left(\partial_\tau \hat{A}_{\mathrm{I}}\right)(\tau, \tau_0) = \left[\hat{H}_{0,\mu}, \hat{A}_{\mathrm{I}}(\tau, \tau_0)\right], \qquad \hat{A}_{\mathrm{I}}(\tau_0) = \hat{A}, \tag{5.147}$$

in the interacting picture.

In the Schrödinger picture, states at the imaginary time τ_0 are related to states at the imaginary time τ by multiplication of the former state from the left with the imaginary-time evolution operator

$$\hat{U}_{\mathrm{S}}(\tau, \tau_0) = e^{-(\tau-\tau_0)\hat{H}_\mu}, \tag{5.148a}$$

as their imaginary-time evolution is governed by the imaginary-time Schrödinger equation

$$\partial_\tau \hat{U}_{\mathrm{S}}(\tau, \tau_0) = -\hat{H}_\mu\, \hat{U}_{\mathrm{S}}(\tau, \tau_0) \Longleftrightarrow \partial_\tau \Psi_{\mathrm{S}}(\tau) = -\hat{H}_\mu \Psi_{\mathrm{S}}(\tau), \qquad \Psi_{\mathrm{S}}(\tau_0) \text{ given.} \tag{5.148b}$$

In the interacting picture, states at imaginary time τ_0 are related to states at the imaginary time τ by multiplication of the former state from the left with the operator

$$\begin{aligned}\hat{U}_{\mathrm{I}}(\tau, \tau_0) \equiv& T_\tau \exp\left(-\int\limits_{\tau_0}^{\tau} \mathrm{d}\tau'\, \hat{H}_{1\mathrm{I}}(\tau', \tau_0)\right)\\ :=& 1 + \sum_{n=1}^{\infty}(-1)^n \int\limits_{\tau_0}^{\tau} \mathrm{d}\tau_n \cdots \int\limits_{\tau_0}^{\tau_2} \mathrm{d}\tau_1\, \hat{H}_{1\mathrm{I}}(\tau_n, \tau_0) \cdots \hat{H}_{1\mathrm{I}}(\tau_2, \tau_0) \hat{H}_{1\mathrm{I}}(\tau_1, \tau_0),\end{aligned} \tag{5.149a}$$

as their imaginary-time evolution is governed by the imaginary-time first-order differential equation

$$\begin{aligned}\partial_\tau \hat{U}_{\mathrm{I}}(\tau, \tau_0) = -\hat{H}_{1\mathrm{I}}(\tau, \tau_0)\, \hat{U}_{\mathrm{I}}(\tau, \tau_0) \Longleftrightarrow&\\ \partial_\tau \Psi_{\mathrm{I}}(\tau) = -\hat{H}_{1\mathrm{I}}(\tau, \tau_0)\, \Psi_{\mathrm{I}}(\tau), \quad& \Psi_{\mathrm{I}}(\tau_0) \text{ given.}\end{aligned} \tag{5.149b}$$

The operation of imaginary-time ordering used in Eqs. (5.148a) and (5.149a) is defined for any pair of operators

$$\begin{aligned} T_\tau\left(\hat{A}(\tau_1,\tau_0)\,\hat{B}(\tau_2,\tau_0)\right) :=& \hat{A}(\tau_1,\tau_0)\,\hat{B}(\tau_2,\tau_0)\Theta\left(\tau_1-\tau_2\right) \\ &\pm \hat{B}(\tau_2,\tau_0)\,\hat{A}(\tau_1,\tau_0)\Theta\left(\tau_2-\tau_1\right) \\ \equiv& \begin{cases} \hat{A}(\tau_1,\tau_0)\,\hat{B}(\tau_2,\tau_0), & \text{when } \tau_1>\tau_2, \\ (\pm)\hat{B}(\tau_2,\tau_0)\,\hat{A}(\tau_1,\tau_0), & \text{when } \tau_2>\tau_1, \end{cases} \end{aligned} \tag{5.150}$$

irrespective of how the imaginary-time evolution is implemented. The sign $+$ holds for a pair of bosonic operators or for a mixed pair of bosonic and fermionic operators. The sign $-$ holds for a pair of fermionic operators.

Because the interaction does not commute with the kinetic energy,

$$\hat{H}_{1\mathrm{I}}(\tau,\tau_0) = e^{+(\tau-\tau_0)\hat{H}_{0,\mu}}\,\hat{H}_{1\mathrm{I}}(\tau_0)\,e^{-(\tau-\tau_0)\hat{H}_{0,\mu}} \tag{5.151}$$

depends explicitly on imaginary time in the Schrödinger-like equation (5.149b). Hence, the integration over imaginary time cannot be performed explicitly in Eq. (5.149a). Neither $\hat{U}_{\mathrm{S}}(\tau,\tau_0)$ nor $\hat{U}_{\mathrm{I}}(\tau,\tau_0)$ are unitary, but they share the composition law

$$\hat{U}_{\mathrm{S}}(\tau,\tau')\,\hat{U}_{\mathrm{S}}(\tau',\tau_0) = \hat{U}_{\mathrm{S}}(\tau,\tau_0) \Longrightarrow \hat{U}_{\mathrm{S}}^{-1}(\tau,\tau') = \hat{U}_{\mathrm{S}}(\tau',\tau), \tag{5.152}$$

and

$$\hat{U}_{\mathrm{I}}(\tau,\tau')\,\hat{U}_{\mathrm{I}}(\tau',\tau_0) = \hat{U}_{\mathrm{I}}(\tau,\tau_0) \Longrightarrow \hat{U}_{\mathrm{I}}^{-1}(\tau,\tau') = \hat{U}_{\mathrm{I}}(\tau',\tau), \tag{5.153}$$

for all triplets (τ,τ',τ_0), respectively. Either $\hat{U}_{\mathrm{S}}(\tau,\tau_0)$ or $\hat{U}_{\mathrm{I}}(\tau,\tau_0)$ becomes unitary under the analytical continuation

$$\tau\in\mathbb{R}\to+\mathrm{i}t, \qquad t\in\mathbb{R}. \tag{5.154}$$

The relation between the imaginary-time evolution in the Schrödinger and interaction pictures is

$$\begin{aligned} \hat{U}_{\mathrm{I}}(\tau_1,\tau_2) = e^{+(\tau_1-\tau_0)\hat{H}_{0,\mu}}\,\hat{U}_{\mathrm{S}}(\tau_1,\tau_2)\,e^{-(\tau_2-\tau_0)\hat{H}_{0,\mu}} \iff& \\ \Psi_{\mathrm{S}}(\tau) = e^{-(\tau-\tau_0)\hat{H}_{0,\mu}}\,\Psi_{\mathrm{I}}(\tau),& \end{aligned} \tag{5.155}$$

where τ_0 is the time at which $\Psi_{\mathrm{S}}(\tau_0)=\Psi_{\mathrm{H}}(\tau_0)=\Psi_{\mathrm{I}}(\tau_0)$.

Let $\hat{A}$ and $\hat{B}$ be any pair of operators with the degrees $\deg(\hat{A})$ and $\deg(\hat{B})$, respectively, acting on the Fock space (5.132) on which the grand-canonical partition function (5.135) is defined. The *time-ordered correlation function* in imaginary time between $\hat{A}$ and $\hat{B}$ is the expectation value

$$\begin{aligned} C_{\beta,\mu;\hat{A},\hat{B}}(\tau_1,\tau_2) :=& -\frac{\mathrm{Tr}_{\mathcal{F}}\left[e^{-\beta\hat{H}_\mu}\,T_\tau\left(\hat{A}_{\mathrm{H}}(\tau_1,\tau_0)\,\hat{B}_{\mathrm{H}}(\tau_2,\tau_0)\right)\right]}{\mathrm{Tr}_{\mathcal{F}}\left(e^{-\beta\hat{H}_\mu}\right)} \\ \equiv& -\left\langle T_\tau\left(\hat{A}_{\mathrm{H}}(\tau_1,\tau_0)\,\hat{B}_{\mathrm{H}}(\tau_2,\tau_0)\right)\right\rangle_{\beta,\mu}. \end{aligned} \tag{5.156}$$

The sign on the right-hand side is convention and it is implicitly assumed that this correlation function does not depend on τ_0. In fact, we are going to prove that:

(i) Translation invariance in imaginary time holds for the correlation function (5.156) as

$$C_{\beta,\mu;\hat{A},\hat{B}}(\tau_1,\tau_2)=C_{\beta,\mu;\hat{A},\hat{B}}(\tau_1-\tau_2). \tag{5.157a}$$

(ii) The correlation function (5.156) decays exponentially fast with $|\tau_1-\tau_2|$ only if

$$|\tau_1-\tau_2|<\beta. \tag{5.157b}$$

It grows exponentially fast with $|\tau_1-\tau_2|$ otherwise.

(iii) If $-\beta<\tau_1-\tau_2<0$, it then follows that

$$C_{\beta,\mu;\hat{A},\hat{B}}(\tau_1+\beta,\tau_2)=\pm C_{\beta,\mu;\hat{A},\hat{B}}(\tau_1,\tau_2), \tag{5.157c}$$

where periodicity holds if $\hat{A}$ and $\hat{B}$ commute while antiperiodicity holds if $\hat{A}$ and $\hat{B}$ anticommute under the operation of imaginary-time ordering.

(iv) If $\hat{A}$ and $\hat{B}$ are bosonic, then

$$\begin{aligned}C_{\beta,\mu;\hat{A}\hat{B}}(\tau)=\frac{1}{\beta}\sum_{l\in\mathbb{Z}}e^{-\mathrm{i}\varpi_l\tau}\,C_{\beta,\mu;\hat{A}\hat{B},\mathrm{i}\varpi_l}&\iff\\ C_{\beta,\mu;\hat{A}\hat{B},\mathrm{i}\varpi_l}&=\int\limits_{0+}^{\beta^-}\mathrm{d}\tau\,e^{+\mathrm{i}\varpi_l\tau}\,C_{\beta,\mu;\hat{A}\hat{B}}(\tau)\end{aligned} \tag{5.157d}$$

with the *bosonic Matsubara frequency* $\varpi_l=2l\pi/\beta$. If $\hat{A}$ and $\hat{B}$ are fermionic, then

$$\begin{aligned}C_{\beta,\mu;\hat{A}\hat{B}}(\tau)=\frac{1}{\beta}\sum_{n\in\mathbb{Z}}e^{-\mathrm{i}\omega_n\tau}\,C_{\beta,\mu;\hat{A}\hat{B},\mathrm{i}\omega_n}&\iff\\ C_{\beta,\mu;\hat{A}\hat{B},\mathrm{i}\omega_n}&=\int\limits_{0+}^{\beta^-}\mathrm{d}\tau\,e^{+\mathrm{i}\omega_n\tau}\,C_{\beta,\mu;\hat{A}\hat{B}}(\tau)\end{aligned} \tag{5.157e}$$

with the *fermionic Matsubara frequency* $\omega_n=(2n+1)\pi/\beta$. The asymmetric convention for the normalization by β is the same as in Eq. (5.143).

Proof. Cyclicity of the trace with the definition (5.144) implies

$$\begin{aligned}C_{\beta,\mu;\hat{A},\hat{B}}(\tau_1,\tau_2)=&-\Theta(\tau_1-\tau_2)\frac{\mathrm{Tr}_{\mathcal{F}}\left[e^{-\beta\hat{H}_\mu}\,e^{+(\tau_1-\tau_2)\hat{H}_\mu}\,\hat{A}\,e^{-(\tau_1-\tau_2)\hat{H}_\mu}\,\hat{B}\right]}{\mathrm{Tr}_{\mathcal{F}}\left(e^{-\beta\hat{H}_\mu}\right)}\\ &\mp\Theta(\tau_2-\tau_1)\frac{\mathrm{Tr}_{\mathcal{F}}\left[e^{-\beta\hat{H}_\mu}\,e^{+(\tau_2-\tau_1)\hat{H}_\mu}\,\hat{B}\,e^{-(\tau_2-\tau_1)\hat{H}_\mu}\,\hat{A}\right]}{\mathrm{Tr}_{\mathcal{F}}\left(e^{-\beta\hat{H}_\mu}\right)}\end{aligned} \tag{5.158}$$

from which (i) and (iii) follow. Insertion of a complete basis of eigenstates of $\hat{H}_\mu$ in Eq. (5.158), where, without loss of generality, the many-body ground state energy is taken to be positive, implies that the support of $C_{\beta,\mu;\hat{A},\hat{B}}(\tau)$ for which it is a decaying function of τ is Eq. (5.157b). The Fourier transforms (5.157d) and (5.157e) follow from the periodicity (iii). □

The correlation function (5.156) cannot be evaluated exactly in practice. For a systematic perturbation theory, a better suited representation of Eq. (5.156) is

$$C_{\beta,\mu;\hat{A},\hat{B}}(\tau_1-\tau_2) = -\frac{\mathrm{Tr}_{\mathcal{F}}\left[e^{-\beta\hat{H}_{0,\mu}}\,\hat{U}_{\mathrm{I}}(\tau_0+\beta,\tau_0)\,T_\tau\left(\hat{A}_{\mathrm{I}}(\tau_1,\tau_0)\,\hat{B}_{\mathrm{I}}(\tau_2,\tau_0)\right)\right]}{\mathrm{Tr}_{\mathcal{F}}\left[e^{-\beta\hat{H}_{0,\mu}}\,\hat{U}_{\mathrm{I}}(\tau_0+\beta,\tau_0)\right]}. \tag{5.159}$$

Proof. Equation (5.159) follows from Eq. (5.156) with the help of Eq. (5.155). First, we observe that

$$\begin{aligned} e^{-\beta\hat{H}_\mu} &= U_{\mathrm{S}}(\tau_0+\beta,\tau_0) \\ &= e^{-\beta\hat{H}_{0,\mu}}\,U_{\mathrm{I}}(\tau_0+\beta,\tau_0). \end{aligned} \tag{5.160}$$

Second, we observe that

$$\begin{aligned} \hat{A}_{\mathrm{H}}(\tau_1,\tau_0)\,\hat{B}_{\mathrm{H}}(\tau_2,\tau_0) &= e^{+(\tau_1-\tau_0)\hat{H}_\mu}\,\hat{A}(\tau_0)\,e^{-(\tau_1-\tau_2)\hat{H}_\mu}\,\hat{B}(\tau_0)\,e^{-(\tau_2-\tau_0)\hat{H}_\mu} \\ &= e^{+(\tau_1-\tau_0)\hat{H}_\mu}\,\hat{A}(\tau_0)\,\hat{U}_{\mathrm{S}}(\tau_1,\tau_2)\,\hat{B}(\tau_0)\,e^{-(\tau_2-\tau_0)\hat{H}_\mu} \\ \text{Eq. (5.145)}\quad &= e^{+(\tau_1-\tau_0)\hat{H}_\mu}\,e^{-(\tau_1-\tau_0)\hat{H}_{0,\mu}}\,\hat{A}_{\mathrm{I}}(\tau_1,\tau_0)\,e^{+(\tau_1-\tau_0)\hat{H}_{0,\mu}}\,\hat{U}_{\mathrm{S}}(\tau_1,\tau_2) \\ &\quad\times e^{-(\tau_2-\tau_0)\hat{H}_{0,\mu}}\,\hat{B}_{\mathrm{I}}(\tau_2,\tau_0)\,e^{+(\tau_2-\tau_0)\hat{H}_{0,\mu}}e^{-(\tau_2-\tau_0)\hat{H}_\mu} \\ &= \hat{U}_{\mathrm{I}}(\tau_0,\tau_1)\,\hat{A}_{\mathrm{I}}(\tau_1,\tau_0)\,\hat{U}_{\mathrm{I}}(\tau_1,\tau_2)\,\hat{B}_{\mathrm{I}}(\tau_2,\tau_0)\,\hat{U}_{\mathrm{I}}(\tau_2,\tau_0). \end{aligned} \tag{5.161}$$

Third, we recall that bosonic (fermionic) operators behave like complex (Grassmann) numbers under the operation of τ-ordering. Hence, we may move the bosonic operator $\hat{U}_{\mathrm{I}}(\tau_0,\tau_1)$ to the right of $\hat{A}_{\mathrm{I}}(\tau_1,\tau_0)$, while we may move the bosonic operator $\hat{U}_{\mathrm{I}}(\tau_2,\tau_0)$ to the left of $\hat{B}_{\mathrm{I}}(\tau_2,\tau_0)$ in

$$T_\tau\left(\hat{U}_{\mathrm{I}}(\tau_0,\tau_1)\,\hat{A}_{\mathrm{I}}(\tau_1,\tau_0)\,\hat{U}_{\mathrm{I}}(\tau_1,\tau_2)\,\hat{B}_{\mathrm{I}}(\tau_2,\tau_0)\,\hat{U}_{\mathrm{I}}(\tau_2,\tau_0)\right). \tag{5.162}$$

Equation (5.159) then follows from the identity $\hat{U}_{\mathrm{I}}(\tau_0,\tau_1)\,\hat{U}_{\mathrm{I}}(\tau_1,\tau_2)\,\hat{U}_{\mathrm{I}}(\tau_2,\tau_0)=1$.

□

Another useful tool to evaluate the correlation function (5.156) is the equation of motion

$$\begin{aligned} -\partial_{\tau_1}C_{\beta,\mu;\hat{A},\hat{B}}(\tau_1-\tau_2) &= \delta(\tau_1-\tau_2)\Big\langle\hat{A}_{\mathrm{H}}(\tau_1,\tau_0)\hat{B}_{\mathrm{H}}(\tau_1,\tau_0) \\ &\quad\mp\hat{B}_{\mathrm{H}}(\tau_1,\tau_0)\hat{A}_{\mathrm{H}}(\tau_1,\tau_0)\Big\rangle_{\beta,\mu} \\ &\quad+\left\langle T_\tau\left\{\left[\hat{H}_\mu,\hat{A}_{\mathrm{H}}(\tau_1,\tau_0)\right]\hat{B}_{\mathrm{H}}(\tau_2,\tau_0)\right\}\right\rangle_{\beta,\mu}, \end{aligned} \tag{5.163}$$

which is obeyed for any unequal imaginary times τ_1 and τ_2. In general, the commutator $\left[\hat{H}_\mu,\hat{A}_{\mathrm{H}}(\tau_1,\tau_0)\right]$ is not proportional to $\hat{A}_{\mathrm{H}}(\tau_1,\tau_0)$ so that this equation does not close on its own. In fact, a closed set of equations of motion is generically infinite.

The definition (5.156) readily generalizes to the $2n$-*point time-ordered correlation function* in imaginary time between operators $\hat{A}_1, \cdots, \hat{A}_n$ and $\hat{B}_1, \cdots, \hat{B}_n$ acting on the Fock space (5.132) on which the grand-canonical partition function (5.135) is defined. It is

$$\begin{aligned} & C_{\beta,\mu;\hat{A}_1,\cdots,\hat{A}_n|\hat{B}_1,\cdots,\hat{B}_n}(\tau_1,\cdots,\tau_n|\tau'_1,\cdots,\tau'_n) \\ & := (-1)^n \left\langle T_\tau\left(\hat{A}_{\mathrm{H}}(\tau_1,\tau_0)\times\cdots\times\hat{A}_{\mathrm{H}}(\tau_n,\tau_0)\times\hat{B}_{\mathrm{H}}(\tau'_1,\tau_0)\times\cdots\times\hat{B}_{\mathrm{H}}(\tau'_n,\tau_0)\right)\right\rangle_{\beta,\mu}. \end{aligned} \tag{5.164}$$

Next, we are going to compute explicitly 2 and 4 points time-ordered Green functions for the non-interacting jellium model, whereby we shall make the identifications

$$\hat{A}\to\hat{c}_{\sigma,\boldsymbol{k}}, \qquad \hat{B}\to\hat{c}^\dagger_{\sigma,\boldsymbol{k}}, \qquad \varepsilon_\iota\to\varepsilon_{\sigma,\boldsymbol{k}}\equiv\frac{\hbar^2\boldsymbol{k}^2}{2m}. \tag{5.165}$$

5.4.3 *Time-ordered Green functions in real time*

Imaginary time τ and real time t are related by the analytical continuation

$$\tau = \mathrm{i}t, \qquad t\in\mathbb{R}. \tag{5.166}$$

As before, let $\hat{A}$ and $\hat{B}$ be any pair of operator with the degrees $\deg(\hat{A})$ and $\deg(\hat{B})$, respectively, acting on the Fock space (5.132) on which the grand-canonical partition function (5.135) is defined. The *time-ordered correlation function* in real time between $\hat{A}$ and $\hat{B}$ is the expectation value

$$\begin{aligned} C_{\beta,\mu;\hat{A},\hat{B}}(t_1,t_2) &= -\frac{\mathrm{Tr}_{\mathcal{F}}\left[e^{-\beta\hat{H}_\mu}\,T_t\left(\hat{A}_{\mathrm{H}}(t_1,t_0)\,\hat{B}_{\mathrm{H}}(t_2,t_0)\right)\right]}{\mathrm{Tr}_{\mathcal{F}}\left(e^{-\beta\hat{H}_\mu}\right)} \\ &\equiv -\left\langle T_t\left(\hat{A}_{\mathrm{H}}(t_1,t_0)\,\hat{B}_{\mathrm{H}}(t_2,t_0)\right)\right\rangle_{\beta,\mu}, \end{aligned} \tag{5.167a}$$

where

$$\hat{A}_{\mathrm{H}}(t,t_0) := e^{+\mathrm{i}(t-t_0)\hat{H}_\mu}\hat{A}(t_0)e^{-\mathrm{i}(t-t_0)\hat{H}_\mu} \tag{5.167b}$$

and

$$\begin{aligned} T_t\left(\hat{A}(t_1,t_0)\hat{B}(t_2,t_0)\right) &:= \hat{A}(t_1,t_0)\hat{B}(t_2,t_0)\Theta(t_1-t_2) \\ &\quad + (-1)^{\deg(\hat{A})\deg(\hat{B})}\hat{B}(t_2,t_0)\hat{A}(t_1,t_0)\Theta(t_2-t_1) \\ &\equiv \begin{cases} \hat{A}(t_1,t_0)\hat{B}(t_2,t_0), & \text{when } t_1>t_2, \\ \\ (-1)^{\deg(\hat{A})\deg(\hat{B})}\hat{B}(t_2,t_0)\hat{A}(t_1,t_0), & \text{when } t_2>t_1. \end{cases} \end{aligned} \tag{5.167c}$$

The sign on the right-hand side of Eq. (5.167a) is here convention, as is the imaginary factor on the left-hand side of Eq. (5.167a).

5.4.4 *Application to the non-interacting jellium model*

The grand-canonical partition function for the non-interacting jellium model is defined in Eq. (5.24). We shall use the more compact notation

$$\hat{H}_\mu := \hat{H} - \mu\hat{N}, \qquad \xi_{\sigma,\boldsymbol{k}} \equiv \varepsilon_{\sigma,\boldsymbol{k}} - \mu. \tag{5.168}$$

5.4.4.1 *Momentum-space representation*

The imaginary-time-ordered single-particle Green function in momentum space is defined by Eq. (5.156) with the identifications

$$\hat{A} \to \hat{c}_{\sigma_1,\boldsymbol{k}_1}, \qquad \hat{B} \to \hat{c}^\dagger_{\sigma_2,\boldsymbol{k}_2}. \tag{5.169}$$

For any $|\tau_1 - \tau_2| < \beta$, it is given by

$$\begin{aligned}
C&_{\beta,\mu;\hat{c}_{\sigma_1,\boldsymbol{k}_1},\hat{c}^\dagger_{\sigma_2,\boldsymbol{k}_2}}(\tau_1-\tau_2)\\
&= -\left\langle T_\tau\left(\hat{c}_{\mathrm{H}\,\sigma_1,\boldsymbol{k}_1}(\tau_1,\tau_0)\hat{c}^\dagger_{\mathrm{H}\,\sigma_2,\boldsymbol{k}_2}(\tau_2,\tau_0)\right)\right\rangle_{\beta,\mu}\\
&= -\Theta(\tau_1-\tau_2)\,\frac{\mathrm{Tr}_{\mathcal{F}}\left[e^{-\beta\hat{H}_\mu}e^{+(\tau_1-\tau_2)\hat{H}_\mu}\hat{c}_{\sigma_1,\boldsymbol{k}_1}e^{+(\tau_2-\tau_1)\hat{H}_\mu}\hat{c}^\dagger_{\sigma_2,\boldsymbol{k}_2}\right]}{\mathrm{Tr}_{\mathcal{F}}\left(e^{-\beta\hat{H}_\mu}\right)}\\
&\quad+\Theta(\tau_2-\tau_1)\,\frac{\mathrm{Tr}_{\mathcal{F}}\left[e^{-\beta\hat{H}_\mu}e^{+(\tau_2-\tau_1)\hat{H}_\mu}\hat{c}^\dagger_{\sigma_2,\boldsymbol{k}_2}e^{+(\tau_1-\tau_2)\hat{H}_\mu}\hat{c}_{\sigma_1,\boldsymbol{k}_1}\right]}{\mathrm{Tr}_{\mathcal{F}}\left(e^{-\beta\hat{H}_\mu}\right)}\\
&\approx \delta_{\sigma_1,\sigma_2}\delta_{\boldsymbol{k}_1,\boldsymbol{k}_2}G_{\beta,\mu}(\tau_1-\tau_2,\boldsymbol{k}_1),
\end{aligned} \tag{5.170a}$$

where

$$G_{\beta,\mu}(\tau,\boldsymbol{k}) = -\left\{\Theta(+\tau)\left[1-\widetilde{f}_{\mathrm{FD}}(\xi_{\boldsymbol{k}})\right]e^{-\tau\xi_{\boldsymbol{k}}} - \Theta(-\tau)\widetilde{f}_{\mathrm{FD}}(\xi_{\boldsymbol{k}})\,e^{-\tau\xi_{\boldsymbol{k}}}\right\}, \tag{5.170b}$$

and [compare with the definition (5.27a) of the Fermi-Dirac distribution that depends explicitly on the chemical potential]

$$\widetilde{f}_{\mathrm{FD}}(\xi_{\boldsymbol{k}}) := f_{\mathrm{FD}}(\varepsilon_{\boldsymbol{k}}) = \frac{1}{e^{\beta\,\xi_{\boldsymbol{k}}}+1}. \tag{5.170c}$$

It is extended to $|\tau_1 - \tau_2| > \beta$ by antiperiodicity. To reach the last line, we made a small error that vanishes in the thermodynamic limit by which the volume $V = L^3$ of the system goes to infinity, while the average number of electrons per unit volume,

$$n_{\mathrm{e}} := \beta^{-1}\frac{\partial \ln Z(V,\beta,\mu)}{\partial\mu}, \tag{5.171}$$

is held fixed. This is so because we need to introduce once the resolution of the identity in terms of the exact many-body energy eigenstates

$$|\iota\rangle^{(N_{\mathrm{e}})}, \qquad E^{(N_{\mathrm{e}})}_\iota \text{ the energy, } N_{\mathrm{e}} \text{ the electron number,} \tag{5.172}$$

of $\hat{H}_\mu$ between the creation and annihilation operators to go from the second equality to the third equality of Eq. (5.170a). This brings the exponentials

$$e^{-(\tau_1-\tau_2)\left(E^{(N_{\mathrm{e}}+1)}_{\iota,\sigma,\boldsymbol{k}} - E^{(N_{\mathrm{e}})}_\iota\right)} \approx e^{-(\tau_1-\tau_2)\xi_{\boldsymbol{k}}}, \tag{5.173}$$

where the many-body energy eigenstate $|\iota,\sigma,\boldsymbol{k}\rangle^{(N_{\mathrm{e}}+1)}$ has one additional occupied single-particle level compared to the many-body energy eigenstate $|\iota\rangle^{(N_{\mathrm{e}})}$, to be thermal averaged when $\tau_1>\tau_2$, and the exponentials

$$e^{+(\tau_1-\tau_2)\left(E^{(N_{\mathrm{e}}-1)}_{\iota,\sigma,\boldsymbol{k}}-E^{(N_{\mathrm{e}})}_{\iota}\right)}\approx e^{-(\tau_1-\tau_2)\xi_{\boldsymbol{k}}}, \tag{5.174}$$

where the many-body energy eigenstate $|\iota,\sigma,\boldsymbol{k}\rangle^{(N_{\mathrm{e}}-1)}$ has one less occupied single-particle level compared to the many-body energy eigenstate $|\iota\rangle^{(N_{\mathrm{e}})}$, to be thermal averaged when $\tau_2>\tau_1$.

Owing to the fact that

$$\lim_{\beta\to\infty}\left[1-\widetilde{f}_{\mathrm{FD}}(\xi_{\boldsymbol{k}})\right]=\Theta(+\xi_{\boldsymbol{k}}),\qquad \lim_{\beta\to\infty}\widetilde{f}_{\mathrm{FD}}(\xi_{\boldsymbol{k}})=\Theta(-\xi_{\boldsymbol{k}}), \tag{5.175}$$

Eq. (5.170b) tells us that, at zero temperature, the imaginary-time-ordered Green function is non-vanishing at positive (negative) time if and only if the single-particle level $\xi_{\boldsymbol{k}}$ is unoccupied (occupied), i.e.,

$$\lim_{\beta\to\infty}G_{\beta,\mu}(\tau,\boldsymbol{k})=-\left[\Theta(+\tau)\,\Theta(+\xi_{\boldsymbol{k}})\,e^{-\tau\xi_{\boldsymbol{k}}}-\Theta(-\tau)\,\Theta(-\xi_{\boldsymbol{k}})\,e^{-\tau\xi_{\boldsymbol{k}}}\right]. \tag{5.176}$$

Owing to the antiperiodic dependence (5.157e), for any fermionic Matsubara frequency $\omega_n=(2n+1)\pi/\beta$,

$$\begin{aligned}G_{\beta,\mu}(\omega_n,\boldsymbol{k}):=&\int\limits_{0+}^{\beta-}\mathrm{d}\tau\,e^{+\mathrm{i}\omega_n\tau}\,G_{\beta,\mu}(\tau,\boldsymbol{k})\\ =&\int\limits_{-\beta+}^{0-}\mathrm{d}\tau\,e^{+\mathrm{i}\omega_n\tau}\,G_{\beta,\mu}(\tau,\boldsymbol{k})\\ \stackrel{T\to0}{=}&\frac{1}{\mathrm{i}\omega_n-\xi_{\boldsymbol{k}}}.\end{aligned} \tag{5.177}$$

The real-time-ordered single-particle Green function in momentum space follows from Eq. (5.170) with the analytical continuation (5.166), i.e., it is

$$\begin{aligned}&C_{\beta,\mu;\hat{c}_{\sigma_1,\boldsymbol{k}_1},\hat{c}^{\dagger}_{\sigma_2,\boldsymbol{k}_2}}(t_1-t_2)\\ &=-\left\langle T_t\left(\hat{c}_{\mathrm{H}\,\sigma_1,\boldsymbol{k}_1}(t_1,t_0)\hat{c}^{\dagger}_{\mathrm{H}\,\sigma_2,\boldsymbol{k}_2}(t_2,t_0)\right)\right\rangle_{\beta,\mu}\\ &=-\Theta(t_1-t_2)\,\frac{\mathrm{Tr}_{\mathcal{F}}\left[e^{-\beta\hat{H}_\mu}e^{+\mathrm{i}(t_1-t_2)\hat{H}_\mu}\hat{c}_{\sigma_1,\boldsymbol{k}_1}e^{+\mathrm{i}(t_2-t_1)\hat{H}_\mu}\hat{c}^{\dagger}_{\sigma_2,\boldsymbol{k}_2}\right]}{\mathrm{Tr}_{\mathcal{F}}\left(e^{-\beta\hat{H}_\mu}\right)}\\ &\quad+\Theta(t_2-t_1)\,\frac{\mathrm{Tr}_{\mathcal{F}}\left[e^{-\beta\hat{H}_\mu}e^{+\mathrm{i}(t_2-t_1)\hat{H}_\mu}\hat{c}^{\dagger}_{\sigma_2,\boldsymbol{k}_2}e^{+\mathrm{i}(t_1-t_2)\hat{H}_\mu}\hat{c}_{\sigma_1,\boldsymbol{k}_1}\right]}{\mathrm{Tr}_{\mathcal{F}}\left(e^{-\beta\hat{H}_\mu}\right)}\\ &\approx\delta_{\sigma_1,\sigma_2}\delta_{\boldsymbol{k}_1,\boldsymbol{k}_2}G_{\beta,\mu}(t_1-t_2,\boldsymbol{k}_1),\end{aligned} \tag{5.178a}$$

where

$$G_{\beta,\mu}(t,\boldsymbol{k})=-\left\{\Theta(+t)\left[1-\widetilde{f}_{\mathrm{FD}}(\xi_{\boldsymbol{k}})\right]e^{-\mathrm{i}t\xi_{\boldsymbol{k}}}-\Theta(-t)\,\widetilde{f}_{\mathrm{FD}}(\xi_{\boldsymbol{k}})\,e^{-\mathrm{i}t\xi_{\boldsymbol{k}}}\right\}. \tag{5.178b}$$

At zero temperature,

$$\lim_{\beta\to\infty} G_{\beta,\mu}(t,\boldsymbol{k}) = -\left\{\Theta(+t)\Theta(+\xi_{\boldsymbol{k}})\, e^{-\mathrm{i}t\xi_{\boldsymbol{k}}} - \Theta(-t)\,\Theta(-\xi_{\boldsymbol{k}})\, e^{-\mathrm{i}t\xi_{\boldsymbol{k}}}\right\}. \tag{5.179}$$

In real-frequency space,

$$\begin{aligned} G_{\beta,\mu}(\omega,\boldsymbol{k}) &:= \int_{\mathbb{R}} \mathrm{d}t\, e^{+\mathrm{i}\omega t} G_{\beta,\mu}(t,\boldsymbol{k}) \\ &= -\left(\frac{1-\widetilde{f}_{\mathrm{FD}}(\xi_{\boldsymbol{k}})}{\omega-\xi_{\boldsymbol{k}}+\mathrm{i}0^{+}} + \frac{\widetilde{f}_{\mathrm{FD}}(\xi_{\boldsymbol{k}})}{\omega-\xi_{\boldsymbol{k}}-\mathrm{i}0^{+}}\right), \end{aligned} \tag{5.180}$$

with the zero-temperature limit

$$\lim_{\beta\to\infty} G_{\beta,\mu}(\omega,\boldsymbol{k}) = \frac{-1}{\omega-\xi_{\boldsymbol{k}}+\mathrm{i}\,0^{+}\,\mathrm{sgn}(\xi_{\boldsymbol{k}})}. \tag{5.181}$$

5.4.4.2 *Position-space representation*

By combining Eqs. (5.142), (5.170), and (5.178), we obtain the position-space representation

$$\begin{aligned} C_{\beta,\mu;\hat{\psi}_{\sigma_1}(\boldsymbol{r}_1),\hat{\psi}^{\dagger}_{\sigma_2}(\boldsymbol{r}_2)}(\tau_1-\tau_2) &= \frac{1}{V}\sum_{\boldsymbol{k}_1}\sum_{\boldsymbol{k}_2} e^{-\mathrm{i}\boldsymbol{k}_1\cdot\boldsymbol{r}_1+\mathrm{i}\boldsymbol{k}_2\cdot\boldsymbol{r}_2}\, C_{\beta,\mu;\hat{c}_{\sigma_1,\boldsymbol{k}_1},\hat{c}^{\dagger}_{\sigma_2,\boldsymbol{k}_2}}(\tau_1-\tau_2) \\ &\approx \delta_{\sigma_1,\sigma_2}\frac{1}{V}\sum_{\boldsymbol{k}} e^{-\mathrm{i}\boldsymbol{k}\cdot(\boldsymbol{r}_1-\boldsymbol{r}_2)}\, G_{\beta,\mu}(\tau_1-\tau_2,\boldsymbol{k}) \\ &\equiv \delta_{\sigma_1,\sigma_2}\, G_{\beta,\mu}(\tau_1-\tau_2,\boldsymbol{r}_1-\boldsymbol{r}_2) \end{aligned} \tag{5.182}$$

and

$$\begin{aligned} C_{\beta,\mu;\hat{\psi}_{\sigma_1}(\boldsymbol{r}_1),\hat{\psi}^{\dagger}_{\sigma_2}(\boldsymbol{r}_2)}(t_1-t_2) &= \frac{1}{V}\sum_{\boldsymbol{k}_1}\sum_{\boldsymbol{k}_2} e^{-\mathrm{i}\boldsymbol{k}_1\cdot\boldsymbol{r}_1+\mathrm{i}\boldsymbol{k}_2\cdot\boldsymbol{r}_2}\, C_{\beta,\mu;\hat{c}_{\sigma_1,\boldsymbol{k}_1},\hat{c}^{\dagger}_{\sigma_2,\boldsymbol{k}_2}}(t_1-t_2) \\ &\approx \delta_{\sigma_1,\sigma_2}\frac{1}{V}\sum_{\boldsymbol{k}} e^{-\mathrm{i}\boldsymbol{k}\cdot(\boldsymbol{r}_1-\boldsymbol{r}_2)}\, G_{\beta,\mu}(t_1-t_2,\boldsymbol{k}) \\ &\equiv \delta_{\sigma_1,\sigma_2}\, G_{\beta,\mu}(t_1-t_2,\boldsymbol{r}_1-\boldsymbol{r}_2) \end{aligned} \tag{5.183}$$

of the single-particle Green function in imaginary and real times, respectively.

At equal points in space, it is useful to introduce the density of states per spin

$$\widetilde{\nu}(\xi) := \nu(\varepsilon) = \frac{1}{V}\sum_{\boldsymbol{k}} \delta(\xi_{\boldsymbol{k}}-\xi), \tag{5.184}$$

in terms of which

$$G_{\beta,\mu}(\tau,\boldsymbol{r}=\boldsymbol{0}) = -\int \mathrm{d}\xi\, \widetilde{\nu}(\xi)\left\{\Theta(+\tau)\left[1-\widetilde{f}_{\mathrm{FD}}(\xi)\right] e^{-\tau\xi} - \Theta(-\tau)\widetilde{f}_{\mathrm{FD}}(\xi)\, e^{-\tau\xi}\right\}, \tag{5.185}$$

and

$$G_{\beta,\mu}(t,\boldsymbol{r}=\boldsymbol{0}) = -\int \mathrm{d}\xi\, \widetilde{\nu}(\xi) \left\{\Theta(+t)\left[1-\widetilde{f}_{\mathrm{FD}}(\xi)\right] e^{-\mathrm{i}t\xi} - \Theta(-t)\widetilde{f}_{\mathrm{FD}}(\xi)\, e^{-\mathrm{i}t\xi}\right\}, \tag{5.186}$$

respectively. For the parabolic spectrum of the jellium model,

$$\nu(\varepsilon) = \begin{cases} \sqrt{\frac{m}{2\pi^2\,\varepsilon}}, & \text{for } d=1,\\ \frac{m}{2\pi}, & \text{for } d=2,\\ \frac{m\sqrt{2m\,\varepsilon}}{2\pi^2}, & \text{for } d=3, \end{cases} \qquad \varepsilon := \xi+\mu, \tag{5.187}$$

so that the density of states per spin can be taken to be the constant ν_{F} around the Fermi energy at very low temperatures.

At zero temperature and assuming a constant density of states per spin, the imaginary-time single-particle Green function at equal points is, up to a proportionality constant, the Laplace transform of the sign function, i.e.,

$$\begin{aligned} G_{\beta=\infty,\mu}(\tau,\boldsymbol{r}=\boldsymbol{0}) &\approx -\nu_{\mathrm{F}} \int \mathrm{d}\xi\, \left[\Theta(+\tau)\Theta(+\xi)e^{-\tau\xi} - \Theta(-\tau)\Theta(-\xi)e^{-\tau\xi}\right] \\ &= -\frac{\nu_{\mathrm{F}}}{\tau}. \end{aligned} \tag{5.188}$$

Analytical continuation to real time gives

$$\begin{aligned} G_{\beta=\infty,\mu}(t,\boldsymbol{r}=\boldsymbol{0}) &\approx -\nu_{\mathrm{F}} \int \mathrm{d}\xi\, \left[\Theta(+t)\,\Theta(+\xi)\,e^{-\mathrm{i}t\xi} - \Theta(-t)\,\Theta(-\xi)\,e^{-\mathrm{i}t\xi}\right] \\ &= +\frac{\mathrm{i}\nu_{\mathrm{F}}}{t-\mathrm{i}0^{+}\,\mathrm{sgn}(t)}. \end{aligned} \tag{5.189}$$

The approximation by which the density of states per spin is assumed to be constant becomes exact in the limits $\tau\to\pm\infty$ $(t\to\pm\infty)$. In other words, Eqs. (5.188) and 5.189) become exact in the limits for which the integrals on the right-hand sides are dominated by the contributions around the Fermi energy $\xi=0$. The algebraic decay on the right-hand sides of Eqs. (5.188) and 5.189) is caused by the discontinuity at the Fermi energy of the Fermi-Dirac distribution at zero temperature. If the density of states per spin tames the discontinuity at the Fermi energy of the Fermi-Dirac distribution at zero temperature, say because it vanishes in a power law fashion at the Fermi energy, $\widetilde{\nu}(\xi)\sim|\xi|^{g}$ with $g>0$, the long-time correlation probed by the single-particle Green function at equal points decay faster, e.g.,

$$G_{\beta=\infty,\mu}(t,\boldsymbol{r}=\boldsymbol{0}) \sim +\mathrm{i}\Gamma(g+1)\,e^{-\mathrm{i}\pi(1+g)/2}\,\frac{\mathrm{sgn}(t)}{|t|^{1+g}}. \tag{5.190}$$

Tunneling experiments give access to the asymptotic time dependence of the single-particle Green function at equal points in space. Thus, they could signal whenever perturbations to the non-interacting limit are sufficiently strong to change the exponent g from the value $g=0$ to $g>0$.

5.4.4.3 *At equal times*

We now combine Eqs. (5.183) and (5.178b) to study

$$\begin{aligned} G_{\beta,\mu}(t,\boldsymbol{r}) :=& \frac{1}{V}\sum_{\boldsymbol{k}} e^{-\mathrm{i}\boldsymbol{k}\cdot\boldsymbol{r}}\, G_{\beta,\mu}(t,\boldsymbol{k}) \\ =& -\frac{1}{V}\sum_{\boldsymbol{k}} e^{-\mathrm{i}\boldsymbol{k}\cdot\boldsymbol{r}} \left\{\Theta(+t)\left[1-\widetilde{f}_{\mathrm{FD}}(\xi_{\boldsymbol{k}})\right] e^{-\mathrm{i}t\xi_{\boldsymbol{k}}} - \Theta(-t)\widetilde{f}_{\mathrm{FD}}(\xi_{\boldsymbol{k}})e^{-\mathrm{i}t\xi_{\boldsymbol{k}}}\right\} \end{aligned} \tag{5.191}$$

at equal times, i.e., in the limit $t \to -0^{+}$ (without loss of generality). In this limit, the equal-time single-particle Green function in position-space is the Fourier transform of the Fermi-Dirac distribution,

$$G_{\beta,\mu}(t=-0^{+},\boldsymbol{r}) = \frac{1}{V}\sum_{\boldsymbol{k}} e^{-\mathrm{i}\boldsymbol{k}\cdot\boldsymbol{r}}\, \widetilde{f}_{\mathrm{FD}}(\xi_{\boldsymbol{k}}). \tag{5.192}$$

At zero temperature and in the thermodynamic limit, Eq. (5.192) reduces to the Fourier transform over the Heaviside step function

$$\begin{aligned} \lim_{\beta\to\infty} G_{\beta,\mu}(t=-0^{+},\boldsymbol{r}) =& \int \frac{\mathrm{d}^{d}\boldsymbol{k}}{(2\pi)^{d}} e^{-\mathrm{i}\boldsymbol{k}\cdot\boldsymbol{r}}\,\Theta(-\xi_{\boldsymbol{k}}) \\ =& \int \frac{\mathrm{d}^{d}\boldsymbol{k}}{(2\pi)^{d}} e^{-\mathrm{i}\boldsymbol{k}\cdot\boldsymbol{r}}\,\Theta\left(\frac{\hbar^{2}k_{\mathrm{F}}^{2}}{2m} - \frac{\hbar^{2}\boldsymbol{k}^{2}}{2m}\right) \\ =& \int \frac{\mathrm{d}^{d}\boldsymbol{k}}{(2\pi)^{d}} e^{-\mathrm{i}\boldsymbol{k}\cdot\boldsymbol{r}}\,\Theta\left(k_{\mathrm{F}} - |\boldsymbol{k}|\right). \end{aligned} \tag{5.193}$$

In $d = 1$,

$$\begin{aligned} \lim_{\beta\to\infty} G_{\beta,\mu}(t=-0^{+},r) =& \int_{-k_{\mathrm{F}}}^{+k_{\mathrm{F}}} \frac{\mathrm{d}k}{2\pi} e^{-\mathrm{i}kr} \\ =& \frac{\sin k_{\mathrm{F}}\, r}{\pi\, r}. \end{aligned} \tag{5.194}$$

For $d > 1$,

$$\begin{aligned} \lim_{\beta\to\infty} G_{\beta,\mu}(t=-0^{+},\boldsymbol{r}) =& \int \frac{\mathrm{d}^{d}\boldsymbol{k}}{(2\pi)^{d}} e^{-\mathrm{i}\boldsymbol{k}\cdot\boldsymbol{r}}\,\Theta\left(k_{\mathrm{F}} - |\boldsymbol{k}|\right) \\ =& \frac{1}{|\boldsymbol{r}|^{d}} \int \frac{\mathrm{d}\omega_{d}}{(2\pi)^{d}} \int_{0}^{k_{\mathrm{F}}|\boldsymbol{r}|} \mathrm{d}p\, p^{d-1}\, e^{-\mathrm{i}p\cos\theta_{1}} \\ =& \frac{1}{|\boldsymbol{r}|^{d}} \int \frac{\mathrm{d}\widehat{\Omega}_{d-1}}{(2\pi)^{d}} \int_{0}^{k_{\mathrm{F}}|\boldsymbol{r}|} \mathrm{d}p\, p^{d-1} \int_{0}^{2\pi\delta_{d,2}+(1-\delta_{d,2})\pi} \mathrm{d}\theta_{1}\, \sin^{d-2}\theta_{1}\, e^{-\mathrm{i}p\cos\theta_{1}}, \end{aligned} \tag{5.195a}$$

where $0 \leq \theta_n < \pi$ for $n = 1, \cdots, d-2$ and $0 \leq \theta_{d-1} < 2\pi$ with

$$\begin{aligned}
\mathrm{d}\omega_d &:= \sin^{d-2}\theta_1\,\mathrm{d}\theta_1\ \sin^{d-3}\theta_2\,\mathrm{d}\theta_2\ \cdots \sin^{d-1-i}\theta_i\,\mathrm{d}\theta_i\ \cdots \sin\theta_{d-2}\,\mathrm{d}\theta_{d-2}\,\mathrm{d}\theta_{d-1},\\
\mathrm{d}\widehat{\Omega}_{d-1} &:= \sin^{d-3}\theta_2\,\mathrm{d}\theta_2\ \cdots \sin^{d-1-i}\theta_i\,\mathrm{d}\theta_i\ \cdots \sin\theta_{d-2}\,\mathrm{d}\theta_{d-2}\,\mathrm{d}\theta_{d-1}.
\end{aligned} \tag{5.195b}$$

(i) Example $d = 2$,

$$\begin{aligned}
\lim_{\beta\to\infty} G_{\beta,\mu}(t=-0^+,\boldsymbol{r}) &= \frac{1/(2\pi)^2}{|\boldsymbol{r}|^2} \int\limits_0^{k_{\mathrm{F}}|\boldsymbol{r}|} \mathrm{d}p\,p \int\limits_0^{2\pi} \mathrm{d}\theta\, e^{-\mathrm{i}p\cos\theta}\\
\text{Eq. (13.6.22) from Ref. [56]} &= \frac{1/(2\pi)^2}{|\boldsymbol{r}|^2} \int\limits_0^{k_{\mathrm{F}}|\boldsymbol{r}|} \mathrm{d}p\,p \sum_{n\in\mathbb{Z}} J_n(p) \int\limits_0^{2\pi} \mathrm{d}\theta\, e^{-\mathrm{i}n\left(\frac{\pi}{2}-\theta\right)}\\
&= \frac{1/(2\pi)}{|\boldsymbol{r}|^2} \int\limits_0^{k_{\mathrm{F}}|\boldsymbol{r}|} \mathrm{d}p\,p\,J_0(p)\\
\text{8.472.3 from Ref. [57]} &= \frac{1/(2\pi)}{|\boldsymbol{r}|^2} \int\limits_0^{k_{\mathrm{F}}|\boldsymbol{r}|} \mathrm{d}p\,\frac{\mathrm{d}}{\mathrm{d}p}\Big(p\,J_1(p)\Big)\\
&= \frac{k_{\mathrm{F}}}{2\pi|\boldsymbol{r}|} J_1(k_{\mathrm{F}}|\boldsymbol{r}|).
\end{aligned} \tag{5.196a}$$

Hence [see Eq. (4.4.5) from Ref. [56]],

$$\lim_{\substack{|\boldsymbol{r}|\to\infty\\ \beta\to\infty}} G_{\beta,\mu}(t=-0^+,\boldsymbol{r}) \sim +\frac{k_{\mathrm{F}}}{2\pi|\boldsymbol{r}|}\sqrt{\frac{2}{\pi k_{\mathrm{F}}|\boldsymbol{r}|}}\,\sin\left(k_{\mathrm{F}}|\boldsymbol{r}| - \frac{\pi}{4}\right), \tag{5.196b}$$

if $d = 2$.

(ii) Example $d = 3$,

$$\begin{aligned}
\lim_{\beta\to\infty} G_{\beta,\mu}(t=-0^+,\boldsymbol{r}) &= \frac{1/(2\pi)^3}{|\boldsymbol{r}|^3} \int\limits_0^{k_{\mathrm{F}}|\boldsymbol{r}|} \mathrm{d}p\,p^2 \int\limits_0^{2\pi} \mathrm{d}\varphi \int\limits_0^{\pi} \mathrm{d}\theta\,\sin\theta\; e^{-\mathrm{i}p\cos\theta}\\
&= \frac{1/(2\pi)^2}{|\boldsymbol{r}|^3} \int\limits_0^{k_{\mathrm{F}}|\boldsymbol{r}|} \mathrm{d}p\,p^2 \int\limits_{-1}^{+1} \mathrm{d}x\, e^{-\mathrm{i}p\,x}\\
&= \frac{e^{-\mathrm{i}\pi/2}/(2\pi)^2}{|\boldsymbol{r}|^3} \int\limits_0^{k_{\mathrm{F}}|\boldsymbol{r}|} \mathrm{d}p\,p\;\left(e^{+\mathrm{i}p} - e^{-\mathrm{i}p}\right)\\
&= \frac{e^{-\mathrm{i}\pi}/(2\pi)^2}{|\boldsymbol{r}|^3} \int\limits_0^{k_{\mathrm{F}}|\boldsymbol{r}|} \mathrm{d}p\,\left\{\frac{\mathrm{d}}{\mathrm{d}p}\left[p\;\left(e^{+\mathrm{i}p}+e^{-\mathrm{i}p}\right)\right] - \left(e^{+\mathrm{i}p}+e^{-\mathrm{i}p}\right)\right\}\\
&= \frac{e^{-\mathrm{i}\pi}/(2\pi)^2}{|\boldsymbol{r}|^3}\Big[k_{\mathrm{F}}|\boldsymbol{r}|\left(e^{+\mathrm{i}k_{\mathrm{F}}|\boldsymbol{r}|} + e^{-\mathrm{i}k_{\mathrm{F}}|\boldsymbol{r}|}\right)\\
&\quad +\mathrm{i}\left(e^{+\mathrm{i}k_{\mathrm{F}}|\boldsymbol{r}|} - e^{-\mathrm{i}k_{\mathrm{F}}|\boldsymbol{r}|}\right)\Big].
\end{aligned} \tag{5.197a}$$

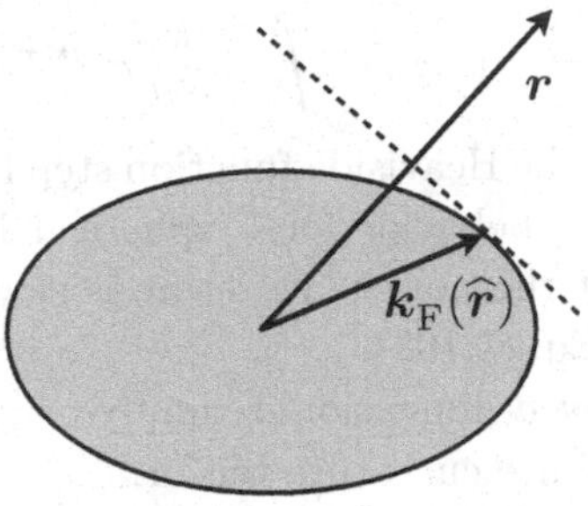

Fig. 5.6 The Fermi vector $k_{\text{F}}(\widehat{r})$ is constructed as follows. Any fixed vector r in real space is determined by its direction $\widehat{r}$ and magnitude $|r|$. The dashed line represents a hyperplane orthogonal to $\widehat{r}$. The vector $k_{\text{F}}(\widehat{r})$ is defined as the point where the hyperplane $[(d-1)$-dimensional$]$ touches the $[(d-1)$-dimensional$]$ Fermi surface.

Hence,

$$\lim_{\substack{|r|\to\infty\\ \beta\to\infty}} G_{\beta,\mu}(t=-0^+,r) \sim \frac{k_{\text{F}}}{(2\pi)^2|r|^2}\left(e^{+\text{i}k_{\text{F}}|r|-\text{i}\pi}+e^{-\text{i}k_{\text{F}}|r|+\text{i}\pi}\right), \tag{5.197b}$$

if $d=3$. Examples (5.194), (5.196b), and (5.197b) illustrate the power-law decay of the equal-time single-particle Green function for large separations in the non-interacting jellium model. This decay is slower the lower the dimensionality. This slow decay reflects the discontinuity of the Fermi-Dirac distribution at zero temperature. More generally, it can be shown that, for any d-dimensional simply-connected closed Fermi surface with a strictly positive-definite curvature tensor,

$$\begin{aligned}\lim_{\substack{|r|\to\infty\\ \beta\to\infty}} G_{\beta,\mu}(t=-0^+,r) \sim\; & A_+(\widehat{r})\,\frac{e^{-\text{i}k_{\text{F}}(-\widehat{r})\cdot r-\text{i}\pi(d+1)/4}}{|r|^{(d+1)/2}}\\ &+A_-(\widehat{r})\,\frac{e^{-\text{i}k_{\text{F}}(+\widehat{r})\cdot r+\text{i}\pi(d+1)/4}}{|r|^{(d+1)/2}},\end{aligned} \tag{5.198}$$

where $A_\pm(\widehat{r})$ are dimensionful [same dimension as $|k_{\text{F}}(\pm\widehat{r})|^{(d-1)/2}$], $\widehat{r}\equiv r/|r|$, and, given a coordinate system in momentum space with the center of gravity of the Fermi surface as origin, $k_{\text{F}}(\pm\widehat{r}) = \pm k_{\text{F}}(\widehat{r})$ are the two Fermi points for which the hyperplanes normal to $\widehat{r}$ are tangent in these points to the Fermi surface, see Fig. 5.6.

5.5 Problems

5.5.1 *Equal-time non-interacting two-point Green function for a Fermi gas*

Introduction

We introduced in Eq. (5.192) the equal-time Green function in position space $G_{\beta,\mu}(t=-0^+,r)$. At zero temperature ($\beta\to\infty$) and in the thermodynamic limit, we found

$$G_{\beta,\mu}(t=-0^{+},\boldsymbol{r})=\int\frac{\mathrm{d}^{d}\boldsymbol{k}}{(2\pi)^{d}}\,e^{-\mathrm{i}\boldsymbol{k}\cdot\boldsymbol{r}}\,\Theta\left(-\xi_{\boldsymbol{k}}\right). \tag{5.199}$$

As usual, the function Θ is the Heaviside function step function. We then proceeded to compute the asymptotic behavior for a spherical Fermi surface in $d=1,2,3$ dimensions and noted that the power-law decay is slower the lower the dimension d. We are going to prove Eq. (5.198).

Exercise 1.1: For any d-dimensional simply-connected closed Fermi surface with a strictly positive-definite curvature tensor:

(a) Convince yourself that $G_{\beta,\mu}(-0^{+},\boldsymbol{r})$ can be written as

$$G_{\beta,\mu}(-0^{+},\boldsymbol{r})=\int\limits_{-\infty}^{+\infty}\mathrm{d}k'\,N(k',\widehat{\boldsymbol{r}})\,e^{-\mathrm{i}k'|\boldsymbol{r}|}, \tag{5.200a}$$

where

$$\begin{aligned}N(k',\widehat{\boldsymbol{r}})=&\int\frac{\mathrm{d}^{d}\boldsymbol{k}}{(2\pi)^{d}}\,\delta\left(k'-\boldsymbol{k}\cdot\widehat{\boldsymbol{r}}\right)\,\Theta\left(-\xi_{\boldsymbol{k}}\right)\\ \approx&\,\tilde{A}_{-}(\widehat{\boldsymbol{r}})\,\Theta\Big(+\boldsymbol{k}_{\mathrm{F}}(+\widehat{\boldsymbol{r}})\cdot\widehat{\boldsymbol{r}}-k'\Big)\,\Theta(+k')\,|+\boldsymbol{k}_{\mathrm{F}}(+\widehat{\boldsymbol{r}})\cdot\widehat{\boldsymbol{r}}-k'|^{\alpha}\\ &+\tilde{A}_{+}(\widehat{\boldsymbol{r}})\,\Theta\Big(-\boldsymbol{k}_{\mathrm{F}}(-\widehat{\boldsymbol{r}})\cdot\widehat{\boldsymbol{r}}+k'\Big)\,\Theta(-k')\,|-\boldsymbol{k}_{\mathrm{F}}(-\widehat{\boldsymbol{r}})\cdot\widehat{\boldsymbol{r}}+k'|^{\alpha},\end{aligned} \tag{5.200b}$$

where $\tilde{A}_{\pm}(\widehat{\boldsymbol{r}})$ are dimensionful, $\widehat{\boldsymbol{r}}\equiv\boldsymbol{r}/|\boldsymbol{r}|$ and, given a coordinate system in momentum space with the center of gravity of the Fermi surface as origin, $\boldsymbol{k}_{\mathrm{F}}(\pm\widehat{\boldsymbol{r}})$ are the two Fermi points for which the hyperplanes normal to $\widehat{\boldsymbol{r}}$ are tangent in these points to the Fermi surface, see Fig. 5.6. For a given $\widehat{\boldsymbol{r}}$, this approximation is good for either $k'\approx\boldsymbol{k}_{\mathrm{F}}(+\widehat{\boldsymbol{r}})\cdot\widehat{\boldsymbol{r}}$ or $k'\approx\boldsymbol{k}_{\mathrm{F}}(-\widehat{\boldsymbol{r}})\cdot\widehat{\boldsymbol{r}}$. Determine the exponent α as a function of the dimension d. What are the coefficients $\tilde{A}_{+}$ and $\tilde{A}_{-}$ in the case of a spherical Fermi surface?

(b) Do the k' integral in Eq. (5.200) to derive Eq. (5.198). Why is the approximation in Eq. (5.200b) valid in the limit of large distance $|\boldsymbol{r}|$?

(c) Compare Eq. (5.198) with the results (5.194), (5.196b), and (5.197b) for the spherical Fermi surface in $d=1$, $d=2$, and $d=3$, respectively.

(d) What happens if the Fermi surface has a flat piece?

5.5.2 *Application of the Kubo formula to the Hall conductivity in the integer quantum Hall effect*

Introduction

The spectrum of a two-dimensional gas of free electrons is strongly reorganized when the electrons are subject to a perpendicular magnetic field. The parabolic dispersion, whose density of states is constant as a function of energy, turns into a sequence of flat bands with an equidistant separation in energy, the so-called Landau levels. Each of these Landau levels comprises an extensive number of degenerate single-particle states. Whenever an integer number $\tilde{n} = 1, 2, \cdots$, of Landau levels is completely filled with electrons and the next-higher Landau level is empty, the single-Slater-determinant ground state is incompressible and insulating as far as longitudinal charge transport is concerned. However, this incompressible state features a non-vanishing and quantized transverse conductivity $\sigma_{\mathrm{H}} = \tilde{n}\, e^2/h$. This is the integer quantum Hall effect (IQHE) [58].

We are going to derive these features of the IQHE. We will see that the non-vanishing Hall conductivity is intimately related to the fact that the electrons experience *non-commutative quantum geometry.* This means that the two in-plane components of the electron's position operator do not commute, when projected to the degrees of freedom from one Landau level. This is in sharp contrast to the case of free electrons, whose position operator components commute.

Diagonalizing the Landau Hamiltonian

Non-interacting electrons confined to two-dimensional position space under a perpendicular uniform magnetic field $\boldsymbol{B} = B\, \boldsymbol{e}_3$ of magnitude $B > 0$ are governed by the Hamiltonian

$$\hat{H} := \frac{1}{2m}\, \hat{\boldsymbol{\pi}}^2, \tag{5.201a}$$

where the gauge-invariant momentum $(-e < 0)$

$$\hat{\boldsymbol{\pi}} := \hat{\boldsymbol{p}} - (-e)\, \boldsymbol{A}(\hat{\boldsymbol{r}}) \tag{5.201b}$$

is given in terms of the vector potential

$$\partial_1 A_2(\boldsymbol{r}) - \partial_2 A_1(\boldsymbol{r}) = B, \tag{5.201c}$$

and the two components of the position operator $\hat{\boldsymbol{r}}$ satisfy canonical commutation relations with the two components of the momentum operator $\hat{\boldsymbol{p}}$

$$[\hat{r}_i, \hat{p}_i] = \mathrm{i}\delta_{ij}, \qquad i, j = 1, 2, \tag{5.201d}$$

in units where the speed of light and Planck's constant are unity. However, the components of $\hat{\boldsymbol{\pi}}$ are not the generators of (magnetic) translations. These are instead given by the so-called guiding center momenta

$$\hat{\boldsymbol{K}} := \hat{\boldsymbol{\pi}} - \frac{1}{\ell^2}\, \boldsymbol{e}_3 \wedge \hat{\boldsymbol{r}}, \tag{5.202a}$$

where

$$\ell := \frac{1}{\sqrt{e B}} \tag{5.202b}$$

is the magnetic length.

Exercise 1.1: Show that

$$\hat{\boldsymbol{\pi}} = \mathrm{i}m\, [\hat{H}, \hat{\boldsymbol{r}}]. \tag{5.203}$$

Exercise 1.2: Show that the components of the guiding center momenta (5.202a) and the components of the gauge-invariant momenta (5.201b) satisfy the commutation relations

$$[\hat{\pi}_i, \hat{\pi}_j] = -\mathrm{i}\frac{\epsilon_{ij}}{\ell^2}, \qquad [\hat{K}_i, \hat{K}_j] = +\mathrm{i}\frac{\epsilon_{ij}}{\ell^2}, \qquad [\hat{K}_i, \hat{\pi}_j] = 0, \qquad i,j = 1,2. \tag{5.204}$$

Due to the different sign of their commutators, $\hat{\boldsymbol{\pi}}$ and $\hat{\boldsymbol{K}}$ are sometimes referred to as the left-handed and right-handed degree of freedom of the Landau level electrons, respectively.

The corresponding position operators for the guiding center are

$$\begin{aligned}\hat{\boldsymbol{X}} &:= \hat{\boldsymbol{r}} + \ell^2\, \boldsymbol{e}_3 \wedge \hat{\boldsymbol{\pi}} \\ &= +\,\ell^2\, \boldsymbol{e}_3 \wedge \hat{\boldsymbol{K}}\end{aligned} \tag{5.205a}$$

and also satisfy the right-handed algebra

$$[\hat{X}_i, \hat{X}_j] = +\mathrm{i}\epsilon_{ij}\,\ell^2, \qquad i,j = 1,2. \tag{5.205b}$$

In order to diagonalize the Hamiltonian (5.201a), it is convenient to introduce the ladder operators

$$\hat{a}^\dagger := \frac{\ell}{\sqrt{2}}\,(\hat{\pi}_1 + \mathrm{i}\hat{\pi}_2), \qquad \hat{a} := \frac{\ell}{\sqrt{2}}\,(\hat{\pi}_1 - \mathrm{i}\hat{\pi}_2), \tag{5.206a}$$

and

$$\hat{b}^\dagger := \frac{1}{\sqrt{2}\,\ell}\left(\hat{X}_1 - \mathrm{i}\hat{X}_2\right), \qquad \hat{b} := \frac{1}{\sqrt{2}\,\ell}\left(\hat{X}_1 + \mathrm{i}\hat{X}_2\right). \tag{5.206b}$$

Exercise 1.3: Show that the ladder operators satisfy the bosonic algebra

$$[\hat{a}, \hat{a}^\dagger] = 1, \qquad [\hat{b}, \hat{b}^\dagger] = 1, \tag{5.206c}$$

with all other commutators vanishing.

Exercise 1.4: After expressing the Hamiltonian (5.201a) in terms of the operators $\hat{a}^\dagger$, $\hat{a}$, $\hat{b}^\dagger$, and $\hat{b}$, show that it has the discrete spectrum of Landau levels indexed by $n = 0, 1, 2, \cdots$,

$$\varepsilon_n = \omega_{\mathrm{c}}\left(n + \frac{1}{2}\right), \qquad \omega_{\mathrm{c}} := \frac{e\,B}{m}, \tag{5.207}$$

and that a basis for the eigenstates of each Landau level n is given by

$$|n, m\rangle := \frac{1}{\sqrt{n!\,m!}}\,\hat{a}^{\dagger\,n}\,\hat{b}^{\dagger\,m}|0\rangle, \qquad \hat{a}\,|0\rangle = \hat{b}\,|0\rangle = 0, \tag{5.208}$$

where $m = 0, 1, 2, \cdots$.

Non-commutative geometry and Hall conductivity

The projector on the states in the n-th Landau level can be represented as

$$\hat{\mathcal{P}}_n := \sum_m |n,m\rangle\langle n,m|. \tag{5.209}$$

Exercise 2.1: Show that the guiding center position $\hat{\boldsymbol{X}}$ is nothing but the projection of the position operator $\hat{\boldsymbol{r}}$ onto any given Landau level, i.e.,

$$\hat{\boldsymbol{X}} = \hat{\mathcal{P}}_n\, \hat{\boldsymbol{r}}\, \hat{\mathcal{P}}_n. \tag{5.210}$$

In this sense, the position operators projected to any given Landau level furnish a non-commutative geometry. The commutation relations (5.205b) say that their components $\hat{X}_1$ and $\hat{X}_2$ are canonically conjugate variables, in the same way as the momentum and position operators of a free electron are canonically conjugate.

This non-commutative geometry is at the heart of both the IQHE and the FQHE. For example, it is intimately related to the quantized Hall conductivity σ_{H}. The Kubo formula for the contribution of the n-th Landau level ($n = 0, 1, 2, \cdots$) to the Hall conductivity is

$$\sigma^{(n)}_{\mathrm{H}} := -\frac{e^2\,\hbar}{\mathrm{i}m^2}\frac{1}{A}\sum_{n'\neq n}\sum_{m}\frac{\langle n,m|\,\hat{\pi}_1\,\hat{\mathcal{P}}_{n'}\,\hat{\pi}_2\,|n,m\rangle - (1\leftrightarrow 2)}{(\varepsilon_n - \varepsilon_{n'})^2}, \tag{5.211}$$

where $A = 2\pi\sum_m \ell^2$ is the area of the Hall droplet, and we reinstated h.

Exercise 2.2: Use Eqs. (5.201b) and (5.203) to show that the Hall conductivity (5.211) is given by

$$\begin{aligned}\sigma^{(n)}_{\mathrm{H}} &= -\frac{\mathrm{i}e^2}{A\,\hbar}\sum_m \left\langle n,m\left|\left[\hat{X}_1,\hat{X}_2\right]\right|n,m\right\rangle \\ &= \frac{e^2}{h},\end{aligned} \tag{5.212}$$

where the commutation relations (5.205b) were used to obtain the last line. The role of the non-commutative position-operator algebra is apparent in the penultimate line. If the components of the position operator were to commute, as they do for free electrons, the Hall conductivity is bound to vanish. If the lowest $\tilde{n}$ Landau levels are filled, each of them contributes the same Hall conductivity (5.212) and the total Hall conductivity is

$$\begin{aligned}\sigma_{\mathrm{H}} &= \sum_{n\leq\tilde{n}}\frac{e^2}{h} \\ &= \tilde{n}\frac{e^2}{h}.\end{aligned} \tag{5.213}$$

5.5.3 *The Hall conductivity and gauge invariance*

Introduction

Shortly after the discovery of the integer quantum Hall effect (IQHE) [58], Laughlin produced a beautiful argument, Laughlin flux insertion argument [59], that explains under what conditions the Hall conductivity in two-dimensional position space must necessarily take universal (i.e., independent of the shape of the Hall bar, independent of the precise value of the applied magnetic field on the Hall bar, robust to changes in the mobility of the Hall bar, etc.) and rational values in units of e^2/h, where $-e < 0$ is the electron charge. We shall adapt his argument to the situation when assumptions L1, L2, and L3 that shortly follow hold. We denote the many-body Hamiltonian for identical electrons by $\hat{H}$. The necessary (but not sufficient) conditions for the Hall conductivity at vanishing temperature to take *rational* values (not only integer values as in the original argument of Laughlin) in units of e^2/h are the following.

L1: The total number (charge) operator commutes with $\hat{H}$ and this symmetry is not broken spontaneously by the ground state. This condition implies that the Hall conductivity of a two-dimensional superconductor need not take rational values in units of e^2/h, see section 7.9.4.
L2: If two-dimensional Euclidean space is compactified so as to be topologically equivalent (homeomorphic) to the two-sphere S^2, then the energy spectrum of $\hat{H}$ displays a gap between its ground state and all other energy eigenstates. A Fermi liquid fails to satisfy this condition. A band insulator meets this requirement, as does an integer number of filled Landau levels.
L3: Galilean invariance is broken by $\hat{H}$. [See Eq. (7.149) in footnote 7 of section 7.7.1 for the definition of a Galilean transformation.] This condition implies that the Hall conductivity of a two-dimensional gas of non-interacting electrons free to propagate in the two-dimensional Euclidean space perpendicular to a uniform and static magnetic field need not take rational values in units of e^2/h. Galilean symmetry is always broken in the laboratory, say by the ionic periodic potential hosting the electrons or by crystalline defects. To appreciate condition L3, we need Kohn theorem [60].

Kohn theorem

Exercise 1.1: Consider N_{e} spinless fermions, each carrying the electron charge $-e < 0$ and the mass m, in the presence of static and uniform magnetic and electric fields $\boldsymbol{B} = B\,\boldsymbol{e}_z$ and $\boldsymbol{E} = E\,\boldsymbol{e}_x$, respectively. Assume that any two spinless fermions separated by the distance $\boldsymbol{r}$ interact through the two-body translation-invariant potential $V(\boldsymbol{r})$.

(a) Write down the classical Lagrangian for these N_{e} spinless fermions.

(b) Derive the classical equations of motion for these N_{e} spinless fermions.
(c) Go to the center-of-mass and relative coordinates for these N_{e} spinless fermions and show that the electric field decouples from the equations of motion for the relative coordinates, while the equations of motion for the center of mass do not depend on the two-body interaction potential.
(d) Show that the center of mass is drifting with the drift velocity $\boldsymbol{v} = c\,\boldsymbol{E}\wedge\boldsymbol{B}/B^2$.
(e) Use the drift velocity of the center of mass and the density n_{e} of electron per unit area to show that the classical electric current per unit time and per unit length is given by

$$\boldsymbol{j} = \begin{pmatrix} 0 \\ n_{\mathrm{e}}\,e\,c\,\frac{E}{B} \end{pmatrix} = \begin{pmatrix} 0 & -\sigma_{\mathrm{H}} \\ +\sigma_{\mathrm{H}} & 0 \end{pmatrix} \begin{pmatrix} E \\ 0 \end{pmatrix}, \tag{5.214}$$

i.e., the Hall conductance is

$$\sigma_{\mathrm{H}} = \frac{n_{\mathrm{e}}\,e\,c}{B}. \tag{5.215}$$

It is independent of the two-body interaction and depends continuously on n_{e}. This is a manifestation of the classical version of Kohn theorem [60].

Exercise 1.2: The decoupling of the center-of-mass coordinate from the two-body interaction at the classical level survives quantization, i.e., the many-body quantum Hamiltonian $\hat{H}$ is the sum of the Hamiltonians $\hat{H}_{\mathrm{cm}}$ and $\hat{H}_{\mathrm{re}}$ that depend solely on the center of mass and relative coordinates, respectively. The quantum dynamics obeyed by the center-of-mass position operator $\hat{\boldsymbol{R}}$ and the center-of-mass momentum operator $\hat{\boldsymbol{P}}$ is governed by

$$\hat{H}_{\mathrm{cm}} = \frac{1}{2M}\left[\hat{P}_x^2 + (\hat{P}_y + \frac{e}{c}\,B\,\hat{R}_x)^2\right] + e\,E\,\hat{R}_x, \tag{5.216}$$

in the Landau gauge $\boldsymbol{A} = (0, B\,R_x, 0)^{\mathrm{T}}$. The total mass is here denoted by M.

(a) Solve for the eigenstates and eigenvalues of $\hat{H}_{\mathrm{cm}}$. *Hint:* Take advantage of section 5.3.4.
(b) Define the center-of-mass (drift) velocity operator

$$\hat{V}_\mu := \frac{\mathrm{i}}{\hbar}\,[\hat{H}_{\mathrm{cm}}, \hat{R}_\mu], \qquad \mu = x, y. \tag{5.217}$$

Compute the expectation value of the center-of-mass (drift) velocity operator in any eigenstate of Hamiltonian (5.216) to deduce that the classical result (5.215) also holds at the quantum level. Explain why this conclusion could have been reached without any calculation. We have established a quantum manifestation of Kohn theorem [60].
(c) Fill $\tilde{n}$ Landau levels and show that Eq. (5.213) follows. What is the Hall conductivity if a Landau level is partially filled? (This is why condition L3 is needed.)

(d) In his paper [60], Kohn was only considering the case of a uniform magnetic field, i.e., $\boldsymbol{E} = \boldsymbol{0}$. He showed that for any exact eigenstate $|E_\iota\rangle$ that is not the ground state of the many-body Hamiltonian $\hat{H}$, there exists a pair of exact eigenstates with the energies $|E_\iota \pm \hbar\omega_{\mathrm{c}}\rangle$, where ω_{c} is the cyclotron frequency. Construct this pair of eigenstates out of $|E_\iota\rangle$ and the center-of-mass momentum operator. What is the lowest excited state of this kind? *Hint:* Derive the equations of motion obeyed by the components of the center-of-mass momentum operator orthogonal to the uniform applied magnetic field.

Laughlin flux insertion argument

We assume that spinless fermions are confined to a ring embedded in three-dimensional Euclidean space spanned by the basis vector $\boldsymbol{e}_\mu$ with $\mu = x, y, z \equiv 1, 2, 3$. We take the ring to be coplanar with $\boldsymbol{e}_1$ and $\boldsymbol{e}_2$. In the context of the IQHE, the many-body interactions between electrons can be neglected, while one-body interactions such as the confining potentials at the edges, impurity potentials, and a uniform magnetic field $\boldsymbol{B} = B\,\boldsymbol{e}_3$ are present [61]. Here, we assume the presence of generic one-body and many-body interactions that meet conditions L1, L2, and L3. The quantum dynamics in the ring considered as a closed system is governed by the Hamiltonian $\hat{H}(r, R)$, where r is the inner radius of the ring and R is the outer radius.

We may model the circles with radius r and R to be the inner and outer edges (boundaries) of the ring, respectively. The interior (bulk) of the ring is then the set of rings with radius strictly larger than r and strictly smaller than R. The limit $r \to 0$ and $R \to \infty$ is always understood as excluding the origin of $\mathbb{R}^2$. Topologically, a ring is homeomorphic to the punctured plane $\mathbb{R}^2 \setminus \{0\}$. Any two points from the punctured plane can be connected by a smooth path. The set of all closed path of the punctured plane decomposes into a set of equivalence classes. Two closed path are equivalent if they wind around the origin the same number of times. The algebraic structure of the Abelian group $\mathbb{Z}$ can be attached to this set of equivalence classes through the so-called fundamental homotopy group $\pi_1(U(1)) = \mathbb{Z}$, see footnote 27 in section F.3.2. The experimental realization of this geometry is called a Corbino disk.

By assumption L1, charge is a good quantum number in the ring, i.e., $\hat{H}(r, R)$ is Hermitian with a global $U(1)$ symmetry that is not spontaneously broken.

By assumption L2, all excited eigenstates of $\hat{H}(r, R)$ are separated from the ground state of the ring by an energy gap $\Delta(r, R)$ that remains non-vanishing in the limit $r \to 0$ and $R \to \infty$ with all the points at $R = \infty$ identified, i.e.,

$$\lim_{\substack{r \to 0 \\ R \to \infty}} \Delta(r, R) > 0 \qquad \text{on the punctured two-sphere.} \tag{5.218}$$

Exercise 2.1: Argue that this implies that all excited states whose wave functions have support in the bulk (interior) of the ring are gapped if open boundary conditions are imposed on a ring of finite width.

Condition L3 is satisfied for any finite value of R. The boundary condition at the origin implementing the removal of the origin in $\mathbb{R}^2 \setminus \{0\}$ implies that condition L3 is also met in the limit $r \to 0$ and $R \to \infty$, irrespectively of the boundary conditions at infinity.

Exercise 2.2: Argue that excited states with support on the boundaries of the ring (inner or outer edges) show a gap bounded from below by a term of order $1/R$ and that assumption L2 does not prevent this gap from disappearing in the limit $r \to 0$ and $R \to \infty$ with open boundary conditions for which finite size effects disappear, i.e.,

$$\lim_{\substack{r\to 0\\ R\to\infty}} \Delta(r, R) = 0 \tag{5.219}$$

is permissible with open boundary conditions at infinity.

Imagine attaching to the axis of symmetry of the ring an infinitely long and infinitesimally thin solenoid. A slowly varying magnetic flux present in the solenoid induces a vector potential tangential to any circle coplanar to the ring with a time-dependent amplitude.

Exercise 2.3: Show that this time-dependent vector potential does not generate a magnetic field in the ring, but it does generate a tangential electric field that can transfer electric charge from the inner to the outer edges of the ring, or vice versa, through the off-diagonal components of the conductivity tensor, i.e., the Hall conductivity.

Exercise 2.4: Argue that assumption L2 prevents dissipation, i.e., the conductivity tensor must be off-diagonal in polar coordinates. *Hint:* Invoke adiabatic continuity.

To discuss the possible values that the Hall conductivity can take, we recall that the Hall conductivity of a Hall bar as shown in Fig. 5.3 is the linear response between an external applied electric field and the charge current it induces in the circuit to which it weakly couples. To reproduce the Hall setup shown in Fig. 5.3 for the Corbino geometry, we imagine connecting the inner and outer edges of the ring to conducting wires connected to a voltmeter so that the difference in chemical potential between the inner and outer edges is the electrostatic potential V_{H}. We also imagine cycling adiabatically the magnetic flux in the solenoid from the value 0 to the value of q times the unit of flux quantum hc/e. More precisely, we prepare the ring in a ground state of $\hat{H}(r, R)$. We then adiabatically couple the ring to the solenoid and the conducting wires during the adiabatic pumping of the flux in the solenoid. This coupling is removed adiabatically after q pumping cycles by which the magnetic flux q times the unit of flux quantum has been transferred to the ring. The question we want to address is what is the final state of the ring after q pumping cycles.

We make the Ansatz

$$\sigma_{\mathrm{H}} = \frac{p}{q}\frac{e^2}{h}, \tag{5.220}$$

with $p < q$ mutually coprime integers, for the Hall conductivity.

Exercise 2.5: Explain why a charge equal to p times the electron charge is transferred from one edge to the other by q pumping cycles.

The energy that has been pumped through the solenoid into the ring is removed once the charge transferred from one edge to the other edge is brought back to its original edge through the conducting wires. In doing so the initial state has been recovered. Hence, the rational Ansatz (5.220) for the Hall conductivity does not contradict conditions L1, L2, and L3.

Instead of the rational Ansatz (5.220) for the Hall conductivity, we make the irrational Ansatz

$$\sigma_{\rm H} = \xi \frac{e^2}{h}, \tag{5.221}$$

with $0 < \xi$ an irrational number. We are going to show that this Ansatz contradicts assumption L2. To this end, we need the following theorem from number theory.

Theorem (Hurwitz): For any irrational number ξ, there are infinitely many pairs of integers p and q such that

$$\left|\xi - \frac{p}{q}\right| < \frac{1}{\sqrt{5}\,q^2}. \tag{5.222}$$

Hence, it is always possible to choose a pair p and q of integers such that

$$q\left(\xi - \frac{p}{q}\right) \times e\,V_{\rm H} < \Delta(r,R), \tag{5.223}$$

where $\Delta(r,R)$ is the gap in the ring. As was the case for the rational Ansatz (5.220) for the Hall conductivity, cycling adiabatically the magnetic flux in the solenoid from the value 0 to the value of q times the unit of flux quantum transfers a charge equal to $q\,\xi$ times the electron charge from one edge to the other. Of this charge, only the integer part p times the electron charge can be brought back to its original edge through the conducting wires, for we assume that only charge in units of the electron charge can be transported along these wires. The final state is thus not equal to the initial state since a charge $q\left(\xi - \frac{p}{q}\right)$ is left over on the "wrong" edge. The final state must then be an excited state of $\hat{H}(r,R)$, the energy of which is nothing but the left-hand side of the inequality (5.223). However, we have constructed a state of $\hat{H}(r,R)$ with an energy below the energy gap $\Delta(r,R)$ that we had assumed between the ground state and all excited states. This is a contradiction with assumption L2. The irrational Ansatz (5.221) is thus not permissible.

The conclusion of this thought experiment is that the Hall conductivity for a two-dimensional Hamiltonian satisfying conditions L1, L2, and L3 must take rational values in units of e^2/h. Which rational value is selected goes beyond Laughlin flux insertion argument, i.e., the ground state and low-lying excited states of $\hat{H}(r,R)$ must be computed. The constraint (5.220) is nevertheless a powerful one that severely limits the admissible effective low-energy field theories for the Hamiltonian $\lim_{\substack{r\to 0\\ R\to\infty}} \hat{H}(r,R)$ supporting a non-vanishing Hall conductance. The discovery of the fractional quantum Hall effect (FQHE) showed that strong many-body effects can stabilize a phase of matter with a non-integer but rational value of $h\,\sigma_{\mathrm{H}}/e^2$ [62].

Exercise 2.6: Convince yourself that the same conclusion would hold if we replace condition L2 by condition L2′.

> **L2′**: There exists a mobility gap above the many-body ground state of the bulk many-body eigenstates.

The notion of a mobility gap covers the case when translation symmetry is broken by disorder in such a way that the spectral gap of condition L2 (that would apply in the ideal limit when the total crystal momentum is a good quantum number) has been filled by impurity states, but all these impurity states are localized, i.e., insensitive to any change in the boundary conditions. The role of disorder is essential to explain the observation of plateaus of the Hall conductivity at rational values in units of e^2/h [59], [61].

Hint: Consider the Hilbert space of smooth functions with support on a circle of radius one obeying periodic boundary conditions. Use the polar angle $-\pi \leq \phi < \pi$. Verify that the wave function $\psi_{\mathrm{pw}}(\phi) := e^{\mathrm{i}\phi}$ and the smooth wave function $\psi_{\mathrm{loc}}(\phi) := 1$ if $|\phi| < (\Delta\phi - \epsilon)/2 \ll \pi$ and $\psi_{\mathrm{loc}}(0) := 0$ if $(\Delta\phi + \epsilon)/2 < |\phi| < \pi$ obey periodic boundary conditions. Here, the support $\Delta\phi + \epsilon \approx \Delta\phi$ of ψ_{loc} is a non-vanishing interval of the circle while ϵ is the small positive number over which range ψ_{loc} changes in magnitude from 1 to zero. Compute the current density carried by ψ_{pw} and ψ_{loc}, respectively, assuming that these wave functions represent a point particle of mass m moving freely around the unit circle. Connect this exercise to the value taken by the polar components of the conductivity tensor in the Corbino geometry if all states in the bulk are localized.

Chapter 6

Jellium model for electrons in a solid

Outline

The jellium model for a three-dimensional Coulomb gas is defined. The path-integral representation of its grand-canonical partition function is presented. Collective degrees of freedom are introduced through a Hubbard-Stratonovich transformation. The low-energy and long-wavelength limit of the effective theory for the collective degrees of freedom that results from integration over the fermions is derived within the random-phase approximation (RPA). A diagrammatic interpretation to the RPA approximation is given. The ground-state energy in the RPA approximation is calculated. The dependence on momenta and frequencies of the RPA polarization function is studied. A qualitative argument is given for the existence of a particle hole continuum and for a branch of sharp excitations called plasmons. The quasi-static and dynamic limits of the polarization function are studied. The quasi-static limit is characterized by screening, Kohn effect, and Friedel oscillations. The dynamic limit is characterized by plasmons and Landau damping. The physical content of the RPA approximation for a repulsive short-range interaction is derived. The physics of zero-sound is discussed. The feedback effect of phonons on the RPA effective interaction between electrons is sketched.

6.1 Introduction

The so-called *jellium model* is a very naive model for interacting electrons hosted in a three-dimensional solid. Electronic interactions are taken to be of the Coulomb type. The ions making up the solid are treated in the simplest possible way, namely as an inert positive background of charges that insures overall charge neutrality. In spite of its simplicity the jellium model is very instructive as it displays interesting many-body effects. We shall see that *screening* of the Coulomb interaction takes place, there are *Friedel oscillations*, and there are collective excitations called *plasmons*.

The method that we employ to derive these phenomena is very general and powerful although it is not "elementary". The idea is to introduce a collective degree of freedom for which a low-energy and long-wavelength-effective theory is derived. The effective theory cannot be obtained exactly. Instead, the effective theory is derived within the *random-phase approximation* (RPA).

6.2 Definition of the Coulomb gas in the Schrödinger picture

6.2.1 *The classical three-dimensional Coulomb gas*

The definition of the classical and three-dimensional jellium model is given by the Hamiltonian

$$\mathfrak{H} := \sum_{i=1}^{N} \frac{\boldsymbol{p}_i^2}{2m} + \sum_{1\leq i<j\leq N} e^2\, V_{\mathrm{cb}}(\boldsymbol{r}_i - \boldsymbol{r}_j), \tag{6.1a}$$

which describes N identical electrons of charge e and mass m. The electronic interaction is mediated by the electrostatic potential $V_{\mathrm{cb}}(\boldsymbol{r})$ that satisfies the Poisson equation [$\hbar = 1$ and Δ is Laplace operator in three-dimensional Euclidean space],

$$-\Delta V_{\mathrm{cb}}(\boldsymbol{r}) = 4\pi\left(\delta(\boldsymbol{r}) - \frac{1}{V}\right). \tag{6.1b}$$

All N electrons are confined to a box of linear size L and volume $V = L^3$. The condition of total (electronic and ionic) charge neutrality, i.e.,

$$0 = \int\limits_V \mathrm{d}^3\boldsymbol{r}\left(\rho(\boldsymbol{r}) - \frac{N}{V}\right), \tag{6.2}$$

where $\rho(\boldsymbol{r})$ is the local electronic density and N/V is the local ionic density, is the only remnant of the underlying ions in the solid. If periodic boundary conditions are imposed on any solution to Eq. (6.1b), one verifies that

$$\frac{1}{V}\sum_{\boldsymbol{q},\,\frac{L}{2\pi}\boldsymbol{q}\in\mathbb{Z}^3} \frac{4\pi}{\boldsymbol{q}^2}\, e^{\mathrm{i}\boldsymbol{q}\cdot\boldsymbol{r}} \equiv \frac{1}{V}\sum_{\boldsymbol{q}} \frac{4\pi}{\boldsymbol{q}^2}\, e^{\mathrm{i}\boldsymbol{q}\cdot\boldsymbol{r}} \tag{6.3}$$

is a solution to the Poisson equation

$$-\Delta\varphi(\boldsymbol{r}) = 4\pi\delta(\boldsymbol{r}), \tag{6.4}$$

and that

$$\begin{aligned} V_{\mathrm{cb}}(\boldsymbol{r}) &= \frac{1}{V}\sum_{\boldsymbol{q}\neq\boldsymbol{0}} \frac{4\pi}{\boldsymbol{q}^2}\, e^{\mathrm{i}\boldsymbol{q}\cdot\boldsymbol{r}} \\ &= \frac{1}{V}\sum_{\boldsymbol{q}} \frac{4\pi}{\boldsymbol{q}^2}\left(1 - \delta_{\boldsymbol{q},\boldsymbol{0}}\right) e^{\mathrm{i}\boldsymbol{q}\cdot\boldsymbol{r}} \\ &\to \frac{1}{L^3}\int \frac{\mathrm{d}^3\boldsymbol{q}}{(2\pi/L)^3}\frac{4\pi}{\boldsymbol{q}^2}\left[1 - \left(\frac{2\pi}{L}\right)^3 \delta(\boldsymbol{q})\right] e^{\mathrm{i}\boldsymbol{q}\cdot\boldsymbol{r}}, \qquad \text{as } L\to\infty, \end{aligned} \tag{6.5}$$

is a solution to Poisson equation (6.1b). Because of translation invariance, it is more advantageous to use the representation

$$\begin{aligned}
\sum_{1\leq i<j\leq N} e^2 V_{\mathrm{cb}}(\boldsymbol{r}_i-\boldsymbol{r}_j) &= \sum_{1\leq i<j\leq N} \frac{1}{V}\sum_{\boldsymbol{q}\neq 0} \frac{4\pi e^2}{\boldsymbol{q}^2} e^{\mathrm{i}\boldsymbol{q}\cdot(\boldsymbol{r}_i-\boldsymbol{r}_j)} \\
&= \frac{1}{V}\sum_{\boldsymbol{q}\neq 0} \frac{4\pi e^2}{\boldsymbol{q}^2}\frac{1}{2}\left[\left(\sum_{j=1}^{N} e^{-\mathrm{i}\boldsymbol{q}\cdot\boldsymbol{r}_j}\right)\left(\sum_{i=1}^{N} e^{+\mathrm{i}\boldsymbol{q}\cdot\boldsymbol{r}_i}\right) - N\right] \\
&= \frac{1}{V}\sum_{\boldsymbol{q}\neq 0} \frac{2\pi e^2}{\boldsymbol{q}^2}\left(\rho_{+\boldsymbol{q}}\,\rho_{-\boldsymbol{q}} - N\right), \qquad (6.6)
\end{aligned}$$

whereby we have introduced the Fourier transform in reciprocal space of the local electronic density

$$\begin{aligned}
\rho_{\boldsymbol{q}} &:= \sum_{j=1}^{N} e^{-\mathrm{i}\boldsymbol{q}\cdot\boldsymbol{r}_j} \\
&= \int_V \mathrm{d}^3\boldsymbol{r}\, e^{-\mathrm{i}\boldsymbol{q}\cdot\boldsymbol{r}} \sum_{j=1}^{N} \delta(\boldsymbol{r}-\boldsymbol{r}_j) \qquad (6.7) \\
&=: \int_V \mathrm{d}^3\boldsymbol{r}\, e^{-\mathrm{i}\boldsymbol{q}\cdot\boldsymbol{r}} \rho(\boldsymbol{r}).
\end{aligned}$$

Hence,

$$\mathfrak{H} = \sum_{i=1}^{N} \frac{\boldsymbol{p}_i^2}{2m} + \frac{1}{V}\sum_{\boldsymbol{q}\neq 0} \frac{2\pi e^2}{\boldsymbol{q}^2}\left(\rho_{+\boldsymbol{q}}\,\rho_{-\boldsymbol{q}} - N\right). \qquad (6.8)$$

6.2.2 *The quantum three-dimensional Coulomb gas*

The *quantum* jellium model in three-dimensional position space is defined by the Schrödinger equation

$$\begin{aligned}
&\mathrm{i}\hbar\,\partial_t \Psi = \hat{H}_N \Psi, \\
&\hat{H}_N = \sum_{j=1}^{N} \frac{\hat{\boldsymbol{p}}_j^2}{2m} + \frac{1}{V}\sum_{\boldsymbol{q}\neq 0} \frac{2\pi e^2}{\boldsymbol{q}^2}\left(\hat{\rho}_{+\boldsymbol{q}}\hat{\rho}_{-\boldsymbol{q}} - N\right), \\
&\hat{\rho}_{\boldsymbol{q}} := \sum_{j=1}^{N} \exp\left(-\mathrm{i}\boldsymbol{q}\cdot\hat{\boldsymbol{r}}_j\right), \qquad \boldsymbol{q} = \frac{2\pi}{L}\boldsymbol{l}, \qquad \boldsymbol{l}\in\mathbb{Z}^3, \\
&\left[\hat{r}_i^a, \hat{p}_j^b\right] = \mathrm{i}\hbar\,\delta_{ij}\,\delta^{ab}, \qquad i,j=1,\cdots,N, \qquad a,b=1,2,3.
\end{aligned} \qquad (6.9)$$

The N-electrons wave functions Ψ spanning the Hilbert space $\mathcal{H}^{(N)}$ are antisymmetric under exchange of any two electrons and obey periodic boundary conditions in the box of volume $V = L^3$ that defines the solid. Equilibrium properties at inverse temperature β are obtained from the canonical partition function

$$Z_{\beta,N} := \mathrm{Tr}_{\mathcal{H}^{(N)}}\left[\exp\left(-\beta\hat{H}_N\right)\right]. \qquad (6.10)$$

What makes the task of solving for the eigenvalues of the Hamiltonian (6.9), that defines the jellium model, difficult is the *competition* between the kinetic energy and the potential energy. The kinetic energy favors extended or delocalized states in position space. The potential energy favors localized states in position space. Accordingly, the kinetic energy is local in momentum space whereas the potential energy is non-local in momentum space.

The choice to work in the momentum space representation in Eq. (6.9) is motivated by the strategy to solve the non-interacting problem first and then to treat the Coulomb interaction by a perturbative expansion. If so, what could be the small expansion parameter that would, a priori, justify treating the Coulomb interaction perturbatively? We can try the following estimates. Identify the microscopic length scale

$$\mathfrak{a} \sim \left(\frac{N}{V}\right)^{-1/3}, \qquad \text{the classical interparticle separation.} \tag{6.11}$$

With this characteristic microscopic length scale in hand, the following estimates of the characteristic kinetic and Coulomb energies

$$E_{\mathrm{kin}} \sim \frac{\hbar^2}{m\,\mathfrak{a}^2} \text{ and } E_{\mathrm{cb}} \sim \frac{e^2}{\mathfrak{a}}, \tag{6.12}$$

respectively, can be made. The ratio of the characteristic Coulomb and kinetic energies defines a dimensionless number r_{s},

$$\begin{aligned} \frac{E_{\mathrm{cb}}}{E_{\mathrm{kin}}} &\sim \frac{m\,e^2}{\hbar^2}\,\mathfrak{a} \\ &= \frac{\mathfrak{a}}{a_{\mathrm{B}}} \qquad a_{\mathrm{B}} := \frac{\hbar^2}{m\,e^2} \text{ being the Bohr radius} \\ &=: r_{\mathrm{s}}. \end{aligned} \tag{6.13}$$

When $r_{\mathrm{s}} \ll 1$, the kinetic energy is the largest energy scale. When $r_{\mathrm{s}} \gg 1$, the potential energy is the largest energy scale. We will only treat the jellium model in the limit $r_{\mathrm{s}} \ll 1$. The two limits $r_{\mathrm{s}} \ll 1$ and $r_{\mathrm{s}} \gg 1$ are not smoothly connected. In the limit $r_{\mathrm{s}} \gg 1$, the electronic ground state is known to break spontaneously translation invariance and is called a *Wigner crystal.* Although it is believed that Wigner crystals could be realized in some regimes of the quantum Hall effect, the limit $r_{\mathrm{s}} \ll 1$ seems to be the relevant one for a majority of materials. Discussion of the transition to and from the Wigner crystal is beyond the scope of this book. The physics of screening, Friedel oscillations, and plasmons that emerge from the so-called RPA in the regime $r_{\mathrm{s}} \ll 1$ will be seen to depend in an *non-analytic* way on r_{s}. By this measure, the RPA is a highly sophisticated approximation. In fact, the first calculation of the ground-state energy within the RPA relied on perturbation theory and was a tour de force in diagrammatics. Fortunately, it is nowadays possible to reproduce the RPA in a more economical, although perhaps less "elementary", way. The "modern" method that we will follow here present the advantage of being of more general use than the pioneering methods of the 50's (see chapters 5 and 6 of Refs. [11] and [9], respectively, for a historical perspective.) The price to be paid is that some machinery to reformulate quantum mechanics as a path integral has to be introduced.

In order to apply the rules for second quantization to the jellium model, we introduce the single-particle momenta and energies ($V = L^3$)

$$\boldsymbol{k} = \frac{2\pi}{L}\boldsymbol{n}, \qquad \boldsymbol{n} \in \mathbb{Z}, \qquad \varepsilon_{\boldsymbol{k}} = \frac{\boldsymbol{k}^2}{2m}. \tag{6.14a}$$

We then postulate the equal-time fermionic algebra

$$\{\hat{c}_{\boldsymbol{k},\sigma}, \hat{c}^{\dagger}_{\boldsymbol{k}',\sigma'}\} = \delta_{\boldsymbol{k},\boldsymbol{k}'}\delta_{\sigma,\sigma'}, \qquad \{\hat{c}_{\boldsymbol{k},\sigma}, \hat{c}_{\boldsymbol{k}',\sigma'}\} = \{\hat{c}^{\dagger}_{\boldsymbol{k},\sigma}, \hat{c}^{\dagger}_{\boldsymbol{k}',\sigma'}\} = 0. \tag{6.14b}$$

If we define the Fourier transforms

$$\hat{\psi}^{\dagger}_{\sigma}(\boldsymbol{r},t) = \frac{1}{\sqrt{V}}\sum_{\boldsymbol{k}} \hat{c}^{\dagger}_{\boldsymbol{k},\sigma}\, e^{-i\boldsymbol{k}\cdot\boldsymbol{r}+i\varepsilon_{\boldsymbol{k}}t}, \qquad \hat{\psi}_{\sigma}(\boldsymbol{r},t) = \frac{1}{\sqrt{V}}\sum_{\boldsymbol{k}} \hat{c}_{\boldsymbol{k},\sigma}\, e^{+i\boldsymbol{k}\cdot\boldsymbol{r}-i\varepsilon_{\boldsymbol{k}}t}, \tag{6.14c}$$

there follows the equal-time fermionic algebra

$$\begin{aligned} &\left\{\hat{\psi}_{\sigma}(\boldsymbol{r},t), \hat{\psi}^{\dagger}_{\sigma'}(\boldsymbol{r}',t)\right\} = \delta_{\sigma,\sigma'}\delta(\boldsymbol{r}-\boldsymbol{r}'), \\ &\left\{\hat{\psi}_{\sigma}(\boldsymbol{r},t), \hat{\psi}_{\sigma'}(\boldsymbol{r}',t)\right\} = \left\{\hat{\psi}^{\dagger}_{\sigma}(\boldsymbol{r},t), \hat{\psi}^{\dagger}_{\sigma'}(\boldsymbol{r}',t)\right\} = 0. \end{aligned} \tag{6.14d}$$

The Fock space $\mathcal{F}$ on which these operators act is

$$\mathcal{F} := \operatorname{span}\left\{\prod_{j}\left(\hat{c}^{\dagger}_{\boldsymbol{k}_j,\sigma_j}\right)^{m_j}|0\rangle \,\middle|\, m_j = 0,1, \quad \hat{c}_{\boldsymbol{k},\sigma}|0\rangle = 0, \quad \frac{L}{2\pi}\boldsymbol{k} \in \mathbb{Z}^3, \quad \sigma = \uparrow,\downarrow\right\}. \tag{6.14e}$$

Define the (total number) operator

$$\hat{Q} := \sum_{\sigma=\uparrow,\downarrow}\sum_{\boldsymbol{k}} \hat{c}^{\dagger}_{\boldsymbol{k},\sigma}\,\hat{c}_{\boldsymbol{k},\sigma}. \tag{6.15}$$

The Hilbert space $\mathcal{H}^{(N)}$ defined in Eq. (6.9) is the subspace of $\mathcal{F}$ for which

$$\hat{Q} = N \tag{6.16}$$

holds. Application of the rules for second quantization to the jellium model yields the identifications ($\hbar = 1$)

$$\hat{\rho}(\boldsymbol{r},t) \to \sum_{\sigma=\uparrow,\downarrow} \hat{\psi}^{\dagger}_{\sigma}(\boldsymbol{r},t)\,\hat{\psi}_{\sigma}(\boldsymbol{r},t), \tag{6.17a}$$

$$\hat{\rho}(\boldsymbol{r},0) \to \frac{1}{V}\sum_{\boldsymbol{q}} e^{+i\boldsymbol{q}\cdot\boldsymbol{r}}\,\hat{\rho}_{\boldsymbol{q}} \equiv \hat{\rho}(\boldsymbol{r}), \tag{6.17b}$$

$$\hat{\rho}_{\boldsymbol{q}} \to \int_V d^3\boldsymbol{r}\, e^{-i\boldsymbol{q}\cdot\boldsymbol{r}}\,\hat{\rho}(\boldsymbol{r}) = \sum_{\sigma=\uparrow,\downarrow}\sum_{\boldsymbol{k}} \hat{c}^{\dagger}_{\boldsymbol{k},\sigma}\,\hat{c}_{\boldsymbol{k}+\boldsymbol{q},\sigma}, \tag{6.17c}$$

and

$$\hat{H}_N \text{ in Eq. (6.9)} \longrightarrow \sum_{\boldsymbol{k},\sigma} \frac{\boldsymbol{k}^2}{2m} \hat{c}^\dagger_{\boldsymbol{k}\sigma} \hat{c}_{\boldsymbol{k}\sigma} + \frac{1}{V} \sum_{\boldsymbol{q}\neq\boldsymbol{0}} \frac{2\pi e^2}{\boldsymbol{q}^2} \left(\hat{\rho}_{+\boldsymbol{q}} \hat{\rho}_{-\boldsymbol{q}} - N\right)$$
$$= \sum_{\boldsymbol{k},\sigma} \frac{\boldsymbol{k}^2}{2m} \hat{c}^\dagger_{\boldsymbol{k}\sigma} \hat{c}_{\boldsymbol{k}\sigma} + \frac{1}{V} \sum_{\boldsymbol{q}\neq\boldsymbol{0}} \frac{2\pi e^2}{\boldsymbol{q}^2} \sum_{\boldsymbol{k},\boldsymbol{k}'} \sum_{\sigma,\sigma'} \hat{c}^\dagger_{\boldsymbol{k}+\boldsymbol{q},\sigma} \hat{c}^\dagger_{\boldsymbol{k}'-\boldsymbol{q},\sigma'} \hat{c}_{\boldsymbol{k}',\sigma'} \hat{c}_{\boldsymbol{k},\sigma}. \tag{6.17d}$$

Observe that the total particle number operator $\hat{Q}$ can only be replaced by the $\mathbb{C}$-number N in the subspace $\mathcal{H}^{(N)}$ of $\mathcal{F}$.

6.3 Path-integral representation of the Coulomb gas

The exact number of electrons is not accessible experimentally in a macroscopic piece of metal. Hence, we can choose the grand-canonical ensemble instead of the canonical ensemble to describe the jellium model. In the grand-canonical ensemble, the uniform density of electrons is allowed to fluctuate around its average,

$$\rho_0 \equiv N/V, \tag{6.18}$$

held fixed by the choice of the chemical potential μ as the temperature β is varied or as the thermodynamic limit $N, V \to \infty$ is taken. In mathematical terms, the transition from the canonical to the grand-canonical ensemble is encoded by the identification of the pair of triplets (Hamiltonian, Hilbert space, and partition function)

$$\hat{H}_N,\ \mathcal{H}^{(N)},\ Z_{\beta,N} \longrightarrow \hat{H}_\mu,\ \mathcal{F},\ Z_{\beta,\mu}, \tag{6.19a}$$

where the Hamiltonian entering the grand-canonical partition function is

$$\hat{H}_\mu := \sum_{\boldsymbol{k},\sigma} \left(\frac{\boldsymbol{k}^2}{2m} - \mu\right) \hat{c}^\dagger_{\boldsymbol{k},\sigma} \hat{c}_{\boldsymbol{k},\sigma} + \frac{1}{V} \sum_{\boldsymbol{q}\neq 0} \frac{2\pi e^2}{\boldsymbol{q}^2} \sum_{\boldsymbol{k},\boldsymbol{k}'} \sum_{\sigma,\sigma'} \hat{c}^\dagger_{\boldsymbol{k}+\boldsymbol{q},\sigma} \hat{c}^\dagger_{\boldsymbol{k}'-\boldsymbol{q},\sigma'} \hat{c}_{\boldsymbol{k}',\sigma'} \hat{c}_{\boldsymbol{k},\sigma}, \tag{6.19b}$$

the Hilbert space over which the trace entering the grand-canonical partition function is to be performed is the Fock space (6.14e), and the grand-canonical partition function is

$$Z_{\beta,\mu} := \mathrm{Tr}_{\mathcal{F}} \left[\exp\left(-\beta \hat{H}_\mu\right)\right]. \tag{6.19c}$$

Of course, one must demand that

$$\frac{1}{V} \left\langle \sum_{\boldsymbol{k}} \sum_{\sigma=\uparrow,\downarrow} \hat{c}^\dagger_{\boldsymbol{k},\sigma} \hat{c}_{\boldsymbol{k},\sigma} \right\rangle_{Z_{\beta,\mu}} := \frac{1}{\beta V} \frac{\partial \ln Z_{\beta,\mu}}{\partial \mu} \equiv \rho_0 \tag{6.19d}$$

is held fixed in order to specify the dependence of μ on the inverse temperature β.

Observe that the Coulomb interaction is normal ordered in Eq. (6.19a), i.e., with all creation to the left of annihilation operators. With the definition of $\hat{H}_N$ on the first line of Eq. (6.17d) and the definition (6.19b) of $\hat{H}_\mu$, the expectation value in the empty state $|0\rangle$ of their difference differ by

$$\left\langle 0\left|\left(\hat{H}_N-\hat{H}_\mu\right)\right|0\right\rangle=-\frac{N}{V}\sum_{\boldsymbol{q}\neq\boldsymbol{0}}\frac{2\pi e^2}{\boldsymbol{q}^2}. \tag{6.20}$$

This $\mathbb{C}$-number does not enter observables obtained from taking logarithmic derivatives of the partition function, but it has to be accounted for when evaluating the logarithm of the partition function, say, as is the case for the ground-state energy.

Rather than calculating the grand-canonical partition function or its logarithmic derivatives by performing the trace over the fermionic Fock space $\mathcal{F}$, we choose to represent the grand-canonical partition function as a path integral over Grassmann coherent states. As is shown in appendices E.1 and E.2, the creation $\hat{c}^\dagger_{\boldsymbol{k},\sigma}(t=-\mathrm{i}\tau)$ and annihilation $\hat{c}_{\boldsymbol{k},\sigma}(t=-\mathrm{i}\tau)$ operators in the Heisenberg picture are replaced, at non-vanishing temperature, by the imaginary-time dependent Grassmann fields $\psi^*_{\boldsymbol{k},\sigma}(\tau)$ and $\psi_{\boldsymbol{k},\sigma}(\tau)$, respectively. Whereas $\hat{c}^\dagger_{\boldsymbol{k},\sigma}(t=-\mathrm{i}\tau)$ is the adjoint of $\hat{c}_{\boldsymbol{k},\sigma}(t=-\mathrm{i}\tau)$, the two Grassmann fields $\psi^*_{\boldsymbol{k},\sigma}(\tau)$ and $\psi_{\boldsymbol{k},\sigma}(\tau)$, are *independent* of each other. The symbol * here should not be construed as implying some dependence, as occurs for complex conjugation say. With the help of the Grassmann integration rules defined in appendices E.1 and E.2, the partition function in the grand-canonical ensemble is given by

$$Z_{\beta,\mu}=\int\mathcal{D}[\psi^*]\,\mathcal{D}[\psi]\,\exp\left(-S_{\beta,\mu}\right). \tag{6.21a}$$

Here, the Boltzmann weight is the exponential of the Euclidean action

$$S_{\beta,\mu}=\int_0^\beta\mathrm{d}\tau\left[\sum_{\boldsymbol{k},\sigma}\psi^*_{\boldsymbol{k},\sigma}(\tau)\left(\frac{\partial}{\partial\tau}+\frac{\boldsymbol{k}^2}{2m}-\mu\right)\psi_{\boldsymbol{k},\sigma}(\tau)+\frac{1}{V}\sum_{\boldsymbol{q}\neq\boldsymbol{0}}\frac{2\pi e^2}{\boldsymbol{q}^2}\rho_{+\boldsymbol{q}}(\tau)\,\rho_{-\boldsymbol{q}}(\tau)\right], \tag{6.21b}$$

where the notation

$$\rho_{+\boldsymbol{q}}(\tau):=\sum_{\boldsymbol{k},\sigma}\psi^*_{\boldsymbol{k},\sigma}(\tau)\,\psi_{\boldsymbol{k}+\boldsymbol{q},\sigma}(\tau) \tag{6.21c}$$

is used for the counterpart to the Fourier component (6.17c) of the electronic density operator, and the Grassmann integration variables obey antiperiodic boundary conditions in imaginary time,

$$\psi^*_{\boldsymbol{k},\sigma}(\tau+\beta)=-\psi^*_{\boldsymbol{k},\sigma}(\tau),\qquad\psi_{\boldsymbol{k},\sigma}(\tau+\beta)=-\psi_{\boldsymbol{k},\sigma}(\tau). \tag{6.21d}$$

Of course, one must demand that

$$\frac{1}{V}\left\langle\rho_{\boldsymbol{q}=\boldsymbol{0}}(\tau)\right\rangle_{Z_{\beta,\mu}}:=\frac{1}{\beta V}\frac{\partial\ln Z_{\beta,\mu}}{\partial\mu}\equiv\rho_0 \tag{6.21e}$$

is held fixed in order to specify the dependence of μ on the inverse temperature β. Finally, the Grassmann measure $\mathcal{D}[\psi^*]\,\mathcal{D}[\psi]$ for the two *independent* Grassmann fields

$$\psi^*_{\boldsymbol{k},\sigma}(\tau)=\frac{1}{\sqrt{\beta}}\sum_{n\in\mathbb{Z}}\psi^*_{\boldsymbol{k},\omega_n,\sigma}\,e^{+\mathrm{i}\omega_n\tau} \tag{6.21f}$$

and

$$\psi_{\boldsymbol{k},\sigma}(\tau)=\frac{1}{\sqrt{\beta}}\sum_{n\in\mathbb{Z}}\psi_{\boldsymbol{k},\omega_n,\sigma}\,e^{-\mathrm{i}\,\omega_n\tau}, \tag{6.21g}$$

where

$$\omega_n=\frac{\pi}{\beta}(2n+1),\qquad n\in\mathbb{Z}, \tag{6.21h}$$

are the discrete (fermionic) Matsubara frequencies, is to be understood as the product measure $\prod_{\boldsymbol{k},n,\sigma}\mathrm{d}\psi^{*}_{\boldsymbol{k},\omega_n,\sigma}\,\mathrm{d}\psi_{\boldsymbol{k},\omega_n,\sigma}$.

Observe that the Fourier series

$$\psi^{*}_{\sigma}(\boldsymbol{r},\tau)=\frac{1}{\sqrt{\beta}}\sum_{n\in\mathbb{Z}}\frac{1}{\sqrt{V}}\sum_{\boldsymbol{k}}\psi^{*}_{\boldsymbol{k},\omega_n,\sigma}\,e^{-\mathrm{i}(\boldsymbol{k}\cdot\boldsymbol{r}-\omega_n\tau)} \tag{6.22a}$$

and

$$\psi_{\sigma}(\boldsymbol{r},\tau)=\frac{1}{\sqrt{\beta}}\sum_{n\in\mathbb{Z}}\frac{1}{\sqrt{V}}\sum_{\boldsymbol{k}}\psi_{\boldsymbol{k},\omega_n,\sigma}\,e^{+\mathrm{i}(\boldsymbol{k}\cdot\boldsymbol{r}-\omega_n\tau)}, \tag{6.22b}$$

can be inverted to yield the position-space and imaginary-time representation of the Euclidean action $S_{\beta,\mu}$,

$$\begin{aligned}S_{\beta,\mu}=&\int\limits_{0}^{\beta}\mathrm{d}\tau\int\limits_{V}\mathrm{d}^3\boldsymbol{r}\left[\sum_{\sigma}\psi^{*}_{\sigma}(\boldsymbol{r},\tau)\left(\partial_\tau-\frac{\Delta}{2m}-\mu\right)\psi_{\sigma}(\boldsymbol{r},\tau)\right]\\&+\frac{1}{2}\int\limits_{0}^{\beta}\mathrm{d}\tau\int\limits_{V}\mathrm{d}^3\boldsymbol{r}\int\limits_{V}\mathrm{d}^3\boldsymbol{r}'\,\left[\rho(\boldsymbol{r},\tau)-\rho_0\right]\frac{e^2}{|\boldsymbol{r}-\boldsymbol{r}'|}\left[\rho(\boldsymbol{r}',\tau)-\rho_0\right].\end{aligned} \tag{6.22c}$$

6.4 The random-phase approximation

6.4.1 *Hubbard-Stratonovich transformation*

We begin with the extension to infinite-product measures of the Gaussian identity

$$\sqrt{\frac{2\pi}{a}}e^{+\frac{1}{2a}z^2}=\int\limits_{-\infty}^{+\infty}\mathrm{d}x\,e^{-\frac{1}{2}ax^2+x\,z},\qquad\forall z\in\mathbb{C}, \tag{6.23}$$

valid for any $a>0$. Second, we introduce the real-valued scalar field

$$\varphi(\boldsymbol{r},\tau)=\frac{1}{\sqrt{\beta V}}\sum_{l\in\mathbb{Z}}\sum_{\boldsymbol{q}}\varphi_{\boldsymbol{q},\varpi_l}\,e^{\mathrm{i}(\boldsymbol{q}\cdot\boldsymbol{r}-\varpi_l\tau)}, \tag{6.24a}$$

where

$$\frac{L}{2\pi}\boldsymbol{q}\in\mathbb{Z}^3,\qquad\varphi_{\boldsymbol{q}=\boldsymbol{0},\varpi_l}=0,\qquad\frac{\beta}{2\pi}\varpi_l\in\mathbb{Z}. \tag{6.24b}$$

By construction, this scalar field obeys periodic boundary condition in the imaginary-time direction and in position space. In addition, it obeys the charge-neutrality condition. This scalar field is the field conjugate to the local electronic density

$$\rho(\boldsymbol{r},\tau)=\sum_{\sigma}(\psi^{*}_{\sigma}\,\psi_{\sigma})(\boldsymbol{r},\tau) \tag{6.25}$$

through the Hubbard-Stratonovich identity

$$\begin{aligned} e^{-\int_0^\beta \mathrm{d}\tau \frac{1}{V}\sum_{\boldsymbol{q}\neq 0}\frac{2\pi e^2}{\boldsymbol{q}^2}\rho_{+\boldsymbol{q}}(\tau)\,\rho_{-\boldsymbol{q}}(\tau)} \propto& \int \mathcal{D}[\varphi]\, e^{-\frac{1}{8\pi}\int_0^\beta \mathrm{d}\tau \sum_{\boldsymbol{q}\neq 0}\boldsymbol{q}^2\varphi_{+\boldsymbol{q}}(\tau)\,\varphi_{-\boldsymbol{q}}(\tau)} \\ &\times e^{-\int_0^\beta \mathrm{d}\tau \frac{\mathrm{ie}}{2\sqrt{V}}\sum_{\boldsymbol{q}\neq 0}\left[\varphi_{+\boldsymbol{q}}(\tau)\,\rho_{-\boldsymbol{q}}(\tau)+\rho_{+\boldsymbol{q}}(\tau)\,\varphi_{-\boldsymbol{q}}(\tau)\right]} \\ {\scriptstyle \text{by charge neutrality } \varphi_{+\boldsymbol{q}=0}(\tau)\,=\,0} =& \int \mathcal{D}[\varphi]\, e^{-\frac{1}{8\pi}\int_0^\beta \mathrm{d}\tau \sum_{\boldsymbol{q}}\boldsymbol{q}^2\,\varphi_{+\boldsymbol{q}}(\tau)\,\varphi_{-\boldsymbol{q}}(\tau)} \\ &\times e^{-\int_0^\beta \mathrm{d}\tau \frac{\mathrm{ie}}{2\sqrt{V}}\sum_{\boldsymbol{q}}\left[\varphi_{+\boldsymbol{q}}(\tau)\,\rho_{-\boldsymbol{q}}(\tau)+\rho_{+\boldsymbol{q}}(\tau)\,\varphi_{-\boldsymbol{q}}(\tau)\right]}. \end{aligned} \tag{6.26}$$

The physical interpretation of φ is that of the scalar potential associated to charge fluctuations. Charge neutrality is here implemented by the condition

$$0=\varphi_{+\boldsymbol{q}=\boldsymbol{0}}(\tau)=\frac{1}{\sqrt{V}}\int \mathrm{d}^3 r\,\varphi(\boldsymbol{r},\tau)=0, \qquad \forall\tau. \tag{6.27}$$

In this way, the grand-canonical partition function becomes

$$Z_{\beta,\mu}=\left[\mathrm{Det}\left(-\frac{\Delta}{4\pi}\right)\right]^{+1/2}\times\int\mathcal{D}[\varphi]\int\mathcal{D}[\psi^*]\,\mathcal{D}[\psi]\,e^{-S'_{\beta,\mu}}, \tag{6.28a}$$

where

$$\begin{aligned} S'_{\beta,\mu}=&\int_0^\beta \mathrm{d}\tau\int_V \mathrm{d}^3\boldsymbol{r}\left[\frac{1}{8\pi}(-\varphi\,\Delta\varphi)+\sum_\sigma \psi^*_\sigma\left(\partial_\tau-\frac{\Delta}{2m}-\mu+\mathrm{i}e\,\varphi\right)\psi_\sigma\right](\boldsymbol{r},\tau) \\ =&\sum_{\boldsymbol{q},l}\Bigg\{\varphi_{+\boldsymbol{q},+\varpi_l}\frac{\boldsymbol{q}^2}{8\pi}\varphi_{-\boldsymbol{q},-\varpi_l}+\sum_{\boldsymbol{k},n,\sigma}\psi^*_{\boldsymbol{k}+\boldsymbol{q},\omega_n+\varpi_l,\sigma} \\ &\times\left[\left(-\mathrm{i}\omega_n+\frac{\boldsymbol{k}^2}{2m}-\mu\right)\delta_{\boldsymbol{q},\boldsymbol{0}}\delta_{\varpi_l,0}+\frac{\mathrm{ie}}{\sqrt{\beta V}}\varphi_{+\boldsymbol{q},+\varpi_l}\right]\psi_{\boldsymbol{k},\omega_n,\sigma}\Bigg\}. \end{aligned} \tag{6.28b}$$

By analogy to the Gaussian integration over Grassmann numbers, the product measure $\mathcal{D}[\varphi]$ is normalized so that $\int\mathcal{D}[\varphi]\,\exp\left(-\frac{1}{2}\varphi A\varphi\right)=\frac{1}{\sqrt{\mathrm{Det}\,A}}$ for any symmetric bilinear form (symmetric kernel) A.

6.4.2 *Integration of the electrons*

With the extension of the Gaussian identity for Grassmann integration,

$$\int \mathrm{d}\psi^* \mathrm{d}\psi \, e^{-\psi^* a \psi} = a, \qquad a \in \mathbb{C}, \tag{6.29}$$

to infinite-product Grassmann measures, integration over the fermions in the background of the real-valued scalar field φ yields

$$\begin{aligned} Z_{\beta,\mu} &= \left[\mathrm{Det}\left(-\frac{\Delta}{4\pi}\right)\right]^{+1/2} \times \int \mathcal{D}[\varphi]\, e^{-S''_{\beta,\mu}}, \\ S''_{\beta,\mu} &= \int_0^\beta \mathrm{d}\tau \int_V \mathrm{d}^3\boldsymbol{r} \frac{1}{8\pi}\left(-\varphi\,\Delta\varphi\right)(\boldsymbol{r},\tau) - 2\,\mathrm{Tr}\,\ln\left(\partial_\tau - \frac{\Delta}{2m} - \mu + \mathrm{i}e\,\varphi\right). \end{aligned} \tag{6.30a}$$

The functional trace Tr $\cdot$ appears when exponentiating the fermionic determinant,

$$\begin{aligned} \left[\mathrm{Det}\left(\partial_\tau - \frac{\Delta}{2m} - \mu + \mathrm{i}e\,\varphi\right)\right]^2 &= \exp\left(2\ln\left[\mathrm{Det}\left(\partial_\tau - \frac{\Delta}{2m} - \mu + \mathrm{i}e\,\varphi\right)\right]\right) \\ &= \exp\left(2\,\mathrm{Tr}\,\ln\left(\partial_\tau - \frac{\Delta}{2m} - \mu + \mathrm{i}e\,\varphi\right)\right). \end{aligned} \tag{6.30b}$$

The prefactor of 2 multiplying the trace is due to the spin degeneracy.

6.4.3 *Gaussian expansion of the fermionic determinant*

We need to approximate

$$\begin{aligned} \mathrm{Tr}\,\ln M &:= \mathrm{Tr}\,\ln(M_0 + M_1) \\ &= \mathrm{Tr}\,\ln\left[M_0\left(1 + M_0^{-1}M_1\right)\right] \\ &= \mathrm{Tr}\,\ln M_0 + \mathrm{Tr}\,\ln\left(1 + M_0^{-1}M_1\right), \end{aligned} \tag{6.31a}$$

where

$$M := \partial_\tau - \frac{\Delta}{2m} - \mu + \mathrm{i}e\,\varphi, \tag{6.31b}$$

$$M_0 := \partial_\tau - \frac{\Delta}{2m} - \mu, \tag{6.31c}$$

$$M_1 := \mathrm{i}e\,\varphi. \tag{6.31d}$$

Define the unperturbed Green function

$$G_0 := -M_0^{-1} = -\frac{1}{\partial_\tau - \frac{\Delta}{2m} - \mu}. \tag{6.32}$$

The sign is convention. Perform the expansion

$$\begin{aligned} \mathrm{Tr}\,\ln M &= \mathrm{Tr}\,\ln(-G_0^{-1} + M_1) \\ &= \mathrm{Tr}\,\ln\left(-G_0^{-1}\right) + \mathrm{Tr}\,\ln\left(1 - G_0 M_1\right) \\ &= \mathrm{Tr}\,\ln(-G_0^{-1}) - \sum_{n=1}^{\infty} \frac{1}{n} \mathrm{Tr}\,\left(G_0\,M_1\right)^n \end{aligned} \tag{6.33}$$

to the desired order. The RPA truncates this expansion to second order.

Use the short-hand notations

$$k := (\boldsymbol{k}, \omega_n) \tag{6.34a}$$

for the fermionic momenta and fermionic Matsubara frequencies and

$$(M_0)_{kk'} := \left(-\mathrm{i}\omega_n + \frac{\boldsymbol{k}^2}{2m} - \mu\right) \delta_{k,k'}, \tag{6.34b}$$

$$(G_0)_{kk'} := \frac{1}{\mathrm{i}\omega_n - \frac{\boldsymbol{k}^2}{2m} + \mu} \delta_{k,k'} \equiv \frac{1}{\mathrm{i}\omega_n - \xi_{\boldsymbol{k}}} \delta_{k,k'} \equiv G_{0k}\, \delta_{k,k'}, \tag{6.34c}$$

$$(M_1)_{kk'} := \frac{\mathrm{i}e}{\sqrt{\beta V}} \varphi_{k-k'}. \tag{6.34d}$$

To first order,

$$\begin{aligned} -\mathrm{Tr}\,(G_0\, M_1) &= -\sum_{k,k'} (G_0)_{kk'}\, (M_1)_{k'k} \\ &= -\sum_{k,k'} G_{0k}\, \delta_{k,k'}\, (M_1)_{k'k} \\ &= -\sum_{k} G_{0k}\, (M_1)_{kk}\,. \end{aligned} \tag{6.35}$$

To second order,

$$\begin{aligned} -\frac{1}{2}\mathrm{Tr}\,(G_0\, M_1)^2 &= -\frac{1}{2}\sum_{k,k'} G_{0k}\, (M_1)_{kk'}\, G_{0k'}\, (M_1)_{k'k} \\ &= -\frac{1}{2}\sum_{k,q} G_{0k}\, (M_1)_{k(k+q)}\, G_{0(k+q)}\, (M_1)_{(k+q)k}\,. \end{aligned} \tag{6.36}$$

We see that charge neutrality insures that the first-order contribution vanishes. The first non-vanishing contribution to the expansion of the fermionic determinant in powers of $e/\sqrt{\beta V}$ is Gaussian and given by

$$-\frac{1}{2}\mathrm{Tr}\,(G_0\, M_1)^2 = (-1)^2 \frac{1}{2} \sum_{q} \frac{e^2}{\beta V} \left(\sum_{k} G_{0k}\, G_{0(k+q)} \right) \varphi_{+q} \varphi_{-q}. \tag{6.37}$$

In summary, expansion of the fermionic logarithm in $S''_{\beta,\mu}$ in powers of $\mathrm{i}e\,\varphi/[\partial_\tau - (\Delta/2m) - \mu]$ yields the first non-vanishing contribution to second order only, owing to the charge-neutrality condition $\varphi_{q=0}(\tau) = 0$. Truncation of this expansion to this order yields the RPA to the jellium model. The RPA partition function in the grand-canonical ensemble is

$$Z_{\beta,\mu} \approx Z^{\mathrm{RPA}}_{\beta,\mu}, \tag{6.38a}$$

where

$$Z^{\mathrm{RPA}}_{\beta,\mu} = \left[\mathrm{Det}\left(-\frac{\Delta}{4\pi}\right)\right]^{+1/2} \times \left[\mathrm{Det}\left(\partial_\tau - \frac{\Delta}{2m} - \mu\right)\right]^2 \times \int \mathcal{D}[\varphi]\, \exp\left(-S^{\mathrm{RPA}}_{\beta,\mu}\right). \tag{6.38b}$$

The RPA action in imaginary time is

$$S^{\mathrm{RPA}}_{\beta,\mu} = \sum_{q=(\boldsymbol{q},\varpi_l)} \frac{1}{2}\varphi_{+q}\left(\frac{\boldsymbol{q}^2}{4\pi} - e^2\Pi^{\mathrm{RPA}}_q\right)\varphi_{-q}, \qquad \frac{L}{2\pi}\boldsymbol{q}\in\mathbb{Z}^3, \qquad \frac{\beta}{2\pi}\varpi_l = l \in \mathbb{Z}, \tag{6.38c}$$

where we have introduced the RPA kernel

$$\Pi^{\mathrm{RPA}}_q := +\frac{2}{\beta V}\sum_k G_{0k}G_{0(k+q)}, \tag{6.38d}$$

and the single-particle Green function

$$G_{0k} := \frac{1}{\mathrm{i}\omega_n - \frac{\boldsymbol{k}^2}{2m} + \mu} \equiv \frac{1}{\mathrm{i}\omega_n - \xi_{\boldsymbol{k}}}, \qquad \frac{L}{2\pi}\boldsymbol{k}\in\mathbb{Z}^3, \qquad \omega_n = \frac{\pi}{\beta}(2n+1),\ n\in\mathbb{Z}. \tag{6.38e}$$

Comments:

- The kernel Π^{RPA}_q that results from the integration over the fermions within the RPA approximation is called the *polarization function.* It endows the field φ with a non-trivial dynamics, i.e., φ is no longer instantaneous. This kernel is a property of the occupied states of a Fermi gas at temperature $T = \beta^{-1}$, i.e., of all states within a small window of energy $T/\varepsilon_{\mathrm{F}} \ll 1$ around the Fermi energy $\varepsilon_{\mathrm{F}} = k^2_{\mathrm{F}}/2m$, $k_{\mathrm{F}} = (3\pi^2 N/V)^{1/3}$ [recall Eq. (5.38)] at sufficiently low temperatures relative to the Fermi energy.
- It is important to stress that the exact effective action $S''_{\beta,\mu}$ in Eq. (6.30) for the order parameter φ has been expanded up to quadratic order in φ within the RPA approximation. However, since this expansion takes place in the *argument* of an exponential, *this is clearly not second-order perturbation theory in the electric charge e.*
- We are using the terminology of an order parameter for the field φ to stress that our treatment of the Coulomb interaction is nothing but a mean-field theory with the inclusion of Gaussian fluctuations around the mean-field value of the order parameter. Indeed, assume that $S''_{\beta,\mu}$ can be "Taylor expanded" around some field configuration φ_0,

$$S''_{\beta,\mu}[\varphi_0+\delta\varphi] = \sum_{m=0}^{\infty}\frac{1}{m!}\left.\frac{\delta^m S''_{\beta,\mu}}{\delta\varphi^m}\right|_{\varphi=\varphi_0}(\delta\varphi)^m, \tag{6.39}$$

and impose the "self-consistency" condition that φ_0 be a local extrema of $S''_{\beta,\mu}$. Then, the mean-field value of the order parameter is the one solving

$$0 = \frac{1}{8\pi}(-1)(\Delta\varphi_0)(\boldsymbol{r},\tau) - 2\left\langle \boldsymbol{r}'\tau'\left|\frac{+\mathrm{i}e\delta(\boldsymbol{r}-\boldsymbol{r}')\delta(\tau-\tau')}{\partial_{\tau'} - \frac{\Delta_{\boldsymbol{r}'}}{2m} - \mu + \mathrm{i}e\,\varphi_0(\boldsymbol{r}',\tau')}\right|\boldsymbol{r}'\tau'\right\rangle. \tag{6.40}$$

The Ansatz that we made to solve this condition is $\varphi_0(\boldsymbol{r},\tau) = 0$ for all $\boldsymbol{r}$ and τ, i.e., we extended the charge-neutrality condition $\varphi_{q=0} = 0$ to all q's. Since mean-field theory is here nothing but a *saddle-point approximation,*[1] it should be exact in the $\beta\to\infty$ limit, provided the saddle-point is an absolute minimum.

[1] The terminology of *steepest descent* or *stationary phase approximation* can also be found in the mathematics literature.

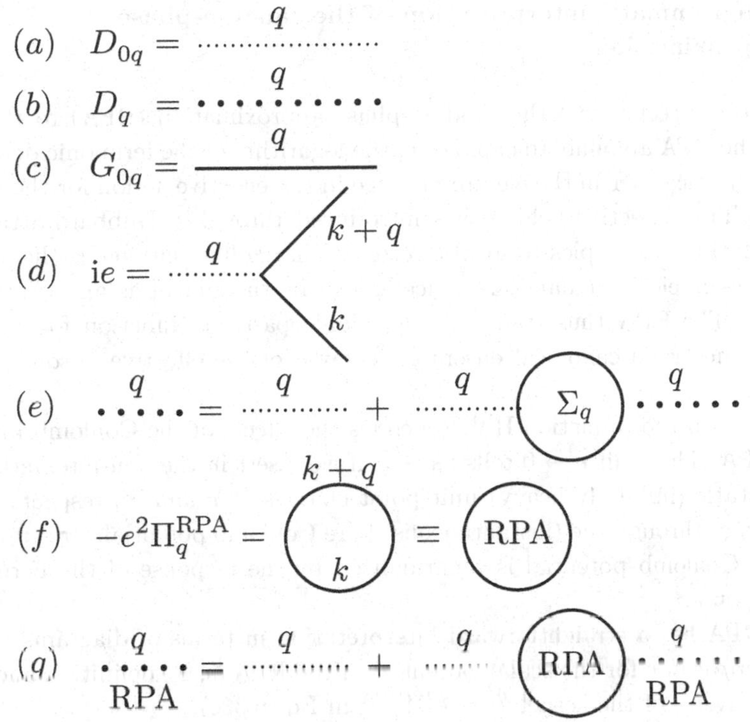

Fig. 6.1 $(a-d)$ Rules to construct (Feynman) diagrams. (e) Dyson's equation. (f) RPA electron-hole bubble. (g) RPA propagator follows from (e) with the substitution of the self-energy Σ_q by the electron-hole bubble $-e^2\Pi_q^{\mathrm{RPA}}$.

- One *a posteriori* justification for the truncation of $S''_{\beta,\mu}$ in Eq. (6.30) up to second order in φ consists in verifying that $\boldsymbol{q}^2/(4\pi) > e^2\,\Pi_q^{\mathrm{RPA}}$. If so, the mean-field configuration $\varphi = 0$ is, at the very least, a local minimum of the exact effective action $S''_{\beta,\mu}$ in Eq. (6.30). This is not to say that $\varphi = 0$ is a global minimum of $S''_{\beta,\mu}$. Unfortunately, short of an exact solution of the problem or the identification of an instability, for example another mean-field Ansatz with lower energy than the energy of the Fermi gas (the ground state when $\varphi = 0$), it is impossible to prove that $\varphi = 0$ is an absolute minimum of $S''_{\beta,\mu}$.
- Mean-field theory with eventual inclusion of fluctuations within the RPA approximation or to more than Gaussian order should not be thought of as a systematic method to solve an interacting many-body problem. Rather, it is a practical method based on physical intuition or prejudice that can only be justified a posteriori by comparison with experiments.

6.5 Diagrammatic interpretation of the random-phase approximation

We derived in section 6.4 the random-phase approximation (RPA) to the jellium model. The RPA amounts to expanding the logarithm of the fermionic determinant up to quadratic order in the electron charge in the effective action for the collective field φ . The collective field φ was introduced through a Hubbard-Stratonovich transformation. It couples to local electronic charge fluctuations as the scalar potential does in electrodynamics. Hence, φ can be thought of as an *effective scalar potential.* The RPA thus trades the fermionic partition function for the jellium model in the grand-canonical ensemble in favor of the effective bosonic partition function (6.38).

The polarization function Π_q^{RPA} encodes the effects of the Coulomb interaction within RPA. The limit $e \to 0$ tells us that if we insert in the non-interacting Fermi gas two static (infinitely heavy) unit point charges at $\boldsymbol{r}$ and $\boldsymbol{r}'$, respectively, then they interact through the (instantaneous) bare Coulomb potential $\delta(\tau-\tau')/|\boldsymbol{r}-\boldsymbol{r}'|$. The bare Coulomb potential is renormalized by the response of the Fermi sea to switching on e.

The RPA has a straightforward interpretation in terms of diagrams. The *Euclidean propagator* for the scalar potential in the RPA is, by definition and up to a sign, the inverse of the kernel $\frac{\boldsymbol{q}^2}{4\pi} - e^2\Pi_q^{\mathrm{RPA}}$ in Eq (6.38c),

$$D_q^{\mathrm{RPA}} := -\frac{1}{\frac{\boldsymbol{q}^2}{4\pi} - e^2\Pi_q^{\mathrm{RPA}}}, \qquad q \equiv (\boldsymbol{q}, \varpi_l). \tag{6.41}$$

The sign is convention. In the absence of Coulomb interaction, $e = 0$, the Euclidean propagator $D_{0\,q}$ is instantaneous as it is independent of the Matsubara frequency ϖ_l,

$$D_{0\,q} := -\frac{4\pi}{\boldsymbol{q}^2}. \tag{6.42}$$

This is nothing but the bare Coulomb potential in Fourier space. We thus have

$$\begin{aligned} D_q^{\mathrm{RPA}} &= \frac{1}{(D_{0\,q})^{-1} + e^2\,\Pi_q^{\mathrm{RPA}}} \\ &= D_{0\,q}\frac{1}{1 + e^2\,D_{0\,q}\Pi_q^{\mathrm{RPA}}} \\ &= D_{0\,q}\sum_{n=0}^{\infty}(-1)^n\left(e^2\,D_{0\,q}\,\Pi_q^{\mathrm{RPA}}\right)^n \\ &= D_{0\,q}\sum_{n=0}^{\infty}\left[(\mathrm{i}e)^2\,\Pi_q^{\mathrm{RPA}}\,D_{0\,q}\right]^n. \end{aligned} \tag{6.43}$$

Equation (6.43) is the approximate solution to *Dyson's equation*,

$$D_q = D_{0\,q} + D_{0\,q}\,\Sigma_q\,D_q \Longleftrightarrow D_q = \frac{1}{1 - D_{0\,q}\,\Sigma_q}\,D_{0\,q}, \tag{6.44a}$$

where D_q is the exact propagator, i.e., (the sign is convention)

$$D_q := -\frac{\int \mathcal{D}[\varphi]\, \varphi_{+q}\varphi_{-q} \exp\left(-S''_{\beta,\mu}\right)}{\int \mathcal{D}[\varphi] \qquad \exp\left(-S''_{\beta,\mu}\right)}, \tag{6.44b}$$

with

$$S''_{\beta,\mu} = \int_0^\beta \mathrm{d}\tau \int_V \mathrm{d}^3\boldsymbol{r} \frac{1}{8\pi}(-\varphi\Delta\varphi)(\boldsymbol{r},\tau) - 2\mathrm{Tr}\ \ln\left(\partial_\tau - \frac{\Delta}{2m} - \mu + \mathrm{i}e\,\varphi\right), \tag{6.44c}$$

and the right-hand side *defines* the so-called *self-energy* Σ_q of the collective field φ. The RPA replaces the self-energy Σ_q of φ by the RPA polarization function,

$$\Sigma_q \to (\mathrm{i}e)^2 \Pi_q^{\mathrm{RPA}}. \tag{6.45}$$

The diagrammatic or perturbative definition of the self-energy of φ goes as follows.

- Draw a dotted thin line for the unperturbed bosonic propagator $D_{0\,q}$ [Fig. 6.1(a)].
- Draw a dotted thick line for the exact bosonic propagator D_q [Fig. 6.1(b)].
- Draw a thin line for the unperturbed fermionic propagator $G_{0\,q}$ [Fig. 6.1(c)].
- The rule for connecting dotted and non-dotted lines is that a dotted (thin or thick) line can only be connected to two distinct non-dotted lines. The connection point is a vertex. It is to be associated with the factor ie [Fig. 6.1(d)]. Momenta and energies on the three lines meeting at a vertex must obey momentum and energy conservation. The electronic spin index is a by-stander.
- Each perturbative contribution to the exact propagator D_q is related to a *connected diagram* that has been built from dotted and non-dotted lines according to the preceding rules with no more and no less than two open ended lines. These two lines are dotted thin lines that are called *external legs*. External legs carry the momentum and energy $q = (\boldsymbol{q}, \varpi_l)$. All other lines are called *internal lines*. Momenta and energies on the internal lines that differ from $q = (\boldsymbol{q}, \varpi_l)$ are said to be *virtual* and are to be summed over. One must also sum over the spin degrees of freedom of the fermionic (non-dotted) internal lines. This gives an extra degeneracy factor of 2 for all diagrams.
- An *irreducible* diagram contributing to the exact propagator is a connected diagram that cannot be divided into two sub-diagrams joined solely by a single dotted line.
- An *irreducible self-energy diagram* is an *irreducible* diagram with the two external legs removed (amputated).
- The *irreducible self-energy* Σ_q is the sum of all irreducible self-energy diagrams.

The diagrammatic counterparts to Eqs. (6.44a) and (6.43) are given in Figs. 6.1(e) and 6.1(g), respectively.

In terms of the original electrons, D_q is closely related to the density-density correlation function

$$\left\langle \rho_{+q}\rho_{-q}\right\rangle_{S_{\beta,\mu}} := \frac{\int\mathcal{D}[\psi^*]\,\mathcal{D}[\psi]\,\rho_{+q}\,\rho_{-q}\,\exp\left(-S_{\beta,\mu}\right)}{\int\mathcal{D}[\psi^*]\,\mathcal{D}[\psi]\qquad\quad \exp\left(-S_{\beta,\mu}\right)}, \tag{6.46a}$$

with

$$S_{\beta,\mu} = \sum_{k,\sigma}\psi^*_{k,\sigma}\left(-\mathrm{i}\omega_n + \frac{\boldsymbol{k}^2}{2m} - \mu\right)\psi_{k,\sigma} + \frac{1}{\beta V}\sum_{\boldsymbol{q}\neq 0,\varpi_l}\frac{2\pi e^2}{\boldsymbol{q}^2}\rho_{+q}\,\rho_{-q}, \tag{6.46b}$$

and

$$\rho_{+q} := \sum_{k,\sigma}\psi^*_{k,\sigma}\,\psi_{k+q,\sigma}, \qquad k=(\boldsymbol{k},\omega_n), \qquad \frac{L}{2\pi}\boldsymbol{k}\in\mathbb{Z}^3, \qquad \omega_n = \frac{\pi}{\beta}(2n+1), \tag{6.46c}$$

for $n\in\mathbb{Z}$. To see this, note that the saddle-point equation

$$0 = \frac{\delta S'_{\beta,\mu}}{\delta\varphi_q}, \tag{6.47}$$

applied to the exact partition function

$$Z_{\beta,\mu} = \left[\mathrm{Det}\left(-\frac{\Delta}{4\pi}\right)\right]^{+1/2} \times \int\mathcal{D}[\varphi]\int\mathcal{D}[\psi^*]\,\mathcal{D}[\psi]\,\exp\left(-S'_{\beta,\mu}\right), \tag{6.48a}$$

where

$$\begin{aligned} S'_{\beta,\mu} =& \int\limits_0^\beta \mathrm{d}\tau\int\limits_V \mathrm{d}^3\boldsymbol{r}\left[\frac{1}{8\pi}\left(-\varphi\,\Delta\varphi\right) + \sum_\sigma\psi^*_\sigma\left(\partial_\tau - \frac{\Delta}{2m} - \mu + \mathrm{i}e\,\varphi\right)\psi_\sigma\right](\boldsymbol{r},\tau)\\ =& \sum_{\boldsymbol{q},l}\Bigg\{\varphi_{+\boldsymbol{q},+\varpi_l}\,\frac{\boldsymbol{q}^2}{8\pi}\,\varphi_{-\boldsymbol{q},-\varpi_l} + \sum_{\boldsymbol{k},n,\sigma}\psi^*_{\boldsymbol{k}+\boldsymbol{q},\omega_n+\varpi_l,\sigma}\\ &\times\left[\left(-\mathrm{i}\omega_n + \frac{\boldsymbol{k}^2}{2m} - \mu\right)\delta_{\boldsymbol{q},\boldsymbol{0}}\,\delta_{\varpi_l,0} + \frac{\mathrm{i}e}{\sqrt{\beta V}}\,\varphi_{+\boldsymbol{q},+\varpi_l}\right]\psi_{\boldsymbol{k},\omega_n,\sigma}\Bigg\}, \end{aligned} \tag{6.48b}$$

yields

$$\begin{aligned} 0 =& -\frac{\delta\ln Z_{\beta,\mu}}{\delta\varphi_q}\\ =& \frac{1}{Z_{\beta,\mu}}\int\mathcal{D}[\varphi]\int\mathcal{D}[\psi^*]\,\mathcal{D}[\psi]\,\frac{\delta S'_{\beta,\mu}}{\delta\varphi_q}\,e^{-S'_{\beta,\mu}}\\ =& \left\langle\frac{\boldsymbol{q}^2}{4\pi}\,\varphi_{-q} + \frac{\mathrm{i}e}{\sqrt{\beta V}}\,\rho_{-q}\right\rangle_{S'_{\beta,\mu}}. \end{aligned} \tag{6.49}$$

Equation (6.49) suggests the identification

$$D_q \longrightarrow \left(\frac{4\pi}{\boldsymbol{q}^2}\right)^2\frac{(\mathrm{i}e)^2}{\beta V}\left\langle\rho_{+q}\,\rho_{-q}\right\rangle_{S_{\beta,\mu}}. \tag{6.50}$$

More generally, any m-point correlation function for the scalar potential φ corresponds to a $(n = 2m)$-point correlation function for electrons. The converse is not true. Not all n-point and fermionic correlation functions can be written as m-point correlation functions for the scalar field φ. For example, the two-point fermionic correlation function (electron propagator)

$$G_k = \frac{\int \mathcal{D}[\psi^*]\,\mathcal{D}[\psi]\,\psi_k \psi_k^* \exp\left(-S_{\beta,\mu}\right)}{\int \mathcal{D}[\psi^*]\,\mathcal{D}[\psi] \qquad \exp\left(-S_{\beta,\mu}\right)} \tag{6.51}$$

has no simple expression in terms of correlation functions for the scalar field φ. The electronic density-density correlation function can be measured by inelastic X-ray scattering. The electronic two-point function can be measured by angular resolved photoemission scattering (ARPES). At zero temperature and as a function of the Matsubara frequency analytically continued to the imaginary axis, poles of the propagator D_q are interpreted as *collective* excitations of the underlying jellium model. Similarly, poles of the two-point fermionic Green function G_k are called *quasiparticle* excitations.

6.6 Ground-state energy in the random-phase approximation

The ground-state energy follows from the partition function by taking the zero temperature limit $\beta \to \infty$

$$\lim_{\beta\to\infty}\left(-\frac{1}{\beta}\ln Z_{\beta,\mu}\right) =: E_{\mathrm{GS}}. \tag{6.52}$$

Remember that we chose to define $\hat{H}_\mu$ to be normal ordered from the outset in section 6.3, i.e., that

$$\left\langle 0\left|\left(\hat{H}_N - \hat{H}_\mu\right)\right|0\right\rangle = -\frac{N}{V}\sum_{\boldsymbol{q}\neq\boldsymbol{0}}\frac{2\pi e^2}{\boldsymbol{q}^2}. \tag{6.53}$$

We need to account for the shift in the energy due this choice. The RPA provides an upper bound to the exact ground-state energy. After evaluating the partition function (6.38), a Gaussian integral in the RPA approximation, we infer that

$$\begin{aligned}
E^{\mathrm{RPA}}_{\mathrm{GS}}\Big|_{e\neq0} - E^{\mathrm{RPA}}_{\mathrm{GS}}\Big|_{e=0} &= -\frac{N}{V}\sum_{\boldsymbol{q}\neq\boldsymbol{0}}\frac{2\pi e^2}{\boldsymbol{q}^2} + \lim_{\beta\to\infty}(-1)\beta^{-1}\ln\left[\prod_{\boldsymbol{q}\neq\boldsymbol{0},\varpi_l}\frac{\frac{\boldsymbol{q}^2}{8\pi} - \frac{e^2}{2}\Pi^{\mathrm{RPA}}_{\boldsymbol{q},\varpi_l}}{\boldsymbol{q}^2/(8\pi)}\right]^{-1/2} \\
&= -\frac{N}{V}\sum_{\boldsymbol{q}\neq\boldsymbol{0}}\frac{2\pi e^2}{\boldsymbol{q}^2} + \lim_{\beta\to\infty}\frac{1}{2\beta}\sum_{\boldsymbol{q}\neq\boldsymbol{0}}\sum_{\varpi_l}\ln\left[\frac{\frac{\boldsymbol{q}^2}{8\pi} - \frac{e^2}{2}\Pi^{\mathrm{RPA}}_{\boldsymbol{q},\varpi_l}}{\boldsymbol{q}^2/(8\pi)}\right] \\
&= \sum_{\boldsymbol{q}\neq\boldsymbol{0}}\left\{-\frac{N}{V}\frac{2\pi e^2}{\boldsymbol{q}^2} + \int_{-\infty}^{+\infty}\frac{\mathrm{d}\varpi}{4\pi}\ln\left[1 - \frac{4\pi e^2}{\boldsymbol{q}^2}\Pi^{\mathrm{RPA}}_{\boldsymbol{q}}(\varpi)\right]\right\}.
\end{aligned} \tag{6.54}$$

In the limit

$$a_{\rm s} := \left(\frac{3}{4\pi}\frac{V}{N}\right)^{1/3} \ll a_{\rm B} := \frac{\hbar^2}{m\,e^2}, \tag{6.55}$$

it can be shown that (see Ref. [63])

$$\frac{E^{\rm RPA}_{\rm GS}}{N} = \left\{\frac{2.21}{(a_{\rm s}/a_{\rm B})^2} - \frac{0.916}{(a_{\rm s}/a_{\rm B})} + 0.062\ln(a_{\rm s}/a_{\rm B}) - 0.096 + \mathcal{O}\left[(a_{\rm s}/a_{\rm B})\ln(a_{\rm s}/a_{\rm B})\right]\right\}{\rm Ry}, \tag{6.56}$$

where

$${\rm Ry} := \frac{\hbar^2}{2ma_{\rm B}^2} = \frac{e^2}{2a_{\rm B}}. \tag{6.57}$$

The first term is called the *Hartree* term. The first two terms are the *Hartree-Fock* terms (see chapter 17 of Ref. [64]). In conventional metals $a_{\rm s}/a_{\rm B}$ range from 2 to 6 which indicates that electronic interactions need to be accounted for to calculate the energy of a metal with any hope of precision. The RPA gives the next two leading corrections in an expansion in powers of $a_{\rm s}/a_{\rm B}$ [63]. Evidently, it is doubtful that such an expansion is of relevance to metals in a computational sense. The RPA is, however, instructive conceptually and was, historically, the first attempt to calculate systematically the effects of electron interactions in a metal. Our next task is to identify the excitation spectrum above the RPA ground state.

6.7 Lindhard response function

It is time to evaluate the polarization function

$$\Pi^{\rm RPA}_{\boldsymbol{q},\varpi_l} := 2\frac{1}{V}\sum_{\boldsymbol{k}}\frac{1}{\beta}\sum_{\omega_n}\frac{1}{({\rm i}\omega_n - \xi_{\boldsymbol{k}})\left({\rm i}\omega_n + {\rm i}\varpi_l - \xi_{\boldsymbol{k}+\boldsymbol{q}}\right)}, \qquad \boldsymbol{q}\neq\boldsymbol{0}. \tag{6.58}$$

The first step consists in performing the summation over fermionic Matsubara frequencies $\omega_n = \pi(2n+1)/\beta$, $n\in\mathbb{Z}$ for any given $\boldsymbol{k} = 2\pi\boldsymbol{l}/L$, $\boldsymbol{l}\in\mathbb{Z}^3$. As an intermediary step, observe that the Fermi-Dirac distribution function

$$\tilde{f}_{\rm FD}(z) := \frac{1}{e^{\beta z}+1}, \qquad z\in\mathbb{C}, \tag{6.59a}$$

has equidistant first-order poles at

$$z_n = {\rm i}\omega_n, \qquad n\in\mathbb{Z}, \tag{6.59b}$$

with residues

$${\rm Res}\,\tilde{f}_{\rm FD}(z)\Big|_{{\rm i}\omega_n} = -\frac{1}{\beta}. \tag{6.59c}$$

For any given $\boldsymbol{k}$, let $\Gamma_{\boldsymbol{k}}$ be the path running antiparallel to the imaginary axis infinitesimally close to its left and parallel to the imaginary axis infinitesimally close

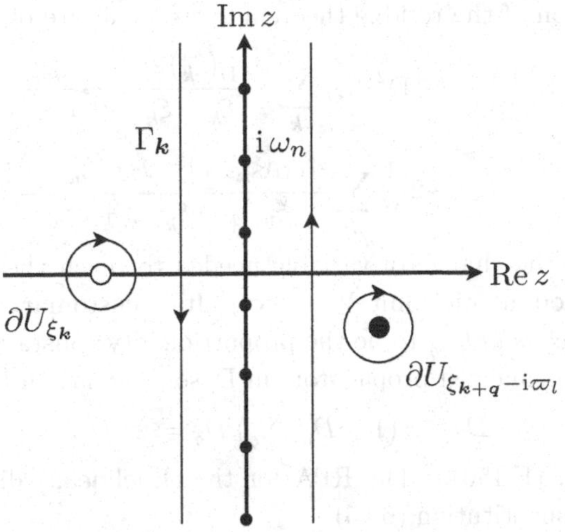

Fig. 6.2 Poles of Euclidean polarization function on and off imaginary z-axis in the representation of Eq. (6.60) arising from an arbitrarily chosen $\boldsymbol{k}$ contribution. Hole- (particle-) like poles are off the imaginary axis and denoted by an empty (filled) circle. Poles on the imaginary axis at the Matsubara frequencies ω_n are denoted by smaller filled circles. Closed integration paths $\Gamma_{\boldsymbol{k}}$, $\partial U_{\xi_{\boldsymbol{k}+\boldsymbol{q}}-\mathrm{i}\varpi_l}$ and $\partial U_{\xi_{\boldsymbol{k}}}$ are also drawn.

to its right, i.e., it goes around the imaginary axis in a counterclockwise fashion. By the residue theorem,

$$\Pi^{\mathrm{RPA}}_{\boldsymbol{q},\varpi_l} = -2\frac{1}{V}\sum_{\boldsymbol{k}}\int\limits_{\Gamma_{\boldsymbol{k}}}\frac{\mathrm{d}z}{2\pi\mathrm{i}}\frac{\tilde{f}_{\mathrm{FD}}(z)}{(z-\xi_{\boldsymbol{k}})\left(z+\mathrm{i}\varpi_l-\xi_{\boldsymbol{k}+\boldsymbol{q}}\right)}. \tag{6.60}$$

Since $\boldsymbol{q}\neq\boldsymbol{0}$, the integrand with $\boldsymbol{k}$ fixed has, aside from first-order poles along the imaginary axis, two isolated and first-order poles at

$$z_{\boldsymbol{k}} := \xi_{\boldsymbol{k}}, \qquad z_{\boldsymbol{k}+\boldsymbol{q},\varpi_l} := \xi_{\boldsymbol{k}+\boldsymbol{q}} - \mathrm{i}\varpi_l, \tag{6.61}$$

with residues

$$\frac{1}{2\pi\mathrm{i}}\frac{\tilde{f}_{\mathrm{FD}}(\xi_{\boldsymbol{k}})}{\xi_{\boldsymbol{k}}-\xi_{\boldsymbol{k}+\boldsymbol{q}}+\mathrm{i}\varpi_l} \tag{6.62}$$

and

$$\frac{1}{2\pi\mathrm{i}}\frac{\tilde{f}_{\mathrm{FD}}(\xi_{\boldsymbol{k}+\boldsymbol{q}}-\mathrm{i}\varpi_l)}{\xi_{\boldsymbol{k}+\boldsymbol{q}}-\xi_{\boldsymbol{k}}-\mathrm{i}\varpi_l} = -\frac{1}{2\pi\mathrm{i}}\frac{\tilde{f}_{\mathrm{FD}}(\xi_{\boldsymbol{k}+\boldsymbol{q}})}{\xi_{\boldsymbol{k}}-\xi_{\boldsymbol{k}+\boldsymbol{q}}+\mathrm{i}\varpi_l}, \tag{6.63}$$

respectively. These two first-order poles merge into a second-order pole when $\varpi_l = 0$ and $\boldsymbol{q}\to\boldsymbol{0}$. By Cauchy theorem, the contour of integration can be deformed into two small circles $\partial U_{z_{\boldsymbol{k}}}$ and $\partial U_{z_{\boldsymbol{k}+\boldsymbol{q},\varpi_l}}$ encircling $z_{\boldsymbol{k}}$ and $z_{\boldsymbol{k}+\boldsymbol{q},\varpi_l}$, respectively, in a *clockwise* fashion (see Fig. 6.2),

$$\Gamma_{\boldsymbol{k}} \to \partial U_{z_{\boldsymbol{k}}} \cup \partial U_{z_{\boldsymbol{k}+\boldsymbol{q},\varpi_l}}. \tag{6.64}$$

A second application of the residue theorem gives [be aware of the extra (-1)]

$$\begin{aligned}\Pi^{\mathrm{RPA}}_{\boldsymbol{q},\varpi_l} &= (-1)^2\,2\,\frac{1}{V}\sum_{\boldsymbol{k}}\frac{\tilde{f}_{\mathrm{FD}}(\xi_{\boldsymbol{k}})-\tilde{f}_{\mathrm{FD}}(\xi_{\boldsymbol{k}+\boldsymbol{q}})}{\xi_{\boldsymbol{k}}-\xi_{\boldsymbol{k}+\boldsymbol{q}}+\mathrm{i}\varpi_l}\\ &= +2\,\frac{1}{V}\sum_{\boldsymbol{k}}\frac{\tilde{f}_{\mathrm{FD}}(\xi_{\boldsymbol{k}-\frac{\boldsymbol{q}}{2}})-\tilde{f}_{\mathrm{FD}}(\xi_{\boldsymbol{k}+\frac{\boldsymbol{q}}{2}})}{\xi_{\boldsymbol{k}-\frac{\boldsymbol{q}}{2}}-\xi_{\boldsymbol{k}+\frac{\boldsymbol{q}}{2}}+\mathrm{i}\varpi_l}.\end{aligned} \tag{6.65}$$

(Strictly speaking, the change of variable needed to reach the second line is only legal in the thermodynamic limit $V\to\infty$.) It is customary to define the (Euclidean) *dielectric constant* ε_q to be the proportionality constant between the bare, Eq. (6.42), and renormalized propagators in Dyson's equation (6.44a),

$$D_{0\,q} = \left(1 - D_{0\,q}\,\Sigma_q\right) D_q =: \varepsilon_q\, D_q. \tag{6.66}$$

[Compare with Eq. (E.150).] The RPA for the (Euclidean) dielectric constant is obtained with the substitution (6.45),

$$\begin{aligned}\varepsilon^{\mathrm{RPA}}_{\boldsymbol{q},\varpi_l} &= 1-\frac{4\pi e^2}{\boldsymbol{q}^2}\,\Pi^{\mathrm{RPA}}_{\boldsymbol{q},\varpi_l}\\ &= 1-2\frac{4\pi e^2}{\boldsymbol{q}^2}\frac{1}{V}\sum_{\boldsymbol{k}}\frac{\tilde{f}_{\mathrm{FD}}(\xi_{\boldsymbol{k}-\frac{\boldsymbol{q}}{2}})-\tilde{f}_{\mathrm{FD}}(\xi_{\boldsymbol{k}+\frac{\boldsymbol{q}}{2}})}{\xi_{\boldsymbol{k}-\frac{\boldsymbol{q}}{2}}-\xi_{\boldsymbol{k}+\frac{\boldsymbol{q}}{2}}+\mathrm{i}\varpi_l}.\end{aligned} \tag{6.67}$$

Equation (6.67) is known as the *Lindhard dielectric constant.* Equation (6.67) was first derived in the static limit $\varpi_l = 0$ [65]. The static limit $\varpi_l \to 0$ of the (Euclidean) polarization function is called the *Lindhard function.*

A useful property of the Euclidean dielectric function at zero temperature is that an excitation in the jellium model with momentum $\boldsymbol{q}$ and real-time frequency $\widetilde{\varpi}_{\boldsymbol{q}}$ shows up as a zero of the analytic continuation of the Euclidean dielectric function to the negative imaginary axis[2]

$$\lim_{\varpi\to-\mathrm{i}\widetilde{\varpi}_{\boldsymbol{q}}+0^+}\varepsilon_{\boldsymbol{q},\varpi} = 0. \tag{6.69}$$

Indeed, a pole of D_q at some $\boldsymbol{q}\neq\boldsymbol{0}$ implies a zero of ε_q at some $\boldsymbol{q}\neq\boldsymbol{0}$ in Eq. (6.66), as the left-hand side of Eq. (6.66) is a non-vanishing and finite number for any $\boldsymbol{q}\neq\boldsymbol{0}$. Alternatively, the physical interpretation of Eqs. (6.69) and (6.66) is that a harmonic perturbation *with arbitrarily small amplitude* induces a *non-vanishing* response of the Fermi sea in the form of a non-vanishing renormalization of the Coulomb potential.[3] The jellium model supports free modes of oscillations with the

[2] Remember that real time t is related to imaginary time τ by $\tau = \mathrm{i}t$. A Matsubara frequency ϖ_l that enters as $\varpi_l\tau$ in the imaginary-time Fourier expansion of fields is related to the real-time frequency $\widetilde{\varpi}_l$ by

$$\varpi_l\tau = (+\mathrm{i}\varpi_l)(-\mathrm{i}\tau) \equiv \widetilde{\varpi}_l t, \qquad \widetilde{\varpi}_l := +\mathrm{i}\varpi_l, \qquad t := -\mathrm{i}\tau. \tag{6.68}$$

[3] The harmonic perturbation can be imposed, for example, by forcing a charge fluctuation in the electron gas that varies periodically in space and time with wave vector $\boldsymbol{q}$ and real-time frequency $\widetilde{\varpi}_{\boldsymbol{q}}$, respectively. The infinitesimal frequency 0^+ in Eq. (6.69) ensures that the perturbation on the jellium model is switched on adiabatically slowly.

dispersion $\widetilde{\varpi}_{\boldsymbol{q}}$ since these oscillations need not be forced by an external probe to the electronic system. The excitation spectrum within the RPA is obtained from solving

$$\begin{aligned} 0 =& \lim_{\varpi \to -\mathrm{i}\widetilde{\varpi}_{\boldsymbol{q}}+0^+} \varepsilon^{\mathrm{RPA}}_{\boldsymbol{q},\varpi} \\ =& 1 - 2\frac{4\pi e^2}{\boldsymbol{q}^2}\frac{1}{V}\sum_{\boldsymbol{k}} \frac{\tilde{f}_{\mathrm{FD}}(\xi_{\boldsymbol{k}}) - \tilde{f}_{\mathrm{FD}}(\xi_{\boldsymbol{k}+\boldsymbol{q}})}{\widetilde{\varpi}_{\boldsymbol{q}} - \left(\xi_{\boldsymbol{k}+\boldsymbol{q}} - \xi_{\boldsymbol{k}} - \mathrm{i}0^+\right)}. \end{aligned} \tag{6.70}$$

Figure 6.3 displays a graphical solution to Eq. (6.70). One distinguishes two types of excitations. There is a continuum of particle-hole excitations when, for $\varpi \geq 0$ and zero temperature say,

$$\begin{aligned} &\widetilde{\varpi}_{\boldsymbol{q}\mathrm{min}} \leq \widetilde{\varpi} \leq \widetilde{\varpi}_{\boldsymbol{q}\mathrm{max}}, \\ &\widetilde{\varpi}_{\boldsymbol{q}\mathrm{min}} := \frac{|\boldsymbol{q}|^2}{2m} - \frac{k_{\mathrm{F}}|\boldsymbol{q}|}{m}, \qquad |\boldsymbol{q}| > 2k_{\mathrm{F}} \\ &\qquad = \inf_{0\leq\theta<2\pi} \left(\frac{|\boldsymbol{q}|^2}{2m} + \cos\theta\frac{k_{\mathrm{F}}|\boldsymbol{q}|}{m}\right), \qquad |\boldsymbol{q}| > 2k_{\mathrm{F}} \\ &\qquad = \inf_{|\boldsymbol{k}|<k_{\mathrm{F}},|\boldsymbol{k}+\boldsymbol{q}|>k_{\mathrm{F}}} \left(\xi_{\boldsymbol{k}+\boldsymbol{q}} - \xi_{\boldsymbol{k}}\right), \qquad |\boldsymbol{q}| > 2k_{\mathrm{F}}, \\ &\widetilde{\varpi}_{\boldsymbol{q}\mathrm{max}} := \frac{|\boldsymbol{q}|^2}{2m} + \frac{k_{\mathrm{F}}|\boldsymbol{q}|}{m} \\ &\qquad = \sup_{0\leq\theta<2\pi} \left(\frac{|\boldsymbol{q}|^2}{2m} + \cos\theta\frac{k_{\mathrm{F}}|\boldsymbol{q}|}{m}\right) \\ &\qquad = \sup_{|\boldsymbol{k}|<k_{\mathrm{F}},|\boldsymbol{k}+\boldsymbol{q}|>k_{\mathrm{F}}} \left(\xi_{\boldsymbol{k}+\boldsymbol{q}} - \xi_{\boldsymbol{k}}\right). \end{aligned} \tag{6.71}$$

There is another branch of excitations called plasmons that merges into the continuum for sufficiently large momentum transfer. Figure 6.4 sketches the excitation spectrum for the jellium model within the RPA.

References have been made to the Fermi sea and Fermi wave vector k_{F}, see section 5.3. We recall that at zero temperature, the Fermi-Dirac distribution becomes the Heaviside step function,

$$\lim_{\beta\to\infty} \tilde{f}_{\mathrm{FD}}(\xi) = \Theta(-\xi) = \begin{cases} 0, & \text{if } \xi > 0, \\ 1, & \text{otherwise,} \end{cases} \tag{6.72}$$

and

$$\lim_{\beta\to\infty} \left[\tilde{f}_{\mathrm{FD}}(\xi_{\boldsymbol{k}-\frac{\boldsymbol{q}}{2}}) - \tilde{f}_{\mathrm{FD}}(\xi_{\boldsymbol{k}+\frac{\boldsymbol{q}}{2}})\right] \neq 0 \Longleftrightarrow \xi_{\boldsymbol{k}-\frac{\boldsymbol{q}}{2}} \times \xi_{\boldsymbol{k}+\frac{\boldsymbol{q}}{2}} < 0. \tag{6.73}$$

We also recall that the interpretation of Eq. (6.72) is that all single-particle states with an energy smaller than the chemical potential (Fermi energy) are occupied in the Fermi sea at zero temperature. The interpretation of Eq. (6.73) is that for the difference of the Fermi-Dirac distributions at two single-particle energies to be non-vanishing at zero temperature, one of the two single-particle states must be

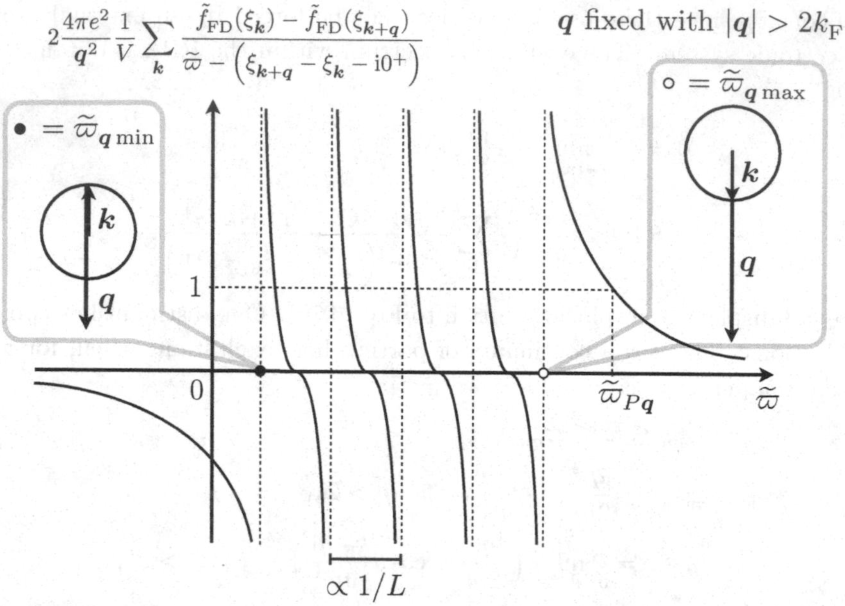

Fig. 6.3 Qualitative plot of $(4\pi e^2/q^2)\Pi^{\rm RPA}_{q,-\mathrm{i}\widetilde{\varpi}+0^+}$ as a function of the real-time frequency $\widetilde{\varpi}$ at fixed momentum q, $|q|>2k_{\rm F}$. The polarization function decays like $1/\widetilde{\varpi}$ when $|\widetilde{\varpi}|\gg|\widetilde{\varpi}_{q\min}|,|\widetilde{\varpi}_{q\max}|$. The number of intercepts between $(4\pi e^2/q^2)\Pi^{\rm RPA}_{q,-\mathrm{i}\widetilde{\varpi}+0^+}$ and the constant line at 1 for $\widetilde{\varpi}_{q\min}\leq\widetilde{\varpi}\leq\widetilde{\varpi}_{q\max}$ scales like the inverse of the level spacing $1/L$, i.e., like $L=V^{1/3}$. There can be one more intercept between $(4\pi e^2/q^2)\Pi^{\rm RPA}_{q,-\mathrm{i}\widetilde{\varpi}+0^+}$ and the constant line at 1 for $\widetilde{\varpi}_{q\max}<\widetilde{\varpi}$. This intercept takes place at the plasma frequency $\widetilde{\varpi}_{Pq}$.

above the Fermi energy while the other single-particle state must be below the Fermi energy. The difference of the Fermi-Dirac distributions in Eq. (6.73) selects an electron-hole pair that is represented by the bubble diagram in Fig. 6.1(e).

At low temperatures, the polarization function (6.65) is thus controlled by the geometrical properties of the *Fermi sea*, the unperturbed ground state of the Fermi gas. We need some characteristic scales of the Fermi sea. The Fermi wave vector $k_{\rm F}$ is defined by filling up all available single-particle energy levels,

$$\begin{aligned}
\frac{N}{V} &= V^{-1}\sum_{\sigma=\uparrow,\downarrow}\sum_{k}\Theta(-\xi_k)\\
&=: 2V^{-1}\sum_{k}\Theta\left(\frac{k_{\rm F}^2}{2m}-\frac{k^2}{2m}\right)\\
&= 2\times\frac{1}{(2\pi)^3}\frac{4\pi}{3}(k_{\rm F})^3\\
&= \frac{(k_{\rm F})^3}{3\pi^2}.
\end{aligned}\tag{6.74}$$

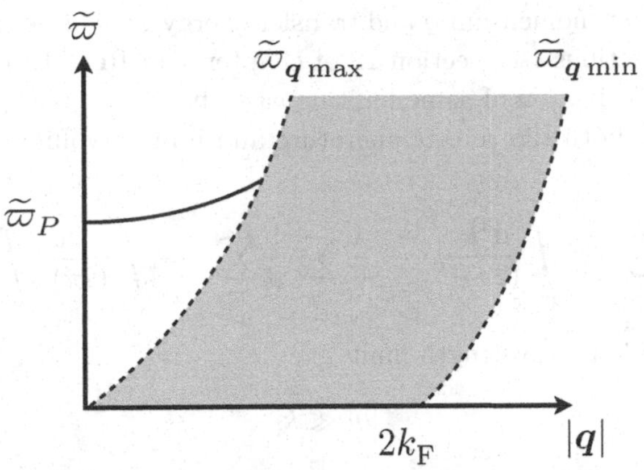

Fig. 6.4 Qualitative excitation spectrum for the jellium model within the RPA approximation. The dashed region represents the particle-hole continuum. The line emanating from $(|\boldsymbol{q}| = 0, \varpi_{\rm P})$ is the plasmon branch of collective excitations.

The Fermi velocity and Fermi energy are

$$v_{\rm F} := \frac{k_{\rm F}}{m} \propto \left(\frac{N}{V}\right)^{1/3}, \qquad \varepsilon_{\rm F} := \frac{(k_{\rm F})^2}{2m} \propto \left(\frac{N}{V}\right)^{2/3}, \tag{6.75}$$

respectively. The Fermi velocity appears naturally when expanding the numerator in powers of $|\boldsymbol{q}|/k_{\rm F}$,

$$\begin{aligned} \lim_{\beta\to\infty} \frac{\partial \tilde{f}_{\rm FD}(\xi_{\boldsymbol{k}})}{\partial \xi_{\boldsymbol{k}}} &= -\delta(\xi_{\boldsymbol{k}}) \\ &= -\delta\left(\frac{\boldsymbol{k}^2}{2m} - \frac{k_{\rm F}^2}{2m}\right) \\ &= -\frac{1}{v_{\rm F}}\delta(|\boldsymbol{k}| - k_{\rm F}). \end{aligned} \tag{6.76}$$

At temperatures much smaller than the Fermi energy, only those single-particle electronic states within a distance β^{-1} of the Fermi surface contribute to the polarization function. At zero temperature and in the infinite-volume limit, the Lindhard function can be calculated explicitly, (we reinstate $\hbar$)

$$\lim_{\beta\to\infty} \lim_{\varpi_l\to 0} \Pi^{\rm RPA}_{\boldsymbol{q},\varpi_l} = -\left(\frac{mk_{\rm F}}{\hbar^2\pi^2}\right)\left(\frac{1}{2} + \frac{1-x^2}{4x}\ln\left|\frac{1+x}{1-x}\right|\right), \qquad x = \frac{|\boldsymbol{q}|}{2k_{\rm F}}. \tag{6.77}$$

Note the presence of logarithmic singularities when the magnitude of the momentum transfer $|\boldsymbol{q}|$ is twice the Fermi wave vector. These logarithmic singularities are responsible for the so-called *Friedel or Ruderman-Kittel-Kasuya-Yosida oscillations*. Note also that (we reinstate $\hbar$)

$$\lim_{\beta\to\infty} \lim_{\boldsymbol{q}\to\boldsymbol{0}} \left(\lim_{\varpi_l\to 0} \Pi^{\rm RPA}_{\boldsymbol{q},\varpi_l}\right) = -\left(\frac{mk_{\rm F}}{\hbar^2\pi^2}\right) \times 1 \equiv -\nu_{\rm F}, \tag{6.78}$$

where $\nu_{\rm F}$ is the single-particle density of states per unit energy and per unit volume at the Fermi energy [see Eq. (5.33)]. In fact the full dependence of the polarization function on transfer momentum $\boldsymbol{q}$ and transfer energy ϖ_l can be expressed in terms of elementary functions (see section 12 of chapter 4 in Ref. [12]). We will restrict ourselves to the derivation of some limiting cases below.

From now on, both the zero-temperature and infinite-volume limits are understood,

$$\frac{1}{V}\sum_{\boldsymbol{k}} \longrightarrow \int\limits_{\mathbb{R}^3} \frac{\mathrm{d}^3\boldsymbol{k}}{(2\pi)^3}, \qquad \frac{1}{V}\sum_{\boldsymbol{q}}\frac{1}{\beta}\sum_{\varpi} \longrightarrow \int\limits_{\mathbb{R}^3} \frac{\mathrm{d}^3\boldsymbol{q}}{(2\pi)^3}\int\limits_{\mathbb{R}} \frac{\mathrm{d}\varpi}{2\pi}. \tag{6.79}$$

Furthermore, the long-wavelength limit

$$|\boldsymbol{q}| \ll k_{\mathrm{F}} \tag{6.80}$$

will also be assumed for some given transfer momentum $\boldsymbol{q}$. Choose a spherical coordinate system in $\boldsymbol{k}$-space with the angle between $\boldsymbol{q}$ and $\boldsymbol{k}$ the polar angle θ:

$$\begin{aligned} &\boldsymbol{k}\cdot\boldsymbol{q} = |\boldsymbol{k}||\boldsymbol{q}|\cos\theta \equiv |\boldsymbol{k}||\boldsymbol{q}|\nu, \\ &\int\limits_{\mathbb{R}^3}\mathrm{d}^3\boldsymbol{k} = \int\limits_0^{+\infty}\mathrm{d}|\boldsymbol{k}||\boldsymbol{k}|^2\int\limits_0^{2\pi}\mathrm{d}\phi\int\limits_0^{\pi}\mathrm{d}\theta\sin\theta \equiv \int\limits_0^{+\infty}\mathrm{d}|\boldsymbol{k}||\boldsymbol{k}|^2\int\limits_0^{2\pi}\mathrm{d}\phi\int\limits_{-1}^{+1}\mathrm{d}\nu. \end{aligned} \tag{6.81}$$

Insertion of

$$\begin{aligned} \xi_{\boldsymbol{k}+\frac{\boldsymbol{q}}{2}} - \xi_{\boldsymbol{k}-\frac{\boldsymbol{q}}{2}} &= \frac{\boldsymbol{k}\cdot\boldsymbol{q}}{m}, \\ \tilde{f}_{\mathrm{FD}}(\xi_{\boldsymbol{k}+\frac{\boldsymbol{q}}{2}}) - \tilde{f}_{\mathrm{FD}}(\xi_{\boldsymbol{k}-\frac{\boldsymbol{q}}{2}}) &= \frac{\partial\tilde{f}_{\mathrm{FD}}(\xi_{\boldsymbol{k}})}{\partial\xi_{\boldsymbol{k}}}\,\frac{\boldsymbol{k}\cdot\boldsymbol{q}}{m} + \mathcal{O}\left(|\boldsymbol{q}|^3\right) \\ &= -\frac{1}{v_{\mathrm{F}}}\,\delta(|\boldsymbol{k}| - k_{\mathrm{F}})\,\frac{\boldsymbol{k}\cdot\boldsymbol{q}}{m} + \mathcal{O}\left(|\boldsymbol{q}|^3\right), \end{aligned} \tag{6.82}$$

into the polarization function (6.65) yields

$$\begin{aligned} \Pi^{\mathrm{RPA}}_{\boldsymbol{q},\varpi} &= +2\frac{1}{V}\sum_{\boldsymbol{k}}\frac{\tilde{f}_{\mathrm{FD}}(\xi_{\boldsymbol{k}-\frac{\boldsymbol{q}}{2}}) - \tilde{f}_{\mathrm{FD}}(\xi_{\boldsymbol{k}+\frac{\boldsymbol{q}}{2}})}{\xi_{\boldsymbol{k}-\frac{\boldsymbol{q}}{2}} - \xi_{\boldsymbol{k}+\frac{\boldsymbol{q}}{2}} + \mathrm{i}\varpi} \\ &= +2\int\limits_0^{\infty}\frac{d|\boldsymbol{k}|}{(2\pi)^3}|\boldsymbol{k}|^2\int\limits_0^{2\pi}\mathrm{d}\phi\int\limits_{-1}^{+1}\mathrm{d}\nu\frac{\frac{1}{k_{\mathrm{F}}}\,\delta(|\boldsymbol{k}| - k_{\mathrm{F}})\,|\boldsymbol{k}|\,|\boldsymbol{q}|\,\nu + \mathcal{O}\left(|\boldsymbol{q}|^3\right)}{\mathrm{i}\varpi - \frac{|\boldsymbol{k}|\,|\boldsymbol{q}|}{m}\,\nu} \\ &= +2\frac{m\,k_{\mathrm{F}}}{(2\pi)^2}\int\limits_{-1}^{+1}\mathrm{d}\nu\frac{v_{\mathrm{F}}\,|\boldsymbol{q}|\,\nu + \mathcal{O}\left(|\boldsymbol{q}|^3/(v_{\mathrm{F}})^3\right)}{\mathrm{i}\varpi - v_{\mathrm{F}}\,|\boldsymbol{q}|\,\nu} \\ &= -\frac{4m\,k_{\mathrm{F}}}{(2\pi)^2}\left[1 - \frac{\varpi}{v_{\mathrm{F}}\,|\boldsymbol{q}|}\arctan\left(\frac{v_{\mathrm{F}}\,|\boldsymbol{q}|}{\varpi}\right)\right] + \mathcal{O}\left[\left(\frac{|\boldsymbol{q}|}{v_{\mathrm{F}}}\right)^2\right], \end{aligned} \tag{6.83}$$

with the help of Eqs. (1.622) and (2.112) from Ref. [57] to reach the last equality. Next, two limits of Eq. (6.83) are considered:

- The *quasi-static* limit,

$$|\varpi| \ll v_{\mathrm{F}}\,|\boldsymbol{q}|, \qquad |\boldsymbol{q}| \ll m\,v_{\mathrm{F}}. \tag{6.84}$$

  In this limit, the RPA encodes the physics of screening.
- The *dynamic* limit,

$$v_{\mathrm{F}}\,|\boldsymbol{q}| \ll |\varpi|, \qquad |\boldsymbol{q}| \ll m\,v_{\mathrm{F}}. \tag{6.85}$$

  In this limit, the RPA encodes the physics of plasma oscillations.

6.7.1 *Long-wavelength and quasi-static limit at* $T = 0$

The regime

$$|\varpi| \ll v_{\mathrm{F}}|\boldsymbol{q}|, \qquad |\boldsymbol{q}| \ll m\,v_{\mathrm{F}}, \tag{6.86}$$

suggests using the expansion

$$\arctan z = \frac{\pi}{2} - \frac{1}{z} + \frac{1}{3z^3} + \cdots, \qquad |z| > 1, \qquad z \to \frac{v_{\mathrm{F}}|\boldsymbol{q}|}{\varpi}, \tag{6.87}$$

in Eq. (6.83). To leading order in this expansion,

$$\Pi^{\mathrm{RPA}}_{\boldsymbol{q},\varpi} = -\frac{4mk_{\mathrm{F}}}{(2\pi)^2}\left(1 - \frac{\pi}{2}\frac{\varpi}{v_{\mathrm{F}}|\boldsymbol{q}|}\right) + \mathcal{O}\left[\left(\frac{|\boldsymbol{q}|}{v_{\mathrm{F}}}\right)^2, \left(\frac{\varpi}{v_{\mathrm{F}}|\boldsymbol{q}|}\right)^2\right]. \tag{6.88}$$

In turn, the RPA propagator in Eq. (6.41) is approximated by

$$\begin{aligned} D^{\mathrm{RPA}}_{\boldsymbol{q},\varpi} &= -\frac{1}{\frac{\boldsymbol{q}^2}{4\pi} - e^2\Pi^{\mathrm{RPA}}_{\boldsymbol{q},\varpi}} \\ &= -\frac{4\pi}{|\boldsymbol{q}|^2 + 8\pi e^2\frac{2mk_{\mathrm{F}}}{(2\pi)^2}\left(1 - \frac{\pi}{2}\frac{\varpi}{v_{\mathrm{F}}|\boldsymbol{q}|}\right)} + \mathcal{O}\left[\left(\frac{|\boldsymbol{q}|}{v_{\mathrm{F}}}\right)^2, \left(\frac{\varpi}{v_{\mathrm{F}}|\boldsymbol{q}|}\right)^2\right]. \end{aligned} \tag{6.89}$$

6.7.1.1 *The physics of screening*

Analytical continuation of Eq. (6.89) onto the negative imaginary axis yields

$$\begin{aligned} \lim_{\varpi\to -\mathrm{i}\tilde{\varpi}+0^+} D^{\mathrm{RPA}}_{\boldsymbol{q},\varpi} = &-\frac{4\pi}{|\boldsymbol{q}|^2 + 8\pi e^2\frac{2mk_{\mathrm{F}}}{(2\pi)^2}\left(1 + \mathrm{i}\frac{\pi}{2}\frac{\tilde{\varpi}}{v_{\mathrm{F}}|\boldsymbol{q}|}\right)} \\ &+ \mathcal{O}\left[\left(\frac{|\boldsymbol{q}|}{v_{\mathrm{F}}}\right)^2, \left(\frac{\varpi}{v_{\mathrm{F}}|\boldsymbol{q}|}\right)^2\right]. \end{aligned} \tag{6.90}$$

The imaginary part of the denominator indicates that the "lifetime" of the field φ is non-vanishing in the quasi-static limit. The bare Coulomb interaction is thus profoundly modified by the Fermi sea. The Fermi sea is characterized by a continuum of particle-hole excitations causing a non-vanishing lifetime of φ at non-vanishing frequencies and screening in the static limit. As we shall see, screening is non-perturbative in powers of e.

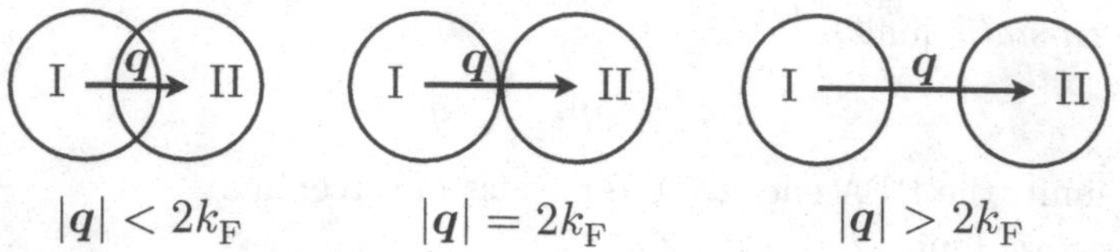

Fig. 6.5 Two Fermi spheres are drawn to represent pictorially the Kohn effect at zero temperature. The center of the Fermi spheres are shifted in reciprocal space by the transfer momentum $-\boldsymbol{q}$. Contributions to the dielectric constant are only possible when $\tilde{f}_{\mathrm{FD}}(\xi_{\boldsymbol{k}}) = 1$ and $\tilde{f}_{\mathrm{FD}}(\xi_{\boldsymbol{k}-\boldsymbol{q}}) = 0$ or $\tilde{f}_{\mathrm{FD}}(\xi_{\boldsymbol{k}}) = 0$ and $\tilde{f}_{\mathrm{FD}}(\xi_{\boldsymbol{k}-\boldsymbol{q}}) = 1$. The condition $\tilde{f}_{\mathrm{FD}}(\xi_{\boldsymbol{k}}) = 1$ defines the interior of the Fermi sphere centered at the origin, the condition $\tilde{f}_{\mathrm{FD}}(\xi_{\boldsymbol{k}-\boldsymbol{q}}) = 0$ defines the outside of the Fermi sphere centered at $+\boldsymbol{q}$, combining those two conditions yields region I. The condition $\tilde{f}_{\mathrm{FD}}(\xi_{\boldsymbol{k}}) = 0$ defines the outside of the Fermi sphere centered at the origin, the condition $\tilde{f}_{\mathrm{FD}}(\xi_{\boldsymbol{k}-\boldsymbol{q}}) = 0$ defines the inside of the Fermi sphere centered at $+\boldsymbol{q}$, combining those two conditions yields region II. The union of regions I and II is the Fermi surface $\xi_{\boldsymbol{k}} = 0$ to a very good approximation for very small momentum transfer $\boldsymbol{q}$. As the momentum transfer increases in magnitude so does the volume of the union of region I and II. The volume of the union of region I and II saturates to twice the Fermi volume at and beyond the value $|\boldsymbol{q}| = 2k_{\mathrm{F}}$.

6.7.1.2 *Thomas-Fermi approximation*

In the static limit, $\varpi = 0$, the field φ acquires an infinite lifetime,

$$\lim_{\varpi\to 0} D^{\mathrm{RPA}}_{\boldsymbol{q},\varpi} = -\frac{4\pi}{|\boldsymbol{q}|^2 + (\lambda_{\mathrm{TF}})^{-2}} + \mathcal{O}\left[\left(\frac{|\boldsymbol{q}|}{v_{\mathrm{F}}}\right)^2, \left(\frac{\varpi}{v_{\mathrm{F}}|\boldsymbol{q}|}\right)^2\right], \tag{6.91}$$

where we have introduced the *Thomas-Fermi screening length*

$$\begin{aligned}\lambda_{\mathrm{TF}} &:= \left[8\pi\, e^2 \frac{2\, m\, k_{\mathrm{F}}}{(2\pi)^2}\right]^{-1/2} \\ &= \left(4e^2 \frac{m\, k_{\mathrm{F}}}{\pi}\right)^{-1/2}.\end{aligned} \tag{6.92}$$

The dependence on e^2 of the Thomas-Fermi screening length is non-analytic in the vicinity of $e^2 = 0$. A position-space Fourier transformation of the right-hand side yields the Yukawa potential

$$\frac{e^{-\frac{|\boldsymbol{r}|}{\lambda_{\mathrm{TF}}}}}{|\boldsymbol{r}|}. \tag{6.93}$$

However, this Fourier transform extends the range of validity of Eq. (6.91) beyond the long-wavelength limit. Fourier transform to position space of the Lindhard function amounts to the replacement

$$\frac{e^{-\frac{|\boldsymbol{r}|}{\lambda_{\mathrm{TF}}}}}{|\boldsymbol{r}|} \longrightarrow |\boldsymbol{r}|^{-3} \cos(2\, k_{\mathrm{F}}\, |\boldsymbol{r}|). \tag{6.94}$$

This oscillatory behavior is known as a *Friedel oscillation.* We will rederive this result by "elementary" means below.

6.7.1.3 *Kohn effect*

We would like to revisit the Thomas-Fermi approximation to the RPA propagator $D^{\mathrm{RPA}}_{\boldsymbol{q}}$ and the condition under which it breaks down. We have seen that in the static limit,

$$\xi_{\boldsymbol{k}+\boldsymbol{q}} - \xi_{\boldsymbol{k}} = +(\boldsymbol{q}\cdot\nabla_{\boldsymbol{k}}\xi_{\boldsymbol{k}}) + \mathcal{O}(\boldsymbol{q}^2), \tag{6.95a}$$

$$\tilde{f}_{\mathrm{FD}}(\xi_{\boldsymbol{k}+\boldsymbol{q}}) - \tilde{f}_{\mathrm{FD}}(\xi_{\boldsymbol{k}}) = \left(\frac{\partial \tilde{f}_{\mathrm{FD}}}{\partial \xi}\right)(\boldsymbol{q}\cdot\nabla_{\boldsymbol{k}}\xi_{\boldsymbol{k}}) + \mathcal{O}(\boldsymbol{q}^2), \tag{6.95b}$$

$$\begin{aligned}\varepsilon^{\mathrm{RPA}}_{\boldsymbol{q},\varpi=0} &= 1 + 2\frac{4\pi e^2}{\boldsymbol{q}^2}\int\frac{\mathrm{d}^3\boldsymbol{k}}{(2\pi)^3}\left[\left(-\frac{\partial \tilde{f}_{\mathrm{FD}}}{\partial \xi}\right) + \mathcal{O}(\boldsymbol{q}^2)\right]\\ &= 1 + \frac{(k_{\mathrm{TF}})^2}{\boldsymbol{q}^2} + \mathcal{O}(\boldsymbol{q}^0),\end{aligned} \tag{6.95c}$$

where the Thomas-Fermi wave vector k_{TF} is proportional to the density of states at the Fermi energy ν_{F} defined in Eq. (6.78). What happens for larger $|\boldsymbol{q}|$'s? We can use the Lindhard function (6.77), (we reinstate $\hbar$)

$$\varepsilon^{\mathrm{RPA}}_{\boldsymbol{q},\varpi=0} = 1 + \frac{4\pi e^2}{\boldsymbol{q}^2}\left(\frac{mk_{\mathrm{F}}}{\hbar^2\pi^2}\right)\left[\frac{1}{2} + \frac{4(k_{\mathrm{F}})^2 - \boldsymbol{q}^2}{8k_{\mathrm{F}}|\boldsymbol{q}|}\ln\left|\frac{2k_{\mathrm{F}} + |\boldsymbol{q}|}{2k_{\mathrm{F}} - |\boldsymbol{q}|}\right|\right], \tag{6.96}$$

to infer that the effective screening length increases with the momentum transfer $|\boldsymbol{q}|$. It is becoming more and more difficult to make electrons screen out potentials on shorter wavelengths. Moreover, when $|\boldsymbol{q}| = 2k_{\mathrm{F}}$, the dielectric constant becomes singular. This singularity comes about from the fact that the summand in the polarization function is proportional to

$$\tilde{f}_{\mathrm{FD}}(\xi_{\boldsymbol{k}+\boldsymbol{q}}) - \tilde{f}_{\mathrm{FD}}(\xi_{\boldsymbol{k}}). \tag{6.97}$$

Only those single-particle states with momenta $\boldsymbol{k}$ and $\boldsymbol{k}+\boldsymbol{q}$ contribute to the sum in the polarization function provided either of one is occupied but not both simultaneously. For small values of $|\boldsymbol{q}|$ the pairs of single-particle states $\boldsymbol{k}$ and $\boldsymbol{k}+\boldsymbol{q}$ contributing to

$$\tilde{f}_{\mathrm{FD}}(\xi_{\boldsymbol{k}+\boldsymbol{q}}) - \tilde{f}_{\mathrm{FD}}(\xi_{\boldsymbol{k}}) \tag{6.98}$$

belong to two regions I and II that are essentially equal to the *surface* of the Fermi sea (see Fig. 6.5). As $|\boldsymbol{q}|$ is increased, regions I and II increase in size and converge smoothly to the Fermi sea. There thus exists a functional change of $\sum_{\boldsymbol{k}}\left[\tilde{f}_{\mathrm{FD}}(\xi_{\boldsymbol{k}+\boldsymbol{q}}) - \tilde{f}_{\mathrm{FD}}(\xi_{\boldsymbol{k}})\right]$ upon a small variation $\delta\boldsymbol{q}$ of $\boldsymbol{q}$,

$$|\boldsymbol{q}| < 2k_{\mathrm{F}} \Longrightarrow \frac{\delta}{\delta\boldsymbol{q}}\sum_{\boldsymbol{k}}\left[\tilde{f}_{\mathrm{FD}}(\xi_{\boldsymbol{k}+\boldsymbol{q}}) - \tilde{f}_{\mathrm{FD}}(\xi_{\boldsymbol{k}})\right] \neq 0. \tag{6.99}$$

However, as soon as $\boldsymbol{q}$ equals in magnitude twice the Fermi wave vector and beyond, there is no functional change of $\sum_{\boldsymbol{k}}\left[\tilde{f}_{\mathrm{FD}}(\xi_{\boldsymbol{k}+\boldsymbol{q}})-\tilde{f}_{\mathrm{FD}}(\xi_{\boldsymbol{k}})\right]$ anymore upon a small variation $\delta\boldsymbol{q}$ of $\boldsymbol{q}$,

$$|\boldsymbol{q}|\geq 2k_{\mathrm{F}}\Longrightarrow\frac{\delta}{\delta\boldsymbol{q}}\sum_{\boldsymbol{k}}\left[\tilde{f}_{\mathrm{FD}}(\xi_{\boldsymbol{k}+\boldsymbol{q}})-\tilde{f}_{\mathrm{FD}}(\xi_{\boldsymbol{k}})\right]=0. \tag{6.100}$$

Hence, transfer momenta obeying $|\boldsymbol{q}|=2k_{\mathrm{F}}$ must be singular points.

This argument does not depend on the shape of the Fermi surface. It also has consequences for the ability of electrons to screen out the electrostatic potential set up by collective modes propagating through the solids. For example, a phonon of wave vector $\boldsymbol{K}$ sets up an external potential due to the coherent motion of ions. The electrons respond by screening the electric field induced by the phonon. Evidently, screening of the ions by the much more mobile electrons changes the effective interaction between the ions in a nearly instantaneous way. This last change should thus be encoded by the electronic dielectric constant in the static limit. Moreover, any singularity in the electronic dielectric constant should show up in the phonon spectrum thereby opening the possibility to measure the Fermi wave vector by inspection of the phonon spectrum. This phenomenon is called the *Kohn effect*.

6.7.1.4 *Friedel or Ruderman-Kittel-Kasuya-Yosida oscillations*

The dependence on position $\boldsymbol{r}$ of the RPA propagator in the static limit is given by

$$\lim_{\varpi\to 0}D^{\mathrm{RPA}}_{\varpi}(\boldsymbol{r})=-\int\mathrm{d}^3\boldsymbol{q}\,e^{+\mathrm{i}\boldsymbol{q}\cdot\boldsymbol{r}}\left(\frac{\boldsymbol{q}^2}{4\pi}+e^2\chi_{\boldsymbol{q}}\right)^{-1}, \tag{6.101}$$

where $\chi_{\boldsymbol{q}}$ is, up to a sign, the static limit of the polarization function, i.e., ($\hbar=1$)

$$\chi_{\boldsymbol{q}}=2\times\left(\frac{mk_{\mathrm{F}}}{2\pi^2}\right)\left(\frac{1}{2}+\frac{1-x^2}{4x}\ln\left|\frac{1+x}{1-x}\right|\right),\qquad x=\frac{|\boldsymbol{q}|}{2k_{\mathrm{F}}}. \tag{6.102}$$

We have explicitly factorized a factor of 2 arising from the two-fold spin degeneracy. As already noted in Eq. (6.77), the logarithmic singularity of $\chi_{\boldsymbol{q}}$ when $|\boldsymbol{q}|=2k_{\mathrm{F}}$ shows up as an oscillatory behavior at long distances. This oscillatory behavior is known as a *Friedel oscillation* in the context of the jellium model. It is also known as the *Ruderman-Kittel-Kasuya-Yosida (RKKY) oscillation* of the static spin susceptibility induced by a magnetic impurity in a free electron gas.

We are going to sketch an alternative derivation of the Friedel oscillations. In this derivation, the emphasis is on the response of the electron gas to a s-wave static charge impurity. Consider the Schrödinger equation

$$\mathrm{i}\partial_t\Psi=\left(\frac{\boldsymbol{p}^2}{2m}+V(|\boldsymbol{r}|)\right)\Psi \tag{6.103}$$

for a (spinless) particle subjected to a spherically symmetric potential $V(|\boldsymbol{r}|)$ that decays faster than $1/|\boldsymbol{r}|$ for large $|\boldsymbol{r}|$. The boundary conditions

$$\Psi(\boldsymbol{r}=\boldsymbol{0},t)\text{ finite},\qquad\Psi\sim\text{ outgoing plane wave for large }|\boldsymbol{r}|, \tag{6.104}$$

are imposed. Stationary states have an energy spectrum $\{\varepsilon_{|\boldsymbol{k}|,l}\}$ that depends on the angular momentum quantum number l and on the magnitude of the momentum $\boldsymbol{k}$ of the outgoing wave. For large $|\boldsymbol{r}|$, we do the expansion in terms of the spherical waves

$$\Psi_{\boldsymbol{k}}(\boldsymbol{r},t)=\sum_{l=0}^{\infty}A_{|\boldsymbol{k}|,l}(t)\,\psi_{|\boldsymbol{k}|,l}(\boldsymbol{r}), \tag{6.105a}$$

where the stationary states behave at large $|\boldsymbol{r}|$ as

$$\psi_{|\boldsymbol{k}|,l}(\boldsymbol{r})\sim\frac{1}{|\boldsymbol{r}|}\sin\left(|\boldsymbol{k}||\boldsymbol{r}|-\frac{\pi}{2}l+\eta_l(|\boldsymbol{k}|)\right)P_l(\cos\theta),\qquad l=0,1,\cdots. \tag{6.105b}$$

Here, θ is the angle between the momentum $\boldsymbol{k}$ and $\boldsymbol{r}$ and P_l is a Legendre polynomial. The phase shifts $\eta_l(|\boldsymbol{k}|)$, which are functions of $|\boldsymbol{k}|$, encode all information on the impurity potential $V(|\boldsymbol{r}|)$. For $V=0$, $\eta_l(|\boldsymbol{k}|)=0$ and stationary states behave at large $|\boldsymbol{r}|$ as

$$\psi^{(0)}_{|\boldsymbol{k}|,l}(\boldsymbol{r})\sim\frac{1}{|\boldsymbol{r}|}\sin\left(|\boldsymbol{k}||\boldsymbol{r}|-\frac{\pi}{2}l\right)P_l(\cos\theta),\qquad l=0,1,\cdots. \tag{6.106}$$

The energy spectrum is discrete if we impose the hard wall boundary condition

$$\lim_{|\boldsymbol{r}|\to R}\psi_{|\boldsymbol{k}|,l}(\boldsymbol{r})=0, \tag{6.107}$$

where R is the radius of a large sphere centered about the origin, i.e., about the impurity, since

$$|\boldsymbol{k}|=\frac{1}{R}\left(n\pi+\frac{\pi}{2}l-\eta_l(|\boldsymbol{k}|)\right),\qquad n\in\mathbb{Z},\qquad l=0,1,\cdots, \tag{6.108}$$

must then hold. Notice that in the absence of the impurity, i.e., when $\eta_l(|\boldsymbol{k}|)=0$ with $l=0,1,\cdots$, the quantization condition

$$|\boldsymbol{k}|=\frac{1}{R}\left(n\pi+\frac{\pi}{2}l\right),\qquad n\in\mathbb{Z},\qquad l=0,1,\cdots, \tag{6.109}$$

yields the same number of energy eigenstates below the Fermi energy ε_{F} as if we had chosen to impose periodic boundary conditions in a box of volume $4\pi R^3/3$ instead. We will denote solutions to Eq. (6.108) by $|\boldsymbol{k}_{n,l}|$ and solutions to Eq. (6.109) by $|\boldsymbol{k}^{(0)}_{n,l}|$.

Consider the momentum range $|\boldsymbol{k}_1|\leq|\boldsymbol{k}|\leq|\boldsymbol{k}_2|$ as is depicted in Fig. 6.6. In the absence of the impurity at the origin we can enumerate all eigenfunctions that decay like

$$\frac{1}{|\boldsymbol{r}|}\sin\left(|\boldsymbol{k}^{(0)}_{n,l}||\boldsymbol{r}|-\frac{\pi}{2}l\right)P_l(\cos\theta) \tag{6.110}$$

by the integers l and n allowed by the hard wall boundary condition on the very large sphere of radius R and for which

$$|\boldsymbol{k}_1|\leq|\boldsymbol{k}^{(0)}_{n,l}|=\left(n+\frac{l}{2}\right)\frac{\pi}{R}\leq|\boldsymbol{k}_2| \tag{6.111}$$

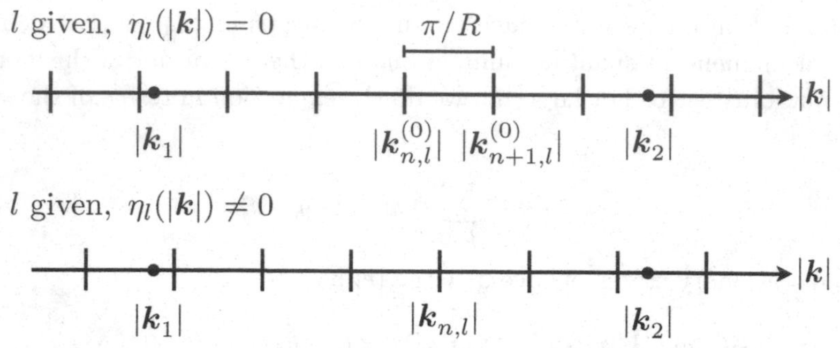

Fig. 6.6 By switching on the spherical impurity potential V, eigenvalues are shifted along the momentum quantization axis that characterizes the large $|\boldsymbol{r}|$ asymptotic behavior of energy eigenstates. This shift of the spectrum can cause a net change in the number of eigenvalues in the fixed interval $|\boldsymbol{k}_1| \leq |\boldsymbol{k}| \leq |\boldsymbol{k}_2|$. The shift induced by the spherical impurity potential between $|\boldsymbol{k}^{(0)}_{n,l}|$ and $|\boldsymbol{k}_{n,l}|$ is $\eta_l(|\boldsymbol{k}^{(0)}_{n,l}|)/R$.

must hold. If we fix l, the spacing in momentum space between neighboring eigenvalues is π/R. Under the terminology of adiabatic switching of the s-wave impurity potential one understands the hypothesis that there exists a one-to-one correspondence between the eigenfunctions (6.110) with the quantization condition (6.111) and all eigenfunctions that decay like

$$\frac{1}{|\boldsymbol{r}|}\sin\left(|\boldsymbol{k}_{n,l}||\boldsymbol{r}| - \frac{\pi}{2}l + \eta_l(|\boldsymbol{k}_{n,l}|)\right) P_l(\cos\theta), \tag{6.112}$$

with the quantization condition

$$|\boldsymbol{k}_1| \leq |\boldsymbol{k}_{n,l}| = \left(n + \frac{l}{2} - \frac{\eta_l(|\boldsymbol{k}_{n,l}|)}{\pi}\right)\frac{\pi}{R} \leq |\boldsymbol{k}_2|, \tag{6.113}$$

up to few states with wave vectors in the vicinity of $|\boldsymbol{k}_1|$ and $|\boldsymbol{k}_2|$. In the spirit of adiabaticity, the phase shift $\eta_l(|\boldsymbol{k}_{n,l}|)$ should be thought of as a function of $|\boldsymbol{k}^{(0)}_{n,l}|$. Moreover, it is worthwhile to keep in mind that the shift

$$|\boldsymbol{k}_{n,l}| - |\boldsymbol{k}^{(0)}_{n,l}| = -\frac{\eta_l(|\boldsymbol{k}_{n,l}|)}{R} \tag{6.114}$$

vanishes in the thermodynamic limit $R \to \infty$.

In the thermodynamic limit $R \to \infty$, the change in the number of energy eigenvalues with fixed l in the range $|\boldsymbol{k}_1| \leq |\boldsymbol{k}| \leq |\boldsymbol{k}_2|$ before and after adiabatically switching the s-wave impurity potential $V(|\boldsymbol{r}|)$ is

$$\frac{1}{\pi}\left[\eta_l(|\boldsymbol{k}_2|) - \eta_l(|\boldsymbol{k}_1|)\right]. \tag{6.115a}$$

If $|\boldsymbol{k}_1|$ and $|\boldsymbol{k}_2|$ are chosen to be infinitesimally far apart, i.e., $|\boldsymbol{k}_1| \to |\boldsymbol{k}|$ and $|\boldsymbol{k}_2| \to |\boldsymbol{k}| + \mathrm{d}|\boldsymbol{k}|$, then the number (6.115a) takes the differential form

$$\frac{1}{\pi} \times \frac{\mathrm{d}\eta_l}{\mathrm{d}|\boldsymbol{k}|} \times \mathrm{d}|\boldsymbol{k}|. \tag{6.115b}$$

Let us further assume that:

(1) First,

$$\lim_{|\boldsymbol{k}|\to 0} \eta_l(|\boldsymbol{k}|) = 0. \tag{6.116}$$

(2) Second, the Fermi momentum k_{F} or, more generally, the volume of the Fermi sea, is left unchanged by switching on $V(|\boldsymbol{r}|)$.

We can then integrate Eq. (6.115b) to obtain the total number

$$2 \times \frac{1}{\pi} \sum_{l=0}^{\infty} (2l+1)\, \eta_l(k_{\mathrm{F}}) \tag{6.117}$$

of *new electrons* required to fill up all single-particle energy levels up to the Fermi energy after switching on the s-wave impurity potential $V(|\boldsymbol{r}|)$. (The factor of 2 accounts for the two-fold spin degeneracy. The factor of $(2l+1)$ accounts for the spherical geometry of the impurity potential.) If we further require that the electric charge of a s-wave impurity must be neutralized by an excess of electrons within a non-vanishing distance R, then the difference Z of the valency between the impurity and the metallic host is given by

$$Z = 2 \times \frac{1}{\pi} \sum_{l=0}^{\infty} (2l+1)\, \eta_l(k_{\mathrm{F}}). \tag{6.118}$$

Equation (6.118) is known as *Friedel sum rule*.

Associated to the phase shifts η_l there are changes in the local electronic density. In the thermodynamic limit and at large distances from the s-wave impurity, the excess charge is given by

$$\begin{aligned} \delta\rho(|\boldsymbol{r}|) &\propto \lim_{R\to\infty} 2 \times e \sum_{l=0}^{\infty} (2l+1) \int_0^{k_{\mathrm{F}}} \frac{\mathrm{d}k}{\pi/R} \left[|\psi_{k,l;\eta_l\neq 0}(|\boldsymbol{r}|)|^2 - |\psi_{k,l;\eta_l=0}(|\boldsymbol{r}|)|^2 \right] \\ &\propto e \sum_{l=0}^{\infty} (2l+1) \frac{1}{r^2} \int_0^{k_{\mathrm{F}}} \mathrm{d}k \left[\sin^2\left(k\,r - \frac{\pi}{2} l + \eta_l(|\boldsymbol{k}|)\right) - \sin^2\left(k\,r - \frac{\pi}{2} l\right) \right] \\ &\propto e \sum_{l=0}^{\infty} \frac{(2l+1)(-1)^l \sin\eta_l(k_{\mathrm{F}}) \left[\cos\left(\eta_l(k_{\mathrm{F}})\right) - \cos\left(2k_{\mathrm{F}} r + \eta_l(k_{\mathrm{F}})\right)\right]}{r^3}, \end{aligned} \tag{6.119}$$

in the static limit.[4] The Yukawa decay predicted by the Thomas-Fermi approximation is replaced by the slower algebraic decay with superimposed periodic oscillations (quantum interferences) with periodicity of twice the Fermi wave vector.

[4] The normalization of a wave function decaying like $r^{-1}\sin(k\,r)$ in a sphere of radius R is proportional to $R^{-1/2}$.

6.7.2 *Long-wavelength and dynamic limit at $T = 0$*

The regime

$$v_{\mathrm{F}}|\boldsymbol{q}| \ll |\varpi|, \qquad |\boldsymbol{q}| \ll m\, v_{\mathrm{F}}, \tag{6.120}$$

suggests using the expansion

$$\arctan z = z - \frac{z^3}{3} + \frac{z^5}{5} - \cdots, \qquad |z| < 1, \qquad z \to \frac{v_{\mathrm{F}}|\boldsymbol{q}|}{\varpi}, \tag{6.121}$$

in Eq. (6.83). To leading order in this expansion,

$$\Pi^{\mathrm{RPA}}_{\boldsymbol{q},\varpi} = -\frac{4mk_{\mathrm{F}}}{3(2\pi)^2}\left(\frac{v_{\mathrm{F}}|\boldsymbol{q}|}{\varpi}\right)^2 + \mathcal{O}\left[\left(\frac{|\boldsymbol{q}|}{v_{\mathrm{F}}}\right)^2, \left(\frac{v_{\mathrm{F}}|\boldsymbol{q}|}{\varpi}\right)^4\right]. \tag{6.122}$$

In turn, the RPA propagator in Eq. (6.41) is approximated by

$$\begin{aligned} D^{\mathrm{RPA}}_{\boldsymbol{q},\varpi} &= -\frac{1}{\frac{\boldsymbol{q}^2}{4\pi} - e^2\Pi^{\mathrm{RPA}}_{\boldsymbol{q},\varpi}} \\ &= -\frac{4\pi}{|\boldsymbol{q}|^2 + 4\pi e^2\frac{4mk_{\mathrm{F}}}{3(2\pi)^2}\left(\frac{v_{\mathrm{F}}|\boldsymbol{q}|}{\varpi}\right)^2} + \mathcal{O}\left[\left(\frac{|\boldsymbol{q}|}{v_{\mathrm{F}}}\right)^2, \left(\frac{v_{\mathrm{F}}|\boldsymbol{q}|}{\varpi}\right)^4\right] \\ &= -\frac{4\pi}{|\boldsymbol{q}|^2\left[1 + \left(\frac{\varpi_{\mathrm{P}}}{\varpi}\right)^2\right]} + \mathcal{O}\left[\left(\frac{|\boldsymbol{q}|}{v_{\mathrm{F}}}\right)^2, \left(\frac{v_{\mathrm{F}}|\boldsymbol{q}|}{\varpi}\right)^4\right], \end{aligned} \tag{6.123}$$

whereby the so-called *plasma frequency* is

$$(\varpi_{\mathrm{P}})^2 := 4\pi e^2\frac{4mk_{\mathrm{F}}}{3(2\pi)^2}(v_{\mathrm{F}})^2 = \frac{4}{3\pi}(k_{\mathrm{F}})^3\frac{e^2}{m} = \frac{4}{3\pi}3\pi^2\frac{N}{V}\frac{e^2}{m} = 4\pi\frac{N}{V}\frac{e^2}{m}. \tag{6.124}$$

Observe that the factorization of $\boldsymbol{q}^2$ in the denominator of Eq. (6.123) is special to the Coulomb interaction.

6.7.2.1 *The physics of plasmons*

Analytical continuation of Eq. (6.123) onto the negative imaginary axis yields

$$\lim_{\varpi\to -\mathrm{i}\tilde{\varpi}+0^+} D^{\mathrm{RPA}}_{\boldsymbol{q},\varpi} = \frac{4\pi}{\boldsymbol{q}^2\left[1 - \left(\frac{\varpi_{\mathrm{P}}}{\tilde{\varpi}}\right)^2\right]} + \mathcal{O}\left[\left(\frac{|\boldsymbol{q}|}{v_{\mathrm{F}}}\right)^2, \left(\frac{v_{\mathrm{F}}|\boldsymbol{q}|}{\tilde{\varpi}}\right)^4\right]. \tag{6.125}$$

After this analytical continuation, we find poles whenever

$$\tilde{\varpi}_{\boldsymbol{q}} = \varpi_{\mathrm{P}}, \qquad \forall \boldsymbol{q}. \tag{6.126}$$

Of course the independence on the momentum transfer $\boldsymbol{q}$ is an artifact of truncating the gradient expansion to leading order. Including higher-order contributions in the gradient expansion gives, up to some numerical constant #, the so-called *plasmon branch of excitations*

$$\tilde{\varpi}_{\boldsymbol{q}} = \varpi_{\mathrm{P}}\left[1 + \#\left(\frac{v_{\mathrm{F}}}{\varpi_{\mathrm{P}}}\boldsymbol{q}\right)^2 + \cdots\right], \tag{6.127}$$

provided the momentum transfer is not too large.

6.7.2.2 *Landau damping*

Once the dispersion curve of plasmons enters in the particle-hole continuum, plasmons become unstable to decay into an electron-hole pair. This phenomenon is signaled by

$$\lim_{\varpi\to -\mathrm{i}\tilde{\varpi}+0+} \left(D^{\mathrm{RPA}}_{\boldsymbol{q},\varpi}\right)^{-1} \tag{6.128}$$

acquiring an imaginary part and thus a non-vanishing lifetime, [use $(x-\mathrm{i}0^+)^{-1} = P_{1/x} + \mathrm{i}\pi\delta(x)$]

$$\mathrm{Im}\left[\lim_{\varpi\to -\mathrm{i}\tilde{\varpi}+0+} \left(D^{\mathrm{RPA}}_{\boldsymbol{q},\varpi}\right)^{-1}\right] \propto \int \frac{\mathrm{d}^3\boldsymbol{k}}{(2\pi)^3}\delta\left(\tilde{\varpi}_{\boldsymbol{q}} - \frac{\boldsymbol{k}\cdot\boldsymbol{q}}{m}\right)\left[\boldsymbol{q}\cdot\nabla_{\boldsymbol{k}}\tilde{f}_{\mathrm{FD}}(\xi_{\boldsymbol{k}})\right]. \tag{6.129}$$

The factor

$$\delta\left(\tilde{\varpi}_{\boldsymbol{q}} - \frac{\boldsymbol{k}\cdot\boldsymbol{q}}{m}\right) \tag{6.130}$$

select electrons whose velocities $|\boldsymbol{k}|/m$ are close to the phase velocity $\tilde{\varpi}_{\boldsymbol{q}}/|\boldsymbol{q}|$ of the plasmon density wave in that $\boldsymbol{k}\cdot\boldsymbol{q}/m = \tilde{\varpi}_{\boldsymbol{q}}$. There is thus a small range of electron velocities for which the electrons are able to surf the plasmon wave. Electrons moving initially slightly more slowly than the plasmon wave will pump energy from the plasmon wave as they are accelerated up to the wave speed by the wave leading edge. Conversely, electrons moving initially faster than the plasmon wave will give up energy to the plasmon wave as they are decelerated up to the wave speed by the wave trailing edge. Because the velocity distribution of electrons

$$\left[\boldsymbol{q}\cdot\nabla_{\boldsymbol{k}}\tilde{f}_{\mathrm{FD}}(\xi_{\boldsymbol{k}})\right] \tag{6.131}$$

is skewed in favor of low energy electrons, the net effect is to damp the wave. This damping is called *Landau damping.*

6.8 Random-phase approximation for a short-range interaction

So far, we have been dealing exclusively with the two-body repulsive potential

$$V_{\mathrm{cb}}(\boldsymbol{r}_1 - \boldsymbol{r}_2) = +\frac{e^2}{|\boldsymbol{r}_1 - \boldsymbol{r}_2|}. \tag{6.132}$$

(The coupling constant e^2 has units of *energy* × *length.*) What if we work instead with a short-range repulsive potential, say

$$V_\lambda(\boldsymbol{r}_1 - \boldsymbol{r}_2) = +\lambda\delta(\boldsymbol{r}_1 - \boldsymbol{r}_2)? \tag{6.133}$$

(The coupling constant λ has units of *energy* × *volume.*) This type of modeling of a two-body interaction is made, for example, to describe the interaction between ^{3}He atoms in liquid ^{3}He. Since

$$\delta(\boldsymbol{r}) = \frac{1}{V}\sum_{\boldsymbol{q}} e^{+\mathrm{i}\boldsymbol{q}\cdot\boldsymbol{r}}, \tag{6.134}$$

in a box of volume V with the imposition of periodic boundary conditions, we have the Fourier transforms

$$V_{\mathrm{cb}\,\boldsymbol{q}}=\frac{4\pi e^2}{\boldsymbol{q}^2}, \tag{6.135a}$$

and

$$V_{\lambda;\boldsymbol{q}}=+\lambda, \tag{6.135b}$$

for the Coulomb and contact repulsive interactions, respectively. RPA fluctuations of the order parameter φ around the mean field $\varphi_{\mathrm{mf}}=0$ is now encoded by the effective action

$$S_{\lambda}^{\mathrm{RPA}}=\sum_{q=(\boldsymbol{q},\varpi_l)}\frac{1}{2}\varphi_{+q}\left(\frac{1}{\lambda}-\Pi_q^{\mathrm{RPA}}\right)\varphi_{-q}, \tag{6.136a}$$

instead of [compare with Eq. (6.38c) and note that the convention for the (engineering) dimension of φ has been changed]

$$S_{\mathrm{cb}}^{\mathrm{RPA}}=\sum_{q=(\boldsymbol{q},\varpi_l)}\frac{1}{2}\varphi_{+q}\left(\frac{\boldsymbol{q}^2}{4\pi e^2}-\Pi_q^{\mathrm{RPA}}\right)\varphi_{-q}. \tag{6.136b}$$

The locations of the poles of

$$D_q^{\mathrm{RPA}}:=-\left[(V_{\boldsymbol{q}})^{-1}-\Pi_q^{\mathrm{RPA}}\right]^{-1} \tag{6.137}$$

depend dramatically on the short distance behavior of $V_{\boldsymbol{q}}$ in the dynamic limit $|\varpi_l|\gg v_{\mathrm{F}}|\boldsymbol{q}|$, $|\boldsymbol{q}|\ll k_{\mathrm{F}}$. Indeed, the plasma dispersion (6.127) becomes gapless for our naive modeling of ^{3}He as

$$\lim_{\varpi\to-\mathrm{i}\tilde{\varpi}+0^+}D_{\lambda;\boldsymbol{q},\varpi}^{\mathrm{RPA}}=\frac{-1}{\left[1-\left(\frac{c|\boldsymbol{q}|}{\tilde{\varpi}}\right)^2\right]}+\mathcal{O}\left[\left(\frac{|\boldsymbol{q}|}{v_{\mathrm{F}}}\right)^2,\left(\frac{v_{\mathrm{F}}|\boldsymbol{q}|}{\tilde{\varpi}}\right)^4\right], \tag{6.138}$$

where

$$\begin{aligned}c^2 &:= \frac{\lambda}{4\pi e^2}\times(\varpi_{\mathrm{P}})^2\\ &\underset{\text{Eq. (6.124)}}{=}\frac{N}{V}\frac{\lambda}{m}.\end{aligned} \tag{6.139}$$

These excitations are just above the particle-hole continuum and are called *zero-sound*. Collective modes whose energies go to zero at large wavelength are the general rule. A non-vanishing energy mode at large wavelengths such as the plasmon is the exception as it is associated with *long-range* forces. Long-range forces are very special. In the context of phase transitions, they cause the breakdown of Goldstone theorem, i.e., of the existence of excitations with arbitrary small energies when a continuous symmetry is spontaneously broken.

Coming back to zero-sound, one can show that zero-sound is a coherent superposition of particle-hole excitations near the Fermi surface tantamount to some $\boldsymbol{q}$-resolved periodic oscillation of the local (in space) Fermi surface (see section 5.4

in Ref. [16]). Zero-sound is thus completely different from thermodynamic sound in a Fermi gas. Thermodynamic sound is a classical phenomenon that can only be observed on time scales much larger than the typical time scale τ (smallest between microscopic time scale and inverse temperature) for particles to interact. Indeed the adjective "thermodynamic" requires thermodynamic equilibrium. In turn, thermodynamic equilibrium can only be achieved due to interactions, i.e., interactions are needed to relax any initial (non-interacting, say) state into thermodynamic equilibrium. Conventional (i.e., thermodynamic) sound results from a time-dependent perturbation whose characteristic time $1/\omega$ is much larger than τ,

$$\omega\tau \ll 1. \tag{6.140}$$

On such time scales, quasiparticle and collective modes have already decayed at non-vanishing temperatures and thus are unrelated to thermodynamic sound. From a geometrical point of view, thermodynamic sound can be viewed as an isotropic pulsating local (in position space) Fermi sphere (see section 5.4 in Ref. [16]). Zero-sound is the opposite extreme to thermodynamic sound. Zero-sound is built out of quasiparticles. At zero temperature, the coherent superposition of quasiparticles responsible for zero-sound acquires an infinite lifetime. Hence, zero-sound can propagate at non-vanishing frequencies. At non-vanishing temperatures, a necessary condition for the observation of zero-sound is that the frequency ω of the laboratory probe be large enough for the characteristic observation time $1/\omega$ to be smaller than the lifetime of quasiparticles,

$$\omega\tau \gg 1. \tag{6.141}$$

6.9 Feedback effect on and by phonons

We now consider a jellium model for ions. Ions are point charges of mass M immersed in an (initially) uniform electron gas of density $\rho_0 = N/V$. The electric charge per ion is denoted $Z\,e$. The averaged number of ions per unit volume is denoted

$$\bar{\rho}_{\text{ion}} \equiv N_{\text{ion}}/V. \tag{6.142a}$$

The number of ions per unit volume

$$\rho_{\text{ion}} = \bar{\rho}_{\text{ion}} + \delta\rho_{\text{ion}} \tag{6.142b}$$

is allowed to weakly fluctuate in space and time through $\delta\rho_{\text{ion}}$. Charge neutrality reads

$$Z\,N_{\text{ion}} = N, \qquad Z\,\bar{\rho}_{\text{ion}} = \rho_0. \tag{6.143}$$

In the absence of any electronic motion but allowing the ionic density to fluctuate in space and time according to

$$\begin{aligned}
&M\,\dot{\boldsymbol{v}}_{\text{ion}} = (Z\,e)\,\boldsymbol{E},\\
&\boldsymbol{\nabla}\cdot\boldsymbol{E} = 4\pi\,e\,(Z\,\rho_{\text{ion}} - \rho_0)\,,\\
&0 = \dot{\rho}_{\text{ion}} + \boldsymbol{\nabla}\cdot\boldsymbol{J}_{\text{ion}}, \qquad \boldsymbol{J}_{\text{ion}} := \rho_{\text{ion}}\,\boldsymbol{v}_{\text{ion}},
\end{aligned} \tag{6.144}$$

we find, up to linear order in $\delta\rho_{\text{ion}}$, the plasma oscillation

$$0 = \delta\ddot{\rho}_{\text{ion}} + \Omega_{\text{P}}^2\,\delta\rho_{\text{ion}}, \tag{6.145a}$$

with the characteristic plasma frequency [compare with Eq. (6.124)]

$$\Omega_{\text{P}}^2 := 4\pi\frac{N_{\text{ion}}}{V}\frac{(Z\,e)^2}{M}. \tag{6.145b}$$

The ionic plasma frequency Ω_{P} is much lower then the electronic one since

$$\left(\frac{\Omega_{\text{P}}}{\varpi_{\text{P}}}\right)^2 = \frac{4\pi\frac{N_{\text{ion}}}{V}\frac{(Z\,e)^2}{M}}{4\pi\frac{N}{V}\frac{(-e)^2}{m}} = \frac{Z\,m}{M} \ll 1. \tag{6.146}$$

Ions move much more slowly that electrons. Electrons can thus adapt to the motion of ions. In particular, any (infinitesimal) local excess of ionic charge $\delta\rho_{\text{ion}}$ induced by a collective motion of the ions that solves Eq. (6.144) is screened by the electrons. To account for this physics, we may set up the following model for the coupled system of ions and electrons in the presence of an external density ρ_{ext}. The dynamical response of the ions with electrons providing screening to an external charge density $e\,\rho_{\text{ext}}$ is governed by the classical model

$$\begin{aligned}
M\,\delta\ddot{\boldsymbol{r}}_{\text{ion}} &= -(Z\,e)\,\boldsymbol{\nabla}\phi,\\
-\Delta\phi + k_0^2\,\phi &= 4\pi\,(Z\,e)\,\delta\rho_{\text{ion}} + 4\pi\,e\,\rho_{\text{ext}},\\
(Z\,e)\,\delta\rho_{\text{ion}} &= -\frac{N_{\text{ion}}}{V}\,(Z\,e)\,\boldsymbol{\nabla}\cdot\delta\boldsymbol{r}_{\text{ion}}.
\end{aligned} \tag{6.147}$$

Here, $\delta\boldsymbol{r}_{\text{ion}}$ is the deviation of the position of an ion with regard to its equilibrium position, the divergence $\boldsymbol{\nabla}\cdot\delta\boldsymbol{r}_{\text{ion}}$ is proportional to the small deviation $\delta\rho_{\text{ion}}$ in the number of ions per unit volume relative to the uniform density $\bar{\rho}_{\text{ion}}$ of ions at equilibrium, and $k_0 = \sqrt{4e^2\,m\,k_{\text{F}}/\pi} = \lambda_{\text{TF}}^{-1}$ is the inverse Thomas-Fermi screening length. We have assumed that the characteristic frequency Ω_{P} that enters the electronic polarization function is, for all intent and purposes, so small that we can use the Thomas-Fermi approximation to account for the screening by the electrons. Fourier transformation of

$$-\Delta\ddot{\phi} + k_0^2\,\ddot{\phi} = 4\pi\,\frac{N_{\text{ion}}}{V}\,\frac{(Z\,e)^2}{M}\Delta\phi + 4\pi\,e\,\ddot{\rho}_{\text{ext}} \tag{6.148}$$

with respect to position space and time gives

$$\boldsymbol{q}^2\,(-\varpi^2)\,\phi_{\boldsymbol{q},\varpi} + k_0^2\,(-\varpi^2)\,\phi_{\boldsymbol{q},\varpi} = \Omega_{\text{P}}^2\,(-\boldsymbol{q}^2)\,\phi_{\boldsymbol{q},\varpi} + 4\pi\,e\,(-\varpi^2)\,\rho_{\text{ext}\,\boldsymbol{q},\varpi}, \tag{6.149}$$

i.e.,

$$\begin{aligned}
\phi_{\boldsymbol{q},\varpi} &= \frac{1}{\varepsilon_{\boldsymbol{q},\varpi}}\,\frac{4\pi\,e}{\boldsymbol{q}^2}\,\rho_{\text{ext}\,\boldsymbol{q},\varpi} \equiv \frac{1}{\varepsilon_{\boldsymbol{q},\varpi}}\,\phi_{\text{ext}\,\boldsymbol{q},\varpi},\\
\frac{1}{\varepsilon_{\boldsymbol{q},\varpi}} &= \frac{1}{1+(k_0^2/\boldsymbol{q}^2)}\,\frac{\varpi^2}{\varpi^2-\varpi_{\boldsymbol{q}}^2},\\
\varpi_{\boldsymbol{q}}^2 &= \frac{\Omega_{\text{P}}^2}{1+(k_0^2/\boldsymbol{q}^2)}.
\end{aligned} \tag{6.150}$$

We conclude that

- The response to an external test charge diverges when $\varpi^2 = (\varpi_{\boldsymbol{q}})^2$. A longitudinal density fluctuation can thus propagate at this frequency. For small $|\boldsymbol{q}|$,

$$\varpi \approx c\,|\boldsymbol{q}|, \tag{6.151a}$$

where the sound velocity is given by

$$\begin{aligned} c^2 &= (\Omega_{\mathrm{P}}/k_0)^2 \\ &= \frac{1}{3}\, Z\, \frac{m}{M}\, v_{\mathrm{F}}^2. \end{aligned} \tag{6.151b}$$

This approximate relation between the speed of sound c and the Fermi velocity v_{F} is called the *Bohm-Staver* relation.

- The effective Coulomb propagator mediating the interaction between electrons is modified by the slow motion of the ions. It becomes [recall Eqs. (6.42), (6.44a), and (E.150)]

$$\begin{aligned} D_{\boldsymbol{q},\varpi} &= \frac{1}{\varepsilon_{\boldsymbol{q},\varpi}}\, D_{0\,\boldsymbol{q},\varpi}, \\ \frac{1}{\varepsilon_{\boldsymbol{q},\varpi}} &\equiv \frac{1}{1+(k_0^2/\boldsymbol{q}^2)}\, \frac{\varpi^2}{\varpi^2-\varpi_{\boldsymbol{q}}^2}, \\ D_{0\,\boldsymbol{q},\varpi} &= -\frac{4\pi}{\boldsymbol{q}^2}. \end{aligned} \tag{6.152}$$

This propagator is frequency dependent, i.e., the force between two electrons is not instantaneous anymore. More importantly, whenever

$$\varpi^2 < \varpi_{\boldsymbol{q}}^2, \tag{6.153}$$

the force between electrons has effectively changed sign; *it has become attractive.* This arises because the passage of an electron nearby an ion draws the ion to the electron. However, in view of the difference in the characteristic energy scales $\Omega_{\mathrm{P}}/\varpi_{\mathrm{P}} \ll 1$, the ion relaxes to its equilibrium position on time scales much larger than the time needed for the electron to be far away. In the mean time, another electron can take advantage of the gain in potential energy caused by moving in the wake of the positive charge induced by the displaced ion. As both Ω_{P}, or, more generally, the Debye energy, are small compared with the Fermi energy ε_{F}, only electrons near the Fermi surface can take advantage of the gain in potential energy induced by the ionic motion.

6.10 Problems

6.10.1 *Static Lindhard function in one-dimensional position space*

The retarded density-density correlation function for an electron gas is defined by [recall Eq. (E.102c)]

$$\chi(q,t-t') := -\frac{\mathrm{i}}{\hbar}\,\Theta(t-t')\,e^2\,\Big\langle\,[\hat{\rho}_{\mathrm{I}}(+q,t),\hat{\rho}_{\mathrm{I}}(-q,t')]\,\Big\rangle. \tag{6.154}$$

Here, $-e < 0$ is the charge of the electron and $\hat{\rho}_{\mathrm{I}}(q,t)$ is the Fourier transform of the local electronic density operator evaluated at the (real) time t and wave vector q in the interaction picture. The infinite volume limit $V \to \infty$ has been taken and dimensionality of position space is $d = 1$, i.e., $q \in \mathbb{R}$. The angular brackets refer to averaging in the grand canonical statistical ensemble and the function Θ is the Heaviside step function.

The function (6.154) gives a linear response of the electron gas. It is related to the dielectric response function by Eq. (E.150b). The static limit $\omega \to 0$ of the time Fourier transform of the function (6.154) reduces to the momentum integral

$$\chi(q,\omega=0) \propto +2\int\frac{\mathrm{d}k}{2\pi}\frac{f_{\mathrm{FD}}(\varepsilon_{k-q/2})-f_{\mathrm{FD}}(\varepsilon_{k+q/2})}{\varepsilon_{k-q/2}-\varepsilon_{k+q/2}+\mathrm{i}0^{+}}, \tag{6.155a}$$

in the Random Phase Approximation (RPA). This function is known in the literature as the Lindhard function. Here, f_{FD} denotes the Fermi function

$$f_{\mathrm{FD}}(\varepsilon) := \frac{1}{e^{\beta(\varepsilon-\mu)}+1}, \qquad \beta := \frac{1}{k_{\mathrm{B}}\,T}. \tag{6.155b}$$

Exercise 1.1:

(a) Show that at zero temperature and for a linearized dispersion, the Fermi function is given by

$$f_{\mathrm{FD}}(\varepsilon_k) = \Theta(k+k_{\mathrm{F}}) - \Theta(k-k_{\mathrm{F}}), \tag{6.156}$$

where k_{F} is the magnitude of the Fermi wave number.

(b) Show that at $T=0$ and for q in the vicinity of $\pm 2k_{\mathrm{F}}$,

$$\chi(q) \approx -\frac{1}{\pi\,\hbar\,v_{\mathrm{F}}}\ln\left|\frac{2\,k_{\mathrm{F}}+q}{2\,k_{\mathrm{F}}-q}\right|\operatorname{sgn}(q), \tag{6.157}$$

where $v_{\mathrm{F}} = \hbar\,k_{\mathrm{F}}/m$ denotes the Fermi velocity. Conclude that the static Lindhard function $\chi(q)$ diverges at $q = \pm 2\,k_{\mathrm{F}}$.

(c) To evaluate the static Lindhard function at a finite temperature we approximate the Fermi-Dirac function by

$$f_{\mathrm{FD}}(\varepsilon) = \begin{cases} 1, & \text{for } \varepsilon < \varepsilon_{\mathrm{F}} - 2\,k_{\mathrm{B}}\,T, \\ g_{\mathrm{FD}}(\varepsilon), & \text{for } |\varepsilon - \varepsilon_{\mathrm{F}}| \le 2\,k_{\mathrm{B}}\,T, \\ 0, & \text{for } \varepsilon > \varepsilon_{\mathrm{F}} + 2\,k_{\mathrm{B}}\,T, \end{cases} \tag{6.158a}$$

where

$$g_{\mathrm{FD}}(\varepsilon) := f'_{\mathrm{FD}}(\varepsilon_{\mathrm{F}})(\varepsilon-\varepsilon_{\mathrm{F}}) + f_{\mathrm{FD}}(\varepsilon_{\mathrm{F}}), \tag{6.158b}$$

with f'_{FD} the derivative of f_{FD}. Use approximation (6.158) to show that the static Lindhard function at a non-vanishing but low (compared to the Fermi energy) temperature and at $q = 2\,k_{\mathrm{F}}$ is given by

$$\chi(2\,k_{\mathrm{F}}) \approx -\frac{2}{\pi\,\hbar\,v_{\mathrm{F}}}\ln\left(\frac{2\varepsilon_{\mathrm{F}} - k_{\mathrm{B}}\,T}{k_{\mathrm{B}}\,T}\right) + \mathcal{O}\left(\frac{k_{\mathrm{B}}\,T}{\varepsilon_{\mathrm{F}}}\right), \tag{6.159}$$

to leading order in $k_{\mathrm{B}}\,T/\varepsilon_{\mathrm{F}}$.

(d) How would approximation (6.159) change had one chosen a different slope for $g_{\mathrm{FD}}(\varepsilon)$?

(e) How are the singularities $q = \pm 2\,k_{\mathrm{F}}$ affected by a finite temperature?

6.10.2 *Luttinger theorem revisited: Adiabatic flux insertion*

Introduction

For a Fermi liquid as defined by the effective Hamiltonian Eq. (F.15), Luttinger theorem holds in the form of Eq. (F.14) [66]. The proof of Luttinger theorem for spinless fermions involves the following ingredients. Let Λ be a finite lattice made of $N_\Lambda = L^3/\mathfrak{a}^3$ sites, where $\mathfrak{a}^3$ is the volume of the elementary unit cell of the lattice. Let $\hat{H}$ be the many-body Hamiltonian acting on the Fock space $\mathcal{F}$ of dimension 2^{N_Λ} for identical spinless fermions. We impose periodic boundary conditions and assume that translation invariance holds so that the total momentum $\hat{\boldsymbol{P}}$ is conserved. The eigenvalues of $\hat{\boldsymbol{P}}$ are countable because of the periodic boundary conditions. The total number operator $\hat{N}$ is also conserved by assumption. The partition function

$$Z^{(N_\Lambda)}(\beta,\mu,\lambda) := \mathrm{Tr}_{\mathcal{F}}\,[e^{-\beta\,(\hat{H}-\mu\hat{N})}] \tag{6.160}$$

in the grand-canonical ensemble is the sum of 2^{N_Λ} analytic functions of the inverse temperature β (the Boltzmann constant $k_{\mathrm{B}} = 1$), the chemical potential μ, and all intrinsic coupling constants of $\hat{H}$ that we denote collectively by the symbol λ. For any given N_Λ, the expectation value of the total number operator divided by the volume, i.e.,

$$n^{(N_\Lambda)}(\beta,\mu,\lambda) := (L^3\,\beta)^{-1}\,\partial_\mu \ln Z^{(N_\Lambda)}(\beta,\mu,\lambda), \tag{6.161}$$

is for the same reason an analytic function of β, μ, and λ. At zero temperature, assuming that translation symmetry is not spontaneously broken and that the ground state is non-degenerate, the expectation value of $\hat{N}$ is an integer, i.e., an analytic function of μ and λ taking discrete values. Consequently, for any given N_Λ, $n^{(N_\Lambda)}(\beta=\infty,\cdot,\cdot)$ is constant as a function of μ and λ, while holding $\beta = \infty$. Assume that

$$\bar{n} = n^{(N_\Lambda)}(\beta=\infty,\mu,\lambda), \tag{6.162a}$$

for some given density

$$\bar{n} := \frac{N_{\mathrm{f}}}{N_\Lambda}\,\frac{1}{\mathfrak{a}^3}, \qquad N_{\mathrm{f}} = 1,2,\cdots,N_\Lambda, \tag{6.162b}$$

of spinless fermions, defines implicitly the real value $\mu(\lambda)$. A mathematically rigorous proof of Luttinger theorem is achieved once the existence of the limiting function

$$\bar{n} = \lim_{\substack{N_{\mathfrak{f}}\to\infty \\ N_\Lambda\to\infty}} n^{(N_\Lambda)}(\beta=\infty, \mu(\cdot), \cdot) \tag{6.163}$$

in an open neighborhood of $\lambda = 0$ is proved.

The proof of Eq. (6.163) is a formidable task because interactions distort the shape of the Fermi surface. Consequently, uniform convergence of perturbation theory for the two-point Green function in powers of the two-point Green function for non-interacting fermions breaks down in the thermodynamic limit very much in the same way as uniform convergence of perturbation theory for the two-point function of the $O(3)$ NLσM was shown to break down in section 3.5. As noted by Luttinger [66], the cure to this problem demands a non-perturbative definition of the Fermi surface, which he defines by the location in momentum space at which the dependence on the single-particle momentum $\boldsymbol{p}$ of the occupation number

$$\begin{aligned} n_{\beta=\infty,\mu}(\cdot,\lambda): \mathbb{R}^3 &\longrightarrow [0,1], \\ \boldsymbol{p} &\longmapsto n_{\beta=\infty,\mu}(\boldsymbol{p},\lambda), \end{aligned} \tag{6.164}$$

[i.e., the ground state expectation value of the number operator $\hat{c}^\dagger(\boldsymbol{p})\,\hat{c}(\boldsymbol{p})$ for the spinless fermions] is discontinuous. The strategy is here similar to the one used in section 3.6 when defining non-perturbatively the renormalization point of the $O(3)$ NLσM. If the Fermi surface for interacting fermions exists, this definition allows to estimate its distortion relative to the non-interacting Fermi surface to any given order in perturbation theory. The thermodynamic limit can then be taken order by order in perturbation theory in powers of two-point functions with poles on the (perturbatively) renormalized Fermi surface, so as to prove Luttinger theorem as can be found in Ref. [67].

Although the mathematically rigorous proof of Luttinger theorem is challenging, there is a tautological flavor to it once the existence of the Fermi surface for interacting fermions is proved (or assumed). This suggests that assuming adiabatic continuity between the Fermi surface for non-interacting fermions and the one for interacting fermions should allow to rationalize Luttinger theorem by elementary means. Indeed, it is possible to relate Eq. (F.14) to spectral flows under an adiabatic insertion of magnetic fluxes without invoking perturbation theory if the Fermi surface exists in the thermodynamic limit, following an argument developed by Oshikawa in Ref. [68].

We shall make the following assumptions:

- Identical spinless fermions are confined to a hypercube of volume

$$V = L_1 \times \cdots \times L_d \subset \mathbb{R}^d. \tag{6.165a}$$

The Cartesian basis of $\mathbb{R}^d$ will be denoted $\boldsymbol{e}_\mu$ with $\mu = 1, \cdots, d$.

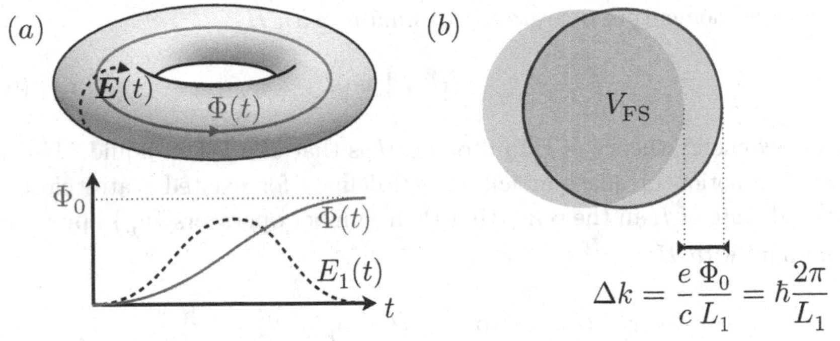

Fig. 6.7 (a) A two-torus T^2 is a surface generated by revolving a circle in three-dimensional space about an axis coplanar with this circle. Topologically, a two-torus T^2 is homeomorphic to the Cartesian product of two circles, $T^2 \sim S^1 \times S^1$. Assign the radius r to the revolving circle. Assign the radius R to the circle traced by the center of the revolving circle. These are the radii of the circles in the homeomorphism $T^2 \sim S^1 \times S^1$. By taking the radius R to infinity holding the radius r fixed, one obtains locally a cylinder whose symmetry axis is the limit of the circle with radius $R \to \infty$. If the circle with radius R is identified with an infinitesimal solenoid in which a magnetic field varies in time, there follows a time-dependent magnetic flux $\Phi(t)$ that generates a time-dependent electric field $\boldsymbol{E}(t)$ tangent to the surface of revolution. (b) Spectral flow of a circular Fermi surface induced by twisting boundary conditions along the $\boldsymbol{e}_1$ direction in Cartesian coordinates.

- Their many-body quantum dynamics is governed by the conserved (i.e., Hermitian) Hamiltonian $\hat{H}$, an operator-valued function of the position operator $\hat{\boldsymbol{r}}_i$ and momentum operator $\hat{\boldsymbol{p}}_i$ obeying the canonical algebra
$$[\hat{r}^{\mu}_{i}, \hat{p}^{\nu}_{j}] = \delta_{ij}\, \delta^{\mu\nu}\, \mathrm{i}\hbar \tag{6.165b}$$
for any pair of fermions labeled by i and j and for any $\mu, \nu = 1, \cdots, d$.
- The total number $\hat{N}$ of spinless fermions is conserved,
$$[\hat{H}, \hat{N}] = 0. \tag{6.165c}$$
- Periodic boundary conditions are assumed, i.e., for any many-body state $|\Psi\rangle$ from the Hilbert space of $\hat{H}$ with a given number of spinless fermions,
$$\hat{T}_{\mu}\, |\Psi\rangle = |\Psi\rangle, \tag{6.165d}$$
where
$$\hat{T}_{\mu} := e^{\mathrm{i}\, L_{\mu}\, \hat{\boldsymbol{P}}\cdot\boldsymbol{e}_{\mu}/\hbar}, \qquad \mu = 1, \cdots, d, \tag{6.165e}$$
is the translation operator across the volume V along the direction $\boldsymbol{e}_{\mu}$ generated by the total momentum operator
$$\hat{\boldsymbol{P}} := \sum_{i} \hat{\boldsymbol{p}}_i. \tag{6.165f}$$
The geometry of position space is thus, in the thermodynamic limit, that of a d-dimensional torus T^d as is illustrated in Fig. 6.7(a) for the two-torus T^2 embedded in $\mathbb{R}^3$.

- The total momentum operator $\hat{\boldsymbol{P}}$ commutes with $\hat{H}$,

$$[\hat{H}, \hat{\boldsymbol{P}}] = 0. \tag{6.165g}$$

- The low-energy theory of Hamiltonian $\hat{H}$ is that of a Fermi liquid. This means that the notion of quasiparticle is well defined for excited states close to the ground state of $\hat{H}$ in the sense that their number operators $\{\hat{\mathsf{n}}_{\boldsymbol{p}}\}$ approximately commute with $\hat{H}$,

$$[\hat{H}, \hat{\mathsf{n}}_{\boldsymbol{p}}] \sim [\hat{H}_{\mathrm{FL}}, \hat{\mathsf{n}}_{\boldsymbol{p}}] = 0, \tag{6.165h}$$

where $\hat{H}_{\mathrm{FL}}$ is defined by [compare with Eq. (F.15)]

$$\hat{H}_{\mathrm{FL}} := \sum_{\boldsymbol{p}} \varepsilon_{\boldsymbol{p}}\, \hat{\mathsf{n}}_{\boldsymbol{p}} + \frac{1}{2L^3} \sum_{\boldsymbol{p}} \sum_{\boldsymbol{p}'} \mathsf{f}_{\boldsymbol{p},\boldsymbol{p}'}\, \hat{\mathsf{n}}_{\boldsymbol{p}}\, \hat{\mathsf{n}}_{\boldsymbol{p}'}. \tag{6.165i}$$

The interpretation of the operator $\hat{\mathsf{n}}_{\boldsymbol{p}}$ is that it measures the occupancy of the quasiparticle state labeled by the single-particle momentum $\boldsymbol{p}$ in a Fermi liquid. Hence, the eigenvalue $\mathsf{n}_{\boldsymbol{p}}$ of $\hat{\mathsf{n}}_{\boldsymbol{p}}$ is either 0 or 1. This operator should not be confused with the number operator $\hat{n}_{\boldsymbol{p}}$ for the original spinless fermions. The latter operator does not commute with $\hat{H}_{\mathrm{FL}}$, whereas Landau postulates that $\hat{\mathsf{n}}_{\boldsymbol{p}}$ does. The many-body eigenstates of $\hat{H}_{\mathrm{FL}}$ take the form $|\cdots, \mathsf{n}_{\boldsymbol{p}}, \cdots\rangle$. Their eigenenergies depend on the phenomenological single-particle dispersion $\varepsilon_{\boldsymbol{p}}$ and on the residual fermion-fermion interaction encoded by the Landau function $\mathsf{f}_{\boldsymbol{p},\boldsymbol{p}'}$. The ground state of $\hat{H}_{\mathrm{FL}}$ is the Fermi sea defined to be the state of the form $|\cdots, \mathsf{n}_{\boldsymbol{p}}, \cdots\rangle$ with the lowest energy eigenvalue. It is characterized by a Fermi surface, a $(d-1)$-dimensional surface in momentum space that bounds a volume, the Fermi sea. In the Fermi sea of $\hat{H}_{\mathrm{FL}}$, $\mathsf{n}_{\boldsymbol{p}}$ takes the value 1 if $\boldsymbol{p}$ belongs to the Fermi sea and zero otherwise.

The assumption that the fermions are spinless is not necessary. It allows to simplify notations.

The notion of Fermi liquid is rooted in the notion of an adiabatic response to switching on a many-body interaction, as explained in appendix F. A Fermi liquid demands the existence of a Fermi surface and of quasiparticles whose lifetimes diverge upon approaching the Fermi surface. If we assume that a Fermi surface exists and that quasiparticles are well defined in its neighborhood, the Fermi surface and the quasiparticles must respond adiabatically to a perturbation. The idea of Oshikawa is to choose a periodic adiabatic perturbation and track the spectral flow undergone by the Fermi surface and the quasiparticles during the period that takes the Fermi liquid back to itself. This periodic and adiabatic perturbation is a one-body perturbation by which a unit of magnetic flux is threaded through the torus. It can be implemented by twisting the boundary conditions as we now explain.

Magnetic flux as twisted boundary conditions

We momentarily work in three-dimensional position space, $d = 3$. We recall that the homogeneous Maxwell equations are

$$\boldsymbol{\nabla} \cdot \boldsymbol{B} = 0, \qquad \boldsymbol{\nabla} \wedge \boldsymbol{E} + \frac{1}{c}\partial_t \boldsymbol{B} = 0, \tag{6.166}$$

while the inhomogeneous Maxwell equations are

$$\boldsymbol{\nabla} \cdot \boldsymbol{E} = 4\pi\rho, \qquad \boldsymbol{\nabla} \wedge \boldsymbol{B} - \frac{1}{c}\partial_t \boldsymbol{E} = \frac{4\pi}{c}\boldsymbol{j}, \tag{6.167}$$

Gaussian units are used.

We consider a cylinder $\Omega \subset \mathbb{R}^3$ in position space with the radius $L_1/(2\pi)$, height L_2, and symmetry axis parallel to $\boldsymbol{e}_3$,

$$\Omega := \left\{ \boldsymbol{r} = \frac{L_1}{2\pi}\cos\varphi\, \boldsymbol{e}_1 + \frac{L_1}{2\pi}\sin\varphi\, \boldsymbol{e}_2 + r_3\, \boldsymbol{e}_3 \in \mathbb{R}^3 \,\middle|\, 0 \le \varphi < 2\pi, \quad 0 \le r_3 \le L_2 \right\}. \tag{6.168}$$

Here, we are using the cylindrical coordinates

$$\boldsymbol{r} = \rho\, \cos\varphi\; \boldsymbol{e}_1 + \rho\, \sin\varphi\; \boldsymbol{e}_2 + r_3\; \boldsymbol{e}_3, \tag{6.169}$$

where

$$\rho := +\sqrt{r_1^2 + r_2^2}, \qquad \varphi := \arctan\frac{r_2}{r_1}, \tag{6.170}$$

and the orthonormal cylindrical basis vectors at $\boldsymbol{r}$ are given by

$$\boldsymbol{e}_\rho = +\cos\varphi\; \boldsymbol{e}_1 + \sin\varphi\; \boldsymbol{e}_2, \qquad \boldsymbol{e}_\varphi = -\sin\varphi\; \boldsymbol{e}_1 + \cos\varphi\; \boldsymbol{e}_2, \qquad \boldsymbol{e}_3. \tag{6.171}$$

We are going to construct time-dependent electric and magnetic fields such that the electric field is pointing along the direction $\boldsymbol{e}_\varphi$ and is constant in magnitude on the surface of the cylinder, whereas the magnetic field vanishes everywhere in space except along the symmetry axis of the cylinder, where it is singular.

Exercise 1.1:

(a) Show that the homogeneous Maxwell equations are satisfied if the magnetic and electric fields are expressed according to

$$\boldsymbol{B} = \boldsymbol{\nabla} \wedge \boldsymbol{A}, \qquad \boldsymbol{E} = -\boldsymbol{\nabla} \cdot A^0 - \frac{1}{c}\partial_t \boldsymbol{A}, \tag{6.172}$$

in terms of the vector $\boldsymbol{A}$ and scalar A^0 potentials.

(b) Show that

$$\Phi_0 := \frac{h\,c}{e}, \qquad h = 2\pi\,\hbar, \tag{6.173}$$

where $-e < 0$ is the electron charge has the units of a magnetic flux.

(c) What are the units of the vector potential expressed in terms of units of length and magnetic flux?

(d) Assume the time-dependent magnetic and electric fields

$$\boldsymbol{B}(\boldsymbol{r},t) = -\frac{\Phi_0}{2\pi\,\rho}\,\phi(t)\,\delta(\rho)\;\boldsymbol{e}_3, \qquad \boldsymbol{E}(\boldsymbol{r},t) = +\frac{\Phi_0}{2\pi\,c\,\rho}\,(\partial_t\phi)(t)\;\boldsymbol{e}_{\varphi}, \tag{6.174}$$

where $\phi:\mathbb{R}\to\mathbb{R}, t\mapsto\phi(t)$ is some dimensionless real-valued function of time. Show that these magnetic and electric fields follow from choosing

$$A^0 = 0, \qquad \boldsymbol{A}(\boldsymbol{r},t) = -\frac{\Phi_0}{2\pi\,\rho}\,\phi(t)\,\boldsymbol{e}_{\varphi}. \tag{6.175}$$

We now return to the case of d-dimensional position space with the assumptions (6.165). We modify the many-body Hamiltonian $\hat{H}$ as follows. Motivated by Eq. (6.175), we couple the vector gauge field

$$\boldsymbol{A}(\phi) := -\frac{\Phi_0}{L_1}\,\phi\;\boldsymbol{e}_1 = -2\pi\,\frac{(\hbar c/e)}{L_1}\,\phi\;\boldsymbol{e}_1, \tag{6.176}$$

where ϕ is a dimensionless number, by the minimal coupling, i.e., through the substitution

$$\hat{\boldsymbol{p}}_i \to \hat{\boldsymbol{p}}_i - \frac{(-e)}{c}\,\boldsymbol{A}(\phi), \tag{6.177}$$

to all electrons (with the label i) carrying the electric charge $-e<0$. The resulting many-body Hamiltonian is denoted $\hat{H}(\phi)$. It acts on a Hilbert space with given number of spinless fermions.

Exercise 1.2:

(a) Verify that, as was the case with Eq. (6.165d), any many-body state $|\Psi\rangle$ from the Hilbert space with given number of spinless fermions of $\hat{H}(\phi)$ obeys the periodic boundary conditions

$$\hat{T}_{\mu}\,|\Psi\rangle = |\Psi\rangle, \tag{6.178}$$

with $\hat{T}_{\mu}$ defined by Eq. (6.165e) for $\mu = 1,\cdots,d$.

(b) Verify that, as was the case with Eq. (6.165g),

$$[\hat{H}(\phi), \hat{\boldsymbol{P}}] = 0. \tag{6.179}$$

(c) How does $\hat{H}(\phi)$ change under the local gauge transformation

$$\begin{aligned}|\Psi\rangle &=: e^{-\mathrm{i}\frac{e}{\hbar c}\,\boldsymbol{A}(\phi)\cdot\sum_i\hat{\boldsymbol{r}}_i}\;|\Theta(\phi)\rangle\\ &= e^{+\mathrm{i}\frac{2\pi\,\phi}{L_1}\,\sum_i\hat{\boldsymbol{r}}_i\cdot\boldsymbol{e}_1}\;|\Theta(\phi)\rangle?\end{aligned} \tag{6.180}$$

(d) What twisted boundary conditions obeys the transformed state $|\Theta(\phi)\rangle$?

The local gauge transformation (6.180) is called a large gauge transformation, for it changes the boundary conditions. We have shown that we can trade the one-body coupling between the fermions and the vector potential (6.176) for twisted boundary conditions.

Spectral flows

Define the vector from $\mathbb{R}^d$

$$\boldsymbol{\phi} = \phi_\mu\, \boldsymbol{e}_\mu, \tag{6.181}$$

and the vector potential

$$\boldsymbol{A}(\boldsymbol{\phi}) := -\frac{\Phi_0}{L_\mu}\, \phi_\mu\, \boldsymbol{e}_\mu = -2\pi\, \frac{(\hbar c/e)}{L_\mu}\, \phi_\mu\, \boldsymbol{e}_\mu, \tag{6.182}$$

with the summation convention implied on the repeated indices $\mu = 1, \cdots, d$. Define $\hat{H}(\boldsymbol{\phi})$ through the minimal coupling

$$\hat{\boldsymbol{p}}_i \to \hat{\boldsymbol{p}}_i - \frac{(-e)}{c}\, \boldsymbol{A}(\boldsymbol{\phi}). \tag{6.183}$$

Any many-body wave function from the Hilbert space with given number of spinless fermions of $\hat{H}(\boldsymbol{\phi})$ obeys periodic boundary conditions. Define the local gauge transformation

$$\hat{U}(\boldsymbol{\phi}) := e^{-\mathrm{i}\frac{e}{\hbar c}\boldsymbol{A}(\boldsymbol{\phi})\cdot\sum_i \hat{\boldsymbol{r}}_i} = e^{+\mathrm{i}\frac{2\pi\,\phi_\mu}{L_\mu}\sum_i \hat{\boldsymbol{r}}_i\cdot\boldsymbol{e}_\mu}, \tag{6.184a}$$

with the summation convention implied on the repeated indices $\mu = 1, \cdots, d$. We can then rewrite Eq. (6.180) as

$$|\Psi\rangle = \hat{U}(\boldsymbol{\phi})\, |\Theta(\boldsymbol{\phi})\rangle. \tag{6.184b}$$

Exercise 2.1:

(a) Show that the transformation law of the operator $\hat{O}$ in the $|\Psi\rangle$ basis under the unitary transformation (6.184) is

$$\hat{O} \to \hat{U}^\dagger(\boldsymbol{\phi})\, \hat{O}\, \hat{U}(\boldsymbol{\phi}). \tag{6.185}$$

(b) With the help of Eq. (6.165e) and the Baker-Campbell-Hausdorff formula, show that

$$\hat{U}^\dagger(\boldsymbol{\phi})\, \hat{T}_\mu\, \hat{U}(\boldsymbol{\phi}) = e^{+\mathrm{i}\,2\pi\,\phi_\mu\,\hat{N}}\, \hat{T}_\mu, \qquad \mu = 1, \cdots, d, \tag{6.186}$$

where $\hat{N}$ is the total number operator for the spinless fermions.

(c) Show that

$$T_\mu\, |\Theta(\boldsymbol{\phi})\rangle = e^{-\mathrm{i}\,2\pi\,\phi_\mu\,\hat{N}}\, |\Theta(\boldsymbol{\phi})\rangle, \qquad \mu = 1, \cdots, d. \tag{6.187}$$

Compare this result with the one obtained in exercise 1.2(d).

(d) Show that

$$\hat{U}^\dagger(\boldsymbol{\phi}')\, \hat{H}(\boldsymbol{\phi} + \boldsymbol{\phi}')\, \hat{U}(\boldsymbol{\phi}') = \hat{H}(\boldsymbol{\phi}) \tag{6.188}$$

for any $\boldsymbol{\phi}$ and $\boldsymbol{\phi}'$ of the form (6.181).

Exercise 2.2: Let $|\Theta_0(\boldsymbol{\phi})\rangle$ be the ground state of $\hat{H}$ obeying the twisted boundary condition implied by Eq. (6.186). By assumption, this ground state is non-degenerate. Hence, it can be chosen to be an eigenstate of the total momentum operator $\hat{\boldsymbol{P}}$ with the eigenvalue $\boldsymbol{P}_0(\boldsymbol{\phi})$. Assume adiabatic continuity and show the spectral flow

$$\boldsymbol{P}_0(\boldsymbol{\phi}+\boldsymbol{\phi}') - \boldsymbol{P}_0(\boldsymbol{\phi}) = -2\pi\hbar\, N_{\mathrm{f}}\, \frac{\phi'_\mu}{L_\mu}\, \boldsymbol{e}_\mu, \tag{6.189}$$

with the summation convention implied on the repeated indices $\mu = 1, \cdots, d$ and where N_{f}, the number of spinless fermions in the Hilbert space of $\hat{H}$, was defined in Eq. (6.162b). *Hint:* Relate $|\Theta_0(\boldsymbol{\phi}+\boldsymbol{\phi}')\rangle$ to $|\Theta_0(\boldsymbol{\phi})\rangle$ with the help of Eq. (6.188). For any position vector $\boldsymbol{r}$, apply the translation operator $\hat{T}(\boldsymbol{r}) := e^{+\mathrm{i}\hat{\boldsymbol{P}}\cdot\boldsymbol{r}/\hbar}$ to this relation and make use of the proper generalization of Eq. (6.186).

Spectral flow of the Fermi sea

The relation (6.189) makes no reference to a Fermi sea. It applies to the non-degenerate ground state of any Hamiltonian that commutes with the total number operator and the total momentum operator, and for which adiabatic continuity with respect to twisting boundary conditions holds. The number of spinless electrons N_{f} is fixed by the choice of the filling fraction of the lattice Λ made of N_Λ unit cells in position space.

To derive Luttinger theorem as stated by Eq. (F.14), we are going to compute the change in the total momentum of the Fermi sea of a Fermi liquid under the assumption of adiabatic continuity under the parametric change of the single-particle momenta given by

$$\hat{\boldsymbol{p}}_i \to \hat{\boldsymbol{p}}_i - 2\pi\hbar\, \frac{\phi_\mu}{L_\mu}\, \boldsymbol{e}_\mu, \tag{6.190}$$

as follows from Eqs. (6.183) and (6.182). [The summation convention is implied on the repeated indices $\mu = 1, \cdots, d$.] Let $\boldsymbol{P}_{\mathrm{FS}}(\boldsymbol{\phi})$ denote the total momentum of the Fermi sea, with the latter *defined* to be an eigenstate of all the *quasiparticle* number operators $\hat{\mathsf{n}}_{\boldsymbol{p}}$, see Eqs. (6.165h) and (6.165i), with the eigenvalues 1 for N_{FS} single-particle momenta and 0 for all $N_\Lambda - N_{\mathrm{FS}}$ remaining single-particle momenta. Which of the single-particle momenta are occupied may change as the interaction is changed or as the adiabatic parameter $\boldsymbol{\phi}$ is changed, but not N_{FS}.

Exercise 3.1:

(a) Find the transformation on the number operators in Eq. (6.165i) that brings $\hat{H}_{\mathrm{FL}}$ to the form given in Eq. (F.15).
(b) Show that

$$\boldsymbol{P}_{\mathrm{FS}}(\boldsymbol{\phi}+\boldsymbol{\phi}') - \boldsymbol{P}_{\mathrm{FS}}(\boldsymbol{\phi}) = -2\pi\hbar\, N_{\mathrm{FS}}\, \frac{\phi'_\mu}{L_\mu}\, \boldsymbol{e}_\mu. \tag{6.191}$$

The summation convention is implied on the repeated indices $\mu = 1, \cdots, d$.

(c) After comparing Eq. (6.189) to Eq. (6.191), deduce that

$$\frac{N_{\mathrm{f}}}{L_\mu} = \frac{N_{\mathrm{FS}}}{L_\mu} + n_\mu, \qquad \mu = 1, \cdots, d, \tag{6.192}$$

where $n_1, \cdots, n_d$ are integers.

(d) What is the origin of the integers n_μ with $\mu = 1, \cdots, d$?

Thermodynamic limit

We assume that the thermodynamic limit is well defined, i.e., that the limit $N_\Lambda, N_{\mathrm{f}}, N_{\mathrm{FS}} \to \infty$, holding the ratios N_{f}/N_Λ and $N_{\mathrm{FS}}/N_\Lambda$ fixed, exists and is unique.

Exercise 4.1:

(a) To simplify notation, we set the unit of length $\mathfrak{a}$ and hence the volume of the unit cell $\mathfrak{a}^d$ to unity. If so, the macroscopic side lengths $L_1, \cdots, L_d$ are integers. Take advantage of the existence of the thermodynamic limit by choosing $L_1, \cdots, L_d$ to be pairwise mutually prime positive integers (two integers are mutually prime if the only positive integer that evenly divides both of them is 1). Show that

$$N_{\mathrm{f}} = N_{\mathrm{FS}} + n \prod_{\mu=1}^{d} L_\mu \iff \bar{n} = \frac{V_{\mathrm{FS}}}{(2\pi)^d} + n, \tag{6.193a}$$

with some integer n, where the density $\bar{n}$ was defined in Eq. (6.162b) and

$$\frac{V_{\mathrm{FS}}}{(2\pi)^d} := \frac{N_{\mathrm{FS}}}{L_1 \times \cdots \times L_d}, \tag{6.193b}$$

where V_{FS} is the volume of the Fermi sea in d-dimensional momentum space. We have recovered Luttinger theorem, as stated by Eq. (F.14), for a Fermi liquid.

(b) What is the interpretation of the integer n on the right-hand side of Eq. (6.193a)?

6.10.3 *Fermionic slave particles*

In strongly correlated physics, one is often confronted with quantum Hamiltonians expressed in terms of local operators that obey an algebra that is different from that obeyed by bosons or fermions. Wick theorem (or Leibniz theorem of differentiation within the path integral formalism) does not apply for such operators, i.e., perturbation theory is very complicated. The simplest physical example of local operators whose algebra differs from the ones bosons or fermions obey is that of quantum spin-S operators. To circumvent this difficulty, we used Holstein-Primakoff bosons in section 2.6.1 to represent the spin operators in quantum Heisenberg magnets. However, there is no unique choice of auxiliary operators to represent spin operators.

The choice made to represent the spin operators is of no consequences as long as no approximation is performed. However, different choices can deliver qualitatively different predictions as soon as approximations are made. The mean-field approximations based on the Holstein-Primakoff bosons representation of quantum spins become exact in the classical limit $S \to \infty$ for the quantum spin number. It has been used successfully to describe collinear magnetic long-range order. It is doubtful that an approximation based on Holstein-Primakoff bosons would be useful to describe a putative magnetic ground state without magnetic long-range order (a so-called spin liquid) in the extreme quantum limit by which the quantum spin number $S = 1/2$ and the lattice has a low coordination number or is geometrically frustrated.

The "slave-particle" method was introduced in condensed matter physics by Read [69] and Coleman [70] in order to study a local quantum spin-1/2 immersed in a sea of conduction electrons (Kondo problem). Later the method was used to treat the large-U Hubbard model in Refs. [71] and [72], which is believed to describe the physics of high-temperature superconductors and has become the cornerstone of the so-called Resonating-Valence-Bond (RVB) approach to high-temperature superconductors.

Both with the Kondo model or with the large-U Hubbard model, it is possible to find a representation of the Hamiltonian purely in terms of bosons or fermions, in which case the applicability of Wick theorem is restored. A price must however be paid as the underlying Hilbert space for the slave particles has been enlarged by the introduction of unphysical degrees of freedom, namely gauge degrees of freedom. In effect, the original problem is traded for a problem in lattice gauge theory at infinite bare gauge coupling.

Here, we are going to present the fermionic "slave-particle" method for the problem of two quantum spin-1/2 particles interacting through the Heisenberg exchange interaction. The following exercises go step by step through the exact calculation of the partition function using a representation of the spin-1/2 algebra in terms of slave-fermions. First, the spin problem is mapped onto a fermion problem. Second, a fermionic path integral representation of the partition function is derived. Third, this path integral is explicitly computed using a diagrammatic method.

Slave-fermion representation of the Heisenberg exchange interaction

Let $\hat{\boldsymbol{S}}_1$ and $\hat{\boldsymbol{S}}_2$ be *two* spin-1/2 operators satisfying the commutation relations ($\hbar = 1$)

$$\left[\hat{S}^a_i, \hat{S}^b_j\right] = \mathrm{i}\,\delta_{ij}\,\epsilon^{abc}\,\hat{S}^c_j, \qquad a,b,c = x,y,z \equiv 1,2,3, \qquad i,j = 1,2. \tag{6.194a}$$

The quantum dynamics for these two spins is governed by the Heisenberg exchange Hamiltonian

$$\hat{H} = \frac{J}{2}\,\hat{\boldsymbol{S}}_1 \cdot \hat{\boldsymbol{S}}_2. \tag{6.194b}$$

Exercise 1.1:

(a) Show that the Hilbert space $\mathcal{H}$ on which $\hat{H}$ is defined is four-dimensional and decomposes (irreducibly) into the singlet and triplet sectors.

(b) Show that the partition function

$$Z := \mathrm{Tr}_{\mathcal{H}}\left(e^{-\beta\hat{H}}\right) \tag{6.195a}$$

can be written as

$$Z = e^{+\beta\frac{3J}{8}} + 3\,e^{-\beta\frac{J}{8}}, \tag{6.195b}$$

where β is the inverse temperature in units for which the Boltzmann constant is unity.

In order to map this spin problem into a fermion problem, we use the following fermion representation of the spins ($\hbar = 1$)[5]

$$\boldsymbol{S}_i \to \frac{1}{2}\,\hat{c}^{\dagger}_{i\alpha}\,\boldsymbol{\sigma}_{\alpha\beta}\,\hat{c}_{i\beta}, \qquad i = 1, 2. \tag{6.196}$$

The summation convention over repeated labels is assumed throughout this section from now on. Here, the $\hat{c}$'s are operators labeled by a site (i) and spin (α) index. They satisfy the anticommuting relations

$$\{\hat{c}_{i\alpha}, \hat{c}^{\dagger}_{j\beta}\} = \delta_{ij}\,\delta_{\alpha\beta}, \qquad \{\hat{c}_{i\alpha}, \hat{c}_{j\beta}\} = \{\hat{c}^{\dagger}_{i\alpha}, \hat{c}^{\dagger}_{j\beta}\} = 0. \tag{6.197}$$

The three Pauli matrices are denoted by the vector $\boldsymbol{\sigma}$. Equation (6.196) must be handled with care, for the Hilbert space of the spins $\mathcal{H}$ and the Fock space $\mathcal{F}$ spanned by the c-operators do not share the same dimension.

Exercise 1.2:

(a) Show that the Hilbert space $\mathcal{H}$ and the Fock space $\mathcal{F}$ do not have the same dimension.

(b) Construct explicitly an isomorphism between $\mathcal{H}$ and the restricted Fock space

$$\mathcal{F}_{\mathrm{phys}} := \left\{ |\psi\rangle \in \mathcal{F} \,\middle|\, \sum_{\alpha=1,2} \hat{c}^{\dagger}_{i\alpha}\,\hat{c}_{i\alpha}\,|\psi\rangle = |\psi\rangle\,, \qquad i = 1, 2 \right\}. \tag{6.198}$$

(c) Verify that the right-hand side of Eq. (6.196) reproduces the angular momentum algebra for spin $s = 1/2$ in the restricted Fock space $\mathcal{F}_{\mathrm{phys}}$.

As a side note, the restriction to $\mathcal{F}_{\mathrm{phys}}$ implies that each site $i = 1, 2$ in the fermion representation is occupied by a *single* fermion with either up or down spin.

Exercise 1.3:

(a) Prove the useful identity

$$\boldsymbol{\sigma}_{ab} \cdot \boldsymbol{\sigma}_{cd} = 2\delta_{ad}\,\delta_{bc} - \delta_{ab}\,\delta_{cd}, \tag{6.199}$$

and show that, in the fermion representation, the Heisenberg exchange Hamiltonian (6.194b) becomes the *quartic* fermion interaction

$$\hat{H}_{\mathrm{f}} = -\frac{J}{4}\,\hat{c}^{\dagger}_{1\alpha}\,\hat{c}^{\dagger}_{2\beta}\,\hat{c}_{1\beta}\,\hat{c}_{2\alpha} - \frac{J}{8}\,\hat{c}^{\dagger}_{1\alpha}\hat{c}_{1\alpha}\hat{c}^{\dagger}_{2\beta}\hat{c}_{2\beta}. \tag{6.200}$$

[5] Repeated indices are to be summed over.

(b) Show that

$$\hat{H}'_{\mathrm{f}} := -\frac{J}{4}\,\hat{c}^{\dagger}_{1\alpha}\,\hat{c}^{\dagger}_{2\beta}\,\hat{c}_{1\beta}\,\hat{c}_{2\alpha} \tag{6.201}$$

commutes with the local fermionic density

$$\hat{n}_i := \hat{c}^{\dagger}_{i\alpha}\,\hat{c}_{i\alpha}. \tag{6.202}$$

Infer from this observation that the partition function (6.195) for the two interacting spins is proportional to

$$Z^{\mathrm{phys}}_{\mathrm{f}} := \mathrm{Tr}_{\mathcal{F}_{\mathrm{phys}}}\left(e^{-\beta\,\hat{H}'_{\mathrm{f}}}\right), \tag{6.203}$$

the proportionality factor being $\exp\left(+\beta\frac{J}{8}\right)$.

Grassmann-path-integral representation

In order to compute explicitly $Z^{\mathrm{phys}}_{\mathrm{f}}$ with the use of a Grassmann path integral, we must replace the trace over the physical subspace by the trace over the entire Fock space. This can be done by integration over two local Lagrange multipliers φ_i, for each site $i = 1, 2$.

Exercise 2.1:

(a) Show that

$$Z^{\mathrm{phys}}_{\mathrm{f}} = \int \mathcal{D}\mu_{\varphi}\;\mathrm{Tr}_{\mathcal{F}}\left(e^{-\beta\,\hat{H}^{\mathrm{phys}}_{\mathrm{f}}}\right), \tag{6.204a}$$

where

$$\hat{H}^{\mathrm{phys}}_{\mathrm{f}} := \hat{H}'_{\mathrm{f}} + \mathrm{i}\sum_{i=1}^{2}\varphi_i\,\hat{n}_i, \tag{6.204b}$$

and the integration measure $\mathcal{D}\mu_{\varphi}$ is given by (Gutzwiller projection)

$$\mathcal{D}\mu_{\varphi} := \prod_{i=1}^{2}\frac{\beta}{2\pi}\,\mathrm{d}\varphi_i\,e^{+\mathrm{i}\beta\,\varphi_i}. \tag{6.204c}$$

(b) What is the range of integration for the Lagrange multipliers? Why?

Exercise 2.2: Now that the trace is over the entire Fock space and that the Hamiltonian is normal-ordered, we can construct the Grassmann path integral following sections E.1 and E.2.

(a) By use of the completeness relation for the coherent states $|\eta\rangle$

$$\int\left(\prod_{i\in I}\mathrm{d}\eta^{*}_i\,\mathrm{d}\eta_i\right)\,e^{-\sum\limits_{j\in I}\eta^{*}_j\,\eta_j}\;|\eta\rangle\,\langle\eta| = \mathbb{1}, \tag{6.205}$$

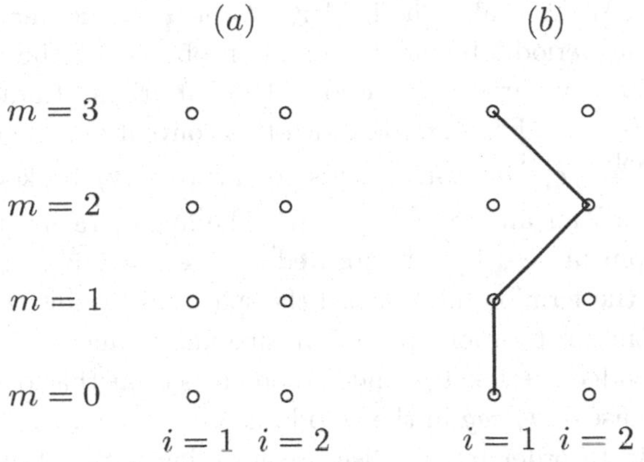

Fig. 6.8 (a) A ladder with $M = 3$ rungs representing position space and imaginary time. (b) A possible world line of a fermion with spin up initially located on site $i = 1$.

where the set I runs over the two sites $i = 1, 2$ and the spin index $\alpha =\uparrow,\downarrow\equiv 1, 2$, show that

$$\begin{aligned}\mathrm{Tr}_{\mathcal{F}}\left(e^{-\beta\,\hat{H}_{\mathrm{f}}^{\mathrm{phys}}}\right) =& \lim_{\substack{M\to\infty\\ \epsilon\to 0}} \int \mathcal{D}\mu_{\eta^*\eta} \left(\prod_{\substack{m,n=1\\ i,j,\alpha=1,2}}^{M} e^{-\eta^*_{im\alpha}\, A^{(\alpha)}_{\substack{im\\ jn}}\, \eta_{jn\alpha}} \right)\\ &\times \left(\prod_{\substack{m=1\\ \alpha,\beta=1,2}}^{M} e^{+\frac{J\epsilon}{4}\,\eta^*_{1m\alpha}\,\eta^*_{2m\beta}\,\eta_{1(m-1)\beta}\,\eta_{2(m-1)\alpha}} \right),\end{aligned} \tag{6.206a}$$

where $M\epsilon = \beta$ is kept fixed, and the Grassmann valued fields $\eta^*_{im\alpha}$ and $\eta_{im\alpha}$ obey antiperiodic boundary conditions in the imaginary-time direction. Here, we have introduced the notation

$$\begin{aligned}\mathcal{D}\mu_{\eta^*\eta} \equiv& \prod_{m=1}^{M} \mathrm{d}\eta^*_m\, \mathrm{d}\eta_m \equiv \prod_{i=1,2}\prod_{m=1}^{M} \mathrm{d}\eta^*_{im}\, \mathrm{d}\eta_{im} \equiv \prod_{i=1,2}\prod_{m=1}^{M}\prod_{\alpha=\uparrow,\downarrow} \mathrm{d}\eta^*_{im\alpha}\, \mathrm{d}\eta_{im\alpha},\\ A^{(\alpha)}_{\substack{im\\ jn}} =& \delta_{ij}\left[\delta_{mn} - \delta_{(m-1)n}\left(1 - \mathrm{i}\epsilon\varphi_j\right)\right], \quad 1 < m \leq M,\ 1 \leq n \leq M,\\ A^{(\alpha)}_{\substack{i1\\ jn}} =& \delta_{ij}\left[\delta_{1n} - (-1)\delta_{Mn}(1 - \mathrm{i}\epsilon\varphi_j)\right], \quad 1 \leq n \leq M.\end{aligned} \tag{6.206b}$$

The representation (6.206) of the partition function $\mathrm{Tr}_{\mathcal{F}}\left(e^{-\beta\,\hat{H}_{\mathrm{f}}^{\mathrm{phys}}}\right)$ can be visualized as follows (see Fig. 6.8). One can think of a ladder whose rungs are labeled by the index m running from 1 to M. A vertex of the ladder corresponds to a point (i, m) in position space and imaginary time with the spatial coordinate taking two possible values $i = 1, 2$ and the imaginary-time coordinate taking the M

values $m=1,\cdots,M$. Actually, the ladder is *toroidal* in the imaginary-time direction due to the antiperiodic boundary condition obeyed by the Grassmann fields, as it should be for any partition function. The left (right) frame of the ladder is labeled by $i=1$ ($i=2$). For example, with this convention, the expectation value $\left\langle 0\left|\hat{c}_{1\uparrow}\exp(-\beta\,\hat{H}^{\mathrm{phys}}_{\mathrm{f}})\,\hat{c}^{\dagger}_{1\uparrow}\right|0\right\rangle$ corresponds to a (possibly) broken line that obeys $(1,m+M)\equiv(1,m)$ for any $m=1,\cdots,M$. The line or rather the *world line* for a fermion with spin up and initially located on site $i=1$ is constructed with the following rule. If the fermion has reached the space-time point (i,m), $1\leq m\leq M$, then the next point in position space and imaginary time is either $(i,m+1)$ or $(i+1,m+1)$ provided $2+1\equiv 1$ is understood. It is clear that only the Heisenberg interaction can cause a *zig zag* in the world line of a fermion.

Exercise 2.3: In order to establish a connection with a lattice gauge theory, we now introduce an additional auxiliary complex-valued field, $Q_{\substack{2m\\1(m-1)}}$, for each *diagonal* link between the sites (i,m) and $(i+1,m+1)$ of the ladder. This is achieved by a Hubbard-Stratonovich transformation on the quartic contribution of $\hat{H}^{\mathrm{phys}}_{\mathrm{f}}$. Recall that the identity

$$\int\frac{\mathrm{d}q^{*}\mathrm{d}q}{2\pi\mathrm{i}}\,e^{-q^{*}q+\sqrt{A}w^{*}q+\sqrt{A}zq^{*}}=\int\frac{\mathrm{d}q^{*}\mathrm{d}q}{2\pi\mathrm{i}}\,e^{-(q-\sqrt{A}w)^{*}(q-\sqrt{A}z)+w^{*}Az}\tag{6.207}$$

holds for any positive number A and for any pair of complex numbers z and w with their complex conjugate denoted by z^{*} and w^{*}, respectively. Here, $\mathrm{d}q^{*}\mathrm{d}q/(2\pi\mathrm{i})$ is nothing but the usual (Riemann) infinitesimal integration area in the complex q plane. If we choose $z=w$, we can do the shift $q\to q+\sqrt{A}z$ of the integration variable q and then perform the integration over the complex plane. Because this integration is over a normalized Gaussian, the multiplicative integration constant is unity and one finds the desired identity

$$e^{z^{*}A\,z}=\int\frac{\mathrm{d}q^{*}\mathrm{d}q}{2\pi\mathrm{i}}\,e^{-q^{*}q+\sqrt{A}z^{*}q+\sqrt{A}zq^{*}}.\tag{6.208}$$

We now decree that the identity

$$\begin{aligned}\exp\left(\frac{J\epsilon}{4}\,\eta^{*}_{1m\alpha}\,\eta_{2(m-1)\alpha}\,\eta^{*}_{2m\beta}\,\eta_{1(m-1)\beta}\right)=&\,\frac{1}{2\pi\mathrm{i}}\int\mathrm{d}Q^{*}_{\substack{2m\\1(m-1)}}\,\mathrm{d}Q_{\substack{2m\\1(m-1)}}\\&\times\exp\left(-\left|Q_{\substack{2m\\1(m-1)}}\right|^{2}+\sqrt{\frac{J\epsilon}{4}}\,\eta^{*}_{1m\alpha}\,\eta_{2(m-1)\alpha}\,Q_{\substack{2m\\1(m-1)}}\right.\\&\left.+\sqrt{\frac{J\epsilon}{4}}\,\eta^{*}_{2m\beta}\,\eta_{1(m-1)\beta}\,Q^{*}_{\substack{2m\\1(m-1)}}\right)\end{aligned}\tag{6.209}$$

holds. Here, $Q_{\substack{2m\\1(m-1)}}$ is treated as if it were a complex number (the auxiliary Hubbard-Stratonovich field). The justification of Eq. (6.209) will be that we can reproduce the partition functions (6.195b) and (6.203) using this identity.

(a) Show that there is an ambiguity in decoupling the quartic interaction in that the decoupling is not unique. For the choice we have taken in Eq. (6.209), the auxiliary field $Q_{\substack{2m\\1(m-1)}}$ is called the Affleck-Marston order parameter.

(b) Show that the final version of the Grassmann path integral for the partition function is

$$Z_{\mathrm{f}}^{\mathrm{phys}} = \int_0^{2\pi/\beta} \mathcal{D}\mu_{\varphi} \lim_{\substack{M\to\infty\\ \epsilon\to 0}} \int_0^{\infty}\int_0^{2\pi} \mathcal{D}\mu_{Q^*Q} \int \mathcal{D}\mu_{\eta^*\eta} \prod_{\alpha=\uparrow,\downarrow} \left[\prod_{\substack{i,j=1,2\\ m,n=1}}^{M} e^{-\eta^*_{im\alpha} \tilde{A}^{(\alpha)}_{\substack{im\\jn}} \eta_{jn\alpha}} \right], \tag{6.210a}$$

where

$$\mathcal{D}\mu_{\varphi} = \prod_{i=1}^{2} \frac{\mathrm{d}\varphi_i}{(2\pi/\beta)}\, e^{+\mathrm{i}\beta\,\varphi_i} \tag{6.210b}$$

is the Riemann measure of the Lagrange multiplier,

$$\mathcal{D}\mu_{Q^*Q} = \prod_{m=1}^{M} \mathrm{d}K^2_{\substack{2m\\1(m-1)}} \exp\left(-K^2_{\substack{2m\\1(m-1)}}\right) \frac{1}{2\pi}\, \mathrm{d}\phi_{\substack{2m\\1(m-1)}} \tag{6.210c}$$

is the Riemann measure of the Hubbard-Stratonovich field,

$$\begin{aligned} Q_{\substack{2m\\1(m-1)}} &\equiv K_{\substack{2m\\1(m-1)}} \exp\left(-\mathrm{i}\phi_{\substack{2m\\1(m-1)}}\right) \\ &\equiv K_{\substack{2m\\1(m-1)}} \exp\left(+\mathrm{i}\phi_{\substack{1m\\2(m-1)}}\right) \\ &\equiv Q^*_{\substack{1m\\2(m-1)}} \end{aligned} \tag{6.210d}$$

is the polar decomposition of the Affleck-Marston auxiliary field Q (there is an amplitude $0 \leq K < \infty$ and a phase $0 \leq \phi < 2\pi$),

$$\mathcal{D}\mu_{\eta^*\eta} = \prod_{i=1}^{2} \prod_{m=1}^{M} \prod_{\alpha=\uparrow,\downarrow} \mathrm{d}\eta^*_{im\alpha}\, \mathrm{d}\eta_{im\alpha} \tag{6.210e}$$

is the Grassmann measure, and

$$\begin{aligned} \tilde{A}^{(\alpha)}_{\substack{im\\jn}} &= \delta_{ij}\left[\delta_{mn} - (1 - \mathrm{i}\epsilon\varphi_j)\,\delta_{(m-1)n}\right] - \sqrt{\frac{J\epsilon}{4}}\, Q^*_{\substack{im\\j(m-1)}}\, \delta_{(m-1)n}, \\ \tilde{A}^{(\alpha)}_{\substack{i1\\jn}} &= \delta_{ij}\left[\delta_{1n} - (-1)\,(1 - \mathrm{i}\epsilon\varphi_j)\,\delta_{Mn}\right] - (-1)\sqrt{\frac{J\epsilon}{4}}\, Q^*_{\substack{i1\\jM}}\, \delta_{Mn}. \end{aligned} \tag{6.210f}$$

To shorten the notation, it is understood that $Q^*_{\substack{im\\j(m-1)}}$ vanishes when $i = j$.

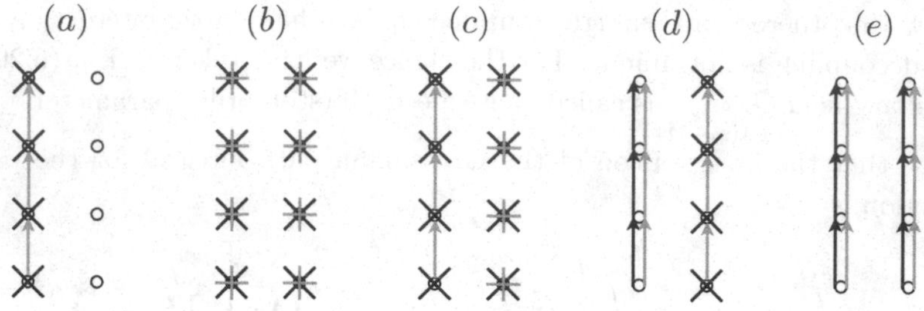

Fig. 6.9 (a) Example of a diagram which is not saturated. (b) Saturated diagram in the sector of total occupation number 0. (c) Example of a saturated diagram in the sector of total occupation number 1. (d) Example of a saturated diagram in the sector of total occupation number 3. (e) Example of a saturated diagram in the sector of total occupation number 4.

Diagrammatic interpretation of the Grassmann path integral

Exercise 3.1: The integrand of Eq. (6.210) is a polynomial of degree $8M$ in the η's and η^*'s with monomials weighted by the phase φ and complex number Q. Only the coefficient of $\prod_{im\alpha}\eta_{im\alpha}\,\eta^*_{im\alpha}$, a monomial of degree $8M$ in the η's and η^*'s, contributes to the integral over the Grassmann variables. This coefficient does not depend on the phase of the Affleck-Marston order parameter and is proportional to the factor $\exp\Big(-\mathrm{i}\beta(\varphi_1+\varphi_2)\Big)$. In the following we use a diagrammatic method to evaluate this coefficient. Verify that the integrand in Eq. (6.210) can be rewritten as the product over the M factors

$$P_m = \prod_{\substack{\alpha=\uparrow,\downarrow\\ i=1,2}}\left[1 - I^{(\alpha)}_{\substack{im\\ im}} + (1-\mathrm{i}\epsilon\varphi_i)\,T^{(\alpha)}_{\substack{im\\ i(m-1)}} + \sqrt{\frac{J\epsilon}{4}}\,Q^*_{\substack{im\\ (i-1)(m-1)}}\;T^{(\alpha)}_{\substack{im\\ (i-1)(m-1)}}\right], \tag{6.211a}$$

with the following Grassmann bilinears

$$\begin{aligned} I^{(\alpha)}_{\substack{im\\ im}} &= \eta^*_{im\alpha}\,\eta_{im\alpha} = e^{-\eta^*_m\eta_m}\,\langle\eta_m|\hat{c}^\dagger_{i\alpha}\,\hat{c}_{i\alpha}\,|\eta_m\rangle,\\ T^{(\alpha)}_{\substack{im\\ i(m-1)}} &= \eta^*_{im\alpha}\,\eta_{i(m-1)\alpha} = e^{-\eta^*_m\eta_{m-1}}\,\langle\eta_m|\hat{c}^\dagger_{i\alpha}\,\hat{c}_{i\alpha}|\eta_{m-1}\rangle,\\ T^{(\alpha)}_{\substack{im\\ (i-1)(m-1)}} &= \eta^*_{im\alpha}\,\eta_{(i-1)(m-1)\alpha} = e^{-\eta^*_m\eta_{m-1}}\,\langle\eta_m|\hat{c}^\dagger_{i\alpha}\,\hat{c}_{(i-1)\alpha}\,|\eta_{m-1}\rangle. \end{aligned} \tag{6.211b}$$

The factor P_m connects the four vertices on the plaquette $\{(1,m-1);(2,m-1);(2,m);(1,m)\}$ of the ladder. The Grassmann bilinears can be interpreted as follows:

(i) $I^{(\alpha)}_{\substack{im\\ im}}$ counts the number of Grassmann degrees of freedom with spin α on site (i,m).

(ii) $T^{(\alpha)}_{\substack{im\\ i(m-1)}}$ transfers one Grassmann degree of freedom with spin α from site

$(i, m-1)$ to site (i, m), i.e., in the imaginary-time direction of the mth plaquette.

(iii) $T^{(\alpha)}{}_{(i-1)(m-1)}^{im}$ transfers one Grassmann degree of freedom with spin α from site $(i-1, m-1)$ to site (i, m), i.e., along the diagonal of the mth plaquette and thereby rounding the corner $(i, m-1)$.

We are now ready to describe diagrammatically the many contributions to the integrand. We keep track of the $I^{(\uparrow(\downarrow))}$'s by assigning the cross $\times$ to the appropriate site. To each $T^{(\uparrow(\downarrow))}$, we assign an arrow $\uparrow$, or $\nearrow$, or $\nwarrow$ linking appropriate sites. The spin index of the I's and T's is fixed by coloring appropriately the crosses and the arrows (say green for spin up and red for spin down). The only non-vanishing contributions to $\int \mathcal{D}\mu_{\eta^*\eta} \prod_m P_m$ are the *fully saturated diagrams*, i.e., those for which any given site of the ladder is such that either (i) two $\times$ of different colors are present, or (ii) one $\times$ of a given color is present together with one arriving and one departing arrow of the other color or (iii) two arrows with different colors arrive and two arrows with different colors depart (see Fig. 6.9). The boundary condition forces the set of arrows to be closed in the imaginary-time direction. We interpret any closed sequence of arrows as the *world line of a fermion of a given spin.*

Exercise 3.2:

(a) Convince yourself that fully saturated diagrams with the initial condition that n $(n = 0, 1, \ldots, 4)$ arrows depart from the rung $m = 1$ contribute to the sector of the Fock space with total occupation number n (see Fig. 6.9).

(b) Show that integration over $\mathcal{D}_{\mu_\varphi}$ in Eq. (6.210) cancels the contribution of all diagrams to the physical partition function, except of some diagrams that correspond to the Fock space with total occupation number 2.

(c) Show that integration over the Affleck-Marston order parameter Q selects those fully saturated diagrams such that any zig from site $(1, m)$ to site $(2, m+1)$ takes place together with a zag from site $(2, m)$ to site $(1, m+1)$. Conclude that the *physical diagrams* (i.e., those diagrams that survive the integration over φ_i and Q) are the fully saturated diagrams with the *initial condition* that one and only one arrow departs from each vertices $(1, 1)$ and $(2, 1)$ and the *dynamical constraint* that at a given imaginary time any zig takes place with a zag (see Fig. 6.10).

If the world lines of a given physical diagram are of the same color, then there can be exchange in the sense that the number d of plaquettes covered by two diagonal arrows is odd, in which case the ladder is covered by a single world line. When the number d of plaquettes covered by two diagonal arrows is even, the ladder is covered by two world lines of the same color. The fact that any $d = 0, \cdots, M$ is allowed reflects the indistinguishability of two fermions of the same spin. There are $\binom{M}{d}$ different ways to cover d plaquettes with two diagonal arrows of the same color.

On the other hand, when the two word lines are of different colors, then the number of plaquettes covered with diagonal arrows is always even.

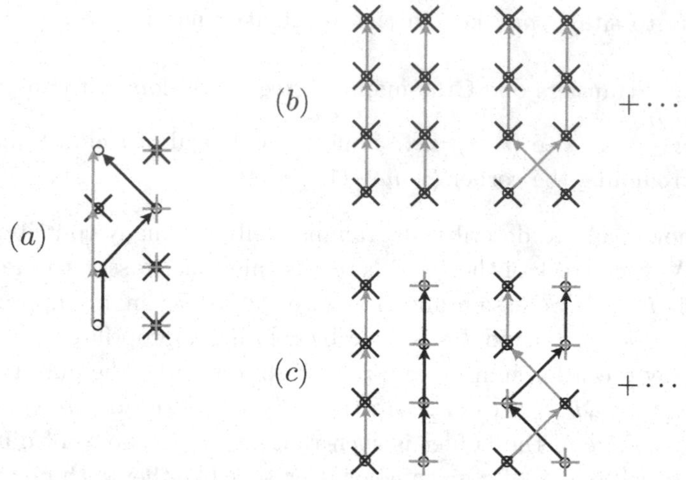

Fig. 6.10 (a) Example of a saturated diagram in the sector of total occupation number 2 which is unphysical. (b) Examples of physical diagrams for indistinguishable world lines. (c) Examples of physical diagrams for distinguishable world lines.

Exercise 3.3: Convince yourself that the physical diagrams can be separated into two different classes. The first class consists of all physical diagrams for which any ladder site is the arrival and departure of one and only one arrow of a given color. The second class consists of all physical diagrams with two world lines which can be distinguished by their color (see Fig. 6.10).

Exercise 3.4: Before calculating the contribution to the partition function of the two classes of physical diagrams we need some few identities that greatly simplify performing the Grassmann integral.

(a) Show that

$$\prod_{m=1}^{M} \eta^{*}_{im\alpha}\eta_{im\alpha} = (-1)^{M} \prod_{m=1}^{M} \eta_{im\alpha}\eta^{*}_{im\alpha}, \qquad i = 1,2, \qquad \alpha = \uparrow,\downarrow\,. \tag{6.212}$$

(b) Show that

$$\prod_{m=1}^{M} \eta^{*}_{im\alpha}\,\eta_{i(m-1)\alpha} = \prod_{m=1}^{M} \eta_{im\alpha}\,\eta^{*}_{im\alpha}, \qquad i = 1,2, \qquad \alpha = \uparrow,\downarrow\,. \tag{6.213}$$

(c) Show that

$$T^{(\alpha)}{}^{\;1m}_{1(m-1)}\; T^{(\alpha)}{}^{\;2m}_{2(m-1)} = -T^{(\alpha)}{}^{\;1m}_{2(m-1)}\; T^{(\alpha)}{}^{\;2m}_{1(m-1)}\,, \qquad \alpha = \uparrow,\downarrow\,. \tag{6.214}$$

(d) What is the effect of the boundary conditions in imaginary time in the physical sector?

Consequently, the only physical diagrams which contribute negatively to the path integral are those with an odd number of plaquettes supporting two diagonal arrows of the same color. We will account for this sign by assigning to each diagonal arrow the imaginary factor i. The full diagrammatic prescription is thus:

- Construct all physical diagrams by covering exactly once all the ladder sites by either one or two closed word lines of the same color or two world lines of different colors and add to each ladder site a cross colored differently from the world line already there.
- Every cross from a physical diagram corresponds to the factor -1.
- Every vertical arrow from a physical diagram which links $(i, m-1)$ to (i, m) corresponds to the factor $[1 - \mathrm{i}\epsilon\varphi_i]$.
- Every diagonal arrow from an physical diagram which links site $(i-1, m-1)$ to (i, m) corresponds to the factor $\mathrm{i}\sqrt{\frac{J\epsilon}{4}}\, Q^{*\ \ im}_{(i-1)(m-1)}$.

Exercise 3.5: Compute the Grassmann path integral (6.210) by explicitly summing over all the physical diagrams. Compare the result with Eq. (6.195b) or Eq. (6.203).

Chapter 7

Superconductivity in the mean-field and random-phase approximations

Outline

The notion of a pairing order is introduced. A repulsive interaction is decoupled through the pairing-order parameter. It is shown that the coupling of a repulsive interaction decreases upon momentum shell integration, whereas the coupling of an attractive interaction increases. An effective action for a uniform and static pairing-order parameter is calculated and analyzed perturbatively as well as non-perturbatively. The BCS mean-field theory for a uniform and static pairing-order parameter is described. Effective theories for the superconducting order parameter are derived in the vicinity of $T = 0$ and $T = T_{\mathrm{c}}$, respectively. First, an effective action is derived that describes in the vicinity of $T = 0$ the long-wavelength and low frequency fluctuations of the phase of the superconducting order parameter. Second, the Gross-Pitaevskii non-linear Schrödinger equation, from which follows the Meissner effect, is derived at $T = 0$. Third, the polarization tensor in the pairing state is calculated to quadratic order in the expansion of the fermionic determinant and shown to encode the Anderson-Higgs mechanism by which the photon acquires an effective mass in a superconductor. Finally, the space-dependent Ginzburg-Landau functional is calculated in the vicinity of T_{c}.

7.1 Pairing-order parameter

In the context of the jellium model in the canonical ensemble with the uniform electronic density N/V, we opted to decouple the four-fermion interaction in

$$\begin{aligned}
&\frac{1}{2}\int\limits_V \mathrm{d}^3\boldsymbol{r}_1 \int\limits_V \mathrm{d}^3\boldsymbol{r}_2\, V_{\mathrm{cb}}(\boldsymbol{r}_1-\boldsymbol{r}_2)\left[\hat{\rho}(\boldsymbol{r}_1)\,\hat{\rho}(\boldsymbol{r}_2)-\delta(\boldsymbol{r}_1-\boldsymbol{r}_2)\frac{N}{V}\right],\\
&\hat{\rho}(\boldsymbol{r}) := \sum_{\sigma=\uparrow,\downarrow}\hat{c}^{\dagger}_{\sigma}(\boldsymbol{r})\hat{c}_{\sigma}(\boldsymbol{r}),
\end{aligned}\tag{7.1}$$

by trading the local electronic density operator $\hat{\rho}(\boldsymbol{r})$ for the fluctuations of the order parameter $\varphi(\boldsymbol{r})$ (interpreted as an effective scalar potential) about the vanishing mean-field value (charge neutrality)

$$0 = V^{-1/2}\varphi_{\boldsymbol{q}=\boldsymbol{0}} = \frac{1}{V}\int\limits_V \mathrm{d}^3\boldsymbol{r}\,\varphi(\boldsymbol{r}). \tag{7.2}$$

However, this choice for decoupling the four-fermion interaction is not unique.

For example, if the interaction potential is spin dependent,

$$\frac{1}{2}\int\limits_V \mathrm{d}^3\boldsymbol{r}_1 \int\limits_V \mathrm{d}^3\boldsymbol{r}_2 \sum_{\sigma_1,\sigma_2} V_{\mathrm{int}\sigma_1,\sigma_2}(\boldsymbol{r}_1-\boldsymbol{r}_2)\left[\hat{\rho}_{\sigma_1}(\boldsymbol{r}_1)\hat{\rho}_{\sigma_2}(\boldsymbol{r}_2) - \delta_{\sigma_1,\sigma_2}\delta(\boldsymbol{r}_1-\boldsymbol{r}_2)\frac{N}{2V}\right],$$

$$\hat{\rho}_\sigma(\boldsymbol{r}) := \hat{c}^\dagger_\sigma(\boldsymbol{r})\hat{c}_\sigma(\boldsymbol{r}), \qquad \sigma = \uparrow,\downarrow, \tag{7.3}$$

it might be better to introduce an order parameter that is sensitive to ordering of the spins at low temperatures. Even for an interaction isotropic in spin space, there are many other possible ways to decouple the four-fermion interaction.

One could imagine an order parameter that corresponds to a ground state in which the Fourier transform $\hat{\rho}_{\boldsymbol{q}}$ of the local electronic density $\hat{\rho}(\boldsymbol{r})$ acquires a non-vanishing expectation value for some special value $\boldsymbol{q} = \boldsymbol{Q}$, i.e., translation invariance could be spontaneously broken if the interaction favors a charge-density wave with momentum $\boldsymbol{Q}$ over the Fermi-liquid ground state of section 6.4.

Another possible instability of the Fermi-liquid ground state of section 6.4 is to a superconducting ground state. The superconducting order parameter is associated with the development of a ground-state expectation value for the pair of adjoint operators

$$\hat{\Phi}^\dagger_{\sigma_1\sigma_2}(\boldsymbol{r}_1,\boldsymbol{r}_2) := \hat{c}^\dagger_{\sigma_1}(\boldsymbol{r}_1)\,\hat{c}^\dagger_{\sigma_2}(\boldsymbol{r}_2), \qquad \hat{\Phi}_{\sigma_1\sigma_2}(\boldsymbol{r}_1,\boldsymbol{r}_2) := \hat{c}_{\sigma_2}(\boldsymbol{r}_2)\,\hat{c}_{\sigma_1}(\boldsymbol{r}_1). \tag{7.4}$$

The operator $\hat{\Phi}_{\sigma_1\sigma_2}(\boldsymbol{r}_1,\boldsymbol{r}_2)$ is quadratic in the electronic annihilation operator. Its adjoint $\hat{\Phi}^\dagger_{\sigma_1\sigma_2}(\boldsymbol{r}_1,\boldsymbol{r}_2)$ creates out of the empty state $|0\rangle$ a pair of electrons at $\boldsymbol{r}_1$ and $\boldsymbol{r}_2$ with the spin quantum numbers σ_1 and σ_2, respectively. The operator $\hat{\Phi}_{\sigma_1\sigma_2}(\boldsymbol{r}_1,\boldsymbol{r}_2)$ is thus called a *pairing operator* and its (complex-valued) expectation value is called a *pairing-order parameter.* Choosing the pairing order parameter is made plausible by normal ordering the interaction,

$$\begin{aligned}
&\frac{1}{2}\int\limits_V \mathrm{d}^3\boldsymbol{r}_1 \int\limits_V \mathrm{d}^3\boldsymbol{r}_2 \sum_{\sigma_1,\sigma_2} V_{\mathrm{eff}\sigma_1,\sigma_2}(\boldsymbol{r}_1-\boldsymbol{r}_2)\left[\hat{\rho}_{\sigma_1}(\boldsymbol{r}_1)\hat{\rho}_{\sigma_2}(\boldsymbol{r}_2) - \delta_{\sigma_1,\sigma_2}\delta(\boldsymbol{r}_1-\boldsymbol{r}_2)\frac{N}{2V}\right] = \\
&\qquad \frac{1}{2}\int\limits_V \mathrm{d}^3\boldsymbol{r}_1 \int\limits_V \mathrm{d}^3\boldsymbol{r}_2 \sum_{\sigma_1,\sigma_2=\uparrow,\downarrow} V_{\mathrm{eff}\sigma_1,\sigma_2}(\boldsymbol{r}_1-\boldsymbol{r}_2)\;\hat{\Phi}^\dagger_{\sigma_1\sigma_2}(\boldsymbol{r}_1,\boldsymbol{r}_2)\hat{\Phi}_{\sigma_1\sigma_2}(\boldsymbol{r}_1,\boldsymbol{r}_2), \\
&\hat{\rho}_\sigma(\boldsymbol{r}) \equiv \hat{c}^\dagger_\sigma(\boldsymbol{r})\hat{c}_\sigma(\boldsymbol{r}), \qquad \int\limits_V \mathrm{d}^3\boldsymbol{r}\,\hat{\rho}_\sigma(\boldsymbol{r}) = \frac{N}{2}, \qquad \sigma = \uparrow,\downarrow,
\end{aligned} \tag{7.5}$$

when the *effective* potential

$$V_{\mathrm{eff}\sigma_1,\sigma_2}(\boldsymbol{r}_1-\boldsymbol{r}_2) \tag{7.6}$$

is attractive in some channel. Development of an expectation value for (condensation of) $\hat{\Phi}^{\dagger}_{\sigma_1\sigma_2}(\boldsymbol{r}_1,\boldsymbol{r}_2)\,\hat{\Phi}_{\sigma_1\sigma_2}(\boldsymbol{r}_1,\boldsymbol{r}_2)$ in the attractive channel would then lower the interaction energy. Thus, one decouples the effective interaction by introducing the order parameters

$$\Delta_{\sigma_1\sigma_2}(\boldsymbol{r}_1,\boldsymbol{r}_2), \qquad [\Delta^{*}_{\sigma_1\sigma_2}(\boldsymbol{r}_1,\boldsymbol{r}_2)], \tag{7.7}$$

for

$$\hat{\Phi}^{\dagger}_{\sigma_1\sigma_2}(\boldsymbol{r}_1,\boldsymbol{r}_2), \qquad [\hat{\Phi}_{\sigma_1\sigma_2}(\boldsymbol{r}_1,\boldsymbol{r}_2)], \tag{7.8}$$

respectively.

The strategy of this chapter consists in constructing an effective theory for the order parameter $\Delta_{\sigma_1\sigma_2}(\boldsymbol{r}_1,\boldsymbol{r}_2)$ conjugate to $\hat{\Phi}^{\dagger}_{\sigma_1\sigma_2}(\boldsymbol{r}_1,\boldsymbol{r}_2)$ by integrating out electrons, once the four-fermion decoupling has been performed, and to verify that the order parameter indeed develops long-range order below some transition temperature. This strategy is identical to the one applied to the jellium model. As before, the effective theory for the order parameter is an approximate one. The difference with the RPA on the jellium model will be the physical content of this approximation, namely the phenomenon of *superconductivity*. As with section 6.4, the approximation to be performed is uncontrolled. This approximation is only to be justified by comparison with experiments. We begin this chapter with an interpretation of any non-vanishing expectation value for the operators in Eq. (7.4) as a signature of *phase ordering or phase stiffness*.

7.1.1 *Phase operator*

Consider the Fock space

$$\mathcal{F} := \text{span}\left\{ |n\rangle := \frac{(\hat{a}^{\dagger})^{n}}{\sqrt{n!}}|0\rangle \,\middle|\, n=0,1,2,\cdots, \qquad \hat{a}|0\rangle = 0 \right\}, \tag{7.9a}$$

which is generated by the bosonic algebra

$$[\hat{a},\hat{a}^{\dagger}] = 1, \qquad [\hat{a},\hat{a}] = [\hat{a}^{\dagger},\hat{a}^{\dagger}] = 0. \tag{7.9b}$$

We are already familiar with the number operator

$$\hat{N} := \hat{a}^{\dagger}\hat{a}. \tag{7.10}$$

Define the operator $\hat{\theta}$ by taking the "square root" of $\hat{N}$, i.e., define on the Fock space $\mathcal{F}\setminus\{\lambda|0\rangle, \lambda\in\mathbb{C}\}$

$$e^{-i\hat{\theta}} := (\hat{N})^{-1/2}\,\hat{a}^{\dagger}, \qquad e^{+i\hat{\theta}} := \hat{a}\,(\hat{N})^{-1/2}, \tag{7.11a}$$

or, upon inversion,

$$\hat{a}^{\dagger} = (\hat{N})^{+1/2}\,e^{-i\hat{\theta}}, \qquad \hat{a} = e^{+i\hat{\theta}}\,(\hat{N})^{+1/2}. \tag{7.11b}$$

Equation (7.11b) resembles the polar representation of complex numbers. Computation of the matrix elements

$$\begin{aligned}\langle m|e^{+\mathrm{i}\hat{\theta}}\,\hat{N}\,e^{-\mathrm{i}\hat{\theta}}|n\rangle &= \langle m|\hat{a}\,(\hat{N})^{-1/2}\,\hat{a}^{\dagger}\hat{a}\,(\hat{N})^{-1/2}\,\hat{a}^{\dagger}|n\rangle \\ {\scriptstyle a^{\dagger}|n\rangle=\sqrt{n+1}|n+1\rangle} &= \langle m+1|\hat{a}^{\dagger}\hat{a}\,|n+1\rangle \\ &= (n+1)\delta_{m,n}, \qquad m,n=1,2,\cdots, \end{aligned} \tag{7.12a}$$

implies that

$$e^{+\mathrm{i}\hat{\theta}}\,\hat{N}\,e^{-\mathrm{i}\hat{\theta}} = \hat{N}+1 \tag{7.12b}$$

holds on $\mathcal{F}\setminus\{\lambda|0\rangle, \lambda\in\mathbb{C}\}$. Extend the definition of $\hat{\theta}$ to all of $\mathcal{F}$ by demanding that

$$e^{-\mathrm{i}\hat{\theta}}\,|0\rangle = |1\rangle, \qquad e^{+\mathrm{i}\hat{\theta}}\,|0\rangle = 0. \tag{7.13}$$

Hence, the "phase" operator $\exp(-\mathrm{i}\hat{\theta})$ turns an eigenstate of the number operator $\hat{N}$ with eigenvalue n into an eigenstate of the number operator $\hat{N}$ with eigenvalue $n+1$. Conversely, the "phase" operator $\exp(+\mathrm{i}\hat{\theta})$ turns an eigenstate of the number operator $\hat{N}$ with eigenvalue n into an eigenstate of the number operator $\hat{N}$ with eigenvalue $n-1$. The phase operator $\exp(-\mathrm{i}\hat{\theta})$ is not quite unitary because of the vacuum state $|0\rangle$ that makes it not norm preserving for *all* states in the Fock space. The commutator between $\hat{N}$ and $\hat{\theta}$ is obtained from

$$\begin{aligned} 1 &= [\hat{a},\hat{a}^{\dagger}] \\ &= \left[e^{+\mathrm{i}\hat{\theta}}\,(\hat{N})^{+1/2}\right]\left[(\hat{N})^{+1/2}\,e^{-\mathrm{i}\hat{\theta}}\right] - \left[(\hat{N})^{+1/2}\,e^{-\mathrm{i}\hat{\theta}}\right]\left[e^{+\mathrm{i}\hat{\theta}}\,(\hat{N})^{+1/2}\right] \\ &= e^{+\mathrm{i}\hat{\theta}}\hat{N}e^{-\mathrm{i}\hat{\theta}} - \hat{N} \\ &= \int\limits_0^1 \mathrm{d}\alpha\,\frac{\mathrm{d}}{\mathrm{d}\alpha}\left(e^{+\mathrm{i}\alpha\hat{\theta}}\hat{N}e^{-\mathrm{i}\alpha\hat{\theta}}\right) \\ &= -\mathrm{i}\int\limits_0^1 \mathrm{d}\alpha\,e^{+\mathrm{i}\alpha\hat{\theta}}\,[\hat{N},\hat{\theta}]\,e^{-\mathrm{i}\alpha\hat{\theta}}, \end{aligned} \tag{7.14a}$$

i.e.,

$$[\hat{N},\hat{\theta}] = +\mathrm{i}, \qquad [\hat{N},(\alpha\,\hat{\theta})] = +\mathrm{i}\alpha, \qquad \forall\alpha\in\mathbb{C}. \tag{7.14b}$$

This is the same algebra as the one obeyed by the position, $\hat{x}$, and momentum, $\hat{p}$, operators except for the important caveat that the eigenvalues of $\hat{\theta}$ are defined on the circle as opposed to the real line for $\hat{p}$. Correspondingly, the eigenvalues of $\hat{N}$ are discrete rather than continuous for $\hat{x}$. The operator $\hat{\theta}$ is called the *phase operator*. The operator $\hat{\theta}$ is canonically conjugate to the number operator $\hat{N}$.

What is the counterpart of $\hat{\theta}$ for the fermionic algebra

$$\{\hat{c}_{\sigma},\hat{c}^{\dagger}_{\sigma'}\} = \delta_{\sigma\sigma'}, \qquad \{\hat{c}_{\sigma},\hat{c}_{\sigma'}\} = \{\hat{c}^{\dagger}_{\sigma},\hat{c}^{\dagger}_{\sigma'}\} = 0? \tag{7.15}$$

The fermionic counterparts are [recall Eq. (7.4)]

$$\hat{a}^{\dagger}\hat{a} \longrightarrow \hat{c}^{\dagger}_{\uparrow}\hat{c}_{\uparrow} + \hat{c}^{\dagger}_{\downarrow}\hat{c}_{\downarrow}, \qquad \exp(-2\mathrm{i}\hat{\theta}) \longrightarrow \hat{c}^{\dagger}_{\uparrow}\hat{c}^{\dagger}_{\downarrow}, \qquad \exp(+2\mathrm{i}\hat{\theta}) \longrightarrow \hat{c}_{\downarrow}\hat{c}_{\uparrow}. \tag{7.16}$$

Indeed, note that

$$\hat{N} \equiv \hat{c}^{\dagger}_{\uparrow}\hat{c}_{\uparrow} + \hat{c}^{\dagger}_{\downarrow}\hat{c}_{\downarrow}, \qquad \hat{N}^2 = \hat{N} + 2\hat{c}^{\dagger}_{\uparrow}\hat{c}^{\dagger}_{\downarrow}\hat{c}_{\downarrow}\hat{c}_{\uparrow}, \tag{7.17a}$$

so that

$$\begin{aligned}
\hat{c}_{\downarrow}\hat{c}_{\uparrow}\left(\hat{c}^{\dagger}_{\uparrow}\hat{c}_{\uparrow} + \hat{c}^{\dagger}_{\downarrow}\hat{c}_{\downarrow}\right)\hat{c}^{\dagger}_{\uparrow}\hat{c}^{\dagger}_{\downarrow} &= \hat{c}_{\downarrow}\hat{c}_{\uparrow}\hat{c}^{\dagger}_{\uparrow}\hat{c}_{\uparrow}\hat{c}^{\dagger}_{\uparrow}\hat{c}^{\dagger}_{\downarrow} + (-1)^2\hat{c}_{\uparrow}\hat{c}_{\downarrow}\hat{c}^{\dagger}_{\downarrow}\hat{c}_{\downarrow}\hat{c}^{\dagger}_{\downarrow}\hat{c}^{\dagger}_{\uparrow} \\
&= \hat{c}_{\downarrow}\hat{c}_{\uparrow}\hat{c}^{\dagger}_{\uparrow}\hat{c}^{\dagger}_{\downarrow} + (\uparrow\leftrightarrow\downarrow) \\
&= \hat{c}_{\downarrow}\hat{c}^{\dagger}_{\downarrow} - \hat{c}_{\downarrow}\hat{c}^{\dagger}_{\uparrow}\hat{c}_{\uparrow}\hat{c}^{\dagger}_{\downarrow} + (\uparrow\leftrightarrow\downarrow) \\
&= 1 - \hat{c}^{\dagger}_{\downarrow}\hat{c}_{\downarrow} - (-1)^2\hat{c}^{\dagger}_{\uparrow}\hat{c}_{\downarrow}\hat{c}^{\dagger}_{\downarrow}\hat{c}_{\uparrow} + (\uparrow\leftrightarrow\downarrow) \\
&= 1 - \hat{c}^{\dagger}_{\downarrow}\hat{c}_{\downarrow} - \hat{c}^{\dagger}_{\uparrow}\hat{c}_{\uparrow} + \hat{c}^{\dagger}_{\uparrow}\hat{c}^{\dagger}_{\downarrow}\hat{c}_{\downarrow}\hat{c}_{\uparrow} + (\uparrow\leftrightarrow\downarrow) \\
\text{By Eq. (7.17a)} &= 2 - 2\hat{N} + \hat{N}(\hat{N}-1) \\
&= (\hat{N}-1)(\hat{N}-2).
\end{aligned} \tag{7.17b}$$

Thus, as it should be, the only non-vanishing matrix element of Eq. (7.17b) in the fermionic Fock space

$$\mathcal{F} := \operatorname{span}\left\{|0\rangle,\ \hat{c}^{\dagger}_{\uparrow}|0\rangle,\ \hat{c}^{\dagger}_{\downarrow}|0\rangle,\ \hat{c}^{\dagger}_{\uparrow}\hat{c}^{\dagger}_{\downarrow}|0\rangle \,\Big|\, \hat{c}_{\uparrow}|0\rangle = \hat{c}_{\downarrow}|0\rangle = 0\right\} \tag{7.18}$$

is the expectation value in the vacuum $|0\rangle$.

At the level of quantum field theory, we deduce from the identifications (Schrödinger picture)

$$\hat{\Phi}^{\dagger}_{\uparrow\downarrow}(\boldsymbol{r},\boldsymbol{r}) =: e^{-2\mathrm{i}\hat{\theta}_{\uparrow\downarrow}(\boldsymbol{r},\boldsymbol{r})}, \qquad \hat{\Phi}_{\uparrow\downarrow}(\boldsymbol{r},\boldsymbol{r}) =: e^{+2\mathrm{i}\hat{\theta}_{\uparrow\downarrow}(\boldsymbol{r},\boldsymbol{r})}, \tag{7.19a}$$

the equal-time (Heisenberg picture)

$$\begin{aligned}
[\hat{\rho}(\boldsymbol{r},t), \hat{\Phi}^{\dagger}_{\uparrow\downarrow}(\boldsymbol{r}',\boldsymbol{r}',t)] &= +2\hat{\Phi}^{\dagger}_{\uparrow\downarrow}(\boldsymbol{r}',\boldsymbol{r}',t)\,\delta(\boldsymbol{r}-\boldsymbol{r}'), \\
[\hat{\rho}(\boldsymbol{r},t), \hat{\Phi}_{\uparrow\downarrow}(\boldsymbol{r}',\boldsymbol{r}',t)] &= -2\hat{\Phi}_{\uparrow\downarrow}(\boldsymbol{r}',\boldsymbol{r}',t)\,\delta(\boldsymbol{r}-\boldsymbol{r}'),
\end{aligned} \tag{7.19b}$$

commutators. By expanding the pairing operators in Eq. (7.19a) to linear order in $\hat{\theta}_{\uparrow\downarrow}(\boldsymbol{r},\boldsymbol{r})$ in the commutators (7.19b), the commutator

$$\left[\hat{\rho}(\boldsymbol{r},t), \hat{\theta}_{\uparrow\downarrow}(\boldsymbol{r}',\boldsymbol{r}',t)\right] = \mathrm{i}\delta(\boldsymbol{r}-\boldsymbol{r}') \tag{7.19c}$$

follows. Hence, a sharp ground-state expectation value of the density operator $\hat{\rho}(\boldsymbol{r},t)$ implies a broad uncertainty in the expectation value of the pairing operator $\hat{\Phi}^{\dagger}_{\uparrow\downarrow}(\boldsymbol{r},\boldsymbol{r},t)$.[1] This scenario is the one realized in section 6.4. In this and chapter 8, we realize the superconducting scenario in which a sharp ground-state expectation value of the Fourier transform of the pairing operator $\hat{\Phi}^{\dagger}_{\uparrow\downarrow}(\boldsymbol{r}_1,\boldsymbol{r}_2,t)$ implies maximum uncertainty in the expectation value of the density operator $\hat{\rho}(\boldsymbol{r},t)$.

[1] By sharp or broad expectation values, we mean small or large mean square root deviations about expectation values, respectively.

7.1.2 Center-of-mass and relative coordinates

In this subsection, we want to argue that the characteristic separation $\boldsymbol{r}_1 - \boldsymbol{r}_2$ is of the order of the inverse Fermi momentum in the pairing operator $\hat{\Phi}^{\dagger}_{\uparrow\downarrow}(\boldsymbol{r}_1, \boldsymbol{r}_2)$, if the pairing operator signals the instability of the non-interacting Fermi sea induced by some interaction to a state that preserves translation invariance. To see this trade $\boldsymbol{r}_1$ and $\boldsymbol{r}_2$ for the relative coordinate $\boldsymbol{r}$ and the center of mass coordinate $\boldsymbol{R}$,

$$\begin{aligned} \boldsymbol{r} &:= \boldsymbol{r}_1 - \boldsymbol{r}_2, & \boldsymbol{r}_1 &= \boldsymbol{R} + \frac{\boldsymbol{r}}{2}, \\ \boldsymbol{R} &:= \frac{\boldsymbol{r}_1 + \boldsymbol{r}_2}{2}, & \boldsymbol{r}_2 &= \boldsymbol{R} - \frac{\boldsymbol{r}}{2}, \end{aligned} \tag{7.20}$$

respectively. Fourier transformation gives

$$\begin{aligned} \hat{\Phi}^{\dagger}_{\sigma_1\sigma_2}(\boldsymbol{r}_1, \boldsymbol{r}_2) &= \left(\frac{1}{\sqrt{V}}\right)^2 \sum_{\boldsymbol{k}_1,\boldsymbol{k}_2} e^{-\mathrm{i}\boldsymbol{k}_1\cdot\boldsymbol{r}_1} e^{-\mathrm{i}\boldsymbol{k}_2\cdot\boldsymbol{r}_2}\, \hat{c}^{\dagger}_{\sigma_1\boldsymbol{k}_1} \hat{c}^{\dagger}_{\sigma_2\boldsymbol{k}_2} \\ &= \frac{1}{V} \sum_{\boldsymbol{k}_1,\boldsymbol{k}_2} e^{-\mathrm{i}\boldsymbol{k}_1\cdot(\boldsymbol{R}+\frac{\boldsymbol{r}}{2})} e^{-\mathrm{i}\boldsymbol{k}_2\cdot(\boldsymbol{R}-\frac{\boldsymbol{r}}{2})}\, \hat{c}^{\dagger}_{\sigma_1\boldsymbol{k}_1} \hat{c}^{\dagger}_{\sigma_2\boldsymbol{k}_2} \\ &= \frac{1}{V} \sum_{\boldsymbol{k}_1,\boldsymbol{k}_2} e^{-\mathrm{i}(\boldsymbol{k}_1+\boldsymbol{k}_2)\cdot\boldsymbol{R}} e^{-\frac{\mathrm{i}}{2}(\boldsymbol{k}_1-\boldsymbol{k}_2)\cdot\boldsymbol{r}}\, \hat{c}^{\dagger}_{\sigma_1\boldsymbol{k}_1} \hat{c}^{\dagger}_{\sigma_2\boldsymbol{k}_2} \\ &= \frac{1}{V} \sum_{\boldsymbol{q},\boldsymbol{Q}} e^{-\mathrm{i}\boldsymbol{Q}\cdot\boldsymbol{R}} e^{-\mathrm{i}\boldsymbol{q}\cdot\boldsymbol{r}}\, \hat{c}^{\dagger}_{\sigma_1(\frac{\boldsymbol{Q}}{2}+\boldsymbol{q})} \hat{c}^{\dagger}_{\sigma_2(\frac{\boldsymbol{Q}}{2}-\boldsymbol{q})}, \end{aligned} \tag{7.21a}$$

whereby

$$\begin{aligned} \boldsymbol{q} &:= \frac{\boldsymbol{k}_1 - \boldsymbol{k}_2}{2}, & \boldsymbol{k}_1 &= \frac{\boldsymbol{Q}}{2} + \boldsymbol{q}, \\ \boldsymbol{Q} &:= \boldsymbol{k}_1 + \boldsymbol{k}_2, & \boldsymbol{k}_2 &= \frac{\boldsymbol{Q}}{2} - \boldsymbol{q}. \end{aligned} \tag{7.21b}$$

Define $\hat{\Phi}^{\dagger}_{\sigma_1\sigma_2\boldsymbol{Q}}(\boldsymbol{r})$ by

$$\begin{aligned} \hat{\Phi}^{\dagger}_{\sigma_1\sigma_2}\left(\boldsymbol{R}+\frac{\boldsymbol{r}}{2}, \boldsymbol{R}-\frac{\boldsymbol{r}}{2}\right) &= \frac{1}{\sqrt{V}} \sum_{\boldsymbol{Q}} e^{-\mathrm{i}\boldsymbol{Q}\cdot\boldsymbol{R}}\, \hat{\Phi}^{\dagger}_{\sigma_1\sigma_2\boldsymbol{Q}}(\boldsymbol{r}), \\ \hat{\Phi}^{\dagger}_{\sigma_1\sigma_2\boldsymbol{Q}}(\boldsymbol{r}) &:= \frac{1}{\sqrt{V}} \sum_{\boldsymbol{q}} e^{-\mathrm{i}\boldsymbol{q}\cdot\boldsymbol{r}}\, \hat{c}^{\dagger}_{\sigma_1(\frac{\boldsymbol{Q}}{2}+\boldsymbol{q})} \hat{c}^{\dagger}_{\sigma_2(\frac{\boldsymbol{Q}}{2}-\boldsymbol{q})}. \end{aligned} \tag{7.22}$$

We are only interested in a temperature range well below the Fermi energy ε_{F},

$$T \ll \varepsilon_{\mathrm{F}}, \qquad (k_{\mathrm{B}} = 1). \tag{7.23}$$

All relevant energy, time, and length scales are controlled by the Fermi momentum k_{F} of the Fermi sea in this range of temperature and in the non-interacting limit. For example, relevant momenta $\boldsymbol{k}_1$ and $\boldsymbol{k}_2$ entering the Fourier expansion of $\hat{\Phi}^{\dagger}_{\sigma_1\sigma_2}(\boldsymbol{r}_1, \boldsymbol{r}_2)$ must obey

$$|\boldsymbol{k}_1| \geq k_{\mathrm{F}}, \qquad |\boldsymbol{k}_2| \geq k_{\mathrm{F}}, \tag{7.24}$$

if $\hat{\Phi}^{\dagger}_{\sigma_1\sigma_2}(\boldsymbol{r}_1,\boldsymbol{r}_2)$ applied to the Fermi sea does not annihilate it at zero temperature. If we presume that the Fermi sea is made unstable by interactions to a many-body state in which some Fourier components of $\hat{\Phi}^{\dagger}_{\sigma_1\sigma_2}(\boldsymbol{r}_1,\boldsymbol{r}_2)$ acquire a non-vanishing expectation value, it is reasonable to assume that these Fourier components have a vanishing center-of-mass momentum $\boldsymbol{Q}=0$, for translation symmetry would be spontaneously broken otherwise. If so, the relative momenta in the Fourier expansion (7.21a) must obey

$$|\boldsymbol{q}|\geq k_{\mathrm{F}}. \tag{7.25}$$

If the single-particle momenta $\boldsymbol{k}_1$ and $\boldsymbol{k}_2$ entering the Fourier expansion (7.21a) are close to the Fermi surface, it then follows that the characteristic size of the relative coordinate $\boldsymbol{r}$ in $\hat{\Phi}^{\dagger}_{\sigma_1\sigma_2\boldsymbol{Q}}(\boldsymbol{r})$ is

$$|\boldsymbol{r}|\sim\frac{1}{k_{\mathrm{F}}}. \tag{7.26}$$

For a good metal,

$$\frac{1}{k_{\mathrm{F}}}\sim\mathfrak{a}, \tag{7.27}$$

$\mathfrak{a}$ the lattice spacing. Hence, the characteristic size of the relative coordinate $\boldsymbol{r}$ in $\hat{\Phi}^{\dagger}_{\sigma_1\sigma_2}(\boldsymbol{R}+\frac{\boldsymbol{r}}{2},\boldsymbol{R}-\frac{\boldsymbol{r}}{2})$ is the lattice spacing for a good metal made unstable by some interaction to a many-body state in which $\hat{\Phi}^{\dagger}_{\sigma_1\sigma_2}(\boldsymbol{R}+\frac{\boldsymbol{r}}{2},\boldsymbol{R}-\frac{\boldsymbol{r}}{2})$ acquires an expectation value that does not break translation invariance.

7.2 Scaling of electronic interactions

7.2.1 *Case of a repulsive interaction*

Consider the electronic Coulomb interaction

$$\begin{aligned}\hat{H}_{\mathrm{int}}:=&\frac{1}{2}\int\limits_{V}\mathrm{d}^3\boldsymbol{r}_1\int\limits_{V}\mathrm{d}^3\boldsymbol{r}_2\sum_{\sigma_1,\sigma_2=\uparrow,\downarrow}V_{\mathrm{cb}}(\boldsymbol{r}_1-\boldsymbol{r}_2)\,\hat{\Phi}^{\dagger}_{\sigma_1\sigma_2}(\boldsymbol{r}_1,\boldsymbol{r}_2)\,\hat{\Phi}_{\sigma_1\sigma_2}(\boldsymbol{r}_1,\boldsymbol{r}_2)\\ =&\frac{1}{2}\int\limits_{2V}\mathrm{d}^3\boldsymbol{r}\int\limits_{V}\mathrm{d}^3\boldsymbol{R}\sum_{\sigma_1,\sigma_2}V_{\mathrm{cb}}(\boldsymbol{r})\,\hat{\Phi}^{\dagger}_{\sigma_1\sigma_2}\left(\boldsymbol{R}+\frac{\boldsymbol{r}}{2},\boldsymbol{R}-\frac{\boldsymbol{r}}{2}\right)\hat{\Phi}_{\sigma_1\sigma_2}\left(\boldsymbol{R}+\frac{\boldsymbol{r}}{2},\boldsymbol{R}-\frac{\boldsymbol{r}}{2}\right).\end{aligned} \tag{7.28}$$

We have seen in section 6.4 that the response of the Fermi sea to the Coulomb interaction is to screen the algebraic tails of the Coulomb interaction in its position-space representation. Moreover, as we have argued in section 7.1.2, we expect that the response of the Fermi sea will be dominated by pairs of electrons (holes) whose relative coordinates $\boldsymbol{r}$ are of the order of the inverse Fermi momentum, which, for a good metal, is the lattice spacing, i.e., $\boldsymbol{r}=\boldsymbol{0}$ on macroscopic length scales. On the lattice scale, the diverging short-distance Coulomb interaction might as well be approximated by a repulsive delta function,

$$V_{\mathrm{cb}}(\boldsymbol{r}) \longrightarrow U\delta(\boldsymbol{r}), \qquad U>0. \tag{7.29}$$

We thus arrive at the approximate interacting Hamiltonian

$$\begin{aligned}\hat{H}_{\mathrm{int}} &\approx \frac{U}{2}\int\limits_{2V}\mathrm{d}^3\boldsymbol{r}\int\limits_{V}\mathrm{d}^3\boldsymbol{R}\sum_{\sigma_1,\sigma_2=\uparrow,\downarrow}\delta(\boldsymbol{r})\,\hat{\Phi}^{\dagger}_{\sigma_1\sigma_2}\left(\boldsymbol{R}+\frac{\boldsymbol{r}}{2},\boldsymbol{R}-\frac{\boldsymbol{r}}{2}\right)\hat{\Phi}_{\sigma_1\sigma_2}\left(\boldsymbol{R}+\frac{\boldsymbol{r}}{2},\boldsymbol{R}-\frac{\boldsymbol{r}}{2}\right)\\ &= U\int\limits_{V}\mathrm{d}^3\boldsymbol{R}\,\hat{\Phi}^{\dagger}_{\uparrow\downarrow}(\boldsymbol{R},\boldsymbol{R})\,\hat{\Phi}_{\uparrow\downarrow}(\boldsymbol{R},\boldsymbol{R}).\end{aligned} \tag{7.30}$$

What is the fate of the repulsive residual interaction $U\delta(\boldsymbol{r})$ upon *partial* integration over electrons, whereby *only* high-energy electrons in a thin shell around the Fermi surface are successively integrated out? We shall see that integration of high-energy electrons induces a renormalization of the interaction strength U such that U decreases with lower energy cut-off!

From now on, we will rely on the Grassmann-path-integral representation of the grand-canonical partition function. The only rules that need to be kept in mind are the substitutions

$$\begin{aligned}\hat{c}^{\dagger}_{\sigma}(\boldsymbol{r}) &\longrightarrow \psi^{*}_{\sigma}(\boldsymbol{r},\tau) = -\psi^{*}_{\sigma}(\boldsymbol{r},\tau+\beta),\\ \hat{c}_{\sigma}(\boldsymbol{r}) &\longrightarrow \psi_{\sigma}(\boldsymbol{r},\tau) = -\psi_{\sigma}(\boldsymbol{r},\tau+\beta),\\ \hat{\Phi}^{\dagger}_{\sigma_1\sigma_2}(\boldsymbol{R},\boldsymbol{R}) &\longrightarrow \Phi^{*}_{\sigma_1\sigma_2}(\boldsymbol{R},\tau) := \psi^{*}_{\sigma_1}(\boldsymbol{R},\tau)\psi^{*}_{\sigma_2}(\boldsymbol{R},\tau),\\ \hat{\Phi}_{\sigma_1\sigma_2}(\boldsymbol{R},\boldsymbol{R}) &\longrightarrow \Phi_{\sigma_1\sigma_2}(\boldsymbol{R},\tau) := \psi_{\sigma_2}(\boldsymbol{R},\tau)\psi_{\sigma_1}(\boldsymbol{R},\tau),\\ Z_{\beta,\mu} := \mathrm{Tr}_{\mathcal{F}}\left(e^{-\beta\hat{H}_{\beta,\mu}}\right) &\longrightarrow \int\mathcal{D}[\psi^{*}]\mathcal{D}[\psi]\,e^{-S_{\beta,\mu}},\end{aligned} \tag{7.31a}$$

where we decompose additively the Euclidean action

$$S_{\beta,\mu} = S_0 + S_U \tag{7.31b}$$

into the non-interacting action

$$S_0 := \int\limits_0^{\beta}\mathrm{d}\tau\int\limits_V\mathrm{d}^3\boldsymbol{r}\sum_{\sigma=\uparrow,\downarrow}\psi^{*}_{\sigma}(\boldsymbol{r},\tau)\left(\partial_{\tau}-\frac{\boldsymbol{\nabla}^2}{2m}-\mu\right)\psi_{\sigma}(\boldsymbol{r},\tau) \tag{7.31c}$$

and the interacting action

$$S_U := U\int\limits_0^{\beta}\mathrm{d}\tau\int\limits_V\mathrm{d}^3\boldsymbol{R}\,\Phi^{*}_{\uparrow\downarrow}(\boldsymbol{R},\tau)\,\Phi_{\uparrow\downarrow}(\boldsymbol{R},\tau). \tag{7.31d}$$

Here, $\psi^{*}_{\sigma}(\boldsymbol{R},\tau)$ and $\psi_{\sigma}(\boldsymbol{R},\tau)$ are two *independent* Grassmann numbers, i.e., they are anticommuting numbers. The Grassmann integral is defined so that it is invariant under a unitary transformation of the ψ's whereas it changes by the *inverse* of the

Jacobian of a non-unitary transformation of the ψ's instead of the usual Jacobian of Riemann integrals.[2]

We can make use of the simplification brought upon by dealing with Grassmann numbers as opposed to operators in the integrand of the grand-canonical partition function. For example, we can freely write

$$e^{-S_0-S_U}=e^{-S_0}\times e^{-S_U}, \tag{7.32}$$

in the path-integral representation of $Z_{\beta,\mu}$ whereas this step is illegal in the trace over the Fock space representation of $Z_{\beta,\mu}$. In turn, we introduce, through the Hubbard-Stratonovich transformation[3]

$$\begin{aligned}\exp(-S_U)&=\exp\left(-U\int\limits_0^\beta \mathrm{d}\tau\int\limits_V \mathrm{d}^3\boldsymbol{R}\,\Phi^*_{\uparrow\downarrow}(\boldsymbol{R},\tau)\,\Phi_{\uparrow\downarrow}(\boldsymbol{R},\tau)\right)\\&\propto\int\mathcal{D}[\Delta^*,\Delta]\exp\left(-\int\limits_0^\beta \mathrm{d}\tau\int\limits_V \mathrm{d}^3\boldsymbol{R}\,\frac{1}{U}\Delta^*(\boldsymbol{R},\tau)\,\Delta(\boldsymbol{R},\tau)\right)\\&\quad\times\exp\left(-\int\limits_0^\beta \mathrm{d}\tau\int\limits_V \mathrm{d}^3\boldsymbol{R}\Big(\mathrm{i}\Delta^*(\boldsymbol{R},\tau)\,\Phi_{\uparrow\downarrow}(\boldsymbol{R},\tau)+\mathrm{i}\Delta(\boldsymbol{R},\tau)\,\Phi^*_{\uparrow\downarrow}(\boldsymbol{R},\tau)\Big)\right),\end{aligned} \tag{7.33}$$

the complex-valued order parameters

$$\Delta^*(\boldsymbol{R},\tau)=|\Delta(\boldsymbol{R},\tau)|\,e^{-\mathrm{i}\phi(\boldsymbol{R},\tau)} \tag{7.34a}$$

and

$$\Delta(\boldsymbol{R},\tau)=|\Delta(\boldsymbol{R},\tau)|\,e^{+\mathrm{i}\phi(\boldsymbol{R},\tau)}. \tag{7.34b}$$

These order parameters will shortly be interpreted as being closely related to the expectation values

$$\langle\Phi^*_{\uparrow\downarrow}(\boldsymbol{R},\tau)\rangle_{Z_{\beta,\mu}}:=\frac{\int\mathcal{D}[\psi^*]\mathcal{D}[\psi]\;\Phi^*_{\uparrow\downarrow}(\boldsymbol{R},\tau)\;e^{-S_0-S_U}}{\int\mathcal{D}[\psi^*]\mathcal{D}[\psi]\qquad\qquad e^{-S_0-S_U}} \tag{7.35a}$$

and

$$\langle\Phi_{\uparrow\downarrow}(\boldsymbol{R},\tau)\rangle_{Z_{\beta,\mu}}:=\frac{\int\mathcal{D}[\psi^*]\mathcal{D}[\psi]\;\Phi_{\uparrow\downarrow}(\boldsymbol{R},\tau)\;e^{-S_0-S_U}}{\int\mathcal{D}[\psi^*]\mathcal{D}[\psi]\qquad\qquad e^{-S_0-S_U}}, \tag{7.35b}$$

respectively. Observe the presence of the imaginary number i in the second exponential on the right-hand side of Eq. (7.33). This imaginary number originates in

[2] This is so because the Grassmann integral is constructed such that $\int\mathrm{d}\psi^*\,\mathrm{d}\psi\,\exp(-\psi^*\,A\,\psi)=A$, i.e., $A\int\frac{\mathrm{d}\psi^*}{\sqrt{A}}\frac{\mathrm{d}\psi}{\sqrt{A}}\exp\big(-(\sqrt{A}\psi^*)(\sqrt{A}\psi)\big)=A\int\mathrm{d}\zeta^*\,\mathrm{d}\zeta\,\exp(-\zeta^*\,\zeta)=A$.

[3] The measure for the auxiliary fields Δ^* and Δ is defined so as to insure convergence. This implies that Δ^* is the complex conjugate to Δ.

the interaction being repulsive. No imaginary number would be needed for an attractive interaction. An upper cut-off in momentum space is imposed in the Fourier expansions

$$\begin{aligned}
\psi^{*}_{\sigma}(\boldsymbol{R},\tau) &=: \frac{1}{\sqrt{\beta V}}\sum_{\omega_n}\sum_{\boldsymbol{K}}^{|\boldsymbol{K}|<\Lambda} e^{-\mathrm{i}(\boldsymbol{K}\cdot\boldsymbol{R}-\omega_n\tau)}\psi^{*}_{\sigma\boldsymbol{K}\omega_n},\\
\psi_{\sigma}(\boldsymbol{R},\tau) &=: \frac{1}{\sqrt{\beta V}}\sum_{\omega_n}\sum_{\boldsymbol{K}}^{|\boldsymbol{K}|<\Lambda} e^{+\mathrm{i}(\boldsymbol{K}\cdot\boldsymbol{R}-\omega_n\tau)}\psi_{\sigma\boldsymbol{K}\omega_n},\\
\Delta^{*}(\boldsymbol{R},\tau) &=: \frac{1}{\sqrt{\beta V}}\sum_{\varpi_l}\sum_{\boldsymbol{Q}}^{|\boldsymbol{Q}|<2\Lambda} e^{-\mathrm{i}(\boldsymbol{Q}\cdot\boldsymbol{R}-\varpi_l\tau)}\,\Delta^{*}_{\boldsymbol{Q}\varpi_l},\\
\Delta(\boldsymbol{R},\tau) &=: \frac{1}{\sqrt{\beta V}}\sum_{\varpi_l}\sum_{\boldsymbol{Q}}^{|\boldsymbol{Q}|<2\Lambda} e^{+\mathrm{i}(\boldsymbol{Q}\cdot\boldsymbol{R}-\varpi_l\tau)}\,\Delta_{\boldsymbol{Q}\varpi_l}.
\end{aligned} \tag{7.36}$$

Remember that the Matsubara frequencies are

$$\omega_n = \frac{\pi}{\beta}(2n+1), \qquad n \in \mathbb{Z}, \tag{7.37}$$

and

$$\varpi_l = \frac{2\pi}{\beta}l, \qquad l \in \mathbb{Z}, \tag{7.38}$$

respectively.

In summary, the full partition function has the Grassmann-path-integral representation

$$Z_{\beta,\mu} \propto \int \mathcal{D}[\psi^{*}]\mathcal{D}[\psi]\int \mathcal{D}[\Delta^{*},\Delta]\; e^{-S'_{\beta,\mu}}, \tag{7.39a}$$

with the additive decomposition of the action

$$S'_{\beta,\mu} = S_{\mathrm{cond}} + S_0 + S'_U \tag{7.39b}$$

into a quadratic action for the order parameter,

$$\begin{aligned}
S_{\mathrm{cond}} &= \sum_{\varpi_l}\sum_{\boldsymbol{Q}} \frac{1}{U}\Delta^{*}_{\boldsymbol{Q}\varpi_l}\Delta_{\boldsymbol{Q}\varpi_l}\\
&= \int_0^{\beta}\mathrm{d}\tau\int_V \mathrm{d}^3\boldsymbol{R}\;\frac{1}{U}\Delta^{*}(\boldsymbol{R},\tau)\Delta(\boldsymbol{R},\tau),
\end{aligned} \tag{7.39c}$$

a quadratic action for the Grassmann variables (the fermions),

$$\begin{aligned}
S_0 &= \sum_{\omega_n}\sum_{\boldsymbol{K}}^{|\boldsymbol{K}|<\Lambda}\sum_{\sigma=\uparrow,\downarrow}(-\mathrm{i}\omega_n+\xi_{\boldsymbol{K}})\,\psi^{*}_{\sigma\boldsymbol{K}\omega_n}\psi_{\sigma\boldsymbol{K}\omega_n}\\
&= \int_0^{\beta}\mathrm{d}\tau\int_V \mathrm{d}^3\boldsymbol{R}\sum_{\sigma=\uparrow,\downarrow}\psi^{*}_{\sigma}(\boldsymbol{R},\tau)\left(\partial_\tau-\frac{\boldsymbol{\nabla}^2}{2m}-\mu\right)\psi_{\sigma}(\boldsymbol{R},\tau),
\end{aligned} \tag{7.39d}$$

and a coupling between the order parameter and bilinears in the Grassmann variables (the fermions),

$$
\begin{aligned}
S'_U =& \frac{\mathrm{i}}{\sqrt{\beta V}} \sum_{\varpi_l} \sum_{\boldsymbol{Q}}^{|\boldsymbol{Q}-\boldsymbol{K}|<\Lambda} \sum_{\omega_n} \sum_{\boldsymbol{K}}^{|\boldsymbol{K}|<\Lambda} \Delta^*_{\boldsymbol{Q}\varpi_l} \psi_{\downarrow(\boldsymbol{Q}-\boldsymbol{K})(\varpi_l-\omega_n)} \psi_{\uparrow\boldsymbol{K}\omega_n} \\
&+ \frac{\mathrm{i}}{\sqrt{\beta V}} \sum_{\varpi_l} \sum_{\boldsymbol{Q}}^{|\boldsymbol{Q}-\boldsymbol{K}|<\Lambda} \sum_{\omega_n} \sum_{\boldsymbol{K}}^{|\boldsymbol{K}|<\Lambda} \Delta_{\boldsymbol{Q}\varpi_l} \psi^*_{\uparrow\boldsymbol{K}\omega_n} \psi^*_{\downarrow(\boldsymbol{Q}-\boldsymbol{K})(\varpi_l-\omega_n)} \\
=& \mathrm{i} \int_0^\beta \mathrm{d}\tau \int_V \mathrm{d}^3\boldsymbol{R} \, \left[\Delta^*(\boldsymbol{R},\tau) \left(\psi_\downarrow\psi_\uparrow\right)(\boldsymbol{R},\tau) + \Delta(\boldsymbol{R},\tau) \left(\psi^*_\uparrow\psi^*_\downarrow\right)(\boldsymbol{R},\tau)\right]. \quad (7.39\mathrm{e})
\end{aligned}
$$

As always, the average number of electrons is $N = \beta^{-1}\partial_\mu \ln Z_{\beta,\mu}$. Finally,

$$
\xi_{\boldsymbol{K}} = \frac{\boldsymbol{K}^2}{2m} - \mu, \qquad \frac{L}{2\pi}\boldsymbol{K} \in \mathbb{Z}^3, \qquad \frac{L}{2\pi}\boldsymbol{Q} \in \mathbb{Z}^3. \tag{7.39f}
$$

The classical equations of motion for the auxiliary fields $\Delta^*(\boldsymbol{R},\tau)$ and $\Delta(\boldsymbol{R},\tau)$ are

$$
\begin{aligned}
0 &= \frac{\partial S'_{\beta,\mu}}{\partial \Delta^*(\boldsymbol{R},\tau)} = \frac{1}{U}\Delta(\boldsymbol{R},\tau) + \mathrm{i}\left(\psi_\downarrow\psi_\uparrow\right)(\boldsymbol{R},\tau), \\
0 &= \frac{\partial S'_{\beta,\mu}}{\partial \Delta(\boldsymbol{R},\tau)} = \frac{1}{U}\Delta^*(\boldsymbol{R},\tau) + \mathrm{i}\left(\psi^*_\uparrow\psi^*_\downarrow\right)(\boldsymbol{R},\tau).
\end{aligned} \tag{7.40}
$$

Hence, the following physical interpretation of the auxiliary fields $\Delta^*(\boldsymbol{R},\tau)$ and $\Delta(\boldsymbol{R},\tau)$ follows. If we compute the expectation value of Eq. (7.40) with the partition function $\int \mathcal{D}[\psi^*]\mathcal{D}[\psi]\, e^{-S_0-S'_U}$, we find that the auxiliary fields $\Delta^*(\boldsymbol{R},\tau)$ and $\Delta(\boldsymbol{R},\tau)$ are a mean-field approximation to $-\mathrm{i}U$ times the expectation values in Eqs. (7.35a) and (7.35b), respectively.

The poor man's scaling procedure that we follow consists in integrating out electrons with momenta $\boldsymbol{K}$ belonging to the momentum shell

$$
\Lambda - \mathrm{d}\Lambda < |\boldsymbol{K}| < \Lambda. \tag{7.41}
$$

In doing so, the bare action

$$
S'_{\beta,\mu} = S_{\mathrm{cond}} + S_0 + S'_U \tag{7.42}
$$

will be modified. The *renormalization-group* (RG) method applies when it is possible to absorb all these modifications through a renormalization of length scales and coupling constants in a way that preserves the form of the action.

Here, we limit ourselves to deriving the changes, induced by a momentum-shell integration for the contribution

$$
\begin{aligned}
S'_{\mathbf{0}0;\Lambda} :=& \frac{1}{U}\Delta^{*}_{\mathbf{0}0}\Delta_{\mathbf{0}0} \\
&+\sum_{\omega_n}\sum_{\boldsymbol{K}}^{|\boldsymbol{K}|<\Lambda}\sum_{\sigma=\uparrow,\downarrow}\left(-\mathrm{i}\omega_n+\xi_{\boldsymbol{K}}\right)\psi^{*}_{\sigma\boldsymbol{K}\omega_n}\psi_{\sigma\boldsymbol{K}\omega_n} \\
&+\frac{\mathrm{i}\Delta^{*}_{\mathbf{0}0}}{\sqrt{\beta V}}\sum_{\omega_n}\sum_{\boldsymbol{K}}^{|\boldsymbol{K}|<\Lambda}\psi_{\downarrow(-\boldsymbol{K})(-\omega_n)}\psi_{\uparrow\boldsymbol{K}\omega_n} \\
&+\frac{\mathrm{i}\Delta_{\mathbf{0}0}}{\sqrt{\beta V}}\sum_{\omega_n}\sum_{\boldsymbol{K}}^{|\boldsymbol{K}|<\Lambda}\psi^{*}_{\uparrow\boldsymbol{K}\omega_n}\psi^{*}_{\downarrow(-\boldsymbol{K})(-\omega_n)},
\end{aligned}
\tag{7.43}
$$

to the action coming from momentum and energy transfer

$$
\boldsymbol{Q}=\mathbf{0},\qquad \varpi_l=0, \tag{7.44}
$$

respectively. This is the reduced action for space- and time-independent configurations of the order parameter.

Observe that $S'_{\mathbf{0}0;\Lambda}$ can be rewritten as

$$
\begin{aligned}
S'_{\mathbf{0}0;\Lambda} =& \frac{1}{U}\Delta^{*}_{\mathbf{0}0}\Delta_{\mathbf{0}0}+\sum_{\omega_n}\sum_{\boldsymbol{K}}^{|\boldsymbol{K}|<\Lambda}\left(\psi^{*}_{\uparrow\boldsymbol{K}\omega_n}\ \psi_{\downarrow(-\boldsymbol{K})(-\omega_n)}\right) \\
&\times\begin{pmatrix}-\mathrm{i}\omega_n+\xi_{\boldsymbol{K}} & \frac{\mathrm{i}\Delta_{\mathbf{0}0}}{\sqrt{\beta V}} \\ \frac{\mathrm{i}\Delta^{*}_{\mathbf{0}0}}{\sqrt{\beta V}} & -\mathrm{i}\omega_n-\xi_{\boldsymbol{K}}\end{pmatrix}\begin{pmatrix}\psi_{\uparrow\boldsymbol{K}\omega_n} \\ \psi^{*}_{\downarrow(-\boldsymbol{K})(-\omega_n)}\end{pmatrix}.
\end{aligned}
\tag{7.45}
$$

Here, the property of inversion symmetry

$$
\xi_{\boldsymbol{K}}=\xi_{-\boldsymbol{K}} \tag{7.46}
$$

of the single-particle dispersion was used. The 2×2 grading that has been introduced is called the particle-hole grading. It plays a very important role in the mean-field theory of superconductivity and for fluctuations about it.

Integration over the fermions within the momentum shell defines the new action $S'_{\mathbf{0}0;\Lambda-\mathrm{d}\Lambda}$,

$$
\exp\left(-S'_{\mathbf{0}0;\Lambda-\mathrm{d}\Lambda}\right):=\int\limits_{\Lambda-\mathrm{d}\Lambda<|\boldsymbol{K}|<\Lambda}\mathcal{D}[\psi^{*}]\mathcal{D}[\psi]\exp\left(-S'_{\mathbf{0}0;\Lambda}\right). \tag{7.47}
$$

But integration over the fermions is the functional determinant

$$
\operatorname{Det}\begin{pmatrix}+\partial_\tau-\frac{\boldsymbol{\nabla}^2}{2m}-\mu & \frac{\mathrm{i}\Delta_{\mathbf{0}0}}{\sqrt{\beta V}} \\ \frac{\mathrm{i}\Delta^{*}_{\mathbf{0}0}}{\sqrt{\beta V}} & +\partial_\tau+\frac{\boldsymbol{\nabla}^2}{2m}+\mu\end{pmatrix} \tag{7.48}
$$

owing to the rules of Grassmann integration. Moreover, since a determinant can always be re-exponentiated, we can write

$$
\begin{aligned}
(S_{\text{cond}})_{\mathbf{0}\mathbf{0};\Lambda-\mathrm{d}\Lambda} &:= \frac{1}{U}\Delta^{*}_{\mathbf{0}\mathbf{0}}\Delta_{\mathbf{0}\mathbf{0}} - \sum_{\omega_n}\sum_{\boldsymbol{K}}^{\Lambda-\mathrm{d}\Lambda<|\boldsymbol{K}|<\Lambda} \log\left(\det\begin{pmatrix} -\mathrm{i}\omega_n+\xi_{\boldsymbol{K}} & \frac{\mathrm{i}\Delta_{\mathbf{0}\mathbf{0}}}{\sqrt{\beta V}} \\ \frac{\mathrm{i}\Delta^{*}_{\mathbf{0}\mathbf{0}}}{\sqrt{\beta V}} & -\mathrm{i}\omega_n-\xi_{\boldsymbol{K}}\end{pmatrix}\right) \\
&= \frac{1}{U}\Delta^{*}_{\mathbf{0}\mathbf{0}}\Delta_{\mathbf{0}\mathbf{0}} - \sum_{\omega_n}\sum_{\boldsymbol{K}}^{\Lambda-\mathrm{d}\Lambda<|\boldsymbol{K}|<\Lambda} \log\left(-\omega_n^2-\xi_{\boldsymbol{K}}^2-\mathrm{i}^2\frac{\Delta_{\mathbf{0}\mathbf{0}}\Delta^{*}_{\mathbf{0}\mathbf{0}}}{\beta V}\right).
\end{aligned}
\tag{7.49}
$$

With the help of the Taylor expansion $\ln(1-x) = -\sum_{j=1}^{+\infty}\frac{1}{j}x^j$, the right-hand side becomes

$$
\begin{aligned}
(S_{\text{cond}})_{\mathbf{0}\mathbf{0};\Lambda-\mathrm{d}\Lambda} =& -\sum_{\omega_n}\sum_{\boldsymbol{K}}^{\Lambda-\mathrm{d}\Lambda<|\boldsymbol{K}|<\Lambda}\log\left(-\omega_n^2-\xi_{\boldsymbol{K}}^2\right) \\
&+\left[\frac{1}{U}+(-\mathrm{i}^2)\frac{1}{\beta V}\sum_{\omega_n}\sum_{\boldsymbol{K}}^{\Lambda-\mathrm{d}\Lambda<|\boldsymbol{K}|<\Lambda}\frac{1}{\omega_n^2+\xi_{\boldsymbol{K}}^2}\right]\Delta^{*}_{\mathbf{0}\mathbf{0}}\Delta_{\mathbf{0}\mathbf{0}} \\
&+\sum_{j=2}^{+\infty}\frac{1}{j}\sum_{\omega_n}\sum_{\boldsymbol{K}}^{\Lambda-\mathrm{d}\Lambda<|\boldsymbol{K}|<\Lambda}\left[(-\mathrm{i}^2)\frac{\Delta^{*}_{\mathbf{0}\mathbf{0}}\Delta_{\mathbf{0}\mathbf{0}}}{\beta V}\frac{1}{\omega_n^2+\xi_{\boldsymbol{K}}^2}\right]^j.
\end{aligned}
\tag{7.50}
$$

If we reinterpret $(S_{\text{cond}})_{\mathbf{0}\mathbf{0};\Lambda}$ as the infinite series

$$
\begin{aligned}
&(S_{\text{cond}})_{\mathbf{0}\mathbf{0};\Lambda} := \sum_{j=0}^{+\infty} a_{\mathbf{0}\mathbf{0};\Lambda,j}\left(\Delta_{\mathbf{0}\mathbf{0}}\Delta^{*}_{\mathbf{0}\mathbf{0}}\right)^j, \\
&a_{\mathbf{0}\mathbf{0};\Lambda,0}=0, \qquad a_{\mathbf{0}\mathbf{0};\Lambda,1}=\frac{1}{U}, \qquad a_{\mathbf{0}\mathbf{0};\Lambda,j}=0, \qquad j=2,3,\cdots,
\end{aligned}
\tag{7.51}
$$

we see that the momentum-shell integration is encoded by a renormalization of the coefficients $a_{\mathbf{0}\mathbf{0};\Lambda,j}$, $j=0,1,2,\cdots$,

$$
\begin{aligned}
(S_{\text{cond}})_{\mathbf{0}\mathbf{0};\Lambda-\mathrm{d}\Lambda} &= \sum_{j=0}^{+\infty} a_{\mathbf{0}\mathbf{0};\Lambda-\mathrm{d}\Lambda,j}\left(\Delta_{\mathbf{0}\mathbf{0}}\Delta^{*}_{\mathbf{0}\mathbf{0}}\right)^j, \\
a_{\mathbf{0}\mathbf{0};\Lambda-\mathrm{d}\Lambda,0} &= -\sum_{\omega_n}\sum_{\boldsymbol{K}}^{\Lambda-\mathrm{d}\Lambda<|\boldsymbol{K}|<\Lambda}\log\left(-\omega_n^2-\xi_{\boldsymbol{K}}^2\right), \\
a_{\mathbf{0}\mathbf{0};\Lambda-\mathrm{d}\Lambda,j} &= \frac{\delta_{j,1}}{U}+\frac{1}{j}\left(\frac{1}{\beta V}\right)^j\sum_{\omega_n}\sum_{\boldsymbol{K}}^{\Lambda-\mathrm{d}\Lambda<|\boldsymbol{K}|<\Lambda}\left(\frac{1}{\omega_n^2+\xi_{\boldsymbol{K}}^2}\right)^j,
\end{aligned}
\tag{7.52}
$$

where $j=1,2,\cdots$.

The coefficient $a_{\mathbf{0}\mathbf{0};\Lambda-\mathrm{d}\Lambda,0}$ is a $\mathbb{C}$ number. It does not enter in any correlation function. We need not worry about it anymore. A very good estimate of

$a_{00;\Lambda-\mathrm{d}\Lambda,j+1}$ can be done at low temperatures. When $\beta \to \infty$ the summation over frequencies can be replaced by an integral,

$$\begin{aligned}
a_{00;\Lambda-\mathrm{d}\Lambda,j+1} &= \frac{\delta_{j+1,1}}{U} + \frac{1}{j+1}\beta\left(\frac{1}{\beta V}\right)^{j+1} \sum_{\boldsymbol{K}}^{\Lambda-\mathrm{d}\Lambda<|\boldsymbol{K}|<\Lambda} \int\limits_{-\infty}^{+\infty} \frac{\mathrm{d}\omega}{2\pi}\left(\frac{1}{\omega^2+\xi^2_{\boldsymbol{K}}}\right)^{j+1} \\
&= \frac{\delta_{j+1,1}}{U} + \frac{1}{j+1}\left(\frac{1}{\beta V}\right)^{j} I_{j+1}\,\frac{1}{V} \sum_{\boldsymbol{K}}^{\Lambda-\mathrm{d}\Lambda<|\boldsymbol{K}|<\Lambda} \left(\frac{1}{|\xi_{\boldsymbol{K}}|}\right)^{2j+1}, \qquad (7.53\text{a})
\end{aligned}$$

where

$$\begin{aligned}
I_{j+1} &:= \int\limits_{-\infty}^{+\infty} \frac{\mathrm{d}x}{2\pi}\left(\frac{1}{x^2+1}\right)^{j+1} \\
&\overset{\text{page 254, section 4.8 from Ref. [73]}}{=} \frac{1}{2\pi} \times \frac{(2j)!}{(j!)^2} 2^{-2j}\pi \\
&= \frac{(2j)!}{(j!)^2} 2^{-2j-1}, \qquad j=0,1,2,\cdots . \qquad (7.53\text{b})
\end{aligned}$$

With the help of the density of states per unit volume

$$\tilde{\nu}(\xi) := \frac{1}{V}\sum_{\sigma=\uparrow,\downarrow}\sum_{\boldsymbol{K}}\delta(\xi-\xi_{\boldsymbol{K}}) = \frac{1}{V}\sum_{\sigma=\uparrow,\downarrow}\sum_{\boldsymbol{K}}\delta(\xi-\varepsilon_{\boldsymbol{K}}+\mu), \qquad (7.54)$$

the momentum summation can be rewritten as an energy integral,

$$\begin{aligned}
\frac{1}{V}\sum_{\boldsymbol{K}}^{\Lambda-\mathrm{d}\Lambda<|\boldsymbol{K}|<\Lambda} |\xi_{\boldsymbol{K}}|^{-(2j+1)} &= \int\limits_{(\Lambda-\mathrm{d}\Lambda)^2/2m}^{\Lambda^2/2m} \mathrm{d}\varepsilon \frac{\tilde{\nu}(\varepsilon-\mu)}{2}|\varepsilon-\mu|^{-(2j+1)} \\
&\overset{\varepsilon_\Lambda > \varepsilon_{\mathrm{F}} \equiv \mu}{=} \frac{\tilde{\nu}\left(\varepsilon_\Lambda-\varepsilon_{\mathrm{F}}\right)}{2}\left(\varepsilon_\Lambda-\varepsilon_{\mathrm{F}}\right)^{-(2j+1)}\mathrm{d}\varepsilon_\Lambda + \mathcal{O}\left[(\mathrm{d}\Lambda)^2\right] \\
&= \frac{\nu_{\mathrm{F}}}{2}\left(\varepsilon_\Lambda\right)^{-2j}\mathrm{d}\ln\varepsilon_\Lambda + \mathcal{O}\left(\frac{\varepsilon_{\mathrm{F}}}{\varepsilon_\Lambda}\right) + \mathcal{O}\left[(\mathrm{d}\Lambda)^2\right]. \qquad (7.55)
\end{aligned}$$

Here, the assumption that the density of states per unit volume varies very slowly from the Fermi energy ε_{F} to the upper energy cut-off ε_Λ has been made,

$$\begin{aligned}
\tilde{\nu}(\varepsilon_\Lambda-\varepsilon_{\mathrm{F}}) &= \tilde{\nu}(0) + \left.\frac{\mathrm{d}\nu}{\mathrm{d}\varepsilon}\right|_{\varepsilon_{\mathrm{F}}}(\varepsilon_\Lambda-\varepsilon_{\mathrm{F}}) + \cdots \\
&\equiv \nu_{\mathrm{F}} + \nu'_{\mathrm{F}}\left(\varepsilon_\Lambda-\varepsilon_{\mathrm{F}}\right) + \cdots \\
&\approx \nu_{\mathrm{F}}. \qquad (7.56)
\end{aligned}$$

Specializing henceforth to the scaling of U, the coefficient $a_{00;\Lambda-\mathrm{d}\Lambda,1}$, we have found that

$$\frac{1}{U_{\Lambda-\mathrm{d}\Lambda}} \approx \frac{1}{U_\Lambda} + \frac{1}{2}\frac{\nu_{\mathrm{F}}}{2}\mathrm{d}\ln\varepsilon_\Lambda \left(\approx \frac{1}{U_\Lambda} - \mathrm{i}^2\frac{\nu_{\mathrm{F}}}{2}\mathrm{d}\ln\Lambda\right). \qquad (7.57)$$

If $\mathrm{d}\Lambda$ is chosen positive, $\Lambda - \mathrm{d}\Lambda$ decreases and so does $U_{\Lambda-\mathrm{d}\Lambda}$ according to Eq. (7.57). For $\mathrm{d}\Lambda$ infinitesimal, the poor man's scaling differential equation

$$\frac{\mathrm{d}\left(U_{\Lambda}^{-1}\right)}{\mathrm{d}\ln\Lambda} = -\frac{\nu_{\mathrm{F}}}{2} \left(= +\mathrm{i}^2\frac{\nu_{\mathrm{F}}}{2}\right) \tag{7.58}$$

equates the logarithmic derivative of the inverse, bare repulsive interaction strength with the *negative* of the density of states at the Fermi energy. Integration of high-energy electrons *decreases* the strength of the repulsive contact interaction. This suggests that at a low enough electronic energy scale, an attractive force between electrons mediated by phonons might overcome the repulsive Coulomb interaction. The coupling constant $\left(U_{\Lambda}^{-1}\right)$ increases with decreasing cut-off Λ as integration of Eq. (7.58) between $0 < \Lambda_1 < \Lambda_2$ yields

$$U_{\Lambda_2}^{-1} - U_{\Lambda_1}^{-1} = -\frac{\nu_{\mathrm{F}}}{2}\ln\left(\frac{\Lambda_2}{\Lambda_1}\right) \Longleftrightarrow U_{\Lambda_1} = \frac{U_{\Lambda_2}}{1 + U_{\Lambda_2}\frac{\nu_{\mathrm{F}}}{2}\ln\left(\frac{\Lambda_2}{\Lambda_1}\right)}. \tag{7.59}$$

In summary, neither does coupling electrons to phonons or coupling electrons to the superconducting order parameter favor repulsion at low energies. In view of this, what can we say if the effective electronic interaction is attractive?

7.2.2 *Case of an attractive interaction*

We reverse the sign of the contact interaction in Eq. (7.29) to make it attractive, i.e., the interacting Hamiltonian in the canonical ensemble is given by

$$\begin{aligned}\hat{H}_{\mathrm{int}} := & -\frac{U}{2}\int\limits_{2V}\mathrm{d}^3\boldsymbol{r}\int\limits_{V}\mathrm{d}^3\boldsymbol{R}\sum_{\sigma_1,\sigma_2=\uparrow,\downarrow}\delta(\boldsymbol{r})\,\hat{\Phi}^{\dagger}_{\sigma_1\sigma_2}\left(\boldsymbol{R}+\frac{\boldsymbol{r}}{2},\boldsymbol{R}-\frac{\boldsymbol{r}}{2}\right)\hat{\Phi}_{\sigma_1\sigma_2}\left(\boldsymbol{R}+\frac{\boldsymbol{r}}{2},\boldsymbol{R}-\frac{\boldsymbol{r}}{2}\right)\\ = & -U\int\limits_{V}\mathrm{d}^3\boldsymbol{R}\,\hat{\Phi}^{\dagger}_{\uparrow\downarrow}(\boldsymbol{R},\boldsymbol{R})\hat{\Phi}_{\uparrow\downarrow}(\boldsymbol{R},\boldsymbol{R}). \qquad U \geq 0.\end{aligned} \tag{7.60}$$

The only modification to the repulsive case is the necessity to remove the imaginary number i in the Hubbard-Stratonovich transformation (7.33). Carrying through this change leads to changing the sign on the right-hand side of the poor man's scaling differential equation (7.58),

$$\frac{\mathrm{d}\left(U_{\Lambda}^{-1}\right)}{\mathrm{d}\ln\Lambda} = +\frac{\nu_{\mathrm{F}}}{2}. \tag{7.61}$$

Equation (7.59) turns into

$$U_{\Lambda_1}^{-1} - U_{\Lambda_2}^{-1} = +\frac{\nu_{\mathrm{F}}}{2}\ln\left(\frac{\Lambda_1}{\Lambda_2}\right) \Longleftrightarrow U_{\Lambda_2} = \frac{U_{\Lambda_1}}{1 - U_{\Lambda_1}\frac{\nu_{\mathrm{F}}}{2}\ln\left(\frac{\Lambda_1}{\Lambda_2}\right)}. \tag{7.62}$$

Taken at face value, Eq. (7.62) predicts a divergence of the attractive interaction when

$$\Lambda_2 = \Lambda_1 \exp\left(-\frac{2}{U_{\Lambda_1}\nu_{\mathrm{F}}}\right). \tag{7.63}$$

However, this conclusion is incorrect since self-consistency of the initial assumption that U is small is violated. Inconsistencies of this type are often the signal that the non-interacting ground state, here the Fermi sea, is not the true ground state of the interacting system.

7.3 Time- and space-independent Landau-Ginzburg action

From now on we will assume an attractive contact interaction. The partition function in the grand-canonical ensemble is given by

$$Z_{\beta,\mu} \propto \int \mathcal{D}[\Delta^*,\Delta] \int \mathcal{D}[\psi^*]\mathcal{D}[\psi]\, e^{-S'_{\beta,\mu}}, \tag{7.64a}$$

with the additive decomposition of the action

$$S'_{\beta,\mu} = S_{\rm cond} + S_0 + S'_U \tag{7.64b}$$

into a quadratic action for the order parameter,

$$\begin{aligned} S_{\rm cond} &= \sum_{\varpi_l}\sum_{\boldsymbol{Q}} \frac{1}{U} \Delta^*_{\boldsymbol{Q}\varpi_l} \Delta_{\boldsymbol{Q}\varpi_l} \\ &= \int\limits_0^\beta \mathrm{d}\tau \int\limits_V \mathrm{d}^3\boldsymbol{R}\, \frac{1}{U} \Delta^*(\boldsymbol{R},\tau)\Delta(\boldsymbol{R},\tau), \end{aligned} \tag{7.64c}$$

a quadratic action for the Grassmann variables (the fermions),

$$\begin{aligned} S_0 &= \sum_{\omega_n}\sum_{\boldsymbol{K}}\sum_{\sigma} (-\mathrm{i}\omega_n + \xi_{\boldsymbol{K}})\, \psi^*_{\sigma\boldsymbol{K}\omega_n} \psi_{\sigma\boldsymbol{K}\omega_n} \\ &= \int\limits_0^\beta \mathrm{d}\tau \int\limits_V \mathrm{d}^3\boldsymbol{R} \sum_\sigma \psi^*_\sigma(\boldsymbol{R},\tau) \left(\partial_\tau - \frac{\boldsymbol{\nabla}^2}{2m} - \mu\right) \psi_\sigma(\boldsymbol{R},\tau), \end{aligned} \tag{7.64d}$$

and a coupling between the order parameter and bilinears in the Grassmann variables (the fermions),

$$\begin{aligned} S'_U = &\frac{1}{\sqrt{\beta V}} \sum_{\varpi_l}\sum_{\boldsymbol{Q}}\sum_{\omega_n}\sum_{\boldsymbol{K}} \Delta^*_{\boldsymbol{Q}\varpi_l} \psi_{\downarrow(\boldsymbol{Q}-\boldsymbol{K})(\varpi_l-\omega_n)} \psi_{\uparrow\boldsymbol{K}\omega_n} \\ &+ \frac{1}{\sqrt{\beta V}} \sum_{\varpi_l}\sum_{\boldsymbol{Q}}\sum_{\omega_n}\sum_{\boldsymbol{K}} \Delta_{\boldsymbol{Q}\varpi_l} \psi^*_{\uparrow\boldsymbol{K}\omega_n} \psi^*_{\downarrow(\boldsymbol{Q}-\boldsymbol{K})(\varpi_l-\omega_n)} \\ = &\int\limits_0^\beta \mathrm{d}\tau \int\limits_V \mathrm{d}^3\boldsymbol{R}\, \left[\Delta^*(\boldsymbol{R},\tau) \left(\psi_\downarrow\psi_\uparrow\right)(\boldsymbol{R},\tau) + \Delta(\boldsymbol{R},\tau) \left(\psi^*_\uparrow\psi^*_\downarrow\right)(\boldsymbol{R},\tau)\right]. \end{aligned} \tag{7.64e}$$

As always, the average number of electrons is $N = \beta^{-1}\partial_\mu \ln Z_{\beta,\mu}$. Finally,

$$\xi_{\boldsymbol{K}} = \frac{\boldsymbol{K}^2}{2m} - \mu, \qquad \frac{L}{2\pi}\boldsymbol{K} \in \mathbb{Z}^3, \qquad \frac{L}{2\pi}\boldsymbol{Q} \in \mathbb{Z}^3. \tag{7.64f}$$

The only important difference with the repulsive case, see Eq. (7.39), is the absence of the imaginary-valued multiplicative factors in S'_U, Eq. (7.64e). The order of path-integral integrations has been exchanged relative to Eq. (7.39a) and the restriction on the summation over fermionic momenta removed in Eqs. (7.64d) and (7.64e).

The classical equations of motion for the auxiliary fields $\Delta^*(\boldsymbol{R},\tau)$ and $\Delta(\boldsymbol{R},\tau)$ are

$$\begin{aligned} 0 &= \frac{\partial S'_{\beta,\mu}}{\partial \Delta^*(\boldsymbol{R},\tau)} = \frac{1}{U}\Delta(\boldsymbol{R},\tau) + \left(\psi_\downarrow\psi_\uparrow\right)(\boldsymbol{R},\tau),\\ 0 &= \frac{\partial S'_{\beta,\mu}}{\partial \Delta(\boldsymbol{R},\tau)} = \frac{1}{U}\Delta^*(\boldsymbol{R},\tau) + \left(\psi^*_\uparrow\psi^*_\downarrow\right)(\boldsymbol{R},\tau). \end{aligned} \tag{7.65}$$

Hence, the following physical interpretation of the auxiliary fields $\Delta^*(\boldsymbol{R},\tau)$ and $\Delta(\boldsymbol{R},\tau)$ follows. If we compute the expectation value of Eq. (7.65) with the partition function $\int \mathcal{D}[\psi^*]\mathcal{D}[\psi]\, e^{-S_0-S'_U}$, we find that the auxiliary fields $\Delta^*(\boldsymbol{R},\tau)$ and $\Delta(\boldsymbol{R},\tau)$ are a mean-field approximation to $-U$ times the expectation values in Eqs. (7.35a) and (7.35b), respectively.

Below, we focus exclusively on the intensive grand canonical potential $F_{\beta,\mu}(\Delta^*,\Delta)$ obtained after integrating over *all* fermions in the background of a space- and time-independent order parameter

$$\Delta^*(\boldsymbol{R},\tau) = \frac{1}{\sqrt{\beta V}}\Delta^*_{0,0} \equiv \Delta^*, \qquad \Delta(\boldsymbol{R},\tau) = \frac{1}{\sqrt{\beta V}}\Delta_{0,0} \equiv \Delta, \qquad \forall \boldsymbol{R},\tau. \tag{7.66}$$

More precisely,

$$\exp\Big(-\beta V F_{\beta,\mu}(\Delta^*,\Delta)\Big) := \int \mathcal{D}[\psi^*]\mathcal{D}[\psi]\exp\left(-S'_{\beta,\mu}\right)\Big|_{\Delta(\boldsymbol{R},\tau)=\Delta}^{\Delta^*(\boldsymbol{R},\tau)=\Delta^*}. \tag{7.67}$$

To compute Eq. (7.67), we can borrow Eq. (7.49), keeping in mind that i^2 together with the restriction on the momentum summation must be removed. In a first application of Eq. (7.49), we calculate

$$V_{\text{eff}}(\Delta^*,\Delta) := \lim_{\beta,V\to\infty} F_{\beta,\mu}(\Delta^*,\Delta) \tag{7.68}$$

in closed form and derive the so-called BCS gap equation at zero temperature. In a second application of Eq. (7.49), we expand the fermionic determinant to study the intensive grand canonical potential in the vicinity of the transition temperature below which Δ^* and Δ acquire expectation values. In the vicinity of the transition temperature, $F_{\beta,\mu}(\Delta^*,\Delta)$ is called the *time-independent and space-independent Landau-Ginzburg free energy.*

7.3.1 *Effective potential at* $T = 0$

The second line of Eq. (7.49) yields, in the limit of infinite volume and vanishing temperature,

$$\begin{aligned} V_{\mathrm{eff}}(\Delta^*,\Delta) &= \frac{1}{U}\Delta^*\Delta - \int\limits_{\mathbb{R}} \frac{\mathrm{d}\omega}{2\pi} \int\limits_{\mathbb{R}^3} \frac{\mathrm{d}^3\boldsymbol{K}}{(2\pi)^3} \log\left(-\omega^2 - \xi^2_{\boldsymbol{K}} - \Delta\Delta^*\right) \\ &= \frac{1}{U}\Delta^*\Delta - \int\limits_{\mathbb{R}} \frac{\mathrm{d}\omega}{2\pi} \int\limits_{\mathbb{R}^3} \frac{\mathrm{d}^3\boldsymbol{K}}{(2\pi)^3} \left\{\ln\left(\omega^2 + \xi^2_{\boldsymbol{K}} + \Delta\Delta^*\right) + \mathrm{i}\pi\right\}. \end{aligned} \tag{7.69}$$

Assuming that $V_{\mathrm{eff}}(\Delta^*,\Delta)$ is differentiable,

$$\begin{aligned} \frac{\partial V_{\mathrm{eff}}(\Delta^*,\Delta)}{\partial\Delta^*} &= U^{-1}\Delta - \int\limits_{\mathbb{R}} \mathrm{d}\xi \frac{\tilde{\nu}(\xi)}{2} \int\limits_{\mathbb{R}} \frac{\mathrm{d}\omega}{2\pi} \frac{\Delta}{\omega^2 + \xi^2 + \Delta\Delta^*}, \\ \frac{\partial V_{\mathrm{eff}}(\Delta^*,\Delta)}{\partial\Delta} &= U^{-1}\Delta^* - \int\limits_{\mathbb{R}} \mathrm{d}\xi \frac{\tilde{\nu}(\xi)}{2} \int\limits_{\mathbb{R}} \frac{\mathrm{d}\omega}{2\pi} \frac{\Delta^*}{\omega^2 + \xi^2 + \Delta\Delta^*}, \end{aligned} \tag{7.70a}$$

where

$$\tilde{\nu}(\xi) := 2 \times \frac{1}{V} \sum_{\boldsymbol{K}} \delta(\xi - \xi_{\boldsymbol{K}}) \tag{7.70b}$$

is the density of states per unit volume. Integration over frequencies is done with the help of the residue theorem,

$$\begin{aligned} \frac{\partial V_{\mathrm{eff}}(\Delta^*,\Delta)}{\partial\Delta^*} &= U^{-1}\Delta - \frac{1}{2} \int\limits_{\mathbb{R}} \mathrm{d}\xi \, \frac{\tilde{\nu}(\xi)}{2} \frac{\Delta}{\sqrt{\xi^2 + \Delta\Delta^*}}, \\ \frac{\partial V_{\mathrm{eff}}(\Delta^*,\Delta)}{\partial\Delta} &= U^{-1}\Delta^* - \frac{1}{2} \int\limits_{\mathbb{R}} \mathrm{d}\xi \, \frac{\tilde{\nu}(\xi)}{2} \frac{\Delta^*}{\sqrt{\xi^2 + \Delta\Delta^*}}. \end{aligned} \tag{7.71}$$

Integration over energies is potentially divergent as it stands. A high-energy cut-off must be introduced. This cut-off could be the bandwidth of some lattice regularization or it could be the Debye frequency above which repulsive Coulomb interactions dominate over the attractive interacting channel mediated by phonons. Historically, the Debye frequency was used. Thus,

$$\begin{aligned} \frac{\partial V_{\mathrm{eff}}(\Delta^*,\Delta)}{\partial\Delta^*} &\to U^{-1}\Delta - \frac{1}{2} \int\limits_{-\omega_{\mathrm{D}}}^{+\omega_{\mathrm{D}}} \mathrm{d}\xi \, \frac{\tilde{\nu}(\xi)}{2} \frac{\Delta}{\sqrt{\xi^2 + \Delta\Delta^*}}, \\ \frac{\partial V_{\mathrm{eff}}(\Delta^*,\Delta)}{\partial\Delta} &\to U^{-1}\Delta^* - \frac{1}{2} \int\limits_{-\omega_{\mathrm{D}}}^{+\omega_{\mathrm{D}}} \mathrm{d}\xi \, \frac{\tilde{\nu}(\xi)}{2} \frac{\Delta^*}{\sqrt{\xi^2 + \Delta\Delta^*}}. \end{aligned} \tag{7.72}$$

The density of states can be Taylor expanded around the Fermi energy and to lowest order in the ratio $\omega_{\mathrm{D}}/\varepsilon_{\mathrm{F}}$,

$$\begin{aligned}\frac{\partial V_{\mathrm{eff}}(\Delta^*,\Delta)}{\partial\Delta^*} &\approx U^{-1}\Delta-\frac{1}{2}\frac{\nu_{\mathrm{F}}}{2}\int\limits_{-\omega_{\mathrm{D}}}^{+\omega_{\mathrm{D}}}\mathrm{d}\xi\frac{\Delta}{\sqrt{\xi^2+\Delta\Delta^*}},\\ \frac{\partial V_{\mathrm{eff}}(\Delta^*,\Delta)}{\partial\Delta} &\approx U^{-1}\Delta^*-\frac{1}{2}\frac{\nu_{\mathrm{F}}}{2}\int\limits_{-\omega_{\mathrm{D}}}^{+\omega_{\mathrm{D}}}\mathrm{d}\xi\frac{\Delta^*}{\sqrt{\xi^2+\Delta\Delta^*}}.\end{aligned} \tag{7.73}$$

To leading order in $|\Delta|/\omega_{\mathrm{D}}$,

$$\begin{aligned}\int\limits_{-\omega_{\mathrm{D}}}^{+\omega_{\mathrm{D}}}\mathrm{d}\xi\frac{1}{\sqrt{\xi^2+\Delta\Delta^*}} &= 2\int\limits_{0}^{+\frac{\omega_{\mathrm{D}}}{|\Delta|}}\mathrm{d}x\frac{1}{\sqrt{x^2+1}}\\ &= 2\mathrm{arcsinh}\left(\frac{\omega_{\mathrm{D}}}{|\Delta|}\right)\\ \text{see Eq. (4.6.31) from Ref. [74]}\qquad &= \ln\left(\frac{4\omega_{\mathrm{D}}^2}{\Delta\Delta^*}\right)+\mathcal{O}\left(\left(\frac{\omega_{\mathrm{D}}}{|\Delta|}\right)^{-2}\right).\end{aligned} \tag{7.74}$$

Thus,

$$\begin{aligned}\frac{\partial V_{\mathrm{eff}}(\Delta^*,\Delta)}{\partial\Delta^*} &\approx U^{-1}\Delta+\frac{1}{2}\frac{\nu_{\mathrm{F}}}{2}\Delta\ln\left(\frac{\Delta^*\Delta}{4\omega_{\mathrm{D}}^2}\right)+\mathcal{O}\left(\frac{\omega_{\mathrm{D}}}{\varepsilon_{\mathrm{F}}},\left(\frac{\omega_{\mathrm{D}}}{|\Delta|}\right)^{-2}\right),\\ \frac{\partial V_{\mathrm{eff}}(\Delta^*,\Delta)}{\partial\Delta} &\approx U^{-1}\Delta^*+\frac{1}{2}\frac{\nu_{\mathrm{F}}}{2}\Delta^*\ln\left(\frac{\Delta^*\Delta}{4\omega_{\mathrm{D}}^2}\right)+\mathcal{O}\left(\frac{\omega_{\mathrm{D}}}{\varepsilon_{\mathrm{F}}},\left(\frac{\omega_{\mathrm{D}}}{|\Delta|}\right)^{-2}\right).\end{aligned} \tag{7.75}$$

With $F(x)=\frac{1}{2}x^2\ln x-\frac{1}{4}x^2$ the primitive of $f(x)=x\ln x$,

$$V_{\mathrm{eff}}(\Delta^*,\Delta)\approx U^{-1}\Delta^*\Delta+\frac{1}{2}\frac{\nu_{\mathrm{F}}}{2}\left[\Delta^*\Delta\ln\left(\frac{\Delta^*\Delta}{4\omega_{\mathrm{D}}^2}\right)-\Delta^*\Delta\right]+A, \tag{7.76a}$$

up to an integration constant $A\in\mathbb{C}$, provided the hierarchy of energy scales

$$|\Delta|\ll\omega_{\mathrm{D}}\ll\varepsilon_{\mathrm{F}} \tag{7.76b}$$

holds. The effective potential only depends on the combination $|\Delta|^2$ of the order parameters [recall Eqs. (7.34b-7.34a)]. In particular, the effective potential does not depend on the phase ϕ of the order parameter Δ. The classical equation of motion for the effective potential amounts to minimization, i.e.,

$$0\approx U^{-1}+\frac{\nu_{\mathrm{F}}}{2}\ln\left(\frac{|\Delta|}{2\omega_{\mathrm{D}}}\right),\qquad |\Delta|\ll\omega_{\mathrm{D}}\ll\varepsilon_{\mathrm{F}}, \tag{7.77a}$$

with the mean-field or saddle-point solution

$$\overline{|\Delta|} \approx 2\omega_{\mathrm{D}} \exp\left(-\frac{2}{U\nu_{\mathrm{F}}}\right), \qquad \overline{|\Delta|} \ll \omega_{\mathrm{D}} \ll \varepsilon_{\mathrm{F}}. \tag{7.77b}$$

Equation (7.77b) is called the BCS gap equation. For Eq. (7.77b) to hold, the weak coupling condition

$$U\,\nu_{\mathrm{F}} \ll 1 \tag{7.78}$$

must be satisfied. The mean-field solution only fixes the magnitude of the order parameter Δ, not its phase.

7.3.2 *Effective free energy in the vicinity of T_{c}*

To probe what happens when Δ^* and Δ become very small as a result of thermal fluctuations, Eq. (7.50) is used,

$$\begin{aligned} F_{\beta,\mu}(\Delta^*,\Delta) =& -\frac{1}{\beta V}\sum_{\omega_n}\sum_{\boldsymbol{K}} \log\left(-\omega_n^2 - \xi_{\boldsymbol{K}}^2\right) \\ &+ \left(\frac{1}{U} - \frac{1}{\beta V}\sum_{\omega_n}\sum_{\boldsymbol{K}}\frac{1}{\omega_n^2+\xi_{\boldsymbol{K}}^2}\right)\Delta^*\Delta \\ &+ \sum_{j=2}^{+\infty}\frac{(-1)^j}{j}\frac{1}{\beta V}\sum_{\omega_n}\sum_{\boldsymbol{K}}\left(\frac{\Delta^*\Delta}{\omega_n^2+\xi_{\boldsymbol{K}}^2}\right)^j. \end{aligned} \tag{7.79}$$

As we did for the repulsive case, we can introduce the density of states per unit volume

$$\tilde{\nu}(\xi) := 2\times\frac{1}{V}\sum_{\boldsymbol{K}}\delta(\xi-\xi_{\boldsymbol{K}}), \tag{7.80a}$$

in terms of which [compare with Eq. (7.52)]

$$F_{\beta,\mu}(\Delta^*,\Delta) = \sum_{j=0}^{+\infty} f_j\,(\Delta^*\Delta)^j, \tag{7.80b}$$

where the zeroth-order expansion coefficient is

$$f_0 = -\frac{1}{2\beta}\sum_{\omega_n}\int\limits_{-\infty}^{+\infty}\mathrm{d}\xi\;\tilde{\nu}(\xi)\,\log\left(-\omega_n^2-\xi^2\right), \tag{7.80c}$$

while the $(j=1,2,\cdots)$-th-order expansion coefficient is

$$\begin{aligned} f_j =& \frac{\delta_{j,1}}{U} + \frac{(-1)^j}{2j}\frac{1}{\beta}\sum_{\omega_n}\int\limits_{-\infty}^{+\infty}\mathrm{d}\xi\;\frac{\tilde{\nu}(\xi)}{(\omega_n^2+\xi^2)^j} \\ =& \frac{\delta_{j,1}}{U} + \frac{(-1)^j}{2j}\frac{1}{\beta}\sum_{\omega_n}|\omega_n|^{-2j+1}\int\limits_{-\infty}^{+\infty}\mathrm{d}x\;\frac{\tilde{\nu}_n(x)}{(1+x^2)^j}. \end{aligned} \tag{7.80d}$$

Here,

$$\tilde{\nu}_n(x) := \tilde{\nu}(x\,\omega_n). \tag{7.80e}$$

The important consequence of dealing with an attractive interaction is that the coefficient f_j alternates in sign with $j = 2, 3, \cdots$, i.e., f_{2j} is positive while f_{2j+1} is negative for $j = 1, 2, \cdots$. The coefficient f_0 is only important insofar one is interested in the absolute scale of $F_{\beta,\mu}(\Delta^*, \Delta)$. The coefficient f_1 is the most interesting one, since its sign has the potential to change from positive to negative as the temperature is decreased. The putative change in the sign of f_1 is yet another signature of the instability of the Fermi-liquid ground state. Above the transition temperature at which f_1 vanishes, the Fermi sea is a reasonable candidate for the ground state. Below the transition temperature, the free energy $F_{\beta,\mu}(\Delta^*, \Delta)$ favors condensation of the order parameters Δ^* and Δ. Since the Fermi sea is not compatible with non-vanishing values Δ^* and Δ (the Fermi sea is built out of a given number of electrons) this indicates that the ground state must be fundamentally different from the Fermi sea.

An estimate of the transition temperature follows from application of the residue theorem to compute f_1,

$$\begin{aligned} f_1 &= \frac{1}{U} - \frac{1}{2\beta}\sum_{\omega_n} |\omega_n|^{-1} \int\limits_{-\infty}^{+\infty} \mathrm{d}x \, \frac{\tilde{\nu}_n(x)}{(1+x^2)} \\ &= \frac{1}{U} - \frac{1}{2\beta} \times \frac{2\pi \mathrm{i}}{2\mathrm{i}} \times \frac{\beta}{\pi} \sum_{n\in\mathbb{Z}} |2n+1|^{-1}\, \tilde{\nu}_n(\mathrm{i}). \end{aligned} \tag{7.81}$$

The summation over fermionic Matsubara frequencies is divergent. To regulate this divergence, introduce as a cut-off the Debye frequency $\omega_{\mathrm{D}} > 0$ above which the effective interaction is expected to become repulsive. Let n_{D} be the smallest positive integer with $0 < \omega_{\mathrm{D}} < \omega_{n_{\mathrm{D}}}$,

$$n_{\mathrm{D}} := \inf_n \left\{ n = 0, 1, 2, \cdots \middle| \omega_{\mathrm{D}} < \frac{\pi}{\beta}(2n+1) \right\}, \tag{7.82}$$

and substitute

$$\begin{aligned} f_1 &\to \frac{1}{U} - \frac{\nu_{\mathrm{F}}}{2} \sum_{n\in\mathbb{Z}}^{|n|<n_{\mathrm{D}}} |2n+1|^{-1} \\ &= \frac{1}{U} - \nu_{\mathrm{F}} \sum_{n=0}^{n_{\mathrm{D}}-1} |2n+1|^{-1}. \end{aligned} \tag{7.83}$$

Here, we used the Debye cut-off and assumed that the density of states is a slowly varying function of energy on the scale of the Debye energy whose Taylor expansion at the Fermi energy starts from a non-vanishing value $\tilde{\nu}(0) \equiv \nu_{\mathrm{F}}$, to replace the density of state $\tilde{\nu}_n(\mathrm{i}) = \tilde{\nu}(\mathrm{i}\omega_n)$ by its value at the Fermi energy,

$$\tilde{\nu}(\xi) = \tilde{\nu}(0) + \left.\frac{\partial \tilde{\nu}(\xi)}{\partial \xi}\right|_0 \xi + \cdots. \tag{7.84}$$

Since

$$n_{\mathrm{D}} \sim \frac{\beta\,\omega_{\mathrm{D}}}{2\pi} + \mathcal{O}(\beta^0), \tag{7.85}$$

we can use the asymptotic formula from Eq. (0.132) of Ref. [57] where $\gamma = 0.5772\ldots$ is Euler's constant and $\gamma' := e^{\gamma}$ to write

$$\sum_{k=1}^{j} (2k-1)^{-1} = \frac{1}{2}\left[\ln j + \ln(4\gamma')\right] + \mathcal{O}(j^{-2}) \tag{7.86}$$

at sufficiently low temperatures. Hence, ($k_{\mathrm{B}} = 1$)

$$\begin{aligned} f_1 \to \frac{1}{U} - \frac{\nu_{\mathrm{F}}}{2}\ln\left(\frac{\beta\,\omega_{\mathrm{D}}}{2\pi}\right) - \frac{\nu_{\mathrm{F}}}{2}\ln(4\gamma') = 0 &\Longleftrightarrow \ln\left(\frac{\beta\,\omega_{\mathrm{D}}}{2\pi}\right) = \frac{2}{U\,\nu_{\mathrm{F}}} - \ln(4\gamma') \\ &\Longleftrightarrow \beta_{\mathrm{c}} = \left(\frac{\pi}{2\gamma'\,\omega_{\mathrm{D}}}\right) e^{+2/(\nu_{\mathrm{F}}\,U)} \\ &\Longleftrightarrow T_{\mathrm{c}} = \left(\frac{2\gamma'\,\omega_{\mathrm{D}}}{\pi}\right) e^{-2/(\nu_{\mathrm{F}}\,U)}, \end{aligned} \tag{7.87a}$$

i.e.,

$$f_1 \to \left(\frac{\nu_{\mathrm{F}}}{2}\right)\ln\left(\frac{T}{T_{\mathrm{c}}}\right). \tag{7.87b}$$

The transition temperature obtained from Eq. (7.87a) will be seen to agree with the transition temperature (7.108).

The temperature dependence of f_j, $j = 2, 3, \cdots$, follows from writing

$$\begin{aligned} \beta^{-1}\sum_{n=0}^{\infty} \omega_n^{\,-2j+1} &= \beta^{-1}\left(\frac{\beta}{\pi}\right)^{2j-1} \sum_{n=0}^{\infty} (2n+1)^{-2j+1} \\ &= \frac{1}{\pi} \times \left(\frac{\beta}{\pi}\right)^{2(j-1)} 2^{-2j+1}\zeta(2j-1, 1/2). \end{aligned} \tag{7.88}$$

Here, the Riemann zeta function is defined by (see section 9.5 of Ref. [57])

$$\zeta(z,q) := \sum_{n=0}^{+\infty} (n+q)^{-z}, \qquad \operatorname{Re} z > 1, \qquad q \neq 0, -1, -2, -3, \cdots. \tag{7.89}$$

Since $f_{j+1}/f_j \propto \beta^2$, the expansion parameter is

$$\beta^2 \Delta^* \Delta. \tag{7.90}$$

For the expansion of $F_{\beta,\mu}(\Delta^*, \Delta)$ to be good,

$$\frac{\Delta^* \Delta}{T^2} \tag{7.91}$$

must be small. This is trivially so when

$$0 = \Delta^* = \Delta. \tag{7.92}$$

We know that the order parameters Δ^* and Δ saturate to non-vanishing values upon approaching $T = 0$. Hence, the expansion that we chose to perform on $F_{\beta,\mu}(\Delta^*, \Delta)$ must break down arbitrarily close to $T = 0$.

The full temperature dependence of Δ^* and Δ follows instead from requiring that $F_{\beta,\mu}(\Delta^*, \Delta)$ be an extremum, i.e.,

$$0 = \frac{\partial F_{\beta,\mu}(\Delta^*, \Delta)}{\partial \Delta^*}, \qquad 0 = \frac{\partial F_{\beta,\mu}(\Delta^*, \Delta)}{\partial \Delta}. \tag{7.93}$$

These two (mean-field) equations can be linearized in the vicinity of T_{c} if $F_{\beta,\mu}(\Delta^*, \Delta)$ is truncated up to quadratic order in $\Delta^*\Delta$, for $\Delta^*\Delta/T^2$ is small near T_{c}. A solution for $T \lesssim T_{\mathrm{c}}$ to

$$0 \approx f_1 \Delta + 2 f_2 \Delta^* \Delta^2, \qquad 0 \approx f_1 \Delta^* + 2 f_2 (\Delta^*)^2 \Delta, \tag{7.94a}$$

is

$$(\overline{\Delta}^* \overline{\Delta})(T) \approx -\frac{f_1}{2 f_2} \propto -T^2 \ln\left(\frac{T}{T_{\mathrm{c}}}\right). \tag{7.94b}$$

Thus,

$$T^{-2} (\overline{\Delta}^* \overline{\Delta})(T) \approx \begin{cases} -(T - T_{\mathrm{c}})/T_{\mathrm{c}}, & T < T_{\mathrm{c}}, \\ 0, & T > T_{\mathrm{c}}, \end{cases} \tag{7.95}$$

in the vicinity of T_{c}. Away from T_{c}, one cannot truncate the expansion of $F_{\beta,\mu}(\Delta^*, \Delta)$ in powers of $T^{-2}(\Delta^*\Delta)$ to extract the dependence on T of the order parameters.

7.4 Mean-field theory of superconductivity

In this section, we derive the full temperature dependence of the order parameters Δ^* and Δ that minimize the intensive grand canonical potential defined by Eq. (7.67). Good references for this material can be found in Refs. [9] and [75]. Starting point is the first line of Eq. (7.49) with the substitution

$$\frac{\mathrm{i}\Delta^*_{00}}{\sqrt{\beta V}} \longrightarrow \Delta^*, \qquad \frac{\mathrm{i}\Delta_{00}}{\sqrt{\beta V}} \longrightarrow \Delta, \tag{7.96}$$

and without the restriction on the summation over fermionic momenta,

$$F_{\beta,\mu}(\Delta^*, \Delta) = \frac{1}{U} \Delta^* \Delta - \frac{1}{\beta V} \sum_{\omega_n} \sum_{\boldsymbol{K}} \log \left[\det \begin{pmatrix} -\mathrm{i}\omega_n + \xi_{\boldsymbol{K}} & \Delta \\ \Delta^* & -\mathrm{i}\omega_n - \xi_{\boldsymbol{K}} \end{pmatrix} \right]. \tag{7.97}$$

The dependence on temperature of the uniform and static order parameters Δ^* and Δ is obtained by requiring that $F_{\beta,\mu}(\Delta^*, \Delta)$ is an extremum at the *mean-field* values $\overline{\Delta}^*$ and $\overline{\Delta}$ (below, we omit the overline for notational simplicity). Thus, one must solve the *saddle or mean-field* equations (7.93), i.e.,

$$\begin{aligned}\Delta &= \frac{U}{\beta V}\sum_{\omega_n}\sum_{\boldsymbol{K}}\frac{\Delta}{\omega_n^2+\xi_{\boldsymbol{K}}^2+\Delta^*\Delta},\\ \Delta^* &= \frac{U}{\beta V}\sum_{\omega_n}\sum_{\boldsymbol{K}}\frac{\Delta^*}{\omega_n^2+\xi_{\boldsymbol{K}}^2+\Delta^*\Delta}.\end{aligned} \tag{7.98}$$

Define the *quasiparticle excitation spectrum*

$$E_{\boldsymbol{K}}^2 := \xi_{\boldsymbol{K}}^2 + \Delta^*\Delta. \tag{7.99}$$

The 2×2 particle-hole grading of the kernel entering the fermionic determinant results in the existence of two branches of quasiparticle excitations,

$$E_{\boldsymbol{K},\pm} := \pm\sqrt{\xi_{\boldsymbol{K}}^2+\Delta^*\Delta} \equiv \pm E_{\boldsymbol{K}}. \tag{7.100}$$

In terms of the quasiparticle excitation spectrum, the saddle-point equations reduce to the single equation

$$\begin{aligned}1 &= (-1)^2\frac{U}{V}\sum_{\boldsymbol{K}}\int\limits_{\Gamma_{\boldsymbol{K}}}\frac{\mathrm{d}z}{2\pi\mathrm{i}}\,\frac{\tilde{f}_{\mathrm{FD}}(z)}{z^2-E_{\boldsymbol{K}}^2}\\ &= \frac{U}{V}\sum_{\boldsymbol{K}}\int\limits_{\Gamma_{\boldsymbol{K}}}\frac{\mathrm{d}z}{2\pi\mathrm{i}}\,\frac{\tilde{f}_{\mathrm{FD}}(z)}{(z-E_{\boldsymbol{K}})(z+E_{\boldsymbol{K}})}.\end{aligned} \tag{7.101a}$$

Here, the Fermi-Dirac distribution function

$$\tilde{f}_{\mathrm{FD}}(z) = \frac{1}{e^{\beta z}+1}, \qquad z\in\mathbb{C}, \tag{7.101b}$$

with its equidistant first-order poles at

$$z_n = \mathrm{i}\omega_n, \qquad n\in\mathbb{Z}, \tag{7.101c}$$

with residues

$$\operatorname{Res}\tilde{f}_{\mathrm{FD}}(z)\Big|_{\mathrm{i}\omega_n} = -\frac{1}{\beta} \tag{7.101d}$$

was introduced (see Fig. 7.1). For any given $\boldsymbol{K}$, $\Gamma_{\boldsymbol{K}}$ is the path running antiparallel to the imaginary axis infinitesimally close to its left and parallel to the imaginary axis infinitesimally close to its right, i.e., it goes around the imaginary axis in a counterclockwise fashion. Let $\partial U_{\pm E_{\boldsymbol{K}}}$ be two small circles centered about the quasiparticle excitation energies $\pm E_{\boldsymbol{K}}$, respectively. The path $\Gamma_{\boldsymbol{K}}$ can be deformed into the two closed path $\partial U_{\pm E_{\boldsymbol{K}}}$ of clockwise orientation without crossing any pole

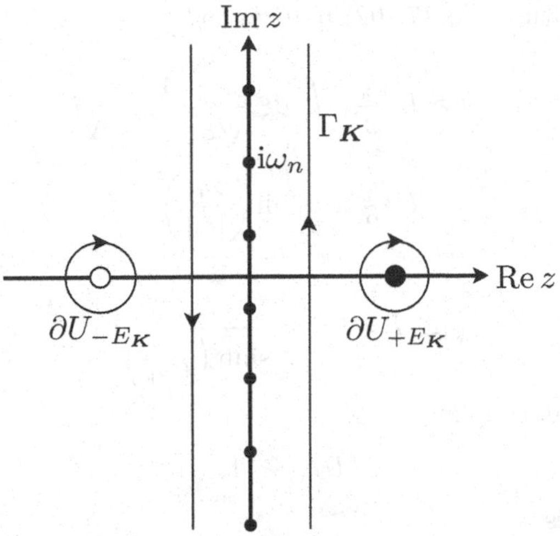

Fig. 7.1 At non-vanishing temperatures, a summation over discrete fermionic Matsubara frequencies is converted into and integral along path $\Gamma_{\boldsymbol{K}}$ in the complex plane with the Fermi-Dirac distribution multiplying the summand. Integration along $\Gamma_{\boldsymbol{K}}$ can be performed by deforming $\Gamma_{\boldsymbol{K}}$ into $\partial U_{-E_{\boldsymbol{K}}} \cup \partial U_{+E_{\boldsymbol{K}}}$ if the summand has poles at the quasiparticle energies $\pm E_{\boldsymbol{K}}$.

in the complex plane $z \in \mathbb{C}$. The saddle-point condition thus becomes

$$
\begin{aligned}
1 &- \frac{U}{V}\sum_{\boldsymbol{K}}\sum_{\pm}\int\limits_{\partial U_{\pm E_{\boldsymbol{K}}}} \frac{\mathrm{d}z}{2\pi\mathrm{i}}\,\frac{\tilde{f}_{\mathrm{FD}}(z)}{(z-E_{\boldsymbol{K}})\,(z+E_{\boldsymbol{K}})}\\
&= -\frac{U}{V}\sum_{\boldsymbol{K}}\sum_{\pm}\left(\frac{\tilde{f}_{\mathrm{FD}}(z)}{z\pm E_{\boldsymbol{K}}}\right)_{z=\pm E_{\boldsymbol{K}}}\\
&= +\frac{U}{V}\sum_{\boldsymbol{K}}\frac{\tilde{f}_{\mathrm{FD}}(-E_{\boldsymbol{K}})-\tilde{f}_{\mathrm{FD}}(+E_{\boldsymbol{K}})}{2E_{\boldsymbol{K}}}\\
&= +\frac{U}{V}\sum_{\boldsymbol{K}}\frac{\tanh\left(\beta E_{\boldsymbol{K}}/2\right)}{2E_{\boldsymbol{K}}}\\
&= \frac{U}{2}\int\limits_{-\omega_{\mathrm{D}}}^{+\omega_{\mathrm{D}}}\mathrm{d}\xi\,\frac{\tilde{\nu}(\xi)}{2}\,\frac{\tanh\left(\beta\sqrt{\xi^2+\Delta^*\Delta}/2\right)}{\sqrt{\xi^2+\Delta^*\Delta}}\\
&\approx U\frac{\nu_{\mathrm{F}}}{2}\int\limits_{0}^{+\omega_{\mathrm{D}}}\mathrm{d}\xi\,\frac{\tanh\left(\beta\sqrt{\xi^2+\Delta^*\Delta}/2\right)}{\sqrt{\xi^2+\Delta^*\Delta}}.
\end{aligned}
\tag{7.102}
$$

At zero temperature, Eq. (7.102) reduces to

$$1 \approx U\,\frac{\nu_{\mathrm{F}}}{2}\int\limits_{0}^{+\omega_{\mathrm{D}}} \mathrm{d}\xi\,\frac{1}{\sqrt{\xi^2+\Delta^*\Delta}} = U\,\frac{\nu_{\mathrm{F}}}{2}\,\mathrm{arcsinh}\left(\frac{\omega_{\mathrm{D}}}{|\Delta|}\right), \tag{7.103a}$$

i.e.,

$$|\Delta(T=0)| \approx \frac{\omega_{\mathrm{D}}}{\sinh\left(\frac{2}{U\,\nu_{\mathrm{F}}}\right)}. \tag{7.103b}$$

In the so-called weak coupling limit,

$$U\nu_{\mathrm{F}} \ll 1, \tag{7.104a}$$

Eq. (7.103b) reduces to

$$|\Delta(T=0)| \approx 2\omega_{\mathrm{D}}\exp\left(-\frac{2}{U\,\nu_{\mathrm{F}}}\right). \tag{7.104b}$$

This expression coincides with Eqs. (7.63) and (7.77b) and resembles the equation (7.87a) for T_{c}.[4]

The self-consistent critical temperature at which

$$0 = \Delta^*(T_{\mathrm{c}}) = \Delta(T_{\mathrm{c}}) \tag{7.106}$$

is defined by

$$\begin{aligned} 1 &= \frac{U}{4}\int\limits_{-\omega_{\mathrm{D}}}^{+\omega_{\mathrm{D}}} \mathrm{d}\xi\;\tilde{\nu}(\xi)\frac{\tanh\left(\beta_{\mathrm{c}}\sqrt{\xi^2}/2\right)}{\sqrt{\xi^2}} \\ &\approx \frac{U\,\nu_{\mathrm{F}}}{2}\int\limits_{0}^{+\omega_{\mathrm{D}}} \mathrm{d}\xi\;\frac{\tanh\left(\beta_{\mathrm{c}}\sqrt{\xi^2}/2\right)}{\sqrt{\xi^2}} \\ &= \frac{U\,\nu_{\mathrm{F}}}{2}\int\limits_{0}^{+\beta_{\mathrm{c}}\omega_{\mathrm{D}}/2} \mathrm{d}x\;\frac{\tanh x}{x} \\ &\approx \frac{U\,\nu_{\mathrm{F}}}{2}\ln\left(\frac{2\gamma'\,\beta_{\mathrm{c}}\,\omega_{\mathrm{D}}}{\pi}\right), \qquad \gamma' = e^{\gamma},\ \gamma\ \text{Euler's constant and}\ \frac{2\gamma'}{\pi}\approx 1.13, \end{aligned} \tag{7.107}$$

[4] The factor of 2 in the gap equation (7.104b) would not be present in Eq. (7.77b) had we used the expansion (see Eq. 4.6.31 of [74])

$$\begin{aligned} \mathrm{arcsinh}(z) &= \log(2z) + \frac{1}{2\times 2z^2} - \frac{1\times 3}{2\times 4\times 4z^4} + \frac{1\times 3\times 5}{2\times 4\times 6\times 6z^6} + \cdots, \qquad |z|\gg 1, \\ &= \log(\ z) + \mathcal{O}\left(z^0\right), \end{aligned} \tag{7.105}$$

to derive Eq. (7.77b). Clearly, the numerical prefactor to the exponential is not to be taken seriously at this level (logarithmic accuracy) of approximation.

in the limit $\beta_{\mathrm{c}}\omega_{\mathrm{D}}\gg 1$. Thus [compare with Eq. (7.87a)],

$$T_{\mathrm{c}}\approx\left(\frac{2\gamma'\,\omega_{\mathrm{D}}}{\pi}\right)\,e^{-2/(U\,\nu_{\mathrm{F}})}. \tag{7.108}$$

For $0\leq T\leq T_{\mathrm{c}}$, the saddle-point equation (7.102) must be solved numerically. Approaching T_{c} from below, Eq. (7.95) implies that

$$\left|\frac{\Delta(T)}{\Delta(T=0)}\right|\propto\left(1-\frac{T}{T_{\mathrm{c}}}\right)^{1/2}. \tag{7.109}$$

The exponent 1/2 is a trademark of the mean-field approximation. For example, the dependence on temperature of the magnetization in the mean-field approximation to the Ising model also behaves like $\sqrt{1-\frac{T}{T_{\mathrm{c}}}}$ in the vicinity of the transition temperature from the ferromagnetic to paramagnetic state. Above T_{c} no non-vanishing solution to the "gap equation" (7.102) exists.

The single-particle Hamiltonian

$$H_{\mathrm{BdG}}:=\begin{pmatrix}-\left(\frac{\nabla^2}{2m}+\mu\right)\otimes\sigma_0 & \Delta(\boldsymbol{R})\otimes(+\mathrm{i}\sigma_2)\\ \Delta^*(\boldsymbol{R})\otimes(-\mathrm{i}\sigma_2) & +\left(\frac{\nabla^2}{2m}+\mu\right)\otimes\sigma_0\end{pmatrix}, \tag{7.110}$$

which enters the fermionic determinant (7.97) through

$$S''_{\beta,\mu}(\Delta^*,\Delta):=\int\limits_0^\beta\mathrm{d}\tau\int\limits_V\mathrm{d}^3\boldsymbol{R}\frac{(\Delta^*\Delta)(\boldsymbol{R},\tau)}{U}-\frac{1}{2}\log\left[\mathrm{Det}\left(\gamma_0\otimes\sigma_0\,\partial_\tau+H_{\mathrm{BdG}}\right)\right], \tag{7.111}$$

is called the Bogoliubov-de-Gennes Hamiltonian. Here, two gradings are displayed explicitly. There is a particle-hole grading generated by the Pauli matrices γ_1, γ_2, and γ_3 together with the identity 2×2 matrix γ_0, and the spin-1/2 grading generated by the Pauli matrices σ_1, σ_2, and σ_3 together with the identity 2×2 matrix σ_0. This is a redundant representation. Correspondingly, there is a factor of 1/2 in front of the logarithm of the determinant. The Bogoliubov-de-Gennes Hamiltonian H_{BdG} is of the general form

$$\mathbb{H}_{\mathrm{BdG}}:=\begin{pmatrix}+\mathbb{K} & +\mathbb{D}\\ +\mathbb{D}^\dagger & -\mathbb{K}^{\mathrm{T}}\end{pmatrix},\qquad \mathbb{K}=\mathbb{K}^\dagger,\qquad \mathbb{D}=-\mathbb{D}^{\mathrm{T}}. \tag{7.112}$$

When the superconducting order parameters $\Delta^*(\boldsymbol{R})$ and $\Delta(\boldsymbol{R})$ that enter in $\mathbb{H}_{\mathrm{BdG}}$ are chosen so as to minimize the effective action $S''_{\beta,\mu}(\Delta^*,\Delta)$, the single-particle eigenstates of $\mathbb{H}_{\mathrm{BdG}}$ are used to construct the mean-field superconducting ground state as well as excitations above it. The single-particle eigenstates of $\mathbb{H}_{\mathrm{BdG}}$ are called quasiparticles since they are not created by the original creation electron operators, but by linear combinations of the original creation and annihilation electron operators (in the same spirit as with the Bogoliubov transformation from section 2.4.1).

The spectrum of $\mathbb{H}_{\text{BdG}}$ is characterized by the presence of a gap. For an order parameter uniform or homogeneous in space,

$$\Delta^*(\boldsymbol{R}) = \Delta^*, \qquad \Delta(\boldsymbol{R}) = \Delta, \tag{7.113}$$

the gap is the same around the entire Fermi surface. For order parameters that vary in momentum space, the gap varies in magnitude and even in sign around the Fermi surface.

Another unique property of $\mathbb{H}_{\text{BdG}}$ is that its eigenvalues occur in pairs of opposite sign. This is because $\mathbb{H}_{\text{BdG}}$, in addition to being Hermitian,

$$\mathbb{H}_{\text{BdG}} = (\mathbb{H}_{\text{BdG}})^\dagger , \tag{7.114}$$

also obeys the transformation law

$$\gamma_1 (\mathbb{H}_{\text{BdG}})^{\text{T}} \gamma_1 = -\mathbb{H}_{\text{BdG}}. \tag{7.115}$$

The antiunitary transformation law on the left-hand side of Eq. (7.115) defines a particle-hole transformation. This transformation implies a spectral symmetry of the spectrum of $\mathbb{H}_{\text{BdG}}$ by which the application of the particle-hole transformation on any eigenstate of $\mathbb{H}_{\text{BdG}}$ with the non-vanishing single-particle energy ε delivers an eigenstate of $\mathbb{H}_{\text{BdG}}$ with the non-vanishing single-particle energy $-\varepsilon$.

7.5 Nambu-Gorkov representation

Presuming an instability of the Fermi sea towards a ground state acquiring a non-vanishing expectation value for the pairing fields

$$\Phi^*_{\uparrow\downarrow}(\boldsymbol{R},\tau) := \lim_{\boldsymbol{r}_1\to\boldsymbol{r}_2} \psi^*_\uparrow(\boldsymbol{r}_1,\tau)\psi^*_\downarrow(\boldsymbol{r}_2,\tau), \quad \Phi_{\uparrow\downarrow}(\boldsymbol{R},\tau) := \lim_{\boldsymbol{r}_1\to\boldsymbol{r}_2} \psi_\downarrow(\boldsymbol{r}_2,\tau)\psi_\uparrow(\boldsymbol{r}_1,\tau), \tag{7.116a}$$

whereby

$$\boldsymbol{R} := \frac{\boldsymbol{r}_1+\boldsymbol{r}_2}{2}, \qquad \boldsymbol{r} := \boldsymbol{r}_1 - \boldsymbol{r}_2, \tag{7.116b}$$

i.e., a *superconducting instability*, we chose in section 7.3 to decouple an instantaneous and attractive contact two-body interaction

$$-U\delta(\boldsymbol{r}), \qquad U>0, \tag{7.117}$$

by presenting the partition function in the grand-canonical ensemble according to Eq. (7.64). Taking note of the fact that Grassmann numbers $\psi^*_{\sigma\boldsymbol{K}\omega_n}$ and $\psi_{\sigma'\boldsymbol{K}'\omega'_n}$ anticommute for all σ, $\boldsymbol{K}$, ω_n, σ', $\boldsymbol{K}'$, and ω'_n, the fermionic contribution $S_0 + S'_U$ to the action $S'_{\beta,\mu}$ can equally well be written

$$S_0 + S'_U \equiv \int_0^\beta \mathrm{d}\tau \int_V \mathrm{d}^3\boldsymbol{R}\,(\mathcal{L}_0 + \mathcal{L}_U), \tag{7.118a}$$

with the Lagrangian densities[5]

$$\mathcal{L}_0+\mathcal{L}_U \equiv \left(\psi^*_\uparrow(\boldsymbol{R},\tau)\ \psi_\downarrow(\boldsymbol{R},\tau)\right)\begin{pmatrix}\partial_\tau-\frac{\boldsymbol{\nabla}^2}{2m}-\mu & \Delta(\boldsymbol{R},\tau)\\ \Delta^*(\boldsymbol{R},\tau) & \partial_\tau+\frac{\boldsymbol{\nabla}^2}{2m}+\mu\end{pmatrix}\begin{pmatrix}\psi_\uparrow(\boldsymbol{R},\tau)\\ \psi^*_\downarrow(\boldsymbol{R},\tau)\end{pmatrix}. \tag{7.118b}$$

The 2×2 grading introduced in Eq. (7.118b) is called the *particle-hole grading.* If one introduces the two independent *Nambu spinors*

$$\Psi^\dagger(\boldsymbol{R},\tau)\equiv\left(\psi^*_\uparrow(\boldsymbol{R},\tau)\ \psi_\downarrow(\boldsymbol{R},\tau)\right), \tag{7.119a}$$

and

$$\Psi(\boldsymbol{R},\tau)\equiv\begin{pmatrix}\psi_\uparrow(\boldsymbol{R},\tau)\\ \psi^*_\downarrow(\boldsymbol{R},\tau)\end{pmatrix}, \tag{7.119b}$$

then Eq. (7.118b) takes the compact form

$$\mathcal{L}_0+\mathcal{L}_U=\Psi^\dagger(\boldsymbol{R},\tau)\ \mathcal{K}_{\Delta^*,\Delta}\ \Psi(\boldsymbol{R},\tau). \tag{7.119c}$$

Recalling that the polar decompositions of the pairing-order parameters Δ^* and Δ are

$$\Delta^*(\boldsymbol{R},\tau)=|\Delta(\boldsymbol{R},\tau)|e^{-\mathrm{i}\phi(\boldsymbol{R},\tau)},\qquad \Delta(\boldsymbol{R},\tau)=|\Delta(\boldsymbol{R},\tau)|e^{\mathrm{i}\phi(\boldsymbol{R},\tau)}, \tag{7.120a}$$

respectively, the kernel $\mathcal{K}_{\Delta^*,\Delta}$ can be represented by

$$\begin{aligned}\mathcal{K}_{\Delta^*,\Delta}=&\ \gamma_0\partial_\tau+\gamma_3\left(-\frac{\boldsymbol{\nabla}^2}{2m}-\mu\right)+\frac{\gamma_1+\mathrm{i}\gamma_2}{2}|\Delta(\boldsymbol{R},\tau)|e^{\mathrm{i}\phi(\boldsymbol{R},\tau)}\\ &+\frac{\gamma_1-\mathrm{i}\gamma_2}{2}|\Delta(\boldsymbol{R},\tau)|e^{-\mathrm{i}\phi(\boldsymbol{R},\tau)},\end{aligned} \tag{7.120b}$$

in imaginary time τ, center-of-mass coordinates $\boldsymbol{R}$, and with the matrices

$$\gamma_0:=\begin{pmatrix}1&0\\0&1\end{pmatrix},\qquad \gamma_1:=\begin{pmatrix}0&1\\1&0\end{pmatrix},\qquad \gamma_2:=\begin{pmatrix}0&-\mathrm{i}\\+\mathrm{i}&0\end{pmatrix},\qquad \gamma_3:=\begin{pmatrix}+1&0\\0&-1\end{pmatrix}, \tag{7.120c}$$

in the particle-hole grading.

The advantage of the Nambu representation, aside from its compactness, is that it displays explicitly the important property that all the non-vanishing eigenvalues of the single-particle Hermitian Hamiltonian

$$H_{\Delta^*,\Delta}:=\mathcal{K}_{\Delta^*,\Delta}-\gamma_0\partial_\tau \tag{7.121}$$

come in pairs of opposite sign, because of Eq. (7.115). Moreover, because of the algebra obeyed by the Pauli matrices

$$\boldsymbol{\gamma}=(\gamma_1\ \gamma_2\ \gamma_3),\qquad \gamma_i\,\gamma_j=\mathrm{i}\epsilon_{ijk}\gamma_k,\qquad \{\gamma_i,\gamma_j\}=2\delta_{ij}\,\gamma_0,\qquad i,j,k=1,2,3, \tag{7.122}$$

[5] The total derivatives that arise from the use of partial integration drop out with the choice made for the boundary conditions.

$$S_{\beta,\mu} \xrightarrow[\text{Stratonovich}]{\text{Hubbard-}} S'_{\beta,\mu} \xrightarrow[\text{integration}]{\text{fermionic}} S''_{\beta,\mu}$$

Fig. 7.2 Strategy used to construct effective action for order parameter: (i) Start from pure fermionic action with quartic and fermionic interaction. (ii) Introduce order parameter by decoupling four-fermion interaction through Hubbard-Stratonovich transformation. (iii) Integrate fermions in background of order parameter field.

the square of the Hamiltonian $H_{\Delta^*,\Delta}$ takes the form

$$\left(H_{\Delta^*,\Delta}\right)^2 \equiv \left(\mathcal{K}_{\Delta^*,\Delta} - \gamma_0\partial_\tau\right)^2 = \gamma_0\left[\left(-\frac{\boldsymbol{\nabla}^2}{2m} - \mu\right)^2 + |\Delta(\boldsymbol{R},\tau)|^2\right]. \tag{7.123}$$

Since the square of an Hermitian operator is a positive operator, $H^2 = H^\dagger H = H H^\dagger$ (although not necessarily a positive definite one), the explicit representation (7.123) of $\left(H_{\Delta^*,\Delta}\right)^2$ allows to solve for the eigenvalues of $H_{\Delta^*,\Delta}$ when the order parameter is independent of $\boldsymbol{R}$ and τ. Indeed, if

$$\Delta^*(\boldsymbol{R},\tau) = |\overline{\Delta}|\, e^{-\mathrm{i}\overline{\phi}}, \qquad \Delta(\boldsymbol{R},\tau) = |\overline{\Delta}|\, e^{+\mathrm{i}\overline{\phi}}, \qquad \forall \boldsymbol{R},\tau, \tag{7.124}$$

we then immediately recover the mean-field quasiparticle spectrum

$$E_{\boldsymbol{K},\pm} := \pm\sqrt{\xi^2_{\boldsymbol{K}} + |\overline{\Delta}|^2} \equiv \pm E_{\boldsymbol{K}} \tag{7.125}$$

of $H_{\overline{\Delta}^*,\overline{\Delta}}$ that we derived in Eq. (7.100).

7.6 Effective action for the pairing-order parameter

We now repeat the strategy of section 6.4 that consists in the computation of the effective action $S''_{\beta,\mu}$ for the order parameter obtained by integrating out all the fermions (see Fig. 7.2), i.e., we are after the action S_{fred} in the effective action

$$S''_{\beta,\mu} = S_{\text{cond}} + S_{\text{fred}}, \tag{7.126a}$$

from the partition function

$$Z_{\beta,\mu} \propto \int \mathcal{D}[\Delta^*,\Delta] \exp\left(-S''_{\beta,\mu}\right). \tag{7.126b}$$

The fermionic determinant

$$\begin{aligned}\exp\left(-S_{\text{fred}}\right) :&= \operatorname{Det}\left(\mathcal{K}_{\Delta^*,\Delta}\right)\\ &= \operatorname{Det}\begin{pmatrix}\partial_\tau - \frac{\boldsymbol{\nabla}^2}{2m} - \mu & \Delta(\boldsymbol{R},\tau)\\ \Delta^*(\boldsymbol{R},\tau) & \partial_\tau + \frac{\boldsymbol{\nabla}^2}{2m} + \mu\end{pmatrix}\end{aligned} \tag{7.127}$$

is known as a Fredholm determinant in mathematics.

We have calculated $S''_{\beta,\mu}$ exactly when the order parameter is space and time independent [as in Eq. (7.124)] and provided the limit of infinite volume and vanishing temperature has been taken, see Eqs. (7.67) and (7.76). Provided $|\overline{\Delta}| \ll \omega_{\mathrm{D}} \ll \varepsilon_{\mathrm{F}}$ holds, we found that $S''_{\beta,\mu}$ only depends on the product $\Delta^*\Delta$,

$$\lim_{\beta,V\to\infty} S''_{\beta,\mu} \approx \lim_{\beta,V\to\infty} \beta V \left\{ \frac{1}{U} + \frac{1}{2}\frac{\nu_{\mathrm{F}}}{2} \left[2\ln\left(\frac{|\overline{\Delta}|}{2\omega_{\mathrm{D}}}\right) - 1 \right] \right\} |\overline{\Delta}|^2. \tag{7.128}$$

Consequently, $S''_{\beta,\mu}$ is independent of the phase $\overline{\phi}$ of $\overline{\Delta}$ in this limit. The classical equation of motion for the action $S''_{\beta,\mu}$ yield the gap equation,

$$|\overline{\Delta}| \approx 2\,\omega_{\mathrm{D}}\, e^{-2/(U\,\nu_{\mathrm{F}})}, \tag{7.129}$$

in the weak coupling limit $U\nu_{\mathrm{F}} \ll 1$. At non-vanishing temperature, we solved for the saddle-point of $S''_{\beta,\mu}$ and found a transition temperature

$$T_{\mathrm{c}} \approx 1.13\,\omega_{\mathrm{D}}\, e^{-2/(U\,\nu_{\mathrm{F}})}, \tag{7.130}$$

above which the magnitude of the pairing-order parameter vanishes. Approaching the transition temperature from below, we expanded $S''_{\beta,\mu}$ in powers of the squared magnitude of the pairing-order parameter, thereby deriving the so-called Ginzburg-Landau free energy,

$$\lim_{V\to\infty} S''_{\beta,\mu} = \lim_{V\to\infty} \beta V \left(\sum_{j=0}^{+\infty} f_j\, |\Delta|^{2j} \right), \tag{7.131}$$

where $f_1 \sim \ln(T/T_{\mathrm{c}})$, $f_j \geq 0$, $j = 2,4,6,\cdots$, and $f_j \leq 0$, $j = 3,5,7,\cdots$. In the following, we are going to expand $S''_{\beta,\mu}$ about the mean-field solution $|\overline{\Delta}|$ to account for fluctuations of the pairing-order parameter that vary in space and time in the vicinity of $T = 0$ and $T = T_{\mathrm{c}}$, respectively.

7.7 Effective theory in the vicinity of $T = 0$

Needed is the evaluation of the determinant (7.127) in the vicinity of $T = 0$. The first question to address is how should we parametrize the pairing-order parameter? To answer this question we shall rely on the fact that Eqs. (7.128) and (7.129) do not depend on the phase ϕ of the pairing-order parameter $\Delta = |\Delta|e^{\mathrm{i}\phi}$. We shall thus choose the non-linear parametrization [compare with Eq. (2.119)]

$$\Delta^*(\boldsymbol{R},\tau) = |\overline{\Delta}|\, e^{-2\mathrm{i}\theta(\boldsymbol{R},\tau)}, \qquad \Delta(\boldsymbol{R},\tau) = |\overline{\Delta}| e^{+2\mathrm{i}\theta(\boldsymbol{R},\tau)}, \tag{7.132}$$

of the manifold for the pairing-order parameter that minimizes Eq. (7.128). By freezing the magnitude of the pairing-order parameter and only allowing space-time fluctuations $\theta(\boldsymbol{R},\tau)$ of the phase of the pairing-order parameters around the mean-field value

$$\theta(\boldsymbol{R},\tau) = \overline{\theta} = 0, \qquad \forall \boldsymbol{R},\tau, \tag{7.133}$$

it is insured that the contribution S_{cond} is minimized. Indeed, the probability weight for configuration (7.132) is only suppressed by the factor

$$\frac{\operatorname{Det}\left(\mathcal{K}_{-2\theta,+2\theta}\right)}{\operatorname{Det}\left(\mathcal{K}_{-2\theta=0,+2\theta=0}\right)} \equiv \frac{\operatorname{Det}\begin{pmatrix} \partial_\tau - \frac{\boldsymbol{\nabla}^2}{2m} - \mu & |\overline{\Delta}|e^{+2\mathrm{i}\theta(\boldsymbol{R},\tau)} \\ |\overline{\Delta}|e^{-2\mathrm{i}\theta(\boldsymbol{R},\tau)} & \partial_\tau + \frac{\boldsymbol{\nabla}^2}{2m} + \mu \end{pmatrix}}{\operatorname{Det}\begin{pmatrix} \partial_\tau - \frac{\boldsymbol{\nabla}^2}{2m} - \mu & |\overline{\Delta}| \\ |\overline{\Delta}| & \partial_\tau + \frac{\boldsymbol{\nabla}^2}{2m} + \mu \end{pmatrix}}, \tag{7.134}$$

compared to the probability weight of the saddle-point configuration. To put it differently, collective excitations associated to space-time fluctuations of the magnitude of the pairing-order parameter cannot have energies below the mean-field gap, whereas collective excitations associated to space-time fluctuations of the phase of the pairing-order parameter can. At energies well below the zero-temperature gap opened by the superconducting order, we can neglect space-time fluctuations in the magnitude of the pairing-order parameter, but we must account for the space-time fluctuations of its phase. Before undertaking a direct evaluation of Eq. (7.134), we look at some simpler limiting cases.

7.7.1 *Spatial twist around* $|\overline{\boldsymbol{\Delta}}|$

Choose in Eq. (7.132)

$$\theta(\boldsymbol{R},\tau) = \boldsymbol{Q}_s \cdot \boldsymbol{R}, \qquad \boldsymbol{Q}_s = |\boldsymbol{Q}_s|\,\widehat{\boldsymbol{Q}}_s \equiv m\,V_s\,\widehat{\boldsymbol{Q}}_s. \tag{7.135}$$

To evaluate

$$\operatorname{Det}\left(\mathcal{K}_{-2\theta,+2\theta}\big|_{\theta=\boldsymbol{Q}_s\cdot\boldsymbol{R}}\right) := \operatorname{Det}\begin{pmatrix} \partial_\tau - \frac{\boldsymbol{\nabla}^2}{2m} - \mu & |\overline{\Delta}|\exp\left(+2\mathrm{i}\boldsymbol{Q}_s\cdot\boldsymbol{R}\right) \\ |\overline{\Delta}|\exp\left(-2\mathrm{i}\boldsymbol{Q}_s\cdot\boldsymbol{R}\right) & \partial_\tau + \frac{\boldsymbol{\nabla}^2}{2m} + \mu \end{pmatrix}, \tag{7.136}$$

perform the unitary (gauge) transformation

$$\mathcal{K}_{-2\theta,+2\theta}\big|_{\theta=\boldsymbol{Q}_s\cdot\boldsymbol{R}} \to \mathcal{K}'_{-2\theta,+2\theta}\big|_{\theta=\boldsymbol{Q}_s\cdot\boldsymbol{R}}, \tag{7.137a}$$

where

$$\begin{aligned} \mathcal{K}'_{-2\theta,+2\theta} :&= \left(\mathcal{U}_{-\theta,+\theta}\right)^{-1}\,\mathcal{K}_{-2\theta,+2\theta}\,\mathcal{U}_{-\theta,+\theta}\Big|_{\theta=\boldsymbol{Q}_s\cdot\boldsymbol{R}} \\ &= \begin{pmatrix} \partial_\tau - \frac{(\boldsymbol{\nabla}+\mathrm{i}\boldsymbol{Q}_s)^2}{2m} - \mu & |\overline{\Delta}| \\ |\overline{\Delta}| & \partial_\tau + \frac{(\boldsymbol{\nabla}-\mathrm{i}\boldsymbol{Q}_s)^2}{2m} + \mu \end{pmatrix}, \end{aligned} \tag{7.137b}$$

and

$$\mathcal{U}_{-\theta,+\theta}\big|_{\theta=\boldsymbol{Q}_s\cdot\boldsymbol{R}} = \begin{pmatrix} e^{+\mathrm{i}\boldsymbol{Q}_s\cdot\boldsymbol{R}} & 0 \\ 0 & e^{-\mathrm{i}\boldsymbol{Q}_s\cdot\boldsymbol{R}} \end{pmatrix}. \tag{7.137c}$$

The identity

$$\mathrm{Det}\left(\mathcal{K}_{-2\theta,+2\theta}\big|_{\theta=\boldsymbol{Q}_s\cdot\boldsymbol{R}}\right)=\mathrm{Det}\left(\mathcal{K}'_{-2\theta,+2\theta}\big|_{\theta=\boldsymbol{Q}_s\cdot\boldsymbol{R}}\right) \tag{7.138}$$

has very important physical consequences. One might be tempted to believe that this identity is a straightforward generalization of the invariance under a unitary transformation of the determinant of a matrix. This is not so however. It is one feature of field theory, from which new physics becomes possible, that an identity such as Eq. (7.138) is highly non-trivial. Proving Eq. (7.138) amounts to proving the absence of the *chiral anomaly in a non-relativistic quantum field theory.* We do not provide a proof of Eq. (7.138) in this book. Instead, we assume that Eq. (7.138) holds.

To proceed, we first make use of

$$\partial_\tau \mp \frac{(\boldsymbol{\nabla}\pm\mathrm{i}\boldsymbol{Q}_s)^2}{2m}\mp\mu=\partial_\tau\mp\frac{\boldsymbol{\nabla}^2}{2m}-\mathrm{i}\frac{\boldsymbol{Q}_s\cdot\boldsymbol{\nabla}}{m}\pm\frac{\boldsymbol{Q}_s^2}{2m}\mp\mu. \tag{7.139}$$

Second, we make use of

$$\begin{aligned}
e^{-\mathrm{i}(\boldsymbol{K}\cdot\boldsymbol{R}-\omega_n\tau)}&\left[\partial_\tau\mp\frac{(\boldsymbol{\nabla}\pm\mathrm{i}\boldsymbol{Q}_s)^2}{2m}\mp\mu\right]e^{+\mathrm{i}(\boldsymbol{K}\cdot\boldsymbol{R}-\omega_n\tau)}\\
&=-\mathrm{i}\omega_n\pm\frac{\boldsymbol{K}^2}{2m}+\frac{\boldsymbol{Q}_s\cdot\boldsymbol{K}}{m}\pm\frac{\boldsymbol{Q}_s^2}{2m}\mp\mu.
\end{aligned} \tag{7.140}$$

Third, we infer that, in a very large volume and at very low temperatures,

$$\begin{aligned}
&\log\left(\mathrm{Det}\left(\mathcal{K}'_{-2\theta,+2\theta}\right)\right)\\
&\approx\beta V\int\limits_V\frac{\mathrm{d}^3\boldsymbol{K}}{(2\pi)^3}\int\limits_{\mathbb{R}}\frac{\mathrm{d}\omega}{2\pi}\log\left(\det\begin{pmatrix}-\mathrm{i}\omega'+\frac{\boldsymbol{K}^2}{2m}-\mu' & |\Delta|\\ |\Delta| & -\mathrm{i}\omega'-\frac{\boldsymbol{K}^2}{2m}+\mu'\end{pmatrix}\right),
\end{aligned} \tag{7.141a}$$

where we have introduced

$$-\mathrm{i}\omega'=-\mathrm{i}\omega+\frac{\boldsymbol{Q}_s\cdot\boldsymbol{K}}{m},\qquad -\mu'=-\mu+\frac{\boldsymbol{Q}_s^2}{2m}. \tag{7.141b}$$

Fourth, we assume that the integration over the fermionic Matsubara frequency ω, whereby

$$\mathrm{i}\omega=\mathrm{i}\omega'+\frac{\boldsymbol{Q}_s\cdot\boldsymbol{K}}{m}\Longleftrightarrow\omega=\omega'-\mathrm{i}\frac{\boldsymbol{Q}_s\cdot\boldsymbol{K}}{m}, \tag{7.142}$$

can be done by shifting the path of integration in the complex-Matsubara-frequency plane from the real axis $\mathbb{R}$ to the horizontal line

$$\gamma^*_{\boldsymbol{K}}:=\mathbb{R}-\mathrm{i}\frac{\boldsymbol{Q}_s\cdot\boldsymbol{K}}{m}, \tag{7.143}$$

without encountering any singularity of the integrand (branch cuts). This assumption is verified when (Λ is a momentum cut-off[6])

$$\left|\frac{\boldsymbol{Q}_s\cdot\boldsymbol{K}}{m}\right|<V_s\Lambda\ll|\Delta|, \tag{7.144a}$$

[6] Remember that one should always impose a momentum cut-off, say Λ, that arises from the band width or from the Debye frequency. Hence, $|\boldsymbol{K}|$ can be replaced by Λ in Eqs. (7.144a) and (7.144b), in which case the validity of shifting the Matsubara frequency integral into the complex plane holds uniformly in the fermionic momentum $\boldsymbol{K}$.

but breaks down, as we shall see after Eq. (7.152) in more details, when

$$\left|\frac{\boldsymbol{Q}_s\cdot\boldsymbol{K}}{m}\right| < V_s\Lambda \lessapprox |\overline{\Delta}|. \tag{7.144b}$$

Here,

$$V_s \equiv \frac{|\boldsymbol{Q}_s|}{m}. \tag{7.144c}$$

If so, Eqs. (7.127) and (7.138) can be used to deduce

$$S_{\mathrm{fred}}\left[|\overline{\Delta}|e^{-2\mathrm{i}\boldsymbol{Q}_s\cdot\boldsymbol{R}}, |\overline{\Delta}|e^{+2\mathrm{i}\boldsymbol{Q}_s\cdot\boldsymbol{R}}\right] = S'_{\mathrm{fred}}\left[|\overline{\Delta}|, |\overline{\Delta}|\right], \tag{7.145a}$$

whereby

$$\begin{aligned}S'_{\mathrm{fred}}\left[|\overline{\Delta}|, |\overline{\Delta}|\right] :=& -\log\left(\mathrm{Det}\left(\mathcal{K}'_{-2\theta,+2\theta}\right)\right)\big|_{\theta=\boldsymbol{Q}_s\cdot\boldsymbol{R}}\\ \approx& -\beta V\int\limits_V\frac{\mathrm{d}^3\boldsymbol{K}}{(2\pi)^3}\int\limits_{\mathbb{R}}\frac{\mathrm{d}\omega}{2\pi}\log\left(-\omega'^2-\left(\frac{\boldsymbol{K}^2}{2m}-\mu'\right)^2-|\overline{\Delta}|^2\right)\end{aligned} \tag{7.145b}$$

for large volumes and low temperatures. Furthermore,

$$\begin{aligned}S'_{\mathrm{fred}}\left[|\overline{\Delta}|, |\overline{\Delta}|\right] \approx& -\beta V\int\limits_V\frac{\mathrm{d}^3\boldsymbol{K}}{(2\pi)^3}\int\limits_{\gamma^*_{\boldsymbol{K}}}\frac{\mathrm{d}\omega}{2\pi}\log\left(-\omega'^2-\left(\frac{\boldsymbol{K}^2}{2m}-\mu'\right)^2-|\overline{\Delta}|^2\right)\\ =& -\beta V\int\limits_V\frac{\mathrm{d}^3\boldsymbol{K}}{(2\pi)^3}\int\limits_{\mathbb{R}}\frac{\mathrm{d}\omega'}{2\pi}\log\left(-\omega'^2-\left(\frac{\boldsymbol{K}^2}{2m}-\mu'\right)^2-|\overline{\Delta}|^2\right),\end{aligned} \tag{7.146}$$

since we have assumed that no branch cuts have been crossed when changing the integration variable in the complex plane of frequencies. Taylor expansion in powers of $\mu'-\mu=-\boldsymbol{Q}_s^2/2m$ then delivers

$$S'_{\mathrm{fred}}\left[|\overline{\Delta}|, |\overline{\Delta}|\right] \approx S_{\mathrm{fred}}\left[|\overline{\Delta}|, |\overline{\Delta}|\right] + \frac{\boldsymbol{Q}_s^2}{2m}\left(-\frac{\partial S_{\mathrm{fred}}}{\partial\mu}\right)\bigg|_{\boldsymbol{Q}_s=\boldsymbol{0}} + \mathcal{O}\left[\left(\frac{\boldsymbol{Q}_s^2}{2m}\right)^2\right]. \tag{7.147}$$

If the definition of $N=\beta^{-1}\partial_\mu\ln Z_{\beta,\mu}$ is used, the final result for Eq. (7.145) becomes

$$S'_{\mathrm{fred}}\left[|\overline{\Delta}|, |\overline{\Delta}|\right] \approx S_{\mathrm{fred}}\left[|\overline{\Delta}|, |\overline{\Delta}|\right] + \beta V\frac{N}{V}\frac{\boldsymbol{Q}_s^2}{2m} + \mathcal{O}\left[\left(\frac{\boldsymbol{Q}_s^2}{2m}\right)^2\right]. \tag{7.148}$$

This is what is expected for a steady uniform flow with velocity $V_s=|\boldsymbol{Q}_s|/m$ of the entire *pairing or superconducting condensate.*

The physical interpretation of a spatial twist of the order parameter is the following. Remember that the partition function $Z_{\beta,\mu}$ describes a system in statistical equilibrium with a reservoir, i.e., the system exchanges energy and particle number with the reservoir whereby β and μ determines the steady average energy and the steady average particle number stored in the system, respectively. Imagine the reservoir as being the walls of a container within which the electrons are interacting. Electrons can go in and out of the container but on average there are N electrons in the container at all times. Equations (7.132) and (7.135) say that the center of mass of all paired electrons participating to the condensate has momentum $2\boldsymbol{Q}_s$, i.e., paired electrons have momenta $\boldsymbol{Q}_s \pm \boldsymbol{K}$, respectively. The unitary transformation (7.137c) is nothing but a Galilean boost into the rest frame of the condensate that is moving with respect to the walls of the reservoir.[7] Whereas unpaired electrons have Matsubara frequencies ω and energies $\boldsymbol{K}^2/2m$ in the frame of reference of the container, Eq. (7.142) tells that, in the rest frame of the condensate, mean-field quasiparticles have Doppler-shifted Matsubara frequencies $\omega + \mathrm{i}V_s\widehat{\boldsymbol{Q}}_s \cdot \boldsymbol{K}$ and Eq. (7.146) tells that quasiparticles have energies

$$E'_{\boldsymbol{K},\pm} := \pm\sqrt{(\boldsymbol{K}^2/2m - \mu')^2 + |\overline{\Delta}|^2} \equiv \pm E'_{\boldsymbol{K}}, \tag{7.150}$$

respectively. The second term on the right-hand side of Eq. (7.147), i.e., the use of the chemical potential μ, says that statistical equilibrium is defined with respect to the stationary walls of the container and that it is the quasiparticle states with negative energies as seen from the container that are occupied [recall Eq. (7.141b) and use Eq. (7.153) with $\boldsymbol{Q}_s = \boldsymbol{0}$].

As soon as the speed V_s of the condensate with respect to the walls of the container reaches the critical value set by the quasiparticle gap $|\overline{\Delta}|$, breaking of some electron pairs in the moving condensate takes place. These unpaired electrons are said to be "normal" and they do not flow along with the condensate. Phenomenologically, the second term on the right-hand side of Eq. (7.148) should then become

$$\beta V \frac{N}{V}\frac{\boldsymbol{Q}_s^2}{2m} \longrightarrow \beta V \frac{N_s}{V}\frac{\boldsymbol{Q}_s^2}{2m} < \beta V \frac{N}{V}\frac{\boldsymbol{Q}_s^2}{2m}. \tag{7.151}$$

The fermionic density N/V has been replaced by the smaller density N_s/V with $N_s < N$ the depleted number of electrons that still participate to the pairing condensate.

The justification of Eq. (7.151) is based on the observation that, as soon as $|\boldsymbol{Q}_s|\Lambda/m$ is larger than the quasiparticle gap $|\overline{\Delta}|$, the transition (7.146) is not legal anymore as the shift $\mathbb{R} \to \gamma^*_{\boldsymbol{K}}$ of path of integration in the complex-valued frequency plane encounters branch cuts of the logarithm in the integrand. A related difficulty also arises if we represent the sum over the fermionic Matsubara frequencies [which at low temperatures becomes an integral as in Eq. (7.145b)] as an integral in the

---

[7] A Galilean transformation is a transformation to a new frame of reference that leaves the time difference $|t_1 - t_2|$ and space separation $|\boldsymbol{x}_1 - \boldsymbol{x}_2|$ unchanged as well as the equation of motion $m\ddot{\boldsymbol{x}} = 0$ of a free particle form invariant. It is given by

$$t' = \pm t + a, \qquad \boldsymbol{x}' = \mathbb{O}\boldsymbol{x} + \boldsymbol{v}t + \boldsymbol{w}, \qquad a \in \mathbb{R}, \qquad \boldsymbol{v}, \boldsymbol{w} \in \mathbb{R}^3, \qquad \mathbb{O} \in O(3). \tag{7.149a}$$

The transformation law of momentum $\boldsymbol{p} \equiv m\dot{\boldsymbol{x}}$ and kinetic energy $E_{\text{kin}} \equiv (1/2)m\dot{\boldsymbol{x}}^2$ under a Galilean transformation are

$$\boldsymbol{p}' = \mathbb{O}\boldsymbol{p} + m\boldsymbol{v}, \qquad E'_{\text{kin}} = E_{\text{kin}} + m\,(\mathbb{O}\dot{\boldsymbol{x}}) \cdot \boldsymbol{v} + (1/2)m\boldsymbol{v}^2. \tag{7.149b}$$

complex-frequency plane that picks up the residues of the Fermi-Dirac distribution, see Fig. 7.1. Indeed, in doing so, the right-hand side of Eq. (7.145b) becomes

$$
\begin{aligned}
-\sum_{\boldsymbol{K}}\sum_{\omega_n}\ln\left(\omega_n'^2+E_{\boldsymbol{K}}'^2\right)=&\sum_{\boldsymbol{K}}\beta\int\limits_{\Gamma_{\boldsymbol{K}}}\frac{\mathrm{d}z}{2\pi\mathrm{i}}\log\left((-\mathrm{i}z)^2+E_{\boldsymbol{K}}'^2\right)\tilde{f}_{\mathrm{FD}}(z)\\
=&\,\beta\sum_{\boldsymbol{K}}\int\limits_{\Gamma_{\boldsymbol{K}}}\frac{\mathrm{d}z}{2\pi\mathrm{i}}\left[\log\left(E_{\boldsymbol{K}}'-z\right)+\log\left(E_{\boldsymbol{K}}'+z\right)\right]\tilde{f}_{\mathrm{FD}}(z).
\end{aligned}
\tag{7.152}
$$

For a given $\boldsymbol{K}$ in the sum on the right-hand side, the integrand in the complex z-plane has branch cuts whenever $|\mathrm{Re}\,z|>E_{\boldsymbol{K}}'$. Hence, it is not permissible to deform the path $\Gamma_{\boldsymbol{K}}$ into semi-circles enclosing isolated poles as is done in Fig. 7.1. What can be done, however, is to perform a Taylor expansion in powers of $\boldsymbol{Q}_s^2/(2m)$ of the integrand that converts the branch cuts into isolated poles. In doing so, one replaces the term $\left.\left(-\frac{\partial S_{\mathrm{fred}}}{\partial\mu}\right)\right|_{\boldsymbol{Q}_s=\boldsymbol{0}}$ in Eq. (7.147) by (remember that $E_{\boldsymbol{K}}^2=\xi_{\boldsymbol{K}}^2+|\overline{\Delta}|^2,\quad \xi_{\boldsymbol{K}}=\frac{\boldsymbol{K}^2}{2m}-\mu$)

$$
\begin{aligned}
&\sum_{\boldsymbol{K}}\sum_{\omega_n}\partial_\mu\ln\left(\omega_n'^2+E_{\boldsymbol{K}}^2\right)\\
&=\sum_{\boldsymbol{K}}\sum_{\omega_n}\frac{\partial_\mu E_{\boldsymbol{K}}^2}{\omega_n'^2+E_{\boldsymbol{K}}^2}\\
&=\sum_{\boldsymbol{K}}\sum_{\omega_n}\frac{(-2)\xi_{\boldsymbol{K}}}{\left(\omega_n+\mathrm{i}\frac{\boldsymbol{Q}_s\cdot\boldsymbol{K}}{m}\right)^2+E_{\boldsymbol{K}}^2}\\
&=\sum_{\boldsymbol{K}}\beta\int\limits_{\Gamma_{\boldsymbol{K}}}\frac{\mathrm{d}z}{2\pi\mathrm{i}}\frac{(-1)^2 2\xi_{\boldsymbol{K}}\tilde{f}_{\mathrm{FD}}(z)}{\left(-\mathrm{i}z+\mathrm{i}\frac{\boldsymbol{Q}_s\cdot\boldsymbol{K}}{m}\right)^2+E_{\boldsymbol{K}}^2}\\
&=\sum_{\boldsymbol{K}}\beta\int\limits_{\Gamma_{\boldsymbol{K}}}\frac{\mathrm{d}z}{2\pi\mathrm{i}}\frac{(-1)^3 2\xi_{\boldsymbol{K}}\tilde{f}_{\mathrm{FD}}(z)}{\left(z-\frac{\boldsymbol{Q}_s\cdot\boldsymbol{K}}{m}\right)^2-E_{\boldsymbol{K}}^2}\\
&=\beta\sum_{\boldsymbol{K}}(-1)^3 2\xi_{\boldsymbol{K}}\int\limits_{\Gamma_{\boldsymbol{K}}}\frac{\mathrm{d}z}{2\pi\mathrm{i}}\frac{\tilde{f}_{\mathrm{FD}}(z)}{\left(z-\frac{\boldsymbol{Q}_s\cdot\boldsymbol{K}}{m}-|E_{\boldsymbol{K}}|\right)\left(z-\frac{\boldsymbol{Q}_s\cdot\boldsymbol{K}}{m}+|E_{\boldsymbol{K}}|\right)}\\
&=\beta\sum_{\boldsymbol{K}}(-1)^4\xi_{\boldsymbol{K}}\sum_{\pm}\frac{\tilde{f}_{\mathrm{FD}}\left(\frac{\boldsymbol{Q}_s\cdot\boldsymbol{K}}{m}\pm|E_{\boldsymbol{K}}|\right)}{(\pm)|E_{\boldsymbol{K}}|}\\
&\equiv\beta N_s.
\end{aligned}
\tag{7.153}
$$

Relative to the limit $\boldsymbol{Q}_s\to\boldsymbol{0}$, the Fermi-Dirac distribution in Eq. (7.153) removes from the summation over $\boldsymbol{K}$ all those contributions such that

$$
\frac{\boldsymbol{Q}_s\cdot\boldsymbol{K}}{m}-|E_{\boldsymbol{K}}|>0,
\tag{7.154a}
$$

while it adds to the summation over $\boldsymbol{K}$ all those contributions such that

$$\frac{\boldsymbol{Q}_s \cdot \boldsymbol{K}}{m} + |E_{\boldsymbol{K}}| < 0. \tag{7.154b}$$

The net result when $|\boldsymbol{V}_s| = |\boldsymbol{Q}_s|/m$ is above the critical velocity set by the mean-field gap is the depletion $N - N_s > 0$ in the number of electrons participating to the condensate.

7.7.2 *Time twist around* $|\overline{\Delta}|$

Choose in Eq. (7.132)

$$\theta(\boldsymbol{R}, \tau) = -\Omega\tau. \tag{7.155}$$

To evaluate

$$\mathrm{Det}\left(\mathcal{K}_{-2\theta,+2\theta}\big|_{\theta=-\Omega\tau}\right) := \mathrm{Det}\begin{pmatrix} \partial_\tau - \frac{\boldsymbol{\nabla}^2}{2m} - \mu & |\overline{\Delta}|\exp(-2\mathrm{i}\Omega\tau) \\ |\overline{\Delta}|\exp(+2\mathrm{i}\Omega\tau) & \partial_\tau + \frac{\boldsymbol{\nabla}^2}{2m} + \mu \end{pmatrix}, \tag{7.156}$$

perform the unitary (gauge) transformation

$$\mathcal{K}_{-2\theta,+2\theta}\big|_{\theta=-\Omega\tau} \to \mathcal{K}'_{-2\theta,+2\theta}\big|_{\theta=-\Omega\tau}, \tag{7.157a}$$

where

$$\begin{aligned} \mathcal{K}'_{-2\theta,+2\theta}\big|_{\theta=-\Omega\tau} &:= \left(\mathcal{U}_{-\theta,+\theta}\right)^{-1}\mathcal{K}_{-2\theta,+2\theta}\,\mathcal{U}_{-\theta,+\theta}\big|_{\theta=-\Omega\tau} \\ &= \begin{pmatrix} \partial_\tau - \mathrm{i}\Omega - \frac{\boldsymbol{\nabla}^2}{2m} - \mu & |\overline{\Delta}| \\ |\overline{\Delta}| & \partial_\tau + \mathrm{i}\Omega + \frac{\boldsymbol{\nabla}^2}{2m} + \mu \end{pmatrix}, \end{aligned} \tag{7.157b}$$

and

$$\mathcal{U}_{-\theta,+\theta}\big|_{\theta=-\Omega\tau} = \begin{pmatrix} e^{-\mathrm{i}\Omega\tau} & 0 \\ 0 & e^{+\mathrm{i}\Omega\tau} \end{pmatrix}. \tag{7.157c}$$

As before, it can be shown that

$$\mathrm{Det}\left(\mathcal{K}_{-2\theta,+2\theta}\big|_{\theta=-\Omega\tau}\right) = \mathrm{Det}\left(\mathcal{K}'_{-2\theta,+2\theta}\big|_{\theta=-\Omega\tau}\right) \tag{7.158}$$

holds. Thus, in a very large volume and at very low temperatures

$$\begin{aligned} \log\left(\mathrm{Det}\left(\mathcal{K}_{-2\theta,+2\theta}\right)\right) &= \log\left(\mathrm{Det}\left(\mathcal{K}'_{-2\theta,+2\theta}\right)\right) \\ &\approx \beta V \int\limits_V \frac{\mathrm{d}^3\boldsymbol{K}}{(2\pi)^3}\int\limits_{\mathbb{R}} \frac{\mathrm{d}\omega}{2\pi}\log\left(\det\begin{pmatrix} -\mathrm{i}\omega + \frac{\boldsymbol{K}^2}{2m} - \mu' & |\overline{\Delta}| \\ |\overline{\Delta}| & -\mathrm{i}\omega - \frac{\boldsymbol{K}^2}{2m} + \mu' \end{pmatrix}\right), \end{aligned} \tag{7.159a}$$

where we have introduced

$$\mu' = \mu + \mathrm{i}\Omega. \tag{7.159b}$$

If so, $-\log\left(\mathrm{Det}\left(\mathcal{K}_{-2\theta,+2\theta}\Big|_{\theta=-\Omega\tau}\right)\right) = S'_{\mathrm{fred}}\left[|\overline{\Delta}|,|\overline{\Delta}|\right]$ with

$$\begin{aligned}
S'_{\mathrm{fred}}\left[|\overline{\Delta}|,|\overline{\Delta}|\right] :=&\, -\log\left(\mathrm{Det}\left(\mathcal{K}'_{-2\theta,+2\theta}\right)\right)\Big|_{\theta=-\Omega\tau}\\
\approx&\, -\beta V\int_V\frac{\mathrm{d}^3\boldsymbol{K}}{(2\pi)^3}\int_{\mathbb{R}}\frac{\mathrm{d}\omega}{2\pi}\log\left(-\omega^2-\left(\frac{\boldsymbol{K}^2}{2m}-\mu'\right)^2-|\overline{\Delta}|^2\right)\\
=&\, S_{\mathrm{fred}}\left[|\overline{\Delta}|,|\overline{\Delta}|\right]-\mathrm{i}\Omega\left(-\frac{\partial S_{\mathrm{fred}}}{\partial\mu}\right)\Bigg|_{\theta=0}+\mathcal{O}\left(\Omega^2\right)\\
=&\, S_{\mathrm{fred}}\left[|\overline{\Delta}|,|\overline{\Delta}|\right]-\mathrm{i}\beta V\frac{N}{V}\Omega+\mathcal{O}\left(\Omega^2\right).
\end{aligned}\tag{7.160}$$

The last equality follows from the definition $N=\beta^{-1}\partial_\mu\ln Z_{\beta,\mu}$.

7.7.3 *Conjectured low-energy action for the phase of the condensate*

Choose in Eq. (7.132)

$$\theta(\boldsymbol{R},\tau)=\frac{1}{2}\phi(\boldsymbol{R},\tau).\tag{7.161}$$

Combination of Eqs. (7.148) and (7.160) yields at low temperatures and in a very large volume

$$\begin{aligned}
&S_{\mathrm{fred}}\left[|\overline{\Delta}|e^{+\mathrm{i}\phi(\boldsymbol{R},\tau)},|\overline{\Delta}|e^{-\mathrm{i}\phi(\boldsymbol{R},\tau)}\right]-S_{\mathrm{fred}}\left[|\overline{\Delta}|,|\overline{\Delta}|\right]\\
&\approx\int_V\mathrm{d}^3\boldsymbol{R}\int_0^\beta\mathrm{d}\tau\left\{\mathrm{i}\frac{N}{V}\left(\frac{\partial_\tau\phi}{2}\right)+\frac{1}{2m}\frac{N}{V}\left(\frac{\boldsymbol{\nabla}\phi}{2}\right)^2+\mathcal{O}\left[(\partial_\tau\phi)^2,(\boldsymbol{\nabla}\phi\cdot\boldsymbol{\nabla}\phi)^2\right]\right\}.
\end{aligned}\tag{7.162}$$

Analytical continuation

$$\tau=\mathrm{i}t,\qquad t\in\mathbb{R},\tag{7.163}$$

gives

$$\begin{aligned}
&-\lim_{\tau\to\mathrm{i}t}\left\{S_{\mathrm{fred}}\left[|\overline{\Delta}|e^{+\mathrm{i}\phi(\boldsymbol{R},\tau)},|\overline{\Delta}|e^{-\mathrm{i}\phi(\boldsymbol{R},\tau)}\right]-S_{\mathrm{fred}}\left[|\overline{\Delta}|,|\overline{\Delta}|\right]\right\}\\
&\approx\mathrm{i}\int_V\mathrm{d}^3\boldsymbol{R}\int_0^{-\mathrm{i}\beta}\mathrm{d}t\left\{-\frac{N}{V}\left(\frac{\partial_t\phi}{2}\right)-\frac{1}{2m}\frac{N}{V}\left(\frac{\boldsymbol{\nabla}\phi}{2}\right)^2+\mathcal{O}\left[(\partial_t\phi)^2,(\boldsymbol{\nabla}\phi\cdot\boldsymbol{\nabla}\phi)^2\right]\right\}.
\end{aligned}\tag{7.164}$$

Observe that Eq. (7.164) is left unchanged if

$$\phi(\boldsymbol{R},\mathrm{i}t)\longrightarrow\phi(\boldsymbol{R},\mathrm{i}t)+2\left(\boldsymbol{Q}_s\cdot\boldsymbol{R}-\frac{1}{2m}\boldsymbol{Q}_s^2\,t\right).\tag{7.165}$$

(The boundary term vanishes because of the periodic boundary conditions.) This transformation is a quantum counterpart to the classical Galilean transformation in footnote 7 [see Eqs. (7.149a) and (7.149b)].

Let us relax the assumption that the electronic density is frozen to the value N/V. It is then tempting to conjecture the Madelung action defined by

$$\begin{aligned} S_{\mathrm{Mad}}\left[\phi,\rho|A_0,\boldsymbol{A}\right] &:= \int\limits_V \mathrm{d}^3\boldsymbol{R} \int\limits_{\mathbb{R}} \mathrm{d}t\, \mathcal{L}_{\mathrm{Mad}}, \\ \mathcal{L}_{\mathrm{Mad}} &= -\rho\left(\frac{\partial_t\phi}{2} - e\,A_0\right) - \frac{1}{2m}\rho\left(\frac{\boldsymbol{\nabla}\phi}{2} + e\boldsymbol{A}\right)^2 - \frac{\lambda}{8}\left(\rho - \frac{N}{V}\right)^2, \end{aligned} \tag{7.166}$$

for the effective low-energy action describing space- and time-dependent fluctuations of the phase $\phi(\boldsymbol{R},t)$ of the pairing-order parameter $\Delta(\boldsymbol{R},t) = |\overline{\Delta}|e^{+\mathrm{i}\phi(\boldsymbol{R},t)}$, as well as space- and time-dependent fluctuations of the electronic density $\rho(\boldsymbol{R},t)$. Here, $A_0(\boldsymbol{R},t)$ and $\boldsymbol{A}(\boldsymbol{R},t)$ are the scalar and vector potentials of electrodynamics, respectively. They should be understood as playing the role of classical external sources obeying Maxwell equations at this stage. The convention $e > 0$ is chosen for the electric charge. The positive coupling λ with the dimension of energy times volume (in units with $\hbar = c = 1$) freezes the density $\rho(\boldsymbol{R},t)$ to the value N/V in the limit $\lambda \to \infty$.

By *defining*[8]

$$\Xi(\boldsymbol{R},t) := \sqrt{\frac{\rho(\boldsymbol{R},t)}{2}}\,\exp\left(\mathrm{i}\phi(\boldsymbol{R},t)\right), \qquad e^* := 2e, \qquad m^* = 2m, \tag{7.167}$$

Eq. (7.166) can be brought to (we have reinstated Planck's constant $\hbar$ and the speed of light c)

$$\begin{aligned} S\left[\Xi|A_0',\boldsymbol{A}'\right] &= \int\limits_V \mathrm{d}^3\boldsymbol{R} \int\limits_{\mathbb{R}} \mathrm{d}t\, \mathcal{L}, \\ \mathcal{L} &= \Xi^*\left(\mathrm{i}\hbar\,\partial_t - (-e^*)\,A_0'\right)\Xi - \frac{1}{2m^*}\left|\left(\frac{\hbar}{\mathrm{i}}\boldsymbol{\nabla} - \frac{(-e^*)}{c}\boldsymbol{A}'\right)\Xi\right|^2 - \frac{\lambda}{2}\left(|\Xi|^2 - \frac{1}{2}\frac{N}{V}\right)^2, \\ A_0' &= A_0 - \frac{\mathrm{i}\hbar}{2e^*}\,\partial_t \ln\rho, \qquad \boldsymbol{A}' = \boldsymbol{A} + \frac{\mathrm{i}\hbar c}{2e^*}\,\boldsymbol{\nabla}\ln\rho. \end{aligned} \tag{7.168}$$

If we ignore the dependence of the gauge fields A_0' and $\boldsymbol{A}'$ on ρ, the classical equation of motion for Ξ that follows from Eq. (7.168),

$$\mathrm{i}\hbar\left(\partial_t - \frac{\mathrm{i}e^*}{\hbar}\,A_0\right)\Xi = -\frac{\hbar^2}{2m^*}\left(\boldsymbol{\nabla} + \frac{\mathrm{i}e^*}{\hbar c}\boldsymbol{A}\right)^2\Xi + \lambda\left(|\Xi|^2 - \frac{1}{2}\frac{N}{V}\right)\Xi, \tag{7.169}$$

is known as the *Gross-Pitaevskii* non-linear Schrödinger equation for Ξ. The Gross-Pitaevskii equation is often used as a model for the motion of the superconductor (superfluid) condensate. Revival in the interest for the Gross-Pitaevskii equation

[8] It is important to stress that Ξ is *not the same as* the pairing-order parameter Δ.

has ensued the experimental observation of Bose-Einstein condensation in vapors of rubidium. A recent review on the Gross-Pitaevskii equation can be found in Ref. [76]. Our "derivation" of the Gross-Pitaevskii equation could be misleading in that we have only been considering space-time dependent fluctuations of the phase of the superconducting order parameter whereas the Gross-Pitaevskii equation allows for fluctuations of the amplitude of Ξ as well. A rigorous justification for the Gross-Pitaevskii equation in the context of superfluidity in three- (two-) dimensional position space can be found in Ref. [77] ([78]).

Invariance of Eq. (7.168) under any global gauge transformation

$$\Xi(\boldsymbol{R},t) \to \exp(\mathrm{i}\alpha)\,\Xi(\boldsymbol{R},t), \qquad \alpha \in \mathbb{R}, \tag{7.170}$$

implies the continuity equation

$$0 = \partial_t J_0 + \boldsymbol{\nabla}\cdot\boldsymbol{J}, \tag{7.171a}$$

whereby the density and currents are

$$J_0 := \Xi^*\,\Xi, \tag{7.171b}$$

and

$$\boldsymbol{J} := \frac{\hbar}{2m^*\mathrm{i}}\left[\Xi^*\left(\boldsymbol{\nabla} + \frac{\mathrm{i}e^*}{\hbar c}\boldsymbol{A}'\right)\Xi - \text{c.c.}\right], \tag{7.171c}$$

respectively. Alternatively, the density J_0 and current $\boldsymbol{J}$ are

$$J_0 = \frac{\rho}{2}, \qquad \boldsymbol{J} = \frac{\rho}{2}\boldsymbol{V}, \qquad \boldsymbol{V} := \frac{\hbar}{m^*}\left(\boldsymbol{\nabla}\phi + \frac{e^*}{\hbar c}\boldsymbol{A}\right), \tag{7.172}$$

if the so-called *Madelung transformation* (7.167) is undone.[9]

When the vector field $\boldsymbol{\nabla}\phi$ is rotation free, one says that there are no vortex singularities in Ξ. If so, the vorticity of the velocity field $\boldsymbol{V}$ is entirely determined by the magnetic field

$$\boldsymbol{B} = \boldsymbol{\nabla}\wedge\boldsymbol{A}' = \boldsymbol{\nabla}\wedge\boldsymbol{A} \tag{7.173}$$

through

$$\boldsymbol{\nabla}\wedge\boldsymbol{V} = +\frac{e^*}{m^*\,c}\boldsymbol{B} \Longleftrightarrow \boldsymbol{\nabla}\wedge\boldsymbol{V} - \frac{e^*}{m^*\,c}\boldsymbol{B} = 0. \tag{7.174}$$

The condition (7.174) for the absence of vortices supplemented with[10]

$$(\boldsymbol{\nabla}\rho)\wedge\boldsymbol{V} = 0 \tag{7.179}$$

---

[9] Alternatively, variation of action (7.166) with respect to ϕ gives the continuity equation.

[10] In a classical fluid, conservation of particle number implies the continuity equation

$$0 = \partial_t\rho + \boldsymbol{\nabla}\cdot(\rho\boldsymbol{v}) = [\partial_t\rho + (\boldsymbol{\nabla}\rho)\cdot\boldsymbol{v}] + \rho\,(\boldsymbol{\nabla}\cdot\boldsymbol{v}) \equiv \frac{\mathrm{d}\rho}{\mathrm{d}t} + \rho\,(\boldsymbol{\nabla}\cdot\boldsymbol{v}). \tag{7.175}$$

In a *steady fluid* all ∂_t vanish, i.e., the continuity equation reduces to

$$0 = \boldsymbol{\nabla}\cdot(\rho\boldsymbol{v}). \tag{7.176}$$

An *incompressible* fluid is defined by the conditions that

$$0 = \partial_t\rho, \qquad 0 = \frac{\mathrm{d}\rho}{\mathrm{d}t}, \tag{7.177}$$

in which case the continuity equation reduces to

$$0 = \boldsymbol{\nabla}\cdot\boldsymbol{v}. \tag{7.178}$$

Neglecting fluctuations in the magnitude of the pairing-order parameter as was done to derive Eq. (7.164) implies that the condensate is taken as incompressible, i.e., ρ is constant.

can be combined with Maxwell's equations

$$\boldsymbol{\nabla}\wedge\boldsymbol{B}=\frac{4\pi e^{*}}{c}\boldsymbol{J},\qquad \boldsymbol{\nabla}\cdot\boldsymbol{B}=0, \tag{7.180}$$

and the definition (7.172) of the current to give the equation of motion

$$\begin{aligned}0&=\boldsymbol{\nabla}\wedge\left(\boldsymbol{\nabla}\wedge\boldsymbol{B}-\frac{4\pi e^{*}}{c}\boldsymbol{J}\right)\\&=\left[\boldsymbol{\nabla}\left(\boldsymbol{\nabla}\cdot\boldsymbol{B}\right)-\boldsymbol{\nabla}^{2}\boldsymbol{B}-\frac{4\pi e^{*}}{c}\boldsymbol{\nabla}\wedge\left(\frac{\rho}{2}\boldsymbol{V}\right)\right]\\&=-\left(\boldsymbol{\nabla}^{2}-\frac{4\pi e^{*2}(\rho/2)}{m^{*}c^{2}}\right)\boldsymbol{B}\end{aligned}\tag{7.181}$$

obeyed by a static magnetic field. The length

$$\lambda_{\mathrm{London}}:=\left(\frac{4\pi e^{*2}(\rho/2)}{m^{*}c^{2}}\right)^{-1/2}\tag{7.182}$$

is called the *London penetration depth.* The London penetration depth does not depend on $\hbar$. Quantum mechanics enters through $\hbar$ only if vortices are present as is implied by the modification to (7.174) brought upon by vortices. The London penetration depth controls the exponential decay of a solution to Eq. (7.181). This property that a static magnetic field becomes massive inside a superconductor is called the *Meissner effect.* We are now going to see that the combined effects of the Meissner effect and of the ability of the condensate to screen static charges is to provide the photon with an effective mass inside a superconductor.

7.7.4 *Polarization tensor for a BCS superconductor*

The goal of this section is to further substantiate a low-energy effective theory for the phase $\phi=2\theta$ of the pairing-order parameter. To this end, we modify the partition function (7.64) by coupling the fermions to the classical gauge fields of electromagnetism $(\varphi,\boldsymbol{A})$ through the minimal coupling (in imaginary time)

$$-\mu\to-\mu+(-e)\,\varphi(\boldsymbol{R},\tau),\qquad \frac{\boldsymbol{\nabla}}{\mathrm{i}}\to\frac{\boldsymbol{\nabla}}{\mathrm{i}}-(-e)\,\boldsymbol{A}(\boldsymbol{R},\tau).\tag{7.183}$$

We choose the conventions $e>0$ for the unit of electric charge, and $\hbar=1$ and $c=1$ for Planck's constant and the speed of light, respectively. We are going to use the Nambu-Gorkov representation introduced in section 7.5. This is to say that we need the Bogoliubov-de-Gennes Hamiltonian (7.112), whereby we make the identification

$$\begin{aligned}+\mathbb{K}&\to+\left[\frac{1}{2m}\left(\frac{\boldsymbol{\nabla}}{\mathrm{i}}+e\,\boldsymbol{A}(\boldsymbol{R},\tau)\right)^{2}-\mu-e\varphi(\boldsymbol{R},\tau)\right]\sigma_{0},\\-\mathbb{K}^{\mathsf{T}}&\to-\left[\frac{1}{2m}\left(\frac{\boldsymbol{\nabla}}{\mathrm{i}}-e\,\boldsymbol{A}(\boldsymbol{R},\tau)\right)^{2}-\mu-e\varphi(\boldsymbol{R},\tau)\right]\sigma_{0},\\+\mathbb{D}&\to|\overline{\Delta}|e^{+2\mathrm{i}\theta(\boldsymbol{R},\tau)}\,(\mathrm{i}\sigma_{2})\,.\end{aligned}\tag{7.184}$$

Because we chose a pairing-order parameter that is a singlet in $SU(2)$ spin space, we might as well ignore the spin-1/2 grading to work with the single-particle Bogoliubov-de-Gennes Hamiltonian [compare with Eq. (7.110)]

$$H_{\mathrm{BdG}} := \begin{pmatrix} -\frac{[\boldsymbol{\nabla}+\mathrm{i}e\boldsymbol{A}(\boldsymbol{R},\tau)]^2}{2m} - \mu - e\,\varphi(\boldsymbol{R},\tau) & |\overline{\Delta}|e^{+2\mathrm{i}\theta(\boldsymbol{R},\tau)} \\ |\overline{\Delta}|e^{-2\mathrm{i}\theta(\boldsymbol{R},\tau)} & +\frac{[\boldsymbol{\nabla}-\mathrm{i}e\boldsymbol{A}(\boldsymbol{R},\tau)]^2}{2m} + \mu + e\,\varphi(\boldsymbol{R},\tau) \end{pmatrix}. \tag{7.185}$$

Needed is the evaluation to Gaussian order in θ, φ, and $\boldsymbol{A}$ in the zero temperature limit $T \to 0$ of the fermionic determinant

$$\begin{aligned} e^{-S_{\mathrm{fred}}[\theta,\varphi,\boldsymbol{A}]} &\equiv \int \mathcal{D}[\psi^*,\psi]\, e^{-(S_0+S'_U)[\theta,\varphi,\boldsymbol{A}]} \\ &:= \mathrm{Det}\,(\partial_\tau \gamma_0 + H_{\mathrm{BdG}})\,. \end{aligned} \tag{7.186}$$

However, because we would like to compare the Gaussian approximation to the effective theory $S_{\mathrm{fred}}[\theta,\varphi,\boldsymbol{A}]$ to the effective action $S^{\mathrm{RPA}}_{\beta,\mu}$ defined in Eq. (6.38), we shall perform the analytical continuation

$$\varphi(\boldsymbol{R},\tau) \in \mathbb{R} \to -\mathrm{i}\varphi(\boldsymbol{R},\tau) \in \mathrm{i}\mathbb{R}, \tag{7.187}$$

which allows us to interpret $\varphi(\boldsymbol{R},\tau)$ as the Hubbard-Stratonovich field that decouples an instantaneous Coulomb interaction. Thus, we shall perform, at vanishing temperature, the Gaussian approximation of

$$\mathrm{Det}\begin{pmatrix} \partial_\tau - \frac{[\boldsymbol{\nabla}+\mathrm{i}e\,\boldsymbol{A}(\boldsymbol{R},\tau)]^2}{2m} - \mu + \mathrm{i}e\,\varphi(\boldsymbol{R},\tau) & |\overline{\Delta}|\,e^{+2\mathrm{i}\theta(\boldsymbol{R},\tau)} \\ |\overline{\Delta}|\,e^{-2\mathrm{i}\theta(\boldsymbol{R},\tau)} & \partial_\tau + \frac{[\boldsymbol{\nabla}-\mathrm{i}e\,\boldsymbol{A}(\boldsymbol{R},\tau)]^2}{2m} + \mu - \mathrm{i}e\,\varphi(\boldsymbol{R},\tau) \end{pmatrix} \tag{7.188a}$$

in the background of the pairing-order parameter

$$\Delta^*(\boldsymbol{R},\tau) = |\overline{\Delta}|e^{-2\mathrm{i}\theta(\boldsymbol{R},\tau)}, \qquad \Delta(\boldsymbol{R},\tau) = |\overline{\Delta}|e^{+2\mathrm{i}\theta(\boldsymbol{R},\tau)}, \tag{7.188b}$$

and of the Euclidean electromagnetic potentials $-\mathrm{i}\varphi(\boldsymbol{R},\tau)$ and $\boldsymbol{A}(\boldsymbol{R},\tau)$. The short-hand notation

$$x \equiv (\boldsymbol{R},\tau), \qquad K \equiv (\boldsymbol{K},\omega_n), \qquad \omega_n = \frac{\pi}{\beta}(2n+1), \qquad n \in \mathbb{Z}, \tag{7.188c}$$

will be used from now on.

The lesson learned in sections 7.7.1 and 7.7.2 is that the *local gauge transformation*

$$\psi^*_\sigma(x) = \psi^{*\prime}_\sigma(x)\, e^{-\mathrm{i}\theta(x)}, \qquad \psi_\sigma(x) = e^{+\mathrm{i}\theta(x)}\, \psi'_\sigma(x), \tag{7.189}$$

is advantageous. It turns the action $S_0 + S'_U$ into the action $(S_0 + S'_U)'$ whereby

$$
\begin{aligned}
(S_0 + S'_U)' &= \int \mathrm{d}^4x \left(\mathcal{L}_0 + \mathcal{L}_{1,1} + \mathcal{L}_{1,2} + \mathcal{L}_2\right), \\
\mathcal{L}_0 &= \Psi^{\dagger\prime} \begin{pmatrix} \partial_\tau - \frac{\boldsymbol{\nabla}^2}{2m} - \mu & |\Delta| \\ |\Delta| & \partial_\tau + \frac{\boldsymbol{\nabla}^2}{2m} + \mu \end{pmatrix} \Psi', \\
\mathcal{L}_{1,1} &= \Psi^{\dagger\prime} \begin{pmatrix} +(\mathrm{i}\partial_\tau\theta) + \mathrm{i}e\,\varphi & 0 \\ 0 & -(\mathrm{i}\partial_\tau\theta) - \mathrm{i}e\,\varphi \end{pmatrix} \Psi', \\
\mathcal{L}_{1,2} &= \Psi^{\dagger\prime} \begin{pmatrix} +\left(\frac{(\boldsymbol{\nabla}\theta)+e\boldsymbol{A}}{2m\mathrm{i}} \cdot \boldsymbol{\nabla}\right)^{\leftrightharpoons} & 0 \\ 0 & +\left(\frac{(\boldsymbol{\nabla}\theta)+e\boldsymbol{A}}{2m\mathrm{i}} \cdot \boldsymbol{\nabla}\right)^{\leftrightharpoons} \end{pmatrix} \Psi', \\
\mathcal{L}_2 &= \Psi^{\dagger\prime} \begin{pmatrix} +\frac{1}{2m}\left[(\boldsymbol{\nabla}\theta) + e\boldsymbol{A}\right]^2 & 0 \\ 0 & -\frac{1}{2m}\left[(\boldsymbol{\nabla}\theta) + e\boldsymbol{A}\right]^2 \end{pmatrix} \Psi'.
\end{aligned}
\tag{7.190}
$$

Here, we are using the notation

$$
f\,(\cdots\boldsymbol{\nabla})^{\leftrightharpoons}\,g \equiv f(\cdots\boldsymbol{\nabla}g) - (\cdots\boldsymbol{\nabla}f)g \tag{7.191}
$$

and we made use of the following identities. First, if δ denotes a differential while f, $g = e^{\mathrm{i}\theta}$, and h denote three functions, then

$$
\begin{aligned}
h^* g^{-1}(\delta + \mathrm{i}f)^2(gh) = {} & \underbrace{\mathrm{i}h^*(\delta^2\theta)h}_{\#1,2} - \underbrace{h^*(\delta\theta)^2 h}_{\#2} + \underbrace{2\mathrm{i}h^*(\delta\theta)(\delta h)}_{\#1,2} + \underbrace{h^*(\delta^2 h)}_{\#0} \\
& + \underbrace{\mathrm{i}h^*(\delta f)h}_{\#1,2} - \underbrace{2h^*(\delta\theta)fh}_{\#2} + \underbrace{2\mathrm{i}h^* f(\delta h)}_{\#1,2} - \underbrace{h^* f^2 h}_{\#2},
\end{aligned}
\tag{7.192}
$$

as

$$
\begin{aligned}
(\delta + \mathrm{i}f)^2(gh) &= \left(\delta^2 + \mathrm{i}(\delta f) + 2\mathrm{i}f\delta - f^2\right)(gh), \\
\delta(gh) &= (\delta g)h + g(\delta h) \\
&= e^{\mathrm{i}\theta}\left[\mathrm{i}(\delta\theta)h + (\delta h)\right], \\
\delta^2(gh) &= (\delta^2 g)h + 2(\delta g)(\delta h) + g(\delta^2 h) \\
&= e^{\mathrm{i}\theta}\left[\mathrm{i}(\delta^2\theta)h - (\delta\theta)^2 h + 2\mathrm{i}(\delta\theta)(\delta h) + (\delta^2 h)\right].
\end{aligned}
\tag{7.193}
$$

Second, we converted the terms #1, 2 into the total differentials

$$
\mathrm{i}h^*(\delta^2\theta)h + 2\mathrm{i}h^*(\delta\theta)(\delta h) = \mathrm{i}\delta\left(h^*(\delta\theta)h\right) - \mathrm{i}(\delta h^*)(\delta\theta)h + \mathrm{i}h^*(\delta\theta)(\delta h), \tag{7.194}
$$

and

$$
\mathrm{i}h^*(\delta f)h + 2\mathrm{i}h^* f(\delta h) = \mathrm{i}\delta\left(h^* f h\right) - \mathrm{i}(\delta h^*)fh + \mathrm{i}h^* f(\delta h), \tag{7.195}
$$

respectively. Third, the periodic boundary conditions obeyed by the functions f, g, and h allow to drop any total derivatives after integration.

The kernel in $\mathcal{L}_0$ defines the unperturbed Green function

$$
G_0 := -\left[\gamma_0\,\partial_\tau + \gamma_3\left(-\frac{\boldsymbol{\nabla}^2}{2m} - \mu\right) + \gamma_1\,|\Delta|\right]^{-1}. \tag{7.196}
$$

To first order in θ, φ, and $\boldsymbol{A}$, there are two contributions

$$V_{1,1} := \gamma_3 \left[(\mathrm{i}\partial_\tau \theta) + \mathrm{i}e\,\varphi\right] \tag{7.197}$$

and

$$V_{1,2} := \gamma_0 \frac{1}{2m\mathrm{i}} \left[(\boldsymbol{\nabla}\theta) + e\boldsymbol{A}\right] \cdot \overleftrightarrow{\boldsymbol{\nabla}} \tag{7.198}$$

in the kernels of $\mathcal{L}_{1,1}$ and $\mathcal{L}_{1,2}$, respectively. Observe that $(\mathrm{i}\partial_\tau\theta)$ couples to the electronic density in the same way as the scalar potential does and thus is proportional to γ_3 in the particle-hole grading. On the other hand, $(\boldsymbol{\nabla}\theta)$ couples to the *paramagnetic current*

$$\boldsymbol{J}_{\mathrm{p}} := \frac{1}{2m\mathrm{i}} \sum_{\sigma=\uparrow,\downarrow} \left[\psi^*_\sigma\,(\boldsymbol{\nabla}\psi_\sigma) - (\boldsymbol{\nabla}\psi^*_\sigma)\,\psi_\sigma\right] \tag{7.199}$$

as the vector potential does and is thus proportional to γ_0 in the particle-hole grading. To second order in θ or $\boldsymbol{A}$, there is a single contribution

$$V_2 := \gamma_3 \frac{1}{2m} \left[(\boldsymbol{\nabla}\theta) + e\boldsymbol{A}\right]^2 \tag{7.200}$$

in the kernel of $\mathcal{L}_2$, i.e., $\boldsymbol{\nabla}\theta$ contributes to the *diamagnetic current*

$$\boldsymbol{J}_{\mathrm{d}} := -\frac{1}{m} \left(\sum_{\sigma=\uparrow,\downarrow} \psi^*_\sigma \psi_\sigma\right) \left[(\boldsymbol{\nabla}\theta) + e\boldsymbol{A}\right] \tag{7.201}$$

as the vector potential does. The matrix elements of G_0, $V_{1,1}$, $V_{1,2}$, and V_2 in reciprocal space are, given the notation $\xi_{\boldsymbol{K}} = \frac{\boldsymbol{K}^2}{2m} - \mu$,

$$\begin{aligned}
G_{0K}\delta_{K',K} &= -\left(-\mathrm{i}\gamma_0\,\omega_n + \gamma_3\,\xi_{\boldsymbol{K}} + \gamma_1|\overline{\Delta}|\right)^{-1}\delta_{K',K} \\
&= -\frac{1}{\omega_n^2 + \xi_{\boldsymbol{K}}^2 + |\overline{\Delta}|^2}\left(+\mathrm{i}\gamma_0\,\omega_n + \gamma_3\,\xi_{\boldsymbol{K}} + \gamma_1|\overline{\Delta}|\right)\delta_{K',K}, \\
V_{1,1\,K',K} &= \gamma_3 \frac{1}{\sqrt{\beta V}}\left[(\omega_n - \omega_{n'})\,\theta_{K-K'} + \mathrm{i}e\,\varphi_{K-K'}\right], \\
V_{1,2\,K',K} &= \gamma_0 \frac{1}{\sqrt{\beta V}}\left[-\frac{(\boldsymbol{K}-\boldsymbol{K}')\cdot(\boldsymbol{K}+\boldsymbol{K}')}{2m\mathrm{i}}\theta_{K-K'} + \mathrm{i}\frac{(\boldsymbol{K}+\boldsymbol{K}')}{2m\mathrm{i}}\cdot e\boldsymbol{A}_{K-K'}\right], \\
V_{2\,K',K} &= \gamma_3 \frac{1}{\sqrt{\beta V}}\frac{1}{2m}\left\{\left[(\boldsymbol{\nabla}\theta) + e\boldsymbol{A}\right]^2\right\}_{K-K'},
\end{aligned} \tag{7.202}$$

respectively. Here, we are using the symmetric convention $[x \equiv (\boldsymbol{R},\tau)$, $K \equiv (\boldsymbol{K},\omega_n)$, and $Q \equiv (\boldsymbol{Q},\varpi_l)]$

$$\begin{aligned}
f(x) &= \frac{1}{\sqrt{\beta V}}\sum_K f_K\, e^{+\mathrm{i}K\,x}, \qquad & f_K &= \frac{1}{\sqrt{\beta V}}\int\limits_{\beta V} \mathrm{d}^4K\, f(x)\, e^{-\mathrm{i}K\,x}, \\
g(x) &= \frac{1}{\sqrt{\beta V}}\sum_Q g_Q\, e^{+\mathrm{i}Q\,x}, \qquad & g_Q &= \frac{1}{\sqrt{\beta V}}\int\limits_{\beta V} \mathrm{d}^4Q\, g(x)\, e^{-\mathrm{i}Q\,x}, \\
(g_1g_2)(x) &= \frac{1}{\sqrt{\beta V}}\sum_Q (g_1g_2)_Q\, e^{+\mathrm{i}Q\,x}, \qquad & (g_1g_2)_Q &= \frac{1}{\sqrt{\beta V}}\sum_K (g_1)_{+K}(g_2)_{Q-K},
\end{aligned} \tag{7.203}$$

for the Fourier transforms of the Grassmann-valued f and the complex-valued g and $g_1 g_2$, respectively.

As in section 6.4, we need to approximate a fermionic determinant of the form

$$\begin{aligned}\mathrm{Tr}\ln M := {} & \mathrm{Tr}\ln(M_0 + M_1) \\ = {} & \mathrm{Tr}\ln\left[M_0\left(1 + M_0^{-1}M_1\right)\right] \\ = {} & \mathrm{Tr}\ln M_0 + \mathrm{Tr}\ln\left(1 + M_0^{-1}M_1\right).\end{aligned} \tag{7.204}$$

We thus perform the expansion

$$\begin{aligned}\mathrm{Tr}\ln M = {} & \mathrm{Tr}\ln(-G_0^{-1} + M_1) \\ = {} & \mathrm{Tr}\ln\left(-G_0^{-1}\right) + \mathrm{Tr}\ln\left(1 - G_0 M_1\right) \\ = {} & \mathrm{Tr}\ln(-G_0^{-1}) - \sum_{n=1}^{\infty}\frac{1}{n}\mathrm{Tr}\,(G_0 M_1)^n\end{aligned} \tag{7.205}$$

to the desired order. The unperturbed Green function $G_0 = -(M_0)^{-1}$ and the perturbation $M_1 = V_{1,1} + V_{1,2} + V_2$ are defined in Eq. (7.202). To quadratic order in the fields θ, φ, and $\boldsymbol{A}$, we must thus evaluate

$$\begin{aligned}\mathrm{Tr}\ln M = {} & \mathrm{Tr}\ln(-G_0^{-1}) \\ & -\mathrm{Tr}\,(G_0 V_{1,1}) - \mathrm{Tr}\,(G_0 V_{1,2}) \\ & -\mathrm{Tr}\,(G_0 V_2) - \frac{1}{2}\mathrm{Tr}\,(G_0 V_{1,1} G_0 V_{1,1}) - \frac{1}{2}\mathrm{Tr}\,(G_0 V_{1,2} G_0 V_{1,2}) \\ & -\mathrm{Tr}\,(G_0 V_{1,1} G_0 V_{1,2}) + \cdots.\end{aligned} \tag{7.206}$$

7.7.4.1 *First-order contributions*

If we impose the condition for charge neutrality, i.e., $\varphi_q\big|_{q=\mathbf{0}} = 0$ and $\boldsymbol{A}_q\big|_{q=\mathbf{0}} = 0$, there is no first-order contributions to the expansion of the fermionic determinant.

7.7.4.2 *Second-order contributions*

There are two contributions, one to order $n = 1$ and one to order $n = 2$ in the expansion of the series (7.205). These two contributions are given by

$$\begin{aligned}-\mathrm{Tr}\,(G_0 V_2) = {} & -\sum_{K,K'}\mathrm{tr}\left(G_{0K}\,\delta_{K,K'}\,V_{2\,K',K}\right) \\ = {} & -\frac{\bar{\rho}_0}{2m}\sum_{Q}\left[(+\mathrm{i}\boldsymbol{Q}\,\theta_{+Q}) + e\,\boldsymbol{A}_{+Q}\right]\cdot\left[(-\mathrm{i}\boldsymbol{Q}\,\theta_{-Q}) + e\,\boldsymbol{A}_{-Q}\right]\end{aligned} \tag{7.207}$$

and

$$-\frac{1}{2}\mathrm{Tr}\,\left(G_0\,V_{1,1}\,G_0\,V_{1,1}\right) - \frac{1}{2}\mathrm{Tr}\,\left(G_0\,V_{1,2}\,G_0\,V_{1,2}\right) - \mathrm{Tr}\,\left(G_0\,V_{1,1}\,G_0\,V_{1,2}\right), \tag{7.208}$$

respectively. We have used Eqs. (7.202) and (7.203) and introduced the mean-field electronic density [compare with Eq. (6.21e)]

$$\begin{aligned}\overline{\rho}_0 :=& \frac{1}{\beta V}\int\limits_0^\beta \mathrm{d}\tau \int\limits_V \mathrm{d}^3\boldsymbol{r} \sum_{\sigma=\uparrow,\downarrow} \int \mathcal{D}[\psi^*,\psi]\, \frac{e^{-\int \mathrm{d}^4x\, \mathcal{L}_0}}{Z_0}\, (\psi^*_\sigma\, \psi_\sigma)(\boldsymbol{r},\tau) \\ =& \frac{1}{\beta V}\sum_K \mathrm{tr}\,(G_{0K}\,\gamma_3)\end{aligned} \tag{7.209}$$

to reach the second line of Eq. (7.207). The cross term in Eq. (7.208) vanishes since the ground state, here an isotropic gapped Fermi sphere, preserves the rotational invariance of the Hamiltonian. In other words,

$$\left\langle \rho(x)\, \boldsymbol{J}_{\mathrm{p}}(x)\right\rangle_{\text{Gapped FS}} = 0. \tag{7.210}$$

We now introduce the Fourier components $\Pi_{\mu\nu q}$ of the polarization tensor $\Pi_{\mu\nu}$, $\mu,\nu = 0,1,2,3$ through

$$\begin{aligned}\Pi_{00q} \equiv& \Pi^{\parallel}_{00q} \\ :=& \frac{1}{\beta V}\sum_k \mathrm{tr}\left[G_{0(k+q)}\,\gamma_3\, G_{0k}\,\gamma_3\right], \\ \Pi_{ijq} \equiv& \Pi^{\perp}_{ijq} \\ :=& \frac{1}{\beta V}\sum_k \frac{(k_i+q_i/2)\,(k_j+q_j/2)}{m^2}\mathrm{tr}\left[G_{0(k+q)}\,\gamma_0\, G_{0k}\,\gamma_0\right], \quad i,j=1,2,3, \\ \Pi_{\mu\nu q} \equiv& 0, \qquad \text{for } \mu=0,\ \nu=1,2,3 \text{ or } \mu=1,2,3,\ \nu=0.\end{aligned} \tag{7.211}$$

In terms of the polarization tensor, the contributions (7.208) to the expansion of the fermionic determinant to second order in the background fields are given by

$$\begin{aligned}&-\frac{1}{2}\mathrm{Tr}\,\left(G_0\,V_{1,1}\,G_0\,V_{1,1}\right) \\ &= -\frac{1}{2}\sum_{q=(\varpi_l,\boldsymbol{q})} \left(+\varpi_l\,\theta_{+q}+\mathrm{i}e\,\varphi_{+q}\right)\Pi^{\parallel}_{00q}\left(-\varpi_l\,\theta_{-q}+\mathrm{i}e\,\varphi_{-q}\right)\end{aligned} \tag{7.212}$$

and

$$\begin{aligned}&-\frac{1}{2}\mathrm{Tr}\,\left(G_0\,V_{1,2}\,G_0\,V_{1,2}\right) \\ &= (-1)^2\frac{1}{2}\sum_{q=(\varpi_l,\boldsymbol{q})}\sum_{i,j=1}^{3}\left(-q_i\,\theta_{+q}+\mathrm{i}e\,A_{i(+q)}\right)\Pi^{\perp}_{ijq}\left(+q_j\,\theta_{-q}+\mathrm{i}e\,A_{j(-q)}\right),\end{aligned} \tag{7.213}$$

respectively. Collecting all terms in the expansion to Gaussian order of the fermionic determinant (7.186) gives the RPA partition function

$$
\begin{aligned}
Z^{\mathrm{RPA}}_{\beta,\mu} &\propto \int \mathcal{D}[\theta]\, e^{-S^{\mathrm{RPA}}_{\beta,\mu}}, \\
S^{\mathrm{RPA}}_{\beta,\mu} &= \frac{1}{2} \sum_{Q=(\Omega_l,\boldsymbol{Q})} \left(+\Omega_l\, \theta_{+Q} + \mathrm{i} e\, \varphi_{+Q}\right) \Pi^{\parallel}_{00Q} \left(-\Omega_l\, \theta_{-Q} + \mathrm{i} e\, \varphi_{-Q}\right) \\
&+ \frac{1}{2} \sum_{Q=(\Omega_l,\boldsymbol{Q})} \sum_{i,j=1}^{3} \Big(Q_i\, \theta_{+Q} - \mathrm{i} e\, A_{i(+Q)}\Big)\, \Pi^{\perp}_{ijQ} \Big(Q_j\, \theta_{-Q} + \mathrm{i} e\, A_{j(-Q)}\Big) \\
&+ \frac{\bar{\rho}_0}{2m} \sum_{Q=(\Omega_l,\boldsymbol{Q})} \left[(+\mathrm{i}\boldsymbol{Q}\, \theta_{+Q}) + e\boldsymbol{A}_{+Q}\right] \cdot \left[(-\mathrm{i}\boldsymbol{Q}\, \theta_{-Q}) + e\, \boldsymbol{A}_{-Q}\right].
\end{aligned}
\tag{7.214}
$$

The RPA action should be compared with action (7.166).

It is time to evaluate the BCS polarization tensor by performing a Taylor expansion in powers of the four-momentum transfer q about $q = 0$. This expansion is well-defined owing to the presence of a gap that removes any potential infrared singularities. For simplicity, we work at vanishing temperature. We begin with the longitudinal component $\Pi^{\parallel}_{00Q}$ defined by

$$
\begin{aligned}
\Pi^{\parallel}_{00q} &= \frac{1}{\beta V} \sum_{k} \mathrm{tr}\left(G_{0(k+q)}\gamma_3 G_{0k}\gamma_3\right) \\
&= -\frac{2}{\beta V} \sum_{\boldsymbol{k}} \sum_{\omega_n} \frac{|\overline{\Delta}|^2 - \xi_{\boldsymbol{k}}\, \xi_{\boldsymbol{k}+\boldsymbol{q}} + \omega_n\, (\omega_n + \varpi_l)}{(\omega_n^2 + E^2_{\boldsymbol{k}})\, [(\omega_n + \varpi_l)^2 + E^2_{\boldsymbol{k}+\boldsymbol{q}}]},
\end{aligned}
\tag{7.215}
$$

where we recall that $E^2_{\boldsymbol{k}} := \xi^2_{\boldsymbol{k}} + |\overline{\Delta}|^2$. When $\beta = \infty$ and $q = (\boldsymbol{q}, \varpi_l) = (\boldsymbol{0}, 0)$,

$$
\begin{aligned}
\Pi^{\parallel}_{00q} &= -\frac{2}{\beta V} \int \frac{\mathrm{d}^3 k}{(2\pi)^3/V} \int\limits_{\mathbb{R}} \frac{\mathrm{d}\omega}{2\pi/\beta}\, \frac{|\overline{\Delta}|^2 - \xi^2_{\boldsymbol{k}} + \omega^2}{(\omega^2 + E^2_{\boldsymbol{k}})^2} \\
&\approx -\nu_{\mathrm{F}} \int\limits_{\mathbb{R}} \mathrm{d}\varepsilon \int\limits_{\mathbb{R}} \frac{\mathrm{d}\omega}{2\pi}\, \frac{|\overline{\Delta}|^2 - \varepsilon^2 + \omega^2}{(\omega^2 + \varepsilon^2 + |\overline{\Delta}|^2)^2}.
\end{aligned}
\tag{7.216}
$$

To reach the last line, we have extended the range of the integration over the single-particle energies to all real numbers, as the integral remains well defined. In doing so the error can be estimated. The frequency integration with the measure $\frac{\mathrm{d}\omega}{2\pi}$ gives the residue

$$
\mathrm{i}\mathrm{Res}\, \left. \frac{|\overline{\Delta}|^2 - \varepsilon^2 + \omega^2}{\left(\omega - \mathrm{i}\sqrt{\varepsilon^2 + |\overline{\Delta}|^2}\right)^2 \left(\omega + \mathrm{i}\sqrt{\varepsilon^2 + |\overline{\Delta}|^2}\right)^2} \right|_{\omega = \mathrm{i}\sqrt{\varepsilon^2 + |\overline{\Delta}|^2}} .
\tag{7.217}
$$

With the help of $\mathrm{Res}_{z=a}\frac{f(z)}{(z-a)^n}=\frac{1}{(n-1)!}\left(\frac{\mathrm{d}^{n-1}f}{dz^{n-1}}\right)_{z=a}$, we conclude that

$$\begin{aligned}
\Pi^{\parallel}_{00q} &\approx -\nu_{\mathrm{F}}\int\limits_{\mathbb{R}} \mathrm{d}\varepsilon\,\mathrm{i}\frac{|\overline{\Delta}|^2-\varepsilon^2+\varepsilon^2+|\overline{\Delta}|^2}{4\left(\sqrt{\varepsilon^2+|\overline{\Delta}|^2}\right)^3\mathrm{i}}\\
&= -\frac{\nu_{\mathrm{F}}}{2}\int\limits_{\mathbb{R}}\mathrm{d}\varepsilon\frac{|\overline{\Delta}|^2}{\left(\varepsilon^2+|\overline{\Delta}|^2\right)^{3/2}}\\
&= -\frac{\nu_{\mathrm{F}}}{2}\int\limits_{\mathbb{R}}\mathrm{d}x\frac{1}{(1+x^2)^{3/2}}\\
&= -\frac{\nu_{\mathrm{F}}}{2}\left.\frac{x}{\sqrt{1+x^2}}\right|_{-\infty}^{+\infty}\\
&= -\nu_{\mathrm{F}}
\end{aligned} \tag{7.218}$$

when $\beta=\infty$ and $q=(\boldsymbol{q},\varpi_l)=(\boldsymbol{0},0)$. This is the same result as obtained from evaluating the RPA polarization function at $q=0$ in the jellium model, see Eqs. (6.58), (6.65), and (6.78).

Next, we turn our attention to the transversal components $\Pi^{\perp}_{ijq}$ of the BCS polarization tensor, which are defined by

$$\begin{aligned}
\Pi^{\perp}_{ijq} &:= \frac{1}{\beta V}\sum_{k}\frac{(k_i+q_i/2)\,(k_j+q_j/2)}{m^2}\mathrm{tr}\left(G_{0(k+q)}\,\gamma_0\,G_{0k}\,\gamma_0\right)\\
&= \frac{2}{\beta V}\sum_{k}\frac{(k_i+q_i/2)\,(k_j+q_j/2)}{m^2}\frac{|\overline{\Delta}|^2+\xi_{\boldsymbol{k}}\,\xi_{\boldsymbol{k}+\boldsymbol{q}}-\omega_n\,(\omega_n+\varpi_l)}{(\omega_n^2+E^2_{\boldsymbol{k}})[(\omega_n+\varpi_l)^2+E^2_{\boldsymbol{k}+\boldsymbol{q}}]},
\end{aligned} \tag{7.219}$$

where we recall that $E^2_{\boldsymbol{k}}:=\xi^2_{\boldsymbol{k}}+|\overline{\Delta}|^2$. When $\beta=\infty$ and $q=(\boldsymbol{q},\varpi_l)=(\boldsymbol{0},0)$,

$$\begin{aligned}
\Pi^{\perp}_{ijq} &= +2\int\frac{\mathrm{d}^3k}{(2\pi)^3}\frac{k_i\,k_j}{m^2}\int\limits_{\mathbb{R}}\frac{\mathrm{d}\omega}{2\pi}\frac{|\overline{\Delta}|^2+\xi^2_{\boldsymbol{k}}-\omega^2}{(\omega^2+E^2_{\boldsymbol{k}})^2}\\
&= +2\int\frac{\mathrm{d}^3k}{(2\pi)^3}\frac{k_i\,k_j}{m^2}\frac{1}{2\pi}\,2\pi\mathrm{i}\frac{|\overline{\Delta}|^2+\xi^2_{\boldsymbol{k}}-\xi^2_{\boldsymbol{k}}-|\overline{\Delta}|^2}{4\left(\sqrt{\xi^2_{\boldsymbol{k}}+|\overline{\Delta}|^2}\right)^3\mathrm{i}}\\
&= 0.
\end{aligned} \tag{7.220}$$

Comparison of Eqs. (7.215) and (7.219) shows that the Matsubara summations agree when $\overline{\Delta}=0$ but differ as soon as $\overline{\Delta}\neq 0$. In other words, the *transversal* components of the Fermi-liquid polarization tensor do not vanish when $T=0$ and $q=0$. The fact that the Matsubara summation vanishes when $T=0$ and $q=0$ in the *transversal* components of the BCS polarization function can thus be ascribed unambiguously to a macroscopic property of the superconducting ground state. The phase of the superconducting ground state is so "stiff" that application of an external perturbation of the vector-gauge type does not induce a *paramagnetic current response* at very low energies and very long wavelengths. The superconducting ground state is said to be incompressible with respect to a vector-gauge perturbation. The only non-vanishing response to the external vector-gauge perturbation comes from the *diamagnetic current response* [term proportional to $\overline{\rho}_0$ in Eq. (7.214)] at very low energies and very long wavelengths in the superconducting ground state. At non-vanishing temperature, particle-hole excitations with energies larger than the single-particle gap $2|\overline{\Delta}|$ induce a non-vanishing paramagnetic response, i.e.,

$$\lim_{q\to 0} \Pi^{\perp}_{ijq} \neq 0 \qquad \text{for } T > 0. \tag{7.221}$$

To gain more insights into the response of the superconducting order to the insertion of a test charge, define the partition function

$$\begin{aligned}
Z_{\beta,\mu} :=& \int \mathcal{D}[\theta,\varphi]\, e^{-S},\\
S :=& \sum_{Q} \frac{\boldsymbol{Q}^2}{8\pi} \varphi_{+Q}\, \varphi_{-Q}\\
&+ \frac{1}{2} \sum_{Q} \left(+\Omega_l\, \theta_{+Q} + \mathrm{i}e\, \varphi_{+Q}\right) \Pi^{\parallel}_{00Q} \left(-\Omega_l\, \theta_{-Q} + \mathrm{i}e\, \varphi_{-Q}\right)\\
&+ \frac{1}{2} \sum_{Q} \sum_{i,j=1}^{3} \left(Q_i\, \theta_{+Q}\right) \Pi^{\perp}_{ijQ} \left(Q_j\, \theta_{-Q}\right)\\
&+ \frac{\overline{\rho}_0}{2m} \sum_{Q} \left(+\mathrm{i}\boldsymbol{Q}\, \theta_{+Q}\right) \cdot \left(-\mathrm{i}\boldsymbol{Q}\, \theta_{-Q}\right).
\end{aligned} \tag{7.222}$$

Here, we have used the RPA approximation (7.214) after switching off the external vector potential. We also added a kinetic term to the scalar potential, i.e., endowed the scalar potential with its own dynamics. We chose the kinetic term corresponding to a Coulomb interaction, although we could have chosen a short-range potential instead. To lowest order in an expansion of the BCS polarization tensor in powers of $\Omega_l/|\overline{\Delta}|$ and $|\boldsymbol{Q}|/|\overline{\Delta}|$, the effective action in Eq. (7.222) simplifies to

$$\begin{aligned}
S \approx \sum_{Q} \Bigg(& \frac{\boldsymbol{Q}^2}{8\pi} \varphi_{+Q}\, \varphi_{-Q} + \frac{\nu_{\mathrm{F}}}{2} \left(e\, \varphi_{+Q} - \mathrm{i}\Omega_l\, \theta_{+Q}\right) \left(e\, \varphi_{-Q} + \mathrm{i}\Omega_l\, \theta_{-Q}\right)\\
&+ \frac{\overline{\rho}_0}{2m} \boldsymbol{Q}^2\, \theta_{+Q}\, \theta_{-Q} \Bigg).
\end{aligned} \tag{7.223}$$

Integration over the scalar potential φ then gives the approximate low-energy effective action

$$\begin{aligned}
S \approx& \sum_{Q} \left(+ \frac{\left[\mathrm{i}(\nu_{\mathrm{F}}/2)\, e\, \Omega_l\right]^2}{\frac{\boldsymbol{Q}^2}{8\pi} + (\nu_{\mathrm{F}}/2)\, e^2} + (\nu_{\mathrm{F}}/2)\, \Omega_l^2 + \frac{\overline{\rho}_0}{2m} \boldsymbol{Q}^2 \right) \theta_{+Q}\, \theta_{-Q}\\
=& \sum_{Q} \left(\frac{(\nu_{\mathrm{F}}/2)\, \frac{\boldsymbol{Q}^2}{8\pi}}{\frac{\boldsymbol{Q}^2}{8\pi} + (\nu_{\mathrm{F}}/2)\, e^2} \Omega_l^2 + \frac{\overline{\rho}_0}{2m} \boldsymbol{Q}^2 \right) \theta_{+Q}\, \theta_{-Q}.
\end{aligned} \tag{7.224}$$

After analytical continuation to real time $\tau = \mathrm{i}t$, the zero's of the kernel for the phase θ of the superconducting order parameter give access to some collective excitations in the superconducting state. The kernel does not possess zero's for arbitrary small frequencies

$$\widetilde{\Omega}_l := \mathrm{i}\Omega_l, \tag{7.225}$$

since $\boldsymbol{Q}^2$ can be factorized from the kernel. Had we replaced the Coulomb potential

$$V_{\mathrm{CB}\,\boldsymbol{Q}} := \frac{4\pi}{\boldsymbol{Q}^2} \tag{7.226}$$

by a short-range potential with a non-diverging and non-vanishing limit as $\boldsymbol{Q} \to \boldsymbol{0}$, we would have found a branch of collective excitations with a dispersion relation

$$\widetilde{\Omega}(\boldsymbol{Q}) \propto |\boldsymbol{Q}|. \tag{7.227}$$

The presence of a such a branch of excitation is an example of *Goldstone theorem* and its absence as a result of the long-range nature of the Coulomb interaction is an example of the *Anderson-Higgs mechanism* by which "would be Goldstone bosons" are eaten up by a gauge degree of freedom and photons acquire an effective mass.

The "would be Goldstone bosons" can be found at the energy scale of the plasma frequency. At this energy scale approximation (7.224) is not reliable anymore. To access the dynamics at the energy scale of the plasma frequency, we must integrate over the superconducting phase in Eq. (7.222), whereby we can ignore the contributions from the transversal components of the BCS polarization tensor. (We are taking the limit $\Omega_l \to 0$ before taking the limit $\boldsymbol{Q} \to 0$.) This gives the effective action for the scalar potential

$$\begin{aligned} S &\approx \sum_{\boldsymbol{Q}} \left(\frac{\boldsymbol{Q}^2}{8\pi} + \frac{1}{2}(\mathrm{i}e)^2\,\Pi^{\parallel}_{00\boldsymbol{Q}} + \frac{\left(\frac{1}{2}\,\Omega_l\,(\mathrm{i}e)\,\Pi^{\parallel}_{00\boldsymbol{Q}}\right)^2}{-\frac{1}{2}\,\Pi^{\parallel}_{00\boldsymbol{Q}}\,\Omega_l^2 + \frac{\overline{\rho}_0}{2m}\,\boldsymbol{Q}^2} \right) \varphi_{+\boldsymbol{Q}}\,\varphi_{-\boldsymbol{Q}} \\ &= \sum_{\boldsymbol{Q}} \left(\frac{-\frac{1}{2}\,\Pi^{\parallel}_{00\boldsymbol{Q}}\left(\frac{\overline{\rho}_0\,e^2}{2m} + \frac{\Omega_l^2}{8\pi}\right) + \frac{\overline{\rho}_0}{16\pi m}\,\boldsymbol{Q}^2}{-\frac{1}{2}\,\Pi^{\parallel}_{00\boldsymbol{Q}}\,\Omega_l^2 + \frac{\overline{\rho}_0}{2m}\,\boldsymbol{Q}^2} \right) \boldsymbol{Q}^2\,\varphi_{+\boldsymbol{Q}}\,\varphi_{-\boldsymbol{Q}}. \end{aligned} \tag{7.228}$$

To leading order in the transfer momentum $\boldsymbol{Q}$, the kernel for the scalar potential is given by

$$\frac{\boldsymbol{Q}^2}{8\pi}\left(1 + \frac{\Omega_{\mathrm{P}}^2}{\Omega_l^2}\right)\varphi_{+\boldsymbol{Q}}\,\varphi_{-\boldsymbol{Q}}, \qquad \Omega_{\mathrm{P}} := \sqrt{\frac{4\pi\,\overline{\rho}_0\,e^2}{m}}. \tag{7.229}$$

This kernel only differs from the one for the jellium model by the replacement of the electron density by $\overline{\rho}_0$ defined in Eq. (7.209). Thus, the physics at high energies ($\sim \Omega_{\mathrm{P}} \gg |\overline{\Delta}|$) is largely unaffected by the superconducting ground state.

7.8 Effective theory in the vicinity of $T = T_c$

Define the dimensionless grand canonical potential, which is also called the Landau-Ginzburg free energy [be aware that we keep the same notation as for the intensive grand canonical potential (7.67) and work in units for which $\hbar = c = k_B = 1$],

$$F[\Delta^*, \Delta] := \int_0^\beta d\tau \int_V d^3\boldsymbol{r}\, \frac{1}{U}\, |\Delta|^2(\boldsymbol{r},\tau) - \log \operatorname{Det} \begin{pmatrix} \partial_\tau - \frac{\boldsymbol{\nabla}^2}{2m} - \mu & \Delta(\boldsymbol{r},\tau) \\ \Delta^*(\boldsymbol{r},\tau) & \partial_\tau + \frac{\boldsymbol{\nabla}^2}{2m} + \mu \end{pmatrix}. \tag{7.230}$$

We have performed an expansion of this functional with the amplitude $|\Delta(\boldsymbol{r},\tau)|$ of the superconducting order parameter frozen to a uniform value $|\overline{\Delta}|$ in powers of a space-time fluctuating phase $\theta(\boldsymbol{r},\tau) = \arg\Delta(\boldsymbol{r},\tau)$ in section 7.7. In the vicinity of T_c, it makes no sense to distinguish between amplitude and phase fluctuations of the order parameter and it is more natural to perform the expansion

$$F[\Delta^*, \Delta] := \sum_{j=0}^{\infty} \int_0^\beta d\tau \int_V d^3\boldsymbol{r}\, \mathcal{F}^{(j)}\left(\mathrm{i}\frac{\partial}{\partial \boldsymbol{r}}, -\mathrm{i}\frac{\partial}{\partial\tau}\right) |\Delta|^{2j}(\boldsymbol{r},\tau). \tag{7.231}$$

In the theory of classical continuous phase transitions, an expansion in powers of the order parameter is called a Landau-Ginzburg theory. Symmetry dictates which powers of the order parameter enter the expansion (7.231). Here, the $U(1)$ gauge symmetry only allows even powers $2j$.

In this section, we give a microscopic derivation to the *time-independent Landau-Ginzburg functional*

$$F[\Delta^*, \Delta] = \beta V \times \text{constant} + \beta \times \int_V d^3\boldsymbol{r} \Big[a(T)\, |\Delta|^2(\boldsymbol{r}) + b(T)\, |\Delta|^4(\boldsymbol{r}) + c(T)\, (\boldsymbol{\nabla}\Delta^*) \cdot (\boldsymbol{\nabla}\Delta)(\boldsymbol{r}) + \cdots \Big], \tag{7.232}$$

by relating the temperature dependent coefficients $a(T)$, $b(T)$, and $c(T)$ to the microscopic coupling constants (m, μ, U) entering the BCS Hamiltonian. We show that

$$a(T) = (\nu_F/2) \ln\left(\frac{T}{T_c}\right), \tag{7.233a}$$

$$b(T) = \frac{1}{16\pi^2}\, \zeta(3, 1/2)\, \frac{(\nu_F/2)}{T^2}, \tag{7.233b}$$

$$c(T) = \frac{1}{24\pi^2}\, \zeta(3, 1/2)\, \frac{\mu(\nu_F/2)}{m\, T^2} = \frac{1}{48\pi^2}\, \zeta(3, 1/2)\, \frac{(v_F)^2 (\nu_F/2)}{T^2}, \tag{7.233c}$$

the ζ-function being defined by

$$\zeta(x, y) := \sum_{n=0}^{+\infty} \frac{1}{(n+y)^x}. \tag{7.233d}$$

Before deriving Eqs. (7.232) and (7.233) observe that by replacing the gradient term $\boldsymbol{\nabla}$ in Eq. (7.230) by the covariant derivative $\boldsymbol{\nabla}+\mathrm{i}e\boldsymbol{A}$ that couples fermions carrying the electric charge $-e<0$ to an external static vector potential $\boldsymbol{A}(\boldsymbol{r})$, one can derive the time-independent Landau-Ginzburg functional

$$F[\Delta^*,\Delta,\boldsymbol{A}] = \beta V\times\text{constant} + \beta\times\int\limits_V \mathrm{d}^3\boldsymbol{r}\Big[a(T)\,|\Delta|^2(\boldsymbol{r}) + b(T)\,|\Delta|^4(\boldsymbol{r}) \\ + c(T)\,(\boldsymbol{D}\Delta)^*\cdot(\boldsymbol{D}\Delta)\,(\boldsymbol{r}) + \cdots\Big], \tag{7.234a}$$

where

$$\boldsymbol{D} := \boldsymbol{\nabla} + \mathrm{i}(2e)\boldsymbol{A}, \tag{7.234b}$$

by a straightforward extension of the computation to follow. For a recent review on the use of Landau-Ginzburg functional for superconductors see Ref. [79].

We organize the expansion of the fermionic determinant around the unperturbed Green function

$$G_0 := -\left[\gamma_0\partial_\tau + \gamma_3\left(-\frac{\boldsymbol{\nabla}^2}{2m}-\mu\right)\right]^{-1}, \tag{7.235}$$

i.e., we need to perform the expansion

$$\begin{aligned}&\mathrm{Tr}\,\log\Big(-(G_0)^{-1} + \Delta(\boldsymbol{r},\tau)\gamma_+ + \Delta^*(\boldsymbol{r},\tau)\gamma_-\Big)\\ &= \mathrm{Tr}\,\log\Big(-(G_0)^{-1}\Big) - \sum_{j=1}^{+\infty}\frac{1}{j}\mathrm{Tr}\,\Big(\{G_0\left[\Delta(\boldsymbol{r},\tau)\gamma_+ + \Delta^*(\boldsymbol{r},\tau)\gamma_-\right]\}^j\Big),\end{aligned} \tag{7.236}$$

where the matrices

$$\gamma_+ := \begin{pmatrix}0&1\\0&0\end{pmatrix},\qquad \gamma_- := \begin{pmatrix}0&0\\1&0\end{pmatrix} \tag{7.237}$$

in the particle-hole grading have been introduced. The contribution

$$\text{const} := \mathrm{Tr}\,\log\left[-(G_0)^{-1}\right] \tag{7.238}$$

induces a renormalization of the Landau-Ginzburg free energy through a mere constant.

First-order contributions ($j=1$)

The contributions to first order in the superconducting order parameter vanish because of the algebra obeyed by the Pauli matrices.

Second-order contributions ($j=2$)

The four traces

$$\mathrm{tr}\begin{pmatrix}a&0\\0&b\end{pmatrix}\begin{pmatrix}0&1\\0&0\end{pmatrix}\begin{pmatrix}a'&0\\0&b'\end{pmatrix}\begin{pmatrix}0&1\\0&0\end{pmatrix} = \mathrm{tr}\begin{pmatrix}0&a\\0&0\end{pmatrix}\begin{pmatrix}0&a'\\0&0\end{pmatrix} = 0, \tag{7.239a}$$

$$\operatorname{tr}\begin{pmatrix} a & 0 \\ 0 & b \end{pmatrix}\begin{pmatrix} 0 & 0 \\ 1 & 0 \end{pmatrix}\begin{pmatrix} a' & 0 \\ 0 & b' \end{pmatrix}\begin{pmatrix} 0 & 0 \\ 1 & 0 \end{pmatrix} = \operatorname{tr}\begin{pmatrix} 0 & 0 \\ b & 0 \end{pmatrix}\begin{pmatrix} 0 & 0 \\ b' & 0 \end{pmatrix} = 0, \tag{7.239b}$$

$$\operatorname{tr}\begin{pmatrix} a & 0 \\ 0 & b \end{pmatrix}\begin{pmatrix} 0 & 1 \\ 0 & 0 \end{pmatrix}\begin{pmatrix} a' & 0 \\ 0 & b' \end{pmatrix}\begin{pmatrix} 0 & 0 \\ 1 & 0 \end{pmatrix} = \operatorname{tr}\begin{pmatrix} 0 & a \\ 0 & 0 \end{pmatrix}\begin{pmatrix} 0 & 0 \\ b' & 0 \end{pmatrix} = ab', \tag{7.239c}$$

$$\operatorname{tr}\begin{pmatrix} a & 0 \\ 0 & b \end{pmatrix}\begin{pmatrix} 0 & 0 \\ 1 & 0 \end{pmatrix}\begin{pmatrix} a' & 0 \\ 0 & b' \end{pmatrix}\begin{pmatrix} 0 & 1 \\ 0 & 0 \end{pmatrix} = \operatorname{tr}\begin{pmatrix} 0 & 0 \\ b & 0 \end{pmatrix}\begin{pmatrix} 0 & a' \\ 0 & 0 \end{pmatrix} = ba', \tag{7.239d}$$

are needed to evaluate all contributions to the expansion of the Fredholm determinant that are of quadratic order in the superconducting order parameter.

Introduce the short-hand notation $[q = (\boldsymbol{q}, \varpi_l)]$

$$F^{(2)}[\Delta^*, \Delta] := \sum_q \frac{1}{U}\Delta^*_q\Delta_q + \frac{1}{2}\operatorname{Tr}\left(\{G_0\left[\Delta(\boldsymbol{r},\tau)\gamma_+ + \Delta^*(\boldsymbol{r},\tau)\gamma_-\right]\}^2\right). \tag{7.240}$$

In reciprocal space, the trace

$$\begin{aligned} F^{(2)}[\Delta^*, \Delta] = & \sum_q \frac{1}{U}\Delta^*_q\Delta_q \\ & + \frac{1}{2\beta V}\sum_q\sum_{\omega_n}\sum_{\boldsymbol{k}} \operatorname{tr}\left[(G_0)_{k+q}\gamma_+(G_0)_k\gamma_-\right]\Delta_{+q}\Delta^*_{+q} \\ & + \frac{1}{2\beta V}\sum_q\sum_{\omega_n}\sum_{\boldsymbol{k}} \operatorname{tr}\left[(G_0)_{k+q}\gamma_-(G_0)_k\gamma_+\right]\Delta^*_{-q}\Delta_{-q} \end{aligned} \tag{7.241}$$

simplifies greatly, for

$$\begin{aligned} F^{(2)}[\Delta^*, \Delta] = & \sum_q \frac{1}{U}\Delta^*_q\Delta_q \\ & + \frac{1}{2\beta V}\sum_q\sum_{\omega_n}\sum_{\boldsymbol{k}} \frac{1}{\left(-\mathrm{i}\omega_n - \mathrm{i}\varpi_l + \xi_{\boldsymbol{k}+\boldsymbol{q}}\right)(-\mathrm{i}\omega_n - \xi_{\boldsymbol{k}})}\Delta_{+q}\Delta^*_{+q} \\ & + \frac{1}{2\beta V}\sum_q\sum_{\omega_n}\sum_{\boldsymbol{k}} \frac{1}{\left(-\mathrm{i}\omega_n - \mathrm{i}\varpi_l - \xi_{\boldsymbol{k}+\boldsymbol{q}}\right)(-\mathrm{i}\omega_n + \xi_{\boldsymbol{k}})}\Delta^*_{-q}\Delta_{-q}. \end{aligned} \tag{7.242}$$

It is convenient to change k to $k - q/2$

$$\begin{aligned} F^{(2)}[\Delta^*, \Delta] = & \sum_q \frac{1}{U}\Delta^*_q\Delta_q \\ & + \frac{1}{2\beta V}\sum_{q,\omega_n,\boldsymbol{k}} \frac{\Delta_{+q}\Delta^*_{+q}}{\left(-\mathrm{i}\omega_n - \mathrm{i}\varpi_l/2 + \xi_{\boldsymbol{k}+\boldsymbol{q}/2}\right)\left(-\mathrm{i}\omega_n + \mathrm{i}\varpi_l/2 - \xi_{\boldsymbol{k}-\boldsymbol{q}/2}\right)} \\ & + \frac{1}{2\beta V}\sum_{q,\omega_n,\boldsymbol{k}} \frac{\Delta^*_{-q}\Delta_{-q}}{\left(-\mathrm{i}\omega_n - \mathrm{i}\varpi_l/2 - \xi_{\boldsymbol{k}+\boldsymbol{q}/2}\right)\left(-\mathrm{i}\omega_n + \mathrm{i}\varpi_l/2 + \xi_{\boldsymbol{k}-\boldsymbol{q}/2}\right)}. \end{aligned} \tag{7.243}$$

Finally, if we change q to $-q$ on the last line, we obtain

$$F^{(2)}[\Delta^{*},\Delta] \equiv \sum_{q} K^{(2)}_{q} \Delta^{*}_{q} \Delta_{q}, \tag{7.244a}$$

where

$$K^{(2)}_{q} := \frac{1}{U} + \frac{1}{\beta V} \sum_{\omega_n, \boldsymbol{k}} \frac{1}{\left(-\mathrm{i}\omega_n - \mathrm{i}\varpi_l/2 + \xi_{\boldsymbol{k}+\boldsymbol{q}/2}\right)\left(-\mathrm{i}\omega_n + \mathrm{i}\varpi_l/2 - \xi_{\boldsymbol{k}-\boldsymbol{q}/2}\right)}. \tag{7.244b}$$

Contribution (7.244) resembles the contribution (6.58) for the polarization of the Jellium model except for one important difference. The Jellium model has no particle-hole grading. Energy eigenvalues of the non-interacting Fermi gas are always subtracted from the Matsubara frequencies in the denominator. Here, the particle-hole grading implies that the energy eigenvalues of the non-interacting Fermi gas are subtracted in the particle-particle channel [factor a in Eq. (7.239c)] and subtracted in the hole-hole channel [factor b' in Eq. (7.239c)].

The fermionic Matsubara sum can be replaced by an integral over the Fermi-Dirac distribution $\tilde{f}_{\mathrm{FD}}(z) = 1/[\exp(\beta z)+1] = 1 - \tilde{f}_{\mathrm{FD}}(-z)$, see Fig. 7.1,

$$\begin{aligned}
K^{(2)}_{q} &= \frac{1}{U} - \frac{1}{V}\sum_{\boldsymbol{k}} \int_{\Gamma_{\boldsymbol{k}}} \frac{\mathrm{d}z}{2\pi\mathrm{i}} \frac{\tilde{f}_{\mathrm{FD}}(z)}{\left(z + \mathrm{i}\varpi_l/2 - \xi_{\boldsymbol{k}+\boldsymbol{q}/2}\right)\left(z - \mathrm{i}\varpi_l/2 + \xi_{\boldsymbol{k}-\boldsymbol{q}/2}\right)} \\
&= \frac{1}{U} + (-1)^2 \frac{1}{V} \sum_{\boldsymbol{k}} \left(\frac{\tilde{f}_{\mathrm{FD}}(+\mathrm{i}\varpi_l/2 - \xi_{\boldsymbol{k}-\boldsymbol{q}/2})}{+\mathrm{i}\varpi_l - \xi_{\boldsymbol{k}-\boldsymbol{q}/2} - \xi_{\boldsymbol{k}+\boldsymbol{q}/2}} + \frac{\tilde{f}_{\mathrm{FD}}(-\mathrm{i}\varpi_l/2 + \xi_{\boldsymbol{k}+\boldsymbol{q}/2})}{-\mathrm{i}\varpi_l + \xi_{\boldsymbol{k}+\boldsymbol{q}/2} + \xi_{\boldsymbol{k}-\boldsymbol{q}/2}} \right) \\
&= \frac{1}{U} + (-1)^2 \frac{1}{V} \sum_{\boldsymbol{k}} \frac{\tilde{f}_{\mathrm{FD}}(+\mathrm{i}\varpi_l/2 - \xi_{\boldsymbol{k}-\boldsymbol{q}/2}) - \tilde{f}_{\mathrm{FD}}(-\mathrm{i}\varpi_l/2 + \xi_{\boldsymbol{k}+\boldsymbol{q}/2})}{+\mathrm{i}\varpi_l - \xi_{\boldsymbol{k}-\boldsymbol{q}/2} - \xi_{\boldsymbol{k}+\boldsymbol{q}/2}} \\
&= \frac{1}{U} + (-1)^2 \frac{1}{V} \sum_{\boldsymbol{k}} \frac{1 - \tilde{f}_{\mathrm{FD}}(-\mathrm{i}\varpi_l/2 + \xi_{\boldsymbol{k}-\boldsymbol{q}/2}) - \tilde{f}_{\mathrm{FD}}(-\mathrm{i}\varpi_l/2 + \xi_{\boldsymbol{k}+\boldsymbol{q}/2})}{+\mathrm{i}\varpi_l - \xi_{\boldsymbol{k}-\boldsymbol{q}/2} - \xi_{\boldsymbol{k}+\boldsymbol{q}/2}}.
\end{aligned} \tag{7.245}$$

The denominator on the right-hand side of Eq. (7.245) differs from the denominator (6.65) in the polarization function for the Jellium model, as it is the sum rather than the difference of the energy eigenvalues of the non-interacting Fermi gas that now appear.

In the static limit $q = (\boldsymbol{q}, \varpi_l) \to (\boldsymbol{q}, 0)$, the kernel $K^{(2)}_{q}$ simplifies to

$$K^{(2)}_{\boldsymbol{q},0} = \frac{1}{U} + \frac{1}{\beta V} \sum_{\omega_n} \sum_{\boldsymbol{k}} \frac{1}{(\mathrm{i}\omega_n + \xi_{\boldsymbol{k}})\left(\mathrm{i}\omega_n - \xi_{\boldsymbol{k}+\boldsymbol{q}}\right)} \tag{7.246a}$$

or, equivalently,

$$K^{(2)}_{\boldsymbol{q},0} = \frac{1}{U} - \frac{1}{V} \sum_{\boldsymbol{k}} \frac{1 - \tilde{f}_{\mathrm{FD}}(+\xi_{\boldsymbol{k}}) - \tilde{f}_{\mathrm{FD}}(+\xi_{\boldsymbol{k}+\boldsymbol{q}})}{\xi_{\boldsymbol{k}} + \xi_{\boldsymbol{k}+\boldsymbol{q}}}. \tag{7.246b}$$

Representation (7.246a) is the one that we choose to perform the *gradient expansion*

$$
\begin{aligned}
K^{(2)}_{\boldsymbol{q},0} &= \frac{1}{U} + \frac{1}{\beta V}\sum_{\omega_n}\sum_{\boldsymbol{k}} \frac{1}{\mathrm{i}\omega_n + \xi_{\boldsymbol{k}}} e^{\boldsymbol{q}\cdot\partial_{\boldsymbol{k}}} \frac{1}{\mathrm{i}\omega_n - \xi_{\boldsymbol{k}}} \\
&= \frac{1}{U} - \frac{1}{\beta V}\sum_{\omega_n}\sum_{\boldsymbol{k}} \frac{1}{\omega_n^2 + \xi_{\boldsymbol{k}}^2} \\
&+ \frac{1}{\beta V}\sum_{\omega_n}\sum_{\boldsymbol{k}} \frac{1}{\mathrm{i}\omega_n + \xi_{\boldsymbol{k}}} \frac{1}{(\mathrm{i}\omega_n - \xi_{\boldsymbol{k}})^2} (-1)^2 \boldsymbol{q}\cdot\left(\frac{\partial \xi_{\boldsymbol{k}}}{\partial \boldsymbol{k}}\right) \\
&+ \frac{1}{\beta V}\sum_{\omega_n}\sum_{\boldsymbol{k}} \frac{1}{\mathrm{i}\omega_n + \xi_{\boldsymbol{k}}} \frac{1}{2}(\boldsymbol{q}\cdot\partial_{\boldsymbol{k}})^2 \frac{1}{\mathrm{i}\omega_n - \xi_{\boldsymbol{k}}} \\
&+ \mathcal{O}(\boldsymbol{q}^3).
\end{aligned} \tag{7.247}
$$

The first line on the last equality of the right-hand side was evaluated in Eq. (7.87b). It is given by $(\nu_{\mathrm{F}}/2)\ln(T/T_{\mathrm{c}})$. The second line vanishes under the assumption that the inversion symmetry

$$
\xi_{+\boldsymbol{k}} = \xi_{-\boldsymbol{k}} \tag{7.248}
$$

holds for the non-interacting fermion gas. The third line can be evaluated with the help of the identities

$$
\frac{1}{2}\left(\boldsymbol{q}\cdot\frac{\partial}{\partial \boldsymbol{k}}\right)\left(\boldsymbol{q}\cdot\frac{\partial}{\partial \boldsymbol{k}}\right)\boldsymbol{k}^2 = \left(\boldsymbol{q}\cdot\frac{\partial}{\partial \boldsymbol{k}}\right)(\boldsymbol{q}\cdot\boldsymbol{k}) = \boldsymbol{q}^2, \tag{7.249a}
$$

$$
\left(\frac{\partial}{\partial \boldsymbol{k}}\cdot\frac{\partial}{\partial \boldsymbol{k}}\right)\boldsymbol{k}^2 = 2\times 3 = 6, \tag{7.249b}
$$

$$
\left(\frac{\partial \xi_{\boldsymbol{k}}}{\partial \boldsymbol{k}}\right)^2 = \frac{\boldsymbol{k}^2}{m^2} = \frac{2}{m}(\xi_{\boldsymbol{k}} + \mu). \tag{7.249c}
$$

Thus,

$$
\begin{aligned}
K^{(2)}_{\boldsymbol{q},0} &= \frac{\nu_{\mathrm{F}}}{2}\ln\left(\frac{T}{T_{\mathrm{c}}}\right) + \frac{\boldsymbol{q}^2}{6\beta V}\sum_{\omega_n}\sum_{\boldsymbol{k}} \frac{1}{\mathrm{i}\omega_n + \xi_{\boldsymbol{k}}} (\partial_{\boldsymbol{k}})^2 \frac{1}{\mathrm{i}\omega_n - \xi_{\boldsymbol{k}}} + \mathcal{O}(\boldsymbol{q}^3) \\
&= \frac{\nu_{\mathrm{F}}}{2}\ln\left(\frac{T}{T_{\mathrm{c}}}\right) - \frac{\boldsymbol{q}^2}{6\beta V}\sum_{\omega_n}\sum_{\boldsymbol{k}} \left(\partial_{\boldsymbol{k}}\frac{1}{\mathrm{i}\omega_n + \xi_{\boldsymbol{k}}}\right)\cdot\left(\partial_{\boldsymbol{k}}\frac{1}{\mathrm{i}\omega_n - \xi_{\boldsymbol{k}}}\right) + \mathcal{O}(\boldsymbol{q}^3) \\
&= \frac{\nu_{\mathrm{F}}}{2}\ln\left(\frac{T}{T_{\mathrm{c}}}\right) + \frac{\boldsymbol{q}^2}{3m\beta V}\sum_{\omega_n}\sum_{\boldsymbol{k}} \frac{\xi_{\boldsymbol{k}} + \mu}{(\mathrm{i}\omega_n + \xi_{\boldsymbol{k}})^2(\mathrm{i}\omega_n - \xi_{\boldsymbol{k}})^2} + \mathcal{O}(\boldsymbol{q}^3).
\end{aligned} \tag{7.250}
$$

The summation over momenta is converted into an integral over the density of states $\tilde{\nu}(\xi)$, whereby the definition of the density of states accounts for the spin-1/2 degree of freedom,

$$
\begin{aligned}
K^{(2)}_{\boldsymbol{q},0} &= \frac{\nu_{\mathrm{F}}}{2}\ln\left(\frac{T}{T_{\mathrm{c}}}\right) + \frac{\boldsymbol{q}^2}{3m\beta}\sum_{\omega_n}\int_{\mathbb{R}} \mathrm{d}\xi\, \frac{\tilde{\nu}(\xi)}{2}\frac{\xi + \mu}{(\omega_n^2 + \xi^2)^2} + \mathcal{O}(\boldsymbol{q}^3) \\
&= \frac{\nu_{\mathrm{F}}}{2}\ln\left(\frac{T}{T_{\mathrm{c}}}\right) + \frac{\boldsymbol{q}^2}{3m\beta}\sum_{\omega_n}\int_{\mathbb{R}} \mathrm{d}\xi\, \frac{\tilde{\nu}(\xi)}{2}\frac{\mu}{(\omega_n^2 + \xi^2)^2} + \mathcal{O}(\boldsymbol{q}^3).
\end{aligned} \tag{7.251}
$$

The last equality follows for any density of states that obeys $\tilde{\nu}(\xi)=\tilde{\nu}(-\xi)$. For a density of state that is not even under $\xi\to-\xi$ but is non-vanishing at the Fermi energy, one has the approximation

$$
\begin{aligned}
K^{(2)}_{\boldsymbol{q},0} &\approx \frac{\nu_{\mathrm{F}}}{2}\ln\left(\frac{T}{T_{\mathrm{c}}}\right)+\frac{\boldsymbol{q}^2\mu}{3m\beta}\frac{\nu_{\mathrm{F}}}{2}\sum_{\omega_n}\int_{\mathbb{R}}\mathrm{d}\xi\frac{1}{(\omega_n^2+\xi^2)^2}+\mathcal{O}(\boldsymbol{q}^3)\\
&= \frac{\nu_{\mathrm{F}}}{2}\ln\left(\frac{T}{T_{\mathrm{c}}}\right)+\frac{\boldsymbol{q}^2\mu}{3m\beta}\frac{\nu_{\mathrm{F}}}{2}\sum_{\omega_n}2\pi\mathrm{i}\mathrm{Res}\left.\frac{1}{(\omega_n^2+\xi^2)^2}\right|_{\xi=+\mathrm{i}|\omega_n|}+\mathcal{O}(\boldsymbol{q}^3).
\end{aligned}
\tag{7.252}
$$

The residue of $\frac{1}{(\omega_n^2+\xi^2)^2}$ at $\xi=+\mathrm{i}|\omega_n|$ is the expansion coefficient

$$
(-2)(2\mathrm{i}|\omega_n|)^{-3}=(-1)^2\frac{1}{4\mathrm{i}|\omega_n|^{+3}}=\frac{1}{4\mathrm{i}}\frac{1}{\left[\frac{\pi}{\beta}(2n+1)\right]^{+3}}
\tag{7.253}
$$

that enters the simple pole in the Laurent series expansion

$$
\frac{1}{(\omega_n^2+\xi^2)^2}=\frac{(\xi+\mathrm{i}|\omega_n|)^{-2}}{(\xi-\mathrm{i}|\omega_n|)^{+2}}=\frac{(2\mathrm{i}|\omega_n|)^{-2}}{(\xi-\mathrm{i}|\omega_n|)^{+2}}+\frac{(-2)(2\mathrm{i}|\omega_n|)^{-3}}{(\xi-\mathrm{i}|\omega_n|)^{+1}}+\cdots.
\tag{7.254}
$$

This gives

$$
\begin{aligned}
K^{(2)}_{\boldsymbol{q},0} &\approx \frac{\nu_{\mathrm{F}}}{2}\ln\left(\frac{T}{T_{\mathrm{c}}}\right)+\frac{\boldsymbol{q}^2\mu\beta^2}{24\pi^2m}\frac{\nu_{\mathrm{F}}}{2}\sum_{n=0}^{+\infty}\frac{1}{\left(n+\frac{1}{2}\right)^3}+\mathcal{O}(\boldsymbol{q}^3)\\
&= \frac{\nu_{\mathrm{F}}}{2}\ln\left(\frac{T}{T_{\mathrm{c}}}\right)+\frac{\boldsymbol{q}^2\mu\beta^2}{24\pi^2m}\frac{\nu_{\mathrm{F}}}{2}\zeta(3,1/2)+\mathcal{O}(\boldsymbol{q}^3).
\end{aligned}
\tag{7.255}
$$

The first term in the kernel gives the coefficient (7.233a) whereas the second term gives the coefficient (7.233c).

Third-order contribution ($j=3$)
They vanish because of the Pauli algebra.
Fourth-order contribution ($j=4$)
The coefficient (7.233b) [as was the coefficient (7.233a)] can be read from the expansion of the Fredholm determinant about a space- and time-independent $|\Delta|^2$ performed in section 7.3.2. In the ratio

$$
\frac{c(T)}{b(T)}=\frac{2}{3}\times\frac{\mu}{m},
\tag{7.256}
$$

the factor $1/3$ comes from Eq. (7.249b), the factor $2\mu/m$ from Eq. (7.249c).

7.9 Problems

7.9.1 *BCS variational method to superconductivity*

Introduction

The mean-field and RPA approximations to superconductivity were treated using path-integral techniques. It is also valuable to derive the mean-field approximation following Bardeen, Cooper, and Schrieffer in their seminal papers [80–82].

Definitions

Consider the interacting Hamiltonian

$$\hat{H} := \hat{H}_0 + \hat{H}_1, \tag{7.257a}$$

$$\hat{H}_0 := \sum_{\boldsymbol{k}} \sum_{\sigma=\uparrow,\downarrow} \xi_{\boldsymbol{k}}\, \hat{c}^{\dagger}_{\boldsymbol{k}\sigma}\, \hat{c}_{\boldsymbol{k}\sigma}, \tag{7.257b}$$

$$\hat{H}_1 := \sum_{\boldsymbol{k},\boldsymbol{k}'} V_{\boldsymbol{k},\boldsymbol{k}'}\, \hat{c}^{\dagger}_{+\boldsymbol{k}\uparrow}\, \hat{c}^{\dagger}_{-\boldsymbol{k}\downarrow}\, \hat{c}_{-\boldsymbol{k}'\downarrow}\, \hat{c}_{+\boldsymbol{k}'\uparrow}, \tag{7.257c}$$

$$\xi_{\boldsymbol{k}} := \frac{\hbar^2 \boldsymbol{k}}{2m} - \mu. \tag{7.257d}$$

The interaction matrix elements obey $V_{\boldsymbol{k},\boldsymbol{k}'} = V^{*}_{\boldsymbol{k}',\boldsymbol{k}}$. The Bardeen-Cooper-Schrieffer (BCS) variational wave function is defined by

$$|\varphi\rangle := \prod_{\boldsymbol{k}} \left(u_{\boldsymbol{k}} + v_{\boldsymbol{k}}\, e^{\mathrm{i}\,\varphi}\, \hat{c}^{\dagger}_{+\boldsymbol{k}\uparrow}\, \hat{c}^{\dagger}_{-\boldsymbol{k}\downarrow} \right) |0\rangle, \tag{7.258}$$

in terms of the electron creation and annihilation operators. This BCS wave function can be thought of as a coherent state for Cooper pairs (see Ref. [80]) with $v_{\boldsymbol{k}}$ ($u_{\boldsymbol{k}}$) the amplitude (not) to have a Cooper pair with relative momentum $\boldsymbol{k}$. The numbers $u_{\boldsymbol{k}}, v_{\boldsymbol{k}} \in \mathbb{R}$ obey the normalization conditions

$$u_{\boldsymbol{k}}^2 + v_{\boldsymbol{k}}^2 = 1 \tag{7.259}$$

and $\varphi \in [0, 2\pi[$ is a global phase.

Exercise 1.1:

(a) Express the expectation value in the variational state $|\varphi\rangle$ of the kinetic energy $\langle\varphi|\,\hat{H}_0\,|\varphi\rangle$ in terms of the parameters $u_{\boldsymbol{k}}$, $v_{\boldsymbol{k}}$, and φ.
(b) Express the expectation value in the variational state $|\varphi\rangle$ of the interacting energy $\langle\varphi|\,\hat{H}_1\,|\varphi\rangle$ in terms of the parameters $u_{\boldsymbol{k}}$, $v_{\boldsymbol{k}}$, and φ.
(c) Does $\langle\varphi|\,\hat{H}\,|\varphi\rangle$ depend on the global phase φ?

From now on, we assume that the matrix elements of the interaction potential take the reduced form

$$V_{\boldsymbol{k},\boldsymbol{k}'} = -V. \tag{7.260}$$

Define the complex-valued parameter

$$\Delta := V \sum_{\boldsymbol{k}} u_{\boldsymbol{k}}\, v_{\boldsymbol{k}}. \tag{7.261}$$

Exercise 1.2:

(a) Express $\langle\varphi|\,\hat{H}\,|\varphi\rangle$ in terms of $v_{\boldsymbol{k}}$ only.
(b) Minimize $\langle\varphi|\,\hat{H}\,|\varphi\rangle$ with respect to $v_{\boldsymbol{k}}$ to show that

$$v_{\boldsymbol{k}}^2 = \frac{1}{2}\left(1 - \frac{\xi_{\boldsymbol{k}}}{E_{\boldsymbol{k}}}\right), \qquad u_{\boldsymbol{k}}^2 = \frac{1}{2}\left(1 + \frac{\xi_{\boldsymbol{k}}}{E_{\boldsymbol{k}}}\right), \tag{7.262a}$$

where

$$E_{\boldsymbol{k}} = +\sqrt{\xi_{\boldsymbol{k}}^2 + \Delta^2}. \tag{7.262b}$$

(c) Express $\langle\varphi|\,\hat{H}\,|\varphi\rangle$ and Eq. (7.261) in terms of $\xi_{\boldsymbol{k}}$ and $E_{\boldsymbol{k}}$.

(d) Consider the subspaces with the quantum numbers $\boldsymbol{k}\uparrow$ and $-\boldsymbol{k}\downarrow$. Show that the states

$$\left(u_{\boldsymbol{k}}+v_{\boldsymbol{k}}\,\hat{c}^{\dagger}_{\boldsymbol{k}\uparrow}\,\hat{c}^{\dagger}_{-\boldsymbol{k}\downarrow}\right)|0\rangle\,,\qquad \hat{c}^{\dagger}_{\boldsymbol{k}\uparrow}\,|0\rangle\,,\qquad \hat{c}^{\dagger}_{-\boldsymbol{k}\downarrow}\,|0\rangle\,,\qquad \left(v_{\boldsymbol{k}}-u_{\boldsymbol{k}}\,\hat{c}^{\dagger}_{\boldsymbol{k}\uparrow}\,\hat{c}^{\dagger}_{-\boldsymbol{k}\downarrow}\right)|0\rangle \tag{7.263}$$

are orthogonal to each other and normalized to one.

(e) Consider the state $|2,\boldsymbol{k}\rangle$ which is defined as

$$|2,\boldsymbol{k}\rangle:=\left(v_{\boldsymbol{k}}-u_{\boldsymbol{k}}\,\hat{c}^{\dagger}_{\boldsymbol{k}\uparrow}\,\hat{c}^{\dagger}_{-\boldsymbol{k}\downarrow}\right)\prod_{\boldsymbol{k}'}^{\boldsymbol{k}'\neq\boldsymbol{k}}\left(u_{\boldsymbol{k}'}+v_{\boldsymbol{k}'}\,\hat{c}^{\dagger}_{\boldsymbol{k}'\uparrow}\,\hat{c}^{\dagger}_{-\boldsymbol{k}'\downarrow}\right)|0\rangle\,. \tag{7.264}$$

Show that

$$\langle 2,\boldsymbol{k}|\,\hat{H}\,|2,\boldsymbol{k}\rangle-\langle\varphi|\,\hat{H}\,|\varphi\rangle\approx 2\,E_{\boldsymbol{k}}, \tag{7.265}$$

where $E_{\boldsymbol{k}}:=\sqrt{\Delta^2+\xi^2_{\boldsymbol{k}}}$. What terms have been dropped here?

Exercise 1.3: We now go back to the Hamiltonian (7.257). Show that the BCS wave function (7.258) is obtained from minimizing the energy $\langle\varphi_{\boldsymbol{k}},\varphi|\,\hat{H}\,|\varphi_{\boldsymbol{k}},\varphi\rangle$ of the trial wave function

$$|\varphi_{\boldsymbol{k}},\varphi\rangle=\left(u_{\boldsymbol{k}}+v_{\boldsymbol{k}}e^{\mathrm{i}\varphi_{\boldsymbol{k}}}\,\hat{c}^{\dagger}_{\boldsymbol{k}\uparrow}\,\hat{c}^{\dagger}_{-\boldsymbol{k}\downarrow}\right)\prod_{\boldsymbol{k}'}^{\boldsymbol{k}'\neq\boldsymbol{k}}\left(u_{\boldsymbol{k}'}+v_{\boldsymbol{k}'}\,e^{\mathrm{i}\varphi}\,\hat{c}^{\dagger}_{\boldsymbol{k}'\uparrow}\,\hat{c}^{\dagger}_{-\boldsymbol{k}'\downarrow}\right)|0\rangle\,, \tag{7.266}$$

where $\varphi,\varphi_{\boldsymbol{k}}\in\mathbb{R}$ and under the condition

$$V_{\boldsymbol{k},\boldsymbol{k}'}=V_{\boldsymbol{k}',\boldsymbol{k}} \tag{7.267}$$

in Eq. (7.257c).

7.9.2 *Flux quantization in a superconductor*

Consider a superconducting hollow cylinder. Assume that the superconducting current density $\boldsymbol{j}_{\mathrm{s}}(\boldsymbol{x},t)$ is defined by

$$\boldsymbol{j}_{\mathrm{s}}(\boldsymbol{x},t):=-e^{*}\,n_{\mathrm{s}}\,\boldsymbol{v}(\boldsymbol{x},t), \tag{7.268}$$

where n_{s} is the superfluid density, $\boldsymbol{v}$ the average speed of the charge carriers, and $-e^{*}<0$ the charge of one charge carrier.

Exercise 1.1:

(a) Show that, in the classical limit, the electric field $\boldsymbol{E}$ and the current density $\boldsymbol{j}_{\mathrm{s}}$ are related by

$$\frac{e^{*}}{m}\boldsymbol{E}=\frac{1}{n_{\mathrm{s}}\,e^{*}}\frac{\mathrm{d}\boldsymbol{j}}{\mathrm{d}t}. \tag{7.269}$$

Fig. 7.3 A hollow cylindrical superconductor.

Consider a closed path C in the x–y plane surrounding the hole shown in Fig. 7.3. The fluxoid ϕ_M is defined by

$$\phi_M := \int_{\Omega} \mathrm{d}\boldsymbol{s} \cdot \boldsymbol{B} + \frac{m\,c}{n_{\mathrm{s}} e^{*2}} \oint_{C} \mathrm{d}\boldsymbol{l} \cdot \boldsymbol{j}. \tag{7.270}$$

Note that $\phi = \int_{\Omega} \mathrm{d}\boldsymbol{s} \cdot \boldsymbol{B}$ is just the ordinary magnetic flux through the area Ω with the oriented (anticlockwise) boundary $C = \partial\Omega$.

(b) Use Maxwell equations and Eq. (7.269) to show that the fluxoid ϕ_M is conserved in time

$$\frac{\partial \phi_M}{\partial t} = 0. \tag{7.271}$$

(c) Using the identity $m\,\boldsymbol{v} = \boldsymbol{p} + \frac{e^*}{c}\,\boldsymbol{A}$, show that the fluxoid can be expressed as

$$\phi_M = -\frac{c}{e^*} \oint_{C} \mathrm{d}\boldsymbol{l} \cdot \boldsymbol{p}. \tag{7.272}$$

(d) Applying the Bohr-Sommerfeld quantum condition conclude that the fluxoid is quantized

$$\phi_M = \frac{h\,c}{e^*}\,n, \qquad n \in \mathbb{Z}. \tag{7.273}$$

Early experiments on superconductivity indicated that $e^* = 2\,e$, which was a big puzzle at that time. Today we know that e^* is in fact the charge of a Cooper pair.

7.9.3 *Collective excitations within the RPA approximation*

Introduction

Equation (7.214) is the main result of chapter 7. It is a dynamical effective theory for the low-energy and long-wavelength degrees of freedom in a superconducting phase with a nodeless mean-field gap that preserves the symmetries under spin rotations and reversal of time. These low-energies and long-wavelength degrees of freedom are represented by a real-valued scalar field θ, the phase of the superconducting order parameter, that couples to external (source) electromagnetic gauge fields A with the component A_0 in imaginary time ($\mu = 0$) and the three components $\boldsymbol{A}$ in three-dimensional position space ($\mu = 1, 2, 3$). The dynamical effective theory is an approximate one, for only terms quadratic in θ and the gauge fields A have been kept in a gradient expansion (an expansion in powers of the mean-field Green function). It is the existence of the mean-field superconducting gap $\overline{\Delta}$ that justifies this expansion. It applies to fields that vary slowly on the characteristic length scale

$$\xi = \frac{\hbar v_{\mathrm{F}}}{\overline{\Delta}}, \tag{7.274}$$

where v_{F} is the Fermi velocity. The length ξ is called the superconducting coherence length. It diverges upon approaching the transition at which the mean-field gap vanishes. Hence, the effective field theory (7.214) applies deep in the superconducting phase. Equation (7.214) is of the generic form

$$Z[A] := \int \mathcal{D}[\theta]\, e^{-S[\theta,A]}, \tag{7.275a}$$

with the Euclidean action

$$\begin{aligned} S[\theta,A] &:= +\frac{1}{2}\int \mathrm{d}^4x \int \mathrm{d}^4y \,\left(\partial_\mu\theta - q\,A_\mu\right)(x)\,\Pi^{\mu\nu}(x-y)\,\left(\partial_\nu\theta - q\,A_\nu\right)(y) \\ &= +\frac{1}{2}\sum_{Q}\left(\mathrm{i}Q_\mu\,\theta_Q - q\,A_{\mu\,Q}\right)^*\,\Pi^{\mu\nu}_{Q}\,\left(\mathrm{i}Q_\nu\,\theta_Q - q\,A_{\nu\,Q}\right), \end{aligned} \tag{7.275b}$$

where q is the electromagnetic gauge charge (not necessarily the charge $-e < 0$ of the electron), summation is implied over repeated labels $\mu, \nu = 0, 1, 2, 3$ in position space and imaginary time, the kernel $\Pi^{\mu\nu}(x-y) = [\Pi^{\nu\mu}(y-x)]^*$ is Hermitian, and the Fourier conventions are defined by Eq. (7.203).

Gauge invariance

Exercise 1.1: Verify that the Euclidean action and the partition function (7.275) are invariant under the local gauge transformation

$$\theta \to \theta + q\,\varphi, \qquad A_\mu \to A_\mu + \partial_\mu\varphi. \tag{7.276}$$

Current-current correlation functions

Define the current functional

$$j^\mu[\theta,A] := +\left(\frac{\delta S}{\delta A_\mu}\right)[\theta,A] \tag{7.277}$$

and the susceptibility functional

$$\upsilon^{\mu\nu}[\theta,A] := +\left(\frac{\delta^2 S}{\delta A_\mu\,\delta A_\nu}\right)[\theta,A], \tag{7.278}$$

for $\mu, \nu = 0, \cdots, 3$. They represent the slope and the curvature of the action S at $[\theta, A]$, respectively. Define the current functional

$$J^{\mu}[A] := \left(\frac{\delta \ln Z}{\delta A_{\mu}}\right)[A] \tag{7.279}$$

and the susceptibility functional

$$\Upsilon^{\mu\nu}[A] := \left(\frac{\delta^2 \ln Z}{\delta A_{\mu}\, \delta A_{\nu}}\right)[A], \tag{7.280}$$

for $\mu, \nu = 0, \cdots, 3$. They represent the slope and the curvature of the action $-F := \ln Z$ at A, respectively.

Exercise 2.1:

(a) Show that

$$J^{\mu}[A] = -\langle j^{\mu}\rangle\,[A], \tag{7.281a}$$

and

$$\Upsilon^{\mu\nu}[A] = +\langle j^{\mu}\, j^{\nu}\rangle\,[A] - (J^{\mu}\, J^{\nu})\,[A] - \langle v^{\mu\nu}\rangle\,[A], \tag{7.281b}$$

where

$$\langle(\cdots)\rangle\,[A] := \frac{\int \mathcal{D}[\theta]\, e^{-S[\theta,A]}\,(\cdots)}{\int \mathcal{D}[\theta]\, e^{-S[\theta,A]}}. \tag{7.281c}$$

(b) Compute $j^{\mu}[\theta, A]$ and $v^{\mu\nu}[\theta, A]$ for the generic action (7.275) both in (position) space and (imaginary) time and in four momentum space.

Collective excitations without gauge invariance

We break the gauge invariance under the transformation (7.276) by defining the effective low-energy and long-wavelength theory of the generic form

$$Z := \int \mathcal{D}[\theta]\, e^{-S[\theta]} \tag{7.282a}$$

with the Euclidean action

$$\begin{aligned} S[\theta] :&= +\frac{1}{2}\int \mathrm{d}^4x \int \mathrm{d}^4y\, (\partial_{\mu}\theta)\,(x)\, \Pi^{\mu\nu}(x-y)\, (\partial_{\nu}\theta)\,(y) \\ &= +\frac{1}{2}\sum_{Q} \left(Q_{\mu}\, \theta_{Q}\right)^{*}\, \Pi^{\mu\nu}_{Q}\, \left(Q_{\nu}\, \theta_{Q}\right). \end{aligned} \tag{7.282b}$$

Exercise 3.1:

(a) Under what generic conditions does the phase field θ support excitations that disperse at vanishing temperature? *Hint:* Do a Taylor expansion of $\Pi^{\mu\nu}_{Q}$ in powers of the four-momentum Q at vanishing temperature.

(b) Write down the kernel for the phase field θ to leading order in a gradient expansion of $\Pi^{\mu\nu}$ for the polarization tensor in section 7.7.4 at zero temperature and show that there are gapless excitations. Compare this conclusion with the discussion that follows Eq. (7.227).

7.9.4 *The Hall conductivity in a superconductor and gauge invariance*

Introduction

What is the Hall conductivity of a superconductor at vanishing temperature? At the mean-field level, we immediately encounter a difficulty with the fact that charge is not anymore a good quantum number. Instances of quantization of the Hall conductivity can only make sense if charge is a good quantum number. This suggests that the Hall conductivity cannot be quantized in a superconductor. Whatever value it takes, it cannot be universal. However, one might object that this conclusion is an artifact of the mean-field approximation. Hence, it would be desirable to reach this conclusion using a more general argument. This we try by revisiting Laughlin flux insertion argument from section 5.5.3.

Definition

We imagine a two-dimensional superconductor with a mean-field nodeless gap confined to the Corbino geometry of section 5.5.3.

Exercise 1.1: Does assumption L1 from section 5.5.3 hold? Answer this question in the mean-field approximation first and then in the RPA approximation presented in section 7.7.4.

Exercise 1.2: Does assumption L2 from section 5.5.3 hold? Answer this question in the mean-field approximation first and then in the RPA approximation presented in section 7.7.4. Discuss the role played by gauge invariance or lack thereof. *Hint:* Explain why it suffices to use the effective action defined by Eq. (7.224) in order to answer this question.

Exercise 1.3: How does flux quantization affect the reasoning from section 5.5.3 that constrains the Hall conductivity to be a rational number in units of e^2/h.

Chapter 8

A single dissipative Josephson junction

Outline

A phenomenological model for a Josephson junction is presented. The DC- and AC-Josephson effects are described. A model for a dissipative Josephson junction is given both at the classical and quantum levels. At the quantum level, a dissipative Josephson junction is shown to realize the Caldeira-Leggett model of dissipative quantum mechanics. The method of instantons in quantum mechanics is reviewed. The phase diagram of a dissipative Josephson junction is discussed using renormalization-group methods and a duality transformation. The existence of lines of weak- and strong-coupling fixed points is established.

8.1 Phenomenological model of a Josephson junction

We have derived in section 7.7 a low-energy effective action for space and time fluctuations of the superconducting order parameter. We have argued that the most important (collective) degree of freedom that needs to be accounted for is the phase $\phi(x) = 2\theta(x)$, $x \equiv (\mathbf{R}, \tau)$ of the superconducting order parameter,

$$\Delta(x) = |\overline{\Delta}| \exp\big(+ 2\mathrm{i}\theta(x)\big) \equiv |\overline{\Delta}| \exp\big(+ \mathrm{i}\phi(x)\big). \tag{8.1}$$

Fluctuations in the magnitude of the pairing-order parameter about the mean-field value $|\overline{\Delta}|$ were argued to account for collective excitations with characteristic energy of the order of the mean-field gap $\propto |\overline{\Delta}|$. This is why such fluctuations can be neglected at temperatures well below the mean-field gap. On the one hand, we argued that the imaginary-time derivative of $\theta(x)$ couples to electrons with the electrical charge $-e < 0$ through the electronic density

$$\rho(x) = \sum_{\sigma=\uparrow,\downarrow} \psi^*_\sigma(x)\, \psi_\sigma(x) \tag{8.2}$$

in the same way as the scalar potential $\varphi(x)$ that conveys the electronic Coulomb interaction does,

$$\mathrm{i}\int\limits_{\beta V}\mathrm{d}^4x\;\rho(x)\left[(\partial_\tau\theta)(x)+e\,\varphi(x)\right]. \tag{8.3}$$

On the other hand, we argued that the space-derivative of $\theta(x)$ couples to electrons through the paramagnetic and diamagnetic currents

$$\begin{aligned}\mathbf{J}_{\mathrm{p}}(x)&=\frac{1}{2\,m\mathrm{i}}\sum_{\sigma=\uparrow,\downarrow}\left[\psi^*_\sigma(x)\,(\boldsymbol{\nabla}\psi_\sigma)(x)-(\boldsymbol{\nabla}\psi^*_\sigma)(x)\,\psi_\sigma(x)\right],\\ \mathbf{J}_{\mathrm{d}}(x)&=+\frac{1}{m}\left(\sum_{\sigma=\uparrow,\downarrow}\psi^*_\sigma(x)\,\psi_\sigma(x)\right)\left[(\boldsymbol{\nabla}\theta)\,(x)+e\,\mathbf{A}(x)\right],\end{aligned} \tag{8.4}$$

respectively, in the same way as the vector potential $\mathbf{A}(x)$ does [recall Eq. (7.190)],

$$\int\limits_{\beta V}\mathrm{d}^4x\left\{\mathbf{J}_p(x)\cdot\left[(\boldsymbol{\nabla}\theta)(x)+e\,\mathbf{A}(x)\right]+\frac{1}{2}\mathbf{J}_d(x)\cdot\left[(\boldsymbol{\nabla}\theta)(x)+e\,\mathbf{A}(x)\right]\right\}. \tag{8.5}$$

Absent from our perturbative calculations of the effective action for θ is the fact that θ is defined modulo π, i.e., that the effective action for θ must be periodic in θ and that, in particular, θ can support singularities called vortices.

The phenomenological model of a Josephson junction attempts to describe the coupling between two superconducting metals in close proximity by postulating the validity of Eqs. (8.3) and (8.5) for each superconductor and by proposing an interaction between the two superconductors that is periodic in the phase mismatch between the phase θ_1 of the pairing-order parameter in superconductor 1 and the phase θ_2 of the pairing-order parameter in superconductor 2. The microscopic mechanism that motivates this choice of a coupling is coherent tunneling of paired electrons between the two superconductors, a highly controversial idea at the time (see Ref. [83] for a historical perspective). A first simplifying assumption is that the phases θ_1 and θ_2 of each superconductor are taken to be constant in space. Correspondingly, the vector potentials $\mathbf{A}_\alpha$ are taken to be vanishing whereas the scalar potentials φ_α only vary in imaginary time, $\alpha=1,2$. In other words, the non-interacting contribution to the action describing a Josephson junction is simply

$$S_0:=\mathrm{i}\sum_{\alpha=1,2}\int\limits_0^\beta\mathrm{d}\tau N_\alpha(\tau)\left[(\partial_\tau\theta_\alpha)\,(\tau)+e\,\varphi_\alpha(\tau)\right], \tag{8.6}$$

whereby $N_\alpha(\tau)$, $\phi_\alpha(\tau)=2\theta_\alpha(\tau)$, and $\varphi_\alpha(\tau)$, are the number of electrons, the phase of the pairing-order parameter, and the applied potential in superconductor α at imaginary time τ, respectively. The effective action (8.6) follows from Eq. (7.190) by ignoring the quasiparticle Lagrangian density $\mathcal{L}_0$ while assuming that $\mathcal{L}_{1,2}$ and $\mathcal{L}_2$ vanish, leaving $\mathcal{L}_{1,1}$ as the sole contribution.

It is customary to perform the change of variables

$$N_\pm:=N_1\pm N_2,\qquad\theta_\pm:=\theta_1\pm\theta_2,\qquad\varphi_\pm:=\varphi_1\pm\varphi_2, \tag{8.7a}$$

when coupling two "levels", in which case

$$S_0 = \frac{\mathrm{i}}{2} \sum_{\alpha=\pm} \int_0^\beta \mathrm{d}\tau N_\alpha(\tau) \left[(\partial_\tau \theta_\alpha)(\tau) + e\, \varphi_\alpha(\tau) \right]. \tag{8.7b}$$

A second assumption consists in considering the case when N_+ is time independent, i.e., both superconductors form a closed system in which the total number of electrons N_+ is conserved at all imaginary times. Consequently,

$$\begin{aligned} S_0 &= \frac{\mathrm{i}}{2} N_+ \left[\theta_+(\beta) - \theta_+(0) + e\,\beta\,\overline{\varphi}_+ \right] + \frac{\mathrm{i}}{2} \int_0^\beta \mathrm{d}\tau\, N_-(\tau) \left[(\partial_\tau \theta_-)(\tau) + e\,\varphi_-(\tau) \right] \\ &= \frac{\mathrm{i}}{2} \int_0^\beta \mathrm{d}\tau\, N_-(\tau) \left[(\partial_\tau \theta_-)(\tau) + e\,\varphi_-(\tau) \right], \end{aligned} \tag{8.8a}$$

if $l = 0$ and $\overline{\varphi}_+ = 0$ are chosen in

$$\theta_+(\beta) - \theta_+(0) = \pi\, l, \qquad l \in \mathbb{Z}, \qquad \overline{\varphi}_+ := \frac{1}{\beta} \int_0^\beta \mathrm{d}\tau\ \varphi_+(\tau). \tag{8.8b}$$

The contribution to the Josephson junction action arising from the interaction between the two superconductors is taken as the simplest function of $\phi_- := \phi_1 - \phi_2$ that is periodic with periodicity 2π and that penalizes a non-vanishing phase difference between the two superconductors,[1]

$$S_{U_\mathrm{J}} := -2\,U_\mathrm{J} \int_0^\beta \mathrm{d}\tau\ \cos\phi_- = -2\,U_\mathrm{J} \int_0^\beta \mathrm{d}\tau\ \cos(2\theta_-), \qquad U_\mathrm{J} > 0. \tag{8.10}$$

The quantum model for a Josephson junction is then defined by the partition function

$$\begin{aligned} Z_\beta &:= \int \mathcal{D}[N_-, \phi_-]\ \exp\left(-S_\beta\right), \\ S_\beta &:= \int_0^\beta \mathrm{d}\tau\ \left(L_0 + L_{U_\mathrm{J}} \right), \\ L_0 &:= \frac{1}{4} N_- \left[(\mathrm{i}\partial_\tau \phi_-) + \mathrm{i}e^* \varphi_- \right], \qquad e^* = 2\,e, \\ L_{U_\mathrm{J}} &:= -2\,U_\mathrm{J} \cos\left(\phi_-\right), \\ \phi_-(\tau) &= \phi_-(\tau + \beta), \qquad \varphi_-(\tau) = \varphi_-(\tau + \beta), \end{aligned} \tag{8.11}$$

[1] Expansion in powers of ϕ_- of the interacting Lagrangian

$$L_{U_\mathrm{J}} := -2\,U_\mathrm{J} \cos\phi_- = -2\,U_\mathrm{J} + U_\mathrm{J}\,\phi_-^2 + \mathcal{O}\left(\phi_-^4\right), \tag{8.9}$$

shows that our Gaussian approximation for the effective action obeyed by ϕ in chapter 7, say Eq. (7.166), is simply obtained by identifying ϕ_- with $\boldsymbol{\nabla}\phi$.

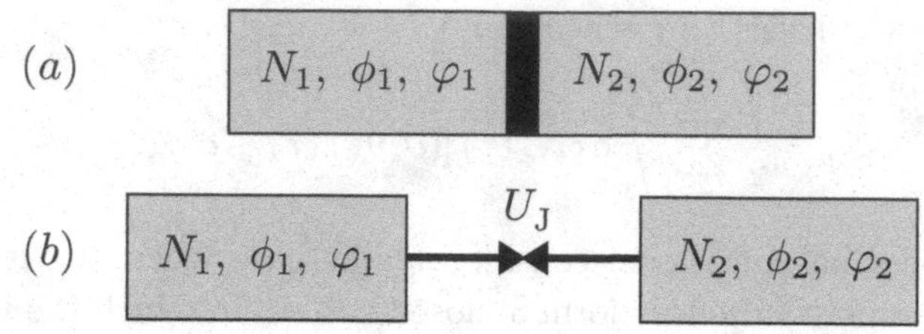

Fig. 8.1 (a) Two superconductors are separated by a thin non-superconducting layer. (b) At zero temperature the thin non-superconducting layer acts like a tunnel barrier to paired electrons (Cooper pairs). Quantum tunneling of Cooper pairs can be driven by application of a voltage difference $V_{\rm J} \equiv \varphi_1 - \varphi_2$ between superconductors 1 and 2. The Josephson equations model macroscopically the current flow driven by quantum tunneling of Cooper pairs across the non-superconducting layer.

in the background of the scalar potential φ_-. Quantum mechanics comes about from the integration over the measure $\mathcal{D}[N_-,\phi_-]$ for the bosonic coherent states ϕ_- and N_- (see appendix A.1.2).

The classical equations of motion for a Josephson junction follow from

$$\begin{aligned} 0 =&\ \delta S_\beta \\ =&\ \frac{\mathrm{i}}{4}\int\limits_0^\beta \mathrm{d}\tau\,(\delta N_-)\left[(\partial_\tau \phi_-) + e^*\varphi_-\right] + \frac{\mathrm{i}}{4}\int\limits_0^\beta \mathrm{d}\tau\,(-\partial_\tau N_-)\,(\delta\phi_-) \\ &+2\,U_{\rm J}\int\limits_0^\beta \mathrm{d}\tau\ \ (\delta\phi_-)\sin\phi_-. \end{aligned} \tag{8.12a}$$

They are

$$\begin{aligned} \mathrm{i}\,(\partial_\tau \phi_-) &= -\mathrm{i}e^*\,\varphi_-, \\ \mathrm{i}\,(\partial_\tau N_-) &= +8\,U_{\rm J}\,\sin\phi_-, \end{aligned} \tag{8.12b}$$

in imaginary time. Analytical continuation to real time[2]

$$\tau = \mathrm{i}t, \qquad \varphi_-(\tau) = +\mathrm{i}\varphi_-(t), \tag{8.13a}$$

yields

$$\begin{aligned} (\partial_t \phi_-)\,(t) &= +e^*\,\varphi_-(t), \\ (\partial_t N_-)\,(t) &= +8\,U_{\rm J}\,\sin\left(\phi_-(t)\right). \end{aligned} \tag{8.13b}$$

Had we chosen canonical quantization instead of a path integral formulation, we would have elevated N_- and ϕ_- to the level of operators $\hat{N}_-$ and $\hat{\phi}_-$ obeying the equal-time commutation relation

$$[\hat{N}_-,\hat{\phi}_-] = \mathrm{i}, \qquad [\hat{N}_-,\hat{N}_-] = [\hat{\phi}_-,\hat{\phi}_-] = 0, \tag{8.14}$$

[2] We are undoing Eq. (7.187).

and used the representation

$$Z_\beta \propto \mathrm{Tr}\left(e^{-\beta\hat{H}}\right),$$
$$\hat{H} := -e^*\,\hat{N}_-\,\varphi_- - 8\,U_\mathrm{J}\,\cos\hat{\phi}_-, \tag{8.15}$$

of the partition function together with the equations of motion

$$\begin{aligned}\mathrm{i}\partial_t\hat{\phi}_- &= [\hat{\phi}_-,\hat{H}],\\ \mathrm{i}\partial_t\hat{N}_- &= [\hat{N}_-,\hat{H}],\end{aligned} \tag{8.16}$$

to recover Eq. (8.13).

To bring Eq. (8.13) to the canonical representation of the Josephson equations,

$$\varphi_-(t) = \varphi_1(t) - \varphi_2(t) =: V_\mathrm{J}(t) \tag{8.17}$$

is reinterpreted as the voltage difference $V_\mathrm{J}(t)$ between superconductors 1 and 2, respectively, at real time t. Correspondingly, the electrical current at real time t that flows between superconductors 2 and 1, owing to the negative charge $-e<0$ of the electron, is

$$I_\mathrm{J}(t) := (-e)\,(\partial_t N_1)\,(t) = \frac{1}{2}(-e)\,(\partial_t N_-)\,(t). \tag{8.18}$$

Here, conservation of the total charge was used,

$$0 = \partial_t N_+ = \partial_t N_1 + \partial_t N_2 \Longrightarrow \partial_t N_- = +2\partial_t N_1. \tag{8.19}$$

Now, the classical equations of motion (8.13) in real time are rewritten

$$\begin{aligned}(\partial_t\phi_-)\,(t) &= +\frac{e^*\,V_\mathrm{J}(t)}{\hbar},\\ I_\mathrm{J}(t) &= -\frac{2\,e^*\,U_\mathrm{J}}{\hbar}\,\sin\left(\phi_-(t)\right).\end{aligned} \tag{8.20}$$

Here, $\hbar$ has been reinstated and it is customary to define the unit of current[3]

$$I_0 := \frac{2\,e^*\,U_\mathrm{J}}{\hbar} = \frac{4\,e\,U_\mathrm{J}}{\hbar}. \tag{8.21}$$

Equations (8.20) and (8.21) define the "classical" Josephson equations of motion (see Fig. 8.1). These equations are "classical" in the sense that they follow from a variational Ansatz on the action of the full quantum mechanical partition function

$$\begin{aligned}Z_{-\mathrm{i}\beta} &:= \int\mathcal{D}[N_-,\phi_-]\,\exp\left(+\mathrm{i}S_{-\mathrm{i}\beta}\right),\\ S_{-\mathrm{i}\beta} &:= \int_0^{-\mathrm{i}\beta}\mathrm{d}t\,\left(L_0 + L_{U_J}\right),\\ L_0 &:= \frac{1}{4}N_-\left(-\left(\partial_t\phi_-\right) + \frac{e^*\,V_\mathrm{J}}{\hbar}\right), \qquad e^* = 2\,e,\\ L_{U_\mathrm{J}} &:= +\frac{2\,U_\mathrm{J}}{\hbar}\,\cos\left(\phi_-\right),\\ \phi_-(t) &= \phi_-(t-\mathrm{i}\beta), \qquad \varphi_-(t) = \varphi_-(t-\mathrm{i}\beta).\end{aligned} \tag{8.22}$$

[3] As a check of units, the potential difference V has the units of energy per unit charge and the coupling constant U_J has the units of energy. Hence, $e^*\,V/\hbar$ has the units of inverse time and $I_0 = 2\,e^*\,U_J/\hbar$ has the units of charge per unit time.

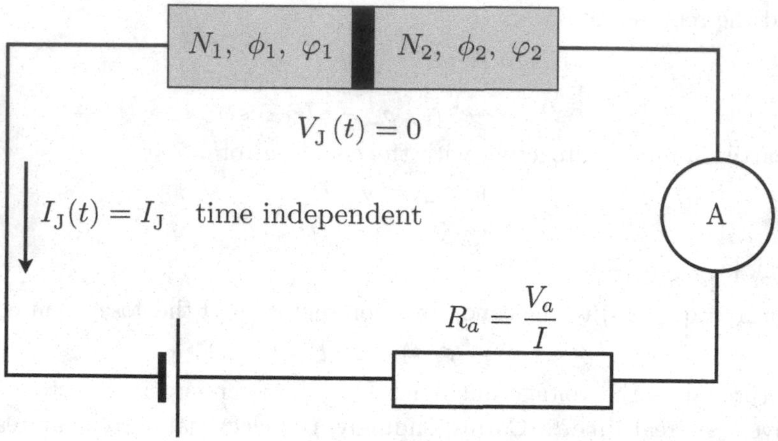

Fig. 8.2 Setup for the DC-Josephson effect. The Josephson current I_{J} is time independent. The Josephson voltage $V_{\mathrm{J}} \equiv \varphi_1 - \varphi_2$ vanishes.

8.2 DC-Josephson effect

The so-called *DC-Josephson effect* is derived from Eq. (8.20) by assuming that the current $I_{\mathrm{J}}(t) = I_{\mathrm{J}}$ in the circuit of Fig. 8.2 is constant in time, in which case

$$\sin\left(\phi_-(t)\right) = -\frac{I_{\mathrm{J}}}{I_0} \tag{8.23a}$$

is time independent and

$$V_{\mathrm{J}}(t) = +\frac{\hbar}{e^*}\left(\partial_t \phi_-\right)(t) = 0. \tag{8.23b}$$

The resistance

$$R_{\mathrm{J}}(t) := \frac{V_{\mathrm{J}}(t)}{I_{\mathrm{J}}(t)} = 0 \tag{8.24}$$

of the Josephson junction vanishes when the current passing through the circuit of Fig. 8.2 is time independent. This is the DC-Josephson effect.

8.3 AC-Josephson effect

The so-called *AC-Josephson effect* is derived from Eq. (8.20) by assuming that the voltage $V_{\mathrm{J}}(t) = V_{\mathrm{J}} \neq 0$ in the circuit of Fig. 8.3 is constant in time and non-vanishing, in which case

$$\phi_-(t) = \phi_-(t=0) + \frac{e^*\,V_{\mathrm{J}}}{\hbar}\,t, \tag{8.25a}$$

and

$$I_{\mathrm{J}}(t) = -I_0 \sin\left(\phi_-(t=0) + \frac{e^*\,V_{\mathrm{J}}}{\hbar}\,t\right). \tag{8.25b}$$

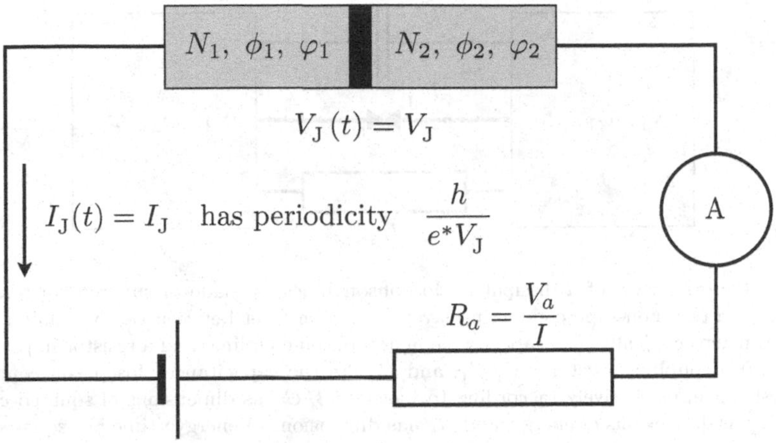

Fig. 8.3 Setup for the AC-Josephson effect. The Josephson voltage $V_J \equiv \varphi_1 - \varphi_2$ is time independent and non-vanishing. The Josephson current J_J is periodic in time with period $h/(e^* V_J)$.

When the voltage drop at the Josephson junction is time independent and non-vanishing, the current in the circuit of Fig. 8.3 is periodic with period

$$\frac{h}{e^* V_J}. \tag{8.26}$$

This is the AC-Josephson effect that allows a measurement of the charge of the Cooper pair

$$e^* = 2e. \tag{8.27}$$

8.4 Dissipative Josephson junction

8.4.1 *Classical*

Consider the setup in Fig. 8.4 that defines a "classical" dissipative Josephson junction. There are three additive contributions to the current $I(t)$ flowing from superconductor 2 to superconductor 1:

(1) A capacitive current

$$I_C(t) := -C\,(\partial_t V_J)(t), \qquad (C \text{ has units of squared charge per energy}), \tag{8.28}$$

where $V_J(t) = \varphi_-(t)$ is the voltage difference between superconductors 1 and 2 and the proportionality constant C is time independent and called the capacitance.

(2) A Josephson current

$$I_J(t) = -\frac{2\,e^*\,U_J}{\hbar}\sin\left(\phi_-(t)\right) \equiv -I_0\sin\left(\phi_-(t)\right), \tag{8.29}$$

where U_J is the characteristic interaction strength between superconductors 1 and 2.

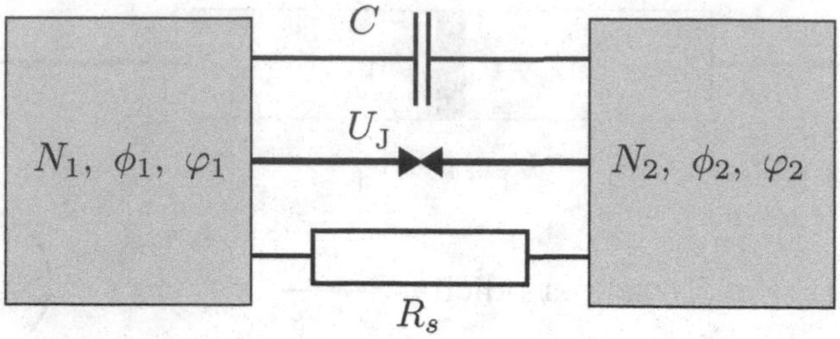

Fig. 8.4 Pictorial view of a dissipative Josephson junction made of superconductors 1 and 2 separated by a thin non-superconducting layer. The thin layer between the two superconductors is modeled macroscopically by a capacitance, a Josephson coupling, and a resistor in parallel. The (dimensionful) coupling constants C, U_{J}, and R_s are the capacitance, Josephson coupling, and Ohmic resistance, respectively (according to footnote 3, C has dimensions of squared charge per energy, U_{J} has dimensions of energy, and R_s has dimensions of energy×time per squared charge).

(3) An Ohmic current

$$I_s(t) := -\frac{V_{\mathrm{J}}(t)}{R_s} \qquad (R_s \text{ has units of energy} \times \text{time per squared charge}). \tag{8.30}$$

We shall shortly see that this current is dissipative. Here, the time-independent Ohmic conductance $1/R_s$ measures the strength of the dissipation. Dissipation is maximal in the limit of an infinite Ohmic conductance (a vanishing Ohmic resistance $R_s = 0$) by which the entire current from superconductor 2 to superconductor 1 is carried by $I_s(t)$, i.e., in the limit $I_C(t)/I_s(t) = I_{\mathrm{J}}(t)/I_s(t) = 0$. In the opposite limit of a vanishing Ohmic conductance (an infinite Ohmic resistance $R_s = \infty$) by which the entire current from superconductor 2 to superconductor 1 is carried by $I_C(t) + I_{\mathrm{J}}(t)$, i.e., in the limit $I_s(t)/I_C(t) = I_s(t)/I_{\mathrm{J}}(t) = 0$, there is no dissipative contribution to the current from superconductor 2 to superconductor 1.

The "classical" equations of motion defining this dissipative Josephson junction are

$$\left(\partial_t \phi_-\right)(t) = +\frac{e^*}{\hbar}\, V_{\mathrm{J}}(t), \tag{8.31a}$$

$$\begin{aligned} I(t) &= I_C(t) + I_{\mathrm{J}}(t) + I_s(t) \\ &= -C\,\left(\partial_t V_{\mathrm{J}}\right)(t) - I_0\,\sin\left(\phi_-(t)\right) - \frac{V_{\mathrm{J}}(t)}{R_s}. \end{aligned} \tag{8.31b}$$

Insertion of (8.31a) into (8.31b) yields the second order differential equation

$$0 = C\,\frac{\hbar}{e^*}\,\ddot{\phi}_- + \frac{1}{R_s}\,\frac{\hbar}{e^*}\,\dot{\phi}_- + I_0\,\sin\phi_- + I, \qquad \dot{\phi}_- \equiv \partial_t \phi_-. \tag{8.32}$$

We will take Eq. (8.32) as our mathematical definition of a classical dissipative Josephson junction. This equation can be reinterpreted as follows. Equation (8.32) describes a classical spinless point particle with the mass

$$C\,\frac{\hbar}{e^{*}}\qquad (e^{*}=2\,e), \tag{8.33}$$

moving on a circle of unit radius with coordinate ϕ_{-} subjected to:[4]

- A driving time-dependent force
$$-\,I(t). \tag{8.35}$$
- A time-independent (conservative) potential
$$-\,I_0\,\cos\phi_{-}. \tag{8.36}$$
- A damping force
$$\frac{1}{R_s}\,\frac{\hbar}{e^{*}}\,\dot{\phi}_{-}. \tag{8.37}$$

When the forcing term $I(t)$ is time independent, the potential

$$V(\phi_{-}):=-I_0\cos\phi_{-}+I\phi_{-},\qquad -V'(\phi_{-})=-I_0\sin\phi_{-}-I \tag{8.38}$$

is sometimes called a washboard potential.

8.4.2 *Caldeira-Leggett model*

In their ground breaking paper [84], Caldeira and Leggett motivated, defined, and resolved the question: *What is the effect of dissipation on quantum tunneling?* As we shall see, their analysis can be recast in the context of a dissipative Josephson junction as the motivation, definition, and resolution to the question: *What is the effect of dissipation on quantum coherence?* To stress the conceptual difference between quantum tunneling and quantum coherence, we will first review the starting point of Caldeira and Leggett.

Caldeira and Leggett consider first an isolated and non-dissipative quantum system that has been initially prepared to be in the close vicinity of the *unique* metastable minimum of the cubic potential[5]

$$\begin{aligned} V(q) &= \frac{1}{2}\,M\,\varpi_0^2\,q^2-\frac{1}{3}\,\lambda^2\,q^3\\ &= \frac{27}{4}\,V_0\left(\frac{q}{q_0}\right)^2\left(1-\frac{q}{q_0}\right),\\ V_0\equiv V(q)\big|_{q=\frac{M\varpi_0^2}{\lambda^2}}&,\qquad q_0\equiv\frac{3}{2}\,\frac{M\,\varpi_0^2}{\lambda^2},\qquad 0<\varpi_0,\lambda\in\mathbb{R}. \end{aligned} \tag{8.39}$$

[4] A particle of mass m obeys Newton equation

$$m\,\ddot{x}=-c\,\dot{x}-V'(x)+F_{\mathrm{ext}}(t), \tag{8.34}$$

when subjected to a damping proportional to its speed $\dot{x}$, an energy conserving potential $V(x)$, and an external force $F_{\mathrm{ext}}(t)$.

[5] The assumption that $V(q)$ is cubic is done without loss of generality as long as $V(q)$ is sufficiently smooth and has the general shape of a cubic potential, i.e., has a single local minimum.

Here, q can be interpreted as the coordinate on the real line $\mathbb{R}$ of a spinless point particle of mass M with the classical Lagrangian[6]

$$L_0 = \frac{1}{2} M \dot{q}^2 - V(q). \tag{8.40}$$

Although a classical particle sitting in the metastable minimum of $V(q)$ cannot escape this local minimum, a quantum particle can decay through the potential barrier V_0 into a continuum of states. Within the WKB approximation, the probability per unit time Γ_0 for the particle to escape a generic metastable potential well is given by

$$\begin{aligned}
\Gamma_0 &= A_0\, e^{-\frac{B_0}{\hbar}} \left[1 + \mathcal{O}(\hbar \varpi_0 / V_0)\right], \\
A_0 &= C_0\, \varpi_0 \sqrt{\frac{B_0}{2\pi\hbar}}, \\
B_0 &= 2 \int_0^{q_0} \mathrm{d}q \sqrt{2 M V(q)},
\end{aligned} \tag{8.41}$$

whereby, for our cubic potential,

$$B_0 = \frac{36}{5} \frac{V_0}{\varpi_0}, \qquad C_0 = \sqrt{60}. \tag{8.42}$$

To model dissipation at the classical level, Caldeira and Leggett choose the simplest possible model in which the classical equations of motion of the isolated system

$$0 = M \ddot{q} + V'(q), \qquad V'(q) \equiv \left(\frac{\mathrm{d}V}{\mathrm{d}q}\right)(q), \tag{8.43}$$

are modified by the addition of damping and forcing terms,

$$0 = M \ddot{q} + \eta \dot{q} + V'(q) - F_{\mathrm{ext}}(t), \qquad \eta \in \mathbb{R}. \tag{8.44}$$

In practice, Eq. (8.44) should be thought of as an equation whose validity is empirical, i.e., the phenomenological parameters M and η together with the potential $V(q)$ and the external force $F_{\mathrm{ext}}(t)$ have been measured experimentally.[7] Caldeira and Leggett then ask:

---

[6] In most applications of quantum tunneling, however, q is not a geometrical coordinate as would be the case if one wants to describe the tunneling of an alpha particle out of a nucleus or of an electron out of an atom in a strong electric field, but a macroscopic or collective degree of freedom. For example, in the case of a SQUID ring, q is the magnetic flux trapped in the ring. SQUID stands for (rf) superconducting quantum interference device, i.e., a superconducting ring interrupted by a Josephson junction (see section 6.3 of Ref. [75] and chapter 7 of Ref. [85]).

[7] Of course, dissipation can manifest itself in much more complicated ways. (i) Fourier transformation of Eq. (8.44) with respect to time holds only for a restricted range of frequencies. (ii) The dissipative term $\eta\,\dot{q}$ is replaced by $f(q)\,\dot{q}$ with the function $f(q)$ not constant in q. (iii) The dissipative term $\eta\,\dot{q}$ is replaced by higher order time derivatives $f(q)\,\partial_t^n q$, n odd. (iv) The dissipative term $\eta\,\dot{q}$ is replaced by the time convolution $\eta * \dot{q}$. (v) Dissipation is non-linear, i.e., $\dot{q}$ enters in a non-linear way in the Lagrangian. Such generalizations are discussed in Ref. [84].

(1) How to construct a quantum mechanical system that yields the equation of motion (8.44) in the classical limit?
(2) Is the effect of dissipation on the tunneling probability (8.41) uniquely determined for a given potential $V(q)$ by the friction coefficient η or is it model dependent?
(3) Does dissipation increase or decrease the tunneling probability (8.41)?

The answers given by Caldeira and Leggett to questions 3 and 2 are that dissipation decreases the tunneling probability and that this effect is indeed uniquely determined by the macroscopic parameter η. We now turn to the answers to questions 1 and 2 that will be directly relevant to the issue of quantum coherence in a dissipative Josephson junction.

8.4.2.1 *Modeling dissipation in quantum mechanics*

To model dissipation at the quantum level, one can imagine coupling the isolated system to some environment. There are then two possible routes. In the first one, the environment is modeled by turning the deterministic Hamiltonian of the isolated system into a statistical ensemble of random Hamiltonians. An alternative approach followed by Caldeira and Leggett is to envision the "universe" made up of the environment and the system of interest as a deterministic one, i.e., endowed with Hamiltonian dynamics, to assume that the environment is made of infinitely many degrees of freedom, and to assume that the coupling between the environment and the system of interest is weak. Assuming that the environment is in the thermodynamic limit allows to consider the limit of strong dissipation without relaxing the condition that the coupling between the environment and the system is weak. A weak coupling between the environment and the system makes plausible the modeling of the environment as a collection or bath of non-interacting harmonic oscillators. The classical Lagrangian of the "universe" is taken by Caldeira and Leggett to be

$$L = L_0 + L_{\mathrm{bath}} + L_{\mathrm{int}} + L_{\mathrm{adia}} + L_{\mathrm{ext}}(t), \tag{8.45a}$$

where the Lagrangian L_0 describes a classical spinless point particle of mass $M > 0$ in a conservative potential V,

$$L_0 := \frac{1}{2} M \dot{q}^2 - V(q), \tag{8.45b}$$

the Lagrangian L_{bath} describes a family of classical harmonic oscillators with the masses $m_\iota > 0$ and harmonic frequencies $\varpi_\iota > 0$ labeled by the index ι,

$$L_{\mathrm{bath}} := \frac{1}{2} \sum_\iota m_\iota \left(\dot{x}_\iota^2 - \varpi_\iota^2 x_\iota^2 \right), \tag{8.45c}$$

the Lagrangian L_{int} describes the linear coupling (with the coupling constants c_ι) between the point particle with mass M and coordinate q and the (harmonic) bath,

$$L_{\mathrm{int}} := +q \sum_\iota c_\iota x_\iota, \tag{8.45d}$$

the Lagrangian L_{adia} is included for convenience,

$$L_{\text{adia}} := -\frac{1}{2} M \left(\varpi_{\text{adia}}\right)^2 q^2, \qquad \frac{1}{2} M \left(\varpi_{\text{adia}}\right)^2 := \sum_{\iota} \frac{c_\iota^2}{2\, m_\iota\, \varpi_\iota^2}, \tag{8.45e}$$

and $L_{\text{ext}}(t)$ describes a non-conservative force,

$$L_{\text{ext}}(t) := +F_{\text{ext}}(t)\, q(t). \tag{8.45f}$$

The first question to address is how to choose the masses $m_\iota > 0$ and frequencies $\varpi_\iota > 0$ for the bath and how to choose the (linear) coupling constants $c_\iota \in \mathbb{R}$ between the bath and the coordinate q so as to reproduce the phenomenological equation of motion (8.44).

To answer this question one must compare the Fourier transform with respect to time of (8.44), i.e.,

$$0 = -M\, \varpi^2\, q_\varpi - \mathrm{i}\eta\, \varpi\, q_\varpi + [V'(q)]_\varpi - F_{\text{ext}\,\varpi}, \tag{8.46}$$

with the Fourier transform with respect to time of the coupled equations of motion for q and the bath

$$\begin{aligned} 0 &= M\, \ddot{q} + V'(q) - \sum_{\iota} c_\iota\, x_\iota + M \left(\varpi_{\text{adia}}\right)^2 q - F_{\text{ext}}(t), \\ 0 &= m_\iota\, \ddot{x}_\iota + m_\iota\, \varpi_\iota^2\, x_\iota - c_\iota\, q, \qquad \forall \iota, \end{aligned} \tag{8.47}$$

i.e.,

$$0 = -M\, \varpi^2\, q_\varpi + [V'(q)]_\varpi - \sum_{\iota} c_\iota\, x_{\iota\varpi} + \left(\sum_{\iota} \frac{c_\iota^2}{m_\iota} \frac{1}{\varpi_\iota^2} \right) q_\varpi - F_{\text{ext}\,\varpi}, \tag{8.48a}$$

and

$$0 = -m_\iota\, \varpi^2\, x_{\iota\varpi} + m_\iota\, \varpi_\iota^2\, x_{\iota\varpi} - c_\iota\, q_\varpi, \qquad \forall \iota. \tag{8.48b}$$

Insertion of Eq. (8.48b),

$$x_{\iota\varpi} = \frac{c_\iota}{m_\iota} \frac{1}{\varpi_\iota^2 - \varpi^2}\, q_\varpi, \qquad \varpi \neq \varpi_\iota\ \forall \iota, \tag{8.49}$$

into Eq. (8.48a) yields

$$\begin{aligned} 0 &= -M\, \varpi^2\, q_\varpi + [V'(q)]_\varpi - K_\varpi\, q_\varpi - F_{\text{ext}\,\varpi}, \\ K_\varpi &:= -\sum_{\iota} \frac{c_\iota^2}{m_\iota} \left(\frac{1}{\varpi_\iota^2} - \frac{1}{\varpi_\iota^2 - \varpi^2} \right), \qquad \varpi \neq \varpi_\iota\ \forall \iota. \end{aligned} \tag{8.50}$$

Equation (8.50) provides two insights. First, the equation of motion obeyed by the coefficient $q_{\varpi=0}$ is unaffected by the coupling to the bath. This is so because the contribution L_{adia} is precisely chosen to prevent a renormalization of the coefficient $[V'(q)]_{\varpi=0}$ in the Fourier expansion with respect of time of $V'(q)$. Second, the bath parameters $\{m_\iota > 0, \varpi_\iota > 0\}$ and the linear coupling constants $\{c_\iota \in \mathbb{R}\}$ to the bath must be chosen in the following way. First, we do the algebraic manipulation

$$\begin{aligned}\operatorname{Im} K_\varpi &= \operatorname{Im}\left(-\sum_\iota \frac{c_\iota^2}{m_\iota}\left(\frac{1}{\varpi_\iota^2} - \frac{1}{\varpi_\iota^2 - \varpi^2}\right)\right) \\ &= \operatorname{Im}\left((-1)^2 \sum_\iota \frac{c_\iota^2}{m_\iota} \frac{\varpi^2}{\varpi_\iota^2\,(\varpi_\iota^2 - \varpi^2)}\right).\end{aligned} \tag{8.51}$$

Second, we do the analytic and algebraic manipulations

$$\begin{aligned}\operatorname{Im} K_\varpi &= \operatorname{Im}\left(\varpi^2 \sum_\iota \frac{c_\iota^2}{m_\iota}\int_0^\infty \mathrm{d}\tilde{\varpi}\,\delta(\tilde{\varpi} - \varpi_\iota)\,\frac{1}{\tilde{\varpi}^2\,(\tilde{\varpi}^2 - \varpi^2)}\right) \\ &= \operatorname{Im}\left(\varpi^2 \int_0^\infty \mathrm{d}\tilde{\varpi}\left(\sum_\iota \frac{c_\iota^2}{m_\iota\,\tilde{\varpi}}\,\delta(\tilde{\varpi} - \varpi_\iota)\right)\frac{1}{\tilde{\varpi}\,(\tilde{\varpi}^2 - \varpi^2)}\right) \\ &= \operatorname{Im}\left(\frac{2\varpi^2}{\pi}\int_0^\infty \mathrm{d}\tilde{\varpi}\,\underline{\left(\frac{\pi}{2}\sum_\iota \frac{c_\iota^2}{m_\iota\,\tilde{\varpi}}\,\delta(\tilde{\varpi} - \varpi_\iota)\right)}\,\frac{1}{\tilde{\varpi}\,(\tilde{\varpi}^2 - \varpi^2)}\right).\end{aligned} \tag{8.52}$$

Third, we introduce the spectral function

$$J(\varpi) := \frac{\pi}{2}\sum_\iota \frac{c_\iota^2}{m_\iota\,\varpi}\,\delta(\varpi - \varpi_\iota) \tag{8.53}$$

to absorb the underlined factor,

$$\begin{aligned}\operatorname{Im} K_\varpi =& \operatorname{Im}\left(\frac{2\varpi^2}{\pi}\int_0^\infty \mathrm{d}\tilde{\varpi}\,J(\tilde{\varpi})\,\frac{1}{\tilde{\varpi}\,(\tilde{\varpi}^2 - \varpi^2)}\right) \\ =& \operatorname{Im}\left(\frac{2\varpi^2}{\pi}\int_0^\infty \mathrm{d}\tilde{\varpi}\,\frac{J(\tilde{\varpi})}{\tilde{\varpi}}\,\frac{1}{\tilde{\varpi} + |\varpi|}\left[\mathcal{P}_{\frac{1}{\tilde{\varpi} - |\varpi|}} + \operatorname{sgn}(\varpi)\,\mathrm{i}\pi\,\delta(\tilde{\varpi} - |\varpi|)\right]\right).\end{aligned} \tag{8.54}$$

The last line defines how to regularize the first-order pole when $\varpi = \varpi_\iota > 0$. To this end, the principal-value distribution $\mathcal{P}$ has been introduced and we have chosen to move the pole away from the real axis in the $\tilde{\varpi}$-complex plane according to the rule $\varpi \to \varpi + \mathrm{i}0^+$. Finally, we impose the condition

$$\operatorname{Im} K_\varpi = \eta\,\varpi. \tag{8.55}$$

Hence, we infer that the choice

$$J(\varpi) = \eta\,\varpi\,\Theta(\varpi), \tag{8.56}$$

where $\Theta(\varpi)$ denotes the Heaviside step function, satisfies Eq. (8.55). This choice requires that the eigenfrequencies $\{\varpi_\iota > 0\}$ are densely distributed on the positive real axis, for the function $J(\varpi)$ would not be continuous had there been a discrete component to the spectrum $\{\varpi_\iota > 0\}$. The real part of K_ϖ is then of the order

$$\eta\,\varpi \times \frac{\varpi}{\varpi_{\text{bath}}}, \tag{8.57}$$

where ϖ_{bath} is some characteristic frequency in the bath. If $\varpi/\varpi_{\text{bath}}$ is typically small, then one can work with Eq. (8.46). Otherwise we must allow for a complex friction coefficient (admittance) in Eq. (8.46).[8] One might naively expect that the characteristic frequency entering the tunneling rate is [see Eq. (8.39) for the definition of ϖ_0]

$$\varpi_0, \qquad \text{if } M\,\varpi_0^2 \gg \eta\,\varpi_0\,, \tag{8.58}$$

in the *lightly damped regime.* Similarly, one might naively expect that the characteristic frequency entering the tunneling rate is

$$\varpi_0 \times \frac{\varpi_0}{\eta/M} \ll \varpi_0, \qquad \text{if } M\,\varpi_0^2 \ll \eta\,\varpi_0\,, \tag{8.59}$$

in the *heavily damped regime.* If so, the approximation of neglecting the real part of the kernel K_ϖ will remain good in the heavily damped regime if it is a good approximation in the lightly damped regime.

With these preliminary considerations in hand, we are ready to define the quantum dissipative model through the partition function ($\hbar = 1$)

$$Z_\beta := \int \mathcal{D}[q] \int \mathcal{D}[x]\;\exp\left(-S_\beta\right), \tag{8.60a}$$

with the action

$$S_\beta = S_0 + S_{\text{bath}} + S_{\text{int}} + S_{\text{adia}} + S_{\text{ext}}, \tag{8.60b}$$

and the Lagrangian

$$= \int_0^\beta \mathrm{d}\tau\,\left(L_0 + L_{\text{bath}} + L_{\text{int}} + L_{\text{adia}} + L_{\text{ext}}\right), \tag{8.60c}$$

whereby

$$L_0 = \frac{M}{2}\,(\partial_\tau q)^2 + V(q) \tag{8.60d}$$

is the Lagrangian for a spinless point particle of mass M in the conservative potential V,

$$L_{\text{bath}} = \sum_\iota \frac{m_\iota}{2}\left((\partial_\tau x_\iota)^2 + \varpi_\iota^2\,x_\iota^2\right) \tag{8.60e}$$

[8] Starting from Ohm's law $V(t) = (R{*}I)(t)$ ($*$ denotes a convolution), Fourier transformation with respect to time defines the complex impedance $V_\varpi = z_\varpi\,I_\varpi$ whereby $\operatorname{Re} z_\varpi$ is called the *resistance* and $\operatorname{Im} z_\varpi$ is called the *reactance*. The *admittance* is $1/z_\varpi$ with $\operatorname{Re} 1/z_\varpi$ the *conductance* and $\operatorname{Im} 1/z_\varpi$ the *susceptance*.

is the Lagrangian for a family of harmonic oscillators,

$$L_{\rm int} = -q \sum_{\iota} c_{\iota}\, x_{\iota} \tag{8.60f}$$

is the Lagrangian that couples the spinless point particle of mass M in the conservative potential V to the bath,

$$L_{\rm adia} = +\frac{1}{2} M\, (\varpi_{\rm adia})^2\, q^2, \qquad \frac{1}{2} M\, (\varpi_{\rm adia})^2 = \sum_{\iota} \frac{c_{\iota}^2}{2\, m_{\iota}\, \varpi_{\iota}^2}, \tag{8.60g}$$

is included for convenience, and

$$L_{\rm ext}(\tau) := -F_{\rm ext}(\tau)\, q(\tau) \tag{8.60h}$$

is the Lagrangian for a driving force. We are imposing the periodic boundary conditions

$$q(\tau) = +q(\tau + \beta), \qquad x_{\iota}(\tau) = +x_{\iota}(\tau + \beta), \qquad \forall \iota. \tag{8.60i}$$

Analytical continuation $\tau = it$ has been performed to go from the Lagrangians in Eq. (8.45) to those in Eq. (8.60). The bosonic measures are best defined after performing a Fourier transformation with respect to imaginary time. In the bath,

$$\begin{aligned} x_{\iota}(\tau) &= \frac{1}{\sqrt{\beta}} \sum_{\varpi_l} x_{\iota\, \varpi_l}\, e^{-\mathrm{i}\varpi_l \tau}, \\ x_{\iota\, \varpi_l} &= \frac{1}{\sqrt{\beta}} \int\limits_0^{\beta} \mathrm{d}\tau\; x_{\iota}(\tau)\, e^{+\mathrm{i}\varpi_l \tau}, \qquad \varpi_l = \frac{2\,\pi}{\beta} l, \qquad l \in \mathbb{Z}, \end{aligned} \tag{8.61}$$

and

$$\int \mathcal{D}[x] \equiv \prod_{\iota} \prod_{l=0}^{\infty} \int\limits_{\mathbb{R}} \frac{\mathrm{dRe}\, x_{\iota\, \varpi_l}}{\sqrt{2\,\pi}} \int\limits_{\mathbb{R}} \frac{\mathrm{dIm}\, x_{\iota\, \varpi_l}}{\sqrt{2\,\pi}}. \tag{8.62}$$

Fourier transform with respect to imaginary time and the measure of q are defined similarly. Fourier transformation with respect to imaginary time gives the representation

$$\begin{aligned} S_{\beta} &= S_0 + S_{\rm bath} + S_{\rm int} + S_{\rm adia} + S_{\rm ext} \\ &= \sum_{\varpi_l} \left(L_0 + L_{\rm bath} + L_{\rm int} + L_{\rm adia} + L_{\rm ext}\right), \\ L_0 &= \frac{M}{2}\, \varpi_l^2\, q_{(+\varpi_l)}\, q_{(-\varpi_l)} + \sqrt{\beta}\, \left[V(q)\right]_{\varpi_l}\, \delta_{\varpi_l, 0}, \\ L_{\rm bath} &= \sum_{\iota} \frac{m_{\iota}}{2} \left(\varpi_l^2 + \varpi_{\iota}^2\right) x_{\iota\,(+\varpi_l)}\, x_{\iota\,(-\varpi_l)}, \\ L_{\rm int} &= -\sum_{\iota} c_{\iota}\, \frac{1}{2} \left(q_{(+\varpi_l)}\, x_{\iota\,(-\varpi_l)} + q_{(-\varpi_l)}\, x_{\iota\,(+\varpi_l)}\right), \\ L_{\rm adia} &= +\frac{1}{2} M (\varpi_{\rm adia})^2\, q_{(+\varpi_l)}\, q_{(-\varpi_l)}, \qquad \frac{1}{2} M (\varpi_{\rm adia})^2 = \sum_{\iota} \frac{c_{\iota}^2}{2\, m_{\iota}\, \varpi_{\iota}^2}, \\ L_{\rm ext} &= -\sqrt{\beta}\, \left[F_{\rm ext}\, q\right]_{\varpi_l}\, \delta_{\varpi_l, 0}, \end{aligned} \tag{8.63}$$

of the action (8.60b).

The strategy that we are going to follow is to integrate the degrees of freedom from the bath. To this end, observe that completing the square of $L_{\mathrm{bath}}+L_{\mathrm{int}}$ with respect to $q_{\pm\varpi_l}$ is achieved by adding and subtracting to $L_{\mathrm{bath}}+L_{\mathrm{int}}$

$$\sum_{\iota}\frac{c_{\iota}^{2}}{2\,m_{\iota}\,(\varpi_l^2+\varpi_{\iota}^2)}\,q_{(+\varpi_l)}\,q_{(-\varpi_l)}. \tag{8.64}$$

Hence,

$$\begin{aligned} L_{\mathrm{bath}}+L_{\mathrm{int}} =& \sum_{\iota}\frac{m_{\iota}}{2}\left(\varpi_l^2+\varpi_{\iota}^2\right)\left(x_{\iota\,(+\varpi_l)}-\frac{c_{\iota}}{m_{\iota}\,(\varpi_l^2+\varpi_{\iota}^2)}\,q_{(+\varpi_l)}\right)\\ &\times\left(x_{\iota\,(-\varpi_l)}-\frac{c_{\iota}}{m_{\iota}\,(\varpi_l^2+\varpi_{\iota}^2)}\,q_{(-\varpi_l)}\right)\\ &-\sum_{\iota}\frac{c_{\iota}^{2}}{2\,m_{\iota}\,(\varpi_l^2+\varpi_{\iota}^2)}\,q_{(+\varpi_l)}\,q_{(-\varpi_l)}. \end{aligned} \tag{8.65}$$

By changing bath integration variables to

$$x_{\iota\,(+\varpi_l)}=\tilde{x}_{\iota\,(+\varpi_l)}+\frac{c_{\iota}}{m_{\iota}\,(\varpi_l^2+\varpi_{\iota}^2)}\,q_{(+\varpi_l)},\qquad \forall\iota,\varpi_l, \tag{8.66}$$

one can decouple the bath from q. Integration over the bath degrees of freedom $\{\tilde{x}_{\iota\,(+\varpi_l)}\}$ produces an overall constant, the bosonic determinant

$$\mathcal{N}_{\mathrm{bath}}:=\prod_{\iota}\prod_{l=0}^{\infty}\frac{1}{m_{\iota}\,(\varpi_l^2+\varpi_{\iota}^2)}, \tag{8.67}$$

while turning the partition function without damping

$$Z_{\beta\,0}:=\int\mathcal{D}[q]\,\exp\left(-S_0\right) \tag{8.68}$$

into the partition function (we set $F_{\mathrm{ext}}=0$ for simplicity)

$$\begin{aligned} Z_{\beta}&=\mathcal{N}_{\mathrm{bath}}\int\mathcal{D}[q]\exp\left(-S_0-S_1\right),\\ S_0+S_1&=\sum_{\varpi_l}\left(L_0+L_1\right),\\ L_0&=\frac{M}{2}\,\varpi_l^2\,q_{(+\varpi_l)}\,q_{(-\varpi_l)}+\sqrt{\beta}\,\left[V(q)\right]_{\varpi_l}\,\delta_{\varpi_l,0},\\ L_1&=\mathcal{K}_{\varpi_l}\,q_{(+\varpi_l)}\,q_{(-\varpi_l)},\\ \mathcal{K}_{\varpi}&=\sum_{\iota}\frac{c_{\iota}^{2}}{2\,m_{\iota}}\left(\frac{1}{\varpi_{\iota}^{2}}-\frac{1}{\varpi_{\iota}^{2}+\varpi^{2}}\right), \end{aligned} \tag{8.69}$$

in the presence of damping. The kernel $\mathcal{K}_{\varpi}$ is related to the kernel in Eq. (8.50) by analytical continuation

$$\mathcal{K}_{\varpi}=-\frac{1}{2}\times\lim_{\omega\to+\mathrm{i}\varpi}K_{\omega}. \tag{8.70}$$

In terms of the spectral function (8.53),

$$\mathcal{K}_{\varpi} = \frac{1}{\pi} \int\limits_{0}^{\infty} \mathrm{d}\tilde{\varpi}\, J(\tilde{\varpi}) \left(\frac{1}{\tilde{\varpi}} - \frac{\tilde{\varpi}}{\tilde{\varpi}^2 + \varpi^2} \right). \tag{8.71}$$

By undoing the Fourier transformation to Matsubara frequencies, the effective action

$$S'_{\beta} = S_0 + S_1 \tag{8.72}$$

induced by integrating over the degrees of freedom from the bath is represented by

$$\begin{aligned} S_0 &= \int\limits_{0}^{\beta} \mathrm{d}\tau \left(\frac{1}{2} M \left(\partial_{\tau} q\right)^2 + V(q) \right), \\ S_1 &= \int\limits_{0}^{\beta} \mathrm{d}\tau \int\limits_{0}^{\beta} \mathrm{d}\tau'\, q(\tau)\, \mathcal{K}(\tau - \tau')\, q(\tau'), \end{aligned} \tag{8.73}$$

where the non-local kernel in imaginary time induced by damping is defined by

$$\mathcal{K}(\tau) := \frac{1}{\beta} \sum_{\varpi_l} \mathcal{K}_{\varpi_l}\, e^{-\mathrm{i}\varpi_l \tau}, \qquad \mathcal{K}_{\varpi_l} = \int\limits_{0}^{\beta} \mathrm{d}\tau\, \mathcal{K}(\tau)\, e^{+\mathrm{i}\varpi_l \tau}. \tag{8.74}$$

It is shown in appendix H.1 that, when the spectral function is chosen as in Eq. (8.56), the kernel $\mathcal{K}(\tau)$ can be written in the form

$$\begin{aligned} S_1 &= \sum_{\varpi_l} \frac{\eta}{2} |\varpi_l|\, q_{(+\varpi_l)}\, q_{(-\varpi_l)} \\ &= \int\limits_{-\infty}^{+\infty} \mathrm{d}\tau \int\limits_{0}^{\beta} \mathrm{d}\tau'\, \frac{\eta}{4\pi} \left(\frac{q(\tau) - q(\tau')}{\tau - \tau'} \right)^2 \end{aligned} \tag{8.75}$$

by which it is explicitly seen to be positive definite.

Equation (8.75) is the central result of this section. Whereas Eq. (8.69) answers question 1, Eq. (8.75) answers question 2. Insofar as all of what is needed of the environment is that the spectral function (8.53) satisfies Eq. (8.56) in some appropriate range of frequencies,[9] the phenomenology of linear damping encoded by Eq. (8.46) is independent of the microscopic details defining the environment,

[9] For example,

$$J_{\mathrm{R}}(\varpi) := \frac{\eta\, \varpi}{1 + \varpi^2\, \tau_{\mathrm{R}}^2} \tag{8.76}$$

reduces to Eq. (8.56) when ϖ is smaller than the characteristic frequency ϖ_0 of the cubic potential which itself is much smaller than the inverse of the relaxation time τ_{R},

$$\varpi^2\, \tau_{\mathrm{R}}^2 < \varpi_0^2\, \tau_{\mathrm{R}}^2 \ll 1. \tag{8.77}$$

say the choice $m_\iota > 0$, $\varpi_\iota > 0$, and $c_\iota \in \mathbb{R}$. Moreover, Eq. (8.75) gives us a strong hint to the answer to question 3. Indeed, Eq. (8.75) tells us that

$$S_0 + S_1 > S_0. \tag{8.78}$$

It is then very suggestive to conjecture that a WKB-like estimate would replace the tunneling rate in Eq. (8.41) by

$$\Gamma' = A' \, e^{-\frac{B'}{\hbar}} \left[1 + \mathcal{O}(\hbar \varpi_0 / V_0)\right], \tag{8.79}$$

whereby

$$S_0 + S_1 > S_0 \Longrightarrow B' > B_0, \tag{8.80}$$

i.e., dissipation would *decrease* the tunneling rate of a particle initially prepared in a metastable minimum of the potential $V(q)$.

It is essential to observe that Eqs. (8.69) and (8.75) hold not only for a cubic potential but for any smooth potential which is a $\mathbb{C}$-number function of q, say the washboard potential (8.38). However, asking about the tunneling rate out of a metastable minimum only makes sense for a potential $V(q)$ of the cubic type. For a potential $V(q)$ with two or more degenerate absolute minima the issue of tunneling rate out of these minima is meaningless as such since the quantum particle can always tunnel back to its initial position at the bottom of one of the wells. For a potential with degenerate minima, we know that quantum eigenstates, as opposed to their classical counterparts, are *delocalized* in the absence of damping. For a symmetric double well potential, the probability to find the particle in the ground state at the bottom of the left well equals the probability to find the particle at the bottom of the right well if the system is isolated. For a cosine potential, eigenstates are Bloch waves, i.e., plane waves with wave vector commensurate with the periodicity of the potential, if the system is isolated. The meaningful question to ask when the potential $V(q)$ has several degenerate absolute minima, say is periodic, is if delocalized states of $Z_{\beta\,0}$ remain delocalized in the presence of damping, i.e., if a state prepared initially to be delocalized remains for ever delocalized as time evolves. The terminology of quantum coherence is also used in the literature as a synonymous to delocalization. The notion of quantum coherence emphasizes the wave-like nature of delocalized states. Quantum coherence can be detected by interference effects. Whether we are after the effects of dissipation on quantum tunneling or dissipation on quantum coherence, a tool is needed to evaluate Z_β in some approximation. Instantons techniques will be the tools that we choose.

8.5 Instantons in quantum mechanics

8.5.1 *Introduction*

Consider the classical Lagrangian ($\dot{}$ denotes t derivative)

$$L := \frac{1}{2}\,\dot{x}^2 - V(x;g), \qquad V(x;g) = \frac{1}{g^2}\,F(g\,x). \tag{8.81}$$

Here, the mass of the particle moving on the real line with the coordinate x has been set to one and the analytic function F has a zero of order 2 at the origin. The classical equation of motion

$$\ddot{x} = -\frac{1}{g^2}\frac{\mathrm{d}F(g\,x)}{\mathrm{d}x} \Longleftrightarrow g\ddot{x} = -\frac{\mathrm{d}F(g\,x)}{\mathrm{d}(g\,x)} \tag{8.82}$$

is independent of the coupling constant g since the coupling constant g factorizes under the rescaling $y := g\,x$:

$$L = \frac{1}{g^2}\left(\frac{1}{2}\dot{y}^2 - F(y)\right) \Longrightarrow \ddot{y} = -\frac{\mathrm{d}\,F(y)}{\mathrm{d}y}. \tag{8.83}$$

If one can solve the classical equation of motion for $g = 1$, we know the solution for all g's. This, however, is not true anymore after quantization as we know that $\hbar$ (or $\hbar\,g^2$ after rescaling) plays a crucial role in the combination $L/\hbar$ [or $L/(\hbar\,g^2)$ after rescaling] that appears in the path integral description of the quantum theory. For example, the amplitude $|T(E)|$ for transmission through a potential barrier of an incoming plane wave with energy E is given by [compare with Eq. (8.41) where $\Gamma_0 \propto |T(E)|^2$]

$$|T(E)| = \exp\left(-\frac{1}{\hbar}\int\limits_{x_1}^{x_2}\mathrm{d}x\sqrt{2(V-E)}\right)[1+\mathcal{O}(\hbar)], \tag{8.84}$$

where x_1 and x_2 are the classical turning points from the left and right, respectively.

What about performing perturbation theory for small g? Hereto, the classical and quantum theory differ. For the classical theory, one would expect perturbation theory around $g = 0$ to be valid. However, this is certainly not true for the quantum theory as is illustrated by Eq. (8.84). A result like Eq. (8.84) that is non-perturbative in $\hbar$ is usually derived by matching solutions of Schrödinger equation in different regions of space (WKB method). This method is difficult to extend beyond one dimension and/or one particle. The method of instantons that relies on the path integral representation of the quantum theory can also reproduce Eq. (8.84). Moreover, it has the advantage of extending to higher dimensions and/or field theory. As with one-dimensional quantum mechanics, instantons techniques give access to phenomena that are intrinsically non-perturbative in the interaction potential of the field theory.

8.5.2 *Semi-classical approximation within the Euclidean-path-integral representation of quantum mechanics*

Consider the quantum Hamiltonian

$$\hat{H} = \frac{\hat{p}^2}{2} + V(\hat{x}), \qquad [\hat{x},\hat{p}] = \mathrm{i}\hbar, \tag{8.85}$$

that describes the motion of a spinless point particle of unit mass on the real line with the position operator $\hat{x}$. Instanton techniques rely not on the operator representation of the quantum theory but on the Euclidean path integral representation,

$$\langle x_f | e^{-\hat{H}\mathcal{T}/\hbar} | x_i \rangle \propto \int \mathcal{D}[x]\, e^{-S/\hbar}. \tag{8.86}$$

On the left-hand side of Eq. (8.86), $|x_i\rangle$ and $|x_f\rangle$ are the initial and final position eigenstates and $\mathcal{T}$ is a positive number with dimension of time. The left-hand side is of interest since it can be expanded in terms of the exact eigenstates $|n\rangle$ of $\hat{H}$,

$$\langle x_f | e^{-\hat{H}\mathcal{T}/\hbar} | x_i \rangle = \sum_n e^{-\varepsilon_n \mathcal{T}/\hbar} \langle x_f | n \rangle \langle n | x_i \rangle, \tag{8.87}$$

so that, for large $\mathcal{T}$, the leading term in this expansion gives the ground state and its energy.

On the right-hand side of Eq. (8.86), S stands for the Euclidean action ($\dot{}$ denotes τ derivative)

$$S := \int_{-\mathcal{T}/2}^{+\mathcal{T}/2} \mathrm{d}\tau \left(\frac{\dot{x}^2}{2} + V(x) \right), \tag{8.88}$$

and $\mathcal{D}[x]$ denotes the measure over all functions x that obey the boundary conditions

$$x(-\mathcal{T}/2) = x_i, \qquad x(+\mathcal{T}/2) = x_f. \tag{8.89}$$

Explicit construction of the measure $\mathcal{D}[x]$ proceeds as follows. If $\bar{x}(\tau)$ is *some arbitrary* function obeying the boundary conditions, then any given function $x(\tau)$ that obeys the same boundary conditions can be expanded in terms of *some chosen* set of complete, real, and orthonormal functions $x_n(\tau)$ that vanish at $\pm\mathcal{T}/2$,

$$x(\tau) = \bar{x}(\tau) + \sum_n c_n\, x_n(\tau), \qquad \int_{-\mathcal{T}/2}^{+\mathcal{T}/2} \mathrm{d}\tau\, x_m(\tau)\, x_n(\tau) = \delta_{m,n}, \qquad x_n(\pm\mathcal{T}/2) = 0, \tag{8.90}$$

and the (normalized) measure is now given by

$$\mathcal{D}[x] = \prod_n \frac{\mathrm{d}c_n}{\sqrt{2\pi\,\hbar}}. \tag{8.91}$$

Observe that the measure does not depend on $\bar{x}(\tau)$. The right-hand side is of interest because it can readily be evaluated in the semi-classical (small $\hbar$) limit through a saddle-point approximation of the argument in the exponential (Boltzmann) weight.

The idea behind the saddle-point approximation relies on the assumption that if the prefactor of the action [here $1/\hbar$ or $1/(\hbar\, g^2)$ when V has the form given in Eq. (8.81)] is extremely large the dominant contribution to the path integral will come from all the paths (there might be more than one) that are global minima of the action. Assume that S has a *minimum* $\bar{x}(\tau)$ that obeys the boundary condition.

Taylor expansion around this minimum of the action yields ($\dot{}$ and $'$ denote τ and x derivatives, respectively)

$$\begin{aligned}
S[\bar{x}+y] = & \int\limits_{-\mathcal{T}/2}^{+\mathcal{T}/2} \mathrm{d}\tau \left(\frac{1}{2}\,\dot{\bar{x}}^2 + V(\bar{x})\right) \\
& + \int\limits_{-\mathcal{T}/2}^{+\mathcal{T}/2} \mathrm{d}\tau \left[-\ddot{\bar{x}} + V'(\bar{x})\right] y \\
& + \int\limits_{-\mathcal{T}/2}^{+\mathcal{T}/2} \mathrm{d}\tau \, \frac{1}{2}\, y \left[-\ddot{y} + V''(\bar{x})\, y\right] \\
& + \cdots . \qquad (8.92)
\end{aligned}$$

Here, $y(\pm\mathcal{T}/2) = 0$ to accommodate the boundary conditions. This is the reason for which all boundary terms vanish after partial integration. By assumption, the second line vanishes and if we truncate the Taylor expansion up to second order in y, we find

$$\begin{aligned}
S[\bar{x}+y] = & \int\limits_{-\mathcal{T}/2}^{+\mathcal{T}/2} \mathrm{d}\tau \left(\frac{\dot{\bar{x}}^2}{2} + V(\bar{x})\right) \\
& + \frac{1}{2} \int\limits_{-\mathcal{T}/2}^{+\mathcal{T}/2} \mathrm{d}\tau \, y \left(-\frac{\mathrm{d}^2}{\mathrm{d}\tau^2} + \left.\frac{\mathrm{d}^2 V}{\mathrm{d}x^2}\right|_{\bar{x}}\right) y \\
& + \cdots , \qquad (8.93a)
\end{aligned}$$

whereby $\bar{x}$ is the solution to the differential equation

$$-\ddot{\bar{x}} + V'(\bar{x}) = 0, \qquad \bar{x}(-\mathcal{T}/2) = x_i, \qquad \bar{x}(+\mathcal{T}/2) = x_f. \qquad (8.93b)$$

The Taylor expansion Eq. (8.93a) suggests that the path y be expanded in terms of the orthonormal eigenfunctions x_n of the Hermitian operator $-\partial_\tau^2 + V''(\bar{x})$

$$y(\tau) = \sum_n c_n \, x_n(\tau), \qquad \left[-\partial_\tau^2 + V''(\bar{x})\right] x_n = \lambda_n \, x_n, \qquad x_n(\pm\mathcal{T}/2) = 0. \qquad (8.94)$$

If $\bar{x}$ is truly a minimum, all eigenvalues λ_n must be larger or equal to zero. Thus, if we insert the expansion (8.94) in terms of the orthonormal modes of the kernel on the second line of the right-hand side of Eq. (8.93a) into Eq. (8.86), we obtain

$$\begin{aligned}
\langle x_f | e^{-\hat{H}\mathcal{T}/\hbar} | x_i \rangle &\propto \int \mathcal{D}[x]\, e^{-S[x]/\hbar} \\
&= e^{-\frac{1}{\hbar}\int\limits_{-\mathcal{T}/2}^{+\mathcal{T}/2} \mathrm{d}\tau \left(\frac{\dot{\bar{x}}^2}{2} + V(\bar{x})\right)} \times \left(\prod_n \int\limits_{-\infty}^{+\infty} \frac{\mathrm{d}c_n}{\sqrt{2\pi\hbar}} e^{-\frac{1}{2} c_n \frac{\lambda_n}{\hbar} c_n} \right) \\
&\quad \times [1 + \mathcal{O}(\hbar)] \\
&= e^{-\frac{S[\bar{x}]}{\hbar}} \times \left(\prod_n \frac{1}{\sqrt{\lambda_n}} \right) \times [1 + \mathcal{O}(\hbar)] \\
&= \frac{\exp\left(-S[\bar{x}]/\hbar\right)}{\sqrt{\mathrm{Det}\left[-\partial_\tau^2 + V''(\bar{x})\right]}} [1 + \mathcal{O}(\hbar)] \, .
\end{aligned} \tag{8.95}$$

In going to the second line, we made use of the orthonormality of the eigenmodes x_n and we assumed that the non-Gaussian contributions are of order $\hbar$. In going to the third line, we assumed that $\lambda_n > 0$, in which case it is reassuring to observe that the Gaussian contribution that results from integrating over c_n is of order zero in powers of $\hbar$. Equation (8.95) encodes the semi-classical approximation to quantum mechanics within the path integral formalism.

From a technical point of view, the semi-classical approximation of quantum mechanics is reduced to:

(1) Solving the differential equation

$$\ddot{\bar{x}} = -[-V'(\bar{x})], \qquad \bar{x}(-\mathcal{T}/2) = x_i, \qquad \bar{x}(+\mathcal{T}/2) = x_f, \tag{8.96}$$

that represents Newton's equation in the potential $-V$. Observe that

$$E := \frac{1}{2}\left(\dot{\bar{x}}\right)^2 - V(\bar{x}) \tag{8.97}$$

is a constant of the motion.

(2) Calculating the determinant

$$\mathrm{Det}\left[-\partial_\tau^2 + V''(\bar{x})\right], \qquad -\mathcal{T}/2 \leq \tau \leq +\mathcal{T}/2, \tag{8.98}$$

with hard-wall boundary conditions.

8.5.3 *Application to a parabolic potential well*

As a first example, we apply the semi-classical approximation to the case when the potential V in Eq. (8.85) has a non-degenerate absolute minimum at the origin (see Fig. 8.5), i.e.,

$$V(x) = \frac{1}{2}\omega^2 x^2 + \mathcal{O}(x^4). \tag{8.99}$$

Initial and final positions are chosen to be

$$x_i = x_f = 0. \tag{8.100}$$

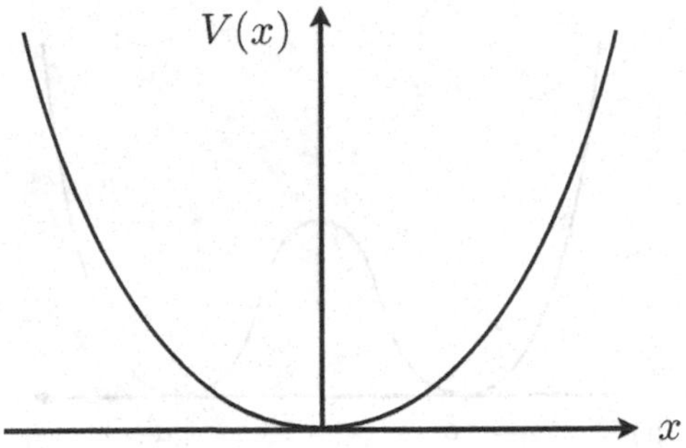

Fig. 8.5 Potential well $V(x)$ with a single non-degenerate minimum at $x = 0$.

The unique solution to Eq. (8.93b) is $\bar{x} = 0$ for which $S[\bar{x}] = 0$ and $\mathrm{Det}\left[-\partial_\tau^2 + V''(\bar{x})\right] = \mathrm{Det}\left(-\partial_\tau^2 + \omega^2\right)$. The amplitude for the particle to remain at the origin after "time" $\mathcal{T}$ is

$$\langle x = 0 | e^{-\hat{H}\mathcal{T}/\hbar} | x = 0 \rangle \propto \frac{1}{\sqrt{\mathrm{Det}\left(-\partial_\tau^2 + \omega^2\right)}} \left[1 + \mathcal{O}(\hbar)\right]. \tag{8.101}$$

Needed is the determinant of the Hermitian operator $-\partial_\tau^2 + \omega^2$. Observe that the wave function

$$\psi_0(x) = \frac{1}{\omega} \sinh\Big(\omega\left[\tau + (\mathcal{T}/2)\right]\Big) \tag{8.102}$$

obeys

$$\left(-\partial_\tau^2 + \omega^2\right)\psi_0(\tau) = 0, \qquad \psi_0(-\mathcal{T}/2) = 0, \qquad (\partial_\tau \psi_0)(-\mathcal{T}/2) = 1. \tag{8.103}$$

Furthermore, it can be shown that[10]

$$\mathrm{Det}\left(-\partial_\tau^2 + \omega^2\right) \propto \psi_0(+\mathcal{T}/2), \tag{8.104}$$

whereby the proportionality constant is independent of ω. Hence,

$$\begin{aligned} \langle x = 0 | e^{-\hat{H}\mathcal{T}/\hbar} | x = 0 \rangle &\propto \frac{1}{\sqrt{\mathrm{Det}\left(-\partial_\tau^2 + \omega^2\right)}} \left[1 + \mathcal{O}(\hbar)\right] \\ &\propto \sqrt{\frac{\omega}{\sinh(\omega\mathcal{T})}} \left[1 + \mathcal{O}(\hbar)\right]. \end{aligned} \tag{8.105}$$

From the asymptotic limit $\mathcal{T} \to \infty$,

$$\begin{aligned} \langle x = 0 | e^{-\hat{H}\mathcal{T}/\hbar} | x = 0 \rangle &= \sum_n |\langle n | x = 0 \rangle|^2 \, e^{-\varepsilon_n \mathcal{T}/\hbar} \\ &\propto \sqrt{\omega}\, e^{-\omega\mathcal{T}/2} \left[1 + \mathcal{O}(e^{-2\omega\mathcal{T}})\right] \left[1 + \mathcal{O}(\hbar)\right], \end{aligned} \tag{8.106}$$

we conclude that the energy of the ground state $n = 0$ is

$$\varepsilon_0 = \frac{\hbar\omega}{2}, \tag{8.107}$$

whereas the probability for the particle to be at the origin is

$$|\langle n = 0 | x = 0 \rangle|^2 \propto \sqrt{\omega}\left[1 + \mathcal{O}(\hbar)\right]. \tag{8.108}$$

---

[10] Appendix 1 in chapter 7 of Ref. [86].

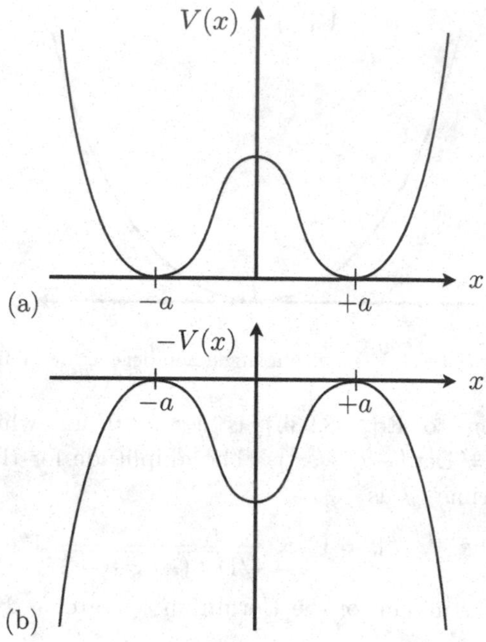

Fig. 8.6 (a) Double potential well. (b) Inverted double potential well.

These are properties of a quantum harmonic oscillator whose ground state wave function is (after reinstating the mass m of the particle)

$$\left(\frac{m\,\omega}{\pi\hbar}\right)^{\frac{1}{4}} e^{-\frac{m\,\omega}{2\hbar}x^2}. \tag{8.109}$$

8.5.4 *Application to the double well potential*

Next, we apply the semi-classical approximation to the case when the potential V in Eq. (8.85) has doubly-degenerate absolute minima at $\pm a$ (see Fig. 8.6), i.e.,

$$V(x) = V(-x), \qquad V(a+x) = \frac{1}{2}\,\omega^2 x^2 + \mathcal{O}(x^4). \tag{8.110}$$

Initial and final positions are:

- Case I : $x_i = +x_f = \mp a$.
- Case II: $x_i = -x_f = \mp a$.

A solution to Eq. (8.93b) in case I is $\bar{x} = \mp a$ for which $S[\bar{x}] = 0$ and $\mathrm{Det}\left[-\partial_\tau^2 + V''(\bar{x})\right] = \mathrm{Det}\left(-\partial_\tau^2 + \omega^2\right)$. The amplitude for the particle to remain at $\mp a$ after "time" $\mathcal{T}$ in the "trivial background" $\bar{x} = \mp a$ is

$$\langle \mp a | e^{-\hat{H}\mathcal{T}/\hbar} | \mp a \rangle \propto \frac{1}{\sqrt{\mathrm{Det}\left(-\partial_\tau^2 + \omega^2\right)}} \left[1 + \mathcal{O}(\hbar)\right]. \tag{8.111}$$

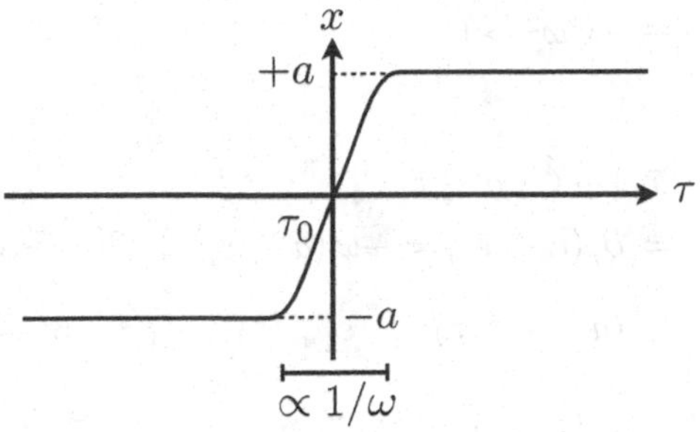

Fig. 8.7 Sketch of a single-instanton profile.

This solution is unique if $\mathcal{T}$ is finite. However, in the limit $\mathcal{T} \to \infty$, whereby the particle initially starts at $-a$, say, reaches $+a$ at some intermediate time, only to come back to its initial position at $+\mathcal{T}/2$. Evidently this process can repeat itself an arbitrary number of times. We thus expect an infinity of non-equivalent solutions $\bar{x}_{2n}$ labeled by the even integer $2n = 2, 4, \cdots$, to Newton's equation in the inverted double well potential. Such solutions $\bar{x}_{2n}$, when they exist, are called *instantons* whenever n is non-vanishing. The "trivial" solution $\bar{x} = \mp a$ is denoted $\bar{x}_0$.

To construct $\bar{x}_{2n}$ as well as to evaluate $S[\bar{x}_{2n}]$ and the Gaussian determinant $\mathrm{Det}[-\partial_\tau^2 + V''(\bar{x}_{2n})]$, we turn our attention to case II and consider the solution $\bar{x}_1$ with an infinitesimally small constant of motion (energy)

$$0^+ = \frac{1}{2}(\dot{\bar{x}}_1)^2 - V(\bar{x}_1). \tag{8.112}$$

The trajectory $\bar{x}_1$ describes the particle starting at $x = -a$, say, and reaching the top of the opposite hill at $+a$ with an infinitesimally small velocity. Hence, $\bar{x}_1$ is a strictly increasing function of $-\mathcal{T}/2 < \tau < +\mathcal{T}/2$, $\dot{\bar{x}}_1 > 0$. This solution is called a single instanton and it satisfies

$$\dot{\bar{x}}_1 = +\sqrt{2V} \Longleftrightarrow \tau = \tau_0 + \int\limits_0^{\bar{x}_1} \frac{\mathrm{d}x}{\sqrt{2V(x)}}. \tag{8.113}$$

The integration constant τ_0 is the time at which $\bar{x}_1 = 0$. As we shall see shortly, τ_0 can be interpreted as the position of the instanton. The first important property of Eq. (8.113) is that the solution $\bar{x}_1$ is a function of the combination $\tau - \tau_0$. The solution obtained by reversing time $\tau \to -\tau$ is called a single *anti-instanton* and is denoted $\bar{x}_{-1}$. The trajectory $\bar{x}_{-1}$ now describes the particle starting at $x = +a$ and reaching the top of the opposite hill at $-a$ with an infinitesimally small velocity after time $\mathcal{T}$. Strictly speaking, an (anti-)instanton can only be constructed in the asymptotic limit $\mathcal{T} \to \infty$ if it is to have a smooth velocity.

The second important property of $\bar{x}_1$ follows from the fact that Eq. (8.113) is, to a very good approximation for large times $\tau \gg 1/\omega$ (remember that $\dot{\bar{x}}_1$ is strictly positive), given by ($\omega \equiv +\sqrt{\omega^2} > 0$)

$$
\begin{aligned}
\dot{\bar{x}}_1 &= \omega\,(a-\bar{x}_1) + \mathcal{O}[(a-\bar{x}_1)^3] \\
&\Longleftrightarrow \partial_\tau (a-\bar{x}_1) = -\omega\,(a-\bar{x}_1) + \mathcal{O}[(a-\bar{x}_1)^3] \\
&\Longleftrightarrow (a-\bar{x}_1)(\tau) \propto c\,e^{-\omega\,\tau} + \cdots, \qquad \tau \gg \frac{1}{\omega} \\
&\Longleftrightarrow \bar{x}_1(\tau) \propto a - c\,e^{-\omega\,\tau} + \cdots, \qquad \tau \gg \frac{1}{\omega},
\end{aligned}
\tag{8.114}
$$

where the constant c is fixed by the boundary condition on $\bar{x}_1$ as $\tau \to \infty$. The larger the curvature ω^2 of V at the two degenerate minima $\mp a$, the steeper the valley between the two degenerate maxima of $-V$ and the longer the time a spinless point particle with unit mass on the real line will be close to a for $\tau > 0$ relative to the time it will be close to the origin. The characteristic time scale of the instanton $1/\omega$ tells us when a spinless point particle with unit mass on the real line has a non-negligible speed. The interpretation of the exponential dependence on time in Eq. (8.114) is that a single instanton is well localized in time around τ_0 (see Fig. 8.7). There are field theories in which instantons can be constructed but for which it is not possible to assign a characteristic scale, say if the instanton is scale invariant. When this happens the semi-classical method outlined here fails to extend to a field theoretical context.

The fact that a single instanton is exponentially localized in time around its "center" τ_0 suggests that it behaves *like* a point particle located at τ_0 in the limit $\omega^{-1}/\mathcal{T} \to 0$, ω^{-1} held fixed. Consequently, the single instanton should really be denoted $\bar{x}_{1;\tau_0}$. However, invariance under time translation as $\mathcal{T} \to \infty$ of $\hat{H}$ and of the boundary conditions implies that τ_0 can be arbitrarily chosen, i.e., $S[\bar{x}_{1;\tau_0}]$ is independent of τ_0.[11] Taken together, these properties suggest that a saddle point $\bar{x}_n$ describing a trajectory that starts at $-a$ at time $-\mathcal{T}/2$, crosses the origin n times at the successive times $\tau_1 \ll \tau_2 \ll \cdots \ll \tau_{n-1} \ll \tau_n$, and reaches $+a$ when n is odd (or $-a$ when n is even) at time $\mathcal{T}/2 \gg \omega^{-1}$, can be construed as a string of single instantons and anti-instantons beginning with a single instanton $\bar{x}_{1;\tau_1}$, followed by a single anti-instanton $\bar{x}_{-1;\tau_2}$, and so on. Except for the ordering of the instantons center, $\tau_1 \ll \tau_2 \ll \cdots \ll \tau_{n-1} \ll \tau_n$, the action in this instanton background is independent of the centers due to invariance under time translation.

In summary, motivated by the existence of single instantons, we have assumed the existence of instanton configurations $\bar{x}_n$, $n \in \mathbb{Z}$, in terms of which the transition amplitude between the initial states $x_i = \pm a$ and the final states $x_f = \pm a$ can be

[11] Technically, one can always write $\bar{x}_{1;\tau_0} = f(\tau - \tau_0)$ and perform the change of variable $\tau - \tau_0 = \tau'$ in the limit $\mathcal{T} \to \infty$.

formally written in the semi-classical approximation as

$$\begin{aligned}\langle -a|e^{-\hat{H}\mathcal{T}/\hbar}|-a\rangle \propto & \left(\sum_{n=0}^{\infty}\int\mathcal{D}[\bar{x}_{2n}]\,e^{-S[\bar{x}_{2n}]/\hbar}\left\{\mathrm{Det}'\left[-\partial_\tau^2+V''(\bar{x}_{2n})\right]\right\}^{-1/2}\right)\\ &\times\left[1+\mathcal{O}(\hbar)\right],\\ \langle +a|e^{-\hat{H}\mathcal{T}/\hbar}|-a\rangle \propto & \left(\sum_{n=0}^{\infty}\int\mathcal{D}[\bar{x}_{2n+1}]e^{-S[\bar{x}_{2n+1}]/\hbar}\left\{\mathrm{Det}'\left[-\partial_\tau^2+V''(\bar{x}_{2n+1})\right]\right\}^{-1/2}\right)\\ &\times\left[1+\mathcal{O}(\hbar)\right].\end{aligned} \tag{8.115}$$

This expression is formal as neither $\bar{x}_n$ nor the measure $\mathcal{D}[\bar{x}_n]$ were explicitly constructed. The meaning of the prime over the functional determinants also needs explanation. The construction of the instanton configurations $\bar{x}_n$ and their measure $\mathcal{D}[\bar{x}_n]$ is performed in the limit $\mathcal{T}\gg\omega^{-1}$ whereby $\omega^2 := V''(\pm a)$. In this limit, the instanton configuration $\bar{x}_n$ is thought of as an ordered string of single instanton/anti-instanton located at $\tau_1\ll\tau_2\ll\cdots\ll\tau_{n-1}\ll\tau_n$. Pictorially, the trajectory $\bar{x}_n$ is a sequence of sharp jumps (on the scale of $\mathcal{T}\gg\omega^{-1}$) between the values $-a$ and $+a$ at time τ_{2i+1} for a single instanton and between the values $+a$ and $-a$ at time $\tau_{2(i+1)}$ for a single anti-instanton. The width $\sim\mathcal{T}/(n+1)$ of the plateaus at $\pm a$ is much larger than the width ω^{-1} of the jumps. This is tantamount to assuming that the single instantons behave like a dilute gas of hardcore point-like particles. If so, it is reasonable to write

$$\begin{aligned}\int\mathcal{D}[\bar{x}_n]\approx & \int\limits_{-\mathcal{T}/2}^{+\mathcal{T}/2}\mathrm{d}\tau_n\int\limits_{-\mathcal{T}/2}^{\tau_n}\mathrm{d}\tau_{n-1}\int\limits_{-\mathcal{T}/2}^{\tau_{n-1}}\mathrm{d}\tau_{n-2}\cdots\int\limits_{-\mathcal{T}/2}^{\tau_3}\mathrm{d}\tau_2\int\limits_{-\mathcal{T}/2}^{\tau_2}\mathrm{d}\tau_1\\ \to & \frac{\mathcal{T}^n}{n!},\qquad\text{by time translation invariance of the integrand,}\end{aligned} \tag{8.116a}$$

for the integral over the measure of the instanton $\bar{x}_n$,

$$\begin{aligned}e^{-S[\bar{x}_n]/\hbar}\approx & e^{-n\,S[\bar{x}_1]/\hbar}\\ \equiv & e^{-n\,S_1/\hbar},\qquad S_1=\int\limits_{-\mathcal{T}/2}^{+\mathcal{T}/2}\mathrm{d}\tau\,(\dot{\bar{x}}_1)^2=\int\limits_{-a}^{+a}\mathrm{d}\bar{x}_1\sqrt{2\,V(\bar{x}_1)},\end{aligned} \tag{8.116b}$$

for the Boltzmann weight of the instanton $\bar{x}_n$, and

$$\left(\prod_{j=1}^{n}\frac{\mathrm{d}c_j^{\text{zero-mode}}}{\sqrt{2\pi\hbar}}\right)\frac{1}{\sqrt{\mathrm{Det}'\left[-\partial_\tau^2+V''(\bar{x}_n)\right]}}=:\left(\prod_{j=1}^{n}\mathrm{d}\tau_j\right)\frac{K^n}{\sqrt{\mathrm{Det}\left(-\partial_\tau^2+\omega^2\right)}}, \tag{8.116c}$$

for the measure of the zero modes from the kernel $\left[-\partial_\tau^2+V''(\bar{x}_n)\right]$ of the instanton $\bar{x}_n$. The right-hand side of Eq. (8.116c) defines the number K that measures the effect on the determinant of "how much" (or rather how little) $V''(\bar{x}_n)$ deviates from

ω^2 as is explained in Eqs. (2.26) and (2.27) from Ref. [86]. Also, K is implicitly assumed to be non-vanishing, positive, and independent of the center of the instanton. To see this requires a careful definition and evaluation of Det $\left[-\partial_\tau^2 + V''(\bar{x}_1)\right]$ in account of time translation invariance that we postpone for the time being. We will see that there always exist an eigenvalue zero of $-\partial_\tau^2 + V''(\bar{x}_1)$ and that this eigenvalue must be removed from Det $\left[-\partial_\tau^2 + V''(\bar{x}_n)\right]$ as is indicated by the prime in Det$'$ $\left[-\partial_\tau^2 + V''(\bar{x}_n)\right]$. We will then see that the integration over the instantons coordinates in Eq. (8.116a) is just what is needed to account for the integration over the measure of the zero modes (the eigenfunctions with vanishing eigenvalue). Finally, observe that although it had been assumed that all instantons center are very (infinitely) far from each others, Eq. (8.116a) breaks this assumption. We will verify below that the error thus committed is negligible in the limit $\mathcal{T} \to \infty$.

The outcome of this discussion is that, within the dilute instanton gas approximation, the semi-classical approximation for the amplitude of a spinless point particle with unit mass on the real line to propagate between the minima of a double well potential V is

$$\begin{aligned}\langle -a|e^{-\hat{H}\mathcal{T}/\hbar}|-a\rangle &\propto \frac{1}{\sqrt{\mathrm{Det}\left(-\partial_\tau^2+\omega^2\right)}}\sum_{n=0}^{\infty}\frac{\left(K\,e^{-S_1/\hbar}\,\mathcal{T}\right)^{2n}}{(2n)!}\times\left[1+\mathcal{O}(\hbar)\right]\\ &\propto \sqrt{\omega}\,e^{-\omega\mathcal{T}/2}\cosh\left(K\,e^{-S_1/\hbar}\,\mathcal{T}\right)\times\left[1+\mathcal{O}(\hbar)\right],\end{aligned} \tag{8.117a}$$

for the case when initial and final positions are the position $-a$ at which V is minimal and

$$\begin{aligned}\langle +a|e^{-\hat{H}\mathcal{T}/\hbar}|-a\rangle &\propto \frac{1}{\sqrt{\mathrm{Det}\left(-\partial_\tau^2+\omega^2\right)}}\sum_{n=0}^{\infty}\frac{\left(K\,e^{-S_1/\hbar}\,\mathcal{T}\right)^{2n+1}}{(2n+1)!}\times\left[1+\mathcal{O}(\hbar)\right]\\ &\propto \sqrt{\omega}\,e^{-\omega\mathcal{T}/2}\sinh\left(K\,e^{-S_1/\hbar}\,\mathcal{T}\right)\times\left[1+\mathcal{O}(\hbar)\right],\end{aligned} \tag{8.117b}$$

for the case when the initial position is the position $-a$ at which V is minimal while the final position is the position $+a$ at which V is also minimal. By comparison with the exact eigenstate expansion

$$\langle x_f|e^{-\hat{H}\,\mathcal{T}/\hbar}|x_i\rangle = \sum_m\sum_{\sigma=\pm}\langle x_f|m;\sigma\rangle\,\langle m;\sigma|x_i\rangle\,e^{-\varepsilon_{m;\sigma}\,\mathcal{T}/\hbar}, \tag{8.118}$$

where m labels all the energy eigenstates in a single potential well whereas $\sigma = \pm$ labels (naively) the "bonding" and "anti-bonding" linear combinations, we conclude that, within the semi-classical approximation,

$$\varepsilon_{0;-} = \frac{1}{2}\,\hbar\omega - \hbar K\,e^{-S_1/\hbar} \tag{8.119a}$$

is the (lowest) energy of the bonding state with the squared amplitude

$$|\langle x=-a|0;-\rangle|^2 \propto +\frac{1}{2}\,\sqrt{\omega}, \tag{8.119b}$$

and the overlap

$$\langle x=+a|0;-\rangle\,\langle 0;-|x=-a\rangle \propto +\frac{1}{2}\,\sqrt{\omega}, \tag{8.119c}$$

on the one hand, while

$$\varepsilon_{0;+} = \frac{1}{2}\,\hbar\omega + \hbar K\, e^{-S_1/\hbar} \tag{8.120a}$$

is the (first excited) energy of the anti-bonding state with the squared amplitude

$$|\langle x = -a|0;+\rangle|^2 \propto +\frac{1}{2}\,\sqrt{\omega}, \tag{8.120b}$$

and the overlap

$$\langle x = +a|0;+\rangle\,\langle 0;+|x = -a\rangle \propto -\frac{1}{2}\sqrt{\omega}, \tag{8.120c}$$

on the other hand. The difference in the energy of the bonding and anti-bonding states is proportional to $\exp(-S_1/\hbar)$. It vanishes as the surface $S_1 = \int_{-a}^{+a} d\bar{x}_1\,\sqrt{2\,V(\bar{x}_1)}$ underneath the "tunneling barrier $\sqrt{2\,V}$" diverges. In this limit of an infinitely high potential barrier, bonding and anti-bonding states are degenerate in energy and the single potential well result is recovered up to a degeneracy of two. This degeneracy is broken by barrier penetration which is exponentially small [strictly speaking only $\varepsilon_{0;-} - \varepsilon_{0;+}$ should be expanded semi-classically since the correction of order $\hbar$ beyond the semi-classical approximation of the individual energies $\varepsilon_{0;\mp}$ is already much larger than the exponentially small lifting of the degeneracy in the limit $S_1 \to \infty$ resulting from the symmetry under $x \to -x$ of $V(\pm a + x) \approx \frac{1}{2}\,\omega^2\,x^2$].

Is our assumption of a dilute gas of instanton self consistent? For any fixed value of

$$z := K\,e^{\,S_1/\hbar}\,\mathcal{T}, \tag{8.121}$$

the terms in the exponential series $\sum_m z^m/m!$ grow with m until m is of order of z. After this point the terms will decrease rapidly with m. The important terms in the instanton gas expansion are thus those for which

$$m \le K\,e^{-S_1/\hbar}\,\mathcal{T} \Longleftrightarrow \frac{m}{\mathcal{T}} \le K\,e^{-S_1/\hbar}. \tag{8.122}$$

For small $\hbar$, the important terms in the dilute gas expansion are those for which the gas density $m/\mathcal{T}$ is exponentially small. The average separation $\mathcal{T}/m$ between instantons is therefore exponentially large and independent of $\mathcal{T}$ for sufficiently large $\mathcal{T}$ (as the dependence of S_1 on $\mathcal{T}$ becomes negligible for $\mathcal{T} \to \infty$) provided we can show that K is also independent of $\mathcal{T}$ as $\mathcal{T} \to \infty$. We conclude that the error committed in Eq. (8.116a) is inconsequential.

It is time to return to the evaluation of the determinant $\mathrm{Det}\left[-\partial_\tau^2 + V''(\bar{x}_1)\right]$. Recall that we are seeking the eigenfunctions and eigenvalues of $-\partial_\tau^2 + V''(\bar{x}_1)$ obeying hard-wall boundary conditions when $\tau = \pm\mathcal{T}/2$. We also recall that $\bar{x}_1$ is a solution to Newton's equation $-\ddot{x} + V'(x) = 0$ with vanishing constant of motion (energy) $\frac{1}{2}(\dot{\bar{x}}_1)^2 - V(\bar{x}_1) = 0$ that, without loss of generality, represents a single instanton centered at τ_0 (the time at which $\bar{x}_1$ vanishes). In other words, the trajectory $\bar{x}_1$ is a strictly increasing function of $\tau - \tau_0$ that interpolates between $-a$ at $-\mathcal{T}/2$ and $+a$ at $+\mathcal{T}/2$ with a velocity $\dot{\bar{x}}_1 = \sqrt{2\,V(\bar{x}_1)}$ which is strictly positive if $-\mathcal{T}/2 < \tau < +\mathcal{T}/2$ and vanishes at $\mp\mathcal{T}/2$.

The first important observation is that the velocity $\dot{\bar{x}}_1$ is itself an eigenfunction of $-\partial_\tau^2 + V''(\bar{x}_1)$ with vanishing eigenvalue that obeys the hard-wall boundary conditions. To see this, we note that $\dot{\bar{x}}_1$ does vanish at the initial and final times $\mp\mathcal{T}/2$. Moreover,

$$\left[-\partial_\tau^2 + V''(\bar{x}_1)\right]\dot{\bar{x}}_1 = \partial_\tau\left[-\ddot{\bar{x}}_1 + V'(\bar{x}_1)\right] = 0, \tag{8.123}$$

since $\bar{x}_1$ was constructed to obey Newton's equation in the inverted potential $-V$. The normalized eigenfunction of $-\partial_\tau^2 + V''(\bar{x}_1)$ with vanishing eigenvalue that obeys hard-wall boundary conditions will be denoted [compare with Eq. (8.94) and make use of Eq. (8.116b)]

$$x_1 := \frac{1}{\sqrt{S_1}}\dot{\bar{x}}_1, \qquad \int\limits_{-\mathcal{T}/2}^{+\mathcal{T}/2} \mathrm{d}\tau\,(\dot{\bar{x}}_1)^2 = \int\limits_{-a}^{+a} \mathrm{d}\bar{x}_1\,\dot{\bar{x}}_1 = S_1. \tag{8.124}$$

Since $-\partial_\tau^2 + V''(\bar{x}_1)$ with hard-wall boundary conditions defines a Hermitian Hamiltonian for a spinless point particle of unit mass in one dimension, and since the eigenfunction x_1 is nodeless on $-\mathcal{T}/2 < \tau < +\mathcal{T}/2$, the eigenvalue $\lambda_1 = 0$ of x_1 must be the lowest in the spectrum: All remaining eigenfunctions x_n, $n = 2, 3, \cdots$, must have nodes on $-\mathcal{T}/2 < \tau < +\mathcal{T}/2$, i.e., strictly positive eigenvalues λ_n. Assuming the expansion

$$y = c_1\,x_1 + \sum_{n=2}^{\infty} c_n\,x_n \tag{8.125}$$

for a small deviation y around $\bar{x}_1$, we define the restricted functional determinant $\mathrm{Det}'[-\partial_\tau^2 + V''(\bar{x}_1)]$ to be the functional determinant of $[-\partial_\tau^2 + V''(\bar{x}_1)]$ with the omission of its vanishing eigenvalue λ_1

$$\frac{1}{\sqrt{\mathrm{Det}'[-\partial_\tau^2 + V''(\bar{x}_1)]}} := \prod_{n=2}^{\infty}\frac{1}{\sqrt{\lambda_n}}. \tag{8.126}$$

With the help of the Gaussian identity

$$\frac{1}{\sqrt{\lambda_n}} = \int\limits_{-\infty}^{+\infty}\frac{\mathrm{d}c_n}{\sqrt{2\,\pi\,\hbar}}\,e^{-\frac{1}{2\hbar}\,c_n\,\lambda_n\,c_n}, \tag{8.127}$$

and the orthonormality of the eigenfunctions x_n in the mode expansion (8.125), we may write

$$\prod_{n=2}^{\infty}\frac{1}{\sqrt{\lambda_n}} = \left(\prod_{n=2}^{\infty}\int\limits_{-\infty}^{+\infty}\frac{\mathrm{d}c_n}{\sqrt{2\,\pi\,\hbar}}\right) e^{-\frac{1}{2\hbar}\sum\limits_{k,l=2}^{\infty} c_k\,c_l \int\limits_{-\mathcal{T}/2}^{+\mathcal{T}/2}\mathrm{d}\tau\,x_k\,[-\partial_\tau^2 + V''(\bar{x}_1)]\,x_l}. \tag{8.128}$$

Since $\lambda_1 = 0$, we can extend the lower bound on the sum in the argument of the exponential to include $k, l = 1$,

$$\prod_{n=2}^{\infty} \frac{1}{\sqrt{\lambda_n}} = \left(\prod_{n=2}^{\infty} \int_{-\infty}^{+\infty} \frac{\mathrm{d}c_n}{\sqrt{2\pi\hbar}} \right) e^{-\frac{1}{2\hbar} \sum_{k,l=1}^{\infty} c_k c_l \int_{-\mathcal{T}/2}^{+\mathcal{T}/2} \mathrm{d}\tau \, x_k [-\partial_\tau^2 + V''(\bar{x}_1)] x_l} . \tag{8.129}$$

Finally, we make another use of the mode expansion (8.125) [see also Eqs. (8.93a) and (8.95)],

$$\frac{1}{\sqrt{\mathrm{Det}'[-\partial_\tau^2 + V''(\bar{x}_1)]}} = \left(\prod_{n=2}^{\infty} \int_{-\infty}^{+\infty} \frac{\mathrm{d}c_n}{\sqrt{2\pi\hbar}} \right) e^{-\frac{1}{2\hbar} \int_{-\mathcal{T}/2}^{+\mathcal{T}/2} \mathrm{d}\tau \, y [-\partial_\tau^2 + V''(\bar{x}_1)] y} . \tag{8.130}$$

The existence of the eigenvalue $\lambda_1 = 0$ has the disastrous consequence that $\mathrm{Det}[-\partial_\tau^2 + V''(\bar{x}_1)]$ vanishes. The eigenfunction x_1 is called a *zero mode*. It originates in the fact that the center τ_0 of $\bar{x}_1$ can always be chosen to be zero with the help of the change of variable $\tau - \tau_0 = \tau'$ in the limit $\mathcal{T} \to \infty$. As a corollary $\mathrm{Det}'[-\partial_\tau^2 + V''(\bar{x}_1)]$ is independent of τ_0 and so is K. The association of instantons to zero modes is not particular to this example but is a generic feature of instanton physics. It is crucial to avoid integrating too early over the measure of zero modes: Had we formally integrated over the measure $\mathrm{d}c_1$ in Eq. (8.130) we would have encountered a divergence. The strategy that we will use instead is to treat the zero mode separately from all other eigenmodes.

Treating the zero modes separately requires an explicit construction of the measure $\mathrm{d}c_1/\sqrt{2\pi\hbar}$ in terms of the instanton $\bar{x}_1$. Fortunately, this can be done without any detailed knowledge on the potential other than the existence of the two degenerate minima of V. On the one hand, a small change $\mathrm{d}\tau_0$ in the center τ_0 of the instanton induces the small change $\dot{\bar{x}}_1 \mathrm{d}\tau_0$ in the instanton. Since this small change vanishes when $\tau = \mp\mathcal{T}/2$, we can ascribe it to a change

$$\mathrm{d}y = \dot{\bar{x}}_1 \, \mathrm{d}\tau_0 \tag{8.131}$$

of the small Gaussian fluctuations obeying hard-wall boundary conditions about the instanton. On the other hand, a small change $\mathrm{d}c_1$ in the expansion coefficient of the zero-mode contributes a change

$$\mathrm{d}y = x_1 \, \mathrm{d}c_1 \tag{8.132}$$

to the same small Gaussian fluctuations obeying hard-wall boundary conditions about the instanton. We have thus derived a relation between the measure $\mathrm{d}c_1/\sqrt{2\pi\hbar}$ of the zero mode and the arbitrariness in the choice of the instanton center,

$$\begin{aligned}
\frac{\mathrm{d}c_1}{\sqrt{2\pi\hbar}} &= \frac{1}{\sqrt{2\pi\hbar}} \frac{\mathrm{d}y}{x_1} && \text{By Eq. (8.132)} \\
\text{By Eq. (8.124)} \quad &= \sqrt{\frac{S_1}{2\pi\hbar}} \frac{\mathrm{d}y}{\dot{\bar{x}}_1} \\
\text{By Eq. (8.131)} \quad &= \sqrt{\frac{S_1}{2\pi\hbar}} \mathrm{d}\tau_0 .
\end{aligned} \tag{8.133}$$

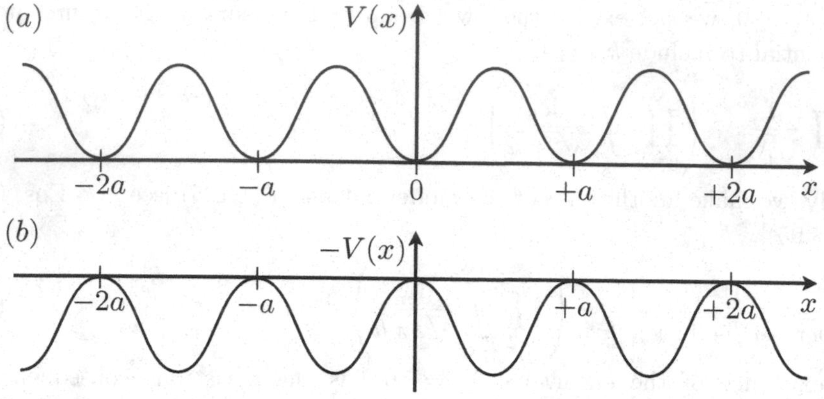

Fig. 8.8 (a) Periodic potential. (b) Inverted periodic potential.

We finally arrive at the result for K defined in Eq. (8.116c)

$$K := \frac{\frac{\mathrm{d}c_1}{\sqrt{2\pi\hbar}}}{\mathrm{d}\tau_0} \times \sqrt{\frac{\mathrm{Det}\,(-\partial_\tau^2 + \omega^2)}{\mathrm{Det}'\,[-\partial_\tau^2 + V''(\bar{x}_1)]}} = \sqrt{\frac{S_1}{2\pi\hbar}} \sqrt{\frac{\mathrm{Det}\,(-\partial_\tau^2 + \omega^2)}{\mathrm{Det}'\,[-\partial_\tau^2 + V''(\bar{x}_1)]}}. \tag{8.134}$$

Observe that K is proportional to $1/\sqrt{\hbar}$.

8.5.5 *Application to the periodic potential*

We apply the semi-classical approximation to the case when the potential V in Eq. (8.85) is periodic, i.e., has infinitely many degenerate minima at $n\,a$, $n \in \mathbb{Z}$ [see Figs. 8.8(a) and 8.8(b)],

$$V(x) = V(x + n\,a), \qquad V(n\,a + x) = \frac{1}{2}\,\omega^2\,x^2 + \mathcal{O}(x^4). \tag{8.135}$$

Initial and final states are

$$x_i = j_i\,a, \qquad x_f = j_f\,a. \tag{8.136}$$

We can borrow the complete analysis of the double well potential except for one restriction present before and absent here. Owing to the periodicity of the minima, it is not necessary anymore to alternate instantons and anti-instantons in time. To see this draw all the minima of the potential on the line, thus defining a one-dimensional lattice with lattice spacing a. For the double well potential, the lattice is made of two sites. For the periodic potential, the lattice is made of infinitely many sites. Picture all the instanton configurations as a time ordered sequence of nearest-neighbor jumps taking place at times $\tau_1 \ll \tau_2 \ll \cdots$, in such a way that initial and final states are reached after a time $\mathcal{T}$. A jump to the right (left) at time τ_m represents a single (anti-) instanton centered at τ_m. For the double well problem, jumping can only take place between the same two sites and thus a jump to the right is necessarily followed by a jump to the left. In contrast, for the periodic potential, the only condition on the sequence of nearest neighbor jumps at the ordered times $\tau_1 \ll \tau_2 \ll \cdots \ll \tau_{n-1} \ll \tau_n$ is that the number n_r of nearest-neighbor jumps to the right minus the number n_l of nearest neighbor-jumps to the left equals $j_f - j_i$. In particular, for n_r and n_l given, it does not matter whether jumps to the right alternate with jumps to the left. This implies that the integration over the instanton centers is

$$\frac{\mathcal{T}^{n_r+n_l}}{n_r!\,n_l!} = \left(\int\limits_{-\mathcal{T}/2}^{+\mathcal{T}/2} \mathrm{d}\tau_{n_r} \int\limits_{-\mathcal{T}/2}^{\tau_{n_r}} \mathrm{d}\tau_{n_r-1} \cdots \int\limits_{-\mathcal{T}/2}^{\tau_3} \mathrm{d}\tau_2 \int\limits_{-\mathcal{T}/2}^{\tau_2} \mathrm{d}\tau_1\right) \times \left(\int\limits_{-\mathcal{T}/2}^{+\mathcal{T}/2} \mathrm{d}\bar{\tau}_{n_l} \int\limits_{-\mathcal{T}/2}^{\bar{\tau}_{n_l}} \mathrm{d}\bar{\tau}_{n_l-1} \cdots \int\limits_{-\mathcal{T}/2}^{\bar{\tau}_3} \mathrm{d}\bar{\tau}_2 \int\limits_{-\mathcal{T}/2}^{\bar{\tau}_2} \mathrm{d}\bar{\tau}_1\right). \tag{8.137}$$

We conclude that the amplitude for the initial state $x_i = j_i\,a$ to evolve into the final state $x_f = j_f\,a$ is, as $\mathcal{T} \to \infty$, given by the semi-classical approximation

$$\langle j_f\,a|e^{-\hat{H}\mathcal{T}/\hbar}|j_i\,a\rangle \approx \sqrt{\omega}\,e^{-\omega\mathcal{T}/2} \sum_{n,\bar{n}=0}^{\infty} \frac{(K\,e^{-S_1/\hbar}\,\mathcal{T})^{n+\bar{n}}}{n!\,\bar{n}!}\,\delta_{n-\bar{n},j_f-j_i}. \tag{8.138}$$

If we represent the Kronecker delta on the right-hand side by an integral, we may then write

$$\langle j_f\,a|e^{-\hat{H}\mathcal{T}/\hbar}|j_i\,a\rangle \approx \sqrt{\omega}\,e^{-\omega\mathcal{T}/2} \sum_{n,\bar{n}=0}^{\infty} \frac{(K\,e^{-S_1/\hbar}\,\mathcal{T})^{n+\bar{n}}}{n!\,\bar{n}!} \int\limits_{0}^{2\pi} \frac{\mathrm{d}\theta}{2\pi}\,e^{+\mathrm{i}\theta(n-\bar{n}-j_f+j_i)}. \tag{8.139}$$

After interchanging the order between the summations over the integers n and $\bar{n}$ and the integral over θ, we reach the desired representation

$$\begin{aligned}\langle j_f\,a|e^{-\hat{H}\mathcal{T}/\hbar}|j_i\,a\rangle &\approx \sqrt{\omega}\,e^{-\omega\mathcal{T}/2} \int\limits_{0}^{2\pi} \frac{\mathrm{d}\theta}{2\pi}\,e^{-\mathrm{i}\theta(j_f-j_i)} \\ &\quad\times \sum_{n,\bar{n}=0}^{\infty} \frac{1}{n!}\left(K\,e^{-S_1/\hbar}\,\mathcal{T}\,e^{+\mathrm{i}\theta}\right)^{n} \frac{1}{\bar{n}!}\left(K\,e^{-S_1/\hbar}\,\mathcal{T}\,e^{-\mathrm{i}\theta}\right)^{\bar{n}} \\ &= \sqrt{\omega}\,e^{-\omega\mathcal{T}/2} \int\limits_{0}^{2\pi} \frac{\mathrm{d}\theta}{2\pi}\,e^{-\mathrm{i}\theta(j_f-j_i)} \exp\left(2\,K\,e^{-S_1/\hbar}\,\mathcal{T}\,\cos\theta\right).\end{aligned} \tag{8.140}$$

The interpretation of this result is that the infinite degeneracy of the harmonic oscillator mode in the limit of infinite potential barrier has been lifted by tunneling. A band of low-lying states has emerged with the energy

$$\varepsilon(\theta) = +\frac{1}{2}\,\hbar\omega - 2\hbar\,K\,e^{-S_1/\hbar}\,\cos\theta, \qquad 0 \le \theta \le 2\pi, \tag{8.141}$$

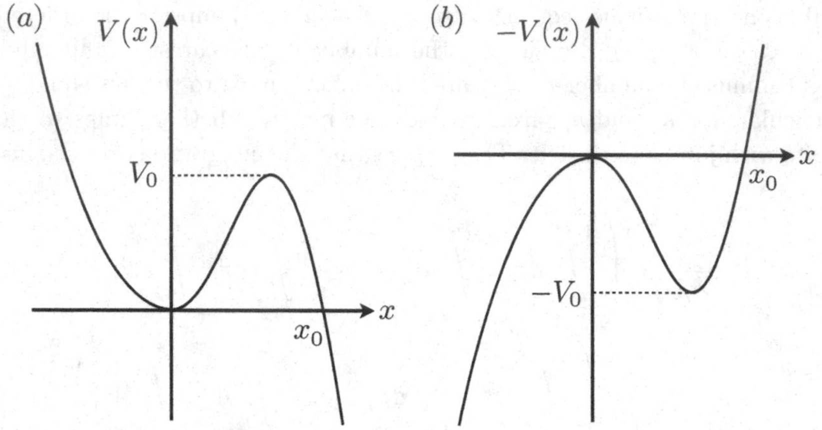

Fig. 8.9 (a) Potential well $V(x)$ with a metastable minimum at $x = 0$. (b) Inverted potential well $V(x)$ with a metastable minimum at $x = 0$.

and the overlap

$$\langle\theta|x = j_i\, a\rangle \propto \frac{1}{\sqrt{2\,\pi}}\,\omega^{1/4}\,e^{+\mathrm{i}\theta j_i}. \tag{8.142}$$

Within the semi-classical approximation, we have recovered the tight-binding band of states of the Hamiltonian

$$\hat{H}_{\mathrm{tb}} := \frac{1}{2}\,\hbar\omega \sum_{j\in\mathbb{Z}} \hat{c}^{\dagger}_{j}\,\hat{c}_{j} + \hbar\,K\,e^{-S_1/\hbar} \sum_{j\in\mathbb{Z}} \left(\hat{c}^{\dagger}_{j}\,\hat{c}_{j+1} + \mathrm{h.c.}\right). \tag{8.143}$$

The band width $4\,\hbar\,K\,e^{-S_1/\hbar}$ is twice that for the double well potential.

8.5.6 *The case of an unbounded potential of the cubic type*

So far, we only considered potentials $V(x)$ that have absolute minima. When the potential has a unique global minimum, the classical path $\bar{x}$ describing a spinless point particle with unit mass on the real line stuck at the bottom of the potential minimizes the Euclidean action. Indeed, for any trajectory $x(\tau)$,

$$\begin{aligned} S[x] &= \int\limits_{-\mathcal{T}/2}^{+\mathcal{T}/2} \mathrm{d}\tau \left(\frac{1}{2}\,\dot{x}^2 + V(x)\right) \\ &\geq \mathcal{T}\,V(\bar{x}) \\ &= S[\bar{x}]. \end{aligned} \tag{8.144}$$

By construction, $\bar{x}$ obeys Newton equation in the inverted potential $-V$, i.e.,

$$\ddot{\bar{x}} = V'(\bar{x}), \tag{8.145}$$

since $\ddot{\bar{x}} = 0$ and $V'(\bar{x}) = 0$. The trajectory $\bar{x}$ is also characterized by the constant of motion

$$E[\bar{x}] = \frac{1}{2}\,\dot{\bar{x}}^2 - V(\bar{x}) \\ = -V(\bar{x}). \tag{8.146}$$

When the potential has several global minima ι, all trajectories $\bar{x}_\iota$ for which a spinless point particle with unit mass on the real line is stuck at the ι-th bottom of the potential are global minima of the Euclidean action $S[x]$ with the action

$$S[\bar{x}_\iota] = \mathcal{T}\, V(\bar{x}_\iota), \tag{8.147}$$

and the constant of motion

$$E[\bar{x}_\iota] = -V(\bar{x}_\iota). \tag{8.148}$$

In addition, we constructed instanton trajectories $\bar{x}_n$ that are *local minima* of the action $S[x]$. Instantons can be visualized as trajectories by which a spinless point particle with unit mass on the real line rolls n-times away from the absolute maxima of the inverted potential $-V(x)$ with the constant of motion [compare with Eq. (8.112) for which $V(\bar{x}_\iota) = 0^+$]

$$E[\bar{x}_n] := \frac{1}{2}\,\dot{\bar{x}}_n^2 - V(\bar{x}_n) = -V(\bar{x}_\iota) = E[\bar{x}_\iota], \qquad \forall\iota. \tag{8.149}$$

Correspondingly, instanton $\bar{x}_n$ has the action

$$\begin{aligned} S[\bar{x}_n] &= \int\limits_{-\mathcal{T}/2}^{+\mathcal{T}/2} \mathrm{d}\tau \left(\frac{1}{2}\,\dot{\bar{x}}_n^2 + V(\bar{x}_n)\right) \\ &= \int\limits_{-\mathcal{T}/2}^{+\mathcal{T}/2} \mathrm{d}\tau \left(\frac{1}{2}\,\dot{\bar{x}}_n^2 - V(\bar{x}_n) + 2\,V(\bar{x}_n)\right) \\ &= \int\limits_{-\mathcal{T}/2}^{+\mathcal{T}/2} \mathrm{d}\tau \left[-V(\bar{x}_\iota) + 2\,V(\bar{x}_n)\right] \\ &\geq \mathcal{T}\, V(\bar{x}_\iota) \\ &= S[\bar{x}_\iota], \qquad \forall\iota. \end{aligned} \tag{8.150}$$

Potentials with degenerate absolute minima are relevant to a dissipative Josephson junction but not to the Caldeira-Leggett model of quantum tunneling for which the potential is of the cubic type [see Eq. (8.39)], i.e., unbounded from below with a single metastable minimum (see Fig. 8.9). For concreteness, we take the classical potential to be

$$\begin{aligned} V(x) &= \frac{1}{2}\,\omega^2\,x^2 - \frac{1}{3}\,\lambda^2\,x^3 \\ &= \frac{27}{4}\,V_0 \left(\frac{x}{x_0}\right)^2 \left(1 - \frac{x}{x_0}\right), \end{aligned} \tag{8.151}$$

$$V_0 \equiv V(x)\big|_{x=\frac{\omega^2}{\lambda^2}}, \qquad x_0 \equiv \frac{3}{2}\,\frac{\omega^2}{\lambda^2}, \qquad 0 < \omega, \qquad \lambda \in \mathbb{R}.$$

Classically, a spinless point particle with unit mass on the real line that initially sits at the origin $x = 0$ will remain forever in the metastable minimum of the cubic potential (8.151). If $\bar{x}$ denotes this trajectory,

$$S[\bar{x}] = \int\limits_{-\mathcal{T}/2}^{+\mathcal{T}/2} \mathrm{d}\tau \left(\frac{1}{2} \dot{\bar{x}}^2 + V(\bar{x}) \right) = \mathcal{T}\, V(0) = 0 \tag{8.152}$$

is a local minimum of the action with the constant of the motion

$$E[\bar{x}] = \frac{1}{2} \dot{\bar{x}}^2 - V(\bar{x}) = -V(0) = 0. \tag{8.153}$$

There are trajectories called *bounces* and denoted $\bar{x}_n(\tau)$ that share the same energy (8.153) with $\bar{x}(\tau) \equiv 0$, are local extrema of the action, and are built out of the single bounce $\bar{x}_1(\tau)$ by which a spinless point particle with unit mass on the real line rolls along the constant energy curve

$$0 = \frac{1}{2} (\dot{\bar{x}}_1)^2 - V(\bar{x}_1) \tag{8.154}$$

from the top of the hill at $x = 0$ to the classical turning point x_0 of the inverted potential $-V(x)$ to come back to the top of the hill at $x = 0$. The action of a single bounce $\bar{x}_1(\tau)$ is larger than that of $\bar{x}(\tau)$,

$$\begin{aligned}
S[\bar{x}_1] &= \int\limits_{-\mathcal{T}/2}^{+\mathcal{T}/2} \mathrm{d}\tau \left(\frac{1}{2} (\dot{\bar{x}}_1)^2 + V(\bar{x}_1) \right) \\
&= \int\limits_{-\mathcal{T}/2}^{+\mathcal{T}/2} \mathrm{d}\tau\, 2\, V(\bar{x}_1) \\
&= \int\limits_{0}^{x_0} \mathrm{d}\bar{x}_1 \frac{\mathrm{d}\tau}{\mathrm{d}\bar{x}_1}\, 2\, V(\bar{x}_1) + \int\limits_{x_0}^{0} \mathrm{d}\bar{x}_1 \frac{\mathrm{d}\tau}{\mathrm{d}\bar{x}_1}\, 2\, V(\bar{x}_1) \\
&= 2 \int\limits_{0}^{x_0} \mathrm{d}\bar{x}_1 \frac{1}{\sqrt{2\, V(\bar{x}_1)}}\, 2\, V(\bar{x}_1) \\
&= 2 \int\limits_{0}^{x_0} \mathrm{d}\bar{x}_1 \sqrt{2\, V(\bar{x}_1)} \\
&\equiv S_1 > 0.
\end{aligned} \tag{8.155}$$

One difference with having a potential with an absolute minimum is that the bounce is not a local minimum but a saddle point. To see this, for any positive energy E consider a classical path $\bar{x}_E(\tau)$ obeying the classical equation of motion

$$\ddot{\bar{x}}_E = V'(\bar{x}_E), \qquad -\frac{\mathcal{T}}{2} \leq \tau \leq +\frac{\mathcal{T}}{2}, \tag{8.156}$$

with the constant energy

$$0 < E = \frac{1}{2}(\dot{\bar{x}}_E)^2 - V(\bar{x}_E), \tag{8.157}$$

whereby the particle sits at $x = 0$ at time $-\mathcal{T}/2$ with the positive kinetic energy E, reaches a turning point x_E *larger* than the classical turning point x_0, and return to $x = 0$ at time $+\mathcal{T}/2$. The dependence on the energy $E > 0$ of the action $S[\bar{x}_E]$ must be unbounded from below as $E \to \infty$ since this limit corresponds to a particle provided with enough kinetic energy to spend more and more time into the classically forbidden region to the right of the turning point x_0 in Fig. 8.10. In mathematical terms,

$$\begin{aligned}
S[\bar{x}_E] =& \int\limits_{-\mathcal{T}/2}^{+\mathcal{T}/2} \mathrm{d}\tau \left(\frac{1}{2}(\dot{\bar{x}}_E)^2 + V(\bar{x}_E)\right) \\
=& \int\limits_{-\mathcal{T}/2}^{+\mathcal{T}/2} \mathrm{d}\tau \left(\frac{1}{2}(\dot{\bar{x}}_E)^2 - V(\bar{x}_E) + 2\,V(\bar{x}_E)\right) \\
=& \int\limits_{-\mathcal{T}/2}^{+\mathcal{T}/2} \mathrm{d}\tau \, [E + 2\,V(\bar{x}_E)] \\
=& \, \mathcal{T}E + 2 \int\limits_{-\mathcal{T}/2}^{+\mathcal{T}/2} \mathrm{d}\tau \, V(\bar{x}_E) \\
=& \, \mathcal{T}E + 4 \int\limits_{0}^{x_0} \mathrm{d}\bar{x}_E \frac{\mathrm{d}\tau}{\mathrm{d}\bar{x}_E} V(\bar{x}_E) + 4 \int\limits_{x_0}^{x_E} \mathrm{d}\bar{x}_E \frac{\mathrm{d}\tau}{\mathrm{d}\bar{x}_E} V(\bar{x}_E) \\
=& \, \mathcal{T}E + 4 \int\limits_{0}^{x_0} \mathrm{d}\bar{x}_E \frac{V(\bar{x}_E)}{\sqrt{2\,[E + V(\bar{x}_E)]}} + 4 \int\limits_{x_0}^{x_E} \mathrm{d}\bar{x}_E \frac{V(\bar{x}_E)}{\sqrt{2\,[E + V(\bar{x}_E)]}}.
\end{aligned} \tag{8.158}$$

The first and second terms are both positive and non-vanishing. The last term is negative and grows as the area underneath the (negative) curve $4\frac{\mathrm{d}\tau}{\mathrm{d}\bar{x}_E}V(\bar{x}_E)$ between x_0 and x_E. As $\lim_{E\to\infty} x_E = \infty$, the last contribution always dominates over the first two contributions to the right-hand side of Eq. (8.158). A single bounce is thus a saddle point as it is a local maximum in the "direction" made of the submanifold of classical path $\bar{x}_E$ whereas it is a local minimum in the remaining orthogonal "directions" in the space of all paths entering the path integral representation of the partition function.

A corollary to the unboundness of $S[\bar{x}_E]$ is the fact that $-\partial_\tau^2 + V''(\bar{x}_1)$ must have a negative eigenvalue. Indeed, one observes that the velocity of a single bounce is an eigenfunction of $[-\partial_\tau^2 + V''(\bar{x}_1)]$ with vanishing eigenvalue that obeys the hard-wall boundary conditions and supports one node. Since the velocity of a single

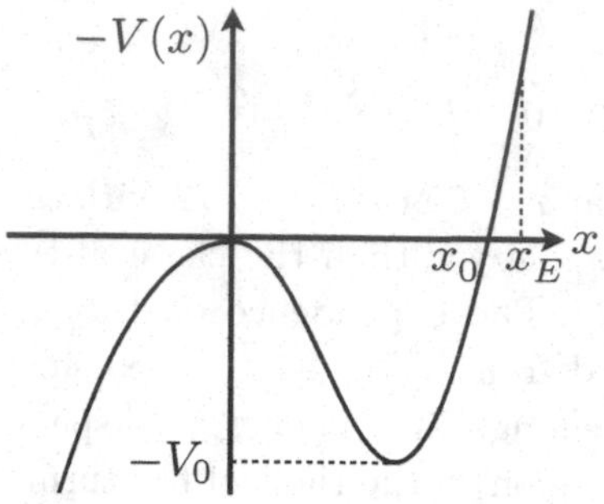

Fig. 8.10 A classical path for an inverted potential well $-V(x)$ with a metastable maximum at $x=0$ by which the particle starts at the origin $x=0$ with kinetic energy E, reaches the turning point x_E past the classical turning point x_0, and "rolls" back to the origin $x=0$ after time $\mathcal{T}$.

bounce is an eigenstate of $[-\partial_\tau^2 + V''(\bar{x}_1)]$ with vanishing eigenvalue and supports one and only one node, there must exist one and only one nodeless eigenfunction of $-\partial_\tau^2 + V''(\bar{x}_1)$ with negative eigenvalue. The counterpart to the constant K in Eq. (8.116c) must then be imaginary for a single bounce.

It is now possible to salvage a physical interpretation of the semi-classical expansions (8.92), (8.95), and (8.106) around multi-bounce trajectories. The fact that the counterpart to the constant K in Eq. (8.116c) is imaginary for bounces means that the ground state in the expansion (8.106) is not the true ground state of the Hamiltonian but must be interpreted as an unstable ground state with the complex energy $\varepsilon_0 = \mathrm{Re}\,\varepsilon_0 + \mathrm{i}\mathrm{Im}\,\varepsilon_0$ and the inverse lifetime

$$\Gamma_0 := \left|\frac{2\,\mathrm{Im}\,\varepsilon_0}{\hbar}\right| = |K|\,e^{-S_1/\hbar}, \tag{8.159}$$

whereby S_1 given by Eq. (8.155). The tunneling rate (8.159) should be compared with Eq. (8.41). To reach this conclusion it is sufficient to replace Eq. (8.117a) by

$$\begin{aligned}\langle x=0|e^{-\hat{H}\mathcal{T}/\hbar}|x=0\rangle \;&\propto\; \frac{1}{\sqrt{\mathrm{Det}\,(-\partial_\tau^2+\omega^2)}}\sum_{n=0}^{\infty}\frac{\left[\mathrm{i}(K/2)\,e^{-S_1/\hbar}\,\mathcal{T}\right]^n}{n!}\times[1+\mathcal{O}(\hbar)]\\ &\propto \sqrt{\omega}\,e^{-\omega\mathcal{T}/2}\,e^{\mathrm{i}(K/2)\,e^{-S_1/\hbar}\,\mathcal{T}}\times[1+\mathcal{O}(\hbar)]\,.\end{aligned} \tag{8.160}$$

Upon the analytical continuation $\mathcal{T}=\mathrm{i}T$, Eq. (8.160) implies that the amplitude for a spinless point particle with unit mass on the real line to remain at the metastable minimum $x=0$ of a cubic-like potential decays exponentially fast with T.

There is a subtlety with the factor 1/2 appearing in $(K/2)$ from Eq. (8.160). The factor 1/2 arises from the full Gaussian path integration about the single-bounce saddle-point. The Gaussian path integral is not convergent since it contains the divergent Riemann integral

$$\int\limits_0^\infty \frac{\mathrm{d}E}{\sqrt{2\pi\,\hbar}}\,e^{-(S_1/\hbar)-\frac{1}{2}\,(S_1''\,E^2/\hbar)+\cdots}. \tag{8.161}$$

Here, E is a "deviation" about the single-bounce saddle-point $E=0$ in the unstable direction corresponding to Fig. 8.10. The unstable direction is encoded by the fact that S_1'' is negative. Hence, the Gaussian Riemann integral over E is divergent. To make sense of this divergent integral one replaces the path of integration $0 \leq E < \infty$ along the half line by the path of integration $E = \mathrm{i}z$ where $0 \leq z < \infty$ along half of the imaginary axis,

$$\int\limits_0^\infty \frac{\mathrm{d}z}{\sqrt{2\pi\hbar}}\, e^{-(S_1/\hbar)-\frac{1}{2}(|S_1''|\,z^2/\hbar)+\cdots} = \frac{1}{2}\frac{1}{\sqrt{|S_1''|}} e^{-S_1/\hbar}. \tag{8.162}$$

The replacement $K \to (K/2)$ originates from the factor 1/2 in Eq. (8.162).

8.6 The quantum-dissipative Josephson junction

In section 8.4.2.1, we introduced the Caldeira-Leggett (CL) model [84] of dissipative quantum mechanics, which is defined by the partition function ($\dot{}$ denotes τ derivative, $'$ denotes q derivative)

$$Z_\beta = \mathcal{N}_{\mathrm{bath}} \int \mathcal{D}[q]\, e^{-S'_\beta/\hbar}, \tag{8.163a}$$

with the additive decomposition of the action

$$S'_\beta := S_0 + S_1 + S_{\mathrm{ext}} \tag{8.163b}$$

into three contributions. There is the dissipative-free action

$$\begin{aligned} S_0 &:= \int\limits_0^\beta \mathrm{d}\tau \left(\frac{M}{2}\dot{q}^2 + V(q)\right) \\ &= \sum_{\varpi_l} \left(\frac{M}{2}\varpi_l^2\, q_{(+\varpi_l)}\, q_{(-\varpi_l)} + \sqrt{\beta}\,[V(q)]_{\varpi_l}\, \delta_{\varpi_l,0}\right). \end{aligned} \tag{8.163c}$$

There is the dissipative action

$$\begin{aligned} S_1 &:= \int\limits_{-\infty}^{+\infty} \mathrm{d}\tau \int\limits_0^\beta \mathrm{d}\tau'\, \frac{\eta}{4\pi} \left(\frac{q(\tau)-q(\tau')}{|\tau-\tau'|}\right)^2 \\ &= \sum_{\varpi_l} \frac{\eta}{2}\, |\varpi_l|\, q_{(+\varpi_l)}\, q_{(-\varpi_l)}. \end{aligned} \tag{8.163d}$$

Finally, there is the driving term

$$\begin{aligned} S_{\mathrm{ext}} &:= -\int\limits_0^\beta \mathrm{d}\tau F_{\mathrm{ext}}(\tau)\, q(\tau) \\ &= -\sum_{\varpi_l} F_{\mathrm{ext}\,(+\varpi_l)}\, q_{(-\varpi_l)}. \end{aligned} \tag{8.163e}$$

Here, all trajectories $q(\tau)$ entering the path integral are periodic in imaginary time,

$$q(\tau) = q(\tau+\beta) = \frac{1}{\sqrt{\beta}}\sum_{\varpi_l} q_{\varpi_l}\, e^{-\mathrm{i}\varpi_l\tau}, \qquad \varpi_l = \frac{2\pi}{\beta}l, \qquad l\in\mathbb{Z}. \tag{8.164}$$

The CL model is constructed so that, after analytical continuation $\tau = it$ to real time t, the particle of mass M with coordinate $q(t)$ along the real line is subjected to a force $-V'(q)$ arising from a potential $V(q)$ of the cubic type with a single metastable minimum, to a frictional force $-\eta\,\dot{q}(\tau)$, and, finally, to an external force (or source term) $F_{\mathrm{ext}}(t)$ in the classical limit $\hbar\to 0$. The kernel of S_1, which is non-local in imaginary (Matsubara) time, and the proportionality constant $\mathcal{N}_{\mathrm{bath}}$ are the remnants of the interaction between the particle of mass M and a bath made of infinitely many independent harmonic oscillators.

A semi-classical estimate of the probability per unit time Γ' for the particle to tunnel out from the metastable minimum of $V(q)$ is [see Eqs. (8.159) and (8.134)], in the absence of an external force,

$$\begin{aligned}
\Gamma' &= A'\, e^{-\frac{B'}{\hbar}}\left[1+\mathcal{O}(\hbar)\right],\\
B' &= S_0[\bar{q}_1] + S_1[\bar{q}_1],\\
A' &= \sqrt{\frac{B'}{2\pi\hbar}}\left|\frac{\mathrm{Det}\ \mathcal{D}_0}{\mathrm{Det}'\ \mathcal{D}_1}\right|^{1/2},
\end{aligned} \tag{8.165a}$$

and

$$\begin{aligned}
\mathcal{D}_0 q(\tau) &= \left(-\partial_\tau{}^2 + \varpi_0{}^2\right) q(\tau) + \frac{\eta}{\pi M}\int\limits_{-\infty}^{+\infty}\mathrm{d}\tau'\,\frac{q(\tau)-q(\tau')}{(\tau-\tau')^2},\\
\mathcal{D}_1 q(\tau) &= \left(-\partial_\tau{}^2 + \frac{1}{M}V''(\bar{q}_1)\right) q(\tau) + \frac{\eta}{\pi M}\int\limits_{-\infty}^{+\infty}\mathrm{d}\tau'\,\frac{q(\tau)-q(\tau')}{(\tau-\tau')^2}.
\end{aligned} \tag{8.165b}$$

This estimates relies on saddle-point approximations to the path integral about *multi-bounce* trajectories $\bar{q}_n$, $n=1,2,\cdots$, that are assumed to behave like n *non-interacting* single-bounce trajectories $\bar{q}_1$. The prime over the functional determinant of the non-local propagator $\mathcal{D}_1$ says that the zero eigenvalue of $\mathcal{D}_1$ (corresponding to a uniform translation of the bounce along Matsubara time) is to be omitted. The absolute value in A' is needed as bounces are not local minima of the action but saddle-points, i.e., there exists one negative eigenvalue of $\mathcal{D}_1$. This is not so for the propagator $\mathcal{D}_0$ which is evaluated at the single local minimum of $V(q)$, $\omega_0^2 \equiv V''(q=q_{\mathrm{min}})$. Equation (8.165) holds for all values of the damping coefficient η.

The quantum-dissipative Josephson junction is related to the CL model by the identifications

$$q(\tau) \longrightarrow \phi_-(\tau) \tag{8.166a}$$

between the particle position in the CL model and the Josephson phase,

$$M \longrightarrow C \left(\frac{\hbar}{e^*}\right)^2 \quad \text{[units are: (charge}^2\text{/energy) (energy}^2\times\text{time}^2\text{/charge}^2\text{)= energy}\times\text{time}^2\text{]} \tag{8.166b}$$

between the particle mass in the CL model and the capacity,

$$V(q) \longrightarrow -\frac{\hbar}{e^*} I_0 \cos\phi_-, \qquad I_0 = \frac{2\,e^*\,U_{\mathrm{J}}}{\hbar} \tag{8.166c}$$

between the metastable potential in the CL model and the Josephson coupling,

$$\eta \longrightarrow \frac{1}{R_s}\left(\frac{\hbar}{e^*}\right)^2 \quad \text{[units are: (energy}\times\text{time/charge}^2)^{-1}\text{ (energy}^2\times\text{time}^2\text{/charge}^2\text{)= energy}\times\text{time]} \tag{8.166d}$$

between the friction in the CL model and the Ohmic resistance,

$$F_{\mathrm{ext}}(\tau) \longrightarrow -\frac{\hbar}{e^*} I(\tau) \tag{8.166e}$$

between the driving forces, and

$$S'_\beta \longrightarrow S_{0\,\mathrm{kin}} + S_{0\,\mathrm{int}} + S_1 + S_{\mathrm{ext}}, \tag{8.166f}$$

whereby we have split the dissipative-free action into the dissipative-free kinetic action

$$S_{0\,\mathrm{kin}} = \frac{\hbar}{e^*}\int\limits_0^\beta \mathrm{d}\tau\, \frac{1}{2}\, C\, \frac{\hbar}{e^*}\, (\dot{\phi}_-)^2, \tag{8.166g}$$

and the dissipative-free interacting action

$$S_{0\,\mathrm{int}} = \frac{\hbar}{e^*}\int\limits_0^\beta \mathrm{d}\tau\, (-I_0)\cos\phi_-, \tag{8.166h}$$

while the dissipative action is

$$\begin{aligned} S_1 &= \frac{\hbar}{e^*}\int\limits_{-\infty}^{+\infty}\mathrm{d}\tau \int\limits_0^\beta \mathrm{d}\tau'\, \frac{1}{4\pi}\,\frac{1}{R_s}\,\frac{\hbar}{e^*}\left(\frac{\phi_-(\tau)-\phi_-(\tau')}{|\tau-\tau'|}\right)^2 \\ &= \frac{\hbar}{e^*}\sum_{\varpi_l}\frac{1}{2}\,\frac{1}{R_s}\,\frac{\hbar}{e^*}\,|\varpi_l|\,\phi_{-(+\varpi_l)}\,\phi_{-(-\varpi_l)}, \end{aligned} \tag{8.166i}$$

and the driving action has become

$$S_{\mathrm{ext}} = \frac{\hbar}{e^*}\int\limits_0^\beta \mathrm{d}\tau\, I(\tau)\,\varphi_-(\tau). \tag{8.166j}$$

There is one essential difference between the original CL and the dissipative Josephson junction, namely the fact that the Josephson angle ϕ_- is a *compact* degree of freedom (it is defined modulo $2\,\pi$, i.e., on the unit circle). This difference invalidates the assumption that multi-bounces (for $I \neq 0$) or multi-instantons configurations (for $I = 0$) form a dilute non-interacting gas. As a result, we will see that the dissipationless regime $1/R_s \to 0$ and strong dissipation regime $1/R_s \to \infty$ are not smoothly connected, to the contrary of the CL model (8.165) for which the limits $\eta \to 0$ and $\eta \to \infty$ are smoothly connected.

With the introduction of

- the quantum resistors for Cooper pairs

$$\frac{\hbar}{e^{*2}} =: R_{\hbar}, \qquad 2\pi\, R_{\hbar} = \frac{2\pi\,\hbar}{e^{*2}} = \frac{h}{e^{*2}} =: R_{h}, \tag{8.167a}$$

- the ratio of the quantum resistor to the Ohmic resistance

$$\frac{1}{R_s}\,\frac{\hbar}{e^{*2}} = \frac{R_{\hbar}}{R_s} = \frac{1}{2\pi}\,\frac{R_h}{R_s} =: \frac{1}{2\pi}\,\alpha, \tag{8.167b}$$

- the Josephson potential

$$-\frac{\hbar}{e^{*}}\,I_0\,\cos\phi_- = -2\,U_{\mathrm{J}}\,\cos\phi_-, \tag{8.167c}$$

the partition function for the dissipative Josephson junction becomes

$$Z_\beta = \mathcal{N}_{\mathrm{bath}} \int \mathcal{D}[\phi_-]\, e^{-S'_\beta/\hbar}, \tag{8.168a}$$

with the additive decomposition of the action

$$S'_\beta := S_{0\,\mathrm{kin}} + S_{0\,\mathrm{int}} + S_1 + S_{\mathrm{ext}} \tag{8.168b}$$

in terms of four contributions: the dissipation-free kinetic action

$$\begin{aligned} S_{0\,\mathrm{kin}} &= \int_0^\beta \mathrm{d}\tau\, \frac{\hbar}{4\pi}\, C\, R_h \left(\partial_\tau \phi_-\right)^2 \\ &= \sum_{\varpi_l} \frac{\hbar}{4\pi}\, C\, R_h\, \varpi_l^2\, \phi_{-(+\varpi_l)}\, \phi_{-(-\varpi_l)}, \end{aligned} \tag{8.168c}$$

the dissipation-free interacting action

$$\begin{aligned} S_{0\,\mathrm{int}} &= \int_0^\beta \mathrm{d}\tau\, (-2\,U_{\mathrm{J}})\, \cos\phi_- \\ &= \sum_{\varpi_l} \sqrt{\beta}\,(-2\,U_{\mathrm{J}})\, \left(\cos\phi_-\right)_{\varpi_l}\, \delta_{\varpi_l,0}, \end{aligned} \tag{8.168d}$$

the dissipative action

$$\begin{aligned} S_1 &= \int_{-\infty}^{+\infty} \mathrm{d}\tau \int_0^\beta \mathrm{d}\tau'\, \frac{\hbar}{4\pi}\,\frac{\alpha}{2\,\pi} \left(\frac{\phi_-(\tau) - \phi_-(\tau')}{|\tau - \tau'|}\right)^2 \\ &= \sum_{\varpi_l} \frac{\hbar}{4\pi}\, \alpha\, |\varpi_l|\, \phi_{-(+\varpi_l)}\, \phi_{-(-\varpi_l)}, \end{aligned} \tag{8.168e}$$

and the driving action

$$\begin{aligned} S_{\mathrm{ext}} &= \int_0^\beta \mathrm{d}\tau\, \frac{\hbar}{e^{*}}\, I(\tau)\, \phi_-(\tau) \\ &= \sum_{\varpi_l} \frac{\hbar}{e^{*}}\, I_{(+\varpi_l)}\, \phi_{-(-\varpi_l)}. \end{aligned} \tag{8.168f}$$

Whereas the Josephson potential $-2\,U_{\mathrm{J}}\cos(\phi_{-})$ is local in Matsubara time, S_1 only becomes local after performing a Fourier transformation to Matsubara frequencies.

From now on, we only consider the case when the time-independent external current (bias) vanishes,

$$I = 0. \tag{8.169}$$

Comparison of $S_{0\,\mathrm{kin}}$ and S_1 suggests that there are two distinct regimes of frequencies:

(1) The regime of weak dissipation

$$C\,R_h\,|\varpi_l| \gg \alpha \Longleftrightarrow |\varpi_l| \gg \frac{1}{C\,R_h}\alpha = \frac{1/R_s}{C}. \tag{8.170a}$$

In this regime, the propagator defined by

$$\begin{aligned}\mathcal{D}_{\varpi_l} :=& \frac{2\,\pi/\hbar}{C\,R_h\,\varpi_l^2 + \alpha\,|\varpi_l|}\\ =& \frac{2\,\pi/\hbar}{C\,R_h\,\varpi_l^2} + \mathcal{O}\left(\frac{\alpha}{C\,R_h\,|\varpi_l|}\right)\end{aligned} \tag{8.170b}$$

decays quadratically fast with large frequencies to a good approximation.

(2) The regime of strong dissipation

$$C\,R_h\,|\varpi_l| \ll \alpha \Longleftrightarrow |\varpi_l| \ll \frac{1}{C\,R_h}\alpha = \frac{1/R_s}{C}. \tag{8.170c}$$

In this regime,

$$\begin{aligned}\mathcal{D}_{\varpi_l} =& \frac{2\,\pi/\hbar}{C\,R_h\,\varpi_l^2 + \alpha\,|\varpi_l|}\\ =& \frac{2\,\pi/\hbar}{\alpha\,|\varpi_l|} + \mathcal{O}\left(\frac{C\,R_h\,|\varpi_l|}{\alpha}\right)\end{aligned} \tag{8.170d}$$

is inversely proportional to small frequencies to a good approximation.

Comparison of $S_{0\,\mathrm{kin}} + S_1$ and $S_{0\,\mathrm{int}}$ suggests that there are two distinct regimes of Josephson coupling:

(1) A regime

$$\frac{C\,R_h \times U_{\mathrm{J}}}{\hbar} \ll 1, \qquad \frac{C\,R_h \times U_{\mathrm{J}}}{\hbar} \ll \alpha, \tag{8.171a}$$

in which perturbation theory in powers of the Josephson coupling might be sensible.

(2) A regime

$$\frac{C\,R_h \times U_{\mathrm{J}}}{\hbar} \gg 1, \qquad \frac{C\,R_h \times U_{\mathrm{J}}}{\hbar} \gg \alpha, \tag{8.171b}$$

in which a good semi-classical approximation when $\alpha = 0$ might remain sensible for finite α.

8.7 Duality in a dissipative Josephson junction

To simplify notation, we rewrite Eq. (8.168) as

$$\begin{aligned}
Z_\beta &= \mathcal{N}_{\text{bath}} \int \mathcal{D}[\phi]\, e^{-S'_\beta/\hbar}, \\
S'_\beta &= S_{0\,\text{kin}} + S_{0\,\text{int}} + S_1 + S_{\text{ext}}, \\
S_{0\,\text{kin}} &= \int_0^\beta \mathrm{d}\tau\, \frac{1}{2} m\, (\partial_\tau \phi)^2 = \sum_{\varpi_l} \frac{1}{2} m\, \varpi_l^2\, \phi_{(+\varpi_l)}\, \phi_{(-\varpi_l)}, \\
S_{0\,\text{int}} &= \int_0^\beta \mathrm{d}\tau\, (-y)\, \cos\phi = \sum_{\varpi_l} \sqrt{\beta}\, (-y)\, (\cos\phi)_{\varpi_l}\, \delta_{\varpi_l,0}, \qquad (8.172a) \\
S_1 &= \int_{-\infty}^{+\infty} \mathrm{d}\tau \int_0^\beta \mathrm{d}\tau'\, \frac{\eta}{4\pi} \left(\frac{\phi(\tau) - \phi(\tau')}{|\tau - \tau'|} \right)^2 = \sum_{\varpi_l} \frac{\eta}{2}\, |\varpi_l|\, \phi_{(+\varpi_l)}\, \phi_{(-\varpi_l)}, \\
S_{\text{ext}} &= \int_0^\beta \mathrm{d}\tau\, J(\tau)\, \phi(\tau) = \sum_{\varpi_l} J_{(+\varpi_l)}\, \phi_{(-\varpi_l)},
\end{aligned}$$

whereby $0 \leq \phi \leq 2\pi$ is the angular coordinate of a "particle" on a circle of unit radius and

$$m \equiv \frac{\hbar}{2\pi} C R_h, \qquad y \equiv 2 U_{\mathrm{J}}, \qquad \eta \equiv \frac{\hbar}{2\pi} \alpha, \qquad J(\tau) \equiv \frac{\hbar}{e^*} I(\tau). \qquad (8.172b)$$

In the absence of an external force (or source), $J(\tau) = 0$, correlation (Green) functions are obtained from

$$\left\langle \prod_{j=1}^{n} \phi(\tau_j) \right\rangle_{Z_\beta} := (-\hbar)^n \frac{1}{Z_\beta} \frac{\partial^n Z_\beta}{\partial J(\tau_1) \cdots \partial J(\tau_n)} \bigg|_{J(\tau)=0}. \qquad (8.172c)$$

From now on, $\hbar = 1$. Our goal is to approximate the partition function in regimes (8.171a) and (8.171b), respectively.

8.7.1 *Regime $m\,y \ll 1$ and $m\,y \ll \eta$*

When the characteristic energy scale y is the smallest in the problem (aside from the temperature β^{-1}), it appears natural to expand formally the partition function in powers of y. In this context we are going to reinterpret $y/2$ as the fugacity of a neutral plasma of classical point-like particles with coordinates $0 \leq \tau \leq \beta$

interacting through the propagator

$$
\begin{aligned}
\mathcal{D}(\tau) &:= \frac{1}{\beta} \sum_{\varpi_l} \mathcal{D}_{\varpi_l}\, e^{-\mathrm{i}\varpi_l \tau} \\
\text{See Eqs. (8.170b) and (8.170d)} \qquad &= \frac{1}{\beta} \sum_{\varpi_l} \frac{1}{m\,\varpi_l^2 + \eta\,|\varpi_l|}\, e^{-\mathrm{i}\varpi_l \tau} \\
\text{In the limit } \beta^{-1} \ll y \qquad &\approx \int\limits_{-\infty}^{+\infty} \frac{d\varpi}{2\pi} \frac{1}{m\,\varpi^2 + \eta\,|\varpi|}\, e^{-\mathrm{i}\varpi \tau},
\end{aligned}
\tag{8.173a}
$$

and Z_β as the grand-canonical partition function of this classical plasma. Here, the divergence due to the pole at the origin of the integral on the right-hand side needs to be regulated. This is achieved by taking the difference between $\mathcal{D}(\tau)$ and $\mathcal{D}(\tau')$ to extract

$$
\mathcal{D}_{\mathrm{reg}}(\tau) \approx
\begin{cases}
-\frac{1}{2\eta}\, r, & \text{if } r := \frac{\eta}{m}\,|\tau| \ll 1, \\
-\frac{1}{\pi\,\eta}\, \ln r, & \text{if } r := \frac{\eta}{m}\,|\tau| \gg 1.
\end{cases}
\tag{8.173b}
$$

To justify this reinterpretation, we first write

$$
Z_\beta = \mathcal{N}_{\mathrm{bath}} \int \mathcal{D}[\phi]\; e^{-(S_{0\,\mathrm{kin}} + S_1) - (S_{0\,\mathrm{int}} + S_{\mathrm{ext}})}.
\tag{8.174}
$$

Observe that

$$
\begin{aligned}
e^{-S_{0\,\mathrm{int}}} &= \exp\left(+y \int\limits_0^\beta \mathrm{d}\tau\, \cos\phi(\tau) \right) \\
&= \sum_{n=0}^{\infty} \frac{1}{n!}\, y^n \prod_{j=1}^{n} \int\limits_0^\beta \mathrm{d}\tau_j\, \cos\phi(\tau_j) \\
&= \sum_{n=0}^{\infty} \frac{1}{n!} \left(\frac{y}{2}\right)^n \prod_{j=1}^{n} \int\limits_0^\beta \mathrm{d}\tau_j \left(e^{+\mathrm{i}\phi(\tau_j)} + e^{-\mathrm{i}\phi(\tau_j)} \right) \\
&= \sum_{n=0}^{\infty} \frac{1}{n!} \left(\frac{y}{2}\right)^n \prod_{j=1}^{n} \int\limits_0^\beta \mathrm{d}\tau_j \left(e^{+\mathrm{i}\int_0^\beta \mathrm{d}\tau\, \delta(\tau-\tau_j)\phi(\tau)} + e^{-\mathrm{i}\int_0^\beta \mathrm{d}\tau\, \delta(\tau-\tau_j)\phi(\tau)} \right) \\
&= \sum_{n=0}^{\infty} \frac{1}{n!} \left(\frac{y}{2}\right)^n \int\limits_0^\beta \mathrm{d}\tau_1 \cdots \int\limits_0^\beta \mathrm{d}\tau_n \sum_{m=0}^{n} \frac{n!}{(n-m)!\,m!}\, e^{-\int_0^\beta \mathrm{d}\tau\, J_{n,m}(\tau)\,\phi(\tau)},
\end{aligned}
\tag{8.175}
$$

whereby the fact that the integrand on the penultimate line is independent of the ordering of "charges" of the same sign has been used, and

$$\begin{aligned} J_{n,m}(\tau) &:= -\mathrm{i}\left(\sum_{j=1}^{n-m}\delta(\tau-\tau_j)-\sum_{l=1}^{m}\delta(\tau-\tau_l)\right)\\ &\equiv -\mathrm{i}\sum_{j=1}^{n} e_j\,\delta(\tau-\tau_j),\\ 1 =: e_1 =: \cdots &=: e_{n-m} =: -e_{n-m+1}\cdots =: -e_n. \end{aligned} \tag{8.176}$$

Insertion of Eq. (8.175) in the partition function gives

$$\begin{aligned} Z_\beta = &\sum_{n=0}^{\infty}\frac{1}{n!}\left(\frac{y}{2}\right)^n\int_0^\beta \mathrm{d}\tau_1\cdots\int_0^\beta \mathrm{d}\tau_n\sum_{m=0}^{n}\frac{n!}{(n-m)!\,m!}\\ &\times\left\langle e^{-\int_0^\beta \mathrm{d}\tau\,(J_{n,m}+J)(\tau)\,\phi(\tau)}\right\rangle_{S_{0\,\mathrm{kin}}+S_1}, \end{aligned} \tag{8.177a}$$

where

$$\langle\cdots\rangle_{S_{0\,\mathrm{kin}}+S_1} := \mathcal{N}_{\mathrm{bath}}\int\mathcal{D}[\phi]\,(\cdots)\,e^{-S_{0\,\mathrm{kin}}-S_1}. \tag{8.177b}$$

Since the argument of the exponential in $\cdots$ is linear in the integration variable ϕ and since the argument $-S_{0\,\mathrm{kin}}-S_1$ entering the Boltzmann weight over which averaging $\langle\cdots\rangle_{S_{0\,\mathrm{kin}}+S_1}$ is to be performed with is quadratic in ϕ, we are dealing with a (bosonic) Gaussian path integral for given n and given m,

$$\begin{aligned} &\left\langle e^{-\int_0^\beta \mathrm{d}\tau(J_{n,m}+J)(\tau)\,\phi(\tau)}\right\rangle_{S_{0\,\mathrm{kin}}+S_1}\\ &\quad=\frac{\mathcal{N}_{\mathrm{bath}}}{\sqrt{\mathrm{Det}'\mathcal{D}}}\,e^{+\frac{1}{2}\int_0^\beta \mathrm{d}\tau\int_0^\beta \mathrm{d}\tau'\,(J_{n,m}+J)(\tau)\,\mathcal{D}(\tau-\tau')\,(J_{n,m}+J)(\tau')}\,\delta_{n,2m}. \end{aligned} \tag{8.178}$$

As we have explained above Eq. (8.173b), $D(\tau-\tau')$ contains the singular term $D(\tau-\tau')-D_{\mathrm{reg}}(\tau-\tau')$. The reason for which the neutrality condition

$$n = 2\,m \tag{8.179}$$

must hold is that

$$\lim_{\varpi\to 0}\frac{1}{\mathcal{D}_\varpi} = 0, \tag{8.180a}$$

i.e.,

$$\lim_{\varpi\to 0}\mathcal{D}_\varpi \tag{8.180b}$$

diverges. Thus, integration over $\phi_{\varpi=0}$ is only well defined if

$$0 = \int_0^\beta \mathrm{d}\tau\,(J_{n,m}+J)(\tau) = -\mathrm{i}(n-2\,m)+\int_0^\beta \mathrm{d}\tau\,J(\tau), \qquad m,n = 0,1,2,\cdots, \tag{8.181}$$

i.e., $n = 2m$ and $\int\limits_0^\beta \mathrm{d}\tau J(\tau) = 0$. Hence, the vanishing eigenvalue (zero mode) of $\mathcal{D}^{-1}$ must be omitted in $(\mathrm{Det}\,\mathcal{D})^{-1/2}$ as is implied by the use of $(\mathrm{Det}'\,\mathcal{D})^{-1/2}$. As a corollary, we can also replace $\mathcal{D}(\tau-\tau')$ by $\mathcal{D}_{\mathrm{reg}}(\tau-\tau')$ on the right-hand side of Eq. (8.178) as the τ-independent divergent contribution $\mathcal{D}(\tau-\tau') - \mathcal{D}_{\mathrm{reg}}(\tau-\tau')$ drops out from the argument of the exponential in the integrand as a consequence of the neutrality condition.

In summary, when the external source $J(\tau)$ is set to zero,

$$Z_\beta = \frac{\mathcal{N}_{\mathrm{bath}}}{\sqrt{\mathrm{Det}'\mathcal{D}}} \sum_{n=0}^{\infty} \left(\frac{1}{n!}\right)^2 \left(\frac{y}{2}\right)^{2n} \int\limits_0^\beta \mathrm{d}\tau_1 \cdots \int\limits_0^\beta \mathrm{d}\tau_{2n}\, e^{-\frac{1}{2}\sum\limits_{j,k=1}^{2n} e_j\,\mathcal{D}_{jk}\,e_k},$$

$$\mathcal{D}_{jk} \equiv \mathcal{D}(\tau_j-\tau_k) \sim \begin{cases} -\frac{1}{2\eta}\, r_{jk}, & \text{if } r_{jk} := \frac{\eta}{m}\,|\tau_j-\tau_k| \ll 1, \\ -\frac{1}{\pi\eta}\,\ln r_{jk}, & \text{if } r_{jk} := \frac{\eta}{m}\,|\tau_j-\tau_k| \gg 1, \end{cases} \tag{8.182a}$$

defines the grand-canonical partition function at the temperature $\beta^{-1} \ll y$ and fugacity $y/2$ of a neutral plasma made of classical point-like particles with coordinates $0 \leq \tau \leq \beta$ interacting through the two-body potential

$$\mathcal{D}(\tau_j-\tau_k) := \frac{1}{\beta} \sum_{\varpi_l} \mathcal{D}_{\varpi_l}\, e^{-\mathrm{i}\varpi_l|\tau_j-\tau_k|}, \qquad \mathcal{D}_{\varpi_l} := \frac{1}{m\,\varpi_l^2 + \eta\,|\varpi_l|}. \tag{8.182b}$$

In the presence of the source term $J(\tau)$, the term

$$\begin{aligned} S_{\mathrm{ext}} &= \int\limits_0^\beta \mathrm{d}\tau \int\limits_0^\beta \mathrm{d}\tau' \left(-\mathrm{i}\sum_{j=1}^{2n} e_j\,\delta(\tau-\tau_j)\right) \mathcal{D}(\tau-\tau')\,J(\tau') \\ &\quad + \frac{1}{2}\int\limits_0^\beta \mathrm{d}\tau \int\limits_0^\beta \mathrm{d}\tau'\,J(\tau)\,\mathcal{D}(\tau-\tau')\,J(\tau') \\ &= \sum_{\varpi_l}\left(-\mathrm{i}\rho_{2n\,(-\varpi_l)} + \frac{1}{2}J_{(-\varpi_l)}\right)\mathcal{D}_{\varpi_l}\,J_{(+\varpi_l)} \end{aligned} \tag{8.183}$$

must be added to the action $\frac{1}{2}\sum_{j,k=1}^{2n} e_j\,\mathcal{D}_{jk}\,e_k$. The plasma density $\rho(\tau)$ is defined to be

$$\rho_{2n}(\tau) := \sum_{j=1}^{2n} e_j\,\delta(\tau-\tau_j), \quad \rho_{2n\,\varpi_l} := \frac{1}{\sqrt{\beta}}\int\limits_0^\beta \mathrm{d}\tau\,\rho_{2n}(\tau)\,e^{+\mathrm{i}\varpi_l\tau} = \frac{1}{\sqrt{\beta}}\sum_{j=1}^{2n} e_j\,e^{+\mathrm{i}\varpi_l\tau_j}, \tag{8.184}$$

in the sector with $2n$ particles.

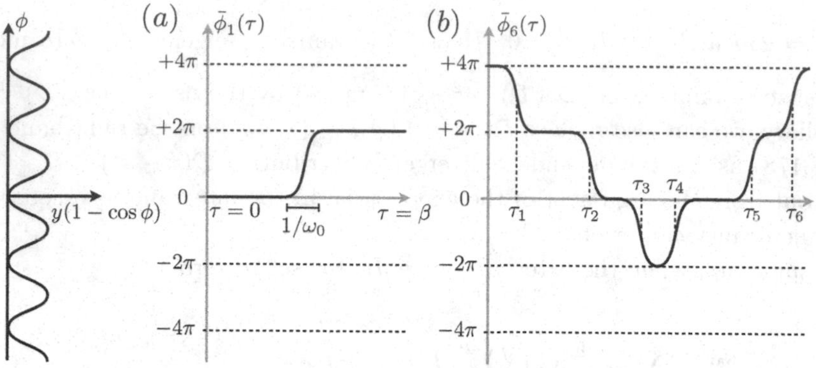

Fig. 8.11 Instanton configurations. (a) A single-instanton configuration $\overline{\phi}_1$ with $\omega_0 \times \beta := \sqrt{y/m}\times\beta$ finite and located at $\beta/2$. (b) A 6-instantons configuration $\overline{\phi}_6$ in the limit $(1/\omega_0)/\beta \to 0$. Anti-instantons are located at τ_1, τ_2, and τ_3, instantons are located at τ_4, τ_5, and τ_6.

8.7.2 *Regime* $m\,y \gg 1$ *and* $m\,y \gg \eta$

When the geometrical mean $\sqrt{m\,y}$ of the characteristic scales in $S_{0\,\text{kin}}$ and $S_{0\,\text{int}}$ becomes arbitrarily large relative to 1 or η, the partition function Z_β reduces to a summation over all the periodic trajectories $\overline{\phi}(\tau)$ that are local minima of the action $S_{0\,\text{kin}} + S_{0\,\text{int}}$, i.e., satisfy

$$0 = m\ddot{\overline{\phi}} - y\sin\overline{\phi}, \qquad \overline{\phi}(\tau) = \overline{\phi}(\tau+\beta). \tag{8.185}$$

Approximate solutions to Eq. (8.185) can be constructed from the instanton $\overline{\phi}_1$ and anti-instanton $\overline{\phi}_{-1}$ solutions

$$\overline{\phi}_1(\tau) = 4\arctan\Big(\exp(\omega_0\tau)\Big), \qquad \omega_0 := \sqrt{\frac{y}{m}}, \tag{8.186a}$$

and

$$\overline{\phi}_{-1}(\tau) = -\overline{\phi}_1(\tau) \tag{8.186b}$$

respectively, to

$$0 = m\ddot{\overline{\phi}} - y\,\sin\overline{\phi} \tag{8.186c}$$

by writing

$$\overline{\phi}(\tau) \approx \sum_{j=1}^{2n} e_j\,\overline{\phi}_1(\tau-\tau_j) \equiv \overline{\phi}_{2n}(\tau), \quad \sum_{j=1}^{2n} e_j = 0, \quad e_j = \pm 1, \quad j = 1,\cdots,2n. \tag{8.186d}$$

The characteristic time $1/\omega_0 = \sqrt{m/y}$ is the time needed for $\overline{\phi}_1(\tau)$ to change by $\pm 2\,\pi$. Such trajectories made of a sequence of kinks when $e_j = +1$ and anti-kinks when $e_j = -1$ become exact solutions of Eq. (8.185) in the limit when the width of the kink $1/\omega_0$ is much smaller than the separation between the kinks,

$$\frac{1/\omega_0}{|\tau_j - \tau_k|} \to 0, \qquad j,k = 1,\cdots,2n. \tag{8.187}$$

An example with 6 instantons is depicted in Fig. 8.11. We have seen in section 8.5.4 that the mean density of kinks is of the order

$$
\begin{aligned}
&K[\overline{\phi}_1] \times e^{-\left(S_{0\,\mathrm{kin}}[\overline{\phi}_1]+\beta\, y+S_{0\,\mathrm{int}}[\overline{\phi}_1]\right)},\\
&K[\overline{\phi}_1] = \sqrt{\frac{S_{0\,\mathrm{kin}}[\overline{\phi}_1]+\beta\, y+S_{0\,\mathrm{int}}[\overline{\phi}_1]}{2\,\pi}}\sqrt{\frac{\mathrm{Det}\,(-\partial_\tau^2+\omega_0^2)}{\mathrm{Det}'\left[-\partial_\tau^2+y\cos(\overline{\phi}_1)\right]}},\\
&S_{0\,\mathrm{kin}}[\overline{\phi}_1]+\beta\, y+S_{0\,\mathrm{int}}[\overline{\phi}_1] = +4\,(m\, y)^{1/2}+\mathcal{O}(e^{-2\,\omega_0\,\beta}),
\end{aligned}
\tag{8.188}
$$

which is indeed negligible for very large $\sqrt{m\, y}$ (β fixed).[12] Integrating fluctuating trajectories about each local minimum $\overline{\phi}_{2n}$ up to Gaussian order yields the partition function

$$
Z_\beta = \frac{\mathcal{N}_{\mathrm{bath}} \times e^{+y\beta}}{\sqrt{\mathrm{Det}\,(-\partial_\tau^2+\omega_0^2)}} \sum_{n=0}^{\infty} \left(\frac{1}{n!}\right)^2 z^{2n} \int_0^\beta \mathrm{d}\tau_1 \cdots \int_0^\beta \mathrm{d}\tau_{2n}\, e^{-S_1[\overline{\phi}_{2n}]}, \tag{8.189a}
$$

where the fugacity z is

$$
\begin{aligned}
z &= K[\overline{\phi}_1]\, e^{-\left(S_{0\,\mathrm{kin}}[\overline{\phi}_1]+\beta\, y+S_{0\,\mathrm{int}}[\overline{\phi}_1]\right)}\\
&= K[\overline{\phi}_1]\, e^{-4\,(m\, y)^{1/2}+\mathcal{O}(e^{-2\,\omega_0\,\beta})}.
\end{aligned}
\tag{8.189b}
$$

The combinatorial factor $1/(n!)^2$ follows from the property that the centers of instantons of a given charge are time ordered. Time ordering can then be removed at the price of the combinatorial factor $1/(n!)^2$ since the interaction $S_1[\overline{\phi}_{2n}]$ is a two-body interaction that is translation invariant as we shall verify explicitly shortly. The new feature brought by the dissipation compared to the dissipationless periodic Hamiltonian of section 8.5.5 is that instanton now interact through $S_1[\overline{\phi}_{2n}]$. Hence, it is not possible to integrate freely over the instanton centers $\tau_1, \cdots, \tau_{2n}$.

We now turn to the evaluation of the interaction between instantons induced by dissipation, i.e., by the coupling to the bath. Let h_{ϖ_l} be the Fourier transform with respect to Matsubara time of $(\partial_\tau\overline{\phi}_1)(\tau)$,

$$
(\partial_\tau\overline{\phi}_1)\,(\tau) = \frac{1}{\sqrt{\beta}}\sum_{\varpi_l} h_{\varpi_l}\, e^{-\mathrm{i}\varpi_l\tau}, \qquad h_{\varpi_l} = \frac{1}{\sqrt{\beta}}\int_0^\beta \mathrm{d}\tau\, (\partial_\tau\overline{\phi}_1)\,(\tau)\, e^{+\mathrm{i}\varpi_l\tau}, \tag{8.190}
$$

---

[12] The contribution $\beta\, y$ comes about by rewriting the interaction as $-y\cos\phi = y(1-\cos\phi)-y$ and working with the potential $y(1-\cos\phi)$ as opposed to the potential $-y\cos\phi$ as is indicated in Fig. 8.11. An overall multiplicative factor $\exp\left((-1)^2\beta\, y\right)$ must then be accounted for in the partition function. The $'$ in Det$'$ means removal of the zero mode. Finally, we make use of the constant of motion $m\,(\dot{\overline{\phi}}_1)^2 - 2\times(1-\cos\overline{\phi})\, y = 0$ to express, with the explicit form of $\overline{\phi}_1$, $S_{0\,\mathrm{int}}[\overline{\phi}_1]$ in terms of $S_{0\,\mathrm{kin}}[\overline{\phi}_1]$. There follows $S_{0\,\mathrm{kin}}[\overline{\phi}_1]+\beta\, y+S_{0\,\mathrm{int}}[\overline{\phi}_1] = 4\,m\,\omega_0^2\int_0^\beta\frac{\mathrm{d}\tau}{\cosh^2(\omega_0\,\tau)}$. Integration over τ gives $S_{0\,\mathrm{kin}}[\overline{\phi}_1]+\beta\, y+S_{0\,\mathrm{int}}[\overline{\phi}_1] = 4\,m\,\omega_0^2\,\frac{\tanh(\omega_0\,\beta)}{\omega_0}$. In the limit, $\omega_0\,\beta \gg 1$, the right-hand side becomes $4\,m\,\omega_0\,\left(1+\mathcal{O}(e^{-2\,\omega_0\,\beta})\right)$.

where $\varpi_l = \frac{2\pi}{\beta} l$ and $l \in \mathbb{Z}$. As we shall see shortly, the most important property of h_{ϖ_l} is

$$h_{\varpi_l=0} = \frac{2\pi}{\sqrt{\beta}}. \tag{8.191}$$

Making use of the neutrality condition, we can express the Fourier transform

$$\overline{\phi}_{2n\,\varpi_l} = \frac{1}{\sqrt{\beta}} \int\limits_0^\beta \mathrm{d}\tau\, \overline{\phi}_{2n}(\tau)\, e^{+\mathrm{i}\varpi_l \tau} \tag{8.192}$$

of $\overline{\phi}_{2n}(\tau)$ in terms of h_{ϖ_l} by taking the τ derivative of Eq. (8.186d) and Fourier transforming it,

$$-\mathrm{i}\varpi_l\, \overline{\phi}_{2n\,\varpi_l} = \sum_{j=1}^{2n} e_j\, h_{\varpi_l}\, e^{+\mathrm{i}\varpi_l \tau_j}, \qquad \sum_{j=1}^{2n} e_j = 0 \Longrightarrow \overline{\phi}_{2n\,\varpi_l} = +\mathrm{i}\frac{h_{\varpi_l}}{\varpi_l} \sum_{j=1}^{2n} e_j\, e^{+\mathrm{i}\varpi_l \tau_j}, \tag{8.193}$$

for all $l \in \mathbb{Z}$. Here, the neutrality condition can be used together with Eq. (8.191) to avoid an inconsistency when $l = 0$. Evaluation of $S_1[\overline{\phi}_{2n}]$ is now straightforward,

$$\begin{aligned}
S_1[\overline{\phi}_{2n}] &= \sum_{\varpi_l} \frac{\eta}{2} |\varpi_l|\, \overline{\phi}_{2n\,(+\varpi_l)}\, \overline{\phi}_{2n\,(-\varpi_l)} \\
&= \sum_{\varpi_l} \frac{\eta}{2} |\varpi_l| \frac{h_{(+\varpi_l)}\, h_{(-\varpi_l)}}{\varpi_l^2} \sum_{j=1}^{2n}\sum_{k=1}^{2n} e_j\, e_k\, e^{+\mathrm{i}\varpi_l(\tau_j - \tau_k)} \\
&= \frac{1}{2} \sum_{j=1}^{2n}\sum_{k=1}^{2n} \left(\underline{\sum_{\varpi_l} \frac{\eta}{|\varpi_l|}\, h_{(+\varpi_l)}\, h_{(-\varpi_l)}\, e^{+\mathrm{i}\varpi_l(\tau_j - \tau_k)}} \right) e_j\, e_k.
\end{aligned} \tag{8.194a}$$

The underlined term defines the two-body interaction potential

$$\begin{aligned}
\Delta(\tau - \tau') &:= \frac{\eta}{\beta} \sum_{\varpi_l} \frac{\beta}{|\varpi_l|}\, h_{(+\varpi_l)}\, h_{(-\varpi_l)}\, e^{+\mathrm{i}\varpi_l(\tau - \tau')} \\
&= \frac{\eta}{\beta} \left(\sum_{|\varpi_l| < \omega_0} + \sum_{|\varpi_l| \geq \omega_0} \right) \frac{\beta}{|\varpi_l|}\, h_{(+\varpi_l)}\, h_{(-\varpi_l)}\, e^{+\mathrm{i}\varpi_l(\tau - \tau')} \\
&\sim \begin{cases} -\text{constant} \times \eta\, \rho^2, & \text{if } \rho := \omega_0\, |\tau - \tau'| \ll 1, \\[2ex] -4\pi\eta \ln \rho, & \text{if } \rho := \omega_0\, |\tau - \tau'| \gg 1, \end{cases}
\end{aligned} \tag{8.194b}$$

(constant is a numerical constant of order unity) in terms of which

$$S_1[\overline{\phi}_{2n}] = \frac{1}{2} \sum_{j=1}^{2n}\sum_{k=1}^{2n} e_j\, \Delta(\tau_j - \tau_k)\, e_k. \tag{8.194c}$$

The limiting form of the two-body interaction (8.194b)

$$\lim_{\tau\to 0}\Delta(\tau)\sim -\text{const}\times\eta\,\rho^2 \tag{8.195}$$

follows from expanding $\exp\left(+\mathrm{i}\varpi_l\,\tau\right)$ in powers of $\varpi_l\,\tau$ up to second order owing to the limit $\omega_0\,|\tau|\ll 1$. The integration $\int^{\omega_0}\mathrm{d}\varpi$ is then independent (divergent) of τ to zero-th order, vanishes to first order in τ, and gives to second order in τ Eq. (8.195). As before, the condition of charge neutrality allows us to ignore the diverging constant. The limiting form of the two-body interaction (8.194b)

$$\lim_{\tau\to\infty}\Delta(\tau)\sim -4\,\pi\,\eta\,\ln\left(\omega_0\,|\tau|\right), \tag{8.196}$$

follows from observing that, in the limit $\omega_0\,|\tau|\gg 1$, the sum over ϖ_l is dominated by the contribution from ϖ_l near zero. One may then take advantage of the fact that $\int_{\omega_0}\mathrm{d}\varpi\,|\varpi|^{-1}$ is invariant under rescaling of ϖ after insertion of Eq. (8.191) into Eq. (8.194b). Equation (8.196) also follows from

$$-\int\limits_{x}^{\infty}\mathrm{d}t\,\frac{\cos t}{t}=\gamma+\ln x+\int\limits_{0}^{x}\mathrm{d}t\,\frac{\cos t-1}{t},\qquad \gamma\text{ denoting Euler's constant.} \tag{8.197}$$

(See Eq. 8.230.2 from Ref. [57]). Correspondingly, $S_1[\overline{\phi}_1]$ is scale invariant below a frequency cut-off. Charge neutrality can then be understood as following from the fact that it costs an infinite action to create a net charge in the "thermodynamic limit" $\beta\to\infty$.

In summary, when the external source $J(\tau)$ is set to zero,

$$\begin{aligned}
Z_\beta &= \frac{\mathcal{N}_{\text{bath}}\times e^{+y\beta}}{\sqrt{\operatorname{Det}\left(-\partial_\tau^2+\omega_0^2\right)}}\sum_{n=0}^{\infty}\left(\frac{1}{n!}\right)^2 z^{2n}\int\limits_0^\beta \mathrm{d}\tau_1\cdots\int\limits_0^\beta \mathrm{d}\tau_{2n}\,e^{-\frac{1}{2}\sum_{j,k=1}^{2n}e_j\,\Delta_{jk}\,e_k},\\
z &= K[\overline{\phi}_1]\,e^{-4\,(m\,y)^{1/2}+\mathcal{O}(e^{-2\,\omega_0\,\beta})},\\
\Delta_{jk} &\equiv \Delta(\tau_j-\tau_k)\\
&\equiv \frac{1}{\beta}\sum_{\varpi_l}\Delta_{\varpi_l}\,e^{+\mathrm{i}\varpi_l\,(\tau_j-\tau_k)}\\
&\sim \begin{cases}
-\text{constant}\times\eta\,\rho_{jk}^2, & \text{if } \rho_{jk}:=\omega_0\,|\tau_j-\tau_k|\ll 1,\\[2ex]
-4\,\pi\,\eta\,\ln\rho_{jk}, & \text{if } \rho_{jk}:=\omega_0\,|\tau_j-\tau_k|\gg 1,
\end{cases}
\end{aligned} \tag{8.198a}$$

with

$$\Delta_{\varpi_l}:=\frac{\eta\,\beta}{|\varpi_l|}\,h_{(+\varpi_l)}\,h_{(-\varpi_l)}, \tag{8.198b}$$

defines the grand-canonical partition function at the temperature $\beta^{-1}\ll y$ and fugacity z of a neutral plasma made of classical point-like particles with coordinates

$0 \leq \tau \leq \beta$ interacting through the two-body potential $\Delta(\tau_j - \tau_k)$. The instanton expansion converges best with small fugacity z, i.e., with large $\sqrt{my}$. In the presence of the source term $J(\tau)$, the term

$$S_{\text{ext}}[\overline{\phi}_{2n}] = -\mathrm{i} \sum_{\varpi_l} J_{(+\varpi_l)} \frac{h_{(-\varpi_l)}}{\varpi_l} \sum_{j=1}^{2n} e_j \, e^{-\mathrm{i}\varpi_l \tau_j} \tag{8.199}$$

must be added to the action $S_1[\overline{\phi}_{2n}] = \frac{1}{2} \sum_{j,k=1}^{2n} e_j \, \Delta_{jk} \, e_k$. Here, the source couples linearly as opposed to Eq. (8.178) where the source couples quadratically.

8.7.3 *Duality*

Duality is the observation [87] that Eqs. (8.182a) and (8.198a) are related by the substitutions

$$\frac{y}{2} \leftrightarrow z, \qquad \frac{\eta}{m} \leftrightarrow \omega_0, \qquad \frac{1}{\pi \, \eta} \leftrightarrow 4\pi \, \eta, \tag{8.200}$$

in the absence of sources and if one is allowed to neglect the difference in the core regions of the interaction potentials $\mathcal{D}_{jk}$ and Δ_{jk}. Duality implies that there is a one to one correspondence between the asymptotic behavior at very low Matsubara frequencies or, equivalently, at very large separations of Matsubara times, of correlation functions in the regimes (8.171a) and (8.171b), respectively. If one knows the asymptotic behavior of one correlation function, say in regime (8.171a), one can use the duality relations (8.200) to reconstruct the corresponding correlation function in the regime (8.171b). To do so, one must carefully account for the source term that allows to derive the correlation function. For example, if one is after the two-point correlation function

$$\langle \phi_{(+\varpi_l)} \, \phi_{(-\varpi_l)} \rangle_{Z_\beta} = \frac{1}{Z_\beta} \frac{\partial^2 Z_\beta}{\partial J_{(-\varpi_l)} \, \partial J_{(+\varpi_l)}} \Bigg|_{J=0}, \tag{8.201a}$$

one finds that

$$\langle \phi_{(+\varpi_l)} \phi_{(-\varpi_l)} \rangle_{Z_\beta} = \begin{cases} \mathcal{D}_{\varpi_l} \left[-1 - \mathcal{D}_{\varpi_l} \langle \rho_{(+\varpi_l)} \, \rho_{(-\varpi_l)} \rangle_{Z_\beta} \right], & \sqrt{my} \ll 1, \ \sqrt{my} \ll \eta, \\ \frac{1}{\eta} \frac{1}{|\varpi_l|} \Delta_{\varpi_l} \langle \rho_{(+\varpi_l)} \, \rho_{(-\varpi_l)} \rangle_{Z_\beta}, & \sqrt{my} \gg 1, \ \sqrt{my} \gg \eta, \end{cases} \tag{8.201b}$$

where the plasma density $\rho(\tau)$ is defined to be

$$\rho(\tau) := \sum_{j=1}^{2n} e_j \, \delta(\tau - \tau_j), \qquad \rho_{\varpi_l} := \frac{1}{\sqrt{\beta}} \int_0^\beta \mathrm{d}\tau \, \rho(\tau) \, e^{+\mathrm{i}\varpi_l \tau} = \frac{1}{\sqrt{\beta}} \sum_{j=1}^{2n} e_j \, e^{+\mathrm{i}\varpi_l \tau_j}, \tag{8.201c}$$

in the sector with $2n$ particles. Here, we also made use of Eqs. (8.183) and (8.199), respectively, as well as the fact that $\mathcal{D}_{+\varpi_l} = \mathcal{D}_{-\varpi_l}$. Establishing duality between two regimes is useful insofar as computations can be carried out in either of one of the regimes. In the next section, a renormalization-group calculation for the flow of the potential height y in the regime (8.171a) is performed. As a by product of duality, the corresponding flow can be obtained in the regime (8.171b).

8.8 Renormalization-group methods

In this section we are, following Ref. [88], going to focus our efforts on a renormalization-group (RG) approach whenever the Josephson coupling is the smallest energy scale in the problem (aside from the temperature β^{-1}) and frequencies are sufficiently small for the motion of the quantum particle to be diffusive (i.e., dissipation S_1 dominates over the kinetic energy $S_{0\,\mathrm{kin}}$). We will then apply duality to investigate the regime when the Josephson coupling is the largest energy scale in the problem and frequencies are sufficiently small for the motion of the quantum particle to be diffusive. The RG approach will allow us to decide whether the diffusive limit is stable or unstable upon elimination (integration) of high frequency modes. We shall argue that the answer to this question depends on how large the Josephson coupling is.

8.8.1 *Diffusive regime when $m\,y \ll 1$ and $m\,y \ll \eta$*

In section 8.7.1 we performed a formal expansion of the partition function in powers of y as y is the smallest energy scale aside from the temperature. Although this expansion is essential in establishing duality with the regime $m\,y \gg 1$ and $m\,y \gg \eta$, it is of limited advantage to the naive evaluation of correlation functions in the static limit. Indeed, perturbation theory in powers of y breaks down in the diffusive regime due to the fact that the free propagator is then proportional to $1/|\varpi_l|$ in Matsubara frequency space to a very high accuracy as is indicated by Eq. (8.170d). To tame the divergences plaguing perturbation theory in powers of y, one relies on a scaling approach. Having identified the upper frequency cut-off

$$\Lambda \sim \frac{\alpha}{C\,R_h} \equiv \frac{1/R_s}{C} \sim \frac{\eta}{m}, \tag{8.202}$$

which defines the diffusive regime from Eq. (8.170c), the idea behind the scaling approach is to distinguish between "slow" and "fast" Fourier components ϕ_{ϖ_l} in the Fourier expansion of the Josephson phase difference $\phi(\tau)$ and to integrate over the "fast" components in the partition function (8.172a). One thus writes

$$\phi_{\varpi_l} = \begin{cases} \phi^{(1)}_{\varpi_l}, & \text{if } |\varpi_l| < \Lambda - \mathrm{d}\Lambda, \\ \\ \phi^{(2)}_{\varpi_l}, & \text{if } \Lambda - \mathrm{d}\Lambda \le |\varpi_l| \le \Lambda, \end{cases} \tag{8.203a}$$

and

$$\begin{aligned}\phi^{(1)}(\tau) &:= \frac{1}{\sqrt{\beta}} \sum_{\varpi_l}^{|\varpi_l|<\Lambda-\mathrm{d}\Lambda} \phi^{(1)}_{\varpi_l}\, e^{-\mathrm{i}\varpi_l\tau},\\ \phi^{(2)}(\tau) &:= \frac{1}{\sqrt{\beta}} \sum_{\varpi_l}^{\Lambda-\mathrm{d}\Lambda\leq|\varpi_l|\leq\Lambda} \phi^{(2)}_{\varpi_l}\, e^{-\mathrm{i}\varpi_l\tau},\end{aligned} \tag{8.203b}$$

for the decomposition into slow and fast Fourier modes which, in turn, is inserted into the partition function (8.172a):

$$\begin{aligned}Z_\beta &= \mathcal{N}_{\text{bath}} \int \mathcal{D}[\phi^{(1)},\phi^{(2)}]\, e^{-S''_\beta+\mathcal{O}(S_{0\,\text{kin}})},\\ S''_\beta &= S_1 + S_{0\,\text{int}} + S_{\text{ext}},\\ S_1 &= \sum_{\varpi_l}^{|\varpi_l|<\Lambda-\mathrm{d}\Lambda} \frac{\eta}{2}\,|\varpi_l|\,\phi^{(1)}_{(+\varpi_l)}\,\phi^{(1)}_{(-\varpi_l)} + \sum_{\varpi_l}^{\Lambda-\mathrm{d}\Lambda\leq|\varpi_l|\leq\Lambda} \frac{\eta}{2}\,|\varpi_l|\,\phi^{(2)}_{(+\varpi_l)}\,\phi^{(2)}_{(-\varpi_l)},\\ S_{0\,\text{int}} &= \int_0^\beta \mathrm{d}\tau\,(-y)\,\cos\left(\phi^{(1)}(\tau)+\phi^{(2)}(\tau)\right),\\ S_{\text{ext}} &= \sum_{\varpi_l}^{|\varpi_l|<\Lambda-\mathrm{d}\Lambda} J^{(1)}_{(+\varpi_l)}\,\phi^{(1)}_{(-\varpi_l)} + \sum_{\varpi_l}^{\Lambda-\mathrm{d}\Lambda\leq|\varpi_l|\leq\Lambda} J^{(2)}_{(+\varpi_l)}\,\phi^{(2)}_{(-\varpi_l)}.\end{aligned} \tag{8.203c}$$

Here, we are neglecting the contribution

$$S_{0\,\text{kin}} = \sum_{\varpi_l}^{|\varpi_l|<\Lambda-\mathrm{d}\Lambda} \frac{1}{2}\,m\,\varpi_l^2\,\phi^{(1)}_{(+\varpi_l)}\,\phi^{(1)}_{(-\varpi_l)} + \sum_{\varpi_l}^{\Lambda-\mathrm{d}\Lambda\leq|\varpi_l|\leq\Lambda} \frac{1}{2}\,m\,\varpi_l^2\,\phi^{(2)}_{(+\varpi_l)}\,\phi^{(2)}_{(-\varpi_l)} \tag{8.203d}$$

to the exact action in Eq. (8.172a).

Integration over the fast modes $\phi^{(2)}_{\varpi_l}$ is free from divergences in view of the lower frequency cut-off $\Lambda - \mathrm{d}\Lambda$ and yields an effective or renormalized action $S^{(1)\,\prime\prime}_\beta$ given by

$$\begin{aligned}Z_\beta &\approx \mathcal{N}_{\text{bath}} \times \mathcal{N} \int \mathcal{D}[\phi^{(1)}]\, e^{-S^{(1)\,\prime\prime}_\beta},\\ S^{(1)\,\prime\prime}_\beta &:= S^{(1)}_1 + S^{(1)}_{0\,\text{int}},\\ S^{(1)}_1 &:= \sum_{\varpi_l}^{|\varpi_l|<\Lambda-\mathrm{d}\Lambda} \frac{\eta}{2}\,|\varpi_l|\,\phi^{(1)}_{(+\varpi_l)}\,\phi^{(1)}_{(-\varpi_l)},\\ S^{(1)}_{0\,\text{int}} &:= -\ln\left\langle \exp\left((-1)^2 \int_0^\beta \mathrm{d}\tau\, y\,\cos\left(\phi^{(1)}(\tau)+\phi^{(2)}(\tau)\right)\right)\right\rangle_2,\end{aligned} \tag{8.204a}$$

in the absence of a source term. The notation $\langle(\cdots)\rangle_2$ refers to

$$\langle(\cdots)\rangle_2 := \frac{\int\mathcal{D}[\phi^{(2)}]\,\exp\left(-\sum_{\varpi_l}^{\Lambda-\mathrm{d}\Lambda\leq|\varpi_l|\leq\Lambda}\frac{\eta}{2}|\varpi_l|\,\phi^{(2)}_{(+\varpi_l)}\,\phi^{(2)}_{(-\varpi_l)}\right)(\cdots)}{\int\mathcal{D}[\phi^{(2)}]\,\exp\left(-\sum_{\varpi_l}^{\Lambda-\mathrm{d}\Lambda\leq|\varpi_l|\leq\Lambda}\frac{\eta}{2}|\varpi_l|\,\phi^{(2)}_{(+\varpi_l)}\,\phi^{(2)}_{(-\varpi_l)}\right)}. \qquad (8.204b)$$

In practice, the computation of $S^{(1)}_{0\,\mathrm{int}}$ cannot be performed exactly but relies on a perturbative expansion in powers of y:

$$\begin{aligned} S^{(1)}_{0\,\mathrm{int}} &= -\ln\left\langle 1 + y\int_0^\beta \mathrm{d}\tau\,\cos\left(\phi^{(1)}(\tau)+\phi^{(2)}(\tau)\right)+\mathcal{O}(y^2)\right\rangle_2 \\ &= -y\left\langle\int_0^\beta \mathrm{d}\tau\,\cos\left(\phi^{(1)}(\tau)+\phi^{(2)}(\tau)\right)\right\rangle_2 + \mathcal{O}(y^2). \end{aligned} \qquad (8.205)$$

To calculate the expectation value of the cosine on the right-hand side of Eq. (8.205), use the identity

$$\begin{aligned} \langle\cos(x'+x)\rangle_x &:= \frac{\int\mathrm{d}x\,e^{-a\,x^2/2}\,\cos(x'+x)}{\int\mathrm{d}x\,e^{-a\,x^2/2}} \\ &= \frac{\int\mathrm{d}x\,e^{-a\,x^2/2}\,\frac{1}{2}\left[e^{+\mathrm{i}(x'+x)}+e^{-\mathrm{i}(x'+x)}\right]}{\int\mathrm{d}x\,e^{-a\,x^2/2}} \\ &= \frac{1}{2}e^{+\mathrm{i}x'-\frac{1}{2}\langle x^2\rangle_x}+\frac{1}{2}e^{-\mathrm{i}x'-\frac{1}{2}\langle x^2\rangle_x} \\ &= e^{-\frac{1}{2}\langle x^2\rangle_x}\cos x', \qquad \langle x^2\rangle_x = 1/a, \qquad a>0, \end{aligned} \qquad (8.206a)$$

with the identifications

$$\begin{aligned} x' \to \phi^{(1)}(\tau) &:= \frac{1}{\sqrt{\beta}}\sum_{\varpi_l}^{|\varpi_l|<\Lambda-\mathrm{d}\Lambda}\phi^{(1)}_{\varpi_l}\,e^{-\mathrm{i}\varpi_l\tau}, \\ x \to \phi^{(2)}(\tau) &:= \frac{1}{\sqrt{\beta}}\sum_{\varpi_l}^{\Lambda-\mathrm{d}\Lambda\leq|\varpi_l|\leq\Lambda}\phi^{(2)}_{\varpi_l}\,e^{-\mathrm{i}\varpi_l\tau}, \\ x^2 \to \left[\phi^{(2)}(\tau)\right]^2 &= \frac{1}{\beta}\sum_{\varpi_l}^{\Lambda-\mathrm{d}\Lambda\leq|\varpi_l|\leq\Lambda}\phi^{(2)}_{(+\varpi_l)}\,\phi^{(2)}_{(-\varpi_l)} \\ &+\frac{1}{\beta}\sum_{\varpi_l\neq-\Omega_l}^{\Lambda-\mathrm{d}\Lambda\leq|\varpi_l|,|\Omega_l|\leq\Lambda}\phi^{(2)}_{(+\varpi_l)}\,\phi^{(2)}_{(+\Omega_l)}e^{-\mathrm{i}\varpi_l\tau-\mathrm{i}\Omega_l\tau}, \\ \langle x^2\rangle_x \to \left\langle\left[\phi^{(2)}(\tau)\right]^2\right\rangle_2 &= \frac{1}{\beta}\sum_{\varpi_l}^{\Lambda-\mathrm{d}\Lambda\leq|\varpi_l|\leq\Lambda}(\eta|\varpi_l|)^{-1} \\ &\approx 2\times\frac{1}{\beta}\frac{1}{\eta\Lambda}\frac{\mathrm{d}\Lambda}{2\pi/\beta} \qquad \text{In the limit } \beta\to\infty \\ &= \frac{1}{\pi\eta}\frac{\mathrm{d}\Lambda}{\Lambda}. \end{aligned} \qquad (8.206b)$$

Thus,

$$S^{(1)}_{0\,\mathrm{int}} = -y\left(1-\frac{1}{2\pi\eta}\frac{\mathrm{d}\Lambda}{\Lambda}+\mathcal{O}\left[\left(\frac{\mathrm{d}\Lambda}{\Lambda}\right)^2\right]\right)\int_0^\beta \mathrm{d}\tau\,\cos\left(\phi^{(1)}(\tau)\right)+\mathcal{O}(y^2)$$

$$\equiv \int_0^\beta \mathrm{d}\tau\,(-y^{(1)})\,\cos\left(\phi^{(1)}(\tau)\right)+\mathcal{O}(y^2), \tag{8.207a}$$

where

$$y^{(1)} := y\left(1-\frac{1}{2\pi\eta}\frac{\mathrm{d}\Lambda}{\Lambda}+\mathcal{O}\left[\left(\frac{\mathrm{d}\Lambda}{\Lambda}\right)^2\right]\right). \tag{8.207b}$$

The action thus transforms covariantly upon integration over the fast degrees of freedom, i.e., the changes induced to the action by integration over the fast degrees of freedom can be absorbed by assigning a scale dependence to the coupling constants η and y according to the transformation laws

$$\eta(\Lambda-\mathrm{d}\Lambda) := \eta(\Lambda)+\mathcal{O}(y^2), \tag{8.208a}$$

and

$$y(\Lambda-\mathrm{d}\Lambda) := y(\Lambda)\left\{1-\frac{1}{2\pi\eta}\frac{\mathrm{d}\Lambda}{\Lambda}+\mathcal{O}\left[\left(\frac{\mathrm{d}\Lambda}{\Lambda}\right)^2\right]\right\}+\mathcal{O}(y^2). \tag{8.208b}$$

In units in which $\hbar=1$, the dissipation strength η is dimensionless whereas the fugacity y has dimensions of inverse time. It is desirable to distinguish in the transformation law (8.208b) the components induced by the dimension carried by y from an "intrinsic" component. To this end, define the dimensionless dissipation strength

$$\widetilde{y}(1) := \frac{y(\Lambda)}{\Lambda} \tag{8.209}$$

in terms of which the RG equations (8.208) become

$$\eta(\Lambda-\mathrm{d}\Lambda) = \eta(\Lambda)+\mathcal{O}(\widetilde{y}^2), \tag{8.210a}$$

and

$$\begin{aligned}
\widetilde{y}(1-\mathrm{d}\ln\Lambda) &= \frac{y(\Lambda-\mathrm{d}\Lambda)}{\Lambda-\mathrm{d}\Lambda}\\
&= \frac{y(\Lambda)}{\Lambda}\left\{1-\frac{1}{2\pi\eta}\frac{\mathrm{d}\Lambda}{\Lambda}+\mathcal{O}\left[\left(\frac{\mathrm{d}\Lambda}{\Lambda}\right)^2\right]\right\}\left\{1+\frac{\mathrm{d}\Lambda}{\Lambda}+\mathcal{O}\left[\left(\frac{\mathrm{d}\Lambda}{\Lambda}\right)^2\right]\right\}\\
&\quad+\mathcal{O}(\widetilde{y}^2)\\
&= \widetilde{y}(1)\left\{1+\left(1-\frac{1}{2\pi\eta}\right)\frac{\mathrm{d}\Lambda}{\Lambda}+\mathcal{O}\left[\left(\frac{\mathrm{d}\Lambda}{\Lambda}\right)^2\right]\right\}+\mathcal{O}(\widetilde{y}^2).
\end{aligned} \tag{8.210b}$$

In the limit $\mathrm{d}\Lambda \to 0$, we obtain the pair of differential equations

$$\begin{aligned}\frac{\mathrm{d}\eta}{\mathrm{d}\ln(\Lambda^{-1})} &= 0 + \mathcal{O}(\widetilde{y}^2),\\ \frac{\mathrm{d}\widetilde{y}}{\mathrm{d}\ln(\Lambda^{-1})} &= \left(1 - \frac{1/(2\pi)}{\eta}\right)\widetilde{y} + \mathcal{O}(\widetilde{y}^2).\end{aligned} \tag{8.211}$$

We are ready to answer the question: How does $\widetilde{y}$ change as the frequency cut-off Λ is *decreased* (or, equivalently, as Λ^{-1} is *increased*)? We must distinguish three cases:

(1) For sufficiently small dissipation

$$\eta < \frac{1}{2\pi}, \tag{8.212}$$

$\widetilde{y}$ decreases with decreasing Λ (or, equivalently, increasing Λ^{-1}). The initial assumption that y is the smallest energy scale in the problem aside from the temperature is consistent and we can trust the RG approach. It is said that the Josephson interaction is irrelevant.

(2) For the critical value

$$\eta = \frac{1}{2\pi}, \tag{8.213}$$

$\widetilde{y}$ does not change with decreasing Λ (or, equivalently, increasing Λ^{-1}). It is said that the Josephson interaction is marginal.

(3) For sufficiently large dissipation

$$\eta > \frac{1}{2\pi}, \tag{8.214}$$

$\widetilde{y}$ increases with decreasing Λ (or, equivalently, increasing Λ^{-1}). The initial assumption that y is the smallest energy scale in the problem aside from the temperature is not consistent and we cannot trust the RG approach. It is said that the Josephson interaction is relevant.

To access the regime in which the RG perturbative approach breaks down we take advantage of duality.

8.8.2 *Diffusive regime when $m\,y \gg 1$ and $m\,y \gg \eta$*

The same RG approach can be used on Eq. (8.198a) when the fugacity $z \propto \exp\left(-4\,(m\,y)^{1/2}\right)$ is small. The counterparts to Eq. (8.211) can be obtained with the help of the duality relation (8.200). They are given by

$$\begin{aligned}\frac{\mathrm{d}\eta^{-1}}{\mathrm{d}\ln(\Lambda^{-1})} &= 0 + \mathcal{O}(\widetilde{z}^2),\\ \frac{\mathrm{d}\widetilde{z}}{\mathrm{d}\ln(\Lambda^{-1})} &= \left(1 - \frac{\eta}{1/(2\pi)}\right)\widetilde{z} + \mathcal{O}(\widetilde{z}^2).\end{aligned} \tag{8.215}$$

The answer to the question how does $\widetilde{z}$ change as the frequency cut-off Λ is *decreased* (or, equivalently, as Λ^{-1} is *increased*) is:

(1) For sufficiently small dissipation

$$\eta < \frac{1}{2\pi}, \tag{8.216}$$

$\widetilde{z}$ increases with decreasing Λ (or, equivalently, increasing Λ^{-1}). The initial assumption that $z \propto \exp\left(-4\,(m\,y)^{1/2}\right)$ is the smallest energy scale in the problem aside from the temperature is not consistent and we cannot trust the RG approach. It is said that the instanton fugacity is relevant.

(2) For the critical value

$$\eta = \frac{1}{2\pi}, \tag{8.217}$$

$\widetilde{z}$ does not change with decreasing Λ (or, equivalently, increasing Λ^{-1}). It is said that the instanton fugacity is marginal.

(3) For sufficiently large dissipation

$$\eta > \frac{1}{2\pi}, \tag{8.218}$$

$\widetilde{z}$ decreases with decreasing Λ (or, equivalently, increasing Λ^{-1}). The initial assumption that $z \propto \exp\left(-4\,(m\,y)^{1/2}\right)$ is the smallest energy scale in the problem aside from the temperature is consistent and we can trust the RG approach. It is said that the instanton fugacity is irrelevant.

8.9 Conjectured phase diagram for a dissipative Josephson junction

We have modeled the dissipative Josephson junction pictured in Fig. 8.4 by the quantum theory defined in Eq. (8.168) or, equivalently, Eq. (8.172a). In this quantum model, the phase difference ϕ_- between two superconductors can be used to decide if the dissipative Josephson junction is in its coherent or decoherent state. An oscillatory and periodic time dependence of ϕ_- in the asymptotic limit $t \to \infty$ characterizes the coherent state of the dissipative Josephson junction. A (time independent) sharp value of ϕ_- in the asymptotic limit $t \to \infty$ characterizes the decoherent state of the dissipative Josephson junction.

In the classical limit, the time dependence of ϕ_- is governed by the second order differential equation

$$0 = C\,\frac{\hbar}{e^*}\,\ddot{\phi}_- + \frac{1}{R_s}\,\frac{\hbar}{e^*}\,\dot{\phi}_- + I_0\,\sin\phi_- + I. \tag{8.219}$$

Without the flow of a resistive current between the two superconductors, i.e., when the dissipation vanishes

$$\eta = \hbar\,\frac{R_\hbar}{R_s} = 0 \tag{8.220}$$

due to an infinite Ohmic resistance $R_s = \infty$, the relative phase ϕ_- oscillates periodically in time for any non-vanishing capacitance $C > 0$. The difference in the

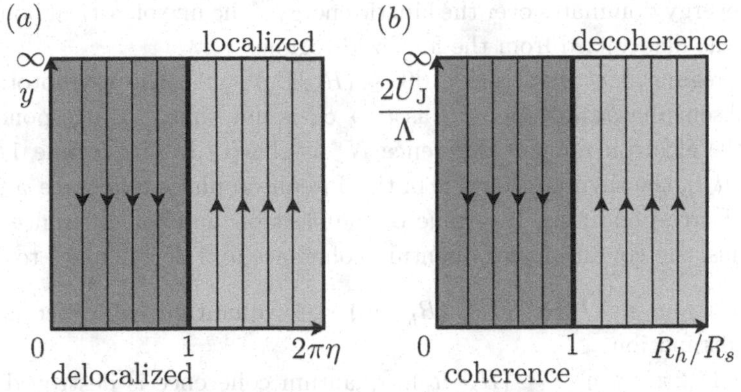

Fig. 8.12 (a) Conjectured phase diagram for a dissipative Josephson junction based on the perturbative RG equations obeyed by the dimensionless dissipation and Josephson coupling constants η and $\widetilde{y}$, respectively. To first order in $\widetilde{y}$, the flow is vertical. Points on the segment $0 \leq 2\,\pi\,\eta < 1$ at $\widetilde{y} = 0$ are attractive fixed points and realize a regime of delocalization of the Josephson phase difference. Points on the segment $1 < 2\,\pi\,\eta < \infty$ at $\widetilde{y} = \infty$ are attractive fixed points and realize a regime of localization of the Josephson phase difference. The thick line at $2\,\pi\,\eta = 1$ represents a line of critical points. (b) Same figure as before but with axis corresponding to a dissipative Josephson junction.

number of electrons between the two supercondutors is then sharp, this is the coherent state of the dissipative Josephson junction that is selected by the capacitance. The effect introduced by a finite Ohmic resistance $0 < R_s < \infty$ is to damp the oscillatory time dependence of the phase difference ϕ_- between the two superconductors. In this classical limit, increasing the strength of the dissipation has no other effect than to smoothly change the time dependence of the phase difference ϕ_- from the under-damped regime, in which an oscillatory time dependence survives, to the over-damped regime, in which the time dependence is purely exponentially damped. For any non-vanishing dissipation, the phase difference ϕ_- becomes asymptotically sharp in the limit $t \gg C \times R_s$, i.e., the dissipative Josephson is in its decoherent state.

Quantum fluctuations change this classical picture dramatically in that they can overcome the effect of weak dissipation. The condition for quantum fluctuations to favor localization over delocalization of ϕ_- is that the Ohmic resistance R_s is sufficiently small (smaller than the quantum resistance R_h).

Indeed, the RG equations (8.211) and (8.215) suggest a two-dimensional phase diagram with the scale dependent coupling constants η and $\widetilde{y}$ as horizontal and vertical axis, respectively, made of vertical flows and separated by a vertical line of fixed points as depicted in Fig. 8.12(a). When the bare dissipation strength η is smaller than $1/(2\,\pi)$, the Josephson junction coupling constant y is irrelevant and the ground state is delocalized as the kinetic energy dominates over the potential energy. When the bare dissipation strength η is larger than $1/(2\,\pi)$, the Josephson junction coupling constant $\widetilde{y}$ is relevant and the ground state is localized as the potential energy dominates over the kinetic energy. The line of critical points $\eta = 2\pi$ separates the delocalized from the localized regime.

In the regime $0 \leq \eta < 1/(2\pi)$ $[0 \leq (R_h/R_s) < 1]$, the asymptotic value of the Josephson phase difference ϕ_- as $t \to \infty$ is not sharp. Correspondingly, the value of the electron number difference N_- is sharp. In the regime $1/(2\pi) < \eta$ $[1 < (R_h/R_s)]$, the asymptotic value of the Josephson phase difference ϕ_- as $t \to \infty$ is sharp. Correspondingly, the value of the electron number difference N_- is not sharp. Thus, the conditions for quantum coherence and decoherence to hold are:

- When $0 \leq \eta < 1/(2\pi)$ $[0 \leq (R_h/R_s) < 1]$, quantum coherence is robust to weak dissipation.
- When $1/(2\pi) < \eta$ $(1 < (R_h/R_s)]$, quantum coherence is destroyed by strong dissipation, quantum decoherence rules.

This interpretation of the phase diagram in Fig. 8.12(a) is given in Fig. 8.12(b). The interplay of dissipation, through an Ohmic current flow between two superconductors coupled capacitively and through a Josephson (periodic) interaction, with quantum fluctuations arising from the uncertainty relation between the relative phase and the difference in the electron number of electrons on superconductors 1 and 2 making up a dissipative Josephson junction has brought about a sharp distinction between the weak and strong dissipative regimes.

8.10 Problems

8.10.1 *The Kondo effect: A perturbative approach*

Introduction

Dissipation is ubiquitous in physics, for the notion of an isolated system for which conservation laws hold is an idealization. Electrons in condensed matter physics originate from atoms with which they exchange quantum numbers such as energy, momentum, angular momentum, etc. In a metal, the couplings between the electrons close to the Fermi energy and the atoms change the ability of the electrons to carry an electrical current as compared to the idealized limit by which the electrons define a closed system. The diagonal contributions to the conductivity tensor are a direct measure of the dissipative effects of these couplings. Electrical resistance is an example of dissipation and its character increasingly becomes quantum mechanical as temperature is lowered. Anomalies in the electrical resistance at low temperature might thus reveal subtle manifestation of quantum mechanics through the presence or absence of dissipation. Superconductivity and the fractional quantum Hall effect have perhaps been historically the most dramatic examples of anomalous dissipation (by its absence), whose explanations have brought paradigm changes in the understanding of many-body quantum mechanics. Another example of a many-body quantum-mechanical effect revealed by an anomalous dissipation is provided by the Kondo effect [89].

In a semi-classical treatment of electrical transport, we may approximate the resistivity of a metal at zero temperature by [recall Eq. (5.71)]

$$\rho \approx \frac{m}{n_{\rm e}\, e^2} \times \frac{1}{\tau_{\rm el}}. \tag{8.221}$$

The lifetime $\tau_{\rm el}$ accounts for the elastic scattering of a single-particle Bloch state to any other Bloch state that is brought about by a static perturbation that breaks the space group symmetry of the lattice (the periodic potential of the ions in their crystalline ground state).[13] This lifetime is calculated to the first non-vanishing order in perturbation theory, whereby the perturbation is a deviation from perfect crystalline order, for every wave vector on the Fermi surface and then averaged over the Fermi surface. In the limit of perfect crystalline order, $\tau_{\rm el} \to \infty$ and the resistivity vanishes. Otherwise, the resistivity is non-vanishing but finite and non-universal.

At any non-vanishing temperature, we may approximate the resistivity of a metal by [recall Eq. (5.71)]

$$\rho \approx \left(\frac{m}{n_{\rm e}\, e^2}\right) \times \left(\frac{1}{\tau_{\rm el}} + \frac{1}{\tau_{\text{e-e}}(T)} + \frac{1}{\tau_{\text{e-pho}}(T)} + \cdots\right). \tag{8.222}$$

The inverse lifetime of the Fermi-liquid quasiparticles in a window of energy of order $k_{\rm B}\,T$ centered about the Fermi energy acquires additional additive channels. A Fermi-liquid quasiparticle can decay through (inelastic) electron-electron interactions, in which case it can be shown that

$$\frac{1}{\tau_{\text{e-e}}(T)} \sim T^2. \tag{8.223}$$

A Fermi-liquid quasiparticle can also decay through electron-phonon interactions (the dynamical counterpart to the static deviation from perfect crystalline symmetry captured by $1/\tau_{\rm el}$), in which case it can be shown that

$$\frac{1}{\tau_{\text{e-pho}}(T)} \sim T^5 \tag{8.224}$$

if only forward scattering is accounted for. There follows the temperature dependence

$$\rho(T) \approx \rho_0 + \rho_2 \left(\frac{k_{\rm B}\,T}{\varepsilon_{\rm F}}\right)^2 + \rho_5 \left(\frac{k_{\rm B}\,T}{\hbar\omega_{\rm D}}\right)^5 + \cdots, \qquad \frac{k_{\rm B}\,T}{\varepsilon_{\rm F}} \ll 1. \tag{8.225}$$

The non-universal numbers ρ_0, ρ_2, and ρ_5 are positive and carry the dimension of the resistivity. The second characteristic energy scale besides the Fermi energy $\varepsilon_{\rm F} > 0$ is here the Debye energy $\hbar\omega_{\rm D} > 0$. In this approximation, the resistivity is a monotonous increasing function of temperature that is dominated by the contribution from phonon scattering at sufficiently large temperatures.

[13] This perturbative expansion might break down. The theory of Anderson localization aims at solving the cases (low dimensionality or strong perturbations) when this perturbation expansion breaks down.

However, it has been known since 1934 that the resistance of gold shows a minimum as a function of temperature when measured between 1 and 21 Kelvins [90]. Evidently, the right-hand side of Eq. (8.225) must be augmented by a scattering channel that increases the resistivity with decreasing temperature. Such a term was found by Kondo in 1964, see Ref. [91], who realized that an effective antiferromagnetic coupling between the spin of the conduction electrons and localized magnetic impurities would produce the temperature dependence

$$\rho(T) \approx \rho_0' + \rho_{\mathrm{K}} \ln\left(\frac{D-\varepsilon_{\mathrm{F}}}{k_{\mathrm{B}}\,T}\right) + \rho_5 \left(\frac{k_{\mathrm{B}}\,T}{\hbar\omega_{\mathrm{D}}}\right)^5 + \cdots, \qquad \frac{k_{\mathrm{B}}\,T_{\mathrm{K}}}{\varepsilon_{\mathrm{F}}} \lesssim \frac{k_{\mathrm{B}}\,T}{\varepsilon_{\mathrm{F}}} \ll 1, \tag{8.226}$$

with a resistivity minimum at the temperature

$$k_{\mathrm{B}}\,T_{\mathrm{K}}^{\mathrm{min}} = \left(\frac{\rho_{\mathrm{K}}}{5\,\rho_5}\right)^{1/5} \hbar\omega_{\mathrm{D}}, \tag{8.227}$$

if the electron-electron interaction is neglected. Here, this magnetic scattering channel changes ρ_0 to $\rho_0' > 0$, while it produces the multiplicative constant ρ_{K} which is again positive, non-universal, and carries the dimension of the resistivity. The band width D is an upper bound to the Fermi energy $0 < \varepsilon_{\mathrm{F}} < D$.

The logarithmic growth

$$\rho_{\mathrm{K}} \ln\left(\frac{D-\varepsilon_{\mathrm{F}}}{k_{\mathrm{B}}\,T}\right) \tag{8.228}$$

with decreasing temperature cannot continue all the way to zero temperature. Upon lowering temperature, perturbation theory must break down when

$$\rho_{\mathrm{K}} \ln\left(\frac{D-\varepsilon_{\mathrm{F}}}{k_{\mathrm{B}}\,T}\right) \tag{8.229}$$

is of the same order as ρ_0', i.e., when $k_{\mathrm{B}}\,T$ is of the order

$$k_{\mathrm{B}}\,T_{\mathrm{K}} := (D-\varepsilon_{\mathrm{F}})\, e^{-\rho_0'/\rho_{\mathrm{K}}}. \tag{8.230}$$

Kondo conjectured a crossover to the contribution

$$\rho_{\mathrm{K}} \ln\left(\frac{D-\varepsilon_{\mathrm{F}}}{k_{\mathrm{B}}\,T_{\mathrm{K}}}\right), \qquad T \ll T_{\mathrm{K}}, \tag{8.231}$$

at temperatures well below the Kondo temperature.

The failure of perturbation theory could signal two scenarios. On the one hand, magnetic impurities are gradually screened by the conduction electrons on the way down to zero temperature. This screening process is non-perturbative in the bare coupling between the electrons and the magnetic impurities very much in the same way as screening of a test point charge is non-perturbative in the electron charge for the jellium model. Nevertheless, the low temperature phase remains a Fermi liquid one, as was conjectured by Kondo, i.e., with the low-temperature dependence of all susceptibilities found for the non-interacting jellium model of section 5.3. On the other hand, screening of the magnetic impurities is not achieved, a non-Fermi liquid phase is selected at low temperatures with deviations from the power laws found for the non-interacting jellium model of section 5.3.

Theoretical methods are thus needed to overcome the singular nature of perturbation theory for temperatures lower than the Kondo temperature, a challenge known as the Kondo problem, in order to establish under what conditions the conjecture (8.231) and its implication that a Fermi liquid phase is recovered for $T \ll T_{\mathrm{K}}$ holds.

As we have seen in sections 3.6, 4.6, and 8.8, and alluded to when discussing the ingredients entering a rigorous proof of Luttinger theorem in section 6.10.2, one option to deal with systemic logarithmic divergences that invalidate perturbation theory is to do a resummation of perturbation theory as dictated by a renormalization group calculation. This program was first carried out by Anderson, Yuval, and Hamann in the spirit of a one-loop renormalization group calculation [92], and then by Wilson who used a non-perturbative renormalization group scheme requiring a numerical implementation [93]. The conclusion is the same by either methods. A single band of non-interacting electrons coupled antiferromagnetically to a dilute concentration of spin-1/2 localized impurities has a Fermi liquid ground state, thereby confirming the conjecture of Kondo.

Perturbative estimates of transition probabilities

Exercise 1.1: We are going to substantiate Eqs. (8.223) and (8.228).

(a) Justify Eq. (8.223) with the help of the zero-temperature estimate for the decay rate of a Fermi liquid quasiparticle derived in footnote 24 from section F.2.

(b) Assume the additive decomposition

$$\hat{H}(t) = \hat{H}_0 + e^{\eta\, t}\,\hat{V} \tag{8.232}$$

into two conserved Hermitian operators $\hat{H}_0$ and $\hat{V}$. The infinitesimal number $\eta > 0$ with the dimension of frequency implements adiabatic switching on of the perturbation $\hat{V}$ at $t = -\infty$. Fermi's golden rule states that the transition probability per unit time from the initial energy eigenstate $|E^{(0)}_a\rangle$ of $\hat{H}_0$ to the final energy eigenstate $|E^{(0)}_b\rangle$ of $\hat{H}_0$ is

$$\begin{aligned} W_{b\leftarrow a} :=& \lim_{t\to\infty}\frac{\mathrm{d}}{\mathrm{d}\,t}\left|\left\langle E^{(0)}_b \left| e^{-\frac{\mathrm{i}}{\hbar}\int_0^t \mathrm{d}t'\,\hat{H}(t')}\right| E^{(0)}_a\right\rangle\right|^2 \\ =& \frac{2\pi}{\hbar}\,\delta\left(E^{(0)}_a - E^{(0)}_b\right) V_{a\,b}\,V_{b\,a}, \end{aligned} \tag{8.233a}$$

to lowest order in an expansion in powers of the matrix elements

$$V_{dc} := \left\langle E^{(0)}_d \left|\hat{V}\right| E^{(0)}_c\right\rangle, \tag{8.233b}$$

in the basis made of the eigenstates of $\hat{H}_0$. Apply Fermi's golden rule to the Fermi-liquid Hamiltonian (F.15) to derive Eq. (8.223).

(c) Show that

$$W_{b\leftarrow a} \equiv \frac{2\pi}{\hbar}\,\delta\left(E^{(0)}_a - E^{(0)}_b\right)|\mathcal{M}_{b\leftarrow a}|^2$$
$$= \frac{2\pi}{\hbar}\,\delta\left(E^{(0)}_a - E^{(0)}_b\right)\left(V_{a\,b}\,V_{b\,a} + \sum_{c\neq a}\frac{V_{a\,b}\,V_{b\,c}\,V_{c\,a} + \text{c.c.}}{E^{(0)}_a - E^{(0)}_c}\right) \tag{8.234a}$$

with the amplitude

$$\mathcal{M}_{b\leftarrow a} := V_{b\,a} + \sum_{c\neq a}\frac{V_{b\,c}\,V_{c\,a}}{E^{(0)}_a - E^{(0)}_c}, \tag{8.234b}$$

up to second order in an expansion in powers of $\hat{V}$.

(d) Kondo chooses[14]

$$\hat{H}_0 := \sum_{\boldsymbol{k}\in\mathrm{BZ}}\sum_{\sigma=\uparrow,\downarrow}\varepsilon_{\boldsymbol{k}}\,\hat{c}^\dagger_{\boldsymbol{k},\sigma}\,\hat{c}_{\boldsymbol{k},\sigma} \tag{8.235a}$$

to encode a single-band of conduction electrons dispersing with the single-particle energies $\varepsilon_{\boldsymbol{k}}$ and the perturbation

$$\hat{V} := \frac{2\,J}{\hbar}\,\frac{1}{N}\sum_{i=1}^{N_{\mathrm{s}}}\sum_{\boldsymbol{k},\boldsymbol{k}'\in\mathrm{BZ}} e^{+\mathrm{i}(\boldsymbol{k}-\boldsymbol{k}')\cdot\boldsymbol{r}_i}\sum_{\alpha,\beta=\uparrow,\downarrow}\hat{c}^\dagger_{\boldsymbol{k}',\alpha}\,\boldsymbol{\sigma}^{\alpha\beta}\,\hat{c}_{\boldsymbol{k},\beta}\cdot\hat{\boldsymbol{S}}_i, \tag{8.235b}$$

by which the conduction electrons couple through an antiferromagnetic Heisenberg exchange coupling $J>0$ to a dilute concentration $N_{\mathrm{s}}/N \ll 1$ of localized spin-1/2 degrees of freedom $\hat{\boldsymbol{S}}_i$ at the site $\boldsymbol{r}_i$ from the lattice Λ obeying the algebra

$$\left[\hat{S}^\mu_i,\hat{S}^\nu_j\right] = \delta_{ij}\,\mathrm{i}\hbar\sum_{\gamma=1,2,3}\epsilon^{\mu\nu\gamma}\,\hat{S}^\gamma_i, \qquad \mu,\nu=1,2,3\equiv x,y,z, \qquad \hat{\boldsymbol{S}}^2_i = \frac{3}{4}\hbar^2\,\mathbb{1}. \tag{8.235c}$$

The lattice Λ is here made of N sites.

We introduce the notation

$$\hat{\boldsymbol{s}}_{\boldsymbol{k}',\boldsymbol{k}} := \frac{\hbar}{2}\sum_{\alpha,\beta=\uparrow,\downarrow}\hat{c}^\dagger_{\boldsymbol{k}',\alpha}\,\boldsymbol{\sigma}^{\alpha\beta}\,\hat{c}_{\boldsymbol{k},\beta} \tag{8.236}$$

and define

$$\hat{V}_{J_\perp,J_\parallel} = \left(\frac{2}{\hbar}\right)^2\frac{1}{N}\sum_{i=1}^{N_{\mathrm{s}}}\sum_{\boldsymbol{k},\boldsymbol{k}'\in\mathrm{BZ}} e^{+\mathrm{i}(\boldsymbol{k}-\boldsymbol{k}')\cdot\boldsymbol{r}_i}$$
$$\times\left[\frac{J_\perp}{2}\left(\hat{s}^+_{\boldsymbol{k}',\boldsymbol{k}}\,\hat{S}^-_i + \hat{s}^-_{\boldsymbol{k}',\boldsymbol{k}}\hat{S}^+_i\right) + J_\parallel\,\hat{s}^z_{\boldsymbol{k}',\boldsymbol{k}}\,\hat{S}^z_i\right]. \tag{8.237}$$

---

[14] Our conventions are

$$\{\hat{c}_{\boldsymbol{k},\sigma},\hat{c}^\dagger_{\boldsymbol{k}',\sigma'}\} = \delta_{\boldsymbol{k},\boldsymbol{k}'}\delta_{\sigma,\sigma'}, \qquad \{\hat{c}_{\boldsymbol{k},\sigma},\hat{c}_{\boldsymbol{k}',\sigma'}\} = \{\hat{c}^\dagger_{\boldsymbol{k},\sigma},\hat{c}^\dagger_{\boldsymbol{k}',\sigma'}\} = 0,$$

and

$$\{\hat{c}_{\boldsymbol{r},\sigma},\hat{c}^\dagger_{\boldsymbol{r}',\sigma'}\} = \delta_{\boldsymbol{r},\boldsymbol{r}'}\delta_{\sigma,\sigma'}, \qquad \{\hat{c}_{\boldsymbol{r},\sigma},\hat{c}_{\boldsymbol{r}',\sigma'}\} = \{\hat{c}^\dagger_{\boldsymbol{r},\sigma},\hat{c}^\dagger_{\boldsymbol{r}',\sigma'}\} = 0,$$

whereby

$$\hat{c}^\dagger_{\boldsymbol{r},\sigma} = \frac{1}{\sqrt{N}}\sum_{\boldsymbol{k}\in\mathrm{BZ}}\hat{c}^\dagger_{\boldsymbol{k},\sigma}\,e^{-\mathrm{i}\boldsymbol{k}\cdot\boldsymbol{r}}, \qquad \hat{c}_{\boldsymbol{r},\sigma} = \frac{1}{\sqrt{N}}\sum_{\boldsymbol{k}\in\mathrm{BZ}}\hat{c}_{\boldsymbol{k},\sigma}\,e^{+\mathrm{i}\boldsymbol{k}\cdot\boldsymbol{r}}.$$

Consider the direct product $|\boldsymbol{k},\alpha\rangle \otimes |i,\beta\rangle$ of a single-particle electronic state with momentum $\boldsymbol{k}$ and projection α of its spin along the quantization axis with a localized spin at site i with the projection β of its spin along the quantization axis. We shall also denote with $\bar{\sigma}$ the reversal of the projection σ along the quantization axis. Compute the sixteen matrix elements

$$\left\langle \boldsymbol{k}'',\alpha'' \right| \otimes \left\langle i'',\beta'' \left| \hat{V}_{J_\perp,J_\parallel} \right| \boldsymbol{k},\alpha \right\rangle \otimes \left| i,\beta \right\rangle, \qquad \alpha,\beta,\alpha'',\beta'' = \uparrow,\downarrow \equiv +,-, \tag{8.238}$$

for given initial $\boldsymbol{k}$, i and final $\boldsymbol{k}''$, i''.

(e) Draw the Feynman diagram describing the amplitude for the following process. An initial state $|\boldsymbol{k},\sigma\rangle \otimes |i,\bar{\sigma}\rangle$ with $\boldsymbol{k}$ just above the Fermi sea overlaps with the virtual state $|\boldsymbol{k}'',\bar{\sigma}\rangle \otimes |i,\sigma\rangle$ with $\boldsymbol{k}''$ outside the Fermi sea owing to the perturbation $\hat{V}$. This virtual state overlaps with the final state $|\boldsymbol{k}',\sigma\rangle \otimes |i,\bar{\sigma}\rangle$ with $\boldsymbol{k}'$ chosen just above the Fermi sea owing to the perturbation $\hat{V}$. Show that the amplitude of this Feynman diagram is, at zero temperature and when the density of state at the Fermi energy $\nu_{\rm F}$ is non-vanishing, proportional to

$$\begin{aligned} J^2 \sum_{\boldsymbol{k}''\in{\rm BZ}} \frac{1-f_{\rm FD}(\varepsilon_{\boldsymbol{k}''})}{\varepsilon_{\boldsymbol{k}}-\varepsilon_{\boldsymbol{k}''}} &= J^2 \int\limits_{\varepsilon_{\rm F}}^{D} {\rm d}\varepsilon \, \frac{\nu(\varepsilon)}{\varepsilon_{\boldsymbol{k}}-\varepsilon} \\ &\approx J^2\,\nu_{\rm F} \ln\left(\frac{\varepsilon_{\boldsymbol{k}}-\varepsilon_{\rm F}}{D-\varepsilon_{\boldsymbol{k}}}\right), \quad 0<\varepsilon_{\rm F}\lesssim\varepsilon_{\boldsymbol{k}}<D. \end{aligned} \tag{8.239}$$

For comparison, estimate the order of magnitude of the integral

$$\int\limits_{0}^{D} {\rm d}\varepsilon \, \frac{\nu(\varepsilon)}{\varepsilon_{\boldsymbol{k}}-\varepsilon}. \tag{8.240}$$

Hint: Assume that $\varepsilon_{\boldsymbol{k}}$ is just above the Fermi energy. Do the change of variable $\xi := \varepsilon - \varepsilon_{\rm F}$ and decompose the density of state $\tilde{\nu}(\xi)$ into the sum of a function of ξ that is even about the Fermi energy $\xi = 0$ and one that is odd.

(f) Usually, there is no need to extend perturbation theory beyond the first non-vanishing order if the expansion is convergent and if the goal of perturbation theory is not merely to increase the precision of the expansion. However, perturbation theory in many-body physics is often not convergent as the function to be expanded is singular in the expansion parameter. The dependence on the squared electric charge of the Thomas-Fermi screening length (6.92) is a case at hand. Keeping this in mind, we are now ready to finish the steps that lead to Eq. (8.228). We need to evaluate Eq. (8.234a). To this end, the initial state $|a\rangle$, the final state $|b\rangle$, and the virtual states $|c\rangle$ must be supplied. We denote with

$$|\cdots;\boldsymbol{k},\sigma_{\boldsymbol{k}};\cdots\rangle := \prod_{\{\boldsymbol{k},\sigma_{\boldsymbol{k}}\}} \hat{c}^{\dagger}_{\boldsymbol{k},\sigma_{\boldsymbol{k}}}|0\rangle \tag{8.241}$$

a Slater determinant for the N conduction electrons. The Fermi sea, a special case of such a Slater determinant, is denoted $|{\rm FS}\rangle$. We work in the approximation with one impurity, i.e., $N_{\rm s} = 1$, and multiply intensive quantities calculated for one impurity by the impurity concentration. Our initial state contains one quasiparticle above the Fermi surface. The quasiparticle has the momentum $\boldsymbol{k}$, the single-particle energy $\varepsilon_{\boldsymbol{k}} > \varepsilon_{\mathrm{F}}$, and the spin-1/2 quantum number α. The spin-1/2 impurity at the lattice site $\boldsymbol{r}_i$ has the spin-1/2 quantum number σ. Thus,

$$|a\rangle := \left(\hat{c}^{\dagger}_{\boldsymbol{k},\alpha}\,|\mathrm{FS}\rangle\right) \otimes |\boldsymbol{r}_i,\sigma\rangle. \tag{8.242}$$

The final state has one quasiparticle above the Fermi surface with the momentum $\boldsymbol{k}'$, the single-particle energy $\varepsilon_{\boldsymbol{k}'} = \varepsilon_{\boldsymbol{k}} > \varepsilon_{\mathrm{F}}$, and the spin-1/2 quantum number α. The spin-1/2 impurity has the spin-1/2 quantum number σ. Thus,

$$|b\rangle := \left(\hat{c}^{\dagger}_{\boldsymbol{k}',\alpha}\,|\mathrm{FS}\rangle\right) \otimes |\boldsymbol{r}_i,\sigma\rangle. \tag{8.243}$$

There are four possible families of virtual states that we arrange pairwise. The first pair of virtual states is constructed by applying $\hat{V}_{J_{\perp}=0,J=J_{\parallel}}$ on either the initial or final state. The virtual state is then either labeled by the momentum $\boldsymbol{p}_{>}$ above the Fermi surface with the spin-1/2 quantum number α, in which case

$$\begin{aligned} |c\rangle &:= \hat{c}^{\dagger}_{\boldsymbol{p}_{>},\alpha}\,\hat{c}_{\boldsymbol{k},\alpha}\,|a\rangle, \qquad && \varepsilon_{\boldsymbol{p}_{>}} > \varepsilon_{\mathrm{F}}, \\ |b\rangle &= \hat{c}^{\dagger}_{\boldsymbol{k}',\alpha}\,\hat{c}_{\boldsymbol{p}_{>},\alpha}\,|c\rangle, && \varepsilon_{\boldsymbol{p}_{>}} > \varepsilon_{\mathrm{F}}, \end{aligned} \tag{8.244a}$$

or it is labeled by the momentum $\boldsymbol{p}_{<}$ of a quasihole below the Fermi surface with the spin-1/2 quantum number α, in which case

$$\begin{aligned} |c\rangle &:= \hat{c}^{\dagger}_{\boldsymbol{k}',\alpha}\,\hat{c}_{\boldsymbol{p}_{<},\alpha}\,|a\rangle, \qquad && \varepsilon_{\boldsymbol{p}_{<}} < \varepsilon_{\mathrm{F}}, \\ |b\rangle &= \hat{c}^{\dagger}_{\boldsymbol{p}_{<},\alpha}\,\hat{c}_{\boldsymbol{k},\alpha}\,|c\rangle, && \varepsilon_{\boldsymbol{p}_{<}} < \varepsilon_{\mathrm{F}}. \end{aligned} \tag{8.244b}$$

The second pair of virtual states is constructed by applying $\hat{V}_{J=J_{\perp},J_{\parallel}=0}$ on either the initial or final state. The virtual state is then either

$$\begin{aligned} |c\rangle &:= \hat{c}^{\dagger}_{\boldsymbol{p}_{>},\bar{\alpha}}\,\hat{c}_{\boldsymbol{k},\alpha}\,\hat{S}^{\alpha}\,|a\rangle, \qquad && \varepsilon_{\boldsymbol{p}_{>}} > \varepsilon_{\mathrm{F}}, \\ |b\rangle &= \hat{c}^{\dagger}_{\boldsymbol{k}',\alpha}\,\hat{c}_{\boldsymbol{p}_{>},\bar{\alpha}}\,\hat{S}^{\bar{\alpha}}\,|c\rangle, && \varepsilon_{\boldsymbol{p}_{>}} > \varepsilon_{\mathrm{F}}, \end{aligned} \tag{8.245a}$$

or

$$\begin{aligned} |c\rangle &:= \hat{c}^{\dagger}_{\boldsymbol{k}',\alpha}\,\hat{c}_{\boldsymbol{p}_{<},\bar{\alpha}}\,\hat{S}^{\bar{\alpha}}\,|a\rangle, \qquad && \varepsilon_{\boldsymbol{p}_{<}} < \varepsilon_{\mathrm{F}}, \\ |b\rangle &= \hat{c}^{\dagger}_{\boldsymbol{p}_{<},\bar{\alpha}}\,\hat{c}_{\boldsymbol{k},\alpha}\,\hat{S}^{\alpha}\,|c\rangle, && \varepsilon_{\boldsymbol{p}_{<}} < \varepsilon_{\mathrm{F}}. \end{aligned} \tag{8.245b}$$

Show that the third-order term on the right-hand side of Eq. (8.234a) is proportional to the integral (8.240) for the first pair (8.244) of virtual states. Show that the third-order term on the right-hand side of Eq. (8.234a) delivers the sum

$$\sum_{\boldsymbol{p}\in\mathrm{BZ}} \frac{f_{\mathrm{FD}}(\varepsilon_{\boldsymbol{p}})}{\varepsilon_{\boldsymbol{k}} - \varepsilon_{\boldsymbol{p}}} \approx \nu_{\mathrm{F}} \ln\left(\frac{\varepsilon_{\boldsymbol{k}}}{\varepsilon_{\boldsymbol{k}} - \varepsilon_{\mathrm{F}}}\right) \tag{8.246}$$

for the second pair (8.245) of virtual states. Which one of the four families of virtual states gives the single-particle result (8.239)? Which ones of the steps of this computation are intrinsically many-body ones, i.e., would not be possible within single-particle physics?

8.10.2 *The Kondo effect: A non-perturbative approach*

Introduction

The model (8.235) studied by Kondo has a very small concentration $N_s/N \ll 1$ of spin-1/2 impurities, where N_s is the number of spin-1/2 impurities and N is the number of sites from the lattice Λ on which the non-interacting electrons are hopping with the single-particle dispersion $\varepsilon_{\boldsymbol{k}}$. In this limit, each spin-1/2 impurity can be treated independently of the others. This is why the model (8.235) with $N_s = 1$ is known as the Kondo model.

Although Kondo solved the mystery of the resistivity minimum in dilute magnetic alloys such as the metals Cu, Ag, Au, Mg, or Zn with Cr, Mn, Fe, Mo, Re, or Os as impurities, he left open the (Kondo) problem of how to prove his conjecture that a spin-1/2 impurity is screened by one band of conduction electrons. This task can be achieved in three different ways.

In the spirit of Anderson, Yuval, and Hamann [92], an effective partition function for the spin-1/2 impurity is obtained after approximately integrating out the conduction electrons. The effective action is defined in imaginary time where it is non-local, a reflection of the fact that the spectrum of the conduction electrons is gapless. It can be studied with the help of the renormalization group. This approach is very similar to that employed for a dissipative Josephson junction in section 8.8.

In the spirit of Wilson [93], a single-particle basis for the conduction electrons is chosen such that their hybridizations with the magnetic impurity is maximized and their energies are as close as possible to the Fermi energy. In other words, electronic states that are far away from the impurity in position space and have a single-particle energy close to the bottom or top of the conduction band can be neglected. This basis selection is similar to the one made in section 6.7.1.4 to derive the Friedel oscillations of a classical point scatterer. One may then integrate those single-particle basis states that reside in a small window of large energies relative to the Fermi energy, as we did to derive the one-loop flow of a repulsive interaction in a superconductor in section 7.2.1 for example. In doing so, the Kondo coupling changes slightly. By iterating this procedure numerically, the flow of the Kondo coupling can be traced in a non-perturbative way all the way from the regime of a free magnetic impurity (high temperature) down to vanishing temperature [94]. For a bare ferromagnetic Kondo coupling, the Kondo coupling flows to zero in magnitude, i.e., the magnetic impurity is essentially free as the temperature approaches zero. For a bare antiferromagnetic Kondo coupling, the Kondo coupling flows to infinity in magnitude. In this limit, the Kondo hybridization term $\hat{V}$ is minimized by forming a singlet between the electronic spin density and the spin-1/2 impurity with an infinitely large gap to the triplet excitations. In effect, the spin-1/2 impurity has been screened.

The conclusions of both approaches have been confirmed by non-perturbative analytical means with the help of the seminal observation made by Andrei in Ref. [95]

and by Wiegmann in Ref. [96] that a mild approximation to the Kondo model brings it to a class of exactly soluble models in (1+1)-dimensional position space and time.

We are going to construct a mean-field approximation of the Kondo model that solves the Kondo problem. This mean-field approximation is uncontrolled. However, it reproduces (by design) qualitatively the crossover to the strong Kondo coupling regime with decreasing temperature uncovered by the aforementioned approaches. Before we do so however, we are going to list some lattice models related to the Kondo model.

Definitions

The Kondo model (8.235) has one magnetic spin-1/2 impurity ($N_{\mathrm{s}} = 1$). In the opposite limit, the Kondo lattice model has one spin-1/2 impurity on each site of the lattice Λ. Conduction electrons from a single band hop between the sites of Λ. Hence, the number of lattice sites N equals the number of magnetic impurities N_{s} and we may write

$$\hat{H} := \hat{H}_0 + \hat{V}_{\mathrm{KL}}, \tag{8.247a}$$

where (footnote 14 defines our normalization convention for Fourier transformations)

$$\hat{H}_0 := \sum_{i,j\in\Lambda}\sum_{\sigma=\uparrow,\downarrow} \hat{c}^{\dagger}_{i\,\sigma}\,t_{ij}\,\hat{c}_{j\,\sigma}, \qquad t_{ij} := \frac{1}{N}\sum_{\boldsymbol{k}\in\mathrm{BZ}} e^{-\mathrm{i}\boldsymbol{k}\cdot(\boldsymbol{r}_i-\boldsymbol{r}_j)}\varepsilon_{\boldsymbol{k}} = t^{*}_{ji}, \tag{8.247b}$$

is the tight-binding representation of Hamiltonian (8.235a), while

$$\hat{V}_{\mathrm{KL}} := \frac{4\,J}{\hbar^2}\sum_{i\in\Lambda}\hat{\boldsymbol{s}}_i\cdot\hat{\boldsymbol{S}}_i, \qquad \hat{\boldsymbol{s}}_i = \frac{\hbar}{2}\sum_{\alpha,\beta=\uparrow,\downarrow}\hat{c}^{\dagger}_{i,\alpha}\,\boldsymbol{\sigma}^{\alpha\beta}\,\hat{c}_{i\,\beta}, \tag{8.247c}$$

is the tight-binding representation of Hamiltonian (8.235b) for $N_{\mathrm{s}} = N$. The Kondo lattice model is more complicated than the Kondo model [97]. The hopping of the electrons is strongly affected by constructive or destructive quantum interferences arising from the localized spins. Conversely, if one imagine integrating out the fermions, there follows a Heisenberg model with long-range oscillatory exchange couplings. The Kondo lattice model is often taken as a starting point to understand a class of materials called heavy fermions [98].

The mean-field method that we shall apply to the Kondo model takes advantage of a representation of the localized spin-1/2 magnetic impurities in terms of auxiliary bosons and fermions. This is a mere trick devised to circumvent the lack of Wick theorem for operators satisfying the $SU(2)$ spin algebra. We already used this trick in section 6.10.3. Alternatively, we may ask the following question. Is there a tight-binding model for more than one band of fermions that is related to the Kondo model? The answer is affirmative as we now show.

The model in question is due to Anderson in 1961 [99]. This model and its variants are now called the Anderson model with various qualifiers, such as single impurity, periodic, $SU(4)$ symmetric, etc. Its relationship to the Kondo model was established by Schrieffer and Wolff [100].

Anderson wanted to construct a simple model such that it captures the competition between the tendency for localized fermions to develop a local magnetic moment by way of a generalization of Hund's coupling from atomic physics and the preference for these localized fermions to lower their energy by hybridization with a conducting band of electrons. The localized electrons often originate from $3d$ or $4f$ atomic orbitals. The letter c is conventionally reserved for the creation and annihilation operators of conduction electrons. The letter f is conventionally reserved for the creation and annihilation operators of the non-dispersing electrons. The two-band model known as the Anderson model is then[15]

$$\hat{H} := \hat{H}_c + \hat{H}_f + \hat{H}_{c-f} \tag{8.248a}$$

with

$$\hat{H}_c := \sum_{\boldsymbol{k}\in \mathrm{BZ}} \sum_{\sigma=\uparrow,\downarrow} \varepsilon_{\boldsymbol{k}}\, \hat{c}^{\dagger}_{\boldsymbol{k},\sigma} \hat{c}_{\boldsymbol{k},\sigma} \tag{8.248b}$$

the non-interacting Hamiltonian for the conduction (c) electrons in the Bloch representation,

$$\hat{H}_f := \varepsilon_f \sum_{i=1}^{N_f} \sum_{\sigma=\uparrow,\downarrow} \hat{f}^{\dagger}_{\boldsymbol{r}_i\,\sigma} \hat{f}_{\boldsymbol{r}_i\,\sigma} + U \sum_{i=1}^{N_f} \left(\hat{f}^{\dagger}_{\boldsymbol{r}_i\uparrow} \hat{f}_{\boldsymbol{r}_i\uparrow}\right) \left(\hat{f}^{\dagger}_{\boldsymbol{r}_i\downarrow} \hat{f}_{\boldsymbol{r}_i\downarrow}\right) \tag{8.248c}$$

the interacting Hamiltonian for the localized (f) electrons in the position representation, and

$$\hat{H}_{c-f} := \sum_{\boldsymbol{k}\in \mathrm{BZ}} \sum_{\sigma=\uparrow,\downarrow} \sum_{i=1}^{N_f} \left(V_{\boldsymbol{k}}\, e^{-\mathrm{i}\boldsymbol{k}\cdot\boldsymbol{r}_i}\, \hat{c}^{\dagger}_{\boldsymbol{k},\sigma} \hat{f}_{\boldsymbol{r}_i\,\sigma} + V^{*}_{\boldsymbol{k}}\, e^{+\mathrm{i}\boldsymbol{k}\cdot\boldsymbol{r}_i}\, \hat{f}^{\dagger}_{\boldsymbol{r}_i\,\sigma} \hat{c}_{\boldsymbol{k},\sigma} \right) \tag{8.248d}$$

the coupling between the conduction band of c electrons and the localized f electrons. The f electrons are not dispersing on their own, they are localized on a subset made of N_f sites from the lattice Λ. Occupation of any one of these sites with two f electrons of opposite spins cost the on-site energy $U > 0$. The coupling between $\hat{c}^{\dagger}_{\boldsymbol{k},\sigma}$ and $\hat{f}_{\boldsymbol{r}_i\,\sigma}$ is mediated by a single-body term with the matrix elements $V_{\boldsymbol{k}} \in \mathbb{C}$ that we assume independent of the spin quantum number σ and of the impurity site i (in magnitude for the latter). The choice for the normalization of $V_{\boldsymbol{k}}$ is motivated by requesting that $\hat{H}_{c-f}$ is proportional to N when $N_f = 1$. We could have made the choice $V_{\boldsymbol{k}} \to V_{\boldsymbol{k}}/\sqrt{N}$. With this choice, the coupling between the f electrons and the c electrons becomes weaker as the number of conduction electrons increases. This is the spirit of the Caldeira-Leggett model that takes advantage of the macroscopic size of the bath to couple each normal mode of the bath

[15] All creation and annihilation operators obey the usual anticommutation relations appropriate to fermions. Creation and annihilation operators with different symbols or labels have a vanishing anticommutator. The anticommutator of an annihilation operator with its adjoint is unity.

to the finite number (N_{s}) of degrees of freedom constituting the "impurity". In fact, this convention is required if we demand that Hamiltonian (8.271b) scales like N in the thermodynamic limit [see Eq. (8.328) at which stage we will need to use Eq. (8.271b)]. Another possibility is to choose $V_{\boldsymbol{k}} \to \sqrt{N_{\mathrm{s}}/N}\, V_{\boldsymbol{k}}$ and to take the thermodynamic limit $N \to \infty$ keeping the ratio N_{s}/N fixed, say $N_{\mathrm{s}}/N \ll 1$ in the dilute limit. In thermodynamic equilibrium the total number of electrons, the sum of c electrons and of f electrons, is fixed by the chemical potential.

The phase diagram of the Anderson model at vanishing temperature depends on the dispersion of the c electrons, on the cost ε_f to fill an impurity site by a single $4f$ electron, on the on-site repulsive interaction U of f electrons, on the hybridization matrix elements $V_{\boldsymbol{k}}$, and on the chemical potential μ that determines the total number N_{e} of electrons.

Exercise 1.1:

(a) When the number of impurity sites N_f is smaller than the number N of lattice sites, the translation symmetry group of the lattice Λ is broken in the Anderson model. The symmetry group of the lattice Λ is recovered when $N_f = N$, in which case the periodic Anderson model follows. Write down the periodic Anderson model with the f creation and annihilation operators expressed in terms of the crystal momentum $\boldsymbol{k} \in \mathrm{BZ}$ (the on-site interaction term should be written in terms of the occupation number operator of f electrons in the first Brillouin zone).

(b) Write down the Anderson model (8.248) with the c electron creation and annihilation operators and their hybridization matrix elements to the f electron creation and annihilation operators expressed in terms of the site index i of the lattice Λ.

When the number of impurity sites $N_f = 1$, the single-impurity Anderson model is obtained. Without loss of generality, we declare that the impurity site is at the origin of the coordinate system in position space. This is the model we want to relate to the Kondo model. If we use a path-integral representation, it is defined by the partition function

$$Z := \int \mathcal{D}[c^*, c] \int \mathcal{D}[f^*, f]\, e^{-S}, \tag{8.249a}$$

with the action

$$S := S_c + S_f + S_{c-f} \tag{8.249b}$$

decomposing into the Euclidean action for the c electrons

$$S_c := \int\limits_0^\beta \mathrm{d}\tau \sum_{\boldsymbol{k}\in\mathrm{BZ}} \sum_{\sigma=\uparrow,\downarrow} \left[c^*_{\boldsymbol{k},\sigma} \left(\partial_\tau + \xi_{\boldsymbol{k}}\right) c_{\boldsymbol{k},\sigma}\right], \tag{8.249c}$$

the Euclidean action for the f electrons

$$S_f := \int\limits_0^\beta \mathrm{d}\tau \left[\sum_{\sigma=\uparrow,\downarrow} f^*_\sigma \left(\partial_\tau + \xi_f\right) f_\sigma + U\, f^*_\uparrow\, f_\uparrow\, f^*_\downarrow\, f_\downarrow\right], \tag{8.249d}$$

and the Euclidean action that couples the c electrons and the f electrons

$$S_{c-f} := \int_0^\beta \mathrm{d}\tau \sum_{\boldsymbol{k}\in\mathrm{BZ}} \sum_{\sigma=\uparrow,\downarrow} \left(V_{\boldsymbol{k}}\, c^*_{\boldsymbol{k},\sigma}\, f_\sigma + V^*_{\boldsymbol{k}}\, f^*_\sigma\, c_{\boldsymbol{k},\sigma}\right). \tag{8.249e}$$

As usual, all Grassmann-valued fields obey antiperiodic boundary conditions in imaginary time. If we opt to work with the Hamiltonian formalism, we deduce from S the Hamiltonian

$$\hat{H} := \hat{H}_c + \hat{H}_f + \hat{H}_{c-f}, \tag{8.250a}$$

from S_c the Hamiltonian

$$\hat{H}_c := \sum_{\boldsymbol{k}\in\mathrm{BZ}} \sum_{\sigma=\uparrow,\downarrow} \xi_{\boldsymbol{k}}\, \hat{n}^c_{\boldsymbol{k},\sigma}, \qquad \hat{n}^c_{\boldsymbol{k},\sigma} := \hat{c}^\dagger_{\boldsymbol{k},\sigma}\, \hat{c}_{\boldsymbol{k},\sigma}, \tag{8.250b}$$

from S_f the Hamiltonian

$$\hat{H}_f := \xi_f \left(\hat{n}^f_\uparrow + \hat{n}^f_\downarrow\right) + U \hat{n}^f_\uparrow\, \hat{n}^f_\downarrow, \qquad \hat{n}^f_\sigma := \hat{f}^\dagger_\sigma\, \hat{f}_\sigma, \tag{8.250c}$$

and from S_{c-f} the Hamiltonian

$$\hat{H}_{c-f} := \sum_{\boldsymbol{k}\in\mathrm{BZ}} \sum_{\sigma=\uparrow,\downarrow} \left(V_{\boldsymbol{k}}\, \hat{c}^\dagger_{\boldsymbol{k},\sigma}\, \hat{f}_\sigma + V^*_{\boldsymbol{k}}\, \hat{f}^\dagger_\sigma\, \hat{c}_{\boldsymbol{k},\sigma}\right). \tag{8.250d}$$

All single-particle energies are now measured relative to the chemical potential μ. This is why we changed the symbol from $\varepsilon_\cdot$ to $\xi_\cdot$ to denote single-particle energies.

The single-impurity Anderson Hamiltonian (8.250) is constructed to be exactly soluble in the limit of no hybridization, i.e., for $V_{\boldsymbol{k}} = 0$ for all wave vectors from the first Brillouin zone. The spectrum of Hamiltonian (8.250a) is the "addition" of the spectrum of Hamiltonian (8.250b) and of the spectrum of Hamiltonian (8.250c).

The spectrum of Hamiltonian (8.250b) is that of a non-interacting gas of electrons, i.e., the Fermi sea ground state $|\mathrm{FS}\rangle_c$ with all possible particle-hole excitations as excited states. The level spacing of Hamiltonian (8.250b) above the ground state is of order $\hbar v_{\mathrm{F}}\, 2\pi/L$.[16] It becomes a continuum in the thermodynamic limit $L \to \infty$ holding the filling fraction N_{e}/N of the lattice fixed.

The spectrum of Hamiltonian (8.250c) consists of four orthonormal energy eigenstates. There is the vacuum with the eigenenergy 0, i.e., the state $|0\rangle_f$ annihilated by $\hat{f}_\sigma$ for both projections $\sigma = \uparrow, \downarrow$ of the spin along the quantization axis. There are two degenerate states with eigenenergy ξ_f, the states

$$|\uparrow\rangle_f := \hat{f}^\dagger_\uparrow\, |0\rangle_f, \qquad |\downarrow\rangle_f := \hat{f}^\dagger_\downarrow\, |0\rangle_f. \tag{8.251}$$

There is one state with the eigenenergy $2\xi_f + U$,

$$|\uparrow\downarrow\rangle_f := \hat{f}^\dagger_\uparrow\, \hat{f}^\dagger_\downarrow\, |0\rangle_f. \tag{8.252}$$

Since we are measuring energies relative to the Fermi energy, at the Fermi energy the f electrons are in their vacuum state $|0\rangle_f$. If we choose ξ_f to be positive,

[16] Here, v_{F} is the Fermi velocity and L the linear size of the lattice.

the many-body eigenstates with one or two f electrons present have an excitation energy bounded from below by

$$\xi_f > 0 \tag{8.253}$$

and

$$2\xi_f + U > 0, \tag{8.254}$$

respectively. Thus, they are inoperative to stabilize a localized magnetic moment at temperatures below the threshold ξ_f/k_{B}.

The regime of parameter space for Hamiltonian $\hat{H}_c + \hat{H}_f$ that is the most favorable to the formation of a localized magnetic moment in the ground state of Hamiltonian $\hat{H}_c + \hat{H}_f$ is when the chain of inequalities

$$\xi_f < 0 < 2\xi_f + U \tag{8.255}$$

holds, for the many-body ground state manifold of $\hat{H}_c + \hat{H}_f$ is then two-fold degenerate, given by

$$\mathrm{span}\left\{|\mathrm{FS}\rangle_c \otimes |\uparrow\rangle_f, |\mathrm{FS}\rangle_c \otimes |\downarrow\rangle_f\right\}, \tag{8.256}$$

and separated from all excited many-body energy eigenstates by the gap of order $\hbar v_{\mathrm{F}}\, 2\pi/L$ in the sector of the Hilbert space for the conduction electrons and the gap $|\xi_f|$ in the sector of the Hilbert space for the f electrons. The chain of inequalities (8.255) is always met for sufficiently large U.

Exercise 1.2:

(a) Calculate the magnetic susceptibility of the ground state manifold (8.256) and show that it obeys the Curie law for temperatures satisfying $k_{\mathrm{B}}T \ll |\xi_f|$.

(b) Assume the inequalities

$$\xi_f < 2\xi_f + U < 0 \tag{8.257}$$

and compute the magnetic susceptibility of $\hat{H}_c + \hat{H}_f$ for temperatures satisfying $k_{\mathrm{B}}T \ll |2\xi_f + U|$.

(c) Assume the inequalities

$$2\xi_f + U < \xi_f < 0 \tag{8.258}$$

and compute the magnetic susceptibility of $\hat{H}_c + \hat{H}_f$ for temperatures satisfying $k_{\mathrm{B}}T \ll |\xi_f|$.

The question to be addressed is what happens to Curie's law obeyed by the ground state manifold (8.256) upon approaching zero temperature once the average value

$$\langle |V_{\boldsymbol{k}}|^2\rangle_{\mathrm{Fs}} := \frac{\sum\limits_{\boldsymbol{k}\in\mathrm{Fs}} |V_{\boldsymbol{k}}|^2}{\sum\limits_{\boldsymbol{k}\in\mathrm{Fs}} 1} \tag{8.259}$$

of the hybridization over the unperturbed Fermi surface (Fs as opposed to FS for the Fermi sea) of the c electrons is switched on adiabatically. Perturbation theory in the hybridization cannot address this issue for the same reason as perturbation theory in the Kondo coupling fails. This can be seen by brute force calculation when reaching the fourth order of perturbation theory. Alternatively, we can map the single-impurity Anderson model to the Kondo model when conditions (8.274) are met, as first shown by Schrieffer and Wolff in Ref. [100].

The Schrieffer-Wolff transformation

We rewrite Hamiltonian (8.250) as

$$\hat{H}(\lambda) = \hat{H}_0 + \lambda\,\hat{H}_1, \tag{8.260}$$

where the real-valued dimensionless number λ is introduced for bookkeeping. Hamiltonian

$$\hat{H}_0 := \hat{H}_c + \hat{H}_f \tag{8.261}$$

is the unperturbed Hamiltonian of which we know the eigenstates and eigenvalues. In particular, we can compute the magnetic susceptibility $\chi_0(T)$ of $\hat{H}_0$. In fact we know that $\chi_0(T)$ is singular (Curie singularity) at $T = 0$. Hamiltonian

$$\hat{H}_1 := \hat{H}_{c-f} \tag{8.262}$$

is the perturbation, which can be shown to be singular when computing the corrections it brings about to the magnetic susceptibility $\chi_0(T)$ of $\hat{H}_0$ at temperature T, i.e., the sequence of functions χ_n defined for any positive integer n by computing the correction to χ_0 up to n-th order in the perturbation $\hat{H}_1$ does not converge uniformly with decreasing temperature to χ_0. This breakdown of perturbation theory signals that the dependence on temperature of $\chi(T)$ could still be singular or it could be regular in the limit $T \to 0$, but it cannot be captured by doing perturbation theory in powers of $\hat{H}_1$. If the manipulation

$$\chi(T) = \chi_0(T) + [\chi(T) - \chi_0(T)] \tag{8.263}$$

seems innocuous at any non-vanishing temperature, it is not at $T = 0$ where $\chi_0(T)$ is singular, and it is not to be expected that perturbation theory in the operator that encodes the difference $\chi(T) - \chi_0(T)$ is a wise approach to capture the dependence of $\chi(T)$ in the limit $T \to 0$.

It would be wiser to seek a function $\chi_0^\star$ that shares the same leading dependence on T for $T \to 0$ as χ and then do the decomposition

$$\chi(T) = \chi_0^\star(T) + [\chi(T) - \chi_0^\star(T)]\,. \tag{8.264}$$

This function $\chi_0^\star$ should be a property of a fixed point that is perturbatively close to the fixed point governing the dependence of $\chi(T)$ on T as $T \to 0$. With the privilege of hindsight, this fixed point should be governed by some strong coupling limit, but it had not been identified prior to the work of Schrieffer and Wolff.

With no such a priori knowledge, as was the case for Schrieffer and Wolff, we may keep $\hat{H}_0$ in the additive decomposition of $\hat{H}$ but opt to use a perhaps simpler term than $\hat{H}_1$, one which would still be singular if treated as a perturbation to $\hat{H}_0$, but one that would offer new insights on how to identify a more reasonable fixed point Hamiltonian $\hat{H}^\star_0$ compared to $\hat{H}_0$.

Hence, we seek a similarity transformation generated by $\hat{S} = -\hat{S}^\dagger$ that acts on the space of creation and annihilation operators and such that the transformed Hamiltonian

$$\hat{H}^\star(\lambda) := e^{+\lambda\,\hat{S}}\,\hat{H}(\lambda)\,e^{-\lambda\,\hat{S}} \tag{8.265}$$

obeys

$$\left(\frac{\mathrm{d}\hat{H}^\star}{\mathrm{d}\lambda}\right)(\lambda=0)=0, \tag{8.266}$$

in the hope that the perturbation $\hat{H}^\star(\lambda)-\hat{H}_0$ leads to useful insights or simplifications. To this end, we may use the following expansion of the similarity transformation (8.265),

$$\hat{H}^\star(\lambda)=\hat{H}(\lambda)+\frac{\lambda}{1!}\,[\hat{S},\hat{H}(\lambda)]+\frac{\lambda^2}{2!}\left[\hat{S},[\hat{S},\hat{H}(\lambda)]\right]+\frac{\lambda^3}{3!}\left[\hat{S},\left[\hat{S},[\hat{S},\hat{H}(\lambda)]\right]\right]+\cdots \tag{8.267a}$$

and demand that

$$[\hat{S},\hat{H}_0] := -\hat{H}_1. \tag{8.267b}$$

Insertion of the condition (8.267b) that defines the similarity transformation $\hat{S}$ into the expansion (8.267a) gives

$$\begin{aligned}\hat{H}^\star(\lambda)=&+\hat{H}_0+\frac{(2-1)\,\lambda^2}{2!}\,[\hat{S},\hat{H}_1(\lambda)]\\&+\frac{(3-1)\,\lambda^3}{3!}\left[\hat{S},[\hat{S},\hat{H}_1(\lambda)]\right]+\frac{(4-1)\,\lambda^4}{4!}\left[\hat{S},\left[\hat{S},[\hat{S},\hat{H}_1(\lambda)]\right]\right]+\cdots,\end{aligned} \tag{8.268}$$

as desired. Computing $\hat{H}^\star$ to all orders is equivalent to diagonalizing $\hat{H}$, which we do not know how to do in the first place. However, the expansion (8.268) becomes useful if there is a range of parameters for which it may be truncated, say by dropping all terms of order λ^3 or higher.

Exercise 2.1:

(a) Verify Eqs. (8.267a) and (8.268). *Hint:* Equation (8.267a) is closely related to the Baker-Campbell-Hausdorff formula on the one hand, and to the equations of motion obeyed by operators in the Heisenberg picture in imaginary time on the other hand [recall Eq. (5.144)].

(b) Verify that

$$\hat{S}=\sum_{\boldsymbol{k}\in\mathrm{BZ}}\sum_{\sigma=\uparrow,\downarrow}\left[\frac{V_{\boldsymbol{k}}\,\hat{n}^f_{\bar{\sigma}}}{\xi_{\boldsymbol{k}}-\xi_f-U}\,\hat{c}^\dagger_{\boldsymbol{k},\sigma}\,\hat{f}_\sigma+\frac{V_{\boldsymbol{k}}\,(1-\hat{n}^f_{\bar{\sigma}})}{\xi_{\boldsymbol{k}}-\xi_f}\,\hat{c}^\dagger_{\boldsymbol{k},\sigma}\,\hat{f}_\sigma-\mathrm{H.c.}\right] \tag{8.269}$$

solves Eq. (8.267b). Use this result to justify identifying the two dimensionless ratios

$$r_J := \frac{\tilde{\tilde{\nu}}_{\mathrm{F}} \times \langle |V_{\boldsymbol{k}}|^2 \rangle_{\mathrm{Fs}}}{|\xi_f + U|}, \qquad r_W := \frac{\tilde{\tilde{\nu}}_{\mathrm{F}} \times \langle |V_{\boldsymbol{k}}|^2 \rangle_{\mathrm{Fs}}}{|\xi_f|}, \tag{8.270}$$

as the ones that control how good the expansion (8.268) is. Here, $\tilde{\tilde{\nu}}_{\mathrm{F}}$ is the density of states per unit energy of the unperturbed c electrons at the Fermi energy, and we made use of the definition (8.259) for the average over their unperturbed Fermi surface.

(c) We use a more compact notation by which $\hat{c}_{\boldsymbol{k}}$ is a column vector with the two components $\hat{c}_{\boldsymbol{k}\uparrow}$ and $\hat{c}_{\boldsymbol{k}\downarrow}$ and similarly for $\hat{f}$. We also use the three-component vector $\boldsymbol{\sigma}$ to present the three Pauli matrices. Verify that

$$\frac{1}{2}\left[\hat{S}, \hat{H}_1\right] = H_1^{(1)} + H_1^{(2)} + H_1^{(3)} + H_1^{(4)}, \tag{8.271a}$$

where

$$H_1^{(1)} := -\frac{1}{4} \sum_{\boldsymbol{k},\boldsymbol{k}' \in \mathrm{BZ}} J_{\boldsymbol{k}',\boldsymbol{k}} \left(\hat{c}^{\dagger}_{\boldsymbol{k}'}\, \boldsymbol{\sigma}\, \hat{c}_{\boldsymbol{k}}\right) \cdot \left(\hat{f}^{\dagger}\, \boldsymbol{\sigma}\, \hat{f}\right),$$

$$J_{\boldsymbol{k}',\boldsymbol{k}} := V_{\boldsymbol{k}'}\, V^{*}_{\boldsymbol{k}} \left(\frac{1}{\xi_{\boldsymbol{k}} - \xi_f - U} + \frac{1}{\xi_{\boldsymbol{k}'} - \xi_f - U} - \frac{1}{\xi_{\boldsymbol{k}} - \xi_f} - \frac{1}{\xi_{\boldsymbol{k}'} - \xi_f}\right), \tag{8.271b}$$

would be a local Heisenberg exchange interaction in the first Brillouin zone if $J_{\boldsymbol{k}',\boldsymbol{k}}$ was proportional to $\delta_{\boldsymbol{k}',\boldsymbol{k}}$,

$$\begin{aligned} H_1^{(2)} &:= \sum_{\boldsymbol{k},\boldsymbol{k}' \in \mathrm{BZ}} \left[W_{\boldsymbol{k}',\boldsymbol{k}} + \frac{1}{4} J_{\boldsymbol{k}',\boldsymbol{k}} \left(\hat{f}^{\dagger}\, \hat{f}\right)\right] \left(\hat{c}^{\dagger}_{\boldsymbol{k}'}\, \hat{c}_{\boldsymbol{k}}\right), \\ W_{\boldsymbol{k}',\boldsymbol{k}} &:= \frac{V_{\boldsymbol{k}'}\, V^{*}_{\boldsymbol{k}}}{2} \left(\frac{1}{\xi_{\boldsymbol{k}} - \xi_f} + \frac{1}{\xi_{\boldsymbol{k}'} - \xi_f}\right), \end{aligned} \tag{8.271c}$$

would be the sum of a local density of c electrons in the first Brillouin zone with a local density-density interaction between c and f electrons in the first Brillouin zone if $W_{\boldsymbol{k}',\boldsymbol{k}}$ and $J_{\boldsymbol{k}',\boldsymbol{k}}$ were both proportional to $\delta_{\boldsymbol{k}',\boldsymbol{k}}$,

$$H_1^{(3)} := -\sum_{\boldsymbol{k} \in \mathrm{BZ}} \sum_{\sigma=\uparrow,\downarrow} \left(W_{\boldsymbol{k},\boldsymbol{k}} + \frac{1}{2} J_{\boldsymbol{k},\boldsymbol{k}}\, \hat{n}^{f}_{\bar{\sigma}}\right) \hat{n}^{f}_{\sigma} \tag{8.271d}$$

shifts the chemical potential of the f electrons, and

$$H_1^{(4)} := \frac{1}{4} \sum_{\boldsymbol{k},\boldsymbol{k}' \in \mathrm{BZ}} \sum_{\sigma=\uparrow,\downarrow} \left(J_{\boldsymbol{k}',\boldsymbol{k}}\, \hat{c}^{\dagger}_{\boldsymbol{k}',\bar{\sigma}}\, \hat{c}^{\dagger}_{\boldsymbol{k},\sigma}\, \hat{f}_{\sigma}\, \hat{f}_{\bar{\sigma}} + \mathrm{H.c.}\right) \tag{8.271e}$$

would hybridize a pair of f electrons with a local pair (in the first Brillouin zone) of c electrons if $J_{\boldsymbol{k}',\boldsymbol{k}}$ was proportional to $\delta_{\boldsymbol{k}',\boldsymbol{k}}$.

(d) Verify that if $\boldsymbol{k}' = \boldsymbol{k}$ is on the unperturbed Fermi surface of the c electrons, i.e., if $\xi_{\boldsymbol{k}'} = \xi_{\boldsymbol{k}} = 0$, then

$$J_{\boldsymbol{k}',\boldsymbol{k}} = 2\,|V_{\boldsymbol{k}}|^2\,\frac{U}{\xi_f\,(\xi_f + U)}, \qquad W_{\boldsymbol{k}',\boldsymbol{k}} = -|V_{\boldsymbol{k}}|^2\,\frac{1}{\xi_f}. \tag{8.272}$$

(e) Equation (8.272) indicates that $J_{\boldsymbol{k}',\boldsymbol{k}}$ enters as an antiferromagnetic coupling for $\boldsymbol{k}' = \boldsymbol{k}$ on the unperturbed Fermi surface if and only if

$$\xi_f\,(\xi_f + U) < 0, \tag{8.273}$$

in which case the unperturbed ground state of $\hat{H}_f$ has one f electron present. The condition (8.273) is met for any negative value of ξ_f provided $\xi_f + U > 0$. In the subspace of the Hilbert space with one f electron present, $\hat{H}_1^{(2)}$ simplifies to a one-body Hamiltonian that changes the dispersion of the c electrons, $\hat{H}_1^{(3)}$ shifts the value of ξ_f by $-\sum_{\boldsymbol{k}\in\mathrm{BZ}} W_{\boldsymbol{k},\boldsymbol{k}}$, and $\hat{H}_1^{(4)}$ is inoperative, for it only has non-vanishing matrix elements between the subspaces with no and two f electrons present. Explain why the two independent conditions

$$r_J \ll 1 \tag{8.274a}$$

and

$$r_W \ll 1 \tag{8.274b}$$

imply that

$$\tilde{\tilde{\nu}}_{\mathrm{F}} \times \left\langle J_{\boldsymbol{k},\boldsymbol{k}} \right\rangle_{\mathrm{Fs}} \ll 1, \tag{8.274c}$$

and explain why, at sufficiently low temperatures, the approximation (the book-keeping parameter λ has been set to the number 1)

$$\hat{H}^{\star} \approx \hat{H}_0 + \hat{H}_1^{(1)} \tag{8.275}$$

is a good one.

Having rewritten the single-impurity Anderson model as a Kondo model shows that the Kondo problem is also present in the single-impurity Anderson model. It also shows that the coupling that drives the crossover from high to low temperature is $\left\langle J_{\boldsymbol{k},\boldsymbol{k}} \right\rangle_{\mathrm{Fs}}$ in the single-impurity Anderson model. Finally, the solution for the crossover from high to low temperatures in either one of the two models can be applied to the other model.

Mean-field approximation for the single-impurity Anderson model

We are going to work with the representation (8.249) of the single-impurity Anderson model. We are going to reproduce a mean-field approximation done by Anderson in Ref. [99] that will give us a complementary perspective to the Schrieffer-Wolff one.

Exercise 3.1: Using a Hubbard-Stratonovich transformation verify that the Euclidean action (8.249d) for the f electrons can be written as

$$S_f = \int_0^\beta d\tau \left[\sum_{\sigma=\uparrow,\downarrow} f^*_\sigma \left(\partial_\tau + \xi_f\right) f_\sigma + \frac{U}{2} \left(f^*_\uparrow f_\uparrow - f^*_\downarrow f_\downarrow\right) \varphi + \frac{U}{4} \varphi^2 \right]. \tag{8.276}$$

Hint: Make use of $4\,\hat{f}^\dagger_\uparrow \hat{f}_\uparrow \hat{f}^\dagger_\downarrow \hat{f}_\downarrow = \left(\hat{f}^\dagger_\uparrow \hat{f}_\uparrow + \hat{f}^\dagger_\downarrow \hat{f}_\downarrow\right)^2 - \left(\hat{f}^\dagger_\uparrow \hat{f}_\uparrow - \hat{f}^\dagger_\downarrow \hat{f}_\downarrow\right)^2$ and explain why and how the first term on the right-hand side of this equality can be ignored. Observe that the Hubbard-Stratonovich field φ couples linearly to the magnetic moment $\hat{f}^\dagger \frac{\sigma^z}{2} \hat{f}$ of the f electron.

Exercise 3.2: The sum of the Euclidean actions (8.249c) and (8.249e) can be written as

$$\begin{aligned} S_c + S_{c-f} = \int_0^\beta d\tau \sum_{\boldsymbol{k}\in \mathrm{BZ}} \sum_{\sigma=\uparrow,\downarrow} \Big[& \left(c^*_{\boldsymbol{k},\sigma} - f^*_\sigma V^*_{\boldsymbol{k}} G^c_{\boldsymbol{k}}\right) \left(-G^c_{\boldsymbol{k}}\right)^{-1} \\ & \times \left(c_{\boldsymbol{k},\sigma} - G^c_{\boldsymbol{k}} V_{\boldsymbol{k}} f_\sigma\right) + f^*_\sigma \left|V_{\boldsymbol{k}}\right|^2 G^c_{\boldsymbol{k}} f_\sigma \Big], \end{aligned} \tag{8.277a}$$

where we have introduced the single-particle Green function

$$\left(G^c_{\boldsymbol{k}}\right)^{-1} := -\left(\partial_\tau + \xi_{\boldsymbol{k}}\right) \tag{8.277b}$$

for the unperturbed c electrons. After going to fermionic Matsubara frequency space, we can independently shift the c^*'s and c's Grassmann integration variables and do their Grassmann integrations. Verify that the integration over the c Grassmann variables delivers

$$Z \propto \int \mathcal{D}[\varphi] \int \mathcal{D}[f^*, f]\, e^{-S'_0 - S'_1} \tag{8.278a}$$

with the effective action

$$S'_0 = \int_0^\beta d\tau \sum_{\sigma=\uparrow,\downarrow} f^*_\sigma \left(\partial_\tau + \xi_f + \hat{\Sigma}\right) f_\sigma \tag{8.278b}$$

and

$$S'_1 = \int_0^\beta d\tau \left[\frac{U}{2} \left(f^*_\uparrow f_\uparrow - f^*_\downarrow f_\downarrow\right) \varphi + \frac{U}{4} \varphi^2 \right]. \tag{8.278c}$$

The f electrons have acquired the self-energy $\hat{\Sigma}$, a non-diagonal operator in the imaginary-time representation that becomes diagonal in the Matsubara frequency representation with the $\mathbb{C}$-valued matrix elements

$$\Sigma_{\mathrm{i}\omega_n} = \sum_{\boldsymbol{k}\in \mathrm{BZ}} \frac{\left|V_{\boldsymbol{k}}\right|^2}{\mathrm{i}\omega_n - \xi_{\boldsymbol{k}}}, \qquad \omega_n = \frac{\pi}{\beta}(2n+1), \qquad n \in \mathbb{Z}, \tag{8.278d}$$

from their hybridization with the c electrons.[17] The Fourier conventions of Eq. (6.22) have been used here. The action $S_0' + S_1'$ is quadratic in the f electrons. Verify that their integration delivers the partition function

$$Z \propto \int \mathcal{D}[\varphi]\, e^{-S''[\varphi]} \tag{8.279a}$$

with the effective action

$$S''[\varphi] = \int\limits_0^\beta \mathrm{d}\tau \left[\frac{U}{4}\varphi^2 - \sum_{\sigma=+,-} \mathrm{Tr}\log\left(\partial_\tau + \xi_f + \hat{\Sigma} + \sigma\frac{U}{2}\varphi\right)\right]. \tag{8.279b}$$

The trace must be done with functions of τ that obey antiperiodic boundary conditions over the interval $[0,\beta]$. Show that

$$S''[+\varphi] = S''[-\varphi]. \tag{8.280}$$

So far no approximations were made and we have access to the projection on the quantization axis of the magnetization

$$\left\langle \hat{f}^\dagger \frac{\hbar\sigma^z}{2} \hat{f} \right\rangle_\beta := \frac{1}{\beta}\left(\frac{\delta \ln Z}{\delta B}\right)[B=0] \tag{8.281}$$

and the dynamical susceptibility

$$\chi(\tau,\tau') := \frac{1}{\beta}\left(\frac{\delta^2 \ln Z}{\delta B(\tau)\,\delta B(\tau')}\right)[B=0] \tag{8.282}$$

on the impurity site. Here, $B(\tau)$ is a source field that enters as the additive term $B(\tau)\left(\hat{f}^\dagger \frac{\hbar\sigma^z}{2}\hat{f}\right)(\tau)$ in the logarithm of S''.

Exercise 3.3: The first approximation that we are going to do is on the self-energy (8.278d). Show that

$$\mathrm{Re}\,\Sigma_{\mathrm{i}\omega_n} = -\sum_{\boldsymbol{k}\in\mathrm{BZ}} \frac{|V_{\boldsymbol{k}}|^2\,\xi_{\boldsymbol{k}}}{\omega_n^2 + \xi_{\boldsymbol{k}}^2} \tag{8.283a}$$

and

$$\mathrm{Im}\,\Sigma_{\mathrm{i}\omega_n} = -\sum_{\boldsymbol{k}\in\mathrm{BZ}} \frac{|V_{\boldsymbol{k}}|^2\,\omega_n}{\omega_n^2 + \xi_{\boldsymbol{k}}^2}. \tag{8.283b}$$

Justify the approximations

$$\mathrm{Re}\,\Sigma_{\mathrm{i}\omega_n} \approx -\sum_{\boldsymbol{k}\in\mathrm{BZ}} \frac{|V_{\boldsymbol{k}}|^2\,\xi_{\boldsymbol{k}}}{\xi_{\boldsymbol{k}}^2} \equiv \delta\xi_f \tag{8.284a}$$

and

$$\mathrm{Im}\,\Sigma_{\mathrm{i}\omega_n} \approx -\Delta\,\mathrm{sgn}\,\omega_n \tag{8.284b}$$

[17] With the convention we chose for the normalization of $V_{\boldsymbol{k}}$ in Eq. (8.248d) or in Eq. (8.250d) the self-energy is extensive, i.e., grows like N in the thermodynamic limit.

with [recall Eq. (8.259)]

$$0 \le \Delta := \pi \, \tilde{\nu}_{\mathrm{F}} \, \langle |V_{\boldsymbol{k}}|^2 \rangle_{\mathrm{Fs}}, \tag{8.284c}$$

where

$$\tilde{\nu}_{\mathrm{F}} := \sum_{\boldsymbol{k} \in \mathrm{BZ}} \delta(\xi_{\boldsymbol{k}}) \tag{8.284d}$$

is the density of states per unit energy at the unperturbed Fermi energy.

Exercise 3.4: The second approximation that we make consists in evaluating the functional derivative of the action (8.279) with the approximation (8.284) for the self-energy and assuming that φ is independent of imaginary time. Neglecting fluctuations in imaginary time ignores quantum fluctuations. Show that, at $\beta = \infty$,

$$\frac{1}{\beta} \left(\frac{\mathrm{d}S''}{\mathrm{d}\varphi} \right) (\varphi) = \frac{U}{2} \left(1 - \frac{U}{\bar{\Delta}} \right) \varphi + \mathcal{O}\left(\varphi^2\right), \tag{8.285a}$$

where

$$\bar{\Delta} := \pi \frac{\Delta^2 + \left(\xi_f + \delta\xi_f\right)^2}{\Delta}. \tag{8.285b}$$

Hint: You will need the integral

$$\int \mathrm{d}x \, \frac{1}{(x+a)^2 + b^2} = \frac{1}{b} \arctan \frac{x+a}{b} \tag{8.286}$$

and the expansion

$$\arctan x = \frac{\pi}{2} - \frac{1}{x} + \cdots \tag{8.287}$$

for $x \gg 1$.

Exercise 3.5: Show that by combining Eq. (8.280) with Eq. (8.285), the condition

$$\frac{U}{\bar{\Delta}} > 1 \tag{8.288}$$

guarantees that $\varphi = 0$ is a local maximum of $S''(\cdot)$ and that there exists at least two minima of $S''(\cdot)$ at $\pm\varphi_0$. Each minimum corresponds to one out of the two possible orientations along the quantization axis of the local magnetic moment. Comment on the similarities and differences between the criterion (8.288) which is known as the Anderson criterion for the emergence of a local magnetic moment in the single-impurity Anderson model, and the Schrieffer-Wolff criterion (8.274) relating the single-impurity Anderson model to the Kondo model.

Assume that the function $S''(\cdot)$ has two absolute minima at $\pm\varphi_0$ when $U > \bar{\Delta}$ and that $\varphi = 0$ is the absolute minimum of $S''(\cdot)$ otherwise. Increasing the dimensionless and positive parameter $U/\bar{\Delta}$ drives a continuous mean-field transition at the value $U/\bar{\Delta} = 1$ from the regime $0 \le U/\bar{\Delta} \le 1$ with one absolute minimum at $\varphi = 0$ of $S''(\cdot)$ to the regime $1 \le U/\bar{\Delta} < \infty$ with two absolute minima at $\pm\varphi_0$ of

$S''(\cdot)$. Can we trust this mean-field prediction? The short answer is negative. Short of a calculation, we can list the following arguments supporting this conclusion.

We know that quantum fluctuations wipe out the semi-classical quantum phase transition of a $(0+1)$-dimensional quantum field theory that is local in time. This is the case of a quantum particle with the kinetic energy $(\partial_\tau\varphi)^2$ and with the double-well potential energy $V(+\varphi)=V(-\varphi)$, whose ground state wave function is an even function of φ with equal probability to have the quantum particle at any one of the two minima of the potential well.

However, the difficulty with the functional $S''[\cdot]$ in Eq. (8.279) to be integrated over in the path integral (8.279a) is that it is not local in imaginary time, for the argument of the logarithm on the right-hand side of Eq. (8.279b) encodes the coupling to a dissipative bath. Non-locality in time opens the possibility that elimination of high-energy quantum fluctuations (large values of $\partial_\tau\varphi$) modify the quantum dynamics of the remaining quantum fluctuations, i.e., the separation into a kinetic and potential contribution to the action. In the case of a local action in imaginary time with a double-well potential, integration of high-energy quantum fluctuations preserves the double-well shape at all energies. Anderson, Yuval, and Hamann on the one hand and Wilson on the other hand tell us that this is not the case for the quantum problem (8.279).

We may imagine the following scenario against spontaneous symmetry breaking of the symmetry under $\varphi\to-\varphi$ of the quantum problem (8.279). At high energies set by the energy scale U the potential has two absolute minima representing the two possible orientations of localized spin-1/2 degree of freedom on the impurity site. At low energies this potential has one absolute minimum representing a localized spin-1/2 degree of freedom that has been screened by the conduction electrons through the formation of a singlet bound state. The flow of this potential from high to low energies corresponds to the crossover from Curie's law $\chi(T)\propto(1/T)$ at high temperatures to the Pauli susceptibility $\chi(T)\propto(1/T_{\mathrm{K}})$ at temperatures well below the Kondo temperature T_{K}.

Large N_{c} expansion for the single-impurity Anderson model

The mean-field approximation

$$Z\propto\int\mathcal{D}[\varphi]\,e^{-S''[\varphi]}\approx e^{-S''(\bar\varphi)},\qquad\left(\frac{\mathrm{d}S''}{\mathrm{d}\varphi}\right)[\bar\varphi]=0,\tag{8.289}$$

dictates that Curie's law can only be obeyed if condition (8.288) holds. However, this mean-field approximation is an uncontrolled approximation. Following Coleman in Ref. [101], we are going to introduce a model for which (i) there exists an exact mean-field theory, and such that (ii) the qualitative behavior expected from exact solutions of the Kondo and related models is captured.

The quantum impurity problem is defined by the partition function

$$Z:=\lim_{U\to\infty}\int\mathcal{D}[c^*,c]\int\mathcal{D}[f^*,f]\,e^{-S}\tag{8.290a}$$

with the action

$$S := S_{c} + S_{f} + S_{c-f} \tag{8.290b}$$

decomposing into the Euclidean action for the c electrons

$$S_{c} := \int\limits_{0}^{\beta} \mathrm{d}\tau \sum_{\boldsymbol{k}\in\mathrm{BZ}} \sum_{\sigma=1}^{N_{\mathrm{c}}} \left[c^{*}_{\boldsymbol{k},\sigma}\left(\partial_{\tau}+\xi_{\boldsymbol{k}}\right) c_{\boldsymbol{k},\sigma}\right], \tag{8.290c}$$

the Euclidean action for the f electrons

$$S_{f} := \int\limits_{0}^{\beta} \mathrm{d}\tau \left[\sum_{\sigma=1}^{N_{\mathrm{c}}} f^{*}_{\sigma}\left(\partial_{\tau}+\xi_{f}\right) f_{\sigma} + U\left(\sum_{\sigma=1}^{N_{\mathrm{c}}} f^{*}_{\sigma} f_{\sigma} - 1\right)\left(\sum_{\sigma=1}^{N_{\mathrm{c}}} f^{*}_{\sigma} f_{\sigma}\right)\right], \tag{8.290d}$$

and the Euclidean action that couples the c electrons to the f electrons

$$S_{c-f} := \int\limits_{0}^{\beta} \mathrm{d}\tau \sum_{\boldsymbol{k}\in\mathrm{BZ}} \sum_{\sigma=1}^{N_{\mathrm{c}}} \frac{1}{\sqrt{N_{\mathrm{c}}}}\left(V_{\boldsymbol{k}}\, c^{*}_{\boldsymbol{k},\sigma} f_{\sigma} + V^{*}_{\boldsymbol{k}}\, f^{*}_{\sigma} c_{\boldsymbol{k},\sigma}\right). \tag{8.290e}$$

As usual, all Grassmann-valued fields obey antiperiodic boundary conditions in imaginary time.

This model differs from the single-site impurity Anderson model (8.249) in two ways. First, the spin index is now a color index σ that runs from 1 to N_{c}. The limit of infinitely many colors, $N_{\mathrm{c}} \to \infty$, is the one for which the mean-field approximation to come is exact. Second, we have replaced the Hubbard on-site repulsive interaction by an on-site repulsive interaction that penalizes with the positive energy $(n-1)\,n\,U$ the occupation of the impurity site by $1 < n \leq N_{\mathrm{c}}$ electrons. The limit $U \to \infty$ is taken from the outset, i.e., a counterpart to Anderson criterion (8.288) would be met if it exists. In the limit $U \to \infty$, the Fock space of the f electrons

$$\begin{aligned}\mathfrak{F}^{\mathrm{phys}}_{f} := \mathrm{span}\Big\{|0\rangle_{f}, \hat{f}^{\dagger}_{\sigma}\,|0\rangle_{f} \,\Big|\, \hat{f}_{\sigma}\,|0\rangle_{f} = 0, \qquad &\{\hat{f}_{\sigma}, \hat{f}^{\dagger}_{\sigma'}\} = \delta_{\sigma,\sigma'}, \\ &\{\hat{f}^{\dagger}_{\sigma}, \hat{f}^{\dagger}_{\sigma'}\} = \{\hat{f}_{\sigma}, \hat{f}_{\sigma'}\} = 0, \qquad \sigma, \sigma' = 1, \cdots, N_{\mathrm{c}}\Big\}\end{aligned} \tag{8.291a}$$

is $(N_{\mathrm{c}}+1)$-dimensional. In contrast, the Fock space for the conduction electrons

$$\begin{aligned}\mathfrak{F}_{c} = \mathrm{span}\Bigg\{\prod_{\boldsymbol{k}\in\mathrm{BZ}} \prod_{\sigma=1}^{N_{\mathrm{c}}} \left(\hat{c}^{\dagger}_{\boldsymbol{k},\sigma}\right)^{n_{\boldsymbol{k},\sigma}} |0\rangle_{c} \,\Bigg|\, \hat{c}_{\boldsymbol{k},\sigma}\,|0\rangle_{c} = 0, \qquad \{\hat{c}_{\boldsymbol{k},\sigma}, \hat{c}^{\dagger}_{\boldsymbol{k}',\sigma'}\} = \delta_{\boldsymbol{k},\boldsymbol{k}'}\,\delta_{\sigma,\sigma'}, \\ \{\hat{c}^{\dagger}_{\boldsymbol{k},\sigma}, \hat{c}^{\dagger}_{\boldsymbol{k}',\sigma'}\} = \{\hat{c}_{\boldsymbol{k},\sigma}, \hat{c}_{\boldsymbol{k}',\sigma'}\} = 0, \quad n_{\boldsymbol{k},\sigma} = 0,1, \quad \boldsymbol{k}, \boldsymbol{k}' \in \mathrm{BZ}, \quad \sigma, \sigma' = 1, \cdots, N_{\mathrm{c}}\Bigg\}\end{aligned} \tag{8.291b}$$

is $2^{N \times N_{\mathrm{c}}}$-dimensional. The Fock space over which the partition function Z is to be traced is

$$\mathfrak{F} := \mathfrak{F}_{c} \otimes \mathfrak{F}^{\mathrm{phys}}_{f}. \tag{8.291c}$$

Implementing constraints is inherently a strong coupling problem as the limit $U \to \infty$ makes explicit. The constraint

$$\sum_{\sigma=1}^{N_{\mathrm{c}}} \hat{f}^{\dagger}_{\sigma} \hat{f}_{\sigma} \leq 1 \tag{8.292}$$

is difficult to implement analytically because it is an inequality. The slave boson method was devised to turn this constraint into an operator equality on a different Hilbert space [69,70,101]. We introduce the infinite-dimensional bosonic Fock space

$$\mathfrak{F}_{b} = \mathrm{span}\left\{ \prod_{n=0}^{\infty} \frac{(\hat{b}^{\dagger})^{n}}{\sqrt{n!}} |0\rangle_{b} \middle| \hat{b}\,|0\rangle_{b} = 0, \quad [\hat{b},\hat{b}^{\dagger}] = 1, \quad [\hat{b}^{\dagger},\hat{b}^{\dagger}] = [\hat{b},\hat{b}] = 0 \right\} \tag{8.293a}$$

and the $2^{N_{\mathrm{c}}}$-dimensional fermionic Hilbert space

$$\begin{aligned}\mathfrak{F}_{s} = \mathrm{span}\Bigg\{ \prod_{\sigma=1}^{N_{\mathrm{c}}} (\hat{s}^{\dagger}_{\sigma})^{n_{\sigma}} |0\rangle_{s} \Bigg| \hat{s}_{\sigma}\,|0\rangle_{s} = 0, \qquad \{\hat{s}_{\sigma},\hat{s}^{\dagger}_{\sigma'}\} = \delta_{\sigma,\sigma'}, \\ \{\hat{s}^{\dagger}_{\sigma},\hat{s}^{\dagger}_{\sigma'}\} = \{\hat{s}_{\sigma},\hat{s}_{\sigma'}\} = 0, \quad n_{\sigma} = 0,1, \quad \sigma,\sigma' = 1,\cdots,N_{\mathrm{c}} \Bigg\}.\end{aligned} \tag{8.293b}$$

The $(N_{\mathrm{c}}+1)$-dimensional physical Fock space is

$$\mathfrak{F}^{\mathrm{phys}}_{b\times s} := \left\{ |b\rangle \otimes |s\rangle \middle| |b\rangle \in \mathfrak{F}_{b}, \quad |s\rangle \in \mathfrak{F}_{s}, \quad \left(\hat{b}^{\dagger}\hat{b} + \sum_{\sigma=1}^{N_{\mathrm{c}}} \hat{s}^{\dagger}_{\sigma}\hat{s}_{\sigma} \right) |b\rangle \otimes |s\rangle = |b\rangle \otimes |s\rangle \right\}. \tag{8.293c}$$

Exercise 4.1: Show that an isomorphism between $\mathfrak{F}^{\mathrm{phys}}_{f}$ and $\mathfrak{F}^{\mathrm{phys}}_{b\times s}$ is established by the maps

$$\begin{aligned} |0\rangle_{f} &\longmapsto \hat{b}^{\dagger}\,|0\rangle_{b}, & \\ \hat{f}^{\dagger}_{\sigma}\,|0\rangle_{f} &\longmapsto \hat{s}^{\dagger}_{\sigma}|0\rangle_{s}, & \sigma = 1,\cdots,N_{\mathrm{c}}, \\ \hat{f}^{\dagger}_{\sigma} &\longmapsto \hat{s}^{\dagger}_{\sigma}\,\hat{b}, & \sigma = 1,\cdots,N_{\mathrm{c}}, \\ \hat{f}_{\sigma} &\longmapsto \hat{b}^{\dagger}\,\hat{s}_{\sigma}, & \sigma = 1,\cdots,N_{\mathrm{c}}. \end{aligned} \tag{8.294}$$

Hint: Verify first that the fermionic algebra of the $\hat{f}$'s is preserved under this mapping. Verify then that the matrix elements of the $\hat{f}$'s in $\mathfrak{F}^{\mathrm{phys}}_{f}$ are in one-to-one correspondence with those of the $\hat{b}^{\dagger}\,\hat{s}$'s in $\mathfrak{F}^{\mathrm{phys}}_{b\times s}$.

Exercise 4.2: Convince yourself that there are uncountably many distinct ways of representing the $(N_{\mathrm{c}}+1)$-dimensional physical Fock space $\mathfrak{F}^{\mathrm{phys}}_{f}$ with auxiliary bosonic and fermionic operators. *Hint:* Do this by way of three examples. Define a transformation by which the phases of the $\hat{s}$ and $\hat{b}$ operators are changed without changing their algebra and leaving the $\hat{f}$ operators unchanged. Second, allow the b bosons to acquire the same index σ as the s fermions. Third, choose the $\hat{b}$ operators to be spinless fermions and the $\hat{s}$ operators to be bosons. This representation is called the slave-fermion representation.

Exercise 4.3: Verify that the partition function (8.290) can be presented as the partition function

$$Z := \int \mathcal{D}[c^*, c] \int \mathcal{D}[\lambda] \int \mathcal{D}[s^*, s] \int \mathcal{D}[b^*, b]\, e^{-S}, \tag{8.295a}$$

where the action

$$S := S_c + S_s + S_b + S_{c-b\times s} - \mathrm{i} \int_0^\beta \mathrm{d}\tau\, \lambda \tag{8.295b}$$

is here decomposed into the bilinear Euclidean action for the c electrons

$$S_c := \int_0^\beta \mathrm{d}\tau \sum_{\boldsymbol{k}\in \mathrm{BZ}} \sum_{\sigma=1}^{N_c} c^*_{\boldsymbol{k},\sigma} \left(\partial_\tau + \xi_{\boldsymbol{k}}\right) c_{\boldsymbol{k},\sigma}, \tag{8.295c}$$

the bilinear Euclidean action for the s electrons

$$S_s := \int_0^\beta \mathrm{d}\tau \sum_{\sigma=1}^{N_c} s^*_\sigma \left(\partial_\tau + \xi_f + \mathrm{i}\lambda\right) s_\sigma, \tag{8.295d}$$

the bilinear Euclidean action for the b bosons

$$S_b := \int_0^\beta \mathrm{d}\tau\, b^* \left(\partial_\tau + \mathrm{i}\lambda\right) b, \tag{8.295e}$$

and the Euclidean action that couples the c electrons to the s electrons and b bosons

$$S_{c-b\times s} := \int_0^\beta \mathrm{d}\tau \sum_{\boldsymbol{k}\in \mathrm{BZ}} \sum_{\sigma=1}^{N_c} \frac{1}{\sqrt{N_c}} \left(V_{\boldsymbol{k}}\, c^*_{\boldsymbol{k},\sigma} s_\sigma\, b^* + V^*_{\boldsymbol{k}}\, s^*_\sigma\, b\, c_{\boldsymbol{k},\sigma}\right). \tag{8.295f}$$

All the Grassmann integration variables c^*, c, s^*, and s are independent and obey antiperiodic boundary conditions in imaginary time. The complex-valued field b^* is the complex conjugate of b, the latter obeying periodic boundary conditions in imaginary time. The real-valued field λ enforces the projection onto the physical Hilbert space (8.293c). It obeys periodic boundary conditions in imaginary time.

Exercise 4.4:

(a) Verify that the partition function (8.295) is invariant under the $U(1)$ local gauge transformation

$$\begin{aligned} &s^*_\sigma \to s^*_\sigma\, e^{+\mathrm{i}\phi}, && s_\sigma \to e^{-\mathrm{i}\phi}\, s_\sigma, && \sigma = 1, \cdots, N_c, \\ &b^* \to b^*\, e^{+\mathrm{i}\phi}, && b \to e^{-\mathrm{i}\phi}\, b, && \\ &\lambda \to \lambda + \partial_\tau \phi, && && \end{aligned} \tag{8.296}$$

for any real-valued and smooth function $\phi : [0, \beta] \to [0, 2\pi[, \tau \mapsto \phi(\tau)$ that obeys periodic boundary conditions in imaginary time.

(b) Use the notation

$$\hat{n}_{b\times s} := \hat{b}^\dagger\,\hat{b} + \sum_{\sigma=1}^{N_c} \hat{s}^\dagger_\sigma\,\hat{s}_\sigma, \qquad n_{b\times s} := b^*\,b + \sum_{\sigma=1}^{N_c} s^*_\sigma\,s_\sigma. \tag{8.297}$$

Show that

$$\begin{aligned} 0 =& \left\langle \left[\hat{n}_{b\times s}(\tau_1) - 1\right] \cdots \left[\hat{n}_{b\times s}(\tau_n) - 1\right] \right\rangle_Z \\ :=& \frac{\operatorname{Tr}\left\{e^{-\beta\hat{H}}\left[\hat{n}_{b\times s}(\tau_1) - 1\right] \cdots \left[\hat{n}_{b\times s}(\tau_n) - 1\right]\right\}}{\operatorname{Tr}\left\{e^{-\beta\hat{H}}\right\}}. \end{aligned} \tag{8.298}$$

Hint: Add a source field to the action S that couples to $n_{b\times s} - 1$. Express the correlation function (8.298) as a functional derivative of the partition with source field. Use the $U(1)$ local gauge symmetry (8.296) to show that the partition function with source field equals the partition function without source field.

Exercise 4.5: Choose the parametrization

$$b^*(\tau) = \sqrt{N_\mathrm{c}}\,\rho(\tau)\,e^{-\mathrm{i}\theta(\tau)}, \qquad b\,(\tau) = \sqrt{N_\mathrm{c}}\,\rho(\tau)\,e^{+\mathrm{i}\theta(\tau)}, \tag{8.299}$$

in terms of the real-valued amplitude field $\rho : [0,\beta] \to [0,\infty[, \tau \mapsto \rho(\tau)$ and the real-valued phase field $\theta : [0,\beta] \to [0,2\pi[, \tau \mapsto \theta(\tau)$. The transformation law of the measure is

$$\mathcal{D}[b^*, b] = \mathcal{D}[\rho,\theta], \qquad \mathrm{d}b^*(\tau)\,\mathrm{d}b(\tau) = \frac{N_\mathrm{c}}{2}\,\mathrm{d}[\rho^2(\tau)]\,\mathrm{d}\theta(\tau). \tag{8.300}$$

Verify that the partition function (8.295a) does not depend on the phase field θ and is given by

$$Z \propto \int \mathcal{D}[c^*,c] \int \mathcal{D}[\lambda] \int \mathcal{D}[s^*,s] \int \mathcal{D}[\rho]\,e^{-S}, \tag{8.301a}$$

where the action

$$S := S_c + S_s + S_\rho + S_{c-\rho\times s} - \mathrm{i}N_\mathrm{c}\int_0^\beta \mathrm{d}\tau\,\frac{\lambda}{N_\mathrm{c}} \tag{8.301b}$$

is here decomposed into the bilinear Euclidean action for the c electrons

$$S_c := \int_0^\beta \mathrm{d}\tau \sum_{\boldsymbol{k}\in\mathrm{BZ}} \sum_{\sigma=1}^{N_\mathrm{c}} c^*_{\boldsymbol{k},\sigma}\left(\partial_\tau + \xi_{\boldsymbol{k}}\right) c_{\boldsymbol{k},\sigma}, \tag{8.301c}$$

the bilinear Euclidean action for the s electrons

$$S_s := \int_0^\beta \mathrm{d}\tau \sum_{\sigma=1}^{N_\mathrm{c}} s^*_\sigma\left(\partial_\tau + \xi_f + \mathrm{i}\lambda\right) s_\sigma, \tag{8.301d}$$

the bilinear Euclidean action for the amplitude of the b bosons

$$S_{\rho} := N_{\mathrm{c}} \int\limits_{0}^{\beta} \mathrm{d}\tau\, \rho\,(\partial_{\tau} + \mathrm{i}\lambda)\,\rho, \tag{8.301e}$$

and the Euclidean action that couples the c electrons to the s electrons and b bosons

$$S_{c-\rho\times s} := \int\limits_{0}^{\beta} \mathrm{d}\tau \sum_{\boldsymbol{k}\in\mathrm{BZ}} \sum_{\sigma=1}^{N_{\mathrm{c}}} \rho\,\left(V_{\boldsymbol{k}}\, c^{*}_{\boldsymbol{k},\sigma} s_{\sigma} + V^{*}_{\boldsymbol{k}}\, s^{*}_{\sigma}\, c_{\boldsymbol{k},\sigma}\right). \tag{8.301f}$$

Observe that the replacements

$$V_{\boldsymbol{k}} \to \sqrt{\langle |V_{\boldsymbol{k}}|^2\rangle_{\mathrm{Fs}}}, \qquad S_{\rho} \to N_{\mathrm{c}} \int\limits_{0}^{\beta} \mathrm{d}\tau \left(\frac{\langle |V_{\boldsymbol{k}}|^2\rangle_{\mathrm{Fs}}}{J_0}\rho^2 - \frac{\xi_f}{N_{\mathrm{c}}}\right), \qquad J_0 > 0, \tag{8.302}$$

define the large N_{c} limit of the Coqblin-Schrieffer model solved by Read and Newns in Ref. [69].[18]

Exercise 4.6: Verify that integration over the c Grassmann variables followed by integration over the s Grassmann variables gives the partition function

$$Z = \int \mathcal{D}[\lambda] \int \mathcal{D}[\rho]\, e^{-N_{\mathrm{c}}\,\beta\, F[\lambda,\rho]}, \tag{8.303a}$$

$$F[\lambda,\rho] := -\frac{1}{\beta}\,\mathrm{Tr}\,\log\left(\partial_{\tau} + \xi_f + \mathrm{i}\lambda + \rho\,\hat{\Sigma}\,\rho\right) + \frac{1}{\beta}\int\limits_{0}^{\beta} \mathrm{d}\tau \left[\rho\,(\partial_{\tau} + \mathrm{i}\lambda)\,\rho - \frac{\mathrm{i}\lambda}{N_{\mathrm{c}}}\right]. \tag{8.303b}$$

So far no approximations were made. *Hint:* All the preparatory work has been done when deriving Eqs. (8.278) and (8.279) from which we borrow the definition of the self-energy operator $\hat{\Sigma}$.

Exercise 4.7: Convince yourself that, in the limit $N_{\mathrm{c}} \to \infty$, the partition function (8.303a) simplifies to

$$Z \propto \sum_{\{\bar{\lambda},\bar{\rho}\}} \lim_{N_{\mathrm{c}}\to\infty} e^{-N_{\mathrm{c}}\,\beta\, F(\bar{\lambda},\bar{\rho})}, \tag{8.304a}$$

where a pair $\{\bar{\lambda},\bar{\rho}\}$ is a saddle-point solution to the equations

$$0 = \left(\frac{\delta F}{\delta\rho(\tau)}\right)[\lambda,\rho] \tag{8.304b}$$

and

$$0 = \left(\frac{\delta F}{\delta\lambda(\tau)}\right)[\lambda,\rho], \tag{8.304c}$$

[18] In the Coqblin-Schrieffer model [102], the field b is not needed to enforce a constraint, it is a Hubbard-Stratonovich field introduced to decouple an interaction between the conduction and impurity electrons. The field λ retains its role as a $U(1)$ temporal gauge field enforcing the constraint that there be no less and no more than one electron at the impurity site. For this reason, the partition function of the Coqblin-Schrieffer model has no separate dependences on ξ_f and $\mathrm{i}\lambda$. It only depends on the linear combination $\xi_f + \mathrm{i}\lambda$.

that is an absolute minimum (possibly degenerate) of the functional (8.303b). We seek all pairs of solutions $\{\bar{\lambda},\bar{\rho}\}$ that are static in imaginary time, as is expected in thermodynamic equilibrium given a conserved Hamiltonian.

Exercise 4.8: From now on, we are going to make the analytical continuation $\lambda\to-\mathrm{i}\lambda$. For ρ and λ independent of imaginary time, show that the functional (8.303b) becomes

$$\begin{aligned}F(-\mathrm{i}\lambda,\rho)=&-\frac{1}{\beta}\sum_{\omega_n}\log\left(-\mathrm{i}\omega_n+\xi_f+\lambda+\rho^2\,\Sigma_{\mathrm{i}\omega_n}\right)+\lambda\left(\rho^2-\frac{1}{N_{\mathrm{c}}}\right)\\=&\oint\limits_{\Gamma}\frac{\mathrm{d}z}{2\pi\mathrm{i}}\,\tilde{f}_{\mathrm{FD}}(z)\,\log\left(-z+\xi_f+\lambda+\rho^2\,\Sigma_z\right)+\lambda\left(\rho^2-\frac{1}{N_{\mathrm{c}}}\right).\end{aligned}\qquad(8.305)$$

The closed path Γ in the z-complex plane is oriented counterclockwise and runs parallel and infinitesimally close to the left and right sides of the imaginary axis. The self-energy $\Sigma_{\mathrm{i}\omega_n}$ acquired by the s electrons as a result of their hybridization to the c electrons was defined in Eq. (8.278d) and approximated by

$$\Sigma_{\mathrm{i}\omega_n}\approx\langle|V_{\boldsymbol{k}}|^2\rangle_{\mathrm{Fs}}\int\mathrm{d}\xi\,\frac{\tilde{\tilde{\nu}}(\xi)}{\mathrm{i}\omega_n-\xi},\qquad\tilde{\tilde{\nu}}(\xi):=\sum_{\boldsymbol{k}\in\mathrm{BZ}}\delta(\xi-\xi_{\boldsymbol{k}}).\qquad(8.306)$$

Hint: Use Eq. (6.59).

Exercise 4.9: It is time to make some assumptions about the nature of the conduction band. We need to distinguish two cases. For the standard case, $\tilde{\tilde{\nu}}_{\mathrm{F}}\equiv\tilde{\tilde{\nu}}(\xi=0)$ is non-vanishing. We then make the simplifying assumption

$$\tilde{\tilde{\nu}}(\xi)=\tilde{\tilde{\nu}}_{\mathrm{F}}\left[\Theta\left(\frac{\xi}{D}+1\right)-\Theta\left(\frac{\xi}{D}-1\right)\right],\qquad(8.307)$$

where Θ is the Heaviside function and the positive energy D is half the bandwidth. This is a mild assumption with no qualitative, only quantitative, consequences. The less common case is that of the power-law dependence

$$\tilde{\tilde{\nu}}(\xi)=\frac{C_r}{D^{1+r}}\times|\xi|^r\left[\Theta\left(\frac{\xi}{D}+1\right)-\Theta\left(\frac{\xi}{D}-1\right)\right]\qquad(8.308)$$

with $r>-1$ a real number and C_r a positive dimensionless number (for $r=0$, $C_0=D\,\tilde{\tilde{\nu}}_{\mathrm{F}}$) was first explored by Withoff and Fradkin in Ref. [103]. (On a square lattice with nearest-neighbor hopping and at half-filling, the density of states is logarithmically divergent. This would be another instance deviating from the standard Fermi liquid scenario.) The choices $r=1$ and $r=2$ apply to graphene and bilayer graphene, respectively. Verify that the approximation (8.306) has the two analytical continuations

$$\begin{aligned}\Sigma_{\omega\mp\mathrm{i}0^+}\approx&\ \langle|V_{\boldsymbol{k}}|^2\rangle_{\mathrm{Fs}}\int\mathrm{d}\xi\,\tilde{\tilde{\nu}}(\xi)\left[\frac{\mathrm{PV}}{\omega-\xi}\pm\mathrm{i}\pi\,\delta(\omega-\xi)\right]\\=&\ C_r\,\frac{\langle|V_{\boldsymbol{k}}|^2\rangle_{\mathrm{Fs}}}{D}\int\limits_{-D}^{+D}\mathrm{d}\xi\,\left(\frac{\xi}{D}\right)^r\left[\frac{\mathrm{PV}}{\omega-\xi}\pm\mathrm{i}\pi\,\delta(\omega-\xi)\right],\end{aligned}\qquad(8.309a)$$

where $\frac{\mathrm{PV}}{x}$ denotes the principal value of $1/x$. Hence,

$$\mathrm{Re}\,\Sigma_{\omega\mp\mathrm{i}0^+} \approx C_r\,\frac{\langle|V_{\boldsymbol{k}}|^2\rangle_{\mathrm{Fs}}}{D}\begin{cases}\int\limits_{-D}^{+D}\mathrm{d}\xi\,\frac{(\xi/D)^r}{\omega-\xi}, & \text{if } |\omega|>D,\\[2ex] \left(\int\limits_{-D}^{\omega-0^+}+\int\limits_{\omega+0^+}^{+D}\right)\mathrm{d}\xi\,\frac{(\xi/D)^r}{\omega-\xi}, & \text{otherwise,}\end{cases} \tag{8.309b}$$

while

$$\mathrm{Im}\,\Sigma_{\omega\mp\mathrm{i}0^+} \approx \pm\pi\,C_r\,\frac{\langle|V_{\boldsymbol{k}}|^2\rangle_{\mathrm{Fs}}}{D}\left(\frac{\omega}{D}\right)^r\left[\Theta\left(\frac{\omega}{D}+1\right)-\Theta\left(\frac{\omega}{D}-1\right)\right]. \tag{8.309c}$$

Compute $\mathrm{Re}\,\Sigma_{\omega\mp\mathrm{i}0^+}$ as it will be needed when solving the saddle-point Eqs. (8.340).

Exercise 4.10: We are going to solve the single-impurity Anderson model in the limit $N_{\mathrm{c}}\to\infty$ when the density of states per unit energy of the conduction electrons is non-vanishing at the Fermi energy and smooth across the band.

(a) Assumption (8.307) allows to define [recall Eq. (8.284c)]

$$\Delta := \pi\,\tilde{\nu}_{\mathrm{F}}\,\langle|V_{\boldsymbol{k}}|^2\rangle_{\mathrm{Fs}}. \tag{8.310}$$

Show that Eq. (8.305) can be written[19]

$$F(-\mathrm{i}\lambda,\rho)=\int\limits_{-D}^{+D}\frac{\mathrm{d}\omega}{\pi}\,\tilde{f}_{\mathrm{FD}}(\omega)\arctan\left(\frac{\Delta\,\rho^2}{\omega-\xi_f-\lambda}\right)+\lambda\left(\rho^2-\frac{1}{N_{\mathrm{c}}}\right). \tag{8.311}$$

Hint: Explain why you may deform the integration path Γ according to the rule

$$\oint\limits_{\Gamma}\mathrm{d}z\to\int\limits_{-\infty}^{+\infty}\mathrm{d}(\omega+\mathrm{i}0^+)+\int\limits_{+\infty}^{-\infty}\mathrm{d}(\omega-\mathrm{i}0^+). \tag{8.312}$$

You may then use

$$\log w-\log w^* = \ln|w|+\mathrm{i}\arg w-\ln|w^*|-\mathrm{i}\arg w^* = 2\mathrm{i}\arg w \tag{8.313}$$

for any complex valued w, and in particular for the case when $w=\omega-\xi_f-\lambda+\mathrm{i}\Delta\,\rho^2$.

At zero temperature, the Fermi-Dirac distribution becomes the Heaviside function $\Theta(-\omega)$. The integral over the bandwidth on the right-hand side of Eq. (8.311) can be done in closed form with the help of

$$\int\mathrm{d}x\arctan\left(\frac{a}{x-b}\right)=\frac{a}{2}\ln\left(a^2+(b-x)^2\right)-b\arctan\left(\frac{b-x}{a}\right)+x\arctan\left(\frac{a}{x-b}\right) \tag{8.314}$$

and

$$\arctan(x)+\arctan(1/x)=\frac{\pi}{2}. \tag{8.315}$$

[19] If non-vanishing, the real part of the self-energy renormalizes ξ_f. We use the same symbol ξ_f for the renormalized value.

(b) Verify that, to leading order in ξ_f/D, λ/D, or $\Delta\,\rho^2/D$,

$$F(-\mathrm{i}(\lambda-\xi_f),\rho) = -\frac{\Delta\,\rho^2}{\pi} + \frac{\lambda}{\pi}\arctan\frac{\Delta\,\rho^2}{\lambda} + \frac{\Delta\,\rho^2}{2\pi}\ln\frac{\lambda^2+\Delta^2\,\rho^4}{D^2} + (\lambda-\xi_f)\left(\rho^2-\frac{1}{N_{\mathrm{c}}}\right). \tag{8.316}$$

The large N_{c} limit of the Coqblin-Schrieffer model that we defined by the replacement (8.302) is, to leading order in ξ_f/D, λ/D, or $\Delta\,\rho^2/D$,

$$F(-\mathrm{i}(\lambda-\xi_f),\rho) = -\frac{\Delta\,\rho^2}{\pi} + \frac{\lambda}{\pi}\arctan\frac{\Delta\,\rho^2}{\lambda} + \frac{\Delta\,\rho^2}{2\pi}\ln\frac{\lambda^2+\Delta^2\,\rho^4}{D^2} + \left(\frac{\Delta\,\rho^2}{\pi\tilde{\tilde{\nu}}_{\mathrm{F}}\,J_0}-\frac{\lambda}{N_{\mathrm{c}}}\right). \tag{8.317}$$

Observe that the free energy of the single-impurity Anderson model in the large N_{c} limit depends on three microscopic energy scales within the approximation that we made. There is the energy scale ξ_f to occupy the impurity site with one electron. There is the hybridization energy scale Δ, the product of the squared coupling between the conduction and impurity electrons with the density of states per unit energy of the conduction electrons averaged over the Fermi surface of the conduction electrons. There is the band width $2\,D$ of the conduction electrons. The argument of the logarithm on the right-hand side of both Eqs. (8.316) and (8.317) defines the new energy scale

$$k_{\mathrm{B}}\,T_{\mathrm{K}}(\lambda,\rho) := \sqrt{(\xi_f+\lambda)^2+\Delta^2\,\rho^4}, \tag{8.318}$$

the Kondo energy scale in the limit $N_{\mathrm{c}}\to\infty$. The Kondo energy scale is the natural unit in which the bandwidth of the conduction electrons is to be measured on a logarithmic scale.

(c) Show that the saddle-point equations

$$\left(\frac{\mathrm{d}F}{\mathrm{d}\rho}\right)(-\mathrm{i}\lambda,\rho) = 0 \tag{8.319a}$$

and

$$\left(\frac{\mathrm{d}F}{\mathrm{d}\lambda}\right)(-\mathrm{i}\lambda,\rho) = 0 \tag{8.319b}$$

have the solutions

$$\sqrt{\left(\bar{\lambda}+\xi_f\right)^2+\Delta^2\,\bar{\rho}^4} = D\,e^{-\pi\,\bar{\lambda}/\Delta} \tag{8.320a}$$

and

$$\frac{1}{\pi}\arctan\frac{\Delta\,\bar{\rho}^2}{\bar{\lambda}+\xi_f} = \frac{1}{N_{\mathrm{c}}} - \bar{\rho}^2, \tag{8.320b}$$

respectively. The term $1/N_{\mathrm{c}}$ on the right-hand side of Eq. (8.320b) is the only term present in the large N_{c} saddle-point solutions

$$\sqrt{\left(\bar{\lambda}+\xi_f\right)^2+\Delta^2\,\bar{\rho}^4}=D\,e^{-1/(\tilde{\nu}_{\mathrm{F}}\,J_0)} \tag{8.321a}$$

and

$$\frac{1}{\pi}\arctan\frac{\Delta\,\bar{\rho}^2}{\bar{\lambda}+\xi_f}=\frac{1}{N_{\mathrm{c}}} \tag{8.321b}$$

to the Coqblin-Schrieffer model. Here, it can be safely dropped in the limit $N_{\mathrm{c}}\to\infty$ of Eq. (8.320). The solution

$$\bar{\rho}^2=0 \tag{8.322}$$

to the saddle-point equation (8.320b) in the limit $N_{\mathrm{c}}\to\infty$ decouples the conduction electrons from the impurity. Show that a non-vanishing solution

$$0<\bar{\rho}^2\leq\frac{1}{2} \tag{8.323a}$$

to Eq. (8.320b) is then possible if and only if

$$\frac{\Delta}{\pi(\bar{\lambda}+\xi_f)}<-1 \tag{8.323b}$$

in the limit $N_{\mathrm{c}}\to\infty$.

Equations (8.320) and (8.321) supply the dependence of the Kondo energy scale (8.318) on the microscopic parameters of the single-impurity Anderson model and the Coqblin-Schrieffer model in the limit $N_{\mathrm{c}}\to\infty$, respectively. The limit $N_{\mathrm{c}}\to\infty$ of the Coqblin-Schrieffer model is simpler than that of the single-impurity Anderson model in the following sense. Both Eqs. (8.316) and (8.317) depend on the bandwidth of the conduction electrons through the term

$$\frac{\Delta\,\rho^2}{\pi}\ln\frac{k_{\mathrm{B}}\,T_{\mathrm{K}}(\lambda,\rho)}{D}=\frac{\Delta\,\rho^2}{\pi}\ln\frac{k_{\mathrm{B}}\,T_{\mathrm{K}}(\lambda,\rho)}{D'}+\frac{\Delta\,\rho^2}{\pi}\ln\frac{D'}{D}, \tag{8.324}$$

where $2\,D'>0$ is the new band width. If the free energies (8.316) and (8.317) are independent on the choice of the band width, i.e., invariant under the infinitesimal transformation $D\to D'$, there must follow that the microscopic parameters of the theory obey RG equations, as we are going to verify for the Coqblin-Schrieffer model.

(d) Show that the infinitesimal transformation $D\to D'$ on the free energy (8.317) can be absorbed by an infinitesimal change of the microscopic coupling J_0. Derive the first-order differential equation obeyed by J_0 that guarantees form invariance of the free energy (8.317) under the infinitesimal transformation $D\to D'$. If $\mu:=\ln D'/D$, show that $J_0(\mu)$ flows to 0 as $\mu\to\infty$. This is an example of asymptotic freedom by which the running coupling constant vanishes at high energies. Show that $J_0(\mu)$ flows to ∞ as $D'\to k_{\mathrm{B}}\,T_{\mathrm{K}}$. Explain why asymptotic freedom implies that the impurity spin susceptibility at high temperatures obeys Curie's law, if we assume that the analysis at $N_{\mathrm{c}} \to \infty$ remains qualitatively correct at $N_{\mathrm{c}} = 2$. Explain why asymptotic freedom implies that the impurity spin susceptibility at temperatures below the Kondo temperature obey the Pauli law, if we assume that the analysis at $N_{\mathrm{c}} \to \infty$ remains qualitatively correct at $N_{\mathrm{c}} = 2$. The specific heat coefficient [recall Eq. (5.67)], the zero-temperature spin susceptibility [recall Eq. (5.100b)], and the charge susceptibility at the impurity site and at zero temperature were computed for the $N_{\mathrm{c}} \to \infty$ limit of the Coqblin-Schrieffer model by Read and Newns in Ref. [69]. They behave as would be expected in a Fermi liquid.

(e) Verify that if we assume a non-vanishing solution to Eq. (8.320b) in the limit $N_{\mathrm{c}} \to \infty$ with $\left(\bar{\lambda} + \xi_f\right)^2 \ll \Delta^2\,\bar{\rho}^4$ then Eq. (8.320a) gives

$$k_{\mathrm{B}}\,T_K(\bar{\lambda},\bar{\rho}) \approx D\,e^{-\pi|\xi_f|/\Delta}. \tag{8.325}$$

The parametric regime for which Eq. (8.325) holds is the Kondo limit of the single-impurity Anderson model. In this Kondo limit, we may borrow the results from exercise 4.10(d) on the $N_{\mathrm{c}} \to \infty$ limit of the Coqblin-Schrieffer model provided we do the identification

$$\frac{\pi\,|\xi_f|}{\Delta} \to \frac{1}{\tilde{\tilde{\nu}}_{\mathrm{F}}\,J_0}. \tag{8.326}$$

(f) Verify that, provided $\bar{\rho} \neq 0$,

$$G_f(\mathrm{i}\omega_n) := \frac{\bar{\rho}^2}{\mathrm{i}\omega_n - \xi_f - \bar{\lambda} + \mathrm{i}\bar{\rho}^2\,\Delta\,\mathrm{sgn}\,\omega_n} \tag{8.327}$$

is the effective Green function at the impurity site for both the single-impurity Anderson model and the Coqblin-Schrieffer model in the limit $N_{\mathrm{c}} \to \infty$. Verify that, upon analytical continuation $\mathrm{i}\omega_n \to \omega_n \pm \mathrm{i}0^+$, the imaginary part of the single-particle Green function displays a peak at the energy $\omega = \xi_f + \bar{\lambda}$ with a width $\bar{\rho}^2\,\Delta$ and weight $\bar{\rho}^2$. Verify that $\bar{\rho}^2 < 1/2$ for the single-impurity Anderson model is a consequence of the saddle-point equation. Is there any constraint on $\bar{\rho}^2$ for the Coqblin-Schrieffer model arising from the saddle-point equation or on physical grounds? Explain why the weight $1 - \bar{\rho}^2$ of the single-particle Green function (8.327) should be distributed on physical grounds over two broad peaks at ξ_f and $\xi_f + U$ for the single-impurity Anderson model and only one broad peak at ξ_f for the Coqblin-Schrieffer model if we assume that we may apply the results of the limit $N_{\mathrm{c}} \to \infty$ to the case of $N_{\mathrm{c}} = 2$ and a large but finite U.

Exercise 4.11: We have argued that any model that couples one Bloch band of non-interacting conduction electrons to a single point-like impurity with internal degrees of freedom is characterized by no less and no more than two fixed points, provided the following two conditions hold. First, the density of states per unit energy of the unperturbed conduction electrons is constant. Second, the coupling is effectively that of an antiferromagnetic exchange interaction with an effective spin-1/2 in the Kondo regime.

There is a fixed point at high energy (high temperature) at which the conduction electrons decouple from the impurity. Perturbation theory in the coupling to the impurity is valid in the vicinity of this fixed point. There is a fixed point at low energy (low temperature) at which the conduction electrons are strongly coupled locally to the impurity so as to screen it. The Kondo energy scale (temperature) signals the crossover between the two fixed points. Although the impurity breaks translation symmetry, the strong-coupling fixed point is nevertheless thought of as a local Fermi liquid. This interpretation is justified by the fact that the two-point Green function at the impurity site has the analytical structure expected from a Fermi liquid.

A scenario with a flow between a decoupled fixed point and a strong coupling fixed point without any intervening quantum critical point is not always true. The zero-temperature phase diagram becomes much richer if we relax two assumptions that we made so far. The multichannel Kondo effect allows for a mismatch between twice the number of conducting bands and the number of internal degrees of freedom at the impurity site [104]. The condition of a non-vanishing density of states per unit energy at the Fermi energy can also be relaxed [103].

We shall now study the situation of a density of states per unit energy that vanishes in a power-law fashion at the Fermi energy that was first investigated by Withoff and Fradkin in Ref. [103].

Equations (8.275) and (8.271b) play a central role in what follows. We consider the Coqblin-Schrieffer model on a lattice made of N lattice sites. The density of states per unit energy $\tilde{\nu}(\xi)$ for the conduction electrons is given by Eq. (8.308). Hence, it is supported on the interval $[-D,+D]$ and singular at the Fermi energy $\xi=0$ if $r\neq 0$. Moreover, it is extensive in the number N of lattice sites with the convention we made for the normalization of the Hubbard-Stratonovich field of the Coqblin-Schrieffer model. The coupling of the conduction electrons to the f electrons on the single impurity site subject to the constraint it be occupied by exactly one of them is denoted by $J(D)$. We opt for the convention that $J(D)$ carries units of energy and scales like $1/N$ with the number of lattice sites. This choice corresponds to having Hamiltonian (8.271b) scales like N [Recall the discussion that followed the definition of Hamiltonian (8.248d)]. We also choose the convention for which $J(D)>0$ corresponds to an antiferromagnetic Heisenberg exchange coupling. This convention for the sign of $J(D)$ is opposite to the convention leading to Eq. (8.271b). Correspondingly,

$$g(D) := \tilde{\nu}(D)\,J(D) \tag{8.328}$$

is a dimensionless coupling constant that is intensive with respect to its scaling relative to the number of lattice sites N.

Rather than doing perturbation theory as Schrieffer and Wolff did to reach Eq. (8.275), we imagine that we integrate out all electrons in a thin shell below the bandwidth D under the assumption that $J(D)$ is small. The resulting bandwidth is D' where $D - D' > 0$ is very small, i.e.,

$$\frac{D - D'}{D} = \mathrm{d}\ell \Longleftrightarrow \frac{D'}{D} = e^{-\mathrm{d}\ell} \tag{8.329}$$

with $\mathrm{d}\ell$ infinitesimal. We still use Eq. (8.271b) with the assumption that all the changes resulting from the integration over the electrons with energies between D' and D are a mere small additive shift that can be absorbed covariantly by changing $J(D)$ to $J'(D')$. In other words, the action with the smaller bandwidth D' and the dimensionful coupling constant $J'(D')$ takes the same form as the action with the original bandwidth D and the dimensionful coupling $J(D)$. Accordingly,

$$J'(D') := J(D) + \tilde{\tilde{\nu}}(D)\, J^2(D)\, \mathrm{d}\ell. \tag{8.330}$$

We seek to express the left-hand side in terms of the original bandwidth D. To this end, we express Eq. (8.330) in terms of the dimensionless couplings $g'(D')$ and $g(D)$ and the density of states per unit energy $\tilde{\tilde{\nu}}(D')$ and $\tilde{\tilde{\nu}}(D)$, respectively,

$$\frac{g'(D')}{\tilde{\tilde{\nu}}(D')} = \frac{g(D)}{\tilde{\tilde{\nu}}(D)} + \frac{g^2(D)}{\tilde{\tilde{\nu}}(D)}\, \mathrm{d}\ell. \tag{8.331}$$

If we assume that the homogeneity relation

$$\tilde{\tilde{\nu}}(D') = \tilde{\tilde{\nu}}(D) \left(\frac{D'}{D}\right)^{+r} = \tilde{\tilde{\nu}}(D)\, e^{-r\,\mathrm{d}\ell} \tag{8.332}$$

holds for $D'/D \leq 1$ and if we do the expansion

$$g'(D')\, e^{+r\,\mathrm{d}\ell} = g(D) + \mathrm{d}g(D) + r\, g(D)\, \mathrm{d}\ell \tag{8.333}$$

on the left-hand side of Eq. (8.331), there follows the one-loop beta function

$$\left(\frac{\mathrm{d}g}{\mathrm{d}\,\ell}\right)(\ell) = -r\, g(\ell) + g^2(\ell). \tag{8.334}$$

(a) Draw the mean-field phase diagram implied by Eq. (8.334). Convince yourself of the following. For negative $-1 < r < 0$ and positive initial g, the flow is to strong coupling away from the fixed point $g = 0$. For $r = 0$, the linear term on the right-hand side drops out, while the positive sign of the quadratic term indicates that a positive g is marginally relevant, it also flows to strong coupling, though much more slowly than when $-1 < r < 0$. The case of $r = 0$ is of course that of a constant density of states per unit energy. For $r > 0$, the flow is to the stable fixed point $g = 0$ as long as the initial condition is compatible with $r\, g(\ell) > g^2(\ell)$. The one-loop beta function vanishes if the condition $g(\ell) = r$ is met. This critical point is unstable, for the flow is to strong coupling as soon as the initial condition satisfies the condition $0 < r\, g(\ell) < g^2(\ell)$.

The crude scaling analysis that we made relies on the homogeneity relation (8.332).

(b) Verify that the density of states per unit energy $\tilde{\nu}(\xi)$ given by Eq. (8.308) is homogeneous of degree r, i.e.,

$$\nu(\kappa\,\xi) = \kappa^r\,\nu(\xi) \tag{8.335}$$

when $0 < |\kappa| \leq 1$.

(c) Show that the number of lattice sites

$$N(D) := \int\limits_{-D}^{+D} \mathrm{d}\xi\,\tilde{\nu}(\xi) \tag{8.336}$$

is homogeneous of order $r+1$, i.e.,

$$N(\kappa\,D) = \kappa^{r+1}\,N(D) \tag{8.337}$$

when $0 < \kappa \leq 1$.

Exercise 4.12: Equation (8.334) is the poor man's one-loop beta function for any impurity problem in the Kondo regime. This is a perturbative calculation. The new feature brought about by a density of states that vanishes in a power-law fashion at the Fermi energy is the possibility of a new fixed point. It is desirable to confirm this perturbative RG result in a non-perturbative fashion and to decide if this critical point represents a non-Fermi liquid critical point. We are going to establish the existence of an unstable and non-Fermi-liquid critical point that intervenes between the stable fixed points at vanishing and infinite coupling in the limit $N_{\mathrm{c}} \to \infty$ of the Coqblin-Schrieffer model for the case $0 < r < 1/2$. We refer the reader to Ref. [103] for a more detailed analysis.

We use the approximation (8.309) to the self-energy of the f electrons acquired after integrating out the conduction electrons. Instead of the single energy scale (8.310) that applies when $r = 0$, we need to introduce the two energy scales,

$$\Delta''_r := \frac{1}{\pi}\,\Delta'_r := \pi\,\frac{\langle |V_{\boldsymbol{k}}|^2\rangle_{\mathrm{Fs}}}{D}\,C_r, \tag{8.338}$$

for the imaginary and real parts of the self-energy, respectively.

(a) Show that the free energy at zero temperature for the Coqblin-Schrieffer model is given by

$$\begin{aligned} F_r(-\mathrm{i}\lambda,\rho) = {} & \int\limits_{-D}^{+D} \frac{\mathrm{d}\omega}{\pi}\,\tilde{f}_{\mathrm{FD}}(\omega)\arctan\left(\frac{\Delta''_r\,\rho^2\,\left(\frac{\omega}{D}\right)^r}{\omega - \xi_f - \lambda - \Delta'_r\,\rho^2 \int\limits_{-D}^{+D} \mathrm{d}\varepsilon\,\left(\frac{\varepsilon}{D}\right)^r \frac{\mathrm{PV}}{\omega-\varepsilon}}\right) \\ & + \left(\frac{\Delta''_r\,\rho^2}{\pi\,C_r\,J_0/D} - \frac{\xi_f + \lambda}{N_{\mathrm{c}}}\right) \end{aligned} \tag{8.339}$$

in the limit $N_{\mathrm{c}} \to \infty$ and to leading order in ξ_f/D, λ/D, $\Delta'\,\rho^2/D$, or $\Delta''\,\rho^2/D$. *Hint:* Modify the derivation of Eq. (8.311) as demanded.

(b) Write the explicit form of the saddle-point equations

$$\left(\frac{\mathrm{d}F_r}{\mathrm{d}\rho}\right)(-\mathrm{i}\lambda,\rho)=0 \tag{8.340a}$$

and

$$\left(\frac{\mathrm{d}F_r}{\mathrm{d}\lambda}\right)(-\mathrm{i}\lambda,\rho)=0 \tag{8.340b}$$

at zero temperature, i.e., when $\tilde{f}_{\mathrm{FD}}(\omega)=\Theta(-\omega)$.

(c) Verify that $\bar{\rho}=0$ is a solution to the saddle-point equation (8.340a).

(d) Show that the saddle-point equation (8.340a) also admits a solution to

$$\frac{1}{\rho}\left(\frac{\mathrm{d}F_r}{\mathrm{d}\rho}\right)(-\mathrm{i}\lambda,\rho)=0 \tag{8.341}$$

for $\bar{\rho}=\bar{\lambda}=\xi_f=0$ that defines a critical value

$$J_0^{(\mathrm{c})}(r)\propto r \tag{8.342}$$

for any $r>0$. This critical value is not predicated on being small, as consistency demands when using poor man's scaling. It thus applies to graphene ($r=1$) or to some stackings of n consecutive layers of graphene ($r=n$).

(e) Argue why this solution can be interpreted as an unstable critical point between two non-Fermi-liquid phases of the Coqblin-Schrieffer model for any $r>0$. *Hint:* Assume (or better show) that $J_0^{(\mathrm{c})}(r)$ is such that any $J_0>J_0^{(\mathrm{c})}(r)$ admits a solution to the saddle-point equations with $\bar{\rho}>0$ while any $0\leq J_0<J_0^{(\mathrm{c})}(r)$ admits only the trivial solution $\bar{\rho}=0$.

Why should we trust the large N_{c} expansion? After all, the saddle-point approximation (8.285) predicts a spurious phase transition. Kondo-like problems can be thought of as realizations of boundary critical phenomena in $(1+1)$-dimensional conformal field theories [105]. For example, observables at the impurity site in the Kondo problem (with $r=0$) such as the spin susceptibility, the decay of the two-point Green function in time, etc., obey scaling laws that can be computed approximately and then compared to the power laws of boundary critical phenomena in $(1+1)$-dimensional conformal field theories. In particular, the critical exponents that can be extracted from a large N_{c} expansion of the Kondo-like problem (with $r=0$) can be compared to the exact boundary critical exponents in $(1+1)$-dimensional conformal field theories. Their agreement is a measure of the quality of the large N_{c} expansion. Cox and Ruckenstein in Ref. [106] showed that the extension of the large N_{c} expansion to the multichannel Kondo problem with a non-singular density of states at the Fermi level already reproduces to leading order the corresponding boundary critical exponents in $(1+1)$-dimensional conformal field theories and that these exponents are not modified to each order in the $1/N_{\mathrm{c}}$ expansion. We refer the reader to the review by Vojta in Ref. [107] of the developments, mostly numerical, beyond the large N_{c} limit of Withoff and Fradkin.

Chapter 9

Abelian bosonization in two-dimensional space and time

Outline

It is shown that the two-dimensional massive Thirring model is related to the two-dimensional Sine-Gordon model through Abelian bosonization. The bosonized solution to the quantum-xxz spin-1/2 chain and to the single-impurity problem in a Luttinger liquid are given.

9.1 Introduction

Quantum field theory in $(1+1)$-dimensional (position) space and time is more tractable than in higher-dimensional space and time. It is also of relevance to both classical statistical mechanics in two-dimensional (position) space and to condensed matter physics in one-dimensional (position) space. The trademark of quantum field theory in $(1+1)$-dimensional space and time is the "equivalence" between Bose and Fermi fields. [108–112] This equivalence is dubbed Abelian bosonization and has been extremely fruitful in applications to condensed matter physics. By way of a comparison between the two-dimensional Sine-Gordon model and the two-dimensional massive Thirring model, we shall illustrate the physical concepts at the root of Abelian bosonization. This will also allow us to connect a model of classical statistical mechanics such as the two-dimensional XY model to the quantum physics of interacting electrons constrained to one-dimensional (position) space.

The two-dimensional Sine-Gordon model is defined by the partition function

$$Z_{\mathrm{SG};t,h} := \int \mathcal{D}[\phi]\, e^{-S_{\mathrm{SG};t,h}} \tag{9.1a}$$

with the action

$$S_{\mathrm{SG};t,h} := \int \mathrm{d}^2x\, \mathcal{L}_{\mathrm{SG};t,h} \tag{9.1b}$$

and the Lagrangian density

$$\mathcal{L}_{\mathrm{SG};t,h} := \frac{1}{2t}(\partial_\mu\phi)^2 - \frac{h}{t}\cos\phi \tag{9.1c}$$

for the real-valued scalar field ϕ. As usual, we are working in two-dimensional Euclidean space, i.e., time is imaginary.

The two-dimensional massive Thirring model describes a massive and interacting quantum field theory for a spinor in $(1+1)$-dimensional Euclidean space and time (i.e., time is imaginary). It is defined by the partition function

$$Z_{\mathrm{Th};m,g} := \int \mathcal{D}[\bar{\psi},\psi]\, e^{-S_{\mathrm{Th};m,g}}, \tag{9.2a}$$

with the action

$$S_{\mathrm{Th};m,g} := \int \mathrm{d}^2x\, \mathcal{L}_{\mathrm{Th};m,g}, \tag{9.2b}$$

and the Lagrangian density

$$\mathcal{L}_{\mathrm{Th};m,g} := \bar{\psi}\left(\mathrm{i}\sigma_\mu\partial_\mu - m\right)\psi - \frac{g}{2}\left(\bar{\psi}\,\sigma_\mu\,\psi\right)^2. \tag{9.2c}$$

The spinor field $\bar{\psi}$ and ψ are Grassmann valued. They each transform according to the spin-1/2 representation of the two-dimensional Euclidean Poincaré group, i.e., they have two components on which the two-vector of Pauli matrices $\sigma_\mu = (\sigma_1, \sigma_2)$ act. We are using the summation convention over repeated indices, i.e., $\sigma_\mu\partial_\mu \equiv \sigma_1\partial_1 + \sigma_2\partial_2$. The dimensionless coupling constant g measures the strength of the current-current interaction $(j_\mu)^2 \equiv j_1^2 + j_2^2$ with the conserved current

$$j_\mu := \bar{\psi}\,\sigma_\mu\,\psi, \qquad \mu = 1,2, \tag{9.3}$$

that results from the invariance of the partition function under the global $U(1)$ gauge transformation

$$\bar{\psi} = \bar{\psi}'\, e^{+\mathrm{i}\alpha}, \qquad \psi = e^{-\mathrm{i}\alpha}\,\psi', \qquad \alpha \in \mathbb{R}. \tag{9.4}$$

In the massless limit $m = 0$, $\mathcal{L}_{\mathrm{Th};m=0,g}$ is also invariant under the global $U(1)$ axial gauge transformation

$$\bar{\psi} = \bar{\psi}'\, e^{-\mathrm{i}\alpha_5\,\gamma_5}, \qquad \psi = e^{-\mathrm{i}\alpha_5\,\gamma_5}\,\psi', \qquad \gamma_5 \equiv -\mathrm{i}\sigma_1\sigma_2, \qquad \alpha_5 \in \mathbb{R}. \tag{9.5}$$

For any non-vanishing g, the symmetry of $\mathcal{L}_{\mathrm{Th};m=0,g}$ under the transformation (9.5) is broken by the measure of the partition function $Z_{\mathrm{Th};m=0,g}$ as is shown in section 9.2.2. The mass term $m\,\bar{\psi}\,\psi$ is symmetric under the transformation (9.4), but it breaks explicitly the symmetry of $\mathcal{L}_{\mathrm{Th};m=0,g}$ under the transformation (9.5), since $\bar{\psi}\,\psi = \bar{\psi}'\, e^{-2\mathrm{i}\alpha_5\,\gamma_5}\,\psi'$.

Abelian-bosonization rules encode the fact that the two-dimensional Sine-Gordon and two-dimensional Thirring models are equivalent in the sense that some local fields share the *same* correlation functions in both theories. This correspondence is summarized in Table 9.1.

Table 9.1 Abelian bosonization rules in two-dimensional Euclidean space.

Sine-Gordon model	Thirring model
$\frac{1}{8\pi}(\partial_\mu\phi)^2$	$\bar{\psi}\,\mathrm{i}\sigma_\mu\,\partial_\mu\,\psi$
$\frac{1}{2\pi}\,\epsilon_{\mu\nu}\,\partial_\nu\,\phi$	$\bar{\psi}\,\mathrm{i}\,\sigma_\mu\,\psi$
$\frac{h}{t}$	$\frac{-\mathrm{i}m}{\pi}$

9.2 Abelian bosonization of the Thirring model

9.2.1 *Free-field fixed point in the massive Thirring model*

Before undertaking a justification of the Abelian bosonization rules, it is useful to gain some familiarity with the two-dimensional Thirring model by studying the massless free-field fixed point $m = g = 0$,

$$\begin{aligned} Z_{\mathrm{Th};0,0} &:= \int \mathcal{D}[\bar{\psi},\psi]\, e^{-S_{\mathrm{Th};0,0}}, \\ S_{\mathrm{Th};0;0} &:= \int \mathrm{d}^2x\, \mathcal{L}_{\mathrm{Th};0,0}, \\ \mathcal{L}_{\mathrm{Th};0,0} &:= \bar{\psi}\,\mathrm{i}\sigma_\mu\,\partial_\mu\,\psi. \end{aligned} \tag{9.6}$$

At the massless free-field fixed point, engineering and scaling dimensions of $\bar{\psi}$ and ψ are equal and given by 1/2 ($[\bar{\psi}] = [\psi] = \text{length}^{-1/2}$). Scale invariance of $S_{\mathrm{Th};0,0}$ under simultaneous rescaling of the coordinates and fields imply[1]

$$\langle \psi(x)\bar{\psi}(y)\rangle_{Z_{\mathrm{Th};0,0}} \sim \frac{1}{|x-y|}, \tag{9.7}$$

if we are to neglect the spinor structure all together. To account for the spinor structure, observe that (σ_0 is the unit 2×2 matrix in spinor space)

$$\begin{aligned} \sigma_\mu\partial_\mu &= \begin{pmatrix} 0 & \partial_1 - \mathrm{i}\partial_2 \\ \partial_1 + \mathrm{i}\partial_2 & 0 \end{pmatrix}, \\ \left(\sigma_\mu\partial_\mu\right)^2 &= \sigma_0\left(\partial_1^2 + \partial_2^2\right) \equiv \sigma_0\,\Delta, \\ \Delta^{-1}(x) &= +\frac{1}{4\pi}\ln\left(\frac{x^2}{\mathfrak{a}^2}\right), \end{aligned} \tag{9.8}$$

[1] We are using the convention

$$\int \mathrm{d}\psi^*\mathrm{d}\psi\, e^{-\psi^*\psi} = \int \mathrm{d}\psi^*\mathrm{d}\psi\,(-)\psi^*\,\psi = 1, \qquad \int \mathrm{d}\psi^*\mathrm{d}\psi\,\psi\,\psi^*\, e^{-\psi^*\psi} = \int \mathrm{d}\psi^*\mathrm{d}\psi\,\psi\,\psi^*\,1 = 1,$$

for the Grassmann integration.

where $\mathfrak{a}$ is the short distance cut-off, i.e., the lattice spacing. Hence,

$$\begin{aligned}\langle\psi(x)\bar{\psi}(0)\rangle_{Z_{\mathrm{Th};0,0}} &= \left(\mathrm{i}\sigma_\mu\partial_\mu\right)^{-1}(x)\\ &= -\mathrm{i}\sigma_\mu\partial_\mu\,\sigma_0\Delta^{-1}(x)\\ &= -\mathrm{i}\sigma_\mu\partial_\mu\left[\frac{1}{4\pi}\ln\left(\frac{x^2}{\mathfrak{a}^2}\right)\right].\end{aligned} \tag{9.9}$$

It is convenient at this stage to rotate the Cartesian coordinates of two-dimensional Euclidean space by $\pi/4$. If

$$x_\pm := x_1 \pm \mathrm{i}x_2, \tag{9.10a}$$

then

$$\partial_\mp \equiv \partial_1 \mp \mathrm{i}\partial_2 = 2\frac{\partial}{\partial x_\pm} \tag{9.10b}$$

and

$$\begin{aligned}\langle\psi(x)\bar{\psi}(0)\rangle_{Z_{\mathrm{Th};0,0}} &= \left(-\frac{\mathrm{i}}{2\pi}\right)\frac{1}{x^2}\begin{pmatrix}0 & x_1-\mathrm{i}x_2\\ x_1+\mathrm{i}x_2 & 0\end{pmatrix}\\ &= \left(-\frac{\mathrm{i}}{2\pi}\right)\begin{pmatrix}0 & \frac{1}{x_1+\mathrm{i}x_2}\\ \frac{1}{x_1-\mathrm{i}x_2} & 0\end{pmatrix}.\end{aligned} \tag{9.10c}$$

The representation in terms of the Pauli matrices chosen here defines the so-called *Dirac representation* of the Thirring model. In the Dirac representation, the *Euclidean Feynman propagator* (9.10c) is off-diagonal. From now on, we choose units in which $\mathfrak{a} = 1$.

Whereas the Feynman propagator (9.10c) is not diagonal, the propagator $\langle\psi(x)\,\psi^*(y)\rangle_{Z_{\mathrm{Th};0,0}}$, where

$$\psi^* := \bar{\psi}\,\sigma_1 \Longleftrightarrow \bar{\psi} =: \psi^*\,\sigma_1 \tag{9.11}$$

is diagonal. To see this, introduce first the third Pauli matrix

$$\gamma_5 \equiv -\mathrm{i}\sigma_1\sigma_2 = \sigma_3. \tag{9.12}$$

Observing that $P_\pm := (1/2)(\sigma_0 \pm \gamma_5)$ form a complete set of projectors onto the spinor subspace, the chiral representation,

$$\psi^* := \begin{pmatrix}\psi^*_- & \psi^*_+\end{pmatrix}, \qquad \psi := \begin{pmatrix}\psi_-\\ \psi_+\end{pmatrix}, \tag{9.13a}$$

is defined by

$$\psi^*_\pm := \psi^*\,\frac{1}{2}\left(\sigma_0 \mp \gamma_5\right), \qquad \psi_\pm := \frac{1}{2}\left(\sigma_0 \mp \gamma_5\right)\psi. \tag{9.13b}$$

With this definition,

$$\begin{aligned}\bar{\psi}\,\mathrm{i}\sigma_\mu\,\partial_\mu\,\psi &= \psi^*\,\sigma_1\mathrm{i}\sigma_\mu\,\partial_\mu\,\psi\\ &= \psi^*\mathrm{i}\left(\sigma_0\,\partial_1 + \mathrm{i}\gamma_5\,\partial_2\right)\psi\\ &= \psi^*_-\mathrm{i}\left(\partial_1 + \mathrm{i}\partial_2\right)\psi_- + \psi^*_+\mathrm{i}\left(\partial_1 - \mathrm{i}\partial_2\right)\psi_+\end{aligned} \tag{9.14}$$

and

$$\begin{aligned}\langle\psi(x)\,\psi^*(0)\rangle_{Z_{\mathrm{Th};0,0}} &= \left(-\frac{\mathrm{i}}{2\pi}\right)\frac{1}{x^2}\begin{pmatrix} x_- & 0\\ 0 & x_+\end{pmatrix}\\ &= \left(-\frac{\mathrm{i}}{2\pi}\right)\begin{pmatrix}\frac{1}{x_+} & 0\\ 0 & \frac{1}{x_-}\end{pmatrix}.\end{aligned} \tag{9.15}$$

The chiral basis displays explicitly the fact that $\mathcal{L}_{\mathrm{Th};0,0}$ has more than the global $U(1)$ gauge invariance of Eq. (9.4). Indeed, Eq. (9.14) is invariant under the local $U(1)\times U(1)$ gauge transformations defined by[2]

$$\begin{aligned}\psi^*_- &=: \psi^{*\,\prime}_-\,e^{+\mathrm{i}\alpha(x_+)}, &\quad \psi_- &=: e^{-\mathrm{i}\alpha(x_+)}\,\psi_-{}',\\ \psi^*_+ &=: \psi^{*\,\prime}_+\,e^{+\mathrm{i}\beta(x_-)}, &\quad \psi_+ &=: e^{-\mathrm{i}\beta(x_-)}\,\psi_+{}'.\end{aligned} \tag{9.16}$$

Here, $\alpha(x_+)$ is any holomorphic function of $x_+=x_1+\mathrm{i}x_2$, whereas $\beta(x_-)$ is any antiholomorphic function of $x_-=x_1-\mathrm{i}x_2$.[3] The corresponding conserved (Noether) currents

$$j_- := j_1-\mathrm{i}j_2 = 2\,\psi^*_-\psi_-, \qquad j_+ = j_1+\mathrm{i}j_2 = 2\,\psi^*_+\psi_+, \tag{9.19a}$$

obey

$$0=\partial_+ j_- = \partial_- j_+. \tag{9.19b}$$

The $U(1)\times U(1)$ local gauge invariance (9.16) has dramatic consequences on correlation functions. For example, consider the two bilinears

$$\Psi_{+-}(x) := (\psi^*_+\,\psi_-)(x), \qquad \Psi_{-+}(x) := (\psi^*_-\,\psi_+)(x). \tag{9.20}$$

These bilinears appear in the "standard"

$$\bar\psi\quad\psi = \quad\psi^*\sigma_1\,\psi = +\left(\Psi_{-+}+\Psi_{+-}\right), \tag{9.21}$$

and "axial" mass terms

$$\bar\psi\,\gamma_5\,\psi = -\mathrm{i}\psi^*\,\sigma_2\,\psi = -\left(\Psi_{-+}-\Psi_{+-}\right), \tag{9.22}$$

respectively. Now, the $4n$-point correlation function

$$\left\langle\prod_{j=1}^{n}\Psi_{-+}(x_j)\,\Psi_{+-}(y_j)\right\rangle_{Z_{\mathrm{Th};0,0}} \tag{9.23}$$

---

[2] This local gauge invariance is a manifestation of conformal invariance in two dimensions.

[3] Equation (9.14) is rewritten

$$\bar\psi\,\mathrm{i}\sigma_\mu\,\partial_\mu\,\psi = \psi^*_-\,\mathrm{i}(2\partial_{\bar z})\,\psi_- + \psi^*_+\,\mathrm{i}(2\partial_z)\,\psi_+, \tag{9.17}$$

whereby we have introduced the notation

$$\begin{aligned} x_+ &:= x_1+\mathrm{i}x_2\equiv z, &\quad x_1 &= \frac{1}{2}(x_++x_-), &\quad \partial_+ &:= \partial_1+\mathrm{i}\partial_2\equiv(2\partial_{\bar z}),\\ x_- &:= x_1-\mathrm{i}x_2\equiv \bar z, &\quad x_2 &= -\frac{\mathrm{i}}{2}(x_+-x_-), &\quad \partial_- &:= \partial_1-\mathrm{i}\partial_2\equiv(2\partial_z).\end{aligned} \tag{9.18}$$

factorizes into the product of holomorphic and antiholomorphic functions. For example,

$$\begin{aligned}
\left\langle \Psi_{-+}(x)\,\Psi_{+-}(y) \right\rangle_{Z_{\mathrm{Th};0,0}} =& \left\langle \psi^{*}_{-}(x)\,\psi_{+}(x)\,\psi^{*}_{+}(y)\,\psi_{-}(y) \right\rangle_{Z_{\mathrm{Th};0,0}} \\
=& -\left\langle \psi_{-}(y)\,\psi^{*}_{-}(x)\,\psi_{+}(x)\,\psi^{*}_{+}(y) \right\rangle_{Z_{\mathrm{Th};0,0}} \\
=& -\left\langle \psi_{-}(y)\,\psi^{*}_{-}(x) \right\rangle_{Z_{\mathrm{Th};0,0}} \times \left\langle \psi_{+}(x)\,\psi^{*}_{+}(y) \right\rangle_{Z_{\mathrm{Th};0,0}} \\
=& -\left(-\frac{\mathrm{i}}{2\pi}\right)^{2} \frac{1}{y_{+}-x_{+}} \times \frac{1}{x_{-}-y_{-}} \\
=& +\left(-\frac{\mathrm{i}}{2\pi}\right)^{2} \frac{1}{x_{+}-y_{+}} \times \frac{1}{x_{-}-y_{-}}.
\end{aligned} \tag{9.24}$$

In general,

$$\begin{aligned}
\left\langle \prod_{j=1}^{n} \Psi_{-+}(x_j)\,\Psi_{+-}(y_j) \right\rangle_{Z_{\mathrm{Th};0,0}} =& (-1)^{n} \left\langle \prod_{j=1}^{n} \psi_{-}(y_j)\,\psi^{*}_{-}(x_j) \right\rangle_{Z_{\mathrm{Th};0,0}} \\
&\times \left\langle \prod_{k=1}^{n} \psi_{+}(x_k)\,\psi^{*}_{+}(y_k) \right\rangle_{Z_{\mathrm{Th};0,0}},
\end{aligned} \tag{9.25a}$$

where ($\mathcal{S}_n$ is the permutation group of n elements)

$$\begin{aligned}
\left\langle \prod_{j=1}^{n} \psi_{-}(y_j)\,\psi^{*}_{-}(x_j) \right\rangle_{Z_{\mathrm{Th};0,0}} =& \sum_{\sigma\in\mathcal{S}_n} \mathrm{sgn}(\sigma) \prod_{j=1}^{n} \left\langle \psi_{-}(y_{\sigma j})\,\psi^{*}_{-}(x_j) \right\rangle_{Z_{\mathrm{Th};0,0}} \\
=& \left(-\frac{\mathrm{i}}{2\pi}\right)^{n} \sum_{\sigma\in\mathcal{S}_n} \mathrm{sgn}(\sigma) \prod_{j=1}^{n} \frac{1}{(y_{\sigma j})_{+}-(x_j)_{+}} \\
=& (-1)^{n} \left(-\frac{\mathrm{i}}{2\pi}\right)^{n} \sum_{\sigma\in\mathcal{S}_n} \mathrm{sgn}(\sigma) \prod_{j=1}^{n} \frac{1}{(x_j)_{+}-(y_{\sigma j})_{+}} \\
=& (-1)^{n} \left(-\frac{\mathrm{i}}{2\pi}\right)^{n} \det\left[\frac{1}{(x_j)_{+}-(y_k)_{+}}\right]_{j,k=1,\cdots,n} \\
=& (-1)^{n} \left(-\frac{\mathrm{i}}{2\pi}\right)^{n} (-1)^{n(n-1)/2} \\
&\times \frac{\prod\limits_{1\le j<k\le n} \left[(x_j)_{+}-(x_k)_{+}\right]\left[(y_j)_{+}-(y_k)_{+}\right]}{\prod\limits_{j,k=1}^{n} \left[(x_j)_{+}-(y_k)_{+}\right]},
\end{aligned} \tag{9.25b}$$

and

$$\left\langle \prod_{k=1}^{n} \psi_+(x_k)\, \psi_+^*(y_k) \right\rangle_{Z_{\mathrm{Th};0,0}} = \left(-\frac{\mathrm{i}}{2\pi}\right)^n (-1)^{n(n-1)/2} \times \frac{\prod\limits_{1\le j<k\le n} [(x_j)_- - (x_k)_-]\,[(y_j)_- - (y_k)_-]}{\prod\limits_{j,k=1}^{n} [(x_j)_- - (y_k)_-]}. \tag{9.25c}$$

Thus,

$$\left\langle \prod_{i=1}^{n} \Psi_{-+}(x_i)\, \Psi_{+-}(y_i) \right\rangle_{Z_{\mathrm{Th};0,0}} = \left(-\frac{\mathrm{i}}{2\pi}\right)^{2n} \frac{\prod\limits_{1\le j<k\le n} \left|x_j - x_k\right|^2 \left|y_j - y_k\right|^2}{\prod\limits_{j,k=1}^{n} \left|x_j - y_k\right|^2}. \tag{9.26}$$

Up to an overall multiplicative factor, the same correlation function is obtained in the two-dimensional Sine-Gordon model with $h = 0$ and $t = 4\pi$ provided one identifies

$$\Psi_{-+} \to \frac{1}{2\pi\mathrm{i}} e^{+\mathrm{i}\phi}, \qquad \Psi_{+-} \to \frac{1}{2\pi\mathrm{i}} e^{-\mathrm{i}\phi}. \tag{9.27}$$

This result is consistent with the Abelian bosonization rule $h/t \longleftrightarrow -\mathrm{i}m/\pi$.

Another consistency check of the Abelian bosonization rules amounts to comparing the generating functional

$$\begin{aligned} Z_{\mathrm{Th};0,0}[J_\mu] :&= \int \mathcal{D}[\bar\psi,\psi]\, \exp\left(-\int \mathrm{d}^2x\, \bar\psi\, \sigma_\mu \left(\mathrm{i}\partial_\mu + J_\mu\right) \psi\right) \\ &= \mathrm{Det}\left[\sigma_\mu \left(\mathrm{i}\partial_\mu + J_\mu\right)\right] \end{aligned} \tag{9.28}$$

for the vector current-current correlation function in the two-dimensional Thirring model with the generating functional ($\tilde\partial_\mu \equiv \epsilon_{\mu\nu}\partial_\nu$ implies $\tilde\partial_\mu\tilde\partial_\nu = \delta_{\mu\nu}\Delta - \partial_\mu\partial_\nu$)

$$\begin{aligned} Z_{\mathrm{SG};0,0}[J_\mu] :&= \int \mathcal{D}\phi\, e^{-\int \mathrm{d}^2x \left[\frac{1}{2t}(\partial_\mu\phi)^2 + \beta(\tilde\partial_\mu\phi)J_\mu\right]} \\ &= \int \mathcal{D}\phi\, e^{-\int \mathrm{d}^2x \left[\frac{1}{2t}\phi(-\Delta)\phi - \beta\phi(\tilde\partial_\mu J_\mu)\right]} \\ &= \left[\mathrm{Det}\,(-)\Delta/t\right]^{-1/2} e^{+\frac{\beta^2 t}{2}\int \mathrm{d}^2x \int \mathrm{d}^2y (\tilde\partial_\mu J_\mu)(x)(-\Delta)^{-1}(x-y)(\tilde\partial_\nu J_\nu)(y)} \\ &= \left[\mathrm{Det}\,(-)\Delta/t\right]^{-1/2} e^{+\frac{\beta^2 t}{2}\int \mathrm{d}^2x \int \mathrm{d}^2y (\Delta\xi)(x)(-\Delta)^{-1}(x-y)(\Delta\xi)(y)} \\ &= \left[\mathrm{Det}\,(-)\Delta/t\right]^{-1/2} e^{+\frac{\beta^2 t}{2}\int \mathrm{d}^2x(-\Delta\xi)\xi} \\ &= \left[\mathrm{Det}\,(-)\Delta/t\right]^{-1/2} e^{+\frac{\beta^2 t}{2}\int \mathrm{d}^2x(\partial_\mu\xi)^2} \\ &\equiv \left[\mathrm{Det}\,(-)\Delta/t\right]^{-1/2} e^{+\frac{\beta^2 t}{2}\int \mathrm{d}^2x \int \mathrm{d}^2y J_\mu(x)\left[\delta_{\mu\nu}\,\delta(x-y) - \frac{\partial_\mu\partial_\nu}{\Delta}\right] J_\nu(y)}. \end{aligned} \tag{9.29}$$

Here, we have assumed that the source J_μ is a smooth vector field, i.e., that the decomposition

$$J_\mu = \partial_\mu \chi + \tilde\partial_\mu \xi \tag{9.30}$$

in terms of the pure-gauge $\partial_\mu \chi$ and the transverse component $\tilde{\partial}_\mu \xi$ is valid everywhere in space. Moreover, we assume that we can safely drop all boundary terms when performing partial integrations.

The fermionic determinant on the second line of Eq. (9.28) for an arbitrary source J_μ needs to be computed. This is done below using the method of Fujikawa with the result

$$\begin{aligned}\frac{\operatorname{Det}\left[\sigma_\mu\left(\mathrm{i}\partial_\mu+J_\mu\right)\right]}{\operatorname{Det}\left[\sigma_\mu\left(\mathrm{i}\partial_\mu\right)\right]}&=e^{-\frac{1}{2\pi}\int \mathrm{d}^2x\,(\partial_\mu\xi)^2}\\&\equiv e^{-\frac{1}{2\pi}\int \mathrm{d}^2x\int \mathrm{d}^2y\,J_\mu(x)\left[\delta_{\mu\nu}\,\delta(x-y)-\frac{\partial_\mu\partial_\nu}{\Delta}\right]J_\nu(y)}.\end{aligned} \tag{9.31}$$

The right-hand side on the second line is defined by the right-hand side on the first line. Hence, we must have

$$\beta^2 t=-\frac{1}{\pi}, \tag{9.32}$$

if the two generating functions are to produce the same correlation functions. With the choice $m=0$ and $t\to 4\pi$ in the two-dimensional Sine-Gordon model, we find that

$$\beta=-\frac{\mathrm{i}}{2\pi} \tag{9.33}$$

agrees with the Abelian bosonization rule $\bar{\psi}\,\mathrm{i}\sigma_\mu\,\psi\longleftrightarrow(1/2\pi)\tilde{\partial}_\mu\,\phi$.

9.2.2 *The $U(1)$ axial-gauge anomaly*

We want to compute the fermionic determinant

$$\operatorname{Det}\left[\sigma_\mu\left(\mathrm{i}\partial_\mu+J_\mu\right)\right]=\int\mathcal{D}[\bar{\psi},\psi]\exp\left(-\int \mathrm{d}^2x\,\bar{\psi}\,\sigma_\mu\left(\mathrm{i}\partial_\mu+J_\mu\right)\psi\right). \tag{9.34}$$

By assumption (J_μ is taken sufficiently smooth), we can always decompose the source J_μ into rotation and divergence free contributions,

$$J_\mu=\partial_\mu\,\chi+\tilde{\partial}_\mu\,\xi. \tag{9.35}$$

We can then perform the change of Grassmann variables

$$\bar{\psi}=:\bar{\psi}'\,e^{-\mathrm{i}\chi+\gamma_5\,\xi},\qquad \psi=:e^{+\mathrm{i}\chi+\gamma_5\,\xi}\,\psi', \tag{9.36}$$

under which

$$\begin{aligned}\mathcal{L}&:=\bar{\psi}\,\sigma_\mu\left(\mathrm{i}\partial_\mu+J_\mu\right)\psi\\&=\bar{\psi}'\,e^{-\mathrm{i}\chi+\gamma_5\,\xi}\,\sigma_\mu\left(\mathrm{i}\partial_\mu+\partial_\mu\chi+\tilde{\partial}_\mu\xi\right)e^{+\mathrm{i}\chi+\gamma_5\xi}\,\psi'\\&=\bar{\psi}'\mathrm{i}\sigma_\mu\,\partial_\mu\,\psi'.\end{aligned} \tag{9.37}$$

To reach the last equality, we made use of $\sigma_\mu\gamma_5\partial_\mu=-\mathrm{i}\epsilon_{\mu\nu}\sigma_\nu\partial_\mu=+\mathrm{i}\epsilon_{\mu\nu}\sigma_\mu\partial_\nu=+\mathrm{i}\sigma_\mu\tilde{\partial}_\mu$. The fact that the transformation (9.36) decouples the spinor from the source field is unique to two dimensions. On general grounds, a change of integration variables costs a Jacobian. What is the Jacobian $\mathcal{J}_{\mathrm{f}}$ of the transformation (9.36)? Fujikawa was the first to propose a method to calculate the Jacobian $\mathcal{J}_{\mathrm{f}}$ associated to Eq. (9.36) [113].

The Grassmann measure is defined in terms of the Grassmann-valued expansion coefficients $\bar{a}_m$ and a_n of the fields $\bar{\psi}$ and ψ, respectively, in the basis of the Dirac operator $\sigma_\mu(\mathrm{i}\partial_\mu + J_\mu)$. In other words,

$$\begin{aligned}\mathcal{D}[\bar{\psi},\psi] &= \left(\prod_m \mathrm{d}\bar{a}_m\right)\left(\prod_n \mathrm{d}a_n\right),\\ \bar{\psi}(x) &= \sum_m \langle m|x\rangle\,\bar{a}_m \equiv \sum_m \varphi^\dagger_m(x)\,\bar{a}_m,\\ \psi(x) &= \sum_n a_n\,\langle x|n\rangle \equiv \sum_n a_n\,\varphi_n(x),\end{aligned} \tag{9.38a}$$

where φ_n is the complete set of orthonormal eigenspinors with eigenvalues λ_n of $\sigma_\mu(\mathrm{i}\partial_\mu + J_\mu)$,

$$\begin{aligned}&\sigma_\mu\left(\mathrm{i}\partial_\mu + J_\mu\right)\varphi_n(x) = \lambda_n\,\varphi_n(x),\\ &\sum_n \varphi_n(x)\,\varphi^\dagger_n(y) = \sigma_0\,\delta(x-y), \qquad \int \mathrm{d}^2x\,\varphi^\dagger_m(x)\,\varphi_n(x) = \delta_{m,n}.\end{aligned} \tag{9.38b}$$

Similarly, after performing the change of integration variables (9.36),

$$\begin{aligned}\mathcal{D}[\bar{\psi}',\psi'] &= \left(\prod_{m'} \mathrm{d}\bar{a}'_{m'}\right)\left(\prod_{n'} \mathrm{d}a'_{n'}\right),\\ \bar{\psi}'(x) &= \sum_{m'} \varphi^\dagger_{m'}(x)\,\bar{a}'_{m'} = \sum_m \varphi^\dagger_m(x)\,\bar{a}_m\,e^{+\mathrm{i}\chi(x)-\gamma_5\,\xi(x)},\\ \psi'(x) &= \sum_{n'} a'_{n'}\,\varphi_{n'}(x) = \sum_n a_n\,e^{-\mathrm{i}\chi(x)-\gamma_5\,\xi(x)}\,\varphi_n(x).\end{aligned} \tag{9.39}$$

The relationship between the expansion coefficients $(\bar{a}_m, a_n)$ and $(\bar{a}'_{m'}, a'_{n'})$ is linear,

$$\begin{aligned}\bar{a}'_{m'} &= \sum_m \bar{U}_{m',m}\,\bar{a}_m, \qquad \bar{U}_{m',m} = \int \mathrm{d}^2x\,\varphi^\dagger_m(x)\,e^{+\mathrm{i}\chi(x)-\gamma_5\,\xi(x)}\,\varphi_{m'}(x),\\ a'_{n'} &= \sum_n U_{n',n}\,a_n, \qquad U_{n',n} = \int \mathrm{d}^2x\,\varphi^\dagger_{n'}(x)e^{-\mathrm{i}\chi(x)-\gamma_5\,\xi(x)}\,\varphi_n(x).\end{aligned} \tag{9.40}$$

Under the linear transformation (9.40), the transformation law of the measure is

$$\begin{aligned}\prod_{m'} \mathrm{d}\bar{a}'_{m'} &= \left(\mathrm{Det}\,\bar{U}\right)^{-1}\prod_m \mathrm{d}\bar{a}_m,\\ \prod_{n'} \mathrm{d}a'_{n'} &= \left(\mathrm{Det}\,U\right)^{-1}\prod_n \mathrm{d}a_n.\end{aligned} \tag{9.41}$$

Notice that it is not the determinant of the linear transformation that appears on the right-hand side, as would be the case for Riemann integrals, but the inverse determinant.[4]

[4] This is so because the Grassmann integral is constructed such that $\int \mathrm{d}\psi^*\,\mathrm{d}\psi\,\exp(-\psi^*\,A\,\psi) = A$, i.e., $A\int \frac{\mathrm{d}\psi^*}{\sqrt{A}}\,\frac{\mathrm{d}\psi}{\sqrt{A}}\,\exp\left(-(\sqrt{A}\psi^*)(\sqrt{A}\psi)\right) = A\int \mathrm{d}\zeta^*\,\mathrm{d}\zeta\,\exp(-\zeta^*\,\zeta) = A$.

We first evaluate $\left(\operatorname{Det}\bar{U}\right)^{-1}$ and $(\operatorname{Det}U)^{-1}$ by assuming that J_μ is infinitesimal,

$$J_\mu=\partial_\mu(\delta\chi)+\tilde{\partial}_\mu(\delta\xi), \tag{9.42}$$

where the rotation-free contribution from $\delta\chi$ and the divergence-free contribution from $\delta\xi$ are infinitesimal. We then integrate the result.

On the one hand,

$$\begin{aligned}
\left(\operatorname{Det}\bar{U}\right)^{-1} &= \left(\operatorname{Det}\left\{\delta_{m,m'}+\int \mathrm{d}^2x\,\varphi^\dagger_m(x)\,[+\mathrm{i}(\delta\chi)(x)-\gamma_5(\delta\xi)(x)]\,\varphi_{m'}(x)\right\}\right)^{-1}\\
&= \operatorname{Det}\left\{\delta_{m,m'}-\int \mathrm{d}^2x\,\varphi^\dagger_m(x)\,[+\mathrm{i}(\delta\chi)(x)-\gamma_5(\delta\xi)(x)]\,\varphi_{m'}(x)\right\}\\
&= e^{\operatorname{Tr}\,\ln\left\{\delta_{m,m'}-\int \mathrm{d}^2x\,\varphi^\dagger_m(x)[+\mathrm{i}(\delta\chi)(x)-\gamma_5(\delta\xi)(x)]\varphi_{m'}(x)\right\}}\\
&= e^{-\sum_m\int \mathrm{d}^2x\,\varphi^\dagger_m(x)[+\mathrm{i}(\delta\chi)(x)-\gamma_5(\delta\xi)(x)]\varphi_m(x)}.
\end{aligned} \tag{9.43}$$

To reach the last equality, we made use of $\ln(1-x)=-x-\frac{x^2}{2}-\cdots$. On the other hand,

$$\begin{aligned}
(\operatorname{Det}U)^{-1} &= \left(\operatorname{Det}\left\{\delta_{n',n}+\int \mathrm{d}^2x\,\varphi^\dagger_{n'}(x)\,[-\mathrm{i}(\delta\chi)(x)-\gamma_5(\delta\xi)(x)]\,\varphi_{n}(x)\right\}\right)^{-1}\\
&= \operatorname{Det}\left\{\delta_{n',n}-\int \mathrm{d}^2x\,\varphi^\dagger_{n'}(x)\,[-\mathrm{i}(\delta\chi)(x)-\gamma_5(\delta\xi)(x)]\,\varphi_{n}(x)\right\}\\
&= e^{\operatorname{Tr}\,\ln\left\{\delta_{n',n}-\int \mathrm{d}^2x\,\varphi^\dagger_{n'}(x)[-\mathrm{i}(\delta\chi)(x)-\gamma_5(\delta\xi)(x)]\varphi_{n}(x)\right\}}\\
&= e^{-\sum_n\int \mathrm{d}^2x\,\varphi^\dagger_n(x)[-\mathrm{i}(\delta\chi)(x)-\gamma_5(\delta\xi)(x)]\varphi_n(x)}.
\end{aligned} \tag{9.44}$$

To reach the last equality, we again made use of $\ln(1-x)=-x-\frac{x^2}{2}-\cdots$. We thus find that the infinitesimal Jacobian

$$\delta\mathcal{J}_{\mathrm{f}}:=\frac{\left(\prod_m \mathrm{d}\bar{a}_m\right)\left(\prod_n \mathrm{d}a_n\right)}{\left(\prod_{m'} \mathrm{d}\bar{a}'_{m'}\right)\left(\prod_{n'} \mathrm{d}a'_{n'}\right)}=\left(\operatorname{Det}\bar{U}\right)(\operatorname{Det}U) \tag{9.45}$$

only depends on the infinitesimal generator $(\delta\xi)$ of the divergence-free contribution to J_μ,

$$\begin{aligned}
\delta\mathcal{J}_{\mathrm{f}} &= e^{-\int \mathrm{d}^2x\,(\delta\xi)(x)\,\mathcal{A}_5(x)},\\
\mathcal{A}_5(x) &:= 2\times\sum_n \varphi^\dagger_n(x)\,\gamma_5\,\varphi_n(x).
\end{aligned} \tag{9.46}$$

The function $\mathcal{A}_5$ is called the "axial anomaly" if it is non-vanishing, for it implies that quantum fluctuations encoded by the measure in the path integral break the axial symmetry of the Lagrangian density.

To make sense of the "axial anomaly" $\mathcal{A}_5(x)$ when

$$J_\mu\to J_\mu+(\delta J)_\mu,\qquad (\delta J)_\mu=\tilde{\partial}_\mu(\delta\xi), \tag{9.47}$$

we need to regularize the summation on the right-hand side of Eq. (9.46), for a given background J_μ. This is done by choosing the following *gauge-invariant* regularization,

$$\begin{aligned}\mathcal{A}_5(x) :&= 2\times\lim_{M\to\infty}\sum_n \varphi^\dagger_n(x)\,\gamma_5\,e^{-(\lambda_n)^2/M^2}\,\varphi_n(x)\\ &= 2\times\lim_{M\to\infty}\sum_n \varphi^\dagger_n(x)\,\gamma_5\,e^{-\left[\mathrm{i}\sigma_\mu\,\partial_\mu+\sigma_\mu\,J_\mu(x)\right]^2/M^2}\,\varphi_n(x).\end{aligned} \tag{9.48}$$

Since it is not possible to construct the eigenfunctions $\varphi_n(x)$ explicitly for an arbitrary source J_μ, we trade the summation over n by a summation over the momenta of the plane-wave basis. This is done by insertion of the resolution of the identity twice. More precisely, for any operator $\mathcal{O}$ acting on the spinor subspace with the representation $\mathcal{O}_{ab}$ with $a,b=1,2$, we may write

$$\mathrm{Tr}\,|x\rangle\,\mathcal{O}\,\langle x| \equiv \sum_n\sum_{a,b=1}^{2}\langle n,a|x\rangle\,\mathcal{O}_{ab}\,\langle x|n,b\rangle \tag{9.49}$$

$$\begin{aligned}&=\sum_n\sum_{a,b=1}^{2}\sum_k\sum_{k'}\langle n,a|k\rangle\langle k|x\rangle\,\mathcal{O}_{ab}\,\langle x|k'\rangle\langle k'|n,b\rangle\\ &=\sum_k\sum_{k'}\sum_{a,b=1}^{2}\langle k'|\underbrace{\left(\sum_n|n,b\rangle\langle n,a|\right)}_{=\delta_{a,b}}|k\rangle\langle k|x\rangle\mathcal{O}_{ab}\,\langle x|k'\rangle\\ &=\int\frac{\mathrm{d}^2k}{(2\pi)^2}\langle k|x\rangle\;(\mathrm{tr}\,\mathcal{O})\;\langle x|k\rangle,\end{aligned} \tag{9.50}$$

where the label n refers to the energy eigenstate, the indices $a,b=1,2$ refer to the spinor indices, and tr refers to the trace over the spinor indices ($\langle x|k\rangle$ is a mere $\mathbb{C}$ number). Thus, after having traded the basis that diagonalizes the Dirac operator $\sigma_\mu(\mathrm{i}\partial_\mu+J_\mu)$ for the plane-wave basis, the axial anomaly in the background J_μ becomes

$$\mathcal{A}_5(x) = 2\times\lim_{M\to\infty}\mathrm{tr}\left(\int\frac{\mathrm{d}^2k}{(2\pi)^2}\gamma_5 e^{-\mathrm{i}kx}\,e^{-\left[\mathrm{i}\sigma_\mu\partial_\mu+\sigma_\mu\,J_\mu(x)\right]^2/M^2}\,e^{+\mathrm{i}kx}\right). \tag{9.51}$$

If we introduce the notation

$$D_\mu := \mathrm{i}\partial_\mu + J_\mu, \qquad \mu=1,2, \tag{9.52a}$$

for the covariant derivative and

$$F_{\mu\nu}(x) := \left(\partial_\mu\,J_\nu\right)(x) - \left(\partial_\nu\,J_\mu\right)(x), \qquad \mu,\nu=1,2, \tag{9.52b}$$

for the field strength of the background vector potential, then are needed

$$
\begin{aligned}
\sum_{\mu=1}^{2}\left(\sigma_\mu D_\mu\right)^2 &= \sum_{\mu=1}^{2}\sigma_\mu\sigma_\mu D_\mu D_\mu + \sum_{\mu\neq\nu}\sigma_\mu\sigma_\nu D_\mu D_\nu \\
&= \sum_{\mu=1}^{2}\sigma_0 D_\mu D_\mu + \sum_{\mu<\nu}\sigma_\mu\sigma_\nu\left(D_\mu D_\nu - D_\nu D_\mu\right) \\
&= \sum_{\mu=1}^{2}\sigma_0 D_\mu D_\mu + \frac{1}{2}\sum_{\mu<\nu}\left(\sigma_\mu\sigma_\nu - \sigma_\nu\sigma_\mu\right)\left(D_\mu D_\nu - D_\nu D_\mu\right) \\
&= \sum_{\mu=1}^{2}\sigma_0 D_\mu D_\mu + \frac{1}{4}\sum_{\mu,\nu=1}^{2}[\sigma_\mu,\sigma_\nu][D_\mu, D_\nu] \\
&= \sum_{\mu=1}^{2}\sigma_0 D_\mu D_\mu + \frac{\mathrm{i}}{4}\sum_{\mu,\nu=1}^{2}[\sigma_\mu,\sigma_\nu]\left(\partial_\mu J_\nu - \partial_\nu J_\mu\right) \\
&\equiv \sum_{\mu=1}^{2}\sigma_0 D_\mu D_\mu + \frac{\mathrm{i}}{4}\sum_{\mu,\nu=1}^{2}[\sigma_\mu,\sigma_\nu]F_{\mu\nu},
\end{aligned}
\tag{9.52c}
$$

and (summation convention over repeated indices reinstated)

$$
\left(\sigma_\mu D_\mu\right)^2 e^{+\mathrm{i}kx} = e^{+\mathrm{i}kx}\left[\sigma_0\left(-k_\mu + D_\mu\right)\left(-k_\mu + D_\mu\right) + \frac{\mathrm{i}}{4}[\sigma_\mu,\sigma_\nu]\,F_{\mu\nu}(x)\right]. \tag{9.52d}
$$

The axial anomaly in the background J_μ is now

$$
\begin{aligned}
\mathcal{A}_5(x) &= 2\times\lim_{M\to\infty}\operatorname{tr}\left(\int\frac{\mathrm{d}^2k}{(2\pi)^2}\gamma_5\, e^{-\mathrm{i}kx}\, e^{-\left(\sigma_\mu D_\mu\right)^2/M^2}\, e^{+\mathrm{i}kx}\right) \\
&= 2\times\lim_{M\to\infty}\operatorname{tr}\left(\int\frac{\mathrm{d}^2k}{(2\pi)^2}\gamma_5\, e^{-\left(k^2-2k_\mu D_\mu + D_\mu D_\mu + \frac{\mathrm{i}}{4}[\sigma_\mu,\sigma_\nu]\,F_{\mu\nu}(x)\right)/M^2}\right) \\
&= 2\times\lim_{M\to\infty}M^2\operatorname{tr}\left(\int\frac{\mathrm{d}^2p}{(2\pi)^2}\gamma_5\, e^{-p^2+2p_\mu D_\mu/M - D_\mu D_\mu/M^2 - \frac{\mathrm{i}}{4}[\sigma_\mu,\sigma_\nu]\,F_{\mu\nu}(x)/M^2}\right) \\
&\equiv 2\times\lim_{M\to\infty}M^2\int\frac{\mathrm{d}^2p}{(2\pi)^2}e^{-p^2}\, g_M(x,p),
\end{aligned}
\tag{9.53a}
$$

where we have introduced the auxiliary function

$$
\begin{aligned}
g_M(x,p) := \operatorname{tr}\Big(\ \gamma_5\Big\{1 + 2p_\mu D_\mu/M - D_\mu D_\mu/M^2 - \frac{\mathrm{i}}{4}[\sigma_\mu,\sigma_\nu]F_{\mu\nu}(x)/M^2 \\
+\frac{1}{2}(2p_\mu D_\mu)^2/M^2 + \mathcal{O}(M^{-3})\Big\}\Big),
\end{aligned}
\tag{9.53b}
$$

to shorten the notation. After performing the trace over the 2×2 matrices, the axial anomaly in the background J_μ turns into

$$\begin{aligned}\mathcal{A}_5(x) &= 2\times\left[\int\frac{\mathrm{d}^2p}{(2\pi)^2}e^{-p^2}\right](-)\frac{\mathrm{i}}{4}\mathrm{tr}\ \{\gamma_5[\sigma_\mu,\sigma_\nu]F_{\mu\nu}(x)\}\\ &= 2\times\qquad\frac{1}{4\pi}\qquad(-)\frac{\mathrm{i}}{4}\mathrm{tr}\ \{(+\mathrm{i})\sigma_2\sigma_1\left[4\times\sigma_1\sigma_2F_{12}(x)\right]\}\\ &= \frac{1}{\pi}F_{12}(x)\\ &= \frac{1}{2\pi}\epsilon_{\mu\nu}F_{\mu\nu}(x).\end{aligned}\tag{9.54}$$

Having obtained with Eq. (9.46) the infinitesimal change $\delta\mathcal{J}_{\mathrm{f}}$ under Eq. (9.47) of the Jacobian $\mathcal{J}_{\mathrm{f}}$ in the background J_μ, we can reconstruct the Jacobian $\mathcal{J}_{\mathrm{f}}$ itself by the composition of infinitesimal $\delta\mathcal{J}_{\mathrm{f}}$,

$$\mathcal{J}_{\mathrm{f}} = \prod_\xi\delta\mathcal{J}_{\mathrm{f}} = \exp\left(\ln\prod_\xi\delta\mathcal{J}_{\mathrm{f}}\right) = \exp\left(\sum_\xi\ln\delta\mathcal{J}_{\mathrm{f}}\right).\tag{9.55}$$

To this end, we express $\delta F_{\mu\nu}$ in terms of $\delta\xi$,

$$\begin{aligned}(\delta J_\mu) = \epsilon_{\mu\nu}\partial_\nu(\delta\xi) \Longleftrightarrow -(\delta F_{12}) &= \partial_2(\delta J_1) - \partial_1(\delta J_2)\\ &= \epsilon_{\mu\mu'}\partial_{\mu'}(\delta J_\mu)\\ &= \epsilon_{\mu\mu'}\epsilon_{\mu\nu}\partial_{\mu'}\partial_\nu(\delta\xi)\\ &= -\epsilon_{\mu'\mu}\epsilon_{\mu\nu}\partial_{\mu'}\partial_\nu(\delta\xi)\\ &= (-)^2\delta_{\mu'\nu}\partial_{\mu'}\partial_\nu(\delta\xi)\\ &= \Delta(\delta\xi).\end{aligned}\tag{9.56}$$

In the same way,

$$-F_{12} = \Delta\xi.\tag{9.57}$$

Since $\mathcal{J}_{\mathrm{f}} = \prod_\xi\delta\mathcal{J}_{\mathrm{f}}$ is obtained by exponentiation of $\int_\xi\delta\xi\ln(\delta\mathcal{J}_{\mathrm{f}})$, we deduce that

$$\begin{aligned}\mathcal{J}_{\mathrm{f}} &= \exp\left(-\int\mathrm{d}^2x\int_\xi(\delta\xi)(x)\mathcal{A}_5(x)\right)\\ &= \exp\left(-\int\mathrm{d}^2x\int_\xi(\delta\xi)(x)\frac{1}{\pi}F_{12}(x)\right).\end{aligned}\tag{9.58}$$

With the help of the decomposition (9.30), we conclude that

$$
\begin{aligned}
\mathcal{J}_{\mathrm{f}} &= \exp\left(+\int \mathrm{d}^2x \int\limits_{\xi} (\delta\xi)(x)\frac{1}{\pi}(\Delta\xi)(x)\right)\\
&= \exp\left(+\frac{1}{2\pi}\int \mathrm{d}^2x\,\xi(x)(\Delta\xi)(x)\right)\\
&= \exp\left(-\frac{1}{2\pi}\int \mathrm{d}^2x\,(\partial_\mu\xi)^2(x)\right)\\
&\equiv \exp\left(-\frac{1}{2\pi}\int \mathrm{d}^2x\int \mathrm{d}^2y\,J_\mu(x)\left[\delta_{\mu\nu}\,\delta(x-y)-\frac{\partial_\mu\partial_\nu}{\Delta}\right]J_\nu(y)\right).
\end{aligned}
\tag{9.59}
$$

The first calculation of the axial anomaly goes back to Schwinger's solution of quantum electrodynamics in (1+1)-dimensional Minkowski space and time (QED_2). [114]

It is important to point out that, had we chosen to regularize the axial anomaly in Eq. (9.46) without respecting gauge invariance, say by writing

$$
\mathcal{A}_5(x) := 2\times\lim_{M\to\infty}\sum_n \varphi^\dagger_n(x)\gamma_5\,e^{-(\mathrm{i}\sigma_\mu\partial_\mu)^2/M^2}\,\varphi_n(x),
\tag{9.60}
$$

we would have then found that $\mathcal{A}_5=0$. We are thus faced with the choice between preserving gauge invariance at the expense of the anomaly, or an anomaly free theory at the expense of gauge invariance.

Where does the anomaly come from? On the one hand, the anomaly comes from the zero-mode sector, i.e., the eigenspace of $\sigma_\mu D_\mu$ with vanishing eigenvalues according to Eq. (9.48). Eigenvalues $\lambda_n\neq 0$ are irrelevant since they do not contribute in the limit $M\to\infty$. On the other hand, the anomaly originates from all plane waves, including ones with very large momenta, according to Eq. (9.54) in the plane-wave basis. The transformation (9.50), from the eigenbasis of $\sigma_\mu D_\mu$ to the plane-wave basis, that we might naively believe to be unitary, is thus far from benign for the operator $\mathcal{O}\equiv\gamma_5$ that anticommutes with $\sigma_\mu D_\mu$.

9.2.3 *Abelian bosonization of the massless Thirring model*

The generating function for the current-current correlation functions in the massless Thirring model is

$$
\begin{aligned}
Z_{\mathrm{Th};g}[J_\mu] &:= \int \mathcal{D}[\bar\psi,\psi]\,\exp(-S_{\mathrm{Th};g}),\\
S_{\mathrm{Th};g} &:= \int \mathrm{d}^2x\,\mathcal{L}_{\mathrm{Th};g},\\
\mathcal{L}_{\mathrm{Th};g} &:= \bar\psi\,\sigma_\mu\left(\mathrm{i}\partial_\mu+J_\mu\right)\psi-\frac{g}{2}\left(\bar\psi\,\sigma_\mu\psi\right)^2.
\end{aligned}
\tag{9.61}
$$

With the help of a Hubbard-Stratonovich transformation,

$$
\begin{aligned}
Z_{\mathrm{Th};g}[J_\mu] &\propto \int \mathcal{D}[B_\mu]\, \exp\left(-\frac{1}{2g}\int \mathrm{d}^2x\, B_\mu^2\right) \int \mathcal{D}[\bar\psi,\psi]\, \exp(-S_{\mathrm{cov}}),\\
S_{\mathrm{cov}} &:= \int \mathrm{d}^2x\, \mathcal{L}_{\mathrm{cov}},\\
\mathcal{L}_{\mathrm{cov}} &:= \bar\psi\, \sigma_\mu \left(\mathrm{i}\partial_\mu + J_\mu + B_\mu\right) \psi.
\end{aligned}
\tag{9.62}
$$

By shifting the functional integration over the auxiliary field B_μ to

$$
B_\mu = A_\mu - J_\mu,
\tag{9.63}
$$

the generating function becomes

$$
\begin{aligned}
Z_{\mathrm{Th};g}[J_\mu] &\propto \int \mathcal{D}[A_\mu]\, \exp\left(-\frac{1}{2g}\int \mathrm{d}^2x\, \left(A_\mu - J_\mu\right)^2\right) \int \mathcal{D}[\bar\psi,\psi]\, \exp(-S_{\mathrm{cov}}),\\
S_{\mathrm{cov}} &:= \int \mathrm{d}^2x\, \mathcal{L}_{\mathrm{cov}},\\
\mathcal{L}_{\mathrm{cov}} &:= \bar\psi\, \sigma_\mu \left(\mathrm{i}\partial_\mu + A_\mu\right) \psi.
\end{aligned}
\tag{9.64}
$$

Assuming that the local decomposition

$$
A_\mu = \partial_\mu \chi + \tilde\partial_\mu \xi
\tag{9.65}
$$

holds,[5] the decoupling transformation

$$
\bar\psi =: \bar\psi'\, e^{-\mathrm{i}\chi+\gamma_5\,\xi}, \qquad \psi =: e^{+\mathrm{i}\chi+\gamma_5\,\xi}\, \psi',
\tag{9.67}
$$

results in the following representation of the generating function,

$$
\begin{aligned}
Z_{\mathrm{Th};g}[J_\mu] &\propto \int \mathcal{D}[\chi,\xi]\, e^{-\frac{1}{2g}\int \mathrm{d}^2x\, \left(\partial_\mu\chi + \tilde\partial_\mu\xi - J_\mu\right)^2}\, e^{-\frac{1}{2\pi}\int \mathrm{d}^2x\, \left(\partial_\mu\xi\right)^2}\\
&\quad \times \int \mathcal{D}[\bar\psi',\psi']\, e^{-S_{\mathrm{free}}},\\
S_{\mathrm{free}} &:= \int \mathrm{d}^2x\, \mathcal{L}_{\mathrm{free}},\\
\mathcal{L}_{\mathrm{free}} &:= \bar\psi' \mathrm{i}\sigma_\mu\, \partial_\mu \psi'.
\end{aligned}
\tag{9.68}
$$

Here, we made use of the fact that the bosonic Jacobian for going from A_μ to (χ,ξ) is independent of (χ,ξ). We also made use of Eq. (9.59).

Integration over the pure-gauge contribution $\partial_\mu\chi$ yields [as $\int \mathrm{d}^2x\, (\partial_\mu\chi)\, (\tilde\partial_\mu\xi) = 0$]

---

[5] The gauge field A_μ is topologically trivial, i.e., it has vanishing winding number (magnetic flux)

$$
\frac{1}{2\pi}\int \mathrm{d}^2x\, \epsilon_{\mu\nu}\, F_{\mu\nu} = \frac{1}{\pi}\int \mathrm{d}^2x\, F_{12} = -\frac{1}{\pi}\int \mathrm{d}^2x\, (\Delta\xi) = 0.
\tag{9.66}
$$

$$\begin{aligned}
\int \mathcal{D}[\chi]\, e^{-\frac{1}{2g}\int d^2x\,[(\partial_\mu\chi)^2-2(\partial_\mu\chi)J_\mu]} &= \int \mathcal{D}[\chi]\, e^{-\frac{1}{2g}\int d^2x\,[\chi(-\Delta)\chi+2\chi(\partial_\mu J_\mu)]} \\
&\propto e^{+\frac{1}{2g}\int d^2x \int d^2y\,(\partial_\mu J_\mu)(x)(-\Delta)^{-1}(x-y)(\partial_\nu J_\nu)(y)} \\
\text{By Eq. (9.30) } J_\mu=\partial_\mu\chi'+\tilde\partial_\mu\xi' \qquad &= e^{+\frac{1}{2g}\int d^2x \int d^2y\,(\Delta\chi')(x)(-\Delta)^{-1}(x-y)(\Delta\chi')(y)} \\
&= e^{+\frac{1}{2g}\int d^2x\,(-\Delta\chi')\chi'} \\
&= e^{+\frac{1}{2g}\int d^2x\,(\partial_\mu\chi')^2} \\
&\equiv e^{+\frac{1}{2g}\int d^2x \int d^2y\, J_\mu(x)\frac{\partial_\mu\partial_\nu}{\Delta}J_\nu(y)}.
\end{aligned} \tag{9.69}$$

Note that the argument on the right-hand side of Eq. (9.69) has the opposite sign to the corresponding (longitudinal) term in Eq. (9.29). After integration over the pure-gauge component $\partial_\mu\chi$ of A_μ, we conclude that the generating function for current-current correlation functions reads

$$\begin{aligned}
Z_{\mathrm{Th};g}[J_\mu] &\propto \int \mathcal{D}[\xi] e^{-\frac{1}{2g}\int d^2x\left\{(1+\frac{g}{\pi})(\partial_\mu\xi)^2-2(\tilde\partial_\mu\xi)J_\mu+\int d^2y\, J_\mu(x)\left[\delta_{\mu\nu}\,\delta(x-y)-\frac{\partial_\mu\partial_\nu}{\Delta}\right]J_\nu(y)\right\}} \\
&\quad\times \int \mathcal{D}[\bar\psi',\psi']\, e^{-S_{\mathrm{free}}}, \\
S_{\mathrm{free}} &:= \int d^2x\, \mathcal{L}_{\mathrm{free}}, \\
\mathcal{L}_{\mathrm{free}} &:= \bar\psi'\, \mathrm{i}\sigma_\mu\,\partial_\mu\,\psi'.
\end{aligned} \tag{9.70}$$

This is not quite yet the canonical form for a generating function since the argument of the exponential is quadratic in the source J_μ for the current. Preferred is a linear dependence on J_μ. To remedy this deficiency, we make use of Eq. (9.29) with

$$\phi\to\theta, \qquad t\to\frac{1}{\alpha}, \qquad \beta\to-\mathrm{i}\sqrt{\frac{\alpha}{g}}, \qquad \beta^2 t=-\frac{1}{\pi}\to\frac{\beta^2}{\alpha}=-\frac{1}{g}, \tag{9.71}$$

to dispose of the term quadratic in the source J_μ in the argument of the Boltzmann weight on the right-hand side of Eq. (9.70) by the introduction of an auxiliary real-valued scalar field θ in the path integral. Hence,

$$\begin{aligned}
Z_{\mathrm{Th};g}[J_\mu] &\propto \int \mathcal{D}[\xi,\theta]\, e^{-\frac{1}{2g}\int d^2x\left\{(1+\frac{g}{\pi})(\partial_\mu\xi)^2+\alpha g(\partial_\mu\theta)^2-2[\tilde\partial_\mu(\xi-g\beta\theta)]J_\mu\right\}} \\
&\quad\times \int \mathcal{D}[\bar\psi',\psi']\, e^{-S_{\mathrm{free}}}, \\
S_{\mathrm{free}} &:= \int d^2x\, \mathcal{L}_{\mathrm{free}}, \\
\mathcal{L}_{\mathrm{free}} &:= \bar\psi'\, \mathrm{i}\sigma_\mu\,\partial_\mu\,\psi'.
\end{aligned} \tag{9.72}$$

The penultimate step consists in the linear transformation

$$\begin{pmatrix}\theta_1\\ \theta_2\end{pmatrix} = \begin{pmatrix}1/g & -\beta\\ a & b\end{pmatrix}\begin{pmatrix}\xi\\ \theta\end{pmatrix} \Longleftrightarrow \begin{pmatrix}\xi\\ \theta\end{pmatrix} = \frac{g}{b+a\beta g}\begin{pmatrix}b & \beta\\ -a & 1/g\end{pmatrix}\begin{pmatrix}\theta_1\\ \theta_2\end{pmatrix}, \tag{9.73}$$

with the ratio of the two adjustable complex-valued parameters a and b chosen such that the coefficient

$$\left(\frac{g}{b+\alpha\beta g}\right)^2\left[\frac{1}{g}\left(1+\frac{g}{\pi}\right)b\beta-\frac{\alpha a}{g}\right] \tag{9.74}$$

of the cross term $(\partial_\mu\theta_1)(\partial_\mu\theta_2)$ vanishes. Since $\beta^2/\alpha=-1/g$, we must then have that

$$\begin{aligned}
\frac{a}{b} &= \left(1+\frac{g}{\pi}\right)\frac{\beta}{\alpha},\\
\begin{pmatrix}\xi\\ \theta\end{pmatrix} &= \frac{g/b}{1+\left(1+\frac{g}{\pi}\right)\frac{\beta^2}{\alpha}g}\begin{pmatrix}b & \beta\\ -a & 1/g\end{pmatrix}\begin{pmatrix}\theta_1\\ \theta_2\end{pmatrix} = -\frac{\pi}{b}\begin{pmatrix}b & \beta\\ -a & 1/g\end{pmatrix}\begin{pmatrix}\theta_1\\ \theta_2\end{pmatrix}.
\end{aligned} \tag{9.75}$$

In terms of the scalar fields θ_1 and θ_2, the generating function reads

$$\begin{aligned}
Z_{\mathrm{Th};g}[J_\mu] &\propto \int\mathcal{D}[\theta_1,\theta_2]\,e^{-\frac{1}{2}\int\mathrm{d}^2x\left[a_1(\partial_\mu\theta_1)^2+a_2(\partial_\mu\theta_2)^2-2(\tilde{\partial}_\mu\theta_1)J_\mu\right]}\\
&\quad\times\int\mathcal{D}[\bar\psi',\psi']\,e^{-S_{\mathrm{free}}},\\
S_{\mathrm{free}} &:= \int\mathrm{d}^2x\,\mathcal{L}_{\mathrm{free}},\\
\mathcal{L}_{\mathrm{free}} &:= \bar\psi'\mathrm{i}\sigma_\mu\,\partial_\mu\psi',
\end{aligned} \tag{9.76a}$$

with the coefficients

$$\begin{aligned}
a_1 :&= \left(\frac{g}{b+a\beta g}\right)^2\left[\frac{1}{g}\left(1+\frac{g}{\pi}\right)b^2+\alpha a^2\right]\\
&= \frac{g}{\left(1+\frac{a}{b}\beta g\right)^2}\left[\left(1+\frac{g}{\pi}\right)+\alpha\frac{a^2}{b^2}g\right]\\
&= \frac{g}{\left[1+\left(1+\frac{g}{\pi}\right)\frac{\beta^2}{\alpha}g\right]^2}\left[\left(1+\frac{g}{\pi}\right)+\alpha\left(1+\frac{g}{\pi}\right)^2\left(\frac{\beta}{\alpha}\right)^2g\right]\\
&= \frac{g}{\left[1-\left(1+\frac{g}{\pi}\right)\right]^2}\left[\left(1+\frac{g}{\pi}\right)-\left(1+\frac{g}{\pi}\right)^2\right]\\
&= \frac{g}{\left(-\frac{g}{\pi}\right)^2}\left(1+\frac{g}{\pi}\right)\left[-\left(\frac{g}{\pi}\right)\right]\\
&= -\pi\left(1+\frac{g}{\pi}\right)
\end{aligned} \tag{9.76b}$$

and

$$\begin{aligned}
a_2 :&= \left(\frac{g}{b+a\beta g}\right)^2\left[\frac{1}{g}\left(1+\frac{g}{\pi}\right)\beta^2+\frac{\alpha}{g^2}\right]\\
&= \frac{1}{b^2}\frac{g}{\left[1+\left(1+\frac{g}{\pi}\right)\frac{\beta^2}{\alpha}g\right]^2}\left[\left(1+\frac{g}{\pi}\right)\beta^2+\frac{\alpha}{g}\right]\\
&= \frac{\alpha}{b^2}\frac{1}{\left[1-\left(1+\frac{g}{\pi}\right)\right]^2}\left[-\left(1+\frac{g}{\pi}\right)+1\right]\\
&= -\pi\frac{\alpha}{b^2}\frac{1}{g}.
\end{aligned} \tag{9.76c}$$

If we choose

$$b = \beta, \tag{9.77a}$$

we find the simple relation

$$\xi = -\pi\,(\theta_1 + \theta_2)\,, \tag{9.77b}$$

as well as the positive coefficient

$$a_2 = +\pi. \tag{9.77c}$$

The real-valued scalar field θ_1 has a negative definite kinetic energy. To bring this field to a canonical form, we perform the final change of variables

$$\theta_1 =: \frac{\mathrm{i}}{2\pi}\phi_1, \qquad \theta_2 =: \frac{1}{2\pi}\phi_2, \tag{9.78}$$

under which we find that

$$\begin{aligned}
Z_{\mathrm{Th};g}[J_\mu] \propto & \int \mathcal{D}[\phi_1,\phi_2]\, e^{-\int \mathrm{d}^2x\,\left[\frac{1}{8\pi}\left(1+\frac{g}{\pi}\right)(\partial_\mu\phi_1)^2 + \frac{1}{8\pi}(\partial_\mu\phi_2)^2 - \frac{\mathrm{i}}{2\pi}(\tilde{\partial}_\mu\phi_1)J_\mu\right]} \\
& \times \int \mathcal{D}[\bar{\psi}',\psi']\, e^{-S_{\mathrm{free}}}, \\
S_{\mathrm{free}} :=& \int \mathrm{d}^2x\, \mathcal{L}_{\mathrm{free}}, \\
\mathcal{L}_{\mathrm{free}} :=& \bar{\psi}'\,\mathrm{i}\sigma_\mu\,\partial_\mu\,\psi', \\
\bar{\psi}\psi =& \bar{\psi}'\, e^{+2\gamma_5\xi}\,\psi', \qquad \xi = -\frac{1}{2}\,(\mathrm{i}\phi_1 + \phi_2)\,.
\end{aligned} \tag{9.79}$$

Observe that ϕ_1 (ϕ_2) enters with (without) an i in $\bar{\psi}\psi = \bar{\psi}'\exp(+2\gamma_5\,\xi)\,\psi' = \bar{\psi}'\exp(-\gamma_5\,(\mathrm{i}\phi_1+\phi_2))\psi'$. This fact is crucial to establish the correspondence between the two-dimensional Sine-Gordon model and the two-dimensional Thirring model. In the mean time, we have established the Abelian bosonization rules of the massless two-dimensional Thirring model

$$\begin{aligned}
\bar{\psi}\,\mathrm{i}\sigma_\mu\,\partial_\mu\,\psi &\longleftrightarrow \frac{1}{8\pi}(\partial_\mu\phi_1)^2, \\
\bar{\psi}\,\sigma_\mu\,\psi &\longleftrightarrow -\frac{\mathrm{i}}{2\pi}(\tilde{\partial}_\mu\phi_1),
\end{aligned} \tag{9.80}$$

that imply

$$\bar{\psi}\,\mathrm{i}\sigma_\mu\,\partial_\mu\,\psi - \frac{g}{2}\left(\bar{\psi}\,\sigma_\mu\,\psi\right)^2 \longleftrightarrow \frac{1}{8\pi}\left(1+\frac{g}{\pi}\right)(\partial_\mu\phi_1)^2. \tag{9.81}$$

Integration over ϕ_2, $\bar{\psi}'$, and ψ' only changes the proportionality factor between the fermionic and bosonic generating functions and is thus of no consequences when calculating current-current correlation functions. Keeping the explicit dependence on $\bar{\psi}'$, ψ', and ϕ_2 as in Eq. (9.79) is needed to establish the equivalence between the two-dimensional massive Thirring model and the two-dimensional Sine-Gordon model.

9.2.4 *Abelian bosonization of the massive Thirring model*

Abelian bosonization of the two-dimensional massive Thirring model follows the steps of Abelian bosonization of the massless two-dimensional Thirring model up to Eq. (9.79), which becomes

$$\begin{aligned}
Z_{\mathrm{Th};m,g}[J_\mu] &\propto \int \mathcal{D}[\phi_1,\phi_2]\, e^{-\int \mathrm{d}^2x \left[\frac{1}{8\pi}\left(1+\frac{g}{\pi}\right)(\partial_\mu \phi_1)^2 + \frac{1}{8\pi}(\partial_\mu \phi_2)^2 - \frac{1}{2\pi}(\tilde{\partial}_\mu \phi_1) J_\mu\right]} \\
&\quad \times \int \mathcal{D}[\bar{\psi}',\psi']\, e^{-S_m}, \\
S_m &:= \int \mathrm{d}^2x\, \mathcal{L}_m, \\
\mathcal{L}_m &:= \bar{\psi}' \left[\mathrm{i}\sigma_\mu \partial_\mu - m\, e^{-\gamma_5(\mathrm{i}\phi_1+\phi_2)}\right] \psi'.
\end{aligned} \tag{9.82}$$

Equivalence between the two-dimensional massive Thirring model (9.82) and the two-dimensional Sine-Gordon model

$$\begin{aligned}
Z_{\mathrm{SG};t,h}[J_\mu] &:= \int \mathcal{D}[\phi]\, e^{-S_{\mathrm{SG};t,h} + \frac{1}{2\pi}\int \mathrm{d}^2x\, (\tilde{\partial}_\mu \phi) J_\mu}, \\
S_{\mathrm{SG};t,h} &:= \int \mathrm{d}^2x\, \mathcal{L}_{\mathrm{SG};t,h}, \\
\mathcal{L}_{\mathrm{SG};t,h} &:= \frac{1}{2t}(\partial_\mu \phi)^2 - \frac{h}{t}\cos\phi, \\
t &= \frac{4\pi}{1+\frac{g}{\pi}}, \qquad \frac{h}{t} \propto -\frac{\mathrm{i}m}{\pi},
\end{aligned} \tag{9.83}$$

is established by comparing term by term the expansions of the two-dimensional Thirring ($Z_{\mathrm{Th};m,g}[J_\mu]$) and two-dimensional Sine-Gordon ($Z_{\mathrm{SG};t,h}[J_\mu]$) current-current generating functions in powers of m and $h/(2t)$, respectively.

A generic charge-neutral term in the two-dimensional Thirring expansion of order $2n$ is of the form (hardcore condition is assumed)

$$\underbrace{m^{2n}\int \mathrm{d}^2x_1 \cdots \int \mathrm{d}^2x_n \int \mathrm{d}^2y_1 \cdots \int \mathrm{d}^2y_n}_{\text{no } x\text{'s or } y\text{'s equal}} \times \left(f_1 \times f_2 \times f\right)(\underbrace{x_1,\cdots,x_n}_{\equiv \boldsymbol{x}_n},\underbrace{y_1,\cdots,y_n}_{\equiv \boldsymbol{y}_n}), \tag{9.84a}$$

where the integrand is the product of three functions. The first function is

$$\begin{aligned}
f_1(\boldsymbol{x}_n,\boldsymbol{y}_n) :&= \int \mathcal{D}[\phi_1]\, e^{-\int \mathrm{d}^2z \left[\frac{1}{8\pi}\left(1+\frac{g}{\pi}\right)(\partial_\mu\phi_1)^2 - \frac{1}{2\pi}(\tilde{\partial}_\mu\phi_1)J_\mu\right]} \\
&\quad \times e^{-\mathrm{i}[\phi_1(x_1)+\cdots+\phi_1(x_n)]+\mathrm{i}[\phi_1(y_1)+\cdots+\phi_1(y_n)]} \\
&= \int \mathcal{D}[\phi]\, e^{-\int \mathrm{d}^2z \left[\frac{1}{2t}(\partial_\mu\phi)^2 - \frac{1}{2\pi}(\tilde{\partial}_\mu\phi)J_\mu\right]} \\
&\quad \times e^{-\mathrm{i}[\phi(x_1)+\cdots+\phi(x_n)]+\mathrm{i}[\phi(y_1)+\cdots+\phi(y_n)]}.
\end{aligned} \tag{9.84b}$$

The second function is

$$f_2(\boldsymbol{x}_n, \boldsymbol{y}_n) := \int \mathcal{D}[\phi_2]\, e^{-\int \mathrm{d}^2 z\, \frac{1}{8\pi}(\partial_\mu \phi_2)^2} \times e^{-[\phi_2(x_1)+\cdots+\phi_2(x_n)]+[\phi_2(y_1)+\cdots+\phi_2(y_n)]}. \tag{9.84c}$$

The third function is

$$\tilde{f}(\boldsymbol{x}_n, \boldsymbol{y}_n) := \int \mathcal{D}[\psi'^*_\pm, \psi'_\pm]\, e^{-\int \mathrm{d}^2 z\, (\psi'^*_- \mathrm{i}\partial_+ \psi'_- + \psi'^*_+ \mathrm{i}\partial_- \psi'_+)} \times (\psi'^*_- \psi'_+)(x_1) \cdots (\psi'^*_- \psi'_+)(x_n) (\psi'^*_+ \psi'_-)(y_1) \cdots (\psi'^*_+ \psi'_-)(y_n). \tag{9.84d}$$

Because there is no i that multiplies ϕ_2 in $e^{\pm\phi_2}$ on the right-hand side of f_2, it is found that

$$f_2(\boldsymbol{x}_n, \boldsymbol{y}_n)\, \tilde{f}(\boldsymbol{x}_n, \boldsymbol{y}_n) = \left(-\frac{\mathrm{i}}{2\pi}\right)^{2n}, \qquad \text{no two points in arguments equal,} \tag{9.85}$$

by Eqs. (9.26) and (9.27). Owing to this cancellation, ϕ_1 can be identified with the real-valued scalar field ϕ in the two-dimensional Sine-Gordon model as represented in Eq. (4.55a), thereby establishing the Abelian bosonization rules for the two-dimensional massive Thirring model in terms of the two-dimensional Sine-Gordon model.

One interesting consequence of the Abelian bosonization rules relating the two-dimensional Thirring model to the two-dimensional Sine-Gordon model is that the Sine-Gordon model at reduced temperature

$$t = 4\pi \Longleftrightarrow g = 0 \qquad \left(\Longleftrightarrow K = \frac{1}{\pi} \text{ in } 2d\text{–}XY \text{ model}\right) \tag{9.86}$$

is equivalent to a free-fermion (relativistic) theory.

9.3 Applications

We present three applications of the Abelian bosonization rules to quantum systems in one-dimensional position space. We limit ourselves to two cases of spinless fermions on a lattice, for simplicity, when dealing with interacting fermions, as it can be shown that the case of fermions carrying spin-1/2 reduces to two copies of bosonized spinless fermions (this is the phenomenon of spin and charge separation in one-dimensional position space). We also consider a quantum magnet and show how it reduces to spinless fermions, before taking advantage of the Abelian bosonization rules to turn this model into one for interacting bosons.

9.3.1 *Spinless fermions with effective Lorentz and global $U(1)$ gauge symmetries*

We consider the Hamiltonian

$$\hat{H}_{\text{kin}} := -t \sum_{j=1}^{N} \left(\hat{c}^\dagger_j\, \hat{c}_{j+1} + \hat{c}^\dagger_{j+1}\, \hat{c}_j\right), \tag{9.87a}$$

where

$$\{\hat{c}_i, \hat{c}_j^\dagger\} = \delta_{ij}, \qquad \{\hat{c}_i^\dagger, \hat{c}_j^\dagger\} = \{\hat{c}_i, \hat{c}_j\} = 0, \qquad i,j = 1,\cdots,N, \tag{9.87b}$$

that acts on the Fock space

$$\mathcal{F} := \text{span}\left\{ \prod_{j=1}^{N} \left(\hat{c}_j^\dagger\right)^{m_j} |0\rangle \,\middle|\, \hat{c}_j|0\rangle = 0, \qquad \hat{c}_j = \hat{c}_{j+N}, \qquad m_j = 0,1 \right\}, \tag{9.87c}$$

subject to the condition that the average number of spinless fermions in the grand-canonical ensemble is N_{f}. The parameter t that sets the energy scale is taken to be positive.

Periodic boundary conditions in a finite system of length $L = N\mathfrak{a}$ have been chosen, as we are interested in the thermodynamic limit, which is defined by the total number of sites $N \to \infty$ and the average number of electrons $N_{\mathrm{f}} \to \infty$ while holding their ratio $N_{\mathrm{f}}/N \leq 1$ fixed. In this limit, our choice of boundary conditions does not affect the conclusions that we draw below. To cover all grounds (see below), we assume that the density of electrons is commensurate to the lattice. For definitiveness, we choose

$$N_{\mathrm{f}} = \frac{N}{2}, \tag{9.88a}$$

i.e., the filling fraction

$$\nu := \frac{N_{\mathrm{f}}}{N} \tag{9.88b}$$

of the lattice is one-half. This choice requires that N is even and that the Fermi wave vector is

$$k_{\mathrm{F}} = \pi/2, \tag{9.88c}$$

in the thermodynamic limit. The many-body ground state is then the non-interacting Fermi sea

$$|\mathrm{FS}\rangle = \prod_{k_m}^{|k_m|<k_{\mathrm{F}}} c_{k_m}^\dagger |0\rangle, \quad c_{k_m}^\dagger := \frac{1}{\sqrt{N}} \sum_{j=1}^{N} e^{\mathrm{i}k_m j} c_j^\dagger, \tag{9.88d}$$

with $k_m = \frac{\pi}{N} m$ and $m = -\frac{N}{2}, \cdots, +\frac{N}{2} - 1$. The Fermi sea obeys the isotropy condition

$$\left\langle \mathrm{FS} \left| c_j^\dagger c_j \right| \mathrm{FS} \right\rangle = \frac{1}{2}, \qquad j = 1, \cdots, N. \tag{9.89}$$

In preparation for taking the continuum limit $\mathfrak{a} \to 0$, we perform the local gauge transformation

$$\hat{c}_{2j} = (-\mathrm{i})^{2j} \hat{f}_{\mathrm{e}i} = e^{-2\mathrm{i}k_{\mathrm{F}} j} \hat{f}_{\mathrm{e}i}, \qquad \hat{c}_{2j+1} = (-\mathrm{i})^{2j+1} \hat{f}_{\mathrm{o}i} = e^{-2\mathrm{i}k_{\mathrm{F}}(j+1/2)} \hat{f}_{\mathrm{o}i}, \tag{9.90a}$$

with $i \equiv 2j$, that leaves the fermionic algebra (9.87b) unchanged and under which

$$\hat{H}_{\mathrm{kin}} = \mathrm{i}t \sum_{i=1}^{N/2} \left[\hat{f}_{\mathrm{e}i}^\dagger \left(\hat{f}_{\mathrm{o}i} - \hat{f}_{\mathrm{o}(i-1)} \right) + \hat{f}_{\mathrm{o}i}^\dagger \left(\hat{f}_{\mathrm{e}(i+1)} - \hat{f}_{\mathrm{e}i} \right) \right]. \tag{9.90b}$$

The (naive) continuum limit of Eq. (9.90b) is the *Dirac* Hamiltonian

$$\begin{aligned}\hat{H}_{\mathrm{D}} &= \int_0^L \mathrm{d}x\, \left(\hat{\eta}_1^\dagger\, \mathrm{i}v_{\mathrm{F}}\, \partial_x \hat{\eta}_2 + \hat{\eta}_2^\dagger\, \mathrm{i}v_{\mathrm{F}}\, \partial_x \hat{\eta}_1\right)\\ &= \int_0^L \mathrm{d}x\, \left(\hat{\eta}_1^\dagger\ \hat{\eta}_2^\dagger\right) \begin{pmatrix} 0 & \mathrm{i}v_{\mathrm{F}}\, \partial_x \\ \mathrm{i}v_{\mathrm{F}}\, \partial_x & 0 \end{pmatrix} \begin{pmatrix} \hat{\eta}_1 \\ \hat{\eta}_2 \end{pmatrix}, \end{aligned} \tag{9.91a}$$

whereby

$$v_{\mathrm{F}} := (2\mathfrak{a})\, t, \qquad \hat{f}_{\mathrm{ei}} \longrightarrow \sqrt{2\mathfrak{a}}\, \hat{\eta}_1(x), \qquad \hat{f}_{\mathrm{oi}} \longrightarrow \sqrt{2\mathfrak{a}}\, \hat{\eta}_2(x), \tag{9.91b}$$

in one-dimensional position space. All derivatives of higher order than one have been dropped here. This approximation should be good in the very close vicinity of the two Fermi points $\pm k_{\mathrm{F}}$. In the chiral basis

$$\begin{aligned} \hat{\eta}_- &:= \sqrt{\tfrac{1}{2}}\, (\hat{\eta}_1 + \hat{\eta}_2) & & \hat{\eta}_1 := \sqrt{\tfrac{1}{2}}\, (\hat{\eta}_- + \hat{\eta}_+)\\ & & \Longleftrightarrow & \\ \hat{\eta}_+ &:= \sqrt{\tfrac{1}{2}}\, (\hat{\eta}_1 - \hat{\eta}_2) & & \hat{\eta}_2 := \sqrt{\tfrac{1}{2}}\, (\hat{\eta}_- - \hat{\eta}_+), \end{aligned} \tag{9.92a}$$

the Dirac Hamiltonian is diagonal

$$\begin{aligned}\hat{H}_{\mathrm{D}} &= \int_0^L \mathrm{d}x\, \left(\hat{\eta}_-^\dagger\, \mathrm{i}v_{\mathrm{F}}\, \partial_x\, \hat{\eta}_- - \hat{\eta}_+^\dagger\, \mathrm{i}v_{\mathrm{F}}\, \partial_x\, \hat{\eta}_+\right)\\ &= \int_0^L \mathrm{d}x\, \left(\hat{\eta}_-^\dagger\ \hat{\eta}_+^\dagger\right) \begin{pmatrix} +\mathrm{i}v_{\mathrm{F}}\partial_x & 0 \\ 0 & -\mathrm{i}v_{\mathrm{F}}\partial_x \end{pmatrix} \begin{pmatrix} \hat{\eta}_- \\ \hat{\eta}_+ \end{pmatrix}. \end{aligned} \tag{9.92b}$$

The partition function at the inverse temperature β (the Boltzmann constant is set to unity) can be represented by a path integral over Grassmann coherent states obeying antiperiodic boundary conditions in the imaginary-time direction,

$$\begin{aligned} Z_{\mathrm{D}} &= \int \mathcal{D}[\eta^*, \eta]\, e^{-S_{\mathrm{D}}},\\ S_{\mathrm{D}} &= \int_0^\beta \mathrm{d}\tau \int_0^L \mathrm{d}x\, \mathcal{L}_{\mathrm{D}},\\ \mathcal{L}_{\mathrm{D}} &= \eta_-^*\, (\partial_\tau + \mathrm{i}v_{\mathrm{F}}\, \partial_x)\, \eta_- + \eta_+^*\, (\partial_\tau - \mathrm{i}v_{\mathrm{F}}\, \partial_x)\, \eta_+. \end{aligned} \tag{9.93}$$

By taking advantage of the fact that $\eta_\mp^*$ is independent of $\eta_\pm$, the change of integration variable

$$\eta_\mp^* =: \mathrm{i}\psi_\mp^*, \qquad \eta_\mp =: \psi_\mp, \tag{9.94}$$

brings the partition function (9.93) to the desired form [see Eq. (9.14)], namely

$$\begin{aligned} Z_{\mathrm{D}} &\propto \int \mathcal{D}[\psi^*, \psi]\, e^{-S_{\mathrm{D}}},\\ S_{\mathrm{D}} &= \int_0^\beta \mathrm{d}\tau \int_0^L \mathrm{d}x\, \mathcal{L}_{\mathrm{D}},\\ \mathcal{L}_{\mathrm{D}} &= \psi_-^*\, \mathrm{i}\, (\partial_\tau + \mathrm{i}v_{\mathrm{F}}\, \partial_x)\, \psi_- + \psi_+^* \mathrm{i}\, (\partial_\tau - \mathrm{i}v_{\mathrm{F}}\, \partial_x)\, \psi_+, \end{aligned} \tag{9.95}$$

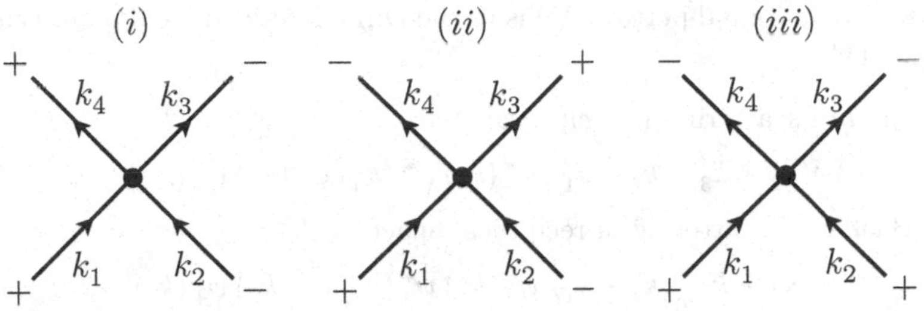

Fig. 9.1 (*i*) Forward, (*ii*) backward, and (*iii*) Umklapp scattering processes in reciprocal space due to a quartic contact interaction in position space.

provided one identifies τ with x_1, x with x_2, and sets the Fermi velocity v_{F} to one,

$$x_1 := \tau, \qquad x_2 := x, \qquad v_{\mathrm{F}} \equiv 1. \tag{9.96}$$

Note that these identifications imply, upon Abelian bosonization, that

$$\begin{aligned}\left(\bar\psi\,\sigma_1\,\psi\right)(\tau,x) =& \left(\psi^*_-\,\psi_- + \psi^*_+\,\psi_+\right)(\tau,x)\\ &\to +\frac{1}{2\pi\mathrm{i}}\left(\partial_x\,\phi\right)(\tau,x),\end{aligned} \tag{9.97a}$$

$$\begin{aligned}\left(\bar\psi\,\sigma_2\,\psi\right)(\tau,x) =& +\mathrm{i}\left(\psi^*_-\,\psi_- - \psi^*_+\,\psi_+\right)(\tau,x)\\ &\to -\frac{1}{2\pi\mathrm{i}}\left(\partial_\tau\,\phi\right)(\tau,x).\end{aligned} \tag{9.97b}$$

Equation (9.97a) tells us that the imaginary-time component $\bar\psi\,\sigma_1\,\psi$ of the (relativistic) two-current $\bar\psi\,\sigma_\mu\,\psi$ becomes the space (x) derivative of a real valued scalar field ϕ upon Abelian bosonization. Equation (9.97b) tells us that the space component $\bar\psi\,\sigma_2\,\psi$ of the (relativistic) two-current $\bar\psi\,\sigma_\mu\,\psi$ becomes the imaginary time (τ) derivative of a real-valued scalar field ϕ upon Abelian bosonization.

The partition function (9.95) defines a free-field fixed point. The engineering dimension of the Dirac field equals its scaling dimension and is given by 1/2 in units where $\hbar = v_{\mathrm{F}} = 1$ as

$$[\psi_\mp] = (\text{length})^{-1/2}. \tag{9.98}$$

The engineering dimension of the local bilinears

$$(\psi^*_-\,\psi_- \pm \psi^*_+\,\psi_+)(x,\tau) \tag{9.99a}$$

and

$$\left(\psi^*_-\,\psi_+ \pm \psi^*_+\,\psi_-\right)(x,\tau) \tag{9.99b}$$

is 1. These are infrared relevant perturbations to the Dirac free-field fixed point. The engineering dimension of the local (contact) quartic interactions

$$(\psi^*_-\,\psi_- \pm \psi^*_+\,\psi_+)^2(x,\tau) \tag{9.100a}$$

and

$$\left(\psi^*_-\,\psi_+ \pm \psi^*_+\,\psi_-\right)^2(x,\tau) \tag{9.100b}$$

is 2. These are marginal perturbations to the Dirac free-field fixed point that encode (see Fig. 9.1):[6]

- (*i*) Forward scattering in reciprocal space,
$$\delta(k_4+k_3-k_2-k_1)\,\psi^*_+(k_4)\,\psi^*_-(k_3)\,\psi_-(k_2)\,\psi_+(k_1). \tag{9.103}$$
- (*ii*) Backward scattering in reciprocal space,
$$\delta(k_4+k_3-k_2-k_1)\,\psi^*_-(k_4)\,\psi^*_+(k_3)\,\psi_-(k_2)\,\psi_+(k_1). \tag{9.104}$$
- (*iii*) Umklapp scattering in reciprocal space,
$$\delta(k_4+k_3-k_2-k_1+K)\,\psi^*_-(k_4)\,\psi^*_-(k_3)\,\psi_+(k_2)\,\psi_+(k_1), \tag{9.105a}$$
where K is any vector from the reciprocal lattice, i.e.,
$$K=\frac{2\pi}{\mathfrak{a}}n, \qquad n\in\mathbb{Z}. \tag{9.105b}$$
Umklapp scattering demands a filling fraction (9.88b) that is commensurate with the lattice in order to satisfy momentum conservation up to momenta from the reciprocal lattice.

Higher powers of Eqs. (9.100a) and (9.100b) are infrared irrelevant at the free-field fixed point.

By demanding that any quartic interaction satisfies the effective Lorentz symmetry and the global $U(1)$ gauge symmetry (9.4) of the free-field Dirac theory (9.95), we are lead to consider the generic quartic interaction

$$\begin{aligned}\mathcal{L}_{\rm int}=&-\frac{\lambda_1}{2}\left[\left(\psi^*_-\psi_-+\psi^*_+\psi_+\right)^2-\left(\psi^*_-\psi_--\psi^*_+\psi_+\right)^2\right]\\&+\frac{\lambda_2}{2}\left(\psi^*_-\psi_++\psi^*_+\psi_-\right)^2+\frac{\lambda_3}{2}\left(\psi^*_-\psi_+-\psi^*_+\psi_-\right)^2,\end{aligned} \tag{9.106}$$

which is parametrized by the three dimensionless coupling constants $\lambda_1,\lambda_2,\lambda_3\in\mathbb{R}$. The Abelian bosonization rules

$$\bar\psi\,\mathrm{i}\sigma_\mu\,\partial_\mu\,\psi\to\frac{1}{8\pi}\left(\frac{\partial\phi}{\partial x_1}\right)^2+\frac{1}{8\pi}\left(\frac{\partial\phi}{\partial x_2}\right)^2 \tag{9.107a}$$

for the Dirac kinetic energy,

$$\bar\psi\,\sigma_1\,\psi\to+\frac{1}{2\pi\mathrm{i}}\left(\frac{\partial\phi}{\partial x_2}\right), \qquad \bar\psi\,\sigma_2\,\psi\to-\frac{1}{2\pi\mathrm{i}}\left(\frac{\partial\phi}{\partial x_1}\right), \tag{9.107b}$$

[6] It might seem peculiar to worry about interactions induced by taking powers of Eq. (9.99) in that they appear to vanish in the continuum limit due to the fermionic algebra
$$0=\left(\psi^*_-\,\psi^*_-\,\psi_+\,\psi_+\right)(x,\tau). \tag{9.101}$$
However, these fields are highly singular at short distances. To deal with such ambiguities one must rely on the regularization procedure known as *point-splitting* by which
$$\left(\psi^*_-\,\psi^*_-\,\psi_+\,\psi_+\right)(x,\tau)\to\psi^*_-(x+\epsilon_4,\tau)\,\psi^*_-(x+\epsilon_3,\tau)\,\psi_+(x+\epsilon_2,\tau)\,\psi_+(x+\epsilon_1,\tau), \tag{9.102}$$
and the limit $\epsilon_{1,2,3,4}\to0$ is carefully taken so as to extract any singular $\mathbb{C}$-number that appears in the expectation value of the right-hand side due to short distance singularities of the free-field Green functions.

for the Dirac current,

$$\bar{\psi}\,\psi \to +\frac{1}{\pi \mathrm{i}}\frac{\cos\phi}{2\mathfrak{a}}, \qquad \bar{\psi}\,\gamma_5\,\psi \to -\frac{1}{\pi}\frac{\sin\phi}{2\mathfrak{a}}, \tag{9.107c}$$

for the Dirac masses, and

$$\psi^*_-\,\psi_+ \to \frac{1}{2\pi \mathrm{i}}\frac{e^{+\mathrm{i}\phi}}{2\mathfrak{a}}, \qquad \psi^*_+\,\psi_- \to \frac{1}{2\pi \mathrm{i}}\frac{e^{-\mathrm{i}\phi}}{2\mathfrak{a}}, \tag{9.107d}$$

for the chiral masses imply that the fermionic partition function

$$Z = \int \mathcal{D}[\psi^*,\psi]\, e^{-\int\limits_{-\infty}^{+\infty} \mathrm{d}^2 x\,(\mathcal{L}_{\mathrm{D}}+\mathcal{L}_{\mathrm{int}})} \tag{9.108a}$$

can be represented by the bosonic partition function

$$\begin{aligned} Z &= \int \mathcal{D}[\phi]\, e^{-\int\limits_{-\infty}^{+\infty} \mathrm{d}^2 x\,\mathcal{L}}, \\ \mathcal{L} &= \frac{1}{8\pi}\left(1+\frac{\lambda_1}{\pi}\right)(\partial_\mu\phi)^2 - \frac{\lambda_2}{2\pi^2}\frac{\cos^2\phi}{(2\mathfrak{a})^2} + \frac{\lambda_3}{2\pi^2}\frac{\sin^2\phi}{(2\mathfrak{a})^2}. \end{aligned} \tag{9.108b}$$

When using the Abelian bosonization rules for the Dirac mass $\bar{\psi}\,\psi$ and the axial mass $\bar{\psi}\,\gamma_5\,\psi$, we have divided the cosine and the sine of the bosonic field ϕ by the lattice spacing $(2\mathfrak{a})$ of the sublattice made of even sites to insure that the bosonic interaction has the units of $(\text{length})^{-2}$ and with a bias towards the microscopic model constructed from Eq. (9.87a) by the addition of some short-range interaction between lattice fermions. The choice of the length scale $(2\mathfrak{a})$ is arbitrary, as we could equally have chosen a length scale that differs from $(2\mathfrak{a})$ by a numerical factor of order 1, say if the hopping took place only between nearest-neighbor even sites. The length scale entering the bosonic interaction cannot be fixed by field theory alone, as to do so would require a detailed knowledge of the physics at short distances. To put it differently, many different microscopic models could be described at long distances and low energies by the field theory (9.108b), although they would differ at short distances and high energies. This ambiguity is reflected by the fact that the ratio m/h of dimensionful couplings cannot be fixed from the sole data provided by the fermionic and bosonic field theories in the bosonization Table 9.1.

The one-loop renormalization-group (RG) flows obeyed by the three dimensionless coupling constants λ_1, λ_2, and λ_3 can be deduced along the same lines as we derived the Kosterlitz-Thouless-RG flows in section 4.6.

9.3.2 *Quantum xxz spin-1/2 chain*

The quantum Hamiltonian for the quantum xxz spin chain is defined by

$$\begin{aligned} \hat{H}^{xxz} := {} & J_\perp \sum_{j=1}^{N} \left(\hat{S}^x_j\,\hat{S}^x_{j+1} + \hat{S}^y_j\,\hat{S}^y_{j+1}\right) + J_z \sum_{j=1}^{N} \hat{S}^z_j\,\hat{S}^z_{j+1} \\ = {} & \frac{J_\perp}{2} \sum_{j=1}^{N} \left(\hat{S}^+_j\,\hat{S}^-_{j+1} + \hat{S}^-_j\,\hat{S}^+_{j+1}\right) + J_z \sum_{j=1}^{N} \hat{S}^z_j\,\hat{S}^z_{j+1}, \end{aligned} \tag{9.109a}$$

whereby

$$\hat{S}^{+}_{j} := \hat{S}^{x}_{j} + \mathrm{i}\hat{S}^{y}_{j}, \qquad \hat{S}^{-}_{j} := \hat{S}^{x}_{j} - \mathrm{i}\hat{S}^{y}_{j}. \tag{9.109b}$$

The local spin operators obey the $SU(2)$ (spin) algebra

$$\left[\hat{S}^{\alpha}_{k}, \hat{S}^{\beta}_{l}\right] = \mathrm{i}\epsilon^{\alpha\beta\gamma}\,\hat{S}^{\gamma}_{k}\,\delta_{kl}, \qquad k,l = 1,\cdots,N, \qquad \alpha,\beta,\gamma = x,y,z, \tag{9.109c}$$

and the periodic boundary conditions

$$\hat{\boldsymbol{S}}_{j+N} = \hat{\boldsymbol{S}}_{j}, \qquad j = 1,\cdots,N. \tag{9.109d}$$

The dimensionful coupling constants $J_{\perp}$ and J_{z} are real-valued. Without loss of generality, we may choose $J_{\perp}$ to be positive, for a rotation about the z axis in spin space by π for every other spins renders $J_{\perp}$ positive.

Before defining the Hilbert space on which $\hat{H}^{xxz}$ is defined, we recall that the spin algebra (9.109c) can be rewritten

$$[\hat{S}^{z}_{j}, \hat{S}^{+}_{k}] = +\hat{S}^{+}_{j}\,\delta_{jk}, \qquad [\hat{S}^{z}_{j}, \hat{S}^{-}_{k}] = -\hat{S}^{-}_{j}\,\delta_{jk}, \qquad [\hat{S}^{+}_{j}, \hat{S}^{-}_{k}] = 2\hat{S}^{z}_{j}\,\delta_{jk}, \tag{9.110a}$$

for $j,k = 1,\cdots,N$. For any s such that $2s$ is a non-vanishing and positive integer and for any site $j = 1,\cdots,N$, the local Hilbert space can be constructed from the highest-weight state $|s\rangle_{j}$, here defined by the condition

$$\hat{S}^{+}_{j}|s\rangle_{j} = 0, \tag{9.110b}$$

by repeated application of the ladder operator $\hat{S}^{-}_{j}$,

$$\mathcal{H}_{j} := \mathrm{span}\left\{|s\rangle_{j}, \hat{S}^{-}_{j}|s\rangle_{j}, \cdots, \left(\hat{S}^{-}_{j}\right)^{2s}|s\rangle_{j}\right\}. \tag{9.110c}$$

The Hilbert space for the quantum xxz spin-s chain is then

$$\mathcal{H}^{xxz} := \bigotimes_{j=1}^{N}\mathcal{H}_{j}. \tag{9.110d}$$

The representation $s = 1/2$ of the spin algebra (9.109c) is specified by demanding that the Casimir operator

$$\begin{aligned}\hat{\boldsymbol{S}}^{2}_{j} :&= (\hat{S}^{x}_{j})^{2} + (\hat{S}^{y}_{j})^{2} + (\hat{S}^{z}_{j})^{2}\\ &= \frac{1}{2}\left(\hat{S}^{+}_{j}\hat{S}^{-}_{j} + \hat{S}^{-}_{j}\hat{S}^{+}_{j}\right) + (\hat{S}^{z}_{j})^{2}\\ &= \frac{1}{2}\left(\frac{1}{2}+1\right)\end{aligned} \tag{9.111}$$

holds locally, i.e., for any site $j = 1,\cdots,N$. In the spin-1/2 representation,

$$\begin{aligned}\hat{S}^{+}_{j}\hat{S}^{-}_{j} &= \hat{\boldsymbol{S}}^{2}_{j} - \hat{S}^{z}_{j}\left(\hat{S}^{z}_{j} - 1\right)\\ &= \frac{3}{4} - \frac{1}{4} + \hat{S}^{z}_{j}\\ &= \hat{S}^{z}_{j} + \frac{1}{2}, \qquad j = 1,\cdots,N.\end{aligned} \tag{9.112}$$

From now on, we assume the spin-1/2 representation, in which case the Hilbert space $\mathcal{H}^{xxz}$ is 2^N-dimensional and it is customary to label states in $\mathcal{H}^{xxz}$ by the Casimir operator for the total spin

$$\hat{\boldsymbol{S}}^2_{\text{tot}} := \left(\sum_{j=1}^{N} \hat{\boldsymbol{S}}_j\right)^2, \tag{9.113}$$

and the z-component $\hat{S}^z_{\text{tot}}$ of the total spin

$$\hat{\boldsymbol{S}}_{\text{tot}} := \sum_{j=1}^{N} \hat{\boldsymbol{S}}_j, \tag{9.114}$$

as one verifies that $\hat{\boldsymbol{S}}^2_{\text{tot}}$ and $\hat{S}^z_{\text{tot}}$ commute with $\hat{H}^{xxz}$. (The choice of the quantization axis along the z direction in spin space is of course a matter of convention.)

Hamiltonian $\hat{H}^{xxz}$ can be diagonalized when $|J_z/J_\perp| = \infty$ and $|J_z/J_\perp| = 0$.

9.3.2.1 *Ising limit*

When $|J_z/J_\perp| = \infty$, $\hat{H}^{xxz}$ reduces to the Ising model along a ring,

$$\hat{H}^z := J_z \sum_{j=1}^{N} \hat{S}^z_j \hat{S}^z_{j+1} \equiv H^{\text{Ising}}, \tag{9.115a}$$

where

$$H^{\text{Ising}} := J_z \sum_{j=1}^{N} s^z_j s^z_{j+1}, \qquad s_j = \pm\frac{1}{2} \text{ eigenvalues of } \hat{S}^z_j. \tag{9.115b}$$

When $J_z > 0$, this is the nearest-neighbor antiferromagnetic Ising model along a ring (periodic boundary conditions). When $J_z < 0$, this is the nearest-neighbor ferromagnetic Ising model along a ring.

The ferromagnetic Ising model (9.115) is related to the antiferromagnetic Ising model (9.115) by the transformation

$$\left(J_z, \hat{S}^z_j\right) \to \left(-J_z, (-1)^j \hat{S}^z_j\right), \tag{9.116}$$

for all sites $j = 1, \cdots, N$. Hence, all properties in thermodynamic equilibrium of the antiferromagnetic Ising model (9.115) are in one-to-one correspondence with those of the ferromagnetic Ising model (9.115) through the transformation (9.116).

The antiferromagnetic Ising model (9.115) supports long-range order (LRO) in the form of two degenerate ground states having all spins parallel to the z-axis in spin space and nearest-neighbor spins antiparallel. Any one of these two ground states is also called the one-dimensional Néel state. Any one of these two Néel states breaks spontaneously the translation symmetry by one lattice spacing of $\hat{H}^{xxz}$. Any one of these two Néel states also breaks spontaneously time-reversal symmetry, defined by reversing locally all the spin directions, of $\hat{H}^{xxz}$. It is essential to realize that the symmetries of $\hat{H}^{xxz}$ that are spontaneously broken by any one of its two Néel states are discrete in the Ising limit $J_z/J_\perp \to \infty$. For this reason, a Néel state is separated from all excited states by a finite-energy gap in the thermodynamic limit. At any finite temperature, the Néel-long-range order is destroyed by the finite-energy excitations owing to a celebrated argument by Peierls. Correspondingly, Ising spin correlation functions decay exponentially fast with large separations in position space at any non-vanishing temperature.

9.3.2.2 *Quantum xy limit*

When $|J_z/J_\perp| = 0$, $\hat{H}^{xxz}$ reduces to

$$\hat{H}^{xy} := J_\perp \sum_{j=1}^{N} \left(\hat{S}^x_j \, \hat{S}^x_{j+1} + \hat{S}^y_j \, \hat{S}^y_{j+1} \right). \tag{9.117}$$

This limit, called the quantum xy limit, represents another point in the coupling-space $0 \leq |J_z/J_\perp| \leq \infty$ which is exactly soluble, as we shall see shortly with the help of the Jordan-Wigner transformation. The ground state of $\hat{H}^{xy}$ is featureless and the finite-size gap to all excitations above this ground state collapses to 0 in the thermodynamic limit $N \to \infty$. Spin-spin correlation functions for the x or y components of the spins obey isotropic power laws for large separations in position space and in the thermodynamic limit when $J_z/J_\perp = 0$. Furthermore, the quantum xy critical point $J_z/J_\perp = 0$ is the lower critical end point of a finite segment of critical points with the upper critical value $0 < (J_z/J_\perp)_c$ as upper critical end point along the parametric line $0 \leq |J_z/J_\perp| \leq \infty$. Above $(J_z/J_\perp)_c$ a zero-temperature gap opens up. This gap evolves smoothly to the one in the antiferromagnetic Ising limit $J_z/J_\perp \to \infty$. This is the antiferromagnetic Ising regime of the zero-temperature phase diagram. The critical point $(J_z/J_\perp)_c$ belongs to the universality class of the Kosterlitz-Thouless transition. The segment $-\infty < J_z/J_\perp < 0$ supports a gap. This gap is smoothly connected to the one in the ferromagnetic Ising limit $J_z/J_\perp \to -\infty$. The half-line $-\infty \leq J_z/J_\perp < 0$ can thus be identified with the ferromagnetic regime. The existence of the lower and upper critical values $J_z/J_\perp = 0$ and $(J_z/J_\perp)_c$, respectively, can be inferred from quantum-field-theoretical arguments, although the numerical value of $(J_z/J_\perp)_c$ is beyond quantum field theory. The key step towards a quantum field theory is the Jordan-Wigner transformation. But before taking advantage of the Jordan-Wigner transformation (which is thoroughly described in appendix E.3), we identify the origin of the quantum fluctuations along $0 \leq |J_z/J_\perp| < \infty$.

9.3.2.3 *Quantum fluctuations*

When $|J_z/J_\perp| = \infty$, $\hat{H}^{xxz}$ becomes the classical Ising model $\hat{H}^z$, for any local $\hat{S}^z_j$ commutes with $\hat{H}^{xxz}$. Quantum fluctuations are restored for any non-vanishing $J_\perp$, since any local $\hat{S}^z_j$ fails to commute with $\hat{H}^{xxz}$ for any $J_\perp \neq 0$.

To appreciate further the role played by quantum fluctuations, observe first that the physics is invariant under transformation (9.116) in the Ising limit $J_{\perp} = 0$. This is not true anymore when $J_{\perp} \neq 0$, as can be verified from the fact that the two ferromagnetic states, which are defined by the condition that they are the eigenstates with eigenvalues $\pm N/2$ of the uniform magnetization

$$\hat{S}^z_{\text{tot}} := \sum_{j=1}^{N} \hat{S}^z_j, \tag{9.118}$$

are always eigenstates of Eq. (9.109), whereas the two (Néel) states, which are defined by the condition that they are the eigenstates with eigenvalues $\pm N/2$ of the staggered magnetization

$$\hat{S}^z_{\text{stag}} := \sum_{j=1}^{N} (-)^j \, \hat{S}^z_j, \tag{9.119}$$

are not eigenstates of Eq. (9.109) when $J_{\perp} \neq 0$.

Second, consider for simplicity the two-sites problem at the Heisenberg point defined by the Hamiltonian

$$\begin{aligned}
\hat{H}^{xxz}_{N=2} :=&\; J\,\hat{\boldsymbol{S}}_1 \cdot \hat{\boldsymbol{S}}_2 \\
=&\; \frac{J}{2}\left(\hat{\boldsymbol{S}}_1 + \hat{\boldsymbol{S}}_2\right)^2 - \frac{3}{4}J \\
=&\; -\frac{3}{4}J|0;0\rangle\langle 0;0| \\
&+\frac{1}{4}J\left(|1,-1\rangle\langle 1;-1| + |1;0\rangle\langle 1;0| + |1;+1\rangle\langle 1;+1|\right),
\end{aligned} \tag{9.120}$$

with open boundary conditions. Here, eigenstates of $\hat{\boldsymbol{S}}^2_{\text{tot}} = (\hat{\boldsymbol{S}}_1 + \hat{\boldsymbol{S}}_2)^2$ and $\hat{S}^z_{\text{tot}} = \hat{S}^z_1 + \hat{S}^z_2$ with eigenvalues $s(s+1)$ and s^z, respectively, are denoted $|s; s^z\rangle$. This basis can also be represented in terms of the tensorial basis $|1/2; s^z_1\rangle \otimes |1/2; s^z_2\rangle \equiv |s^z_1; s^z_2\rangle_{\frac{1}{2}}$ as

$$|0;0\rangle := \frac{1}{\sqrt{2}}\left(|\uparrow\downarrow\rangle_{\frac{1}{2}} - |\downarrow\uparrow\rangle_{\frac{1}{2}}\right), \tag{9.121a}$$

$$|1;+1\rangle := |\uparrow\uparrow\rangle_{\frac{1}{2}}, \tag{9.121b}$$

$$|1;0\rangle := \frac{1}{\sqrt{2}}\left(|\uparrow\downarrow\rangle_{\frac{1}{2}} + |\downarrow\uparrow\rangle_{\frac{1}{2}}\right), \tag{9.121c}$$

$$|1;-1\rangle := |\downarrow\downarrow\rangle_{\frac{1}{2}}. \tag{9.121d}$$

The ferromagnetic states are given by Eqs. (9.121b) and (9.121d). Evidently, they are eigenstates of $\hat{H}^{xxz}_{N=2}$ and are symmetric under exchange of the two spin labels. The Néel states are given by $\frac{1}{\sqrt{2}}(|0;0\rangle \pm |1;0\rangle)$, i.e., they are linear superpositions of eigenstates belonging to the subspace $\hat{S}^z_{\text{tot}} = 0$, but with different $\hat{\boldsymbol{S}}^2_{\text{tot}}$. Evidently, they are neither eigenstates of $\hat{H}^{xxz}_{N=2}$ nor eigenstates of the operator that interchanges label 1 and 2 of the spins. Conversely, the ground state of the antiferromagnetic $(J > 0)$ $\hat{H}^{xxz}_{N=2}$ is built out of an antisymmetric linear superposition of the classical (Néel) states. By a classical state, we thus understand a many-body state that is the tensorial product of "single-particle" states.

9.3.2.4 *Jordan-Wigner transformation*

With the help of appendix E.3, $\hat{H}^{xxz}$ can be represented solely in terms of spinless fermions, which are called *Jordan-Wigner* fermions. To see this, define the *non-local* operators

$$\hat{K}_j := e^{\mathrm{i}\pi \sum_{k=1}^{j-1} \left(\hat{S}^z_k + \frac{1}{2}\right)} = e^{\mathrm{i}\pi \sum_{k=1}^{j-1} \hat{S}^+_k \hat{S}^-_k}, \qquad j = 1, \cdots, N. \tag{9.122}$$

The non-local operator $\hat{K}_j$ rotates all spins to the left of site j by the angle π around the z axis in spin space. The operator $\hat{K}_j$ is called a kink operator. It is shown in appendix E.3 that the operators

$$\hat{c}^\dagger_j := \hat{S}^+_j \hat{K}_j, \qquad \hat{c}_j := \hat{K}^\dagger_j \hat{S}^-_j, \tag{9.123a}$$

realize the fermion algebra

$$\{\hat{c}_k, \hat{c}^\dagger_l\} = \delta_{k,l}, \qquad 0 = \{\hat{c}_k, \hat{c}_l\} = \{\hat{c}^\dagger_k, \hat{c}^\dagger_l\}, \qquad k, l = 1, \cdots, N. \tag{9.123b}$$

These operators create and destroy Jordan-Wigner fermions. As shown in appendix E.3, the quantum Hamiltonian (9.109) becomes

$$\begin{aligned}
\hat{H}^{xxz} &= +\frac{J_\perp}{2} \sum_{j=1}^{N} \left(\hat{c}^\dagger_j \hat{c}_{j+1} + \hat{c}^\dagger_{j+1} \hat{c}_j\right) + J_z \sum_{j=1}^{N} \left(\hat{c}^\dagger_j \hat{c}_j - \frac{1}{2}\right) \left(\hat{c}^\dagger_{j+1} \hat{c}_{j+1} - \frac{1}{2}\right) \\
&\to -\frac{J_\perp}{2} \sum_{j=1}^{N} \left(\hat{c}^\dagger_j \hat{c}_{j+1} + \hat{c}^\dagger_{j+1} \hat{c}_j\right) + J_z \sum_{j=1}^{N} \left(\hat{c}^\dagger_j \hat{c}_j - \frac{1}{2}\right) \left(\hat{c}^\dagger_{j+1} \hat{c}_{j+1} - \frac{1}{2}\right),
\end{aligned} \tag{9.124a}$$

in the Jordan-Wigner representation. We have performed the local gauge transformation

$$\hat{c}_j \to (-)^j \, \hat{c}_j \tag{9.124b}$$

to reach the second equality.

The total number-operator

$$\hat{N}_{\mathrm{tot}} := \sum_{j=1}^{N} \hat{c}^\dagger_j \, \hat{c}_j \tag{9.125}$$

for Jordan-Wigner fermions commutes with $\hat{H}^{xxz}$. It is related to the total spin operator (9.114) by

$$\hat{N}_{\mathrm{tot}} = \hat{S}^z_{\mathrm{tot}} + \frac{N}{2}. \tag{9.126}$$

The boundary conditions obeyed by the Jordan-Wigner fermions depend on the eigenvalue $s^z_{\rm tot}$ of $\hat{S}^z_{\rm tot}$ through [see Eq. (E.49)]

$$\hat{c}_{j+N} = (-)^{\hat{N}_{\rm tot}+1}\,\hat{c}_j. \tag{9.127}$$

The unitary transformation

$$\hat{c}_j \to (-)^j\,\hat{c}_j \tag{9.128}$$

of the fermions corresponds to the unitary transformation

$$\hat{S}^x_j \to (-)^j\,\hat{S}^x_j, \qquad \hat{S}^y_j \to (-)^j\,\hat{S}^y_j, \qquad \hat{S}^z_j \to \hat{S}^z_j, \tag{9.129}$$

of the spins, i.e., to a local rotation by the angle π around the z axis in spin space.

9.3.2.5 *Path-integral representation*

A path-integral representation of

$$Z^{xxz} := \operatorname{Tr}\left(e^{-\beta\hat{H}^{xxz}}\right), \tag{9.130}$$

with $\hat{H}^{xxz}$ defined in Eq. (9.109) and the trace performed over the Hilbert space (9.110d), is

$$Z^{xxz} = \int\mathcal{D}[c^*,c]\,e^{-S_{xxz}}, \tag{9.131a}$$

where the Euclidean action reads

$$S^{xxz} := \int\limits_0^\beta \mathrm{d}\tau \sum_{j=1}^{N} \mathcal{L}^{xxz}_{j,j+1}, \tag{9.131b}$$

with the Lagrangian density

$$\mathcal{L}^{xxz}_{j,j+1} := c^*_j\,\partial_\tau\,c_j - \frac{J_\perp}{2}\left(c^*_j c_{j+1} + c^*_{j+1}c_j\right) + J_z\left(c^*_j c_j - \frac{1}{2}\right)\left(c^*_{j+1}c_{j+1} - \frac{1}{2}\right), \tag{9.131c}$$

and the boundary conditions

$$c^*_{j+N}(\tau+\beta) = (-)^{\hat{S}^z_{\rm tot}+\frac{N}{2}}\,c^*_j(\tau), \qquad c_{j+N}(\tau+\beta) = (-)^{\hat{S}^z_{\rm tot}+\frac{N}{2}}\,c_j(\tau). \tag{9.131d}$$

9.3.2.6 *Field theory*

When $|J_z/J_\perp| = 0$, the Hamiltonian (9.117) for interacting spin-1/2 is nothing but the non-interacting Hamiltonian (9.87a) for spinless fermions provided one makes the identification

$$t \to +\frac{J_\perp}{2}. \tag{9.132}$$

The many-body ground state is represented by a Fermi sea for the Jordan-Wigner fermions (see appendix E.3.2). Thus, it is featureless. Correlation functions at long distances are controlled by the two Fermi points and their immediate vicinity.

In others words, linearization of the spectrum around the two Fermi points gives the Dirac Hamiltonian in the non-diagonal or diagonal representations (9.91a) and (9.92b), respectively, from which the asymptotic decays of spin-spin correlation functions can be calculated and shown to be power laws.

A naive continuum limit of the Ising contribution (9.115) to the spin Hamiltonian (9.109) replaces

$$\begin{aligned}
\hat{H}^z &= +J_z \sum_{j=1}^{N} \left(\hat{c}_j^\dagger \hat{c}_j - \frac{1}{2}\right) \left(\hat{c}_{j+1}^\dagger \hat{c}_{j+1} - \frac{1}{2}\right) \\
&= -\frac{J_z}{2} \sum_{j=1}^{N} \left(\hat{c}_j^\dagger \hat{c}_j - \hat{c}_{j+1}^\dagger \hat{c}_{j+1}\right)^2 + \frac{J_z N}{4} \\
&= -\frac{J_z}{2} \sum_{i=1}^{N/2} \left[\left(\hat{f}_{ei}^\dagger \hat{f}_{ei} - \hat{f}_{oi}^\dagger \hat{f}_{oi}\right)^2 + \left(\hat{f}_{oi}^\dagger \hat{f}_{oi} - \hat{f}_{e(i+1)}^\dagger \hat{f}_{e(i+1)}\right)^2\right] + \frac{J_z N}{4} \\
&= -\frac{J_z}{2} \sum_{i=1}^{N/2} \left[\left(\hat{f}_{ei}^\dagger \hat{f}_{ei} - \hat{f}_{oi}^\dagger \hat{f}_{oi}\right)^2 + \left(\hat{f}_{oi}^\dagger \hat{f}_{oi} - \hat{f}_{ei}^\dagger \hat{f}_{ei} + \cdots\right)^2\right] + \frac{J_z N}{4}
\end{aligned} \tag{9.133}$$

by

$$\begin{aligned}
\hat{H}^z &\approx -J_z \sum_{i=1}^{N/2} \left(\hat{f}_{ei}^\dagger \hat{f}_{ei} - \hat{f}_{oi}^\dagger \hat{f}_{oi}\right)^2 + \frac{J_z N}{4} \\
\text{By Eq. (9.91b)} &\rightarrow -v_{\mathrm{F}} \frac{2J_z}{J_\perp} \int_0^L \mathrm{d}x \left(\hat{\eta}_1^\dagger \hat{\eta}_1 - \hat{\eta}_2^\dagger \hat{\eta}_2\right)^2 + \frac{J_z N}{4} \\
\text{By Eq. (9.92a)} &= -v_{\mathrm{F}} \frac{2J_z}{J_\perp} \int_0^L \mathrm{d}x \left(\hat{\eta}_-^\dagger \hat{\eta}_+ + \hat{\eta}_+^\dagger \hat{\eta}_-\right)^2 + \frac{J_z N}{4},
\end{aligned} \tag{9.134}$$

whereby it is assumed that

$$|J_z| \ll J_\perp, \tag{9.135}$$

for linearization of the kinetic dispersion relation to make sense.

To pass to the Grassmann representation of the partition function (9.130), we need to normal order Eq. (9.134). At the operator level, normal ordering is a highly non-trivial step as it requires a careful regularization through point-splitting of the product of fields at the same position in space (see footnote 6). We gloss over these subtleties and assume that we can replace the operators on the last line of Eq. (9.134) by Grassmann-valued fields in the path-integral representation of the partition function. The Ising interaction then becomes

$$\begin{aligned}
(\eta^*_- \eta_+ + \eta^*_+ \eta_-)^2 &= +2\eta^*_- \eta_+ \eta^*_+ \eta_- + (\eta^*_- \eta_+)^2 + (\eta^*_+ \eta_-)^2 \\
&= -2\eta^*_- \eta_- \eta^*_+ \eta_+ + (\eta^*_- \eta_+)^2 + (\eta^*_+ \eta_-)^2 \\
&= -\frac{1}{2} \underbrace{\left[(\eta^*_- \eta_- + \eta^*_+ \eta_+)^2 - (\eta^*_- \eta_- - \eta^*_+ \eta_+)^2\right]}_{\text{Current}-\text{Current interaction}} \\
&\quad + \frac{1}{2} \underbrace{\left[(\eta^*_- \eta_+ + \eta^*_+ \eta_-)^2 + (\eta^*_- \eta_+ - \eta^*_+ \eta_-)^2\right]}_{\text{Umklapp interaction}} \\
\text{By Eq. (9.12)} &= -\frac{1}{2} \underbrace{\left[(\bar\eta\, \sigma_1\, \eta)^2 - (\bar\eta\, \sigma_2\, \eta)^2\right]}_{\text{Current}-\text{Current interaction}} + \frac{1}{2} \underbrace{\left[(\bar\eta\, \eta)^2 + (\bar\eta\, \gamma_5\, \eta)^2\right]}_{\text{Umklapp interaction}} .
\end{aligned} \tag{9.136}$$

With the help of the transformation (9.94), setting the Fermi velocity [recall Eqs. (9.132) and (9.91b)]

$$v_{\mathrm{F}} = (2\mathfrak{a}) \frac{J_\perp}{2} = \mathfrak{a}\, J_\perp \tag{9.137}$$

to one, and ignoring the constant on the right-hand side of Eq. (9.134), we end up with the path-integral representation of the partition function at vanishing temperature and in the thermodynamic limit $L \to \infty$ [recall transformation (9.94)]

$$\begin{aligned}
Z^{xxz} &\propto \int \mathcal{D}[\bar\psi, \psi]\, e^{\,S^{xxz}}, \\
S^{xxz} &= \int\limits_{-\infty}^{+\infty} \mathrm{d}x_1 \int\limits_{-\infty}^{+\infty} \mathrm{d}x_2\, \mathcal{L}^{xxz}, \\
\mathcal{L}^{xxz} &= \bar\psi \mathrm{i} \sigma_\mu\, \partial_\mu\, \psi - \frac{J_z}{J_\perp} (\bar\psi\, \sigma_1\, \psi)^2 - \frac{J_z}{J_\perp} (\bar\psi\, \sigma_2\, \psi)^2 + \frac{J_z}{J_\perp} \left[(\bar\psi\, \psi)^2 + (\bar\psi\, \gamma_5\, \psi)^2\right].
\end{aligned} \tag{9.138}$$

The Abelian bosonization rules (9.107) give[7] the bosonic representation of the partition function

$$\begin{aligned}
Z &= \int \mathcal{D}[\phi]\, e^{-\int\limits_{-\infty}^{+\infty} \mathrm{d}\tau \int\limits_{-\infty}^{+\infty} \mathrm{d}x\, \mathcal{L}}, \\
\mathcal{L} &= \left(\frac{1}{8\pi} + \frac{J_z}{4\pi^2 J_\perp}\right) \left(\frac{\partial \phi}{\partial \tau}\right)^2 + \left(\frac{1}{8\pi} + \frac{J_z}{4\pi^2 J_\perp}\right) \left(\frac{\partial \phi}{\partial x}\right)^2 - \frac{J_z}{\pi^2 J_\perp} \frac{\cos(2\phi)}{(2\mathfrak{a})^2} \\
&= +\frac{1}{8\pi} \left(1 + \frac{2 J_z}{\pi J_\perp}\right) \left[\left(\frac{\partial \phi}{\partial \tau}\right)^2 + \left(\frac{\partial \phi}{\partial x}\right)^2\right] - \frac{J_z}{\pi^2 J_\perp} \frac{\cos(2\phi)}{(2\mathfrak{a})^2}.
\end{aligned} \tag{9.139}$$

[7] Use $\cos^2 \alpha = \frac{1}{2}\left[1 + \cos(2\alpha)\right]$, $\sin^2 \alpha = \frac{1}{2}\left[1 - \cos(2\alpha)\right]$, and $\cos^2 \alpha - \sin^2 \alpha = \cos(2\alpha)$.

Equation (9.139) is the Sine-Gordon model (9.1) at the inverse temperature

$$t = \frac{4\pi}{1 + \frac{2J_z}{\pi J_\perp}}, \tag{9.140}$$

and in the magnetic field

$$\frac{h}{t} = \frac{J_z}{\pi^2 J_\perp} \times \frac{1}{(2\mathfrak{a})^2}, \tag{9.141}$$

corresponding to vortices of charge 2 if the Sine-Gordon model is interpreted as the classical two-dimensional XY model.

The cosine interaction in Eq. (9.139) breaks the symmetry

$$\phi \to \phi + \text{const} \tag{9.142}$$

of the bosonic kinetic energy down to the discrete subgroup

$$\text{const} = \pm\pi, \tag{9.143}$$

which is isomorphic to the multiplicative group

$$\mathbb{Z}_2 := \{+1, -1\}. \tag{9.144}$$

The transformation (9.142) corresponds to the global $U(1)$ axial transformation (9.5) in the fermionic representation. It is a symmetry of the Lagrangian density in the massless two-dimensional Thirring model. The transformation (9.142) with the choice (9.143) corresponds to the global $U(1)$ axial transformation (9.5) with the choice $\alpha_5 = \pm\pi/2$. The latter transformation changes the sign of the Dirac and axial masses $\bar\psi\psi$ and $\bar\psi\gamma_5\psi$, respectively. Hence, adding a squared Dirac mass or a squared axial mass to the Lagrangian density in the two-dimensional massless Thirring model breaks the global $U(1)$ axial symmetry down to the discrete axial subgroup $\mathbb{Z}_2$. From the point of view of lattice fermions, this discrete remnant of the continuous axial symmetry is nothing but a manifestation of the bipartite nature of the underlying lattice in the tight-binding microscopic model. From this microscopic perspective, the cosine interaction in Eq. (9.139) corresponds to Umklapp processes by which the net change of momentum in a scattering event is twice the Fermi momentum and equals the reciprocal vector spanning the first Brillouin zone. It should be noted that linearization of the lattice spectrum induces a much larger symmetry group (the chiral symmetry) at the level of the field theory than the original (microscopic) sublattice symmetry. Destruction of the continuous chiral symmetry of the field theory with forward and backward scattering only is not generic at the microscopic level, for it requires commensuration between the Fermi wave vector and the lattice.

The lessons from the Kosterlitz-Thouless transition is that theory (9.139) has a "low-temperature" spin-wave phase and a "high-temperature" paramagnetic phase. The spin-wave phase corresponds here to a line of critical points characterized by power laws obeyed by correlation functions when $0 \leq J_z/|J_\perp| < (J_z/|J_\perp|)_c$. The paramagnetic phase when $(J_z/|J_\perp|)_\mathrm{c} \le J_z/|J_\perp| \le \infty$ is here characterized by long-range order, a gap, and exponentially decaying correlation functions. The difference with the Kosterlitz-Thouless transition studied in the context of the classical two-dimensional XY model is that it is the vortices of charge 2 and not vortices of charge 1 that trigger the continuous phase transition from a quasi-long-range-ordered phase (quantum xy regime) to a massive phase (Ising regime). Moreover, the terminology of long-range order in the Ising regime refers to long-range order for Ising degrees of freedom and should not be confused with the long-range order of the ground state of the classical $O(2)$ Heisenberg model. The latter order refers to a continuous symmetry, while the former order refers to a discrete symmetry.

The critical value $(J_z/|J_\perp|)_\mathrm{c}$ cannot be predicted reliably from the naive continuum limit that we took. One needs to resort to a matching of field theory and exact methods such as the Bethe Ansatz solution to the xxz lattice model to extract the true dependence of the couplings $\lambda_{1,2,3}$ on the lattice coupling constants from which the critical value $(J_z/|J_\perp|)_\mathrm{c}$ follows by demanding that the cosine interaction is marginal. It turns out that the critical value $(J_z/|J_\perp|)_\mathrm{c}$ corresponds to the so-called Heisenberg point

$$(J_z/|J_\perp|)_\mathrm{c} = 1. \tag{9.145}$$

9.3.3 *Single impurity of the mass type*

As a third example of interacting fermions, we consider the two-dimensional Thirring model (9.2) with the mass term

$$m(x,\tau) := -\mathrm{i}(2\pi\mathfrak{a})\, V_0\, \delta(x), \qquad \forall\tau\in\mathbb{R}. \tag{9.146}$$

This mass term varies only in the space direction with a delta-function profile. It can be interpreted as a *static* impurity located at the origin that scatters incoming right and left movers through a delta function potential. The strength of the impurity is measured by the dimensionful coupling constant V_0, whereby $[V_0] = (\text{length})^{-1}$. Upon Abelian bosonization, the partition function at zero temperature becomes [see Eq. (9.83)]

$$\begin{aligned}
Z &= \int \mathcal{D}[\phi]\, e^{-S},\\
S &= \int\limits_{-\infty}^{+\infty} \mathrm{d}\tau \int\limits_{-\infty}^{+\infty} \mathrm{d}x\; \mathcal{L}, \qquad\qquad (9.147\mathrm{a})\\
\mathcal{L} &= \frac{1}{4\pi\,\eta}(\partial_\mu\phi)^2 + V_0\,\delta(x)\,\cos\phi,
\end{aligned}$$

where

$$\eta = \frac{2}{1+\frac{g}{\pi}}. \tag{9.147b}$$

Our strategy is to integrate over all the components $\phi(x\neq 0,\tau)$ of the field in the path integral so as to induce an effective action for the field

$$\theta(\tau)\equiv\phi(x=0,\tau). \tag{9.148}$$

The first step towards this goal is to rewrite the integration measure as

$$\begin{aligned}\mathcal{D}[\phi] &= \mathcal{D}[\phi]\,\mathcal{D}[\theta]\,\delta\left[\theta(\tau)-\phi(0,\tau)\right]\\ &= \mathcal{D}[\phi]\,\mathcal{D}[\theta]\int\mathcal{D}[\lambda(\tau)]\,e^{\mathrm{i}\int\limits_{-\infty}^{+\infty}\mathrm{d}\tau\lambda(\tau)[\theta(\tau)-\phi(0,\tau)]}.\end{aligned} \tag{9.149}$$

On the second line, the Lagrange multiplier $\lambda(\tau)$ is introduced at each imaginary time τ to enforce the delta-function constraint on the scalar field θ by way of a modification of the Lagrangian density.

Second, we may perform the path integrals in the following order. We begin by integrating over $\phi(x,\tau)$ for all x and τ. Then, we integrate over $\lambda(\tau)$ for all τ. In this way we obtain an effective action in $(0+1)$-dimensional space and (imaginary) time for θ,

$$\begin{aligned}Z_{\mathrm{eff}} &= \int\mathcal{D}[\theta]\;e^{-S_{\mathrm{eff}}[\theta]},\\ e^{-S_{\mathrm{eff}}[\theta]} &\propto \int\mathcal{D}[\lambda]e^{-S_{\mathrm{int}}[\theta,\lambda]},\\ e^{-S_{\mathrm{int}}[\theta,\lambda]} &\propto e^{-\int\limits_{-\infty}^{+\infty}\mathrm{d}\tau[V_0\cos\theta(\tau)-\mathrm{i}\theta(\tau)\,\lambda(\tau)]}\\ &\quad\times\int\mathcal{D}[\phi]\,e^{-\int\limits_{-\infty}^{+\infty}\mathrm{d}\tau\int\limits_{-\infty}^{+\infty}\mathrm{d}x\left[\frac{1}{4\pi\eta}(\partial_\mu\phi)^2(x,\tau)+\mathrm{i}\delta(x)\phi(x,\tau)\lambda(\tau)\right]}.\end{aligned} \tag{9.150}$$

The path integral over ϕ on the last line of Eq. (9.150) is Gaussian and is most easily performed in Fourier space,

$$\begin{aligned}&\int\mathcal{D}[\phi]\mathrm{e}^{-\int\limits_{-\infty}^{+\infty}\mathrm{d}\tau\int\limits_{-\infty}^{+\infty}\mathrm{d}x\left[\frac{1}{4\pi\eta}(\partial_\mu\phi)^2(x,\tau)+\mathrm{i}\delta(x)\phi(x,\tau)\lambda(\tau)\right]}\\ &=\int\mathcal{D}[\phi]\mathrm{e}^{-\int\limits_{-\infty}^{+\infty}\mathrm{d}\varpi\int\limits_{-\infty}^{+\infty}\mathrm{d}q\left\{\frac{\varpi^2+q^2}{4\pi\eta}\phi(+q,+\varpi)\phi(-q,-\varpi)+\mathrm{i}\frac{1}{2\sqrt{2\pi}}[\phi(+\varpi,+q)\lambda(-\varpi)+\phi(-\varpi,-q)\lambda(+\varpi)]\right\}}\\ &\propto e^{-\int\limits_{-\infty}^{+\infty}\mathrm{d}\varpi\left[\int\limits_{-\infty}^{+\infty}\frac{\mathrm{d}q}{2\pi}\frac{\pi\eta}{\varpi^2+q^2}\right]\lambda(+\varpi)\lambda(-\varpi)}\\ &=e^{-\int\limits_{-\infty}^{+\infty}\mathrm{d}\varpi\frac{\pi\eta}{2|\varpi|}\lambda(+\varpi)\lambda(-\varpi)}.\end{aligned} \tag{9.151}$$

The path integral on the second line of Eq. (9.150) is again Gaussian and is also most easily performed in Fourier space,

$$\begin{aligned}&\int\mathcal{D}[\lambda]\mathrm{e}^{-\int\limits_{-\infty}^{+\infty}\mathrm{d}\varpi\left\{\frac{\pi\eta}{2|\varpi|}\lambda(+\varpi)\lambda(-\varpi)-\frac{\mathrm{i}}{2}[\lambda(+\varpi)\theta(-\varpi)+\lambda(-\varpi)\theta(+\varpi)]\right\}}\\ &\propto e^{-\int\limits_{-\infty}^{+\infty}\mathrm{d}\varpi\frac{|\varpi|}{2\pi\eta}\theta(+\varpi)\theta(-\varpi)}.\end{aligned} \tag{9.152}$$

We thus conclude that

$$S_{\text{eff}}[\theta] = \int\limits_{-\infty}^{+\infty} \mathrm{d}\varpi\, \frac{|\varpi|}{2\pi\eta}\, \theta(+\varpi)\, \theta(-\varpi) + V_0 \int\limits_{-\infty}^{+\infty} \mathrm{d}\tau\, \cos\theta(\tau). \tag{9.153}$$

This is nothing but the effective action $S_1 + S_{0\,\text{int}}$ from Eq. (8.172a) for a dissipative Josephson junction. It can be interpreted as a single particle moving on the circle subject to the periodic potential $V_0 \cos\theta$ and to a dissipation with friction coefficient

$$\gamma = \frac{1}{\pi\eta}. \tag{9.154}$$

We saw that the single-particle motion was either delocalized or localized depending on the strength of the friction coefficient γ. The critical value γ_{c} at which the cosine interaction is marginal, to first non-trivial order in a perturbative RG analysis, is given by [recall Eq. (9.147b)]

$$\gamma_{\text{c}} = \frac{1}{2\pi} \Longleftrightarrow \frac{1}{4\pi}\left(1 + \frac{g_{\text{c}}}{\pi}\right) = \frac{1}{4\pi} \Longleftrightarrow g_{\text{c}} = 0. \tag{9.155}$$

When the friction coefficient γ is larger than the critical value γ_{c}, the cosine interaction is relevant, the particle is localized in a minimum of the impurity potential, and translation symmetry is broken. When the friction coefficient γ is smaller than the critical value γ_{c}, the cosine interaction is irrelevant, and the particle is in a Bloch state that preserves the periodicity of the action. From the point of view of the Thirring model (9.2) (and of the underlying tight-binding electronic model), the critical value γ_{c} corresponds to the free fermionic point $g_{\text{c}} = 0$, $\gamma > \gamma_{\text{c}}$ to a repulsive current-current interaction, and $\gamma < \gamma_{\text{c}}$ to an attractive current-current interaction. Here, the interpretation of $g > 0$ ($g < 0$) in terms of a repulsive (attractive) interaction follows from the identification of g with $4J_z/|J_\perp|$ in the fermionized lattice xxz chain model [see Eqs. (9.2), (9.109), (9.138), and (9.139)]. Finally, we can use the Abelian bosonization rules (9.107) to infer that a change of $\theta(\tau)$ by 2π corresponds to the transmission of one electron (in its spinor incarnation) through the impurity site. The localized nature of $\theta(\tau)$ when $\gamma > \gamma_{\text{c}}$ means that all incoming electrons on the impurity site are reflected, i.e., total reflection by the impurity. The delocalized nature of $\theta(\tau)$ when $\gamma < \gamma_{\text{c}}$ means that all incoming electrons on the impurity site are transmitted, i.e., total transmission by the impurity. A deviation from total reflection or total transmission can only occur at the critical value of the friction coefficient γ_{c} when the impurity potential is exactly marginal. In other words, partial transmission and partial reflection by the impurity can only occur when the electrons are non-interacting. The transmission and reflection probabilities depend on V_0 when $\gamma = \gamma_{\text{c}}$ and can be computed by elementary means. This is a unique feature of one-dimensional physics.

9.4 Problems

9.4.1 *Quantum chiral edge theory*

Introduction

We have shown in section 9.2.1 that the massless Thirring model in two-dimensional Euclidean space realizes a line of critical points labeled by the dimensionless coupling constant of the current-current interaction. At each critical point, the partition function factorizes into two sectors, a holomorphic and an antiholomorphic sector, respectively. We then showed in section 9.2.3 that the massless Thirring model could be rewritten as a free real-valued scalar field theory in two-dimensional Euclidean space through the equivalence (9.81). We left open the question of how to demonstrate factorization of this free real-valued scalar field theory into holomorphic and antiholomorphic sectors. We are going to provide a constructive answer to this question that goes beyond the equivalence (9.81).

We are going to abandon imaginary time and work within the Hamiltonian formalism of quantum field theory in one-dimensional position space. The factorization into a holomorphic and an antiholomorphic sector becomes a factorization into a right-moving and left-moving sector. In the right-moving sector, the quantum fields depend exclusively on the linear combination $x - v_{\mathrm{F}}\, t$ of the coordinate x in position space and the coordinate t in time. In the left-moving sector, the quantum fields depend exclusively on the linear combination $x + v_{\mathrm{F}}\, t$. Whereas in a Lorentz-invariant quantum field theory the identification of the velocity v_{F} with the speed of light c would hold, as an emerging low-energy and long-wavelength theory we interpret v_{F} as the Fermi velocity from an underlying lattice model.

Right-moving and left-moving quantum fields in one-dimensional position space differ by their chirality, the choice of sign in their dependence on $x \pm v_{\mathrm{F}}\, t$. As we have shown for the Dirac Lagrangian on the left-hand side of the equivalence (9.81), the massless and neutral Klein-Gordon Lagrangian on the right-hand side of the equivalence (9.81) will be shown to decompose additively into a right-moving and a left-moving sector.

However, we will show that it is also possible to define quantum field theories in one-dimensional position space with unequal numbers of fields in the left-moving and right-moving sectors. These so-called chiral quantum field theories cannot emerge from lattice fermions defined on a one-dimensional lattice due to the fermion-doubling obstruction (the Nielsen-Ninomiya no-go theorem from Ref. [115]). They can emerge on the boundary of a two-dimensional lattice model and play an essential role in the quantum Hall effect as emphasized by Halperin in Ref. [61] for the IQHE and by Wen in Ref. [116] for the FQHE. As a byproduct, we will obtain a representation of right-moving and left-moving fermions in terms of right-moving and left-moving bosons, respectively, that goes back to Mandelstam [109].

Definition

Define the quantum Hamiltonian (in units with the positron charge $e > 0$, the speed of light c, and $\hbar$ set to one)

$$\hat{H}[A_\mu(t,x)] := \int_0^L \mathrm{d}x \left[\frac{1}{4\pi} V_{ij} \left(D_x \hat{u}_i\right) \left(D_x \hat{u}_j\right) + A_0 \left(\frac{q_i}{2\pi} K^{-1}_{ij} \left(D_x \hat{u}_j\right)\right)\right](t,x),$$

$$D_x \hat{u}_i := \left(\partial_x \hat{u}_i + q_i A_1\right), \qquad i = 1, \cdots, N. \tag{9.156a}$$

The summation convention over repeated indices is implied throughout. The N Hermitian quantum fields $\hat{u}_i(t,x)$ are postulated to obey the equal-time commutation relations[8]

$$\left[\hat{u}_i(t,x), \hat{u}_j(t,y)\right] := \mathrm{i}\pi \left[K_{ij} \operatorname{sgn}(x-y) + L_{ij}\right], \qquad i,j = 1, \cdots, N. \tag{9.156b}$$

The function $\operatorname{sgn}(x) = -\operatorname{sgn}(-x)$ gives the sign of the real variable x and will be assumed to be periodic with periodicity L. The $N \times N$ matrix K is integer-valued, symmetric, and invertible

$$K_{ij} = K_{ji} \in \mathbb{Z}, \qquad K^{-1}_{ij} = K^{-1}_{ji} \in \mathbb{Q}, \qquad i,j = 1, \cdots, N. \tag{9.156c}$$

The $N \times N$ matrix L is antisymmetric

$$L_{ij} = -L_{ji} = \begin{cases} 0, & \text{if } i = j, \\ \operatorname{sgn}(i-j)\left(K_{ij} + q_i q_j\right), & \text{otherwise}, \end{cases} \tag{9.156d}$$

for $i,j = 1, \cdots, N$. The sign function $\operatorname{sgn}(i)$ of any integer i is here not made periodic and taken to vanish at the origin of $\mathbb{Z}$. The external scalar gauge potential $A_0(t,x)$ and vector gauge potential $A_1(t,x)$ are real-valued functions of time t and space x coordinates. The $N \times N$ matrix V is symmetric and positive definite

$$V_{ij} = V_{ji} \in \mathbb{R}, \qquad v_i V_{ij} v_j > 0, \qquad i,j = 1, \cdots, N, \tag{9.156e}$$

for any non-vanishing vector $v = (v_i) \in \mathbb{R}^N$. The charges q_i are integer-valued and satisfy

$$(-1)^{K_{ii}} = (-1)^{q_i}, \qquad i = 1, \cdots, N. \tag{9.156f}$$

Finally, we shall impose the boundary conditions

$$\hat{u}_i(t, x+L) = \hat{u}_i(t,x) + 2\pi n_i, \qquad n_i \in \mathbb{Z}, \tag{9.156g}$$

and

$$\left(\partial_x \hat{u}_i\right)(t, x+L) = \left(\partial_x \hat{u}_i\right)(t,x), \tag{9.156h}$$

for any $i = 1, \cdots, N$.

[8] As we shall see, this is the algebra to be imposed on the phase operator [recall Eq. (7.11)] of creation and annihilation operators if they are to obey the canonical commutation relations of quantum fields. This interpretation also justifies the definition of the covariant derivative (9.156a).

Chiral equations of motion

Exercise 1.1: We set $A_0 = A_1 = 0$.

(a) Show that, for any $i = 1, \cdots, N$, the equations of motion are

$$\mathrm{i}\left(\partial_t \hat{u}_i\right)(t,x) = -\,\mathrm{i}K_{ij}V_{jk}\left(\partial_x \hat{u}_k\right)(t,x). \tag{9.157}$$

(b) The equation of motion

$$0 = \left(\delta_{ik}\partial_t + K_{ij}V_{jk}\partial_x\right)\hat{u}_k, \qquad i = 1, \cdots, N, \tag{9.158}$$

is chiral. Show that if we define a Hamiltonian of the form Eq. (9.156) with the substitution $\hat{u}_i \to \hat{v}_i$ and if we change the sign of the right-hand side of Eq. (9.156b), we then find the chiral equation

$$0 = \left(\delta_{ik}\partial_t - K_{ij}V_{jk}\partial_x\right)\hat{v}_k, \qquad i = 1, \cdots, N, \tag{9.159}$$

with the opposite chirality.

Gauge invariance

Exercise 2.1: Verify that Hamiltonian (9.156a) is invariant under the local $U(1)$ gauge transformation

$$\begin{aligned} A_0(t,x) &= A_0'(t,x), \\ A_1(t,x) &= A_1'(t,x) - \left(\partial_x \chi\right)(t,x), \\ \hat{u}_i(t,x) &= \hat{u}_i'(t,x) + q_i\chi(t,x), \qquad i = 1, \cdots, N. \end{aligned} \tag{9.160a}$$

for any real-valued function χ that satisfies the periodic boundary conditions

$$\chi(t, x+L) = \chi(t,x). \tag{9.160b}$$

Differentiation of Hamiltonian (9.156a) with respect to the gauge potentials allows defining the gauge-invariant two-current with the components

$$\begin{aligned} \hat{J}_0(t,x) :&= \frac{\delta \hat{H}}{\delta A_0(t,x)} \\ &= \frac{1}{2\pi} q_i K^{-1}_{ij}\left(D_x \hat{u}_j\right)(t,x), \end{aligned} \tag{9.161a}$$

and

$$\begin{aligned} \hat{J}_1(t,x) :&= \frac{\delta \hat{H}}{\delta A_1(t,x)} \\ &= \frac{1}{2\pi} q_i V_{ij}\left(D_x \hat{u}_j\right)(t,x) + \frac{1}{2\pi}\left(q_i K^{-1}_{ij} q_j\right) A_0(t,x). \end{aligned} \tag{9.161b}$$

Exercise 2.2: Verify that the two components of this gauge-invariant two-current reduce to

$$\hat{\rho}(t,x) := \frac{1}{2\pi} q_i K^{-1}_{ij}\left(\partial_x \hat{u}_j\right)(t,x), \tag{9.162a}$$

and

$$\hat{\jmath}(t,x) := \frac{1}{2\pi} q_i V_{ij} \left(\partial_x \hat{u}_j\right)(t,x), \tag{9.162b}$$

when the external gauge fields vanish.

Exercise 2.3: Verify that

$$\begin{aligned}\partial^\mu \hat{J}_\mu \equiv&\ \partial_t \hat{J}_0 + \partial_x \hat{J}_1\\ =&\ \partial_t \hat{\rho} + \partial_x \hat{\jmath} + \frac{1}{2\pi}\left(q_i K^{-1}_{ij} q_j\right)\left(\partial_t A_1\right)\\ &+\frac{1}{2\pi}\left(q_i V_{ij} q_j\right)\left(\partial_x A_1\right) + \frac{1}{2\pi}\left(q_i K^{-1}_{ij} q_j\right)\left(\partial_x A_0\right).\end{aligned} \tag{9.163}$$

We recall that the magnetic and electric fields are related to the gauge fields by (remember that $c=1$)

$$\boldsymbol{B} = \boldsymbol{\nabla}\wedge\boldsymbol{A}, \qquad \boldsymbol{E} = -\boldsymbol{\nabla}A_0 - \partial_t \boldsymbol{A}, \tag{9.164}$$

in d-dimensional position space. We also recall that the constraints

$$\partial_t A_0 + \boldsymbol{\nabla}\cdot\boldsymbol{A} = 0 \tag{9.165}$$

and

$$\boldsymbol{\nabla}\cdot\boldsymbol{A} = 0 \tag{9.166}$$

are called the Lorenz and Coulomb gauges, respectively [117].

Exercise 2.4: Show that, for the one-dimensional chiral edge in the Coulomb gauge,

$$\partial^\mu \hat{J}_\mu = \partial_t \hat{\rho} + \partial_x \hat{\jmath} - \frac{1}{2\pi}\left(q_i K^{-1}_{ij} q_j\right) E, \tag{9.167a}$$

where

$$E(t,x) = -\left(\partial_x A_0\right)(t,x) - \left(\partial_t A_1\right)(t,x), \tag{9.167b}$$

and

$$0 = \left(\partial_x A_1\right)(t,x). \tag{9.167c}$$

Conserved topological charges

We turn off the external gauge potentials

$$A_0(t,x) = A_1(t,x) = 0, \tag{9.168a}$$

and use the short-hand notation

$$\hat{H} \equiv \hat{H}[A_\mu(t,x)=0]. \tag{9.168b}$$

For any $i = 1, \cdots, N$, define the operator

$$\begin{aligned}\hat{\mathcal{N}}_i(t) :=&\ \frac{1}{2\pi}\int_0^L \mathrm{d}x\, \left(\partial_x \hat{u}_i\right)(t,x)\\ =&\ \frac{1}{2\pi}\left[\hat{u}_i(t,L) - \hat{u}_i(t,0)\right].\end{aligned} \tag{9.169}$$

Exercise 3.1:

(a) Show that $\hat{\mathcal{N}}_i(t)$ is conserved (i.e., time independent) if and only if

$$(\partial_x \hat{u}_i)(t,x) = (\partial_x \hat{u}_i)(t,x+L), \qquad 0 \le x \le L. \tag{9.170}$$

(b) Show that, if we demand that there exists an $n_i \in \mathbb{Z}$ such that

$$\hat{u}_i(t,x+L) = \hat{u}_i(t,x) + 2\pi n_i, \tag{9.171}$$

it then follows that

$$\hat{N}_i = n_i. \tag{9.172}$$

(c) Show that the N conserved topological charges $\mathcal{N}_i$ with $i = 1,\cdots,N$ commute pairwise. *Hint:* Make us of the fact that, for any $i,j = 1,\cdots,N$,

$$\left[\hat{\mathcal{N}}_i, \hat{u}_j(y)\right] = \mathrm{i}K_{ij} \tag{9.173}$$

is independent of y.

The local counterpart to the global conservation of the topological charge is

$$\partial_t \hat{\rho}_i^{\text{top}} + \partial_x \hat{j}_i^{\text{top}} = 0, \tag{9.174a}$$

where the local topological density operator is defined by

$$\hat{\rho}_i^{\text{top}}(t,x) := \frac{1}{2\pi}(\partial_x \hat{u}_i)(t,x), \tag{9.174b}$$

and the local topological current operator is defined by

$$\hat{j}_i^{\text{top}}(t,x) := \frac{1}{2\pi} K_{ik}\, V_{kl}\, (\partial_x \hat{u}_l)(t,x), \tag{9.174c}$$

for $i = 1,\cdots,N$.

Exercise 3.2: Verify the equal-time current algebra

$$\left[\hat{\rho}_i^{\text{top}}(t,x), \hat{\rho}_j^{\text{top}}(t,y)\right] = -\frac{\mathrm{i}}{2\pi} K_{ij}\, \partial_x \delta(x-y), \tag{9.175a}$$

$$\left[\hat{j}_i^{\text{top}}(t,x), \hat{j}_j^{\text{top}}(t,y)\right] = -\frac{\mathrm{i}}{2\pi} K_{ik} V_{kl} K_{jk'} V_{k'l'} K_{ll'}\, \partial_x \delta(x-y), \tag{9.175b}$$

$$\left[\hat{\rho}_i^{\text{top}}(t,x), \hat{j}_j^{\text{top}}(t,y)\right] = -\frac{\mathrm{i}}{2\pi} K_{jk} V_{kl} K_{il}\, \partial_x \delta(x-y), \tag{9.175c}$$

for any $i,j = 1,\cdots,N$.

We also introduce the local charges and currents

$$\hat{\rho}_i(t,x) := K_{ij}^{-1} \hat{\rho}_j^{\text{top}}(t,x), \tag{9.176a}$$

and

$$\hat{j}_i(t,x) := K_{ij}^{-1} \hat{j}_j^{\text{top}}(t,x), \tag{9.176b}$$

respectively, for any $i = 1,\cdots,N$. The continuity equation (9.174a) is unchanged under this linear transformation,

$$\partial_t \hat{\rho}_i + \partial_x \hat{j}_i = 0, \tag{9.176c}$$

for any $i = 1, \cdots, N$. The topological current algebra (9.175) transforms into

$$\left[\hat{\rho}_i(t,x), \hat{\rho}_j(t,y)\right] = -\frac{\mathrm{i}}{2\pi} K^{-1}_{ij}\, \partial_x \delta(x-y), \tag{9.177a}$$

$$\left[\hat{j}_i(t,x), \hat{j}_j(t,y)\right] = -\frac{\mathrm{i}}{2\pi} V_{ik} V_{jl} K_{kl}\, \partial_x \delta(x-y), \tag{9.177b}$$

$$\left[\hat{\rho}_i(t,x), \hat{j}_j(t,y)\right] = -\frac{\mathrm{i}}{2\pi} V_{ij}\, \partial_x \delta(x-y), \tag{9.177c}$$

for any $i, j = 1, \cdots, N$.

At last, if we contract the continuity equation (9.176c) with the integer-valued charge vector, we obtain the flavor-global continuity equation [compare with Eq. (9.163)]

$$\partial_t \hat{\rho} + \partial_x \hat{j} = 0, \tag{9.178a}$$

where the local flavor-global charge operator is [compare with Eq. (9.162a)]

$$\hat{\rho}(t,x) := q_i K^{-1}_{ij}\, \hat{\rho}^{\mathrm{top}}_j(t,x) \tag{9.178b}$$

and the local flavor-global current operator is [compare with Eq. (9.162b)]

$$\hat{j}(t,x) := q_i K^{-1}_{ij} \hat{j}^{\mathrm{top}}_j(t,x). \tag{9.178c}$$

The flavor-resolved current algebra (9.177) turns into the flavor-global current algebra

$$\left[\hat{\rho}(t,x), \hat{\rho}(t,y)\right] = -\frac{\mathrm{i}}{2\pi} \left(q_i K^{-1}_{ij} q_j\right) \partial_x \delta(x-y), \tag{9.179a}$$

$$\left[\hat{j}(t,x), \hat{j}(t,y)\right] = -\frac{\mathrm{i}}{2\pi} \left(q_i V_{ik} K_{kl} V_{lj} q_j\right) \partial_x \delta(x-y), \tag{9.179b}$$

$$\left[\hat{\rho}(t,x), \hat{j}(t,y)\right] = -\frac{\mathrm{i}}{2\pi} \left(q_i V_{ij} q_j\right) \partial_x \delta(x-y). \tag{9.179c}$$

Quasiparticle and electronic excitations

When Eq. (9.168a) holds, there exist N conserved global topological (i.e., integer-valued) charges $\hat{\mathcal{N}}_i$ with $i = 1, \cdots, N$ defined in Eq. (9.169) that commute pairwise. Define the N global charges

$$\hat{Q}_i := \int_0^L \mathrm{d}x\, \hat{\rho}_i(t,x) = K^{-1}_{ij} \hat{\mathcal{N}}_j, \qquad i = 1, \cdots, N. \tag{9.180}$$

We shall shortly interpret these charges as the elementary Fermi-Bose charges.

Define for any $i = 1, \cdots, N$ the pair of vertex operators

$$\hat{\Psi}^{\dagger}_{\text{q-p},i}(t,x) := e^{-\mathrm{i} K^{-1}_{ij}\, \hat{u}_j(t,x)} \tag{9.181a}$$

and

$$\hat{\Psi}^{\dagger}_{\text{f-b},i}(t,x) := e^{-\mathrm{i}\delta_{ij}\, \hat{u}_j(t,x)}, \tag{9.181b}$$

respectively. The quasiparticle vertex operator $\hat{\Psi}^{\dagger}_{\text{q-p},i}(t,x)$ is multi-valued under a shift by 2π of all $\hat{u}_j(t,x)$ with $j=1,\cdots,N$. The Fermi-Bose vertex operator $\hat{\Psi}^{\dagger}_{\text{f-b},i}(t,x)$ is single-valued under a shift by 2π of all $\hat{u}_j(t,x)$ with $j=1,\cdots,N$.

Exercise 4.1: Verify that, for any pair $i,j=1,\cdots,N$, the commutator (9.173) delivers the identities

$$\left[\hat{\mathcal{N}}_i,\hat{\Psi}^{\dagger}_{\text{q-p},j}(t,x)\right]=\delta_{ij}\hat{\Psi}^{\dagger}_{\text{q-p},j}(t,x),\qquad \left[\hat{\mathcal{N}}_i,\hat{\Psi}^{\dagger}_{\text{f-b},j}(t,x)\right]=K_{ij}\,\hat{\Psi}^{\dagger}_{\text{f-b},j}(t,x), \tag{9.182a}$$

and

$$\left[\hat{Q}_i,\hat{\Psi}^{\dagger}_{\text{q-p},j}(t,x)\right]=K^{-1}_{ij}\hat{\Psi}^{\dagger}_{\text{q-p},j}(t,x),\qquad \left[\hat{Q}_i,\hat{\Psi}^{\dagger}_{\text{f-b},j}(t,x)\right]=\delta_{ij}\,\hat{\Psi}^{\dagger}_{\text{f-b},j}(t,x), \tag{9.182b}$$

respectively. The quasiparticle vertex operator $\hat{\Psi}^{\dagger}_{\text{q-p},i}(t,x)$ is an eigenstate of the topological number operator $\hat{\mathcal{N}}_i$ with eigenvalue one. The Fermi-Bose vertex operator $\hat{\Psi}^{\dagger}_{\text{f-b},i}(t,x)$ is an eigenstate of the charge number operator $\hat{Q}_i$ with eigenvalue one.

The Baker-Campbell-Hausdorff formula implies that

$$e^{\hat{A}}\,e^{\hat{B}}=e^{\hat{A}+\hat{B}}\,e^{+(1/2)[\hat{A},\hat{B}]}=e^{\hat{B}}\,e^{\hat{A}}\,e^{[\hat{A},\hat{B}]}, \tag{9.183}$$

whenever two operators $\hat{A}$ and $\hat{B}$ have a $\mathbb{C}$-number as their commutator.

Exercise 4.2: Show that a first application of the Baker-Campbell-Hausdorff formula to any pair of quasiparticle vertex operators at equal time t but two distinct space coordinates $x\neq y$ gives

$$\begin{aligned}&\hat{\Psi}^{\dagger}_{\text{q-p},i}(t,x)\,\hat{\Psi}^{\dagger}_{\text{q-p},j}(t,y)\\&=\begin{cases}\hat{\Psi}^{\dagger}_{\text{q-p},i}(t,y)\,\hat{\Psi}^{\dagger}_{\text{q-p},i}(t,x)\,e^{-\mathrm{i}\pi\left[K^{-1}_{ii}\,\mathrm{sgn}(x-y)+\left(K^{-1}_{ik}K^{-1}_{il}K_{kl}+q_kK^{-1}_{ik}K^{-1}_{il}q_l\right)\mathrm{sgn}(k-l)\right]}, & \text{if } i=j,\\ \hat{\Psi}^{\dagger}_{\text{q-p},j}(t,y)\,\hat{\Psi}^{\dagger}_{\text{q-p},i}(t,x)\,e^{-\mathrm{i}\pi\left[K^{-1}_{ji}\,\mathrm{sgn}(x-y)+\left(K^{-1}_{ik}K^{-1}_{jl}K_{kl}+q_kK^{-1}_{ik}K^{-1}_{jl}q_l\right)\mathrm{sgn}(k-l)\right]}, & \text{if } i\neq j.\end{cases}\end{aligned} \tag{9.184}$$

Here and below, it is understood that

$$\mathrm{sgn}(k-l)=0, \tag{9.185}$$

when $k=l=1,\cdots,N$. Argue that the quasiparticle vertex operators obey neither bosonic nor fermionic statistics whenever $\det K\neq\pm1$. *Hint:* $K^{-1}_{ij}\in\mathbb{Q}$ has rational matrix elements.

Exercise 4.3: Show that the same exercise applied to the Fermi-Bose vertex operators yields

$$\hat{\Psi}^{\dagger}_{\text{f-b},i}(t,x)\,\hat{\Psi}^{\dagger}_{\text{f-b},j}(t,y)=\begin{cases}(-1)^{K_{ii}}\,\hat{\Psi}^{\dagger}_{\text{f-b},i}(t,y)\,\hat{\Psi}^{\dagger}_{\text{f-b},i}(t,x), & \text{if } i=j,\\ (-1)^{q_i\,q_j}\,\hat{\Psi}^{\dagger}_{\text{f-b},j}(t,y)\,\hat{\Psi}^{\dagger}_{\text{f-b},i}(t,x), & \text{if } i\neq j,\end{cases} \tag{9.186}$$

when $x \neq y$. The self statistics of the Fermi-Bose vertex operators is carried by the diagonal matrix elements $K_{ii} \in \mathbb{Z}$. The mutual statistics of any pair of Fermi-Bose vertex operators labeled by $i \neq j$ is carried by the product $q_i\, q_j \in \mathbb{Z}$ of the integer-valued charges q_i and q_j. Had we not assumed that K_{ij} with $i \neq j$ are integers, the mutual statistics would not be Fermi-Bose because of the non-local term $K_{ij}\,\mathrm{sgn}\,(x-y)$.

Exercise 4.4: Show that a third application of the Baker-Campbell-Hausdorff formula allows to determine the boundary conditions

$$\hat{\Psi}^{\dagger}_{\text{q-p},i}(t,x+L) = \hat{\Psi}^{\dagger}_{\text{q-p},i}(t,x)\, e^{-2\pi\mathrm{i}\,K^{-1}_{ij}\hat{\mathcal{N}}_i}\, e^{-\pi\mathrm{i}\,K^{-1}_{ii}}, \tag{9.187}$$

and

$$\hat{\Psi}^{\dagger}_{\text{f-b},i}(t,x+L) = \hat{\Psi}^{\dagger}_{\text{f-b},i}(t,x)\, e^{-2\pi\mathrm{i}\hat{\mathcal{N}}_i}\, e^{-\pi\mathrm{i}\,K_{ii}}, \tag{9.188}$$

obeyed by the quasiparticle and Fermi-Bose vertex operators, respectively.

We close this discussion with the following definitions. Introduce the operators

$$\hat{Q} := q_i \hat{Q}_i, \qquad \hat{\Psi}^{\dagger}_{\text{q-p},\boldsymbol{m}} := e^{-\mathrm{i}m_i K^{-1}_{ij}\hat{u}_j(t,x)}, \qquad \hat{\Psi}^{\dagger}_{\text{f-b},\boldsymbol{m}} := e^{-\mathrm{i}m_i \delta_{ij}\hat{u}_j(t,x)}, \tag{9.189}$$

where $\boldsymbol{m} \in \mathbb{Z}^N$ is the vector with the integer-valued components m_i for any $i = 1,\cdots,N$. The N charges q_i with $i=1,\cdots,N$ that enter Hamiltonian (9.156a) can also be viewed as the components of the vector $\boldsymbol{q} \in \mathbb{Z}^N$. Define the functions

$$\begin{aligned} q : \mathbb{Z}^N &\longrightarrow \mathbb{Z}, \\ \boldsymbol{m} &\longmapsto q(\boldsymbol{m}) := q_i m_i \equiv \boldsymbol{q}\cdot\boldsymbol{m}, \end{aligned} \tag{9.190a}$$

and

$$\begin{aligned} K : \mathbb{Z}^N &\longrightarrow \mathbb{Z} \\ \boldsymbol{m} &\longmapsto K(\boldsymbol{m}) := m_i K_{ij} m_j. \end{aligned} \tag{9.190b}$$

On the one hand, for any distinct pair of space coordinates $x \neq y$, we deduce from Eqs. (9.182b), (9.184), and (9.187) that

$$\begin{aligned} &\left[\hat{Q}, \hat{\Psi}^{\dagger}_{\text{q-p},\boldsymbol{m}}(t,x)\right] = \left(q_i K^{-1}_{ij} m_j\right) \hat{\Psi}^{\dagger}_{\text{q-p},\boldsymbol{m}}(t,x), \\ &\hat{\Psi}^{\dagger}_{\text{q-p},\boldsymbol{m}}(t,x)\,\hat{\Psi}^{\dagger}_{\text{q-p},\boldsymbol{n}}(t,y) = \hat{\Psi}^{\dagger}_{\text{q-p},\boldsymbol{n}}(t,y)\,\hat{\Psi}^{\dagger}_{\text{q-p},\boldsymbol{m}}(t,x) \\ &\quad\times e^{-\mathrm{i}\pi\left[m_i K^{-1}_{ij} n_j \mathrm{sgn}(x-y) + \left(m_i K^{-1}_{ik} K_{kl} K^{-1}_{lj} n_j + q_k K^{-1}_{ki} m_i n_j K^{-1}_{jl} q_l\right)\mathrm{sgn}(k-l)\right]}, \\ &\hat{\Psi}^{\dagger}_{\text{q-p},\boldsymbol{m}}(t,x+L) = \hat{\Psi}^{\dagger}_{\text{q-p},\boldsymbol{m}}(t,x)\, e^{-2\pi\mathrm{i}\,m_i K^{-1}_{ij}\hat{\mathcal{N}}_j}\, e^{-\pi\mathrm{i}\,m_i K^{-1}_{ij} m_j}, \end{aligned} \tag{9.191}$$

respectively. On the other hand, for any distinct pair of space coordinates $x \neq y$, we deduce from Eqs. (9.182b), (9.186), and (9.188) that

$$
\begin{aligned}
&\left[\hat{Q},\hat{\Psi}^{\dagger}_{\text{f-b},\boldsymbol{m}}(t,x)\right]=q(\boldsymbol{m})\,\hat{\Psi}^{\dagger}_{\text{f-b},\boldsymbol{m}}(t,x),\\
&\hat{\Psi}^{\dagger}_{\text{f-b},\boldsymbol{m}}(t,x)\,\hat{\Psi}^{\dagger}_{\text{f-b},\boldsymbol{n}}(t,y)=\hat{\Psi}^{\dagger}_{\text{f-b},\boldsymbol{n}}(t,y)\,\hat{\Psi}^{\dagger}_{\text{f-b},\boldsymbol{m}}(t,x)\\
&\qquad\qquad\times e^{-\mathrm{i}\pi\left[m_i\,K_{ij}\,n_j\,\operatorname{sgn}(x-y)+m_i\left(K_{ij}+q_iq_j\right)n_j\,\operatorname{sgn}(i-j)\right]},\\
&\hat{\Psi}^{\dagger}_{\text{f-b},\boldsymbol{m}}(t,x+L)=\hat{\Psi}^{\dagger}_{\text{f-b},\boldsymbol{m}}(t,x)\,e^{-2\pi\mathrm{i}\,m_i\hat{\mathcal{N}}_i}\,e^{-\pi\mathrm{i}\,m_iK_{ij}m_j},
\end{aligned}
\tag{9.192}
$$

respectively.

Exercise 4.5: The integer quadratic form $K(\boldsymbol{m})$ is thus seen to dictate whether the vertex operator $\hat{\Psi}^{\dagger}_{\text{f-b},\boldsymbol{m}}(t,x)$ realizes a fermion or a boson. The vertex operator $\hat{\Psi}^{\dagger}_{\text{f-b},\boldsymbol{m}}(t,x)$ realizes a fermion if and only if

$$K(\boldsymbol{m}) \text{ is an odd integer,} \tag{9.193}$$

or a boson if and only if

$$K(\boldsymbol{m}) \text{ is an even integer.} \tag{9.194}$$

Show that, because of assumption (9.156f),

$$(-1)^{K(\boldsymbol{m})}=(-1)^{q(\boldsymbol{m})}. \tag{9.195}$$

Hence, the vertex operator $\hat{\Psi}^{\dagger}_{\text{f-b},\boldsymbol{m}}(t,x)$ realizes a fermion if and only if

$$q(\boldsymbol{m}) \text{ is an odd integer} \tag{9.196}$$

or a boson if and only if

$$q(\boldsymbol{m}) \text{ is an even integer.} \tag{9.197}$$

From the Hamiltonian to the Lagrangian formalism

What is the Minkowski path integral that is equivalent either to the quantum theory defined by Eq. (9.156) or to the quantum theory defined with the opposite chirality as is explained in exercise 1.1(b)? The label $(+)$ will be associated to the choice of chirality made in Eq. (9.158), the label $(-)$ to the choice of chirality made in Eq. (9.159). In other words, we seek the path integrals

$$Z^{(\pm)}:=\int\mathcal{D}[\phi]\,e^{\mathrm{i}S^{(\pm)}[\phi]}, \tag{9.198a}$$

with the Minkowski action

$$S^{(\pm)}[\phi]:=\int\limits_{-\infty}^{+\infty}\mathrm{d}t\,L^{(\pm)}[\phi]\equiv\int\limits_{-\infty}^{+\infty}\mathrm{d}t\int\limits_{0}^{L}\mathrm{d}x\,\mathcal{L}^{(\pm)}[\phi](t,x), \tag{9.198b}$$

such that one of the two Hamiltonians

$$H^{(\pm)}:=\int\limits_{0}^{L}\mathrm{d}x\left[\Pi^{(\pm)}_i\,(\partial_t\phi_i)-\mathcal{L}^{(\pm)}[\phi]\right] \tag{9.199}$$

can be identified with $\hat{H}$ in Eq. (9.156a) after elevating the classical fields

$$\phi_i(t,x) \tag{9.200a}$$

and

$$\Pi_i^{(\pm)}(t,x) := \frac{\delta \mathcal{L}^{(\pm)}}{\delta(\partial_t \phi_i)(t,x)} \tag{9.200b}$$

entering $\mathcal{L}^{(\pm)}[\phi]$ to the status of quantum fields $\hat{\phi}_i(t,x)$ and $\hat{\Pi}_j^{(\pm)}(t,y)$ upon imposing the equal-time commutation relations

$$\left[\hat{\phi}_i(t,x), \hat{\Pi}_j^{(\pm)}(t,y)\right] = \frac{\mathrm{i}}{2}\,\delta_{ij}\,\delta(x-y), \tag{9.200c}$$

for any $i,j=1,\cdots,N$. The unusual factor 1/2 (instead of 1) on the right-hand side of the commutator between pairs of canonically conjugate fields arises because each real-valued scalar field ϕ_i with $i=1,\cdots,N$ is chiral, i.e., it represents "one-half" of a canonical real-valued scalar field.

Exercise 5.1: We try

$$\mathcal{L}^{(\pm)} := \frac{1}{4\pi}\left[\mp (\partial_x \phi_i)\, K_{ij}^{-1}\, (\partial_t \phi_j) - (\partial_x \phi_i)\, V_{ij}\, (\partial_x \phi_j)\right]. \tag{9.201a}$$

Show that there follows the chiral equations of motion

$$\begin{aligned} 0 &= \partial_\mu \frac{\delta \mathcal{L}^{(\pm)}}{\delta \partial_\mu \phi_i} - \frac{\delta \mathcal{L}^{(\pm)}}{\delta \phi_i} \\ &= \mp \frac{K_{ij}^{-1}}{2\pi} \partial_x \left(\delta_{jl}\partial_t \pm K_{jk} V_{kl} \partial_x\right) \phi_l, \end{aligned} \tag{9.201b}$$

for any $i=1,\cdots,N$.

Exercise 5.2: Show that it is only the term that mixes time t and space x derivatives that becomes imaginary upon analytical continuation from real time t to Euclidean time $\tau = \mathrm{i}t$.

We need to verify that the Hamiltonian density that follows from the Lagrangian density (9.201a) is, upon quantization, Eq. (9.156) with the gauge fields set to zero.

Exercise 5.3: The canonical momentum $\Pi_i^{(\pm)}$ to the field ϕ_i is

$$\Pi_i^{(\pm)}(t,x) := \frac{\delta \mathcal{L}^{(\pm)}}{\delta(\partial_t \phi_i)(t,x)} = \mp \frac{1}{4\pi} K_{ij}^{-1}\left(\partial_x \phi_j\right)(t,x), \tag{9.202}$$

for any $i=1,\cdots,N$ owing to the symmetry of the matrix K. Show that the Legendre transform

$$\mathcal{H}^{(\pm)} := \Pi_i^{(\pm)}\left(\partial_t \phi_i\right) - \mathcal{L}^{(\pm)} \tag{9.203}$$

delivers

$$\mathcal{H}^{(\pm)} = \frac{1}{4\pi}\left(\partial_x \phi_i\right) V_{ij} \left(\partial_x \phi_j\right). \tag{9.204}$$

The right-hand side does not depend on the chiral index $(\pm)$.

Exercise 5.4: We now quantize the theory by elevating the classical fields ϕ_i to the status of operators $\hat{\phi}_i$ obeying either the algebra (9.156b) for the choice of chirality (+) or the one with a minus sign on the right-hand side of Eq. (9.156b) for the choice of chirality (−). Show that this gives a quantum theory that meets all the demands of the quantum chiral edge theory (9.156) in all compatibility with the canonical quantization rules (9.200c), for

$$\left[\hat{\phi}_i(t,x), \hat{\Pi}_j^{(\pm)}(t,y)\right] = \frac{\mathrm{i}}{2}\delta_{ij}\delta(x-y) \tag{9.205}$$

where $i, j = 1, \cdots, N$.

Finally, analytical continuation to Euclidean time

$$\tau = \mathrm{i}t \tag{9.206a}$$

allows to define the finite temperature quantum chiral theory through the path integral

$$\begin{aligned} Z_\beta^{(\pm)} &:= \int \mathcal{D}[\phi]\, e^{-S^{(\pm)}[\phi]}, \\ S^{(\pm)}[\phi] &:= \int_0^\beta \mathrm{d}\tau \int_0^L \mathrm{d}x\, \frac{1}{4\pi}\left[(\pm)\mathrm{i}\,(\partial_x\phi_i)\, K_{ij}^{-1}\,(\partial_\tau\phi_j) + (\partial_x\phi_i)\, V_{ij}\,(\partial_x\phi_j)\right]. \end{aligned} \tag{9.206b}$$

9.4.2 *Two-point correlation function in the massless Thirring model*

Introduction

We are going to derive the two-point function (F.47) for the massless Thirring model, a relativistic quantum field theory in (1+1)-dimensional position space and time, that we choose to represent through Eq. (9.210). We work in units of $\hbar = 1$ and $c = 1$ (or $v_{\mathrm{F}} = 1$ if the Lorentz invariance is an emergent one at low energies and long wavelength of some underlying lattice model). The short-distance cut-off $\mathfrak{a}$ is also set to one.

A pair of freely counterpropagating chiral bosons

Define the bosonic Hamiltonian

$$\hat{H}_0 := \frac{1}{4\pi}\int_0^L \mathrm{d}x \left(\left(\partial_x\hat{\phi}_-\right)\left(\partial_x\hat{\phi}_-\right) + \left(\partial_x\hat{\phi}_+\right)\left(\partial_x\hat{\phi}_+\right) \right)(t,x). \tag{9.207a}$$

The dependence on t on the right-hand side refers to the Heisenberg picture. Of course, $\hat{H}_0$ is conserved in time. The two Hermitian quantum fields $\hat{\phi}_i(t,x)$ with

$i = 1, 2 \equiv -, +$ are postulated to obey the equal-time commutation relations

$$\begin{aligned}
\left[\hat{\phi}_{-}(t,x), \hat{\phi}_{-}(t,y)\right] &:= +\mathrm{i}\pi \,\mathrm{sgn}(x-y), \\
\left[\hat{\phi}_{+}(t,x), \hat{\phi}_{+}(t,y)\right] &:= -\mathrm{i}\pi \,\mathrm{sgn}(x-y), \\
\left[\hat{\phi}_{-}(t,x), \hat{\phi}_{+}(t,y)\right] &= -\left[\hat{\phi}_{+}(t,x), \hat{\phi}_{-}(t,y)\right] := -\mathrm{i}\pi.
\end{aligned} \tag{9.207b}$$

The function $\mathrm{sgn}(x) = -\mathrm{sgn}(-x)$ gives the sign of the real variable x and will be assumed to be periodic with periodicity L. We shall impose the boundary conditions

$$\hat{\phi}_i(t, x+L) = \hat{\phi}_i(t,x) + 2\pi n_i, \qquad n_i \in \mathbb{Z}, \tag{9.207c}$$

and

$$\left(\partial_x \hat{\phi}_i\right)(t, x+L) = \left(\partial_x \hat{\phi}_i\right)(t,x), \tag{9.207d}$$

for any $i = 1, 2 \equiv -, +$.

Exercise 1.1:

(a) What are the matrices K, V, L, and the charge vector $\boldsymbol{q}$ defined in Eq. (9.156) that deliver Eq. (9.207)?

(b) Show that

$$\partial_t \hat{\phi}_{-} = -\partial_x \hat{\phi}_{-}, \tag{9.208a}$$

and

$$\partial_t \hat{\phi}_{+} = +\partial_x \hat{\phi}_{+}. \tag{9.208b}$$

Hence, $\hat{\phi}_{-}$ is a right-moving bosonic field, while $\hat{\phi}_{+}$ is a left-moving bosonic field.

(c) Define the pair of operators

$$\hat{J}_{\mp} := \frac{1}{2\pi}\,\partial_x \hat{\phi}_{\mp}. \tag{9.209a}$$

Show that they obey periodic boundary conditions under $x \to x + L$, satisfy the equal-time (Schwinger) algebra

$$\begin{aligned}
\left[\hat{J}_{-}(t,x), \hat{J}_{-}(t,y)\right] &= -\frac{\mathrm{i}}{2\pi}\,\partial_x \delta(x-y), \\
\left[\hat{J}_{+}(t,x), \hat{J}_{+}(t,y)\right] &= +\frac{\mathrm{i}}{2\pi}\,\partial_x \delta(x-y), \\
\left[\hat{J}_{-}(t,x), \hat{J}_{+}(t,y)\right] &= \left[\hat{J}_{+}(t,x), \hat{J}_{-}(t,y)\right] = 0,
\end{aligned} \tag{9.209b}$$

and

$$\begin{aligned}
\hat{H}_0 &= \pi \int\limits_0^L \mathrm{d}x \left(\hat{J}_{-}\hat{J}_{-} + \hat{J}_{+}\hat{J}_{+}\right)(t,x), \\
&(\partial_t \pm \partial_x)\,\hat{J}_{\mp}(t,x) = 0.
\end{aligned} \tag{9.209c}$$

A pair of interacting counterpropagating chiral bosons

Define the interacting Hamiltonian

$$\hat{H} := \hat{H}_0 + \hat{H}_1(\delta v) + \hat{H}_2(\lambda), \tag{9.210a}$$

where

$$\hat{H}_0 = \pi \int\limits_0^L \mathrm{d}x \left(\hat{J}_- \hat{J}_- + \hat{J}_+ \hat{J}_+ \right)(t,x) \tag{9.210b}$$

was defined in Eq. (9.209),

$$\hat{H}_1(\delta v) = \pi\,\delta v \int\limits_0^L \mathrm{d}x \left(\hat{J}_- \hat{J}_- + \hat{J}_+ \hat{J}_+ \right)(t,x) \tag{9.210c}$$

renormalizes additively the coefficient (the bare energy scale $\pi\,\hbar\,v_{\mathrm{F}}/\mathfrak{a}$ if we reinstate all units) of $\hat{H}_0$ by the dimensionless real-valued number $\pi\,\delta v$, and

$$\hat{H}_2(\lambda) = \lambda \int\limits_0^L \mathrm{d}x \left(\hat{J}_- \hat{J}_+ \right)(t,x) \tag{9.210d}$$

mixes left with right movers for any real-valued and non-vanishing λ.

Exercise 2.1: Define the one-parameter family of currents

$$\begin{aligned} \hat{J}^{\theta}_- &:= \cosh\theta\,\hat{J}_- + \sinh\theta\,\hat{J}_+,\\ \hat{J}^{\theta}_+ &:= \sinh\theta\,\hat{J}_- + \cosh\theta\,\hat{J}_+, \end{aligned} \tag{9.211}$$

with the label $\theta \in \mathbb{R}$. Fix θ to the value

$$\bar{\theta} := \frac{1}{2}\,\mathrm{arctanh}\,\frac{\lambda}{2\pi(1+\delta v)}. \tag{9.212}$$

Define the finite real number

$$\bar{v} := \frac{1+\delta v}{\cosh 2\bar{\theta}}. \tag{9.213}$$

Show that the currents $\hat{J}^{\bar{\theta}}_{\mp}$ satisfy periodic boundary conditions under $x \to x + L$, obey the equal-time (Schwinger) algebra

$$\begin{aligned} \left[\hat{J}^{\bar{\theta}}_-(t,x), \hat{J}^{\bar{\theta}}_-(t,y)\right] &= -\frac{\mathrm{i}}{2\pi}\,\partial_x \delta(x-y),\\ \left[\hat{J}^{\bar{\theta}}_+(t,x), \hat{J}^{\bar{\theta}}_+(t,y)\right] &= +\frac{\mathrm{i}}{2\pi}\,\partial_x \delta(x-y),\\ \left[\hat{J}^{\bar{\theta}}_-(t,x), \hat{J}^{\bar{\theta}}_+(t,y)\right] &= \left[\hat{J}^{\bar{\theta}}_+(t,x), \hat{J}^{\bar{\theta}}_-(t,y)\right] = 0, \end{aligned} \tag{9.214a}$$

and

$$\begin{aligned} \hat{H} &= \pi\,\bar{v} \int\limits_0^L \mathrm{d}x \left(\hat{J}^{\bar{\theta}}_- \hat{J}^{\bar{\theta}}_- + \hat{J}^{\bar{\theta}}_+ \hat{J}^{\bar{\theta}}_+ \right)(t,x),\\ &(\partial_t \pm \bar{v}\,\partial_x)\,\hat{J}^{\bar{\theta}}_{\mp}(t,x) = 0. \end{aligned} \tag{9.214b}$$

Diagonalization of $\hat{H} \equiv \hat{H}^{\bar{\theta}}$

We are going to diagonalize

$$\hat{H}_0^{\bar{\theta}} := \pi \bar{v} \int_0^L \mathrm{d}x \left(\hat{J}_-^{\bar{\theta}} \hat{J}_-^{\bar{\theta}} + \hat{J}_+^{\bar{\theta}} \hat{J}_+^{\bar{\theta}} \right)(t, x), \tag{9.215a}$$

where the currents obey the equal-time (Schwinger) algebra

$$\begin{aligned}
\left[\hat{J}_-^{\bar{\theta}}(t,x), \hat{J}_-^{\bar{\theta}}(t,y)\right] &= -\frac{\mathrm{i}}{2\pi}\,\partial_x \delta(x-y),\\
\left[\hat{J}_+^{\bar{\theta}}(t,x), \hat{J}_+^{\bar{\theta}}(t,y)\right] &= +\frac{\mathrm{i}}{2\pi}\,\partial_x \delta(x-y),\\
\left[\hat{J}_-^{\bar{\theta}}(t,x), \hat{J}_+^{\bar{\theta}}(t,y)\right] &= \left[\hat{J}_+^{\bar{\theta}}(t,x), \hat{J}_-^{\bar{\theta}}(t,y)\right] = 0,
\end{aligned} \tag{9.215b}$$

and periodic boundary conditions under $x \to x + L$. To make contact to the representation (9.207), we also define

$$\hat{J}_\mp^{\bar{\theta}}(t,x) =: \frac{1}{2\pi}\left(\partial_x \hat{\phi}_\mp^{\bar{\theta}}\right)(t,x). \tag{9.215c}$$

We observe that the Schwinger algebra (9.215) is very close to the equal-time algebra obeyed by two Hermitian (field-valued) operators $\hat{\varphi}_-^{\bar{\theta}}$ and $\hat{\varphi}_+^{\bar{\theta}}$ and their canonical momenta $\hat{\Pi}_-^{\bar{\theta}}$ and $\hat{\Pi}_+^{\bar{\theta}}$, namely

$$\begin{aligned}
\left[\hat{\varphi}_-^{\bar{\theta}}(t,x), \hat{\Pi}_-^{\bar{\theta}}(t,y)\right] &= \mathrm{i}\delta(x-y),\\
\left[\hat{\varphi}_+^{\bar{\theta}}(t,x), \hat{\Pi}_+^{\bar{\theta}}(t,y)\right] &= \mathrm{i}\delta(x-y),\\
\left[\hat{\varphi}_-^{\bar{\theta}}(t,x), \hat{\Pi}_+^{\bar{\theta}}(t,y)\right] &= \left[\hat{\varphi}_+^{\bar{\theta}}(t,x), \hat{\Pi}_-^{\bar{\theta}}(t,y)\right] = 0.
\end{aligned} \tag{9.216}$$

We seek to relate the pair $(\hat{J}_i^{\bar{\theta}}, \hat{J}_i^{\bar{\theta}})$ to the pair of raising and lowering operators associated to $(\hat{\varphi}_i^{\bar{\theta}}, \hat{\Pi}_i^{\bar{\theta}})$ for $i = -, +$. This can be done in momentum space.

Exercise 3.1: We do the Fourier expansions

$$\begin{aligned}
\hat{\phi}_i^{\bar{\theta}}(t,x) &= \frac{1}{\sqrt{L}} \sum_{\frac{Lp}{2\pi}\in\mathbb{Z}} e^{+\mathrm{i}p\,x}\, \hat{\phi}_{i\,p}^{\bar{\theta}}(t), &
\hat{\phi}_{i\,p}^{\bar{\theta}}(t) &= \frac{1}{\sqrt{L}} \int_0^L \mathrm{d}x\, e^{-\mathrm{i}p\,x}\, \hat{\phi}_i^{\bar{\theta}}(t,x),\\
\hat{J}_i^{\bar{\theta}}(t,x) &= \frac{1}{\sqrt{L}} \sum_{\frac{Lp}{2\pi}\in\mathbb{Z}} e^{+\mathrm{i}p\,x}\, \hat{J}_{i\,p}^{\bar{\theta}}(t), &
\hat{J}_{i\,p}^{\bar{\theta}}(t) &= \frac{1}{\sqrt{L}} \int_0^L \mathrm{d}x\, e^{-\mathrm{i}p\,x}\, \hat{J}_i^{\bar{\theta}}(t,x),\\
\hat{\varphi}_i^{\bar{\theta}}(t,x) &= \frac{1}{\sqrt{L}} \sum_{\frac{Lp}{2\pi}\in\mathbb{Z}} e^{+\mathrm{i}p\,x}\, \hat{\varphi}_{i\,p}^{\bar{\theta}}(t), &
\hat{\varphi}_{i\,p}^{\bar{\theta}}(t) &= \frac{1}{\sqrt{L}} \int_0^L \mathrm{d}x\, e^{-\mathrm{i}p\,x}\, \hat{\varphi}_i^{\bar{\theta}}(t,x),\\
\hat{\Pi}_i^{\bar{\theta}}(t,x) &= \frac{1}{\sqrt{L}} \sum_{\frac{Lp}{2\pi}\in\mathbb{Z}} e^{+\mathrm{i}p\,x}\, \hat{\Pi}_{i\,p}^{\bar{\theta}}(t), &
\hat{\Pi}_{i\,p}^{\bar{\theta}}(t) &= \frac{1}{\sqrt{L}} \int_0^L \mathrm{d}x\, e^{-\mathrm{i}p\,x}\, \hat{\Pi}_i^{\bar{\theta}}(t,x),
\end{aligned} \tag{9.217}$$

for $i = -, +$. From now on, any summation over the momenta p is understood as the sum over the integer n in $p = 2\pi\, n/L$.

(a) Verify that

$$\begin{aligned}\hat{\phi}^{\bar{\theta}}_{i\,(+p)}(t) &= \left[\hat{\phi}^{\bar{\theta}}_{i\,(-p)}(t)\right]^{\dagger},\\ \hat{J}^{\bar{\theta}}_{i\,(+p)}(t) &= \left[\hat{J}^{\bar{\theta}}_{i\,(-p)}(t)\right]^{\dagger},\\ \hat{\varphi}^{\bar{\theta}}_{i\,(+p)}(t) &= \left[\hat{\varphi}^{\bar{\theta}}_{i\,(-p)}(t)\right]^{\dagger},\\ \hat{\Pi}^{\bar{\theta}}_{i\,(+p)}(t) &= \left[\hat{\Pi}^{\bar{\theta}}_{i\,(-p)}(t)\right]^{\dagger},\end{aligned} \tag{9.218}$$

for $i=-,+$.

(b) Show that

$$\begin{aligned}\left[\hat{J}^{\bar{\theta}}_{-\,(+p)}(t),\hat{J}^{\bar{\theta}}_{-\,(+p')}(t)\right] &= +\frac{p}{2\pi}\,\delta_{p,-p'},\\ \left[\hat{J}^{\bar{\theta}}_{+\,(+p)}(t),\hat{J}^{\bar{\theta}}_{+\,(+p')}(t)\right] &= -\frac{p}{2\pi}\,\delta_{p,-p'},\\ \left[\hat{J}^{\bar{\theta}}_{-\,(+p)}(t),\hat{J}^{\bar{\theta}}_{+\,(+p')}(t)\right] &= \left[\hat{J}^{\bar{\theta}}_{+\,(+p)}(t),\hat{J}^{\bar{\theta}}_{-\,(+p')}(t)\right]=0,\end{aligned} \tag{9.219}$$

whereas

$$\begin{aligned}\left[\hat{\varphi}^{\bar{\theta}}_{i\,p}(t),\hat{\varphi}^{\bar{\theta}}_{j\,p'}(t)\right] &= 0,\\ \left[\hat{\Pi}^{\bar{\theta}}_{i\,p}(t),\hat{\Pi}^{\bar{\theta}}_{j\,p'}(t)\right] &= 0,\\ \left[\hat{\varphi}^{\bar{\theta}}_{i\,p}(t),\hat{\Pi}^{\bar{\theta}}_{j\,p'}(t)\right] &= \mathrm{i}\,\delta_{i,j}\,\delta_{p,-p'},\end{aligned} \tag{9.220}$$

for $i,j=-,+$.

(c) Verify that if we invert the relation

$$\hat{J}^{\bar{\theta}}_{i\,(+p)}(t)=\frac{\mathrm{i}p}{2\pi}\,\hat{\phi}^{\bar{\theta}}_{i\,(+p)}(t) \tag{9.221a}$$

for any $p=2\pi\,n/L$ with $n\in\mathbb{Z}\setminus\{0\}$ according to

$$\begin{aligned}\hat{\phi}^{\bar{\theta}}_{-}(t,x) =& \sqrt{\frac{\pi}{2}}\left(\hat{b}^{\bar{\theta}}_{0}+\mathrm{i}\hat{b}^{\bar{\theta}\dagger}_{0}\right)\\ &+\lim_{\epsilon\to0+}2\pi\,\frac{1}{\sqrt{L}}\sum_{p>0}e^{-\epsilon\,p/2}\,\frac{1}{\mathrm{i}p}\left[e^{+\mathrm{i}p\,x}\,\hat{J}^{\bar{\theta}}_{-\,(+p)}-e^{-\mathrm{i}p\,x}\,\hat{J}^{\bar{\theta}}_{-\,(-p)}\right],\\ \hat{\phi}^{\bar{\theta}}_{+}(t,x) =& \sqrt{\frac{\pi}{2}}\left(\hat{b}^{\bar{\theta}}_{0}-\mathrm{i}\hat{b}^{\bar{\theta}\dagger}_{0}\right)\\ &+\lim_{\epsilon\to0+}2\pi\,\frac{1}{\sqrt{L}}\sum_{p>0}e^{-\epsilon\,p/2}\,\frac{1}{\mathrm{i}p}\left[e^{+\mathrm{i}p\,x}\,\hat{J}^{\bar{\theta}}_{+\,(+p)}-e^{-\mathrm{i}p\,x}\,\hat{J}^{\bar{\theta}}_{+\,(-p)}\right],\end{aligned} \tag{9.221b}$$

for $i=-,+$, we then recover Eq. (9.207b) in the limit $L\to\infty$ provided the only non-vanishing commutator

$$\frac{1}{L}\left[\hat{\phi}^{\bar{\theta}}_{-\,0}(t),\hat{\phi}^{\bar{\theta}}_{+\,0}(t)\right]=-\mathrm{i}\pi \tag{9.221c}$$

of the zero-mode operators

$$\frac{1}{\sqrt{L}}\hat{\phi}^{\bar{\theta}}_{-\,0}(t) =: \sqrt{\frac{\pi}{2}}\left(\hat{b}^{\bar{\theta}}_{0} + \mathrm{i}\hat{b}^{\bar{\theta}\dagger}_{0}\right), \qquad \frac{1}{\sqrt{L}}\hat{\phi}^{\bar{\theta}}_{+\,0}(t) =: \sqrt{\frac{\pi}{2}}\left(\hat{b}^{\bar{\theta}}_{0} - \mathrm{i}\hat{b}^{\bar{\theta}\dagger}_{0}\right), \tag{9.221d}$$

follows from the zero-mode algebra

$$\left[\hat{b}^{\bar{\theta}}_{0}, \hat{b}^{\bar{\theta}\dagger}_{0}\right] = 1, \qquad \left[\hat{b}^{\bar{\theta}\dagger}_{0}, \hat{b}^{\bar{\theta}\dagger}_{0}\right] = \left[\hat{b}^{\bar{\theta}}_{0}, \hat{b}^{\bar{\theta}}_{0}\right] = 0. \tag{9.221e}$$

Hint: Use the integral representation

$$\mathrm{sgn}(x) = \lim_{\epsilon\to 0+}\frac{2}{\pi}\int_{0}^{\infty}\mathrm{d}p\, e^{-\epsilon p}\,\frac{\sin p x}{p}. \tag{9.222}$$

(d) Verify that

$$\hat{H}^{\bar{\theta}}_{0} = \pi\,\bar{v}\sum_{i=-,+}\sum_{p>0}\left(\hat{J}^{\bar{\theta}}_{i\,(+p)}\,\hat{J}^{\bar{\theta}}_{i\,(-p)} + \hat{J}^{\bar{\theta}}_{i\,(-p)}\,\hat{J}^{\bar{\theta}}_{i\,(+p)}\right)(t), \tag{9.223}$$

whereby one must also explain why the contribution $\left(\hat{J}^{\bar{\theta}}_{i\,0}\,\hat{J}^{\bar{\theta}}_{i\,0}\right)(t)$ is not accounted for.

Exercise 3.2: From now on, $p = 2\pi\, n/L$ is strictly positive, i.e., $n = 1, 2, 3, \cdots$. Inspection of exercise 3.1(d) and exercise 3.1(b) suggests that we define the operators

$$\hat{J}^{\bar{\theta}}_{-\,(+p)}(t) =: \sqrt{\frac{p}{2\pi}}\,\hat{b}^{\bar{\theta}}_{-\,(+p)}(t), \qquad \hat{J}^{\bar{\theta}}_{-\,(-p)}(t) =: \sqrt{\frac{p}{2\pi}}\,\hat{b}^{\bar{\theta}\,\dagger}_{-\,(+p)}(t), \tag{9.224a}$$

and

$$\hat{J}^{\bar{\theta}}_{+\,(+p)}(t) =: \sqrt{\frac{p}{2\pi}}\,\hat{b}^{\bar{\theta}\dagger}_{+\,(+p)}(t), \qquad \hat{J}^{\bar{\theta}}_{+\,(-p)}(t) =: \sqrt{\frac{p}{2\pi}}\,\hat{b}^{\bar{\theta}}_{+\,(+p)}(t). \tag{9.224b}$$

(a) Verify that the only possible non-vanishing commutators originate from

$$\left[\hat{b}^{\bar{\theta}}_{i\,(+p)}(t), \hat{b}^{\bar{\theta}\dagger}_{j\,(+p')}(t)\right] = \delta_{i,j}\,\delta_{p,p'}, \qquad i,j = -,+, \qquad \frac{Lp}{2\pi}, \frac{Lp'}{2\pi} = 1, 2, 3, \cdots. \tag{9.225}$$

(b) Verify that

$$\hat{H}^{\bar{\theta}}_{0} = \frac{\bar{v}}{2}\sum_{i=-,+}\sum_{p>0} p\left(\hat{b}^{\bar{\theta}\dagger}_{i\,(+p)}\,\hat{b}^{\bar{\theta}}_{i\,(+p)} + \hat{b}^{\bar{\theta}}_{i\,(+p)}\,\hat{b}^{\bar{\theta}\dagger}_{i\,(+p)}\right)(t) \tag{9.226}$$

and show that the ground state of $\hat{H}^{\bar{\theta}}_{0}$ is the state $|0\rangle$ annihilated by $\hat{b}^{\bar{\theta}}_{i\,(+p)}$ for $i = -, +$ and $\frac{Lp}{2\pi} = 1, 2, 3, \cdots$.

(c) Verify that

$$
\begin{aligned}
\hat{\phi}^{\bar{\theta}}_{-}(t,x) =& \sqrt{\frac{\pi}{2}}\left(\hat{b}^{\bar{\theta}}_{0}+\mathrm{i}\hat{b}^{\bar{\theta}\dagger}_{0}\right)\\
&+\lim_{\epsilon\to 0^{+}}\frac{-\mathrm{i}}{\sqrt{L}}\sum_{p>0}e^{-\epsilon\, p/2}\sqrt{\frac{2\pi}{p}}\left[e^{+\mathrm{i}p\,x}\,\hat{b}^{\bar{\theta}}_{-\,(+p)}-e^{-\mathrm{i}p\,x}\,\hat{b}^{\bar{\theta}\dagger}_{-\,(+p)}\right],\\
\hat{\phi}^{\bar{\theta}}_{+}(t,x) =& \sqrt{\frac{\pi}{2}}\left(\hat{b}^{\bar{\theta}}_{0}-\mathrm{i}\hat{b}^{\bar{\theta}\dagger}_{0}\right)\\
&+\lim_{\epsilon\to 0^{+}}\frac{-\mathrm{i}}{\sqrt{L}}\sum_{p>0}e^{-\epsilon\, p/2}\sqrt{\frac{2\pi}{p}}\left[e^{+\mathrm{i}p\,x}\,\hat{b}^{\bar{\theta}\dagger}_{+\,(+p)}-e^{-\mathrm{i}p\,x}\,\hat{b}^{\bar{\theta}}_{+\,(+p)}\right].
\end{aligned}
\tag{9.227}
$$

Exercise 3.3: We define the additive decomposition

$$
\hat{\phi}^{\bar{\theta}}_{-}(t,x)=\hat{\phi}^{\bar{\theta}-}_{-}(t,x)+\hat{\phi}^{\bar{\theta}+}_{-}(t,x), \tag{9.228a}
$$

where

$$
\begin{aligned}
\hat{\phi}^{\bar{\theta}-}_{-}(t,x) &:= +\sqrt{\frac{\pi}{2}}\,\hat{b}^{\bar{\theta}}_{0}+\lim_{\epsilon\to 0^{+}}\frac{-\mathrm{i}}{\sqrt{L}}\sum_{p>0}e^{-\epsilon\, p/2}\sqrt{\frac{2\pi}{p}}e^{+\mathrm{i}p\,x}\,\hat{b}^{\bar{\theta}}_{-\,(+p)},\\
\hat{\phi}^{\bar{\theta}+}_{-}(t,x) &:= +\sqrt{\frac{\pi}{2}}\,\mathrm{i}\hat{b}^{\bar{\theta}\dagger}_{0}+\lim_{\epsilon\to 0^{+}}\frac{+\mathrm{i}}{\sqrt{L}}\sum_{p>0}e^{-\epsilon\, p/2}\sqrt{\frac{2\pi}{p}}e^{-\mathrm{i}p\,x}\,\hat{b}^{\bar{\theta}\dagger}_{-\,(+p)},
\end{aligned}
\tag{9.228b}
$$

in the right-moving sector. Similarly, we define the additive decomposition

$$
\hat{\phi}^{\bar{\theta}}_{+}(t,x)=\hat{\phi}^{\bar{\theta}-}_{+}(t,x)+\hat{\phi}^{\bar{\theta}+}_{+}(t,x), \tag{9.229}
$$

where

$$
\begin{aligned}
\hat{\phi}^{\bar{\theta}-}_{+}(t,x) &:= +\sqrt{\frac{\pi}{2}}\,\hat{b}^{\bar{\theta}}_{0}+\lim_{\epsilon\to 0^{+}}\frac{+\mathrm{i}}{\sqrt{L}}\sum_{p>0}e^{-\epsilon\, p/2}\sqrt{\frac{2\pi}{p}}e^{-\mathrm{i}p\,x}\,\hat{b}^{\bar{\theta}}_{+\,(+p)},\\
\hat{\phi}^{\bar{\theta}+}_{+}(t,x) &:= -\sqrt{\frac{\pi}{2}}\,\mathrm{i}\hat{b}^{\bar{\theta}\dagger}_{0}+\lim_{\epsilon\to 0^{+}}\frac{-\mathrm{i}}{\sqrt{L}}\sum_{p>0}e^{-\epsilon\, p/2}\sqrt{\frac{2\pi}{p}}e^{+\mathrm{i}p\,x}\,\hat{b}^{\bar{\theta}\dagger}_{+\,(+p)},
\end{aligned}
\tag{9.230}
$$

in the left-moving sector. These additive decompositions are advantageous in that

$$
\langle 0|\hat{\phi}^{\bar{\theta}\dagger}_{i}(t,x)=\langle 0|\hat{\phi}^{\bar{\theta}+\dagger}_{i}(t,x),\qquad \hat{\phi}^{\bar{\theta}}_{i}(t,x)|0\rangle=\hat{\phi}^{\bar{\theta}+}_{i}(t,x)|0\rangle, \tag{9.231}
$$

for $i=-,+$.

(a) Verify that

$$
\left[\hat{\phi}^{\bar{\theta}+}_{-}(t,x),\hat{\phi}^{\bar{\theta}-}_{-}(t,x')\right]=-\frac{\mathrm{i}\pi}{2}-\lim_{\epsilon\to 0^{+}}\frac{2\pi}{L}\sum_{p>0}e^{-\epsilon\, p}\,\frac{1}{p}\,e^{-\mathrm{i}p\,(x-x')}. \tag{9.232}
$$

(b) Verify that

$$
\left[\hat{\phi}^{\bar{\theta}+}_{+}(t,x),\hat{\phi}^{\bar{\theta}-}_{+}(t,x')\right]=+\frac{\mathrm{i}\pi}{2}-\lim_{\epsilon\to 0^{+}}\frac{2\pi}{L}\sum_{p>0}e^{-\epsilon\, p}\,\frac{1}{p}\,e^{+\mathrm{i}p\,(x-x')}. \tag{9.233}
$$

(c) Verify that

$$
\begin{aligned}
\left[\hat{\phi}^{\bar{\theta}-}_{-}(t,x),\hat{\phi}^{\bar{\theta}-}_{+}(t,x')\right] &= \left[\hat{\phi}^{\bar{\theta}+}_{-}(t,x),\hat{\phi}^{\bar{\theta}+}_{+}(t,x')\right] = 0,\\
\left[\hat{\phi}^{\bar{\theta}-}_{-}(t,x),\hat{\phi}^{\bar{\theta}+}_{+}(t,x')\right] &= \left[\hat{\phi}^{\bar{\theta}+}_{-}(t,x),\hat{\phi}^{\bar{\theta}-}_{+}(t,x')\right] = -\mathrm{i}\frac{\pi}{2}.
\end{aligned}
\tag{9.234}
$$

(d) Verify that

$$
\lim_{L\to\infty}\left\{\left[\hat{\phi}^{\bar{\theta}+}_{j}(t,x),\hat{\phi}^{\bar{\theta}-}_{j}(t,x')\right]-\left[\hat{\phi}^{\bar{\theta}+}_{j}(t,0),\hat{\phi}^{\bar{\theta}-}_{j}(t,0)\right]\right\} = \lim_{\epsilon\to0+}\log\frac{\epsilon-\mathrm{i}j\,(x-x')}{\epsilon}
\tag{9.235}
$$

for $j=-,+$.

(e) Verify that

$$
\begin{aligned}
e^{-\mathrm{i}\,j\,\hat{\phi}^{\bar{\theta}}_{j}(t,x)}\,e^{+\mathrm{i}\,j\,\hat{\phi}^{\bar{\theta}}_{j}(t,x')} = {} & e^{+\mathrm{i}\,j\left(-\hat{\phi}^{\bar{\theta}+}_{j}(t,x)+\hat{\phi}^{\bar{\theta}+}_{j}(t,x')\right)}\\
& \times e^{\left[\hat{\phi}^{\bar{\theta}+}_{j}(t,0),\hat{\phi}^{\bar{\theta}-}_{j}(t,0)\right]-\left[\hat{\phi}^{\bar{\theta}+}_{j}(t,x'),\hat{\phi}^{\bar{\theta}-}_{j}(t,x)\right]}\\
& \times e^{+\mathrm{i}\,j\left(-\hat{\phi}^{\bar{\theta}-}_{j}(t,x)+\hat{\phi}^{\bar{\theta}-}_{j}(t,x')\right)},
\end{aligned}
\tag{9.236}
$$

for $j=-,+$. *Hint:* Use twice the Baker-Campbell-Hausdorff formula (9.183).

(f) Verify that, at any unequal two points in position space and in the thermodynamic limit $L\to\infty$,

$$
\lim_{\epsilon\to0+}\frac{1}{\epsilon}\left\langle 0\left|e^{-\mathrm{i}\,j\,\hat{\phi}^{\bar{\theta}}_{j}(t,x)}\,e^{+\mathrm{i}\,j\,\hat{\phi}^{\bar{\theta}}_{j}(t,x')}\right|0\right\rangle = \frac{1}{\mathrm{i}j\,(x-x')},\qquad j=-,+.
\tag{9.237}
$$

Lorentz covariance then dictates that, at any two unequal points in position space and time,

$$
\begin{aligned}
\lim_{\epsilon\to0+}\frac{1}{\epsilon}\left\langle 0\left|e^{+\mathrm{i}\,\hat{\phi}^{\bar{\theta}}_{-}(t,x)}\,e^{-\mathrm{i}\,\hat{\phi}^{\bar{\theta}}_{-}(t',x')}\right|0\right\rangle &= \frac{-\mathrm{i}}{(t-t')-(x-x')},\\
\lim_{\epsilon\to0+}\frac{1}{\epsilon}\left\langle 0\left|e^{-\mathrm{i}\,\hat{\phi}^{\bar{\theta}}_{+}(t,x)}\,e^{+\mathrm{i}\,\hat{\phi}^{\bar{\theta}}_{+}(t',x')}\right|0\right\rangle &= \frac{-\mathrm{i}}{(t-t')+(x-x')}.
\end{aligned}
\tag{9.238}
$$

Exercise 3.4: Let $|\mathrm{FS}\rangle$ be the Fermi sea obtained from the linear dispersion relation $\varepsilon_{-\,(p)}=+p$ for the right movers and $\varepsilon_{+\,(p)}=-p$ for the left movers when the Fermi energy equals zero, i.e., when the Fermi point is $p=0$. Let $n_{-(p)}:=\Theta(-p)$ be the occupation number of all single-particle states with negative momentum and negative energy. Let $n_{+(p)}:=\Theta(+p)$ be the occupation number of all single-particle states with positive momentum and negative energy. Show that, up to the multiplicative prefactor $1/(2\pi)$,

$$
\lim_{\epsilon\to0+}\lim_{L\to\infty}\frac{1}{L}\sum_{L\,p/(2\pi)\in\mathbb{Z}} e^{-\epsilon|p|-\mathrm{i}p\,(x-x')}\,n_{j(p)},\qquad j=-,+,
\tag{9.239}
$$

is nothing but the right-hand side (9.237).

Fermionic two-point functions

Exercise 4.1: We may define the pair of adjoint vertex operators

$$\hat{\psi}^{(\epsilon)\,\dagger}_{\text{q-p}-}(t,x) := \sqrt{\frac{1}{2\pi\,\epsilon}}\, e^{-\mathrm{i}\hat{\phi}_{-}(t,x)}, \qquad \hat{\psi}^{(\epsilon)}_{\text{q-p}-}(t,x) := \sqrt{\frac{1}{2\pi\,\epsilon}}\, e^{+\mathrm{i}\hat{\phi}_{-}(t,x)}, \tag{9.240a}$$

and

$$\hat{\psi}^{(\epsilon)\,\dagger}_{\text{q-p}+}(t,x) := \sqrt{\frac{1}{2\pi\,\epsilon}}\, e^{+\mathrm{i}\hat{\phi}_{+}(t,x)}, \qquad \hat{\psi}^{(\epsilon)}_{\text{q-p}+}(t,x) := \sqrt{\frac{1}{2\pi\,\epsilon}}\, e^{-\mathrm{i}\hat{\phi}_{+}(t,x)}. \tag{9.240b}$$

Verify that these vertex operators obey the fermion algebra.

Exercise 4.2: Alternatively, we may define the pair of adjoint vertex operators

$$\hat{\psi}^{(\epsilon)\,\dagger}_{\text{f}-}(t,x) := \sqrt{\frac{1}{2\pi\,\epsilon}}\, e^{-\mathrm{i}\hat{\phi}_{-}(t,x)}, \qquad \hat{\psi}^{(\epsilon)}_{\text{f}-}(t,x) := \sqrt{\frac{1}{2\pi\,\epsilon}}\, e^{+\mathrm{i}\hat{\phi}_{-}(t,x)}, \tag{9.241a}$$

and

$$\hat{\psi}^{(\epsilon)\,\dagger}_{\text{f}+}(t,x) := \sqrt{\frac{1}{2\pi\,\epsilon}}\, e^{-\mathrm{i}\hat{\phi}_{+}(t,x)}, \qquad \hat{\psi}^{(\epsilon)}_{\text{f}+}(t,x) := \sqrt{\frac{1}{2\pi\,\epsilon}}\, e^{+\mathrm{i}\hat{\phi}_{+}(t,x)}. \tag{9.241b}$$

Verify that these vertex operators also satisfy the fermion algebra.

Exercise 4.3: Define the parameter

$$g := \sinh^2\bar{\theta} \geq 0. \tag{9.242}$$

Show that

$$\begin{aligned}
\left\langle 0 \left| \hat{\psi}^{(\epsilon)}_{\text{f}-}(t,x)\, \hat{\psi}^{(\epsilon)\,\dagger}_{\text{f}-}(t',x') \right| 0 \right\rangle &\propto \frac{-\mathrm{i}}{(t-t')-(x-x')}\, \frac{\epsilon^{2g}}{[(t-t')^2-(x-x')^2]^g}, \\
\left\langle 0 \left| \hat{\psi}^{(\epsilon)\,\dagger}_{\text{f}+}(t,x)\, \hat{\psi}^{(\epsilon)}_{\text{f}+}(t',x') \right| 0 \right\rangle &\propto \frac{-\mathrm{i}}{(t-t')+(x-x')}\, \frac{\epsilon^{2g}}{[(t-t')^2-(x-x')^2]^g},
\end{aligned} \tag{9.243}$$

in the ground state of Hamiltonian (9.210). *Hint*: Invert Eq. (9.211), i.e.,

$$\begin{aligned}
\hat{J}_{-} &= +\cosh\theta\, \hat{J}^{\theta}_{-} - \sinh\theta\, \hat{J}^{\theta}_{+}, \\
\hat{J}_{+} &= -\sinh\theta\, \hat{J}^{\theta}_{-} + \cosh\theta\, \hat{J}^{\theta}_{+},
\end{aligned} \tag{9.244}$$

and make use of Eq. (9.238). Comment on the presence of the factor ϵ^{2g} on the right-hand side of Eq. (9.243). Fourier transformation of Eq. (9.243) to frequency and momentum space delivers Eq. (F.47).

Appendix A

The harmonic-oscillator algebra and its coherent states

A.1 The harmonic-oscillator algebra and its coherent states

A.1.1 *Bosonic algebra*

The quantum Hamiltonian for the harmonic oscillator is

$$\hat{H} = \hbar\omega\left(\hat{a}^\dagger\hat{a} + \frac{1}{2}\right), \tag{A.1}$$

when represented in terms of the lowering (annihilation) and raising (creation) operators $\hat{a}$ and $\hat{a}^\dagger$, respectively. This pair of operators obeys the *bosonic* algebra

$$[\hat{a}, \hat{a}^\dagger] = 1, \qquad [\hat{a}, \hat{a}] = [\hat{a}^\dagger, \hat{a}^\dagger] = 0. \tag{A.2}$$

A complete, orthogonal, and normalized basis of $\hat{H}$ is given by

$$|n\rangle = \frac{(\hat{a}^\dagger)^n}{\sqrt{n!}}|0\rangle, \qquad \hat{H}|n\rangle = \hbar\omega\left(n + \frac{1}{2}\right)|n\rangle, \qquad n = 0, 1, 2, \cdots, \tag{A.3}$$

where the ground state (vacuum) $|0\rangle$ is annihilated by $\hat{a}$,

$$\hat{a}|0\rangle = 0. \tag{A.4}$$

For $\hat{H}$ to be Hermitian, annihilation $\hat{a}$ and creation $\hat{a}^\dagger$ operators must be adjoint to each other, i.e., represented by

$$\begin{aligned} \hat{a}\,|n\rangle &= \sqrt{n}\,|n-1\rangle, & \hat{a}^\dagger\,|n\rangle &= \sqrt{n+1}\,|n+1\rangle, \\ \langle m|\hat{a}|n\rangle &= \sqrt{n}\,\delta_{m+1,n}, & \langle m|\hat{a}^\dagger|n\rangle &= \sqrt{n+1}\,\delta_{m-1,n}. \end{aligned} \tag{A.5}$$

The single-particle Hilbert space $\mathcal{H}^{(1)}$ of twice differentiable and square integrable functions on the real line for the harmonic oscillator can be *reinterpreted* as the Fock space $\mathcal{F}$ for the annihilation and creation operators $\hat{a}$ and $\hat{a}^\dagger$, respectively, since the number operator

$$\hat{N} := \hat{a}^\dagger\hat{a} \tag{A.6}$$

commutes with the Hamiltonian and the Fock space $\mathcal{F}$ is, by definition, the direct sum of the energy eigenspaces:

$$\mathcal{H}^{(1)} \cong \mathcal{F} := \bigoplus_{n=0}^{\infty}\{\lambda|n\rangle|\lambda \in \mathbb{C}\}. \tag{A.7}$$

One possible resolution of the identity $\mathbb{1}$ on $\mathcal{H}^{(1)} \cong \mathcal{F}$ is

$$\mathbb{1} = \sum_{n=0}^{\infty} |n\rangle\langle n|. \tag{A.8}$$

More informations on the harmonic oscillator can be found in chapter V of Ref. [118].

A.1.2 *Coherent states*

Define the uncountable set of coherent states for the harmonic oscillator, in short *bosonic coherent states*, by

$$|\alpha\rangle_{\mathrm{cs}} := e^{\alpha\,\hat{a}^{\dagger}}|0\rangle := \sum_{n=0}^{\infty} \frac{\alpha^n}{\sqrt{n!}}|n\rangle, \qquad \alpha \in \mathbb{C}. \tag{A.9a}$$

The adjoint set is (α^* denotes the complex conjugate of $\alpha \in \mathbb{C}$)

$${}_{\mathrm{cs}}\langle\alpha| := \langle 0|e^{\hat{a}\,\alpha^*} := \sum_{n=0}^{\infty} \langle n|\frac{(\alpha^*)^n}{\sqrt{n!}}, \qquad \alpha \in \mathbb{C}. \tag{A.9b}$$

Properties of bosonic coherent states are:

- Coherent state $|\alpha\rangle_{\mathrm{cs}}$ is a right eigenstate with eigenvalue α of the annihilation operator $\hat{a}$,[9]

$$\begin{aligned}
\hat{a}|\alpha\rangle_{\mathrm{cs}} &= \hat{a}\, e^{\alpha\,\hat{a}^{\dagger}}|0\rangle \\
&= \sum_{n=0}^{\infty} \frac{\alpha^n}{\sqrt{n!}}\hat{a}|n\rangle \\
&= \sum_{n=1}^{\infty} \frac{\alpha^n}{\sqrt{n!}}\sqrt{n}|n-1\rangle \\
&= \alpha \sum_{n=1}^{\infty} \frac{\alpha^{n-1}}{\sqrt{(n-1)!}}|n-1\rangle \\
&= \alpha|\alpha\rangle_{\mathrm{cs}}.
\end{aligned} \tag{A.10}$$

- Coherent state ${}_{\mathrm{cs}}\langle\alpha|$ is a left eigenstate with eigenvalue α^* of the creation operator $\hat{a}^{\dagger}$,

$$\hat{a}|\alpha\rangle_{\mathrm{cs}} = \alpha|\alpha\rangle_{\mathrm{cs}} \Longrightarrow {}_{\mathrm{cs}}\langle\alpha|\hat{a}^{\dagger} = {}_{\mathrm{cs}}\langle\alpha|\alpha^*. \tag{A.11}$$

- The action of creation operator $\hat{a}^{\dagger}$ on coherent state $|\alpha\rangle_{\mathrm{cs}}$ is differentiation with

[9] Non-Hermitian operators need not have the same left and right eigenstates.

respect to α,

$$\begin{aligned}\hat{a}^\dagger|\alpha\rangle_{\rm cs} &= \hat{a}^\dagger\, e^{\alpha\,\hat{a}^\dagger}|0\rangle \\ &= \sum_{n=0}^{\infty}\frac{\alpha^n}{\sqrt{n!}}\hat{a}^\dagger|n\rangle \\ &= \sum_{n=0}^{\infty}\frac{\alpha^n}{\sqrt{n!}}\sqrt{n+1}|n+1\rangle \\ &= \sum_{n=0}^{\infty}\left(\frac{\mathrm{d}}{\mathrm{d}\alpha}\frac{\alpha^{n+1}}{\sqrt{(n+1)!}}\right)|n+1\rangle \\ &= \frac{\mathrm{d}}{\mathrm{d}\alpha}|\alpha\rangle_{\rm cs}.\end{aligned} \tag{A.12}$$

- The action of creation operator $\hat{a}$ on coherent state ${}_{\rm cs}\langle\alpha|$ is differentiation with respect to α^*,

$$\hat{a}^\dagger|\alpha\rangle_{\rm cs} = \frac{\mathrm{d}}{\mathrm{d}\alpha}|\alpha\rangle_{\rm cs} \Longrightarrow {}_{\rm cs}\langle\alpha|\hat{a} = \frac{\mathrm{d}}{\mathrm{d}\alpha^*}\,{}_{\rm cs}\langle\alpha|. \tag{A.13}$$

- The overlap ${}_{\rm cs}\langle\alpha|\beta\rangle_{\rm cs}$ between two coherent states is $\exp(\alpha^*\beta)$,

$$\begin{aligned}{}_{\rm cs}\langle\alpha|\beta\rangle_{\rm cs} &= \sum_{m,n=0}^{\infty}\langle m|\frac{(\alpha^*)^m}{\sqrt{m!}}\frac{\beta^n}{\sqrt{n!}}|n\rangle \\ \overset{\langle m|n\rangle=\delta_{m,n}}{} &= \sum_{n=0}^{\infty}\frac{(\alpha^*\beta)^n}{n!} \\ &= e^{\alpha^*\beta}.\end{aligned} \tag{A.14}$$

- There exists a resolution of the identity in terms of bosonic coherent states,

$$\begin{aligned}\mathbb{1} &= \int\frac{\mathrm{d}z^*\mathrm{d}z}{2\pi\mathrm{i}}\;e^{-z^*z}\;|z\rangle_{\rm cs\,cs}\langle z| \\ &:= \frac{1}{\pi}\int_{-\infty}^{+\infty}\mathrm{dRe}\,z\int_{-\infty}^{+\infty}\mathrm{dIm}\,z\;e^{-z^*z}\;|z\rangle_{\rm cs\,cs}\langle z|.\end{aligned} \tag{A.15}$$

Proof.
Write

$$\hat{O} := \int\frac{\mathrm{d}z^*\mathrm{d}z}{2\pi\mathrm{i}}\;e^{-z^*z}\;|z\rangle_{\rm cs\,cs}\langle z|. \tag{A.16}$$

By construction, $\hat{O}$ belongs to the algebra of operators generated by $\hat{a}$ and $\hat{a}^\dagger$.

Step 1: With the help of Eqs. (A.10) and (A.13),

$$\begin{aligned}[\hat{a},|z\rangle_{\rm cs\,cs}\langle z|] &= \left(\hat{a}|z\rangle_{\rm cs}\right){}_{\rm cs}\langle z| - |z\rangle_{\rm cs}\left({}_{\rm cs}\langle z|\hat{a}\right) \\ &= \left(z|z\rangle_{\rm cs}\right){}_{\rm cs}\langle z| - |z\rangle_{\rm cs}\left(\frac{\mathrm{d}}{\mathrm{d}z^*}\,{}_{\rm cs}\langle z|\right) \\ &= \left(z-\frac{\mathrm{d}}{\mathrm{d}z^*}\right)|z\rangle_{\rm cs\,cs}\langle z|.\end{aligned} \tag{A.17}$$

Hence, after making use of integration by parts,

$$[\hat{a},\hat{O}] = \int \frac{\mathrm{d}z^*\mathrm{d}z}{2\pi\mathrm{i}}\, e^{-z^*z} \left(z - \frac{\mathrm{d}}{\mathrm{d}z^*}\right) |z\rangle_{\mathrm{cs}\,\mathrm{cs}}\langle z| \\ = 0. \tag{A.18}$$

– Step 2: By taking the adjoint of Eq. (A.18), $[\hat{a}^\dagger,\hat{O}] = 0$.
– Step 3:

$$\langle 0|\hat{O}|0\rangle = \int \frac{\mathrm{d}z^*\mathrm{d}z}{2\pi\mathrm{i}}\, e^{-z^*z}\, \langle 0|z\rangle_{\mathrm{cs}\,\mathrm{cs}}\langle z|0\rangle \\ \overset{\langle m|n\rangle = \delta_{m,n}}{=} \int \frac{\mathrm{d}z^*\mathrm{d}z}{2\pi\mathrm{i}}\, e^{-z^*z} \\ = 1. \tag{A.19}$$

– Step 4: Any linear operator from $\mathcal{F}$ to $\mathcal{F}$ belongs to the algebra generated by $\hat{a}$ and $\hat{a}^\dagger$. Since $\hat{O}$ commutes with both $\hat{a}$ and $\hat{a}^\dagger$ by steps 1 and 2, $\hat{O}$ commutes with all linear operators from $\mathcal{F}$ to $\mathcal{F}$. By Schur's lemma, $\hat{O}$ must be proportional to the identity operator. By Step 3, the proportionality factor is 1.

□

- For any operator $\hat{a}:\mathcal{F}\to\mathcal{F}$,

$$\mathrm{Tr}\,\hat{a} := \sum_{n=0}^{\infty}\langle n|\hat{a}|n\rangle \\ \overset{\text{By Eq. (A.15)}}{=} \int \frac{\mathrm{d}z^*\mathrm{d}z}{2\pi\mathrm{i}}\, e^{-z^*z} \sum_{n=0}^{\infty}\langle n|z\rangle_{\mathrm{cs}\,\mathrm{cs}}\langle z|\hat{a}|n\rangle \\ = \int \frac{\mathrm{d}z^*\mathrm{d}z}{2\pi\mathrm{i}}\, e^{-z^*z}\, {}_{\mathrm{cs}}\langle z|\hat{a}\left(\sum_{n=0}^{\infty}|n\rangle\langle n|\right)|z\rangle_{\mathrm{cs}} \\ \overset{\text{By Eq. (A.8)}}{=} \int \frac{\mathrm{d}z^*\mathrm{d}z}{2\pi\mathrm{i}}\, e^{-z^*z}\, {}_{\mathrm{cs}}\langle z|\hat{a}|z\rangle_{\mathrm{cs}}. \tag{A.20}$$

- Any operator $\hat{a}:\mathcal{F}\to\mathcal{F}$ is some linear combination of products of $\hat{a}$'s and $\hat{a}^\dagger$'s. Normal ordering of $\hat{a}$, which is denoted $:\hat{a}:$, is the operation of moving all creation operators to the left of annihilation operators as if all operators were to commute. For example,

$$\hat{a} = \hat{a}^\dagger\hat{a}\hat{a}\hat{a}^\dagger + \hat{a}^\dagger\hat{a}\hat{a}^\dagger \Longrightarrow :\hat{a}: = \hat{a}^\dagger\hat{a}^\dagger\hat{a}\hat{a} + \hat{a}^\dagger\hat{a}^\dagger\hat{a} = \hat{a} - 2\hat{a}^\dagger\hat{a} - \hat{a}^\dagger. \tag{A.21}$$

The matrix element of any normal ordered operator $:\hat{a}(\hat{a}^\dagger,\hat{a}):$ between any two coherent states ${}_{\mathrm{cs}}\langle z|$ and $|z'\rangle_{\mathrm{cs}}$ follows from Eqs. (A.10), (A.11), and (A.14),

$${}_{\mathrm{cs}}\langle z| : \hat{a}(\hat{a}^\dagger,\hat{a}) : |z'\rangle_{\mathrm{cs}} = {}_{\mathrm{cs}}\langle z| : A(z^*,z') : |z'\rangle_{\mathrm{cs}} = e^{z^*z'} : A(z^*,z') : . \tag{A.22}$$

Here, $:A(z^*,z'):$ is the complex-valued function obtained from the normal ordered operator $:\hat{a}(\hat{a}^\dagger,\hat{a}):$ by substituting $\hat{a}^\dagger$ for the complex number z^* and $\hat{a}$ for the complex number z'.

- Define the continuous family of unitary operators

$$D(\alpha) := e^{\alpha \hat{a}^{\dagger} - \alpha^{*} \hat{a}}, \qquad \alpha \in \mathbb{C}. \tag{A.23}$$

From Glauber formula[10]

$$D(\alpha) = e^{-\frac{|\alpha|^2}{2}} \, e^{+\alpha \hat{a}^{\dagger}} \, e^{-\alpha^{*} \hat{a}}, \tag{A.25}$$

which implies that

$$\begin{aligned} D(\alpha)|0\rangle &= e^{-\frac{|\alpha|^2}{2}} \, e^{+\alpha \hat{a}^{\dagger}} \, e^{-\alpha^{*} \hat{a}} |0\rangle \\ &= e^{-\frac{|\alpha|^2}{2}} \, e^{+\alpha \hat{a}^{\dagger}} \, |0\rangle \\ &= e^{-\frac{|\alpha|^2}{2}} \, |\alpha\rangle_{\mathrm{cs}}. \end{aligned} \tag{A.26}$$

Hence, $D(\alpha)$ is the unitary transformation that rotates the vacuum $|0\rangle$ into the coherent state $|\alpha\rangle_{\mathrm{cs}}$, up to a proportionality constant.

More informations on bosonic coherent states can be found in complement $\mathrm{G_V}$ of Ref. [118].

A.2 Path-integral representation of the anharmonic oscillator

Define the anharmonic oscillator of order $n = 2, 3, 4, \cdots$ by

$$\hat{H} = \hat{H}_0 + \hat{H}_n, \qquad \hat{H}_0 := \hbar\omega \left(\hat{a}^{\dagger} \hat{a} + \frac{1}{2} \right), \qquad \hat{H}_n := \sum_{m=3}^{2n} \lambda_m \left(\hat{a}^{\dagger} + \hat{a} \right)^m. \tag{A.27}$$

Of the real-valued parameters λ_m, $m = 3, 4, \cdots, 2n$, it is only required that $\lambda_{2n} > 0$. This insures that there exists a vacuum $|0\rangle$ annihilated by $\hat{a}$. With the help of the bosonic algebra (A.2), it is possible to move all annihilation operators to the right of the creation operators in the interaction $\hat{H}_n$. This action generates many terms that can be grouped by ascending order in the combined number of creation and annihilation operators. The monomials of largest order are all contained in $:\hat{H}_n:$. For example, $:(\hat{a}^{\dagger} + \hat{a})^3: = \left(\hat{a}^{\dagger}\hat{a}^{\dagger}\hat{a}^{\dagger} + 3\hat{a}^{\dagger}\hat{a}^{\dagger}\hat{a} + \mathrm{H.c.} \right)$. Evidently, $:\hat{H}_n:$ cannot be written anymore as a polynomial in $\hat{x} \propto (\hat{a}^{\dagger} + \hat{a})$ of degree $2n$.

After normal ordering of $\hat{H}$, the canonical partition function on the Hilbert space $\mathcal{F}$ in Eq. (A.7) becomes

$$Z := e^{-\beta E_0} \, \mathrm{Tr} \, e^{-\beta :\hat{H}:} = e^{-\beta E_0} \sum_{n=0}^{\infty} \langle n | e^{-\beta :\hat{H}:} | n \rangle, \tag{A.28}$$

---

[10] Let A and B be two operators that both commute with their commutator $[A, B]$. Then,

$$e^{A} \, e^{B} = e^{A+B} \, e^{\frac{1}{2}[A,B]}. \tag{A.24}$$

where E_0 is the normal ordering energy, i.e., the expectation value $\langle 0|\hat{H}|0\rangle$. We will now give an alternative representation of the canonical partition function that relies on the use of coherent states. We begin with the trace formula (A.20)

$$Z = \exp(-\beta E_0) \int \frac{d\varphi_0^* d\varphi_0}{2\pi i} \, e^{-\varphi_0^* \varphi_0} \, {}_{\rm cs}\langle\varphi_0| \exp(-\beta : \hat{H} :)|\varphi_0\rangle_{\rm cs}. \tag{A.29}$$

For M a large positive integer, write

$$\begin{aligned}\exp(-\beta : \hat{H} :) &= \exp\left(-\frac{\beta}{M}\sum_{j=0}^{M-1} : \hat{H} :\right) \\ &= 1 - \frac{\beta}{M}\sum_{j=0}^{M-1} : \hat{H} : + \mathcal{O}\left[\left(\frac{\beta}{M}\right)^2\right].\end{aligned} \tag{A.30}$$

To the same order of accuracy,

$$e^{-\beta:\hat{H}:} = \left[e^{-\beta:\hat{H}:/M}\right]^M. \tag{A.31}$$

Insert the resolution of identity (A.15) $(M-1)$-times,

$$e^{-\beta:\hat{H}:} = e^{-\frac{\beta}{M}:\hat{H}:}\left(\prod_{j=M-1}^{1}\int \frac{d\varphi_j^* d\varphi_j}{2\pi i} \, e^{-\varphi_j^*\varphi_j} \, |\varphi_j\rangle_{\rm cs} \, {}_{\rm cs}\langle\varphi_j| \, e^{-\frac{\beta}{M}:\hat{H}:}\right). \tag{A.32}$$

Equation (A.22) together with Eq. (A.30) gives

$${}_{\rm cs}\langle\varphi_0| \, e^{-\frac{\beta}{M}:\hat{H}:}|\varphi_{M-1}\rangle_{\rm cs} = e^{+\varphi_0^*\varphi_{M-1} - \frac{\beta}{M}:H(\varphi_0^*,\varphi_{M-1}):} + \mathcal{O}\left[\left(\frac{\beta}{M}\right)^2\right], \tag{A.33a}$$

and

$${}_{\rm cs}\langle\varphi_j| \, e^{-\frac{\beta}{M}:\hat{H}:}|\varphi_{j-1}\rangle_{\rm cs} = e^{+\varphi_j^*\varphi_{j-1} - \frac{\beta}{M}:H(\varphi_j^*,\varphi_{j-1}):} + \mathcal{O}\left[\left(\frac{\beta}{M}\right)^2\right], \tag{A.33b}$$

for $j = M-1, M-2, \cdots, 1$. The operator-valued function $: \hat{H} :$ of $\hat{a}$ and $\hat{a}^\dagger$ has been replaced by a complex-valued function $: H :$ of φ and φ^*, respectively. Altogether, a M-dimensional integral representation of the partition function has been found,

$$\begin{aligned}Z = {} & \exp(-\beta E_0) \int \left(\prod_{j=0}^{M-1} \frac{d\varphi_j^* d\varphi_j}{2\pi i}\right) \\ & \times \exp\left(-\sum_{j=1}^{M}\left[\varphi_j^*\left(\varphi_j - \varphi_{j-1}\right) + \frac{\beta}{M} : H(\varphi_j^*, \varphi_{j-1}) :\right]\right) \\ & + \mathcal{O}\left[\left(\frac{\beta}{M}\right)^2\right],\end{aligned} \tag{A.34a}$$

whereby

$$\varphi_M := \varphi_0, \qquad \varphi_M^* := \varphi_0^*. \tag{A.34b}$$

It is customary to write, in the limit $M \to \infty$, the functional path integral representation of the partition function

$$Z = e^{-\beta E_0} \int \mathcal{D}[\varphi^*, \varphi] e^{-S_{\mathrm{E}}[\varphi^*, \varphi]}, \tag{A.35a}$$

where the so-called Euclidean action $S_{\mathrm{E}}[\varphi^*, \varphi]$ is given by

$$S_{\mathrm{E}}[\varphi^*, \varphi] = \int_0^\beta \mathrm{d}\tau \left\{\varphi^*(\tau)\partial_\tau \varphi(\tau) + : H[\varphi^*(\tau), \varphi(\tau)] :\right\}, \tag{A.35b}$$

and the complex-valued fields $\varphi^*(\tau)$ and $\varphi(\tau)$ obey the periodic boundary conditions

$$\varphi^*(\tau) = \varphi^*(\tau + \beta), \qquad \varphi(\tau) = \varphi(\tau + \beta). \tag{A.35c}$$

Hence, their Fourier transform are

$$\varphi^*(\tau) = \frac{1}{\beta} \sum_{l \in \mathbb{Z}} \varphi_l^* \, e^{+\mathrm{i}\varpi_l \tau}, \qquad \varphi(\tau) = \frac{1}{\beta} \sum_{l \in \mathbb{Z}} \varphi_l \, e^{-\mathrm{i}\varpi_l \tau}. \tag{A.36a}$$

The frequencies

$$\varpi_l := \frac{2\pi}{\beta} l, \qquad l \in \mathbb{Z}, \tag{A.36b}$$

are the so-called bosonic *Matsubara* frequencies.

Convergence of the (functional) integral representing the partition function is guaranteed by the contribution $\lambda_{2n}(\varphi^* \, \varphi)^{2n}$ to the interaction $: H_{\mathrm{n}}(\varphi^*, \varphi) :$. Thus, convergence of an integral is the counterpart in a path integral representation to the existence of a ground state in operator language.

Quantum mechanics at zero temperature is recovered from the partition function after performing the analytical continuation (also called a Wick rotation)

$$\tau = +\mathrm{i}t, \qquad \mathrm{d}\tau = +\mathrm{i}\mathrm{d}t, \qquad \partial_\tau = -\mathrm{i}\partial_t, \tag{A.37}$$

under which

$$\begin{aligned} -S_{\mathrm{E}} &\to +\mathrm{i}S \\ &= +\mathrm{i} \int_{-\infty}^{+\infty} \mathrm{d}t \left\{\varphi^*(t)\mathrm{i}\partial_t \varphi(t) - : H[\varphi^*(t), \varphi(t)] :\right\}. \end{aligned} \tag{A.38}$$

The path-integral representation of the anharmonic oscillator relies solely on two properties of bosonic coherent states: Equations (A.15) and (A.22). Raising, $\hat{a}^\dagger$, and lowering, $\hat{a}$, operators are not unique to bosons. As we shall see, one can also associate raising and lowering operators to fermions. Raising and lowering operators are also well known to be involved in the theory of the angular momentum. In general, raising and lowering operators appear whenever a finite (infinite) set of operators obey a finite (infinite) dimensional *Lie algebra*. Coherent states are those states that are eigenstates of lowering operators in the Lie algebra and they obey extensions of Eqs. (A.15) and (A.22). Hence, it is possible to generalize the path-integral representation of the partition function for the anharmonic oscillator to Hamiltonians expressed in terms of operator obeying a fermion, spin, or any type of Lie algebra. Due to the non-vanishing overlap of coherent states, a first-order imaginary-time derivative term always appears in the action. This term is called a *Berry phase* when it yields a pure phase in an otherwise real-valued Euclidean action as is the case, say, when dealing with spin Hamiltonians.[11] It is the first-order imaginary-time derivative term that encodes quantum mechanics in the path-integral representation of the partition function. A reference on generalized coherent states is the book in Ref. [119].

A.3 Higher dimensional generalizations

The path-integral representation of the partition function for a single anharmonic oscillator is a functional integral over the exponential of the Euclidean classical action (A.35b) in $(0+1)$-dimensional (position) space and (imaginary) time. The path-integral representation of the quantum field theory of a d-dimensional continuum of coupled anharmonic oscillators is a functional integral over the exponential of the Euclidean classical action in $(d+1)$-dimensional (position) space and (imaginary) time of the form

$$\begin{aligned} S_{\mathrm{E}}[\varphi^{*},\varphi] \equiv& \int\limits_{0}^{\beta} \mathrm{d}\tau \int \mathrm{d}^{d}\boldsymbol{r} \mathcal{L}_{\mathrm{E}} \\ =& \int\limits_{0}^{\beta} \mathrm{d}\tau \int \mathrm{d}^{d}\boldsymbol{r} \left\{\varphi^{*}(\boldsymbol{r},\tau)\partial_{\tau}\varphi(\boldsymbol{r},\tau) + : H[\varphi^{*}(\boldsymbol{r},\tau),\varphi(\boldsymbol{r},\tau)] :\right\}. \end{aligned} \tag{A.41a}$$

The classical fields $\varphi^{*}(\boldsymbol{r},\tau)$, and $\varphi(\boldsymbol{r},\tau)$ obey periodic boundary conditions in imaginary time τ,

$$\varphi(\boldsymbol{r},\tau) = \varphi(\boldsymbol{r},\tau+\beta), \qquad \varphi^{*}(\boldsymbol{r},\tau) = \varphi^{*}(\boldsymbol{r},\tau+\beta). \tag{A.41b}$$

---

[11] By writing [compare with Eq. (1.62a)]

$$\varphi(\tau) = \sqrt{\frac{1}{2}}\,[x(\tau) + \mathrm{i}p(\tau)]\,, \qquad \varphi^{*}(\tau) = \sqrt{\frac{1}{2}}\,[x(\tau) - \mathrm{i}p(\tau)]\,, \tag{A.39}$$

we can derive the path-integral representation of the (an)harmonic oscillator in terms of the coordinate and momentum of the single particle of unit mass $m=1$, unit characteristic frequency $\omega = 1$, and with $\hbar = 1$. The first-order partial derivative term becomes purely imaginary

$$\int\limits_{0}^{\beta} \mathrm{d}\tau (\varphi^{*}\partial_{\tau}\varphi)(\tau) = \mathrm{i} \int\limits_{0}^{\beta} \mathrm{d}\tau (x\partial_{\tau}p)(\tau). \tag{A.40}$$

At zero temperature, analytical continuation $\tau = +\mathrm{i}t$ of the action yields

$$
\begin{aligned}
S[\varphi^*, \varphi] &\equiv \int\limits_{-\infty}^{+\infty} \mathrm{d}t \int \mathrm{d}^d\boldsymbol{r} \mathcal{L} \\
&= \int\limits_{-\infty}^{+\infty} \mathrm{d}t \int \mathrm{d}^d\boldsymbol{r} \left\{\varphi^*(\boldsymbol{r},\tau)\mathrm{i}\partial_t\varphi(\boldsymbol{r},\tau) - : H[\varphi^*(\boldsymbol{r},\tau), \varphi(\boldsymbol{r},\tau)] :\right\}. \qquad \text{(A.42)}
\end{aligned}
$$

The classical canonical field conjugate to $\varphi(\boldsymbol{r}, t)$ is

$$
\pi(\boldsymbol{r}, t) := \frac{\delta \mathcal{L}}{\delta[\partial_t \varphi(\boldsymbol{r}, t)]} = \mathrm{i}\varphi^*(\boldsymbol{r}, t). \qquad \text{(A.43)}
$$

Canonical quantization is obtained by replacing the classical fields $\varphi(\boldsymbol{r}, t)$ and $\varphi^*(\boldsymbol{r}, t)$ with quantum fields $\hat{\varphi}(\boldsymbol{r}, t)$ and $\hat{\varphi}^\dagger(\boldsymbol{r}, t)$ that obey the equal-time algebra

$$
[\hat{\varphi}(\boldsymbol{r}, t), \hat{\varphi}^\dagger(\boldsymbol{r}', t)] = \delta(\boldsymbol{r} - \boldsymbol{r}'), \qquad [\hat{\varphi}(\boldsymbol{r}, t), \hat{\varphi}(\boldsymbol{r}', t)] = [\hat{\varphi}^\dagger(\boldsymbol{r}, t), \hat{\varphi}^\dagger(\boldsymbol{r}', t)] = 0. \qquad \text{(A.44)}
$$

Appendix B

Some Gaussian integrals

B.1 Generating function

Path integrals are generalizations of multi-dimensional Riemann integrals. Integrands of path integrals for non-interacting bosons are exponentials of quadratic forms. Hence, for any positive real-valued a, their evaluations require generalizations to path integrals of the two Gaussian integrals

$$
\begin{aligned}
Z_a(j^*,j) :=& \int \frac{\mathrm{d}z^*\mathrm{d}z}{2\pi\mathrm{i}}\, e^{-z^* a z + j^* z + j z^*} \\
=& \int \frac{\mathrm{d}z^*\mathrm{d}z}{2\pi\mathrm{i}}\, e^{-z^* a z + j^* z + j z^* - j^* a^{-1} j + j^* a^{-1} j} \\
=& \, e^{+j^* a^{-1} j} \int \frac{\mathrm{d}z^*\mathrm{d}z}{2\pi\mathrm{i}}\, e^{-(z - a^{-1} j)^* a (z - a^{-1} j)} \\
=& \, e^{+j^* a^{-1} j} \int \frac{\mathrm{d}z^*\mathrm{d}z}{2\pi\mathrm{i}}\, e^{-z^* a z} \\
=& \, \frac{e^{+j^* a^{-1} j}}{a} \int \frac{\mathrm{d}z^*\mathrm{d}z}{2\pi\mathrm{i}}\, e^{-z^* z} \\
=& \, \frac{e^{+j^* a^{-1} j}}{a} \int_0^\infty \mathrm{d}r\, r\, \frac{2\pi}{\pi}\, e^{-r^2} \\
=& \, \frac{e^{+j^* a^{-1} j}}{a},
\end{aligned}
\qquad \text{(B.1a)}
$$

and

$$
\begin{aligned}
\langle z^* z \rangle_a :=& \int \frac{\mathrm{d}z^*\mathrm{d}z}{2\pi\mathrm{i}}\, z^* z\, e^{-z^* a z} \Big/ Z_a(j^*,j)\big|_{j^*=j=0} \\
=& \, \frac{1}{Z_a(j^*,j)} \left. \frac{\partial^2 Z_a(j^*,j)}{\partial j \partial j^*} \right|_{j^*=j=0} \\
=& \, \frac{1}{a}.
\end{aligned}
\qquad \text{(B.1b)}
$$

The function $Z_a(j^*, j)$ is called a *generating function.* From it all moments of the form

$$\langle (z^* z)^n \rangle_a := \frac{1}{Z_a(j^*, j)} \left. \frac{\partial^{2n} Z_a(j^*, j)}{\partial j^n \partial j^{*n}} \right|_{j^*=j=0}, \qquad n = 0, 1, 2, \cdots, \tag{B.2}$$

can be calculated.

Generalization of Eqs. (B.1) and (B.2) to N-dimensional Riemann integrals is straightforward. Replace the complex conjugate pair z^* and z by N-dimensional vectors $\boldsymbol{z}^\dagger$ and $\boldsymbol{z}$, respectively. Replace the complex number a with a strictly positive real part by the $N \times N$ positive definite Hermitian matrix $\mathbb{A}$. Define the generating functional

$$Z_{\mathbb{A}}(\boldsymbol{j}^\dagger, \boldsymbol{j}) := \int \frac{\mathrm{d}^N \boldsymbol{z}^\dagger \mathrm{d}^N \boldsymbol{z}}{(2\pi \mathrm{i})^N} e^{-\boldsymbol{z}^\dagger \mathbb{A} \boldsymbol{z} + \boldsymbol{j}^\dagger \cdot \boldsymbol{z} + \boldsymbol{z}^\dagger \cdot \boldsymbol{j}}, \tag{B.3}$$

from which all moments

$$\left\langle \prod_{m=1}^{n} z_m^* z_m \right\rangle_{\mathbb{A}} := \frac{1}{Z_{\mathbb{A}}(\boldsymbol{j}^\dagger, \boldsymbol{j})} \prod_{m=1}^{n} \left. \frac{\partial^2 Z_{\mathbb{A}}(\boldsymbol{j}^\dagger, \boldsymbol{j})}{\partial j_m \partial j_m^*} \right|_{j^*=j=0}, \qquad n = 0, 1, 2, \cdots, \tag{B.4}$$

can be calculated. Since the measure of the generating functional is invariant under any unitary transformation of $\mathbb{C}^N$, we can choose a basis of $\mathbb{C}^N$ that diagonalizes the positive definite Hermitian matrix $\mathbb{A}$, in which case Eq. (B.1) can be used for each independent integration over the N normal modes. Thus,

$$\begin{aligned} Z_{\mathbb{A}}(\boldsymbol{j}^\dagger, \boldsymbol{j}) &= \frac{e^{+\boldsymbol{j}^\dagger \mathbb{A}^{-1} \boldsymbol{j}}}{\det \mathbb{A}}, \\ \langle z_m^* z_n \rangle_{\mathbb{A}} &= \left(\mathbb{A}^{-1}\right)_{mn}, \qquad m, n = 1, \cdots, N. \end{aligned} \tag{B.5}$$

Imposing periodic boundary conditions for continuous systems results in having a countable infinity of normal modes. In this case Eq. (B.5) is generalized by replacing $\mathbb{A}$, whose determinant is made of a finite product of eigenvalues, by a kernel, whose determinant is made of a countable product of eigenvalues. For infinite dimensional vector spaces, we use the notation Det $\cdots$ for the determinant of the kernel $\cdots$. After taking the thermodynamic limit, the number of normal modes is uncountable. The logarithm of the determinant of the kernel becomes an integral instead of a sum.

B.2 Bose-Einstein distribution and the residue theorem

The Bose-Einstein distribution

$$f_{\mathrm{BE}}(z) := \frac{1}{e^{\beta z} - 1} \tag{B.6}$$

is analytic in the complex plane except for the equidistant first-order poles

$$z_l = \frac{2\pi \mathrm{i}}{\beta} l, \qquad l \in \mathbb{Z}, \tag{B.7}$$

on the imaginary axis. Each pole $z_l = (2\pi\mathrm{i}/\beta)l$ of $f_{\mathrm{BE}}(z)$ has the residue $1/\beta$ since

$$\exp\left[\beta(z_l + z)\right] - 1 = \beta z + \mathcal{O}(z^2), \qquad \forall l \in \mathbb{Z}. \tag{B.8}$$

Let $g(z)$ be a complex function such that:

- $g(z)$ decreases sufficiently fast at infinity,

$$\lim_{|z|\to\infty} |z| f_{\mathrm{BE}}(z) g(z) = 0. \tag{B.9}$$

- $g(z)$ is analytic everywhere in the complex planes except for two poles on the real axis away from the origin, say at $z = \pm x \neq 0$.

Let Γ be a closed path infinitesimally close to the imaginary axis and running antiparallel (parallel) to the imaginary axis when $\mathrm{Re}\, z < 0$ ($\mathrm{Re}\, z > 0$). Let $\partial U_{\pm x}$ be circular paths running clockwise and centered about $\pm x$. Then, path Γ can be deformed into path $\partial U_{-x} \cup \partial U_{+x}$ by Cauchy theorem, and the residue theorem yields

$$\begin{aligned}
\sum_l g(z_l) &= +\beta \int_\Gamma \frac{\mathrm{d}z}{2\pi\mathrm{i}} f_{\mathrm{BE}}(z) g(z) \\
&= +\beta \left(\int_{\partial U_{-x}} + \int_{\partial U_{+x}} \right) \frac{\mathrm{d}z}{2\pi\mathrm{i}} f_{\mathrm{BE}}(z) g(z) \\
&= -\beta \left[\mathrm{Res}\,(f_{\mathrm{BE}} g)(-x) + \mathrm{Res}\,(f_{\mathrm{BE}} g)(+x) \right].
\end{aligned} \tag{B.10}$$

Appendix C

Non-Linear Sigma Models (NLσM) on Riemannian manifolds

C.1 Introduction

We have seen in section 3.2.2 that the $O(N)$ NLσM is an example of a NLσM on a Riemannian manifold. The goal of this appendix is to derive the one-loop RG equations obeyed by the metric tensor $\mathfrak{g}_{ab}$ that enters the action of a generic NLσM on a Riemannian manifold. These equations were derived up to two loops by Friedan in Ref. [120]. In terms of the short-distance cut-off $\mathfrak{a} \equiv \Lambda^{-1}$ and up to one loop, they are given by

$$\mathfrak{a}\frac{\partial}{\partial \mathfrak{a}}\mathfrak{g}_{ab} = \epsilon\, \mathfrak{g}_{ab} - \frac{1}{2\pi}R_{ab}, \qquad \epsilon \text{ infinitesimal}, \tag{C.1}$$

when the Euclidean base space is $d = (2+\epsilon)$-dimensional, while R_{ab} represents the Ricci tensor and R_{apqr} represents the curvature tensor of the (Riemannian) target manifold. Summation convention over repeated indices is here implied. Equation (C.1) generalizes Eq. (3.224).

To this end, we shall employ the background-field method [121]. This method dictates how to separate fields into slow and fast modes in such a way that the action can be expanded in a Taylor series in powers of the fast modes which is covariant under reparametrization of the Riemannian manifold. Corrections to the action of the slow modes are then computed to any desired order in a cumulant expansion by integration over the fast modes in $d = (2+\epsilon)$ dimensions.

C.2 A few preliminary definitions

We begin with a collection of mathematical definitions needed to make precise the concept of a Riemannian manifold. This section can be ignored if one is not interested in this level of rigor.

A Riemannian manifold is a smooth manifold endowed with a metric. We thus need to define a smooth manifold and a metric. In turn, a smooth manifold is a special type of topological space.

Topological space: Let X be a set. A topology on X is a set $\underline{\underline{T}}$ of subsets of

X such that $\underline{\underline{T}}$ contains:

(1) The empty set and X itself.
(2) The union of any subset of $\underline{\underline{T}}$.
(3) The intersection of any finite subset of $\underline{\underline{T}}$.

A topological space $(X, \underline{\underline{T}})$ is a set X with a topology $\underline{\underline{T}}$ on X.

Homeomorphism: Let $(X, \underline{\underline{T}})$ and $(X', \underline{\underline{T}}')$ be two topological spaces. A mapping $f : X \longrightarrow X'$ is called a homeomorphism if

(1) f is one-to-one and onto.
(2) $U \in \underline{\underline{T}} \Longrightarrow f(U) \in \underline{\underline{T}}'$.
(3) $U' \in \underline{\underline{T}}' \Longrightarrow f^{-1}(U') \in \underline{\underline{T}}$.

Open sets and neighborhoods: Let $(X, \underline{\underline{T}})$ be a topological space. Elements of the topology $(\underline{\underline{T}})$ are called open sets. A neighborhood of $x \in X$ is a subset of X that includes an open set to which x belongs to.

Hausdorff topological space: A topological space $(X, \underline{\underline{T}})$ is called Hausdorff if any two distinct points possess disjoint neighborhoods.

Topological manifold: A N-dimensional topological manifold is a Hausdorff topological space such that every point has a neighborhood homeomorphic to $\mathbb{R}^N$.

Chart: A chart (U, φ) of a N-dimensional topological manifold $(X, \underline{\underline{T}})$ is an open set U of X, called the domain of the chart, together with a homeomorphism $\varphi : U \longrightarrow V$ onto an open set V in $\mathbb{R}^N$. The coordinates $(x^1, \cdots, x^N)$ of the image $\varphi(x) \in \mathbb{R}^N$ of the point $x \in U \subset X$ are called the coordinates of x in the chart (U, φ) or, in short, local coordinates of x. Here, the N coordinates $(x^1, \cdots, x^N)$ of the point $\varphi(x) \in \mathbb{R}^N$ are short-hand notations for the mappings

$$\begin{aligned} a^i : \mathbb{R}^N &\longrightarrow \mathbb{R}, \\ (x^1, \cdots, x^N) &\longrightarrow a^i(x^1, \cdots, x^N) = x^i. \end{aligned} \tag{C.2}$$

A chart (U, φ) is also called a local coordinate system.

Atlas: An atlas of class C^k of a N-dimensional topological manifold $(X, \underline{\underline{T}})$ is a set of charts $\{(U_\alpha, \varphi_\alpha)\}$ such that:

(1) $\bigcup\limits_\alpha U_\alpha = X$.
(2) The maps $\varphi_\beta \circ \varphi_\alpha^{-1} : \varphi_\alpha(U_\alpha \bigcap U_\beta) \longrightarrow \varphi_\beta(U_\alpha \bigcap U_\beta)$ are maps of open sets of $\mathbb{R}^N$ into $\mathbb{R}^N$ of class C^k, i.e., k-times continuously differentiable.

Equivalent atlases: Two C^k atlases $\{(U_\alpha, \varphi_\alpha)\}$ and $\{(U_{\alpha'}, \varphi_{\alpha'})\}$ are equivalent if and only if the set of domains $\{U_\alpha\} \bigcup \{U_{\alpha'}\}$ and the set of homeomorphisms $\{\varphi_\alpha\} \bigcup \{\varphi_{\alpha'}\}$ is again a C^k atlas.

C^k manifold: A N-dimensional topological manifold $(X, \underline{\underline{T}})$ together with an equivalence class of C^k atlases is a C^k structure on X. It is also said that X is a C^k manifold. When $k = \infty$ the manifold is said to be smooth.

Differentiable functions: Charts make it possible to extend the notion of differentiability of functions $f : \mathbb{R}^N \longrightarrow \mathbb{R}$ to functions whose domain of definitions are C^k manifolds. Let f be a real-valued function with the C^k manifold X as domain of definition. Hence, we associate to any $x \in X$ the image $f(x) \in \mathbb{R}$. Let (U, φ) be a chart at x, i.e., $x \in U$. The function $f \circ \varphi^{-1} : \varphi(U) \longrightarrow \mathbb{R}$ is a mapping from an open set of $\mathbb{R}^N$ into $\mathbb{R}$. Just as the coordinates of $\varphi(x)$ represent x in the local chart (U, φ), the mapping $f \circ \varphi^{-1}$ represents f in the local chart. The function f is of class C^j at x with $j \leq k$ if $f \circ \varphi^{-1}$ is of class C^j at $\varphi(x)$.

Tangent vector $\boldsymbol{v}_x$: A tangent vector to a C^k manifold X at a point $x \in X$ is a function $\boldsymbol{v}_x$ from the space of functions defined and differentiable on some neighborhood of $x \in X$ into $\mathbb{R}$, that satisfies

(1) $\boldsymbol{v}_x(\alpha f + \beta g) = \alpha \boldsymbol{v}_x(f) + \beta \boldsymbol{v}_x(g)$ (linearity),
(2) $\boldsymbol{v}_x(fg) = f \boldsymbol{v}_x(g) + g \boldsymbol{v}_x(f)$ (Leibniz rule),

for all α and β in $\mathbb{R}$ and for all functions f and g on X that are differentiable at x. In the chart (U, φ), the local coordinates (components) of a tangent vector $\boldsymbol{v}_x$ are the N numbers $(v^1, \cdots, v^N)$ where

$$v^i := \boldsymbol{v}_x(\varphi^i), \tag{C.3a}$$

$\varphi^i \equiv a^i \circ \varphi$. Here, the N coordinates functions $(a^1, \cdots, a^N)$ are defined by

$$\begin{aligned} a^i : \mathbb{R}^N &\longrightarrow \mathbb{R}, \\ (u^1, \cdots, u^N) &\longrightarrow a^i(u^1, \cdots, u^N) = u^i. \end{aligned} \tag{C.3b}$$

The tangent vector $\boldsymbol{v}_x$ is also called a derivation.

Tangent vector as a directional derivative: Let f be a function defined on some neighborhood of $x \in X$ into $\mathbb{R}$ that is differentiable. The directional derivative of f along $\boldsymbol{v}_x$ is the image $\boldsymbol{v}_x(f)$ of f. The rational for this terminology follows from the following argument. Define $F : \mathbb{R}^N \longrightarrow \mathbb{R}$ through the composition $F := f \circ \varphi^{-1}$ and assume that f is a C^∞ function in U, the domain of the local chart (U, φ). Taylor expansion gives (for f a C^1 function, one uses the mean value theorem of analysis)

$$f(x) = F\big(\varphi(y)\big) + \sum_{i=1}^{N} \Big(\varphi^i(x) - \varphi^i(y)\Big) \left.\frac{\partial F}{\partial x^i}\right|_{\varphi(y)} + \cdots, \tag{C.4a}$$

for any pair x and y in U. By definition, the directional derivative of f along $\boldsymbol{v}_x$ is

$$\sum_{i=1}^{N} \boldsymbol{v}_x\left(\varphi^i\right) \left.\frac{\partial F}{\partial x^i}\right|_{\varphi(y)} + \cdots, \tag{C.4b}$$

i.e., it reduces to

$$\sum_{i=1}^{N} \boldsymbol{v}_y\left(\varphi^i\right) \left.\frac{\partial F}{\partial x^i}\right|_{\varphi(y)} = \sum_{i=1}^{N} v^i \left.\frac{\partial F}{\partial x^i}\right|_{\varphi(y)}, \tag{C.4c}$$

when $x = y$.

Tangent vector space: The set of all tangent vectors to the C^k manifold X at $x \in X$ together with the addition and scalar multiplication defined by

(1) $(\alpha \boldsymbol{v}_x + \beta \boldsymbol{w}_x)(f) := \alpha \boldsymbol{v}_x(f) + \beta \boldsymbol{w}_x(f)$

is a vector space called the tangent vector space and denoted $T_x X$. According to Eq. (C.4) the vectors of $T_x X$ can be represented as linear combinations of the basis $\{\partial/\partial x^1, \cdots, \partial/\partial x^N\}$, which is also called the natural basis of the tangent vector space. The natural or coordinate basis of $T_x X$ is also denoted by $\{e_a\}$ where $e_a \equiv \partial_a \equiv \partial/\partial x^a$ for $a = 1, \cdots, N$. A chart (U, φ) has thus induced an isomorphism between $T_x X$ and $\mathbb{R}^N$. The basis of $T_x X$ need not be the natural or the coordinate one. We may also choose the basis $\{\hat{e}_{\hat{a}}\}$ defined by

$$\hat{e}_{\hat{a}} := A_{\hat{a}}{}^{a} e_a \tag{C.5}$$

where summation over repeated upper and lower indices is implied and the $N \times N$ matrix $A \equiv (A_{\hat{a}}{}^{a})$ belongs to the group $GL(N, \mathbb{R})$ of $N \times N$ real-valued and invertible matrices. The basis (C.5) is known as the non-coordinate basis of $T_x X$.

Cotangent vector space: The cotangent space $T_x^* X$ is the vector space dual to the tangent space $T_x X$, i.e., it is the vector space of all linear functions $f : T_x X \longrightarrow \mathbb{R}$. The basis $\{e^{*a}\}$ of the cotangent space $T_x^* X$ dual to the basis $\{e_a\}$ of the tangent space $T_x X$ is defined by the condition

$$e^{*a}(e_b) = \delta^a_{\ b}. \tag{C.6}$$

Tensors and tensor fields: A tensor T of a smooth manifold X at $x \in X$ of type $\binom{m}{n}$ is a multilinear mapping that maps m dual vectors and n vectors into $\mathbb{R}$,

$$T : \underbrace{(T_x^* X \times \cdots \times T_x^* X)}_{m\text{-times}} \times \underbrace{(T_x X \times \cdots \times T_x X)}_{n\text{-times}} \longrightarrow \mathbb{R}. \tag{C.7}$$

The set of all tensors of a smooth manifold X at $x \in X$ of type $\binom{m}{n}$ is called the tensor space of a smooth manifold X at $x \in X$ of type $\binom{m}{n}$ and denoted by $T^m_{n,x} X$ where $T^1_{0,x} X \equiv T_x X$ and $T^0_{1,x} X \equiv T^*_x X$. By defining a linear combination of two tensors of the same type by the same linear combination of their point-wise values, the tensor space of a smooth manifold X at $x \in X$ of type $\binom{m}{n}$ is endowed with the structure of a vector space. A smooth assignment of an element of $T^m_{n,x} X$ at each point $x \in X$ defines a smooth tensor field on the smooth manifold X. The set of all tensor fields of type $\binom{m}{n}$ is denoted $T^m_n X$.

Riemannian manifold: A Riemannian manifold is a smooth manifold $\mathfrak{M}$ together with a smooth tensor field $\mathfrak{g} : T\mathfrak{M} \times T\mathfrak{M} \longrightarrow \mathbb{R}$ of type $\binom{0}{2}$ such that:

(1) $\mathfrak{g}$ is symmetric.
(2) For each $\mathfrak{p} \in \mathfrak{M}$, the bilinear form $\mathfrak{g}_{\mathfrak{p}}$ is positive definite.

C.3 Definition of a NLσM on a Riemannian manifold

Consider the NLσM defined by the partition function

$$Z := \int \mathcal{D}[\phi] \exp\left(-S[\phi]\right), \tag{C.8a}$$

the Euclidean action

$$S[\phi] := \int \frac{\mathrm{d}^d x}{\mathfrak{a}^{d-2}}\, \mathcal{L}(\phi), \tag{C.8b}$$

the Lagrangian density

$$\mathcal{L}(\phi) := \frac{1}{2} \mathfrak{g}_{ab}(\phi) \partial_\mu \phi^a \, \partial_\mu \phi^b, \tag{C.8c}$$

and the measure

$$\mathcal{D}[\phi] := \prod_{x\in\mathbb{R}^d} \sqrt{||\mathfrak{g}(\phi)||} \prod_a \mathrm{d}\phi^a(x). \tag{C.8d}$$

At each point $x \in \mathbb{R}^d$, the $N \times N$ ϕ-dependent matrix $\mathfrak{g}(\phi)$ with real-valued matrix elements $\mathfrak{g}_{ab}(\phi)$ is positive definite and symmetric with determinant $||\mathfrak{g}(\phi)||$.[12] Some few words about the conventions we are using in Eq. (C.8). We reserve the Greek alphabet to denote the coordinates of $x \in \mathbb{R}^d$. We reserve the Latin alphabet to denote the $N \times N$ real-valued entries $\mathfrak{g}_{ab}(\phi)$ in the defining representation of the Riemannian metric $\mathfrak{g}$, which, for each point $\mathfrak{p}$ [represented by $\phi(x) \in \mathbb{R}^N$] in the Riemannian manifold $(\mathfrak{M}, \mathfrak{g})$, is the bilinear mapping

$$\begin{aligned} \mathfrak{g}_\mathfrak{p} : T_\mathfrak{p}\mathfrak{M} \times T_\mathfrak{p}\mathfrak{M} &\longrightarrow \mathbb{R} \\ (U,V) &\longrightarrow \mathfrak{g}_\mathfrak{p}(U,V) \end{aligned} \tag{C.9a}$$

($T_\mathfrak{p}\mathfrak{M}$ the tangent space to $\mathfrak{p} \in \mathfrak{M}$) that obeys the condition for symmetry

$$\mathfrak{g}_\mathfrak{p}(U,V) = \mathfrak{g}_\mathfrak{p}(V,U), \qquad \forall U,V \in T_\mathfrak{p}\mathfrak{M}, \tag{C.9b}$$

the condition for positivity

$$\mathfrak{g}_\mathfrak{p}(U,U) \geq 0, \qquad \forall U \in T_\mathfrak{p}\mathfrak{M}, \tag{C.9c}$$

and the condition for non-degeneracy

$$\mathfrak{g}_\mathfrak{p}(U,U) = 0 \Longrightarrow U = 0. \tag{C.9d}$$

There is no distinction between upper and lower Greek indices and we will always choose them to be lower indices. Summation over repeated Greek indices is always implied. There is a distinction between upper and lower Latin indices. Summation over repeated upper and lower Latin indices is implied. Raising and lowering Latin indices is done with the metric

$$\phi_a \equiv \mathfrak{g}_{ab}\phi^b, \tag{C.10a}$$

where the convention

$$\mathfrak{g}^{ab}\, \mathfrak{g}_{bc} \equiv \delta^a_c \tag{C.10b}$$

[12] If A is a matrix, $||A|| := |\det A|$.

is used to denote the matrix with entries $\mathfrak{g}_{ab}$ which is the inverse of the matrix with entries $\mathfrak{g}^{ab}$. A partial derivative only acts on the first object to its right. For example,

$$\begin{aligned}
&\partial_\mu \phi^a \phi^b \equiv \frac{\partial \phi^a}{\partial x_\mu} \phi^b, \qquad \partial_c \phi^a \phi^b \equiv \frac{\partial \phi^a}{\partial \phi^c} \phi^b, \\
&\partial_\mu (\phi^a \phi^b) \equiv (\partial_\mu \phi^a) \phi^b + \phi^a (\partial_\mu \phi^b), \\
&\partial_c (\phi^a \phi^b) \equiv (\partial_c \phi^a) \phi^b + \phi^a (\partial_c \phi^b).
\end{aligned} \tag{C.11}$$

In this appendix, we are going to derive the change in the action (C.8b) evaluated at a solution φ of its equations of motion due to fluctuations arising from the path integral. This will be done perturbatively up to one loop in the so-called loop expansion. For conciseness, we shall abusively call the saddle-point solution a classical solution, while referring to the fluctuations about it as quantum fluctuations.

C.4 Classical equations of motion for NLσM: Christoffel symbol and geodesics

Since the plan of action is to expand about some classical solution of the equations of motion, we need to derive them. First, for any $a = 1, \cdots, N$, we choose arbitrarily small functional variations of the independent fields ϕ^a and $\partial_\mu \phi^a$,[13]

$$\phi^a \longrightarrow \phi^a + \delta\phi^a, \qquad \partial_\mu \phi^a + \delta(\partial_\mu \phi^a) = \partial_\mu \phi^a + \partial_\mu (\delta\phi^a), \tag{C.12a}$$

up to the condition that they vanish when x is at infinity. Variations (C.12a) induce for the action (C.8b) the change

$$\begin{aligned}
\delta S :=& -\int \frac{d^d x}{\mathfrak{a}^{d-2}} \left[\partial_\mu \frac{\delta \mathcal{L}}{\delta(\partial_\mu \phi^a)} - \frac{\delta \mathcal{L}}{\delta \phi^a} \right] \delta\phi^a \\
=& -\int \frac{d^d x}{\mathfrak{a}^{d-2}} \left[\partial_\mu \left(\mathfrak{g}_{ab} \partial_\mu \phi^b \right) - \frac{1}{2} \partial_a \mathfrak{g}_{bc} \partial_\mu \phi^b \partial_\mu \phi^c \right] \delta\phi^a .
\end{aligned} \tag{C.12b}$$

The chain rule for differentiation delivers for the right-hand side

$$\int \frac{d^d x}{\mathfrak{a}^{d-2}} \left(\mathfrak{g}_{ab} \partial_\mu \partial_\mu \phi^b + \partial_c \mathfrak{g}_{ab} \partial_\mu \phi^b \partial_\mu \phi^c - \frac{1}{2} \partial_a \mathfrak{g}_{bc} \partial_\mu \phi^b \partial_\mu \phi^c \right) \delta\phi^a . \tag{C.12c}$$

In turn, relabeling of summation indices delivers for the right-hand side

$$\int \frac{d^d x}{\mathfrak{a}^{d-2}} \left(\mathfrak{g}_{ab} \partial_\mu \partial_\mu \phi^b + \frac{1}{2} \partial_c \mathfrak{g}_{ab} \partial_\mu \phi^b \partial_\mu \phi^c + \frac{1}{2} \partial_b \mathfrak{g}_{ac} \partial_\mu \phi^b \partial_\mu \phi^c - \frac{1}{2} \partial_a \mathfrak{g}_{bc} \partial_\mu \phi^b \partial_\mu \phi^c \right) \delta\phi^a . \tag{C.12d}$$

Since the infinitesimal $\delta\phi^a$ is arbitrarily chosen, there follows, with the help of Eq. (C.10), the N functional derivatives

$$\frac{\delta S}{\delta \phi_a} = - \left(\partial_\mu \partial_\mu \phi^a + \left\{ {a \atop bc} \right\} \partial_\mu \phi^b \partial_\mu \phi^c \right), \tag{C.13a}$$

[13]The equality $\delta(\partial_\mu \phi^a) = \partial_\mu (\delta\phi^a)$ is only true for infinitesimal variations.

where $\left\{ {a \atop bc} \right\}$ is called the Christoffel symbol and defined to be[14]

$$\left\{ {a \atop bc} \right\} := \frac{1}{2} \mathfrak{g}^{ad} \Big(\partial_b \mathfrak{g}_{dc} + \partial_c \mathfrak{g}_{db} - \partial_d \mathfrak{g}_{bc} \Big). \tag{C.13b}$$

The metric is flat whenever it is independent of ϕ, in which case the Christoffel symbol vanishes. The classical equations of motion are obtained by demanding that S be extremal, i.e., are given by the saddle-point equations

$$0 = \partial_\mu \partial_\mu \phi^a + \left\{ {a \atop bc} \right\} \partial_\mu \phi^b \partial_\mu \phi^c. \tag{C.14}$$

For a flat metric they are just the equations of motion of N independent, massless, and free bosonic fields.

Next, we consider the curve C in $\mathbb{R}^d$ parametrized by

$$\begin{aligned} x : [0,1] &\longrightarrow \mathbb{R}^d \\ t &\longrightarrow x(t) \end{aligned} \tag{C.15a}$$

between the end points $x(0)$ and $x(1)$. We then associate to the curve (C.15a) the curve $C_{\mathfrak{M}}$ in $\mathfrak{M}$ between the end points $\lambda(0)$ and $\lambda(1)$ through

$$\lambda^a(t) := \phi^a\big(x(t)\big), \qquad \dot{\lambda}^a \equiv \frac{\mathrm{d}\lambda^a}{\mathrm{d}t}, \qquad a = 1, \cdots, N. \tag{C.15b}$$

The *arclength* $L[C_{\mathfrak{M}}]$ of the curve $C_{\mathfrak{M}}$ is then defined to be

$$L[C_{\mathfrak{M}}] := \int\limits_0^1 \mathrm{d}t \sqrt{\mathfrak{g}_{ab}\, \dot{\lambda}^a\, \dot{\lambda}^b} \equiv \int\limits_{C_{\mathfrak{M}}} \mathrm{d}s, \tag{C.16a}$$

where the *arclength line element* ds is defined by

$$\left(\frac{\mathrm{d}s}{\mathrm{d}t} \right)^2 := \mathfrak{g}_{ab} \dot{\lambda}^a \dot{\lambda}^b \Longleftrightarrow (\mathrm{d}s)^2 := \mathfrak{g}_{ab}\, \mathrm{d}\lambda^a\, \mathrm{d}\lambda^b. \tag{C.16b}$$

The arclength (C.16a) can also be thought of as a functional restricted to any smooth path with given end points, in which case the extremal paths satisfy the N geodesic differential equations

$$0 = \ddot{\lambda}^a + \left\{ {a \atop bc} \right\} \dot{\lambda}^b\, \dot{\lambda}^c, \qquad a = 1, \cdots, N. \tag{C.17}$$

As we shall see next, the arclength $L[C_{\mathfrak{M}}]$ of the curve $C_{\mathfrak{M}}$ is invariant under reparametrization of the manifold $\mathfrak{M}$. In this sense, the arclength $L[C_{\mathfrak{M}}]$ is a geometrical invariant.

[14] Observe that any of these components of the Christoffel symbol is unchanged under $\mathfrak{g}_{ab} \to -\mathfrak{g}_{ab}$.

C.5 Riemann, Ricci, and scalar curvature tensors

In Eq. (C.8), we have chosen a specific parametrization of the Riemannian manifold $\mathfrak{M}$ in terms of the N coordinates ϕ^a. In this section, we are going to investigate the consequences of demanding that the theory (C.8) be invariant under the reparametrization

$$\phi^a = \phi^a(\phi'), \tag{C.18}$$

in terms of N coordinates ϕ'^b.

The transformation law of $\partial_\mu \phi^a$ under the reparametrization (C.18) is

$$\partial_\mu \phi^a = T^a_c\, \partial_\mu \phi'^c, \qquad T^a_c(\phi') := \frac{\partial \phi^a}{\partial \phi'^c}. \tag{C.19}$$

Invariance of the Lagrangian (C.8c) under the reparametrization (C.18),

$$\mathcal{L}(\phi) = \mathcal{L}(\phi'), \tag{C.20}$$

is achieved if and only if the metric transforms as

$$\mathfrak{g}_{ab}(\phi) = \left(T^{-1}\right)^e_a \left(T^{-1}\right)^f_b\, \mathfrak{g}'_{ef}(\phi'), \tag{C.21}$$

where

$$T^a_b(\phi') = \frac{\partial \phi^a}{\partial \phi'^b}, \qquad T^a_b \left(T^{-1}\right)^b_c(\phi') = \delta^a_c \implies \left(T^{-1}\right)^b_c(\phi') = \frac{\partial \phi'^b}{\partial \phi^c}. \tag{C.22}$$

If so, $\sqrt{||\mathfrak{g}(\phi)||}$ transforms as

$$\begin{aligned} \sqrt{||\mathfrak{g}(\phi)||} &= \sqrt{||T^{-1}(\phi')|| \times ||T^{-1}(\phi')|| \times ||\mathfrak{g}'(\phi')||} \\ &= \frac{\sqrt{||\mathfrak{g}'(\phi')||}}{||T(\phi')||}, \end{aligned} \tag{C.23}$$

while $\prod_a \mathrm{d}\phi^a$ transforms as

$$\prod_a \mathrm{d}\phi^a = ||T(\phi')|| \prod_b \mathrm{d}\phi'^b. \tag{C.24}$$

We conclude that the measure (C.8d) transforms as

$$\mathcal{D}[\phi] = \mathcal{D}[\phi'], \tag{C.25}$$

under the reparametrization (C.18). We have thus proved that the transformation laws (C.19) and (C.21) under the reparametrization (C.18) guarantee both the (classical) invariance (C.20) and the (quantum) invariance (C.25). As a byproduct we have also proved that the infinitesimal and finite arclengths (C.16) are invariant under the reparametrization (C.18).

Transformation laws (C.20) and (C.25) define scalar quantities under the reparametrization (C.18). Transformation law (C.19) defines a contravariant vector under the reparametrization (C.18). From a contravariant vector V^a, which transforms as

$$V^a = T^a_{\bar{a}}\, V'^{\bar{a}}, \tag{C.26}$$

under the reparametrization (C.18), one defines a covariant vector

$$V_b := \mathfrak{g}_{bc}\, V^c, \tag{C.27}$$

which must then transform as

$$\begin{aligned} V_b &= \left(T^{-1}\right)^{\bar{b}}_{\ b} \left(T^{-1}\right)^{\bar{c}}_{\ c} T^{c}_{\ \bar{c}}\, \mathfrak{g}'_{\bar{b}\bar{c}}\, V'^{\bar{c}} \\ &= \left(T^{-1}\right)^{\bar{b}}_{\ b} \mathfrak{g}'_{\bar{b}\bar{c}}\, V'^{\bar{c}} \\ &= \left(T^{-1}\right)^{\bar{b}}_{\ b} V'_{\bar{b}} \end{aligned} \tag{C.28}$$

under the reparametrization (C.18). An example of a covariant vector is $\partial_a \mathcal{L}$ since it transforms as

$$\partial_a \mathcal{L} = \left(T^{-1}\right)^{b}_{\ a} \partial'_b \mathcal{L}, \qquad \left(T^{-1}\right)^{b}_{\ a} := \frac{\partial \phi'^b}{\partial \phi^a}, \qquad \partial'_b \mathcal{L} := \frac{\partial \mathcal{L}}{\partial \phi'^b}, \tag{C.29}$$

under the reparametrization (C.18). As a corollary of transformation laws (C.26) and (C.28), it follows that

$$V^a\, W_a = V^a\, \mathfrak{g}_{ab}\, W^b = V_a\, W^a \tag{C.30}$$

is a scalar for any pair V^a and W_a of contravariant and covariant vectors. Transformation law (C.21) defines a covariant tensor of rank 2 under the reparametrization (C.18). An object $V^{a_1\cdots a_m}_{b_1\cdots b_n}$ which transforms like the tensor product of m contravariant vectors and n covariant vectors,

$$V^{a_1\cdots a_m}_{b_1\cdots b_n} = T^{a_1}_{c_1} \cdots T^{a_m}_{c_m} \left(T^{-1}\right)^{d_1}_{b_1} \cdots \left(T^{-1}\right)^{d_n}_{b_n} V'^{c_1\cdots c_m}_{d_1\cdots d_n} \tag{C.31}$$

defines a tensor of rank $\binom{m}{n}$ under the reparametrization (C.18) [see Eq. (C.7)].

The derivative $\partial_a V^b$ of a contravariant vector V^b does not transform as a tensor of rank $\binom{1}{1}$ under the reparametrization (C.18). Rather it transforms as

$$\begin{aligned} \partial_a V^b &= \left(T^{-1}\right)^{\bar{a}}_{\ a} \partial'_{\bar{a}} \left(T^{b}_{\bar{b}}\, V'^{\bar{b}}\right) \\ &= \left(T^{-1}\right)^{\bar{a}}_{\ a} T^{b}_{\bar{b}}\, \partial'_{\bar{a}} V'^{\bar{b}} + \left(T^{-1}\right)^{\bar{a}}_{\ a} \partial'_{\bar{a}} T^{b}_{\bar{b}}\, V'^{\bar{b}}. \end{aligned} \tag{C.32}$$

The *covariant derivative* $\nabla_a V^b$ is defined by the condition that it transforms like a tensor of type $\binom{1}{1}$ under the reparametrization (C.18),

$$\nabla_a V^b := \left(T^{-1}\right)^{\bar{a}}_{\ a} T^{b}_{\bar{b}}\, \nabla'_{\bar{a}} V'^{\bar{b}}. \tag{C.33}$$

To verify that such an object does indeed exist, write

$$\nabla_a V^b \equiv \partial_a V^b + \Gamma^b_{\ ac}\, V^c. \tag{C.34}$$

The object $\Gamma^b_{\ ac}$ is called a *linear or affine connection* when it exists. The transformation law obeyed by the affine connection $\Gamma^b_{\ ac}$ (when it exists) under the reparametrization (C.18) is deduced in two steps. First, Eqs. (C.33) and (C.34) deliver

$$\begin{aligned} \nabla_a V^b &= \left(T^{-1}\right)^{\bar{a}}_{\ a} T^{b}_{\bar{b}}\, \nabla'_{\bar{a}} V'^{\bar{b}} \\ &\equiv \left(T^{-1}\right)^{\bar{a}}_{\ a} T^{b}_{\bar{b}} \left(\partial'_{\bar{a}} V'^{\bar{b}} + \Gamma'^{\bar{b}}_{\ \bar{a}\bar{c}}\, V'^{\bar{c}}\right). \end{aligned} \tag{C.35}$$

Second, Eq. (C.32) delivers

$$\begin{aligned}\nabla_a V^b &\equiv \partial_a V^b + \Gamma^b_{\ ac} V^c \\ &= \left(T^{-1}\right)^{\bar a}_{a} T^b_{\bar b}\, \partial'_{\bar a} V'^{\bar b} + \left(T^{-1}\right)^{\bar a}_{a} \partial'_{\bar a} T^b_{\bar b}\, V'^{\bar b} + \Gamma^b_{\ ac} T^c_{\bar c} V'^{\bar c}. \end{aligned} \tag{C.36}$$

Comparing the right-hand sides of Eqs. (C.35) and (C.36) gives

$$\left(T^{-1}\right)^{\bar a}_{a} T^b_{\bar b}\, \Gamma'^{\bar b}_{\ \bar a \bar c} = \left(T^{-1}\right)^{\bar a}_{a} \partial'_{\bar a} T^b_{\bar c} + \Gamma^b_{\ ac} T^c_{\bar c}, \tag{C.37}$$

since V^b is arbitrary. Multiplication of Eq. (C.37) by $T^a_{\bar{\bar a}} \left(T^{-1}\right)^{\bar{\bar b}}_{b}$ and summation over a and b gives the final transformation law

$$\begin{aligned}\Gamma'^{\bar b}_{\ \bar a \bar c} &= \left(T^{-1}\right)^{\bar b}_{b} \partial'_{\bar a} T^b_{\bar c} + T^a_{\bar a} \left(T^{-1}\right)^{\bar b}_{b} T^c_{\bar c} \Gamma^b_{\ ac} \\ &= \frac{\partial \phi'^{\bar b}}{\partial \phi^b} \frac{\partial^2 \phi^b}{\partial \phi'^{\bar a} \partial \phi'^{\bar c}} + \frac{\partial \phi^a}{\partial \phi'^{\bar a}} \frac{\partial \phi'^{\bar b}}{\partial \phi^b} \frac{\partial \phi^c}{\partial \phi'^{\bar c}} \Gamma^b_{\ ac}, \end{aligned} \tag{C.38}$$

obeyed by the affine connection, provided it exists. The inhomogeneous transformation law (C.38) immediately implies two important properties of affine connections:

(1) If $\Gamma^a_{\ bc}$ and $\widetilde{\Gamma}^a_{\ bc}$ are two affine connections obeying the transformation law (C.38) under the reparametrization (C.18), then their difference is a tensor of type $\binom{1}{2}$.
(2) If $\Gamma^a_{\ bc}$ is an affine connection obeying the transformation law (C.38) under the reparametrization (C.18), and if $t^a_{\ bc}$ is a tensor of type $\binom{1}{2}$, then $\Gamma^a_{\ bc} + t^a_{\ bc}$ is an affine connection obeying the transformation law (C.38) under the reparametrization (C.18).

On the way to proving the existence of the affine connection and thus of the covariant derivative of a contravariant vector we need to extend the definition of the action of the covariant derivative to arbitrary linear combinations of tensors. The action of the covariant derivative on an arbitrary tensor $V^{a_1\cdots a_m}_{b_1\cdots b_n}$ of type $\binom{m}{n}$ is to produce a tensor of type $\binom{m}{n+1}$ given by

$$\begin{aligned}\nabla_a V^{a_1\cdots a_m}_{b_1\cdots b_n} :=& \partial_a V^{a_1\cdots a_m}_{b_1\cdots b_n} \\ &+ \sum_{i=1}^{m} \Gamma^{a_i}_{\ a\bar a} V^{a_1\cdots a_{i-1}\bar a\, a_{i+1}\cdots a_m}_{b_1\cdots b_n} - \sum_{j=1}^{n} \Gamma^{\bar b}_{\ ab_j} V^{a_1\cdots a_m}_{b_1\cdots b_{j-1}\bar b\, b_{j+1}\cdots b_n}, \end{aligned} \tag{C.39a}$$

where it is understood that the covariant derivative is simply the usual partial derivative

$$\nabla_a f = \partial_a f, \tag{C.39b}$$

when applied to a scalar function f. Second, Eq. (C.39) is supplemented by the condition that it remains valid if indices are contracted. Third, Eq. (C.39) is supplemented by the condition that the covariant derivative acts linearly on linear combinations of tensors. The claim that the transformation law of the covariant derivative (C.39) under the reparametrization (C.18), if it exists, is that of a tensor of type $\binom{m}{n+1}$ then follows from the fact that an arbitrary tensor of type $\binom{m}{n}$ is nothing but the direct product of m contravariant vectors and n covariant vectors together with Eq. (C.33) and its counterpart for covariant vectors. Finally, we must make the affine connection compatible with the metric by demanding that the scalar product defined in Eq. (C.56a) transforms covariantly, i.e., as in Eq. (C.56b). This condition is achieved if

$$\begin{aligned} 0 &= \nabla_a \mathfrak{g}_{bc} \\ &= \partial_a \mathfrak{g}_{bc} - \Gamma^{\bar{b}}{}_{ab}\, \mathfrak{g}_{\bar{b}c} - \Gamma^{\bar{c}}{}_{ac}\, \mathfrak{g}_{b\bar{c}}, \end{aligned} \tag{C.40a}$$

$$\begin{aligned} 0 &= \nabla_b \mathfrak{g}_{ca} \\ &= \partial_b \mathfrak{g}_{ca} - \Gamma^{\bar{c}}{}_{bc}\, \mathfrak{g}_{\bar{c}a} - \Gamma^{\bar{a}}{}_{ba}\, \mathfrak{g}_{c\bar{a}}, \end{aligned} \tag{C.40b}$$

$$\begin{aligned} 0 &= \nabla_c \mathfrak{g}_{ab} \\ &= \partial_c \mathfrak{g}_{ab} - \Gamma^{\bar{a}}{}_{ca}\, \mathfrak{g}_{\bar{a}b} - \Gamma^{\bar{b}}{}_{cb}\, \mathfrak{g}_{a\bar{b}}. \end{aligned} \tag{C.40c}$$

All three equations are here related by cyclic permutations of a, b, c.

The compatibility condition (C.40) can be used to construct the affine connection explicitly. The combination −(C.40a) +(C.40b) +(C.40c) yields, owing to the symmetry of the metric tensor,

$$\begin{aligned} 0 = &-\partial_a \mathfrak{g}_{bc} + \underline{\Gamma^{d}{}_{ab}\, \mathfrak{g}_{dc}} + \underline{\underline{\Gamma^{d}{}_{ac}\, \mathfrak{g}_{db}}} \\ &+\partial_b \mathfrak{g}_{ca} - \underline{\underline{\underline{\Gamma^{d}{}_{bc}\, \mathfrak{g}_{da}}}} - \underline{\Gamma^{d}{}_{ba}\, \mathfrak{g}_{dc}} \\ &+\partial_c \mathfrak{g}_{ba} - \underline{\underline{\Gamma^{d}{}_{ca}\, \mathfrak{g}_{db}}} - \underline{\underline{\underline{\Gamma^{d}{}_{cb}\, \mathfrak{g}_{da}}}} \end{aligned} \tag{C.41}$$

Introducing the notation

$$\Gamma^{a}{}_{\{bc\}} := \frac{1}{2}\left(\Gamma^{a}{}_{bc} + \Gamma^{a}{}_{cb}\right), \qquad \Gamma^{a}{}_{[bc]} := \frac{1}{2}\left(\Gamma^{a}{}_{bc} - \Gamma^{a}{}_{cb}\right), \tag{C.42}$$

we can regroup underlined terms in Eq. (C.41) to get

$$0 = -\partial_a \mathfrak{g}_{bc} + \partial_b \mathfrak{g}_{ca} + \partial_c \mathfrak{g}_{ba} + \underline{2\Gamma^{d}{}_{[ab]}\, \mathfrak{g}_{dc}} + \underline{\underline{2\Gamma^{d}{}_{[ac]}\, \mathfrak{g}_{db}}} - \underline{\underline{\underline{2\Gamma^{d}{}_{\{bc\}}\, \mathfrak{g}_{da}}}}. \tag{C.43}$$

Multiplication by $\mathfrak{g}^{a\bar{a}}$ and summation over a turns Eq. (C.43) into

$$\Gamma^{\bar{a}}{}_{\{bc\}} = \left\{ {\bar{a} \atop bc} \right\} + \mathfrak{g}^{\bar{a}a}\left(\Gamma^{d}{}_{[ab]}\, \mathfrak{g}_{dc} + \Gamma^{d}{}_{[ac]}\, \mathfrak{g}_{db}\right), \tag{C.44}$$

where we already encountered the Christoffel symbol $\left\{ {\bar{a} \atop bc} \right\}$ in Eq. (C.13b). We conclude that an affine connection compatible with the metric is given by

$$\begin{aligned} \Gamma^{a}{}_{bc} &= \Gamma^{a}{}_{\{bc\}} + \Gamma^{a}{}_{[bc]} \\ &= \left\{ {a \atop bc} \right\} + \mathfrak{g}^{a\bar{a}}\left(\Gamma^{d}{}_{[\bar{a}b]}\, \mathfrak{g}_{dc} + \Gamma^{d}{}_{[\bar{a}c]}\, \mathfrak{g}_{db}\right) + \Gamma^{a}{}_{[bc]}. \end{aligned} \tag{C.45}$$

The antisymmetric part

$$T^{a}_{bc} := 2\,\Gamma^{a}{}_{[bc]} \tag{C.46}$$

of the affine connection is called the *torsion tensor.* The terminology tensor is here justified, for T^a_{bc} transforms like a tensor of type $\binom{1}{2}$ as antisymmetrization of Eq. (C.38) kills the inhomogeneous term.

An affine connection whose torsion tensor vanishes everywhere on the manifold is called the *Levi-Civita connection,* in which case

$$\begin{aligned} \Gamma^a{}_{bc} &= \left\{ {a \atop bc} \right\} \\ &= \frac{1}{2} \mathfrak{g}^{ad} \Big(\partial_b \mathfrak{g}_{dc} + \partial_c \mathfrak{g}_{db} - \partial_d \mathfrak{g}_{bc} \Big). \end{aligned} \tag{C.47}$$

The term

$$K^a{}_{bc} := \mathfrak{g}^{a\bar{a}} \left(\Gamma^d{}_{[\bar{a}b]}\, \mathfrak{g}_{dc} + \Gamma^d{}_{[\bar{a}c]}\, \mathfrak{g}_{db} \right) + \Gamma^a{}_{[bc]} \tag{C.48}$$

in Eq. (C.45) is called the *contorsion tensor.* With the help of Eq. (C.46), it turns into

$$\begin{aligned} K^a{}_{bc} &= \frac{1}{2} \Big(\mathfrak{g}^{a\bar{a}}\, T^d_{\bar{a}b}\, \mathfrak{g}_{dc} + \mathfrak{g}^{a\bar{a}}\, T^d_{\bar{a}c}\, \mathfrak{g}_{db} + T^a_{bc} \Big) \\ &\equiv \frac{1}{2} \Big(T_c{}^a{}_b + T_b{}^a{}_c + T^a_{bc} \Big). \end{aligned} \tag{C.49}$$

By construction, the contorsion tensor is of type $\binom{1}{2}$ and it vanishes if the torsion tensor vanishes. A *symmetric affine connection* is an affine connection with vanishing torsion tensor.

To sum up, we have proved the *fundamental theorem of Riemannian geometry.* On a Riemannian manifold $(\mathfrak{M}, \mathfrak{g})$, there exists a unique symmetric connection which is compatible with the metric $\mathfrak{g}$. It is given by the Levi-Civita connection (C.47).

We now return to the definition (C.34) of the covariant derivative $\nabla_a V^b$ of an arbitrary contravariant vector V^b. Let W^a be another arbitrary contravariant vector from which we construct

$$W^a\, \nabla_a V^b = W^a \partial_a\, V^b + W^a \Gamma^b{}_{ac}\, V^c. \tag{C.50}$$

Next, we choose the contravariant vector W^a to be the tangent vector to the curve

$$\begin{aligned} \phi : [0,1] &\longrightarrow \mathfrak{M} \\ t &\longrightarrow \phi(t), \end{aligned} \tag{C.51a}$$

i.e.,

$$W^a = \frac{\mathrm{d}\phi^a}{\mathrm{d}t}. \tag{C.51b}$$

With this choice, Eq. (C.50) becomes

$$\begin{aligned} W^a \nabla_a V^b &= \frac{\mathrm{d}\phi^a}{\mathrm{d}t} \frac{\partial V^b}{\partial \phi^a} + \Gamma^b{}_{ac} \frac{\mathrm{d}\phi^a}{\mathrm{d}t} V^c \\ &= \frac{\mathrm{d}V^b}{\mathrm{d}t} + \Gamma^b{}_{ac} \frac{\mathrm{d}\phi^a}{\mathrm{d}t} V^c. \end{aligned} \tag{C.52}$$

The contravariant vector V^a is said to be *parallel transported* along the curve (C.51a) with tangent vector (C.51b) when the N equations

$$\begin{aligned} 0 &= W^a \nabla_a V^b \\ &= \frac{\mathrm{d}V^b}{\mathrm{d}t} + \Gamma^b{}_{ac} \frac{\mathrm{d}\phi^a}{\mathrm{d}t} V^c \end{aligned} \tag{C.53}$$

are satisfied. If it is the tangent vector W^b itself which is parallel transported along the curve $\phi(t)$, it must satisfy

$$\begin{aligned} 0 &= W^a \nabla_a W^b \\ &= \frac{\mathrm{d}W^b}{\mathrm{d}t} + \Gamma^b{}_{ac} \frac{\mathrm{d}\phi^a}{\mathrm{d}t} W^c \\ &= \frac{\mathrm{d}^2\phi^b}{\mathrm{d}t^2} + \Gamma^b{}_{ac} \frac{\mathrm{d}\phi^a}{\mathrm{d}t} \frac{\mathrm{d}\phi^c}{\mathrm{d}t}. \end{aligned} \tag{C.54}$$

These are nothing but the N geodesic equations (C.17) when the affine connection is restricted to the Levi-Civita connection. A geodesic can thus be interpreted as a curve with the property that its tangent vector is parallel transported along itself in accordance with the intuition of a straight line being the shortest path between two points in Euclidean space.

The metric compatibility condition (C.40) is related to parallel transport in the following manner. Let W^a be the tangent vector to the arbitrarily chosen curve (C.51) and let X^b and Y^c be arbitrarily chosen contravariant vectors which are parallel transported with respect to W^a,

$$0 = W^a \nabla_a X^b, \qquad 0 = W^a \nabla_a Y^c. \tag{C.55}$$

We now require that the scalar product

$$X^b Y_b = X^b \mathfrak{g}_{bc} Y^c \tag{C.56a}$$

is covariantly constant as defined by the condition

$$\begin{aligned} 0 &= W^a \nabla_a \left(X^b Y_b\right) \\ &= \mathfrak{g}_{bc} Y^c W^a \nabla_a X^b + W^a X^b Y^c \nabla_a \mathfrak{g}_{bc} + X^b \mathfrak{g}_{bc} W^a \nabla_a Y^c \\ &= W^a X^b Y^c \nabla_a \mathfrak{g}_{bc}. \end{aligned} \tag{C.56b}$$

Condition (C.55) was used to reach the last line, while the penultimate line is a consequence of the definition of the covariant derivative. The metric compatibility condition (C.40) is seen to follow from the arbitrariness of W^a, X^b, and Y^c.

If a vector is parallel transported along different curves between the same initial and final points the resulting vectors are curve dependent in general. This is most evidently seen by considering two antipodal points on the equator of the sphere and connecting them along the parallel or the meridian passing through them. The *Riemann curvature tensor* is a covariant measure of this difference. The Riemann curvature tensor is defined by the action

$$[\nabla_a, \nabla_b] V^c = R^c{}_{dab} V^d - T^d{}_{ab} \nabla_d V^c, \tag{C.57a}$$

on an arbitrary contravariant vector V^c, i.e., in components,

$$R^c{}_{dab} = \partial_a \Gamma^c{}_{bd} - \partial_b \Gamma^c{}_{ad} + \Gamma^c{}_{ae}\,\Gamma^e{}_{bd} - \Gamma^c{}_{be}\,\Gamma^e{}_{ad}. \tag{C.57b}$$

Equation (C.57a) implies that:

(1) The Riemann curvature tensor $R^c{}_{dab}$ is a tensor of type $\binom{1}{3}$.
(2) The Riemann curvature tensor $R^c{}_{dab}$ is antisymmetric in the indices a and b,

$$R^c{}_{dab} = -R^c{}_{dba}. \tag{C.58}$$

(3) For the Levi-Civita connection, the Jacobi identity for commutators $[A,[B,C]]+[B,[C,A]]+[C,[A,B]]=0$ implies the Bianchi identity

$$0 = \nabla_a R^e{}_{dbc} + \nabla_b R^e{}_{dca} + \nabla_c R^e{}_{dab}. \tag{C.59}$$

For the Levi-Civita connection, the Riemann curvature takes the explicit form

$$R^c{}_{dab} = \mathfrak{g}^{c\bar{c}} R_{\bar{c}dab}, \tag{C.60a}$$

with

$$\begin{aligned}
R_{\bar{c}dab} = {}& \frac{1}{2}\left(\partial_a \partial_d \mathfrak{g}_{\bar{c}b} - \partial_a \partial_{\bar{c}} \mathfrak{g}_{bd} - \partial_b \partial_d \mathfrak{g}_{\bar{c}a} + \partial_b \partial_{\bar{c}} \mathfrak{g}_{ad}\right) \\
& -\frac{1}{4}\left(\partial_a \mathfrak{g}_{m\bar{c}} + \partial_{\bar{c}} \mathfrak{g}_{ma} - \partial_m \mathfrak{g}_{a\bar{c}}\right) \mathfrak{g}^{mn} \left(\partial_b \mathfrak{g}_{nd} + \partial_d \mathfrak{g}_{nb} - \partial_n \mathfrak{g}_{bd}\right) \\
& +\frac{1}{4}\left(\partial_b \mathfrak{g}_{m\bar{c}} + \partial_{\bar{c}} \mathfrak{g}_{mb} - \partial_m \mathfrak{g}_{b\bar{c}}\right) \mathfrak{g}^{mn} \left(\partial_a \mathfrak{g}_{nd} + \partial_d \mathfrak{g}_{na} - \partial_n \mathfrak{g}_{ad}\right).
\end{aligned} \tag{C.60b}$$

Contraction of the pair ${}^{c}{}_{\cdot\, a\,\cdot}$ of indices in the Riemann curvature tensor defines the *Ricci tensor*,

$$\begin{aligned}
R_{db} &= \delta^a{}_c\, R^c{}_{dab} \\
&= \partial_a \Gamma^a{}_{bd} - \partial_b \Gamma^a{}_{ad} + \Gamma^a{}_{ae}\,\Gamma^e{}_{bd} - \Gamma^a{}_{be}\,\Gamma^e{}_{ad}.
\end{aligned} \tag{C.61}$$

Of course, we could have equally well chosen to contract the pair ${}^{c}{}_{\cdot\,\cdot\, b}$ of indices in the Riemann curvature tensor to obtain

$$\begin{aligned}
\delta^b{}_c\, R^c{}_{dab} &= R^c{}_{dac} \\
&= \partial_a \Gamma^c{}_{cd} - \partial_c \Gamma^c{}_{ad} + \Gamma^c{}_{ae}\Gamma^e{}_{cd} - \Gamma^c{}_{ce}\Gamma^e{}_{ad} \\
&= -R_{\;da}.
\end{aligned} \tag{C.62}$$

This would give a Ricci tensor with the opposite sign convention. The choice (C.61) is made so that the Ricci tensor of the surface of a unit sphere can be chosen locally to be the unit matrix up to a positive normalization constant.

Contracting the remaining two indices of the Ricci tensor defines the *scalar curvature*

$$R := R_{ab}\,\mathfrak{g}^{ba}. \tag{C.63}$$

One verifies that the Ricci tensor is a symmetric tensor for the Levi-Civita connection. Observe that the Riemann tensor, the Ricci tensor, and the scalar curvature transform like

$$R^c{}_{dab} = +R^c{}_{dab}, \qquad R_{db} \to +R_{db}, \qquad R \to -R, \tag{C.64}$$

respectively, under $\mathfrak{g}_{ab} \to -\mathfrak{g}_{ab}$, see footnote 14.

C.6 Normal coordinates and vielbeins for NLσM

C.6.1 *The background-field method*

Only quantities intrinsic to the Riemannian manifold $(\mathfrak{M}, \mathfrak{g})$ that defines the target space of the NLσM (C.8) are physical. Any choice of local coordinate system can introduce unphysical degrees of freedom since the theory is invariant under reparametrization whereas the coordinates (ϕ^a) are not. The background-field method is aimed at handling this complication.

The background-field method applied to the NLσM consists in decomposing the components $\phi^a(x)$, with $a = 1, \cdots, N$, of the contravariant vector field ϕ in the action (C.8b) into two fields ψ and π according to the additive rule

$$\begin{aligned}
\phi^a(x) &= \int\limits_{|k|<\Lambda} \frac{\mathrm{d}^d k}{(2\pi)^d} e^{+\mathrm{i}kx} \phi^a(k) \\
&= \underbrace{\int\limits_{|k|<\Lambda-\mathrm{d}\Lambda} \frac{\mathrm{d}^d k}{(2\pi)^d} e^{+\mathrm{i}kx} \phi^a(k)}_{=:\psi^a(x)} + \underbrace{\int\limits_{\Lambda-\mathrm{d}\Lambda<|k|<\Lambda} \frac{\mathrm{d}^d k}{(2\pi)^d} e^{+\mathrm{i}kx} \phi^a(k)}_{=:\pi^a(x)} \\
&= \psi^a(x) + \pi^a(x),
\end{aligned} \tag{C.65}$$

whereby $(\psi^a(x))$ is assumed to be a slowly varying solution to the classical equations of motion (C.14) that transforms like a contravariant vector and $(\pi^a(x))$ represents fast degrees of freedom. Here, Λ plays the role of an ultraviolet cut-off. Having identified the contravariant vectors $(\psi^a(x))$ and $(\phi^a(x))$ with two points on the Riemannian manifold $(\mathfrak{M}, \mathfrak{g})$, say $\mathfrak{p}$ and $\mathfrak{q}$, respectively, we cannot in general interpret their difference

$$\pi^a(x) = \phi^a(x) - \psi^a(x), \qquad a = 1, \cdots, N, \tag{C.66}$$

as the coordinates in $\mathbb{R}^N$ of some contravariant vector. The best we can do is to assume that the points $\mathfrak{p}$ and $\mathfrak{q}$ are close enough, i.e., the fields $(\pi^a(x))$ are "small" enough, for there to be a unique geodesic that connects them.

The renormalization program in the background-field method is carried out in two steps. First, the metric, Lagrangian, or, more generally, any function of the fields (ϕ^a) are Taylor expanded in powers of the fast degrees of freedom $(\pi^a(x))$. Second, an integration over the fast degrees of freedom $(\pi^a(x))$ in the partition function or in correlation functions is performed order by order in this expansion.

For example, when carried on the partition function (C.8), this program gives

$$\begin{aligned} Z &= \int \mathcal{D}[\psi] \exp\left(-S[\psi]\right), \\ S[\psi] &= \int \frac{\mathrm{d}^d x}{\mathfrak{a}^{d-2}} \mathcal{L}(\psi), \\ \mathcal{L}(\psi) &= \frac{1}{2}\left[\mathfrak{g}_{ab}(\psi) + T^{(1)}_{ab}(\psi) + T^{(2)}_{ab}(\psi) + \cdots\right] \partial_\mu \psi^a \, \partial_\mu \psi^b, \\ \mathcal{D}[\psi] &= \prod_{x\in\mathbb{R}^d} \sqrt{||\mathfrak{g}(\psi)||} \prod_a \mathrm{d}\psi^a(x). \end{aligned} \tag{C.67}$$

Here, the object

$$T_{ab}(\psi) := \mathfrak{g}_{ab}(\psi) + T^{(1)}_{ab}(\psi) + T^{(2)}_{ab}(\psi) + \cdots, \tag{C.68}$$

is a symmetric tensor of type $\binom{0}{2}$ which, as we shall verify explicitly to first order in the expansion, is an algebraic function of the curvature tensor and the covariant derivative for the Levi-Civita connection. Had we not chosen ψ to satisfy the classical equations of motion, we would need to account for the additional contribution

$$\delta\mathcal{L}(\psi) = \Gamma_a(\psi)\left(\partial_\mu\partial_\mu\psi^a + \left\{{a \atop bc}\right\}\partial_\mu\psi^b\partial_\mu\psi^c\right) \tag{C.69}$$

to the renormalization of the action [see Eq. (C.13a)] where $\Gamma_a(\psi)$ is typically non-covariant under reparametrization and can be expressed in terms of the Christoffel connection (C.13b).

However, integration over the fast fields π quickly becomes very tedious as the expansion of covariant quantities in powers of the components π^a is not manifestly covariant anymore. This difficulty can be overcome by expanding the fast degrees of freedom $\left(\pi^a(x)\right)$ as a power series of fields $\left(\xi^a(x)\right)$ that transform like contravariant vectors. This intermediate step can be done in a unique way once it is guaranteed that there is a unique geodesic connecting $\mathfrak{p} \sim \left(\psi^a(x)\right)$ to $\mathfrak{q} \sim \left(\phi^a(x)\right)$. The existence of a unique geodesic connecting $\mathfrak{p}$ to $\mathfrak{q}$ is closely related to the existence of normal coordinates, which we are going to define from a purely geometrical point of view in the following.

C.6.2 *A mathematical excursion*

Let $\mathfrak{p} \in \mathfrak{M}$ be an arbitrary point of the Riemannian manifold $(\mathfrak{M}, \mathfrak{g})$ defined by Eq. (C.8). Let $\mathfrak{U} \subset \mathfrak{M}$ be an open set of the manifold that contains $\mathfrak{p}$. Let $\xi : \mathfrak{U} \longrightarrow U$ be a smooth homeomorphism between the open set $\mathfrak{U} \subset \mathfrak{M}$ in the manifold and the open set $U \subset \mathbb{R}^N$ such that

$$\xi^{-1}(0) = \mathfrak{p}. \tag{C.70}$$

The pair $(\mathfrak{U}, \xi)$ is said to be a *coordinate system on* $\mathfrak{M}$ *with* $\xi^{-1}(0) = \mathfrak{p}$. This coordinate system is said to be *normal* with respect to $\mathfrak{p}$ if the inverse image under ξ of straight lines through the origin in $\mathbb{R}^N$ are geodesics on $\mathfrak{M}$ with respect to the Levi-Civita connection (C.47).

To explore the usefulness of this definition, let $\boldsymbol{w}$ be an arbitrary vector in $\mathbb{R}^N$. The set of points $t\,\boldsymbol{w}$ with $t \in \mathbb{R}$ defines a straight line through the origin of $\mathbb{R}^N$. We assume that the homeomorphic mapping $\xi : \mathfrak{U} \longrightarrow \mathbb{R}^N$ from the open set $\mathfrak{U}$ that contains $\mathfrak{p} \in \mathfrak{M}$ into $\mathbb{R}^N$ realizes a normal coordinate system. By definition, there must be a $\epsilon > 0$ such that any curve $C_{\boldsymbol{w}} : [-\epsilon, \epsilon] \longrightarrow \mathfrak{M}$ through $\mathfrak{p}$ defined by

$$C_{\boldsymbol{w}}(t) = \xi^{-1} \circ \left(w^1 t, \cdots, w^N t\right) \tag{C.71}$$

is a geodesic whose components in $\mathbb{R}^N$ labeled by $a = 1, \cdots, N$ obey the generic geodesic equations

$$0 = \ddot{C}^a + \Gamma^a{}_{bc}\,\dot{C}^b\,\dot{C}^c, \qquad \Gamma^a{}_{bc} = \Gamma^a{}_{cb} = \left\{ {a \atop bc} \right\}. \tag{C.72}$$

However, since the curve $t\boldsymbol{w}$ in $\mathbb{R}^N$ is linear, insertion of Eq. (C.71) into Eq. (C.72) brings about the simplification

$$0 = \Gamma^a{}_{bc}\big(C_{\boldsymbol{w}}(t)\big)\, w^b\, w^c, \qquad \forall \big(w^1, \cdots, w^N\big) \in \xi(\mathfrak{U}) \subset \mathbb{R}^N. \tag{C.73}$$

Equation (C.73) restricted to $t = 0$ implies

$$0 = \Gamma^a{}_{bc}(\mathfrak{p})\, w^b\, w^c, \qquad \forall \big(w^1, \cdots, w^N\big) \in \xi(\mathfrak{U}) \subset \mathbb{R}^N. \tag{C.74}$$

Since $\boldsymbol{w} \in \mathbb{R}^N$ is arbitrary, there follows

$$0 = \Gamma^a{}_{bc}(\mathfrak{p}), \tag{C.75}$$

for the Levi-Civita connection. In the normal coordinates with respect to $\mathfrak{p}$, the Riemann curvature tensor (C.57b) takes the simpler form

$$R^c{}_{dab}(\mathfrak{p}) = \partial_a \Gamma^c{}_{bd}(\mathfrak{p}) - \partial_b \Gamma^c{}_{ad}(\mathfrak{p}). \tag{C.76}$$

Evidently a non-vanishing curvature tensor at $\mathfrak{p}$ implies a non-vanishing derivative of the Levi-Civita connection at $\mathfrak{p}$ in the normal coordinates with respect to $\mathfrak{p}$. The Levi-Civita connection is, by this argument, not expected to vanish at a generic $\mathfrak{q} \neq \mathfrak{p}$ in $\mathfrak{M}$ when represented in the normal coordinates with respect to $\mathfrak{p}$. The covariant derivative (C.34) also takes the simpler form

$$\nabla_a V^b(\mathfrak{p}) = \partial_a V^b(\mathfrak{p}), \tag{C.77}$$

when evaluated at $\mathfrak{p}$ in the normal coordinates with respect to $\mathfrak{p}$.

Conversely, any geodesic through $\mathfrak{p}$ must be of the form (C.71) in the coordinate system normal with respect to $\mathfrak{p}$. To see this we need the so-called exponential map. For any $\mathfrak{p} \in \mathfrak{M}$ there must be a $\epsilon > 0$ and an open neighborhood $U_{\mathfrak{p}}$ of $0 \in \mathbb{R}^N \cong T_{\mathfrak{p}}\mathfrak{M}$ such that there is, for any $\boldsymbol{w} \in U_{\mathfrak{p}}$, a unique solution to Eq. (C.72)

$$\begin{aligned} C_{\boldsymbol{w}} : [-\epsilon, \epsilon] &\longrightarrow \mathfrak{M}, \\ t &\longmapsto C_{\boldsymbol{w}}(t) \end{aligned} \tag{C.78a}$$

with

$$C_{\boldsymbol{w}}(0) = \mathfrak{p}, \qquad \dot{C}_{\boldsymbol{w}}(0) = \boldsymbol{w}, \tag{C.78b}$$

and a smooth dependence of $C_{\boldsymbol{w}}$ on $\boldsymbol{w}$. The curve $C_{\boldsymbol{w}}$ is the geodesic through $\mathfrak{p}$ with tangent vector $\boldsymbol{w} \in T_{\mathfrak{p}}\mathfrak{M}$ at $\mathfrak{p}$. Since Eq. (C.72) is invariant under the affine (Galilean boost) transformation

$$t = A + B\,t' \qquad \forall A, B \in \mathbb{R}, \tag{C.79}$$

it follows that:

(1) The curve

$$C'_{w}:[-(\epsilon+A)/B,(\epsilon-A)/B]\longrightarrow\mathfrak{M},\qquad t'\longmapsto C'_{w}(t')=C_{w}(t),\tag{C.80a}$$

is a geodesic with

$$C'_{w}(-A/B)=\mathfrak{p},\qquad \left.\frac{\mathrm{d}C'_{w}}{\mathrm{d}t'}\right|_{t'=-A/B}=B\left.\frac{\mathrm{d}C_{w}}{\mathrm{d}t}\right|_{t=0}=Bw=\left.\frac{\mathrm{d}C_{Bw}}{\mathrm{d}t}\right|_{t=0}.\tag{C.80b}$$

(2) The domain of definition of C_{w} can always be extended to the interval $[-1,1]$ after proper rescaling of the open neighborhood $U_{\mathfrak{p}}\subset T_{\mathfrak{p}}\mathfrak{M}$.

Define the exponential mapping by

$$\mathrm{EXP}:U_{\mathfrak{p}}\longrightarrow\mathfrak{M},\qquad w\longmapsto\mathrm{EXP}(w)=C_{w}(1),\tag{C.81a}$$

where we note that, with the help of Eq. (C.80),

$$\mathrm{EXP}(tw)=C_{tw}(1)=C_{w}(t),\qquad \left.\frac{\mathrm{dEXP}(tw)}{\mathrm{d}t}\right|_{t=0}=w,\qquad 0\le t\le 1.\tag{C.81b}$$

The exponential mapping can be used to define a normal coordinate system on an open neighborhood of $\mathfrak{p}\in\mathfrak{M}$ through the definition

$$\xi(\mathrm{EXP}(w))=w.\tag{C.82}$$

By this definition,

$$\xi(C_{w}(t))=t\,w\tag{C.83}$$

parametrizes a straight line passing through the origin in $T_{\mathfrak{p}}\mathfrak{M}$ when $t=0$, i.e., $C_{w}(0)=\mathfrak{p}$. According to Eq. (C.83) the normal coordinates on an open neighborhood of $\mathfrak{p}\in\mathfrak{M}$ of a geodesic passing through $\mathfrak{p}$ are the coordinates of the vector tangent to this geodesic at $\mathfrak{p}$ with a magnitude $t\,|w|$ increasing linearly with t.

Normal coordinates are used in proving the following theorem: Any point of a Riemannian manifold has a neighborhood $\mathfrak{U}$ such that for any two points in $\mathfrak{U}$ there is a unique geodesic that joins the points and lies in $\mathfrak{U}$.

C.6.3 *Normal coordinates for NLσM*

We want to integrate over the fields $\pi^{a}(x)$ with $a=1,\cdots,N$ in the partition function

$$\begin{aligned}
Z&=\int\mathcal{D}[\psi,\pi]\,\exp\left(-S[\psi,\pi]\right),\\
S[\psi,\pi]&=\int\frac{\mathrm{d}^{d}x}{\mathfrak{a}^{d-2}}\,\mathcal{L}(\psi,\pi),\\
\mathcal{L}(\psi,\pi)&=\frac{1}{2}\mathfrak{g}_{ab}(\psi,\pi)\partial_{\mu}(\psi^{a}+\pi^{a})\,\partial_{\mu}(\psi^{b}+\pi^{b}),\\
\mathcal{D}[\psi,\pi]&=\prod_{x\in\mathbb{R}^{d}}\sqrt{||\mathfrak{g}(\psi,\pi)||}\,\prod_{a}\mathrm{d}(\psi^{a}(x)+\pi^{a}(x)),
\end{aligned}\tag{C.84a}$$

where we assume that:

(1) The coordinates (ψ^a) and $(\psi^a + \pi^a)$ describe two points $\mathfrak{p}$ and $\mathfrak{q}$, respectively, from the Riemannian manifold $(\mathfrak{M}, \mathfrak{g})$ which can be connected in a unique way by the geodesic

$$0 = \ddot{\lambda}^a + \Gamma^a_{\ bc}\dot{\lambda}^b\dot{\lambda}^c, \quad \lambda^a(0) = \psi^a, \quad \lambda^a(1) = \psi^a + \pi^a, \quad a = 1, \cdots, N, \tag{C.84b}$$

that lies in some open set $U \subset \mathbb{R}^N$ homeomorphic to the open neighborhood $\mathfrak{U}_\mathfrak{p}$ of $\mathfrak{p}$.

(2) The field $\psi : \mathbb{R}^d \longrightarrow \mathfrak{M}$ defined by $x \longrightarrow \psi(x)$ satisfies the classical equations of motion (C.14).

Given $(\psi^a(x))$ let $(\psi^a(x)+\pi^a(x))$ be some arbitrary point belonging to the open neighborhood $U \subset \mathbb{R}^N$ homeomorphic to $\mathfrak{U}_\mathfrak{p}$ and in which the geodesic (C.84b) lies. We begin by performing a Taylor expansion in powers of $0 \leq t \leq 1$ of Eq. (C.84b),

$$\begin{aligned} &\lambda^a(t) = \lambda^a(0) + \frac{1}{1!}\dot{\lambda}^a(0)\,t + \frac{1}{2!}\ddot{\lambda}^a(0)\,t^2 + \frac{1}{3!}\dddot{\lambda}^a(0)\,t^3 + \cdots, \\ &\lambda^a(1) = \psi^a + \pi^a, \qquad a = 1, \cdots, N. \end{aligned} \tag{C.85}$$

From the geodesic equations of motion (C.84b), we may express all coefficients of order $n > 1$ in terms of linear combinations of the Levi-Civita connection or its derivatives evaluated at $t = 0$. For examples,

$$\begin{aligned} \ddot{\lambda}^a &= -\Gamma^a_{\ bc}\dot{\lambda}^b\dot{\lambda}^c, \\ \dddot{\lambda}^a &= -\dot{\Gamma}^a_{\ bc}\dot{\lambda}^b\dot{\lambda}^c - \Gamma^a_{\ bc}\ddot{\lambda}^b\dot{\lambda}^c - \Gamma^a_{\ bc}\dot{\lambda}^b\ddot{\lambda}^c \\ &= -\frac{\partial\Gamma^a_{\ bc}}{\partial\lambda^d}\dot{\lambda}^d\dot{\lambda}^b\dot{\lambda}^c + \Gamma^a_{\ bc}\Gamma^b_{\ de}\dot{\lambda}^d\dot{\lambda}^e\dot{\lambda}^c + \Gamma^a_{\ bc}\dot{\lambda}^b\Gamma^c_{\ de}\dot{\lambda}^d\dot{\lambda}^e \\ &= -\frac{\partial\Gamma^a_{\ bc}}{\partial\lambda^d}\dot{\lambda}^d\dot{\lambda}^b\dot{\lambda}^c + \Gamma^a_{\ ec}\Gamma^e_{\ db}\dot{\lambda}^d\dot{\lambda}^b\dot{\lambda}^c + \Gamma^a_{\ be}\Gamma^e_{\ dc}\dot{\lambda}^b\dot{\lambda}^d\dot{\lambda}^c \\ &= -\underbrace{\left(\frac{\partial\Gamma^a_{\ bc}}{\partial\lambda^d} - \Gamma^a_{\ ec}\Gamma^e_{\ db} - \Gamma^a_{\ be}\Gamma^e_{\ dc}\right)}_{=:\Gamma^a_{\ dbc}}\dot{\lambda}^d\dot{\lambda}^b\dot{\lambda}^c \\ &= -\Gamma^a_{\ dbc}\dot{\lambda}^d\dot{\lambda}^b\dot{\lambda}^c. \end{aligned} \tag{C.86}$$

If we introduce the tangent vector

$$v^a := \dot{\lambda}^a(0), \tag{C.87a}$$

we have found the Taylor expansion

$$\begin{aligned} \lambda^a(t) &= \psi^a + v^a t - \sum_{n=2}^{\infty}\frac{1}{n!}\Gamma^a_{\ a_1 a_2\cdots a_n}(\psi)v^{a_1}v^{a_2}\cdots v^{a_n}t^n \\ &= \psi^a + v^a t - \sum_{n=2}^{\infty}\frac{1}{n!}\Gamma^a_{\ (a_1 a_2\cdots a_n)}(\psi)v^{a_1}v^{a_2}\cdots v^{a_n}t^n, \\ \lambda^a(1) &= \psi^a + \pi^a, \qquad a = 1, \cdots, N. \end{aligned} \tag{C.87b}$$

The coefficient $\Gamma^{a}{}_{a_1 a_2 \cdots a_n}(\psi)$ on the first line of this Taylor expansion is defined recursively by

$$\begin{aligned}\Gamma^{a}{}_{a_1 a_2 \cdots a_n}(\psi) :&= \nabla_{a_1} \Gamma^{a}{}_{a_2 a_3 \cdots a_n}(\psi)\\ &= \nabla_{a_1} \cdots \nabla_{a_{n-2}} \Gamma^{a}{}_{a_{n-1} a_n}(\psi),\end{aligned} \tag{C.87c}$$

for $n = 2, 3, \cdots$ with the seed $\Gamma^{a}{}_{bc}$ and the rule $\Gamma^{a}{}_{dbc} := \frac{\partial \Gamma^{a}{}_{bc}}{\partial \lambda^{d}} - \Gamma^{a}{}_{ec}\Gamma^{e}{}_{db} - \Gamma^{a}{}_{be}\Gamma^{e}{}_{dc}$. The operation $\nabla_{a_1}\Gamma^{a}{}_{a_2 a_3 \cdots a_n}(\psi)$ resembles the action of the covariant derivative defined in Eq. (C.39a) on tensor fields with, however, the caveat that only the lower indices of the symbols $\Gamma^{a}{}_{a_2 \cdots a_n}$ (they are not tensor fields) are operated on, i.e., the first summation is omitted on the right-hand side of Eq. (C.39a). The coefficient $\Gamma^{a}{}_{(a_1 a_2 \cdots a_n)}(\psi)$ on the second line of this Taylor expansion is defined by the symmetrization

$$\Gamma^{a}{}_{(a_1 a_2 \cdots a_n)}(\psi) := \frac{1}{n!} \sum_{\mathcal{P} \in \mathcal{S}_n} \Gamma^{a}{}_{a_{\mathcal{P}(1)} a_{\mathcal{P}(2)} \cdots a_{\mathcal{P}(n)}}(\psi). \tag{C.87d}$$

The permutation group of n objects is here denoted by $\mathcal{S}_n$. Equation (C.87) for $t = 1$ gives the Taylor expansion

$$\begin{aligned}\partial_{\mu}\left(\psi^{a} + \pi^{a}\right) = {}& \partial_{\mu}\psi^{a} + \partial_{\mu} v^{a}\\ & - \sum_{n=2}^{\infty} \frac{1}{n!} \Big[\partial_{\mu}\psi^{b} \partial_{b} \Gamma^{a}{}_{(a_1 a_2 \cdots a_n)}(\psi) v^{a_1} v^{a_2} \cdots v^{a_n}\\ & \qquad + \Gamma^{a}{}_{(a_1 a_2 \cdots a_n)}(\psi) \partial_{\mu} v^{a_1} v^{a_2} \cdots v^{a_n} + \cdots\\ & \qquad + \Gamma^{a}{}_{(a_1 a_2 \cdots a_n)}(\psi) v^{a_1} v^{a_2} \cdots \partial_{\mu} v^{a_n} \Big],\end{aligned} \tag{C.88}$$

which we will use shortly.

Next, we choose to parametrize the geodesic (C.84b) in terms of the normal coordinates (ξ^{a}) with respect to $\mathfrak{p} \equiv \psi(x)$ as defined in section C.6.3. We also add an overline on the expansion coefficients in Eq. (C.87), Christoffel-symbol, and tensors, etc., when using normal coordinates to represent them. If we introduce the tangent vector

$$\xi^{a} := \dot{\lambda}^{a}(0), \tag{C.89a}$$

then Eq. (C.87), when expressed in the normal coordinates with respect to the classical solution ψ of the equations of motion evaluated at x, becomes[15]

$$\begin{aligned}\overline{\lambda}^{a}(t) &= \overline{\psi}^{a} + \xi^{a} t - \sum_{n=2}^{\infty} \frac{1}{n!} \overline{\Gamma}^{a}{}_{a_1 a_2 \cdots a_n}(\psi) \xi^{a_1} \xi^{a_2} \cdots \xi^{a_n} t^{n}\\ &= \overline{\psi}^{a} + \xi^{a} t - \sum_{n=2}^{\infty} \frac{1}{n!} \overline{\Gamma}^{a}{}_{(a_1 a_2 \cdots a_n)}(\psi) \xi^{a_1} \xi^{a_2} \cdots \xi^{a_n} t^{n},\\ \overline{\lambda}^{a}(1) &= \overline{\psi}^{a} + \overline{\pi}^{a}, \qquad a = 1, \cdots, N,\end{aligned} \tag{C.89b}$$

[15] The point $\psi(x) \in \mathfrak{M}$ is represented by the point $(\psi^{a}(x)) \in \mathbb{R}^{N}$ in an arbitrary local coordinate system. To emphasize that the local coordinate system is chosen to be the normal coordinate system with respect to $\psi(x)$, we use the notation $(\overline{\psi}^{a}(x)) \in \mathbb{R}^{N}$ to represent $\psi(x) \in \mathfrak{M}$.

on the one hand. On the other hand, the defining property of the normal coordinates with respect to a point in the manifold is to represent any geodesics through this point by a straight line in the tangent space to this point. For Eq. (C.89b) to be a straight line,

$$0 = \overline{\Gamma}^{a}{}_{(a_1 a_2 \cdots a_n)}(\psi), \qquad a, a_1, a_2, \cdots a_n = 1, \cdots, N, \qquad n = 2, 3, \cdots, \tag{C.90}$$

must hold. In the normal coordinate system with respect to the point $\psi(x)$ defined by Eq. (C.89b), the covariant derivative $\overline{\nabla}_a$ defined by its action Eq. (C.39a) on tensor fields reduces to a partial derivative

$$\overline{\nabla}_a = \partial_a, \qquad a = 1, \cdots, N, \tag{C.91}$$

at $\psi(x)$. If so, Eq. (C.90) is nothing but

$$0 = \partial_{(a_1} \partial_{a_2} \cdots \partial_{a_{n-2}} \overline{\Gamma}^{a}{}_{a_{n-1} a_n)}(\psi), \quad a, a_1, a_2, \cdots a_n = 1, \cdots, N, \; n = 2, 3, \cdots. \tag{C.92}$$

The Taylor expansion (C.88) is also simplified when represented in terms of the normal coordinates with respect to $\psi(x)$,

$$\begin{aligned}\partial_\mu \left(\overline{\psi}^a + \overline{\pi}^a\right) &= \partial_\mu \overline{\psi}^a + \partial_\mu \xi^a - \partial_\mu \overline{\psi}^b \sum_{n=2}^{\infty} \frac{1}{n!} \partial_b \overline{\Gamma}^{a}{}_{(a_1 a_2 \cdots a_n)}(\psi) \xi^{a_1} \xi^{a_2} \cdots \xi^{a_n} \\ &= \partial_\mu \overline{\psi}^a + \overline{D}_\mu \xi^a - \partial_\mu \overline{\psi}^b \sum_{n=2}^{\infty} \frac{1}{n!} \partial_b \overline{\Gamma}^{a}{}_{(a_1 a_2 \cdots a_n)}(\psi) \xi^{a_1} \xi^{a_2} \cdots \xi^{a_n},\end{aligned} \tag{C.93}$$

where a second covariant derivative defined by

$$\begin{aligned}\overline{D}_\mu \xi^a &:= \partial_\mu \xi^a + \overline{\Gamma}^{a}{}_{bc} \xi^b \partial_\mu \overline{\psi}^c \\ &= \partial_\mu \xi^a, \qquad \mu = 1, \cdots, d, \qquad a = 1, \cdots, N,\end{aligned} \tag{C.94}$$

has been introduced.[16] Expansion (C.93) is not expressed in an optimal way since the right-hand side explicitly depends on derivatives of the Levi-Civita connection and thus is not manifestly covariant under reparametrization of $\psi(x)$. This problem can be fixed by an iterative use of Eqs. (C.92) and (C.76) as we now illustrate with the second order term in the expansion. Condition (C.92) with $n = 3$ gives (see footnote 16)

$$\begin{aligned}0 &= \partial_{(a_1} \overline{\Gamma}^{a}{}_{a_2 a_3)} \\ &= \frac{1}{3!} \left(\partial_{a_1} \overline{\Gamma}^{a}{}_{a_2 a_3} + \partial_{a_2} \overline{\Gamma}^{a}{}_{a_3 a_1} + \partial_{a_3} \overline{\Gamma}^{a}{}_{a_1 a_2} + \partial_{a_1} \overline{\Gamma}^{a}{}_{a_3 a_2} + \partial_{a_2} \overline{\Gamma}^{a}{}_{a_1 a_3} + \partial_{a_3} \overline{\Gamma}^{a}{}_{a_2 a_1} \right) \\ &= \frac{1}{3} \left(\partial_{a_1} \overline{\Gamma}^{a}{}_{a_2 a_3} + \partial_{a_2} \overline{\Gamma}^{a}{}_{a_3 a_1} + \partial_{a_3} \overline{\Gamma}^{a}{}_{a_1 a_2} \right).\end{aligned} \tag{C.95}$$

By Eq. (C.76), the Riemann curvature tensor in the normal coordinates with respect to ψ simplifies to

$$\begin{aligned}\overline{R}^{a}{}_{a_1 a_2 a_3} &= \partial_{a_2} \overline{\Gamma}^{a}{}_{a_1 a_3} - \partial_{a_3} \overline{\Gamma}^{a}{}_{a_1 a_2}, \\ \overline{R}^{a}{}_{a_3 a_2 a_1} &= \partial_{a_2} \overline{\Gamma}^{a}{}_{a_3 a_1} - \partial_{a_1} \overline{\Gamma}^{a}{}_{a_3 a_2}, \\ \overline{R}^{a}{}_{a_1 a_2 a_3} + \overline{R}^{a}{}_{a_3 a_2 a_1} &= 2 \partial_{a_2} \overline{\Gamma}^{a}{}_{a_1 a_3} - \partial_{a_3} \overline{\Gamma}^{a}{}_{a_1 a_2} - \partial_{a_1} \overline{\Gamma}^{a}{}_{a_3 a_2},\end{aligned} \tag{C.96}$$

[16] Recall that $\Gamma^{a}{}_{bc} = \Gamma^{a}{}_{cb}$ holds for the Levi-Civita connection.

at $\psi(x)$. Here, the symmetry of the Levi-Civita connection with respect to interchange of its two lower indices was used. Combining Eqs. (C.95) and (C.96) allows to express the derivative of the Levi-Civita connection in terms of the Riemann curvature tensor,

$$\begin{aligned}\partial_{a_2}\overline{\Gamma}^{a}{}_{a_1 a_3} &= +\frac{1}{3}\left(\overline{R}^{a}{}_{a_1 a_2 a_3} + \overline{R}^{a}{}_{a_3 a_2 a_1}\right)\\ &= -\frac{1}{3}\left(\overline{R}^{a}{}_{a_1 a_3 a_2} + \overline{R}^{a}{}_{a_3 a_1 a_2}\right).\end{aligned} \tag{C.97}$$

The antisymmetry in the interchange of the last two lower indices of the Riemann curvature tensor was used to reach the last line. Since the right-hand side is here symmetric under interchange of the lower indices a_1 and a_3, we conclude that the Taylor expansion (C.93) up to second order in the normal coordinates with respect to $\psi(x)$ is given by

$$\partial_\mu\left(\overline{\psi}^a + \overline{\pi}^a\right) = \partial_\mu\overline{\psi}^a + \overline{D}_\mu\xi^a + \frac{1}{3}\overline{R}^{a}{}_{a_1 a_2 b}\xi^{a_1}\xi^{a_2}\partial_\mu\overline{\psi}^b. \tag{C.98}$$

Although expressed in normal coordinates with respect to $\psi(x)$, Eq. (C.98) for an arbitrary chart containing $\psi(x)$ is simply obtained by removing the overline as Eq. (C.98) is manifestly covariant under reparametrization of $\psi(x)$.

In the Taylor expansion

$$T^{a_1\cdots a_k}_{b_1\cdots b_l}(\psi,\pi) = \sum_{n=0}^{\infty}\frac{1}{n!}\left(\partial_{c_1}\cdots\partial_{c_n}T^{a_1\cdots a_k}_{b_1\cdots b_l}\right)(\psi)\,\pi^{c_1}\cdots\pi^{c_n} \tag{C.99}$$

of the tensor $T^{a_1\cdots a_k}_{b_1\cdots b_l}$ of type $\binom{k}{l}$ neither the expansion coefficients $\left(\partial_{c_1}\cdots\partial_{c_n}T^{a_1\cdots a_k}_{b_1\cdots b_l}\right)(\psi)$ nor the expansion variables (π^c) transform covariantly under reparametrization of (ψ^a). By choosing to parametrize (π^a) in terms of the normal coordinates (ξ^a) with respect to ψ, both the expansion variables (ξ^c) and the expansion coefficients $\left(\partial_{c_1}\cdots\partial_{c_n}\overline{T}^{a_1\cdots a_k}_{b_1\cdots b_l}\right)(\psi)$ transform covariantly under reparametrization of $(\overline{\psi}^a)$ in the Taylor expansion

$$\overline{T}^{a_1\cdots a_k}_{b_1\cdots b_l}(\psi,\pi) = \sum_{n=0}^{\infty}\frac{1}{n!}\left(\partial_{c_1}\cdots\partial_{c_n}\overline{T}^{a_1\cdots a_k}_{b_1\cdots b_l}\right)(\psi)\,\xi^{c_1}\cdots\xi^{c_n}. \tag{C.100}$$

In other words, the expansion coefficients $\left(\partial_{c_1}\cdots\partial_{c_n}\overline{T}^{a_1\cdots a_k}_{b_1\cdots b_l}\right)(\psi)$ can be expressed solely in terms of covariant derivatives of T and the Riemann curvature tensor of the manifold by a direct extension of the method used to reach Eq. (C.97).

As an illustration we prove the identities, valid at $\psi(x)$,

$$\partial_c\overline{T}_{b_1\cdots b_l} = \overline{\nabla}_c\overline{T}_{b_1\cdots b_l}, \tag{C.101a}$$

and

$$\partial_{c_1}\partial_{c_2}\overline{T}_{b_1\cdots b_l} = \overline{\nabla}_{c_1}\overline{\nabla}_{c_2}\overline{T}_{b_1\cdots b_l} - \frac{1}{3}\sum_{i=1}^{l}\left(\overline{R}^{b}{}_{c_2 b_i c_1} + \overline{R}^{b}{}_{b_i c_2 c_1}\right)\overline{T}_{b_1\cdots b_{i-1} b b_{i+1}\cdots b_l}, \tag{C.101b}$$

that we will apply shortly to the covariant Taylor expansion of the metric $\mathfrak{g}_{ab}(\psi,\pi)$. Equation (C.101a) is a direct consequence of definition (C.39) together with Eq. (C.90). The proof of Eq. (C.101b) starts from the observation that $\nabla_{c_2}T_{b_1\cdots b_l}$ is a tensor of type $\binom{0}{l+1}$. We first implement Eq. (C.39) for ∇_{c_1},

$$\begin{aligned}\nabla_{c_1}\left(\nabla_{c_2}T_{b_1\cdots b_l}\right) =& \, \partial_{c_1}\left(\nabla_{c_2}T_{b_1\cdots b_l}\right)\\ &- \Gamma^{a}{}_{c_1 c_2}\nabla_a T_{b_1\cdots b_l} - \sum_{i=1}^{l}\Gamma^{a}{}_{c_1 b_i}\nabla_{c_2}T_{b_1\cdots b_{i-1} a b_{i+1}\cdots b_l},\end{aligned} \tag{C.102}$$

at $\psi(x)$. We follow up by implementing Eq. (C.39) for ∇_{c_2},

$$\begin{aligned}\nabla_{c_1}\left(\nabla_{c_2}T_{b_1\cdots b_l}\right) =& \, \partial_{c_1}\left(\partial_{c_2}T_{b_1\cdots b_l} - \sum_{i=1}^{l}\Gamma^{a}{}_{c_2 b_i}T_{b_1\cdots b_{i-1} a b_{i+1}\cdots b_l}\right)\\ &- \Gamma^{a}{}_{c_1 c_2}\nabla_a T_{b_1\cdots b_l} - \sum_{i=1}^{l}\Gamma^{a}{}_{c_1 b_i}\nabla_{c_2}T_{b_1\cdots b_{i-1} a b_{i+1}\cdots b_l},\end{aligned} \tag{C.103}$$

at $\psi(x)$. The partial derivative ∂_{c_1} is then distributed by the product rule,

$$\begin{aligned}\nabla_{c_1}\left(\nabla_{c_2}T_{b_1\cdots b_l}\right) =& \, \partial_{c_1}\partial_{c_2}T_{b_1\cdots b_l}\\ &- \sum_{i=1}^{l}\partial_{c_1}\Gamma^{a}{}_{c_2 b_i}T_{b_1\cdots b_{i-1} a b_{i+1}\cdots b_l} - \sum_{i=1}^{l}\Gamma^{a}{}_{c_2 b_i}\partial_{c_1}T_{b_1\cdots b_{i-1} a b_{i+1}\cdots b_l}\\ &- \Gamma^{a}{}_{c_1 c_2}\nabla_a T_{b_1\cdots b_l} - \sum_{i=1}^{l}\Gamma^{a}{}_{c_1 b_i}\nabla_{c_2}T_{b_1\cdots b_{i-1} a b_{i+1}\cdots b_l},\end{aligned} \tag{C.104}$$

at $\psi(x)$. The right-hand side, when restricted to the normal coordinates with respect to $\psi(x)$, reduces to

$$\begin{aligned}\overline{\nabla}_{c_1}\overline{\nabla}_{c_2}\overline{T}_{b_1\cdots b_l} &= \partial_{c_1}\partial_{c_2}\overline{T}_{b_1\cdots b_l} - \sum_{i=1}^{l}\partial_{c_1}\overline{\Gamma}^{a}{}_{c_2 b_i}\overline{T}_{b_1\cdots b_{i-1} a b_{i+1}\cdots b_l}\\ &= \partial_{c_1}\partial_{c_2}\overline{T}_{b_1\cdots b_l} + \frac{1}{3}\sum_{i=1}^{l}\left(\overline{R}^{b}{}_{c_2 b_i c_1} + \overline{R}^{b}{}_{b_i c_2 c_1}\right)\overline{T}_{b_1\cdots b_{i-1} b b_{i+1}\cdots b_l},\end{aligned} \tag{C.105}$$

at $\psi(x)$. Equation (C.97) was used to reach the last line. The Taylor expansion (C.99) simplifies for the metric tensor in view of the compatibility condition (C.40),

$$\begin{aligned}
\overline{\mathfrak{g}}_{b_1 b_2}(\psi,\pi) =&\, \overline{\mathfrak{g}}_{b_1 b_2}(\psi) \\
&- \tfrac{1}{2!}\tfrac{1}{3}\left[\left(\overline{R}^{b}{}_{c_2 b_1 c_1} + \overline{R}^{b}{}_{b_1 c_2 c_1}\right)\overline{\mathfrak{g}}_{b b_2} + \left(\overline{R}^{b}{}_{c_2 b_2 c_1} + \overline{R}^{b}{}_{b_2 c_2 c_1}\right)\overline{\mathfrak{g}}_{b b_1}\right](\psi)\xi^{c_1}\xi^{c_2} \\
&+ \cdots \\
=&\, \overline{\mathfrak{g}}_{b_1 b_2}(\psi) \\
&- \tfrac{1}{2!}\tfrac{1}{3}\left[\left(\overline{R}_{b_2 c_2 b_1 c_1} + \underline{\overline{R}_{b_2 b_1 c_2 c_1}}\right) + \left(\overline{R}_{b_1 c_2 b_2 c_1} + \underline{\overline{R}_{b_1 b_2 c_2 c_1}}\right)\right](\psi)\xi^{c_1}\xi^{c_2} \\
&+ \cdots \\
=&\, \overline{\mathfrak{g}}_{b_1 b_2}(\psi) - \frac{1}{3}\overline{R}_{b_1 c_1 b_2 c_2}(\psi)\xi^{c_1}\xi^{c_2} + \cdots .
\end{aligned} \tag{C.106}$$

To reach the last line, Eq. (C.60b) was used to deduce from

$$R_{abcd} = -R_{abdc}, \qquad R_{abcd} = R_{cdab}, \qquad a,b,c,d = 1,\cdots,N, \tag{C.107}$$

that underlined curvature tensors cancel out and that the remaining two curvature tensors, when contracted with $\xi^{c_1}\xi^{c_2}$, are equal.

C.6.4 *Gaussian expansion of the action*

With the help of normal coordinates with respect to a solution ψ of the classical equations of motion of the NLσM, we have found in section C.6.3 the covariant expansions

$$\partial_\mu\left(\psi^a + \pi^a\right) = \partial_\mu\psi^a + D_\mu\xi^a + \frac{1}{3}R^{a}{}_{c_1 c_2 c_3}\xi^{c_1}\xi^{c_2}\partial_\mu\psi^{c_3} + \cdots, \tag{C.108a}$$

and

$$\mathfrak{g}_{ab}(\psi,\pi) = \mathfrak{g}_{ab}(\psi) - \frac{1}{3}R_{a c_1 b c_2}(\psi)\xi^{c_1}\xi^{c_2} + \cdots . \tag{C.108b}$$

The validity of this expansion is conditioned by the existence of a unique geodesic that connects the points $\mathfrak{p} \equiv \psi(x)$ and $\mathfrak{q} \equiv \psi(x) + \pi(x)$ in the Riemannian manifold $(\mathfrak{M},\mathfrak{g})$. Here the geodesic lies in an open neighborhood of the domain $\mathfrak{U}$ from the chart $(\mathfrak{U},\xi)$ at the point $\psi(x)$. We are going to prove that the corresponding expansion for the action is given by

$$\begin{aligned}
S[\psi,\pi] =&\, S[\psi] + \frac{1}{2}\int\frac{\mathrm{d}^d x}{\mathfrak{a}^{d-2}}\left[\mathfrak{g}_{ab}(\psi)D_\mu\xi^a D_\mu\xi^b - \partial_\mu\psi^a\partial_\mu\psi^b R_{acbd}\xi^c\xi^d\right] + \cdots \\
=&\, S[\psi] + \frac{1}{2}\int\frac{\mathrm{d}^d x}{\mathfrak{a}^{d-2}}\left[\mathfrak{g}_{ab}(\psi)D_\mu\xi^a D_\mu\xi^b + \partial_\mu\psi^a\partial_\mu\psi^b R_{acdb}\xi^c\xi^d\right] + \cdots .
\end{aligned} \tag{C.108c}$$

(The second equality was reached with the help of $R_{acbd} = -R_{acdb}$.) Recall that we have introduced the covariant derivative

$$D_\mu\xi^a = \partial_\mu\xi^a + \partial_\mu\psi^b\Gamma^{a}{}_{bc}\xi^c, \qquad \mu = 1,\cdots,d, \qquad a = 1,\cdots,N, \tag{C.108d}$$

where $\Gamma^{a}{}_{bc}$ denotes the Levi-Civita connection and $R_{abcd} \equiv \mathfrak{g}_{a\bar{a}} R^{\bar{a}}{}_{bcd}$ denotes the associated curvature tensor.

Proof.
Step 1: Choose an arbitrary pair (μ, ν) with $\mu, \nu = 1, \cdots, d$,

$$\begin{aligned}\mathfrak{g}_{ij}(\psi,\pi)\partial_\mu(\psi+\pi)^i\partial_\nu(\psi+\pi)^j = &\left(\mathfrak{g}_{ij}(\psi) - \frac{1}{3}R_{il_1jl_2}(\psi)\xi^{l_1}\xi^{l_2} + \cdots\right)\\ &\times\left(\partial_\mu\psi^i + D_\mu\xi^i - \frac{1}{3}\partial_\mu\psi^m R^i{}_{l_1ml_2}(\psi)\xi^{l_1}\xi^{l_2} + \cdots\right)\\ &\times\left(\partial_\nu\psi^j + D_\nu\xi^j - \frac{1}{3}\partial_\nu\psi^n R^j{}_{l_1nl_2}(\psi)\xi^{l_1}\xi^{l_2} + \cdots\right).\end{aligned} \tag{C.109}$$

Distribution of the products gives

$$\begin{aligned}\mathfrak{g}_{ij}(\psi,\pi)\partial_\mu(\psi+\pi)^i\partial_\nu(\psi+\pi)^j =& \,\mathfrak{g}_{ij}(\psi)\partial_\mu\psi^i\partial_\nu\psi^j\\ &+\mathfrak{g}_{ij}(\psi)\partial_\mu\psi^i D_\nu\xi^j + \mathfrak{g}_{ij}(\psi)D_\mu\xi^i\partial_\nu\psi^j\\ &+\mathfrak{g}_{ij}(\psi)D_\mu\xi^i D_\nu\xi^j\\ &-\frac{1}{3}R_{il_1jl_2}(\psi)\partial_\mu\psi^i\partial_\nu\psi^j\xi^{l_1}\xi^{l_2}\\ &-\frac{1}{3}\mathfrak{g}_{ij}(\psi)\partial_\mu\psi^m R^i{}_{l_1ml_2}(\psi)\partial_\nu\psi^j\xi^{l_1}\xi^{l_2}\\ &-\frac{1}{3}\mathfrak{g}_{ij}(\psi)\partial_\mu\psi^i\partial_\nu\psi^n R^j{}_{l_1nl_2}(\psi)\xi^{l_1}\xi^{l_2}\\ &+\cdots,\end{aligned} \tag{C.110}$$

Contraction of the metric tensor lowers indices on the curvature tensor of the last two lines,

$$\begin{aligned}\mathfrak{g}_{ij}(\psi,\pi)\partial_\mu(\psi+\pi)^i\partial_\nu(\psi+\pi)^j =& \,\mathfrak{g}_{ij}(\psi)\partial_\mu\psi^i\partial_\nu\psi^j\\ &+\mathfrak{g}_{ij}(\psi)\partial_\mu\psi^i D_\nu\xi^j + \mathfrak{g}_{ij}(\psi)D_\mu\xi^i\partial_\nu\psi^j\\ &+\mathfrak{g}_{ij}(\psi)D_\mu\xi^i D_\nu\xi^j\\ &-\frac{1}{3}\partial_\mu\psi^i\partial_\nu\psi^j R_{il_1jl_2}(\psi)\xi^{l_1}\xi^{l_2}\\ &-\frac{1}{3}\partial_\mu\psi^i\partial_\nu\psi^j R_{jl_1il_2}(\psi)\xi^{l_1}\xi^{l_2}\\ &-\frac{1}{3}\partial_\mu\psi^i\partial_\nu\psi^j R_{il_1jl_2}(\psi)\xi^{l_1}\xi^{l_2}\\ &+\cdots.\end{aligned} \tag{C.111}$$

With the help of $R_{abcd} = R_{cdab}$, we conclude that

$$\begin{aligned}\mathfrak{g}_{ij}(\psi,\pi)\partial_\mu(\psi+\pi)^i\partial_\nu(\psi+\pi)^j =& \,\mathfrak{g}_{ij}(\psi)\partial_\mu\psi^i\partial_\nu\psi^j + \mathfrak{g}_{ij}(\psi)\partial_\mu\psi^i D_\nu\xi^j\\ &+\mathfrak{g}_{ij}(\psi)D_\mu\xi^i\partial_\nu\psi^j + \mathfrak{g}_{ij}(\psi)D_\mu\xi^i D_\nu\xi^j\\ &-\partial_\mu\psi^i\partial_\nu\psi^j R_{il_1jl_2}(\psi)\xi^{l_1}\xi^{l_2}\\ &+\cdots.\end{aligned} \tag{C.112}$$

When $\mu = 1,\cdots,d$,

$$\begin{aligned}\mathfrak{g}_{ij}(\psi,\pi)\partial_\mu(\psi+\pi)^i\partial_\mu(\psi+\pi)^j =\ & \mathfrak{g}_{ij}(\psi)\partial_\mu\psi^i\partial_\mu\psi^j \\ & +2\mathfrak{g}_{ij}(\psi)\partial_\mu\psi^i D_\mu\xi^j \\ & +\mathfrak{g}_{ij}(\psi)D_\mu\xi^i D_\mu\xi^j - \partial_\mu\psi^i\partial_\mu\psi^j R_{il_1jl_2}(\psi)\xi^{l_1}\xi^{l_2} \\ & +\cdots .\end{aligned} \tag{C.113}$$

Step 2: Summation over $\mu = 1,\cdots,d$, gives

$$\begin{aligned}\mathcal{L}(\psi,\pi) =\ & \frac{1}{2}\mathfrak{g}_{ij}(\psi,\pi)\partial_\mu(\psi+\pi)^i\partial_\mu(\psi+\pi)^j \\ =\ & \frac{1}{2}\mathfrak{g}_{ij}(\psi)\partial_\mu\psi^i\partial_\mu\psi^j \\ & +\mathfrak{g}_{ij}(\psi)\partial_\mu\psi^i D_\mu\xi^j \\ & +\frac{1}{2}\mathfrak{g}_{ij}(\psi)D_\mu\xi^i D_\mu\xi^j - \frac{1}{2}\partial_\mu\psi^i\partial_\mu\psi^j R_{il_1jl_2}(\psi)\xi^{l_1}\xi^{l_2} \\ & +\cdots \\ =\ & \mathcal{L}(\psi) \\ & +\mathfrak{g}_{ij}(\psi)\partial_\mu\psi^i D_\mu\xi^j \\ & +\frac{1}{2}\left[\mathfrak{g}_{ij}(\psi)D_\mu\xi^i D_\mu\xi^j - \partial_\mu\psi^i\partial_\mu\psi^j R_{il_1jl_2}(\psi)\xi^{l_1}\xi^{l_2}\right] \\ & +\cdots .\end{aligned} \tag{C.114}$$

Step 3: By assumption ψ is a solution of the classical equations of motion, i.e., ψ is an extremum of the action. Hence the term linear in ξ^j must vanish in the action constructed from Eq. (C.114),

$$S[\psi,\pi] = S[\psi] + \int\frac{\mathrm{d}^d x}{\mathfrak{a}^{d-2}}\,\frac{1}{2}\left[\mathfrak{g}_{ij}(\psi)D_\mu\xi^i D_\mu\xi^j - \partial_\mu\psi^i\partial_\mu\psi^j R_{il_1jl_2}(\psi)\xi^{l_1}\xi^{l_2}\right] + \cdots . \tag{C.115}$$

□

C.6.5 *Diagonalization of the metric tensor through vielbeins*

The covariant Gaussian expansion (C.108c) about an extremum ψ of the action is still not practical for computations because of the presence of the ψ-dependent metric tensor. It is desirable to perform a $GL(N,\mathbb{R})$ transformation of the tangent space at $\psi(x)$ that diagonalizes the metric tensor at $\psi(x)$. To this end, in any chart $(\mathfrak{U},\xi)$ for which it is possible to find an open neighborhood such that therein lies a unique geodesic that connects $\psi(x)$ to $\psi(x)+\pi(x)$, introduce the fast degree of freedom $\zeta(x)$ by the linear transformation

$$\zeta^{\hat{a}} := \hat{e}^{\hat{a}}{}_{a}(\psi)\xi^{a}, \qquad \xi^{a} = \hat{e}_{\hat{a}}{}^{a}(\psi)\zeta^{\hat{a}}, \tag{C.116a}$$

where $\left(\hat{e}^{\hat{a}}{}_{a}(\psi)\right) \in GL(N,\mathbb{R})$, which is called the *vielbeins*, is the inverse of $\left(\hat{e}_{\hat{a}}{}^{a}(\psi)\right) \in GL(N,\mathbb{R})$,

$$\hat{e}^{\hat{a}}{}_{a}(\psi)\hat{e}_{\hat{b}}{}^{a}(\psi) = \delta^{\hat{a}}_{\hat{b}}, \qquad \hat{e}_{\hat{a}}{}^{a}(\psi)\hat{e}^{\hat{a}}{}_{b}(\psi) = \delta^{a}_{b}, \tag{C.116b}$$

by demanding that

$$\xi^a\, \mathfrak{g}_{ab}(\psi)\, \xi^b = \zeta^{\hat{a}}\, \delta_{\hat{a}\hat{b}}\, \zeta^{\hat{b}} \overset{(C.116a)}{\Longleftrightarrow} \hat{e}_{\hat{a}}{}^{a}(\psi)\, \mathfrak{g}_{ab}(\psi)\; \hat{e}_{\hat{b}}{}^{b}(\psi) = \delta_{\hat{a}\hat{b}}, \qquad \overset{(C.116b)}{\Longleftrightarrow} \hat{e}^{\hat{a}}{}_{a}(\psi)\, \delta_{\hat{a}\hat{b}}\, \hat{e}^{\hat{b}}{}_{b}(\psi) = \mathfrak{g}_{ab}(\psi). \tag{C.116c}$$

From now on, latin letters with a hat refer to the coordinates of the Riemannian manifold in the vielbein basis (C.116). Since the metric tensor is diagonal in the vielbein basis (C.116), we will not distinguish between upper and lower indices in this basis with the exception of the vielbein matrices (C.116a). We now show that under transformation (C.116) the covariant expansion (C.108c) becomes

$$S[\psi,\pi] = S[\psi] + \frac{1}{2}\int \frac{\mathrm{d}^d x}{\mathfrak{a}^{d-2}} \left[\widehat{D}_\mu \zeta^{\hat{a}} \widehat{D}_\mu \zeta^{\hat{a}} + \left(\partial_\mu \psi^a \partial_\mu \psi^b R_{acdb} \hat{e}_{\hat{c}}{}^{c} \hat{e}_{\hat{d}}{}^{d}\right) \zeta^{\hat{c}} \zeta^{\hat{d}}\right] + \cdots, \tag{C.117a}$$

where yet another covariant derivative

$$\widehat{D}_\mu \zeta^{\hat{a}} := \partial_\mu \zeta^{\hat{a}} + A^{\hat{a}\hat{b}}_{\mu}(\psi)\zeta^{\hat{b}}, \qquad \mu = 1,\cdots,d, \qquad \hat{a} = 1,\cdots,N, \tag{C.117b}$$

has been introduced together with the $O(N)$ spin connection

$$\omega^{\hat{a}}_{b\hat{b}}(\psi) := \hat{e}^{\hat{a}}{}_{a}(\psi)\partial_b \hat{e}_{\hat{b}}{}^{a}(\psi) + \hat{e}^{\hat{a}}{}_{a}(\psi)\Gamma^{a}{}_{bc}(\psi)\hat{e}_{\hat{b}}{}^{c}(\psi) \equiv \omega^{\hat{a}\hat{b}}_{b}(\psi), \quad \hat{a},\hat{b},b = 1,\cdots,N, \tag{C.117c}$$

the $O(N)$ gauge field

$$A^{\hat{a}\hat{b}}_{\mu}(\psi) := \partial_\mu \psi^a \omega^{\hat{a}\hat{b}}_{a}(\psi), \qquad \mu = 1,\cdots,d, \qquad \hat{a},\hat{b} = 1,\cdots,N, \tag{C.117d}$$

and the $O(N)$ field strength tensor

$$\begin{aligned} F^{\hat{a}\hat{b}}_{\mu\nu}(\psi) :&= \partial_\mu A^{\hat{a}\hat{b}}_{\nu}(\psi) - \partial_\nu A^{\hat{a}\hat{b}}_{\mu}(\psi) + A^{\hat{a}\hat{c}}_{\mu}(\psi)A^{\hat{c}\hat{b}}_{\nu}(\psi) - A^{\hat{a}\hat{c}}_{\nu}(\psi)A^{\hat{c}\hat{b}}_{\mu}(\psi) \\ &= \partial_\mu \psi^a \partial_\nu \psi^b R_{abcd}(\psi)\hat{e}_{\hat{a}}{}^{c}(\psi)\hat{e}_{\hat{b}}{}^{d}(\psi), \end{aligned} \tag{C.117e}$$

for $\mu,\nu = 1,\cdots,d$, and $\hat{a},\hat{b} = 1,\cdots,N$.

Proof. To shorten the notation, we omit to specify explicitly the ψ-dependence of the metric tensor, the Levi-Civita connection, and of the vielbeins.

Step 1: The transformation law of the covariant derivative $D_\mu \xi^a$ for $\mu = 1,\cdots,d$ and $a = 1,\cdots,N$ is, by Eq. (C.116a),

$$\begin{aligned} D_\mu \xi^a &= \partial_\mu \xi^a + \partial_\mu \psi^b \Gamma^{a}{}_{bc} \xi^c \\ &= \partial_\mu (\hat{e}_{\hat{c}}{}^{a} \zeta^{\hat{c}}) + \partial_\mu \psi^b \Gamma^{a}{}_{bc} (\hat{e}_{\hat{c}}{}^{c} \zeta^{\hat{c}}) \\ &= \hat{e}_{\hat{c}}{}^{a} \partial_\mu \zeta^{\hat{c}} + \underline{\partial_\mu \hat{e}_{\hat{c}}{}^{a}} \zeta^{\hat{c}} + \partial_\mu \psi^b \Gamma^{a}{}_{bc} \hat{e}_{\hat{c}}{}^{c} \zeta^{\hat{c}} \\ &= \hat{e}_{\hat{c}}{}^{a} \partial_\mu \zeta^{\hat{c}} + \underline{\partial_\mu \psi^b \partial_b \hat{e}_{\hat{c}}{}^{a}} \zeta^{\hat{c}} + \partial_\mu \psi^b \Gamma^{a}{}_{bc} \hat{e}_{\hat{c}}{}^{c} \zeta^{\hat{c}} \\ &= \hat{e}_{\hat{c}}{}^{a} \partial_\mu \zeta^{\hat{c}} + \partial_\mu \psi^b \left(\underline{\partial_b \hat{e}_{\hat{c}}{}^{a}} + \Gamma^{a}{}_{bc} \hat{e}_{\hat{c}}{}^{c}\right) \zeta^{\hat{c}}, \end{aligned} \tag{C.118}$$

where we made use of the chain rule to evaluate the underlined term.

Step 2: The definition of the action of the covariant derivative $\widehat{D}_\mu \zeta^{\hat{a}}$ for $\mu = 1,\cdots,d$ and $\hat{a} = 1,\cdots,N$ is given by

$$\begin{aligned}
\widehat{D}_\mu \zeta^{\hat{a}} &:= \hat{e}^{\hat{a}}{}_a D_\mu \xi^a \\
\text{Eq. (C.118)} &= \hat{e}^{\hat{a}}{}_a \left[\hat{e}_{\hat{c}}{}^a \partial_\mu \zeta^{\hat{c}} + \partial_\mu \psi^b \left(\underline{\partial_b \hat{e}_{\hat{c}}{}^a} + \Gamma^a{}_{bc} \hat{e}_{\hat{c}}{}^c \right) \zeta^{\hat{c}} \right] \\
\text{Eq. (C.116b)} &= \delta^{\hat{a}}_{\hat{c}} \partial_\mu \zeta^{\hat{c}} + \partial_\mu \psi^b \left(\hat{e}^{\hat{a}}{}_a \underline{\partial_b \hat{e}_{\hat{c}}{}^a} + \hat{e}^{\hat{a}}{}_a \Gamma^a{}_{bc} \hat{e}_{\hat{c}}{}^c \right) \zeta^{\hat{c}} \\
\text{Eq. (C.116b)} &= \delta^{\hat{a}}_{\hat{c}} \partial_\mu \zeta^{\hat{c}} + \partial_\mu \psi^b \left(-\hat{e}_{\hat{c}}{}^a \underline{\partial_b \hat{e}^{\hat{a}}{}_a} + \hat{e}^{\hat{a}}{}_a \Gamma^a{}_{bc} \hat{e}_{\hat{c}}{}^c \right) \zeta^{\hat{c}}.
\end{aligned} \tag{C.119}$$

Step 3: With the introduction of the $O(N)$ spin connection

$$\begin{aligned}
\omega^{\hat{a}}_{b\hat{b}} &:= +\hat{e}^{\hat{a}}{}_a \underline{\partial_b \hat{e}_{\hat{b}}{}^a} + \hat{e}^{\hat{a}}{}_a \Gamma^a{}_{bc} \hat{e}_{\hat{b}}{}^c \\
&= -\hat{e}_{\hat{b}}{}^a \underline{\partial_b \hat{e}^{\hat{a}}{}_a} + \hat{e}^{\hat{a}}{}_a \Gamma^a{}_{bc} \hat{e}_{\hat{b}}{}^c \\
&\equiv \omega^{\hat{a}\hat{b}}_b, \qquad \hat{a}, \hat{b}, b = 1,\cdots,N,
\end{aligned} \tag{C.120}$$

Eq. (C.119) can be rewritten

$$\widehat{D}_\mu \zeta^{\hat{a}} = \partial_\mu \zeta^{\hat{a}} + \partial_\mu \psi^b \omega^{\hat{a}\hat{b}}_b \zeta^{\hat{b}}, \qquad \mu = 1,\cdots,d, \qquad \hat{a} = 1,\cdots,N. \tag{C.121}$$

It is important to note that

$$\omega^{\hat{a}\hat{b}}_b = -\omega^{\hat{b}\hat{a}}_b, \tag{C.122a}$$

as follows from

$$\begin{aligned}
\omega^{\hat{b}\hat{a}}_b &= +\hat{e}^{\hat{b}}{}_a \partial_b \hat{e}_{\hat{a}}{}^a + \hat{e}^{\hat{b}}{}_a \Gamma^a{}_{bc} \hat{e}_{\hat{a}}{}^c \\
&= -\hat{e}_{\hat{a}}{}^a \partial_b \hat{e}^{\hat{b}}{}_a + \hat{e}_{\hat{a}}{}^c \Gamma^a{}_{bc} \hat{e}^{\hat{b}}{}_a \\
&= -\hat{e}^{\hat{a}a} \partial_b \hat{e}_{\hat{b}a} + \hat{e}^{\hat{a}c} \Gamma^a{}_{bc} \hat{e}_{\hat{b}a} \\
&= -\hat{e}^{\hat{a}}{}_{\bar{a}} \mathfrak{g}^{\bar{a}a} \partial_b \left(\mathfrak{g}_{a\bar{\bar{a}}} \hat{e}_{\hat{b}}{}^{\bar{\bar{a}}} \right) + \hat{e}^{\hat{a}}{}_{\bar{a}} \mathfrak{g}^{\bar{a}c} \Gamma^a{}_{bc} \mathfrak{g}_{a\bar{\bar{a}}} \hat{e}_{\hat{b}}{}^{\bar{\bar{a}}} \\
&= -\hat{e}^{\hat{a}}{}_{\bar{a}} \partial_b \hat{e}_{\hat{b}}{}^{\bar{a}} - \hat{e}^{\hat{a}}{}_{\bar{a}} \left(\mathfrak{g}^{\bar{a}a} \partial_b \mathfrak{g}_{a\bar{\bar{a}}} - \mathfrak{g}^{\bar{a}c} \Gamma^a{}_{bc} \mathfrak{g}_{a\bar{\bar{a}}} \right) \hat{e}_{\hat{b}}{}^{\bar{\bar{a}}} \\
&= -\hat{e}^{\hat{a}}{}_{\bar{a}} \partial_b \hat{e}_{\hat{b}}{}^{\bar{a}} - \hat{e}^{\hat{a}}{}_{\bar{a}} \left[\mathfrak{g}^{\bar{a}a} \partial_b \mathfrak{g}_{a\bar{\bar{a}}} - \mathfrak{g}^{\bar{a}c} \frac{1}{2} \mathfrak{g}^{ad} \left(\partial_b \mathfrak{g}_{dc} + \partial_c \mathfrak{g}_{db} - \partial_d \mathfrak{g}_{bc} \right) \mathfrak{g}_{a\bar{\bar{a}}} \right] \hat{e}_{\hat{b}}{}^{\bar{\bar{a}}} \\
&= -\hat{e}^{\hat{a}}{}_{\bar{a}} \partial_b \hat{e}_{\hat{b}}{}^{\bar{a}} - \hat{e}^{\hat{a}}{}_{\bar{a}} \left[\mathfrak{g}^{\bar{a}a} \partial_b \mathfrak{g}_{a\bar{\bar{a}}} - \frac{1}{2} \mathfrak{g}^{\bar{a}c} \left(\partial_b \mathfrak{g}_{\bar{\bar{a}}c} + \partial_c \mathfrak{g}_{\bar{\bar{a}}b} - \partial_{\bar{\bar{a}}} \mathfrak{g}_{bc} \right) \right] \hat{e}_{\hat{b}}{}^{\bar{\bar{a}}} \\
&= -\hat{e}^{\hat{a}}{}_{\bar{a}} \partial_b \hat{e}_{\hat{b}}{}^{\bar{a}} - \hat{e}^{\hat{a}}{}_{\bar{a}} \left[\mathfrak{g}^{\bar{a}c} \partial_b \mathfrak{g}_{c\bar{\bar{a}}} - \frac{1}{2} \mathfrak{g}^{\bar{a}c} \left(\partial_b \mathfrak{g}_{c\bar{\bar{a}}} + \partial_c \mathfrak{g}_{\bar{\bar{a}}b} - \partial_{\bar{\bar{a}}} \mathfrak{g}_{cb} \right) \right] \hat{e}_{\hat{b}}{}^{\bar{\bar{a}}} \\
&= -\hat{e}^{\hat{a}}{}_{\bar{a}} \partial_b \hat{e}_{\hat{b}}{}^{\bar{a}} - \hat{e}^{\hat{a}}{}_{\bar{a}} \left[\frac{1}{2} \mathfrak{g}^{\bar{a}c} \left(\partial_b \mathfrak{g}_{c\bar{\bar{a}}} - \partial_c \mathfrak{g}_{\bar{\bar{a}}b} + \partial_{\bar{\bar{a}}} \mathfrak{g}_{cb} \right) \right] \hat{e}_{\hat{b}}{}^{\bar{\bar{a}}} \\
&= -\hat{e}^{\hat{a}}{}_{\bar{a}} \partial_b \hat{e}_{\hat{b}}{}^{\bar{a}} - \hat{e}^{\hat{a}}{}_{\bar{a}} \Gamma^{\bar{a}}_{b\bar{\bar{a}}} \hat{e}_{\hat{b}}{}^{\bar{\bar{a}}} \\
&= -\omega^{\hat{a}\hat{b}}_b.
\end{aligned} \tag{C.122b}$$

The object

$$A^{\hat{a}\hat{b}}_\mu(\psi) := \partial_\mu \psi^c \omega^{\hat{a}\hat{b}}_c = -A^{\hat{b}\hat{a}}_\mu(\psi) \tag{C.122c}$$

is thus also antisymmetric under exchange of the vielbeins indices $\hat{a}$ and $\hat{b}$.

Step 4: By combining Eq. (C.116c) and Eq. (C.119) the term quadratic in the covariant derivatives in the covariant expansion (C.108c) becomes

$$\begin{aligned}\mathfrak{g}_{ab} D_\mu \xi^a D_\mu \xi^b &= \hat{e}^{\hat{a}}{}_{a}\, \delta_{\hat{a}\hat{b}}\, \hat{e}^{\hat{b}}{}_{b} D_\mu \xi^a D_\mu \xi^b \\ &= \delta_{\hat{a}\hat{b}} \widehat{D}_\mu \zeta^{\hat{a}} \widehat{D}_\mu \zeta^{\hat{b}} \\ &\equiv \widehat{D}_\mu \zeta^{\hat{a}} \widehat{D}_\mu \zeta^{\hat{a}}. \end{aligned} \tag{C.123}$$

Step 5: Equation (C.117a) follows immediately from Eq. (C.123) and applying Eq. (C.116a) to the term containing the Levi-Civita curvature tensor in the covariant expansion (C.108c),

$$\begin{aligned}\partial_\mu \psi^a \partial_\mu \psi^b R_{acdb} \xi^c \xi^d &= \partial_\mu \psi^a \partial_\mu \psi^b R_{acdb} \big(\hat{e}_{\hat{c}}{}^{c} \zeta^{\hat{c}}\big)\big(\hat{e}_{\hat{d}}{}^{d} \zeta^{\hat{d}}\big) \\ &= \left(\partial_\mu \psi^a \partial_\mu \psi^b R_{acdb} \hat{e}_{\hat{c}}{}^{c} \hat{e}_{\hat{d}}{}^{d}\right) \zeta^{\hat{c}} \zeta^{\hat{d}}. \end{aligned} \tag{C.124}$$

Step 6: The last step in the proof consists in verifying Eq. (C.117e). First, we have that

$$\begin{aligned} F^{\hat{a}\hat{b}}_{\mu\nu}(\psi) &:= \partial_\mu A^{\hat{a}\hat{b}}_\nu - \partial_\nu A^{\hat{a}\hat{b}}_\mu + A^{\hat{a}\hat{c}}_\mu A^{\hat{c}\hat{b}}_\nu - A^{\hat{a}\hat{c}}_\nu A^{\hat{c}\hat{b}}_\mu \\ &= \partial_\mu \left(\partial_\nu \psi^j \omega^{\hat{a}\hat{b}}_j\right) - \partial_\nu \left(\partial_\mu \psi^j \omega^{\hat{a}\hat{b}}_j\right) + \partial_\mu \psi^k \omega^{\hat{a}\hat{c}}_k \partial_\nu \psi^l \omega^{\hat{c}\hat{b}}_l - \partial_\nu \psi^k \omega^{\hat{a}\hat{c}}_k \partial_\mu \psi^l \omega^{\hat{c}\hat{b}}_l, \end{aligned} \tag{C.125}$$

by Eq. (C.117d). Applying the chain rule gives

$$F^{\hat{a}\hat{b}}_{\mu\nu}(\psi) = \partial_\nu \psi^j \partial_\mu \psi^k \partial_k \omega^{\hat{a}\hat{b}}_j - \partial_\mu \psi^j \partial_\nu \psi^l \partial_l \omega^{\hat{a}\hat{b}}_j + \partial_\mu \psi^k \partial_\nu \psi^l \omega^{\hat{a}\hat{c}}_k \omega^{\hat{c}\hat{b}}_l - \partial_\nu \psi^k \partial_\mu \psi^l \omega^{\hat{a}\hat{c}}_k \omega^{\hat{c}\hat{b}}_l. \tag{C.126}$$

Relabeling dummy indices delivers

$$F^{\hat{a}\hat{b}}_{\mu\nu}(\psi) = \partial_\nu \psi^l \partial_\mu \psi^k \partial_k \omega^{\hat{a}\hat{b}}_l - \partial_\mu \psi^k \partial_\nu \psi^l \partial_l \omega^{\hat{a}\hat{b}}_k + \partial_\mu \psi^k \partial_\nu \psi^l \omega^{\hat{a}\hat{c}}_k \omega^{\hat{c}\hat{b}}_l - \partial_\nu \psi^l \partial_\mu \psi^k \omega^{\hat{a}\hat{c}}_l \omega^{\hat{c}\hat{b}}_k. \tag{C.127}$$

With the help of factorization, we conclude that

$$F^{\hat{a}\hat{b}}_{\mu\nu}(\psi) = \partial_\nu \psi^l \partial_\mu \psi^k \left(\partial_k \omega^{\hat{a}\hat{b}}_l - \partial_l \omega^{\hat{a}\hat{b}}_k + \omega^{\hat{a}\hat{c}}_k \omega^{\hat{c}\hat{b}}_l - \omega^{\hat{a}\hat{c}}_l \omega^{\hat{c}\hat{b}}_k\right). \tag{C.128}$$

Second, insertion of the definition (C.117c) of the $O(N)$ spin connection yields

$$\begin{aligned} \partial_k \omega^{\hat{a}\hat{b}}_l - \partial_l \omega^{\hat{a}\hat{b}}_k =&\ \partial_k \left(-\hat{e}_{\hat{b}}{}^{m} \partial_l \hat{e}^{\hat{a}}{}_{m} + \hat{e}^{\hat{a}}{}_{m} \Gamma^{m}_{\ ln} \hat{e}_{\hat{b}}{}^{n}\right) - (k \leftrightarrow l) \\ =&\ -\underline{\partial_k \hat{e}_{\hat{b}}{}^{m} \partial_l \hat{e}^{\hat{a}}{}_{m}} + \underline{\underline{\partial_k \hat{e}^{\hat{a}}{}_{m} \Gamma^{m}_{\ ln} \hat{e}_{\hat{b}}{}^{n}}} + \hat{e}^{\hat{a}}{}_{m} \partial_k \Gamma^{m}_{\ ln} \hat{e}_{\hat{b}}{}^{n} + \underline{\underline{\underline{\hat{e}^{\hat{a}}{}_{m} \Gamma^{m}_{\ ln} \partial_k \hat{e}_{\hat{b}}{}^{n}}}} \\ &-(k \leftrightarrow l), \end{aligned} \tag{C.129}$$

and

$$\begin{aligned} \omega^{\hat{a}\hat{c}}_k \omega^{\hat{c}\hat{b}}_l - \omega^{\hat{a}\hat{c}}_l \omega^{\hat{c}\hat{b}}_k =&\ \left(-\hat{e}_{\hat{c}}{}^{m} \partial_k \hat{e}^{\hat{a}}{}_{m} + \hat{e}^{\hat{a}}{}_{m} \Gamma^{m}_{\ kn} \hat{e}_{\hat{c}}{}^{n}\right)\left(+\hat{e}^{\hat{c}}{}_{\bar{m}} \partial_l \hat{e}_{\hat{b}}{}^{\bar{m}} + \hat{e}^{\hat{c}}{}_{\bar{m}} \Gamma^{\bar{m}}_{\ l\bar{n}} \hat{e}_{\hat{b}}{}^{\bar{n}}\right) - (k \leftrightarrow l) \\ =&\ -\underline{\partial_k \hat{e}^{\hat{a}}{}_{m} \partial_l \hat{e}_{\hat{b}}{}^{m}} - \underline{\underline{\partial_k \hat{e}^{\hat{a}}{}_{m} \Gamma^{m}_{\ ln} \hat{e}_{\hat{b}}{}^{n}}} + \underline{\underline{\underline{\hat{e}^{\hat{a}}{}_{m} \Gamma^{m}_{\ kn} \partial_l \hat{e}_{\hat{b}}{}^{n}}}} + \hat{e}^{\hat{a}}{}_{m} \Gamma^{m}_{\ kp} \Gamma^{p}_{\ ln} \hat{e}_{\hat{b}}{}^{n} \\ &-(k \leftrightarrow l). \end{aligned} \tag{C.130}$$

When adding Eq. (C.129) to Eq. (C.130), all terms underlined with the same number of lines cancel pairwise. We are left with

$$\partial_k \omega_l^{\hat{a}\hat{b}} - \partial_l \omega_k^{\hat{a}\hat{b}} + \omega_k^{\hat{a}\hat{c}} \omega_l^{\hat{c}\hat{b}} - \omega_l^{\hat{a}\hat{c}} \omega_k^{\hat{c}\hat{b}} = \hat{e}^{\hat{a}}{}_m \hat{e}_{\hat{b}}{}^n \left(\partial_k \Gamma^m{}_{ln} - \partial_l \Gamma^m{}_{kn} + \Gamma^m{}_{kp} \Gamma^p{}_{ln} - \Gamma^m{}_{lp} \Gamma^p{}_{kn} \right). \tag{C.131}$$

With the help of Eq. (C.57),

$$\partial_k \omega_l^{\hat{a}\hat{b}} - \partial_l \omega_k^{\hat{a}\hat{b}} + \omega_k^{\hat{a}\hat{c}} \omega_l^{\hat{c}\hat{b}} - \omega_l^{\hat{a}\hat{c}} \omega_k^{\hat{c}\hat{b}} = \hat{e}^{\hat{a}}{}_m \hat{e}_{\hat{b}}{}^n R^m{}_{nkl}, \qquad \hat{a}, \hat{b}, k, l = 1, \cdots, N, \tag{C.132}$$

i.e.,

$$\begin{aligned}
F_{\mu\nu}^{\hat{a}\hat{b}} &= \hat{e}^{\hat{a}}{}_m \hat{e}_{\hat{b}}{}^n \partial_\mu \psi^k \partial_\nu \psi^l R^m{}_{nkl} \\
&= \hat{e}^{\hat{a}}{}_m \hat{e}_{\hat{b}}{}^n \partial_\mu \psi^k \partial_\nu \psi^l \mathfrak{g}^{mp} R_{pnkl} \\
{\scriptstyle \hat{e}^{\hat{c}m} := \mathfrak{g}^{mn} \hat{e}^{\hat{c}}{}_n} \quad &= \hat{e}^{\hat{a}}{}_m \hat{e}_{\hat{b}}{}^n \partial_\mu \psi^k \partial_\nu \psi^l \hat{e}^{\hat{c}m} \hat{e}^{\hat{c}p} R_{pnkl} \\
{\scriptstyle \hat{e}_{\hat{c}}{}^m := \delta_{\hat{c}\hat{d}} \hat{e}^{\hat{d}m}} \quad &= \underline{\hat{e}^{\hat{a}}{}_m} \hat{e}_{\hat{b}}{}^n \partial_\mu \psi^k \partial_\nu \psi^l \underline{\hat{e}_{\hat{c}}{}^m} \hat{e}_{\hat{c}}{}^p R_{pnkl} \\
{\scriptstyle \text{Eq. (C.116b)}} \quad &= \hat{e}_{\hat{a}}{}^p \hat{e}_{\hat{b}}{}^n \partial_\mu \psi^k \partial_\nu \psi^l R_{pnkl},
\end{aligned} \tag{C.133}$$

for $\mu, \nu = 1, \cdots, d$ and $\hat{a}, \hat{b} = 1, \cdots, N$. □

C.6.6 *Renormalization of the action after integration over the fast degrees of freedom*

Integration over the fast mode ζ can be performed with the help of the cumulant expansion, which is defined by the partition function

$$Z := \int \mathcal{D}[\psi] \exp\left(- S[\psi] - \delta S[\psi] \right), \tag{C.134a}$$

with the action

$$\begin{aligned}
\delta S[\psi] := &\left\langle S_2^{(1)}[\psi,\zeta] + S_2^{(0)}[\psi,\zeta] + \cdots \right\rangle_0^\zeta - \frac{1}{2} \left[\left\langle \left(S_2^{(1)}[\psi,\zeta] + S_2^{(0)}[\psi,\zeta] + \cdots \right)^2 \right\rangle_0^\zeta \right. \\
&\left. - \left(\left\langle S_2^{(1)}[\psi,\zeta] + S_2^{(0)}[\psi,\zeta] + \cdots \right\rangle_0^\zeta \right)^2 \right] + \cdots,
\end{aligned} \tag{C.134b}$$

where

$$\langle F[\psi,\zeta] \rangle_0^\zeta := \frac{\int \mathcal{D}[\zeta]\, e^{-S_0[\zeta]} F[\psi,\zeta]}{\int \mathcal{D}[\zeta]\, e^{-S_0[\zeta]}}. \tag{C.134c}$$

Here, within the Gaussian approximation of section C.6.5, the partition function over slow and fast modes is

$$Z_{\text{Gaussian}} := \int \mathcal{D}[\psi] \mathcal{D}[\zeta] \exp\left(- S[\psi,\zeta] \right), \tag{C.135a}$$

with the action

$$S[\psi,\zeta] := S[\psi] + S_0[\zeta] + S_2^{(1)}[\psi,\zeta] + S_2^{(0)}[\psi,\zeta], \tag{C.135b}$$

given by

$$\begin{aligned}
S[\psi] &:= \frac{1}{2}\int \frac{\mathrm{d}^d x}{\mathfrak{a}^{d-2}}\, \mathfrak{g}_{ab}(\psi)\partial_\mu\psi^a\partial_\mu\psi^b,\\
S_0[\zeta] &:= \frac{1}{2}\int \frac{\mathrm{d}^d x}{\mathfrak{a}^{d-2}}\, \partial_\mu\zeta^{\hat{a}}\partial_\mu\zeta^{\hat{a}},\\
S^{(1)}_2[\psi,\zeta] &:= \int \frac{\mathrm{d}^d x}{\mathfrak{a}^{d-2}}\, A^{\hat{a}\hat{b}}_\mu(\psi)\partial_\mu\zeta^{\hat{a}}\zeta^{\hat{b}},\\
S^{(0)}_2[\psi,\zeta] &:= \frac{1}{2}\int \frac{\mathrm{d}^d x}{\mathfrak{a}^{d-2}}\, \left[A^{\hat{a}\hat{b}}_\mu(\psi)A^{\hat{a}\hat{c}}_\mu(\psi) + \partial_\mu\psi^i\partial_\mu\psi^j R_{iklj}(\psi)\,\hat{e}_{\hat{b}}{}^k(\psi)\hat{e}_{\hat{c}}{}^l(\psi)\right]\zeta^{\hat{b}}\zeta^{\hat{c}}.
\end{aligned} \tag{C.135c}$$

(The upper indices in $S^{(0,1,\cdots)}_n[\psi,\zeta]$ refer to the number of gradients acting on the vielbeins coordinates while the lower index n refers to the order n in the expansion in powers of the vielbeins coordinates. Because ψ is an extremum of the action, the second non-vanishing contribution to the expansion in powers of ζ can have a n no less than $n = 2$.) To first order in the cumulant expansion and to Gaussian order in the expansion of the action

$$\delta S[\psi] = \left\langle S^{(1)}_2[\psi,\zeta] + S^{(0)}_2[\psi,\zeta]\right\rangle^{\zeta}_0. \tag{C.136}$$

To evaluate Eq. (C.136), we introduce the Green function

$$\begin{aligned}
G^{\hat{a}\hat{b}}_0(x-y) &:= \left\langle \zeta^{a}(x)\zeta^{\hat{b}}(y)\right\rangle^{\zeta}_0\\
&= \delta^{\hat{a}\hat{b}}\,\mathfrak{a}^{d-2} \int\limits_{\Lambda-\mathrm{d}\Lambda<|k|<\Lambda} \frac{\mathrm{d}^d k}{(2\pi)^d}\frac{e^{\mathrm{i}k(x-y)}}{k^2+M^2}\\
&\equiv \delta^{\hat{a}\hat{b}}\,G_0(x-y), \qquad \hat{a},\hat{b}=1,\cdots,N,
\end{aligned} \tag{C.137}$$

where it is understood that M^2 is an infrared cut-off, which is to be removed at the end of the calculation. Observe that

$$\begin{aligned}
\lim_{y\to x}\partial_\mu G^{\hat{a}\hat{b}}_0(x-y) &:= \lim_{y\to x}\left\langle \partial_\mu\zeta^{\hat{a}}(x)\zeta^{\hat{b}}(y)\right\rangle^{\zeta}_0\\
&= \delta^{\hat{a}\hat{b}}\,\mathfrak{a}^{d-2}\lim_{y\to x}\int\limits_{\Lambda-\mathrm{d}\Lambda<|k|<\Lambda} \frac{\mathrm{d}^d k}{(2\pi)^d}\,\mathrm{i}k_\mu\frac{e^{\mathrm{i}k(x-y)}}{k^2+M^2}\\
&= 0, \qquad \hat{a},\hat{b}=1,\cdots,N, \qquad \mu=1,\cdots,d.
\end{aligned} \tag{C.138}$$

Equation (C.138) implies that

$$\left\langle S^{(1)}_2[\psi,\zeta]\right\rangle^{\zeta}_0 = 0. \tag{C.139}$$

To first order in the cumulant expansion, we are left with the contribution

$$
\begin{aligned}
\delta S^{(1)}[\psi] :=& \left\langle S^{(0)}_{2}[\psi,\varsigma]\right\rangle^{\varsigma}_{0} \\
=& \left\langle \frac{1}{2}\int \frac{\mathrm{d}^{d}x}{\mathfrak{a}^{d-2}} \left[A^{\hat a\hat b}_{\mu}(\psi)A^{\hat a\hat c}_{\mu}(\psi) + \partial_{\mu}\psi^{i}\partial_{\mu}\psi^{j}R_{iklj}(\psi)\,\hat e_{\hat b}{}^{k}(\psi)\hat e_{\hat c}{}^{l}(\psi)\right]\varsigma^{\hat b}\varsigma^{\hat c}\right\rangle^{\varsigma}_{0} \\
=& \frac{1}{2}\int \frac{\mathrm{d}^{d}x}{\mathfrak{a}^{d-2}} \left[A^{\hat a\hat b}_{\mu}(\psi)A^{\hat a\hat c}_{\mu}(\psi) + \partial_{\mu}\psi^{i}\partial_{\mu}\psi^{j}R_{iklj}(\psi)\,\hat e_{\hat b}{}^{k}(\psi)\hat e_{\hat c}{}^{l}(\psi)\right]\left\langle\varsigma^{\hat b}\varsigma^{\hat c}\right\rangle^{\varsigma}_{0} \\
=& G_{0}(0)\frac{1}{2}\int \frac{\mathrm{d}^{d}x}{\mathfrak{a}^{d-2}} \left[A^{\hat a\hat b}_{\mu}(\psi)A^{\hat a\hat b}_{\mu}(\psi) + \partial_{\mu}\psi^{i}\partial_{\mu}\psi^{j}R_{iklj}(\psi)\,\hat e_{\hat b}{}^{k}(\psi)\hat e_{\hat b}{}^{l}(\psi)\right]
\end{aligned}
\tag{C.140}
$$

to $\delta S[\psi]$.

Can there be contributions to $\delta S[\psi]$ proportional to $G_{0}(0)$ originating from carrying the cumulant expansion beyond order one with the Gaussian approximation to the action? To answer this question we limit ourselves to $d=2$ and use dimensional analysis with the rules

$$
\begin{aligned}
&[\psi^{a}] = (\text{length})^{0}, \qquad [\mathfrak{g}_{ab}(\psi)] = (\text{length})^{0}, \qquad [R_{acdb}] = (\text{length})^{0}, \\
&[\varsigma^{\hat a}],\,[\hat e^{\hat a}{}_{a}] = (\text{length})^{0}, \qquad [A^{\hat a\hat b}_{\mu}] = (\text{length})^{-1}, \qquad [F^{\hat a\hat b}_{\mu\nu}] = (\text{length})^{-2},
\end{aligned}
\tag{C.141}
$$

as inputs. By dimensional analysis only the term

$$
\left(\int \mathrm{d}^{2}x\,\partial_{\mu}\varsigma^{\hat a}\varsigma^{\hat b}\right)^{n} \equiv \left(\int\limits_{x} \partial^{x}_{\mu}\varsigma^{\hat a}\varsigma^{\hat b}\right)^{n}
\tag{C.142}
$$

with $n=2$ taken from the cumulant expansion satisfies

$$
\left[\left(\int \mathrm{d}^{2}x\,\partial_{\mu}\varsigma^{\hat a}\varsigma^{\hat b}\right)^{n}\right] = (\text{length})^{2}.
\tag{C.143}
$$

Hence, to second order in the cumulant expansion, we need to consider the contribution

$$
\delta S^{2}[\psi] := \frac{1}{2}\left(\left\langle\left(S^{(1)}_{2}\right)^{2}\right\rangle^{\varsigma}_{0} - \left\langle S^{(1)}_{2}\right\rangle^{\varsigma}_{0}\left\langle S^{(1)}_{2}\right\rangle^{\varsigma}_{0}\right).
\tag{C.144}
$$

With an abbreviated notation for integrals, we proceed with the evaluation of the connected expectation value

$$
\begin{aligned}
\delta S^{2}[\psi] = \frac{1}{2}\Bigg[&\left\langle\left(\int\limits_{x} A^{\hat a\hat b}_{\mu}\partial_{\mu}\varsigma^{\hat a}\varsigma^{\hat b}\right)\left(\int\limits_{y} A^{\hat c\hat d}_{\nu}\partial_{\nu}\varsigma^{\hat c}\varsigma^{\hat d}\right)\right\rangle^{\varsigma}_{0} \\
&- \left\langle\left(\int\limits_{x} A^{\hat a\hat b}_{\mu}\partial_{\mu}\varsigma^{\hat a}\varsigma^{\hat b}\right)\right\rangle^{\varsigma}_{0}\left\langle\left(\int\limits_{y} A^{\hat c\hat d}_{\nu}\partial_{\nu}\varsigma^{\hat c}\varsigma^{\hat d}\right)\right\rangle^{\varsigma}_{0}\Bigg].
\end{aligned}
\tag{C.145}
$$

Wick's theorem implies that the first expectation value delivers three distinct products of two Green functions, one of which cancels upon subtraction of the expectation value squared. Two distinct products of Green functions contribute additively after performing the average over the fast fields,

$$\delta S^2[\psi] = \frac{1}{2}\int\limits_x\int\limits_y A^{\hat{a}\hat{b}}_\mu A^{\hat{c}\hat{d}}_\nu \Big[\delta^{\hat{a}\hat{c}}\delta^{\hat{b}\hat{d}}\partial^x_\mu\partial^y_\nu G_0(x-y)G_0(x-y) + \delta^{\hat{a}\hat{d}}\delta^{\hat{b}\hat{c}}\partial^x_\mu G_0(x-y)\partial^y_\nu G_0(x-y)\Big]. \tag{C.146}$$

We insert the momentum representation of the Green function (C.137) to find

$$\delta S^2[\psi] = \frac{\mathfrak{a}^{2(d-2)}}{2}\int\limits_x\int\limits_y A^{\hat{a}\hat{b}}_\mu A^{\hat{c}\hat{d}}_\nu \Bigg[\delta^{\hat{a}\hat{c}}\delta^{\hat{b}\hat{d}}\int\limits_k\int\limits_{k'}\frac{k_\mu k_\nu}{(k^2+M^2)(k'^2+M^2)}e^{\mathrm{i}(k+k')(x-y)} +\delta^{\hat{a}\hat{d}}\delta^{\hat{b}\hat{c}}\int\limits_k\int\limits_{k'}\frac{k_\mu k'_\nu}{(k^2+M^2)(k'^2+M^2)}e^{\mathrm{i}(k+k')(x-y)}\Bigg]. \tag{C.147}$$

We are ready to perform one of the integrations over position space, say the y integration.[17] There results the delta function $(2\pi)^d\,\delta(k+k')$ in the numerator that multiplies $k_\mu\, k_\nu$ in the first contribution from Wick's theorem and that multiplies $k_\mu\, k'_\nu$ in the second contribution from Wick's theorem. Integration over one of the two momenta, say k', produces a relative sign difference between the two contributions with the common factor

$$\mathfrak{a}^{d-2}\int\frac{\mathrm{d}^dk}{(2\pi)^d}\frac{k_\mu k_\nu}{(k^2+M^2)(k^2+M^2)} = \mathfrak{a}^{d-2}\int\frac{\mathrm{d}^dk}{(2\pi)^d}\frac{(1/d)\,\delta_{\mu\nu}}{k^2+M^2} \tag{C.148}$$

as $M\to 0$. The right-hand side is thus proportional to the Green function (C.137) evaluated at a vanishing argument in position space and we are left with, if we choose $d=2$,

$$\begin{aligned}\delta S^2[\psi] &= \frac{1}{2}\int\limits_x A^{\hat{a}\hat{b}}_\mu A^{\hat{c}\hat{d}}_\nu\left(\delta^{\hat{a}\hat{c}}\delta^{\hat{b}\hat{d}}-\delta^{\hat{a}\hat{d}}\delta^{\hat{b}\hat{c}}\right)G_0(0)\frac{1}{2}\delta_{\mu\nu} + \text{subleading corrections}\\ &= G_0(0)\frac{1}{4}\int\limits_x\left(A^{\hat{a}\hat{b}}_\mu A^{\hat{a}\hat{b}}_\mu - A^{\hat{a}\hat{b}}_\mu A^{\hat{b}\hat{a}}_\mu\right) + \text{subleading corrections}\\ &= G_0(0)\frac{1}{2}\int\limits_x A^{\hat{a}\hat{b}}_\mu A^{\hat{a}\hat{b}}_\mu + \text{subleading corrections}.\end{aligned} \tag{C.149}$$

To reach the final line, we made use of $A^{\hat{b}\hat{a}}_\mu = -A^{\hat{a}\hat{b}}_\mu$ that follows from Eq. (C.122). Subtraction of Eq. (C.149) from Eq. (C.140) cancels the term quadratic in the gauge

---

[17] We here use the fact that $A^{\hat{a}\hat{b}}_\mu(x)$ and $A^{\hat{c}\hat{d}}_\nu(y)$ vary slowly in position space so that we may write $A^{\hat{a}\hat{b}}_\mu(x)\,A^{\hat{c}\hat{d}}_\nu(y) = A^{\hat{a}\hat{b}}_\mu(x)\,A^{\hat{c}\hat{d}}_\nu(x+(y-x))$ and do a Taylor expansion in powers of $(y-x)$ that we truncate to zero-th order.

field. We conclude that

$$\begin{aligned}\delta S[\psi] &= G_0(0)\frac{1}{2}\int \mathrm{d}^2x\,\partial_\mu\psi^i\partial_\mu\psi^j R_{iklj}(\psi)\,\hat{e}_{\hat{b}}{}^k(\psi)\hat{e}_{\hat{b}}{}^l(\psi)\\ \text{Eq. (C.116c)} &= G_0(0)\frac{1}{2}\int \mathrm{d}^2x\,\partial_\mu\psi^i\partial_\mu\psi^j R_{iklj}(\psi)\,\mathfrak{g}^{kl}\\ \text{Eq. (C.107)} &= G_0(0)\frac{1}{2}\int \mathrm{d}^2x\,\partial_\mu\psi^i\partial_\mu\psi^j R_{ljik}(\psi)\mathfrak{g}^{kl}\\ \text{Eq. (C.61)} &= -G_0(0)\frac{1}{2}\int \mathrm{d}^2x\,\partial_\mu\psi^i\partial_\mu\psi^j R_{ij}(\psi) \end{aligned}\tag{C.150}$$

is the leading contribution to the increment of the action of the slow mode ψ resulting from the integration over the fast mode ζ within the second-order cumulant expansion. The sign convention for the Ricci tensor is to be found in Eq. (C.61).

C.6.7 *One-loop scaling flow obeyed by the metric tensor*

In $d = (2+\epsilon)$-dimensional Euclidean space (ϵ infinitesimal), the one-loop scaling flow obeyed by the metric tensor follows from the equality

$$\int \frac{\mathrm{d}^d x}{\mathfrak{a}'^{d-2}}\frac{1}{2}\mathfrak{g}'_{ab}(\psi)\partial_\mu\psi^a\,\partial_\mu\psi^b = \int \frac{\mathrm{d}^d x}{\mathfrak{a}^{d-2}}\frac{1}{2}\left[\mathfrak{g}_{ab}(\psi) - G_0(0)\,R_{ab}(\psi)\right]\partial_\mu\psi^a\,\partial_\mu\psi^b, \tag{C.151a}$$

between the action $S[\psi]$ with the ultraviolet cut-off $\Lambda' = \frac{1}{\mathfrak{a}'}$ and the renormalized action $S[\psi] + \delta S[\psi]$ with the ultraviolet cut-off $\Lambda = \frac{1}{\mathfrak{a}}$, whereby

$$0 < 1 - \frac{\Lambda'}{\Lambda} = 1 - \frac{\mathfrak{a}}{\mathfrak{a}'} \equiv 1 - \frac{\mathfrak{a}}{\mathfrak{a}+\mathrm{d}\mathfrak{a}} = \frac{\mathrm{d}\mathfrak{a}}{\mathfrak{a}} + \cdots \tag{C.151b}$$

is infinitesimal.

To prove Eq. (C.1), it suffices to do

$$\begin{aligned}\frac{\mathfrak{g}'_{ab}(\psi)}{\mathfrak{a}'^{d-2}} &\equiv \frac{\mathfrak{g}_{ab}(\psi)+\mathrm{d}\mathfrak{g}_{ab}(\psi)}{(\mathfrak{a}+\mathrm{d}\mathfrak{a})^{d-2}}\\ &= \frac{1}{\mathfrak{a}^{d-2}}\left[\mathfrak{g}_{ab}(\psi) - (d-2)\frac{\mathrm{d}\mathfrak{a}}{\mathfrak{a}}\mathfrak{g}_{ab}(\psi) + \mathrm{d}\mathfrak{g}_{ab}(\psi) + \cdots\right],\end{aligned}\tag{C.152}$$

on the left-hand side of Eq. (C.151a), and to use the $M \to 0$ limit of Eq. (C.137), i.e.,

$$\begin{aligned}\lim_{M\to 0} G_0(0) &= \mathfrak{a}^{d-2}\,\frac{\Omega(d)}{(2\pi)^d}\,\frac{\Lambda^{d-2}-\Lambda'^{d-2}}{d-2}\\ &= \frac{\Omega(d)}{(2\pi)^d}\frac{\mathrm{d}\mathfrak{a}}{\mathfrak{a}} + \cdots\\ &= \frac{1}{2\pi}\frac{\mathrm{d}\mathfrak{a}}{\mathfrak{a}} + \mathcal{O}(\epsilon) + \cdots,\end{aligned}\tag{C.153}$$

on the right-hand side of Eq. (C.151a).

C.7 How many couplings flow on a NLσM?

The flow encoded by Eq. (C.1) involves N^2 real-valued couplings on a generic N-dimensional Riemannian manifold. This is the largest number of independent running couplings that a N-dimensional Riemannian manifold can support. However, the number of independent running couplings on a N-dimensional Riemannian manifold may be reduced from N^2 if algebraic structures associated to symmetries characterize the Riemannian manifold.

For example, the $O(3)$ NLσM has a two-dimensional Riemannian manifold, the unit sphere, but only one running coupling constant. The unit sphere is an example of a symmetric space. A symmetric space is an homogeneous space with a special property. We have introduced homogeneous spaces through their cosets G/H in Eq. (3.47). The unit sphere is thus to be understood as a $O(3)/O(2)$ coset. Symmetric spaces such as the $O(3)$ NLσM models are special cases of homogeneous spaces in that H is the maximal subgroup of G.

The number of coupling constants for a NLσM on a coset space G/H can be understood as follows.

If H is a maximal subgroup, i.e., the target space is a symmetric space, then the NLσM has precisely one coupling constant. If there exists precisely one intervening subgroup H', i.e.,

$$H \subset H' \subset G, \tag{C.154}$$

then the NLσM with target space G/H turns out to have precisely two independent coupling constants. If, in the latter case, we run the RG into the infrared (i.e., to large length scales), then one of the two coupling constants disappears. Thus, we end up with a NLσM with one coupling constant, whose target space is either G/H' or H'/H (both are, in the current case, by assumption symmetric spaces in the sense of maximal subgroups). The number of coupling constants for a NLσM on a homogeneous space with more intervening subgroups increases accordingly.

If we begin with a NLσM on a general homogeneous space, the RG will select, asymptotically at long length scales, a target space which is a symmetric space G/H where H is a maximal subgroup of G. So, all NLσM on coset spaces become NLσM on symmetric spaces, asymptotically at long length scales. It is for this reason that they are the "stable" large-scale limits, and appear naturally, without fine-tuning.

A NLσM (on a, in general, possibly homogeneous and not necessarily symmetric space) is determined by the very general principle of symmetry breaking. The group G is the global symmetry group of the problem. The subgroup H (not necessarily maximal) characterizes the symmetries which are preserved when the global symmetry G is broken.

The NLσM manifold that encodes the Goldstone modes [the fields $\pi_1, \cdots, \pi_{N-1}$ for the $\mathrm{O}(N)/\mathrm{O}(N-1)$ NLσM] is in general G/H.

Appendix D

The Villain model

Our RG treatment of the KT transition in chapter 4 has relied on the factorization of the spin waves from the vortices. This factorization might be an artifact of the gradient expansion performed on the cosine in Eq. (4.8). To dispel this possibility, it would thus be highly desirable to construct an alternative lattice model for which this factorization is exact and for which the continuum limit delivers the Kosterlitz-Thouless RG flow (4.107). The first task was achieved by Villain in Ref. [44]. The verification that the $2d$–XY model and the Villain model belong to the same universality class is often attributed to Ref. [45]. In this appendix, we shall define the Villain model and show that the factorization of the spin waves from the vortices is exact.

We begin with the partition function

$$Z[K_{ij},A_{ij}] := \left(\prod_{i=1}^{L^2}\int\limits_0^{2\pi}\frac{\mathrm{d}\phi_i}{2\pi}\right)\prod_{\langle ij\rangle}^{2L^2} e^{-\frac{K_{ij}}{2}[1-\cos(\phi_i-\phi_j-A_{ij})]}, \tag{D.1}$$

for the $2d$–XY model on a square lattice made of L^2 sites. Directed links (two per site) on the square lattice are denoted by $\langle ij\rangle$. The reduced spin stiffness $K_{ij}>0$ and the phase $A_{ij}\in\mathbb{R}$ are not required to be uniform throughout the lattice. Furthermore, the phase A_{ij} does not always need to be restricted to $0\le A_{ij}<2\pi$, even though it enters as an argument of a function with the periodicity 2π. An example when A_{ij} is not defined modulo 2π occurs when it is to be interpreted as a random phase whose probability distribution is not periodic with period 2π.

The Villain model consists in defining on the same lattice the partition function

$$\begin{aligned} Z_{\mathrm{V}}[K_{ij},A_{ij}] :&= e^{-\frac{K}{2}2L^2}\left(\prod_{i=1}^{L^2}\int\limits_0^{2\pi}\frac{\mathrm{d}\phi_i}{2\pi}\right)\prod_{\langle ij\rangle}^{2L^2}\sum_{l_{ij}\in\mathbb{Z}} e^{-\frac{K_{ij}}{2}\left(\phi_i-\phi_j-A_{ij}-2\pi l_{ij}\right)^2} \\ &= e^{-KL^2}\left(\prod_{i=1}^{L^2}\int\limits_0^{2\pi}\frac{\mathrm{d}\phi_i}{2\pi}\right)\sum_{\{l_{ij}\}\in\mathbb{Z}^{2L^2}} e^{-\frac{K_{ij}}{2}\sum\limits_{\langle ij\rangle}\left(\phi_i-\phi_j-A_{ij}-2\pi l_{ij}\right)^2}. \end{aligned} \tag{D.2a}$$

Here,

$$K := \frac{1}{2L^2} \sum_{\langle ij \rangle} K_{ij}. \tag{D.2b}$$

The periodicity under a shift of any ϕ_i by 2π is preserved in the Villain action. The non-linearity of the cosine has been removed in the Villain action.

To proceed, we need to introduce some notation. Given the link degree of freedom A_{ij}, we define the longitudinal and transversal components $A^{\parallel}_{ij}$ and $A^{\perp}_{ij}$, respectively, by

$$\begin{aligned}
&A_{ij} := A^{\parallel}_{ij} + A^{\perp}_{ij},\\
&\operatorname{curl}_{\tilde{i}}\, A^{\parallel}_{ij} := A^{\parallel}_{i(i+\hat{\boldsymbol{x}})} + A^{\parallel}_{(i+\hat{\boldsymbol{x}})(i+\hat{\boldsymbol{x}}+\hat{\boldsymbol{y}})} + A^{\parallel}_{(i+\hat{\boldsymbol{x}}+\hat{\boldsymbol{y}})(i+\hat{\boldsymbol{y}})} + A^{\parallel}_{(i+\hat{\boldsymbol{y}})i} = 0,\\
&\operatorname{div}_i\, A^{\perp}_{ij} := A^{\perp}_{i(i+\hat{\boldsymbol{x}})} - A^{\perp}_{(i-\hat{\boldsymbol{x}})i} + A^{\perp}_{i(i+\hat{\boldsymbol{y}})} - A^{\perp}_{(i-\hat{\boldsymbol{y}})i} = 0.
\end{aligned} \tag{D.3}$$

We have introduced the basis vectors $\hat{\boldsymbol{x}}$ and $\hat{\boldsymbol{y}}$ of the square lattice and the dual lattice labeled by the sites $\tilde{i} := i + \frac{1}{2}\hat{\boldsymbol{x}} + \frac{1}{2}\hat{\boldsymbol{y}}$. The important point is that the lattice curl yields a dual lattice scalar, whereas the lattice divergence yields a lattice scalar. We now check that this decomposition into longitudinal and transversal components yields the correct number of degrees of freedom. There are $2L^2$ independent links degrees of freedom A_{ij}. There are only L^2 independent link degrees of freedom $A^{\parallel}_{ij}$ due to L^2 constraints $\operatorname{curl}_{\tilde{i}}\, A^{\parallel}_{ij} = 0$, $\tilde{i}$ in the dual lattice. There are only L^2 independent link degrees of freedom $A^{\perp}_{ij}$ due to L^2 constraints $\operatorname{div}_i\, A^{\perp}_{ij} = 0$, i in the lattice. For completeness, the gradient of a lattice scalar can be defined by

$$\operatorname{grad}_{\hat{\mu}}\, \phi_i := \phi_{i+\hat{\mu}} - \phi_i, \qquad \hat{\mu} = \hat{\boldsymbol{x}}, \hat{\boldsymbol{y}}. \tag{D.4}$$

We are now ready to proceed with the Villain model. Introduce two lattice degrees of freedom $\alpha_i \in \mathbb{R}$ and $\beta_i \in \mathbb{Z}$ in terms of which we can write

$$A^{\parallel}_{ij} =: \alpha_i - \alpha_j, \qquad l^{\parallel}_{ij} =: \beta_i - \beta_j. \tag{D.5}$$

One verifies that both $A^{\parallel}_{ij}$ and $l^{\parallel}_{ij}$ are indeed curl free. The introduction of L^2 lattice scalar degrees of freedom β_i, $i = 1, \cdots, L^2$, allows to extend the range of allowed values that ϕ_i can take,

$$\begin{aligned}
Z_{\mathrm{V}}[K_{ij}, \alpha_i, A^{\perp}_{ij}] &= e^{-KL^2} \left(\prod_{i=1}^{L^2} \delta(\operatorname{div}_i A^{\perp}_{ij}) \right) \left(\prod_{i=1}^{L^2} \int_0^{2\pi} \frac{\mathrm{d}\phi_i}{2\pi} \right) \prod_{\langle ij \rangle}^{2L^2} \sum_{\beta_i \in \mathbb{Z}} \sum_{l^{\perp}_{ij} \in \mathbb{Z}} \delta(\operatorname{div}_i l^{\perp}_{ij})\\
&\quad \times e^{-\frac{K_{ij}}{2} \left[(\phi_i - \alpha_i - 2\pi\beta_i) - (\phi_j - \alpha_j - 2\pi\beta_j) - A^{\perp}_{ij} - 2\pi l^{\perp}_{ij} \right]^2}\\
&= e^{-KL^2} \left(\prod_{i=1}^{L^2} \delta(\operatorname{div}_i A^{\perp}_{ij}) \right) \left(\prod_{i=1}^{L^2} \int_{-\infty}^{+\infty} \frac{\mathrm{d}\phi'_i}{2\pi} \right) \prod_{\langle ij \rangle}^{2L^2} \sum_{l^{\perp}_{ij} \in \mathbb{Z}} \delta(\operatorname{div}_i l^{\perp}_{ij})\\
&\quad \times e^{-\frac{K_{ij}}{2} \left(\phi'_i - \phi'_j - A^{\perp}_{ij} - 2\pi l^{\perp}_{ij} \right)^2},
\end{aligned} \tag{D.6}$$

where we have performed a shift of integration variables $\phi'_i := \phi_i - \alpha_i - 2\pi\beta_i$.

The second step consists in performing a Hubbard-Stratonovich transformation on each link (recall that we have assumed that $K_{ij} > 0$),

$$Z_{\rm V}[K_{ij}, A^{\perp}_{ij}] = e^{-KL^2} \left(\prod_{i=1}^{L^2} \delta(\mathrm{div}_i A^{\perp}_{ij}) \right) \left(\prod_{i=1}^{L^2} \int\limits_{-\infty}^{+\infty} \frac{\mathrm{d}\phi'_i}{2\pi} \right) \prod_{\langle ij \rangle}^{2L^2} \sum_{l^{\perp}_{ij} \in \mathbb{Z}} \delta(\mathrm{div}_i l^{\perp}_{ij})$$
$$\times \frac{1}{\sqrt{2\pi K_{ij}}} \int\limits_{-\infty}^{+\infty} \mathrm{d}\psi_{ij} e^{-\frac{1}{2K_{ij}}\psi^2_{ij} - \mathrm{i}\psi_{ij}\left(\phi'_i - \phi'_j - A^{\perp}_{ij} - 2\pi l^{\perp}_{ij}\right)}. \tag{D.7}$$

The advantage of this transformation is that the integral over ϕ'_i with $i = 1, \cdots, L^2$ can be performed,

$$Z_{\rm V}[K_{ij}, A^{\perp}_{ij}] = e^{-KL^2} \left(\prod_{i=1}^{L^2} \delta(\mathrm{div}_i A^{\perp}_{ij}) \right) \left(\prod_{\langle ij \rangle}^{2L^2} \int\limits_{-\infty}^{+\infty} \frac{\mathrm{d}\psi_{ij}}{\sqrt{2\pi K_{ij}}} \right)$$
$$\times \prod_{\langle ij \rangle}^{2L^2} \sum_{l^{\perp}_{ij} \in \mathbb{Z}} \delta(\mathrm{div}_i l^{\perp}_{ij})\, \delta(\mathrm{div}_i \psi_{ij}) e^{-\frac{1}{2K_{ij}}\psi^2_{ij} + \mathrm{i}\psi_{ij}\left(A^{\perp}_{ij} + 2\pi l^{\perp}_{ij}\right)}. \tag{D.8}$$

Again, not all $2L^2$ ψ_{ij} are independent, only L^2 are, due to the L^2 constraints enforced by the L^2 integrations over all ϕ'_i's. The crucial point here is that the Villain partition function has been expressed solely in terms of *transversal lattice link degrees of freedom.*

The next step is to dispose of all three constraints

$$\mathrm{div}_i A^{\perp}_{ij} = \mathrm{div}_i l^{\perp}_{ij} - \mathrm{div}_i \psi_{ij} = 0, \tag{D.9}$$

by going over to the dual lattice. More precisely, recall that if our original lattice is the set of points

$$\Lambda := \{ i = (m, n) | m, n \in \mathbb{Z} \}, \tag{D.10}$$

then the dual lattice is the set of points (having set the lattice spacing a to unity)

$$\Lambda^{\star} := \left\{ \tilde{i} = \left(\frac{1}{2} + m, \frac{1}{2} + n \right) \middle| m, n \in \mathbb{Z} \right\}. \tag{D.11}$$

In other words, the lattice and its dual are two isomorphic sets Λ and $\Lambda^{\star}$, respectively, with the isomorphism between sites given by the mapping $\Lambda \to \Lambda^{\star}$ defined by

$$(m, n) \to \left(\frac{1}{2} + m, \frac{1}{2} + n \right). \tag{D.12}$$

To complete the isomorphism, we need to construct a one-to-one correspondence between the nearest-neighbor links in Λ and $\Lambda^{\star}$. The isomorphism between directed-nearest-neighbor links is defined as follows,

$$\begin{aligned} \langle i(i + \boldsymbol{x}) \rangle &\to \langle \tilde{i}(\tilde{i} - \boldsymbol{y}) \rangle, \\ \langle i(i + \boldsymbol{y}) \rangle &\to \langle (\tilde{i} - \boldsymbol{x})\tilde{i} \rangle. \end{aligned} \tag{D.13}$$

In other words, the isomorphism is fully specified by the assignment

$$\left[i,\langle i(i+\hat{\boldsymbol{x}})\rangle,\langle i(i+\hat{\boldsymbol{y}})\rangle\right]\to\left[\tilde{i},\langle\tilde{i}(\tilde{i}-\hat{\boldsymbol{y}})\rangle,\langle(\tilde{i}-\hat{\boldsymbol{x}})\tilde{i})\rangle\right]. \tag{D.14}$$

In words, the site dual to i sits in the middle of the plaquette having i as lower left corner. The nearest-neighbor link dual to $\langle ij\rangle$ is the only nearest-neighbor link on the dual lattice intersecting $\langle ij\rangle$. Directed links on the dual lattice are oriented from up-to-down and left-to-right if on the original lattice they correspond to left-to-right and down-to-up, respectively. Clearly, $\hat{\boldsymbol{x}}\to-\hat{\boldsymbol{y}}$ and $\hat{\boldsymbol{y}}\to+\hat{\boldsymbol{x}}$ (a clockwise rotation by $\pi/2$ in this convention) when going over to the dual lattice. According to these rules, one verifies that the links on a plaquette of the original lattice map onto four (nearest-neighbors) dual links sharing the site dual to the one sitting on the lower left corner of the plaquette. Conversely, the four independent nearest-neighbor links on the original lattice sharing the site i map onto the dual links defining a plaquette with upper right corner $\tilde{i}$ on the dual lattice. Moreover, our definition was chosen so that curl and div are interchanged when going over to the dual lattice as can be checked directly with the rule

$$\begin{aligned}\hat{\boldsymbol{x}}\partial_x+\hat{\boldsymbol{y}}\partial_y&\to-\hat{\boldsymbol{y}}\partial_x+\hat{\boldsymbol{x}}\partial_y,\\ \hat{\boldsymbol{x}}\partial_y-\hat{\boldsymbol{y}}\partial_x&\to-\hat{\boldsymbol{y}}\partial_y-\hat{\boldsymbol{x}}\partial_x.\end{aligned} \tag{D.15}$$

We are now ready to take advantage of the relationship between the square lattice and its dual. We take the divergence of a lattice-link degree of freedom, say $\mathrm{div}_i\psi_{ij}$. We define a lattice-link degree of freedom $\psi_{\tilde{i}\tilde{j}}$ on the dual lattice. There are as many independent ψ_{ij} as they are $\psi_{\tilde{i}\tilde{j}}$. Consider the isomorphism between lattice-link degrees of freedom, namely

$$\psi_{i(i+\hat{\boldsymbol{x}})}\to\psi_{\tilde{i}(\tilde{i}-\hat{\boldsymbol{y}})},\qquad \psi_{i(i+\hat{\boldsymbol{y}})}\to\psi_{(\tilde{i}-\hat{\boldsymbol{x}})\tilde{i}}, \tag{D.16}$$

and require that it preserves linearity, say

$$\begin{aligned}\mathrm{div}_i\psi_{ij}\equiv&\,\psi_{i(i+\hat{\boldsymbol{x}})}-\psi_{(i-\hat{\boldsymbol{x}})i}+\psi_{i(i+\hat{\boldsymbol{y}})}-\psi_{(i-\hat{\boldsymbol{y}})i}\\ \to&\,\psi_{\tilde{i}(\tilde{i}-\hat{\boldsymbol{y}})}-\psi_{(\tilde{i}-\hat{\boldsymbol{x}})(\tilde{i}-\hat{\boldsymbol{x}}-\hat{\boldsymbol{y}})}+\psi_{(\tilde{i}-\hat{\boldsymbol{x}})\tilde{i}}-\psi_{(\tilde{i}-\hat{\boldsymbol{y}}-\hat{\boldsymbol{x}})(\tilde{i}-\hat{\boldsymbol{y}})}\\ =&\,\psi_{\tilde{i}(\tilde{i}-\hat{\boldsymbol{y}})}+\psi_{(\tilde{i}-\hat{\boldsymbol{x}}-\hat{\boldsymbol{y}})(\tilde{i}-\hat{\boldsymbol{x}})}+\psi_{(\tilde{i}-\hat{\boldsymbol{x}})\tilde{i}}+\psi_{(\tilde{i}-\hat{\boldsymbol{y}})(\tilde{i}-\hat{\boldsymbol{y}}-\hat{\boldsymbol{x}})}\\ =&\,:\mathrm{curl}_{\tilde{i}}\,\psi_{\tilde{i}\tilde{j}},\end{aligned} \tag{D.17}$$

where we have used the convention that $\psi_{ij}=-\psi_{ji}$ on the lattice (as well as on the dual lattice).

We are now ready to take advantage of the fact that the partition function in Eq. (D.8) is solely expressed in terms of transversal lattice links degrees of freedom. We could have equally well used purely longitudinal link degrees of freedom on the dual lattice,

$$
\begin{aligned}
Z_{\mathrm{V}}[K_{\tilde{i}\tilde{j}}, A^{\perp}_{\tilde{i}\tilde{j}}] = {}& e^{-KL^2}\left[\prod_{\tilde{i}=1}^{L^2}\delta(\mathrm{curl}_i\,A^{\perp}_{\tilde{i}\tilde{j}})\right]\sum_{\left\{l^{\perp}_{\tilde{i}\tilde{j}}\right\}\in\mathbb{Z}^{2L^2}}\delta(\mathrm{curl}_i\,l^{\perp}_{\tilde{i}\tilde{j}})\left(\prod_{\langle ij\rangle}^{2L^2}\int_{-\infty}^{+\infty}\frac{\mathrm{d}\psi_{\tilde{i}\tilde{j}}}{\sqrt{2\pi K_{\tilde{i}\tilde{j}}}}\right)\\
&\times\prod_{\langle\tilde{i}\tilde{j}\rangle}^{2L^2}\delta(\mathrm{curl}_i\,\psi_{\tilde{i}\tilde{j}})e^{-\frac{1}{2K_{\tilde{i}\tilde{j}}}\psi^2_{\tilde{i}\tilde{j}}+\mathrm{i}\psi_{\tilde{i}\tilde{j}}\left(A^{\perp}_{\tilde{i}\tilde{j}}+2\pi l^{\perp}_{\tilde{i}\tilde{j}}\right)},
\end{aligned}
\tag{D.18}
$$

where we have interchanged the sum over all L^2 independent degrees of freedom $l^{\perp}_{\tilde{i}\tilde{j}}$ with integration over $\psi_{\tilde{i}\tilde{j}}$'s. We implement the constraint that $\psi_{\tilde{i}\tilde{j}}$ has vanishing curl by introducing the scalar lattice field $\varphi_{\tilde{i}}$ through

$$
\psi_{\tilde{i}\tilde{j}} =: \varphi_{\tilde{i}} - \varphi_{\tilde{j}}.
\tag{D.19}
$$

We then account for the $2\times L^2$ terms in the product over links so as to obtain

$$
\begin{aligned}
Z_{\mathrm{V}}[K_{\tilde{i}\tilde{j}}, A^{\perp}_{\tilde{i}\tilde{j}}] = {}& e^{-KL^2-\frac{1}{2}\sum\limits_{\langle\tilde{i}\tilde{j}\rangle}^{2L^2}\ln K_{\tilde{i}\tilde{j}}}\left[\prod_{\tilde{i}=1}^{L^2}\delta\left(\mathrm{curl}_i\,A^{\perp}_{\tilde{i}\tilde{j}}\right)\right]\sum_{\left\{l^{\perp}_{\tilde{i}\tilde{j}}\right\}\in\mathbb{Z}^{2L^2}}\delta\left(\mathrm{curl}_i\,l^{\perp}_{\tilde{i}\tilde{j}}\right)\left(\prod_{\tilde{i}=1}^{L^2}\int_{-\infty}^{+\infty}\frac{\mathrm{d}\varphi_{\tilde{i}}}{2\pi}\right)\\
&\times\prod_{\tilde{i}=1}^{L^2}e^{-\frac{1}{2}\sum\limits_{\hat{\mu}=-\hat{x},+\hat{y}}K^{-1}_{\tilde{i}(\tilde{i}+\hat{\mu})}\left[\left(\varphi_{\tilde{i}}-\varphi_{\tilde{i}+\hat{\mu}}\right)^2-2\mathrm{i}\left(\varphi_{\tilde{i}}-\varphi_{\tilde{i}+\hat{\mu}}\right)K_{\tilde{i}(\tilde{i}+\hat{\mu})}\left(A^{\perp}_{\tilde{i}(\tilde{i}+\hat{\mu})}+2\pi l^{\perp}_{\tilde{i}(\tilde{i}+\hat{\mu})}\right)\right]}\\
= {}& e^{-KL^2-\frac{1}{2}\sum\limits_{\langle\tilde{i}\tilde{j}\rangle}^{2L^2}\ln K_{\tilde{i}\tilde{j}}}\left[\prod_{\tilde{i}=1}^{L^2}\delta(\mathrm{curl}_i\,A^{\perp}_{\tilde{i}\tilde{j}})\right]\sum_{\{l^{\perp}_{\tilde{i}\tilde{j}}\}\in\mathbb{Z}^{2L^2}}\delta(\mathrm{curl}_i\,l^{\perp}_{\tilde{i}\tilde{j}})\left(\prod_{\tilde{i}=1}^{L^2}\int_{-\infty}^{+\infty}\frac{\mathrm{d}\varphi_{\tilde{i}}}{2\pi}\right)\\
&\times\prod_{\tilde{i}=1}^{L^2}e^{-\frac{1}{2}\sum\limits_{\hat{\mu}=-\hat{x},+\hat{y}}K^{-1}_{\tilde{i}(\tilde{i}+\hat{\mu})}\left[(\mathrm{grad}_{\hat{\mu}}\varphi_{\tilde{i}})^2-2\mathrm{i}(\mathrm{grad}_{\hat{\mu}}\varphi_{\tilde{i}})K_{\tilde{i}(\tilde{i}+\hat{\mu})}\left(A^{\perp}_{\tilde{i}(\tilde{i}+\hat{\mu})}+2\pi l^{\perp}_{\tilde{i}(\tilde{i}+\hat{\mu})}\right)\right]}.
\end{aligned}
\tag{D.20}
$$

This is our main result. It shows that one can decouple the spin wave from the vortex sector at the price of a shift of spin-wave-integration variable,

$$
\mathrm{grad}_{\hat{\mu}}\varphi'_{\tilde{i}} := \mathrm{grad}_{\hat{\mu}}\varphi_{\tilde{i}} + \mathrm{i}K_{\tilde{i}(\tilde{i}+\hat{\mu})}\left(A^{\perp}_{\tilde{i}(\tilde{i}+\hat{\mu})} + 2\pi l^{\perp}_{\tilde{i}(\tilde{i}+\hat{\mu})}\right).
\tag{D.21}
$$

For simplicity, we assume that K_{ij} is given by the real number K on directed-nearest-neighbor links, while it vanishes otherwise, to illustrate in this limit the factorization between spin waves and vortices. After performing an integration by part to convert the lattice gradient to a lattice divergence and ignoring boundary terms as follows from imposing periodic boundary conditions, we can integrate over all $\varphi_{\tilde{i}}$'s. More precisely, introducing a first set of L^2 degrees of freedom on the dual lattice

$$
\mathrm{div}_{\tilde{i}}A^{\perp}_{\tilde{i}\tilde{j}} =: \sqrt{2\pi}n_{\tilde{i}},
\tag{D.22}
$$

and a second set of L^2 degrees of freedom on the dual lattice

$$
\mathrm{div}_{\tilde{i}}l^{\perp}_{\tilde{i}\tilde{j}} =: \frac{1}{\sqrt{2\pi}}m_{\tilde{i}},
\tag{D.23}
$$

we find

$$
\begin{aligned}
Z_{\mathrm{V}}[K,n_{\tilde{i}}] &= e^{-(K+\ln K)\,L^2} \sum_{\left\{\frac{1}{\sqrt{2\pi}}m_{\tilde{i}}\right\}\in\mathbb{Z}^{L^2}} \left(\prod_{\tilde{i}=1}^{L^2}\int\limits_{-\infty}^{+\infty}\frac{\mathrm{d}\varphi_{\tilde{i}}}{2\pi}\right) \\
&\quad\times\prod_{\tilde{i}=1}^{L^2} e^{-\frac{1}{2K}\sum\limits_{\hat{\mu}=-\hat{\boldsymbol{x}},+\hat{\boldsymbol{y}}}\left[\left(\varphi_{\tilde{i}}-\varphi_{\tilde{i}+\hat{\mu}}\right)^2+2\mathrm{i}\sqrt{2\pi}\,K\left(n_{\tilde{i}}+m_{\tilde{i}}\right)\varphi_{\tilde{i}}\right]} \\
&= e^{-(K+\ln K)\,L^2}\, Z_{\mathrm{sw}} \sum_{\left\{\frac{1}{\sqrt{2\pi}}m_{\tilde{i}}\right\}\in\mathbb{Z}^{L^2}} \prod_{\tilde{i},\tilde{j}=1}^{L^2} e^{-\pi K(m_{\tilde{i}}+n_{\tilde{i}})D_{\tilde{i}\tilde{j}}(m_{\tilde{j}}+n_{\tilde{j}})}.
\end{aligned}
\tag{D.24a}
$$

Here, $D_{\tilde{i}\tilde{j}}$ is the Green function for the Villain model on the square lattice and Z_{SW} is the "spin-wave" partition function, namely the normalization factor induced by the integration over the $\varphi_{\tilde{i}}$'s. The precise definition of the Green function is given by

$$
\Delta_{\tilde{i}\tilde{j}}\; D_{\tilde{j}\tilde{k}} \;=\; -\delta_{\tilde{i}\tilde{k}}, \tag{D.24b}
$$

where Δ is defined by its action on a scalar lattice degree of freedom,

$$
\begin{aligned}
\Delta\phi_i &:= \mathrm{div}_i(\mathrm{grad}_{\hat{\mu}}\phi_i) \\
&= \mathrm{grad}_{\hat{\boldsymbol{x}}}\phi_i - \mathrm{grad}_{\hat{\boldsymbol{x}}}\phi_{i-\hat{\boldsymbol{x}}} + \mathrm{grad}_{\hat{\boldsymbol{y}}}\phi_i - \mathrm{grad}_{\hat{\boldsymbol{y}}}\phi_{i-\hat{\boldsymbol{y}}} \\
&= \phi_{i+\hat{\boldsymbol{x}}} + \phi_{i-\hat{\boldsymbol{x}}} - 2\phi_i + \hat{\boldsymbol{x}} \to \hat{\boldsymbol{y}}.
\end{aligned}
\tag{D.24c}
$$

Appendix E

Coherent states for fermions, Jordan-Wigner fermions, and linear-response theory

E.1 Grassmann coherent states

To simplify notation, we consider the two-dimensional fermionic Fock space $\mathcal{F}$ spanned by the fermionic algebra of operators

$$\{\hat{c}, \hat{c}^{\dagger}\} = 1, \qquad \{\hat{c}, \hat{c}\} = \{\hat{c}^{\dagger}, \hat{c}^{\dagger}\} = 0. \tag{E.1}$$

In other words, $\mathcal{F}$ is spanned by the two orthonormal vectors

$$|0\rangle, \qquad |1\rangle := \hat{c}^{\dagger}|0\rangle, \tag{E.2}$$

where $\hat{c}|0\rangle = 0$ defines the vacuum $|0\rangle$. To be concrete, define the fermionic harmonic oscillator by the Hamiltonian

$$\hat{H} := \hat{c}^{\dagger}\hat{c}. \tag{E.3}$$

There are two eigenvalues, 0 and 1, and the partition function at inverse temperature β is

$$Z(\beta) := \operatorname{tr}|_{\mathcal{F}}\, e^{-\beta \hat{H}} = 1 + e^{-\beta}. \tag{E.4}$$

Is it possible to construct a path-integral representation of this partition function as was done for the bosonic harmonic oscillator in appendix A.2? If the bosonic harmonic oscillator is to be a guide, we need to construct coherent states defined by the conditions that (i) they are eigenstates of the annihilation operator $\hat{c}$, (ii) overcomplete, and (iii) provide a resolution of the identity. The route to fermionic coherent states goes through the introduction of a *Grassmann algebra*.

A four-dimensional Grassmann algebra is defined by considering all possible polynomials with complex-valued coefficients that can be built from monomials in the two *independent* Grassmann numbers η^* and η that obey the multiplication rules

$$\{\eta, \eta^*\} = \{\eta, \eta\} = \{\eta^*, \eta^*\} = 0. \tag{E.5}$$

In other words, η^* and η are anticommuting numbers. A generic element of the Grassmann algebra $\mathfrak{G}$ is written

$$\mathfrak{a} \equiv a_1 + a_2\eta + a_3\eta^* + a_4\eta^*\eta = a_1 + a_2\eta + a_3\eta^* - a_4\eta\eta^*, \qquad a_{1,2,3,4} \in \mathbb{C}. \tag{E.6}$$

Observe that the Fock space $\mathcal{F}$ is a vector space spanned by $|0\rangle$ and $|1\rangle$ over the complex numbers,

$$\mathcal{F} := \left\{ |a_1, a_2\rangle \,\middle|\, |a_1, a_2\rangle = a_1\,|0\rangle + a_2\,|1\rangle, \qquad a_{1,2} \in \mathbb{C} \right\}. \tag{E.7}$$

By analogy, define the Grassmann Fock space $\mathcal{F}_{\mathfrak{G}}$ to be the vector space spanned by $|0\rangle$ and $|1\rangle$ over the Grassmann algebra (E.5-E.6), i.e.,

$$\begin{aligned}\mathcal{F}_{\mathfrak{G}} := \Big\{ |\mathfrak{a}, \mathfrak{b}\rangle \Big| |\mathfrak{a}, \mathfrak{b}\rangle = \; & \Big(a_1 + a_2\eta + a_3\eta^* + a_4\eta^*\eta\Big)|0\rangle \\ & + \Big(b_1 + b_2\eta + b_3\eta^* + b_4\eta^*\eta\Big)\,|1\rangle, \qquad a_i, b_i \in \mathbb{C}\Big\},\end{aligned} \tag{E.8a}$$

whereby the consistency rule that Grassmann numbers η and η^* anticommute with fermion annihilation $\hat{c}$ and creation $\hat{c}^\dagger$ operators,

$$\{\eta, \hat{c}\} = \{\eta, \hat{c}^\dagger\} = \{\eta^*, \hat{c}\} = \{\eta^*, \hat{c}^\dagger\} = 0, \tag{E.8b}$$

must be imposed.

The pair of fermionic coherent states $|a_2\eta\rangle$ and $|a_3^*\eta^*\rangle$ from the Grassmann Fock space $\mathcal{F}_{\mathfrak{G}}$ are defined by

$$|a_2\eta\;\rangle := e^{-a_2\eta\;\hat{c}^\dagger}|0\rangle \equiv \sum_{n=0}^{\infty} \frac{\left(-a_2\eta\;\hat{c}^\dagger\right)^n}{n!}|0\rangle = \left(1 - a_2\eta\;\hat{c}^\dagger\right)|0\rangle, \qquad a_2 \in \mathbb{C}, \tag{E.9a}$$

and

$$|a_3^*\eta^*\rangle := e^{-a_3^*\eta^*\hat{c}^\dagger}|0\rangle \equiv \sum_{n=0}^{\infty} \frac{\left(-a_3^*\eta^*\hat{c}^\dagger\right)^n}{n!}|0\rangle = \left(1 - a_3^*\eta^*\hat{c}^\dagger\right)|0\rangle, \qquad a_3^* \in \mathbb{C}, \tag{E.9b}$$

respectively. The corresponding pair of adjoint fermionic coherent states $\langle a_2\eta|$ and $\langle a_3^*\eta^*|$ from the Grassmann Fock space $\mathcal{F}_{\mathfrak{G}}$ are defined by

$$\langle a_2\eta\;| := \langle 0|e^{-\hat{c}\eta^* a_2^*} \equiv \langle 0|\sum_{n=0}^{\infty} \frac{\left(-\hat{c}\eta^* a_2^*\right)^n}{n!} = \langle 0|\left(1 - \hat{c}\eta^* a_2^*\right), \qquad a_2 \in \mathbb{C}, \tag{E.10a}$$

and

$$\langle a_3^*\eta^*| := \langle 0|e^{-\hat{c}\eta\;a_3} \equiv \langle 0|\sum_{n=0}^{\infty} \frac{\left(-\hat{c}\eta\;a_3\right)^n}{n!} = \langle 0|\left(1 - \hat{c}\eta\;a_3\right), \qquad a_3^* \in \mathbb{C}, \tag{E.10b}$$

respectively. With these definitions in hand, one verifies that $|a_2\eta\rangle$ ($|a_3^*\eta^*\rangle$) is a right eigenstate of $\hat{c}$ with Grassmann eigenvalue $a_2\eta$ ($a_3^*\eta^*$) and that $\langle a_2\eta|$ ($\langle a_3^*\eta^*|$) is a left eigenstate of $\hat{c}^\dagger$ with Grassmann eigenvalue $a_2^*\eta^*$ ($a_3\eta$),

$$\hat{c}|a_2\eta\;\rangle = (-1)^2 a_2\eta\;|0\rangle = a_2\eta\;|a_2\eta\;\rangle, \qquad \langle a_2\eta\;|\hat{c}^\dagger = \langle 0|a_2^*\eta^*(-1)^2 = \langle a_2\eta\;|a_2^*\eta^*, \tag{E.11a}$$

and

$$\hat{c}|a_3^*\eta^*\rangle = (-1)^2 a_3^*\eta^*|0\rangle = a_3^*\eta^*|a_3^*\eta^*\rangle, \qquad \langle a_3^*\eta^*|\hat{c}^\dagger = \langle 0|a_3\eta\;(-1)^2 = \langle a_3^*\eta^*|a_3\eta\;. \tag{E.11b}$$

Fermionic coherent states are neither normalized nor orthogonal,

$$\begin{aligned}
\langle\eta\,|\eta\,\rangle &= \langle 0|\,(1-\hat{c}\eta^*)\,(1-\eta\,\hat{c}^\dagger)\,|0\rangle = 1+(-1)^4\eta^*\eta\; = e^{+\eta^*\eta},\\
\langle\eta^*|\eta^*\rangle &= \langle 0|\,(1-\hat{c}\eta\,)\,(1-\eta^*\hat{c}^\dagger)\,|0\rangle = 1+(-1)^4\eta\;\eta^* = e^{-\eta^*\eta},\\
\langle\eta\,|\eta^*\rangle &= \langle 0|\,(1-\hat{c}\eta^*)\,(1-\eta^*\hat{c}^\dagger)\,|0\rangle = 1+(-1)^4\eta^*\eta^* = 1,\\
\langle\eta^*|\eta\,\rangle &= \langle 0|\,(1-\hat{c}\eta\,)\,(1-\eta\,\hat{c}^\dagger)\,|0\rangle = 1+(-1)^4\eta\;\eta\; = 1.
\end{aligned} \tag{E.12}$$

The expectation value of any normal-ordered operator $\hat{C}(\hat{c}^\dagger,\hat{c})$ in the fermionic coherent state $|\eta\rangle$ is

$$\langle\eta|\hat{C}(\hat{c}^\dagger,\hat{c})|\eta\rangle = e^{+\eta^*\eta}\,C(\eta^*,\eta). \tag{E.13}$$

Here, the Grassmann-valued function $C(\eta^*,\eta)$ is obtained from the operator $\hat{C}(\hat{c}^\dagger,\hat{c})$ by replacing $\hat{c}^\dagger$ with η^* and $\hat{c}$ with η.

Fermionic coherent states are merely a mathematical trick that allows a path-integral representation of partition functions for fermions.[18] To this end, we still need a resolution of the identity, which, in turn, demands the notion of Grassmann integration. Grassmann integrations $\int \mathrm{d}\eta$ and $\int \mathrm{d}\eta^*$ are multilinear mappings from the Grassmann algebra (E.5-E.6) to the complex numbers which are defined by linear extension of the rules

$$\begin{aligned}
&0=\int \mathrm{d}\eta\; 1, \qquad 0=\int \mathrm{d}\eta\;\,\eta^* = -\eta^*\int \mathrm{d}\eta\; 1, \qquad 1=\int \mathrm{d}\eta\;\,\eta\,,\\
&0=\int \mathrm{d}\eta^*\, 1, \qquad 0=\int \mathrm{d}\eta^*\,\eta\; = -\eta\;\int \mathrm{d}\eta^*\, 1, \qquad 1=\int \mathrm{d}\eta^*\,\eta^*,\\
&\int \mathrm{d}\eta^*\int \mathrm{d}\eta\quad\cdots = -\int \mathrm{d}\eta\int \mathrm{d}\eta^*\quad\cdots \equiv \int \mathrm{d}\eta^*\mathrm{d}\eta\quad\cdots.
\end{aligned} \tag{E.14}$$

Thus,

$$\begin{aligned}
&\int \mathrm{d}\eta\;\;(a_1+a_2\eta+a_3\eta^*+a_4\eta^*\eta) = a_2-a_4\eta^*, \qquad \forall a_{1,2,3,4}\in\mathbb{C},\\
&\int \mathrm{d}\eta^*\;(a_1+a_2\eta+a_3\eta^*+a_4\eta^*\eta) = a_3+a_4\eta\;, \qquad \forall a_{1,2,3,4}\in\mathbb{C},\\
&\int \mathrm{d}\eta^*\int \mathrm{d}\eta\;\;(a_1+a_2\eta+a_3\eta^*+a_4\eta^*\eta) = -a_4, \qquad \forall a_{1,2,3,4}\in\mathbb{C}.
\end{aligned} \tag{E.15}$$

Grassmann integration over the Grassmann Fock space $\mathcal{F}_{\mathfrak{G}}$ in Eq. (E.8) is the same as Grassmann integration over the Grassmann algebra with the caveat that $\hat{c}$ and $\hat{c}^\dagger$ anticommute with $\int \mathrm{d}\eta$ and $\int \mathrm{d}\eta^*$.

We are now ready to establish a resolution of the identity for fermionic coherent states. Indeed, with the help of the resolution of the identity

$$\mathbb{1}_{\mathcal{F}} = |0\rangle\langle 0| + |1\rangle\langle 1|, \tag{E.16}$$

[18] A word of caution here. Gauge potentials in classical electrodynamics were also thought to be mathematical curiosities before the advent of quantum mechanics.

in $\mathcal{F}$, we have

$$\begin{aligned}\int \mathrm{d}\eta^* \int \mathrm{d}\eta\, e^{-\eta^*\eta}\, |\eta\rangle\langle\eta| &= \int \mathrm{d}\eta^* \int \mathrm{d}\eta\, (1-\eta^*\eta)\,(1-\eta\hat{c}^\dagger)\,|0\rangle\langle 0|\,(1-\hat{c}\eta^*)\\ &= \int \mathrm{d}\eta^* \int \mathrm{d}\eta \Big(|0\rangle\langle 0| - |0\rangle\langle 1|\eta^* - \eta|1\rangle\langle 0| + \eta|1\rangle\langle 1|\eta^* \\ &\qquad - \eta^*\eta|0\rangle\langle 0|\Big)\\ &= |1\rangle\langle 1| + |0\rangle\langle 0|\\ &= \mathbb{1}_{\mathcal{F}}.\end{aligned} \tag{E.17}$$

Moreover, we have the trace formula

$$\begin{aligned}\int \mathrm{d}\eta^* \int \mathrm{d}\eta\, e^{-\eta^*\eta}\, \langle -\eta|\hat{C}| + \eta\rangle &= \int \mathrm{d}\eta^* \int \mathrm{d}\eta\, (1-\eta^*\eta)\, \Big(\langle 0| - \eta^*\langle 1|\Big)\hat{C}\Big(|0\rangle + |1\rangle\eta\Big)\\ &= \int \mathrm{d}\eta^* \int \mathrm{d}\eta \Big(\langle 0|\hat{C}|0\rangle + \langle 0|\hat{C}|1\rangle\eta - \eta^*\langle 1|\hat{C}|0\rangle \\ &\qquad - \eta^*\langle 1|\hat{C}|1\rangle\eta - \eta^*\eta\langle 0|\hat{C}|0\rangle\Big)\\ &= \langle 1|\hat{C}|1\rangle + \langle 0|\hat{C}|0\rangle\\ &= \mathrm{tr}_{\mathcal{F}}\,\hat{C},\end{aligned} \tag{E.18}$$

for any linear operator $\hat{C}: \mathcal{F} \to \mathcal{F}$. The trace formula (E.18), with its asymmetry in the sign of η entering a bra relative to that entering a ket, should be contrasted with the trace formula (A.20). This asymmetry delivers the Fermi-Dirac distribution function as opposed to the Bose-Einstein distribution.

E.2 Path-integral representation for fermions

With the help of the overlap, resolution of the identity, and trace formula in Eqs. (E.12), (E.17), and (E.18), respectively, it is possible to represent the partition function (E.4) as the Grassmann path integral,

$$\begin{aligned}Z(\beta) &= \lim_{M\to\infty} \left(\prod_{j=0}^{M-1} \int \mathrm{d}\eta_j^* \int \mathrm{d}\eta_j\right) e^{-\sum\limits_{j=1}^{M}\left[\eta_j^*(\eta_j-\eta_{j-1}) + \frac{\beta}{M} H(\eta_j^*,\eta_{j-1})\right]}\\ &\equiv \int \mathcal{D}[\eta^*,\eta]\, e^{-\int\limits_0^\beta \mathrm{d}\tau \left\{\eta_j^*(\tau)\partial_\tau\eta(\tau) + H[\eta^*(\tau),\eta(\tau)]\right\}},\end{aligned} \tag{E.19a}$$

where

$$\begin{aligned}\eta_M &\equiv -\eta_0 \Longrightarrow \eta\ (\tau+\beta) = -\eta\ (\tau),\\ \eta_M^* &\equiv -\eta_0^* \Longrightarrow \eta^*(\tau+\beta) = -\eta^*(\tau).\end{aligned} \tag{E.19b}$$

The manipulations that lead to Eq. (E.19a) are identical to those made for the bosonic harmonic oscillator in Eq. (A.34). The only change comes about due to the asymmetry of the trace formula (E.18). It leads to the integration variables in the path-integral representation obeying antiperiodic boundary conditions in (imaginary) time. Equation (E.19a) holds not only for $\hat{H}$ in Eq. (E.3) but for any normal-ordered operator. However, since Eq. (E.3) is quadratic, direct evaluation of the first line in Eq. (E.19a) can be performed,

$$\begin{aligned} Z(\beta) &= \lim_{M\to\infty} \det \begin{pmatrix} 1 & 0 & 0 \cdots 0 & 0 & 0 & 1-\frac{\beta}{M} \\ \frac{\beta}{M}-1 & 1 & 0 \cdots 0 & 0 & 0 & 0 \\ 0 & \frac{\beta}{M}-1 & 1 \cdots 0 & 0 & 0 & 0 \\ & & & & & \\ \vdots & \vdots & \vdots \vdots \vdots & \vdots & \vdots & \vdots \\ & & & & & \\ 0 & 0 & 0 \cdots 0 & \frac{\beta}{M}-1 & 1 & 0 \\ 0 & 0 & 0 \cdots 0 & 0 & \frac{\beta}{M}-1 & 1 \end{pmatrix} \\ &= \lim_{M\to\infty} \left[1+(-1)^{M-1}\left(1-\frac{\beta}{M}\right)\prod_{j=1}^{M-1}(-1)\left(1-\frac{\beta}{M}\right)\right] \\ &= \lim_{M\to\infty} \left[1+\left(1-\frac{\beta}{M}\right)^{M}\right] \\ &= 1+e^{-\beta}, \end{aligned} \tag{E.20}$$

as it should be. It follows that the internal energy [recall Eq. (5.28a)] $-\partial_\beta \ln Z(\beta)$ in a unit volume and at a vanishing chemical potential is nothing but the Fermi-Dirac distribution function,

$$f_{\mathrm{FD}}(1) = \frac{e^{-\beta}}{1+e^{-\beta}} = \frac{1}{e^{+\beta}+1}, \tag{E.21}$$

of the excited single-particle state $|1\rangle$ [recall Eq. (5.27a) with $\mu=0$, $\iota=0,1$, $\varepsilon_0=0$, and $\varepsilon_1=1$].

Had we imposed a boson algebra rather than the fermion algebra in Eq. (E.1), the path-integral representation (E.19a) would have to be modified in two ways. First, the integration variables would be conventional complex numbers and the path integral would be an infinite product of one-dimensional Riemann integrals. Second, the integration variables would obey periodic boundary conditions. These two changes would turn Eq. (E.20) into

$$Z(\beta) = \left(1-e^{-\beta}\right)^{-1}, \tag{E.22}$$

from which follows that the internal energy $-\partial_\beta \ln Z(\beta)$ in a unit volume and at a vanishing chemical potential is nothing but the Bose-Einstein distribution

$$f_{\mathrm{BE}}(1) = \frac{e^{-\beta}}{1-e^{-\beta}} = \frac{1}{e^{+\beta}-1}. \tag{E.23}$$

E.3 Jordan-Wigner fermions

E.3.1 *Introduction*

To illustrate the physical relevance of spinless fermions, we are going to introduce Jordan-Wigner fermions in this section. This will also allow us to demonstrate how spinless fermions can emerge from a many-body Hamiltonian built out of hard-core bosonic operators.

Modern high-energy physics postulates the existence of two kinds of point-like elementary particle. There are bosons that mediate the strong (through the gluons), weak (through the $W^{\pm}$ and Z bosons), and electromagnetic (through the photon) interactions. There are fermions such as quarks or such as leptons. Quarks make up composite particles known as baryons or mesons as a result of the strong interactions mediated by the gluons. Leptons interact with each others and with quarks through the weak and electromagnetic interactions. Finally, there is the Higgs boson that endows selected fermions with a mass.

All present scattering experiments probing length scales of the order of the weak-interaction range can be understood within the standard model. The standard model is a relativistic quantum field theory built out of quarks and leptons that interact by the exchange of gluons, $W^{\pm}$ and Z bosons, and the photon through gauge-invariant interactions. In the standard model, quarks, leptons, gauge bosons, and the Higgs boson are all point-like particles that are treated on equal footing. However, it cannot be ruled out on logical grounds that the Higgs boson is a mere mathematical abstraction that quantifies a structure of the world below some characteristic length scale that has yet to be observed by modern physics. Since the Higgs boson is the agent responsible for the measured rest masses of leptons in the standard model, it could very well be that leptons in the standard model are not elementary particle at sufficiently small length scale but quasiparticles that emerge from a more fundamental organization principle than the standard model.

Perhaps the most ambitious and radical organization principle for the physics beyond the standard model is string theory. String theory is an attempt to deduce the standard model with its elementary point-like particles and gravity from a more fundamental organization principle based on elementary objects that have a one-dimensional extension in space, i.e., strings. In string theory, fermions and bosons have been replaced by more fundamental objects; strings.

To this date, string theory has not been validated experimentally. It is a speculative theory that has yet to be confronted with experiments. On the other hand, there are some examples in condensed matter physics for which the interacting elementary constituents are neither bosonic nor fermionic although the relevant low-energy degrees of freedom can be.

To illustrate this fact, we consider a condensed matter system made of point-like particles that are (i) static (i.e., of infinite mass), but (ii) neither bosonic nor fermionic, and (iii) interact with each others. These point-like particles are defined on the sites of some lattice Λ (any countable set Λ of cardinality $|\Lambda|$), i.e., a regular arrangement of points in d-dimensional space. We shall denote the sites of the lattice embedded in a space of dimension d larger than one by boldfaced Latin letters, say $\boldsymbol{i}$ and $\boldsymbol{j}$. To each lattice site $\boldsymbol{i}$, we assign the vector-valued operator

$$\boldsymbol{\sigma}_{\boldsymbol{i}} := \begin{pmatrix} \sigma^{x}_{\boldsymbol{i}} \\ \sigma^{y}_{\boldsymbol{i}} \\ \sigma^{z}_{\boldsymbol{i}} \end{pmatrix}, \tag{E.24a}$$

where

$$\sigma^{x}_{\boldsymbol{i}} := \begin{pmatrix} 0 & 1 \\ 1 & 0 \end{pmatrix}, \qquad \sigma^{y}_{\boldsymbol{i}} := \begin{pmatrix} 0 & -\mathrm{i} \\ +\mathrm{i} & 0 \end{pmatrix}, \qquad \sigma^{z}_{\boldsymbol{i}} := \begin{pmatrix} +1 & 0 \\ 0 & -1 \end{pmatrix}, \tag{E.24b}$$

are the usual Pauli matrices. These Pauli matrices act on the local two-dimensional Hilbert space spanned by the eigenstates

$$\mathcal{H}_{\boldsymbol{i}} := \mathrm{span}\left\{ |\uparrow\rangle_{\boldsymbol{i}} := \begin{pmatrix} 1 \\ 0 \end{pmatrix}, \quad |\downarrow\rangle_{\boldsymbol{i}} := \begin{pmatrix} 0 \\ 1 \end{pmatrix} \right\}, \tag{E.24c}$$

of $\sigma^{z}_{\boldsymbol{i}}$. Next, we assign to the lattice Λ the global Hilbert space

$$\mathcal{H}_{\Lambda} := \bigotimes_{\boldsymbol{i}\in\Lambda} \mathcal{H}_{\boldsymbol{i}}, \tag{E.25a}$$

with the operator algebra

$$\left[\sigma^{a}_{\boldsymbol{i}}, \sigma^{b}_{\boldsymbol{j}}\right] = 2\delta_{\boldsymbol{i}\boldsymbol{j}}\mathrm{i}\epsilon^{abc}\sigma^{c}_{\boldsymbol{i}}, \qquad \left(\sigma^{a}_{\boldsymbol{i}}\right)^{2} = \mathbb{1}_{\boldsymbol{i}}, \qquad a,b = x,y,z, \qquad \boldsymbol{i},\boldsymbol{j}\in\Lambda. \tag{E.25b}$$

Finally, given any choice of a $|\Lambda| \times |\Lambda|$ real-valued and symmetric matrix with the dimensionful matrix elements

$$J_{\boldsymbol{i}\boldsymbol{j}} = J^{*}_{\boldsymbol{i}\boldsymbol{j}} = J^{*}_{\boldsymbol{j}\boldsymbol{i}} = J_{\boldsymbol{j}\boldsymbol{i}}, \qquad \boldsymbol{i},\boldsymbol{j}\in\Lambda, \tag{E.26a}$$

together with the real-valued and dimensionless number λ, we define the Hamiltonian

$$\hat{H}^{xxz} := \frac{1}{8}\sum_{\boldsymbol{i},\boldsymbol{j}\in\Lambda} J_{\boldsymbol{i}\boldsymbol{j}}\left(\sigma^{x}_{\boldsymbol{i}}\,\sigma^{x}_{\boldsymbol{j}} + \sigma^{y}_{\boldsymbol{i}}\,\sigma^{y}_{\boldsymbol{j}} + \lambda\,\sigma^{z}_{\boldsymbol{i}}\,\sigma^{z}_{\boldsymbol{j}}\right). \tag{E.26b}$$

The $U(1)$ symmetric case $\lambda = 0$ defines the so-called quantum xy model. The $SU(2)$ symmetric case $\lambda = 1$ defines the Heisenberg model. The $\mathbb{Z}_2$ symmetric cases $\lambda = \pm\infty$ define the classical Ising model. All three cases are of relevance to certain classes of materials in condensed matter physics.

For any value of $\lambda \neq \pm\infty$, the local degrees of freedom (E.24) are neither bosonic nor fermionic according to the global Hilbert space and algebra (E.25). The fact that they commute when localized on different sites is reminiscent of bosons. The fact that the local Hilbert space is finite is reminiscent of hard-core bosons or, alternatively, of fermions.

In fact, it is always possible to reformulate the problem solely in terms of spinless fermions defined by the non-local (Jordan-Wigner) transformation

$$\hat{c}^{\dagger}_{\boldsymbol{i}_{\iota}} := \frac{1}{2}\left(\sigma^{x}_{\boldsymbol{i}_{\iota}} + \mathrm{i}\sigma^{y}_{\boldsymbol{i}_{\iota}}\right)\prod_{\iota'<\iota}\sigma^{z}_{\boldsymbol{i}_{\iota'}}, \qquad \hat{c}_{\boldsymbol{i}_{\iota}} := \frac{1}{2}\left(\sigma^{x}_{\boldsymbol{i}_{\iota}} - \mathrm{i}\sigma^{y}_{\boldsymbol{i}_{\iota}}\right)\prod_{\iota'<\iota}\sigma^{z}_{\boldsymbol{i}_{\iota'}}, \tag{E.27a}$$

where we have chosen the ordering of the lattice

$$\Lambda = \left\{ \boldsymbol{i}_1, \boldsymbol{i}_2, \cdots, \boldsymbol{i}_\iota, \cdots, \boldsymbol{i}_{|\Lambda|} \right\}, \tag{E.27b}$$

for it can be shown (see below) that

$$\left\{ \hat{c}_{\boldsymbol{i}_\iota}, \hat{c}^\dagger_{\boldsymbol{i}_{\iota'}} \right\} = \delta_{\iota\iota'}, \qquad 0 = \left\{ \hat{c}_{\boldsymbol{i}_\iota}, \hat{c}_{\boldsymbol{i}_{\iota'}} \right\} = \left\{ \hat{c}^\dagger_{\boldsymbol{i}_\iota}, \hat{c}^\dagger_{\boldsymbol{i}_{\iota'}} \right\}. \tag{E.27c}$$

Observe however that the transformation (E.27a) does not necessarily simplify a generic Hamiltonian (E.26b), since the quantum xy limit of Hamiltonian (E.26b) is generically non-local when expressed in terms of the Jordan-Wigner fermions (E.27a). A remarkable exception to this rule was discovered by Jordan and Wigner. It occurs when the lattice is one-dimensional and the exchange couplings (E.26a) are only non-vanishing for pairs of nearest-neighbor sites (see below).

Before we consider the case of the nearest-neighbor quantum xy model in one dimension, we should remember that the nearest-neighbor quantum Heisenberg antiferromagnet on a cubic lattice breaks spontaneously the $SU(2)$ spin-rotation symmetry down to its $U(1)$ subgroup so that, by the Goldstone theorem, the low-lying excitations are bosonic modes called magnons. Anticipating the results of the following section, we thus conclude that the low-energy excitations of the xxz-Hamiltonian (E.26b) can either be bosons or fermions depending on the dimensionality of space, the lattice structure, and the sign, range, and symmetries of the exchange couplings.

E.3.2 *Nearest-neighbor and quantum xy limit in one-dimensional position space*

When the lattice Λ is a one-dimensional ring made of $|\Lambda| = N$ sites $i = 1, \cdots, N$ with $i + N \equiv i$, and the exchange couplings are only non-vanishing between the ordered pair i and $i+1$ of nearest-neighbor sites in which case they are given by

$$J_{i(i+1)} := \begin{cases} J_\perp, & i = 1, \cdots, N-1, \\ \\ J_\perp \cos\phi, & i = N, \end{cases} \tag{E.28}$$

the quantum xy Hamiltonian that acts on the Hilbert space (E.25a) then reduces to

$$\hat{H}^{xy} := \frac{J_\perp}{2} \sum_{i=1}^{N} \left[1 - \delta_{iN}\left(1 - \cos\phi\right)\right] \left(\hat{S}^+_i \hat{S}^-_{i+1} + \hat{S}^-_i \hat{S}^+_{i+1} \right), \tag{E.29a}$$

where, in units for which $\hbar = 1$,

$$\hat{S}^\pm_i := \hat{S}^x_i \pm \mathrm{i}\hat{S}^y_i \equiv \frac{1}{2}\left(\sigma^x_i \pm \mathrm{i}\sigma^y_i\right) \equiv \hat{S}^\pm_{i+N}, \qquad \hat{S}^z_i := \frac{1}{2}\sigma^z_i \equiv \hat{S}^z_{i+N}. \tag{E.29b}$$

The parameter $0 \leq \phi < 2\pi$ is the ring parameter. It fixes the boundary conditions at the level of the Hamiltonian. For examples, the choice $\phi = 0$ is equivalent to the replacement $[1-\delta_{iN}(1-\cos\phi)] \to 1$ in the Hamiltonian while imposing the periodic boundary conditions $\hat{\boldsymbol{S}}_i = +\hat{\boldsymbol{S}}_{i+N}$, the choice $\phi = \pi/2$ is equivalent to the replacement $[1-\delta_{iN}(1-\cos\phi)] \to 1$ in the Hamiltonian while imposing open boundary conditions, and the choice $\phi = \pi$ is equivalent to the replacement $[1-\delta_{iN}(1-\cos\phi)] \to 1$ in the Hamiltonian while imposing the antiperiodic boundary conditions $\hat{\boldsymbol{S}}_i = -\hat{\boldsymbol{S}}_{i+N}$.

We are going to show that excitations above the ground state form a continuum. Remarkably, this continuum is not separated by an energy gap from the ground state energy. Correspondingly, spin-spin correlation functions decay algebraically and not exponentially with separation at zero temperature. The quantum xy limit at zero temperature defines a *quantum critical point* in that all correlation functions between local spin operators are algebraic functions of the space arguments when sufficiently far apart. At a quantum critical point there is no characteristic intrinsic length scale, scale invariance rules. The key step in deriving this result is the Jordan-Wigner transformation, a remarkable identity that relates the raising and lowering spin-1/2 operators to spinless fermions through a non-local transformation, but preserves the locality of the xy Hamiltonian when expressed in terms of these spinless fermions.

For any $i \in \Lambda$, define the operators

$$\hat{f}_i^{\dagger} := \hat{K}_i\,\hat{S}_i^{+}, \qquad \hat{f}_i := \hat{S}_i^{-}\,\hat{K}_i^{\dagger}, \tag{E.30a}$$

where the unitary operator

$$\hat{K}_i := \exp\left(\mathrm{i}\pi\sum_{j=1}^{i-1}\left(\hat{S}_j^{z}+\frac{1}{2}\right)\right) \tag{E.30b}$$

has been introduced.

For any $i \in \Lambda$, the non-local operator

$$\begin{aligned}\hat{K}_i &= e^{\mathrm{i}(i-1)\pi/2}\prod_{j=1}^{i-1} e^{\mathrm{i}\pi\hat{S}_j^{z}}\\ &= \mathrm{i}^{i-1}\prod_{j=1}^{i-1}\left(\cos\frac{\pi}{2}\,\hat{\sigma}_j^{0}+\mathrm{i}\sin\frac{\pi}{2}\,\hat{\sigma}_j^{z}\right)\\ &= (-1)^{i-1}\hat{\sigma}_1^{z}\cdots\hat{\sigma}_{i-1}^{z}\end{aligned} \tag{E.31a}$$

rotates by the angle π around the z axis in spin space all the spins left to site i,

$$\begin{aligned}\hat{K}_i\,\hat{S}_j^{x}\hat{K}_i^{\dagger} &= -\Theta(i-j)\hat{S}_j^{x}+\Theta(j-i)\hat{S}_j^{x},\\ \hat{K}_i\,\hat{S}_j^{y}\hat{K}_i^{\dagger} &= -\Theta(i-j)\hat{S}_j^{y}+\Theta(j-i)\hat{S}_j^{y},\\ \hat{K}_i\,\hat{S}_j^{z}\hat{K}_i^{\dagger} &= \hat{S}_j^{z},\end{aligned} \tag{E.31b}$$

where $i,j = 1,\cdots,N$ and $\Theta(x)$ is the Heaviside step function. It is thus the non-local nature of the operator $\hat{K}_i$ that allows it to either anticommute or commute with the raising or lowering operators $\hat{S}_j^{\pm}$ depending on whether $j < i$ or $j \geq i$, respectively. The choice of the phase factor $\exp\left(\mathrm{i}(i-1)\pi/2\right)$ insures that its eigenvalues are ± 1. Finally, since $\hat{K}_i$ is unitary by construction and has only real eigenvalues, it must satisfy

$$\hat{K}_i \hat{K}_i^{\dagger} = \hat{K}_i^{\dagger} \hat{K}_i = 1, \qquad \hat{K}_i = \hat{K}_i^{\dagger}, \qquad i = 1, \cdots, N. \tag{E.31c}$$

Furthermore, we observe that $\hat{K}_i$ is built exclusively from $\hat{S}_j^z$ with $j = 1, \cdots, i-1$. Thus, it obeys the commutation relations

$$\left[\hat{K}_i, \hat{K}_j\right] = 0, \qquad i, j = 1, \cdots, N, \tag{E.31d}$$

and

$$\left[\hat{K}_i, \hat{S}_j^z\right] = 0, \qquad i, j = 1, \cdots, N. \tag{E.31e}$$

Hence, we can rewrite Eq. (E.30) as

$$\hat{f}_i^{\dagger} = \hat{S}_i^{+} \hat{K}_i = \hat{K}_i \hat{S}_i^{+}, \qquad \hat{f}_i = \hat{S}_i^{-} \hat{K}_i = \hat{K}_i \hat{S}_i^{-}, \qquad i = 1, \cdots, N. \tag{E.32}$$

We are now in position to prove that

$$\left\{\hat{f}_i, \hat{f}_j^{\dagger}\right\} = \delta_{ij}, \qquad \left\{\hat{f}_i^{\dagger}, \hat{f}_j^{\dagger}\right\} = \left\{\hat{f}_i, \hat{f}_j\right\} = 0, \qquad i, j = 1, \cdots, N, \tag{E.33}$$

i.e., that we have constructed out of the spin-1/2 operators fermionic operators called Jordan-Wigner fermions.

Proof. Let $i, j \in \Lambda$. When $i < j$, we have,

$$\begin{aligned}
\hat{f}_i \hat{f}_j^{\dagger} + \hat{f}_j^{\dagger} \hat{f}_i &= \hat{S}_i^{-} \hat{K}_i \hat{S}_j^{+} \hat{K}_j + \hat{S}_j^{+} \hat{K}_j \hat{S}_i^{-} \hat{K}_i \\
&= \hat{S}_i^{-} \hat{S}_j^{+} \hat{K}_i \hat{K}_j + \hat{S}_j^{+} \hat{K}_j \hat{S}_i^{-} \hat{K}_j \hat{K}_i \hat{K}_j \\
&= \hat{S}_i^{-} \hat{S}_j^{+} \hat{K}_i \hat{K}_j - \hat{S}_j^{+} \hat{S}_i^{-} \hat{K}_i \hat{K}_j \\
&= \left[\hat{S}_i^{-}, \hat{S}_j^{+}\right] \hat{K}_i \hat{K}_j \\
&= 0,
\end{aligned} \tag{E.34}$$

and

$$\begin{aligned}
\hat{f}_i \hat{f}_j + \hat{f}_j \hat{f}_i &= \hat{S}_i^{-} \hat{K}_i \hat{S}_j^{-} \hat{K}_j + \hat{S}_j^{-} \hat{K}_j \hat{S}_i^{-} \hat{K}_i \\
&= \hat{S}_i^{-} \hat{S}_j^{-} \hat{K}_i \hat{K}_j + \hat{S}_j^{-} \hat{K}_j \hat{S}_i^{-} \hat{K}_j \hat{K}_i \hat{K}_j \\
&= \hat{S}_i^{-} \hat{S}_j^{-} \hat{K}_i \hat{K}_j - \hat{S}_j^{-} \hat{S}_i^{-} \hat{K}_i \hat{K}_j \\
&= \left[\hat{S}_i^{-}, \hat{S}_j^{-}\right] \hat{K}_i \hat{K}_j \\
&= 0.
\end{aligned} \tag{E.35}$$

When $i = j$, we have,

$$\begin{aligned}
\hat{f}_i \hat{f}_i^{\dagger} + \hat{f}_i^{\dagger} \hat{f}_i &= \hat{S}_i^{-} \hat{K}_i \hat{S}_i^{+} \hat{K}_i + \hat{S}_i^{+} \hat{K}_i \hat{S}_i^{-} \hat{K}_i \\
&= \left\{\hat{S}_i^{-}, \hat{S}_i^{+}\right\} \hat{K}_i \hat{K}_i \\
&= 1,
\end{aligned} \tag{E.36}$$

and

$$\begin{aligned} \hat{f}_i \hat{f}_i + \hat{f}_i \hat{f}_i &= \hat{S}_i^- \hat{K}_i \hat{S}_i^- \hat{K}_i + \hat{S}_i^- \hat{K}_i \hat{S}_i^- \hat{K}_i \\ &= \left\{ \hat{S}_i^-, \hat{S}_i^- \right\} \hat{K}_i \hat{K}_i \\ &= 0. \end{aligned} \tag{E.37}$$

Observe that only $\left(\hat{S}_i^-\right)^2 = 0$ is needed to reach the last line. The case of $i > j$ follows by interchanging i and j in the proof for the case of $i < j$. □

The interpretation of the fermion creation operator $\hat{f}_i^\dagger$ is that it creates a defect, a kink, in an ordered state. To see this, consider the states

$$|\mathrm{F}\rangle := \bigotimes_{i=1}^{N} \frac{1}{\sqrt{2}} \left(|\uparrow\rangle_i + |\downarrow\rangle_i\right), \tag{E.38}$$

and

$$|\bar{\mathrm{F}}\rangle := \bigotimes_{i=1}^{N} \frac{1}{\sqrt{2}} \left(|\uparrow\rangle_i - |\downarrow\rangle_i\right), \tag{E.39}$$

which are eigenstates of $\hat{S}_i^x$ with eigenvalues $+1/2$ and $-1/2$, respectively, for all sites $i \in \Lambda$. For any $j \in \Lambda$, the state

$$\hat{f}_j^\dagger |\mathrm{F}\rangle = (-1)^{j-1} \left(\bigotimes_{i=1}^{j-1} \frac{1}{\sqrt{2}} \left(|\uparrow\rangle_i - |\downarrow\rangle_i\right) \right) \frac{1}{\sqrt{2}} |\uparrow\rangle_j \left(\bigotimes_{i=j+1}^{N} \frac{1}{\sqrt{2}} \left(|\uparrow\rangle_i + |\downarrow\rangle_i\right) \right) \tag{E.40}$$

is an eigenstate of $\hat{S}_i^x$ with eigenvalue $-1/2$ for all sites $i = 1, \cdots, j-1$, an eigenstate of $\hat{S}_j^z$ with eigenvalue $+1/2$, and an eigenstate of $\hat{S}_i^x$ with eigenvalue $+1/2$ for all sites $i = j+1, \cdots, N$. The state (E.40) interpolates between $|\mathrm{F}\rangle$ and $|\bar{\mathrm{F}}\rangle$ with a spin up at the boundary $j \in \Lambda$.

Having established that the definition (E.30) yields operators obeying the fermionic algebra, we are going to express the spin operators in terms of these fermions. To this end, we assume the fermionic algebra (E.33). One then verifies that the spin operators defined by

$$\hat{S}_i^+ := \hat{K}_i \hat{f}_i^\dagger, \qquad \hat{S}_i^- := \hat{f}_i \hat{K}_i^\dagger, \qquad \hat{S}_i^z := \hat{f}_i^\dagger \hat{f}_i - \frac{1}{2}, \tag{E.41a}$$

where

$$\hat{K}_i := \exp\left(\mathrm{i}\pi \sum_{j=1}^{i-1} \hat{f}_j^\dagger \hat{f}_j \right) = \prod_{j=1}^{i-1} \left(1 - 2\hat{f}_j^\dagger \hat{f}_j\right) = \hat{K}_i^\dagger \tag{E.41b}$$

satisfy the $su(2)$ Lie algebra

$$\left[\hat{S}_i^+, \hat{S}_j^-\right] = \delta_{ij}\, 2\hat{S}^z, \qquad \left[\hat{S}_i^-, \hat{S}_j^z\right] = \delta_{ij}\, \hat{S}_i^-, \qquad \left[\hat{S}_i^z, \hat{S}_j^+\right] = \delta_{ij}\, \hat{S}_i^+, \tag{E.41c}$$

for all $i, j = 1, \cdots, N$.

The fermion representation (E.41b) of the kink operator (E.30b) is useful to establish the identities

$$
\begin{aligned}
&\hat{K}_i\hat{K}_{i+1} = \hat{K}_{i+1}\hat{K}_i = e^{i\pi \hat{f}_i^\dagger \hat{f}_i} = 1 - 2\hat{f}_i^\dagger \hat{f}_i,\\
&\hat{f}_i^\dagger \hat{K}_i\hat{K}_{i+1} = \hat{f}_i^\dagger \hat{K}_{i+1}\hat{K}_i = \hat{f}_i^\dagger\left(1 - 2\hat{f}_i^\dagger \hat{f}_i\right) = +\hat{f}_i^\dagger,\\
&\hat{f}_i\hat{K}_i\hat{K}_{i+1} = \hat{f}_i\hat{K}_{i+1}\hat{K}_i = \hat{f}_i\left(1 - 2\hat{f}_i^\dagger \hat{f}_i\right) = -\hat{f}_i,\\
&\hat{S}_i^+\hat{S}_{i+1}^- = \hat{f}_i^\dagger\hat{K}_i\hat{K}_{i+1}\hat{f}_{i+1} = +\hat{f}_i^\dagger\hat{f}_{i+1},\\
&\hat{S}_i^-\hat{S}_{i+1}^+ = \hat{f}_i\,\hat{K}_i\hat{K}_{i+1}\hat{f}_{i+1}^\dagger = -\hat{f}_i\hat{f}_{i+1}^\dagger,\\
&\hat{S}_i^+\hat{S}_{i+1}^- + \hat{S}_i^-\hat{S}_{i+1}^+ = \hat{f}_i^\dagger\hat{f}_{i+1} + \hat{f}_{i+1}^\dagger\hat{f}_i,
\end{aligned}
\tag{E.42}
$$

for $i = 1, \cdots, N-1$. The term

$$
\begin{aligned}
\hat{S}_N^+\hat{S}_{N+1}^- + \hat{S}_N^-\hat{S}_{N+1}^+ &\equiv \hat{S}_N^+\hat{S}_1^- + \hat{S}_N^-\hat{S}_1^+\\
&= \hat{K}_N\hat{f}_N^\dagger\hat{f}_1\hat{K}_1 + \hat{f}_N\hat{K}_N\hat{K}_1\hat{f}_1^\dagger,
\end{aligned}
\tag{E.43}
$$

which encodes a ring topology must be treated with care. Since

$$
\hat{K}_1 = 1, \qquad \left[\hat{f}_N, \hat{K}_N\right] = 0,
\tag{E.44}
$$

we can make the simplification

$$
\hat{S}_N^+\hat{S}_1^- + \hat{S}_N^-\hat{S}_1^+ = \hat{K}_N\hat{f}_N^\dagger\hat{f}_1 + \hat{K}_N\hat{f}_N\hat{f}_1^\dagger.
\tag{E.45}
$$

Next, we observe that

$$
\begin{aligned}
\hat{K}_N &= \exp\left(i\pi\sum_{j=1}^{N-1}\hat{f}_j^\dagger\hat{f}_j\right)\\
&= \exp\left(i\pi\sum_{j=1}^{N}\hat{f}_j^\dagger\hat{f}_j\right)\exp\left(-i\pi\hat{f}_N^\dagger\hat{f}_N\right)\\
&= \hat{K}\left(1 - 2\hat{f}_N^\dagger\hat{f}_N\right),
\end{aligned}
\tag{E.46a}
$$

where we have introduced

$$
\hat{N}_f := \sum_{j=1}^{N}\hat{f}_j^\dagger\hat{f}_j, \qquad \hat{K} := \exp\left(i\pi\hat{N}_f\right).
\tag{E.46b}
$$

Now,

$$
\hat{K}_N\,\hat{f}_N^\dagger = -\hat{K}\,\hat{f}_N^\dagger, \qquad \hat{K}_N\,\hat{f}_N = +\hat{K}\,\hat{f}_N,
\tag{E.47}
$$

so that the ring topology is encoded by the relation [recall Eq. (E.29b)]

$$
\begin{aligned}
\hat{S}_N^+\hat{S}_{N+1}^- + \hat{S}_N^-\hat{S}_{N+1}^+ &\equiv \hat{S}_N^+\hat{S}_1^- + \hat{S}_N^-\hat{S}_1^+\\
&= -\hat{K}\left(\hat{f}_N^\dagger\hat{f}_1 + \hat{f}_1^\dagger\hat{f}_N\right).
\end{aligned}
\tag{E.48}
$$

We conclude that, in the sector of the Hilbert space (E.25a) with the given number N_f of fermions [the eigenvalue of the fermion-number operator in Eq. (E.46b)], the ring topology is achieved by the condition

$$\hat{f}^{\dagger}_{i+N} = (-1)^{N_f+1}\hat{f}^{\dagger}_{i}, \qquad \hat{f}_{i+N} = (-1)^{N_f+1}\hat{f}_{i}, \qquad i \in \Lambda. \tag{E.49}$$

Since the fermion-number operator in Eq. (E.46b) is related to the total spin operator

$$\hat{\boldsymbol{S}} := \sum_{i=1}^{N} \hat{\boldsymbol{S}}_i \tag{E.50}$$

by[19]

$$\hat{N}_f = \hat{S}^z + \frac{N}{2}, \qquad \hat{S}^z := \sum_{i=1}^{N} \hat{S}^z_i, \tag{E.51}$$

the sector of the Hilbert space with the quantum number $S^z = 0$ of the total spin operator $\hat{S}^z$ along the quantization axis is equivalent to demanding that the number of fermions N_f is half the number of lattice sites, i.e., the half-filled condition $N_f = N/2$ for spinless fermions.

With the help of the Jordan-Wigner transformation (E.41), the xy-Hamiltonian (E.29a) is fermionized to

$$\hat{H}^{xy} := \operatorname{sgn}(J_{\perp})\,\frac{|J_{\perp}|}{2}\sum_{i=1}^{N}\left[1-\delta_{iN}\left(1-\cos\phi\right)\right]\left(\hat{f}^{\dagger}_{i}\hat{f}_{i+1}+\hat{f}^{\dagger}_{i+1}\hat{f}_{i}\right). \tag{E.52a}$$

Here, the spinless fermions obey the boundary conditions

$$\hat{f}^{\dagger}_{i+N} = (-1)^{N_f+1}\hat{f}^{\dagger}_{i}, \qquad \hat{f}_{i+N} = (-1)^{N_f+1}\hat{f}_{i}, \qquad i \in \Lambda, \tag{E.52b}$$

in the subspace of the Hilbert space with the fermion-number operator constrained to

$$\hat{N}_f = \hat{S}^z + \frac{N}{2} = N_f, \qquad N_f = 0, 1, \cdots, N. \tag{E.52c}$$

For a ferromagnetic coupling, $\operatorname{sgn}(J_{\perp}) = -$. For an antiferromagnetic coupling $\operatorname{sgn}(J_{\perp}) = +$. The case of the antiferromagnetic coupling can be brought to the case of the ferromagnetic coupling through the local gauge transformation

$$\hat{f}^{\dagger}_{i} \to (-1)^{i}\hat{f}^{\dagger}_{i}, \qquad \hat{f}_{i} = (-1)^{i}\hat{f}_{i}, \qquad i \in \Lambda, \tag{E.52d}$$

that leaves the boundary conditions (E.52b) unchanged for N even. Hence, we may adopt the convention

$$\hat{H}^{xy} := -\frac{|J_{\perp}|}{2}\sum_{i=1}^{N}\left[1-\delta_{iN}\left(1-\cos\phi\right)\right]\left(\hat{f}^{\dagger}_{i}\hat{f}_{i+1}+\hat{f}^{\dagger}_{i+1}\hat{f}_{i}\right), \tag{E.52e}$$

without loss of generality.

---

[19] One verifies as an exercise that the right-hand side of Eq. (E.51) can only take integer-valued eigenvalues irrespective of the parity of the integer N.

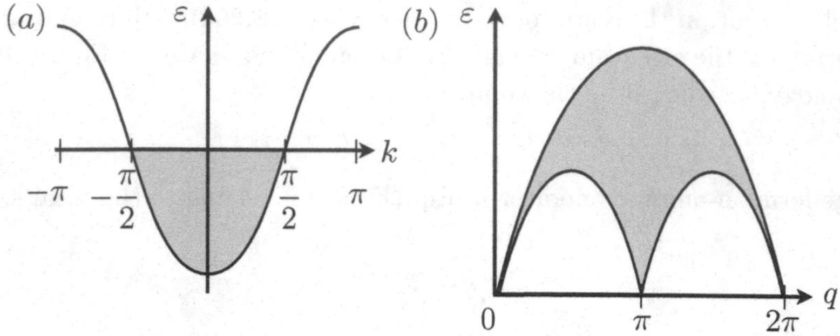

Fig. E.1 (a) Single-particle eigenvalues for Jordan-Wigner fermions and the Fermi sea at half-filling. (b) Jordan-Wigner particle-hole continuum in the nearest-neighbor quantum xy chain with $q = k_p - k_h$ where k_p is the particle and k_h the hole wave number.

When the ring-parameter $\phi = 0$, the single-particle energy spectrum ε_k in

$$\hat{H}^{xy} = \sum_{k\in\Lambda_{\mathrm{BZ}}} \varepsilon_k\, \hat{f}^{\dagger}_k\, \hat{f}_k, \qquad \varepsilon_k = -\,|J_{\perp}|\,\cos k, \tag{E.53a}$$

follows from using the Bloch states ($[x]$ denotes the largest integer part in absolute value of $x \in \mathbb{R}$)

$$\hat{f}^{\dagger}_j = \frac{1}{\sqrt{N}} \sum_{k\in\Lambda_{\mathrm{BZ}}} e^{-\mathrm{i}k\,j}\, \hat{f}^{\dagger}_k, \qquad \hat{f}_j = \frac{1}{\sqrt{N}} \sum_{k\in\Lambda_{\mathrm{BZ}}} e^{+\mathrm{i}k\,j}\, \hat{f}_k, \qquad j \in \Lambda, \tag{E.53b}$$

with the first Brillouin zone

$$\Lambda_{\mathrm{BZ}} = \left\{ k = \frac{2\pi n}{N} \,\middle|\, n = -\left[\frac{N}{2}\right], -\left[\frac{N}{2}\right]+1, \cdots, \left[\frac{N}{2}\right]-1, \left[\frac{N}{2}\right]\delta_{1,N\mathrm{mod}2} \right\}, \tag{E.53c}$$

when periodic boundary condition hold and with the first Brillouin zone

$$\Lambda_{\mathrm{BZ}} = \left\{ k = \frac{\pi(2n+1)}{N} \,\middle|\, n = -\left[\frac{N}{2}\right], -\left[\frac{N}{2}\right]+1, \cdots, \left[\frac{N}{2}\right]-1, \left[\frac{N}{2}\right]\delta_{1,N\mathrm{mod}2} \right\}, \tag{E.53d}$$

when antiperiodic boundary conditions hold.

When N is even, the ground state can be shown to be a singlet. The ground state can be represented by the Fermi sea of Jordan-Wigner fermions at half-filling as shown in Fig. E.1(a). The first excited states of the quantum xy chain have total spin-1 quantum number and can be represented by particle-hole excitations of Jordan-Wigner fermions.

Reinstating a small $J^z/J_{\perp} \equiv \lambda$ does not violate the locality of the representation of the xy Hamiltonian in terms of the Jordan-Wigner fermions,

$$\begin{aligned}
\hat{H}^{xxz} =& -\frac{|J_{\perp}|}{2} \sum_{i=1}^{N} \left[1 - \delta_{iN}\,(1-\cos\phi)\right] \left(\hat{f}^{\dagger}_i \hat{f}_{i+1} + \hat{f}^{\dagger}_{i+1} \hat{f}_i\right) \\
&+ J^z \sum_{i=1}^{N} \left[1 - \delta_{iN}\,(1-\cos\phi)\right] \left(\hat{f}^{\dagger}_i \hat{f}_i - \frac{1}{2}\right) \left(\hat{f}^{\dagger}_{i+1} \hat{f}_{i+1} - \frac{1}{2}\right).
\end{aligned} \tag{E.54}$$

The quartic interaction for the Jordan-Wigner fermions cannot be treated by perturbation theory in one-dimensional position space. Non-perturbative techniques are available and go under the name of Bethe Ansatz or bosonization. For antiferromagnetic exchange couplings, the qualitative nature of the non-interacting xy limit survives until the Heisenberg point $J^z/|J_\perp| = 1$ is reached, i.e., the line $0 \leq J^z/|J_\perp| \leq 1$ realizes a line of critical points with all local spin-correlation functions decaying like power laws with exponents determined by the ratio $J^z/|J_\perp|$, see sections 3.3.2 and 4.2. Beyond the antiferromagnetic Heisenberg point $J^z/|J_\perp| = 1$, an energy gap opens between the ground state and the low-lying excitations as does the onset of long-range order in the ground state. The quantum transition at the antiferromagnetic Heisenberg point $J^z/|J_\perp| = 1$ is, mathematically, closely related to the (classical) Kosterlitz-Thouless transition studied in section 4.6.

E.4 The ground state energy and the single-particle time-ordered Green function

We are going to perturb the non-interacting jellium model with a Coulomb interaction. This is to say that the collective index ι that labels the fermionic generators of the Fock space (5.132) becomes the quantized momenta

$$\boldsymbol{k} = \frac{2\pi}{L}\boldsymbol{n}, \qquad \boldsymbol{n} \in \mathbb{Z}^3, \tag{E.55a}$$

in a cubic box of volume L^3 upon imposing periodic boundary conditions on the one hand, and the spin-1/2 index

$$\sigma = \uparrow, \downarrow, \tag{E.55b}$$

on the other hand. Let $0 \leq \lambda \leq 1$ be a dimensionless coupling that allows us to treat the Coulomb interaction adiabatically, i.e., we define

$$\begin{aligned} \hat{H}_\mu(\lambda) &:= \hat{H}_{0,\mu} + \lambda\, \hat{H}_1, \\ \hat{H}_{0,\mu} &:= \sum_{\sigma,\boldsymbol{k}} \left(\frac{\boldsymbol{k}^2}{2m} - \mu\right) \hat{c}^\dagger_{\sigma,\boldsymbol{k}}\, \hat{c}_{\sigma,\boldsymbol{k}}, \\ \hat{H}_1 &:= \frac{1}{V} \sum_{\boldsymbol{q}\neq\boldsymbol{0}} \frac{2\pi e^2}{\boldsymbol{q}^2} \sum_{\sigma,\sigma'} \sum_{\boldsymbol{k},\boldsymbol{k}'} \hat{c}^\dagger_{\sigma,\boldsymbol{k}+\boldsymbol{q}}\, \hat{c}^\dagger_{\sigma',\boldsymbol{k}'-\boldsymbol{q}}\, \hat{c}_{\sigma',\boldsymbol{k}'}\, \hat{c}_{\sigma,\boldsymbol{k}}. \end{aligned} \tag{E.56}$$

We shall use the notation

$$\hat{H}_\mu \equiv \hat{H}_\mu(\lambda = 1), \tag{E.57}$$

so that $\hat{H}_\mu(\lambda)$ interpolates between $\hat{H}_{0,\mu}$ and $\hat{H}_\mu$ as λ varies between 0 and 1. The restriction to non-vanishing transfer momenta

$$\boldsymbol{q} \neq \boldsymbol{0} \tag{E.58}$$

implements the charge-neutrality condition. We work in the grand-canonical ensemble with the grand-canonical partition function

$$Z(\beta,\mu;\lambda):=\mathrm{Tr}_{\mathcal{F}}\left[e^{-\beta\,\hat{H}_{\mu}(\lambda)}\right]. \tag{E.59}$$

The grand-canonical potential F is defined by [recall Eq. (5.28b)]

$$F(\beta,\mu;\lambda):=U(\beta,\mu;\lambda)-TS(\beta,\mu;\lambda)\equiv-\frac{1}{\beta}\ln\mathrm{Tr}_{\mathcal{F}}\left[e^{-\beta\,\hat{H}_{\mu}(\lambda)}\right]\equiv-\frac{1}{\beta}\ln Z(\beta,\mu;\lambda). \tag{E.60}$$

The thermal expectation value of the Coulomb interaction is

$$\frac{\partial F(\beta,\mu;\lambda)}{\partial\lambda}=\frac{\mathrm{Tr}_{\mathcal{F}}\left[e^{-\beta\,\hat{H}_{\mu}(\lambda)}\,\hat{H}_{1}\right]}{\mathrm{Tr}_{\mathcal{F}}\left[e^{-\beta\,\hat{H}_{\mu}(\lambda)}\right]}\equiv\left\langle\hat{H}_{1}\right\rangle_{\beta,\mu;\lambda}. \tag{E.61}$$

The change in the grand-canonical potential F induced by switching on the Coulomb interaction adiabatically is

$$F(\beta,\mu;1)-F(\beta,\mu;0)=\int\limits_{0}^{1}\mathrm{d}\lambda\,\frac{\partial F(\beta,\mu;\lambda)}{\partial\lambda}=\int\limits_{0}^{1}\frac{\mathrm{d}\lambda}{\lambda}\left\langle\lambda\hat{H}_{1}\right\rangle_{\beta,\mu;\lambda}. \tag{E.62}$$

Our goal is now to relate the grand-canonical expectation value

$$\left\langle\lambda\hat{H}_{1}\right\rangle_{\beta,\mu;\lambda} \tag{E.63}$$

to the so-called single-particle Green function. To this end, we first define the time-ordered Green functions for the Coulomb gas in the grand-canonical ensemble.

Define the *single-particle and time-ordered Green function*

$$G_{\sigma,\boldsymbol{k}}(\tau;\lambda):=-\left\langle T_{\tau}\left(\hat{c}_{\mathrm{H}\sigma,\boldsymbol{k}}(\tau;\lambda)\,\hat{c}^{\dagger}_{\mathrm{H}\sigma,\boldsymbol{k}}(0;\lambda)\right)\right\rangle_{\beta,\mu;\lambda}, \tag{E.64}$$

where the λ dependence comes from the use of Eq. (E.56) to define the grand-canonical partition function in the Heisenberg picture. The terminology single particle stems from the fact that the $\hat{c}^{\dagger}$'s create single-particle-like excitations in the non-interacting limit. According to Eq. (5.163), the single-particle Green function (E.64) obeys the equation of motion

$$-\partial_{\tau}G_{\sigma,\boldsymbol{k}}(\tau;\lambda)=\delta(\tau)+\left\langle T_{\tau}\left\{\left[\hat{H}_{\mu}(\lambda),\hat{c}_{\mathrm{H}\sigma,\boldsymbol{k}}(\tau;\lambda)\right]\hat{c}^{\dagger}_{\mathrm{H}\sigma,\boldsymbol{k}}(0;\lambda)\right\}\right\rangle_{\beta,\mu;\lambda}. \tag{E.65}$$

We take advantage of the fact that $\hat{H}_{\mu}(\lambda)$ is conserved, i.e.,

$$\hat{H}_{\mu}(\lambda)=\hat{H}_{\mathrm{H}\mu}(\tau;\lambda) \tag{E.66}$$

for any imaginary time τ. We can bring the equation of motion (E.65) to the form

$$-\partial_{\tau}G_{\sigma,\boldsymbol{k}}(\tau;\lambda)=\delta(\tau)+\left\langle T_{\tau}\left\{\left[\hat{H}_{\mathrm{H}\mu}(\tau;\lambda),\hat{c}_{\mathrm{H}\sigma,\boldsymbol{k}}(\tau;\lambda)\right]\hat{c}^{\dagger}_{\mathrm{H}\sigma,\boldsymbol{k}}(0;\lambda)\right\}\right\rangle_{\beta,\mu;\lambda} \tag{E.67}$$

in which the commutator on the right-hand side now only involves operators at equal imaginary time. When evaluated at an infinitesimal time $\tau = 0^-$ before $\tau = 0$, Eq. (E.67) can be brought to the form

$$\begin{aligned}\sum_{\sigma,\boldsymbol{k}} \left[\left(\partial_\tau G_{\sigma,\boldsymbol{k}}\right)(0^-;\lambda) + \delta(0^-) \right] \\ = \sum_{\sigma,\boldsymbol{k}} \left\langle T_\tau \left\{ \hat{c}^\dagger_{\mathrm{H}\sigma,\boldsymbol{k}}(0;\lambda) \left[\hat{H}_{\mathrm{H}\mu}(0^-;\lambda), \hat{c}_{\mathrm{H}\sigma,\boldsymbol{k}}(0^-;\lambda) \right] \right\} \right\rangle_{\beta,\mu;\lambda}. \end{aligned} \tag{E.68}$$

Evaluation of the commutator on the right-hand side of Eq. (E.68) yields

$$\sum_{\sigma,\boldsymbol{k}} \left[\left(\partial_\tau G_{\sigma,\boldsymbol{k}}\right)(0^-;\lambda) + \delta(0^-) \right] = -\sum_{\sigma,\boldsymbol{k}} \xi_{\mu\boldsymbol{k}} G_{\sigma,\boldsymbol{k}}(0^-;\lambda) - 2 \times \left\langle \lambda \hat{H}_1 \right\rangle_{\beta,\mu;\lambda}, \tag{E.69a}$$

with

$$\xi_{\mu\boldsymbol{k}} := \frac{\boldsymbol{k}^2}{2m} - \mu, \tag{E.69b}$$

as we now show.

Proof. With the help of the identity

$$\begin{aligned}\left[\hat{A}\hat{B}, \hat{C}\right] &= \hat{A}\hat{B}\hat{C} - \hat{C}\hat{A}\hat{B} = \hat{A}\hat{B}\hat{C} + \hat{A}\hat{C}\hat{B} - \hat{A}\hat{C}\hat{B} - \hat{C}\hat{A}\hat{B} \\ &= \hat{A}\left\{\hat{B}, \hat{C}\right\} - \left\{\hat{A}, \hat{C}\right\}\hat{B},\end{aligned} \tag{E.70}$$

which is valid for any triplet of operators acting on the Fock space, we deduce (for $\lambda = 1$ without loss of generality)

$$\begin{aligned}\left[\hat{H}_{0,\mu}, \hat{c}_{\sigma,\boldsymbol{k}}\right] &= \left[\sum_{\sigma',\boldsymbol{k}'} \xi_{\mu\boldsymbol{k}'}\, \hat{c}^\dagger_{\sigma',\boldsymbol{k}'}\, \hat{c}_{\sigma',\boldsymbol{k}'}, \hat{c}_{\sigma,\boldsymbol{k}}\right] \\ &= -\xi_{\mu\boldsymbol{k}}\, \hat{c}_{\sigma,\boldsymbol{k}},\end{aligned} \tag{E.71}$$

on the one hand, and

$$\begin{aligned}\left[\hat{H}_1, \hat{c}_{\sigma,\boldsymbol{k}}\right] &= \left[\frac{1}{2V}\sum_{\boldsymbol{q}\neq\boldsymbol{0}} V_{\mathrm{cb}\,\boldsymbol{q}} \sum_{\sigma',\sigma''}\sum_{\boldsymbol{k}',\boldsymbol{k}''} \hat{c}^\dagger_{\sigma',\boldsymbol{k}'+\boldsymbol{q}}\, \hat{c}^\dagger_{\sigma'',\boldsymbol{k}''-\boldsymbol{q}}\, \hat{c}_{\sigma'',\boldsymbol{k}''}\, \hat{c}_{\sigma',\boldsymbol{k}'}, \hat{c}_{\sigma,\boldsymbol{k}}\right] \\ &= \frac{1}{2V}\sum_{\boldsymbol{q}\neq\boldsymbol{0}} V_{\mathrm{cb}\,\boldsymbol{q}} \sum_{\sigma',\sigma''}\sum_{\boldsymbol{k}',\boldsymbol{k}''} \Big(\hat{c}^\dagger_{\sigma',\boldsymbol{k}'+\boldsymbol{q}}\, \hat{c}_{\sigma'',\boldsymbol{k}''}\, \hat{c}_{\sigma',\boldsymbol{k}'}\, \delta_{\sigma'',\sigma}\, \delta_{\boldsymbol{k}''-\boldsymbol{q},\boldsymbol{k}} \\ &\quad - \hat{c}^\dagger_{\sigma'',\boldsymbol{k}''-\boldsymbol{q}}\, \hat{c}_{\sigma'',\boldsymbol{k}''}\, \hat{c}_{\sigma',\boldsymbol{k}'}\, \delta_{\sigma',\sigma}\, \delta_{\boldsymbol{k}'+\boldsymbol{q},\boldsymbol{k}} \Big),\end{aligned} \tag{E.72}$$

on the other hand. Here, we have introduced the short-hand notations

$$\xi_{\mu\boldsymbol{k}} := \frac{\boldsymbol{k}^2}{2m} - \mu, \qquad V_{\mathrm{cb}\,\boldsymbol{q}} := \frac{4\pi e^2}{\boldsymbol{q}^2}. \tag{E.73}$$

The summation over the momentum $\boldsymbol{q}$ delivers

$$\left[\hat{H}_1, \hat{c}_{\sigma,\boldsymbol{k}}\right] = \frac{1}{2V} \sum_{\boldsymbol{k}',\boldsymbol{k}''} \Big(V_{\mathrm{cb}\,\boldsymbol{k}''-\boldsymbol{k}\neq\boldsymbol{0}} \sum_{\sigma'} \hat{c}^{\dagger}_{\sigma',\boldsymbol{k}'+\boldsymbol{k}''-\boldsymbol{k}} \underline{\hat{c}_{\sigma,\boldsymbol{k}''}}\, \hat{c}_{\sigma',\boldsymbol{k}'} \\ -V_{\mathrm{cb}\,\boldsymbol{k}-\boldsymbol{k}'\neq\boldsymbol{0}} \sum_{\sigma''} \hat{c}^{\dagger}_{\sigma'',\boldsymbol{k}'+\boldsymbol{k}''-\boldsymbol{k}}\, \hat{c}_{\sigma'',\boldsymbol{k}''}\, \hat{c}_{\sigma,\boldsymbol{k}'} \Big). \tag{E.74}$$

By moving the underlined annihilation operator to the right, we may factorize a minus sign,

$$\left[\hat{H}_1, \hat{c}_{\sigma,\boldsymbol{k}}\right] = -\frac{1}{2V} \sum_{\boldsymbol{k}',\boldsymbol{k}''} \Big(V_{\mathrm{cb}\,\boldsymbol{k}''-\boldsymbol{k}\neq\boldsymbol{0}} \sum_{\sigma'} \hat{c}^{\dagger}_{\sigma',\boldsymbol{k}'+\boldsymbol{k}''-\boldsymbol{k}} \hat{c}_{\sigma',\boldsymbol{k}'} \underline{\hat{c}_{\sigma,\boldsymbol{k}''}} \\ +V_{\mathrm{cb}\,\boldsymbol{k}-\boldsymbol{k}'\neq\boldsymbol{0}} \sum_{\sigma''} \hat{c}^{\dagger}_{\sigma'',\boldsymbol{k}'+\boldsymbol{k}''-\boldsymbol{k}} \hat{c}_{\sigma'',\boldsymbol{k}''} \hat{c}_{\sigma,\boldsymbol{k}'} \Big). \tag{E.75}$$

After relabeling some momenta to be summed over, we finally arrive to

$$\left[\hat{H}_1, \hat{c}_{\sigma,\boldsymbol{k}}\right] = -\frac{1}{2V} \sum_{\boldsymbol{k}',\boldsymbol{k}''} \Big(V_{\mathrm{cb}\,\boldsymbol{k}''-\boldsymbol{k}\neq\boldsymbol{0}} \sum_{\sigma'} \hat{c}^{\dagger}_{\sigma',\boldsymbol{k}'+\boldsymbol{k}''-\boldsymbol{k}}\, \hat{c}_{\sigma',\boldsymbol{k}'}\, \hat{c}_{\sigma,\boldsymbol{k}''} \\ +V_{\mathrm{cb}\,\boldsymbol{k}-\boldsymbol{k}''\neq\boldsymbol{0}} \sum_{\sigma'} \hat{c}^{\dagger}_{\sigma',\boldsymbol{k}'+\boldsymbol{k}''-\boldsymbol{k}}\, \hat{c}_{\sigma',\boldsymbol{k}'}\, \hat{c}_{\sigma,\boldsymbol{k}''} \Big). \tag{E.76}$$

The Coulomb potential is an even function of momentum. It can thus be factorized from Eq. (E.76),

$$\left[\hat{H}_\mu, \hat{c}_{\sigma,\boldsymbol{k}}\right] = -\,\xi_{\mu\boldsymbol{k}} \hat{c}_{\sigma,\boldsymbol{k}} - \frac{1}{V} \sum_{\boldsymbol{k}',\boldsymbol{k}''} V_{\mathrm{cb}\,\boldsymbol{k}''-\boldsymbol{k}\neq\boldsymbol{0}} \sum_{\sigma'} \hat{c}^{\dagger}_{\sigma',\boldsymbol{k}'+\boldsymbol{k}''-\boldsymbol{k}}\, \hat{c}_{\sigma',\boldsymbol{k}'}\, \hat{c}_{\sigma,\boldsymbol{k}''}. \tag{E.77}$$

Observe that

$$\begin{aligned} \sum_{\sigma}\sum_{\boldsymbol{k}} \hat{c}^{\dagger}_{\sigma,\boldsymbol{k}} \left[\hat{H}_\mu, \hat{c}_{\sigma,\boldsymbol{k}}\right] =& -\sum_{\sigma}\sum_{\boldsymbol{k}} \xi_{\mu\boldsymbol{k}}\, \hat{c}^{\dagger}_{\sigma,\boldsymbol{k}}\, \hat{c}_{\sigma,\boldsymbol{k}} \\ & -\frac{1}{V} \sum_{\boldsymbol{k},\boldsymbol{k}',\boldsymbol{k}''} V_{\mathrm{cb}\,\boldsymbol{k}''-\boldsymbol{k}\neq\boldsymbol{0}} \sum_{\sigma,\sigma'} \hat{c}^{\dagger}_{\sigma,\boldsymbol{k}}\, \hat{c}^{\dagger}_{\sigma',\boldsymbol{k}'+\boldsymbol{k}''-\boldsymbol{k}}\, \hat{c}_{\sigma',\boldsymbol{k}'}\, \hat{c}_{\sigma,\boldsymbol{k}''}. \\ =& -\sum_{\sigma}\sum_{\boldsymbol{k}} \xi_{\mu\boldsymbol{k}}\, \hat{c}^{\dagger}_{\sigma,\boldsymbol{k}}\, \hat{c}_{\sigma,\boldsymbol{k}} \\ & -2\times\frac{1}{2V} \sum_{\boldsymbol{q}} V_{\mathrm{cb}\,\boldsymbol{q}\neq\boldsymbol{0}} \sum_{\boldsymbol{k},\boldsymbol{k}'}\sum_{\sigma,\sigma'} \hat{c}^{\dagger}_{\sigma,\boldsymbol{k}}\, \hat{c}^{\dagger}_{\sigma',\boldsymbol{k}'+\boldsymbol{q}}\, \hat{c}_{\sigma',\boldsymbol{k}'}\, \hat{c}_{\sigma,\boldsymbol{k}+\boldsymbol{q}} \\ =& -\hat{H}_{0,\mu} - 2\times\hat{H}_1. \end{aligned} \tag{E.78}$$

Equation (E.69a) follows from inspection of Eqs. (E.64), (E.68), and (E.78).

□

Equation (E.69a) allows to express the expectation value of the Coulomb interaction in terms of single-particle properties according to

$$2\times\left\langle \lambda\,\hat{H}_1\right\rangle_{\beta,\mu;\lambda} = -\sum_{\sigma,\boldsymbol{k}} \left[\left(\partial_\tau G_{\sigma,\boldsymbol{k}}\right)(0^-;\lambda) + \delta(0^-)\right] - \sum_{\sigma,\boldsymbol{k}} \xi_{\mu\boldsymbol{k}} G_{\sigma,\boldsymbol{k}}(0^-;\lambda). \tag{E.79}$$

If we use the Fourier conventions [compare with Eq. (5.157e)]

$$\begin{aligned} G_{\sigma,\boldsymbol{k}}(\tau;\lambda) &= \frac{1}{\beta}\sum_{n\in\mathbb{Z}} e^{-\mathrm{i}\omega_n\tau}\, G_{\sigma,\boldsymbol{k},\mathrm{i}\omega_n}(\lambda), \\ \delta(\tau) &= \frac{1}{\beta}\sum_{n\in\mathbb{Z}} e^{-\mathrm{i}\omega_n\tau}, \end{aligned} \tag{E.80}$$

we arrive at

$$\begin{aligned} \left\langle\lambda\,\hat{H}_1\right\rangle_{\beta,\mu;\lambda} &= -\frac{1}{2\beta}\sum_{n\in\mathbb{Z}}\sum_{\sigma,\boldsymbol{k}} e^{-\mathrm{i}\omega_n 0^-}\left[\left(-\mathrm{i}\omega_n+\xi_{\mu\boldsymbol{k}}\right)G_{\sigma,\boldsymbol{k},\mathrm{i}\omega_n}(\lambda)+1\right] \\ &= +\frac{1}{2\beta}\sum_{n\in\mathbb{Z}}\sum_{\sigma,\boldsymbol{k}} e^{-\mathrm{i}\omega_n 0^-}\left[\left(\mathrm{i}\omega_n-\xi_{\mu\boldsymbol{k}}\right)-\left(G^{-1}\right)_{\sigma,\boldsymbol{k},\mathrm{i}\omega_n}(\lambda)\right]G_{\sigma,\boldsymbol{k},\mathrm{i}\omega_n}(\lambda). \end{aligned} \tag{E.81}$$

Here, the inverse operator G^{-1} to the Green function G is defined by the condition

$$G^{-1}G = G\,G^{-1} = 1, \tag{E.82}$$

which holds as an operator equation on the space of propagators that evolve the many-body wave functions from the Fock space in imaginary time following the dynamics set by $\hat{H}_\mu$. Equation (E.81) suggests that we introduce the non-interacting Green function through the matrix elements

$$\left(G_{0\mu}^{-1}\right)_{\sigma,\boldsymbol{k},\mathrm{i}\omega_n} := \mathrm{i}\omega_n - \xi_{\mu\boldsymbol{k}}, \tag{E.83a}$$

and the self-energy

$$\begin{aligned} \Sigma_{\sigma,\boldsymbol{k},\mathrm{i}\omega_n}(\lambda) &:= \left(G_{0\mu}^{-1}\right)_{\sigma,\boldsymbol{k},\mathrm{i}\omega_n} - \left(G^{-1}\right)_{\sigma,\boldsymbol{k},\mathrm{i}\omega_n}(\lambda) \iff \\ &\left(G^{-1}\right)_{\sigma,\boldsymbol{k},\mathrm{i}\omega_n}(\lambda) = \mathrm{i}\omega_n - \xi_{\mu\boldsymbol{k}} - \Sigma_{\sigma,\boldsymbol{k},\mathrm{i}\omega_n}(\lambda). \end{aligned} \tag{E.83b}$$

This self-energy measures the difference induced by the Coulomb interaction between the non-interacting propagator $\left(G_{0\mu}^{-1}\right)$ and the exact propagator $\left(G_{\mu}^{-1}\right)$. With these definitions, Eqs. (E.81) and (E.62) take the final forms

$$\left\langle\lambda\,\hat{H}_1\right\rangle_{\beta,\mu;\lambda} = +\frac{1}{2\beta}\sum_{n\in\mathbb{Z}}\sum_{\sigma,\boldsymbol{k}} e^{-\mathrm{i}\omega_n 0^-}\,\Sigma_{\sigma,\boldsymbol{k},\mathrm{i}\omega_n}(\lambda)\,G_{\sigma,\boldsymbol{k},\mathrm{i}\omega_n}(\lambda), \tag{E.84}$$

and

$$F(\beta,\mu;1)-F(\beta,\mu;0) = +\frac{1}{2\beta}\sum_{n\in\mathbb{Z}}\sum_{\sigma,\boldsymbol{k}}\int_0^1\frac{\mathrm{d}\lambda}{\lambda}e^{-\mathrm{i}\omega_n 0^-}\,\Sigma_{\sigma,\boldsymbol{k},\mathrm{i}\omega_n}(\lambda)\,G_{\sigma,\boldsymbol{k},\mathrm{i}\omega_n}(\lambda), \tag{E.85}$$

respectively, where the chemical potential μ is fixed by the condition that there is an average number of N_{e} electrons in the cubic box of volume L^3. The change in

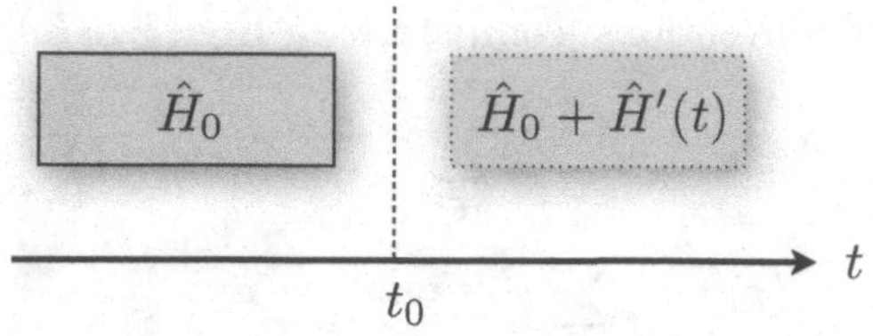

Fig. E.2 A material is described by the conserved Hamiltonian $\hat{H}_0$ at times $t < t_0$. At time t_0 a time-dependent interaction $\hat{H}'(t)$ to the environment is switched on.

the ground state energy $E_{0\mu}$ of the Fermi sea induced by switching on adiabatically the Coulomb potential is

$$
\begin{aligned}
E(\mu;1) - E(\mu;0) &= \lim_{\beta\to\infty} \frac{1}{2\beta} \sum_{n\in\mathbb{Z}} \sum_{\sigma,\boldsymbol{k}} \int_0^1 \frac{d\lambda}{\lambda} e^{-i\omega_n 0^-} \Sigma_{\sigma,\boldsymbol{k},i\omega_n}(\lambda) G_{\sigma,\boldsymbol{k},i\omega_n}(\lambda) \\
&= \lim_{\beta\to\infty} \frac{1}{\beta} \sum_{n\in\mathbb{Z}} \sum_{\boldsymbol{k}} \int_0^1 \frac{d\lambda}{\lambda} e^{-i\omega_n 0^-} \Sigma_{\sigma,\boldsymbol{k},i\omega_n}(\lambda) G_{\sigma,\boldsymbol{k},i\omega_n}(\lambda), \quad \text{(E.86)}
\end{aligned}
$$

for any projection $\sigma =\uparrow,\downarrow$ of the electronic spin along the quantization axis.

E.5 Linear response

E.5.1 *Introduction*

To characterize experimentally the different states or phases in condensed matter physics, a macroscopic sample is subjected to weak external perturbations. A sample can thus be probed in various different ways. Mechanical forces are exerted on it. Electric (thermal) leads are attached to it to drive an electric (thermal) current with varying applied magnetic fields or applied pressures. Beams of light, electrons, neutrons, or muons are aimed to it with detectors recording the outcome of these scattering experiments. As long as these external perturbations preserve the integrity of the sample under study on the relevant length and time scales, one hopes that intrinsic properties of the sample can be deduced at these length and time scales. The mathematical description of such experiments is the following.

Let $\hat{H}_0$ be the Hamiltonian describing the macroscopic sample to be investigated under the assumption that it is well isolated from its environment. Thus, $\hat{H}_0$ is conserved and acts on the Hilbert space $\mathcal{H}_0$. In thermodynamic equilibrium at the inverse temperature β the sample is described by the partition function

$$
Z_0 := \mathrm{Tr}_{\mathcal{H}_0}\left(e^{-\beta \hat{H}_0}\right) \equiv \sum_{\iota} e^{-\beta E_{0\,\iota}}. \quad \text{(E.87a)}
$$

Here, $E_{0\,\iota}$ denotes an exact eigenvalue indexed by ι of $\hat{H}_0$. If the density-matrix is defined by

$$
\hat{\rho}_{0\,\beta} := e^{-\beta \hat{H}_0} \equiv \sum_{\iota} e^{-\beta E_{0\,\iota}} \left|\iota\right\rangle \left\langle\iota\right|, \quad \text{(E.87b)}
$$

then the thermal average of any operator $\hat{A}_0$ acting on $\mathcal{H}_0$ is

$$\left\langle \hat{A}_0 \right\rangle_{0\,\beta} := \frac{\mathrm{Tr}_{\mathcal{H}_0}\left(e^{-\beta\,\hat{H}_0}\right)\hat{A}_0}{\mathrm{Tr}_{\mathcal{H}_0}\left(e^{-\beta\,\hat{H}_0}\right)} \equiv Z_0^{-1}\,\mathrm{Tr}_{\mathcal{H}_0}\left(\hat{\rho}_{0\,\beta}\,\hat{A}_0\right). \tag{E.87c}$$

Let the Hilbert space of the environment be denoted by $\mathcal{H}'$. The Hilbert space of the "universe" is then

$$\mathcal{H} := \mathcal{H}_0 \otimes \mathcal{H}'. \tag{E.88}$$

We now imagine that at time t_0 a weak time-dependent coupling to the environment is switched on, see Fig. E.2. This physics is modeled by the time-dependent Hamiltonian

$$\hat{H}(t) := \hat{H}_0 + \Theta(t-t_0)\,\hat{H}'(t), \tag{E.89}$$

acting on the Hilbert space $\mathcal{H}$. Here, Θ is the Heaviside step function. The question we then pose is the following. What is the change in the expectation value (E.87c) induced by the time-dependent coupling (E.89) to first order in perturbation theory? The answer to this question is the Kubo formula that we now derive.

E.5.2 *The Kubo formula*

Assume that we have solved the time-dependent Schrödinger equation

$$\mathrm{i}\hbar\,\partial_t\,|\iota(t)\rangle = \hat{H}(t)\,|\iota(t)\rangle\,, \tag{E.90a}$$

with the initial condition that

$$|\iota(t_0)\rangle = |\iota\rangle \tag{E.90b}$$

is the eigenstate of $\hat{H}_0$ with eigenvalue $E_{0\,\iota}$. We then *define* the time-dependent (instantaneous) density matrix by

$$\hat{\rho}_{0\,\beta}(t,t_0) := \sum_{\iota} e^{-\beta\,E_{0\,\iota}}\,|\iota(t)\rangle\,\langle\iota(t)|\,, \qquad \hat{\rho}_{0\,\beta}(t_0,t_0) = \hat{\rho}_{0\,\beta}. \tag{E.91a}$$

The time-dependent (instantaneous) density matrix allows to define the expectation value

$$\left\langle \hat{A}(t,t_0)\right\rangle_{0\,\beta} := Z_0^{-1}\mathrm{Tr}_{\mathcal{H}}\left(\hat{\rho}_{0\,\beta}(t,t_0)\,\hat{A}\right), \qquad \left\langle \hat{A}(t_0,t_0)\right\rangle_{0\,\beta} = \left\langle \hat{A}\right\rangle_{0\,\beta}, \tag{E.91b}$$

out of thermodynamic equilibrium. The Kubo formula states that, to first order in time-dependent perturbation theory,

$$\left\langle \hat{A}(t,t_0)\right\rangle_{0\,\beta} = \left\langle \hat{A}\right\rangle_{0\,\beta} + \int_{t_0}^{\infty} \mathrm{d}t'\,C^{\mathrm{R}}_{0\,\beta\,\hat{A},\hat{H}'}(t,t'), \tag{E.92a}$$

where the retarded Green function

$$C^{\mathrm{R}}_{0\,\beta\,\hat{A},\hat{H}'}(t,t') := -\frac{\mathrm{i}}{\hbar}\Theta(t-t')\left\langle\left[\hat{A}_{\mathrm{I}}(t,t_0),\hat{H}'_{\mathrm{I}}(t',t_0)\right]\right\rangle_{0\,\beta} \tag{E.92b}$$

has been introduced together with

$$\hat{A}_{\mathrm{I}}(t,t_0) := e^{+\mathrm{i}\hat{H}_0(t-t_0)/\hbar}\,\hat{A}\,e^{-\mathrm{i}\hat{H}_0(t-t_0)/\hbar}, \tag{E.92c}$$

and

$$\hat{H}'_{\mathrm{I}}(t,t_0) := e^{+\mathrm{i}\hat{H}_0(t-t_0)/\hbar}\,\hat{H}'(t)\,e^{-\mathrm{i}\hat{H}_0(t-t_0)/\hbar}. \tag{E.92d}$$

Proof. Step 1: Make the Ansatz

$$|\iota(t)\rangle = e^{-\mathrm{i}\hat{H}_0(t-t_0)/\hbar}\,\hat{U}_{\mathrm{I}}(t,t_0)\;|\iota(t_0)\rangle\,, \tag{E.93}$$

which defines the unitary time-evolution operator $\hat{U}_{\mathrm{I}}(t,t_0)$ in the interaction picture from the initial time t_0 to the final time $t \geq t_0$. Insertion of Eq. (E.93) into the left-hand side of Schrödinger equation (E.90a) gives

$$\mathrm{i}\hbar\,\partial_t\,|\iota(t)\rangle = e^{-\mathrm{i}\hat{H}_0(t-t_0)/\hbar}\left[\hat{H}_0\,\hat{U}_{\mathrm{I}}(t,t_0) + \mathrm{i}\hbar\left(\partial_t\,\hat{U}_{\mathrm{I}}\right)(t,t_0)\right]|\iota(t_0)\rangle\,, \tag{E.94a}$$

on the one hand. Insertion of Eq. (E.93) into the right-hand side of Schrödinger equation (E.90a) gives

$$\mathrm{i}\hbar\,\partial_t\,|\iota(t)\rangle = \left(\hat{H}_0 + \Theta(t-t_0)\hat{H}'(t)\right)e^{-\mathrm{i}\hat{H}_0(t-t_0)/\hbar}\hat{U}_{\mathrm{I}}(t,t_0)\,|\iota(t_0)\rangle\,, \tag{E.94b}$$

on the other hand. In other words, the unitary operator $\hat{U}_{\mathrm{I}}(t,t_0)$ obeys the differential equation

$$\begin{aligned}\mathrm{i}\hbar\left(\partial_t\,\hat{U}_{\mathrm{I}}\right)(t,t_0) &= e^{+\mathrm{i}\hat{H}_0(t-t_0)/\hbar}\,\Theta(t-t_0)\,\hat{H}'(t)\,e^{-\mathrm{i}\hat{H}_0(t-t_0)/\hbar}\,\hat{U}_{\mathrm{I}}(t,t_0)\\ &\equiv \Theta(t-t_0)\,\hat{H}'_{\mathrm{I}}(t,t_0)\,\hat{U}_{\mathrm{I}}(t,t_0),\end{aligned} \tag{E.95a}$$

with the initial condition

$$\hat{U}_{\mathrm{I}}(t_0,t_0) = 1, \tag{E.95b}$$

i.e.,

$$\hat{U}_{\mathrm{I}}(t,t_0) = T_t \exp\left(-\frac{\mathrm{i}}{\hbar}\int\limits_{t_0}^{t}\mathrm{d}t'\,\hat{H}'_{\mathrm{I}}(t',t_0)\right). \tag{E.95c}$$

We recall that

$$\hat{B}_{\mathrm{I}}(t,t_0) = e^{+\mathrm{i}\hat{H}_0(t-t_0)/\hbar}\,\hat{B}\,e^{-\mathrm{i}\hat{H}_0(t-t_0)/\hbar}. \tag{E.96}$$

Step 2: We insert into

$$\begin{aligned}\left\langle\hat{A}(t,t_0)\right\rangle_{0\,\beta} &= Z_0^{-1}\sum_{\iota} e^{-\beta\,E_{0\,\iota}}\,\langle\iota(t)|\,\hat{A}\,|\iota(t)\rangle\\ &= Z_0^{-1}\sum_{\iota} e^{-\beta\,E_{0\,\iota}}\,\langle\iota(t_0)|\,\hat{U}^{\dagger}_{\mathrm{I}}(t,t_0)\,e^{+\mathrm{i}\hat{H}_0(t-t_0)/\hbar}\,\hat{A}\,e^{-\mathrm{i}\hat{H}_0(t-t_0)/\hbar}\,\hat{U}_{\mathrm{I}}(t,t_0)\;|\iota(t_0)\rangle\\ &= Z_0^{-1}\sum_{\iota} e^{-\beta\,E_{0\,\iota}}\,\langle\iota(t_0)|\,\hat{U}^{\dagger}_{\mathrm{I}}(t,t_0)\,\hat{A}_{\mathrm{I}}(t,t_0)\,\hat{U}_{\mathrm{I}}(t,t_0)\;|\iota(t_0)\rangle\end{aligned} \tag{E.97}$$

the expansion

$$\hat{U}_{\mathrm{I}}(t,t_0) = 1 - \frac{\mathrm{i}}{\hbar}\int\limits_{t_0}^{t} \mathrm{d}t'\, \hat{H}'_{\mathrm{I}}(t',t_0), \tag{E.98}$$

to first order in $\hat{H}'_{\mathrm{I}}$. We find

$$\begin{aligned}\left\langle \hat{A}(t,t_0)\right\rangle_{0\,\beta} =& Z_0^{-1} \sum_{\iota} e^{-\beta E_{0\,\iota}} \left\langle \iota(t_0) \middle| \hat{A}_{\mathrm{I}}(t,t_0) \middle| \iota(t_0)\right\rangle \\ &- \frac{\mathrm{i}}{\hbar}\int\limits_{t_0}^{t} \mathrm{d}t'\, Z_0^{-1} \sum_{\iota} e^{-\beta E_{0\,\iota}} \left\langle \iota(t_0) \middle| \left[\hat{A}_{\mathrm{I}}(t,t_0), \hat{H}'_{\mathrm{I}}(t',t_0)\right] \middle| \iota(t_0)\right\rangle ,\end{aligned} \tag{E.99}$$

to first order in $\hat{H}'_{\mathrm{I}}$. If we undo the unitary transformation to the interaction picture on the first line on the right-hand side of Eq. (E.99) and if we replace the upper limit of the integral over t' by ∞ at the expense of inserting the Heaviside step function $\Theta(t-t')$ in the integrand on the second line on the right-hand side of Eq. (E.99), we may write [we use the short-hand notation (E.90b)]

$$\begin{aligned}\left\langle \hat{A}(t,t_0)\right\rangle_{0\,\beta} =& Z_0^{-1} \sum_{\iota} e^{-\beta E_{0\,\iota}} \left\langle \iota \middle| \hat{A} \middle| \iota\right\rangle \\ &- \frac{\mathrm{i}}{\hbar}\int\limits_{t_0}^{\infty} \mathrm{d}t'\, \Theta(t-t')\, Z_0^{-1} \sum_{\iota} e^{-\beta E_{0\,\iota}} \left\langle \iota \middle| \left[\hat{A}_{\mathrm{I}}(t,t_0), \hat{H}'_{\mathrm{I}}(t',t_0)\right] \middle| \iota\right\rangle .\end{aligned} \tag{E.100}$$

To first order in perturbation theory, we have thus expressed the evolution in time out of thermodynamic equilibrium solely in terms of expectation values in thermodynamic equilibrium,

$$\left\langle \hat{A}(t,t_0)\right\rangle_{0\,\beta} = \left\langle \hat{A}\right\rangle_{0\,\beta} + \int\limits_{t_0}^{\infty} \mathrm{d}t' \left(-\frac{\mathrm{i}}{\hbar}\Theta(t-t') \left\langle\left[\hat{A}_{\mathrm{I}}(t,t_0), \hat{H}'_{\mathrm{I}}(t',t_0)\right]\right\rangle_{0\,\beta}\right). \tag{E.101}$$

□

The Kubo formula in the frequency domain applies to the case when

$$\hat{H}'(t) = f(t)\,\hat{B}, \tag{E.102a}$$

with f any complex-valued function and $\hat{B}$ any operator acting on $\mathcal{H}$. Indeed, Eq. (E.92) becomes

$$\left\langle \hat{A}(t,t_0)\right\rangle_{0\,\beta} = \left\langle \hat{A}\right\rangle_{0\,\beta} + \int\limits_{t_0}^{\infty} \mathrm{d}t'\, C^{\mathrm{R}}_{0\,\beta\,\hat{A},\hat{B}}(t-t')\, f(t'), \tag{E.102b}$$

where

$$C^{\mathrm{R}}_{0\,\beta\,\hat{A},\hat{B}}(t-t') := -\frac{\mathrm{i}}{\hbar}\Theta(t-t')\left\langle\left[\hat{A}_{\mathrm{I}}(t,t_0),\hat{B}_{\mathrm{I}}(t',t_0)\right]\right\rangle_{0\,\beta}. \tag{E.102c}$$

Invariance under time translation of Eq. (E.102c) follows from the fact that both $\hat{A}$ and $\hat{B}$ are time independent. If the Fourier transform of f exists on the open line $\mathbb{R}$ and if one is not interested in transients, then the convolution in time

$$\left\langle\hat{A}(t)\right\rangle_{0\,\beta} := \lim_{t_0\to-\infty}\left\langle\hat{A}(t,t_0)\right\rangle_{0\,\beta} = \left\langle\hat{A}\right\rangle_{0\,\beta} + \int\limits_{\mathbb{R}}\mathrm{d}t'\,C^{\mathrm{R}}_{0\,\beta\,\hat{A},\hat{B}}(t-t')\,f(t') \tag{E.103a}$$

is represented in frequency space by the product

$$\left\langle\hat{A}(\omega)\right\rangle_{0\,\beta} = 2\pi\left\langle\hat{A}\right\rangle_{0\,\beta}\delta(\omega) + C^{\mathrm{R}}_{0\,\beta\,\hat{A},\hat{B}}(\omega)\,f(\omega), \tag{E.103b}$$

where we are using the Fourier convention

$$g(t) = \int\limits_{\mathbb{R}}\frac{\mathrm{d}\omega}{2\pi}e^{-\mathrm{i}\omega t}\,g(\omega) \Longleftrightarrow g(\omega) = \int\limits_{\mathbb{R}}\mathrm{d}t\,e^{+\mathrm{i}\omega t}\,g(t), \tag{E.104}$$

for all three functions in Eq. (E.103).

By construction, the Kubo formula is linear in the perturbation so that

$$\left\langle\hat{A}(t)\right\rangle_{0\,\beta} = \left\langle\hat{A}\right\rangle_{0\,\beta} + \sum_a\int\mathrm{d}^d\boldsymbol{r}\int\limits_{\mathbb{R}}\mathrm{d}t'\,C^{\mathrm{R}}_{0\,\beta\,\hat{A},\hat{B}^a(\boldsymbol{r})}(t-t')\,f^a(\boldsymbol{r},t'), \tag{E.105a}$$

and

$$\left\langle\hat{A}(\omega)\right\rangle_{0\,\beta} = 2\pi\left\langle\hat{A}\right\rangle_{0\,\beta}\delta(\omega) + \sum_a\int\mathrm{d}^d\boldsymbol{r}\,C^{\mathrm{R}}_{0\,\beta\,\hat{A},\hat{B}^a(\boldsymbol{r})}(\omega)\,f^a(\boldsymbol{r},\omega) \tag{E.105b}$$

hold when

$$\hat{H}'(t) = \sum_a\int\mathrm{d}^d\boldsymbol{r}\,f^a(\boldsymbol{r},t)\,\hat{B}^a(\boldsymbol{r}). \tag{E.105c}$$

E.5.3 *Kubo formula for the conductivity*

We start from the time-dependent Hamiltonian (the electron carries the negative charge $-e$ with e positive)

$$\begin{aligned}\hat{H}(t) = &\sum_{\sigma=\uparrow,\downarrow}\int\mathrm{d}^d\boldsymbol{r}\,\hat{\psi}^\dagger_\sigma(\boldsymbol{r})\,\frac{1}{2m}\left(\frac{\hbar}{\mathrm{i}}\boldsymbol{\nabla} - \frac{(-e)}{c}\boldsymbol{A}(\boldsymbol{r},t)\right)^2\hat{\psi}_\sigma(\boldsymbol{r})\\ &+(-e)\int\mathrm{d}^d\boldsymbol{r}\,\hat{n}(\boldsymbol{r})\,\varphi(\boldsymbol{r},t)\\ &+\frac{1}{2}\sum_{\sigma=\uparrow,\downarrow}\sum_{\sigma'=\uparrow,\downarrow}\int\mathrm{d}^d\boldsymbol{r}\int\mathrm{d}^d\boldsymbol{r}'\,\hat{\psi}^\dagger_\sigma(\boldsymbol{r})\,\hat{\psi}^\dagger_{\sigma'}(\boldsymbol{r}')\,\hat{\psi}_{\sigma'}(\boldsymbol{r}')\,\hat{\psi}_\sigma(\boldsymbol{r})\,W_{\mathrm{int}}(\boldsymbol{r}-\boldsymbol{r}'),\end{aligned} \tag{E.106a}$$

where

$$\hat{n}(\boldsymbol{r}) := \sum_{\sigma=\uparrow,\downarrow} \hat{\psi}^{\dagger}_{\sigma}(\boldsymbol{r})\,\hat{\psi}_{\sigma}(\boldsymbol{r}). \tag{E.106b}$$

The time dependence originates from the classical gauge fields in terms of which the classical electromagnetic fields read

$$\boldsymbol{E}(\boldsymbol{r},t) = -\left(\boldsymbol{\nabla}\varphi\right)(\boldsymbol{r},t) - \left(\frac{1}{c}\partial_t \boldsymbol{A}\right)(\boldsymbol{r},t), \tag{E.107a}$$

and

$$\boldsymbol{B}(\boldsymbol{r},t) = \left(\boldsymbol{\nabla}\wedge\boldsymbol{A}\right)(\boldsymbol{r},t). \tag{E.107b}$$

The corresponding classical Lagrangian

$$\mathfrak{L} := \sum_{\sigma=\uparrow,\downarrow} \int \mathrm{d}t \int \mathrm{d}^d\boldsymbol{r}\, \psi^{*}_{\sigma}\, \mathrm{i}\hbar\, \partial_t\, \psi_{\sigma} - \int \mathrm{d}t\, \mathfrak{H} \tag{E.108a}$$

is invariant under the local gauge transformation

$$\psi^{*} \to \psi^{*} e^{+\mathrm{i}e\chi/(\hbar c)}, \qquad \psi \to e^{-\mathrm{i}e\chi/(\hbar c)}\psi, \qquad \boldsymbol{A} \to \boldsymbol{A} + \boldsymbol{\nabla}\chi, \qquad \varphi \to \varphi - \frac{1}{c}\partial_t\chi. \tag{E.108b}$$

At the operator level, this implies that the continuity equation

$$\frac{1}{c}\partial_t \hat{n}_{\mathrm{H}} + \boldsymbol{\nabla}\cdot\hat{\boldsymbol{\jmath}}_{\mathrm{H}} = 0 \tag{E.109a}$$

is obeyed in the Heisenberg picture. Here, the gauge-invariant number-density and current-density operators are

$$\hat{n}_{\mathrm{H}} = \sum_{\sigma=\uparrow,\downarrow} \hat{\psi}^{\dagger}_{\mathrm{H}\,\sigma}\,\hat{\psi}_{\mathrm{H}\,\sigma}, \tag{E.109b}$$

and

$$\hat{\boldsymbol{\jmath}}_{\mathrm{H}} = \frac{1}{2m}\sum_{\sigma=\uparrow,\downarrow}\left[\hat{\psi}^{\dagger}_{\mathrm{H}\,\sigma}\left(\frac{\hbar}{\mathrm{i}}\boldsymbol{\nabla} + \frac{e}{c}\boldsymbol{A}\right)\hat{\psi}_{\mathrm{H}\,\sigma} + \mathrm{H.c.}\right], \tag{E.109c}$$

respectively. We shall assume that the global gauge symmetry responsible for this continuity equation is not spontaneously broken, as happens in a mean-field treatment of superconductivity for example.

We choose a gauge in which

$$\varphi(\boldsymbol{r},t) = 0, \tag{E.110}$$

and make the linear-response Ansatz

$$\boldsymbol{A}(\boldsymbol{r},t) = \boldsymbol{A}_{\mathrm{eq}}(\boldsymbol{r}) + \Theta(t-t_0)\,\boldsymbol{A}'(\boldsymbol{r},t). \tag{E.111}$$

The vector potential $\boldsymbol{A}_{\mathrm{eq}}(\boldsymbol{r})$ is chosen static to insure thermodynamic equilibrium prior to switching on the coupling $\boldsymbol{A}'(\boldsymbol{r},t)$ to the environment at time t_0. Correspondingly, we do the decomposition

$$\hat{H}(t) = \hat{H}_0 + \Theta(t-t_0)\,\hat{H}'(t) + \cdots, \tag{E.112a}$$

where

$$\begin{aligned}\hat{H}_0 = &\sum_{\sigma=\uparrow,\downarrow} \int \mathrm{d}^d\boldsymbol{r}\, \hat{\psi}^\dagger_\sigma(\boldsymbol{r}) \frac{1}{2m} \left(\frac{\hbar}{\mathrm{i}}\boldsymbol{\nabla} + \frac{e}{c}\boldsymbol{A}_{\mathrm{eq}}(\boldsymbol{r})\right)^2 \hat{\psi}_\sigma(\boldsymbol{r}) \\ &+ \frac{1}{2} \sum_{\sigma=\uparrow,\downarrow} \sum_{\sigma'=\uparrow,\downarrow} \int \mathrm{d}^d\boldsymbol{r} \int \mathrm{d}^d\boldsymbol{r}'\, \hat{\psi}^\dagger_\sigma(\boldsymbol{r})\, \hat{\psi}^\dagger_{\sigma'}(\boldsymbol{r}')\, \hat{\psi}_{\sigma'}(\boldsymbol{r}')\, \hat{\psi}_\sigma(\boldsymbol{r})\, W_{\mathrm{int}}\,(\boldsymbol{r}-\boldsymbol{r}'),\end{aligned} \tag{E.112b}$$

and

$$\hat{H}'(t) = \frac{e}{c} \int \mathrm{d}^d\boldsymbol{r}\, \boldsymbol{A}'(\boldsymbol{r},t)\cdot \hat{\boldsymbol{j}}_0(\boldsymbol{r}). \tag{E.112c}$$

The gauge-invariant current-density operator in thermodynamic equilibrium

$$\hat{\boldsymbol{j}}_0(\boldsymbol{r}) = \frac{1}{2m} \sum_{\sigma=\uparrow,\downarrow} \left[\hat{\psi}^\dagger_\sigma \left(\frac{\hbar}{\mathrm{i}}\boldsymbol{\nabla} + \frac{e}{c}\boldsymbol{A}_{\mathrm{eq}}\right) \hat{\psi}_\sigma + \mathrm{H.c.}\right] \tag{E.113a}$$

is sometimes broken up into

$$\hat{\boldsymbol{j}}_0(\boldsymbol{r}) = \hat{\boldsymbol{j}}_{0\,\mathrm{p}}(\boldsymbol{r}) + \hat{\boldsymbol{j}}_{0\,\mathrm{d}}(\boldsymbol{r}), \tag{E.113b}$$

with the paramagnetic contribution

$$\hat{\boldsymbol{j}}_{0\,\mathrm{p}}(\boldsymbol{r}) := \frac{\hbar}{2m\mathrm{i}} \sum_{\sigma=\uparrow,\downarrow} \left(\left(\hat{\psi}^\dagger_\sigma \boldsymbol{\nabla} \hat{\psi}_\sigma\right) - \left(\boldsymbol{\nabla}\hat{\psi}^\dagger_\sigma\right) \hat{\psi}_\sigma \right)(\boldsymbol{r}), \tag{E.113c}$$

and with the diamagnetic contribution

$$\hat{\boldsymbol{j}}_{0\,\mathrm{d}}(\boldsymbol{r}) := \frac{e}{m\,c}\boldsymbol{A}_{\mathrm{eq}}(\boldsymbol{r})\,\hat{n}(\boldsymbol{r}), \qquad \hat{n}(\boldsymbol{r}) := \sum_{\sigma=\uparrow,\downarrow} \left(\hat{\psi}^\dagger_\sigma \hat{\psi}_\sigma\right)(\boldsymbol{r}). \tag{E.113d}$$

Observe that (i) these currents have the dimensions of $(\text{length})^{-(d-1)} \times (\text{time})^{-1}$ and (ii) $\hat{\boldsymbol{j}}_0(\boldsymbol{r}) = \hat{\boldsymbol{j}}_{0\,\mathrm{p}}(\boldsymbol{r})$ when $\boldsymbol{A}_{\mathrm{eq}} = \boldsymbol{0}$. Terms omitted in $\cdots$ from Eq. (E.112a) are total derivatives and the second-order term

$$\int \mathrm{d}^d\boldsymbol{r}\,\hat{n}(\boldsymbol{r})\,\frac{1}{2m}\left(\frac{e}{c}\boldsymbol{A}'(\boldsymbol{r},t)\right)^2. \tag{E.114}$$

We seek the instantaneous statistical average defined by Eq. (E.91b) of the current density $\hat{\boldsymbol{J}}(\boldsymbol{r},t)$ defined by the functional derivative of Hamiltonian (E.112a) with respect to $\boldsymbol{A}'(\boldsymbol{r},t)$,

$$\begin{aligned}\hat{\boldsymbol{J}}(\boldsymbol{r},t) :=& -c\,\frac{\delta \hat{H}(t)}{\delta \boldsymbol{A}'(\boldsymbol{r},t)} \\ =& (-e)\left[\hat{\boldsymbol{j}}_0(\boldsymbol{r}) + \frac{e}{m\,c}\,\boldsymbol{A}'(\boldsymbol{r},t)\,\hat{n}(\boldsymbol{r})\right].\end{aligned} \tag{E.115}$$

We achieve this by evaluating

$$\boldsymbol{J}_{0\,\beta}(\boldsymbol{r},\omega) := (-e)\left[\left\langle \hat{\boldsymbol{j}}_0(\boldsymbol{r},\omega)\right\rangle_{0\,\beta} + \frac{e}{m\,c}\left\langle \boldsymbol{A}'(\boldsymbol{r},\omega)\,\hat{n}(\boldsymbol{r})\right\rangle_{0\,\beta}\right], \tag{E.116}$$

in the frequency domain to linear order in $\boldsymbol{A}'$.[20]

Needed first is

$$\left\langle \hat{\boldsymbol{j}}_0(\boldsymbol{r},\omega)\right\rangle_{0\,\beta} = \left\langle \left(\hat{\boldsymbol{j}}_{0\,\mathrm{p}}(\boldsymbol{r},\omega) + \hat{\boldsymbol{j}}_{0\,\mathrm{d}}(\boldsymbol{r},\omega)\right)\right\rangle_{0\,\beta}, \tag{E.117}$$

to linear order in $\boldsymbol{A}'$. To this end, we observe that $\hat{H}'(t)$ is of the form (E.105c) with the identifications

$$f^b(\boldsymbol{r},t) \to \frac{e}{c} A'^b(\boldsymbol{r},t), \qquad \hat{B}^b(\boldsymbol{r}) \to \hat{j}^b_0(\boldsymbol{r}). \tag{E.118}$$

We can then apply Eq. (E.105b) to the computation of the expectation value of $\hat{\boldsymbol{j}}_0(\boldsymbol{r},\omega)$ out of the thermodynamic equilibrium, provided we make the identification

$$\hat{A} \to \hat{\boldsymbol{j}}_0(\boldsymbol{r}). \tag{E.119}$$

Thus, for any component $\hat{j}^a_0(\boldsymbol{r},\omega)$ of $\hat{\boldsymbol{j}}_0(\boldsymbol{r},\omega)$, we find

$$\left\langle \hat{j}^a_0(\boldsymbol{r},\omega)\right\rangle_{0\,\beta} = 2\pi \left\langle \hat{j}^a_0(\boldsymbol{r})\right\rangle_{0\,\beta} \delta(\omega) + \frac{e}{c}\sum_b \int \mathrm{d}^d\boldsymbol{r}'\, C^{\mathrm{R}}_{\hat{j}^a_0(\boldsymbol{r}),\hat{j}^b_0(\boldsymbol{r}')}(\omega)\, A'^b(\boldsymbol{r}',\omega), \tag{E.120}$$

in the frequency domain and to linear order in $\boldsymbol{A}'$. Furthermore, with our choice of gauge

$$\left(\frac{\mathrm{i}\omega}{c}\right) A'^b(\boldsymbol{r}',\omega) = E'^b(\boldsymbol{r}',\omega), \tag{E.121}$$

so that

$$\left\langle \hat{j}^a_0(\boldsymbol{r},\omega)\right\rangle_{0\,\beta} = 2\pi \left\langle \hat{j}^a_0(\boldsymbol{r})\right\rangle_{0\,\beta} \delta(\omega) + \frac{e}{\mathrm{i}\omega}\sum_b \int \mathrm{d}^d\boldsymbol{r}'\, C^{\mathrm{R}}_{\hat{j}^a_0(\boldsymbol{r}),\hat{j}^b_0(\boldsymbol{r}')}(\omega)\, E'^b(\boldsymbol{r}',\omega), \tag{E.122}$$

in the frequency domain and to linear order in $\boldsymbol{E}'$.

Second, we need

$$\langle \boldsymbol{A}'(\boldsymbol{r},\omega)\,\hat{n}(\boldsymbol{r})\rangle_{0\,\beta} = \boldsymbol{A}'(\boldsymbol{r},\omega)\,\langle \hat{n}(\boldsymbol{r})\rangle_{0\,\beta} = \frac{c}{\mathrm{i}\omega}\boldsymbol{E}'(\boldsymbol{r},\omega)\,\langle \hat{n}(\boldsymbol{r})\rangle_{0\,\beta}, \tag{E.123}$$

to first order in $\boldsymbol{A}'$. Here, the average local number of electrons per unit volume

$$n_{0\,\beta}(\boldsymbol{r}) := \langle \hat{n}(\boldsymbol{r})\rangle_{0\,\beta} \tag{E.124}$$

is time independent, for it is the one in thermodynamic equilibrium prior to switching on $\boldsymbol{A}'$, see footnote 20. As a corollary, the current density

$$\boldsymbol{j}_{0\,\beta}(\boldsymbol{r}) := \left\langle \hat{\boldsymbol{j}}_0(\boldsymbol{r})\right\rangle_{0\,\beta} = 0 \tag{E.125}$$

[20] The operators $\hat{\boldsymbol{j}}_0(\boldsymbol{r})$ and $\hat{n}(\boldsymbol{r})$ are not dependent on time on the right-hand side of Eq. (E.115) in the Schrödinger picture. The operator $\hat{\boldsymbol{j}}_0(\boldsymbol{r})$ acquires a time dependence in the interaction picture as its expectation value out of thermodynamic equilibrium is to be evaluated with Eq. (E.99). This is why we use $\hat{\boldsymbol{j}}_0(\boldsymbol{r},\omega)$ for the first term on the right-hand side of Eq. (E.116). On the other hand, to first order in $\boldsymbol{A}'(\boldsymbol{r},t)$, we use the first term on the right-hand side of Eq. (E.99) to compute the expectation value out of thermodynamic equilibrium of $\boldsymbol{A}'(\boldsymbol{r},t)\,\hat{n}(\boldsymbol{r})$. This is why we use $\hat{n}(\boldsymbol{r})$ for the second term on the right-hand side of Eq. (E.116).

vanishes in thermodynamic equilibrium by the continuity equation so that the first term on the right-hand side of Eq. (E.122) drops out.[21]

By combining Eq. (E.116) with Eqs. (E.122)–(E.125), we conclude that

$$\begin{aligned} J^{a}_{0\,\beta}(\boldsymbol{r},\omega) &= \frac{\mathrm{i}e^2}{\omega}\sum_{b}\int \mathrm{d}^d\boldsymbol{r}'\, C^{\mathrm{R}}_{\hat{j}^a_0(\boldsymbol{r}),\hat{j}^b_0(\boldsymbol{r}')}(\omega)\, E'^{b}(\boldsymbol{r}',\omega) + \frac{\mathrm{i}e^2}{\omega\, m}\, n_{0\,\beta}(\boldsymbol{r})\, E'^{a}(\boldsymbol{r},\omega) \\ &= \frac{\mathrm{i}e^2}{\omega}\sum_{b}\int \mathrm{d}^d\boldsymbol{r}' \left[C^{\mathrm{R}}_{\hat{j}^a_0(\boldsymbol{r}),\hat{j}^b_0(\boldsymbol{r}')}(\omega) + \frac{1}{m}\, n_{0\,\beta}(\boldsymbol{r})\,\delta(\boldsymbol{r}-\boldsymbol{r}')\,\delta^{ab}\right] E'^{b}(\boldsymbol{r}',\omega). \end{aligned} \tag{E.126}$$

The non-local conductivity tensor is defined by

$$J^{a}_{0\,\beta}(\boldsymbol{r},t) =: \sum_{b}\int_{\mathbb{R}} \mathrm{d}t' \int \mathrm{d}^d\boldsymbol{r}'\, \sigma^{ab}_{0\,\beta}(\boldsymbol{r},t;\boldsymbol{r}',t')\, E'^{b}(\boldsymbol{r}',t'). \tag{E.127}$$

In the frequency domain, under the assumption that the non-local conductivity tensor is a function of $t-t'$ only,

$$J^{a}_{0\,\beta}(\boldsymbol{r},\omega) = \sum_{b}\int \mathrm{d}^d\boldsymbol{r}'\, \sigma^{ab}_{0\,\beta}(\boldsymbol{r},\boldsymbol{r}',\omega)\, E'^{b}(\boldsymbol{r}',\omega). \tag{E.128a}$$

With the help of Eq. (E.126), we arrive at the Kubo formula for the conductivity

$$\sigma^{ab}_{0\,\beta}(\boldsymbol{r},\boldsymbol{r}',\omega) = \frac{\mathrm{i}e^2}{\omega}\left[\Pi^{\mathrm{R}\,ab}_{0\,\beta}(\boldsymbol{r},\boldsymbol{r}',\omega) + \frac{1}{m} n_{0\,\beta}(\boldsymbol{r})\,\delta(\boldsymbol{r}-\boldsymbol{r}')\,\delta^{ab}\right] \tag{E.128b}$$

of a non-relativistic (i.e., with a kinetic energy that is quadratic in the momentum) interacting electron gas. Here, we have introduced the notation

$$\Pi^{\mathrm{R}\,ab}_{0\,\beta}(\boldsymbol{r},\boldsymbol{r}',\omega) \equiv C^{\mathrm{R}}_{\hat{j}^a_0(\boldsymbol{r}),\hat{j}^b_0(\boldsymbol{r}')}(\omega) = \int_{\mathbb{R}} \mathrm{d}t\, e^{+\mathrm{i}\omega t}\,\frac{\Theta(t)}{\mathrm{i}\hbar}\left\langle\left[\hat{j}^a_{0\mathrm{I}}(\boldsymbol{r},t),\hat{j}^b_{0\mathrm{I}}(\boldsymbol{r}',0)\right]\right\rangle_{0\,\beta}. \tag{E.128c}$$

As it should be, the Kubo conductivity (E.128b) is gauge invariant.

E.5.4 *Kubo formula for the dc conductance*

The conductivity (E.128b) relates linearly the charge-current density to an applied electric field. In an infinite system (or far away from the boundaries of a macroscopic sample), the conductivity (E.128b) is an intrinsic property of the system (macroscopic sample). On the other hand, Ohm's law,

$$I = G\,U \Longrightarrow [G] = \frac{\text{charge}}{\text{time}} \times \frac{\text{charge}}{\text{energy}} = \left[\frac{e^2}{\hbar}\right], \tag{E.129}$$

[21] Indeed, $(\boldsymbol{\nabla}\cdot\boldsymbol{j}_0)(\boldsymbol{r}) = 0$ for all $\boldsymbol{r}$ with the proper boundary conditions implies Eq. (E.125). The continuity equation breaks down when the global gauge symmetry responsible for the continuity equation is spontaneously broken as is the case in a superconducting phase of matter. In a superconducting phase, the magnetic field $\boldsymbol{B}_0 = \boldsymbol{\nabla}\wedge\boldsymbol{A}_{\mathrm{eq}}$ is screened by the diamagnetic current through the relation $\boldsymbol{\nabla}\wedge\boldsymbol{B}_0 = \frac{4\pi e^*}{c}\boldsymbol{j}_{0\,\mathrm{d}}$, see Eq. (7.180), where we may use Eq. (E.113d) with $e^* = 2\,e$ and $m^* = 2m$, operators replaced by functions, the density $n(\boldsymbol{r})$ replaced by the superfluid density $n_{0\,\beta\,\mathrm{sc}}$ so that $\boldsymbol{j}_{0\,\mathrm{d}} = +\frac{e^*\,n_{0\,\beta\,\mathrm{sc}}}{m^*\,c}\boldsymbol{A}_{\mathrm{eq}}$.

relates linearly the electric current I through a macroscopic sample to the voltage U applied to it, thereby defining the conductance G. The conductance depends on the shape of the material and is thus not an intrinsic property. For three-dimensional materials with a constant local conductivity, the conductance and conductivity σ are related by the aspect ratio

$$G = \frac{W}{L}\sigma, \tag{E.130}$$

where L is the sample length and W the area of its cross-section, as

$$J = \sigma\, E \Longrightarrow [\sigma] = \frac{\text{charge}}{\text{length}^{d-1}\times\text{time}} \times \frac{\text{charge}\times\text{length}}{\text{energy}} = \left[\frac{e^2}{\hbar}\right] \times \text{length}^{2-d}. \tag{E.131}$$

For materials for which the local approximation to the conductivity tensor is not good, a Kubo formula for the conductance is needed. Mesoscopic conductors are characterized by a size smaller than the typical thermalization length so that a local approximation for the conductivity is inadequate. We shall assume that the wavelength of the external perturbation is much longer than the sample, in which case the perturbing frequency is usually very small and can be set to zero. This is the static or dc approximation. We want to compute the evolution in time out of thermodynamic equilibrium of the current

$$I_{0\,\beta\,\mathrm{dc}} := \int \mathrm{d}a\,\boldsymbol{\xi}_{\perp}(\boldsymbol{r})\cdot \mathrm{Re}\,\boldsymbol{J}_{0\,\beta\,\mathrm{dc}}(\boldsymbol{r}), \qquad \boldsymbol{J}_{0\,\beta\,\mathrm{dc}}(\boldsymbol{r}) \equiv \boldsymbol{J}_{0\,\beta}(\boldsymbol{r},\omega=0), \tag{E.132}$$

where the integration is over an oriented $(d-1)$-dimensional cross-section of the d-dimensional volume with the infinitesimal oriented volume $\mathrm{d}a\,\boldsymbol{\xi}_{\perp}(\boldsymbol{r})$ whereby the unit vector $\boldsymbol{\xi}_{\perp}(\boldsymbol{r})$ is orthogonal to the cross-section. Current conservation here selects the choice of the cross-section. The perturbing dc electric field

$$\boldsymbol{E}'_{\mathrm{dc}}(\boldsymbol{r}) \equiv \boldsymbol{E}'(\boldsymbol{r},\omega=0) \tag{E.133}$$

defines equipotential cross-sections by the condition

$$\boldsymbol{a}_{\parallel}(\boldsymbol{r})\cdot \boldsymbol{E}'_{\mathrm{dc}}(\boldsymbol{r}) = 0, \qquad \boldsymbol{a}^2_{\parallel}(\boldsymbol{r}) = 1. \tag{E.134}$$

We thus choose the coordinate system

$$\mathsf{a}_{\parallel}(\boldsymbol{r})\,\boldsymbol{a}_{\parallel}(\boldsymbol{r}) + \xi_{\perp}(\boldsymbol{r})\,\boldsymbol{\xi}_{\perp}(\boldsymbol{r}), \qquad \boldsymbol{\xi}_{\perp}(\boldsymbol{r}) := \frac{\boldsymbol{E}'_{\mathrm{dc}}(\boldsymbol{r})}{|\boldsymbol{E}'_{\mathrm{dc}}(\boldsymbol{r})|}, \tag{E.135}$$

in order to parametrize the oriented cross-section with infinitesimal volume element $\mathrm{d}a\,\boldsymbol{\xi}_{\perp}(\boldsymbol{r})$. In this coordinate system, Eq. (E.132) becomes

$$\begin{aligned} I_{0\,\beta\,\mathrm{dc}} &= \int \mathrm{d}a\,\boldsymbol{\xi}_{\perp}(\boldsymbol{r})\cdot \mathrm{Re}\,\boldsymbol{J}_{0\,\beta\,\mathrm{dc}}(\boldsymbol{r}) \\ &= \int \mathrm{d}\xi_{\perp}(\boldsymbol{r}')\,|\boldsymbol{E}'_{\mathrm{dc}}(\boldsymbol{r}')| \int \mathrm{d}a \int \mathrm{d}a' \sum_{a,b} \left[\mathrm{Re}\,\sigma^{ab}_{0\,\beta\,\mathrm{dc}}(\boldsymbol{r},\boldsymbol{r}')\right]\, \xi^a_{\perp}(\boldsymbol{r})\,\xi^b_{\perp}(\boldsymbol{r}'), \end{aligned} \tag{E.136}$$

where we made use of Eq. (E.128a). The second (diamagnetic) term on the right-hand side of Eq. (E.128b) is purely imaginary. Hence, it drops out from the real part of the conductivity tensor. We are then left with

$$I_{0\,\beta\,\mathrm{dc}}=\int \mathrm{d}\xi_{\perp}(\boldsymbol{r}')\,|\boldsymbol{E}'_{\mathrm{dc}}(\boldsymbol{r}')|\int \mathrm{da}\int \mathrm{da}'\sum_{a,b}\left[\lim_{\omega\to 0}\mathrm{Re}\,\frac{\mathrm{i}e^2}{\omega}\Pi^{\mathrm{R}\,ab}_{0\,\beta\,\mathrm{dc}}(\boldsymbol{r},\boldsymbol{r}',\omega)\right]\xi^{a}_{\perp}(\boldsymbol{r})\,\xi^{b}_{\perp}(\boldsymbol{r}'), \tag{E.137a}$$

where, according to Eq. (E.128c),

$$\lim_{\omega\to 0}\int \mathrm{da}\int \mathrm{da}'\sum_{a,b}\Pi^{\mathrm{R}\,ab}_{0\,\beta\,\mathrm{dc}}(\boldsymbol{r},\boldsymbol{r}',\omega)\,\xi^{a}_{\perp}(\boldsymbol{r})\,\xi^{b}_{\perp}(\boldsymbol{r}')=\int_{\mathbb{R}}\mathrm{d}t\,\frac{\Theta(t)}{\mathrm{i}\hbar}\left\langle\left[\hat{I}^{\mathrm{dc}}_{0\,\mathrm{I}}(t),\hat{I}^{\mathrm{dc}}_{0\,\mathrm{I}}(0)\right]\right\rangle_{0\,\beta}, \tag{E.137b}$$

with the short-hand notation

$$\hat{I}^{\mathrm{dc}}_{0\,\mathrm{I}}(t)\equiv\sum_{a}\int \mathrm{da}\,\xi^{a}_{\perp}(\boldsymbol{r})\,\hat{j}^{a}_{0\mathrm{I}}(\boldsymbol{r},t). \tag{E.137c}$$

The operator $\hat{I}^{\mathrm{dc}}_{0\,\mathrm{I}}(t)$ carries the dimension of inverse time. Current conservation through the choice (E.135) for the coordinate system implies that the operator $\hat{I}^{\mathrm{dc}}_{0\,\mathrm{I}}(t)$ is independent of the coordinate parallel to the electric field. Hence,

$$\begin{aligned}I_{0\,\beta\,\mathrm{dc}}&=\lim_{\omega\to 0}\mathrm{Re}\left[\frac{\mathrm{i}e^2}{\omega}C^{\mathrm{R}}_{0\,\beta\,\hat{I}^{\mathrm{dc}}_{0},\hat{I}^{\mathrm{dc}}_{0}}(\omega)\right]\int \mathrm{d}\xi_{\perp}(\boldsymbol{r}')\,|\boldsymbol{E}'_{\mathrm{dc}}(\boldsymbol{r}')|\\&=\lim_{\omega\to 0}\mathrm{Re}\left[\frac{\mathrm{i}e^2}{\omega}C^{\mathrm{R}}_{0\,\beta\,\hat{I}^{\mathrm{dc}}_{0},\hat{I}^{\mathrm{dc}}_{0}}(\omega)\right]U.\end{aligned} \tag{E.138}$$

The dc conductance is thus given by

$$G=\lim_{\omega\to 0}\mathrm{Re}\left[\frac{\mathrm{i}e^2}{\omega}C^{\mathrm{R}}_{0\,\beta\,\hat{I}^{\mathrm{dc}}_{0},\hat{I}^{\mathrm{dc}}_{0}}(\omega)\right]. \tag{E.139}$$

E.5.5 *Kubo formula for the dielectric function*

What if we had used a time-dependent scalar potential as coupling to the environment instead of a time-dependent vector gauge field in section E.5.3? In other words, we start from the time-dependent Hamiltonian

$$\begin{aligned}\hat{H}(t)=&\sum_{\sigma=\uparrow,\downarrow}\int \mathrm{d}^d\boldsymbol{r}\,\hat{\psi}^{\dagger}_{\sigma}(\boldsymbol{r})\,\frac{1}{2m}\left(\frac{\hbar}{\mathrm{i}}\boldsymbol{\nabla}\right)^2\hat{\psi}_{\sigma}(\boldsymbol{r})\\&-e\int \mathrm{d}^d\boldsymbol{r}\,\hat{n}(\boldsymbol{r})\,\varphi(\boldsymbol{r},t)\\&+\frac{1}{2}\sum_{\sigma=\uparrow,\downarrow}\sum_{\sigma'=\uparrow,\downarrow}\int \mathrm{d}^d\boldsymbol{r}\int \mathrm{d}^d\boldsymbol{r}'\,\hat{\psi}^{\dagger}_{\sigma}(\boldsymbol{r})\,\hat{\psi}^{\dagger}_{\sigma'}(\boldsymbol{r}')\,\hat{\psi}_{\sigma'}(\boldsymbol{r}')\,\hat{\psi}_{\sigma}(\boldsymbol{r})\,W_{\mathrm{int}}(\boldsymbol{r}-\boldsymbol{r}'),\end{aligned} \tag{E.140a}$$

where

$$\hat{n}(\boldsymbol{r}):=\sum_{\sigma=\uparrow,\downarrow}\hat{\psi}^{\dagger}_{\sigma}(\boldsymbol{r})\,\hat{\psi}_{\sigma}(\boldsymbol{r}). \tag{E.140b}$$

We make the linear-response Ansatz

$$\varphi(\boldsymbol{r},t)=\varphi_{\text{eq}}(\boldsymbol{r})+\Theta(t-t_0)\,\varphi'(\boldsymbol{r},t), \tag{E.141}$$

where the scalar potential $\varphi_{\text{eq}}(\boldsymbol{r})$ is chosen static to insure thermodynamic equilibrium prior to switching on the coupling $\varphi'(\boldsymbol{r},t)$ to the environment at time t_0. Correspondingly, we do the decomposition

$$\hat{H}(t)=\hat{H}_0+\Theta(t-t_0)\,\hat{H}'(t), \tag{E.142a}$$

where

$$\begin{aligned}\hat{H}_0=&\sum_{\sigma=\uparrow,\downarrow}\int\mathrm{d}^d\boldsymbol{r}\,\hat{\psi}^{\dagger}_{\sigma}(\boldsymbol{r})\,\frac{1}{2m}\left(\frac{\hbar}{\mathrm{i}}\boldsymbol{\nabla}\right)^2\hat{\psi}_{\sigma}(\boldsymbol{r})\\&-e\int\mathrm{d}^d\boldsymbol{r}\,\hat{n}(\boldsymbol{r})\,\varphi_{\text{eq}}(\boldsymbol{r})\\&+\frac{1}{2}\sum_{\sigma=\uparrow,\downarrow}\sum_{\sigma'=\uparrow,\downarrow}\int\mathrm{d}^d\boldsymbol{r}\int\mathrm{d}^d\boldsymbol{r}'\,\hat{\psi}^{\dagger}_{\sigma}(\boldsymbol{r})\,\hat{\psi}^{\dagger}_{\sigma'}(\boldsymbol{r}')\,\hat{\psi}_{\sigma'}(\boldsymbol{r}')\,\hat{\psi}_{\sigma}(\boldsymbol{r})\,W_{\text{int}}(\boldsymbol{r}-\boldsymbol{r}'),\end{aligned} \tag{E.142b}$$

and

$$\hat{H}'(t)=-e\int\mathrm{d}^d\boldsymbol{r}\,\varphi'(\boldsymbol{r},t)\,\hat{n}(\boldsymbol{r}). \tag{E.142c}$$

We observe that $\hat{H}'(t)$ is of the form (E.105c) with the identifications

$$f^b(\boldsymbol{r},t)\to-e\,\varphi'(\boldsymbol{r},t),\qquad \hat{B}^b(\boldsymbol{r})\to\hat{n}(\boldsymbol{r}). \tag{E.143}$$

We can then apply Eq. (E.105b) to the computation of the out-of-thermodynamic-equilibrium expectation value of $\hat{n}(\boldsymbol{r},t)$ provided we make the identification

$$\hat{A}\to\hat{n}(\boldsymbol{r}). \tag{E.144}$$

Thus, to linear order in φ', we find

$$\langle\hat{n}(\boldsymbol{r},\omega)\rangle_{0\,\beta}=2\pi\,\langle\hat{n}(\boldsymbol{r})\rangle_{0\,\beta}\,\delta(\omega)-e\int\mathrm{d}^d\boldsymbol{r}'\,C^{\mathrm{R}}_{0\,\beta\,\hat{n}(\boldsymbol{r}),\hat{n}(\boldsymbol{r}')}(\omega)\,\varphi'(\boldsymbol{r}',\omega), \tag{E.145}$$

in the frequency domain or

$$\langle\hat{n}(\boldsymbol{r},t)\rangle_{0\,\beta}=\langle\hat{n}(\boldsymbol{r})\rangle_{0\,\beta}-e\int\mathrm{d}^d\boldsymbol{r}'\int\limits_{\mathbb{R}}\mathrm{d}t'\,C^{\mathrm{R}}_{0\,\beta\,\hat{n}(\boldsymbol{r}),\hat{n}(\boldsymbol{r}')}(t-t')\,\varphi'(\boldsymbol{r}',t'), \tag{E.146}$$

in the time domain.

It is customary to denote the retarded density-density correlation function by

$$\chi^{\mathrm{R}}_{0\,\beta}(\boldsymbol{r},t;\boldsymbol{r}',t')\equiv C^{\mathrm{R}}_{0\,\beta\,\hat{n}(\boldsymbol{r}),\hat{n}(\boldsymbol{r}')}(t-t'). \tag{E.147}$$

This correlation function is related to the dielectric constant as follows. Define the classical scalar gauge potential induced in linear response by the solution to Poisson equation

$$\Delta\varphi_{0\,\beta\,\text{tot}}(\boldsymbol{r},t)=-4\pi\rho_{0\,\beta\,\text{tot}}(\boldsymbol{r},t), \tag{E.148a}$$

where

$$\rho_{0\,\beta\,\mathrm{tot}}(\boldsymbol{r},t) = -e\left(n'(\boldsymbol{r},t) + \langle\hat{n}(\boldsymbol{r},t)\rangle_{0\,\beta} - \langle\hat{n}(\boldsymbol{r})\rangle_{0\,\beta}\right), \tag{E.148b}$$

and

$$\Delta\varphi'(\boldsymbol{r},t) = +4\pi e\, n'(\boldsymbol{r},t). \tag{E.148c}$$

Let W_{cb} be the solution to

$$\Delta W_{\mathrm{cb}}(\boldsymbol{r}) = +4\pi e\, \delta(\boldsymbol{r}). \tag{E.149}$$

It then follows that

$$\begin{aligned}
\varphi_{0\,\beta\,\mathrm{tot}}(\boldsymbol{r},t) &= \varphi'(\boldsymbol{r},t) + \int \mathrm{d}^d\boldsymbol{r}'' W_{\mathrm{cb}}(\boldsymbol{r}-\boldsymbol{r}'')\left(\langle\hat{n}(\boldsymbol{r}'',t)\rangle_{0\,\beta} - \langle\hat{n}(\boldsymbol{r}'')\rangle_{0\,\beta}\right)\\
&= \varphi'(\boldsymbol{r},t) - e\int \mathrm{d}^d\boldsymbol{r}'' W_{\mathrm{cb}}(\boldsymbol{r}-\boldsymbol{r}'')\int \mathrm{d}^d\boldsymbol{r}'\int \mathrm{d}t'\chi^{\mathrm{R}}_{0\,\beta}(\boldsymbol{r}'',t;\boldsymbol{r}',t')\varphi'(\boldsymbol{r}',t')\\
&= \int \mathrm{d}^d\boldsymbol{r}'\int \mathrm{d}t'\,\varepsilon^{-1}_{0\,\beta}(\boldsymbol{r},t;\boldsymbol{r}',t')\,\varphi'(\boldsymbol{r}',t'),
\end{aligned} \tag{E.150a}$$

where we have introduced the dielectric kernel through its inverse

$$\varepsilon^{-1}_{0\,\beta}(\boldsymbol{r},t;\boldsymbol{r}',t') := \delta(\boldsymbol{r}-\boldsymbol{r}')\,\delta(t-t') - e\int \mathrm{d}^d\boldsymbol{r}''\, W_{\mathrm{cb}}(\boldsymbol{r}-\boldsymbol{r}'')\,\chi^{\mathrm{R}}_{0\,\beta}(\boldsymbol{r}'',t;\boldsymbol{r}',t'). \tag{E.150b}$$

This result should be compared with the discussion of the dielectric function in section 6.7.

E.5.6 *Fluctuation-dissipation theorem*

The Kubo formula is an example of the fluctuation-dissipation theorem that relates dissipative processes in a system out of thermodynamic equilibrium due to weak couplings to the environment (the dissipation) to its fluctuations in thermodynamic equilibrium when isolated from the environment. To emphasize the notion of fluctuations in the Kubo formula, we are going to relate the retarded correlation function to a *spectral density function.*

We consider the closed system defined in Eq. (E.87). Let $\hat{A}$ and $\hat{B}$ be any pair of operators acting on $\mathcal{H}_0$. It is standard practice to call

$$\begin{aligned}
J_{0\,\beta\,\hat{A},\hat{B}}(\omega) &:= \int\limits_{-\infty}^{+\infty} \mathrm{d}t\, e^{+\mathrm{i}\omega t}\left\langle \hat{A}_{\mathrm{H}}(t)\hat{B}_{\mathrm{H}}(0)\right\rangle_{0\,\beta}\\
&= \int\limits_{-\infty}^{+\infty} \mathrm{d}t\, e^{+\mathrm{i}\omega t}\left\langle \hat{A}_{\mathrm{H}}(0)\hat{B}_{\mathrm{H}}(-t)\right\rangle_{0\,\beta}\\
&\equiv \int\limits_{-\infty}^{+\infty} \mathrm{d}t\, e^{+\mathrm{i}\omega t} J_{0\,\beta\,\hat{A},\hat{B}}(t)
\end{aligned} \tag{E.151a}$$

the *spectral density function* associated with the time correlation (we set $\hbar = 1$)

$$\begin{aligned} J_{0\,\beta\,\hat{A},\hat{B}}(t) &\equiv \left\langle \hat{A}_{\mathrm{H}}(t)\hat{B}_{\mathrm{H}}(0)\right\rangle_{0\,\beta} \\ &:= \frac{\operatorname{Tr}\left(e^{-\beta\hat{H}_0}\, e^{+\mathrm{i}\hat{H}_0 t}\hat{A}e^{-\mathrm{i}\hat{H}_0 t}\hat{B}\right)}{\operatorname{Tr}\left(e^{-\beta\hat{H}_0}\right)} \\ &= \left\langle \hat{A}_{\mathrm{H}}(0)\hat{B}_{\mathrm{H}}(-t)\right\rangle_{0\,\beta}. \end{aligned} \tag{E.151b}$$

The second line in Eq. (E.151a) or the last two lines in Eq. (E.151b) follow from the cyclicity of the trace. In turn, it is translation invariance in time, i.e., energy conservation as the system is assumed closed, that allows to take advantage of the cyclicity of the trace. By construction the spectral density function depends on both the energy transfer $\hbar\omega$ and the inverse temperature β aside from its dependence on the operators $\hat{A}$ and $\hat{B}$. In the limit $\beta = \infty$ corresponding to zero temperature, only one term survives in the trace (E.151b), namely the ground state expectation value of the product $\hat{A}_{\mathrm{H}}(t)\hat{B}_{\mathrm{H}}(0)$.

The desired link between the physics of fluctuations encoded by the spectral density function and the physics of dissipation is made by relating the spectral density function to the retarded Green function for operators $\hat{A}$ and $\hat{B}$ [22]

$$\begin{aligned} C^{\mathrm{R}}_{0\,\beta\,\hat{A},\hat{B}}(t) &:= -\mathrm{i}\Theta(t)\left\langle\left[\hat{A}_{\mathrm{H}}(t)\hat{B}_{\mathrm{H}}(0) - \hat{B}_{\mathrm{H}}(0)\hat{A}_{\mathrm{H}}(t)\right]\right\rangle_{0\,\beta} \\ &\equiv -\mathrm{i}\Theta(t)\left[J_{0\,\beta\,\hat{A},\hat{B}}(+t) - J_{0\,\beta\,\hat{B},\hat{A}}(-t)\right]. \end{aligned} \tag{E.152}$$

We recall that the prefactor $-\mathrm{i}$ is convention and that the function Θ is the Heaviside step function, i.e., the step function taking the value 0 when $t < 0$ and 1 otherwise. The meaning of the terminology retarded is made transparent after performing the following Fourier transformation in time,

$$\begin{aligned} C^{\mathrm{R}}_{0\,\beta\,\hat{A},\hat{B}}(\omega) &:= \int\limits_{-\infty}^{+\infty} \mathrm{d}t\, e^{+\mathrm{i}\omega t}\, C^{\mathrm{R}}_{0\,\beta\,\hat{A},\hat{B}}(t) \\ &= \int\limits_{0}^{+\infty} \mathrm{d}t\, e^{+\mathrm{i}\omega t}\, C^{\mathrm{R}}_{0\,\beta\,\hat{A},\hat{B}}(t) \\ &= -\mathrm{i}\int\limits_{0}^{+\infty} \mathrm{d}t\, e^{+\mathrm{i}\omega t}\left\langle\left[\hat{A}_{\mathrm{H}}(t)\hat{B}_{\mathrm{H}}(0) - \hat{B}_{\mathrm{H}}(0)\hat{A}_{\mathrm{H}}(t)\right]\right\rangle_{0\,\beta} \\ &= -\mathrm{i}\int\limits_{0}^{+\infty} \mathrm{d}t\, e^{+\mathrm{i}\omega t}\left[J_{0\,\beta\,\hat{A},\hat{B}}(+t) - J_{0\,\beta\,\hat{B},\hat{A}}(-t)\right], \end{aligned} \tag{E.153}$$

[22] As we are in a closed system, we use the Heisenberg picture. We nevertheless use the same symbol for the retarded Green function as we did for open systems in Eq. (E.102c).

to which only $C^{\mathrm{R}}_{0\,\beta\,\hat{A},\hat{B}}(t)$ with $t>0$ contributes.

The retarded Green function of operators $\hat{A}$ and $\hat{B}$ as a function of t has the integral representation

$$\begin{aligned}
C^{\mathrm{R}}_{0\,\beta\,\hat{A},\hat{B}}(t) &= -\mathrm{i}\Theta(t)\left\langle\left[\hat{A}_{\mathrm{H}}(t)\hat{B}_{\mathrm{H}}(0)-\hat{B}_{\mathrm{H}}(0)\hat{A}_{\mathrm{H}}(t)\right]\right\rangle_{0\,\beta}\\
&= -\mathrm{i}\Theta(t)\left[J_{0\,\beta\,\hat{A},\hat{B}}(+t)-J_{0\,\beta\,\hat{B},\hat{A}}(-t)\right]\\
&= -\mathrm{i}\Theta(t)\int\limits_{-\infty}^{+\infty}\frac{\mathrm{d}\omega'}{2\pi}e^{-\mathrm{i}\omega' t}\left[J_{0\,\beta\,\hat{A},\hat{B}}(+\omega')-J_{0\,\beta\,\hat{B},\hat{A}}(-\omega')\right],
\end{aligned}\tag{E.154}$$

in terms of the spectral density function of operators $\hat{A}$ and $\hat{B}$. Now, $J_{0\,\beta\,\hat{A},\hat{B}}(+\omega')$ and $J_{0\,\beta\,\hat{B},\hat{A}}(-\omega')$ are related. To see this denote by $\{|\mu\rangle\}$ the exact basis of eigenstates of $\hat{H}_0$ with eigenvalues $\{E_{0\,\mu}\}$. From the definitions (E.151a) and (E.151b),

$$\begin{aligned}
J_{0\,\beta\,\hat{A},\hat{B}}(+\omega) &= \int\limits_{-\infty}^{+\infty}\mathrm{d}t\, e^{+\mathrm{i}\omega t}\langle\hat{A}_{\mathrm{H}}(t)\hat{B}_{\mathrm{H}}(0)\rangle_{0\,\beta}\\
&= Z^{-1}_{0\,\beta}\sum_{\mu,\nu}\int\limits_{-\infty}^{+\infty}\mathrm{d}t\, e^{+\mathrm{i}(\omega+E_{0\,\mu}-E_{0\,\nu})t}e^{-\beta E_{0\,\mu}}\left\langle\mu\left|\hat{A}_{\mathrm{H}}(0)\right|\nu\right\rangle\left\langle\nu\left|\hat{B}_{\mathrm{H}}(0)\right|\mu\right\rangle\\
&= 2\pi Z^{-1}_{0\,\beta}\sum_{\mu,\nu}e^{-\beta E_{0\,\mu}}\left\langle\mu\left|\hat{A}_{\mathrm{H}}(0)\right|\nu\right\rangle\left\langle\nu\left|\hat{B}_{\mathrm{H}}(0)\right|\mu\right\rangle\delta(\omega+E_{0\,\mu}-E_{0\,\nu}),
\end{aligned}\tag{E.155}$$

and

$$\begin{aligned}
J_{0\,\beta\,\hat{B},\hat{A}}(-\omega) &= \int\limits_{-\infty}^{+\infty}\mathrm{d}t\, e^{+\mathrm{i}\omega t}\langle\hat{B}_{\mathrm{H}}(0)\hat{A}_{\mathrm{H}}(t)\rangle_{0\,\beta}\\
&= Z^{-1}_{0\,\beta}\sum_{\mu,\nu}\int\limits_{-\infty}^{+\infty}\mathrm{d}t\, e^{+\mathrm{i}(\omega+E_{0\,\nu}-E_{0\,\mu})t}e^{-\beta E_{0\,\mu}}\left\langle\mu\left|\hat{B}_{\mathrm{H}}(0)\right|\nu\right\rangle\left\langle\nu\left|\hat{A}_{\mathrm{H}}(0)\right|\mu\right\rangle\\
&= 2\pi Z^{-1}_{0\,\beta}\sum_{\mu,\nu}e^{-\beta E_{0\,\mu}}\left\langle\mu\left|\hat{B}_{\mathrm{H}}(0)\right|\nu\right\rangle\left\langle\nu\left|\hat{A}_{\mathrm{H}}(0)\right|\mu\right\rangle\delta(\omega+E_{0\,\nu}-E_{0\,\mu})\\
&= 2\pi Z^{-1}_{0\,\beta}\sum_{\mu,\nu}e^{-\beta E_{0\,\nu}}\left\langle\mu\left|\hat{A}_{\mathrm{H}}(0)\right|\nu\right\rangle\left\langle\nu\left|\hat{B}_{\mathrm{H}}(0)\right|\mu\right\rangle\delta(\omega+E_{0\,\mu}-E_{0\,\nu}),
\end{aligned}\tag{E.156}$$

where the canonical partition function is given by

$$Z_{0\,\beta}:=\sum_{\mu}e^{-\beta E_{0\,\mu}}.\tag{E.157}$$

Making use of the constraint of energy conservation in the so-called Lehmann expansions (E.155) and (E.156), we infer that

$$J_{0\,\beta\,\hat{B},\hat{A}}(-\omega)=e^{-\beta\omega}J_{0\,\beta\,\hat{A},\hat{B}}(+\omega).\tag{E.158}$$

Insertion of Eq. (E.158) into Eq. (E.154) gives

$$\begin{aligned} C^{\mathrm{R}}_{0\,\beta\,\hat{A},\hat{B}}(t) &= -\mathrm{i}\Theta(t) \int\limits_{-\infty}^{+\infty} \frac{\mathrm{d}\omega'}{2\pi} e^{-\mathrm{i}\omega' t} \left[J_{0\,\beta\,\hat{A},\hat{B}}(+\omega') - J_{0\,\beta\,\hat{B},\hat{A}}(-\omega') \right] \\ &= -\mathrm{i}\Theta(t) \int\limits_{-\infty}^{+\infty} \frac{\mathrm{d}\omega'}{2\pi} e^{-\mathrm{i}\omega' t} J_{0\,\beta\,\hat{A},\hat{B}}(+\omega') \left(1 - e^{-\beta\omega'}\right). \end{aligned} \tag{E.159}$$

Time-Fourier transformation of the retarded Green function demands a regularization at long positive time which is implemented by the addition of an infinitesimal convergence factor $|\eta|$ ($\frac{\mathrm{PV}}{x}$ denotes the principal value of $1/x$):

$$\begin{aligned} C^{\mathrm{R}}_{0\,\beta\,\hat{A},\hat{B}}(\omega) &= \lim_{\eta\to 0} \int\limits_{-\infty}^{+\infty} \mathrm{d}t\, e^{+\mathrm{i}\omega t - |\eta| t} C^{\mathrm{R}}_{0\,\beta\,\hat{A},\hat{B}}(t) \\ &= -\mathrm{i} \int\limits_{-\infty}^{+\infty} \frac{\mathrm{d}\omega'}{2\pi} J_{0\,\beta\,\hat{A},\hat{B}}(\omega') \left(1 - e^{-\beta\omega'}\right) \lim_{\eta\to 0} \int\limits_{0}^{+\infty} \mathrm{d}t\, e^{+\mathrm{i}(\omega - \omega' + \mathrm{i}|\eta|)t} \\ &= \lim_{\eta\to 0} \int\limits_{-\infty}^{+\infty} \frac{\mathrm{d}\omega'}{2\pi} J_{0\,\beta\,\hat{A},\hat{B}}(\omega') \left(1 - e^{-\beta\omega'}\right) \frac{1}{\omega - \omega' + \mathrm{i}|\eta|} \\ &= \int\limits_{-\infty}^{+\infty} \frac{\mathrm{d}\omega'}{2\pi} J_{0\,\beta\,\hat{A},\hat{B}}(\omega') \left(1 - e^{-\beta\omega'}\right) \left[\frac{\mathrm{PV}}{\omega - \omega'} - \mathrm{i}\pi\delta(\omega - \omega') \right] \\ &= \int\limits_{-\infty}^{+\infty} \frac{\mathrm{d}\omega'}{2\pi} J_{0\,\beta\,\hat{A},\hat{B}}(\omega') \left(1 - e^{-\beta\omega'}\right) \frac{\mathrm{PV}}{\omega - \omega'} - \frac{\mathrm{i}}{2} J_{0\,\beta\,\hat{A},\hat{B}}(\omega) \left(1 - e^{-\beta\omega}\right). \end{aligned} \tag{E.160}$$

Whenever $J_{0\,\beta\,\hat{A},\hat{B}}(\omega)$ is real valued for all ω, we conclude that

$$J_{0\,\beta\,\hat{A},\hat{B}}(\omega) = -\frac{2}{1 - e^{-\beta\omega}} \operatorname{Im} C^{\mathrm{R}}_{0\,\beta\,\hat{A},\hat{B}}(\omega), \qquad \forall \omega \neq 0. \tag{E.161}$$

This is one form of the fluctuation-dissipation theorem. The condition for $J_{0\,\beta\,\hat{A},\hat{B}}(\omega)$ to be real is that

$$\hat{A} = \hat{B}^{\dagger}. \tag{E.162}$$

Appendix F

Landau theory of Fermi liquids

Introduction

The theory of Fermi liquids was initiated by Landau in 1956 [122]. Landau's theory of interacting fermions ushered a new theoretical paradigm to approach many-body physics. Rather than computing perturbatively properties of a microscopic Hamiltonian, Landau postulated a phase of matter to deduce what interesting properties would then be observable. The central postulate of Landau is encoded by the notion of quasiparticles. Assuming the existence of quasiparticles and postulating their quantum dynamics, it is possible to predict equilibrium and out-of-equilibrium properties (in a semi-classical approximation) that can be tested with experiments. Hence, the theory of Fermi liquids is falsifiable and has turned out to be extremely successful for understanding the properties of metals at sufficiently low temperatures. The terminology "Landau theory of Fermi liquids" is often shortened to the theory of Fermi liquids in the physics literature. We will follow this convention. We refer the reader to Refs. [10], [11], [16], and [123] for a systematic review of the consequences of the theory of Fermi liquids.

The purpose of this appendix is to motivate the notion of quasiparticles (not necessarily of the Landau type). We shall follow the point of view of Volovik who pioneered in Ref. [124] the notion that topological attributes are fundamental to understanding Fermi surfaces and their low-lying excitations (see also Ref. [125]).

The theory of Fermi liquids is built from the Pauli principle obeyed by fermions, energy conservation, and adiabatic continuity. We are first going to illustrate by way of an example the notion of adiabatic continuity.

F.1 Adiabatic continuity

For simplicity, we consider the quantum mechanics of a point particle of mass m restricted to the real line that is governed by the Hamiltonian

$$\hat{H} := \frac{\hat{p}^2}{2\,m} - v_0\,V(\hat{q}). \tag{F.1}$$

Here, the position operator $\hat{q}$ with the eigenvalues $q \in \mathbb{R}$ and the momentum operator $\hat{p}$ with the eigenvalues $p \in \mathbb{R}$ do not commute, $[\hat{q},\hat{p}] = \mathrm{i}\hbar$. We assume a potential well $-v_0\,V$ such that (i) its global minimum is reached at $q=0$ at which it takes the value $-v_0 < 0$, (ii) it is of finite range, (iii) $V(q\to\pm\infty)=0$, and (iv) it is deep enough for the eigenstates of $\hat{H}$ to be made of countably many bound states $\Psi^{(v_0)}_{\varepsilon_n}(q)$ with the energies $\varepsilon_0<\varepsilon_1<\cdots<0$ and uncountably many extended states $\Psi^{(v_0)}_{\varepsilon(p)}(q)$ with the energies $\varepsilon(p)=p^2/(2m)\geq 0$.[23]

We now imagine that the depth (strength) of the potential well varies with the parameter t that we *interpret* as time, i.e., we define the time-dependent Hamiltonian

$$\hat{H}(t) := \frac{\hat{p}^2}{2\,m} - v_0(t)\,V(\hat{q}), \tag{F.2}$$

together with the Schrödinger equation

$$\mathrm{i}\hbar\,(\partial_t\Psi)(q,t) = \left(-\frac{\hbar^2\partial_q^2}{2\,m} - v_0(t)\,V(q)\right)\Psi(q,t), \tag{F.3}$$

given some initial condition. The question we are after is the following. Under what conditions can we use our knowledge of the eigenenergies and eigenfunctions of the time-independent Hamiltonian (F.1) to solve the Schrödinger equation (F.3)? This question is certainly of relevance if the initial state is $\Psi^{(v_0)}_{\varepsilon}(q)$. Does $\Psi^{(v_0)}_{\varepsilon}(q)$ evolve under the Schrödinger equation (F.3) as the instantaneous eigenstate defined by solving

$$\hat{H}(t)\,\Psi^{(v_0(t))}_{\varepsilon(t)}(q) = \varepsilon(t)\,\Psi^{(v_0(t))}_{\varepsilon(t)}(q), \tag{F.4}$$

or does it mix with all eigenstates of the time-independent Hamiltonian (F.1) as a function of time?

A quantitative answer to this question follows from inserting the adiabatic Ansatz

$$\Psi_{\mathrm{adia}}(q,t) := e^{-\mathrm{i}\varepsilon(t)\,t/\hbar}\,\Psi^{(v_0(t))}_{\varepsilon(t)}(q) \tag{F.5}$$

into the Schrödinger equation (F.3). The adiabatic Ansatz (F.5) is a solution to the Schrödinger equation (F.3) if and only if

$$\left[t\left(\frac{\partial\varepsilon}{\partial v_0}\right)(t)\,\Psi^{(v_0(t))}_{\varepsilon(t)}(q) + \mathrm{i}\hbar\frac{\partial\Psi^{(v_0(t))}_{\varepsilon(t)}(q)}{\partial v_0(t)}\right]\left(\frac{\partial v_0}{\partial t}\right)(t) = 0. \tag{F.6}$$

It thus appears that the adiabatic Ansatz becomes exact in the limit

$$\left(\frac{\partial v_0}{\partial t}\right)(t) \to 0, \tag{F.7}$$

[23] If we define the potential square well $-v_0\,V(q)=0$ for $|q|>q_0>0$ and $-v_0\,V(q)=-v_0<0$ for $|q|\leq q_0$, then v_0 and q_0 are the strength and range of the square well, respectively. The same is true for the smooth potential well $-v_0\,V(q)=-v_0\,\exp\left(-q^2/(2\,q_0^2)\right)$. In both cases, the spectrum consists of bound states with discrete negative energies and of plane waves with positive energies.

for which the parametric rate of change is infinitesimal, provided the square bracket remains finite. However, the functional change

$$\mathrm{d}\Psi^{(v_0(t))}_{\varepsilon(t)}(q) = \Psi^{(v_0(t)+\mathrm{d}v_0(t))}_{\varepsilon(t)}(q) - \Psi^{(v_0(t))}_{\varepsilon(t)}(q) \tag{F.8}$$

of the wave function is not always vanishing in the limit $\mathrm{d}v_0(t) \to 0$, in which case the derivative with respect to v_0 of $\Psi^{(v_0(t))}_{\varepsilon(t)}(q)$ is divergent. This situation occurs when level crossing takes place. In the example of a potential well, we may start initially with a bound state. With evolving time, this bound state might merge into the continuum of extended states. Since there is no smooth (perturbative) deformation of a bound state that turns it into an extended state, the level crossing of a bound state into the continuum of extended states cannot be captured by the adiabatic approximation, i.e., Eq. (F.7) cannot be satisfied.

F.2 Quasiparticles

We consider N identical electrons of mass m. For simplicity, they are spinless, propagate freely in a volume L^3, and interact with each others through a two-body potential. Hermiticity of the many-body Hamiltonian implies that the total particle number is conserved. The boundary conditions are chosen so that the total momentum is also conserved. The thermodynamic limit $N, L \to \infty$ with the density N/L^3 held fixed is then taken.

We shall assume that the many-body electron interaction is switched on adiabatically at the initial time $t_0 = -\infty$ and is fully established at time $t \gg t_0$. Any many-body state at time t evolves from an initial state at the initial time $t_0 = -\infty$ through the (unitary) evolution operator $\hat{U}_{\mathrm{I}}(t, t_0)$ in the interaction picture [recall Eq. (E.93)]. Hence, the initial many-body ground state is the Fermi sea [recall Eq. (5.22) and Fig. 5.1]

$$|\Phi_{0,\mathrm{free}}\rangle = \prod_{\boldsymbol{p}}^{|\boldsymbol{p}|<p_{\mathrm{F}}} \hat{c}^{\dagger}_{\boldsymbol{p}} |0\rangle. \tag{F.9}$$

The Fermi sea evolves into the state

$$|\Phi_0(t)\rangle_{\mathrm{I}} = \hat{U}_{\mathrm{I}}(t, t_0)\, |\Phi_{0,\mathrm{free}}\rangle \tag{F.10}$$

at time t in the interaction picture. The theory of Fermi liquids postulates that $|\Phi_0(t)\rangle_{\mathrm{I}}$ remains the ground state. This assumption rules out any long-range order, say superfluidity, superconductivity, crystallization, etc. The theory of Fermi liquids does not attempt to prove this postulate. It seeks the consequences of this hypothesis, such as the fact that the state $|\Phi_0\rangle_{\mathrm{I}}$ shares the same quantum numbers as the Fermi sea $|\Phi_{0,\mathrm{free}}\rangle$.

Next, we consider the non-interacting particle-like excited state

$$|\boldsymbol{p}, -e\rangle := \hat{c}^{\dagger}_{\boldsymbol{p}}\, |\Phi_{0,\mathrm{free}}\rangle, \qquad |\boldsymbol{p}| > p_{\mathrm{F}}, \tag{F.11a}$$

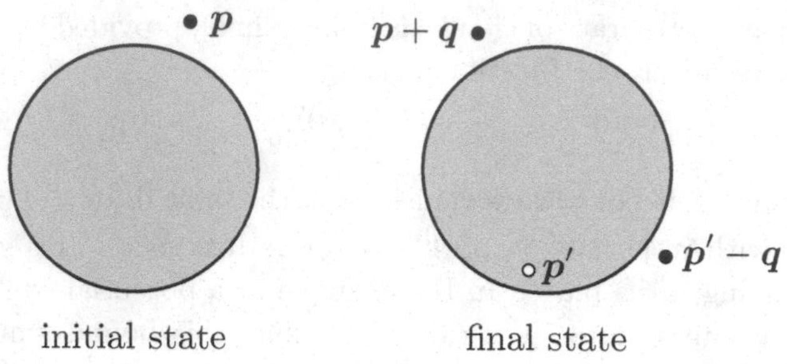

Fig. F.1 Initial and final states for the decay of a quasiparticle just above the Fermi surface.

and the non-interacting hole-like excited state

$$|\boldsymbol{p}',+e\rangle := \hat{c}_{\boldsymbol{p}'}\,|\Phi_{0,\mathrm{free}}\rangle, \qquad |\boldsymbol{p}'|<p_{\mathrm{F}}. \tag{F.11b}$$

Here, $-e<0$ denotes the electron charge. (We work in the grand canonical potential.) At time t, these two eigenstates have evolved into the states

$$|\boldsymbol{p},-e;t\rangle := \hat{U}_{\mathrm{I}}(t,t_0)\,|\boldsymbol{p},-e\rangle, \tag{F.12a}$$

and

$$|\boldsymbol{p}',+e;t\rangle := \hat{U}_{\mathrm{I}}(t,t_0)\,|\boldsymbol{p}',+e\rangle, \tag{F.12b}$$

respectively, neither of which are eigenstates of the many-body Hamiltonian anymore. Indeed, switching on the many-body interaction induces scattering out of the state $|\boldsymbol{p},-e;t\rangle$, i.e., the state $|\boldsymbol{p},-e;t\rangle$ acquires a lifetime. However, if the adiabatic energy of the state $|\boldsymbol{p},-e;t\rangle$ is sufficiently close to that of the Fermi sea during the adiabatic switching-on process of the interaction, then the rate of decay of the state $|\boldsymbol{p},-e;t\rangle$ scales like $[\boldsymbol{p}^2/(2m)-\mu_{\mathrm{F}}]^2$.[24]

The estimate $\propto [\boldsymbol{p}^2/(2m)-\mu_{\mathrm{F}}]^{-2}$ for the lifetime of the state $|\boldsymbol{p},-e;t\rangle$ presumes that the state $|\boldsymbol{p},-e;t\rangle$ exists at the time t_{adia} needed for the adiabatic switching on of the many-body interaction to be completed. If the scaling of the time $t_{\mathrm{adia}}\to\infty$ and the scaling of the energy difference $\boldsymbol{p}^2/(2m)-\mu_{\mathrm{F}}\to 0$ are such that the ratio between the lifetime of the state $|\boldsymbol{p},-e;t\rangle$ and t_{adia} is large, then it is consistent to assume that the state $|\boldsymbol{p},-e;t\rangle$ is yet to decay through the many-body interaction. This condition can always be met if $\boldsymbol{p}^2/(2m)\to\mu_{\mathrm{F}}$. If so, Landau infers that all the

[24] What is the decay rate of state $|\boldsymbol{p},-e;t\rangle$ if its energy is $\boldsymbol{p}^2/(2m)-\mu_{\mathrm{F}}>0$ above the Fermi energy μ_{F} assuming a translation-invariant two-body interaction? The quasiparticle in the state $|\boldsymbol{p},-e;t\rangle$ above the Fermi surface may scatter off a quasiparticle occupying the single-particle state with momentum $\boldsymbol{p}'$ below the Fermi surface by exchanging the momentum $\boldsymbol{q}$ from the interaction. To first order in perturbation theory, the outcome of this scattering event is the final state comprising the two quasiparticles $|\boldsymbol{p}+\boldsymbol{q},-e;t\rangle$ and $|\boldsymbol{p}'-\boldsymbol{q},-e;t\rangle$ and one quasihole $|\boldsymbol{p}',+e;t\rangle$, by the Pauli principle, see Fig. F.1. Energy conservation requires that $\mu_{\mathrm{F}}-\boldsymbol{p}'^2/(2m)>0$ is smaller or equal to $\boldsymbol{p}^2/(2m)-\mu_{\mathrm{F}}>0$. The number of states with momentum $\boldsymbol{p}'$ below the Fermi surface with which the quasiparticle $|\boldsymbol{p},-e;t\rangle$ can exchange the momentum $\boldsymbol{q}$ from the interaction is thus proportional to $\boldsymbol{p}^2/(2m)-\mu_{\mathrm{F}}>0$. The number of final states (labeled by $\boldsymbol{q}$) is also proportional to $\boldsymbol{p}^2/(2m)-\mu_{\mathrm{F}}>0$, by energy conservation. The decay rate of the state $|\boldsymbol{p},-e;t\rangle$ that follows from Fermi golden rule is thus proportional to $[\boldsymbol{p}^2/(2m)-\mu_{\mathrm{F}}]^2$.

low-lying excited states in the non-interacting limit become low-lying excited states after an appropriate adiabatic switching on of the interaction. Moreover, all the quantum numbers in the non-interacting limit that are conserved by the interaction and characterize the low-lying excited states remain good quantum numbers after an appropriate adiabatic switching on of the interaction. These low-lying states are of the generic form

$$|\boldsymbol{p}_1, \boldsymbol{p}_2, \cdots, \boldsymbol{p}_{n_\mathrm{p}}, \boldsymbol{p}'_1, \boldsymbol{p}'_2, \cdots, \boldsymbol{p}'_{n_\mathrm{h}}, (n_\mathrm{h} - n_\mathrm{p})e; t\rangle \\ := \hat{U}_\mathrm{I}(t, t_0) \left(\prod_{j=1}^{n_\mathrm{p}} \hat{c}^\dagger_{\boldsymbol{p}_j}\right) \left(\prod_{j'=1}^{n_\mathrm{h}} \hat{c}_{\boldsymbol{p}'_{j'}}\right) |\Phi_{0,\mathrm{free}}\rangle. \tag{F.13}$$

The case $n_\mathrm{p} = 1$ and $n_\mathrm{h} = 0$ defines a quasiparticle with the momentum $\boldsymbol{p}$ and the charge $-e$ relative to the momentum and charge of the Fermi sea, respectively. The case $n_\mathrm{p} = 0$ and $n_\mathrm{h} = 1$ defines a quasihole with the momentum $-\boldsymbol{p}'$ and the charge $+e$ relative to the momentum and charge of the Fermi sea, respectively.

The concept of quasiparticles and quasiholes presumes implicitly that the volume of the Fermi sea is unchanged under adiabatic switching on of the many-body interactions. A microscopic justification of this assumption is known as Luttinger theorem [66]. The argumentation in Ref. [66] consists of two steps. First, the Fermi surface is defined in the presence of many-body interactions. Second, it is shown that if a Fermi surface exists, then its volume is independent of the interaction strength. The second step is achieved with the help of perturbative techniques introduced in Ref. [126]. Luttinger theorem states that the volume V_FS of the Fermi sea enclosed by the Fermi surface in wave-number space is related to the electron density N/L^3 by

$$\frac{N}{L^3} = \frac{V_\mathrm{FS}}{(2\pi)^3}, \tag{F.14}$$

irrespective of the many-body interactions. Many-body interactions can destroy the Fermi surface, but if the Fermi surface survives the presence of many-body interactions, then they may change its shape relative to that in the non-interacting limit, but not its volume.

The defining property of a Fermi liquid is that, in addition to the many-body ground state evolving adiabatically from the non-interacting Fermi sea, so do the non-interacting low-lying particle-hole excitations. All the low-lying excited states in the non-interacting limit remain low-lying states in the Fermi liquid. The converse statement that all low-lying excited states of a Fermi liquid are low-lying states in the non-interacting limit is incorrect, as shown by Landau in Ref. [127]. The latter claim is established within Fermi liquid theory by postulating the effective low-energy Hamiltonian

$$\hat{H}_\mathrm{FL} := \sum_{\boldsymbol{p}} \xi_{\boldsymbol{p}}\, \delta\hat{n}_{\boldsymbol{p}} + \frac{1}{2L^3} \sum_{\boldsymbol{p}} \sum_{\boldsymbol{p}'} f_{\boldsymbol{p},\boldsymbol{p}'}\, \delta\hat{n}_{\boldsymbol{p}}\, \delta\hat{n}_{\boldsymbol{p}'}, \tag{F.15}$$

from which equilibrium and out-of-equilibrium observables are computed. Here, the real-valued functions $\xi_{\boldsymbol{p}}$ and $f_{\boldsymbol{p},\boldsymbol{p}'}$ are phenomenological functions of the momenta and the operator $\delta\hat{n}_{\boldsymbol{p}}$ carries the eigenvalues 0 and $+1$ if $\boldsymbol{p}$ does not belong to the Fermi sea or the eigenvalues 0 and -1 if $\boldsymbol{p}$ belongs to the Fermi sea. The Fermi surface is defined by the condition $\xi_{\boldsymbol{p}} = 0$. The eigenstate of $\delta\hat{n}_{\boldsymbol{p}}$ with eigenvalue $+1$ is a quasiparticle with the energy $+\xi_{\boldsymbol{p}}$ slightly positive. The eigenstate of $\delta\hat{n}_{\boldsymbol{p}}$ with eigenvalue -1 is a quasihole with the energy $-\xi_{\boldsymbol{p}}$ slightly positive. The quasiparticle density operators $\{\delta\hat{n}_{\boldsymbol{p}}\}$ commute pairwise. We refer the reader to Refs. [10], [11], [16], and [123] for more insights on the consequences of Eq. (F.15) for equilibrium and out-of-equilibrium properties of a Fermi liquid. Instead, we are going to provide a complementary point of view for the notion of quasiparticles.

F.3 Topological stability of the Fermi surface

A Fermi liquid is a phase of matter in d-dimensional position space with two defining properties. There is the ground state, a Fermi sea with its Fermi surface, that is adiabatically connected to the non-interacting Fermi surface. There is a continuum of gapless excitations, the quasiparticles, that are adiabatically connected to the low-lying excitations of the non-interacting Fermi sea.

The notion of a Fermi surface is more fundamental than that of quasiparticles, for the former can exist without the latter as occurs in a (Tomonaga-) Luttinger liquid for example [128–130].

The Fermi surface is defined in momentum space as the boundary of a volume. Hence, its very existence demands the conservation of the total fermion particle number and of the total momentum, the two prerequisites to define a volume in momentum space. Global gauge symmetry and translation invariance must therefore hold, an assumption made throughout this appendix. Following Volovik in Ref. [124], we are going to assign to the Fermi surface a topological invariant. This is first done in the non-interacting limit. We shall then extend this result to include many-body interactions.

F.3.1 *The case of no many-body interactions*

In the non-interacting and thermodynamic limits, the many-body Hamiltonian is reducible because of translation invariance. It decomposes into the direct sum over a family (bundle) of $n \times n$ Hermitian matrices $\mathcal{H}_{\boldsymbol{p}}$ labeled by the single-particle momentum $\boldsymbol{p} \in \mathbb{R}^d$. Diagonalization of each member $\mathcal{H}_{\boldsymbol{p}}$ from the bundle of $n \times n$ Hermitian matrices $\{\mathcal{H}_{\boldsymbol{p}}\}$ delivers n bands. By the Pauli principle, all single-particle eigenstates with the lowest eigenenergies are to be filled up to the Fermi energy μ_{F} in the non-interacting ground state, the Fermi sea. The Fermi surface that encloses the Fermi sea, if it exists (i.e., when there is at least one partially filled band), can be identified with the help of the inverse of the $n \times n$ matrix $z - \mathcal{H}_{\boldsymbol{p}} + \mu_{\mathrm{F}}$, namely the $n \times n$ matrix

$$\mathcal{G}_{\boldsymbol{p}}(z) = \frac{1}{z - \mathcal{H}_{\boldsymbol{p}} + \mu_{\mathrm{F}}}, \tag{F.16}$$

with $z \in \mathbb{C}$ off the real axis in the complex plane. In the mathematics literature, $\mathcal{G}_{\boldsymbol{p}}(z)$ is called the resolvent of $\mathcal{H}_{\boldsymbol{p}} - \mu_{\mathrm{F}}$. In the physics literature, $\mathcal{G}_{\boldsymbol{p}}(z)$ is related through an appropriate analytic continuation to the real axis to the appropriate (time-ordered, retarded, advanced, etc.) single-particle Green function. Eigenvalues of $\mathcal{H}_{\boldsymbol{p}} - \mu_{\mathrm{F}}$ are n discrete poles of $\mathcal{G}_{\boldsymbol{p}}(z)$ on the real axis in the complex z plane. The set of poles at the origin $z = 0$ from the family $\{\mathcal{G}_{\boldsymbol{p}}(z)\}$ labeled by the momenta $\boldsymbol{p} \in \mathbb{R}^d$ defines the Fermi surface. Thus, the Fermi surface is a $(d-1)$-dimensional submanifold of momentum space $\mathbb{R}^d$ made of, discrete points when $d = 1$, lines when $d = 2$, surfaces when $d = 3$, and so on; on which, the mapping $\mathcal{G}_{\boldsymbol{p}} : \mathbb{C} \to GL(n, \mathbb{C})$ that assigns to the complex number z the matrix $\mathcal{G}_{\boldsymbol{p}}(z)$ is singular at the origin in the complex plane.

For simplicity, we momentarily consider the case of one band, i.e., $n = 1$. We *assume* that there exists a momentum $\boldsymbol{p}$ at which $\mathcal{G}_{\boldsymbol{p}}(z)$ has a pole at $z = 0$. This pole is necessarily of first order. Let $\gamma_{\boldsymbol{p}} : [0, 2\pi[\to \mathbb{C},\ l \mapsto z(l)$ be any closed path that winds counterclockwise once around the origin in the z complex plane. By Cauchy theorem,[25]

$$\begin{aligned} 2\pi\mathrm{i} &= \int\limits_{\gamma_{\boldsymbol{p}}} \mathrm{d}z\, \mathcal{G}_{\boldsymbol{p}}(z) \\ &= \int\limits_0^{2\pi} \mathrm{d}l \left(\frac{\mathrm{d}z}{\mathrm{d}l}\right)(l)\, \mathcal{G}_{\boldsymbol{p}}(z(l)) \\ &= \int\limits_0^{2\pi} \mathrm{d}l \left(\frac{\partial\, \mathcal{G}^{-1}_{\boldsymbol{p}}}{\partial l}\right)(l)\, \mathcal{G}_{\boldsymbol{p}}(l). \end{aligned} \tag{F.17}$$

If we relax the condition that the number of bands is unity, i.e., $n \geq 1$, we can state that, for any momentum $\boldsymbol{p} \in \mathbb{R}^d$ (not necessarily on the Fermi surface),

$$N^{(1)}_{\boldsymbol{p}} := \int\limits_0^{2\pi} \frac{\mathrm{d}l}{2\pi\mathrm{i}}\, \mathrm{tr}\, \left[\left(\partial_l\, \mathcal{G}^{-1}_{\boldsymbol{p}}\right)(l)\, \mathcal{G}_{\boldsymbol{p}}(l) \right] = 0, 1, 2, \cdots, \tag{F.18}$$

for any closed path $\gamma_{\boldsymbol{p}} : [0, 2\pi[\to \mathbb{C},\ l \mapsto \gamma_{\boldsymbol{p}}(l) = z(l)$ in a sufficiently small open neighborhood of $z = 0$ that winds counterclockwise once around $z = 0$. When a non-interacting Fermi surface exists, $N^{(1)}_{\boldsymbol{p}}$ is non-vanishing whenever $\boldsymbol{p}$ belongs to the Fermi surface. Otherwise, it vanishes for all $\boldsymbol{p}$.

[25] Cauchy theorem is verified explicitly with the choice $\gamma(l) = \exp(\mathrm{i}l)$ with $0 \leq l < 2\pi$ for which $\int\limits_{\gamma} \frac{\mathrm{d}z}{z} = \int\limits_0^{2\pi} \mathrm{d}l\, \frac{\mathrm{i}e^{\mathrm{i}l}}{e^{\mathrm{i}l}} = 2\pi\mathrm{i}$.

If we denote by $\boldsymbol{p}_{\mathrm{F}}$ a momentum from the Fermi surface, the condition $N^{(1)}_{\boldsymbol{p}_{\mathrm{F}}} = 1$ for a non-degenerate Fermi surface has the following topological character. Assume that the dependence of the bundle of single-particle Hamiltonians

$$\left\{ \mathcal{H}_{\boldsymbol{p}}(\lambda) = \sum_{j=1}^{n} |\varepsilon_{\boldsymbol{p},j}(\lambda)\rangle\langle\varepsilon_{\boldsymbol{p},j}(\lambda)| \right\}, \tag{F.19}$$

labeled by the momentum $\boldsymbol{p} \in \mathbb{R}^d$ on the parameter λ denotes a smooth parametric change of the orthonormal eigenstates $|\varepsilon_{\boldsymbol{p},j}(\lambda)\rangle$ and their eigenenergies $\varepsilon_{\boldsymbol{p},j}(\lambda)$ in the spectral representation of $\mathcal{H}_{\boldsymbol{p}}(\lambda)$. In other words, if $|\varepsilon_{\boldsymbol{p},j}(\lambda)\rangle$ is an orthonormal eigenstate of $\mathcal{H}_{\boldsymbol{p}}(\lambda)$ with the eigenenergy $\varepsilon_{\boldsymbol{p},j}(\lambda)$, then the state $|\varepsilon_{\boldsymbol{p},j}(\lambda + \mathrm{d}\lambda)\rangle$ is an orthonormal eigenstate of $\mathcal{H}_{\boldsymbol{p}}(\lambda + \mathrm{d}\lambda)$ with the eigenenergy $\varepsilon_{\boldsymbol{p},j}(\lambda + \mathrm{d}\lambda)$ that is uniquely (up to a phase) determined by $|\varepsilon_{\boldsymbol{p},j}(\lambda)\rangle$. If $\boldsymbol{p}_{\mathrm{F}}(\lambda)$ denotes a momentum on the Fermi surface belonging to band j_{F}, then the state $|\varepsilon_{\boldsymbol{p}_{\mathrm{F}}(\lambda),j_{\mathrm{F}}}(\lambda)\rangle$ is, by definition, an occupied state from the Fermi sea with the largest eigenenergy μ_{F}. By adiabaticity, the state $|\varepsilon_{\boldsymbol{p}_{\mathrm{F}}(\lambda+\mathrm{d}\lambda),j_{\mathrm{F}}}(\lambda+\mathrm{d}\lambda)\rangle$ remains an occupied state from the Fermi sea with the largest eigenenergy μ_{F}, i.e., $\boldsymbol{p}_{\mathrm{F}}(\lambda + \mathrm{d}\lambda)$ belongs to the Fermi surface. Even though $\boldsymbol{p}_{\mathrm{F}}(\lambda + \mathrm{d}\lambda)$ might not be equal to $\boldsymbol{p}_{\mathrm{F}}$,[26] we must have the invariant

$$N^{(1)}_{\boldsymbol{p}_{\mathrm{F}}(\lambda+\mathrm{d}\lambda)} = N^{(1)}_{\boldsymbol{p}_{\mathrm{F}}(\lambda)} = 1. \tag{F.20}$$

Equation (F.20) is the counterpart to Luttinger theorem restricted to one-body perturbations that change single-particle energy eigenstates and their eigenvalues in a smooth (adiabatic) way. Equation (F.20) states that any smooth change of the family $\{\mathcal{H}_{\boldsymbol{p}}\}$ labeled by the single-particle momentum $\boldsymbol{p} \in \mathbb{R}^d$ cannot destroy the non-interacting Fermi surface, i.e., the existence of a boundary in momentum space between the occupied and the empty single-particle states. The shape (geometry) of the non-interacting Fermi surface can change smoothly under the influence of a smooth parametric change of single-particle energy eigenstates and eigenenergies, but not the very existence of the boundary between occupied and empty single-particle states.

In mathematics, an integral whose value is quantized is called a topological invariant. The integrand of a topological invariant can be smoothly deformed without changing the value of the integral. This allows to classify the set of all integrands into equivalence classes by declaring that two integrands belong to the same equivalence class if they share the same quantized value for their integrals.

The vorticities of the vortices in the XY model from chapter 4, the integer values taken by the Hall conductance (see section 5.5), and the quantization of flux in a superconductor (see section 7.9) are examples of physical observables that are topological invariants. We have just shown that a non-interacting Fermi surface can be assigned the topological invariant (F.20).

[26] The difference $|\boldsymbol{p}_{\mathrm{F}}(\lambda + \mathrm{d}\lambda) - \boldsymbol{p}_{\mathrm{F}}(\lambda)|$ must be of order $|\mathrm{d}\lambda|$ by adiabaticity.

The winding number (F.18) is unity for a non-interacting and non-degenerate Fermi surface. It vanishes for a non-interacting insulating ground state that results from having no partially filled bands and no band touching at the Fermi energy, i.e., a band insulator. A non-interacting Fermi sea with and one without Fermi surface are not topologically equivalent. Hence, the gap of a band insulator that descends from a non-interacting Fermi sea with Fermi surface by the application of a one-body interaction cannot be obtained from a perturbative treatment of this one-body interaction. An example thereof is the Peierls instability by which the electron-phonon coupling opens a gap at commensurate filling fractions of a one-dimensional periodic lattice [131]. The topological character of the Fermi surface implies that the Peierls gap must be a non-analytic function of the one-body coupling strength between the electrons and the phonons. The changes in the eigenstates and eigenenergies induced by the infinitesimal electron-phonon coupling responsible for the Peierls instability are not smooth (adiabatic). This singularity carries through to the dependence of the partition function on the electron-phonon coupling (recall that the thermodynamic limit is assumed), i.e., a quantum phase transition.

F.3.2 *The case with many-body interactions*

In the presence of many-body interactions, we need to diagonalize the many-body Hamiltonian $\hat{H} - \mu_{\mathrm{F}}\,\hat{N}$ with $\hat{N}$ the conserved total number operator. Because of the many-body interactions, $\hat{H}$ does not decompose as a direct sum over the single-particle momenta. Hence, its resolvent $\hat{G}(z) := (z - \hat{H} + \mu_{\mathrm{F}}\,\hat{N})$ is not the proper generalization of the resolvent (F.16), for it does not share the same quantum numbers as (F.16). Needed is a many-body counterpart to the single-particle Green function (F.16).

We seek an alternative definition of the single-particle Green function (F.16) that lends itself to a generalization that can account for many-body interactions. To this end, we define the non-interacting many-body Hamiltonian

$$\hat{H}_0 := \sum_{\boldsymbol{p}} \hat{\Psi}^{\dagger}_{\boldsymbol{p}} \left(\mathcal{H}_{\boldsymbol{p}} - \mu_{\mathrm{F}}\right) \hat{\Psi}_{\boldsymbol{p}}, \tag{F.21a}$$

where the n components $\hat{\Psi}^{\dagger}_{\boldsymbol{p},\alpha}$ of the operator-valued row vector $\hat{\Psi}^{\dagger}_{\boldsymbol{p}}$ and the n components $\hat{\Psi}_{\boldsymbol{p},\alpha}$ of the operator-valued column vector $\hat{\Psi}_{\boldsymbol{p}}$ $(\alpha = 1, \cdots, n)$ obey the usual fermion algebra. We are going to show that the single-particle Green function (F.16) can be represented by

$$\mathcal{G}_{0\,\boldsymbol{p}}(z) := \int\limits_{-\infty}^{+\infty} \mathrm{d}\tau\, e^{z\,\tau}\, \mathcal{G}_{0\,\boldsymbol{p}}(\tau), \tag{F.21b}$$

where

$$\mathcal{G}_{0\,\boldsymbol{p}}(\tau) := -\left\langle T_{\tau}\left(\hat{\Psi}_{\boldsymbol{p}}(\tau)\,\hat{\Psi}^{\dagger}_{\boldsymbol{p}}(0)\right)\right\rangle, \tag{F.21c}$$

with the angular brackets denoting the normalized many-body ground-state expectation value, T_τ denoting the ordering with respect to the imaginary time $\tau \in \mathbb{R}$, and

$$\hat{\Psi}^{\dagger}_{\boldsymbol{p}}(\tau) := e^{+\tau\,\hat{H}_0}\,\hat{\Psi}^{\dagger}_{\boldsymbol{p}}\,e^{-\tau\,\hat{H}_0}, \qquad \hat{\Psi}_{\boldsymbol{p}}(\tau) := e^{+\tau\,\hat{H}_0}\,\hat{\Psi}_{\boldsymbol{p}}\,e^{-\tau\,\hat{H}_0}, \tag{F.21d}$$

denoting the imaginary-time evolution of operators in the Heisenberg picture.

Proof. Because we can always diagonalize the single-particle $n \times n$ Hermitian matrix $\mathcal{H}_{\boldsymbol{p}} - \mu_{\mathrm{F}}$, it suffices to consider the case of one band, $n = 1$. The definition of the imaginary-time ordering for fermions implies that

$$\mathcal{G}_{0\,\boldsymbol{p}}(\tau) = -\left[\Theta(+\tau)\left\langle\hat{\Psi}_{\boldsymbol{p}}(\tau)\,\hat{\Psi}^{\dagger}_{\boldsymbol{p}}(0)\right\rangle - \Theta(-\tau)\left\langle\hat{\Psi}^{\dagger}_{\boldsymbol{p}}(0)\,\hat{\Psi}_{\boldsymbol{p}}(\tau)\right\rangle\right], \tag{F.22}$$

where Θ denotes the Heaviside step function. We use the short-hand notation $\xi_{\boldsymbol{p}} := \mathcal{H}_{\boldsymbol{p}} - \mu_{\mathrm{F}}$ and make use of

$$\hat{\Psi}^{\dagger}_{\boldsymbol{p}}(\tau) = e^{+\tau\,\xi_{\boldsymbol{p}}}\,\hat{\Psi}^{\dagger}_{\boldsymbol{p}}, \qquad \hat{\Psi}_{\boldsymbol{p}}(\tau) = e^{-\tau\,\xi_{\boldsymbol{p}}}\,\hat{\Psi}_{\boldsymbol{p}}, \tag{F.23}$$

to find

$$\mathcal{G}_{0\,\boldsymbol{p}}(\tau) = -e^{-\tau\,\xi_{\boldsymbol{p}}}\left[\Theta(+\tau)\left\langle\hat{\Psi}_{\boldsymbol{p}}(0)\,\hat{\Psi}^{\dagger}_{\boldsymbol{p}}(0)\right\rangle - \Theta(-\tau)\left\langle\hat{\Psi}^{\dagger}_{\boldsymbol{p}}(0)\,\hat{\Psi}_{\boldsymbol{p}}(0)\right\rangle\right]. \tag{F.24}$$

We may replace the ground-state expectation values by the Fermi-Dirac distributions at zero temperature (6.72),

$$\mathcal{G}_{0\,\boldsymbol{p}}(\tau) = -e^{-\tau\,\xi_{\boldsymbol{p}}}\left[\Theta(+\tau)\,\Theta(+\xi_{\boldsymbol{p}}) - \Theta(-\tau)\,\Theta(-\xi_{\boldsymbol{p}})\right]. \tag{F.25}$$

Integration of Eq. (F.25) by $\int\limits_{-\infty}^{+\infty} \mathrm{d}\tau\, e^{+\tau\, z}$ with z on the imaginary axis in the complex plane gives two terms on the right-hand side. The first term is

$$-\,\Theta(+\xi_{\boldsymbol{p}})\int\limits_{0}^{+\infty}\mathrm{d}\tau\, e^{+\tau\,(z-\xi_{\boldsymbol{p}})} = +\Theta(\xi_{\boldsymbol{p}})\,(z-\xi_{\boldsymbol{p}})^{-1}. \tag{F.26a}$$

The second term on the right-hand side reduces to the integral

$$+\,\Theta(-\xi_{\boldsymbol{p}})\int\limits_{-\infty}^{0}\mathrm{d}\tau\, e^{+\tau\,(z-\xi_{\boldsymbol{p}})} = +\Theta(-\xi_{\boldsymbol{p}})\,(z-\xi_{\boldsymbol{p}})^{-1}. \tag{F.26b}$$

We may then analytically continue z to $\mathbb{C}\setminus\mathbb{R}$. Thus, when the number of bands is $n = 1$, we have shown that

$$\mathcal{G}_{0\,\boldsymbol{p}}(z) = \frac{1}{z - \xi_{\boldsymbol{p}}}, \tag{F.27}$$

for any complex number z off the real axis in the complex plane. For $n > 1$ bands, after undoing the unitary transformation that diagonalizes $\mathcal{H}_{\boldsymbol{p}}$, we find the $n \times n$ matrix

$$\mathcal{G}_{0\,\boldsymbol{p}}(z) = \frac{1}{z - \mathcal{H}_{\boldsymbol{p}} + \mu_{\mathrm{F}}}, \tag{F.28}$$

for any complex number z off the real axis in the complex plane. □

For a generic many-body Hamiltonian $\hat{H}$ that commutes with the total fermion number operator $\hat{N}$, say (summation over repeated band index is understood) [recall Eq. (6.19)]

$$\hat{H} := \sum_{\boldsymbol{p}} \hat{\Psi}^{\dagger}_{\boldsymbol{p}}\, \mathcal{H}_{\boldsymbol{p}}\, \hat{\Psi}_{\boldsymbol{p}} + \frac{1}{L^d} \sum_{\boldsymbol{p},\boldsymbol{p}',\boldsymbol{q}\neq 0} \hat{\Psi}^{\dagger}_{\boldsymbol{p}+\boldsymbol{q},\alpha}\, \hat{\Psi}^{\dagger}_{\boldsymbol{p}'-\boldsymbol{q},\gamma}\, V^{\alpha,\beta|\gamma,\delta}_{\boldsymbol{q}}\, \hat{\Psi}_{\boldsymbol{p}',\delta}\, \hat{\Psi}_{\boldsymbol{p},\beta} + \cdots, \tag{F.29a}$$

we define the two-point Green function to be

$$G_{\boldsymbol{p}}(z,z^{*}) := \int\limits_{-\infty}^{+\infty} \mathrm{d}\tau\, e^{z\,\tau}\, G_{\boldsymbol{p}}(\tau), \tag{F.29b}$$

where

$$G_{\boldsymbol{p}}(\tau) := -\left\langle T_{\tau}\left(\hat{\Psi}_{\boldsymbol{p}}(\tau)\, \hat{\Psi}^{\dagger}_{\boldsymbol{p}}(0)\right)\right\rangle, \tag{F.29c}$$

with the angular brackets denoting the normalized many-body ground-state expectation value, T_{τ} denoting the ordering with respect to the imaginary time $\tau \in \mathbb{R}$, and

$$\hat{\Psi}^{\dagger}_{\boldsymbol{p}}(\tau) := e^{+\tau\,\hat{H}_{\mathrm{F}}}\, \hat{\Psi}^{\dagger}_{\boldsymbol{p}}\, e^{-\tau\,\hat{H}_{\mathrm{F}}}, \qquad \hat{\Psi}_{\boldsymbol{p}}(\tau) := e^{+\tau\,\hat{H}_{\mathrm{F}}}\, \hat{\Psi}_{\boldsymbol{p}}\, e^{-\tau\,\hat{H}_{\mathrm{F}}}, \tag{F.29d}$$

denoting the imaginary-time evolution of operators in the Heisenberg picture generated by

$$\hat{H}_{\mathrm{F}} := \hat{H} - \mu_{\mathrm{F}}\, \hat{N}. \tag{F.29e}$$

The chemical potential μ_{F} at zero temperature is fixed by the fermion density N/L^d with L^d the volume in d-dimensional position space. The dependence on the complex conjugate z^{*} to z of the argument of the two-point Green function (F.29b) is there to emphasize that the two-point Green function (F.29b) need not be an analytic function of z with a pole, as is the case in the non-interacting limit and as is postulated in the Landau theory of Fermi liquids.

As was the case with Eq. (F.18), we may, for any momentum $\boldsymbol{p} \in \mathbb{R}^d$ (not necessarily on the Fermi surface), define the integral

$$N^{(1)}_{\boldsymbol{p}} := \int\limits_{0}^{2\pi} \frac{\mathrm{d}l}{2\pi\mathrm{i}}\, \mathrm{tr}\, \left[\left(\partial_l\, G^{-1}_{\boldsymbol{p}}\right)(l)\, G_{\boldsymbol{p}}(l)\right] \tag{F.30}$$

for any closed path $\gamma_{\boldsymbol{p}} : [0, 2\pi[\to \mathbb{C}$, $l \mapsto \gamma_{\boldsymbol{p}}(l) = z(l)$ in a sufficiently small open neighborhood U of $z = 0$ that winds counterclockwise once around $z = 0$. (The punctured neighborhood $U \setminus \{0\}$ of $z = 0$ should be free of singularities of $G_{\boldsymbol{p}}$, such as poles of quasiparticles on the real axis say.) Here, we have anticipated that the right-hand side of Eq. (F.30) is independent of the closed path $\gamma_{\boldsymbol{p}}$ in the open neighborhood U of $z = 0$. This property is not anymore obvious, for we cannot rely on the condition of meromorphy of the integrand, as was the case with the resolvent in Eq. (F.17), to prove independence on the choice of $\gamma_{\boldsymbol{p}}$. Instead, we are going to verify explicitly this property by performing an infinitesimal variation of the path,

$$\gamma_{\boldsymbol{p}} \to \gamma_{\boldsymbol{p}} + \delta\gamma_{\boldsymbol{p}} \tag{F.31}$$

holding the value of $\gamma_{\boldsymbol{p}}(0) = \gamma_{\boldsymbol{p}}(2\pi)$ fixed, and showing that the change in the integral (F.30) vanishes to first order in $\delta\gamma_{\boldsymbol{p}}$. The proof that the infinitesimal variation $\delta N^{(1)}_{\boldsymbol{p}}$ under Eq. (F.31) vanishes goes as follows.

Proof. The infinitesimal variation (F.31) leaves $\boldsymbol{p}$ and $\gamma_{\boldsymbol{p}}(0) = \gamma_{\boldsymbol{p}}(2\pi)$ unchanged. Hence,

$$\delta N^{(1)}_{\boldsymbol{p}} = \int\limits_0^{2\pi} \frac{\mathrm{d}l}{2\pi\mathrm{i}}\, \delta\, \mathrm{tr}\, \left[\left(\partial_l\, G^{-1}_{\boldsymbol{p}}\right) G_{\boldsymbol{p}}\right](l). \tag{F.32}$$

Invertibility of $G_{\boldsymbol{p}}(l)$ implies that

$$\left(\partial_l\, G^{-1}_{\boldsymbol{p}}\right) = -G^{-1}_{\boldsymbol{p}}\, \left(\partial_l\, G_{\boldsymbol{p}}\right)\, G^{-1}_{\boldsymbol{p}}, \tag{F.33}$$

and

$$\left(\delta G^{-1}_{\boldsymbol{p}}\right) = -G^{-1}_{\boldsymbol{p}}\, \left(\delta G_{\boldsymbol{p}}\right)\, G^{-1}_{\boldsymbol{p}}. \tag{F.34}$$

By the product rule of differentiation,

$$\delta N^{(1)}_{\boldsymbol{p}} = -\int\limits_0^{2\pi} \frac{\mathrm{d}l}{2\pi\mathrm{i}}\, \mathrm{tr}\, \left[-G^{-1}_{\boldsymbol{p}}\, \left(\delta G_{\boldsymbol{p}}\right)\, G^{-1}_{\boldsymbol{p}}\, \left(\partial_l G_{\boldsymbol{p}}\right) + G^{-1}_{\boldsymbol{p}}\, \left(\partial_l\, \delta\, G_{\boldsymbol{p}}\right)\right](l). \tag{F.35}$$

Here, the change in the order from $\delta\, \partial_l$ to $\partial_l\, \delta$ for the second term on the right-hand side holds because the variation (F.31) is infinitesimal. If we do the manipulation

$$G^{-1}_{\boldsymbol{p}}\, \left(\partial_l\, \delta\, G_{\boldsymbol{p}}\right) = \partial_l\, \left[G^{-1}_{\boldsymbol{p}}\, \left(\delta\, G_{\boldsymbol{p}}\right)\right] + (-1)^2\, G^{-1}_{\boldsymbol{p}}\, \left(\partial_l\, G_{\boldsymbol{p}}\right) G^{-1}_{\boldsymbol{p}}\, \left(\delta\, G_{\boldsymbol{p}}\right), \tag{F.36}$$

we can rewrite the integrand in Eq. (F.35) as the sum over two terms. There is the trace over the commutator of the $n \times n$ matrices $G^{-1}_{\boldsymbol{p}}\, \left(\delta\, G_{\boldsymbol{p}}\right)$ and $G^{-1}_{\boldsymbol{p}}\, \left(\partial_l\, G_{\boldsymbol{p}}\right)$. There is the total derivative of a trace. Because the trace of the commutator of finite-dimensional matrices vanishes by the cyclicity of the trace, the integrand in Eq. (F.35) is a mere total derivative

$$\delta N^{(1)}_{\boldsymbol{p}} = -\int\limits_0^{2\pi} \frac{\mathrm{d}l}{2\pi\mathrm{i}}\, \partial_l\, \mathrm{tr}\, \left[G^{-1}_{\boldsymbol{p}}\, \left(\delta\, G_{\boldsymbol{p}}\right)\right](l). \tag{F.37}$$

Since $G^{-1}_{\boldsymbol{p}}(0) = G^{-1}_{\boldsymbol{p}}(2\pi)$ and $\delta G_{\boldsymbol{p}}(0) = \delta G_{\boldsymbol{p}}(2\pi) = 0$, this line integral vanishes. □

We already know that the line integral (F.30) is an integer if the two-point Green function is meromorphic. We are going to prove that this is always the case for sufficiently well-behaved two-point Green functions.

To this end, we first observe that the line integral (F.30) is an integer for any two-point Green function of the form

$$G_{\boldsymbol{p}}(z, z^{*}) = \frac{c\, f_{\boldsymbol{p}}(z, z^{*})}{\left(z - A_{\boldsymbol{p}}\right)^{\nu}\, \left(z - B_{\boldsymbol{p}}\right)^{1-\nu}}, \tag{F.38}$$

where c is a real number, $f_{\boldsymbol{p}} : \mathbb{R}^2 \to]0, \infty[$ is strictly positive for any non-vanishing momentum $\boldsymbol{p}$ and at least once differentiable, $0 < \nu \leq 1$ is a number, and the pair of $n \times n$ matrices $A_{\boldsymbol{p}}$ and $B_{\boldsymbol{p}}$ are Hermitian with the property that the functions $\operatorname{tr}\left(z-A_{\boldsymbol{p}}\right)^{-1}$ and $\operatorname{tr}\left(z-B_{\boldsymbol{p}}\right)^{-1}$ share the same residue. Indeed, we then have the identity

$$\begin{aligned}\int\limits_0^{2\pi}\frac{\mathrm{d}l}{2\pi\mathrm{i}}\operatorname{tr}\left[(\partial_l\, G_{\boldsymbol{p}}^{-1})\,(l)\, G_{\boldsymbol{p}}(l)\right\} &= \int\limits_0^{2\pi}\frac{\mathrm{d}l}{2\pi\mathrm{i}}\operatorname{tr}\left\{(\partial_l\, z)(l)\left[\frac{\nu}{z-A_{\boldsymbol{p}}}+\frac{1-\nu}{z-B_{\boldsymbol{p}}}\right]\right.\\ &\qquad\left. -\,\partial_l \ln f_{\boldsymbol{p}}\big(z(l),z^*(l)\big)\right\}\\ &= \int\limits_{\gamma_{\boldsymbol{p}}}\frac{\mathrm{d}z}{2\pi\mathrm{i}}\operatorname{tr}\left(\frac{\nu}{z-A_{\boldsymbol{p}}}+\frac{1-\nu}{z-B_{\boldsymbol{p}}}\right)\\ &= 0,1,2,\cdots.\end{aligned} \tag{F.39}$$

This example suggests that the quantization of the number $N_{\boldsymbol{p}}^{(1)}$ defined by Eq. (F.30) holds on more general grounds.

Proof. The fact that we can interpret the number $N_{\boldsymbol{p}}^{(1)}$ defined by Eq. (F.30) as a winding (i.e., integer-valued) number is a consequence of the following result from algebraic topology,

$$\pi_1\big(GL(n,\mathbb{C})\big) = \mathbb{Z}. \tag{F.40}$$

The meaning of this equation is that the set of equivalence classes of smooth mappings from the pointed circle S^1 into the group of pointed $n\times n$ invertible complex-valued matrices $GL(n,\mathbb{C})$ is a group (an example of a homotopy group[27]) isomorphic to the group of integers $\mathbb{Z}$. As the integral (F.30) involves a closed path in the Lie algebra $gl(n,\mathbb{C})$ of the Lie group $GL(n,\mathbb{C})$, it turns out that the homotopy property

[27] The group $GL(n,\mathbb{C})$ is an example of a topological space, i.e., a space on which the notion of open sets is well defined, recall section C.2. So is the circle $S^1 \sim SO(2)$. Since S^1 and $GL(n,\mathbb{C})$ are topological spaces, it is possible to define continuous maps from S^1 to $GL(n,\mathbb{C})$. Choose a point x from S^1 and a matrix X from $GL(n,\mathbb{C})$. The set of all continuous maps from S^1 to $GL(n,\mathbb{C})$ that map x into X can be studied on its own right. A homotopy is a continuous path between two such continuous maps, i.e., any two maps from (S^1,x) to $\big(GL(n,\mathbb{C}),X\big)$ connected by a homotopy are said to be homotopic. Mathematicians then show that the set of homotopic maps from (S^1,x) to $\big(GL(n,\mathbb{C}),X\big)$, the homotopic classes, form a group, the homotopy group, which is denoted by $\pi_1\big(GL(n,\mathbb{C})\big)$. If the pointed base space (S^1,x) is replaced by the pointed sphere (S^n,x) with $n=1,2,\cdots$ and if the target pointed topological space $\big(GL(n,\mathbb{C}),X\big)$ is replaced by the pointed connected topological space (G,X), then one denotes this homotopy group by $\pi_n(G)$. The fundamental homotopy group of G is $\pi_1(G)$. Homotopy groups record informations about the basic shapes, or holes, of topological spaces. Homotopy groups are therefore a mean to distinguish topological spaces, although two topological spaces sharing the same homotopy group can nevertheless be distinct. (A cup of coffee and a doughnut share the same fundamental homotopy group, but try drinking from a doughnut.) Homotopy groups allow to discard variations that do not affect the outcomes of interest. The residue theorem of complex analysis is an example thereof. The closed paths in the residue theorem surround a selected point of the complex plane. Hence they are nothing but continuous maps from the circle into the punctured complex plane. Two closed paths in the residue theorem produce the same integral result if they are homotopic.

of $GL(n,\mathbb{C})$ is reflected in the quantization of (F.30). The essence of this quantization is that any matrix M from $GL(n,\mathbb{C})$ admits the unique left polar decomposition $M = R\,U$ with the positive definite $n\times n$ matrix $R = \sqrt{M\,M^{\dagger}}$ and the $n\times n$ unitary matrix U. There follows the identity

$$\operatorname{tr}\left(M^{-1}\partial M\right) = \operatorname{tr}\left(U^{-1}\partial U\right) + \operatorname{tr}\left(R^{-1}\partial R\right). \tag{F.41}$$

Because $U \in U(n) = U(1) \times SU(n)$, $U^{-1}\partial U$ belongs to the Lie algebra $u(n) = u(1) \oplus su(n)$ of $U(n)$, and $\operatorname{tr} U^{-1}\partial U = n\,\mathrm{i}\partial\phi$ with ϕ the generator of the $U(1)$ subgroup of $U(n) = U(1) \times SU(n)$. Because R is positive definite, $\operatorname{tr} R^{-1}\partial R = \partial \operatorname{tr}\, \ln R$. If the matrix-valued function M is single-valued, so is R. On the otherhand, ϕ can be multivalued, in which case

$$\oint \operatorname{tr}\left(M^{-1}\mathrm{d}\,M\right) = \mathrm{i}\,n \oint \mathrm{d}\phi \tag{F.42}$$

is quantized in some units.[28] □

Now that we understand why the integral (F.30) is a winding number, we return to its relevance to the stability of the Fermi surface. The momentum $\boldsymbol{p}$ belongs to a Fermi surface if the winding number $N^{(1)}_{\boldsymbol{p}}$ defined by Eq. (F.30) is non-vanishing. The connected set of momenta $\boldsymbol{p}$ for which $N^{(1)}_{\boldsymbol{p}}$ is non-vanishing and constant as a function of $\boldsymbol{p}$ defines a branch of the Fermi surface of the interacting many-body Hamiltonian $\hat{H}$.

Under a parametric change of the many-body interaction during which eigenstates and eigenenergies of $\hat{H}$ change adiabatically, the many-body Fermi surface can change its shape (geometry), but the winding number (F.30) on each branch of the Fermi surface is unchanged, i.e., the many-body version of Eq. (F.20) holds. Gap-opening perturbations must necessarily break the adiabatic response of the eigenstates and eigenenergies of $\hat{H}$, for they induce a quantum phase transition across which the winding number (F.30) changes from a non-vanishing to a vanishing value.

F.4 Quasiparticles in the Landau theory of Fermi liquids as poles of the two-point Green function

In a Fermi liquid with a spherical and non-degenerate Fermi surface of radius p_{F}, the two-point Green function (F.29b) takes the asymptotic form

[28] The following example is illuminating. Let $M := \operatorname{diag}\left(e^{\mathrm{i}\varphi}, 1\right)$. This complex-valued matrix has a non-vanishing determinant. Hence, it belongs to $GL(2,\mathbb{C})$. The polar decomposition of M follows from $M = e^{\mathrm{i}(\varphi\,\sigma_0 + \varphi\,\sigma_3)/2}$, i.e., $R = \sigma_0$ and $U = e^{\mathrm{i}(\varphi\,\sigma_0 + \varphi\,\sigma_3)/2}$, where σ_0 denotes the unit 2×2 matrix and $\boldsymbol{\sigma}$ denotes the three Pauli matrices. The matrix from the Lie algebra associated to M is the matrix whose exponential gives M, i.e., $\left(\varphi\,\sigma_0 + \varphi\,\sigma_3\right)/2$. The angle ϕ in Eq. (F.42) is thus to be identified with the angle $\varphi/2$ in this example. The quantization unit for ϕ in this example is π. Evidently, the quantization unit depends depends on M and becomes explicit through the polar decomposition of M.

$$G_{\boldsymbol{p}}(z,z^{*}) \sim \frac{Z}{z - v_{\mathrm{F}}\left(|\boldsymbol{p}| - p_{\mathrm{F}}\right)}, \tag{F.43}$$

when the momentum $\boldsymbol{p}$ is very close to the Fermi surface and z is very close to the origin in the complex plane. The non-vanishing number $0 < Z < 1$ is the residue of the pole that defines a quasiparticle with the energy

$$\varepsilon_{\boldsymbol{p}} \sim v_{\mathrm{F}}\left(|\boldsymbol{p}| - p_{\mathrm{F}}\right), \tag{F.44}$$

measured relative to the ground state energy, asymptotically close to the Fermi surface. The residue Z is less than unity, the residue in the non-interacting limit. The Fermi velocity v_{F} is no longer the ratio of the Fermi momentum p_{F} to the fermion bare mass m. It gives the effective (renormalized) mass $m^{*} := p_{\mathrm{F}}/v_{\mathrm{F}}$. Neither the changes from unity to Z and m to m^{*} induced in a free gas of spinless fermions by the many-body interactions affect the value one of the winding number (F.30). Conversely, a smooth change of the many-body interaction may change Z and v_{F}, but it leaves the value one of the winding number (F.30) unchanged. In the theory of Fermi liquids with a spherical Fermi surface, p_{F}, Z, and v_{F} are fundamental constants. (For a non-spherical Fermi surface, p_{F} is replaced by the volume enclosed by the Fermi surface.)

If the periodic lattice of the crystal from which the conduction electrons originate is accounted for, we can take the asymptotics

$$G_{\boldsymbol{p}}(z,z^{*}) \sim \frac{Z}{z - \xi_{\boldsymbol{p}}} + \cdots \tag{F.45}$$

of the two-point Green function (F.29b) as the *definition* of a Fermi liquid with a non-degenerate Fermi surface defined by the condition $\xi_{\boldsymbol{p}} = 0$ and the long-lived quasiparticles with the asymptotic dispersion $\xi_{\boldsymbol{p}} \in \mathbb{R}$. Here, $\xi_{\boldsymbol{p}} \in \mathbb{R}$ is a linear function of the deviation in momentum space away from the Fermi surface that respects the point group symmetries of the lattice. The subleading corrections implied by $\cdots$ do not have to be meromorphic functions of z.

F.5 Breakdown of Landau Fermi liquid theory

F.5.1 *Gapped phases*

Phases of fermionic matter for which the winding number (F.30) vanishes are not Fermi liquids. Examples of such phases occur when a gap opens up at the Fermi surface due to a perturbation. A continuous opening of the gap at zero temperature as a function of a coupling strength is to be ascribed to a continuous quantum phase transition.

F.5.2 *Luttinger liquids*

There are phases of matter that support a non-vanishing winding number (F.30), but are nevertheless distinct from a Fermi liquid. For example, the Luttinger liquid supports a Fermi surface with the winding number (F.30) equal to the number 1. However, the Luttinger liquid is distinct from a Fermi liquid through the nature of its quasiparticles.

In one-dimensional position space, the Fermi surface of a non-interacting gas of spinless fermions consists of two Fermi points, labeled by the wave numbers $\pm k_{\mathrm{F}}$ where we are setting $\hbar = 1$ in this discussion (see Fig. E.1). In the close vicinity of the Fermi point $k_{\mathrm{F}} > 0$, the single-particle Green function (F.16) scales like

$$\mathcal{G}_{0\,k_{\mathrm{F}}+k}(z) \sim \frac{1}{z - v_{\mathrm{F}}\,k}. \tag{F.46}$$

We are going to argue that the one-dimensional Fermi surface $\{\pm k_{\mathrm{F}}\}$ survives in the presence of a weak short-range density-density interaction. This will be achieved by studying the analytical properties of the two-point Green function (F.29b). In the process, we shall observe that the quasiparticles in the sense of Landau are absent.

The calculation of the two-point Green function (F.29b) requires the tools from chapter 9, namely Abelian bosonization. With the help of Abelian bosonization, the representation of the time-ordered two-point Green function in position space x and real time t [recall Eq. (5.167)] can be achieved at the expense of a moderate amount of work [some Gaussian integrations as shown in Ref. [132] or in section 9.4.2 with Eq. (9.243)]. Performing a Fourier transformation from (x,t) space to (k,ω) space is demanding as it involves the computation of a convolution [133]. This task was performed in a form that is convenient for our purpose by Wen in Ref. [134],

$$G_{k_{\mathrm{F}}+k}(\omega) \sim \frac{(v_{\mathrm{F}}^2\,k^2 - \omega^2)^g}{\omega - v_{\mathrm{F}}\,k + \mathrm{i}\,0^+\,\mathrm{sgn}\,k}. \tag{F.47}$$

The wave number $k \in \mathbb{R}$ is very small compared to the momentum cut-off ($2\pi/\mathfrak{a}$ with $\mathfrak{a}$ the lattice spacing of an underlying lattice model, see section E.3 for an example with spinless fermions). The exponent $g \geq 0$ is a positive real number. When $g = 0$, the non-interacting limit (5.181) for a linear spectrum in one-dimensional position space is recovered. This number, when strictly positive, $g > 0$, encodes the effect of all short-range many-body interactions between the spinless fermions, provided the filling fraction is not commensurate to the underlying lattice model. As was observed by Volovik in Ref. [124], analytic continuation of Eq. (F.47) from the real frequency $\omega + \mathrm{i}\,0^+\,\mathrm{sgn}\,k$ to the Matsubara frequency $\mathrm{i}\omega \equiv z$ gives the asymptotics[29]

$$G_{k_{\mathrm{F}}+k}(z,z^*) \sim \frac{(v_{\mathrm{F}}\,k - z)^g (v_{\mathrm{F}}\,k - z^*)^g}{z - v_{\mathrm{F}}\,k}, \tag{F.48}$$

for the two-point Green function (F.29b). The non-interacting limit corresponds to $g = 0$ and for a strictly positive $g > 0$, Eq. (F.48) only applies at filling fractions that are not commensurate with the underlying one-dimensional lattice model. The effects of interaction is to make the two-point Green function (F.29b) non-analytic through an explicit dependence on z^*.

[29]According to Eq. (5.166), we need to perform the analytic continuation $\omega + \mathrm{i}\,0^+\,\mathrm{sgn}\,k \to \mathrm{i}\omega \equiv z$. [Compare Eq. (5.181) to Eq. (5.177)].

The two-point Green function (F.48) is of the form (F.38). Hence, its winding number (F.30) is unity at the Fermi point. The Fermi surface is robust to the many-body interaction. On the other hand, analytical continuation $z \to \omega + \mathrm{i}0^{+}\,\mathrm{sgn}\,k$ gives either the function of ω (F.47) with a branch cut at the energies ($\hbar = 1$) $\omega = v_{\mathrm{F}}\,k$ for any strictly positive g that is not an integer, or a function of ω (F.47) that vanishes at the energies $\omega = v_{\mathrm{F}}\,k$ for any positive integer value of $g > 0$. This is to say that for any strictly positive value of g, the interactions in a Luttinger liquid wipe out the poles at non-vanishing energies that signal the existence of quasiparticles in the Landau sense.

The effect of short-range interactions between electrons constrained to a one-dimensional position space are even more dramatic than those for spinless fermions. Interactions cause the phenomenon of spin and charge separation as signaled by the asymptotics [125]

$$G_{k_{\mathrm{F}}+k}(z, z^{*}) \sim \left(\frac{(v_{\mathrm{c}}\,k - z)^{g}(v_{\mathrm{c}}\,k - z^{*})^{g}}{z - v_{\mathrm{c}}\,k}\right)^{1/2} \times \left(\frac{(v_{\mathrm{s}}\,k - z)^{g}(v_{\mathrm{s}}\,k - z^{*})^{g}}{z - v_{\mathrm{s}}\,k}\right)^{1/2}, \tag{F.49}$$

for the two-point Green function (F.29b), provided the filling fraction is not commensurate with the underlying lattice. Hereto, the two-point Green function (F.29b) has factorized into two sectors. There are two distinct Fermi velocities v_{c} and v_{s}. The Fermi velocity v_{c} arises from low-lying excitations that carry the charge quantum number of the electron but are spinless. The Fermi velocity v_{s} arises from low-lying excitations that carry the spin quantum number of the electron but are charge neutral. The non-interacting limit corresponds to $g = 0$ and $v_{\mathrm{s}} = v_{\mathrm{c}}$, in which case Eq. (5.181) with its analytic dependence on z up to poles on mass shell is recovered in each spin sector. Any microscopic many-body interaction is signaled by $g > 0$ with the possibility of $v_{\mathrm{s}} \neq v_{\mathrm{c}}$. Although the value of the winding number (F.30) for the two-point Green function (F.49) is independent of $g \geq 0$ [because of Eq. (F.39)], all the poles that define the existence of dispersing long-lived quasiparticles in the sense of Landau have been wiped out for any strictly positive g. Instead, the low-lying energy sector has been rearranged in a non-adiabatic way to support two independent continua of excitations. There are the holons that carry a charge but no spin quantum numbers. There are the spinons that carry spin-1/2 but no charge quantum numbers. Although both holons and spinons have their own distinct Fermi surfaces (distinct through their Fermi velocities), they are not Landau quasiparticles. At commensurate filling fractions, the Fermi surface of holons or that of spinons can be gapped independently of each others by interactions, very much in the same way as the electron-phonon coupling can open a Peierls gap at a commensurate filling fraction.

Appendix G

First-order phase transitions induced by thermal fluctuations

Outline

We have illustrated in chapters 1-3 the fact that thermal fluctuations wipe out a continuous phase transition predicted by mean-field theory at or below the lower critical dimension.

The purpose of this appendix is to show that thermal fluctuations may turn a phase transition, that is predicted to be continuous within the mean-field approximation, into a discontinuous one for a dimensionality of space in between the lower and upper critical dimensions.

It is sufficient to demonstrate this effect by way of an example. To this end, we shall consider a microscopic model in three-dimensional space with a global $U(1)$ symmetry that is spontaneously broken below some ordering temperature T_{c}. We assume that this transition is captured by a Landau-Ginzburg theory with the Lagrangian density $\mathcal{L}_0$. We first review the connection between the Landau-Ginzburg theory with $U(1)$ global symmetry and the mean-field theory of a continuous phase transition. We then elevate the global $U(1)$ symmetry to a local one by way of the minimal coupling to a static vector gauge field. We define a gauge-invariant partition function by integrating over the corresponding static magnetic field and show under what conditions the fluctuations induced by the integration over the static magnetic field change the order of the transition from a continuous one to a first-order transition.

G.1 Landau-Ginzburg theory and the mean-field theory of continuous phase transitions

Position space is d-dimensional and we consider a very large volume

$$V = L^d, \tag{G.1a}$$

at which boundaries the order parameter is either vanishing or repeating (periodic boundary conditions) so that we may always drop boundary terms upon partial integration. Statistical equilibrium at the temperature T is fulfilled and we denote with

$$\beta = \frac{1}{k_{\mathrm{B}}\,T} \tag{G.1b}$$

the inverse temperature with the dimension of inverse energy. We assume that, upon integration over the microscopic degrees of freedom, the partition function becomes

$$Z_0[j^*, j] := \int \mathcal{D}[\psi^*, \psi]\, e^{-\beta S_0}, \tag{G.1c}$$

where the effective action (free energy) is

$$\begin{aligned} S_0 &:= \int\limits_V \mathrm{d}^d\boldsymbol{r}\, \mathcal{L}_0, \\ \mathcal{L}_0 &:= a_0^2\, t\, |\psi(\boldsymbol{r})|^2 + \frac{1}{2} b_0^2\, |\psi(\boldsymbol{r})|^4 + \gamma_0^2\, |(\boldsymbol{\nabla}\psi)(\boldsymbol{r})|^2 - \psi^*(\boldsymbol{r})\, j(\boldsymbol{r}) - \psi(\boldsymbol{r})\, j^*(\boldsymbol{r}). \end{aligned} \tag{G.1d}$$

The order parameter $\psi : \mathbb{R}^d \to \mathbb{C}$, a complex-valued field that assigns to any point $\boldsymbol{r}$ in space the complex number $\psi(\boldsymbol{r})$, is smooth in the sense that its Fourier transform $\psi(\boldsymbol{k})$ vanishes if $|\boldsymbol{k}| \sim \Lambda$ with Λ the effective ultraviolet cut-off much smaller than the inverse lattice spacing $1/\mathfrak{a}$. The measure $\mathcal{D}[\psi^*, \psi]$ is the bosonic measure for a complex-valued bosonic field (two real bosonic fields). The external source fields j^* and j have been introduced in order to compute correlation functions of the order parameter. The couplings a_0^2, b_0^2, and γ_0^2 are dimensionful and positive.[30] The dimensionless parameter t is the reduced temperature

$$t := \frac{T - T_{\mathrm{c},0}}{T_{\mathrm{c},0}}. \tag{G.1e}$$

It is positive when the temperature T is larger than the critical temperature $T_{\mathrm{c},0}$. It vanishes when $T = T_{\mathrm{c},0}$. It is negative otherwise.

The Lagrangian density $\mathcal{L}_0$ is assumed to be local. This assumption is justified if all connected correlation functions for local operators built out of the microscopic degrees of freedom in the many-body ground state decay exponentially fast with decay lengths bounded from above by a finite zero-temperature correlation length. When the external source fields vanish, $\mathcal{L}_0$ is also invariant under the global gauge transformation

$$\psi(\boldsymbol{r}) =: e^{\mathrm{i}\alpha}\, \psi'(\boldsymbol{r}), \tag{G.2}$$

for any real-valued α, and so is the partition function $Z[j^*, j]$ when the external source fields are absent, i.e., $j^* = j = 0$. This symmetry is inherited from a

[30] Dimensional analysis assigns

$$[\psi] = 1, \qquad [a_0^2] = [b_0^2] = [j] = \frac{\text{energy}}{\text{volume}}, \qquad [\gamma_0^2] = \frac{\text{energy} \times \text{area}}{\text{volume}}.$$

global $U(1)$ symmetry of the Hamiltonian governing the dynamics of the microscopic degrees of freedom.

Mean-field theory is the saddle-point approximation by which the partition function $Z_0[j^*,j]$ is approximated by the Boltzmann weight $e^{-\beta S_0}$ evaluated at the solution $\psi_j(\boldsymbol{r})$ to the saddle-point equation

$$a_0^2\, t\, \psi(\boldsymbol{r}) + b_0^2\, \psi^*(\boldsymbol{r})\, \psi^2(\boldsymbol{r}) - \gamma_0^2\, \left(\boldsymbol{\nabla}^2\psi\right)(\boldsymbol{r}) = j(\boldsymbol{r}). \tag{G.3}$$

When $t > 0$, the saddle-point solution

$$\psi_{j=0}(\boldsymbol{r}) = 0 \tag{G.4a}$$

is unique in the absence of external sources. When $t < 0$, the saddle-point solution is not unique, since

$$\psi_{j=0}(\boldsymbol{r}) = \sqrt{\frac{a_0^2\,|t|}{b_0^2}}\, e^{\mathrm{i}\alpha} \tag{G.4b}$$

is a saddle-point solution for any $0 \leq \alpha < 2\pi$, in the absence of external sources. As the temperature T is lowered from above $T_{\mathrm{c},0}$ to below $T_{\mathrm{c},0}$, t changes smoothly from positive to negative values, while the magnitude of the order parameter that solves the saddle-point equations changes smoothly from 0 to a positive value. The phase transition at $T = T_{\mathrm{c},0}$ within the mean-field approximation is continuous. This phase transition is characterized by the following mean-field thermodynamic critical exponents, as can be inferred from the saddle-point equation (G.3) and the value

$$\mathcal{L}_0 = \begin{cases} 0, & \text{if } T > T_{\mathrm{c},0}, \\ -\frac{1}{2}\,\frac{a_0^4\,t^2}{b_0^2}, & \text{if } T < T_{\mathrm{c},0}, \end{cases} \tag{G.5}$$

taken by the Lagrangian density at the saddle-point in the absence of external sources. More generally, the mean-field approximation in the presence of the source j is the approximation

$$\langle\psi(\boldsymbol{r})\rangle_0 \approx \psi_j(\boldsymbol{r}), \tag{G.6a}$$

where $\psi_j(\boldsymbol{r})$ denotes a solution to the saddle-point equation (G.3) and we use the notation

$$\Big\langle(\cdots)\Big\rangle_0 := \frac{\int\mathcal{D}[\psi^*,\psi]\, e^{-\beta S_0}\,(\cdots)}{\int\mathcal{D}[\psi^*,\psi]\, e^{-\beta S_0}}. \tag{G.6b}$$

The critical exponent β (not to be confused with the inverse temperature $1/(k_{\mathrm{B}}\,T)$ for which the same symbol is used) is defined by the power-law growth of the order parameter below the transition,

$$\langle\psi(\boldsymbol{r})\rangle_0 \sim |t|^\beta, \qquad T < T_{\mathrm{c},0}, \tag{G.7a}$$

in the absence of source terms. The mean-field approximation (G.6a) gives

$$\beta = 1/2. \tag{G.7b}$$

The critical exponent α is defined by the power-law divergence of the heat capacity at constant volume upon approaching the transition. The heat capacity at constant volume is related to the entropy S by $C_V := T\left(\frac{\partial S}{\partial T}\right)_V$. Hence,

$$C_V = -T\,\frac{\partial\langle\mathcal{L}_0\rangle_0}{\partial T^2} \sim |t|^{-\alpha}, \tag{G.8a}$$

in the absence of source terms. The mean-field approximation (G.6a) gives

$$\alpha = 0. \tag{G.8b}$$

The critical exponent γ is defined by the power-law divergence of the uniform susceptibility within linear-response theory upon approaching the transition

$$\chi := \left.\frac{\partial\langle\psi(\boldsymbol{r})\rangle_0}{\partial j}\right|_{j=0} \sim |t|^{-\gamma}. \tag{G.9a}$$

The mean-field approximation (G.6a) gives

$$\gamma = 1. \tag{G.9b}$$

The critical exponent δ is defined by the power-law dependence of the source field on the order parameter on the critical isotherm $t = 0$,

$$j(\boldsymbol{r}) \sim |\langle\psi(\boldsymbol{r})\rangle_0|^{\delta}, \qquad t = 0. \tag{G.10a}$$

The mean-field approximation (G.6a) gives

$$\delta = 3. \tag{G.10b}$$

The critical exponent ν is defined by the power-law divergence

$$\xi \sim |t|^{-\nu} \tag{G.11a}$$

of the finite correlation length ξ defined by

$$\left(-\boldsymbol{\nabla}^2 + \xi^{-2}\right)\chi_0(\boldsymbol{r}-\boldsymbol{r}') = \frac{1}{\gamma_0^2}\delta(\boldsymbol{r}-\boldsymbol{r}'), \tag{G.11b}$$

where the susceptibility χ_0 is defined by

$$\chi_0(\boldsymbol{r}-\boldsymbol{r}') := \left.\frac{\delta\langle\psi(\boldsymbol{r})\rangle_0}{\delta j(\boldsymbol{r}')}\right|_{j=0}. \tag{G.11c}$$

Thus, ξ is defined within linear-response theory whereby translation invariance is assumed.[31] The mean-field approximation (G.6a) gives

$$\nu = 1/2. \tag{G.11d}$$

The critical exponent η is defined at criticality $t = 0$ by the algebraic decay of the susceptibility,

$$\chi_0(\boldsymbol{r}-\boldsymbol{r}') \sim \left(\frac{\mathfrak{a}}{|\boldsymbol{r}-\boldsymbol{r}'|}\right)^{d-2+\eta}. \tag{G.12a}$$

[31] Dimensional analysis assigns

$$\left[\frac{\delta}{\delta j(\boldsymbol{r}')}\right] = [\chi_0(\boldsymbol{r}-\boldsymbol{r}')] = \frac{1}{\text{energy}}.$$

The mean-field approximation (G.6a) gives

$$\eta = 0. \tag{G.12b}$$

The validity of the mean-field approximation (G.6a) is controlled by the dimensionless ratio

$$r_{\mathrm{G}} := \frac{k_{\mathrm{B}}\, T_{\mathrm{c},0} \int\limits_{\xi^d(T)} \mathrm{d}^d \boldsymbol{r}\, |\chi_0(\boldsymbol{r}-\boldsymbol{r}')|}{\int\limits_{\xi^d(T)} \mathrm{d}^d \boldsymbol{r}\, |\langle \psi(\boldsymbol{r}) \rangle_0|^2}. \tag{G.13}$$

(The label G stands for Ginzburg.) The rational for this dimensionless number is the following. At temperature $T < T_{\mathrm{c},0}$ and in a volume V, the correlation function

$$\langle \psi(\boldsymbol{r})\, \psi(\boldsymbol{r}') \rangle_0 = \langle \psi(\boldsymbol{r}) \rangle_0 \langle \psi(\boldsymbol{r}') \rangle_0 + c(V,T) \times e^{-\frac{|\boldsymbol{r}-\boldsymbol{r}'|}{\xi(T)}}. \tag{G.14}$$

The non-vanishing number $c(V,T)$ accounts for both the thermal fluctuations at a non-vanishing temperature T and the residual fluctuations in the large but finite volume V. The former fluctuations vanish at $T = 0$, while the latter ones vanish in the thermodynamic limit $V \to \infty$, i.e., $c(V,T) \to 0$ in the limit $T \to 0$ and $V \to \infty$. The susceptibility is proportional to the connected two-point correlation function

$$\langle \psi(\boldsymbol{r})\, \psi(\boldsymbol{r}') \rangle_0 - \langle \psi(\boldsymbol{r}) \rangle_0 \langle \psi(\boldsymbol{r}') \rangle_0 = c(V,T) \times e^{-\frac{|\boldsymbol{r}-\boldsymbol{r}'|}{\xi(T)}}. \tag{G.15}$$

Hence, the susceptibility vanishes in the limit $V \to \infty$ and $T \to 0$, for which perfect long-range order is established. Now, the numerator of r_{G} is the integrated susceptibility at temperature T over a volume $V = \xi^d(T)$ set by the correlation length at T multiplied with the energy scale $k_{\mathrm{B}}\, T_{\mathrm{c},0}$. The numerator is constructed to be much smaller than the denominator, which is the expectation-value of the squared magnitude of the order parameter at temperature T, as $T \to 0$. Thus,

$$\lim_{T \to 0} r_{\mathrm{G}} \ll 1, \tag{G.16}$$

for large V. Accordingly, the mean-field approximation (G.6a) is expected to hold when $r_{\mathrm{G}} \ll 1$. What happens to this inequality when $T \to T_{\mathrm{c},0}$ in the mean-field approximation (G.6a)?

We make the mean-field estimate

$$\begin{aligned} r_{\mathrm{G}} &\sim \frac{k_{\mathrm{B}}\, T_{\mathrm{c},0}}{a_0^2\, |t|} \times \frac{b_0^2}{a_0^2\, \xi^d(T=0)\, |t|^{(2-d)/2}} \\ &\sim \frac{k_{\mathrm{B}}}{\Delta C_0\, \xi^d(T=0)} \frac{1}{|t|^{(4-d)/2}}, \end{aligned} \tag{G.17}$$

where $\Delta C_0 = a_0^4/(b_0^2\, T_{\mathrm{c},0})$ is the mean-field jump of the heat capacity at constant volume V at the transition. The mean-field approximation is good as long as the Ginzburg criterion

$$r_{\mathrm{G}} \ll 1 \Longleftrightarrow \left(\frac{t_{\mathrm{G}}}{|t|}\right)^{(4-d)/2} \ll 1, \qquad t_{\mathrm{G}} := \left(\frac{k_{\mathrm{B}}}{\Delta C_0\, \xi^d(T=0)}\right)^{2/(4-d)} \tag{G.18a}$$

holds. The Ginzburg criterion is satisfied for any $|t| \ll 1$ above the upper mean-field critical dimension $d = 4$. It is marginally satisfied at the upper mean-field critical dimension $d = 4$ as $|t| \to 0$. It is violated when

$$|t| \lesssim t_{\mathrm{G}} \ll 1, \tag{G.18b}$$

below the upper mean-field critical dimension $d = 4$. The condition (G.18b) defines the critical regime in which fluctuations are dominant and the mean-field approximation breaks down [135].

G.2 Fluctuations induced by a local gauge symmetry

It is time to gauge the theory as follows. We define the partition function

$$Z := \int \mathcal{D}[\psi^*, \psi] \int \frac{\mathcal{D}[\boldsymbol{A}]}{\mathcal{V}} e^{-\beta S}, \tag{G.19a}$$

with the action (free energy)

$$S := \int\limits_V \mathrm{d}^3\boldsymbol{r}\, \mathcal{L}, \tag{G.19b}$$

and the Lagrangian density

$$\mathcal{L} := a_0^2\, t\, |\psi(\boldsymbol{r})|^2 + \frac{1}{2} b_0^2\, |\psi(\boldsymbol{r})|^4 + \gamma_0^2\, |(\boldsymbol{D}\psi)(\boldsymbol{r})|^2 + \frac{1}{8\pi\mu_0^2}\, |\boldsymbol{\nabla} \wedge \boldsymbol{A}(\boldsymbol{r})|^2. \tag{G.19c}$$

Here, the covariant derivative

$$\boldsymbol{D} := \boldsymbol{\nabla} - \mathrm{i}\, q_0\, \boldsymbol{A} \tag{G.19d}$$

couples the real-valued and dimensionless vector gauge field $\boldsymbol{A}$, a map $\boldsymbol{A} : \mathbb{R}^3 \to \mathbb{R}^3$, to the complex-valued and dimensionless order parameter ψ through the real-valued and dimensionful gauge coupling q_0. The dimensionful coupling constant μ_0^2, the magnetic permeability, is positive.[32] The measure for the vector gauge field is divided by the gauge volume

$$\mathcal{V} := \int \mathcal{D}[\alpha], \tag{G.19e}$$

which accounts for the local gauge symmetry under the transformations

$$\psi(\boldsymbol{r}) =: e^{\mathrm{i}\,\alpha(\boldsymbol{r})}\, \psi'(\boldsymbol{r}), \tag{G.19f}$$

and

$$\boldsymbol{A}(\boldsymbol{r}) =: \boldsymbol{A}'(\boldsymbol{r}) + \frac{1}{q_0}\, (\boldsymbol{\nabla}\alpha)(\boldsymbol{r}). \tag{G.19g}$$

The boundary conditions and the remaining couplings were already defined in Eq. (G.1).

[32] The gauge coupling q_0 carries the dimension of $(\text{length})^{-1}$ in order for the gauge field $\boldsymbol{A}(\boldsymbol{r})$ to be dimensionless. The coupling constant μ_0^2 carries the dimension of (length/energy).

The saddle-point solutions are, up to a local gauge transformation, given by

$$\boldsymbol{A}(\boldsymbol{r})=0, \qquad |\psi(\boldsymbol{r})|=\begin{cases}0, & t>0,\\ \sqrt{\frac{a_0^2\,|t|}{b_0^2}}, & t<0.\end{cases} \tag{G.20}$$

We choose to work in any range of temperature that satisfies the Ginzburg criterion (G.18a) for the validity of the mean-field approximation. In this regime and after fixing the Coulomb gauge (the condition $\boldsymbol{\nabla}\cdot\boldsymbol{A}=0$), we assume that the order parameter is constant throughout space (the source term j vanishes) so that the partition function (G.19a) simplifies to

$$\begin{aligned} Z\approx{}&\frac{2\pi}{2}\int\limits_0^\infty \mathrm{d}\left(|\psi|^2\right)\,e^{-\beta V\left(a_0^2\,t\,|\psi|^2+\frac{1}{2}b_0^2\,|\psi|^4\right)}\\ &\times\int\mathcal{D}[\boldsymbol{A}]\,\delta(\boldsymbol{\nabla}\cdot\boldsymbol{A}=0)\,e^{-\beta\int\limits_V \mathrm{d}^3\boldsymbol{r}\,\frac{1}{8\pi\mu_0^2}\left(8\pi\mu_0^2\,\gamma_0^2\,q_0^2\,|\psi|^2\,\boldsymbol{A}^2(\boldsymbol{r})+|\boldsymbol{\nabla}\wedge\boldsymbol{A}(\boldsymbol{r})|^2\right)}. \end{aligned} \tag{G.21}$$

We proceed in three steps. First, we introduce the notation

$$e^{-\beta\,\tilde{S}(|\psi|)}\equiv e^{-\beta V\left(a_0^2\,t\,|\psi|^2+\frac{1}{2}b_0^2\,|\psi|^4\right)}\times\tilde{Z}(|\psi|), \tag{G.22a}$$

where

$$\tilde{Z}(|\psi|):=\int\mathcal{D}[\boldsymbol{A}]\,\delta(\boldsymbol{\nabla}\cdot\boldsymbol{A}=0)\,e^{-\beta V\,\tilde{\mathcal{L}}[|\psi|,\boldsymbol{A}]}, \tag{G.22b}$$

and

$$\tilde{\mathcal{L}}[|\psi|,\boldsymbol{A}]:=\int\limits_V\frac{\mathrm{d}^3\boldsymbol{r}}{V}\,\frac{1}{8\pi\mu_0^2}\left(8\pi\mu_0^2\,\gamma_0^2\,q_0^2\,|\psi|^2\,\boldsymbol{A}^2(\boldsymbol{r})+|\boldsymbol{\nabla}\wedge\boldsymbol{A}(\boldsymbol{r})|^2\right). \tag{G.22c}$$

There follows the first-order differential equation

$$\frac{\mathrm{d}\tilde{S}(|\psi|)}{\mathrm{d}|\psi|}=2\,V\left(a_0^2\,t\,|\psi|+b_0^2\,|\psi|^3\right)+2\,V\,\gamma_0^2\,q_0^2\,|\psi|\left\langle\int\limits_V\frac{\mathrm{d}^3\boldsymbol{r}}{V}\boldsymbol{A}^2(\boldsymbol{r})\right\rangle_{\tilde{Z}(|\psi|)}, \tag{G.23a}$$

with

$$\left\langle\int\limits_V\frac{\mathrm{d}^3\boldsymbol{r}}{V}\boldsymbol{A}^2(\boldsymbol{r})\right\rangle_{\tilde{Z}(|\psi|)}:=\frac{\int\mathcal{D}[\boldsymbol{A}]\,\delta(\boldsymbol{\nabla}\cdot\boldsymbol{A}=0)\,e^{-\beta V\,\tilde{\mathcal{L}}(|\psi|,\boldsymbol{A})}\int\limits_V\frac{\mathrm{d}^3\boldsymbol{r}}{V}\boldsymbol{A}^2(\boldsymbol{r})}{\int\mathcal{D}[\boldsymbol{A}]\,\delta(\boldsymbol{\nabla}\cdot\boldsymbol{A}=0)\,e^{-\beta V\,\tilde{\mathcal{L}}(|\psi|,\boldsymbol{A})}}. \tag{G.23b}$$

Second, we compute the dependence on $|\psi|$ of Eq. (G.23b). Third, we set the right-hand side of Eq. (G.23a) to 0 to solve for the saddle-point equation obeyed by the action $\tilde{S}(|\psi|)$. In this way, a mean-field phase transition is obtained that might differ from the one of section G.1, the latter corresponding to the limit $q_0=0$. This computation is self-consistent if the renormalized transition temperature controlled by the gauge charge q_0 belongs to the regime (G.18b).

To compute the right-hand side of Eq. (G.23b), we choose the Fourier convention

$$\boldsymbol{A}(\boldsymbol{r})=\frac{1}{V}\sum_{|\boldsymbol{k}|<\Lambda}e^{+\mathrm{i}\boldsymbol{k}\cdot\boldsymbol{r}}\,\boldsymbol{A}_{\boldsymbol{k}}, \qquad \boldsymbol{A}_{\boldsymbol{k}}=\int\limits_V\mathrm{d}^3\boldsymbol{r}\,e^{-\mathrm{i}\boldsymbol{k}\cdot\boldsymbol{r}}\,\boldsymbol{A}(\boldsymbol{r}). \tag{G.24}$$

Here, Λ is a momentum cut-off. Since $\boldsymbol{A}(\boldsymbol{r}) \in \mathbb{R}^3$,

$$\int\limits_V \mathrm{d}^3\boldsymbol{r}\, \boldsymbol{A}^2(\boldsymbol{r}) = \frac{1}{V} \sum_{|\boldsymbol{k}|<\Lambda} |\boldsymbol{A}_{\boldsymbol{k}}|^2, \tag{G.25}$$

and, because of the Coulomb gauge condition $\boldsymbol{k} \cdot \boldsymbol{A}_{\boldsymbol{k}} = 0$,

$$\int\limits_V \mathrm{d}^3\boldsymbol{r}\, (\boldsymbol{\nabla} \wedge \boldsymbol{A})^2(\boldsymbol{r}) = \frac{1}{V} \sum_{|\boldsymbol{k}|<\Lambda} \boldsymbol{k}^2\, |\boldsymbol{A}_{\boldsymbol{k}}|^2. \tag{G.26}$$

Because of the Coulomb gauge, $\boldsymbol{A}$ has only two independent components out of three. Hence, we may choose to present the measure for the gauge fields in terms of polar coordinates,

$$\mathcal{D}[\boldsymbol{A}]\,\delta(\boldsymbol{\nabla}\cdot\boldsymbol{A}=0) = \prod_{|\boldsymbol{k}|<\Lambda} \mathrm{d}\varphi_{\boldsymbol{k}}\, \frac{1}{2}\mathrm{d}\left(|\boldsymbol{A}_{\boldsymbol{k}}|^2\right), \qquad 0 \le \varphi_{\boldsymbol{k}} < 2\pi, \qquad 0 \le |\boldsymbol{A}_{\boldsymbol{k}}| < \infty. \tag{G.27}$$

We conclude that[33]

$$\left\langle \int\limits_V \frac{\mathrm{d}^3\boldsymbol{r}}{V} \boldsymbol{A}^2(\boldsymbol{r}) \right\rangle_{\tilde{Z}(|\psi|)} = \frac{8\pi\mu_0^2\, k_{\mathrm{B}}\, T}{V} \sum_{|\boldsymbol{k}|<\Lambda} \frac{1}{\boldsymbol{k}^2 + k_{\mathrm{s}}^2(|\psi|)} \tag{G.28a}$$

with

$$k_{\mathrm{s}}^2(|\psi|) := 8\pi\mu_0^2\,\gamma_0^2\, q_0^2\, |\psi|^2. \tag{G.28b}$$

The length scale $k_{\mathrm{s}}^{-1}(|\psi|)$ is reminiscent of the Thomas-Fermi length scale that arises due to the screening of the scalar potential by the Fermi sea in the jellium model. This length scale is also reminiscent of the London penetration depth in a superconductor [see Eq. (7.182)].

For a very large volume V, we may replace the sum over the discrete wave numbers by an integral (say, assuming periodic boundary conditions) and carry the elementary integration

$$\begin{aligned}\left\langle \int\limits_V \frac{\mathrm{d}^3\boldsymbol{r}}{V} \boldsymbol{A}^2(\boldsymbol{r}) \right\rangle_{\tilde{Z}(|\psi|)} &\approx \frac{8\pi\mu_0^2\, k_{\mathrm{B}}\, T}{(2\pi)^3} \int\limits_{|\boldsymbol{k}|<\Lambda} \mathrm{d}^3\boldsymbol{k}\, \frac{1}{\boldsymbol{k}^2 + k_{\mathrm{s}}^2(|\psi|)} \\ &= \frac{4\mu_0^2\, k_{\mathrm{B}}\, T_{\mathrm{c},0}\,(1+t)}{\pi} \left(\Lambda - k_{\mathrm{s}}(|\psi|)\, \arctan\frac{\Lambda}{k_{\mathrm{s}}(|\psi|)}\right).\end{aligned} \tag{G.29}$$

In the mean-field approximation, we replace in the partition function (G.19a) the integral over the magnitude $|\psi|$ of the order parameter by the saddle-point contribution. This is to say that we do the mean-field substitution

$$k_{\mathrm{s}}(|\psi|) := 2\sqrt{2\,\pi}|\mu_0\,\gamma_0\, q_0\,\psi| \to 2\sqrt{2\,\pi}|\mu_0\,\gamma_0\, q_0\,\psi_{j=0}| =: k_{\mathrm{s},0}, \tag{G.30}$$

[33] Use the pair of integrals $\int\limits_0^\infty \mathrm{d}x\, x^{2n+1}\, e^{-ax^2} = n!/(2a^{n+1})$ with $n=1$ in the numerator and $n=0$ in the denominator and the identifications $x \to |\boldsymbol{A}_{\boldsymbol{k}}|$ and $a \to \frac{1}{8\,\pi\,\mu_0^2\, k_{\mathrm{B}}\, T\, V}[\boldsymbol{k}^2 + k_{\mathrm{s}}^2(|\psi|)]$.

with $|\psi_{j=0}|$ a saddle-point solution to Eq. (G.23). We now assume that

$$k_{\mathrm{s},0} \ll \Lambda. \tag{G.31}$$

If so, we may do the expansion $\arctan x = (\pi/2) - (1/x) + \cdots$ valid for large positive values of x to write

$$\left\langle \int\limits_V \frac{\mathrm{d}^3\boldsymbol{r}}{V} \boldsymbol{A}^2(\boldsymbol{r}) \right\rangle_{\tilde{Z}(|\psi_{j=0}|)} \approx \frac{4\mu_0^2\, k_{\mathrm{B}}\, T_{\mathrm{c},0}\,(1+t)}{\pi} \left(\Lambda - \frac{\pi}{2} k_{\mathrm{s},0} + \cdots\right). \tag{G.32}$$

The leading contribution on the right-hand side is proportional to the effective momentum cut-off Λ. It is positive and independent of the saddle-point $|\psi_{j=0}|$, for we only consider the regime of non-vanishing temperature for which $1+t>0$ holds. The leading correction is independent of the momentum cut-off Λ. It is negative and linear in the saddle-point value $|\psi_{j=0}|$. We conclude that the saddle-point approximation to the partition function (G.19a) is

$$Z \approx e^{-\beta\,\tilde{S}_0}, \tag{G.33a}$$

where the saddle-point value of the action is, after integration of Eq. (G.23),

$$\tilde{S}_0 := \tilde{S}(0) + V \left(a_{\mathrm{r}}^2\, t\, |\psi_{j=0}|^2 - \frac{2}{3}\, c_{\mathrm{r}}^2\, |\psi_{j=0}|^3 + \frac{1}{2}\, b_{\mathrm{r}}^2\, |\psi_{j=0}|^4 \right), \tag{G.33b}$$

with the t-dependent couplings

$$a_{\mathrm{r}}^2 := a_0^2 + \frac{4\,\gamma_0^2\, q_0^2\, \mu_0^2\, k_{\mathrm{B}}\, T_{\mathrm{c},0}\,(1+t)\,\Lambda}{\pi\, t}, \tag{G.33c}$$

$$c_{\mathrm{r}}^2 := 4\,\gamma_0^2\, q_0^2\, \mu_0^2\, k_{\mathrm{B}}\, T_{\mathrm{c},0}\,(1+t)\,\sqrt{2\,\pi\,\mu_0^2\,\gamma_0^2\, q_0^2}, \tag{G.33d}$$

and the t-independent coupling

$$b_{\mathrm{r}}^2 := b_0^2. \tag{G.33e}$$

The effect of the fluctuations of the static vector gauge field on the uniform order parameter is two-fold. First, the quadratic coefficient $a_{\mathrm{r}}^2\, t$ of the order parameter has increased. Of its own, this change in the quadratic coefficient thus amounts to a mere lowering in the mean-field critical temperature of the mean-field continuous phase transition, for the root of $a_{\mathrm{r}}^2\, t = 0$ now occurs at a *non-vanishing negative* value of t. This is not surprising, the coupling to the static vector gauge fields brings about an additional source of fluctuations in addition to thermal fluctuations from section G.1. Second, a new *cubic term* has appeared with a negative coefficient. The appearance of the cubic term has dramatic consequences. It induces a discontinuous phase transition (see Fig. G.1).

To see this, we minimize the effective action (G.33b) with respect to $|\psi_{j=0}|$ by seeking all solutions to the renormalized saddle-point equation

$$|\psi_{j=0}|\left(a_{\mathrm{r}}^2\, t - c_{\mathrm{r}}^2\, |\psi_{j=0}|^2 + b_{\mathrm{r}}^2\, |\psi_{j=0}|^4\right) = 0. \tag{G.34}$$

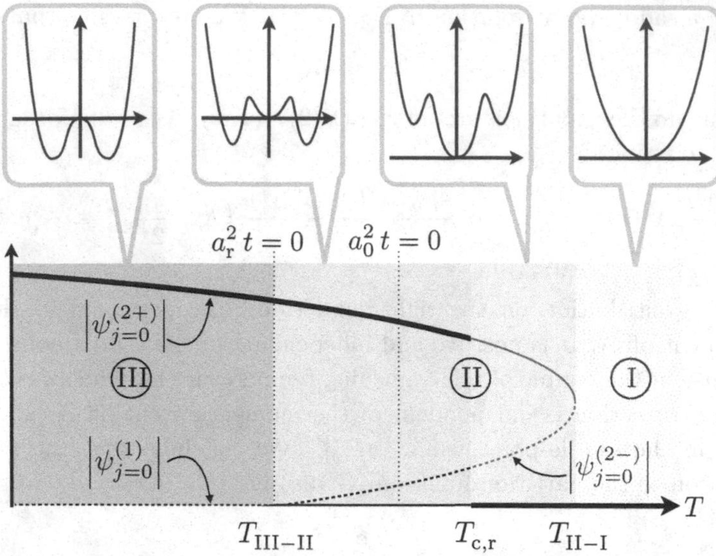

Fig. G.1 Characteristic dependence of $\tilde{S}_0 - \tilde{S}(0)$ defined by the effective action (G.33b) on $|\psi_{j=0}|$ corresponding to regimes I, II, and III. The thick, thin, and dashed lines denote the absolute minima, local minima, and local maxima of the effective action (G.33b) as a function of the reduced temperature t.

We recall here that the renormalized coefficients $a_{\rm r}^2$ and $c_{\rm r}^2$ are both functions of t according to Eq. (G.33). There are three solutions to the renormalized saddle-point equation (G.34). The first renormalized saddle-point solution is the trivial solution

$$|\psi^{(1)}_{j=0}| = 0. \tag{G.35a}$$

The remaining two renormalized saddle-point solutions are

$$|\psi^{(2\pm)}_{j=0}| = \frac{c_{\rm r}^2 \pm \sqrt{c_{\rm r}^4 - 4\,b_{\rm r}^2\,a_{\rm r}^2\,t}}{2\,b_{\rm r}^2}. \tag{G.35b}$$

By selecting which of these solutions are absolute minima of the (renormalized) effective action (G.33b), we identify three ranges of temperature.

In the range of temperatures I that includes the $T\to\infty$ limit, the renormalized (quadratic) discriminant

$$\Delta_{\rm r} := c_{\rm r}^4 - 4\,b_{\rm r}^2\,a_{\rm r}^2\,t \tag{G.36}$$

obeys the condition

$$\mathrm{I} := \{T\,|\,\Delta_{\rm r} < 0\}. \tag{G.37}$$

In the range of temperature I (large T), only the solution $|\psi^{(1)}_{j=0}|$ is an absolute minimum of the effective action (G.33b) (the solutions $|\psi^{(2\pm)}_{j=0}|$ are unphysical for they are complex-valued).

In the range of temperatures II, the temperature is bounded from above by the temperature at which the renormalized (quadratic) discriminant (G.36) vanishes, while it is bounded from below by the temperature at which the quadratic renormalized coefficient $a_{\mathrm{r}}^2\, t$ of the effective action (G.33b) vanishes,

$$\mathrm{II} := \left\{T_{\mathrm{III-II}} < T < T_{\mathrm{II-I}} \,|\, \Delta_{\mathrm{r}} = 0 \text{ if } T = T_{\mathrm{II-I}},\, a_{\mathrm{t}}^2\, t = 0 \text{ if } T = T_{\mathrm{III-II}} \right\}. \quad \text{(G.38)}$$

In the range of temperature II (intermediate T), only the saddle-point solution $|\psi_{j=0}^{(2-)}|$ is a local maximum. The saddle-point solutions $|\psi_{j=0}^{(1)}|$ and $|\psi_{j=0}^{(2+)}|$ are local minima. The two local minima $|\psi_{j=0}^{(1)}|$ and $|\psi_{j=0}^{(2+)}|$ are degenerate when condition (G.40) holds. Condition (G.40) defines the temperature $T_{\mathrm{c,r}}$. When $T_{\mathrm{c,r}} < T < T_{\mathrm{II-I}}$, it is $|\psi_{j=0}^{(1)}|$ that is the absolute minimum of the effective action (G.33b), $|\psi_{j=0}^{(+)}|$ is only a local minimum. When $T_{\mathrm{III-II}} < T < T_{\mathrm{c,r}}$, it is $|\psi_{j=0}^{(2+)}|$ that is the absolute minimum of the effective action (G.33b), $|\psi_{j=0}^{(1)}|$ is only a local minimum.

In the range of temperatures III that includes the $T \to 0$ limit, the quadratic renormalized coefficient $a_{\mathrm{r}}^2\, t$ of the effective action (G.33b) is negative,

$$\mathrm{III} := \left\{T \,|\, a_{\mathrm{t}}^2\, t < 0\right\}. \quad \text{(G.39)}$$

In the range of temperatures III (low T), the saddle-point solution $|\psi_{j=0}^{(2+)}|$ is an absolute minimum, while $|\psi_{j=0}^{(1)}|$ is a local maximum, of the effective action (G.33b) ($|\psi_{j=0}^{(2-)}|$ is unphysical for it is negative).

The condition

$$c_{\mathrm{r}}^4 = \frac{9}{2}\, b_{\mathrm{r}}^2\, a_{\mathrm{r}}^2\, t \Longleftrightarrow \tilde{S}_0(|\psi_{j=0}^{(1)}|) = \tilde{S}_0(|\psi_{j=0}^{(2+)}|) \quad \text{(G.40)}$$

defines implicitly the renormalized critical temperature $T_{\mathrm{c,r}} > T_{\mathrm{c,0}}$. The renormalized critical temperature $T_{\mathrm{c,r}}$ signals a discontinuous phase transition, for the location of the absolute minimum of the effective action (G.33b) jumps discontinuously from $|\psi_{j=0}^{(1)}|$ just above $T_{\mathrm{c,r}}$ to $|\psi_{j=0}^{(2+)}|$ just below $T_{\mathrm{c,r}}$.

It remains to verify the consistency of the solution $T_{\mathrm{c,r}}$ with regard to the assumption that this temperature satisfies the Ginzburg criterion (G.18b). To this end, we perform the estimate (in three-dimensional position space)

$$t_{\mathrm{r}} \approx \left.\frac{c_{\mathrm{r}}^4}{a_0^2\, b_0^2}\right|_{t=0}, \quad \text{(G.41)}$$

for the solution t_{r} of Eq. (G.40). It is convenient to express the dimensionless ratio on the right-hand side of Eq. (G.41) as the Ginzburg reduced temperature t_{G} defined in Eq. (G.18b) multiplied by some dimensionless number. To construct the latter dimensionless number, we take the ratio of two independent characteristic scales at our disposal. There is the correlation length ξ_0 defined after performing the mean-field approximation on Eq. (G.11). There is the screening length of the vector gauge propagator

$$\lambda_{\mathrm{s,0}} := k_{\mathrm{s,0}}^{-1}, \quad \text{(G.42)}$$

where the mean-field approximation $k_{\mathrm{s},0}$ of k_{s} was defined in Eq. (G.30). Equipped with the mean-field estimates for the correlation length associated to the breaking of $U(1)$ gauge invariance and the screening length of the corresponding $U(1)$ vector gauge field, we define their ratio

$$\kappa_0 := \frac{\lambda_{\mathrm{s},0}}{\xi_0}. \tag{G.43}$$

There follows the estimate

$$t_{\mathrm{r}} \approx \kappa_0^{-6} \times t_{\mathrm{G}}. \tag{G.44}$$

This estimate is consistent with the assumption that we may neglect variations in space of the order parameter ψ if

$$t_{\mathrm{r}} \gg t_{\mathrm{G}}, \tag{G.45}$$

i.e., if

$$\kappa_0 \ll 1. \tag{G.46}$$

When the consistency condition (G.46) is violated, renormalization-group techniques must be employed to perform an integration over the vector gauge fields so as to renormalize consistently the action for the field ψ.

G.3 Applications

The first application of section G.2 is to type I superconductors. We identify ψ with the s-wave order parameter of a superconductor and do the identifications

$$\begin{aligned} \gamma_0 &\to \frac{\hbar^2}{2\,m^*}, \qquad && m^* := 2\,m, \\ q_0 &\to \frac{e^*}{\hbar c}, && e^* = 2\,e. \end{aligned} \tag{G.47}$$

The correlation length ξ_0 is the mean-field coherence length [see Eq. (7.274)] and the screening length λ_0 is the mean-field London penetration depth [see Eq. (7.182)]. The ratio κ_0 defines a type I superconductor when $\kappa_0 < 1/\sqrt{2}$.

Type II superconductors have $\kappa_0 > 1/\sqrt{2}$. For type II superconductors, the coupling between the gradient of the order parameter and the vector gauge field cannot be neglected anymore. Halperin, Lubensky, and Ma have computed the one-loop flows of the couplings a_0^2, b_0^2, γ_0^2, $1/\mu_0^2$, and q_0 when the order parameter with two real-valued components is replaced by N real-valued components in a $4-\epsilon$ expansion about the upper critical dimension [136]. They find a non-trivial critical point under the condition that $N \gtrsim 366$ but none otherwise. They interpret the absence of a critical point for the one-loop flow when $N = 2$ as the signature of a first-order phase transition.

Halperin and Lubensky in Ref. [137] have also shown that the transition from the smectic A phase to the nematic phase in a liquid crystal is driven by fluctuations to be of first order. The crucial observation here is again that the order parameter couples to a vector gauge field (the local deviations of the director from the symmetry axis).[34]

Finally, Coleman and Weinberg had anticipated these results from condensed matter physics in their study of quantized gauge fields coupled to quantized scalar fields in a setting preserving the relativistic symmetry [139].

[34] Liquid crystals are composed of rod-like molecules. In the nematic phase of liquid crystals, nearest-neighbor rod-like molecules have their long axes approximately parallel to one another. The order parameter for this anisotropic structure is a dimensionless unit vector $\boldsymbol{n}$ called the director to account for the fact that $\boldsymbol{n}$ and $-\boldsymbol{n}$ are to be identified [138].

Appendix H

Useful identities

This appendix is an exercise in complex analysis allowing to prove Eq. (8.75). We work in units with $\eta = 1$ throughout this appendix.

H.1 Proof of Equation (8.75)

Let [recall Eqs. (8.71) and (8.56)]

$$\mathcal{K}_{\varpi_l} := \int_0^\infty \frac{\mathrm{d}\varpi}{\pi}\, \varpi \left(\frac{1}{\varpi} - \frac{\varpi}{\varpi_l^2 + \varpi^2} \right), \tag{H.1}$$

and let [recall Eq. (8.74)]

$$\mathcal{K}(\tau) := \frac{1}{\beta} \sum_{\varpi_l} \mathcal{K}_{\varpi_l}\, e^{-\mathrm{i}\varpi_l \tau}. \tag{H.2}$$

Observe that

$$\mathcal{K}_{\varpi_l} = +\mathcal{K}_{-\varpi_l} \Longleftrightarrow \mathcal{K}(\tau) = +\mathcal{K}(-\tau). \tag{H.3}$$

We are going to show that

$$\int_0^\beta \mathrm{d}\tau \int_0^\beta \mathrm{d}\tau'\, \mathcal{K}(\tau - \tau')\, f(\tau)\, g(\tau') = \int_{-\infty}^{+\infty} \mathrm{d}\tau \int_0^\beta \mathrm{d}\tau'\, \frac{[f(\tau) - g(\tau')]^2 - [f(\tau') - g(\tau')]^2}{4\pi\, |\tau - \tau'|^2} \tag{H.4}$$

holds for any pair of functions f and g periodic on $0 \leq \tau \leq \beta$.

H.1.1 *Proof of Eq. (H.5)*

Proof of

$$\frac{1}{\beta} \sum_{l \in \mathbb{Z}} e^{-\mathrm{i}\varpi_l \tau} = \sum_{m \in \mathbb{Z}} \delta(\tau - m\,\beta), \tag{H.5}$$

with $\varpi_l = \frac{2\pi}{\beta}\, l$.

Proof. Let $f(\tau)$ be β periodic but arbitrary otherwise. Define the Fourier coefficients f_{ϖ_l} by

$$f(\tau) =: \frac{1}{\sqrt{\beta}} \sum_{l\in\mathbb{Z}} f_{\varpi_l}\, e^{-\mathrm{i}\varpi_l \tau} \iff f_{\varpi_l} := \frac{1}{\sqrt{\beta}} \int\limits_0^\beta \mathrm{d}\tau\, f(\tau)\, e^{+\mathrm{i}\varpi_l\tau}. \tag{H.6}$$

Needed is the identity

$$\int\limits_0^\beta \mathrm{d}\tau\, f(\tau)\, \frac{1}{\beta} \sum_{l\in\mathbb{Z}} e^{-\mathrm{i}\varpi_l\tau} = \int\limits_0^\beta \mathrm{d}\tau\, f(\tau) \sum_{m\in\mathbb{Z}} \delta(\tau - m\,\beta). \tag{H.7}$$

On the one hand,

$$\int\limits_0^\beta \mathrm{d}\tau\, f(\tau)\, \frac{1}{\beta} \sum_{l\in\mathbb{Z}} e^{-\mathrm{i}\varpi_l\tau} = \frac{1}{\sqrt{\beta}} \sum_{l\in\mathbb{Z}} f_{-\varpi_l} = \frac{1}{\sqrt{\beta}} \sum_{l\in\mathbb{Z}} f_{+\varpi_l}. \tag{H.8}$$

On the other hand,

$$\begin{aligned}
\int\limits_0^\beta \mathrm{d}\tau\, f(\tau) \sum_{m\in\mathbb{Z}} \delta(\tau - m\,\beta) &= \sum_{m\in\mathbb{Z}} \int\limits_0^\beta \mathrm{d}\tau\, f(\tau)\, \delta(\tau - m\,\beta) \\
&= \sum_{m\in\mathbb{Z}} \int\limits_{m\,\beta}^{(m+1)\,\beta} \mathrm{d}y\, f(y + m\,\beta)\, \delta(y) \\
&= f(0) + \sum_{0\neq m\in\mathbb{Z}} \int\limits_{m\,\beta}^{(m+1)\,\beta} \mathrm{d}y\, f(y + m\,\beta)\, \delta(y) \\
&= f(0) \\
&= \frac{1}{\sqrt{\beta}} \sum_{l\in\mathbb{Z}} f_{+\varpi_l}.
\end{aligned} \tag{H.9}$$

We have proven

$$\frac{1}{\beta} \sum_{l\in\mathbb{Z}} e^{-\mathrm{i}\varpi_l\tau} = \sum_{m\in\mathbb{Z}} \delta(\tau - m\,\beta). \tag{H.10}$$

□

H.1.2 *Proof of Eq. (H.11)*

Proof of

$$\frac{1}{\beta} \sum_{l\in\mathbb{Z}} \frac{e^{-\mathrm{i}\varpi_l\tau}}{\varpi_l^2 + \varpi^2} = \frac{\cosh\left(\varpi\left[|\tau| - (\beta/2)\right]\right)}{2\,\varpi \sinh\left(\frac{\beta\,\varpi}{2}\right)}, \tag{H.11}$$

with $\varpi_l = \frac{2\,\pi}{\beta} l$.

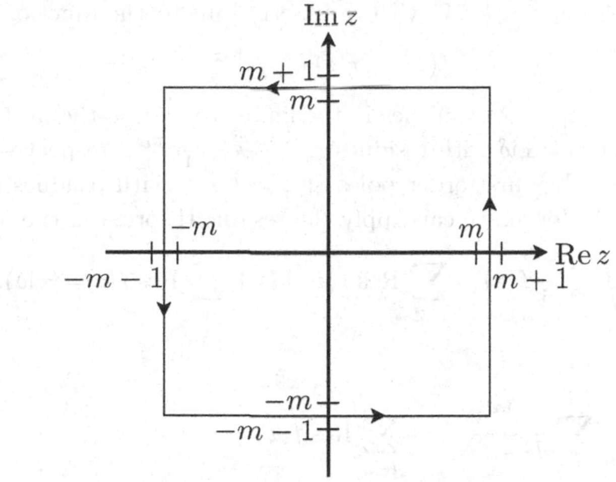

Fig. H.1 Definition of integration path Γ_m.

Proof. With the short-hand notation

$$\alpha := 2\pi\,\frac{|\tau|}{\beta}, \qquad \delta := \frac{\beta\,\varpi}{2\pi}, \tag{H.12}$$

we have

$$\frac{1}{\beta}\sum_{l\in\mathbb{Z}}\frac{e^{-\mathrm{i}\varpi_l\tau}}{\varpi_l^2+\varpi^2} = \frac{1}{\beta}\sum_{l\in\mathbb{Z}}\frac{e^{-\mathrm{i}2\pi l(\tau/\beta)}}{(2\pi/\beta)^2\,l^2+\varpi^2} = \frac{\beta}{(2\pi)^2}\sum_{l\in\mathbb{Z}}\frac{e^{\mathrm{i}\alpha l}}{l^2+\delta^2}. \tag{H.13}$$

Let

$$\Gamma_m = \gamma_m + \gamma'_m + \gamma''_m + \gamma'''_m \tag{H.14}$$

be the square path in the complex plane with the lower left and lower right corners (see Fig. H.1)

$$\left(m+\frac{1}{2}\right)(-1-\mathrm{i}), \qquad \left(m+\frac{1}{2}\right)(+1-\mathrm{i}), \tag{H.15a}$$

and the upper right and upper left corners

$$\left(m+\frac{1}{2}\right)(+1+\mathrm{i}), \quad \left(m+\frac{1}{2}\right)(-1+\mathrm{i}), \tag{H.15b}$$

respectively. The auxiliary function

$$r(z) := \frac{1}{z^2+\delta^2} \tag{H.16}$$

has two first-order poles at $z = \pm\mathrm{i}\delta$ with residues $\pm 1/(2\mathrm{i}\delta)$, respectively. The auxiliary function

$$u(z) := \frac{2\pi\mathrm{i}}{e^{2\pi\mathrm{i}z}-1} \tag{H.17}$$

has first-order poles at $z = l \in \mathbb{Z}$ with residue 1. Finally, the function

$$f(z) := r(z)\, u(z)\, e^{\mathrm{i}\alpha z} \tag{H.18}$$

decays like $1/z^2$ as $|z| \to \infty$, vanishes in the limit $m \to \infty$ on the path Γ_m, has two first-order poles at $z = \pm \mathrm{i}\delta$ with residues $\pm\frac{1}{2\mathrm{i}\delta}\frac{2\pi\mathrm{i}}{e^{\mp 2\pi\delta}-1}e^{\mp\alpha\delta}$, respectively, along the imaginary axis, and has first-order poles at $z = l \in \mathbb{Z}$ with residues $e^{\mathrm{i}\alpha l}/(l^2+\delta^2)$ along the real axis. Hence, we can apply the residue theorem in the form

$$0 = \lim_{m\to\infty} \int\limits_{\Gamma_m} \frac{\mathrm{d}z}{2\pi\mathrm{i}} f(z) = \sum_{l\in\mathbb{Z}} \operatorname{Res} f(z=l) + \sum_{\pm} \operatorname{Res} f(z=\pm\mathrm{i}\delta). \tag{H.19}$$

This gives

$$\begin{aligned}
\sum_{l\in\mathbb{Z}} \frac{e^{\mathrm{i}\alpha l}}{l^2+\delta^2} &= +\sum_{l\in\mathbb{Z}} \operatorname{Res} f(z=l) \\
&= -\sum_{\pm} \operatorname{Res} f(z=\pm\mathrm{i}\delta) \\
&= -\sum_{\pm} \frac{(\pm 1)}{2\mathrm{i}\delta}\,\frac{2\pi\mathrm{i}}{e^{\mp 2\pi\delta}-1}\, e^{\mp\alpha\delta} \\
&= -\sum_{\pm} \frac{(\pm 1)}{2\mathrm{i}\delta}\,\frac{2\pi\mathrm{i}e^{\pm\pi\delta}}{e^{\mp\pi\delta}-e^{\pm\pi\delta}}\, e^{\mp\alpha\delta} \\
&= -\frac{\pi}{\delta}\sum_{\pm} \frac{e^{\pm\pi\delta\mp\alpha\delta}}{(\pm 2)\sinh(\mp\pi\delta)} \\
&= +\frac{\pi}{\delta}\,\frac{\sum\limits_{\pm} e^{\pm\pi\delta\mp\alpha\delta}}{2\sinh(\pi\delta)} \\
&= +\frac{\pi}{\delta}\,\frac{\cosh\big((\pi-\alpha)\,\delta\big)}{\sinh(\pi\delta)}.
\end{aligned} \tag{H.20}$$

Hence, since $\pi\,\delta = \beta\,\varpi/2$ and $\alpha\,\delta = |\tau|\,\varpi$,

$$\begin{aligned}
\frac{1}{\beta}\sum_{l\in\mathbb{Z}} \frac{e^{-\mathrm{i}\varpi_l\tau}}{\varpi_l^2+\varpi^2} &= \frac{\beta}{(2\pi)^2}\sum_{l\in\mathbb{Z}} \frac{e^{\mathrm{i}\alpha l}}{l^2+\delta^2} \\
&= \frac{\beta}{(2\pi)^2}\,\frac{\pi}{\delta}\,\frac{\cosh\big((\pi-\alpha)\,\delta\big)}{\sinh(\pi\delta)} \\
&= \frac{\beta}{(2\pi)^2}\,\frac{2\pi^2}{\beta\varpi}\,\frac{\cosh\big((\beta/2-|\tau|)\,\varpi\big)}{\sinh(\beta\varpi/2)} \\
&= \frac{1}{2\varpi}\,\frac{\cosh\big((\beta/2-|\tau|)\,\varpi\big)}{\sinh(\beta\varpi/2)}.
\end{aligned} \tag{H.21}$$

□

H.1.3 *Proof of Eq. (H.22)*

Proof of

$$\sum_{m\in\mathbb{Z}} e^{-|\tau+m\,\beta|\,\varpi} = \frac{\cosh\Big(\big[|\tau|-(\beta/2)\big]\varpi\Big)}{\sinh\left(\frac{\beta\,\varpi}{2}\right)}, \tag{H.22}$$

with $\varpi > 0$.

Proof. As a first warm up, consider the case when $\tau = 0$,

$$\begin{aligned}\sum_{m\in\mathbb{Z}} e^{-|m\,\beta|\,\varpi} &= 2\sum_{m=0}^{\infty} e^{-m\,\beta\,\varpi} - 1\\ &= \frac{2}{1-e^{-\beta\,\varpi}} - 1\\ &= \frac{2e^{+\beta\,\varpi/2}}{e^{+\beta\,\varpi/2}-e^{-\beta\,\varpi/2}} - 1\\ &= \frac{e^{+\beta\,\varpi/2}+e^{-\beta\,\varpi/2}}{e^{+\beta\,\varpi/2}-e^{-\beta\,\varpi/2}}\\ &= \frac{\cosh(\beta\,\varpi/2)}{\sinh(\beta\,\varpi/2)}.\end{aligned} \tag{H.23}$$

As a second warm up, consider the case when $\tau = \beta$,

$$\begin{aligned}\sum_{m\in\mathbb{Z}} e^{-|(m+1)\,\beta|\,\varpi} &= \sum_{n\in\mathbb{Z}} e^{-|n\,\beta|\,\varpi}\\ &= \frac{\cosh(\beta\,\varpi/2)}{\sinh(\beta\,\varpi/2)}.\end{aligned} \tag{H.24}$$

Next, consider the case $0 < \tau < \beta$ for which

$$\begin{aligned}\sum_{m\in\mathbb{Z}} e^{-|\tau+m\,\beta|\,\varpi} &= e^{-\tau\,\varpi} + e^{-\tau\,\varpi}\sum_{m=1}^{\infty} e^{-m\,\beta\,\varpi} + e^{+\tau\,\varpi}\sum_{m=1}^{\infty} e^{-m\,\beta\,\varpi}\\ &= e^{-\tau\,\varpi} + e^{-\tau\,\varpi}\left(\frac{1}{1-e^{-\beta\,\varpi}}-1\right) + e^{+\tau\,\varpi}\left(\frac{1}{1-e^{-\beta\,\varpi}}-1\right)\\ &= e^{-\tau\,\varpi}\left(1+\frac{e^{-\beta\,\varpi/2}}{e^{+\beta\,\varpi/2}-e^{-\beta\,\varpi/2}}\right) + e^{+\tau\,\varpi}\frac{e^{-\beta\,\varpi/2}}{e^{+\beta\,\varpi/2}-e^{-\beta\,\varpi/2}}\\ &= \frac{e^{-\tau\,\varpi+\beta\,\varpi/2}+e^{+\tau\,\varpi-\beta\,\varpi/2}}{e^{+\beta\,\varpi/2}-e^{-\beta\,\varpi/2}}\\ &= \frac{\cosh(\tau\,\varpi-\beta\,\varpi/2)}{\sinh(\beta\,\varpi/2)}.\end{aligned} \tag{H.25}$$

Finally, consider the case $-\beta < \tau < 0$ for which

$$\begin{aligned}\sum_{m\in\mathbb{Z}} e^{-|\tau+m\,\beta|\,\varpi} &= e^{-|\tau|\,\varpi} + e^{+|\tau|\,\varpi}\sum_{m=1}^{\infty} e^{-m\,\beta\,\varpi} + e^{-|\tau|\,\varpi}\sum_{m=1}^{\infty} e^{-m\,\beta\,\varpi}\\ &= \frac{\cosh(|\tau|\,\varpi-\beta\,\varpi/2)}{\sinh(\beta\,\varpi/2)}.\end{aligned} \tag{H.26}$$

Since the case $|\tau| > \beta$ can always be reduced to the four special cases studied above by a change of summation variable, we have proven that

$$\sum_{m\in\mathbb{Z}} e^{-|\tau+m\,\beta|\,\varpi} = \frac{\cosh\left(\left[|\tau| - (\beta/2)\right]\varpi\right)}{\sinh\left(\frac{\beta\,\varpi}{2}\right)}. \tag{H.27}$$

□

H.1.4 *Proof of Eq. (H.28)*

Proof of

$$\mathcal{K}(\tau) = \int_0^\infty \frac{\mathrm{d}\varpi}{\pi}\,\varpi \sum_{m\in\mathbb{Z}} \left[\frac{1}{\varpi}\delta(\tau - m\,\beta) - \frac{1}{2}e^{-\varpi\,|\tau+m\,\beta|}\right]. \tag{H.28}$$

Proof. By Eqs. (H.2) and (H.1),

$$\begin{aligned}\mathcal{K}(\tau) &\equiv \frac{1}{\beta}\sum_{l\in\mathbb{Z}} \mathcal{K}_{\varpi_l}\, e^{-\mathrm{i}\varpi_l\tau}\\ &= \int_0^\infty \frac{\mathrm{d}\varpi}{\pi}\frac{1}{\beta}\sum_{l\in\mathbb{Z}} e^{-\mathrm{i}\varpi_l\tau} - \int_0^\infty \frac{\mathrm{d}\varpi}{\pi}\,\varpi^2\,\frac{1}{\beta}\sum_{l\in\mathbb{Z}} \frac{e^{-\mathrm{i}\varpi_l\tau}}{\varpi_l^2+\varpi^2}.\end{aligned} \tag{H.29}$$

By Eqs. (H.5), (H.11), and (H.22),

$$\begin{aligned}\mathcal{K}(\tau) &= \int_0^\infty \frac{\mathrm{d}\varpi}{\pi}\sum_{m\in\mathbb{Z}} \delta(\tau - m\,\beta) - \int_0^\infty \frac{\mathrm{d}\varpi}{\pi}\,\varpi^2\,\frac{\cosh\left(\varpi\left[|\tau| - (\beta/2)\right]\right)}{2\varpi\sinh\left(\frac{\beta\,\varpi}{2}\right)}\\ &= \int_0^\infty \frac{\mathrm{d}\varpi}{\pi}\sum_{m\in\mathbb{Z}} \delta(\tau - m\,\beta) - \int_0^\infty \frac{\mathrm{d}\varpi}{2\,\pi}\,\varpi\sum_{m\in\mathbb{Z}} e^{-\varpi|\tau+m\,\beta|}.\end{aligned} \tag{H.30}$$

□

H.1.5 *Proof of Eq. (H.31)*

Proof of

$$\begin{aligned}\int_0^\beta \mathrm{d}\tau \int_0^\beta \mathrm{d}\tau'\,\mathcal{K}(\tau-\tau')\,f(\tau)\,g(\tau') = &\int_{-\infty}^{+\infty} \mathrm{d}\tau \int_0^\beta \mathrm{d}\tau' \left(\int_0^\infty \frac{\mathrm{d}\varpi}{4\,\pi}\,\varpi\, e^{-\varpi\,|\tau-\tau'|}\right)\\ &\times\left\{\left[f(\tau) - g(\tau')\right]^2 - \left[f(\tau') - g(\tau')\right]^2\right\},\end{aligned} \tag{H.31}$$

for any pair of functions f and g periodic on $0 \leq \tau \leq \beta$.

Proof. This identity follows from Eq. (H.28) as we now show. First,

$$
\begin{aligned}
&\int_0^\beta \mathrm{d}\tau \int_0^\beta \mathrm{d}\tau' \, \mathcal{K}(\tau-\tau')\, f(\tau)\, g(\tau') \\
&\quad = \int_0^\beta \mathrm{d}\tau \int_0^\beta \mathrm{d}\tau' \int_0^\infty \frac{\mathrm{d}\varpi}{\pi}\, \varpi \sum_{m\in\mathbb{Z}} \frac{1}{\varpi}\, \delta(\tau-\tau'-m\,\beta)\, f(\tau)\, g(\tau') \\
&\qquad - \int_0^\beta \mathrm{d}\tau \int_0^\beta \mathrm{d}\tau' \int_0^\infty \frac{\mathrm{d}\varpi}{\pi}\, \varpi \sum_{m\in\mathbb{Z}} \frac{1}{2} e^{-\varpi|\tau-\tau'+m\,\beta|}\, f(\tau)\, g(\tau') \\
&\quad = \int_0^\beta \mathrm{d}\tau' \int_0^\infty \frac{\mathrm{d}\varpi}{\pi}\, f(\tau')\, g(\tau') \\
&\qquad - \int_0^\beta \mathrm{d}\tau \int_0^\beta \mathrm{d}\tau' \int_0^\infty \frac{\mathrm{d}\varpi}{\pi}\, \varpi \sum_{m\in\mathbb{Z}} \frac{1}{2} e^{-\varpi|\tau-\tau'+m\,\beta|}\, f(\tau)\, g(\tau').
\end{aligned}
\tag{H.32}
$$

We momentarily concentrate on the second line on the right-hand side of Eq. (H.32). We do the change of integration variable $\tau+m\,\beta \to \tau$ and make use of $f(\tau-m\,\beta)=f(\tau)$ for any integer m to rewrite the second line on the right-hand side of Eq. (H.32) as

$$
\begin{aligned}
&\sum_{m\in\mathbb{Z}} \int_{m\,\beta}^{(m+1)\,\beta} \mathrm{d}\tau \int_0^\beta \mathrm{d}\tau' \int_0^\infty \frac{\mathrm{d}\varpi}{\pi}\, \varpi\, \frac{1}{2}\, e^{-\varpi|\tau-\tau'|}\, f(\tau)\, g(\tau') \\
&\quad = \int_{\mathbb{R}} \mathrm{d}\tau \int_0^\beta \mathrm{d}\tau' \int_0^\infty \frac{\mathrm{d}\varpi}{2\,\pi}\, \varpi\, e^{-\varpi|\tau-\tau'|}\, f(\tau)\, g(\tau').
\end{aligned}
\tag{H.33}
$$

We proceed by bringing the first line on the right-hand side of Eq. (H.32) to a form as close as possible to the right-hand side of Eq. (H.33). To this end, we use the identity

$$
\frac{1}{\varpi} = \int_0^\infty \mathrm{d}\tau\, e^{-\varpi\tau} = \frac{1}{2} \int_{\mathbb{R}} \mathrm{d}\tau\, e^{-\varpi|\tau|}
\tag{H.34}
$$

to rewrite the first line on the right-hand side of Eq. (H.32) as

$$
\int_{\mathbb{R}} \mathrm{d}\tau \int_0^\beta \mathrm{d}\tau' \int_0^\infty \frac{\mathrm{d}\varpi}{2\,\pi}\, \varpi\, e^{-\varpi|\tau-\tau'|}\, f(\tau')\, g(\tau').
\tag{H.35}
$$

Equation (H.32) thus takes the more symmetric form

$$
\int_0^\beta \mathrm{d}\tau \int_0^\beta \mathrm{d}\tau' \, \mathcal{K}(\tau-\tau')\, f(\tau)\, g(\tau') = \int_{\mathbb{R}} \mathrm{d}\tau \int_0^\beta \mathrm{d}\tau' \int_0^\infty \frac{\mathrm{d}\varpi}{2\,\pi}\, \varpi\, e^{-\varpi|\tau-\tau'|} \left[f(\tau')\, g(\tau') - f(\tau)\, g(\tau')\right].
\tag{H.36}
$$

To close the proof, we do the manipulation

$$f(\tau')\,g(\tau') - f(\tau)\,g(\tau') = f(\tau')\,g(\tau') - \frac{1}{2}\left[f^2(\tau) + g^2(\tau')\right] + \frac{1}{2}\left[f(\tau) - g(\tau')\right]^2. \tag{H.37}$$

We may now use the evenness of the kernel $\mathcal{K}(\tau)$, see Eq. (H.3), to deduce that

$$\int\limits_0^\beta \mathrm{d}\tau \int\limits_0^\beta \mathrm{d}\tau'\, \mathcal{K}(\tau-\tau')\, f^2(\tau) = \int\limits_0^\beta \mathrm{d}\tau \int\limits_0^\beta \mathrm{d}\tau'\, \mathcal{K}(\tau-\tau')\, f^2(\tau'). \tag{H.38}$$

Hence, we are allowed to do the substitution

$$\begin{aligned} f(\tau')\,g(\tau') - f(\tau)\,g(\tau') &= f(\tau')\,g(\tau') - \frac{1}{2}\left[f^2(\tau) + g^2(\tau')\right] + \frac{1}{2}\left[f(\tau) - g(\tau')\right]^2 \\ &\rightarrow f(\tau')\,g(\tau') - \frac{1}{2}\left[f^2(\tau') + g^2(\tau')\right] + \frac{1}{2}\left[f(\tau) - g(\tau')\right]^2 \\ &= \frac{1}{2}\left[f(\tau) - g(\tau')\right]^2 - \frac{1}{2}\left[f(\tau') - g(\tau')\right]^2 \end{aligned} \tag{H.39}$$

in the integrand on the right-hand side of Eq. (H.36). Identity (H.31) follows. □

H.1.6 *Proof of Eq. (H.40)*

Proof of

$$\int\limits_0^\infty \frac{\mathrm{d}\varpi}{4\pi}\, \varpi\, e^{-\varpi\,|\tau-\tau'|} = \frac{1}{4\pi}\,\frac{1}{|\tau-\tau'|^2}. \tag{H.40}$$

Proof. This identity is established with the help of partial integration,

$$\begin{aligned} \int\limits_0^\infty \frac{\mathrm{d}\varpi}{4\pi}\, \varpi\, e^{-\varpi\,|\tau|} &= \frac{1}{4\pi}\left(\varpi\, \frac{(-1)}{|\tau|}\, e^{-\varpi\,|\tau|}\Big|_0^\infty - \int\limits_0^\infty \mathrm{d}\varpi\, \frac{(-1)}{|\tau|}\, e^{-\varpi\,|\tau|} \right) \\ &= \frac{1}{4\pi}\left(0 - \frac{1}{|\tau|^2}\, e^{-\varpi\,|\tau|}\Big|_0^\infty\right) \\ &= \frac{1}{4\pi|\tau|^2}. \end{aligned} \tag{H.41}$$

□

Appendix I

Non-Abelian bosonization

I.1 Introduction

We have bosonized the xxz spin-1/2 chain in section 9.3.2. At the Heisenberg point defined by the condition $J_{\perp} = J_z$, Hamiltonian (9.109) is invariant under an arbitrary global rotation of all spin operators. This is the $SU(2)$ spin symmetry. This symmetry is not manifest in the bosonized Lagrangian density (9.139).

More generally, the Abelian bosonization rules (9.107) fail to manifest explicitly any internal continuous symmetry that a Dirac Lagrangian might support. The non-Abelian bosonization rules for free massless Dirac fermions discovered by Witten in Ref. [111] remedy this deficiency.

We are going to derive these non-Abelian bosonization rules by generalizing the functional bosonization method of chapter 9 to Dirac fermions interacting with non-Abelian gauge fields taking values in the Lie algebra $u(N) = u(1) \oplus su(N)$. This method was introduced by Polyakov and Wiegmann in Ref. [112]. But before doing so, we are going to proceed heuristically by drawing analogies with the Abelian bosonization rules.

I.2 Minkowski versus Euclidean spaces

In this appendix, we shall work in $(1+1)$-dimensional Minkowski space as opposed to working in two–dimensional Euclidean space as we did in chapter 9. We briefly review the conventions that we are using to define Minkowski space and how we choose to relate Minkowski and Euclidean spaces.

We shall denote $(d+1)$-dimensional Minkowski space and time by $\mathbb{M}_{1,d}$. Minkowski space $\mathbb{M}_{1,d}$ is a $(d+1)$-dimensional vector space over the field of real numbers $\mathbb{R}$ with the Lorentz scalar product. If we denote the pair $a = (a^{\mu})$ and $b = (b^{\nu})$ of vectors from $\mathbb{M}_{1,d}$ where the labels $\mu, \nu = 0, \cdots, d$, then their Lorentz scalar product is

$$a\,b := a^{\mu}\, g_{\mu\nu}\, b^{\nu}. \tag{I.1}$$

Here, the Lorentz metric is represented by a diagonal $(d+1)\times(d+1)$ matrix $I_{1,d}$ with the components $g_{\mu\nu}$, whereby $\mu,\nu=0,\cdots,d$ and $g_{00}=1$, $g_{11}=\cdots=g_{dd}=-1$. The summation convention over repeated indices is implied. A component with an upper index, say a^{μ}, of a is referred to as a contravariant component. A lower one, say a_{ν}, is referred to as a covariant component. Contravariant components are related to covariant ones by contraction with the Lorentz metric, $a^{\mu}=g^{\mu\nu}\,a_{\nu}$. We may thus write

$$a=(a^0,\boldsymbol{a}),\qquad b=(b^0,\boldsymbol{b}),\qquad a\,b=a_{\mu}\,b^{\mu}=a^0\,b^0-\boldsymbol{a}\cdot\boldsymbol{b},\tag{I.2}$$

where $\boldsymbol{a}$ and $\boldsymbol{b}$ are a pair of vectors from the Euclidean space $\mathbb{R}^d$. We reserve the notation $\boldsymbol{a}\cdot\boldsymbol{b}\equiv a^1\,b^1+\cdots+a^d\,b^d$ to denote the Euclidean scalar product.

The Lorentz metric $I_{1,d}$ of Minkowski space $\mathbb{M}_{1,d}$ defines a bilinear form that is invariant under the group of Lorentz transformations $O(1,d)$, i.e., if the $(d+1)\times(d+1)$-dimensional matrix $L=(L^{\mu}_{\nu})$ with $\mu,\nu=0,\cdots,d$ is a Lorentz transformation, then

$$L^{\mu}_{\mu'}\,g_{\mu\nu}\,L^{\nu}_{\nu'}=g_{\mu'\nu'}\Longleftrightarrow L^{\mathsf{T}}\,I_{1,d}\,L=I_{1,d},\tag{I.3}$$

for any $\mu',\nu'=0,\cdots,d$. In contrast, the Euclidean (Kronecker) metric of Euclidean space $\mathbb{R}^d$ defines a bilinear form that is invariant under the group of orthogonal transformations $O(d)$, i.e., if the $d\times d$-dimensional matrix $O=(O_{ij})$ with $i,j=1,\cdots,d$ is an orthogonal transformation, then

$$O_{ii'}\,\delta_{ij}\,O_{jj'}=\delta_{i'j'}\Longleftrightarrow O^{\mathsf{T}}\,I_d\,O=I_d,\tag{I.4}$$

where I_d is the $d\times d$ unit matrix, for any $i',j'=1,\cdots,d$. We observe that both Lorentz and orthogonal transformations from $O(1,d)$ and $O(d)$, respectively, have determinants that square to unity. Since Lorentz and orthogonal transformations are real-valued matrices, their determinants are either -1 or $+1$.

Let A define a one-to-one linear mapping from $\mathbb{M}_{1,d}$ to $\mathbb{M}_{1,d}$. It follows that A can be represented by a $(d+1)\times(d+1)$-dimensional matrix $(A^{\mu\nu})$ with $\mu,\nu=0,\cdots,d$ with the non-vanishing determinant

$$\det A=\sum_{\sigma\in\mathcal{S}_{d+1}}\operatorname{sgn}(\sigma)\,A^{0\sigma(0)}\cdots A^{d\sigma(d)}.\tag{I.5}$$

Here, $\mathcal{S}_{d+1}$ is the group of permutations of $(d+1)$ objects and $\operatorname{sgn}(\sigma)=\pm1$ is the parity of the permutation σ. Denote with $\epsilon_{\mu_0\cdots\mu_d}$ the fully antisymmetric Levi-Civita covariant tensor by

$$\epsilon_{\mu_0\cdots\mu_d}=\operatorname{sgn}(\sigma),\tag{I.6}$$

where σ is the permutation that takes $0,\cdots,d$ to $\mu_0,\cdots,\mu_d$. We may then write

$$\det A=\epsilon_{\mu_0\cdots\mu_d}A^{0\mu_0}\cdots A^{d\mu_d}.\tag{I.7a}$$

Since $\det L^{\mathsf{T}}=\det L=\pm1$ for any $L\in O(1,d)$. Similarly, it follows that

$$\det\left(L^{\mathsf{T}}\,A\,L\right)=\det A.\tag{I.7b}$$

Moreover, for any choice of $a_0, \cdots, a_d \in \mathbb{M}_{1,d}$,

$$\det\,(a_0,\cdots,a_d) \equiv \epsilon_{\mu_0\cdots\mu_d}\, a_0^{\mu_0}\cdots a_d^{\mu_d} \tag{I.8a}$$

is a Lorentz pseudo-scalar, since

$$\det\,(L\,a_0,\cdots,L\,a_d) = \det L \times \det\,(a_0,\cdots,a_d)\,, \tag{I.8b}$$

for any $L \in O(1,d)$. Finally, we observe that

$$\epsilon^{\mu_0\cdots\mu_d} = g^{\mu_0\nu_0}\cdots g^{\mu_d\nu_d}\,\epsilon_{\nu_0\cdots\nu_d} = (-1)^d\,\epsilon_{\mu_0\cdots\mu_d}. \tag{I.9}$$

There are two possible ways to relate Minkowski $\mathbb{M}_{1,2}$ to Euclidean space $\mathbb{R}^2$. First, if $(x_1,x_2) \equiv (x^1,x^2)$ denote the Euclidean coordinates, then the Wick rotation

$$x_1 \to x^1 = -x_1, \qquad x_2 \to +\mathrm{i}x^0 = +\mathrm{i}x_0, \tag{I.10a}$$

is one possibility to define the Minkowski coordinates. By this Wick rotation from Euclidean space $\mathbb{R}^2$ to Minkowski space $\mathbb{M}_{1,1}$,

$$a_\mu\, b_\mu \to -a_\mu\, b^\mu, \qquad \epsilon_{\mu\nu} a_\mu\, b_\nu \to -\mathrm{i}\epsilon_{\mu\nu} a^\mu\, b^\nu, \qquad a_\mu := \epsilon_{\mu\nu}\partial_\nu\phi \to a^\mu := \mathrm{i}\epsilon^{\mu\nu}\partial_\nu\phi. \tag{I.10b}$$

Second, if $(x_1,x_2) \equiv (x^1,x^2)$ denote the Euclidean coordinates, then the Wick rotation

$$x_1 \to x^0 = +x_0, \qquad x_2 \to +\mathrm{i}x^1 = -\mathrm{i}x_1, \tag{I.11a}$$

is another possibility to define the Minkowski coordinates. By this Wick rotation from Euclidean space $\mathbb{R}^2$ to Minkowski space $\mathbb{M}_{1,1}$,

$$a_\mu\, b_\mu \to +a_\mu\, b^\mu, \qquad \epsilon_{\mu\nu} a_\mu\, b_\nu \to +\mathrm{i}\epsilon_{\mu\nu} a^\mu\, b^\nu, \qquad a_\mu := \epsilon_{\mu\nu}\partial_\nu\phi \to a^\mu := \mathrm{i}\epsilon^{\mu\nu}\partial_\nu\phi. \tag{I.11b}$$

It is this second option that we choose, for the two components of the current (9.3) become

$$j_1 = \bar\psi\,\sigma_1\psi = \psi^\dagger\,\sigma_1\,\sigma_1\psi = \psi^\dagger\,\psi \to j^0 = \bar\psi\,\gamma^0\psi = \psi^\dagger\,\psi \tag{I.12a}$$

and

$$j_2 = \bar\psi\,\sigma_2\psi = \psi^\dagger\,\sigma_1\,\sigma_2\psi = \psi^\dagger\,\mathrm{i}\sigma_3\psi \to +\mathrm{i}\,j^1 = +\mathrm{i}\,\bar\psi\,\gamma^1\psi = +\mathrm{i}\,\psi^\dagger\,\gamma^0\,\gamma^1\psi, \tag{I.12b}$$

respectively, where $\gamma^0 \equiv \sigma^1$ and $\gamma^1 \equiv -\mathrm{i}\sigma^2$. This is the definition for the Abelian conserved Dirac current that we are going to use in Minkowski space $\mathbb{M}_{1,1}$.

I.3 Free massless Dirac fermions and the Wess-Zumino-Witten theory

A massless Dirac fermion that propagates freely in $(1+1)$-dimensional Minkowski space and time is described by the partition function

$$Z_0 := \int \mathcal{D}[\bar{\psi}]\,\mathcal{D}[\psi]\,e^{\mathrm{i}S_0}, \tag{I.13a}$$

with the action

$$S_0 := \int \mathrm{d}x^0\mathrm{d}x^1\,\mathcal{L}_0, \tag{I.13b}$$

and the Lagrangian density

$$\mathcal{L}_0 := \bar{\psi}\,\gamma^\mu\,\mathrm{i}\partial_\mu\,\psi. \tag{I.13c}$$

The independent Grassmann degrees of freedom $\bar{\psi}$ and ψ describe a two-component spinor. The 2×2 matrices $\gamma^\mu = (\gamma^0, \gamma^1)$ obey the anticommutation relations

$$\{\gamma^\mu, \gamma^\nu\} = 2\,g^{\mu\nu}, \qquad \mu,\nu = 0,1. \tag{I.14a}$$

We shall also make use of the 2×2 matrix

$$\gamma^5 := \gamma^0\gamma^1 = -\gamma_0\gamma_1, \tag{I.14b}$$

which squares to the unit 2×2 matrix owing to the Clifford algebra (I.14a). One possible representation of the Clifford algebra (I.14) is the Weyl representation $\gamma^0 = \begin{pmatrix} 0 & 1 \\ 1 & 0 \end{pmatrix} \equiv \sigma^1$, $\gamma^1 = \begin{pmatrix} 0 & -1 \\ +1 & 0 \end{pmatrix} \equiv -\mathrm{i}\sigma^2$, and $\gamma^5 = \begin{pmatrix} +1 & 0 \\ 0 & -1 \end{pmatrix} \equiv \sigma^3$.

We introduce the light-cone coordinates

$$x^\pm := x^0 \pm x^1 = x_0 \mp x_1 =: x_\mp \tag{I.15}$$

that we extend to any contravariant two-vector a^μ or any covariant two-vector b_μ. There follows the useful identities

$$g_{\mu\nu}\,a^\mu\,b^\nu = \frac{1}{2}\left(a_+\,b_- + a_-\,b_+\right) \tag{I.16a}$$

and

$$\epsilon_{\mu\nu}\,a^\mu\,b^\nu = \frac{1}{2}\left(a_+\,b_- - a_-\,b_+\right). \tag{I.16b}$$

We also define

$$\bar{\psi} =: \psi^\dagger\,\gamma^0, \tag{I.17a}$$

together with the chiral basis

$$\psi^\dagger_\pm := \psi^\dagger\,\frac{1}{2}\left(1\mp\gamma^5\right), \qquad \psi_\pm := \frac{1}{2}\left(1\mp\gamma^5\right)\psi, \tag{I.17b}$$

in terms of which the Lagrangian density (I.13c) becomes

$$\begin{aligned}\bar{\psi}\,\mathrm{i}\gamma^\mu\,\partial_\mu\,\psi &= \psi^\dagger\,\mathrm{i}\left(\partial_0 + \gamma^5\,\partial_1\right)\psi \\ &= \psi^\dagger_-\,\mathrm{i}\left(\partial_0+\partial_1\right)\psi_- + \psi^\dagger_+\,\mathrm{i}\left(\partial_0-\partial_1\right)\psi_+ \\ &\equiv \psi^\dagger_-\,\mathrm{i}\partial_+\psi_- + \psi^\dagger_+\,\mathrm{i}\partial_-\psi_+.\end{aligned} \tag{I.17c}$$

The two chiral sectors defined by Eq. (I.17) decouple in the partition function (I.13) of a massless Dirac fermion propagating freely in $(1+1)$-dimensional Minkowski space and time $\mathbb{M}_{1,1}$,

$$\begin{aligned} Z_0 &= Z_0^{(-)} \times Z_0^{(+)}, \\ Z_0^{(-)} &:= \int \mathcal{D}[\psi_-^\dagger]\,\mathcal{D}[\psi_-] e^{-\int \frac{\mathrm{d}x_-\mathrm{d}x_+}{2}\psi_-^\dagger\partial_+\psi_-}, \\ Z_0^{(+)} &:= \int \mathcal{D}[\psi_+^\dagger]\,\mathcal{D}[\psi_+] e^{-\int \frac{\mathrm{d}x_-\mathrm{d}x_+}{2}\psi_+^\dagger\partial_-\psi_+}. \end{aligned} \tag{I.18}$$

Hence, the equations of motion

$$\partial_-\,\psi_+ = 0, \qquad \partial_+\,\psi_- = 0, \tag{I.19}$$

follow. In turn, the currents

$$j_\mp := 2\,\psi_\mp^\dagger\,\psi_\mp \tag{I.20a}$$

obey the conservation laws

$$\partial_-\,j_+ = 0, \qquad \partial_+\,j_- = 0. \tag{I.20b}$$

If we borrow the results of section 9.2.1 and then perform the analytical continuation (I.11) from two-dimensional Euclidean space $\mathbb{R}^2$ to $(1+1)$-dimensional Minkowski space and time $\mathbb{M}_{1,1}$, we infer that the correlation functions of the conserved chiral currents (I.20) are identical with the correlation functions of the conserved currents[35] [recall Table 9.1 or Eq. (9.80) and the choice fixed by Eq. (I.12) to perform the Wick rotation]

$$j_\mu := \frac{1}{2\pi}\epsilon_{\mu\nu}\partial^\nu\phi \Longleftrightarrow j_- := +\frac{1}{2\pi}\partial_-\phi, \qquad j_+ := -\frac{1}{2\pi}\partial_+\phi, \tag{I.21}$$

defined with respect to the partition function

$$Z_0 := \int \mathcal{D}[\phi]\, e^{+\mathrm{i}S_0} \tag{I.22a}$$

with the action

$$S_0 := \int \mathrm{d}x^0\mathrm{d}x^1\mathcal{L}_0 \tag{I.22b}$$

and the Lagrangian density

$$\mathcal{L}_0 := \frac{1}{8\pi}(\partial_\mu\phi)(\partial^\mu\phi) \tag{I.22c}$$

for a real-valued massless scalar field propagating freely in $(1+1)$-dimensional Minkowski space and time $\mathbb{M}_{1,1}$.

Following Witten in Ref. [111], we define the field

$$U := e^{+\mathrm{i}\phi} \tag{I.23}$$

[35] Strictly speaking, we should use different symbols to denote the fermionic currents (I.20a) and the bosonic currents (I.21). The same comment applies to the partition function, action, and Lagrangian density defined in Eqs. (I.13) and (I.22) respectively.

that takes values on the unit circle S^1. The bosonic currents (I.21) become

$$\begin{aligned} j_\mu &:= -\tfrac{\mathrm{i}}{2\pi}\epsilon_{\mu\nu}U^{-1}\partial^\nu U \\ &= -\tfrac{\mathrm{i}}{2\pi}\epsilon_{\mu\nu}\left(\partial^\nu U\right)U^{-1} \end{aligned} \iff \begin{aligned} j_- &:= -\tfrac{\mathrm{i}}{2\pi}U^{-1}\partial_- U = -\tfrac{\mathrm{i}}{2\pi}\left(\partial_- U\right)U^{-1}, \\ j_+ &:= +\tfrac{\mathrm{i}}{2\pi}U^{-1}\partial_+ U = +\tfrac{\mathrm{i}}{2\pi}\left(\partial_+ U\right)U^{-1}. \end{aligned} \tag{I.24}$$

They obey the conservation laws

$$\partial_+ j_- = 0, \qquad \partial_- j_+ = 0, \tag{I.25a}$$

given the partition function

$$Z_0 := \int \mathcal{D}[U]\, e^{+\mathrm{i}S_0}, \tag{I.25b}$$

with the action

$$S_0 := \int \mathrm{d}x^0 \mathrm{d}x^1 \mathcal{L}_0, \tag{I.25c}$$

and the Lagrangian density

$$\mathcal{L}_0 := \frac{1}{8\pi}(\partial_\mu U^{-1})(\partial^\mu U). \tag{I.25d}$$

We have emphasized in Eq. (I.24) that the ordering between U^{-1} and ∂U is irrelevant because U is defined to be an element of the Abelian group $U(1) \sim S^1$. The measure $\mathcal{D}[U] \equiv \mathcal{D}[\phi]$ is called the Haar measure of the group $U(1)$. This Haar measure is invariant under the group multiplication from the left or from the right by an element that may depend on $x = (x^\mu)$.

We now generalize Eq. (I.24) by replacing the group element $U \in S^1$ with the group element $G \in U(N)$

$$J_- := -\frac{\mathrm{i}}{8\pi}\left(\partial_- G\right)G^{-1}, \qquad J_+ := +\frac{\mathrm{i}}{8\pi}G^{-1}\partial_+ G. \tag{I.26}$$

[The choice of the multiplicative factor $1/(8\pi)$ is tied to the choice of the normalization that we shall make below for the generators of the Lie algebra $u(N)$. It is only the relative sign on the right-hand sides of J_- and J_+ that matters. We could have equally well chosen the convention by which the right-hand sides of J_- and J_+ are simultaneously multiplied by -1.] We ask the following two questions. (i) Is there a partition function for the group-valued field G [here $G \in U(N) = U(1) \times SU(N)$] for which there follows the equations of motion

$$\partial_+ J_- = 0, \qquad \partial_- J_+ = 0, \tag{I.27}$$

for the $N \times N$ Lie-algebra-valued currents J_- and J_+ [here the Lie algebra is $u(N) = u(1) \oplus su(N)$]? (ii) With this partition function, are the correlation functions of the conserved currents (I.26) identical to the corresponding correlation functions built from the conserved currents

$$J^{ab}_- := 2\,\psi^{a\dagger}_-\,\psi^b_-, \qquad J^{ab}_+ := 2\,\psi^{a\dagger}_+\,\psi^b_+, \qquad a,b = 1,\cdots,N, \tag{I.28}$$

with the partition function

$$
\begin{aligned}
Z_0 &= Z_0^{(-)} \times Z_0^{(+)}, \\
Z_0^{(-)} &:= \int \mathcal{D}[\psi_-^\dagger]\,\mathcal{D}[\psi_-] e^{-\int \frac{\mathrm{d}x_- \mathrm{d}x_+}{2} \sum_{a=1}^{N} \psi_-^{a\dagger} \partial_+ \psi_-^a}, \\
Z_0^{(+)} &:= \int \mathcal{D}[\psi_+^\dagger]\,\mathcal{D}[\psi_+] e^{-\int \frac{\mathrm{d}x_- \mathrm{d}x_+}{2} \sum_{a=1}^{N} \psi_+^{a\dagger} \partial_- \psi_+^a},
\end{aligned}
\tag{I.29}
$$

for N massless Dirac fermions propagating freely in $(1+1)$-dimensional Minkowski space and time $\mathbb{M}_{1,1}$?

The answers to both questions (i) and (ii) are positive as shown by Witten in Ref. [111]. Before deriving this remarkable result, we first explain what generalizations of Eqs. (I.24) and (I.25) do not work, in order to motivate the positive answer of Witten.

First, had we chosen to order the factor G^{-1} to the left of the factor $\partial_- G$ in Eq. (I.26), the conservation laws (I.27) would not follow. Instead, because of the ordering chosen in Eq. (I.26), we have

$$
\begin{aligned}
0 &= \partial_+ \left[(\partial_- G)\, G^{-1} \right] \\
&= (\partial_+ \partial_- G)\, G^{-1} - (\partial_- G)\, G^{-1}\, (\partial_+ G)\, G^{-1},
\end{aligned}
\tag{I.30}
$$

which, upon multiplication from the left with G^{-1} and from the right with G, is equivalent to

$$
\begin{aligned}
0 &= \partial_- \left(G^{-1} \partial_+ G \right) \\
&- - G^{-1}\, (\partial_- G)\, G^{-1}\, (\partial_+ G) + G^{-1}\, (\partial_- \partial_+ G)\,.
\end{aligned}
\tag{I.31}
$$

(We assume that G is smooth in that $\partial_- \partial_+ G = \partial_+ \partial_- G$.)

Second, had we guessed that the principal chiral NLσM (PCM) is the generalization of the free massless scalar-field theory defined by Eq. (I.25), then we would have a mismatch of symmetries and we would fail to satisfy condition (i), as we now explain.

We begin with the definition of the PCM through the Lagrangian density

$$
\begin{aligned}
\mathcal{L}_{\mathrm{PCM}} :&= +\frac{1}{4\lambda^2} \operatorname{tr} \left(\partial_\mu G^{-1} \right) \left(\partial^\mu G \right) \\
&= -\frac{1}{4\lambda^2} \operatorname{tr} \left(G^{-1} \partial_\mu G \right) \left(G^{-1} \partial^\mu G \right),
\end{aligned}
\tag{I.32a}
$$

the action

$$
S_{\mathrm{PCM}} := \int \mathrm{d}x^0 \mathrm{d}x^1 \mathcal{L}_{\mathrm{PCM}},
\tag{I.32b}
$$

and the partition function

$$
Z_{\mathrm{PCM}} := \int \mathcal{D}[G]\, e^{\mathrm{i} S_{\mathrm{PCM}}}.
\tag{I.32c}
$$

The real-valued coupling constant λ^2 is dimensionless and positive. It should be compared with the real-valued and positive coupling constant g^2 of the $O(3)$ NLσM of chapter 3. As was the case with g^2, we must choose λ^2 to be positive in order to bound from below the value of the action in two-dimensional Euclidean space. The trace is to be performed in the representation chosen for the $U(N)$-valued fields. In the fundamental representation, the trace is performed over $N\times N$ unitary matrices. The convention $1/(4\lambda^2)$ in the PCM Lagrangian density follows from the choice of the normalization (I.39) that we are making for the generators of the Lie algebra.

The continuous symmetries of the PCM are the following. The PCM is endowed with the global (chiral) $U(N)\times U(N)$ symmetry by which the Lagrangian density and the Haar measure $\mathcal{D}[G]$ of the group $U(N)$ are left invariant by the transformation

$$G(x)\to L\,G(x)\,R^{-1}, \tag{I.33}$$

for any given pair of matrix L and R from $U(N)$. This global symmetry is to be compared with the larger symmetry group of the partition function (I.29). In the latter case, the local symmetry operation is defined by performing the local chiral gauge transformation

$$\psi^{a\dagger}_{-}(x)\to\sum_{b=1}^{N}G^{ab*}_{-}(x_{-})\,\psi^{b\dagger}_{-}(x),\qquad \psi^{a}_{-}(x)\to\sum_{c=1}^{N}G^{ac}_{-}(x_{-})\,\psi^{c}_{-}(x), \tag{I.34a}$$

whereby

$$\sum_{c}^{N}G^{ca*}_{-}(x_{-})\,G^{cb}_{-}(x_{-})=\delta^{ab}, \tag{I.34b}$$

for one group factor in $U(N)\times U(N)$ and by performing the local chiral gauge transformation

$$\psi^{a\dagger}_{+}(x)\to\sum_{b=1}^{N}G^{ab*}_{+}(x_{+})\,\psi^{b\dagger}_{+}(x),\qquad \psi^{a}_{+}(x)\to\sum_{c=1}^{N}G^{ac}_{+}(x_{+})\,\psi^{c}_{+}(x), \tag{I.35a}$$

whereby

$$\sum_{c}^{N}G^{ca*}_{+}(x_{+})\,G^{cb}_{+}(x_{+})=\delta^{ab}, \tag{I.35b}$$

for the second group factor in $U(N)\times U(N)$. The independent local $U(N)\times U(N)$ transformations (I.34) and (I.35) leave the Lagrangian density and Grassmann measure in Eq. (I.29) unchanged.

The global continuous symmetry (I.33) is spontaneously broken at the fixed point $\lambda^2=0$ down to the subgroup $U(N)$ generated by the transformations

$$G(x)\to D\,G(x)\,D^{-1}, \tag{I.36}$$

for any $D\in U(N)$ independent of space and time. Indeed, the sole $U(N)$-valued fields that contribute to the partition function when $\lambda^2=0$ are those $U(N)$-valued fields G_0 that are constant in $(1+1)$-dimensional Minkowski space $\mathbb{M}_{1,1}$, say $G(x)=$

I with I the unit $N\times N$ matrix for any $x\equiv(x^{\mu})\in\mathbb{M}_{1,1}$.[36] The pattern of symmetry breaking $U(N)\times U(N)\to U(N)$ that is realized spontaneously when $\lambda^2=0$ is explicitly enforced by the addition of the "mass" term

$$\mathcal{L}_{m^2/\lambda^2}:=\frac{m^2}{4\lambda^2}\,\mathrm{tr}\,(G+G^{-1}) \tag{I.37}$$

to the Lagrangian density $\mathcal{L}_{\mathrm{PCM}}$. Goldstone modes must be associated with this instance of spontaneous symmetry breaking. To uncover the Goldstone modes, we parametrize the matrix-valued field G according to

$$\begin{aligned} G(x) &=: \exp\left(+\mathrm{i}|\lambda|\sum_{\mathsf{a}=1}^{N^2}\pi^{\mathsf{a}}(x)\,T^{\mathsf{a}}\right) \\ &= 1+\mathrm{i}|\lambda|\sum_{\mathsf{a}=1}^{N^2}\pi^{\mathsf{a}}(x)T^{\mathsf{a}}+\mathcal{O}(\lambda^2), \end{aligned} \tag{I.38}$$

where the N^2 real-valued fields π^{a} are labeled by Latin letters in sans-serif fonts and any one of the N^2 Hermitian generators of the Lie algebra $u(N)=u(1)\oplus su(N)$ is denoted by T^{a}. We choose the normalization condition[37]

$$\mathrm{tr}\,(T^{\mathsf{a}}\,T^{\mathsf{b}})=2\,\delta^{\mathsf{ab}},\qquad \mathsf{a},\mathsf{b}=1,\cdots,N^2-1 \tag{I.39}$$

for the generators of $su(N)$. By expanding $\mathcal{L}_{\mathrm{PCM}}+\mathcal{L}_{m^2/\lambda^2}$ to second order in powers of the fields π^{a}, we deduce that these fields are massless when the symmetry-breaking field $m^2/\lambda^2\to0$, while they are gapped for any $m^2/\lambda^2>0$. This is the expected behavior of Goldstone modes when their interactions are neglected.

The Lagrangian density of the PCM (I.32) has the following three discrete symmetries. First, it is left invariant by the operation of matrix transposition

$$G\to G^{\mathsf{T}}. \tag{I.40}$$

Since the Haar measure is also invariant under this operation, so is the PCM. The operation of matrix transposition (I.40) can be interpreted as a manifestation of the operation of charge conjugation on the Goldstone bosons that might emerge at low energies by perturbing the free massless Dirac fermions with the partition function (I.29) through interactions. Those Goldstone modes associated to generators that are symmetric matrices are charge neutral and left invariant by the operation of charge conjugation with respect to the conserved charges of the theory (I.29). Those Goldstone modes associated to generators that are antisymmetric matrices

36 Any space- and time-independent $G(x)=G_0$ can be brought to the $N\times N$ identity matrix I by making use of the global symmetry transformation (I.33) with $L=I$ and $R=G_0$.

37 The $(N^2-1)\times(N^2-1)$ matrix defined by the matrix elements $\mathrm{tr}\,(T^{\mathsf{a}}\,T^{\mathsf{b}})$ with $\mathsf{a},\mathsf{b}=1,\cdots,N^2-1$ is symmetric owing to the cyclicity of the trace. It is also real-valued owing to the Hermiticity of the generators of $su(N)$, the invariance of the trace with respect to the operation of matrix transposition, and the cyclicity of the trace. Hence, this matrix can be diagonalized by an orthogonal transformation. Because the group $SU(N)$ is compact, it can be shown that this diagonal matrix is positive (negative) definite. This diagonal matrix can thus be made proportional to the unit matrix by a rescaling.

carry a charge that is reversed by the operation of charge conjugation with respect to the conserved charges of the theory (I.29). The operation of matrix transposition (I.40) cannot be a symmetry of the free massless Dirac Lagrangian density defined by Eq. (I.29) if we assume that Eqs. (I.26) and (I.27) hold.

Second, the Lagrangian density of the PCM is left invariant by the operation of matrix inversion

$$G \to G^{-1}. \tag{I.41}$$

Since the Haar measure is also invariant under this operation, so is the PCM. The operation of matrix inversion (I.41) assigns a charge to all bosons and reverses it, for matrix inversion of the left-hand side of Eq. (I.38) amounts to changing the sign of all terms of odd powers in $|\lambda|$ in the Taylor expansion of the exponential on the right-hand side of Eq. (I.38). This operation is thus even (odd) on all states with a definite even (odd) number of Goldstone bosons, i.e., this operation can be interpreted as implementing the Goldstone-number operator $\hat{N}_\pi$ modulo 2 through the action of $(-1)^{\hat{N}_\pi}$ on all states from the bosonic Hilbert space. However, the free fermion theory (I.29) does not conserve the parity of the number of Goldstone bosons, i.e., there is no symmetry of the free massless Dirac fermions with the partition function (I.29) corresponding to the matrix inversion (I.41). A fortiori Eqs. (I.26) and (I.27) are not symmetric under the matrix inversion (I.41). In line with this observation, the equations of motion of the PCM,

$$\partial_\mu \left(G^{-1}\,\partial^\mu G\right) = 0 \Longleftrightarrow \partial_\mu \left(G\,\partial^\mu G^{-1}\right) = 0, \tag{I.42}$$

which are of course compatible with the operation of matrix inversion, are not equivalent to the equations of motion (I.27).

The PCM is also invariant under the operation

$$x^1 \to -x^1. \tag{I.43}$$

This is the parity operation. Since the Haar measure is also invariant under this operation, so is the PCM. However, the Lagrangian density of the free massless Dirac fermions theory (I.29) is not symmetric under the parity operation (I.43) under which $x_\pm \to x_\mp$. A fortiori Eqs. (I.26) and (I.27) are not symmetric under the parity operation (I.43).

Although none of the three operations (I.40), (I.41), and (I.43) leave Eqs. (I.26) and (I.27) unchanged, one verifies that the composition of any two of them is consistent with the non-Abelian bosonization formula (I.26).

For example, we observe that by composing parity with matrix inversion, we exchange the chiral currents defined in Eq. (I.26). Hence, if the non-Abelian bosonization formula (I.26) is to make any sense, the symmetry of the PCM under the composed discrete operation

$$G(x^0, x^1) \to G^{-1}(x^0, -x^1) \tag{I.44}$$

should also be a symmetry of the free massless Dirac fermion with the partition function (I.29).

Finally, the PCM fails to satisfy condition (ii) as all the equal-time correlation functions for its conserved currents decay exponentially fast in position space in the limit of large separations. In contrast, all equal-time correlation functions for the conserved currents in the free massless Dirac with the partition function (I.29) decay algebraically fast in position space.

To prove this fact, we may use again the parametrization (I.38), expand the Lagrangian density $\mathcal{L}_{\mathrm{PCM}} + \mathcal{L}_{m^2/\lambda^2}$ about the $N \times N$ unit matrix I to quartic order in the N^2 fields π^{a} very much as we did in chapter 3 when expanding the Lagrangian density of the $O(N)$ NLσM to quartic order in the $N-1$ spin waves, and then integrate the fast modes in favor of the slow modes. The changes incurred to the action of the slow modes when we take the limit $m^2 \to 0$ can then be encoded into a renormalization of the coupling λ^2. Alternatively, we can borrow the Ricci flow of appendix C to derive the infrared RG flow obeyed by λ^2. It is the infrared flow to strong coupling [use Eq. (3.230) with $\epsilon = 0$ and $c_v = N$]

$$\mathfrak{a}\frac{\partial \lambda^2}{\partial \mathfrak{a}} = +\frac{N}{4\pi}\lambda^4 + \mathcal{O}(\lambda^6), \tag{I.45}$$

where $\mathfrak{a}$ is the short-distance cut-off. As was the case of the $O(3)$ NLσM, the long-range-ordered phase at $\lambda^2 = 0$ is unstable to a massive phase characterized by correlation functions that decay exponentially fast. Hence, the PCM (I.32) cannot be equivalent to N massless Dirac fermions propagating freely in $(1+1)$-dimensional Minkowski space and time $\mathbb{M}_{1,1}$.

Nevertheless, the PCM is part of the bosonic quantum field theory that solves conditions (i) and (ii). The PCM is a theory for N^2 interacting real-valued scalar fields. Had we used Abelian bosonization, we would have turned N massless Dirac fermions propagating freely in $(1+1)$-dimensional Minkowski space and time $\mathbb{M}_{1,1}$ into a theory of N real-valued scalar fields. The PCM has thus $N(N-1)$ undesirable degrees of freedom that we would like to eliminate, while retaining N critical modes, if we are to relate the PCM to N massless Dirac fermions propagating freely in $(1+1)$-dimensional Minkowski space and time $\mathbb{M}_{1,1}$. The infrared flow to strong coupling (I.45) can be used for small values of λ^2 to eliminate these $N(N-1)$ unwanted degrees of freedom, provided we can terminate the infrared flow at a critical value of the coupling λ^2 for which the theory reaches the critical point equivalent to N massless Dirac fermions propagating freely in $(1+1)$-dimensional Minkowski space and time $\mathbb{M}_{1,1}$.

We are thus seeking to modify S_{PCM} additively with the action $\kappa\Gamma$, where κ is a dimensionless coupling, in such a way that we may replace the one-loop infrared flow (I.45) by

$$\begin{aligned} \mathfrak{a}\frac{\partial \lambda^2}{\partial \mathfrak{a}} &= +\frac{N}{4\pi}\lambda^4\left[1-\left(\frac{\lambda^2\,\kappa}{4\pi}\right)^2\right] + \cdots, \\ \mathfrak{a}\frac{\partial \kappa}{\partial \mathfrak{a}} &= 0 + \cdots, \end{aligned} \tag{I.46}$$

and we may interpret the partition function at the one-loop fixed point $\lambda^2 = 4\pi/|\kappa|$ as (the bosonic sector of) the theory for N massless Dirac fermions propagating freely in $(1+1)$-dimensional Minkowski space and time $\mathbb{M}_{1,1}$.

Witten solved this challenge by constructing Γ to be a functional of the group-valued field G such that (1) it shares the same continuous chiral symmetry as S_{PCM}, (2) it violates the symmetry of S_{PCM} under either the matrix transposition (I.40), the matrix inversion (I.41), or the parity transformation (I.43), but it preserves the symmetry under the composed operation of parity and matrix inversion of $\mathcal{L}_{\text{PCM}}$, (3) it delivers the local equations of motion (which is left unchanged by the composition of matrix transposition and inversion)

$$\partial_\mu \left(\frac{1}{\lambda^2} g^{\mu\nu} - \frac{\kappa}{4\pi} \epsilon^{\mu\nu} \right) \left(G\, \partial_\nu G^{-1} \right) = 0, \tag{I.47a}$$

or, under the composed operation of parity and matrix inversion (the composition of parity and matrix transposition also does the job),

$$\partial_\mu \left(\frac{1}{\lambda^2} g^{\mu\nu} + \frac{\kappa}{4\pi} \epsilon^{\mu\nu} \right) \left(G^{-1}\, \partial_\nu G \right) = 0, \tag{I.47b}$$

(4) a non-perturbative extension of Eq. (I.46) holds, and (5) the continuity equations (I.27) follow from Eqs. (I.47a) and (I.47b) when $\lambda^2 = 4\pi/|\kappa|$.

The insight of Witten was to construct a functional Γ that associates to any $U(N)$-valued field G a real number such that $S_{\text{PCM}} + \kappa\,\Gamma$ delivers the equation of motion (I.47) and Γ is multivalued. The multivaluedness of Γ is allowed as long as $\exp(\mathrm{i}\kappa\,\Gamma)$ is single valued, for the theory is defined by a partition function with the Boltzmann weight $e^{\mathrm{i}(S_{\text{PCM}}+\kappa\,\Gamma)}$. Single valuedness of $\exp(\mathrm{i}\kappa\,\Gamma)$ can be achieved if and only if the ambiguity in computing the value of Γ associated to the group-valued field G is quantized.[38] If so the normalization of the functional Γ may always be chosen in such a way that the ambiguity in computing the value of Γ associated to the group-valued field G is 2π times an integer. Single valuedness of $\exp(\mathrm{i}\kappa\,\Gamma)$ is then achieved if and only if κ is restricted to be an integer $k \in \mathbb{Z}$. The restriction of κ to the integer k implies that k cannot smoothly change under an infinitesimal change of the short-distance cut-off $\mathfrak{a}$, i.e.,

$$\mathfrak{a}\frac{\partial \kappa}{\partial \mathfrak{a}} = 0 \tag{I.48}$$

is an exact result. For some value of the integer $0 \neq |k| \in \mathbb{Z}$, the one-loop critical point

$$\lambda^2 = \frac{4\pi}{|k|} \tag{I.49}$$

[38] Imagine that we allow the group-valued field G to depend parametrically on the variable t (not to be confused with x^0), i.e., the mapping $x \in \mathbb{M}_{1,1} \to G(x) \in \mathrm{G}$ is extended to the mapping $x, t \in \mathbb{M}_{1,1} \times \mathbb{R} \to G_t(x) \in \mathrm{G}$. This parametric dependence is cyclic if $G_{t=0}(x) = G_{t=1}(x)$ for all x in $(1+1)$-dimensional Minkowski space and time $\mathbb{M}_{1,1}$. The ambiguity in computing the value of Γ associated to the group-valued field G can also be understood by the property that Γ may not be cyclic under some cyclic parametric change of G.

of the infrared flow

$$\mathfrak{a}\frac{\partial\lambda^2}{\partial\mathfrak{a}} = +\frac{N}{4\pi}\lambda^4\left[1-\left(\frac{\lambda^2\kappa}{4\pi}\right)^2\right]+\cdots \tag{I.50}$$

is then a plausible candidate to realize the critical point for a certain number of massless Dirac fermions propagating freely in $(1+1)$-dimensional Minkowski space and time $\mathbb{M}_{1,1}$. To substantiate the conjecture that a critical point (I.49) of the quantum field theory with the action $S_{\text{PCM}} + k\,\Gamma$ is equivalent to a free fermionic quantum field theory and to select the value $|k| = 1$ at which this free fermionic quantum field theory is the quantum field theory (I.13) of N massless Dirac fermions propagating freely in $(1+1)$-dimensional Minkowski space and time $\mathbb{M}_{1,1}$, Witten showed that the so-called current algebra obeyed by the bosonic currents is identical to the current algebra obeyed by the fermionic currents. Generalizations of this current algebra and establishing their intimate relations to conformal field theory in $(1+1)$-dimensional Minkowski space and time $\mathbb{M}_{1,1}$ [140] were explored in a seminal paper by Knizhnik and Zamolodchikov in Ref. [141].

Before writing explicitly the form taken by the functional Γ and studying its properties, a few words about terminology.

The action Γ written down by Witten is an example from the family of actions referred to as Wess-Zumino (WZ) actions in the physics literature [142]. Such actions are rooted in quantum anomalies, as we shall illustrate using a method devised by Polyakov and Wiegmann in the case of interest [112]. The action $S_{\text{PCM}} + \kappa\,\Gamma$ is called the Wess-Zumino-Novikov-Witten (WZNW) action for generic values of λ^2 and κ. At the critical points (I.49), this action is called the Wess-Zumimo-Witten (WZW) action.

I.4 A quantum-mechanical example of a Wess-Zumino action

The construction of the WZ action Γ made by Witten in Ref. [111] applies to the group-valued field $G : x \in \mathbb{M}_{1,1} \to G(x) \in \mathrm{G}$, whereby the group G is any simple compact Lie group. Owing to its generality, this construction relies on powerful mathematics from the second half of the 20th century that will be sketched in section I.5.

Following Witten in Ref. [143], we may simplify the problem to that of the quantum mechanics of a point particle restricted to moving on the surface S^2 of a unit ball B in the presence of a magnetic monopole at the center of the sphere. It is then possible to construct the WZ action Γ using the standard lore of calculus from the 19th century. Moreover, this problem is important in its own right, since it is of relevance to quantum magnetism [35] or to the fractional quantum Hall effect [144] as shown in both cases by Haldane.

We thus consider a point particle of mass m whose classical dynamics is defined by the Lagrangian

$$\mathfrak{L}_{m,U} := \int_0^T \mathrm{d}t \left\{ \frac{1}{2} m\, \dot{\boldsymbol{x}}^2(t) - U \left[\boldsymbol{x}^2(t) - 1\right]^2 \right\} \tag{I.51a}$$

between the initial time $t = 0$ and the final time $t = T$. We shall consider the limit $U \to \infty$, for which the point particle is constrained to move on the surface

$$S^2 := \left\{ \boldsymbol{x}(t) \in \mathbb{R}^3 \middle| \, \boldsymbol{x}^2(t) = 1 \right\} \tag{I.51b}$$

of the unit ball

$$B := \left\{ \boldsymbol{x}(t) \in \mathbb{R}^3 \middle| \, \boldsymbol{x}^2(t) \leq 1 \right\}. \tag{I.51c}$$

Lagrangian (I.51a) is symmetric under any static and proper rotation of $\boldsymbol{x}$. It is also symmetric under two discrete transformations. There is the operation of parity

$$\boldsymbol{x}(t) \to -\boldsymbol{x}(t), \tag{I.52a}$$

that leaves Lagrangian (I.51a) invariant. When combined with the static proper rotations, the symmetry group $O(3) = \mathbb{Z}_2 \times SO(3)$ follows. There also is the operation of time reversal

$$t \to -t, \tag{I.52b}$$

that leaves Lagrangian (I.51a) invariant.

The $SO(3)$-symmetric classical equations of motion

$$\left(\frac{\mathrm{d}}{\mathrm{d}t} \frac{\delta}{\delta \dot{\boldsymbol{x}}} - \frac{\delta}{\delta \boldsymbol{x}} \right) \mathfrak{L}_{m,U=\infty} = 0 \tag{I.53}$$

may be written down using spherical coordinates and imposing the condition that the radius is set to unity in view of the limit $U \to \infty$. These equations of motion may also be written down for an arbitrary local parametrization of S^2 using the equations of motion (C.14) that apply to the spin-wave approximation of the $O(3)$ NLσM in $(0+1)$-dimensional Euclidean space $\mathbb{R}$. However, whatever the choice of local coordinate system on S^2, the $SO(3)$ symmetry is not manifest anymore. The $SO(3)$ symmetry is manifest if we use the redundant parametrization in terms of $\boldsymbol{x} \in \mathbb{R}^3$ supplemented by the constraint $\boldsymbol{x}^2 = 1$ (to be implemented by a Lagrange multiplier, say), in which case we may write[39]

$$\ddot{\boldsymbol{x}} + \dot{\boldsymbol{x}}^2\, \boldsymbol{x} = 0. \tag{I.54}$$

These equations of motion are invariant under the operations of time reversal (I.52b) and parity (I.52a).

[39] If λ denotes a Lagrange multiplier that implements the constraint $\boldsymbol{x}^2 = 1$, then we need to solve the two coupled Euler-Lagrange equations of motion (i) $0 = m\, \ddot{\boldsymbol{x}} + 2\lambda\, \boldsymbol{x}$ and (ii) $\boldsymbol{x}^2 = 1$. To solve for the Lagrange multiplier, we write (i) as (iii) $\ddot{\boldsymbol{x}} = -2\lambda\, \boldsymbol{x}/m$ and differentiate (ii) twice with respect to time to obtain (iv) $\dot{\boldsymbol{x}}^2 + \boldsymbol{x} \cdot \ddot{\boldsymbol{x}} = 0$. Insertion of (iii) into (iv), whereby one makes use of (ii), delivers (v) $\lambda = m\, \dot{\boldsymbol{x}}^2/2$. In turn, insertion of (v) into (i) gives Eq. (I.54).

We now seek a Lagrangian

$$\mathfrak{L}_{m,\kappa} := \mathfrak{L}_{m,U=\infty} + \kappa\,\Gamma, \tag{I.55}$$

for a point particle moving on the sphere S^2 that shares the continuous $SO(3)$ symmetry of $\mathfrak{L}_{m,U=\infty}$ but breaks the $\mathbb{Z}_2 \times \mathbb{Z}_2$ discrete symmetries under the operations of time reversal (I.52b) and parity (I.52a) down to the $\mathbb{Z}_2$ subgroup defined by the operation

$$t \to -t, \qquad \boldsymbol{x}(t) \to -\boldsymbol{x}(-t). \tag{I.56}$$

More precisely, we demand that $\mathfrak{L}_{m,\kappa}$ gives rise to the equations of motion

$$m\left(\ddot{\boldsymbol{x}} + \dot{\boldsymbol{x}}^2\,\boldsymbol{x}\right) = -\kappa\,\boldsymbol{x} \wedge \dot{\boldsymbol{x}} \Longleftrightarrow m\left(\ddot{x}^i + \dot{\boldsymbol{x}}^2\,x^i\right) = -\kappa\,\epsilon^{ijk}\,x^j\,\dot{x}^k, \tag{I.57}$$

for any $i = 1, 2, 3$ and for some real-valued coupling constant κ.

The symmetry breaking pattern

$$\mathbb{Z}_2 \times \mathbb{Z}_2 \to \mathbb{Z}_2, \tag{I.58}$$

which Γ encodes is not enforced by the Ansatz

$$\int\limits_0^T \mathrm{d}t\, f\big(\boldsymbol{x}(t), \dot{\boldsymbol{x}}(t)\big)\,\epsilon^{ijk}\,x^i(t)\,x^j(t)\,\dot{x}^k(t), \tag{I.59}$$

where the function $f : \mathbb{R}^3 \times \mathbb{R}^3 \to \mathbb{R}$ is a scalar, i.e., it is invariant under any static and proper rotation of $\boldsymbol{x}$ and $\dot{\boldsymbol{x}}$ as well as invariant under the operations of time reversal (I.52b) and parity (I.52a). This is so because $\epsilon^{ijk}\,x^i(t)\,x^j(t)\,\dot{x}^k(t)$ necessarily vanishes.

The clue that delivers Γ is the observation that the symmetry-breaking term on the right-hand side of Eq. (I.57) is the Lorentz-force

$$\dot{\boldsymbol{x}}(t) \wedge \boldsymbol{B}\big(\boldsymbol{x}(t)\big), \qquad \boldsymbol{B}(\boldsymbol{r}) := \frac{\boldsymbol{r}}{|\boldsymbol{r}|^3}, \tag{I.60}$$

that a classical particle experiences along its trajectory on $S^2 = \partial B$ when a magnetic monopole is placed at the center of the ball B. If we introduce the vector potential $\boldsymbol{A}$ through

$$\boldsymbol{B} =: \boldsymbol{\nabla} \wedge \boldsymbol{A}, \tag{I.61}$$

one verifies that

$$\mathfrak{L}_{m,\kappa} := \int\limits_0^T \mathrm{d}t \left[\frac{1}{2} m\,\dot{\boldsymbol{x}}^2(t) + \kappa\,\dot{\boldsymbol{x}}(t) \cdot \boldsymbol{A}\big(\boldsymbol{x}(t)\big)\right], \qquad \boldsymbol{x}^2(t) = 1, \tag{I.62}$$

delivers the equations of motion (I.57).

The action

$$\begin{aligned}\Gamma :&= \int\limits_0^T \mathrm{d}t\,\dot{\boldsymbol{x}}(t) \cdot \boldsymbol{A}\big(\boldsymbol{x}(t)\big) \\ &= \int\limits_{\boldsymbol{x}(0)}^{\boldsymbol{x}(T)} \mathrm{d}\boldsymbol{x} \cdot \boldsymbol{A}(\boldsymbol{x})\end{aligned} \tag{I.63}$$

on S^2 is not manifestly symmetric under static proper rotations of $\boldsymbol{x}$ and $\dot{\boldsymbol{x}}$. This loss of explicit symmetry is not due to an internal inconsistency. It is a manifestation of the fact that the vector potential is not unique. Indeed, the static gauge transformation

$$\boldsymbol{A} \to \boldsymbol{A} + \boldsymbol{\nabla}\chi \tag{I.64}$$

leaves the magnetic field unchanged for any sufficiently smooth and static real-valued scalar field χ. The symmetries of the equations of motion (I.57) are implemented at the level of Γ by the fact that the change of Γ by a static proper rotation of $\boldsymbol{x}$ can be compensated by a static gauge transformation of the form (I.64).

The integrand in the action (I.63) is also singular in the following way. The condition $(\boldsymbol{\nabla}\cdot\boldsymbol{B})(\boldsymbol{r}) = 0$ for any $\boldsymbol{r} \neq 0$ that the magnetic field of the magnetic monopole must respect is solved by interpreting it as the end point of an infinitely thin solenoid that extends from the center of the ball B all the way to infinity, say along the polar axis through the south pole. This infinitely thin solenoid is called a Dirac string [145]. Consequently, $\boldsymbol{A}$ must be singular at the intersection between S^2 and the Dirac string, say at the south pole.

At this stage, we impose periodic boundary conditions by which the initial and final positions on S^2 are identical,

$$\boldsymbol{x}(0) = \boldsymbol{x}(T). \tag{I.65}$$

Because all closed path of S^2 are smoothly contractible to a point,[40] we may apply Stokes theorem to convert the line integral along the closed path $\gamma_{\text{acw}} \subset S^2$ when traversed anticlockwise into a surface integral over the cap $\Sigma_{\gamma_{\text{acw}}} \subset S^2$ with the oriented γ_{acw} as its boundary,

$$\begin{aligned}\Gamma(\Sigma_{\text{acw}}) :=& \oint\limits_{\gamma_{\text{acw}}} \mathrm{d}\boldsymbol{x}\cdot\boldsymbol{A}(\boldsymbol{x})\\ =& + \int\limits_{\Sigma_{\gamma_{\text{acw}}}} \mathrm{d}\boldsymbol{\Sigma}\cdot\boldsymbol{B}.\end{aligned} \tag{I.66}$$

Had we applied Stokes theorem to convert the line integral along the closed path $\gamma_{\text{acw}} \subset S^2$ into a surface integral over the cap $S^2 \setminus \Sigma_{\gamma_{\text{acw}}} \subset S^2$ with the oriented γ_{cw} as its boundary traversed clockwise,

$$\begin{aligned}\Gamma(S^2\setminus\Sigma_{\text{acw}}) :=& \oint\limits_{\gamma_{\text{acw}}} \mathrm{d}\boldsymbol{x}\cdot\boldsymbol{A}(\boldsymbol{x})\\ =& - \int\limits_{S^2\setminus\Sigma_{\gamma_{\text{acw}}}} \mathrm{d}\boldsymbol{\Sigma}\cdot\boldsymbol{B}.\end{aligned} \tag{I.67}$$

[40] In the jargon of algebraic topology, $\pi_1(S^2) = 0$.

We quantize the classical theory through the path integral

$$Z := \int \mathcal{D}[\boldsymbol{x}]\, \delta(\boldsymbol{x}^2 - 1)\, \exp\left(+\mathrm{i} \int\limits_0^T \mathrm{d}t\, \mathcal{L}_{m,\kappa} \right) \tag{I.68a}$$

over all closed paths by demanding that

$$e^{\mathrm{i}\kappa\, \Gamma(\Sigma_{\mathrm{acw}})} = e^{\mathrm{i}\kappa\, \Gamma(S^2 \backslash \Sigma_{\mathrm{acw}})} \tag{I.68b}$$

holds. We can then combine Eq. (I.66), Eq. (I.67), and Eq. (I.68b) to deduce that

$$1 = \exp\left(\mathrm{i}\kappa \int\limits_{S^2} \mathrm{d}\boldsymbol{\Sigma} \cdot \boldsymbol{B} \right). \tag{I.69}$$

Since the magnetic field $\boldsymbol{B}$ on S^2 is nothing but the unit vector pointing outwardly, $\int\limits_{S^2} \mathrm{d}\boldsymbol{\Sigma} \cdot \boldsymbol{B}$ is nothing but the area 4π of S^2. We conclude that

$$1 = \exp\left(\mathrm{i}\kappa\, 4\pi\right), \tag{I.70}$$

i.e., the dimensionless coupling κ multiplying the Wess-Zumino action Γ must either be an integer or a half-integer k.

I.5 Wess-Zumino action in (1+1)-dimensional Minkowski space and time

Needed is a functional $\kappa\,\Gamma$ of the $U(N)$-valued field G that is odd under both the operation of matrix inversion (I.41) and the operation of parity (I.43) with a real-valued dimensionless coupling constant κ.

To implement oddness under parity, we cannot use the Lorentz metric tensor $g_{\mu\nu}$ to contract derivatives in space and time. This leaves the Levi-Civita tensor $\epsilon_{\mu\nu}$ at our disposal to contract derivatives in space and time. However, the sole pseudo-scalar functional built from a local monomial of order 2 in $G^{-1}\partial_\mu G$,

$$\begin{aligned} \int \mathrm{d}x^0 \mathrm{d}x^1\, \epsilon_{\mu\nu}\, \mathrm{tr}\, \left(\partial^\mu G^{-1}\right) \left(\partial^\nu G\right) &= -\int \mathrm{d}x^0 \mathrm{d}x^1\, \epsilon_{\mu\nu}\, \mathrm{tr}\, \left(G^{-1}\partial^\mu G\right) \left(G^{-1}\partial^\nu G\right) \\ &= 0, \end{aligned} \tag{I.71}$$

vanishes owing to the cyclicity of the trace.

Moreover, a local monomial in powers of $G^{-1}\partial_\mu G$ is odd under $G \to G^{-1}$ if and only if it is of an odd order. This observation rules out a dimensionless coupling constant κ if we insist on locality in terms of the fields $G^{-1}\partial_\mu G$.

It thus appears hopeless to built out of the $U(N)$-valued field G a local Lagrangian density in (1+1)-dimensional Minkowski space that delivers the equations of motion (I.47).

The solution to this dilemma is to abandon the condition that Γ is to be represented by a local polynomial in the space and time derivatives of the $U(N)$-valued field G, provided locality is maintained if we formulate the theory using the underlying coordinates of the Riemannian manifold.

We recall here that the PCM (I.32) is a symmetric and homogeneous NLσM. According to appendix C, there exists a Riemannian metric tensor $\mathfrak{g}_{\mathfrak{p}}$ in the vicinity of any point $\mathfrak{p}$ pf the Riemannian manifold that can be represented by a $N^2 \times N^2$ real-valued symmetric matrix. Any one of its components, say $\mathfrak{g}_{\mathfrak{a}\mathfrak{b}}(\phi)$, depends on the choice of N^2 coordinates $\phi_{\mathfrak{a}}$ made to parametrize the Riemannian manifold $U(N) \times U(N)/U(N)$ in the vicinity of $\mathfrak{p}$. In terms of these local coordinates, the partition function of the PCM (I.32) becomes

$$Z_{\mathrm{PCM}} = \int \mathcal{D}[\phi]\, e^{\mathrm{i}S_{\mathrm{PCM}}[\phi]}, \tag{I.72a}$$

the Minkowski action of the PCM (I.32) becomes

$$S_{\mathrm{PCM}}[\phi] = \int \mathrm{d}x^0 \mathrm{d}x^1\, \mathcal{L}_{\mathrm{PCM}}(\phi), \tag{I.72b}$$

the Lagrangian density of the PCM (I.32) becomes

$$\mathcal{L}_{\mathrm{PCM}}(\phi) = \frac{1}{2\lambda^2}\, \mathfrak{g}_{\mathfrak{a}\mathfrak{b}}(\phi)\, g^{\mu\nu}\, \partial_\mu \phi^{\mathfrak{a}}\, \partial_\nu \phi^{\mathfrak{b}}, \tag{I.72c}$$

and the measure of the PCM (I.32) becomes

$$\mathcal{D}[\phi] = \prod_x \sqrt{||\mathfrak{g}(\phi)||}\, \prod_{\mathfrak{a}} \mathrm{d}\phi^{\mathfrak{a}}(x). \tag{I.72d}$$

The summation convention over repeated $\mathsf{a} = 1, \cdots, N^2$ is implied. Needed is thus an antisymmetric tensor $\mathfrak{e}_{\mathfrak{p}}$ on the Riemannian manifold that would play the role of the Levi-Civita tensor in $(1+1)$-dimensional Minkowski space and time. Equipped with its representation by the $N^2 \times N^2$ real-valued antisymmetric matrix $\mathfrak{e}_{\mathfrak{a}\mathfrak{b}}(\phi)$, we could try the Ansatz

$$\Gamma \stackrel{?}{=} \int \mathrm{d}x^0 \mathrm{d}x^1\, \mathfrak{e}_{\mathfrak{a}\mathfrak{b}}(\phi)\, \epsilon^{\mu\nu}\, \partial_\mu \phi^{\mathfrak{a}}\, \partial_\nu \phi^{\mathfrak{b}}. \tag{I.73}$$

By construction, the right-hand side is odd under both the operation $x^1 \to -x^1$ and the operation $\mathfrak{a} \leftrightarrow \mathfrak{b}$, but even under the composition of both operations. Locality is now manifest, albeit at the price of loosing the manifest $U(N) \times U(N)$ symmetry.

The question to be answered is what conditions must $\mathfrak{e}_{\mathfrak{a}\mathfrak{b}}(\phi)$ then satisfy? To answer this question, we impose boundary conditions on the fields such that all fields are equal at "infinity". With these boundary conditions, we can identify Minkowski space with the area S^2 of the "sphere".[41] In turn, the area S^2 of a "sphere" can be thought of as the boundary ∂B of a "ball" denoted B. We are now going to sketch the manipulations using the tools of differential geometry and algebraic topology that allow to present (I.73) in a manifestly $U(N) \times U(N)$ symmetric way.

[41] Upon a Wick rotation to two-dimensional Euclidean space, the identification with S^2 is rigorous.

We would like to interpret Eq. (I.73) as the integral

$$\begin{aligned}\Gamma &\stackrel{?}{=} \int\limits_{S^2=\partial B} G^*\,\zeta \\ &\equiv \int\limits_{\partial B} G^*\,\zeta,\end{aligned} \tag{I.74}$$

where G^* is called the pull-back of the map G from the manifold S^2 to the $U(N)\times U(N)/U(N)$ manifold [30]. We now assume that Stoke's theorem holds. Application of Stokes theorem converts the integral over the form $G^*\,\zeta$ within the domain of integration ∂B into the integral over the form $\mathrm{d}\,(G^*\,\zeta)$ within the domain of integration B. In other words,

$$\begin{aligned}\Gamma &\stackrel{?}{=} \int\limits_{\partial B} G^*\,\zeta \\ &= \int\limits_{B} \mathrm{d}\,(G^*\,\zeta).\end{aligned} \tag{I.75}$$

Now, it is a property of pull-backs that $\mathrm{d}\,(G^*\,\zeta) = G^*\,\mathrm{d}\zeta$, i.e.,

$$\begin{aligned}\Gamma &\stackrel{?}{=} \int\limits_{B} \mathrm{d}\,(G^*\,\zeta) \\ &= \int\limits_{B} G^*\,\mathrm{d}\zeta,\end{aligned} \tag{I.76}$$

provided it is possible to extend the pull-back G^* to a pull-back from the manifold B to the $U(N)\times U(N)/U(N)$ manifold. We introduce a symbol for the form $\mathrm{d}\zeta$,

$$\omega := \mathrm{d}\,\zeta. \tag{I.77}$$

We have thus arrived at the guess

$$\Gamma \stackrel{?}{=} \int\limits_{B} G^*\,\omega. \tag{I.78}$$

It is known from differential geometry on manifolds that for any simple and compact Lie group G, there exists a G × G invariant third rank tensor field ω that obeys $\mathrm{d}\omega = 0$ globally, $\omega = \mathrm{d}\zeta$ locally, and its integral over any B in G is an integer multiple of 2π. We can apply this result to $SU(N)$. There follows, in the case of $U(N)$ in the fundamental representation with the convention (I.39), the Wess-Zumino action

$$\Gamma := \frac{1}{24\pi}\int\limits_{B} \mathrm{d}x^0\mathrm{d}x^1\mathrm{d}x^2\,\epsilon_{\bar\mu\bar\nu\bar\lambda}\mathrm{tr}\,(\bar G^{-1}\partial^{\bar\mu}\bar G)(\bar G^{-1}\partial^{\bar\nu}\bar G)(\bar G^{-1}\partial^{\bar\lambda}\bar G), \tag{I.79a}$$

where the overline over the group element refers to the extension of the mapping

$$\begin{aligned}S^2 &\longrightarrow U(N), \\ x &\longmapsto G(x),\end{aligned} \tag{I.79b}$$

where $x = (x^{\mu})$ with $\mu = 0, 1$, to the mapping

$$\begin{aligned} B &\longrightarrow U(N), \\ \bar{x} &\longmapsto \bar{G}(\bar{x}), \end{aligned} \tag{I.79c}$$

where $\bar{x} = (\bar{x}^{\bar{\mu}})$ with $\bar{\mu} = 0, 1, 2$. The extension of the $U(N)$-valued field defined by Eq. (I.79b) to the $U(N)$-valued field defined by Eq. (I.79c) exists because of the global topological property $\pi_2(SU(N)) = 0$. This extension is, however, not unique. Consequently, Γ is multivalued. Nevertheless, $\exp(\mathrm{i}\kappa\,\Gamma)$ is unique when κ is restricted to the integer k as a consequence of the global topological property $\pi_3(SU(N)) = \mathbb{Z}$.

The Wess-Zumino action (I.79) is odd under any one of the three discrete operations (I.40), (I.41), and (I.43), respectively. It is even under the composition of any two of these.

A justification of (I.79) that does not rely on differential geometry and algebraic topology consists in deriving the equations of motion of $S_{\mathrm{PCM}} + k\,\Gamma$ with k any integer, S_{PCM} defined in Eq. (I.32), and Γ defined in Eq. (I.79), respectively. We are going to show that these equations of motion are nothing but Eqs. (I.47) with the choice $\kappa = k$.

I.6 Equations of motion for the WZNW action

In this section, the $SU(N)$-valued field g denotes the mapping

$$\begin{aligned} g : \mathbb{M}_{1,1} &\longrightarrow SU(N), \\ x &\longmapsto g(x), \end{aligned} \tag{I.80a}$$

which obeys the boundary conditions

$$\lim_{x^0 \to \pm\infty} g(x^0, x^1) = 1, \qquad \lim_{x^1 \to \pm\infty} g(x^0, x^1) = 1. \tag{I.80b}$$

The $SU(N)$-valued field $\bar{g}$ denotes the mapping

$$\begin{aligned} \bar{g} : \mathbb{M}_{1,1} \times [0,1] &\longrightarrow SU(N), \\ \bar{x} &\longmapsto \bar{g}(\bar{x}), \end{aligned} \tag{I.81a}$$

which obeys the boundary conditions

$$\lim_{\bar{x}^0 \to \pm\infty} \bar{g}(\bar{x}^0, \bar{x}^1, \bar{x}^2) = 1, \qquad \lim_{\bar{x}^1 \to \pm\infty} \bar{g}(\bar{x}^0, \bar{x}^1, \bar{x}^2) = 1, \tag{I.81b}$$

and

$$\bar{g}(\bar{x}^0, \bar{x}^1, 0) = 1, \qquad \bar{g}(\bar{x}^0, \bar{x}^1, 1) = g(x^0, x^1). \tag{I.81c}$$

We define the functional W_1 to be the mapping from the space of the $SU(N)$-valued fields g defined by Eq. (I.80) to the real numbers through

$$\begin{aligned} W_1[g] :=& \int \mathrm{tr}\, \left[(\partial_+ g) g^{-1} (\partial_- g) g^{-1} \right] \\ =& \int \mathrm{tr}\, \left[(\partial_\mu g) g^{-1} (\partial^\mu g) g^{-1} \right]. \end{aligned} \tag{I.82a}$$

Here, the short-hand notation

$$\int \equiv \int \mathrm{d}x^0 \mathrm{d}x^1 = \int \frac{\mathrm{d}x^- \mathrm{d}x^+}{2} \tag{I.82b}$$

is used and we have combined Eq. (I.15) and [recall Eq. (I.16)]

$$\begin{aligned} a_+\, b_- &= a_\mu (g^{\mu\nu} + \epsilon^{\mu\nu}) b_\nu, \\ a_-\, b_+ &= a_\mu (g^{\mu\nu} - \epsilon^{\mu\nu}) b_\nu, \end{aligned} \tag{I.82c}$$

together with the cyclicity of the trace.

We define the functional W_2 to be the mapping from the space of the $SU(N)$-valued fields $\bar{g}$ defined by Eq. (I.81) to the real numbers through

$$\begin{aligned} W_2[\bar{g}] &:= \overline{\int} \mathrm{tr}\, \left\{ (\partial_2 \bar{g})\bar{g}^{-1} \left[(\partial_+ \bar{g})\bar{g}^{-1}, (\partial_- \bar{g})\bar{g}^{-1} \right] \right\} \\ \text{Eq. (I.9)} \quad &= -\frac{2}{3} \overline{\int} \epsilon^{\bar{\mu}\bar{\nu}\bar{\lambda}}\, \mathrm{tr}\, \left[(\partial_{\bar{\mu}} \bar{g})\bar{g}^{-1} (\partial_{\bar{\nu}} \bar{g})\bar{g}^{-1} (\partial_{\bar{\lambda}} \bar{g})\bar{g}^{-1} \right]. \end{aligned} \tag{I.83a}$$

Here, $\bar{\mu}, \bar{\nu}, \bar{\lambda} = 0, 1, 2$, and the short-hand notation

$$\overline{\int} \equiv \int \mathrm{d}x^0 \mathrm{d}x^1 \int_0^1 \mathrm{d}x^2 = \int \frac{\mathrm{d}x^- \mathrm{d}x^+}{2} \int_0^1 \mathrm{d}x^2 \tag{I.83b}$$

is used.

We are going to show that if we define the WZNW action to be

$$\begin{aligned} S_{\mathrm{WZNW}} &\equiv S_{\mathrm{PCM}} + k\,\Gamma \\ &= -\frac{1}{4} \left(\frac{1}{\lambda^2} W_1 + \frac{k}{4\pi} W_2 \right), \end{aligned} \tag{I.84}$$

then the equations of motion (I.47a) with $\kappa = k$ an integer follow. To this end, it is sufficient to show that the infinitesimal changes of W_1 and W_2 if $g \to g + \delta g$ with δg infinitesimal are

$$\begin{aligned} \delta W_1[g] &= -\int \mathrm{tr}\, \left\{ \delta g g^{-1} \Big[\partial_+ \Big((\partial_- g) g^{-1} \Big) + \partial_- \Big((\partial_+ g) g^{-1} \Big) \Big] \right\} \\ &= -2 \int \mathrm{tr}\, \left\{ \delta g g^{-1} \Big[g^{\mu\nu}\, \partial_\mu \Big((\partial_\nu g) g^{-1} \Big) \Big] \right\}, \end{aligned} \tag{I.85a}$$

and

$$\begin{aligned} \delta W_2[\bar{g}] &= +\int \mathrm{tr}\, \left\{ \delta g g^{-1} \Big[\partial_+ \Big((\partial_- g) g^{-1} \Big) - \partial_- \Big((\partial_+ g) g^{-1} \Big) \Big] \right\} \\ \text{Eq. (I.9)} \quad &= +2 \int \mathrm{tr}\, \left\{ \delta g g^{-1} \Big[\epsilon^{\mu\nu}\, \partial_\mu \Big((\partial_\nu g) g^{-1} \Big) \Big] \right\}, \end{aligned} \tag{I.85b}$$

respectively. Observe that the right-hand side of Eq. (I.85b) only depends on g and not on $\bar{g}$. The equations of motion (I.47b) follow from the substitution $g \to g^{-1}$ in $W_1[g]$ and $W_2[g]$.

Proof. The proof of Eq. (I.85) relies on the following identities. Let $g : \mathbb{M}_{1,1} \to SU(N)$, $f : \mathbb{M}_{1,1} \to su(N)$, and $h : \mathbb{M}_{1,1} \to su(N)$ be smooth fields that obey the boundary conditions

$$\begin{aligned}
&\lim_{x^0\to\pm\infty} g(x^0,x^1) = 1, \qquad &&\lim_{x^1\to\pm\infty} g(x^0,x^1) = 1,\\
&\lim_{x^0\to\pm\infty} f(x^0,x^1) = 0, \qquad &&\lim_{x^1\to\pm\infty} f(x^0,x^1) = 0,\\
&\lim_{x^0\to\pm\infty} h(x^0,x^1) = 0, \qquad &&\lim_{x^1\to\pm\infty} h(x^0,x^1) = 0.
\end{aligned} \tag{I.86a}$$

Define the covariant derivative with respect to g to be the operation

$$\nabla^g_\mu f := \partial_\mu f - \left[(\partial_\mu g)g^{-1}, f\right]. \tag{I.86b}$$

Then, for any $\mu,\nu = 0,1$, the identities

$$\delta\Big((\partial_\mu g)g^{-1}\Big) = \nabla^g_\mu\Big((\delta g)g^{-1}\Big), \tag{I.86c}$$

$$\int \operatorname{tr}\left[(\nabla^g_\mu f)h\right] = -\int \operatorname{tr}\left[f(\nabla^g_\mu h)\right], \tag{I.86d}$$

$$\nabla^g_\mu\Big((\partial_\nu g)g^{-1}\Big) = \partial_\nu\Big((\partial_\mu g)g^{-1}\Big), \tag{I.86e}$$

hold as a consequence of the cyclicity of the trace and the vanishing of the $su(N)$ fields f and h upon performing partial integrations.[42]

The computation of the variation of W_1 goes as follows,

$$\begin{aligned}
\delta W_1[g] &= +\int \operatorname{tr}\left[\delta\Big((\partial_+ g)g^{-1}\Big)(\partial_- g)g^{-1} + (\partial_+ g)g^{-1}\delta\Big((\partial_- g)g^{-1}\Big)\right]\\
&= +\int \operatorname{tr}\left[\Big(\nabla^g_+(\delta g g^{-1})\Big)(\partial_- g)g^{-1} + (\partial_+ g)g^{-1}\Big(\nabla^g_-(\delta g g^{-1})\Big)\right]\\
&= -\int \operatorname{tr}\left[\delta g g^{-1}\nabla^g_+\Big((\partial_- g)g^{-1}\Big) + \delta g g^{-1}\nabla^g_-\Big((\partial_+ g)g^{-1}\Big)\right]\\
&= -\int \operatorname{tr}\left[\delta g g^{-1}\partial_-\Big((\partial_+ g)g^{-1}\Big) + \delta g g^{-1}\partial_+\Big((\partial_- g)g^{-1}\Big)\right]\\
&= -\int \operatorname{tr}\left[\delta g g^{-1}\bigg(\partial_+\Big((\partial_- g)g^{-1}\Big) + \partial_-\Big((\partial_+ g)g^{-1}\Big)\bigg)\right].
\end{aligned} \tag{I.87}$$

[42]Equation (I.86c) follows from

$$\delta\Big((\partial_\mu g)g^{-1}\Big) = \partial_\mu\Big((\delta g)g^{-1}\Big) + (\delta g)g^{-1}(\partial_\mu g)g^{-1} - (\partial_\mu g)g^{-1}\,(\delta g)g^{-1}.$$

Equation (I.86d) follows from partial integrations, the boundary conditions obeyed by f, g, h at infinity, together with

$$\operatorname{tr}\{[g,f]h\} = \operatorname{tr}\{gfh - fgh\} = \operatorname{tr}\{fhg - fgh\} = -\operatorname{tr}\{f[g,h]\}.$$

Equation (I.86e) follows from

$$\begin{aligned}
\partial_\mu\,\Big((\partial_\nu g)g^{-1}\Big) - \partial_\nu\,\Big((\partial_\mu g)g^{-1}\Big) &= (-)(\partial_\nu g)g^{-1}\,(\partial_\mu g)g^{-1}\\
&\quad - (-)(\partial_\mu g)g^{-1}\,(\partial_\nu g)g^{-1}\\
&= +\left[(\partial_\mu g)g^{-1}, (\partial_\nu g)g^{-1}\right].
\end{aligned}$$

The computation of the variation of W_2 is more involved. First,

$$\delta W_2[\bar g] = +\overline{\int} \delta \operatorname{tr}\left\{(\partial_2 \bar g)\bar g^{-1}\left[(\partial_+ \bar g)\bar g^{-1},\ (\partial_- \bar g)\bar g^{-1}\right]\right\}. \tag{I.88}$$

Because δ is an infinitesimal variation, it can be distributed according to

$$\begin{aligned}\delta W_2[\bar g] = &+\overline{\int} \operatorname{tr}\left\{\nabla_2^{\bar g}\left(\delta \bar g \bar g^{-1}\right)\left[(\partial_+ \bar g)\bar g^{-1},\ (\partial_- \bar g)\bar g^{-1}\right]\right\}\\ &+\overline{\int} \operatorname{tr}\left\{(\partial_2 \bar g)\bar g^{-1}\left[\nabla_+^{\bar g}\left(\delta \bar g \bar g^{-1}\right),\ (\partial_- \bar g)\bar g^{-1}\right]\right\}\\ &+\overline{\int} \operatorname{tr}\left\{(\partial_2 \bar g)\bar g^{-1}\left[(\partial_+ \bar g)\bar g^{-1},\ \nabla_-^{\bar g}\left(\delta \bar g \bar g^{-1}\right)\right]\right\}.\end{aligned} \tag{I.89}$$

Here, we made use of Eq. (I.86c). With the help of the cyclicity of the trace, the covariant derivative can be brought to the left within the trace on each line,

$$\begin{aligned}\delta W_2[\bar g] = &+\overline{\int} \operatorname{tr}\left\{\nabla_2^{\bar g}\left(\delta \bar g \bar g^{-1}\right)\left[(\partial_+ \bar g)\bar g^{-1}, (\partial_- \bar g)\bar g^{-1}\right]\right\}\\ &+\overline{\int} \operatorname{tr}\left\{\nabla_+^{\bar g}\left(\delta \bar g \bar g^{-1}\right)\left[(\partial_- \bar g)\bar g^{-1}, (\partial_2 \bar g)\bar g^{-1}\right]\right\}\\ &+\overline{\int} \operatorname{tr}\left\{\nabla_-^{\bar g}\left(\delta \bar g \bar g^{-1}\right)\left[(\partial_2 \bar g)\bar g^{-1}, (\partial_+ \bar g)\bar g^{-1}\right]\right\}.\end{aligned} \tag{I.90}$$

Because of Eq. (I.86d), we can perform three partial integrations that deliver

$$\begin{aligned}\delta W_2[\bar g] = &+\overline{\int} \partial_2 \operatorname{tr}\left\{(\delta g)\bar g^{-1}\left[(\partial_+ \bar g)g^{-1}, (\partial_- \bar g)\bar g^{-1}\right]\right\}\\ &-\overline{\int} \operatorname{tr}\left\{\delta \bar g \bar g^{-1}\, \nabla_2^{\bar g}\left[(\partial_+ \bar g)\bar g^{-1}, (\partial_- \bar g)\bar g^{-1}\right]\right\}\\ &-\overline{\int} \operatorname{tr}\left\{\delta \bar g \bar g^{-1}\, \nabla_+^{\bar g}\left[(\partial_- \bar g)\bar g^{-1}, (\partial_2 \bar g)\bar g^{-1}\right]\right\}\\ &-\overline{\int} \operatorname{tr}\left\{\delta \bar g \bar g^{-1}\, \nabla_-^{\bar g}\left[(\partial_2 \bar g)\bar g^{-1}, (\partial_+ \bar g)\bar g^{-1}\right]\right\}.\end{aligned} \tag{I.91}$$

We may now use the Jacobi identity that applies for any Lie algebra to simplify the last three terms on the right-hand side.[43] We then make the additive decomposition

$$\delta W_2[\bar g] = \delta W_2'[\bar g] + \delta W_2''[\bar g], \tag{I.92a}$$

where

$$\delta W_2'[g] := +\overline{\int} \partial_2 \operatorname{tr}\left\{(\delta \bar g)\bar g^{-1}\left[(\partial_+ \bar g)\bar g^{-1}, (\partial_- \bar g)\bar g^{-1}\right]\right\}, \tag{I.92b}$$

[43] Choose the three elements A, B, and C from a Lie algebra equipped with the Lie bracket $[\cdot,\cdot]$. The Jacobi identity $[A,[B,C]]+[B,[C,A]]+[C,[A,B]]=0$ is one of the defining property of the Lie algebra.

and

$$\begin{aligned}\delta W_2''[\bar{g}] := &-\overline{\int}\mathrm{tr}\,\left\{\delta\bar{g}\bar{g}^{-1}\,\partial_2\left[(\partial_+\bar{g})\bar{g}^{-1},(\partial_-\bar{g})\bar{g}^{-1}\right]\right\}\\ &-\overline{\int}\mathrm{tr}\,\left\{\delta\bar{g}\bar{g}^{-1}\,\partial_+\left[(\partial_-\bar{g})\bar{g}^{-1},(\partial_2\bar{g})\bar{g}^{-1}\right]\right\}\\ &-\overline{\int}\mathrm{tr}\,\left\{\delta\bar{g}\bar{g}^{-1}\,\partial_-\left[(\partial_2\bar{g})\bar{g}^{-1},(\partial_+\bar{g})\bar{g}^{-1}\right]\right\}.\end{aligned}\tag{I.92c}$$

The contribution $\delta W_2'[\bar{g}]$ is very similar to $\delta W_1[g]$ since the identity

$$\begin{aligned}\left[(\partial_+\bar{g})\bar{g}^{-1},(\partial_-\bar{g})\bar{g}^{-1}\right] =&\,(\partial_+\bar{g})\bar{g}^{-1}\,(\partial_-\bar{g})\bar{g}^{-1}-(\partial_-\bar{g})\bar{g}^{-1}\,(\partial_+\bar{g})\bar{g}^{-1}\\ =&-\partial_-\left((\partial_+\bar{g})\bar{g}^{-1}\right)+\partial_+\left((\partial_-\bar{g})\bar{g}^{-1}\right)\\ =&\,\partial_+\left((\partial_-\bar{g})\bar{g}^{-1}\right)-\partial_-\left((\partial_+\bar{g})\bar{g}^{-1}\right)\end{aligned}\tag{I.93}$$

holds. After integration in the direction 2, we find the important relationship

$$\delta W_2'[\bar{g}] = \int\mathrm{tr}\,\left\{\delta g g^{-1}\left[\partial_+\left((\partial_- g)g^{-1}\right)-\partial_-\left((\partial_+ g)g^{-1}\right)\right]\right\}.\tag{I.94}$$

Equation (I.85b) follows by establishing that

$$\delta W_2''[\bar{g}] \;=\; 0.\tag{I.95}$$

Equation (I.95) is another consequence of the Jacobi identity for Lie algebras. To see this, define

$$I(\bar{g}) := \partial_2\left[(\partial_+\bar{g})\bar{g}^{-1},(\partial_-\bar{g})\bar{g}^{-1}\right] + \text{cyclic permutations of } (2,+,-).\tag{I.96}$$

We have that

$$\begin{aligned}I(\bar{g}) =& \left[(\partial_2\partial_+\bar{g})\bar{g}^{-1},(\partial_-\bar{g})\bar{g}^{-1}\right]+\left[(\partial_+\bar{g})\bar{g}^{-1},(\partial_2\partial_-\bar{g})\bar{g}^{-1}\right]\\ &-\left[(\partial_+\bar{g})\bar{g}^{-1}(\partial_2\bar{g})\bar{g}^{-1},(\partial_-\bar{g})\bar{g}^{-1}\right]-\left[(\partial_+\bar{g})\bar{g}^{-1},(\partial_-\bar{g})\bar{g}^{-1}(\partial_2\bar{g})\bar{g}^{-1}\right]\\ &+\text{cyclic permutations of } (2,+,-).\end{aligned}\tag{I.97}$$

By making use of the antisymmetry of the commutator and some reordering, we conclude that

$$\begin{aligned}I(\bar{g}) =& \left[(\partial_2\partial_+\bar{g})\bar{g}^{-1},(\partial_-\bar{g})\bar{g}^{-1}\right]-\left[(\partial_+\partial_2\bar{g})\bar{g}^{-1},(\partial_-\bar{g})\bar{g}^{-1}\right]\\ &-\left[(\partial_+\bar{g})\bar{g}^{-1}(\partial_2\bar{g})\bar{g}^{-1},(\partial_-\bar{g})\bar{g}^{-1}\right]+\left[(\partial_2\bar{g})\bar{g}^{-1}(\partial_+\bar{g})\bar{g}^{-1},(\partial_-\bar{g})\bar{g}^{-1}\right]\\ &+\text{cyclic permutations of } (2,+,-)\\ =& \left[(\partial_-\bar{g})\bar{g}^{-1},\left[(\partial_+\bar{g})\bar{g}^{-1},(\partial_2\bar{g})\bar{g}^{-1}\right]\right]+\text{cyclic permutations of } (-,+,2)\\ =&\,0.\end{aligned}\tag{I.98}$$

□

I.7 One-loop RG flow for the WZNW theory

We continue working with fields taking values in $SU(N)$, although the manipulations that follow apply equally well to fields taking values in any simple and compact Lie group. The notations and boundary conditions from section I.6 carry through. The WZNW partition function is defined by

$$Z_{\text{WZNW}} := \int \mathcal{D}[g]\, e^{\mathrm{i}S_{\text{WZNW}}} \tag{I.99a}$$

with the WZNW action

$$S_{\text{WZNW}} := S_{\text{PCM}} + k\,\Gamma, \tag{I.99b}$$

where[44]

$$S_{\text{PCM}} = -\frac{1}{4\lambda^2}\int g^{\mu\nu}\,\mathrm{tr}\,\left(g^{-1}\partial_\mu g\right)\left(g^{-1}\partial_\nu g\right) \tag{I.99c}$$

is the action for the PCM, while the integer k multiplies the WZ action

$$\begin{aligned}\Gamma &= +\frac{1}{24\pi}\int \epsilon^{\bar\mu\bar\nu\bar\lambda}\,\mathrm{tr}\,\left(\bar g^{-1}\partial_{\bar\mu}\bar g\right)\left(\bar g^{-1}\partial_{\bar\nu}\bar g\right)\left(\bar g^{-1}\partial_{\bar\lambda}\bar g\right)\\ \overset{\text{Eq. (I.9)}}{=} &-\frac{1}{16\pi}\int \epsilon^{\mu\nu}\,\mathrm{tr}\,\left(\bar g^{-1}\partial_2\bar g\right)\left[\left(\bar g^{-1}\partial_\mu\bar g\right),\left(\bar g^{-1}\partial_\nu\bar g\right)\right].\end{aligned} \tag{I.99d}$$

The Haar measure $\mathcal{D}[g]$ on $SU(N)$ exists because $SU(N)$ is compact. It is invariant under left or right multiplication $g(x)\to l(x)\,g(x)\,r(x)$ for any $SU(N)$-valued fields l and r.

To implement an RG scheme, we choose to parametrize the $SU(N)$-valued field g according to the multiplicative rule

$$g(x) = g_0(x)\,u(x), \tag{I.100}$$

where the $SU(N)$-valued field g_0 solves the equations of motion (I.47). The $SU(N)$-valued field u encodes a small deviation away from the unit $N\times N$ matrix,

$$u(x) = e^{\mathrm{i}\lambda T^{\mathsf{a}}\pi^{\mathsf{a}}(x)} \approx 1 + \mathrm{i}\lambda T^{\mathsf{a}}\,\pi^{\mathsf{a}}(x) + \frac{1}{2}\left(\mathrm{i}\lambda T^{\mathsf{a}}\,\pi^{\mathsf{a}}(x)\right)^2 + \cdots, \qquad \lambda\geq 0. \tag{I.101}$$

The N^2-1 Hermitian and traceless $N\times N$ generators T^{a} of $su(N)$ are normalized according to Eq. (I.39). The summation convention over repeated $\mathsf{a}=1,\cdots,N^2-1$ is implied. The fluctuations $\pi^{\mathsf{a}}(x)$ are the fast modes in the terminology of section 3.7, as we shall explain shortly. We are seeking the effects on the partition function of integrating out the fast modes. We are going to sketch how this integration can be summarized by the one-loop RG flows (I.46).

We begin by inserting the expansion

$$\begin{aligned}g^{-1}\partial_\mu g &= u^{-1}\left(g_0^{-1}\partial_\mu g_0\right)u + u^{-1}\partial_\mu u\\ &= g_0^{-1}\partial_\mu g_0 + \sum_{n=1}^{\infty}\lambda^n\,\Delta_\mu^{(n)},\end{aligned} \tag{I.102a}$$

[44] The flat Minkowski metric $g^{\mu\nu}$ should not be confused with the $SU(N)$ valued field g.

where

$$\Delta^{(1)}_{\mu} := \mathrm{i}\left[\left(g_0^{-1}\partial_{\mu}g_0\right), T^{\mathrm{a}}\,\pi^{\mathrm{a}}\right] + \mathrm{i}\left(T^{\mathrm{a}}\partial_{\mu}\pi^{\mathrm{a}}\right),$$

$$\Delta^{(2)}_{\mu} := T^{\mathrm{b}}\,\pi^{\mathrm{b}}\left(g_0^{-1}\partial_{\mu}g_0\right)T^{\mathrm{a}}\,\pi^{\mathrm{a}} + \frac{\mathrm{i}^2}{2}\left\{\left(g_0^{-1}\partial_{\mu}g_0\right), \left(T^{\mathrm{a}}\,\pi^{\mathrm{a}}\right)^2\right\}, \tag{I.102b}$$

$$\vdots$$

into Eqs. (I.99c) and (I.99d). We thereby obtain an expansion in powers of λ of the integrands in Eqs. (I.99c) and (I.99d). By choosing g_0 to be a solution of the equations of motion of S_{WZNW}, we insure that the term of linear order in λ vanishes in this expansion. It can be shown that this expansion of $S_{\mathrm{PCM}}+k\,\Gamma$, up to quadratic order in $\boldsymbol{\pi}$, delivers [111]

$$S_{\mathrm{PCM}} = \int \mathrm{tr}\left\{\frac{1}{4\lambda^2}\partial_{\mu}g_0^{-1}\,\partial^{\mu}g_0 + \frac{1}{2}\partial_{\mu}\pi^{\mathrm{a}}\,\partial^{\mu}\pi^{\mathrm{a}} + \frac{g^{\mu\nu}}{4}\left(g_0^{-1}\partial_{\mu}g_0\right)\left[T^{\mathrm{a}}\,\pi^{\mathrm{a}}, T^{\mathrm{b}}\partial_{\nu}\pi^{\mathrm{b}}\right]\right\} \tag{I.103a}$$

and, making use of Eq. (I.9),

$$\Gamma = -\int \frac{\lambda^2}{16\pi}\mathrm{tr}\left\{\epsilon^{\mu\nu}\left(g_0^{-1}\partial_{\mu}g_0\right)\left[T^{\mathrm{a}}\,\pi^{\mathrm{a}}, T^{\mathrm{b}}\partial_{\nu}\pi^{\mathrm{b}}\right]\right\}. \tag{I.103b}$$

The terminology of slow modes for the $SU(N)$-valued field g_0 and fast modes for the $su(N)$-valued fields π^{a} is justified when $\lambda^2 \ll 1$. The reasoning here is the same as when interpreting Eq. (3.206). The penalty for variations in space of g_0 is of order $\lambda^{-2} \gg 1$. The penalty for variations in space of π^{a} is of order 1. Remarkably, the expansion up to quadratic order in $\boldsymbol{\pi}$ of the WZ action delivers a single term, that couples the slow to the fast modes. [143] This term only differs from the interaction between slow and fast modes arising from the expansion up to quadratic order in $\boldsymbol{\pi}$ of the PCM action through the prefactor $-k\,\lambda^2/(4\pi)$ and the presence of the Levi-Civita antisymmetric tensor instead of the flat Lorentz metric.

Integration over the fast modes in a small momentum shell may be performed as was done in section 3.7. There results an additive change to the Lagrangian density of the slow modes through a term proportional to

$$\left[1 - \left(\frac{k\,\lambda^2}{4\pi}\right)^2\right]\,\mathrm{tr}\,\left(\partial_{\mu}g_0^{-1}\,\partial^{\mu}g_0\right). \tag{I.104}$$

The relative sign that distinguishes the renormalization stemming from the interaction between fast and slow modes in the PCM action from that in the WZ action is dictated by the $\mathbb{Z}_2 \times \mathbb{Z}_2$ discrete symmetries respected by the PCM action rather than the smaller $\mathbb{Z}_2$ symmetry respected by the WZ action. This difference is encoded by the contraction $\epsilon^{\mu\lambda}\,\epsilon^{\lambda\nu} = -\delta^{\mu\nu}$ arising from renormalization induced by the WZ action in contrast to the contraction $g^{\mu\lambda}\,g^{\lambda\nu} = +\delta^{\mu\nu}$ arising from renormalization induced by the PCM action. No contribution involving $g^{\mu\lambda}\,\epsilon^{\lambda\nu}$ is permissible when renormalizing tr $\left(\partial_{\mu}g_0^{-1}\,\partial^{\mu}g_0\right)$, by the discrete symmetries it obeys. This completes the sketch of the computation that delivers the one-loop RG flows (I.46).

I.8 The Polyakov-Wiegmann identity

The notations and boundary conditions from section I.6 carry through. In particular, we keep using the lower case g to denote a $SU(N)$-valued field as opposed to the upper case G to denote a $U(N)$-valued field. We define the WZW action in Minkowski space $\mathbb{M}_{1,1}$ to be

$$\begin{aligned}S_{\rm WZW} := &\frac{1}{16\pi}\int \mathrm{d}x^0\mathrm{d}x^1\,\mathrm{tr}\,\left[(\partial_\mu g)(\partial^\mu g^{-1})\right]\\ &-\frac{1}{24\pi}\int_{\rm B} \mathrm{d}x^0\mathrm{d}x^1\mathrm{d}x^2\,\epsilon^{\bar\mu\bar\nu\bar\lambda}\,\mathrm{tr}\,\left[(\partial_{\bar\mu}\bar g)\bar g^{-1}(\partial_{\bar\nu}\bar g)\bar g^{-1}(\partial_{\bar\lambda}\bar g)\bar g^{-1}\right].\end{aligned}\tag{I.105}$$

The boundary conditions (I.81b) allow to interpret Minkowski space $\mathbb{M}_{1,1}$ as the boundary ∂B of the "ball" B, whereby the (not unique) extension of the fields (I.81c) is assumed. The sign of the WZ term, the second line on the right-hand side of Eq. (I.105), is a matter of convention, for it may be reversed by either $g\to g^{-1}$ or $x^1\to -x^1$. The choice made here is consistent with the choice made in Eq. (I.26). The Polyakov-Wiegmann identity states that

$$S_{\rm WZW}[g^{-1}\,h] = S_{\rm WZW}[g^{-1}] + S_{\rm WZW}[h] + \frac{1}{8\pi}\int\frac{\mathrm{d}x^-\mathrm{d}x^+}{2}\,\mathrm{tr}\,\left[(g\,\partial_+ g^{-1})\,(h\,\partial_- h^{-1})\right].\tag{I.106}$$

The proof of Eq. (I.106) is left to the reader.

I.9 Integration of the anomaly in QCD$_2$

We saw in chapter 9 that the Abelian bosonization rules, that relate the quantum theory of a fermion obeying the Dirac equation in the fundamental representation of the Lorentz group to the quantum theory of a real-valued scalar field obeying the massless Sine-Gordon equation, remain valid in the presence of interactions.

The extension of Witten's non-Abelian bosonization rules to interacting fermions is a more subtle matter, as was shown by Polyakov and Wiegmann in Ref. [112] and as we shall now illustrate.

I.9.1 *General symmetry considerations*

We start from the Grassmann partition function

$$Z[A,A^5] := \int \mathcal{D}[\bar\psi]\,\mathcal{D}[\psi]\,e^{\mathrm{i}S},\tag{I.107a}$$

with the action

$$S := \int \mathrm{d}x^0\mathrm{d}x^1\,\mathcal{L},\tag{I.107b}$$

and the Lagrangian density

$$\mathcal{L} := \bar\psi\,\gamma^\mu\left(\mathrm{i}\partial_\mu + A_\mu + \gamma^5 A^5_\mu\right)\psi.\tag{I.107c}$$

The Grassmann-valued row, $\bar{\psi}$, and column, ψ, vectors have $2N$ components. On the one hand, they represent a Dirac spinor that transforms according to the two-dimensional representation of the Lorentz group acting on $(1+1)$-dimensional Minkowski space and time $\mathbb{M}_{1,1}$. On the other hand, they also represent a Dirac spinor that transforms according to the fundamental representation of the group $U(N) = U(1) \times SU(N)$ acting on N internal degrees of freedom. The internal subgroup $U(1)$ will be associated to charge conservation, while the internal subgroup $SU(N)$ will be associated to color conservation. The background gauge fields A_μ belong to the Lie algebra $u(N)$. The background axial gauge fields A^5_μ also belong to the Lie algebra $u(N)$.

The quantum theory (I.107) is a generalization from the quantum theory (I.29) for N massless Dirac fermions propagating freely in $(1+1)$-dimensional Minkowski space and time $\mathbb{M}_{1,1}$ to a quantum theory for N massless Dirac fermions propagating in the background of the field $A_\mu + \gamma^5 A^5_\mu$. The presence of the background gauge and axial gauge fields will be shown to bring about a dramatic departure from Witten's non-Abelian bosonization rules. When the axial gauge field is vanishing, the coupling between the fermions and the gauge fields is similar to the coupling in quantum chromodynamics (QCD). Since space and time is two-dimensional Minkowski space $\mathbb{M}_{1,1}$, we shall use the acronym QCD_2.

As usual, the symmetries obeyed by the quantum theory (I.107) play a crucial role. We begin with the classical symmetries, namely the symmetries obeyed by the Lagrangian density (I.107c).

For any set of N^2 functions

$$\begin{aligned} \omega^{\mathsf{a}} : \mathbb{M}_{1,1} &\longrightarrow \mathbb{R}, \\ x &\longmapsto \omega^{\mathsf{a}}(x), \end{aligned} \tag{I.108a}$$

with $\mathsf{a} = 1, \cdots, N^2$, we may define the $U(N)$-valued field

$$U(\omega) := e^{\mathrm{i}T^{\mathsf{a}}\,\omega^{\mathsf{a}}}. \tag{I.108b}$$

The N^2 generators T^{a} of $u(N)$ are any set of linearly independent Hermitian matrices in a given representation and the summation convention over repeated $\mathsf{a} = 1, \cdots, N^2$ is implied. The Lagrangian density (I.107c) is invariant under the local continuous transformations

$$\begin{aligned} &\bar{\psi} =: \bar{\psi}(\omega)\, U^{-1}(\omega), \qquad \psi =: U(\omega)\, \psi(\omega), \\ &A_\mu =: U(\omega)\, A_\mu(\omega)\, U^{-1}(\omega) - \Big(\mathrm{i}\partial_\mu U(\omega)\Big) U^{-1}(\omega), \\ &A^5_\mu =: U(\omega)\, A^5_\mu(\omega)\, U^{-1}(\omega). \end{aligned} \tag{I.108c}$$

Similarly, for any set of N^2 functions

$$\begin{aligned} \omega^{\mathsf{a}}_5 : \mathbb{M}_{1,1} &\longrightarrow \mathbb{R}, \\ x &\longmapsto \omega^{\mathsf{a}}_5(x), \end{aligned} \tag{I.109a}$$

with $\mathsf{a} = 1, \cdots, N^2$, we may define the $U(N)$-valued field

$$U(\omega_5) := e^{\mathrm{i}\gamma^5\, T^{\mathsf{a}}\, \omega^{\mathsf{a}}_5} \equiv e^{\mathrm{i}T^{\mathsf{a}}\, \gamma^5\, \omega^{\mathsf{a}}_5}. \tag{I.109b}$$

The Lagrangian density (I.107c) is also invariant under the local continuous transformations

$$\begin{aligned}\bar{\psi} &=: \bar{\psi}(\omega_5)\,U(\omega_5), \qquad \psi =: U(\omega_5)\,\psi(\omega_5),\\ A^5_\mu &=: U(\omega_5)\,A^5_\mu(\omega_5)\,U^{-1}(\omega_5) - \gamma^5\left(\mathrm{i}\partial_\mu U(\omega_5)\right)U^{-1}(\omega_5),\\ A_\mu &=: U(\omega_5)\,A_\mu(\omega_5)\,U^{-1}(\omega_5).\end{aligned} \tag{I.109c}$$

Observe that

$$\bar{\psi}\,\psi = \bar{\psi}(\omega)\,\psi(\omega), \tag{I.110}$$

whereas

$$\bar{\psi}\,\psi = \bar{\psi}(\omega_5)\,U(\omega_5)\,\psi(\omega_5). \tag{I.111}$$

In words, a mass term is gauge invariant but breaks the axial gauge symmetry, classically.

The symmetry of the Lagrangian density (I.107c) under a composition of the gauge and axial gauge transformations also holds, but the transformation laws of the gauge fields are more complicated.

According to Noether's theorem, the symmetry operations (I.108) and (I.109) must be associated to generalized continuity equations. These are constructed from the (Dirac) equations of motion

$$(\partial_\mu\bar{\psi}\,\gamma^\mu) = -\mathrm{i}\bar{\psi}\,\gamma^\mu\left(A_\mu + \gamma^5\,A^5_\mu\right), \tag{I.112a}$$

$$(\gamma^\mu\,\partial_\mu\psi) = +\mathrm{i}\,\gamma^\mu\left(A_\mu + \gamma^5\,A^5_\mu\right)\psi, \tag{I.112b}$$

whereby, to establish Eq. (I.112a), we have assumed boundary conditions on the spinors such that we may drop all boundary terms upon partial integrations.

For any $\mu = 0, 1$, it follows from Eq. (I.112) that the Abelian (charge) currents

$$j^\mu_{\bar{\psi},\psi} := \bar{\psi}\,\gamma^\mu\,\psi \tag{I.113a}$$

and

$$j^{5\,\mu}_{\bar{\psi},\psi} := \bar{\psi}\,\gamma^\mu\,\gamma^5\,\psi \tag{I.113b}$$

obey the continuity equations

$$\partial_\mu\,j^\mu_{\bar{\psi},\psi} = 0, \tag{I.114a}$$

and

$$\partial_\mu\,j^{5\,\mu}_{\bar{\psi},\psi} = 0, \tag{I.114b}$$

respectively.

The Abelian currents (I.113) can be embedded into the multiplets

$$J^{\mathrm{a}\,\mu}_{\bar{\psi},\psi} := \bar{\psi}\,T^{\mathrm{a}}\,\gamma^\mu\,\psi \equiv \bar{\psi}\,\gamma^\mu\,T^{\mathrm{a}}\,\psi, \tag{I.115a}$$

and

$$J^{5\,\mathrm{a}\,\mu}_{\bar{\psi},\psi} := \bar{\psi}\,T^{\mathrm{a}}\,\gamma^\mu\,\gamma^5\,\psi \equiv \bar{\psi}\,\gamma^\mu\,\gamma^5\,T^{\mathrm{a}}\,\psi, \tag{I.115b}$$

with the convention that T^{a} with $\mathsf{a}=1,\cdots,N^2-1$ are the Hermitian and traceless generators of $su(N)$ and T^{N^2} is proportional to the unit $N\times N$ matrix that generates $u(1)$. At this stage, it is convenient to use the notation $\mathcal{A}_\mu$ as a short-hand notation for the two-vector field $A_\mu+\gamma^5 A^5_\mu$,

$$\mathcal{A}_\mu \equiv A_\mu+\gamma^5 A^5_\mu, \qquad \mathcal{A}_\mu \equiv \mathcal{A}^{\mathsf{a}}_\mu\, T^{\mathsf{a}}, \tag{I.116}$$

where the summation convention over the repeated index of the $u(N)$ generators is implied. We shall also use the fully antisymmetric structure constants of $u(N)$ that define the Lie bracket of $u(N)$,

$$[T^{\mathsf{a}},T^{\mathsf{b}}]=\mathrm{i}f^{\mathsf{abc}}\,T^{\mathsf{c}} \tag{I.117}$$

for any pair $\mathsf{a},\mathsf{b}=1,\cdots,N^2$. For any $\mathsf{a}=1,\cdots,N^2$, application of the equations of motion (I.112) to the gauge current (I.115a) delivers

$$\begin{aligned}
\partial_\mu J^{\mathsf{a}\mu}_{\bar\psi,\psi} &= (\partial_\mu\bar\psi\,\gamma^\mu)\,T^{\mathsf{a}}\,\psi+\bar\psi\,T^{\mathsf{a}}(\gamma^\mu\,\partial_\mu\psi)\\
&= -\mathrm{i}\bar\psi\,\gamma^\mu\,\mathcal{A}_\mu\,T^{\mathsf{a}}\,\psi+\mathrm{i}\bar\psi\,\gamma^\mu\,T^{\mathsf{a}}\,\mathcal{A}_\mu\,\psi\\
&= +\mathrm{i}\bar\psi\,\gamma^\mu\,\left[T^{\mathsf{a}},\mathcal{A}_\mu\right]\psi\\
&= -f^{\mathsf{abc}}\,\bar\psi\,\gamma^\mu\,T^{\mathsf{c}}\,\psi\,\mathcal{A}^{\mathsf{b}}_\mu\\
&= +f^{\mathsf{abc}}\,J^{\mathsf{b}\mu}_{\bar\psi,\psi}\,\mathcal{A}^{\mathsf{c}}_\mu .
\end{aligned} \tag{I.118a}$$

For any $\mathsf{a}=1,\cdots,N^2$, application of the equations of motion (I.112) to the gauge current (I.115b) together with the fact that γ^5, by definition, anticommutes with $\gamma^\mu=(\gamma^0,\gamma^1)$, delivers

$$\begin{aligned}
\partial_\mu J^{5\,\mathsf{a}\mu}_{\bar\psi,\psi} &= (\partial_\mu\bar\psi\,\gamma^\mu)\,\gamma^5\,T^{\mathsf{a}}\,\psi-\bar\psi\,T^{\mathsf{a}}\,\gamma^5\,(\gamma^\mu\,\partial_\mu\psi)\\
&= -\mathrm{i}\bar\psi\,\gamma^\mu\,\mathcal{A}_\mu\,\gamma^5\,T^{\mathsf{a}}\,\psi-\mathrm{i}\bar\psi\,\gamma^5\,\gamma^\mu\,T^{\mathsf{a}}\,\mathcal{A}_\mu\,\psi\\
&= +\mathrm{i}\bar\psi\,\gamma^\mu\,\gamma^5\,\left[T^{\mathsf{a}},\mathcal{A}_\mu\right]\psi\\
&= -f^{\mathsf{abc}}\,\bar\psi\,\gamma^\mu\,\gamma^5\,T^{\mathsf{c}}\,\psi\,\mathcal{A}^{\mathsf{b}}_\mu\\
&= +f^{\mathsf{abc}}\,J^{5\,\mathsf{b}\mu}_{\bar\psi,\psi}\,\mathcal{A}^{\mathsf{c}}_\mu .
\end{aligned} \tag{I.118b}$$

In order to recast the equations of motion (I.118) in a form reminiscent of a continuity equation, we first define

$$J^{\mu}_{\bar\psi,\psi}\equiv J^{\mathsf{a}\mu}_{\bar\psi,\psi}\,T^{\mathsf{a}}, \qquad J^{5\,\mu}_{\bar\psi,\psi}\equiv J^{5\,\mathsf{a}\mu}_{\bar\psi,\psi}\,T^{\mathsf{a}}. \tag{I.119}$$

After contracting the equations of motion (I.118) with the generators T^{a} of $u(N)$, we may then use the defining Lie bracket of $u(N)$, recall Eq. (I.117), to write

$$\begin{aligned}
\partial_\mu J^{\mu}_{\bar\psi,\psi} &=(-\mathrm{i})\times(+\mathrm{i})\,T^{\mathsf{a}}\,f^{\mathsf{abc}}\,J^{\mathsf{b}\mu}_{\bar\psi,\psi}\,\mathcal{A}^{\mathsf{c}}_\mu\\
&=(-\mathrm{i})\times[T^{\mathsf{b}},T^{\mathsf{c}}]\,J^{\mathsf{b}\mu}_{\bar\psi,\psi}\,\mathcal{A}^{\mathsf{c}}_\mu\\
&=(+\mathrm{i})[\mathcal{A}_\mu,J^{\mu}_{\bar\psi,\psi}]
\end{aligned} \tag{I.120a}$$

and

$$\begin{aligned}\partial_\mu J^{5\,\mu}_{\bar\psi,\psi} &= (-\mathrm{i}) \times (+\mathrm{i})\, T^{\mathsf{a}} f^{\mathsf{abc}}\, J^{5\,\mathsf{b}\mu}_{\bar\psi,\psi}\, \mathcal{A}^{\mathsf{c}}_\mu \\ &= (-\mathrm{i}) \times [T^{\mathsf{b}}, T^{\mathsf{c}}]\, J^{5\,\mathsf{b}\mu}_{\bar\psi,\psi}\, \mathcal{A}^{\mathsf{c}}_\mu \\ &= (+\mathrm{i})[\mathcal{A}_\mu, J^{5\,\mu}_{\bar\psi,\psi}].\end{aligned} \tag{I.120b}$$

It is convenient to define the covariant derivative in the adjoint representation by

$$\nabla_\mu \cdot := \partial_\mu \cdot - \mathrm{i}\left[\mathcal{A}_\mu, \cdot\right]. \tag{I.121}$$

We can now recast the equations of motion (I.120a) and (I.120b) as the covariant continuity equations

$$\nabla_\mu\, J^{\mu}_{\bar\psi,\psi} = 0 \tag{I.122a}$$

and

$$\nabla_\mu\, J^{5\,\mu}_{\bar\psi,\psi} = 0, \tag{I.122b}$$

respectively. Equations (I.122a) and (I.122b) reduce to Eqs. (I.114a) and (I.114b), respectively, for the component $\mathsf{a} = N^2$. For the $su(N)$ components $\mathsf{a} = 1, \cdots, N^2 - 1$, Eqs. (I.122a) and (I.122b) are explicitly dependent on the background of non-Abelian gauge and axial gauge fields.

The covariant continuity equations (I.122) hold at the classical level, i.e., they are consequences of the equations of motion derived from the action (I.107b). Is there a counterpart to these covariant continuity equations (I.122) at the quantum level? To make this question more precise, we define the $u(N)$-valued quantum gauge currents and quantum axial gauge currents by

$$\begin{aligned}J^{\mu}[A, A^5] &:= -\mathrm{i}\left(\frac{\delta \ln Z}{\delta A_\mu}\right)[A, A^5] \\ &\equiv \left(\frac{1}{Z}\left\langle \frac{\delta S}{\delta A_\mu}\right\rangle_Z\right)[A, A^5],\end{aligned} \tag{I.123a}$$

and

$$\begin{aligned}J^{5\,\mu}[A, A^5] &:= -\mathrm{i}\left(\frac{\delta \ln Z}{\delta A^5_\mu}\right)[A, A^5] \\ &\equiv \left(\frac{1}{Z}\left\langle \frac{\delta S}{\delta A^5_\mu}\right\rangle_Z\right)[A, A^5],\end{aligned} \tag{I.123b}$$

respectively. We recognize in

$$\frac{\delta S}{\delta\left(A_\mu\right)_{ab}} = \bar\psi_a\, \gamma^\mu\, \psi_b, \qquad a, b = 1, \cdots, N, \tag{I.124a}$$

and

$$\frac{\delta S}{\delta\left(A^5_\mu\right)_{ab}} = \bar\psi_a\, \gamma^\mu\, \gamma^5\, \psi_b, \qquad a, b = 1, \cdots, N, \tag{I.124b}$$

when evaluated at the solution of the equations of motion (I.112), the $u(N)$-valued counterparts to the currents (I.115a) and (I.115b), respectively. The quantum currents are not limited to the solution of the equations of motion (I.112) obeyed by the Dirac spinor. They involve a weighted average over all possible configurations of the Dirac spinor. We make the Ansatz that these quantum fluctuations turn Eqs. (I.122a) and (I.122b) into

$$\begin{aligned}\nabla_\mu J^\mu[A,A^5] :=& \partial_\mu J^\mu[A,A^5] - \mathrm{i}\Big[A_\mu + \gamma^5 A^5_\mu, J^\mu[A,A^5]\Big]\\ \stackrel{?}{=}& \mathcal{A}[A,A^5],\end{aligned} \tag{I.125a}$$

and

$$\begin{aligned}\nabla^5_\mu J^{5\,\nu}[A,A^5] :=& \partial_\mu J^{5\,\mu}[A,A^5] - \mathrm{i}\Big[A_\mu + \gamma^5 A^5_\mu, J^{5\,\mu}[A,A^5]\Big]\\ \stackrel{?}{=}& \mathcal{A}^5[A,A^5],\end{aligned} \tag{I.125b}$$

respectively. Here, $\mathcal{A}[A,A^5]$ and $\mathcal{A}^5[A,A^5]$ are $u(N)$-valued and presumed to depend locally on $x \in \mathbb{M}_{1,1}$. The question mark over the equality stresses that we are making an educated guess at this stage.

We proceed by making formal manipulations, i.e., neither do we prove the existence of the objects we manipulate nor do we prove the legality of our manipulations. The formal object is

$$Z[A,A^5] = \mathrm{Det}\,\left[\gamma^\mu \left(\mathrm{i}\partial_\mu + A_\mu + \gamma^5 A^5_\mu\right)\right]. \tag{I.126}$$

This is a formal object because (i) it is the product of an infinite number of eigenvalues $\lambda_\iota \in \mathbb{C}$ labeled by an index ι belonging to an ordered set that need not be countable, (ii) some of the eigenvalues λ_ι might be vanishing (they are called zero modes), and (iii) there is no upper bound on the spectrum $|\lambda_\iota| \in \mathbb{R}$. In short, this is an ill-conditioned product.

Difficulty (i) is overcome by choosing boundary conditions that amount to compactifying $\mathbb{M}_{1,1}$.

To overcome difficulty (ii), we shall assume that

$$\lambda_\iota \neq 0, \qquad \forall \iota, \tag{I.127}$$

for the background field (I.116). Under this assumption, if

$$\gamma^\mu \left(\mathrm{i}\partial_\mu + A_\mu + \gamma^5 A^5_\mu\right) \psi_\iota = \lambda_\iota\, \psi_\iota, \tag{I.128a}$$

then

$$\gamma^\mu \left(\mathrm{i}\partial_\mu + A_\mu + \gamma^5 A^5_\mu\right) \left(\gamma^5\, \psi_\iota\right) = -\lambda_\iota \left(\gamma^5\, \psi_\iota\right). \tag{I.128b}$$

Equations (I.127) and (I.128) imply that all eigenvalues λ_ι of $\gamma^\mu \left(\mathrm{i}\partial_\mu + A_\mu + \gamma^5 A^5_\mu\right)$ come in pairs of opposite sign. This spectral property is called the chiral symmetry. When the operator $\mathrm{i}\partial_\mu + A_\mu + \gamma^5 A^5_\mu$ has a chiral-symmetric spectrum, we avoid the ambiguity of taking the logarithm of 0 when evaluating the logarithm of the partition function. There remains the ambiguity of taking the logarithm of a pair of complex-valued numbers with opposite signs. This difficulty is surmounted by substituting for $Z[A,A^5]$ the ratio $Z[A,A^5]/Z[0,0]$ in Eqs. (I.123a) and (I.123b). This substitution does not change the values of the quantum currents. Nevertheless, the formal expression

$$\ln\frac{Z[A,A^5]}{Z[0,0]} = \operatorname{Tr}\ln\left[\gamma^\mu\left(\mathrm{i}\partial_\mu + A_\mu + \gamma^5 A^5_\mu\right)\right] - \operatorname{Tr}\ln\left(\gamma^\mu\,\mathrm{i}\partial_\mu\right) \tag{I.129}$$

still remains an ill-conditioned sum, whereas

$$\frac{Z[A,A^5]}{Z[0,0]} = \frac{\operatorname{Det}\left[\gamma^\mu\left(\mathrm{i}\partial_\mu + A_\mu + \gamma^5 A^5_\mu\right)\right]}{\operatorname{Det}\left(\gamma^\mu\,\mathrm{i}\partial_\mu\right)} \tag{I.130}$$

still remains an ill-conditioned product, for difficulty (iii) has not yet been overcome.

If we assume that we choose the ordering $|\lambda_\iota| \leq |\lambda_{\iota'}|$ for $\iota < \iota'$, a bound of the type $|\lambda_\iota| < f_\iota$ for some positive function f that imposes a regularization condition as $\iota \to \infty$ must be imposed, with the dramatic consequence that $\mathcal{A}[A,A^5]$ or $\mathcal{A}^5[A,A^5]$ on the right-hand sides of Eqs. (I.125a) and (I.125b) might become non-vanishing depending on how the regularization is imposed. In effect, the integration over the Grassmann measure that implements the evaluation of the fermion determinant (I.126) has less symmetries than the action (I.107b) as soon as the spectrum of eigenvalues of $\gamma^\mu\left(\mathrm{i}\partial_\mu + A_\mu + \gamma^5 A^5_\mu\right)$ has been regularized.

Under what conditions are the covariant continuity equations (I.122) broken by the quantum fluctuations? To answer this question, we define the covariant derivative

$$D_\mu := \partial_\mu - \mathrm{i}A_\mu. \tag{I.131}$$

We observe that

$$\mathrm{i}\left[D_\mu, D_\nu\right] = \partial_\mu A_\nu - \partial_\nu A_\mu - \mathrm{i}\left[A_\mu, A_\nu\right] \tag{I.132}$$

reduces to the field strength of electrodynamics when the gauge fields belong to the algebra $u(1)$. We thus use Eq. (I.132) to define the field strength

$$F_{\mu\nu} := \mathrm{i}\left[D_\mu, D_\nu\right] = \partial_\mu A_\nu - \partial_\nu A_\mu - \mathrm{i}\left[A_\mu, A_\nu\right], \tag{I.133}$$

for an arbitrary compact gauge group. The expansions $A = A^{\mathsf{a}}\,T^{\mathsf{a}}$ and $A^5 = A^{5\,\mathsf{a}}\,T^{\mathsf{a}}$ then delivers

$$F^{\mathsf{a}}_{\mu\nu} := \partial_\mu A^{\mathsf{a}}_\nu - \partial_\nu A^{\mathsf{a}}_\mu + f^{\mathsf{abc}}\,A^{\mathsf{b}}_\mu\,A^{\mathsf{c}}_\nu, \tag{I.134a}$$

and

$$F^{5\,\mathsf{a}}_{\mu\nu} := \partial_\mu A^{5\,\mathsf{a}}_\nu - \partial_\nu A^{5\,\mathsf{a}}_\mu + f^{\mathsf{abc}}\,A^{5\,\mathsf{b}}_\mu\,A^{5\,\mathsf{c}}_\nu, \tag{I.134b}$$

respectively, with $\mathsf{a} = 1,\cdots,N^2$, and where f^{abc} are the (fully antisymmetric) structure constants of $u(N)$, i.e., $[T^{\mathsf{a}}, T^{\mathsf{b}}] = \mathrm{i}f^{\mathsf{abc}}\,T^{\mathsf{c}}$. If we use dimensional analysis and request both Lorentz invariance and locality, we conclude that $\mathcal{A}[A,A^5]$ and $\mathcal{A}^5[A,A^5]$ on the right-hand sides of Eqs. (I.125a) and (I.125b) should be functions of $\epsilon_{\mu\nu}\,F^{\mu\nu}$ and $\epsilon_{\mu\nu}\,F^{5\,\mu\nu}$.

I.9.2 The axial gauge anomaly for QCD_2

We specialize to the partition function of a Dirac fermion in the background of a $U(N)$ gauge field,

$$Z[A] := \int \mathcal{D}[\bar{\psi}]\,\mathcal{D}[\psi]\,e^{\mathrm{i}S}, \tag{I.135a}$$

$$S := \int \mathrm{d}x^0 \mathrm{d}x^1\,\mathcal{L}, \tag{I.135b}$$

$$\mathcal{L} := \bar{\psi}\,\gamma^\mu \left(\mathrm{i}\partial_\mu + A_\mu\right)\psi. \tag{I.135c}$$

We demand that this quantum field theory be invariant under any local gauge transformation (I.108). This means that the quantum gauge current

$$J^\mu := -\mathrm{i}\left(\frac{\delta \ln Z}{\delta A_\mu}\right) \tag{I.136}$$

is conserved in the sense of the classical covariant continuity equation (I.122a), whereas

$$J^{5\,\mu} := \epsilon^{\mu\nu}\,J_\nu \tag{I.137}$$

will be shown to violate the classical covariant continuity equation (I.122b) for any non-vanishing $F_{\mu\nu}$.

We borrow the computation (9.54) of the primary $U(1)$ axial anomaly from chapter 9 in two-dimensional Euclidean space, we perform the Wick rotations (I.10), and we postulate the quantum anomalous covariant continuity equations

$$\begin{aligned}&\partial_\mu J^\mu - \mathrm{i}\left[A_\mu, J^\mu\right] = 0,\\ &\epsilon_{\mu\nu}\left(\partial^\mu J^\nu - \mathrm{i}\left[A^\mu, J^\nu\right]\right) = \frac{1}{4\pi}\epsilon_{\mu\nu}\,F^{\mu\nu}.\end{aligned} \tag{I.138}$$

The factor $1/(4\pi)$ is here fixed by the normalization convention (I.39). Following Polyakov and Wiegmann in Ref. [112], we are going to solve (I.138) as the preliminary step needed for the integration of Eq. (I.136). As a check of the postulate (I.138), we shall recover the Abelian bosonization rules.

Solving the anomaly (I.138) is performed in the light-cone coordinates defined by Eq. (I.15). With the help of the identities (I.16), the anomaly becomes

$$\begin{aligned}&\frac{1}{2}\left(\nabla_+ J_- + \nabla_- J_+\right) = 0,\\ &\frac{1}{2}\left(\nabla_+ J_- - \nabla_- J_+\right) = \frac{1}{8\pi}\left(F_{+-} - F_{-+}\right),\end{aligned} \tag{I.139a}$$

where we have defined

$$\nabla_\pm \cdot := \partial_\pm \cdot - \mathrm{i}[A_\pm, \cdot] \tag{I.139b}$$

and

$$F_{+-} := \partial_+ A_- - \partial_- A_+ - \mathrm{i}[A_+, A_-] =: -F_{-+}. \tag{I.139c}$$

Adding and subtracting the two lines of Eq. (I.139) yields

$$\begin{aligned} \nabla_+ J_- &= \frac{1}{4\pi} F_{+-}, \\ \nabla_- J_+ &= \frac{1}{4\pi} F_{-+}. \end{aligned} \tag{I.140}$$

Polyakov and Wiegmann have shown in Ref. [112] that, up to initial conditions,

$$\begin{aligned} J_- &= -\frac{\mathrm{i}}{4\pi} (\partial_- G_-)\, G_-^{-1} + \frac{\mathrm{i}}{4\pi} (\partial_- G_+)\, G_+^{-1}, \\ J_+ &= -\frac{\mathrm{i}}{4\pi} (\partial_+ G_+)\, G_+^{-1} + \frac{\mathrm{i}}{4\pi} (\partial_+ G_-)\, G_-^{-1}, \end{aligned} \tag{I.141}$$

solve the anomaly equations (I.139). This general solution is however not adequate for the purpose of integrating Eq. (I.136).

The main idea that we use to integrate Eq. (I.136) is to represent the $u(N)$-valued background gauge fields A_- and A_+ by $SU(N)$-valued fields G_- and G_+, respectively. To this end, for any $x \in \mathbb{M}_{1,1}$, we define

$$\begin{aligned} G_-(x^-, x^+) &:= \mathcal{T}_- \exp\left(+\mathrm{i} \int\limits_{x_0^-}^{x^-} \mathrm{d}y\, A_-(y, x^+) \right) G_-(x_0^-, x^+), \\ &\equiv \left(\sum_{n=0}^{\infty} (+\mathrm{i})^n \int\limits_{y_{n-1}}^{x^-} \mathrm{d}y_n\, A_-(y_n, x^+) \cdots \int\limits_{x_0^-}^{y_2} \mathrm{d}y_1\, A_-(y_1, x^+) \right) G_-(x_0^-, x^+), \\ G_+(x^-, x^+) &:= \mathcal{T}_+ \exp\left(+\mathrm{i} \int\limits_{x_0^+}^{x^+} \mathrm{d}y\, A_+(x^-, y) \right) G_+(x^-, x_0^+) \\ &\equiv \left(\sum_{n=0}^{\infty} (+\mathrm{i})^n \int\limits_{y_{n-1}}^{x^+} \mathrm{d}y_n\, A_+(x^-, y_n) \cdots \int\limits_{x_0^+}^{y_2} \mathrm{d}y_1\, A_+(x^-, y_1) \right) G_+(x^-, x_0^+). \end{aligned} \tag{I.142a}$$

The descending ordering with respect to $x^\mp$ when reading from left to right in the power expansion of the exponential is implied by the notation $\mathcal{T}_\mp$. In effect, we are propagating the seed values $G_\mp(x_0^-, x_0^+) \in U(N)$ through the first-order differential equations

$$\begin{aligned} +\mathrm{i}A_- &= (\partial_- G_-)\, G_-^{-1}, \qquad G_- \in U(N), \\ +\mathrm{i}A_+ &= (\partial_+ G_+)\, G_+^{-1}, \qquad G_+ \in U(N), \end{aligned} \tag{I.142b}$$

to their final values $G_\mp(x^-, x^+) \in U(N)$. The field strength (I.139c) then becomes

$$\begin{aligned}
F_{+-} &= -\partial_{+}\Big(\mathrm{i}(\partial_{-}G_{-})\,G_{-}^{-1}\Big) + \mathrm{i}(\partial_{+}G_{+})\,G_{+}^{-1}(\partial_{-}G_{-})\,G_{-}^{-1} - \Big(+\leftrightarrow-\Big) \\
&= -\mathrm{i}(\partial_{+}\partial_{-}G_{-})\,G_{-}^{-1} + \mathrm{i}(\partial_{-}G_{-})\,G_{-}^{-1}(\partial_{+}G_{-})\,G_{-}^{-1} + \mathrm{i}(\partial_{+}G_{+})\,G_{+}^{-1}(\partial_{-}G_{-})\,G_{-}^{-1} \\
&\quad -\Big(+\leftrightarrow-\Big)
\end{aligned} \tag{I.143}$$

when expressed in terms of the Lie group instead of the Lie algebra.

The advantage of the parametrization (I.142) is that the transformation law obeyed by the fields $G_{\mp}$ under the local $U(N)\times U(N)$ transformations that leave the action in Eq. (I.135b) invariant is very simple, namely it amounts to multiplication from the left,

$$\begin{aligned}
G_{-} &=: U_{\omega_{-}}\ {}^{\omega_{-}}G_{-}, & U_{\omega_{-}} &:= e^{\mathrm{i}T^{\mathsf{a}}\,\omega^{\mathsf{a}}_{-}}, & \omega^{\mathsf{a}}_{-} &: \mathbb{M}_{1,1}\to\mathbb{R}, \\
G_{+} &=: U_{\omega_{+}}\ {}^{\omega_{+}}G_{+}, & U_{\omega_{+}} &:= e^{\mathrm{i}T^{\mathsf{a}}\,\omega^{\mathsf{a}}_{+}}, & \omega^{\mathsf{a}}_{+} &: \mathbb{M}_{1,1}\to\mathbb{R}.
\end{aligned} \tag{I.144}$$

The disadvantage is that the relationship between $A_{\mp}\in u(N)$ and $G_{\mp}\in U(N)$ is non-linear for $N>1$.

It is also important to notice that the relationship is not one-to-one if the initial conditions are arbitrary. Indeed, multiplication from the right of any pair of solutions $G_{\mp}$ by a pair of $U(N)$-valued matrices that do not depend on $x^{\mp}$, respectively, is also an appropriate parametrization,

$$\begin{aligned}
\mathrm{i}A_{-} &= (\partial_{-}G_{-})G_{-}^{-1} = \Big(\partial_{-}(G_{-}\,H_{-})\Big)H_{-}^{-1}\,G_{-}^{-1}, \quad G_{-}(x^{+},x^{-}),H_{-}(x^{+})\in U(N), \\
\mathrm{i}A_{+} &= (\partial_{+}G_{+})G_{+}^{-1} = \Big(\partial_{+}(G_{+}\,H_{+})\Big)H_{+}^{-1}\,G_{+}^{-1}, \quad G_{+}(x^{+},x^{-}),H_{+}(x^{-})\in U(N).
\end{aligned} \tag{I.145}$$

The second step towards solving the anomaly equations (I.140) is to fix the gauge. This step relies on the assumption that the theory is invariant under any local gauge transformation (I.108).

For example, we may assume that

$$A_{+}=0, \tag{I.146a}$$

in which case

$$F_{+-} = -\mathrm{i}(\partial_{+}\partial_{-}G_{-})\,G_{-}^{-1} + \mathrm{i}(\partial_{-}G_{-})\,G_{-}^{-1}(\partial_{+}G_{-})\,G_{-}^{-1}. \tag{I.146b}$$

In the gauge (I.146), the anomaly equations (I.140) simplifies to

$$\begin{aligned}
\partial_{+}J_{-} &= +\frac{1}{4\pi}\partial_{+}A_{-}, \\
\nabla_{-}J_{+} &= -\frac{1}{4\pi}\partial_{+}A_{-}.
\end{aligned} \tag{I.147}$$

One then verifies that, up to initial conditions,

$$\begin{aligned}
J_{-} &= -\frac{\mathrm{i}}{4\pi}(\partial_{-}G_{-})\,G_{-}^{-1}, \\
J_{+} &= +\frac{\mathrm{i}}{4\pi}(\partial_{+}G_{-})\,G_{-}^{-1},
\end{aligned} \tag{I.148}$$

solve the anomaly equations (I.147).

Alternatively, we may assume that

$$A_- = 0, \tag{I.149a}$$

in which case

$$F_{+-} = +\mathrm{i}(\partial_-\partial_+ G_+)\, G_+^{-1} - \mathrm{i}(\partial_+ G_+)\, G_+^{-1}(\partial_- G_+)\, G_+^{-1}. \tag{I.149b}$$

In the gauge (I.149), the anomaly equations (I.140) simplifies to

$$\begin{aligned} \nabla_+ J_- &= -\frac{1}{4\pi}\partial_- A_+, \\ \partial_- J_+ &= +\frac{1}{4\pi}\partial_- A_+. \end{aligned} \tag{I.150}$$

One then verifies that, up to initial conditions,

$$\begin{aligned} J_- &= +\frac{\mathrm{i}}{4\pi}(\partial_- G_+)\, G_+^{-1}, \\ J_+ &= -\frac{\mathrm{i}}{4\pi}(\partial_+ G_+)\, G_+^{-1}, \end{aligned} \tag{I.151}$$

solve the anomaly equations (I.150).

To proceed, we rewrite the definition (I.136) of the quantum current as

$$\begin{aligned} \delta \ln Z[A_0, A_1] &= \mathrm{i}\int \mathrm{d}x^0 \mathrm{d}x^1 \mathrm{tr}\,\left(J_\mu\, \delta A^\mu\right)(x^0, x^1) \\ &= \mathrm{i}\int \frac{\mathrm{d}x^- \mathrm{d}x^+}{2}\frac{1}{2}\mathrm{tr}\,\left(J_+\, \delta A_- + J_-\, \delta A_+\right)(x^-, x^+) \\ &= \delta \ln Z[A_-, A_+]. \end{aligned} \tag{I.152}$$

In the gauge $A_+ = 0$,

$$\delta \ln Z[A_-, A_+ = 0] = \mathrm{i}\int \frac{\mathrm{d}x^- \mathrm{d}x^+}{2}\frac{1}{2}\mathrm{tr}\,\left(J_+\, \delta A_-\right)(x^-, x^+), \tag{I.153a}$$

where

$$J_+ = +\frac{\mathrm{i}}{4\pi}(\partial_+ G_-)\, G_-^{-1} \tag{I.153b}$$

and

$$\begin{aligned} \delta A_- &= -\,\delta\Big(\mathrm{i}(\partial_- G_-)\, G_-^{-1}\Big) \\ \overset{\text{Eq. (I.86c)}}{} &= -\,\nabla_-\Big(\mathrm{i}(\delta G_-)\, G_-^{-1}\Big). \end{aligned} \tag{I.153c}$$

We introduce the short-hand notation

$$\delta W_-[G_-] := \frac{1}{8\pi}\int \mathrm{tr}\,\bigg((\partial_+ G_-)\, G_-^{-1}\nabla_-\Big((\delta G_-)\, G_-^{-1}\Big)\bigg) \tag{I.154}$$

to represent the integral on the right-hand side of Eq. (I.153a) expressed solely in terms of fields from the Lie group. The cyclicity of the trace allows to perform a "partial integration" over the covariant derivative ∇_-, recall identity (I.86d),

$$\delta W_-[G_-] = -\frac{1}{8\pi}\int \mathrm{tr}\,\bigg((\delta G_-)\, G_-^{-1}\nabla_-\Big((\partial_+ G_-)\, G_-^{-1}\Big)\bigg). \tag{I.155}$$

Identity (I.86e) allows the manipulation

$$\delta W_-[G_-] = -\frac{1}{8\pi}\int \mathrm{tr}\left((\delta G_-)\,G_-^{-1}\partial_+\left((\partial_- G_-)\,G_-^{-1}\right)\right). \tag{I.156}$$

Finally, we use Eq. (I.85) to obtain

$$\delta W_-[G_-] = \frac{1}{16\pi}\left(\delta\,W_1[G_-] - \delta\,W_2[G_-]\right). \tag{I.157}$$

With the help of Eq. (I.84), we conclude that, up to a (functional) integration constant, the logarithm of the fermion determinant of a Dirac fermion in the fundamental representation of $U(N)$ that is coupled to an $u(N)$ gauge field in the gauge $A_+ = 0$ is nothing but, up to the multiplicative number $-\mathrm{i}$ (note the all important negative sign), the WZW action (I.105),

$$\ln Z[A_-, A_+ = 0] = -\mathrm{i}\,S_{\mathrm{WZW}}[G_-], \qquad \mathrm{i}A_- := \left(\partial_- G_-\right)G_-^{-1}. \tag{I.158}$$

Had we chosen the gauge $A_- = 0$, we would have had to extract the term obtained from Eq. (I.156) by the substitution $(- \leftrightarrow +)$ out of $W_1[G_+] = +W_1[G_+^{-1}]$ and $W_2[G_+] = -W_2[G_+^{-1}]$. We would then have concluded that, up to a (functional) integration constant, the logarithm of the fermion determinant of a Dirac fermion in the fundamental representation of $U(N)$ that is coupled to an $u(N)$ gauge field in the gauge $A_- = 0$ is nothing but, up to the multiplicative number $-\mathrm{i}$ (note the all important negative sign), the WZW action (I.105),

$$\ln Z[A_- = 0, A_+] = -\mathrm{i}\,S_{\mathrm{WZW}}[G_+^{-1}], \qquad \mathrm{i}A_+ := \left(\partial_+ G_+\right)G_+^{-1}. \tag{I.159}$$

The case of both A_- and A_+ not necessarily vanishing follows from performing a pure gauge transformation [recall Eq. (I.144)] and is given by

$$\ln Z[A_-, A_+] = -\mathrm{i}\,S_{\mathrm{WZW}}[G_+^{-1}\,G_-]. \tag{I.160a}$$

This representation is manifestly gauge invariant, for the product $G_+^{-1}\,G_-$ is invariant under any left multiplication of G_- and G_+ by the same $U(N)$-valued field. If we first made use of the Polyakov-Wiegmann identity (I.106) that we follow up using

$$\mathrm{i}A_- := \left(\partial_- G_-\right)G_-^{-1}, \qquad \mathrm{i}A_+ := \left(\partial_+ G_+\right)G_+^{-1}, \tag{I.160b}$$

we obtain the representation

$$\ln Z[A_-, A_+] = -\mathrm{i}\,S_{\mathrm{WZW}}[G_+^{-1}] - \mathrm{i}\,S_{\mathrm{WZW}}[G_-] + \frac{\mathrm{i}}{8\pi}\int \mathrm{tr}\,(A_+\,A_-). \tag{I.160c}$$

This representation is not manifestly gauge invariant anymore.

Observe that the sign with which the WZW actions $S_{\mathrm{WZW}}[G_+^{-1}]$ and $S_{\mathrm{WZW}}[G_-]$ enter is opposite to the sign with which the mass term

$$\frac{1}{8\pi}\int \mathrm{tr}\,(A_+\,A_-) = \int \frac{1}{8\pi}\mathrm{tr}\,\left(A_\mu A^\mu\right) \tag{I.161}$$

enters. Thus, upon an analytical continuation to imaginary time, i.e., a Wick rotation from $\mathbb{M}_{1,1}$ to $\mathbb{R}^2$, it is the mass term that provides the convergence factor

$$e^{-\frac{1}{8\pi}\int \mathrm{d}x^1\mathrm{d}x^2 \mathrm{tr}\left(A_1^2+A_2^2\right)}, \tag{I.162}$$

rather than the "wrong" sign with which the PCM enters the Euclidean action,

$$\begin{aligned} &e^{+\frac{1}{16\pi}\int \mathrm{d}x^1\mathrm{d}x^2 \mathrm{tr}\left[(\partial_1 G_-)(\partial_1 G_-^{-1})+(\partial_2 G_-)(\partial_2 G_-^{-1})\right]} \\ &\times e^{+\frac{1}{16\pi}\int \mathrm{d}x^1\mathrm{d}x^2 \mathrm{tr}\left[(\partial_1 G_+)(\partial_1 G_+^{-1})+(\partial_2 G_+)(\partial_2 G_+^{-1})\right]}, \end{aligned} \tag{I.163}$$

were we to define a path integral by integrating over the gauge fields.

We have shown that there exists the gauge-invariant regularization

$$\begin{aligned} Z[A_+,A_-] \equiv& \int \mathcal{D}[\psi_-^\dagger,\psi_-,\psi_+^\dagger,\psi_+]\, e^{+\mathrm{i}\int\left[\psi_-^\dagger(\mathrm{i}\partial_+ + A_+)\psi_- + \psi_+^\dagger(\mathrm{i}\partial_- + A_-)\psi_+\right]} \\ =& e^{-\mathrm{i}S_{\mathrm{WZW}}[G_+^{-1}G_-]}, \end{aligned} \tag{I.164a}$$

where the relation

$$\mathrm{i}A_- := (\partial_- G_-)\,G_-^{-1}, \qquad \mathrm{i}A_+ := (\partial_+ G_+)\,G_+^{-1}, \tag{I.164b}$$

between fields from the Lie algebra and fields from the Lie group is non-linear and one-to-many. Whereas the symmetry under the local gauge transformation

$$G_- = U\,G'_-, \qquad G_+ = U\,G'_+, \tag{I.165}$$

is not manifest in the definition of the fermionic partition function, owing to the ambiguous definition of the Grassmann path integral, it is manifest in the definition of the fermionic determinant by the exponential $\exp\left(-\mathrm{i}S_{\mathrm{WZW}}[G_+^{-1}G_-]\right)$ of the WZW action. On the other hand, under the axial gauge transformation

$$G_- = U_5\,G'_-, \qquad G_+ = U_5^{-1}\,G'_+, \tag{I.166}$$

the WZW action transforms like

$$S_{\mathrm{WZW}}[G_+^{-1}\,G_-] = S_{\mathrm{WZW}}[G_+'^{-1}\,U_5^2\,G'_-]. \tag{I.167}$$

We are going to take advantage of Eq. (I.167) in two powerful ways. First, we are going to represent the gauge-invariant regularization of the fermion determinant as a path integral over bosons. Second, we are going to bosonize the quantum gauge current (I.136). To this end, the Polyakov-Wiegmann identity (I.106) is crucial.

Let U denote the field $U:\mathbb{M}_{1,1}\to U(N)$. Let $\mathcal{D}[U]$ denote the Haar measure of the field U through

$$\mathcal{D}[U] := \prod_{x\in\mathbb{M}_{1,1}} \mathrm{d}U(x), \tag{I.168}$$

with $\mathrm{d}U(x)$ the infinitesimal volume of integration in the Lie group $U(N)$. Since the Haar measure in the Lie group $U(N)$ is invariant under left or right multiplication, the functional measure $\mathcal{D}[U]$ is invariant under the local chiral gauge transformation

$$U(x) =: L(x)\,U'(x)\,R(x), \tag{I.169}$$

for any pair of sufficiently smooth fields $L : \mathbb{M}_{1,1} \to U(N)$ and $R : \mathbb{M}_{1,1} \to U(N)$. For our purpose, it is useful to represent the number 1 as

$$1 = \frac{\int \mathcal{D}[U]\, e^{+\mathrm{i}S_{\mathrm{WZW}}[U]}}{\int \mathcal{D}[U]\, e^{+\mathrm{i}S_{\mathrm{WZW}}[U]}}. \tag{I.170}$$

We may change the integration variable in the numerator according to Eq. (I.169) with the choices $L \to G_{+}^{-1}$ and $R \to G_{-}$,

$$1 = \frac{\int \mathcal{D}[U]\, e^{+\mathrm{i}S_{\mathrm{WZW}}[G_{+}^{-1}\,U\,G_{-}]}}{\int \mathcal{D}[U]\, e^{+\mathrm{i}S_{\mathrm{WZW}}[U]}}. \tag{I.171}$$

Multiplication on both sides by the phase $\exp\left(-\mathrm{i}S_{\mathrm{WZW}}[G_{+}^{-1}\,G_{-}]\right)$ gives

$$\begin{aligned} Z[A_{+},A_{-}] &= e^{-\mathrm{i}S_{\mathrm{WZW}}[G_{+}^{-1}\,G_{-}]} \\ &= \frac{\int \mathcal{D}[U]\, e^{+\mathrm{i}\left\{S_{\mathrm{WZW}}[G_{+}^{-1}\,U\,G_{-}]-S_{\mathrm{WZW}}[G_{+}^{-1}\,G_{-}]\right\}}}{\int \mathcal{D}[U]\, e^{+\mathrm{i}S_{\mathrm{WZW}}[U]}}. \end{aligned} \tag{I.172}$$

This is the representation of the fermionic determinant as a bosonic path integral that we seek. It is advantageous for two reasons. On the one hand, because of the chiral symmetry of the Haar measure, gauge invariance has become the symmetry of the numerator under

$$G_{+} \to H\,G_{+}, \qquad G_{-} \to H\,G_{-}, \qquad U \to H\,U\,H^{-1}. \tag{I.173}$$

On the other hand, chiral gauge invariance under any

$$G_{+} \to L\,G_{+}, \qquad G_{-} \to R\,G_{-}, \qquad L \neq R, \tag{I.174}$$

is explicitly broken by the Boltzmann weight in the numerator.

The usefulness of the representation (I.172) arises from the fact that the action

$$S_{\mathrm{WZ}}[G_{+}^{-1},U,G_{-}] := S_{\mathrm{WZW}}[G_{+}^{-1}\,U\,G_{-}] - S_{\mathrm{WZW}}[G_{+}^{-1}\,G_{-}] \tag{I.175}$$

is a functional of U, A_{+}, and A_{-} only. Once this property has been explicitly established, it is a mere exercise in functional differentiation to compute the quantum gauge current (I.136). However, before doing so, we reiterate that the invariance

$$S_{\mathrm{WZ}}[G_{+}^{-1}\,H^{-1},H\,U\,H^{-1},H\,G_{-}] = S_{\mathrm{WZ}}[G_{+}^{-1},U,G_{-}] \tag{I.176}$$

encodes the gauge invariance of the fermion determinant, whereas the Wess-Zumino transformation law [142]

$$\begin{aligned} S_{\mathrm{WZ}}[G_{+}^{-1}\,H,H^{-1}\,U\,H^{-1},H\,G_{-}] &= S_{\mathrm{WZW}}[G_{+}^{-1}\,U\,G_{-}] - S_{\mathrm{WZW}}[G_{+}^{-1}\,H^{2}\,G_{-}] \\ &= S_{\mathrm{WZ}}[G_{+}^{-1},U,G_{-}] - S_{\mathrm{WZ}}[G_{+}^{-1},H^{2},G_{-}] \end{aligned} \tag{I.177}$$

encodes the breaking of the axial gauge symmetry by the chosen regularization of the fermion determinant.

It is time to bosonize the quantum gauge current (I.136). Repeated application of the Polyakov-Wiegmann identity (I.106) delivers

$$\begin{aligned} S_{\mathrm{WZ}}[G_+^{-1}, U, G_-] &:= S_{\mathrm{WZW}}[G_+^{-1} U\, G_-] - S_{\mathrm{WZW}}[G_+^{-1} G_-] \\ &= S_{\mathrm{WZW}}[U] + \frac{1}{8\pi} \int \mathrm{tr}\, K[U, A_+, A_-], \end{aligned} \tag{I.178a}$$

where

$$\begin{aligned} K[U, A_+, A_-] :=& [U^{-1} G_+ \partial_+ (G_+^{-1} U)]\, (G_-\, \partial_- G_-^{-1}) + (G_+ \partial_+ G_+^{-1})\, (U\, \partial_- U^{-1}) \\ & - (G_+ \partial_+ G_+^{-1})\, (G_-\, \partial_- G_-^{-1}) \\ =& - U^{-1} A_+ U\, A_- - \mathrm{i}(U^{-1} \partial_+ U)\, A_- - \mathrm{i} A_+\, (U\, \partial_- U^{-1}) + A_+\, A_- . \end{aligned} \tag{I.178b}$$

To summarize, we have established that the fermionic determinant (I.164) can be represented by the bosonic path integral

$$Z[A_+, A_-] = \frac{\int \mathcal{D}[U]\, e^{+\mathrm{i} S_{\mathrm{WZ}}[U, A_+, A_-]}}{\int \mathcal{D}[U]\, e^{+\mathrm{i} S_{\mathrm{WZW}}[U]}}, \tag{I.179a}$$

where the WZW action is defined by Eq. (I.105) and the WZ action is given by

$$\begin{aligned} S_{\mathrm{WZ}}[U, A_+, A_-] = S_{\mathrm{WZW}}[U] + \frac{1}{8\pi} \int \mathrm{tr} \Big[& - \mathrm{i}(U^{-1} \partial_+ U)\, A_- - \mathrm{i} A_+\, (U\, \partial_- U^{-1}) \\ & - U^{-1} A_+ U\, A_- + A_+\, A_- \Big]. \end{aligned} \tag{I.179b}$$

There follows that the light-cone components of the quantum gauge current (I.136) are

$$J_- = \frac{1}{8\pi} \left[-\mathrm{i}(U\, \partial_- U^{-1}) - U\, A_- U^{-1} + A_- \right], \tag{I.180a}$$

and

$$J_+ = \frac{1}{8\pi} \left[-\mathrm{i}(U^{-1}\, \partial_+ U) - U^{-1}\, A_+\, U + A_+ \right]. \tag{I.180b}$$

They depend explicitly on A_+ and A_- in the non-Abelian factor group of the Lie group $U(N)$, here $SU(N)$. As a corollary, they agree with Witten's bosonization rules (I.26) only and only if the limit $A_+ = A_- = 0$ for massless free Dirac fermions is taken. For the Abelian factor group $U(1)$ of $U(N)$, the dependencies on A_+ and A_- cancel out and the bosonization rules

$$j_\mp = \pm \frac{1}{2\pi} \partial_\mp \varphi, \tag{I.181}$$

follow if we use the parametrization $U = e^{-4\mathrm{i}\, I\, \varphi / N}$ with I the $N \times N$ unit matrix. This agrees with Eq. (I.21).

I.10 Bosonization of QCD_2 for infinitely strong gauge coupling

We shall define QCD_2 in the infinitely strong gauge coupling limit by integrating the partition function (I.135a) with equal weight over all gauge configurations such that the determinant of the Dirac kernel, $\mathrm{Det}\left[\gamma^{\mu}(\mathrm{i}\partial_{\mu}+A_{\mu})\right]$, is non-vanishing. We thus write the partition function

$$Z_{\mathrm{QCD}_2} := \mathcal{V}^{-1}\int\mathcal{D}[A]\int\mathcal{D}[\bar{\psi}]\,\mathcal{D}[\psi]\,e^{+\mathrm{i}\int dx^0 dx^1\,\mathcal{L}_{\mathrm{QCD}_2}}, \tag{I.182a}$$

with the Lagrangian density

$$\mathcal{L}_{\mathrm{QCD}_2} := \bar{\psi}\,\gamma^{\mu}\left(\mathrm{i}\partial_{\mu}+A_{\mu}\right)\psi. \tag{I.182b}$$

Here, $\mathcal{D}[A]$ is the measure for the $u(N)$ gauge fields. The normalization factor $\mathcal{V}$ is a symbolic expression for the volume

$$\mathcal{V} := \int\mathcal{D}[\omega], \tag{I.182c}$$

where ω parametrizes a local gauge transformation according to

$$A_{\mu} =: U_{\omega}\,{}^{\omega}A_{\mu}\,U^{-1}_{\omega} - \left(\mathrm{i}\partial_{\mu}U_{\omega}\right)U^{-1}_{\omega}, \qquad U_{\omega} := e^{\mathrm{i}T^{\mathrm{a}}\omega^{\mathrm{a}}} \in U(N). \tag{I.182d}$$

It avoids overcounting gauge configurations that are related by a local gauge transformation. The contribution

$$-\frac{1}{8e^2}\mathrm{tr}\left(F_{\mu\nu}\,F^{\mu\nu}\right) = -\frac{1}{4e^2}F^{\mathrm{a}}_{\mu\nu}\,F^{\mathrm{a}\,\mu\nu} \tag{I.182e}$$

to the Lagrangian density of QCD_2 has been suppressed by taking the limit $e^2\to\infty$ of an infinite gauge charge.

Our strategy, as was the case in chapter 9, consists in decoupling the gauge fields from the Dirac spinors at the classical level, while paying a tribute to the consequence of this decoupling at the level of the bosonic (gauge fields) and fermionic measures.

We start from the representation

$$\mathcal{L}_{\mathrm{QCD}_2} = \psi^{\dagger}_{-}\left(\mathrm{i}\partial_{+}+A_{+}\right)\psi_{-} + \psi^{\dagger}_{+}\left(\mathrm{i}\partial_{-}+A_{-}\right)\psi_{+} \tag{I.183}$$

in the chiral basis (I.17). Next, we insert the parametrization (I.142) of the gauge fields $A_{\pm}$ in terms of the fields $G_{\pm}$ taking values in the Lie group,

$$\mathcal{L}_{\mathrm{QCD}_2} = \psi^{\dagger}_{-}\,\mathrm{i}\Big[\partial_{+} - \left(\partial_{+}G_{+}\right)G^{-1}_{+}\Big]\psi_{-} + \psi^{\dagger}_{+}\,\mathrm{i}\Big[\partial_{-} - \left(\partial_{-}G_{-}\right)G^{-1}_{-}\Big]\psi_{+}. \tag{I.184}$$

We first perform the local gauge transformation

$$\begin{aligned} \psi^{\dagger}_{-} &=: \tilde{\psi}^{\dagger}_{-}\,G^{-1}_{+}, & \psi_{-} &=: G_{+}\,\tilde{\psi}_{-}, \\ \psi^{\dagger}_{+} &=: \tilde{\psi}^{\dagger}_{+}\,G^{-1}_{+}, & \psi_{+} &=: G_{+}\,\tilde{\psi}_{+}, \end{aligned} \tag{I.185a}$$

for which

$$\mathcal{L}_{\mathrm{QCD}_2} = \tilde{\psi}^{\dagger}_{-}\,\mathrm{i}\partial_{+}\,\tilde{\psi}_{-} + \tilde{\psi}^{\dagger}_{+}\,\mathrm{i}\Big[\partial_{-} - \left(\partial_{-}\tilde{G}_{-}\right)\tilde{G}^{-1}_{-}\Big]\tilde{\psi}_{+}, \tag{I.185b}$$

where [recall Eq. (I.144)]

$$\begin{aligned}\tilde{G}_+ &:= G_+^{-1}\, G_+ = 1,\\ \tilde{G}_- &:= G_+^{-1}\, G_-.\end{aligned} \tag{I.185c}$$

We have fixed the gauge to satisfy the condition $A_+ = 0$.

We now want to decouple A_- from the fermions. This is done with the help of the mixed local gauge and local axial gauge transformation defined by

$$\omega^{\mathrm{a}} = -\omega_5^{\mathrm{a}} \Longleftrightarrow \omega_+^{\mathrm{a}} = 2\omega^{\mathrm{a}}, \qquad \omega_-^{\mathrm{a}} = 0, \qquad \mathrm{a} = 1, \cdots, N^2. \tag{I.186}$$

In other words, we do the transformation

$$\begin{aligned}\tilde{\psi}_-^{\dagger} &=: \psi_-^{\prime\dagger}, & \tilde{\psi}_- &=: \psi'_-,\\ \tilde{\psi}_+^{\dagger} &=: \psi_+^{\prime\dagger}\,\tilde{G}_-^{-1}, & \tilde{\psi}_+ &=: \tilde{G}_-\psi'_+,\end{aligned} \tag{I.187a}$$

for which

$$\mathcal{L}_{\mathrm{QCD}_2} = \psi_-^{\prime\dagger}\,\mathrm{i}\partial_+\,\psi'_- + \psi_+^{\prime\dagger}\,\mathrm{i}\partial_-\psi'_+, \tag{I.187b}$$

where [recall Eq. (I.144)]

$$\begin{aligned}G'_+ &:= \tilde{G}_+ = 1,\\ G'_- &:= \tilde{G}_-^{-1}\,\tilde{G}_- = 1.\end{aligned} \tag{I.187c}$$

To recapitulate, the chiral transformation defined by

$$\begin{aligned}\psi_-^{\dagger} &=: \psi_-^{\prime\dagger}\,G_+^{-1}, & \psi_- &=: G_+\,\psi'_-, & G'_+ &:= G_+^{-1}\,G_+ = 1,\\ \psi_+^{\dagger} &=: \psi_+^{\prime\dagger}\,G_-^{-1}, & \psi_+ &=: G_-\,\psi'_+, & G'_- &:= G_-^{-1}\,G_- = 1,\end{aligned} \tag{I.188}$$

decouples the gauge fields from the Dirac spinors in the Lagrangian density of QCD_2.

It is time to turn our attention to the effects of the changes of integration variables $A_\pm \to G_\pm$ and $\psi_\mp^{\dagger}, \psi_\mp \to \psi_\mp^{\prime\dagger}, \psi'_\mp$, i.e., we need to compute the corresponding Jacobians.

We begin with the bosonic Jacobian defined by

$$\mathcal{D}[A_+, A_-] =: \mathcal{D}[G_+, G_-] \times \mathcal{V}_{+,-}^{-1} \times \mathcal{J}[G_+, G_-]. \tag{I.189}$$

The relation between the Lie algebra and the Lie group is one-to-many [recall Eq. (I.145)] so that we need to correct for overcounting elements from the Lie group that yield the same element from the Lie algebra. This correction is achieved with the normalization factor

$$\mathcal{V}_{+,-} \equiv \mathcal{V}_+ \times \mathcal{V}_- := \int \mathcal{D}[\varpi_+] \times \int \mathcal{D}[\varpi_-], \tag{I.190}$$

where ϖ_+ and ϖ_- parametrize the transformations defined in Eq. (I.145). The bosonic Jacobian $\mathcal{J}[G_+, G_-]$ may depend on the fields G_+ and G_-. Because the measure $\mathcal{D}[A_+, A_-]$ factorizes into the product of the measures $\mathcal{D}[A_+]$ and $\mathcal{D}[A_-]$,

$$\mathcal{J}[G_+, G_-] = \mathrm{Det}\left(\frac{\mathcal{D}[A_+]}{\mathcal{D}[G_+]}\right) \times \mathrm{Det}\left(\frac{\mathcal{D}[A_-]}{\mathcal{D}[G_-]}\right), \tag{I.191}$$

and with the help of the identity[45]

$$\delta A_{\pm} = \nabla_{\pm}\Big[\big(-\mathrm{i}\delta G_{\pm}\big)G_{\pm}^{-1}\Big], \tag{I.193}$$

we deduce that the bosonic Jacobian is the product of the determinants of the two light-cone components of the covariant derivative in the adjoint representation,

$$\mathcal{J}[G_{+},G_{-}] = \mathrm{Det}\,\nabla_{+} \times \mathrm{Det}\,\nabla_{-}. \tag{I.194}$$

The measure $\mathcal{D}[G_{+},G_{-}]$ is, at each point of (1+1)-dimensional Minkowski space and time $\mathbb{M}_{1,1}$, the Haar measure of the chiral group $U(N)\times U(N)$. It is thus locally invariant under any left or right multiplication by group elements. In particular, it is invariant under any $U(N)\times U(N)$ local gauge transformation. The local symmetry induced by the one-to-many relation between $A_{\pm}$ and $G_{\pm}$ is also preserved by the measure $\mathcal{D}[G_{+},G_{-}]$.

On the other hand, $\mathcal{J}[G_{+},G_{-}]$ is not invariant under all $U(N)\times U(N)$ local gauge transformation. To see this, we use the identity

$$z = \int \mathrm{d}\beta\,\mathrm{d}\alpha\, e^{-\beta\, z\,\alpha} \tag{I.195}$$

that is valid for any complex number z provided the right-hand side is an integral over the two Grassmann variables β and α. We may convert the determinants of the covariant derivatives into the Grassmann integrals

$$\mathrm{Det}\,\nabla_{+} = \int \mathcal{D}[\beta_{-},\alpha_{-}]\, e^{+\mathrm{i}\int \frac{\mathrm{d}x^{-}\mathrm{d}x^{+}}{2}\frac{2}{\mathrm{y_f}}\mathrm{tr}\left(\beta^{\mathsf{a}}_{-}\,T^{\mathsf{a}}\,\mathrm{i}\nabla_{+}\,T^{\mathsf{b}}\alpha^{\mathsf{b}}_{-}\right)}, \tag{I.196}$$

and

$$\mathrm{Det}\,\nabla_{-} = \int \mathcal{D}[\beta_{+},\alpha_{+}]\, e^{+\mathrm{i}\int \frac{\mathrm{d}x^{-}\mathrm{d}x^{+}}{2}\frac{2}{\mathrm{y_f}}\mathrm{tr}\left(\beta^{\mathsf{a}}_{+}\,T^{\mathsf{a}}\,\mathrm{i}\nabla_{-}\,T^{\mathsf{b}}\,\alpha^{\mathsf{b}}_{+}\right)}, \tag{I.197}$$

respectively, over the ghost fields β_{-}, α_{-}, β_{+}, and α_{+}. Here the constant $\mathrm{y_r}$ depends on the representation chosen for the generators of $su(N)$ according to

$$\mathrm{tr}\,(T^{\mathsf{a}}\,T^{\mathsf{b}}) =: \mathrm{y_r}\,\delta^{\mathsf{ab}}, \qquad \mathsf{a},\mathsf{b}=1,\cdots,N^{2}-1. \tag{I.198}$$

Thus, in our case, it is to be taken in the fundamental representation, for which we have chosen the convention $\mathrm{y_f}=2$.

[45] Without loss of generality, it suffices to consider the case of δA_{+} when $A_{+}(x)\in u(N)$ is expressed in terms of $G_{+}(x)\in U(N)$ with the help of Eq. (I.142). Needed is the infinitesimal relation

$$\begin{aligned}
\delta A_{+} &= \delta\left[(-\mathrm{i}\partial_{+}G_{+})G_{+}^{-1}\right]\\
&= (-\mathrm{i}\partial_{+}\delta G_{+})G_{+}^{-1} - (-\mathrm{i}\partial_{+}G_{+})G_{+}^{-1}\,\delta G_{+}\,G_{+}^{-1}\\
&= \partial_{+}\left[(-\mathrm{i}\delta G_{+})G_{+}^{-1}\right] + \mathrm{i}(-\mathrm{i}\delta G_{+})G_{+}^{-1}\,(-\mathrm{i}\partial_{+}G_{+})\,G_{+}^{-1} - \mathrm{i}(-\mathrm{i}\partial_{+}G_{+})G_{+}^{-1}\,(-\mathrm{i}\delta G_{+})\,G_{+}^{-1}\\
&= \partial_{+}\left[(-\mathrm{i}\delta G_{+})\,G_{+}^{-1}\right] - \mathrm{i}\left[(-\mathrm{i}\partial_{+}G_{+})G_{+}^{-1},(-\mathrm{i}\delta G_{+})\,G_{+}^{-1}\right]\\
&= \partial_{+}\left[(-\mathrm{i}\delta G_{+})\,G_{+}^{-1}\right] - \mathrm{i}\left[A_{+},(-\mathrm{i}\delta G_{+})\,G_{+}^{-1}\right].
\end{aligned} \tag{I.192}$$

The behavior under $U(N) \times U(N)$ local gauge transformations of the bosonic Jacobian $\mathcal{J}[G_+, G_-]$ is now determined by the transformation properties of the Lagrangian density

$$\mathcal{L}_{\mathrm{gh}} := \frac{2}{\mathrm{y_f}}\mathrm{tr}\,\left(\beta^{\mathrm{a}}_{-}\, T^{\mathrm{a}}\, \mathrm{i}\nabla_{+}\, T^{\mathrm{b}}\, \alpha^{\mathrm{b}}_{-}\right) + \frac{2}{\mathrm{y_f}}\mathrm{tr}\,\left(\beta^{\mathrm{a}}_{+}\, T^{\mathrm{a}}\, \mathrm{i}\nabla_{-}\, T^{\mathrm{b}}\, \alpha^{\mathrm{b}}_{+}\right) \tag{I.199}$$

and of the Grassmann measure for the ghost fields β_-, α_-, β_+, and α_+.

The ghost Lagrangian density $\mathcal{L}_{\mathrm{gh}}$ shares the local chiral gauge invariance of the original Lagrangian density $\mathcal{L}_{\mathrm{QCD}_2}$ provided the ghost fields β_-, α_-, β_+, and α_+ transform according to the law

$$\begin{aligned} \beta_- &=: U_{\omega_-}\, {}^{\omega_-}\beta_-\, U^{-1}_{\omega_-}, \qquad & \alpha_- &=: U_{\omega_-}\, {}^{\omega_-}\alpha_-\, U^{-1}_{\omega_-}, \\ \beta_+ &=: U_{\omega_+}\, {}^{\omega_+}\beta_+\, U^{-1}_{\omega_+}, \qquad & \alpha_+ &=: U_{\omega_+}\, {}^{\omega_+}\alpha_+\, U^{-1}_{\omega_+}. \end{aligned} \tag{I.200}$$

Any non-invariance of the bosonic Jacobian under local gauge transformations must therefore be associated with the Grassmann measure. We opt for a regularization of the Grassmann measure that preserves the diagonal subgroup of the $U(N) \times U(N)$ local gauge transformations defined in Eq. (I.165), i.e., by the condition $\omega_- = \omega_+ \equiv \omega$.

Let's pause and sum up our intermediary results. We have rewritten the partition function of QCD_2 in the infinitely strong gauge coupling limit by using the chiral basis for the Dirac fermions and representing the gauge fields in terms of $U(N)$-valued fields,

$$Z_{\mathrm{QCD}_2} = \int \frac{\mathcal{D}[G_+, G_-]}{\mathcal{V} \times \mathcal{V}_{\mathrm{l},-}} \int \mathcal{D}[\beta_-, \alpha_-, \beta_+, \alpha_+] \int \mathcal{D}[\psi^{\dagger}_-, \psi_-, \psi^{\dagger}_+, \psi_+]\, e^{+\mathrm{i}S_{\mathrm{QCD}_2}}, \tag{I.201a}$$

with the action S_{QCD_2} given by [recall Eqs. (I.184) and (I.199)]

$$S_{\mathrm{QCD}_2} := \int \frac{\mathrm{d}x^+ \mathrm{d}x^-}{2} \left(\mathcal{L}_{\mathrm{QCD}_2} + \mathcal{L}_{\mathrm{gh}}\right), \tag{I.201b}$$

and all $U(N)$ valued fields taken in the fundamental representation. The bosonic measure is invariant under local $U(N) \times U(N)$ gauge transformations parametrized by the independent fields ω_+ and ω_-. However, the fermionic measures are only invariant under the diagonal subgroup of $U(N) \times U(N)$ defined by $\omega_- = \omega_+ \equiv \omega$. Finally, the redundancy in the parametrization of the gauge fields in terms of group-valued fields has induced a local symmetry in the bosonic sector.

We now repeat the sequence of the local diagonal gauge transformation (I.185) followed by the local mixed diagonal and axial gauge transformation (I.187) (also to be applied to the ghosts fields). We make use of the fact that the bosonic measure $\mathcal{D}[G_-, G_+]$ is left invariant under any local $U(N) \times U(N)$ gauge transformation parametrized by the independent fields ω_+ and ω_-. The diagonal local gauge transformation, $\omega_+ = \omega_- \equiv \omega$, is constructed so as to eliminate the dependency on G_+ of both the chiral and ghost Lagrangian densities. By construction, the Grassmann measures are invariant under any local diagonal gauge transformation. It then follows that the integrand of the partition function is fully independent of G_+. Thus, performing the integration over the measure $\mathcal{D}\,[G_+]$ precisely cancels the normalization $\mathcal{V}^{-1}$. We are then left with the partition function

$$Z_{\mathrm{QCD}_2} = \int \frac{\mathcal{D}[\tilde{G}]}{\mathcal{V}_{+,-}} \int \mathcal{D}[\tilde{\beta}_-,\tilde{\alpha}_-,\tilde{\beta}_+,\tilde{\alpha}_+] \int \mathcal{D}[\tilde{\psi}^\dagger_-,\tilde{\psi}_-,\tilde{\psi}^\dagger_+,\tilde{\psi}_+]\, e^{+\mathrm{i}S_{\mathrm{QCD}_2}}. \tag{I.202}$$

Next, we perform the local mixed diagonal and axial gauge transformation under which $\tilde{G} = G_+^{-1} G_-$ drops out from the chiral and ghost Lagrangian densities, i.e., $\omega_+ = 2\omega$ and $\omega_- = 0$. Whereas the bosonic measure is unchanged, the chiral and ghost Grassmann measures change by the Jacobians $\mathcal{J}_{\mathrm{f}}[\tilde{G}]$ and $\mathcal{J}_{\mathrm{ad}}[\tilde{G}]$, respectively. We arrive at the representation

$$Z_{\mathrm{QCD}_2} = \int \frac{\mathcal{D}[\tilde{G}]}{\mathcal{V}_{+,-}} \int \mathcal{D}[\beta'_-,\alpha'_-,\beta'_+,\alpha'_+]\,\mathcal{J}_{\mathrm{ad}}[\tilde{G}] \int \mathcal{D}[\psi'^\dagger_-,\psi'_-,\psi'^\dagger_+,\psi'_+]\,\mathcal{J}_{\mathrm{f}}[\tilde{G}]\, e^{+\mathrm{i}S_{\mathrm{QCD}_2}}. \tag{I.203}$$

Here, the Jacobians $\mathcal{J}_{\mathrm{f}}[\tilde{G}]$ and $\mathcal{J}_{\mathrm{ad}}[\tilde{G}]$ differ in that the Grassmann measures correspond to two different representations of the gauge group, namely the fundamental and the adjoint representations, respectively. We have already computed

$$\mathcal{J}_{\mathrm{f}}[\tilde{G}] \equiv \frac{\mathrm{Det}\left[\gamma^\mu\left(\mathrm{i}\partial_\mu + A_\mu\right)\right]}{\mathrm{Det}\left(\gamma^\mu \mathrm{i}\partial_\mu\right)} = e^{-\mathrm{i}Z_{\mathrm{WZW}}[\tilde{G}]}, \qquad \tilde{G} \equiv G_+^{-1} G_-. \tag{I.204}$$

The method of Polyakov and Wiegmann also applies to the Dirac equation in the adjoint representation with the caveat that we need to replace

$$\frac{1}{4\pi}\epsilon_{\mu\nu}\,F^{\mu\nu}, \tag{I.205}$$

by

$$\frac{\mathrm{y}_{\mathrm{ad}}}{\mathrm{y}_{\mathrm{f}}}\frac{1}{4\pi}\epsilon_{\mu\nu}\,F^{\mu\nu}, \tag{I.206}$$

in the anomaly equation (I.138). This gives

$$\mathcal{J}_{\mathrm{ad}}[\tilde{G}] \equiv \frac{\mathrm{Det}\left[\gamma^\mu\left(\mathrm{i}\partial_\mu + [A_\mu,\cdot\,]\right)\right]}{\mathrm{Det}\left(\gamma^\mu \mathrm{i}\partial_\mu\right)} = e^{-\mathrm{i}\frac{\mathrm{y}_{\mathrm{ad}}}{\mathrm{y}_{\mathrm{f}}}Z_{\mathrm{WZW}}[\tilde{G}]}, \qquad \tilde{G} \equiv G_+^{-1} G_-. \tag{I.207}$$

For the group $SU(N)$, the ratio $\mathrm{y}_{\mathrm{ad}}/\mathrm{y}_{\mathrm{f}}$ is particularly simple, namely

$$\frac{\mathrm{y}_{\mathrm{ad}}}{\mathrm{y}_{\mathrm{f}}} = 2\,c_v, \tag{I.208}$$

where c_v is the quadratic Casimir operator.

We are ready to finalize the main result of this appendix. QCD_2 in the infinitely strong gauge coupling limit,

$$\begin{aligned} Z_{\mathrm{QCD}_2} &= \mathcal{V}^{-1}\int \mathcal{D}[A_+,A_-] \int \mathcal{D}[\psi^\dagger_-,\psi_-,\psi^\dagger_+,\psi_+]\, e^{+\mathrm{i}S_{\mathrm{QCD}_2}}, \\ S_{\mathrm{QCD}_2} &= \int \frac{\mathrm{d}x^-\mathrm{d}x^+}{2}\left[\psi^\dagger_-\left(\mathrm{i}\partial_+ + A_+\right)\psi_- + \psi^\dagger_+\left(\mathrm{i}\partial_- + A_-\right)\psi_+\right], \end{aligned} \tag{I.209}$$

is nothing but the quantum *critical* field theory with the partition function

$$Z = \int \frac{\mathcal{D}[\tilde{G}]}{\mathcal{V}_{+,-}} \int \mathcal{D}[\beta'_-, \alpha'_-, \beta'_+, \alpha'_+] \int \mathcal{D}[\psi'^{\dagger}_-, \psi'_-, \psi'^{\dagger}_+, \psi'_+] \, e^{+\mathrm{i}(S_1 - S_2 + S_3)}, \tag{I.210a}$$

where the actions

$$\begin{aligned} S_1 &= \int \frac{\mathrm{d}x^- \mathrm{d}x^+}{2} (\psi'^{\dagger}_- \, \mathrm{i}\partial_+ \, \psi'_- + \psi'^{\dagger}_+ \, \mathrm{i}\partial_- \, \psi'_+), \\ S_2 &= (1 + 2c_v) \, S_{\mathrm{WZW}}[\tilde{G}], \\ S_3 &= \int \frac{\mathrm{d}x^- \mathrm{d}x^+}{2} \left[\frac{2}{\mathrm{y_f}} \mathrm{tr} \, (\beta'^{\mathrm{a}}_- T^{\mathrm{a}} \, \mathrm{i}\partial_+ \, T^{\mathrm{b}} \, \alpha'^{\mathrm{b}}_-) + \frac{2}{\mathrm{y_f}} \mathrm{tr} \, (\beta'^{\mathrm{a}}_+ T^{\mathrm{a}} \, \mathrm{i}\partial_- \, T^{\mathrm{b}} \, \alpha'^{\mathrm{b}}_+) \right], \end{aligned} \tag{I.210b}$$

account for the fermion, boson, and ghost contributions, respectively. The relationship between the ψ''s and the ψ's is

$$\psi^{\dagger}_- = \psi'^{\dagger}_- \, G^{-1}_+, \qquad \psi_- = G_+ \, \psi'_-, \qquad \psi^{\dagger}_+ = \psi'^{\dagger}_+ \, G^{-1}_-, \qquad \psi_+ = G_- \, \psi'_+, \tag{I.210c}$$

where

$$\tilde{G} = G^{-1}_+ \, G_-. \tag{I.210d}$$

The relationship between the $A_\pm$'s and the $G_\pm$'s is

$$\mathrm{i}A_+ = (\partial_+ G_+) G^{-1}_+, \qquad \mathrm{i}A_- = (\partial_- G_-) G^{-1}_-. \tag{I.210e}$$

This is a remarkable result, for a strongly interacting quantum field theory has been solved! The nature of the critical theory requires knowledge from two-dimensional conformal field theory that goes beyond the scope of this book. Suffices here to say that it is possible to construct a number, called the Virasoro central charge, that counts how many degrees of freedom are critical (see Refs. [140] and [141]). The free real-valued scalar field theory has the Virasoro central charge 1. By bosonization a single two-component Dirac spinor also has the Virasoro central charge 1. Because of the factorization of the partition function into three independent sectors, each of which is critical, the Virasoro central charge of the non-Abelian sector $SU(N)$ of the theory is [46]

$$c_{123} = c_1 + c_2 + c_3. \tag{I.211}$$

Since S_1 represents N free two-components Dirac fermions, one has

$$c_1 = N. \tag{I.212}$$

According to Knizhnik and Zamolodchikov, the Virasoro central charge associated to a WZW action of level k is

$$c = \frac{k}{k + c_v} \dim su(N). \tag{I.213}$$

[46] The $U(1)$ sector decouples from the $SU(N)$ sector. The $U(1)$ sector is nothing but the infinitely-strong coupling limit of quantum electrodynamics in (1+1)-dimensional Minkowski space and time $\mathbb{M}_{1,1}$ (QED_2). We can borrow the results of chapter 9 to infer that this theory has a vanishing central charge, i.e., all modes are gapped.

In our case, if we choose $c_v = N$, i.e., $\text{y}_\text{f} = 2$ and $-k = 1 + 2c_v = 1 + 2N$, then

$$c_2 = \frac{(1+2N)(N^2-1)}{1+N}. \tag{I.214}$$

Finally, the Virasoro central charge for the ghost sector is negative and given by

$$c_3 = -2\dim su(N) = -2(N^2-1). \tag{I.215}$$

We thus conclude that

$$c_{123} = N + \frac{(1+2N)(N^2-1)}{1+N} - 2(N^2-1). \tag{I.216}$$

For example, if $N = 2$, then

$$c_{123} = 2 + 5 - 6 = 1. \tag{I.217}$$

We see that the ghost sector of the theory is responsible for removing unphysical degrees of freedom associated to the sector of the Hilbert space that is not invariant under local gauge transformations.

Bibliography

[1] G. Baym, *Lectures On Quantum Mechanics*, W. A. Benjamin, New York (1969).

[2] A. Messiah, *Mécanique quantique, Tome 1*, Dunod, Paris (1999); *Mécanique quantique, Tome 2*, Dunod, Paris (2003).

[3] P. A. M. Dirac, *The Principles of Quantum Mechanics*, Oxford University Press, New York (1958).

[4] R. Becker, *Theorie der Wärme*, Springer, Berlin (1985).

[5] H. B. Callen, *Thermodynamics*, John Wiley, New York (1960).

[6] K. Huang, *Statistical mechanics*, John Wiley, New York (1963).

[7] R. P. Feynman, *Statistical mechanics*, Westview Press (1972).

[8] C. Itzykson and J.-B. Zuber, *Quantum Field Theory*, McGraw-Hill, New York (1980).

[9] C. Kittel, *Quantum theory of solids*, John Wiley, New York (1963).

[10] A. A. Abrikosov, L. P. Gorkov, and I. E. Dzyaloshinskii, *Methods of Quantum Field Theory in Statistical Physics*, Dover, New York (1963).

[11] D. Pines and P. Nozières, *The theory of quantum liquids*, Volume 1 (*Normal Fermi liquids*), W. A. Benjamin, New York (1966).

[12] A. L. Fetter and J. D. Walecka, *Quantum theory of many-particle systems*, McGraw-Hill, New York (1971).

[13] S. Doniach and E. H. Sondheimer, *Green's functions for solid state physicists*, Addison-Wesley, Redwood City (1982).

[14] A. M. Polyakov, *Gauge Fields and Strings*, Harwood Academic Publishers, Chur (1987).

[15] G. Parisi, *Statistical Field Theory*, Addison Wesley, New York (1988).

[16] J. W. Negele and H. Orland, *Quantum many-particle systems*, Addison-Wesley, Redwood City (1988).

[17] E. Fradkin, *Field theory of condensed matter systems* second edition, Cambridge University Press, Cambridge (2013).

[18] A. M. Tsvelik, *Quantum field theory in condensed matter physics*, Cambridge University Press, Cambridge (1995).

[19] A. O. Gogolin, A. A. Nersesyan, and A. M. Tsvelik, *Bosonization and strongly correlated systems*, Cambridge University Press, Cambridge (1999).

[20] N Nagaosa, *Quantum field theory in condensed matter physics* and *Quantum field theory in strongly correlated electronic systems*, Springer, Berlin (1999).

[21] M. Stone, *The physics of quantum fields*, Springer, New York (2000).

[22] S. Sachdev, *Quantum phase transitions* second edition, Cambridge University Press, Cambridge (2013).

[23] N. D. Mermin, *Crystalline Order in Two Dimensions*, Phys. Rev. **176**, 250 (1968); Erratum, Phys. Rev. B **20**, 4762 (1979); Erratum, Phys. Rev. B **74**, 149902 (2006).
[24] D. Serre, *Matrices: Theory and Applications*, Springer, New York (2002).
[25] J. Zinn-Justin, fourth edition, *Quantum Field Theory and Critical Phenomena*, Oxford University press, New York (2002).
[26] J. Glimm and A. Jaffe, *Quantum physics*, Springer, New York (1987).
[27] N. N. Bogoliubov, *On the Theory of Superfluidity*, J. Phys. (USSR) **11**, 23 (1947); [reprinted in D. Pines, *The Many-body Problem*, W. A. Benjamin, New York (1961)].
[28] A. J. Leggett, *Bose-Einstein condensation in the alkali gases: Some fundamental concepts*, Rev. Mod. Phys. **73**, 307 (2001).
[29] T. Holstein and H. Primakoff, *Field dependence of the intrinsic domain magnetization of a ferromagnet*, Phys. Rev. **58**, 1098 (1940).
[30] M. Stone and P. Goldbart, *Mathematics for physics*, Cambridge University Press, Cambridge (2009).
[31] E. Abdalla, M. C. B Abdalla, and K. D. Rothe, *2 dimensional quantum field theory*, World Scientific Publishing, Singapore (1991).
[32] A. M. Polyakov, *Interaction of Goldstone particles in 2 dimensions - applications to ferromagnets and massive Yang-Mills fields*, Phys. Lett. B **59**, 79 (1975).
[33] S. Helgason, *Differential Geometry, Lie Groups, and Symmetric Spaces*, Academic Press, London (1978).
[34] N. D. Mermin and H. Wagner, *Absence of Ferromagnetism or Antiferromagnetism in One- or Two-Dimensional Isotropic Heisenberg Models*, Phys. Rev. Lett. **17**, 1133 (1966).
[35] F. D. M. Haldane, *Continuum dynamics of the 1-D Heisenberg antiferromagnet: Identification with the O(3) nonlinear sigma model*, Phys. Lett. A **93**, 464 (1983), and *Nonlinear Field Theory of Large-Spin Heisenberg Antiferromagnets: Semiclassically Quantized Solitons of the One-Dimensional Easy-Axis Néel State*, Phys. Rev. Lett. **50**, 1153 (1983).
[36] M. V. Berry, *Quantal phase factors accompanying adiabatic changes*, Proceedings of the Royal Society of London A **392**, 45 (1984).
[37] V. L. Berezinskii and A. Ya. Blank, *Thermodynamics of layered isotropic magnets at low temperatures*, Zh. Eksp. Teor. Fiz. **64**, 725 (1973) [Sov. Phys. JETP **37**, 369 (1973)].
[38] A.V. Chubukov, S. Sachdev, and J. Ye, *Theory of two-dimensional quantum Heisenberg antiferromagnets with a nearly critical ground state*, Phys. Rev. B **49**, 11919 (1994).
[39] S. Chakravarty, B. I. Halperin, and D. R. Nelson, *Two-dimensional quantum Heisenberg-antiferromagnet at low temperatures*, Phys. Rev. B **39**, 2344 (1989).
[40] V. L. Berezinskii, *Violation of long range order in one-dimensional and two-dimensional systems with a continuous symmetry group. I. Classical systems*, Zh. Eksp. Teor. Fiz. **59**, 907 (1970) [Sov. Phys. JETP. **32**, 493 (1971)] and *Destruction of long-range order in one-dimensional and two-dimensional systems with a continuous symmetry group. II. Quantum systems*, Zh. Eksp. Teor. Fiz. **61**, 1144 (1971) [Sov. Phys. JETP. **34**, 610 (1972)].
[41] J. M. Kosterlitz and D. J. Thouless, *Ordering, metastability and phase transitions in two-dimensional systems*, J. Phys. C **6**, 1181 (1973).
[42] J. M. Kosterlitz, *The critical properties of the two-dimensional xy model*, J. Phys. C **7**, 1046 (1974).
[43] L. V. Ahlfors, *Complex analysis*, McGraw-Hill, New York (1979).
[44] J. Villain, *Theory of one- and two-dimensional magnets with an easy magnetization*

plane. II. The planar, classical, two-dimensional magnet, J. Physique **36**, 581 (1975).
[45] J. V. José, L. P. Kadanoff, S. Kirkpatrick, and D. R. Nelson, *Renormalization, vortices, and symmetry-breaking perturbations in the two-dimensional planar model*, Phys. Rev. B **16**, 1217 (1977).
[46] D. J. Amit, Y. Y. Goldschmidt, and G. Grinstein, *Renormalisation group analysis of the phase transition in the 2D Coulomb gas, Sine-Gordon theory and XY-model*, J. Phys. A **13**, 585 (1980).
[47] D. R. Nelson and J. M. Kosterlitz, *Universal Jump in the Superfluid Density of Two-Dimensional Superfluids*, Phys. Rev. Lett. **39**, 1201 (1977).
[48] M. Rubinstein, B. Shraiman, and D. R. Nelson, *Two-dimensional XY magnets with random Dzyaloshinskii-Moriya interactions*, Phys. Rev. B **27**, 1800 (1983).
[49] S. E. Korshunov, *Possible destruction of the ordered phase in Josephson-junction arrays with positional disorder*, Phys. Rev. B **48**, 1124 (1993).
[50] T. Nattermann, S. Scheidl, S. E. Korshunov, and M. S. Li, *Absence of Reentrance in the Two-Dimensional XY-Model with Random Phase Shift*, J. Phys. I **5**, 565 (1995).
[51] M. Cha and H. A. Fertig, *Disorder-Induced Phase Transitions in Two-Dimensional Crystals*, Phys. Rev. Lett. **74**, 4867 (1995).
[52] C. Mudry and X.G. Wen, *Does quasi-long-range order in the two-dimensional XY model really survive weak random phase fluctuations*, Nucl. Phys. B **549**, 613 (1999).
[53] D. Carpentier and P. Le Doussal, *Topological transitions and freezing in XY models and Coulomb gases with quenched disorder: renormalization via traveling waves*, Nucl. Phys. B **588**, 565 (2000).
[54] V. Alba, A. Pelissetto, and E. Vicari, *Magnetic and glassy transitions in the square-lattice XY model with random phase shifts*, J. Stat. Mech. **2010**, P03006 (2010).
[55] L. D. Landau and E. M. Lifshitz, *Statistical Physics*, part 1 third edition, Pergamon Press, Oxford (1980).
[56] H. W. Wyld, *Mathematical methods for physics*, third edition, W. A. Benjamin, Reading (1976).
[57] I. S. Gradshteyn and I. M. Ryzhik, *Table of integrals, series, and products*, fifth edition, Academic Press, London (1980).
[58] K. V. Klitzing, G. Dorda, and M. Pepper, *New Method for High-Accuracy Determination of the Fine-Structure Constant Based on Quantized Hall Resistance*, Phys. Rev. Lett. **45**, 494 (1980).
[59] R. B. Laughlin, *Quantized Hall conductivity in two dimensions*, Phys. Rev. B **23**, 5632 (1981).
[60] W. Kohn, *Cyclotron Resonance and de Haas-van Alphen Oscillations of an Interacting Electron Gas*, Phys. Rev. **123**, 1242 (1961).
[61] B. I. Halperin, *Quantized Hall conductance, current-carrying edge states, and the existence of extended states in a two-dimensional disordered potential*, Phys. Rev. B **25**, 2185 (1982).
[62] D. C. Tsui, H. L. Stormer, and A. C. Gossard, *Two-Dimensional Magnetotransport in the Extreme Quantum Limit*, Phys. Rev. Lett. **48**, 1559 (1982).
[63] M. Gell-Mann and K. A. Brueckner, *Correlation Energy of an Electron Gas at High Density*, Phys. Rev. **106**, 364 (1957).
[64] N. W. Ashcroft and N. D. Mermin, *Solid state physics*, Holt-Saunders, Philadelphia (1976).
[65] J. Lindhard, *On the properties of a gas of charged particles*, Kgl. Danske Videnskab. Selskab Mat.-Fys. Medd. **28**, (8) 1 (1954).
[66] J. M. Luttinger, *Fermi Surface and Some Simple Equilibrium Properties of a System of Interacting Fermions*, Phys. Rev. **119**, 1153 (1960).

[67] A. Praz, J. Feldman, H. Knörrer, and E. Trubowitz, *A proof of Luttinger's theorem*, Europhys. Lett. **72**, 49 (2005); A. Praz, *A proof of Luttinger's theorem*, Dissertation ETH No. 15383.
[68] M. Oshikawa, *Topological Approach to Luttinger's Theorem and the Fermi Surface of a Kondo Lattice*, Phys. Rev. Lett. **84**, 3370 (2000).
[69] N. Read and D. M. Newns, *On the solution of the Coqblin-Schrieffer Hamiltonian by the large-N expansion technique*, J. Phys. C **16**, 3273 (1983).
[70] P. Coleman, *New approach to the mixed-valence problem*, Phys. Rev. B **29**, 3035 (1984).
[71] G. Kotliar and A. E. Ruckenstein, *New functional integral approach to strongly correlated Fermi systems: The Gutzwiller approximation as a saddle point*, Phys. Rev. Lett. **57**, 1362 (1986).
[72] G. Baskaran, Z. Zou, and P. W. Anderson, *The resonating valence bond state and high-Tc superconductivity – A mean field theory*, Solid State Communications **63**, 973 (1987).
[73] P. Henrici, *Applied and computational complex analysis*, volume I, John Wiley, New York (1974).
[74] M. Abramowitz and I. A. Stegun, *Handbook of mathematical functions*, Dover, New York (1965).
[75] M. Tinkham, *Introduction to superconductivity*, McGraw-Hill, New York (1980).
[76] F. Dalfovo, S. Giorgini, L. P. Pitaevskii, and S. Stringari, *Theory of Bose-Einstein condensation in trapped gases*, Rev. Mod. Phys. **71**, 463 (1999).
[77] E. H. Lieb, R. Seiringer, and J. Yngvason, *Bosons in a Trap: A Rigorous Derivation of the Gross-Pitaevskii Energy Functional*, Phys. Rev. A **61**, 043602 (2000) and arXiv:9908027[math-ph]. See also: Proceedings of *Quantum Theory and Symmetries* (Goslar, 18-22 July 1999), edited by H.-D. Doebner, V. K. Dobrev, J.-D. Hennig, and W. Luecke, World Scientific (2000) and arXiv:9911026[math-ph].
[78] E. H. Lieb, R. Seiringer, and J. Yngvason, *A Rigorous Derivation of the Gross-Pitaevskii Energy Functional for a Two-Dimensional Bose Gas*, Commun. Math. Phys. **224**, 17 (2001).
[79] A. I. Larkin and A. A. Varlamov, *Fluctuation Phenomena in Superconductors*, arXiv:0109177[cond-mat] and chapter 10 of volume 1 in *Superconductivity: Conventional and Unconventional Superconductors*, edited by K.-H.Bennemann and J. B. Ketterson, Springer (2008).
[80] L. N. Cooper, *Bound electron pairs in a degenerate Fermi gas*, Phys. Rev. **104**, 1189 (1956).
[81] J. Bardeen, L. N. Cooper, and J. R. Schrieffer, *Microscopic theory of superconductivity*, Phys. Rev. **106**, 162 (1957).
[82] J. Bardeen, L. N. Cooper, and J. R. Schrieffer, *Theory of superconductivity*, Phys. Rev. **108**, 1175 (1957).
[83] D. G. McDonald, *Physics Today, The Nobel Laureate Versus the Graduate Student*, **54**, (7) 46 (2001).
[84] A. O. Caldeira and A. J. Leggett, *Quantum tunnelling in a dissipative system*, Ann. Phys. **149**, 374 (1983).
[85] O. V. Lounasmaa, *Experimental principles and methods below 1K*, Academic Press, New York (1974).
[86] S. Coleman, *Aspects of symmetry*, Cambridge University Press, Cambridge (1985).
[87] A. Schmid, *Diffusion and Localization in a Dissipative Quantum System*, Phys. Rev. Lett. **51**, 1506 (1983).
[88] M. P. A. Fisher and W. Zwerger, *Quantum Brownian motion in a periodic potential*,

Phys. Rev. B **32**, 6190 (1985).

[89] A. C. Hewson, *The Kondo Problem to Heavy Fermions*, Cambridge University Press, Cambridge (1997).

[90] W. J. de Haas, J. de Boer, and G. J. van den Berg, *The electrical resistance of gold, copper and lead at low temperatures*, Physica **1**, 1115 (1934).

[91] J. Kondo, *Resistance Minimum in Dilute Magnetic Alloys* Prog. Theor. Phys. **32**, 37 (1964).

[92] P. W. Anderson and G. Yuval, *Exact Results in the Kondo Problem: Equivalence to a Classical One-Dimensional Coulomb Gas*, Phys. Rev. Lett. **23**, 89 (1969); G. Yuval and P. W. Anderson, *Exact Results for the Kondo Problem: One-Body Theory and Extension to Finite Temperature* Phys. Rev. B **1**, 1522 (1970); P. W. Anderson, G. Yuval, and D. R. Hamann, *Exact Results in the Kondo Problem. II. Scaling Theory, Qualitatively Correct Solution, and Some New Results on One-Dimensional Classical Statistical Models*, Phys. Rev. B **1**, 4464 (1970). P. W. Anderson, *A poor man's derivation of scaling laws for the Kondo problem*, J. Phys. C **3**, 2436 (1970).

[93] K. G. Wilson, *The renormalization group: Critical phenomena and the Kondo problem*, Rev. Mod. Phys. **47**, 773 (1975).

[94] R. Bulla, T. A. Costi, and T. Pruschke, *Numerical renormalization group method for quantum impurity systems*, Rev. Mod. Phys. **80**, 395 (2008).

[95] N. Andrei, *Diagonalization of the Kondo Hamiltonian*, Phys. Rev. Lett. **45**, 379 (1980).

[96] P. B. Wiegmann, *Towards an exact solution of the Anderson model*, Phys. Lett. A **80**, 163 (1980); *Tochnoe reshenie s-d obmennoi modeli pri T=0*, Pisma Zh. Eksp. Teor. Fiz. **31**, 392 (1980).

[97] H. Tsunetsugu, M. Sigrist, and K. Ueda, *The ground-state phase diagram of the one-dimensional Kondo lattice model*, Rev. Mod. Phys. **69**, 809 (1997).

[98] G. R. Stewart, *Heavy-fermion systems*, Rev. Mod. Phys. **56**, 755 (1984).

[99] P. W. Anderson, *Localized Magnetic States in Metals*, Phys. Rev. **124**, 41 (1961).

[100] J. R. Schrieffer and P. A. Wolff, *Relation between the Anderson and Kondo Hamiltonians*, Phys. Rev. **149**, 491 (1966).

[101] P. Coleman, *Mixed valence as an almost broken symmetry*, Phys. Rev. B **35**, 5072 (1987).

[102] B. Coqblin and J. R. Schrieffer, *Exchange Interaction in Alloys with Cerium Impurities*, Phys. Rev. **185**, 847 (1969).

[103] D. Withoff and E. Fradkin, *Phase transitions in gapless Fermi systems with magnetic impurities*, Phys. Rev. Lett. **64**, 1835 (1990).

[104] P. Nozières and A. Blandin, *Kondo effect in real metals*, J. Physique **41**, 193 (1980).

[105] I. Affleck and A. W. W. Ludwig, *The Kondo effect, conformal field theory and fusion rules*, Nucl. Phys. B **352**, 849 (1991); *Critical theory of overscreened Kondo fixed points*, Nucl. Phys. B **360**, 641 (1991); A. W. W. Ludwig and I. Affleck, *Exact, asymptotic, three-dimensional, space- and time-dependent, Green's functions in the multichannel Kondo effect*, Phys. Rev. Lett. **67**, 3160 (1991).

[106] D. L. Cox and A. E. Ruckenstein, *Spin-flavor separation and non-Fermi-liquid behavior in the multichannel Kondo problem: A large-N approach*, Phys. Rev. Lett. **71**, 1613 (1993).

[107] M. Vojta, *Impurity quantum phase transitions*, Philos. Mag. **86**, 1807 (2006).

[108] S. Coleman, *Quantum sine-Gordon equation as the massive Thirring model*, Phys. Rev. D **11**, 2088 (1975).

[109] S. Mandelstam, *Soliton operators for the quantized sine-Gordon equation*, Phys. Rev. D **11**, 3026 (1975).

[110] A. Luther and I. Peschel, *Calculation of critical exponents in two dimensions from quantum field theory in one dimension*, Phys. Rev. B **12**, 3908 (1975).

[111] E. Witten, *Non-abelian bosonization in two dimensions*, Comm. Math. Phys. **92**, 455 (1984).

[112] A. Polyakov and P. B. Wiegmann, *Theory of Non-Abelian Goldstone bosons in two-dimensions*, Phys. Lett. B **131**, 121 (1983); *Goldstone fields in 2 dimensions with multivalued actions*, Phys. Lett. B **141**, 223 (1984).

[113] K. Fujikawa, *Path-Integral Measure for Gauge-Invariant Fermion Theories*, Phys. Rev. Lett. **42**, 1195 (1979); K. Fujikawa, *Path integral for gauge theories with fermions*, Phys. Rev. D **21**, 2848 (1980).

[114] J. Schwinger, *Gauge Invariance and Mass. II*, Phys. Rev. **128**, 2425 (1962).

[115] H. B. Nielsen and M. Ninomiya, *A no-go theorem for regularizing chiral fermions*, Phys. Lett. B **105**, 219 (1981); *Absence of neutrinos on a lattice: (I). Proof by homotopy theory*, Nucl. Phys. B **185**, 20 (1981); *Absence of neutrinos on a lattice: (II). Intuitive topological proof*, Nucl. Phys. B **193**, 173 (1981).

[116] X.-G. Wen, *Electrodynamical properties of gapless edge excitations in the fractional quantum Hall states*, Phys. Rev. Lett. **64**, 2206 (1990); *Gapless boundary excitations in the quantum Hall states and in the chiral spin states*, Phys. Rev. B **43**, 11025 (1991); *Edge transport properties of the fractional quantum Hall states and weak-impurity scattering of a one-dimensional charge-density wave*, Phys. Rev. B **44**, 5708 (1991).

[117] L. Lorenz, *On the Identity of the Vibrations of Light with Electrical Currents*, Philos. Mag. **34**, 287 (1867).

[118] C. Cohen-Tannoudji, B. Diu, and F. Laloë, *Quantum mechanics*, Hermann, Paris (1977).

[119] A. M. Perelomov, *Generalized coherent states and their applications*, Springer, Berlin (1986).

[120] D. H. Friedan, *Nonlinear Models in $2+\epsilon$ Dimensions*, Ann. Phys. **163**, 318 (1985).

[121] L. Alvarez-Gaumé, D. Z. Freedman, and S. Mukhi, *The background field method and the ultraviolet structure of the supersymmetric nonlinear σ-model,* Ann. Phys. **134**, 85-109 (1981).

[122] L. D. Landau, *The theory of a Fermi liquid*, Sov. Phys. JETP **3**, 920 (1957) [reprinted in D. Pines, *The Many-body Problem*, W. A. Benjamin, New York (1961)].

[123] H. Bruus and K. Flensberg, *Many-body quantum theory in condensed matter physics*, Oxford University press, New York (2004).

[124] G. E. Volovik, *A new class of normal Fermi liquids*, JETP Lett., **53** 4, 222 (1991).

[125] G. E. Volovik, *The universe in a Helium droplet*, Oxford University press, New York (2003).

[126] J. M. Luttinger and J. C. Ward, *Ground-State Energy of a Many-Fermion System. II*, Phys. Rev. **118**, 1417 (1960).

[127] L. D. Landau, *Oscillations in a Fermi liquid*, Sov. Phys. JETP **5**, 101 (1957) [reprinted in D. Pines, *The Many-body Problem*, W. A. Benjamin, New York (1961)].

[128] S. Tomonaga, *Remarks on Bloch's Method of Sound Waves applied to Many-Fermion Problems*, Prog. Theor. Phys. **5**, 544 (1950).

[129] J. M. Luttinger, *An Exactly Soluble Model of a Many-Fermion System*, J. Math. Phys. **4**, 1154 (1963).

[130] D. C. Mattis and E. H. Lieb, *Exact Solution of a Many-Fermion System and Its Associated Boson Field*, J. Math. Phys. **6**, 304 (1965).

[131] C. Kittel, *Introduction to Solid State Physics*, eighth edition, John Wiley, New York (2004).

[132] A. Luther and I. Peschel, *Single-particle states, Kohn anomaly, and pairing fluctuations in one dimension*, Phys. Rev. B **9**, 2911 (1974).

[133] T. Giamarchi, *Quantum Physics in one dimension*, Oxford University Press, New York (2004).

[134] X. G. Wen, *Metallic non-Fermi-liquid fixed point in two and higher dimensions*, Phys. Rev. B **42**, 6623 (1990).

[135] V. L. Ginzburg, *Some remarks on phase transitions of the 2nd kind and the microscopic theory of ferroelectric materials*, Fiz. Tverd. Tela **2**, 2031 (1960) [Sov. Phys. Solid State **2**, 1824 (1961)].

[136] B. I. Halperin, T. C. Lubensky, and S. K. Ma, *First-Order Phase Transitions in Superconductors and Smectic-A Liquid Crystals*, Phys. Rev. Lett. **32**, 292 (1974).

[137] B. I. Halperin and T. C. Lubensky, *On the analogy between smectic a liquid crystals and superconductors*, Solid State Communications **14**, 997 (1974).

[138] P. G. de Gennes and J. Prost, *The Physics of Liquid Crystals*, second edition, Oxford University Press, New York (1993).

[139] S. Coleman and E. Weinberg, *Radiative Corrections as the Origin of Spontaneous Symmetry Breaking*, Phys. Rev. D **7**, 1888 (1973).

[140] A. A. Belavin, A. M. Polyakov, and A. B. Zamolodchikov, *Infinite conformal symmetry in two-dimensional quantum field theory*, Nucl. Phys. B **241**, 333 (1984).

[141] V. G. Knizhnik, and A. B. Zamolodchikov, *Current Algebra and Wess-Zumino Model in Two Dimensions*, Nucl. Phys. B **247**, 83 (1984).

[142] J. Wess and B. Zumino, *Consequences of anomalous ward identities*, Phys. Lett. B **37**, 95 (1971).

[143] E. Witten, *Global aspects of current algebra*, Nucl. Phys. B **223**, 422 (1983).

[144] F. D. M. Haldane, *Fractional Quantization of the Hall Effect: A Hierarchy of Incompressible Quantum Fluid States*, Phys. Rev. Lett. **51**, 605 (1983).

[145] P. A. M. Dirac, *Quantised Singularities in the Electromagnetic Field*, Proceedings of the Royal Society of London A **133**, 60 (1931)

Index

Printed in the United States
By Bookmasters